中国金属学会　编

第十一届
中国钢铁年会论文集
（摘要）

Proceedings of the 11th CSM Steel Congress

北京

冶金工业出版社

2017

内 容 简 介

本论文集共收录论文 1200 余篇，约 1300 万字。全书内容包括炼铁与原料、炼钢与连铸、轧制与热处理、表面与涂镀、金属材料深加工、钢铁材料、汽车钢、海洋工程用钢、轴承钢、电工钢、粉末冶金、非晶合金、高温合金、耐火材料、能源与环保、分析检测、冶金设备与工程技术、冶金自动化与智能管控、冶金技术经济等方面，全面反映了近两年来我国及世界钢铁行业科研、生产、管理等方面的最新成果，是一本内容全面、新颖，具有较高学术水平的专业论文集。本书可供钢铁行业的科研人员、管理人员、工程技术人员等学习参考。

本论文集以纸质图书和电子版方式出版。纸质图书为全部论文的摘要集（不含大会报告），电子版为所有论文的全文和部分摘要。

图书在版编目(CIP)数据

第十一届中国钢铁年会论文集：摘要／中国金属学会编. —北京：冶金工业出版社，2017.11
ISBN 978-7-5024-7690-8

Ⅰ.①第 … Ⅱ.①中 … Ⅲ.①钢铁工业—学术会议—文集 Ⅳ.①TF4-53

中国版本图书馆 CIP 数据核字 (2017) 第 273136 号

出 版 人　谭学余
地　　址　北京市东城区嵩祝院北巷 39 号　邮编　100009　电话　(010)64027926
网　　址　www.cnmip.com.cn　电子信箱　yjcbs@cnmip.com.cn
责任编辑　李培禄　美术编辑　彭子赫　版式设计　孙跃红
责任校对　王永欣　责任印制　牛晓波
ISBN 978-7-5024-7690-8
冶金工业出版社出版发行；各地新华书店经销；北京京华虎彩印刷有限公司印刷
2017 年 11 月第 1 版，2017 年 11 月第 1 次印刷
210mm×297mm；62.5 印张；1962 千字；913 页
185.00 元

冶金工业出版社　投稿电话　(010)64027932　投稿信箱　tougao@cnmip.com.cn
冶金工业出版社营销中心　电话　(010)64044283　传真　(010)64027893
冶金书店　地址　北京市东四西大街 46 号(100010)　电话　(010)65289081（兼传真）
冶金工业出版社天猫旗舰店　yjgycbs.tmall.com
（本书如有印装质量问题，本社营销中心负责退换）

第十一届中国钢铁年会
组委会

年会主席　干　勇

执行主席　赵　沛

委　　员（按拼音排序）

才　让　干　勇　国文清　马国强

沈　彬　唐复平　王新江　杨　锐

于　勇　张欣欣　赵　继　赵　沛

赵民革

年会秘书长　王新江

年会秘书处　中国金属学会学术部

《第十一届中国钢铁年会论文集》
编 委 会

主 任 赵 沛

副主任 王新江

编 委 （按拼音排序）

鲍 磊	毕中南	柴俊兰	陈 兵	陈 杰	陈 卓
陈其安	崔 毅	戴毅刚	邓秀银	丁 波	董 瀚
杜 涛	胡守天	贾成厂	焦克新	李东迟	李美玲
李亚丽	李志芳	刘 辉	刘 洋	刘国栋	刘剑辉
刘湘江	罗光敏	米振莉	倪伟明	谯朝晖	尚成嘉
沈学静	苏 岚	孙 健	孙文强	孙彦广	唐 荻
童金涛	王 华	王 惠	王 凯	王存宇	王俊莉
王凌晗	王卫卫	王新江	王战民	夏 春	谢英秀
谢振家	徐安军	杨景玲	张 继	张 杰	张建良
张万山	张小会	张英华	赵 沛	赵希超	周少雄

前　言

第十一届中国钢铁年会于 2017 年 11 月 21 日至 22 日在北京召开。中国共产党第十九次全国代表大会报告指出，我国经济已由高速增长阶段转向高质量发展阶段，要以供给侧结构性改革为主线，支持传统产业优化升级，坚持去产能、降成本，坚持人与自然和谐共生。加快建设制造强国，推动互联网、大数据、人工智能和实体经济深度融合。近两年来钢铁行业通过大力清除地条钢，推进钢铁去产能，钢铁总量严重过剩矛盾得到缓解，钢铁产业绿色化、智能化有了不断进步。本届年会以"让钢铁更绿色，更智能"为主题，汇聚产学研用各领域专家和科技人员，交流冶金与材料基础理论、生产工艺、新材料开发及其应用、装备和自动化技术、节能环保技术等方面的最新科技成果，探求钢铁行业绿色化、智能化发展路径。

本届年会的征文工作继续得到钢铁企业、科研院所和高校，以及广大科技人员的积极支持，共收到投稿论文和摘要 1400 余篇，经年会技术委员会专家评审，录用 1200 余篇结集出版。内容包括：炼铁与原料、炼钢与连铸、轧制与热处理、表面与涂镀、金属材料深加工、钢铁材料、汽车钢、海洋工程用钢、轴承钢、电工钢、粉末冶金、非晶合金、高温合金、耐火材料、能源与环保、分析检测、冶金设备与工程技术、冶金自动化与智能管控、冶金技术经济等方面。年会收录论文的全文和部分摘要以电子版方式出版，录用论文摘要以纸质版方式出版。

由于论文集出版、编辑时间较紧，难免有疏漏与错误之处，恳请读者批评指正。

中国金属学会

2017 年 11 月

目　录

1. 炼铁与原料

2. 炼钢与连铸

3．轧制与热处理

4．表面与涂镀

5．金属材料深加工

6．钢铁材料

7. 汽车钢

8. 海洋工程用钢

9. 轴承钢

10. 电工钢

11．粉末冶金

12．非晶合金

13. 高温合金

14. 耐火材料

15. 能源与环保

16. 分析检测

17．冶金设备与工程技术

18. 冶金自动化与智能管控

19．冶金技术经济

1　炼铁与原料

★　炼铁与原料
　　炼钢与连铸
　　轧制与热处理
　　表面与涂镀
　　金属材料深加工
　　钢铁材料
　　汽车钢
　　海洋工程用钢
　　轴承钢
　　电工钢
　　粉末冶金
　　非晶合金
　　高温合金
　　耐火材料
　　能源与环保
　　分析检测
　　冶金设备与工程技术
　　冶金自动化与智能管控
　　冶金技术经济

无返矿炼铁工艺构想

沙永志，王志花，李光森，腾 飞，曹 军

（中国钢研科技集团，北京 100081）

摘 要： 大量返矿循环是当今烧结+高炉炼铁流程的严重内在缺陷，对炼铁成本和污染物排放造成巨大影响。提出的无返矿炼铁工艺，是将返矿除尘后直接用于高炉，使炼铁生产更高效环保。实现无返矿炼铁的关键是提高高炉使用颗粒矿的能力。采取的措施包括：去除粉末，分级入炉、优化布料、煤气流控制、控制烧结矿 RDI，以及智能炼铁等。国内外生产实践已证明无返矿炼铁工艺的工业应用基础和实现的可行性。

关键词： 炼铁，返矿，高炉，烧结，颗粒矿

No Return Fine BF Ironmaking Process

Sha Yongzhi, Wang Zhihua, Li Guangsen, Teng Fei, Cao Jun

(China Iron & Steel Research Institute Group, Beijing 100081, China)

Abstract: A large quantity of return fines recycling is a serious shortage of current sinter + BF ironmaking process, which causes big negative effects on operation cost and pollutant emission. No Return Fine (BF) Ironmaking Process (NRIP) is designed to use return fines in BF directly, after removing dust in it. NRIP could make sinter+BF ironmaking process more efficient and environment friendly. The key point of NRIP application is to enhance capacity of BF granular burden acceptance. The methods are dust removing, separate charging, optimum burden distribution, gas flow control, sinter RDI control, smart ironmaking,etc. Industrial practices around the world show profound basis and the feasibility of NRIP application.

Key words: ironmaking, return fine, BF, sinter, granular ore

宝钢绿色炼铁之实践

朱仁良

（宝钢股份炼铁厂，上海 200941）

摘 要： 阐述了当前炼铁普遍面临的挑战，展示了宝钢股份炼铁厂"绿色炼铁"的理念；重点介绍了宝钢股份炼铁厂为实现"绿色炼铁"目标而在生产、铁前系统改造方面采取的一些节能、减排优化措施，应用和即将应用的新技术、新工艺，绿色炼铁的成效；最后展望了宝钢绿色炼铁的愿景、目标，指明了今后炼铁节能、环保的技术发展方向。

关键词： 炼铁，节能，减排，厂城共融

Practice of Green Ironmaking in Baosteel

Zhu Renliang

(Ironmaking Plant of Baoshan Iron & Steel Co., Shanghai 200941, China)

Abstract: The challenges faced by ironmaking have been expounded and the concept of green ironmaking in Baoshan Iron & Steel company has been demonstracted. The key point this paper emphasized is focusing on some energy-saving and emission reduction optimization measurements adopted in production and transformation of iron front system. It also includes the application and new technology, new process for the goal of green ironmaking in Baoshan Iron & Steel Works and the effectiveness of green ironmaking. The further vision and objective have also been expected which give a technology guidance on ironmaking energy saving and and environmental protection in Baoshan Iron & Steel company.
Key words: ironmaking, energy saving, emission reduction, symbiosis between factory and city

A Critical Review and Future Prospects of Ironmaking Processes
(瞻前顾后话钢铁工艺)

Lu Weikao (卢维高)

(Professor Emeritus, McMaster University, Canada; 北京科技大学　荣誉教授)

Abstract: This article is intended to present a critical review of the evolution of iron-making technology. The contributions of science and engineering to the development of Blast Furnace and Shaft Furnace on one hand and Smelting Reduction and Rotary Hearth on the other will be analyzed. It is written for mature iron-makers and for the limitation of length of the article so that some professional jargons are used without adequate explanation. The success of blast furnace and the role of coke will be examined in detail. The evolution of iron- and steel-making processes must and will continue under major driving forces of the time, currently sustainable development, i.e. broadening supplies of raw materials and minimizing emissions to environment as well as economic efficiency. It is assumed that carbon will continue to serve as reducing agent and the source of processing energy. Efficient use of carbon will be analyzed somewhat similar to Professor Sven Eketorp's work on Smelting Reduction, but with more emphasis on heat transfer. Based on this concept and assumption that coal-ore composites will be used as starting material, then, the goal of elimination of coke-making and sintering operations becomes possible to reach. Suitable reactors, mainly hearth-type furnaces, to process composites will be discussed in some details. The impact of availability of less contaminated crude iron on subsequent refining steps will also be discussed.

炉缸积水与炉缸长寿的探讨

邹忠平[1], 张正好[2], 姜　华[2], 印　民[1]

（1. 中冶赛迪工程技术股份有限公司，重庆　401122; 2. 宝山钢铁股份有限公司，上海　201900）

摘　要：文从水进入炉缸后的演变、水对炉缸炭砖和凝固层的影响分析入手，论述了炉缸积水对炉缸长寿的影响，提出了防止炉缸进水和炉缸定期有效排水的建议，希望引起广大炼铁工作者的关注和思索，为改善我国高炉长寿现状发挥有益的作用。

关键词：高炉，炉缸，长寿，积水，凝固层

首钢烧结高温烟气循环提质节能减排新工艺

赵志星[1,2]，潘　文[1,2]，焦光武[3]，刘征建[4]，裴元东[1,2]，高新洲[3]，赵俊花[3]

（1. 绿色可循环钢铁流程北京市重点实验室，北京　100043；2. 首钢技术研究院，北京　100043；
3. 首钢股份有限公司，河北迁安　064400；4. 北京科技大学，北京　100083）

摘　要：围绕烧结大烟道中废烟气的资源化循环利用，开发了烧结高温烟气循环提质节能减排新工艺。根据大烟道热工测试、烧结料层高温带仿真模拟、实验室试验研究等工作，确定了烧结高温烟气循环工艺关键技术参数，循环烟气上罩温度保持在 250~300℃ 左右，氧含量 17%~20%。该工艺在首钢股份公司 $360m^2$ 成功应用后取得了烧结矿返矿率下降 6.6 个百分点，粉尘排放降低 27.30%，SO_2 减排 15.34% 等综合效果。与国内外主要烧结烟气循环工艺参数对比，首钢高温烟气循环工艺实现了烧结提质、节能、减排协同强化，具有推广价值。

关键词：热风烧结，高温烟气，循环利用，提质，节能减排

A New Process on Quality Improvement, Energy and Emission Reduction by Recycling High Temperature Sintering Flue Gas

Zhao Zhixing[1,2], Pan Wen[1,2], Jiao Guangwu[3], Liu Zhengjian[4], Pei Yuandong[1,2],
Gao Xinzhou[3], Zhao Junhua[3]

(1. Beijing Key Laboratory of Green Recyclable Process for Iron & Steel Production Technology, Beijing 100043, China; 2. Shougang Research Institute of Technology, Beijing 100043, China; 3. Shougang Qian'an Company, Qian'an 064400, China; 4. University of Science and Technology Beijing, Beijing 100083, China)

Abstract: Focused on resource utilization of sintering flue gas, a new process on quality improvement, energy and emission reduction by recycling high temperature sintering flue gas has been developed. Based on thermal testing of sintering flue, computer simulation of sintering combustion zone, laboratory investigation, etc., key technical parameters were established. According to above fundamental research, the recommended temperature and oxygen content of flue gas is 250~300℃ and 17%~20%, respectively. After industrial application in $360m^2$ sintering machine of Shougang Qian'an, the return fine ratio, dust and SO_2 emission, etc. were reduced by 6.6 %, 27.30 %, 15.34 %, respectively. By comparison with domestic and overseas recycling process of sintering flue gas, cooperative reinforcement of quality improvement, energy and emission reduction have been achieved with Shougang recycling process of high temperature sintering flue gas.

Key words: hot gas sintering, high temperature flue gas, cyclic utilization, quality improvement, energy and emission reduction

高铬型钒钛磁铁矿资源综合利用
关键技术——高炉冶炼

薛向欣[1,2]，程功金[1]，姜　涛[1,2]，杨　合[1,2]，段培宁[1,2]

（1. 东北大学冶金学院，辽宁沈阳　110819；

2. 辽宁省冶金资源循环科学重点实验室，辽宁沈阳　110819）

摘　要：高铬型钒钛磁铁矿矿物组成复杂，综合利用难度极大，之前国内外无工业化生产实践。本文分别从高铬型钒钛磁铁矿球团矿制备与性能、烧结矿制备与性能以及高炉炉料结构与操作等方面对高铬型钒钛磁铁矿资源综合利用关键技术进行了研究，实现了从实验室研究到工业化试验与应用的跨越，为高铬型钒钛磁铁矿的大规模利用起到了指导和借鉴作用。

关键词：高铬型钒钛磁铁矿，球团矿，烧结矿，炉料结构，高炉操作

Key Technology of Resource Comprehensive Utilization for
High-Chromium Vanadium-Titanium Magnetite
——Blast Furnace Smelting

Xue Xiangxin[1,2], Cheng Gongjin[1], Jiang Tao[1,2], Yang He[1,2], Duan Peining[1,2]

(1. School of Metallurgy, Northeastern University, Shenyang 110819, China; 2. Liaoning Key Laboratory of Recycling Science for Metallurgical Resources, Shenyang 110819, China)

Abstract: High-chromium vanadium-titanium magnetite has complex mineral compositions and fairly difficult comprehensive utilization technologies, and the industrial production had been elusive before this study. Pellet preparation and properties, sinter preparation and properties, blast furnace burden structure and operation, etc, were investigated for the key technology of resource comprehensive utilization for high-chromium vanadium-titanium magnetite in this paper. The technological leapfrogging from laboratory investigations to industrial tests and applications was realized, providing the guidance and reference for the utilization of high-chromium vanadium-titanium magnetite on a large scale.

Key words: high-chromium vanadium-titanium magnetite, pellet, sinter, furnace burden structure, blast furnace operation

宝钢湛江钢铁熔剂性球团稳定生产实践

梁利生，周　琦，易陆杰

（宝钢湛江钢铁有限公司，广东湛江　524072）

摘　要：本文总结了宝钢湛江钢铁 500 万 t/a 链箅机-回转窑生产线生产碱性球团矿的情况，通过对配矿结构、原料处理、造球控制和链箅机热工制度等不断的探讨和优化，目前已实现熔剂性球团矿的连续稳定生产。

关键词：链箅机-回转窑，熔剂性球团，生产操作

Production Performance of Grate-Kiln Fluxed Pellet in Baosteel Zhanjiang Iron & Steel Plant

Liang Lisheng, Zhou Qi, Yi Lujie

(Baosteel Zhanjiang Iron & Steel Co., Ltd., Zhanjiang 524072, China)

Abstract: There a grate-kiln pellet process in Baosteel Zhanjiang Iron & Steel base, which is designed annual fluxed pellet production of 5 million tons. This paper briefly introduced the adjustment of raw materials blending, pellting control and optimizing of thermal regulation for steady production of fluxed pellet. So far, grate-kiln pellet process in zhanjiang base has kept smooth and steady condition, and high quality of fluxed pellet been achieved.

Key words: grate-kiln, fluxed pellet, production operation

强混对制粒过程的影响研究

毛晓明，熊　林，李　建

（宝山钢铁股份有限公司研究院，上海　201999）

摘　要： 为优化制粒效果，通过对比不同制粒阶段的粒度组成、粘附粉粘附率、粘附层强度、化学成分标准差、原始颗粒行为和显微形貌研究了强混对混合料制粒过程的影响。研究结果表明：强混不加水会改变制粒过程，使原本已粘附在粗颗粒周围的细粉物料被剥落下来，导致粘附粉粘附率和低强度粘附层厚下降，但强混对混合料混匀效果明显，且对最终制粒结果的粘附粉粘附率没有影响，对高强度粘附层厚度影响也不大，在制粒过程中总粘附层厚度的升高主要源于高强度粘附层不断增加。强混对各手筛粒级粉/核比影响程度不同，大粒级和中小粒级受影响较大，中大粒级受影响较小。强混可以改变铁矿石颗粒的粘附层结构，使铁矿石颗粒有机会粘附更多的熔剂和燃料，从而有利于较大颗粒铁矿石，特别是褐铁矿烧结成矿。

关键词： 烧结，强混，制粒，粘附粉

Study on the Effect of Strong Mixing on Granulation Process

Mao Xiaoming, Xiong Lin, Li Jian

(Research Institute, Baoshan Iron & Steel Co., Ltd., Shanghai 201999, China)

Abstract: In order to optimize the effect of granulation, the effects of strong mixing on the granulation process of the mixture were studied by comparing the particle size composition, layering particles adhesion rate, adhesive layer strength, standard deviation of chemical composition, original particle behavior and microstructure at different granulation stages. The result shows that strong mixing without adding water will change the granulation process，stripping the fine powder material that had been adhered to the coarse particles, Cause layering particles adhesion rate and low adhesion strength of the layer thickness decrease. But the effect of mixing on the mixture is obvious, and the adhesion rate of the adhesive powder to the final granulation effect is not affected, and the effect on the thickness of the high-strength adhesive layer is not significant, during the granulation the total thickness of the adhesive layer increases is mainly due to the high strength of the adhesive layer increased. The effect of strong mixing on powder and core ratio of the hand sieve grades is different, Large grain size and medium and small grain size are affected greatly, medium and large grain size is less affected. Strong mixing can change the structure of the iron ore particles adhesion layer, so that iron ore particles have the opportunity to adhere more flux and fuel, which is conducive to larger particles of iron ore, especially limonite sintering mineralization.

Key words: sintering, strong mixing, granulation, layering particles

鞍钢鲅鱼圈近年来烧结生产工艺的技术进步

王宝海，杨熙鹏，孙俊波，尹冬松

（鞍钢股份有限公司鲅鱼圈钢铁分公司，辽宁营口　115007）

摘　要： 鲅鱼圈烧结自开工以来围绕优质、环保、低碳、降本等方面开展了诸多项的技术改造。先后实施了低成本配矿烧结生产技术；冶金工厂含铁废弃物综合处理工艺；烧结烟气脱硫绿色环保项目；烧结环冷废气余热发电和主抽风机采用变频控制降低电耗等节能项目；更换新型烧结机头、尾密封盖板降低系统有害漏风改善烧结矿质量和降低烧结矿电耗的技术改造。实现了优质、低碳，绿色烧结生产。

关键词： 低成本，绿色环保，工业废弃物，密封盖板，工序能耗

Technical Advances for Sintering Production in Bayuquan Branch of Ansteel in Recent Years

Wang Baohai, Yang Xipeng, Sun Junbo, Yin Dongsong

(Bayuquan Iron & Steel Subsidiary Company of Angang Steel Co., Ltd., Yingkou 115007, China)

Abstract: In order to decrease the energy consumption, a series of measures were taken by Bayuquan Iron & Steel Subsidiary Company of Angang Steel Co. in recent years, consisting of getting electric power by use of the waste gas of the circular cooler for preheating the sinter mixture, doing the ore blending for sintering by adding the carbon-bearing industrial waste, upgradingthe enclosing cover plates at both ends of the sintering machine to reduce the detrimental airleakage of the sintering system so that the energy consumption in sintering process was decreasedgreatly and thus the effect of energy saving is obvious.

Key words: sintering, hot waste gas, industrial waste, sealing cover plate, energy consumption in process

球团矿三维结构重构及表征研究

王　炜，郑　恒，徐润生，周进东，欧阳泽林，杨代伟

（武汉科技大学省部共建耐火材料与冶金国家重点实验室，湖北武汉　430081）

摘　要： 本文基于序列切片-三维重建法提出了一种球团矿三维结构表征方法。通过对球团矿微观结构的三维重建，分析了球团矿矿物在三维空间的形貌结构和分布特征。结果表明，球团矿不同区域的矿物分布状态不同。该球团主

要矿相为互连较好的赤铁矿，磁铁矿含量较少。球团矿的良好抗压强度与互连赤铁矿良好的连接程度密切相关。

关键词：球团矿，微观结构，三维重建，形貌

Study on 3D Reconstruction and Characterization of Pellet

Wang Wei, Zheng Heng, Xu Runsheng, Zhou Jindong, Ouyang Zelin, Yang Daiwei

(Key Laboratory for Ferrous Metallurgy and Resources Utilization of Ministry of Education,
Wuhan University of Science and Technology, Wuhan 430081, China)

Abstract: A new analysis method based on the serial sectioning and three-dimensional (3D) reconstruction has been developed for characterization of pellet three dimensional structures. The morphology and spatial distribution of the minerals in the three-dimensional space are analyzed by the 3D reconstruction of the iron ore pellet. The results show the morphology and spatial distributions of different minerals varied with the different positions. The pellet contained a large amount of hematite and a small amount of magnetite. The interconnection state of the hematite was quite good. The good compressive strength of the pellet significantly related to its good interconnection hematite.

块矿与烧结矿混合炉料软融性能测定新方法

刘新亮 [1,2]，Honeyands Tom [1,2]，O'Dea Damien [3]，李国玮 [4]

（1. 纽卡斯尔大学纽卡斯尔能源与资源研究院炼铁原材料研究中心，澳大利亚纽卡斯尔　2308；
2. 纽卡斯尔大学 ARC 澳大利亚铁矿石先进技术研究中心，澳大利亚纽卡斯尔　2308；
3. 必和必拓，澳大利亚布里斯班　4000；4. 必和必拓，中国上海　200021）

摘　要：提升高炉内块矿比例能有效地降低铁水成本、能量消耗及环境污染，但关于块矿软融性能弱于烧结矿的认知限制着块矿的使用比例。研究已表明块矿和烧结矿间交互作用能显著改善混合炉料的软融性能，因此仅采用块矿自身软融性能对其进行评价是不恰当的。本研究表明：（1）交互作用使混合炉料软融温度区间变窄、透气性变好，其软融性能要明显优于单一块矿和烧结矿；（2）块矿比的升高带来烧结矿碱度的增加，这能降低烧结过程燃耗、提升烧结效率、改善烧结矿强度和还原性等，进而能改善整个高炉的表现而不仅仅是软融带；（3）与致密块矿和烧结矿相比，澳洲多孔质块矿（如纽曼混合块矿）与烧结矿间交互作用更强、软融性能更好，因此要更注重混合炉料的性能；（4）气孔率是影响块矿还原性和交互性的关键因素之一，在900℃以下块矿气孔率随还原度增加而线性升高。另外，可采用同步辐射计算机断层扫描技术考察料柱的孔隙结构以解析料柱压差与收缩率间的关系。

关键词：块矿，交互作用，软融性能，气孔率，同步辐射计算机断层扫描技术

New Techniques to Measure Softening and Melting Properties of Mixed Burdens of Lump Ore and Sinter

Liu Xinliang [1,2], Honeyands Tom [1,2], O'Dea Damien [3], Li Guowei [4]

(1. Centre for Ironmaking Materials Research, Newcastle Institute for Energy and Resources, The University of Newcastle, Callaghan, NSW 2308, Australia; 2. ARC Research Hub for Advanced Technologies for Australian Iron

Ore, The University of Newcastle, Callaghan, NSW 2308, Australia; 3. BHP Billiton, Brisbane, QLD 4000, Australia; 4. BHP Billiton China, Shanghai 200021, China)

Abstract: Increasing the lump ore proportion in the blast furnace (BF) burden is an effective way to decrease the overall cost of hot metal, environmental impact and energy consumption of ironmaking. The perception that the softening and melting (SM) properties of lump ore are not as good as sinter is limiting its utilisation in some blast furnaces in China. However, it is not appropriate to evaluate lump ore based only on its individual SM behaviour as studies have shown that the SM properties are significantly improved when lump ore and sinter interact. In this work, SM properties of individual lump ores, sinters and mixed burdens are investigated to examine the influence of this interaction. The SM properties of mixed burdens are usually better than both individual lump ore and sinter due to their favourable interaction, leading to a narrower and more permeable cohesive zone (CZ) in the BF. Adding more lump ore in the burden also allows the sinter basicity to be higher which has the additional benefits of decreasing the fuel rate and increasing productivity in the sinter plant and improving sinter strength and reducibility enabling better performance in the whole BF, not only the CZ. Furthermore, this study found that the performance of mixed porous Australian hematite-goethite lump ore like Newman Blend Lump and sinter appears to be better than mixed dense lump ore and sinter because of the higher interactivity of the former. Thus, more attention should be given to the properties of mixed burdens of lump ore and sinter rather than individual materials. Porosity is a key factor influencing the reducibility and interactivity of lump ore and it increases linearly with reduction degree below 900℃. In addition, synchrotron computer tomography (CT) scanning can be used to study the voidage structure of ferrous material during SM under load testing to interpret the fundamentals of pressure drop and bed contraction.

Key words: lump ore, interaction, softening and melting, porosity, synchrotron computer tomography scanning

铁酸钙系熔体在 TiO_2 上的润湿行为

杨明睿，吕学伟，白晨光，魏瑞瑞

（重庆大学，重庆 400044）

摘　要：铁矿石烧结是一个依靠液相产生和再结晶来实现颗粒粘结的高温物理化学过程。在含钛矿烧结中，TiO_2 会与最初的粘接相铁酸钙熔体反应生成钙钛矿，减少了粘结液相量，从而降低烧结矿品质。本研究采用改进的座滴法探索了铁酸钙系熔体 1250℃时在 TiO_2 上的润湿行为。结果表明：平衡接触角小于 25°，铁酸钙渣系在 TiO_2 基片上具有良好的润湿性。铁酸钙渣系在 TiO_2 表面上的润湿行为属于界面反应驱动润湿，界面上钙钛矿的形成是主要驱动力。熔体中添加 MgO 可以提高铺展速率，添加 Al_2O_3 和 SiO_2 会降低铺展速率。润湿结束后，钙钛矿和钛铁矿在界面处富集。

关键词：润湿，接触角，铁酸钙系熔体，界面反应

Wetting Behavior of Calcium Ferrite Based Slag on TiO_2

Yang Mingrui, Lv Xuewei, Bai Chenguang, Wei Ruirui

(Chongqing University, Chongqing 400044, China)

Abstract: Iron ore sinter is a high-temperature physicochemical process that depends on the formation and crystallization of liquid phase to bond the ore particles. In the titanium-bearing sintering, TiO_2 will react with the initial bonding phase calcium ferrite to produce perovskite, reducing the amount of adhesive liquid phase, thereby reducing the sinter quality. In this study, the wetting behavior of calcium ferrite based slag at 1250℃ was investigated by modified sessile drop method. The results show that the equilibrium contact angle is less than 20°, and the calcium ferrite slag has good wettability on the TiO_2 substrate. The wetting behavior of calcium ferrite slag on the surface of TiO_2 belongs to the interface reaction driven wetting, and the formation of perovskite at the interface is the main driving force. Adding MgO to the melt increases the spreading rate, and the addition of Al_2O_3 and SiO_2 reduces the spreading rate. After the wetting, the perovskite and ilmenite are enriched at the interface.

Key words: wetting, contact angle, calcium ferrite based slag, interfacial reaction

近年来湘钢烧结配矿技术探讨

周云花

（湖南华菱湘潭钢铁有限公司，湖南湘潭 411101）

摘 要： 本文总结了湘钢近十年烧结矿质量改进与配矿进步的关系，分析了矿石价格、高炉长寿、有害元素、配矿技术等因素对配矿进步的促进和影响。探讨了今后配矿的基本思路。

关键词： 烧结矿，质量，配矿

基于焦炉煤气强化烧结的富氧喷吹技术研究

韩凤光 [1,2]，李和平 [2]，许力贤 [3]，毕传光 [2]，谢朝晖 [2]，聂慧远 [2]

（1. 中南大学，湖南长沙 410083；2. 上海梅山钢铁股份有限公司，江苏南京 210039；
3. 东北大学，辽宁沈阳 110819）

摘 要： 在梅钢技术中心对"焦炉煤气强化烧结技术"研究的基础上，进行了基于焦炉煤气强化烧结的富氧喷吹技术研究。通过试验探索，在烧结台车上的一定面积范围内喷洒焦炉煤气的基础上喷吹氧气，观察其对烧结矿质量的影响。试验结果表明，在基于焦炉煤气强化烧结的基础上喷吹氧气对改善烧结矿质量具有一定作用。

关键词： 富氧，焦炉煤气，强化，烧结，新技术

钒钛矿块矿煤基回转窑直接还原过程

秦 晋，柳政根，储满生，唐 珏

（东北大学冶金学院，辽宁沈阳 110819）

摘 要： 钒钛矿块矿属于高钒高钛型钒钛磁铁矿，该矿产资源拥有巨大的利用价值，探索其有价组元的高效利用具有重要意义。回转窑直接还原-熔分工艺是目前处理钒钛矿重要技术。本文在实验室条件下研究了钒钛矿块矿煤基回转窑直接还原过程，考察了还原温度和还原时间对还原产物金属化率和外部形貌的影响及等温还原过程物相迁移

规律。实验结果表明，还原温度 1150℃、还原时间 4 h 时还原产物的金属化率达到 85%以上，且表面未出现熔化现象，满足生产要求；等温还原过程中，钛铁氧化物相变历程：$Fe_9TiO_{15} \rightarrow Fe_{2.75}Ti_{0.25}O_4 \rightarrow Fe_{2.5}Ti_{0.5}O_4 \rightarrow FeTiO_3 \rightarrow Fe$，而钒、铬铁氧化物被还原为氧化物。

关键词：钒钛矿，块矿，还原过程，回转窑，物相转变

Coal-Based Direct Reduction Process of Vanadium Titanium Lump Ore in Rotary Kiln

Qin Jin, Liu Zhenggen, Chu Mansheng, Tang Jue

(School of Metallurgy, Northeastern University, Shenyang 110819, China)

Abstract: The vanadium titanium lump ore, enriched with high vanadium and high titanium, has great value to be utilized. Exploring the high utilization of its valuable components is of great importance. Rotary kiln direct reduction-melting process is an important technology currently to deal with vanadium titanium ore. In this paper, coal-based direct reduction process of vanadium titanium lump ore in rotary kiln was experimentally investigated under laboratory conditions. The effects of reduction temperature and reduction time on the metallization ratio and the morphology of reduction products was researched. Simultaneously, the phase transfer in the isothermal reduction process was conducted. The results showed that when the reduction temperature is 1150℃ and the reduction time is 4 h, the metallization ratio is over 85%, and no melting phenomenon was observed on the surface of the reduction products, which meets the requirements of the production. In isothermal reduction process, the phase transition process of titanium iron oxides is $Fe_9TiO_{15} \rightarrow Fe_{2.75}Ti_{0.25}O_4 \rightarrow Fe_{2.5}Ti_{0.5}O_4 \rightarrow FeTiO_3 \rightarrow Fe$, however, the vanadium and chromium iron oxides are reduced to oxides.

Key words: vanadium titanium ore, lump ore, reduction process, rotary kiln, phase transfer

CO 还原 FeO 过程金属铁析出微观形态原位观察研究

丁智勇，钟怡玮，郭占成

（北京科技大学钢铁冶金国家重点实验室，北京　100083）

摘　要：本文利用超高温共聚焦显微镜原位观察和电子探针扫描，研究 CO 还原 FeO 反应过程中（FeO→Fe）金属铁的析出形态以及结构演变的规律，明确金属铁的析出形态与还原条件之间的关系。结果表明升高温度有利于层状铁的形成，而降低温度有利于铁晶须的生长。这是由于随着温度的升高，反应速率明显加快并且大于界面扩散速率。在 85%CO+15%CO_2 的还原气氛下，700℃时金属铁析出的主要形式是铁晶须，随着温度升高逐渐会有层状铁生成。在 CO 还原 FeO 的过程中，随着反应的进行首先会在 FeO 基底生成一个浓度梯度，浓度梯度厚度大约为 1μm。随着铁原子在浓度梯度中不断扩散，铁晶核逐渐生成。

关键词：非高炉炼铁，气基还原，原位观察，铁晶须

In Situ Observation on Micro-Morphology of the Precipitation of Metallic Iron in the Reduction of FeO by CO

Ding Zhiyong, Zhong Yiwei, Guo Zhancheng

(State Key Laboratory of Advanced Metallurgy, University of Science and Technology Beijing, Beijing 100083, China)

Abstract: In this paper, in situ observation of the precipitation morphology and evolution structure of iron in the reduction of FeO by CO (FeO→Fe) were investigated by high-temperature confocal scanning laser microscopy and electron probe micro-analyzer. The relationship between the precipitation morphology of metallic iron and reduction conditions was established. The results showed that the increase of temperature was favorable to the formation of laminated iron, while the decrease of temperature was favorable to the growth of iron whiskers. The reason was that the reaction rate was obviously accelerated and greater than the interface diffusion rate with increasing temperature. Under the 85% CO+15% CO$_2$ reducing atmosphere, the main form of metallic iron precipitation was iron whisker at 700℃. Laminated iron was gradually generated as the temperature increased. In the reduction of FeO by CO, with the reaction a concentration gradient generated on FeO base at first, and the concentration gradient thickness was about 1μm. As the iron atoms diffused in the concentration gradient, iron crystal nucleus gradually generated.

Key words: non-blast furnace ironmaking, gaseous reduction, in situ observation, iron whisker

宝钢智能化高炉炼铁软件系统研发实践

黄建波，毛晓明

（中国宝武钢铁集团中央研究院，上海　201900）

摘　要：本文介绍了宝钢股份公司近几年来在智能化高炉炼铁软件系统研发方面的实践和进展。我们研发的全高炉瞬态过程模拟软件已具备自动建模，自动网格划分，高效，多功能，全自动数据处理及计算结果图形生成的功能。该软件在普通单 CPU 笔记本电脑上用 3 到 4 小时即可完成高炉一组工况的计算模拟及结果数据可视化。采用了国际公认，技术相对成熟的多相，多流体计算流体动力学模型，包含了从炉喉到炉身，再到炉缸的全高炉流动过程。气态，液态，固态三相。并将煤气流，球团，烧结，焦炭，铁水，熔渣等分别当作不同的流体。数学模型考虑了各相，各流体的运动，传热和相变过程，以及各流体间的相互作用。对不同操作条件下料面变化过程，炉内及炉缸气，固，液三相的速度，温度分布及变化过程进行了详细的描述。通过对铁水相变过程的模拟，预测不同冷却制度下炉缸内壁凝铁层的形成及熔化过程，达到炉内全过程可视化的目的。本系统对优化高炉操作，确保高炉稳定顺行，降低能源消耗及炼铁成本，减少污染物排放具有重要作用。

关键词：智能化炼铁，高炉炼铁，高炉数值模拟，高炉多流体模型，计算流体动力学

On Development of a Smart Iron-making Software System

Huang Jianbo, Mao Xiaoming

(Central Research Institute, China Baowu Iron & Steel Group, Shanghai 201900, China)

Abstract: This paper introduces the R&D work for the development of s smart iron-making software system, taking place at Central Research Institute, Baosteel Co., Ltd. The software system is based on the framework of a transient, multiphase, multi-fluid model for computational simulation of blast furnace processes. It possesses the capacity of automatic model setup, automatic meshing, automatic results sampling and visualization. It is highly efficient to run with multiple functionality. On an ordinary laptop with single CPU, it takes 3 to 4 hours to complete a blast furnace (BF) simulation case, including meshing, simulation and results post-processing. The software is developed on the framework of the internationally-recognized multi-phase, multi-fluid CFD models for blast furnaces. It includes modelling of the whole internal fluid-dynamics process, from BF top to hearth flows. The model involves all three phases, liquid, gas and solid, all the main BF burden types, coke, sinter, lump ore and pellets, iron and slag, are simulated. The software can also predict the formation and disappearance of the protective solid iron layer in the vicinity of hearth walls, under different cooling conditions, thus visualize the whole heat transfer and fluid flow processes taking place in the BF. Application of the software will help BF operators to ensure smooth and stable BF operations, reduce energy consumption and ironmaking costs as well as emission of pollutants, making the BF ironmaking process greener and more economic.

Key words: smart ironmaking, blast furnace ironmaking, blast furnace numerical simulation, computational fluid dynamics

我国大型高炉生产现状分析及展望

张寿荣[1]，姜　曦[2]

（1. 武汉钢铁（集团）公司，湖北武汉　430083；2. 中国钢铁工业协会，北京　100711）

摘　要： 4000m³ 以上大型高炉是高炉炼铁先进技术发展方向的体现，大型高炉相对于容积小的高炉，单位炉容投资经济、能耗低、环境负荷低、劳动生产率高。2000 年以来，随着钢铁需求的迅猛增加，一批 4000m³ 以上大型高炉在我国相继投入运行，在生产管理和操控技术等方面取得了明显的成绩和效果。本文对我国 4000m³ 以上大型高炉的装备水平、原燃料指标、技术经济指标等方面进行了分析，对存在的问题提出了改进建议，并对炼铁的发展进行了展望。

关键词： 大型高炉，生产指标，展望

Production and Development of Large Blast Furnaces in China

Zhang Shourong[1], Jiang Xi[2]

(1. Wuhan Iron and Steel (Group) Corp., Wuhan 430083, China;
2. China Iron and Steel Association, Beijing 100711, China)

Abstract: The large blast furnaces with volume larger than 4000m³ is the concentration represents of new techniques in ironmaking blast furnace. Compared to the blast furnace with small volume, blast furnace with larger volume has lower economic investment, smaller energy consumption and environmental load, higher labor productivity. A large number of 4000m³ blast furnaces were built when the steel demands increased dramatically after the year of 2000. In the efforts of ironmaking operators, the operation indexes of large blast furnace have reached advanced levels. In this paper, the operation situation of large blast furnace was analyzed, while some suggestions for existing problems were proposed. In addition, the tendencies of ironmaking in future were prospected, hoping it could improve the development of ironmaking in China.

Key words: large blast furnace, production indexes, prospection

莱钢 1#1080m³ 高炉排碱操作研究

孙建设，王庆会

（山东莱芜钢铁股份有限公司炼铁厂，山东莱芜　271104）

摘　要：本文主要介绍了莱钢 1#1080m³ 高炉 Zn 元素及碱金属富集炉况表现，通过分析引起 Zn 元素及碱金属富集的原因，探索高炉排碱操作期间高炉操作制度，并对排碱过程进行跟踪，归纳排碱结果得出在保证渣铁热量的前提下，降低炉渣碱度、控制合适炉温，有利于提高排碱效率，为同类型相关操作提供了借鉴案例。
关键词：高炉，碱金属，炉渣碱度

Study and Summarization of Alkali in Laiwu Steel 1# 1080m³ Blast Furnace

Sun Jianshe, Wang Qinghui

(Laiwu Steel Corporation Ironworks, Laiwu 271104, China)

Abstract: The reasons of Zn element and alkali metal enrichment in the 1#1080m³ blast furnace of Laiwu Steel were analyzed. The blast furnace operation system during the operation of blast furnace alkali was explored, and the process of rowing was analyzed , The results of the analysis of the summary of starch, for the future operation of the summed up the experience.
Key words: blast furnace, lalkali metal, slag basicity

宝钢三号高炉炉缸侧壁侵蚀微观剖析和侵蚀机理研究

王训富[1,2]

（1. 上海大学材料科学与工程学院，上海　200444；
2. 宝山钢铁股份有限公司研究院，上海　200941）

摘　要：停炉后的宝钢三号高炉根据工况分成风口区域、铁口区域、铁口上侧壁区域和铁口下侧壁区域，通过化学分析、SEM、XRD、EDS 和其他手段，分析炉缸不同部位残留炭砖的化学成分和微观结构，发现不同区域的侵蚀机理存在显著差异。以 ZnO 形式存在的 Zn 元素沉积在炭砖热面表面区域，没有显著的证据显示 Zn 会对炭砖侵蚀产生影响。除了铁口区域外，K.Na.Fe 元素在炭砖中的含量分布从热面到冷面逐渐降低，这降低了炭砖的气孔率，增加了密度和热传导率。Fe 元素含量越高的区域，其侵蚀越严重，这是因为 Fe 元素的渗入显著的改变了炭砖的性能。在铁口区域，由于液态渣铁的机械冲刷，Si、Al 较为集中，SiO_2 和 Al_2O_3 颗粒由于温度的波动导致与炭砖基体存在显著的热膨胀差异。在侵蚀最为严重的 H4 段发现石墨化 C 的存在表明该处的铁水可能与炭砖发生了直接接触。
关键词：高炉侧壁，化学成分，微观结构，侵蚀

Corrosion Mechanism Research and Microstructure Analysis of Baosteel No.3 Blast Furnace Hearth

Wang Xunfu[1,2]

(1. School of Materials Science and Engineering, Shanghai University, Shanghai 200444, China;
2. Baoshan Iron & Steel Co., Ltd., Shanghai 200941, China)

Abstract: Baosteel No.3 Blast Furnace hearth was divided into the tuyere area, the taphole area, the taphole upper side wall and the taphole lower side wall according to different working situations. Through the chemical composition analysis, SEM, XRD, EDS and other means, the chemical composition and microstructure of different parts of hearth carbon brick were analyzed and it was found that the corrosion mechanism of these areas was remarkably different. Zn element in form of ZnO mainly deposited on the hot side of carbon brick. There was no obvious evidence that Zn permeated into carbon bricks and eroded them. Except for taphole area, K,Na,Fe contents from the hot side to the cold side gradually increased and decreased, which results in apparent decreasing in porosity, increasing in density and higher thermal conductivity than new carbon brick.The higher content of Fe in carbon bricks lead to more serious erosion because Fe has greatly changed the physical properties of carbon bricks. In the taphole area, the contents of Si,Al presented obvious concentration gradient because of the mechanical souring of molten iron and slag. SiO_2 and Al_2O_3 particles that have different thermal expansion factors from carbon bricks damaged the carbon substrate because of temperature fluctuation. Graphitized carbon found on H4 where there is the most serious corrosion indicating that the carbon brick ever directly contacted with molten iron.
Key words: blast furnace side wall, chemical components, microstructure, corrosion

中国高炉铜冷却壁破损机理分析

焦克新 [1,2]，张建良 [1,2]，郭光胜 [1,2]，杨天钧 [1,2]

（1. 北京科技大学冶金与生态工程学院，北京　100083；
2. 北京科技大学钢铁冶金新技术国家重点实验室，北京　100083）

摘　要：高炉炉腰、炉腹及炉身下部应用铜冷却壁是保障高炉中部区域安全长寿的重要手段。而近年来铜冷却壁破损的案例逐渐增多。本文结合湘钢高炉铜冷却壁的破损状况，对铜冷却壁的破损形式、显微形貌进行分析，然后运用 ANSYS 数值模拟软件对铜冷却壁在不同工况条件下的温度场进行了计算，并对铜氧化物在高炉内的还原行为进行了热力学分析。结果表明，铜冷却壁破损主要是由于下降炉料膨胀对炉墙产生较大的压应力，造成渣皮脱落，温度升高的铜冷却壁强度降低，最终使得铜冷却壁磨损。本文还提出了高炉炉型自适应指数，表征炉型可接受炉料的膨胀（收缩）程度，包括炉身膨胀指数和炉腹收缩指数，并对中国部分高炉的炉型自适应指数进行了计算。
关键词：高炉，铜冷却壁，磨损，热力学，炉型自适应指数

湛江钢铁 2 号高炉高产低耗生产实践

王志宇，沙华玮

（宝钢湛江钢铁有限公司炼铁厂，广东湛江 524072）

摘　要： 本文主要对湛江钢铁 2 号高炉一段时间内的高炉生产稳定顺行情况进行介绍。湛江钢铁二号高炉在产量爬坡生产过程中稳定顺行良好，煤比逐步上升，燃料比降低，各项经济技术指标优异，已处于世界大型高炉先进水平。入炉原料采取"适当质量"原则，寻求高性价比，但原燃料质量仍有较大改善空间，严格控制有害元素入炉负荷。高炉操作者精心操业，合理平衡高炉四大制度并合理安排炉前出渣铁作业，确保生产出优质铁水。同时，湛江钢铁注重人员及设备管理，统一各班正确操作思路，提高每个员工的责任感，为 2 号高炉长期稳定、顺行、高产、低耗、环保及长寿等打下坚实的基础。

关键词： 高炉，高产低耗，原燃料，管理

High Production and Low Consumption Practice of No.2 Blast Furnace in Baosteel Zhanjiang Iron and Steel

Wang Zhiyu, Sha Huawei

(Ironmaking Plant, Baosteel Zhanjiang Iron and Steel Co., Ltd., Zhanjiang 524072, China)

Abstract: This article mainly introduces the stable production of blast furnace in No. 2 blast furnace of Zhanjiang Iron & Steel Co., Ltd. for a period of time. The BF condition was well during ramp production with excellent economic and technical indexes, and The ratio of coal increases gradually with the fuel ratio decrease. Raw materials into the furnace to take the "appropriate quality" principle with cost-effective and control harmful elements into the furnace strictly. Operators operate seriously with balance the blast furnace with four major systems and arrange slag out reasonable to ensure the production of high quality molten iron. At the same time, Zhanjiang Iron & Steel focuses on personnel and equipment management with consolidate the correct operation ideas of each class and improve the sense of responsibility of each employee to lay a solid foundation for long-term smooth and stable, high production, low consumption, environmental protection and long campaign of No. 2 blast furnace.

Key words: blast furnace, high production and low consumption, raw materials, management

碳复合砖在高炉炉缸炉底的应用实践

宁晓钧[1]，马洪修[1]，张建良[1]，吴森然[1]，焦克新[1]，郝良元[2]，
武会卿[3]，赵永安[4]

（1. 北京科技大学冶金与生态工程学院，北京 100083；2. 河钢集团钢研总院，河北石家庄 050000；
3. 山西通才炼铁厂，山西临汾 041000；4. 河南五耐集团实业有限公司，河南巩义 451200）

摘　要：高炉炉缸作为高炉长寿的限制性区域，承载着高温铁水和高温炉渣的侵蚀，与其直接接触的砖衬必须在具有高导热性的同时还具有较好的抗铁水溶蚀性、抗水蒸汽的氧化性和抗锌侵蚀性等性能，碳复合砖结合了优质炭砖和陶瓷杯的优点具备这样的性能，综合性能较佳，且已经在高炉炉缸区域得到了广泛的应用，本文对使用碳复合砖的通才3#、高义3#和4#、柳钢5#高炉进行了跟踪调查，分析了使用碳复合砖高炉的原燃料情况以及生产参数，计算了高炉炉缸侧壁砖衬的残余厚度，评估了其安全状态；综合分析后认为：碳复合砖在低品位矿高冶炼强度条件下有较好的抗侵蚀能力，以上跟踪的4座高炉生产稳定顺行，炉缸炉底传热体系稳定通畅，温度梯度分布合理，砖衬侵蚀量极小，证明了碳复合砖在炉缸炉底使用的安全性和长寿性，具有较大的推广使用价值。

关键词：高炉，炉缸，安全长寿，碳复合砖

高球团比生产的操作制度优化探讨

张培峰

（宝山钢铁股份有限公司总部炼铁厂，上海　200941）

摘　要：本文结合球团矿的布料特性及软熔特性，分析了高球团矿比生产时，操作制度的优化途径。关键措施是：（1）调整适宜的风口面积，控制合理的风量，确保炉腹煤气指数在60～65m/min、炉缸回旋区长度≥2.3m、炉身平均煤气流速在2.6～2.9m/s。（2）降低料线，调整布料矩阵，拓宽平台到2.1～2.3m，调整炉料排出顺序和切出量，减少边缘球团矿和块矿量，增加平台球团矿量和焦炭量。（3）控制炉温在中下限，并力求稳定。

关键词：操作制度，高球团比，低燃料比，大型高炉

Research of Optimizing of Operation System with Higher Ratio of Pellet Production

Zhang Peifeng

(Ironmaking Plant, Baoshan Iron & Steel Co., Ltd., Shanghai 200941, China)

Abstract: The optimization approach of operation system with higher ratio of pellet production is analyzed in this paper, according to the characteristics of pellet in the process of burden distribution, soften and melting. The major measures is described: (1) Adjusting suitable tuyere area and controlling reasonable blast volume with the aim of keeping the bosh gas volume index at 60~65 m/min, annular length of hearth raceway greater than or equal to 2.3m, and average gas flow-rate of furnace stack at 2.6~2.9m/s. (2) Reducing stuff line and adjusting burden distribution matrix to form platform width at 2.1~2.3m, adjusting discharge sequence and weight of burden, reducing the weight of pellet and lump ore in the edge, and increasing the weight of pellet and coke in the platform. (3) Controlling pig temperature at lower limit, and ensuring stable.

Key words: operation system, higher ratio of pellet, low fuel ratio, large sized blast furnace

本钢北营新 2 号高炉降低燃料比操作实践

王光亮，陈正林，刘　森

（北营炼铁厂新 2 炉作业区，辽宁本溪　117000）

摘　要：本钢北营新 2 号高炉通过加强原料管理、优化上下部制度及日常操作、组织好出铁，高炉炉况稳定顺行，煤气利用率逐步提高到 49.5%以上，燃料比降低到 510kg/t 以下。

关键词：大型高炉，操作制度，燃料比

The New North 2[#] Blast Furnace to Reduce Fuel Ratio Operation Practice

Wang Guangliang, Chen Zhenglin, Liu Sen

(North 2[#] Blast Furnace Ironmaking Plant of New Operation Area, Benxi 117000, China)

Abstract: The new North 2[#] Blast Furnace by strengthening raw material management, lower system optimization and daily operation, organize tapping, The stable operation of blast furnace, The gas utilization rate gradually increased to more than 49.5%, Fuel ratio decreased to less than 510kg/t.

Key words: large blast furnace, operation system, fuel ratio

湘钢 2580m³ 高炉经济冶炼问题的探讨

但家云

（湖南华菱湘潭钢铁有限公司，湖南湘潭　411101）

摘　要：通过引入炉腹煤气指数的概念，分析煤气体积变化对高炉冶炼的影响，对比湘钢高炉目前生产运行数据，提出对湘钢高炉改善和强化的建议。

马钢 4 号高炉强化冶炼实践

王志堂，聂长果，张　群，赵淑文，陈　军

（马鞍山钢铁股份有限公司，安徽马鞍山　243000）

摘　要：对马钢4号高炉开炉后的强化冶炼实践进行了总结。通过做好精料入炉工作、优化高炉操作、加强炉型管理以及出渣铁管理等措施，使开炉后的高炉冶炼水平不断提高，获得了较好的生产技术经济指标。

关键词：大型高炉，原燃料，强化冶炼

Practice of Intensified Smelting in No. 4 Blast Furnace of Maanshan Iron & Steel Co., Ltd.

Wang Zhitang, Nie Changguo, Zhang Qun, Zhao Shuwen, Chen Jun

(Maanshan Iron & Steel Co.,Ltd., Maanshan 243000, China)

Abstract: In the practice of intensified smelting in Masteel No.4 BF are summarized. By doing fine material into the furnace, optimizing BF operation, strengthen the management and the type of furnace slag and iron management measures, the level of blast furnace smelting in the production technology continues to improve, obtained better economic index.

Key words: large-sized blast furnace, raw material, fuel intensified smelting operation

梅钢2高炉喷吹焦炉煤气风口回旋区的数值模拟

唐　珏[1]，储满生[1]，毕传光[2]，孙俊杰[2]

（1. 东北大学冶金学院，辽宁沈阳　110819；2. 梅钢技术中心，江苏南京　210039）

摘　要：基于梅钢2高炉生产操作参数和原燃料条件，系统研究了不同焦炉煤气喷吹量对理论燃烧温度、炉腹煤气量、炉腹煤气组成、回旋区形状、回旋区焦流量的影响。结果表明，在保持梅钢2#高炉现有操作不变的情况下，从风口喷吹焦炉煤气，回旋区温度降低，炉腹煤气流量增加。为了保持良好的炉缸热状态和维持稳定的风口回旋区条件，需增大富氧对回旋区进行热补偿；热补偿后，当焦炉煤气喷吹量为50m³/thm时，富氧率为9.98%，炉腹煤气中总还原气量为55.29%，回旋区体积为0.78m³，焦炭流量为17.547kg/s。每增加1m³/thm的焦炉煤气喷吹量，富氧率增加0.11%，总还原气量增加0.16%，回旋区体积增大0.0022m³，焦炭流量增大0.0366kg/s。

关键词：梅钢高炉，风口喷吹焦炉煤气，回旋区，数值模拟

Numerical Simulation of Meishan Steel No.2 Blast Furnace Raceway with Coke Oven Gas Injection

Tang Jue[1], Chu Mansheng[1], Bi Chuanguang[2], Sun Junjie[2]

(1. School of Metallurgy, Northeastern University, Shenyang 110819, China;
2. Meishan Steel Technology Center, Nanjing 210039, China)

Abstract: Based on the conditions of the Meishan Steel No.2 blast furnace, the effects of the different coke oven gas (COG) injection on theoretical flame temperature, bosh gas volume, composition of bosh gas, raceway shape, and coke mass flow rate had been researched systematically. It was shown that the theoretical flame temperature tends to decreasing and bosh gas volume tends to increasing, when COG injection volume increasing with the benchmark operation of blast furnace unchanged. The theoretical flame temperature can be constant by increasing oxygen enrichment when blast furnace operating with COG injection. After thermal compensation, when the flow rate of COG injection was 50m³/thm, the oxygen enrichment rate was 9.98%, the total reduction gas in bosh gas was 55.29%, the volume of raceway was 0.78m³, and the coke mass flow rate was 17.547kg/s. With 1m³/thm increasing of COG injection, the oxygen enrichment rate increased by 0.11%, the total reduction gas in bosh gas increased by 0.16%, the volume of raceway increased by 0.0022m³, and the coke mass flow rate increased by 0.0366%.

Key words: Meishan Steel bastfrnace, COG injection from tuyere, raceway, numerical simulation

一种新型铸铁冷却壁综合性能分析

吴振威[1]，张建良[1]，焦克新[1]，沈　猛[2]，铁金艳[2]

（1. 北京科技大学冶金与生态工程学院，北京　100083；

2. 河北天宇高科冶金铸造有限公司，河北肃宁　062350）

摘　要： 冷却壁作为高炉的重要部件，对高炉的长寿起着重要的作用，基于开发的新型椭圆形水管铸铁冷却壁，展开以下研究：基于 FLUENT 以及 ANSYS 软件，建立铸铁冷却壁三维传热模型以及三维热应力模型，比较了不同椭圆度的椭圆形冷却水管的冷却效果，同时与热态试验进行对比，并分析了不同椭圆度的椭圆形冷却水管铸铁冷却壁的应力场；利用数值模拟对铸铁冷却壁结构优化进行研究。结果表明：与圆形截面相比，椭圆形水管能提高冷却壁的换热能力，当椭圆短长轴之比 b/a 取 0.4~0.5 时，冷却壁的冷却效果最好；而椭圆短长轴之比 b/a 取 0.4 时，冷却壁热应力较小；利用热态试验得到的数据对数值模拟结果进行验证，计算结果与试验实测值基本吻合，表明数值模拟能满足工程研究的需要；椭圆形水管的短长轴之比在 0.2~0.3 时水流阻力太大，应取 0.4~0.5。

关键词： 椭圆形水管，冷却壁，温度场，热态试验，应力场，数值模拟，优化

Analysis of Comprehensive Performance of a New Type Cast Iron Cooling Stave

Wu Zhenwei[1], Zhang Jianliang[1], Jiao Kexin[1], Shen Meng[2], Tie Jinyan[2]

(1. Metallurgical and Ecological Engineering School, University of Science and Technology Beijing, Beijing 100083, China; 2. Hebei Tianyu High-Tech. Metallurgical Casting Co., Ltd., Suning 062350, China)

Abstract: The cooling stave is an important part of blast furnace, it plays an important role in the longevity of the blast furnace, and based on the development of a new type elliptical cast iron cooling stave, the following studies are carried out: The three-dimensional steady-state heat transfer model and the three-dimensional thermal stress model of a cast iron stave is built based on FLUENT and ANSYS software, the cooling effects of different kinds of elliptical pipes are compared, at the same time, the numerical simulation results are compared with the thermal test data, and the stress field of different kinds of elliptical pipes are analysed; The optimization of cast iron cooling stave was studied by numerical simulation. The results showed that compared with a circular cross section, the elliptical cross section can increase the cooling ability of the stave, when the elliptical semi-minor axis and semi-major axis ratio b/a takes 0.4~0.5, the stave has the best cooling performance; and when b/a takes 0.4, the thermal stress of the cooling stave is smaller. The numerical simulation results are in good agreement with the thermal test data, indicating that the numerical simulation can meet the requirements of engineering research; when b/a within the range of 0.2~0.3, there is too much resistance, so it is should be kept at 0.4~0.5.

Key words: elliptical pipe, cooling stave, temperature field, thermal test, stress field, numerical simulation, optimization

韶钢 3200m³ 高炉热风炉及预热器运行状态分析与评估

杨国新，廖经文，柏德春，曾宗浪，李鲜明，凌志宏

（宝武集团韶关钢铁公司炼铁厂，广东韶关　512123）

摘　要： 通过对 8 号高炉热风炉及预热器实际运行情况、运行状态分析，得出热风炉本体是从 2015 年 1 月开始衰退，目前热风炉状态对比 2015 年，在同等能耗水平下，风温下降了 6℃，能耗上升了 0.05GJ/t。热管预热器因设计上存在空气、煤气的换热能力偏小，导致换热效率能力受限。预热器投入后的第 4 个月即 2013 年 8 月就出现了严重的质量问题，质量问题长期制约预热器的换热效率。考虑热风炉工艺特性及对风温的影响，提出了先进行预热器更换，后续进行热风炉本体改造，达到在提高 8 号炉风温水平的同时并降低热风炉能耗的改进思路。

关键词： 热风炉，预热器，运行状态，分析，评估

Shaogang 3200m³ Hot Blast Furnace and Preheater Operation State Analysis and Evaluation

Yang Guoxin, Liao Jingwen, Bai Dechun, Zeng Zonglang, Li Xianming, Ling Zhihong

(Baowu Group Shaoguan Iron and Steel Company Ironworks, Shaoguan 512123, China)

Abstract: Based on the analysis of the actual operation and operation status of hot blast stove and preheater of blast furnace No. 8, it is concluded that the hot blast furnace body is declining from January 2015, and the current state of hot stove is compared with 2015. Under the same energy consumption level, The air temperature dropped by 6℃ and the energy consumption increased by 0.05GJ/t. Heat pipe preheater due to the design of air, gas heat transfer capacity is too small, resulting in limited heat transfer efficiency. Preheater after the first four months in August 2013 there have been serious quality problems, quality problems long-term constraints preheater heat transfer efficiency. Considering the characteristics of the hot air stove and the influence on the air temperature, it is put forward that the preheater is replaced and the reform of the hot blast furnace will be carried out to improve the energy consumption of the blast furnace at the same time and reduce the energy consumption of the hot blast furnace.

Key words: hot blast stove, preheater, operating state, analysis, evaluation

八钢 2500m³ 高炉原燃料思想与技术的进步

元婷婷，张文庆

（宝钢集团八钢炼铁分公司第二炼铁分厂，新疆乌鲁木齐　830022）

摘　要： 阐述了为应对市场资源的劣化，八钢原燃料条件发生的变化，通过配煤比的调整，细化原燃料管理，转变炉况管理思路，稳定生产，有效地降低铁水成本。

关键词： 高炉，原燃料，稳定生产

Abstract: In response to the deterioration of marker resources and original fuel condition change,by adjusting the blengding ratio,refine the original fuel management and shift dealted management idea and stable production,reduce the cost of hot metal effectively.

Key words: the blast furnace, raw material, production

宝钢1号高炉增加氧量鼓风冶炼实践

王　波，宋文刚，朱锦明，华建明

（宝钢股份炼铁厂，上海　200941）

摘　要：高炉增加氧量鼓风是高炉的重要技术发展方向，具有强化冶炼、提高利用系数、促进间接还原、利于喷煤等优势，但同时具有理论燃烧温度过高、初始气流分布变化、高温区下降、鼓风带入热量降低等诸多因素影响高炉的冶炼进程。本文主要总结1号高炉增加增加氧气量鼓风情况下高炉操作参数变化特点、指标及成本经济合理性，从而为今后宝钢高炉操作及公司科学决策提供基本依据。

关键词：增加氧量，炉缸侧壁，高炉操作，基本制度

Practice of Improving the Oxygen Enrichment Rate in Baosteel No. 1 Blast Furnace

Wang Bo, Song Wengang, Zhu Jinming, Hua Jianming

(Ironmaking Plant, Baoshan Iron & Steel Co., Ltd., Shanghai 200941, China)

Abstract: Oxygen-enriched blast is an important technology development direction of blast furnace. It has the advantages of strengthening smelting, improving utilization coefficient, promoting indirect reduction and facilitating coal injection. But at the same time, the theoretical combustion temperature is too high, the initial air flow distribution is changed, high temperature area decreased,blast into the heat reduction and many other factors affect the smelting process of blast furnace. This paper summarizes the characteristics, indicators and economic rationality of blast furnace operating parameters under the situation of increasing the oxygen enrichment rate, which will provide the basic basis for Baosteel blast furnace operation and scientific decision-making.

Key words: oxygen enrichment, hearth sidewall, blast furnace operation, fundamental systems

我国钢铁工业能耗现状与节能潜力分析

王维兴

（中国金属学会，北京　100010）

摘　要：文章指出了我国钢铁工业能源利用情况，我国吨钢综合能耗偏高的主要原因是，我国铁钢比高；分析了

2016 年中钢协会员单位有关工序能源利用情况，及相关指标。按主要生产工序分析了其节能潜力，分析有关工序节能技术装备应用情况和指标，指出了钢铁工业节能工作的重点在炼铁系统（占总用能耗的 70%左右），要努力降低炼铁燃料比，提高热风温度，提高入炉矿铁品位等。

关键词：钢铁工业，能耗现状，节能潜力，分析

Abstract: The paper points out the energy utilization of steel industry in China, and the main reason for the high comprehensive energy consumption of tonnage in China is that the iron steel ratio is higher than that in China. This paper analyzes the energy utilization and related indexes of the member units of China steel association in 2016. According to the main production process analyses the energy saving potential of energy saving technology and equipment application and analysis about the working procedure, and points out the focus of the iron and steel industry energy conservation in ironmaking system (about 70% of the total in energy consumption), try to reduce ironmaking fuel ratio, improve the hot blast temperature, improve the grade of iron ore into the and so on.

Key words: iron and steel industry, present situation of energy consumption, energy saving potential, analysis

立式强力混合机及其在烧结工艺中的应用

卢兴福，刘克俭，戴　波

（中冶长天国际工程有限责任公司，湖南长沙　410205）

摘　要：针对具有高效混匀作用的立式强力混合机，仿真研究指出其混匀机理是搅拌桨机械扰动下对流、剪切和扩散三种混合方式共同作用实现高效混匀的过程，在此基础上分析指出桨叶磨损变短是混合机的主要失效形式，并提出了射流防磨和结硬边界控制的防磨降磨技术。分析了立式强力混合机在烧结工艺中的优势，提出在烧结使用立式强力混合机的三段混合制粒、粗细筛分强化制粒以及强化混匀制粒三种工艺技术。

关键词：立式强力混合机，混匀机理，磨损失效，射流防磨，结硬边界控制，烧结

Vertical Intensive Mixer and Its Application in Sintering

Lu Xingfu, Liu Kejian, Dai Bo

(Zhongye Changtian International Engineering Co., Ltd., Changsha 410205, China)

Abstract: In view of the vertical intensive mixer which has high-efficiency mixing effect on bulk powder, the simulation research showed the mixing mechanism of the vertical intensive mixer that under mechanical disturbance of the stirring paddle, is the mixing process which contains three kinds of mixing modes of convection, shearing and diffusion. On this basis, the main failure mode is that the blades get shorter for wear. The jet protection of wear and control of agglomerate boundary technology are put forward to protect and reduce wear. The advantage of using vertical intensive mixer in sintering is analyzed. The three stage mixing granulation, the sizing intensifying granulation and the intensifying mixing granulation technology are proposed to using vertical intensive mixer in sintering.

Key words: vertical intensive mixer, mixing mechanism, wear failure, jet protection of wear, control of agglomerate boundary, sintering

HIsmelt 熔融还原工艺工业化最新进展

曹朝真，孟玉杰，梅丛华，章启夫，李　勇，毛庆武

（北京首钢国际工程技术有限公司，
北京市冶金三维仿真设计工程技术研究中心，北京　100043）

摘　要：我国首座 HIsmelt 熔融还原工业化炼铁厂的建成投产，标志着 HIsmelt 工艺工业化进程步入了新的发展阶段。介绍了中国 HIsmelt 工厂的主要工艺流程和技术指标；结合澳大利亚 kwinana 工厂的生产实践，重点从改善炉缸耐材寿命、提高能源利用效率、提高工艺稳定性和生产效率等几个方面，论述了 HIsmelt 工艺工业化过程中取得的关键技术突破，对 HIsmelt 技术的发展和推广进行了展望。

关键词：HIsmelt，熔融还原，工业化进展，关键技术突破，发展前景

Latest Progress in Industrialization of HIsmelt Process

Cao Chaozhen, Meng Yujie, Mei Conghua, Zhang Qifu, Li Yong, Mao Qingwu

(Beijing Shougang International Engineering Technology Co., Ltd., Beijing Metallurgical 3-D Simulation Design Engineering Technology Research Center, Beijing 100043, China)

Abstract: It indicates that the industrialization of HIsmelt process has entered a new stage of development that the commissioning of the first HIsmelt smelting reduction commercial ironmaking plant in China. The main process flow and technical indexes of Molong plant are introduced. New key technological breakthroughs of HIsmelt process during the industrialization are discussed, from the aspects of prolonging the hearth refractory campaign life, improving energy efficiency, improving process availability and productivity etc.. The development and popularization of HIsmelt technology in future are prospected.

Key words: HIsmelt, smelting reduction, industrialization progress, key technical breakthrough, development prospect

铁矿烧结过程 NOx 减排研究

春铁军[1]，李东升[1]，余正伟[1]，王　臻[1]，龙红明[1,2]

（1. 安徽工业大学冶金工程学院，安徽马鞍山　243002；2. 冶金减排与资源综合利用教育部重点实验室，安徽工业大学，安徽马鞍山　243002）

摘　要：铁矿烧结工序是现代钢铁生产流程中必不可少的环节，也是钢铁行业最大的 NOx 排放源，占总排放量的 55%左右。本文阐述了铁矿烧结过程 NOx 的生成机理及其排放特征，分别从原料控制、过程控制及末端控制三个方面介绍了作者关于烧结过程 NOx 减排的相关研究，并提出了烧结过程 NOx 减排研究的发展方向。

关键词：铁矿烧结，NOx 减排，源头控制，过程控制，末端治理

Study on NOx Emission Reduction in Sintering Process of Iron Ore

Chun Tiejun[1], Li Dongsheng[1], Yu Zhengwei[1], Wang Zhen[1], Long Hongming[1,2]

(1. School of Metallurgical Engineering, Anhui University of Technology, Maanshan 243002, China;

2. Key Laboratory of Metallurgical Emission Reduction & Resources Recycling, Ministry of Education, Anhui University of Technology, Maanshan 243002, China)

Abstract: Iron ore sintering process is an indispensable part of modern steel production process. It is also the largest source of NOx in iron and steel industry, accounting for about 55% of total emissions. In this paper, the generation mechanism and emission characteristics of NOx in the sintering process of iron ore were also carried out. The research progress of NOx emission reduction in sintering process was introduced from three parts: raw material control, process control and end treatment, and the development direction of NOx emission reduction in sintering process was put forward.

Key words: iron ore sintering, NOx emission reduction, source control, process control, end treatment

钢铁企业原料场能耗分析与节能设计研究

毕　琳

（中冶京诚工程技术有限公司，北京　100176）

摘　要：本文首先按照原料场工艺流程介绍了钢铁企业原料场主要能耗作业情况，同时阐述了原料场能耗指标评价项目；其次依据原料场工艺作业的特点和功能配置，对重要耗能指标项目内容等进行了分析；探讨了伴随原料场工程技术的进步，钢铁企业对原料准备工序节能要求的提高，如何研究原料场能耗影响因素及开展节能设计，更好地实现原料场功能性和节能设计先进性。

关键词：能耗，节能，指标，原料场，设计

Analysis for Energy Consumption and Energy Conservation Study of Raw Material Yard of Iron and Steel Plants

Bi Lin

(Capital Engineering & Research Incorporation Limited, Beijing 100176, China)

Abstract: This paper briefly introduces the major energy consumption items and operations of metallurgical raw material yard. According to the features and function of metallurgical raw material yard, makes a analysis and induction of energy consumption items. It is also discussed that conditions of raw material yard process design are the foundation of index and value evaluation. Accompany by the progress of the engineering technology, energy conservation index are built up. Energy conservation design and function design of metallurgical raw material yard is realized.

Key words: energy consumption, energy conservation, index, raw material yard, design

日照钢铁600m²烧结余热直罩式锅炉应用技术

沈玉冰，代俊楠

（日照钢铁有限公司炼铁制造部，山东日照 276800）

摘 要： 烧结余热发电锅炉工艺新技术—直罩式烧结余热锅炉，成功应用于日照钢铁 600 m² 烧结机余热发电项目，并于 2014 年 7 月份投产运行，至今已运行近一年。本文重点介绍直罩式余热锅炉的总体布置、技术优势、运行情况以及发电增效。直罩式余热锅炉投资省，占地小，结构简单紧凑，能降低散热损失和阻力损失，有效提高余热锅炉产汽量和出口蒸汽温度，解决典型烧结余热发电技术的不足之处，大幅度提高烧结冷却机废热的热能利用率，节能效果和经济效果显著。据生产统计，2014 年 7 月～2015 年 5 月该项目累计发电 1.1×10^8 kWh，累计上网 8.3×10^7 kWh，获利 5388 万元，效益十分可观。

关键词： 烧结，余热利用，直罩式锅炉，新技术

Bell Type Sintering Waste Heat Boiler and Its Production of Rizhao Steel 600m² Sinter Machine

Shen Yubing, Dai Junnan

(Rizhao Steel Holding Group Co., Ltd. Ironmaking Department, Rizhao 276800, China)

Abstract: The waste heat generation project of Rizhao steel 600m² sinter machine was commissioned in July 2014, and it has been in operation for about one year.The general conditions of the project and production condition after the commissioning of the project were intro duced, including bell type sintering waste heat boiler system, steam water system , waste gas system and steam turbine system.According to the production statistics, the accum ulated generated electricity from July of 2014 to May of 2015 was 1.078×10^8 kWh, and accumulated electricity on the grid was 8.290×10^7 kWh, the profit was 1.078×10^8 RMB Yuan, and the benefits are substantial In addition, the energy saving and reduction emission effects are very good.

Key words: utiliza tion of sintering waste heat, waste heat power generation, energy saving, bell type sintering waste heat boiler

碳氢铁浴还原浸入式侧吹搅拌数值模拟研究

王东彦，姜伟忠，朱锦明，华建明，王士彬

（宝山钢铁股份有限公司炼铁厂，上海 200941）

摘 要： 运用多相流 VOF 模型，对碳氢铁浴熔融还原炉浸入式侧吹搅拌进行了数值模拟研究。研究结果表明，单层喷枪侧吹搅拌效果随喷吹流量增加而快速增加，而在同样喷吹流量下，双层喷枪侧吹搅拌效果显著高于单层。侧

吹搅拌效果随喷枪层数和气流量的提高程度可由数值模拟计算确定。

关键词：熔融还原炉，侧吹搅拌，多相流模型，数值模拟

Numerical Simulation Study on the Behavior of Submersive Side-blowing Stirring in Hydrocarbon Iron-bath Smelting Reduction Furnace

Wang Dongyan, Jiang Weizhong, Zhu Jinming, Hua Jianming, Wang Shibin

(Ironmaking Plant Baoshan Iron & Steel Co., Ltd., Shanghai 200941, China)

Abstract: The multiphase flow VOF model was used to simulate the behavior of submersive side - blowing stirring in hydrocarbon iron-bath smelting reduction furnace . The results show that the blowing effect of the single - layer spray gun increases rapidly with the increase of the blowing flow rate, while the double - layer spray gun side - blowing stirrer is significantly higher than the single layer under the same jet flow rate. The degree of stirring effect increase determined by the number of spray gun layers and the amount of gas flow can be figured out by numerical simulation.

Key words: side - blowing stirring, smelting reduction, mathematical simulation, VOF model

烟气循环烧结工艺综述及其在宝钢应用的探讨

周茂军，张代华

（宝山钢铁股份有限公司，上海 200941）

摘 要： 本文对烟气循环烧结原理、烧结烟气特点、烧结气流介质变化影响等方面进行了综合阐述，介绍了国内外的烟气循环烧结工程应用业绩，并从工艺特点、应用情况、减排及节能效果、产品质量影响等方面进行了比较，分析了几种典型烟气循环烧结工艺的优缺点。继而结合宝钢烧结系统综合改造项目情况，提出了宝钢烧结应用烟气循环烧结工艺的框架性方案。

关键词： 烧结，烟气循环，烧结烟气，减排

Review of Flue Gas Circulation Sintering Process and Its Application in Baosteel

Zhou Maojun, Zhang Daihua

(Ironmaking Plant Baoshan Iron & Steel Co., Ltd., Shanghai 200941, China)

Abstract: In this paper, the principle and characteristics of the flue gas circulation sintering process were described systematically, as well as the influencing of sintering flue gas medium change. The application performances of flue gas circulation sintering projects with different processes at home and abroad were also introduced, and compared in terms of process feature, in terms of business, in terms of emission reduction and save energy, in terms of production and quality. The advantages and disadvantages of typical flue gas circulation sintering processes were analyzed. Then, based on the comprehensive modification project of Baosteel sintering system, a framework scheme for sintering flue gas sintering process in Baosteel was put forward.

Key words: sintering, flue gas circulation, sintering flue gas, emission reduction

冶金粉尘资源化清洁再利用关键技术研究与示范应用

刘德军[1]，张洪宇[2]，贾　振[1]，郝　博[1]

（1. 鞍钢集团钢铁研究院，辽宁鞍山　114091；2. 鞍钢股份有限公司炼铁总厂，
辽宁鞍山　114000）

摘　要：阐述了冶金粉尘再资源化科学利用的重要性；通过对当前冶金除尘灰各种再利用方式的特点进行详细分析与甄别，阐明了"除尘灰与粉煤同喷"利用方式的科学性；重点就鞍钢首创的冶金粉尘高炉综合喷吹技术对冶金粉尘的种类及量等因素的普适性、含铁除尘灰等冷料下料量的工程级精准配料、高炉喷吹含铁除尘灰回旋区热补偿实现手段、消除喷吹除尘灰输送管道堵塞技术以及整个工艺过程无二次扬尘绿色技术等进行了详细的论述，同时阐明了鞍钢首创的冶金粉尘高炉综合喷吹技术对冶金粉尘具有最佳性。
关键词：冶金除尘灰，再资源化，科学利用，集成技术

Study and Demonstration Application of Key Technology of Recycling Clean Recycle of Metallurgical Dust

Liu Dejun[1], Zhang Hongyu[2], Jia Zhen[1], Hao Bo[1]

Abstract: It's stated that it's important to utilize scientifically metallurgical dusts recycling. After analyzing and distinguishing features of different kinds of metallurgical dusts, we concluded that it's scientific of injecting metallurgical dust and pulverized coal into burst furnace together. In the paper, it's expounded on confirming the kind and amount of dusts injected into BF, the accurately mixture in engineering of the materials such as dusts with iron injected into BF, thermal compensation of dusts with iron in the blast furnace raceway, avoidance of blocking the tunnel as injecting into the BF and the green environment technology without second generate dust in the whole process. Injecting metallurgical dusts into the BF applied firstly in Ansteel and the best technology.
Key words: metallurgical dust, recycling, science utilization, integration technology

高炉用焦炭高温性能探索

胡德生，孙维周，钱　晖，毛晓明

（宝山钢铁股份有限公司研究院，上海　201999）

摘　要：本文简要叙述了高炉用焦炭质量的影响因素，炼焦煤资源及其特性是影响焦炭质量的根本因素，配煤和炼焦工艺也是重要因素，焦炭在高炉内行为认识和评价方法是最终影响因素。简要分析了高炉内焦炭行为及消耗过程，指出了影响消耗的因素，主要是矿石的还原性，反应失重率取决于直接还原度和未燃煤粉率。解析了高炉内焦炭破坏产生粉化的因素，目前焦炭质量指标中只有机械强度能表征焦炭在高炉内机械磨损破坏，传统方法的反应性和反应后强度不能表征高炉内焦炭溶损和气化破坏，其它破坏因素还没有表征指标。介绍了四种不同焦炭高温性能探索

试验，结果表明焦炭高温反应后强度与传统试验方法的反应后强度无关，主要取决于焦炭的冷态强度，炭质结构和灰成分内在影响因素。

关键词：高炉，焦炭，高温性能，探索

The Exploration of Coke's Property in High Temperature for Blast Furnace

Hu Desheng, Sun Weizhou, Qian Hui, Mao Xiaoming

(Research Institute, Baoshan Iron & Steel Co., Ltd., Shanghai 201999, China)

Abstract: This paper briefly describe the influence factor of coke quality in Blast Furnace,coking coal resource and characteristic is the fundamental factor for coke quality, blending coal and coking process are also the important factor,but the recognition and evaluation method of coke in Blast Furnace is the decisive factor.this paper simply analyze the coke's behavior and consumption process in Blast Furnace,and consider the ore's reduction property is the primary factor for coke's consumption ,moreover the coke's reaction loss in Blast Furnace depend on direct reduction and unburned pulverized coal rate.it also analyze the factors of coke degradation process in Blast Furnace,and put forward that only coke mechanical strength index can simulate the coke mechanical destruction process in Blast Furnace among the coke quality index in present, the traditional index of CRI and CSR cannot simulate the degradation due to carbon resolution reaction in Blast Furnace,and other degradation factor also cannot be simulate yet.4 different coke high temperature experiment were carried out in this paper,the results show that the strength after high temperature reaction is unrelated with CSR,it chiefly depend on coke mechanical strength,carbon texture, and ash's influence.

Key words: blast furnace, coke, high temperature property, exploration

铁焦气化反应和反应后强度与钾的关系

毕学工[1]，李　鹏[1,3]，刘　威[1]，张慧轩[1]，周进东[1]，史世庄[2]

（1. 武汉科技大学耐火材料与冶金国家重点实验室，湖北武汉　430081；2. 武汉科技大学煤转化与先进炭素材料湖北省重点实验室，湖北武汉　430081；3. 中冶南方，湖北武汉　430081）

摘　要： 弄清铁焦受碱金属影响的气化行为对其实际应用于高炉非常重要。本文中，通过将试样在 K_2CO_3 水溶液中浸泡和煮沸把钾引入铁焦，焦炭失重随时间的变化用热重法测定，用粒焦法测定焦炭的反应性和反应后强度，并基于收缩核模型（SCM）对气化反应机理进行分析。研究证明钾对铁焦气化有催化作用，表现为使其气化开始温度和反应活化能降低。与传统焦炭比较，当加入 K_2O 时，铁焦的反应性指数增大而反应后强度变小，而且当加入相等数量的 K_2O 时，铁焦的这两个指标的变化趋势和传统焦炭相同，但变化幅度比传统焦炭稍微大一些。与传统焦炭的另一不同之处是，当温度从 950℃提高到 1050℃而且气化率不超过 50%的时候，铁焦的反应始终是化学反应和内扩散的混合控制。

关键词：铁焦，钾，气化反应，收缩核模型，热性能

Relationship between Potassium and Gasification Reaction /Post-Reaction Strength of Ferro-coke

Bi Xuegong[1], Li Peng[1,3], Liu Wei[1], Zhang Huixuan[1], Zhou Jindong[1], Shi Shizhuang[2]

(1. State Key Laboratory for Refractory Materials and Metallurgy, Wuhan University of Science and Technology, Wuhan 430081, China; 2. Key Laboratory for Ferrous Metallurgy and Resources Utilization of Ministry of Education, Wuhan University of Science and Technology, Wuhan 430081, China; 3. Key Laboratory for Coal Conversion and Advanced Carbon Materials of Hubei Province, Wuhan University of Science and Technology, Wuhan 430081, China)

Abstract: To clarify the behavior of its gasification affected by alkalis is very important for its practical use in blast furnace. In this paper, potassium was added to sample by soaking and boiling it in K_2CO_3 aqueous solution, the variation of coke weight loss vs time was measured with the thermogravimetry approach, the Tablets Coke Method was applied for measurement of the reactivity and post-reaction strength of cokes, and the gasification reaction mechanism was analyzed based on the Shrinking-Core Model (SCM). This work demonstrated that potassium had a catalytic effect on ferro-coke gasification in terms of a decrease in the starting reaction temperature and reaction activation energies. In comparison with traditional coke, when no K_2O was added, the reactivity index of ferro-coke was higher and the post-reaction strength index of ferro-coke was lower, while as a same amount of K_2O was added and in respect to these two indices, the direction of variation of ferro-coke was the same and the extent of variation of ferro-coke was slightly higher. Different from traditional coke, when temperature ranges from 950 to 1050℃ and the degree of burning-off is within 50%, the reaction of ferro-coke is always mixed controlled by chemical reaction and inner diffusion.

Key words: ferro-coke, potassium, gasification reaction, shrinking-core model, thermal properties

炼焦煤炭化关联性和焦炭综合热性能评价新方法

汪 琦[1]，梁英华[2]，赵雪飞[1]

（1. 辽宁科技大学辽宁省化学冶金重点实验室，辽宁鞍山　114051；

2. 华北理工大学河北省煤化工工程技术研究中心，河北唐山　063210）

摘　要：介绍了"炼焦煤炭化关联性评价方法"和"焦炭综合热性能评价方法"，分别用于检测单种炼焦煤及配合煤在炭化过程中挥发分析出、胶质层生成和膨胀压力之间的关联性；检测焦炭溶损反应开始温度、溶损速率、等溶损率（25%）反应后强度及其对温度的依赖、反应后焦炭热缩聚指数和热强度。这两种方法可以进一步深入辨识炼焦煤和焦炭的性能，建立新的质量评价方法。

关键词：炼焦煤，焦炭，炭化关联性，综合热性质

New Evaluation Methods for Related Properties during Coking Coal Carbonization and the Comprehensive Thermal Performance of Coke

Wang Qi[1], Liang Yinghua[2], Zhao Xuefei[1]

(1. The Key Laboratory of Chemical Metallurgy Engineering of Liaoning Province, University of Science and

Technology Liaoning, Anshan 114051, China; 2. Hebei Province Coal Chemical-Engineering Technology Research Center, North China University of Science and Technology, Tangshan 063210, China)

Abstract: The evaluation methods for "the related properties during carbonization of coking coal" and "the comprehensive thermal performance of coke" were developed. The former one was used to measure the relevance between the volatile matter escaping, plastic layer forming, and swelling pressure during single and blending coking coal carbonization process; the latter one was used to detect the initial temperature of coke reaction, the average coke reaction rate and the coke strength after reaction at the coke weight loss up to 25%, the dependence of the coke strength after reaction on the reaction temperature, the coke polycondensation index and coke strength after polycondensation for the coke weight loss up to 25%. These two methods can be used to further identify the performances of coking coal and coke, and establish some new evaluation methods of coking coal and coke quality.

Key words: coking coal, coke, related properties during carbonization, comprehensive thermal properties

高炉焦炭行为及其对炼焦新技术的指导

李克江，孙敏敏，张建良，刘征建，王广伟，焦克新，杨天钧

（北京科技大学冶金与生态工程学院，北京　100083）

摘　要： 节约能源和控制 CO_2 排放是未来钢铁产业的发展趋势，焦炭的使用使得炼铁和炼焦也面临着前所未有的压力。在有限的炼焦煤资源和严格的环境制度的双重压力下，基于焦炭行为和近年来高炉内部焦炭质量要求，炼铁和炼焦操作对改善炼焦技术提出了更高的要求。本文为高炉内焦炭的行为做出了系统的概述，揭示了高炉对焦炭性能的具体要求，以及提高焦炭质量的炼焦技术。分析结果表明，设计能够满足高炉生产要求的焦炭对于稳定高炉生产至关重要。

关键词： 炼体，炼焦，焦炭质量，焦炭行为，结构

Coke Behavior Inside Blast Furnace and Its Indication to New Cokemaking Techniques

Li Kejiang, Sun Minmin, Zhang Jianliang, Liu Zhengjian, Wang Guangwei, Jiao Kexin, Yang Tianjun

(School of Metallurgical and Ecological Engineering, University of Science and Technology Beijing, Beijing 100083, China)

Abstract: The future potential and pressure of energy-saving and CO_2 emission in iron and steel industry lie on the ironmaking and cokemaking processes which are connected by the utilization of coke. Under the twin pressure of limited coking coal resources and stringent environmental regulations, ironmaking and cokemaking operations have joined forces to improve the cokemaking techniques based on coke behavior and quality requirement inside blast furnace in recent years. This paper presents a comprehensive overview of the recent advancements in uncovering coke behavior inside blast furnace, understanding specific requirements of blast furnace for coke function, as well as development of cokemaking techniques to improve coke properties. This literature survey has shown that it is essentially important to design the coke for meeting properties required in the blast furnace based on its specific function.

Key words: ironmaking, cokemaking, coke quality, coke behavior, structure

焦炉用燃料燃烧控制分析与改进

王 超，庞克亮，武 吉，朱庆庙

（鞍钢股份有限公司技术中心，辽宁鞍山 114009）

摘 要： 焦炉加热用燃料燃烧主要能量损失与空气过剩系数有关，结合鞍钢炼焦股份炼焦总厂焦炉燃料燃烧过程中空气过剩系数偏大的实际情况，分析空气过剩系数的影响因素和控制关键难点，提出多点测量改进措施并实施。现场测定结果表明：与改进前相比，各焦炉烟气空气过剩系数均由最开始的 1.45~1.75 下降到 1.26~1.33 左右，煤气燃烧更趋充分，燃烧趋于合理；燃烧过程中氮氧化物生成减少，6m 焦炉废气中氮氧化物含量由下降比例达到 17.60% 以上；7m 焦炉废气中氮氧化物下降比例达到 20.22% 以上；降低吨焦耗热量 5.0% 以上，节能降本效果显著。

关键词： 焦炉，空气过剩系数，烟气控制，多点测量

Analysis and Improvement of Combustion Control of Coke Oven Fuel

Wang Chao, Pang Keliang, Wu Ji, Zhu Qingmiao

(Technology Center of Angang Steel Co., Ltd., Anshan 114009, China)

Abstract: The main process of the energy loss in coke oven heating combustion is associated with excess air coefficient, considering the actual situation of excess air coefficient is large in General Cokemaking Works of Angang Steel Co., Ltd., the key elements and factors influencing the excess air coefficient are analyzed and the multi-point measurement were put forward to implement. The testing results show that: compared with unimproved time, The excess air coefficient of 6 meters coke oven and 7 meters coke oven are fell to 1.26~1.33 from 1.45~1.75, the gas burn more fully and the gas burn more reasonable. In the process of combustion, the formation of nitrogen oxides decreased, the content of nitrogen oxides in the exhaust gas of the 6 meters coke oven decreased by more than 17.60% and the content of nitrogen oxides in the exhaust gas of the 7 meters coke oven decreased by more than 20.22%. The unit of coking heat consumption is decreased by more than 5%, the energy saving is remarkable.

Key words: coke oven, exess air coefficient, control of flue gas, multi-point measurement

宝钢焦炉烟气净化运行实践

余国普，张福行，曹银平，许永跃，支晓明

（宝山钢铁股份有限公司炼铁厂，上海 200941）

摘 要： 宝钢焦炉烟气净化系统是一种先进的全新的焦炉烟气净化工艺，采用"SDA 半干法脱硫+袋式除尘系统+前置加热炉烟温调节＋低温选择性催化还原脱硝（SCR）＋热交换器(GGH)热量循环回收"等多种工艺组合，可显著降低焦炉烟气中的 SO_2 和 NO_x 排放，达到国家最新的环保标准，同时成本、能耗控制较好。

关键词： 焦炉，烟气净化，脱硫，脱硝，能耗，成本

Practice of Flue Gas Purification Operation in Baosteel Coke Oven

Yu Guopu, Zhang Fuhang, Cao Yinping, Xu Yongyue, Zhi Xiaoming

(Ironmaking Plant Baoshan Iron & Steel Co., Ltd., Shanghai 200941, China)

Abstract: Baosteel coke oven flue gas purification system is an advanced and new coke oven flue gas purification process, using "SDA semi-dry desulfurization + bag dust removal system + front heating furnace smoke temperature adjustment + low temperature selective catalytic reduction +gas gas heater "and so on, can significantly reduce the coke oven flue gas SO_2 and NO_x emissions, to the latest national environmental standards, while the cost, energy control is better.

Key words: coke oven, flue gas purification, desulfurization, denitrification, energy consumption, cost quenching

微波联合助剂作用下高硫煤中有机硫的变迁行为研究

伍　岳，陶善龙，汪金宇，张生富，朱荣锦，张青云，白晨光

（重庆大学材料科学与工程学院，重庆　400044）

摘　要： 高硫煤中有机硫的变迁行为研究是高硫煤清洁利用技术的关键和脱硫技术的难点。本文以南桐煤（NTC）与莱钢煤（LGC）为高硫煤样，借助 NaOH 与 H_2O_2 助剂进行微波脱硫实验，并采用 X 射线衍射仪（XRD）、电感耦合等离子体质谱仪（ICP）和红外光谱（ATR-FTIR）对高硫煤中有机硫的物相转变、官能团含量变化进行检测分析。结果表明：含硫物相宏观转变过程中，有机硫转化为硫化铁、硫酸盐，制约了脱硫率提升；含硫官能团微观变迁过程中，-S-S-具备良好的微波响应能力，而 S=O 是提升脱硫率的关键要素；此外，助剂中 NaOH 与 H_2O_2 比例为1：4时脱硫效果最佳，为工业脱硫工艺提供技术借鉴。

关键词： 高硫煤，微波，有机硫，含硫官能团，变迁行为

Transformation of Organic Sulfur in Coal under Solvent-assisted Microwave Irradiation

Wu Yue, Tao Shanlong, Wang Jinyu, Zhang Shengfu, Zhu Rongjin,

Zhang Qingyun, Bai Chenguang

(College of Materials Science and Engineering, Chongqing University, Chongqing 400044, China)

Abstract: The transformation of organic sulfur in coal under solvent-assisted microwave irradiation was investigated via X-ray diffraction (XRD), Inductively coupled plasma (ICP) and Fourier transform infrared (ATR-FTIR). It was found that a certain amount of organic sulfur in Nantong coal (NTC) and Laigang coal (LGC) was transformed into iron sulfide, sulfate and sulfur-containing gas while some kinds of sulfur compound dissolved. In the two kinds of coal, the retention proportion of the three typical sulfur-containing functional groups lists as: S=O > -S-H- > -S-S-. Moreover, the 1:4 ratio of NaOH and H_2O_2 was the optimum condition for the removal of organic sulfur in both Nantong coal and Laigang coal.

Key words: high sulfur coal, microwave, organic sulfur, functional group, transformation

干熄焦装入装置技术改造与应用

王永亮，王志森，沈洪春，于进江，郭　飞

（青岛特钢焦化厂，山东青岛　266400）

摘　要：对大型干熄焦装入装置，运行状况及存在问题进行了研究与分析，对装入装置受力进行了重新分析，对驱动机构进行了技术改造，并对装入装置控制程序进行了优化，提高了装入装置运行的稳定性，延长了使用寿命。

关键词：干熄焦，装入装置，分体式，技术改造

炼焦煤预处理工艺现状及改进

Alex Wong[1]，胡德生[2]，牛　虎[1]

（1. 济南戴尔塔干燥技术有限公司，山东济南　250101；2. 宝钢股份研究院，上海　201900）

摘　要：本文简要叙述了炼焦煤预处理的主要工艺，煤调湿、煤预热、煤干燥加成型和 SCOPE21 的发展过程。重点分析综合各种炼焦煤预处理工艺开发的 SCOPE21 工艺，在此基础上，提出了改进的高温流化床工艺。

关键词：炼焦煤，CMC，DAPS，SCOPE21，高温流化床

Current Status and Improvement of Coking Coal Pretreatment Processes

Alex Wong[1], Hu Desheng[2], Niu Hu[1]

(1. Jinan Delta Drying Technology Ltd., Jinan 250101, China; 2. Research Institute of Baosteel Co., Ltd., Shanghai 201900, China)

Abstract: Development process of major coking coal pretreatment processes, CMC, Coal Preheating, DAPS and SCOPE21, was briefly described. The key analysis was focused on the development of SCOPE21, which synthesized various coking coal pretreatment processes, and then high temperature fluidized bed process was put forward on the basis of SCOPE21.

Key words: coking coal, CMC, DAPS, SCOPE21, high temperature fluidized bed

影响干熄焦焦炭烧损率的原因分析及措施

刘智江

（宝钢集团八钢公司炼铁分公司焦化分厂，新疆乌鲁木齐　830022）

摘　要：本文对干熄炉内焦炭烧损原因进行分析，进行了相关试验，并采取措施降低焦炭烧损量。

关键词：干熄焦，焦炭烧损量

The Analysis and Measure of the Causation Rate of Dry Coking Coke

Liu Zhijiang

(Baosteel Group Xinjiang Bayi Steel & Iron Company, Urumchi 830022, China)

Abstract: In this paper, the cause of coke burning loss in dry extinguishing furnace was analyzed, and the related experiments were carried out and measures were taken to reduce the coke loss.

Key words: coke dry quenching, coke burning loss rate

焦炭强度对欧冶炉冶炼的影响分析

陈若平

（宝钢集团八钢公司，新疆乌鲁木齐　830022）

摘　要：本文分析了使用低强度焦炭对八钢欧冶炉的影响，并通过国内外炼铁炉的焦炭强度使用情况进行理论分析，推导出八钢欧冶适宜的焦炭强度。

关键词：欧冶炉，焦炭强度

Effect of Coke Strength on the OY Smelting

Chen Ruoping

(Baosteel Group Xinjiang Bayi Iron and Steel Co., Urumchi 830022, China)

Abstract: In this paper, analysis the effect of low Coke-strength on the OY smelting,and theoretically analyzes the coke-strength utilization of blast furnace at home and abroad, derive the appropriate strength of coke in Bayi OY.

Key words: OY, green iron smelting

提高焦炭热反应准确率分析方法研究

黄宗仁，杜志刚，高　巍，张永丰，潘　彤

（鞍钢股份鲅鱼圈钢铁分公司化检验中心，辽宁营口　115007）

摘　要：通过对焦炭热反应制样设备和分析设备炉体加热系统的改进，使制样系统自动化、规范化。加热系统实施三段控温使恒温区由原来的±5℃控制到±3℃以内，提高了控温精度，同时通过内控样品的制备与监测，保证了设备的稳定性，从而总体提高焦炭热态强度分析的准确率。

关键词：焦炭，反应性，反应后强度，恒温区，控样

Study on the Analysis Method of Improving the Accuracy of Coke Thermal Reaction

Huang Zongren, Du Zhigang, Gao Wei, Zhang Yongfeng, Pan Tong

(Chemical Inspection Center, Bayuquan Iron and Steel Subsidiary, Angang Co.,Ltd., Yingkou 115007, China)

Abstract: The sample preparation system can realize automation and standardization by improving the heating system of sample preparation and analysis equipment of coke thermal reaction system. The heating system has three sections of temperature control, which make the constant temperature zone from the original ±5℃ to ±3℃, the temperature precision is improved. The equipment's stability is ensured by control sample preparation and monitoring and the accuracy rate of coke thermal strength analysis is improved.

Key words: coke, reactivity, post reaction strength, constant temperature zone, control sample

还原气氛和温度对 Fe₂O₃ 还原粘结强度的影响

马凯辉，廖哲晗，孙成烽，徐　扬，徐　健，温良英，白晨光

（重庆大学材料科学与工程学院，重庆　400044）

摘　要： 本文以非高炉炼铁中竖炉炉料还原粘结现象及其定量化表征为应用背景，采用 H₂ 和 CO 还原块状 Fe₂O₃，提出以剪切强度来表征还原过程中粘结的强度的方法。探究了还原气氛和温度对粘结强度的影响规律。实验结果表明：还原温度越高，粘结强度越大，最高可达 79MPa；H₂ 气氛下还原样品的粘结强度大于 CO 气氛下的还原样品，但粘结样品的韧性小于后者。

关键词：间接还原，粘结强度，还原气氛，还原温度

Effect of Reducing Composition and Temperature on Sticking Strength of Ferric Oxides after Reduction

Ma Kaihui, Liao Zhehan, Sun Chengfeng, Xu Yang, Xu Jian,

Wen Liangying, Bai Chenguang

(College of Materials Science and Engineering, Chongqing University, Chongqing 400044, China)

Abstract: In this paper, the application background are the sticking phenomenon in the reduction process of shaft furnace of non-blast furnace ironmaking and its quantitative characterization. The reduction of bulk Fe₂O₃ with H₂ and CO is proposed, and the method of characterizing the strength of sticking in reduction process is proposed. The influence of reducing atmosphere and temperature on the sticking strength is investigated. The experimental results show that the higher the reduction temperature, the greater the sticking strength, the highest up to 79MPa. The sticking strength of the reduced sample in H_2 atmosphere is larger than that in CO atmosphere, but the toughness of the bonded sample is smaller than that of the latter.

Key words: indirect reduction, sticking strength, reducing temperature, reducing composition

首钢京唐低硅烧结试验研究及应用

陈绍国[1]，裴元东[1]，石江山[2]，王同宾[2]，赵志星[1]，赵景军[2]

（1. 首钢集团有限公司技术研究院，北京　100043；2. 首钢京唐钢铁联合有限责任公司，

河北曹妃甸　063200）

摘　要： 采用烧结杯实验方法对首钢京唐铁矿粉低硅烧结及优化措施进行了研究，并进行了工业试验及应用效果分析。试验结果表明：在京唐配料结构下，烧结矿硅由 5.6 降至 5.0，烧结矿转鼓指数、成品率和粒度有所变差。粗化焦粉粒度对提高转鼓指数的效果最明显；提高混合料温度对烧结各项指标基本都有改善，尤其对提高成品率的效果最明显；提高料层厚度对降低燃耗等指标有利。在工业应用中，所采取的提高料层厚度 30mm、提高料温 5~10℃、放宽燃料粒度（<3mm 比例）2% 以及优化配矿等措施，使京唐烧结矿的铁酸钙含量基本得到保证，达到 45% 以上，保证了降硅后烧结矿产质量的稳定。

关键词： 低硅烧结，烧结杯，工业试验，应用效果

Experimental Research and Application on Low Silicon Sintering in Shougang Jingtang

Chen Shaoguo[1], Pei Yuandong[1], Shi Jiangshan[2], Wang Tongbin[2],

Zhao Zhixing[1], Zhao Jingjun[2]

(1. Technology Research Institute of Shougang Group Co., Ltd., Beijing 100043, China;

2. Shougang Jingtang United Iron and Steel Co., Ltd., Caofeidian 063200, China)

Abstract: sintering pot tests have been carried out to study low silicon sintering and optimization measures, and then industrial test and application have been done. The results showed that the sinter of drum index, yield and grain size have deteriorated when silicon in sinter decreased from 5.6 to 5.0 under the match structure of iron ores in Jingtang. It was obvious effect of coarsening coke powder size to improve sinter drum index. Each basic index of sinter were improved by increasing the temperature of blending material, especially to obviously improve the sinter yield. It was beneficial for reducing fuel consumption to increase the sintering material thickness. Moreover, in industrial practice, to improve the material thickness of 30 mm, raise temperature 5~10℃, loosen the fuel particle size (< 3 mm) 2% and optimize ore-blending, etc. That made the content of SFCA up to above 45% and guaranteed the stability of the sinter quality after reducing silicon.

Key words: low silicon sintering, sintering pot, industrial test, application effect

CO 和 H₂ 气体还原氧化铁的微观行为及新生铁晶形貌研究

鲁　峰，温良英，韩　旭，魏　晋，徐　健，张生富，白晨光

（重庆大学材料科学与工程学院，重庆　400030）

摘　要：CO 和 H₂ 及其混合气体还原氧化铁颗粒表面新生铁形貌的差异，与还原气体的吸附反应过程中需要克服的能垒、物质的迁移和混合气体间的竞争吸附行为相关。本文基于密度泛函理论计算和 Langmuir 等温吸附理论，研究了 CO 和 H₂ 混合气体还原氧化铁颗粒过程中表面新生铁形貌差异的机理过程。计算结果表明：CO 还原会对铁离子产生较大的垂直方向的牵引力，而氢气则无明显的取向作用；一个 CO 分子还原 FeO 会释放 0.18eV 的能量，一个 H₂ 分子还原 FeO 会吸收 0.94eV 的能量，而铁离子跃迁到空位所要克服的能垒为 2.41eV；等温吸附实验结果表明 CO 和 H₂ 气体还原片状 Fe₂O₃ 颗粒的热重曲线可以分为 3 个阶段，铁晶须的生长发生在 FeO→Fe 阶段；在温度为 900℃时，CO 分子在 FeO 表面的吸附反应能力比 H₂ 分子强。在此基础上，提出须状铁晶形成的判定方法和条件，并通过 SEM+EDS 分析 CO 和 H₂ 及其混合气体流态化还原片状 Fe₂O₃ 颗粒试样实验进行了验证。

关键词：氧化铁颗粒，竞争吸附，还原反应，密度泛函理论，能垒，微观行为，新生铁晶

Study on the Microscopic Behavior and Metallic Iron Morphology in the Process of CO and H₂ Fluidized Reductions for Iron Oxide Particles

Lu Feng, Wen Liangying, Han Xu, Wei Jin, Xu Jian, Zhang Shengfu, Bai Chenguang

(School of Materials Science and Engineering, Chongqing University, Chongqing 400030, China)

Abstract: The morphology of metallic iron on the surface of the iron oxide particles is different when the particles are reduced by carbon monoxide and hydrogen as well as mixture gas. The differences is related to the microstructure conversion, micro-energy transfer and adsorption behavior of the reducing gases. Based on the Density Functional Theory (DFT) and Langmuir adsorption isotherm, the mechanism of metallic iron morphology in difference are studied in the process of CO and H₂ fluidized reductions for iron oxide particles. The results are shown that the CO molecule has strongly stretched on iron ion of wustite at vertical direction in the process of reduction, but H₂ molecule has no directional force on the structure of wustite. The released energy in the process of CO reduction will be used to overcome the barrier energy of iron ion diffusion, but the energy will be consumed in the process of H₂ reduction to hinder the iron ion diffusion. The released energy using a CO molecule reduction is 0.18eV, the consumed energy using a H₂ molecule reduction is 0.94eV and the barrier energy of a iron ion diffusion is 2.41eV. The TG curves of the reduction of Fe₂O₃ layer particles can be divided into three steps, and the growth of iron whiskers happens in the step of FeO→Fe. The ability of CO molecules adsorption on the FeO surface is higher than the ability of H₂ molecules at 900℃. On the bases, a method is given to estimate the iron whiskers growth and is verified by the SEM and EDS experimental analysis of the fluidized reduction of Fe₂O₃ layer particles in the different ratio CO and H₂.

Key words: iron oxide particles, competitive adsorption, reduction, DFT, microscopic behavior, new metallic iron

红格高铬型钒钛磁铁矿球团氧化行为和固结特性

汤卫东，薛向欣，杨松陶，黄　壮

（东北大学冶金学院资源与环境研究所，辽宁沈阳　110819）

摘　要： 本文通过在不同焙烧温度和时间条件下，对红格高铬钒钛磁铁矿进行了氧化焙烧实验。根据定义的氧化率，研究了不同温度下氧化率和时间的关系。通过对球团显微组织的分析，研究了球团矿的微观结构和矿物组成规律。研究表明：HCVTM 氧化球团在氧化和未氧化分界处的结构存在显著差异。随焙烧温度的提高，低熔点液相的增加和赤铁矿晶粒的生成和再结晶，晶粒间逐渐形成连续的粘结相，空隙数量减少，直径增大。随焙烧时间的增加，赤铁矿相表现为生成，长大和再结晶，低熔点液相促进了赤铁矿晶粒间的粘接和长大。钛铁矿和钛磁铁矿随氧化焙烧时间的增加，生成铁板钛矿和钙钛矿相。

关键词： 红格高铬钒钛磁铁矿，球团，氧化行为，固结特性

Oxidized Behavior and Consolidation Characteristics of Hongge High Chromium Vanadium Titanium Magnetite Pellets

Tang Weidong, Xue Xiangxin, Yang Songtao, Huang Zhuang

(Department of Resources and Environment, School of Metallurgy,
Northeastern University, Shenyang 110819, China)

Abstract: The roasting HCVTM pellet was experimented under different temperature and time. Combing the defined oxidation rate, the relationship of oxidation rate and time with different roasting temperature was researched. The microstructures and mineral compositions of HCVTM pellets were analyzed by microscopic analysis. The results are as follows: the boundary of HCVTM pellet between oxidized and un-oxidized part was obvious. The bonding phase was generated due to increase of low melting point liquid phase, and the formation and recrystallization of hematite grain. The numbers of interspace were decreased and size of interspace was increased. Bonding and generation of hematite particles were generated with formation of liquid phase and recrystallization of hematite. Ilmenite and titanomagnetite were oxidized to pseudobrookite and perovskite with increase of roasting time.

Key words: HCVTM, pellet, oxidation behavior, consolidation characteristic

富氧对赤铁矿球团焙烧性能影响的研究与应用

王代军

（北京首钢国际工程技术有限公司，北京市冶金三维仿真设计工程技术研究中心，北京　100043）

摘　要： 赤铁矿球团富氧焙烧表明：提高助燃风含氧量至 23% 时，燃烧温度达到 1355.10℃，氧化球团的物理性能和冶金性能达到最佳。在相同的焙烧时间下，随着含氧量增加，赤铁矿球团的氧化分数先逐渐增加后逐渐减少。矿

相分析表明：氧浓度适当增加，氧的吸附、扩散速率加快，促进赤铁矿氧化球团焙烧速度加快，氧浓度过高则促使球团表面氧化反应速率过快，导致较早形成致密的外壳阻碍氧向球核内部扩散，致使球团内部来不及完全氧化。链算机-回转窑赤铁矿球团富氧焙烧应用表明：富氧焙烧有利于提高燃料燃烧效率、降低回转窑结圈，每吨氧化球团减少 2 元加工成本，氧化球团的质量及回转窑作业率得到提高。

关键词： 赤铁矿，氧化球团，富氧焙烧，冶金性能，矿相分析

Study and Application of Oxygen Enrichment on Hematite Pellets Roasting Performance

Wang Daijun

(Beijing Shougang International Engineering Technology Co., Ltd., Beijing Metallurgy Three Dimensional Simulation Design Engineering Technology Research Center, Beijing 100043, China)

Abstract: Oxygen-enriched roasting of hematite pellets showed that: when the oxygen content of combustion supporting air was increased to 23%, the burning temperature reached 1355.10℃, the physical and metallurgical properties of the oxidized pellets achieved optimum. At the same roasting time, with the increase of oxygen content, the oxidation fraction of hematite pellets were increased first and then decreased gradually. Mineral phase analysis showed that: the oxygen concentration were increased slightly, the adsorption and diffusion rate of oxygen would be accelerated, and the roasting speed of hematite oxidized pellet was accelerated too. When the oxygen concentration was too high, the surface oxidation rate of pellets would be too fast, which would lead to form a dense shell earlier and hinder the diffusion of oxygen toward the nucleus of the ball, as a result the inside of the pellets there's not enough time to be completely oxidized. The application of oxygen-enriched roasting in grate-kiln hematite pellets showed that: the oxygen-enriched roasting was beneficial to increase fuel combustion efficiency and reduce the ring formation of rotary kiln, the oxidized pellets processing costs was reduced 2 ￥/t, the quality of the oxidized pellets and the operation rate of the rotary kiln were improved.

Key words: hematite, oxidized pellet, oxygen-enriched roasting, metallurgical properties, mineral phase analysis

宝钢四烧结运行改进实践

谢学荣，鲁　健，王跃飞

（宝钢股份炼铁厂，上海　200941）

摘　要： 宝钢四烧结设计进行了不少革新，投产后取得了不少效果。本文针对投产后面临的碱度、FeO 检验数据检测滞后、皮带积料、环保悬臂型棒条筛磨损和返矿跑粗、燃料系统粒度粗和作业率低等问题、改进措施进行了概述。

关键词： 烧结，碱度，FeO

Improvement Practice of Baosteel's No.4 Sintering Operation

Xie Xuerong, Lu Jian, Wang Yuefei

(Baosteel Iron and Steel Plant, Shanghai 200941, China)

Abstract: The design of Baosteel's No.4 sintering has made a lot of innovation, and achieved a lot of results after production. This paper introduces the problems and improvement measures, such as the lagging of the alkalinity and FeO test data, the belt material accumulation, the environmental protection cantilever bar sieve wear and the return run, the roughness of the fuel system and the low operating rate.

Key words: sintering, alkalinity, FeO

Fe$_2$O$_3$ 颗粒表观粘度的表征与测试研究

王浩然，张延玲，安卓卿，赵世强，郭占成

（北京科技大学钢铁冶金新技术国家重点实验室，北京　100083）

摘　要： 借鉴流体粘度的表征方式将表观粘度的概念引入来表征颗粒间的相互作用力，结合能量耗散原理，利用旋转粘度仪测定不同 Fe$_2$O$_3$ 颗粒的粘度，测试了温度，粒径和添加剂的加入对颗粒表观粘度的影响。结果表明：温度是颗粒粘结的主要因素。无论颗粒粒径大小，是否加入添加剂，颗粒表观粘度的总体趋势都是随着温度升高而增大直到颗粒粘结；颗粒粒径越大，颗粒粘结趋势越小。主要原因是颗粒粒径越大，比表面积越小，在高温下颗粒间粘结力越小，粘结越缓慢。加入添加剂后，颗粒表观粘度明显降低。分析可知，纳米 SiO$_2$ 的加入对颗粒形成了包覆，抑制了颗粒间的团聚和烧结。

关键词： 表观粘度，温度，粒径，添加剂，粘结

Characterization and Testing of Apparent Viscosity of Fe$_2$O$_3$ Particles

Wang Haoran, Zhang Yanling, An Zhuoqing, Zhao Shiqiang, Guo Zhancheng

(State Key Laboratory of Advanced Metallurgy, University of Science & Technology Beijing, Beijing 100083, China)

Abstract: Based on the characterization method of fluid viscosity, the concept of viscosity is introduced to characterize the interaction force between particles. Combining with the principle of energy dissipation, the apparent viscosity of different Fe$_2$O$_3$ particles is measured by rotational viscometer. The effects of temperature, particle size and additive addition on the apparent viscosity of the particles are tested. The results show that Temperature is the main factor in particle sticking. Regardless of the particle size, whether or not the additive is added, the overall trend of the apparent viscosity of the particles increases with increasing temperature until the particles are sticking; The larger the particle size, the smaller the grain sticking trend. The main reason is the larger particle size, the smaller the specific surface area, the smaller the cohesive force between the particles at high temperatures, the slower sticking. After adding the additive, the apparent viscosity of the particles is obviously reduced. It is found that the addition of nano-SiO$_2$ form a coating on the particles, which inhibits the agglomeration and sintering of the particles.

Key words: apparent viscosity, temperature, particle size, additive, sticking

基于还原-自粉化的高铁铝土矿 铁铝分离工艺实验研究

柳政根，储满生，王　峥，高立华，王宏涛，赵　伟

（东北大学冶金学院，辽宁沈阳　110819）

摘　要： 高铁铝土矿是一种铁铝复合型矿产资源。针对高铁铝土矿的研究利用难题，本文提出了高铁铝土矿热压块-还原-自粉化铁铝分离新工艺。通过实验研究，考察了还原温度、还原时间、配碳比等工艺参数对高铁铝土矿铁铝的影响。研究表明，当高铁铝土矿热压块配碳比为0.9，C/A为1.71，还原温度1425℃，还原时间25 min时，得到了粒铁和自粉渣，实现了铁回收率97.42%，铝回收率94.62%；得到的粒铁可用于钢铁生产，自粉渣可用于湿法浸出提铝。

关键词： 高铁铝土矿，热压块，还原，自粉化，铁铝分离

Experimental Research on Separating Iron and Alumina from High Iron Bauxite Based on Reduction and Self-disintegrating

Liu Zhenggen, Chu Mansheng, Wang Zheng, Gao Lihua, Wang Hongtao, Zhao Wei

(School of Metallurgy, Northeastern University, Shenyang 110819, China)

Abstract: High iron bauxite is a complex mineral resource bearing iron and alumina. According to the problem of comprehensive utilization of high iron bauxite, this paper presents a new process of recovering iron and alumina from high iron bauxite by reduction and self-disintegrating based on hot briquette. Through experimental research, the effects of reduction temperature, reduction time and carbon ratio on the recovery ratio of iron and alumina are investigated. The results show that, as the carbon ratio of high iron gibbsite hot briquette 1.0, C/A 1.71, reduction temperature 1425℃, reduction time 25 min, the iron nuggets and self-disintegrating slag are obtained. The recovery ratio of iron and alumina are 97.42% and 94.62%, respectively. The iron nuggets could be used in the iron and steel production, and the slag could be used to leach alumina.

Key words: high iron bauxite, hot briquette, reduction, self-disintegrating, separation of iron and alumina

邢钢炼铁优化配矿及稳定原料生产实践

刘玉江

（邢台钢铁有限责任公司炼铁厂，河北邢台　054000）

摘　要： 阐述了邢钢炼铁原料生产的思路和实践以及几种常用的测算模型。介绍邢钢炼铁在原料的选择、使用，配矿的优化以及高炉用料平衡上做的工作。包括原料性价比排序的测算方式，烧结混匀料如何组合优化配矿。烧结生

产如何实现快速响应市场，既实现烧结矿成分稳定又满足烧结混匀料快速调整结构降成本的需要。

关键词： 优化配矿，烧结生产，高炉生产

料场棚化期混匀矿生产实践

刘鹏君，薛凤萍，齐俊茹，郝新文

（河钢集团唐钢公司，河北唐山　063000）

摘　要： 本文主要论述了混匀料场在棚化封闭技术改造施工前后混匀生产工艺的适时调整及优化，混匀堆积方式的改进，混匀矿主要质量影响因素的控制及混匀堆积过程的控制优化，保证了混匀矿产质量，确保炼铁各生产工序的安全、稳定运行。

关键词： 棚化，混匀矿生产工艺，优化，混匀矿质量

The Production Practice of Mixed Ore in the Stock Yard Shed Stage

Liu Pengjun, Xue Fengping, Qi Junru, Hao Xinwen

(HBIS Group Tangsteel Company, Tangshan 063000, China)

Abstract: This paper mainly describes the timely adjustment and optimization of the mixing production process before and after the transformation of the mixed material field in the construction of the closed and closed technology, and the improvement of the mixing and stacking method, the control of the main quality of the mixed ore and the control and optimization of the mixing and stacking process, ensure the quality of the mixed mineral resources, and ensure the safe and stable operation of the production processes of the iron and steel industry.

Key words: canopy, production process of mixed ore, optimize, quality of mixed ore

高纯铁精矿制备工业纯铁还原过程的动力学分析

刘程宏，丁学勇，唐昭辉，董　越，刘向东

（东北大学冶金学院，辽宁沈阳　110819）

摘　要： 本文以普通铁精矿经摇床所得高纯铁精矿为原料，重点考察了还原温度对还原度的影响规律，并对直接还原过程中的动力学条件进行了初步探讨。结果表明：提高还原温度有利于改善还原反应的动力学条件，但这种促进作用随着温度的升高呈减弱趋势。直接还原过程中反应控制环节为界面化学反应，经计算此控制过程的表观活化能为 $85.27\text{kJ}\cdot\text{mol}^{-1}$。

关键词： 高纯铁精矿，工业纯铁，直接还原，动力学

Direct Reduction Dynamics of Industrial Pure Iron Using High Purity Iron Ore Concentrate

Liu Chenghong, Ding Xueyong, Tang Zhaohui, Dong Yue, Liu Xiangdong

(School of Metallurgy, Northeastern University, Shenyang 110819, China)

Abstract: In this paper, the study focused on the influence of reduction degree of reduction temperature and the kinetic conditions in the process of direct reduction were discussed by using high purity iron ore concentrate as raw material. It was shown that increasing the reduction temperature was beneficial to improve the kinetic conditions of the reduction reaction, but the effect of this reduction was decreasing with the increase of temperature. Interface chemical reaction was confirmed as reaction control link and the apparent activation energy was 85.27 kJ·mol^{-1}.

Key words: high purity iron ore concentrate, industrial pure iron, direct reduction, dynamics

铁矿石集约化高效利用关键技术研究

韩跃新，唐志东，高　鹏，李艳军

（东北大学矿物工程系，辽宁沈阳　110819）

摘　要： 针对我国铁矿资源禀赋差、利用效率低、成本高、优质铁精矿匮乏的开发利用现状，提出了我国铁矿资源"劣质能用、优质优用"的发展战略。重点介绍了复杂难选铁矿石分步、分散浮选技术、绿色高效浮选药剂研发技术、深度还原技术以及悬浮焙烧技术研究现状，讨论了该类技术在盘活劣质铁矿资源，提高资源利用效率方面的应用前景；针对优质铁矿资源提出了超级铁精矿及洁净钢基料短流程绿色制备技术路线，以期通过多学科交叉融合，推动我国钢铁材料产品从中低端向中高端迈进。最后指出了强化技术创新，加快产业结构调整，提高铁矿资源利用效率和产品质量是选矿重点发展方向。

关键词： 铁矿石，劣质能用，悬浮焙烧，优质优用，超级铁精矿，发展战略

Study on the Key Technology of Intensive and Efficient Utilization for Iron Ore

Han Yuexin, Tang Zhidong, Gao Peng, Li Yanjun

(Department of Mineral Engineering, Northeastern University, Shenyang 110819, China)

Abstract: Based on the iron ore resources endowment characteristics, such as high processing cost, lack of high quality iron concentrate and low efficiency utilization, the development strategies of available use of the inferior quality ore and optimal use of the high quality ore for the domestic iron ore resources have been put forward. The step-by-step and dispersing flotation technology, green high efficiency flotation reagent research and development technology, coal-based reduction technology and preconcentration-suspension roasting-magnetic separation (PSRM) technology for the complex refractory iron ores was emphatically introduced, and the application prospect of these technologies in reusing of the inferior iron ores and improving the efficiency of resources was discussed. While, the green preparation technology route of super iron concentrate and clean steel base material has been proposed for the high quality iron ore resources, in order to accelerate our country's iron and steel material products upgrading from low-end to high-end through multidisciplinary fusion methods. The key development directions in exploitation and utilization of iron ore resources, such as strengthen technological innovation, speeding up industrial restructuring, and improving utilization efficiency and product quality were prospected.

Key words: iron ore, available use of inferior quality, PSRM, optimal use of high quality, super iron concentrate, development strategy

碱度对印尼钒钛矿烧结矿质量的影响及机理分析

高强健，孟繁明，郑海燕，姜　鑫，沈峰满

（东北大学冶金学院，辽宁沈阳　110819）

摘　要： 以典型钢铁企业矿石资源为基础，进行不同碱度条件下配加印尼钒钛矿的烧结试验，考察了二元碱度（$R=CaO/SiO_2$）对该矿烧结过程及烧结矿质量的影响。研究表明：当碱度 R 由 1.4 逐渐增至 2.3 时，烧结有效利用系数由 1.63t/(m²·h)降至 1.44t/(m²·h)，烧结矿成品率由 78.5%降至 75.3%；同时，垂直烧结速度有所上升；烧结矿中 $w(FeO)$ 不断下降；转鼓指数先下降而后逐渐回升，并在 $R=1.7$ 时出现最小值；随着碱度的逐渐增加，低温还原粉化性能和还原性逐渐改善，这有利于高炉上部的透气性的改善和间接还原的发展。

关键词： 碱度，印尼钒钛矿，烧结矿，低温还原粉化性，还原性

Effect of Basicity on Quality of Indonesia Vanadium Titanium Sinter

Gao Qiangjian, Meng Fanming, Zheng Haiyan, Jiang Xin, Shen Fengman

(School of Metallurgy, Northeastern University, Shenyang 110819, China)

Abstract: Basing on raw materials conditions of typical iron and steel plant, the sinter experiments were arranged for the Indonesia Vanadium Titanium Magnetite. The effects of binary basicity on the sintering process and sinter quality were investigated and discussed in present work. The results have shown that, when the basicity (R) of sinter increases from 1.4 to the 2.3, the sintering utilization coefficient decreases from 1.63t/(m²·h) to 1.44t/(m²·h) and the finished rate of sinters decreases from 78.5% to 75.3%, while the vertical sintering speed increases gradually. Moreover, with the increasing of basicity, the $w(FeO)$ decreases gradually. For the drum index, with the increasing of basicity from 1.4 to the 2.3, it decreases down first and gets up gradually, and at the basicity of 1.7, the drum index is lowest considered here. Finally, with the addition of basicity in present work, both the RDI and RI are improved gradually.

Key words: basicity, indonesia vanadium titanium magnetite, sinter, RDI, RI

配加 Sibelco 镁橄榄石对球团性能影响研究

王榕榕[1]，张建良[1]，Slagnes Steinar[2]，Rødal Tore[2]，王　川[3]，

青格勒[4]，程铮明[5]，刘征建[1]，刘兴乐[1]，王晓哲[1]，姜春鹤[1]

（1. 北京科技大学冶金与生态工程学院，北京　100083；2. Sibelco Nordic AS，AAHEIM，6146；

3. Swerea MEFOS，Luleå，971 25；4. 首钢技术研究院，北京　100043；

5. 首钢京唐公司，河北唐山　063200）

摘　要：本文分别采用 Sibelco 镁橄榄石及国内某种氧化镁粉作为添加剂进行造球试验、预热焙烧试验及冶金性能检测试验，以添加氧化镁粉的球团作为基准球团，与添加镁橄榄石球团性能进行对比。造球结果表明两种添加剂制备的球团均可满足实际生产需求，镁橄榄石添加量较少（1.2%及1.8%）时球团铁品位仍可保持在65%左右。在相同的预热焙烧条件下，各组球团抗压强度由高到低依次为镁橄榄石1.2%>镁橄榄石1.8%>基准球团>镁橄榄石3.0%；在相同的还原条件下，各组球团还原度由高到低依次为基准球团>镁橄榄石1.8%>镁橄榄石1.2%>镁橄榄石3.0%，球团还原膨胀性能由优到劣依次为镁橄榄石1.8%>镁橄榄石3.0%>镁橄榄石1.2%>基准球团。综合以上结果，球团中镁橄榄石添加量为1.2%~1.8%时，既可保证球团含铁品位，又可保证球团具有较高的抗压强度，较好的还原性及良好的还原膨胀性能，因此 Sibelco 镁橄榄石具有良好的应用前景。

关键词：镁橄榄石，抗压强度，还原度，还原膨胀率

Effect of Sibelco Magnesium Olivine on the Performance of Pellet

Wang Rongrong[1], Zhang Jianliang[1], Slagnes Steinar[2], Rødal Tore[2], Wang Chuan[3],
Qing Gele[4], Cheng Zhengming[5], Liu Zhengjian[1], Liu Xingle[1],
Wang Xiaozhe[1], Jiang Chunhe[1]

(1. School of Metallurgical and Ecological Engineering, University of Science and Technology Beijing,
Beijing 100083, China; 2. Sibelco Nordic AS, AAHEIM, 6146; 3. Swerea MEFOS, Luleå, 971 25;
4. Shougang Research Institute of Technology, Beijing 100043, China;
5. Shougang Jingtang Iron & Steel Co., Ltd., Tangshan 063200, China)

Abstract: Magnesium olivine provided by Sibelco and a type of domestic magnesium fines were used as Mg-containing additives to carry out pelletizing experiments, preheating and roasting test as well as metallurgical performance testing experiment. Pellet prepared with magnesium fines were used as the reference pellet to compare with magnesium olivine pellet. Pelletizing results showed that pellet prepared with two additives could meet the actual production requirement, and TFe of pellet could be kept at about 65% when the amount of magnesium olivine was small (1.2% and 1.8%). Under the same preheating and roasting conditions, compressive strength of each type of pellet ranked from high to low was as follow: magnesium olivine 1.2% > magnesium olivine 1.8% > magnesium olivine 3.0% > reference pellet; Under the same reducing conditions, reducibility of each type of pellet ranked from high to low was as follow: reference pellet > magnesium olivine 1.8% > magnesium olivine 1.2% > magnesium olivine 3.0%, as for the swelling performance, magnesium olivine 1.8% showed smallest swelling index and followed by magnesium olivine 3.0%, magnesium olivine 1.2% as well as reference pellet. Based on the above results, pellet prepared with the magnesium olivine addition amount of 1.2% to 1.8% could not only ensure TFe grade, but also could ensure that the pellets have a high compressive strength, good reducibility and reduction swelling performance, so Sibelco magnesium olivine has a great application prospect.

Key words: magnesium olivine, compressive strength, reducibility, swelling performance

高 Al_2O_3 条件下烧结过程铁酸钙生成研究

张诗瀚，王　广，杜亚星，王静松，薛庆国

（北京科技大学钢铁冶金新技术国家重点实验室，北京　100083）

摘　要：以分析纯试剂为原料，采用微型烧结法，系统研究了碱度、Al_2O_3 含量和 MgO 含量对复合铁酸钙（SFCA）生成特性和烧结过程液相生成特性的影响。结果表明：碱度、Al_2O_3 含量和 MgO 含量的增加，促进了铁酸钙的生成，而碱度、Al_2O_3 含量和 MgO 含量过高，会对复合铁酸钙的性能产生不利影响；随碱度增加，铁酸钙形貌由短小的板条状、粒状向相互粘结的细长针状发展，铁酸钙中 Fe、Ca 含量逐渐增加，Si、Al、Mg 的含量逐渐降低；随 Al_2O_3 含量增加，铁酸钙形貌由细长的针状向粗大、尺寸较短的板条状发展，铁酸钙中 Al 元素的含量总体呈增加的趋势，Si、Mg、Ca 元素的含量维持在一个相对固定的水平，Fe 元素的含量先降低后增加；随 MgO 含量增加，铁酸钙形貌由短小针状和片状向细长的针状发展，铁酸钙中各元素的含量没有明显规律性影响。

关键词：烧结，碱度，Al_2O_3，MgO，铁酸钙

Study on the Formation of Calcium Ferrite during Sintering Process under High Al₂O₃

Zhang Shihan, Wang Guang, Du Yaxing, Wang Jingsong, Xue Qingguo

(State Key Laboratory of Advanced Metallurgy, University of Science and Technology Beijing, Beijing 100083, China)

Abstract: The effects of alkalinity, Al_2O_3 content and MgO content on the formation characteristics of c-omposite calcium ferrite (SFCA) and the formation characteristics of liquid phase during sintering process were studied by means of micro sintering method using pure reagent as raw material. The results show-ed that the increase of alkalinity, Al_2O_3 content and MgO content promoted the formation of calcium ferrite, but when the alkalinity, Al_2O_3 content and MgO content were too high, which could adversely affect the properties of calcium ferrite; With the increase of alkalinity, the microscopic morphology of calcium ferrite was developed from short platelets and granular to elongated needle which sticks with each ot-her. The content of Fe and Ca in calcium ferrite gradually increased, and the contents of Si, Al and Mg decreased gradually; With the increase of Al_2O_3 content, the morphology of calcium ferrite was developed from slender needle to coarse and short strip. The content of Al in calcium ferrite increased with the increase of Si, Mg and Ca Maintained at a relatively fixed level, and the content of Fe decreased firstl-y and then increased; With the increase of MgO content, the microscopic morphology of calcium ferrite was developed from short needle-like and flaky to slender acicular, and the content of each element in calcium ferrite was not obvious influences.

Key words: sinter, basicity, alumina, magnesium oxide, calcium ferrite

柔性密封装置在大型烧结机中的优势及应用前景

谭哲湘，臧疆文

（宝武集团新疆八钢炼铁分公司，新疆乌鲁木齐　830022）

摘　要：分析八钢 A265m² 烧结机漏风率偏高的原因，使用柔性密封装置后密封作用改善，点火强度提高，烧结机有效面积扩大，台时产量增加，电耗降低，达到了优质增产降耗的效果。

关键词：柔性密封装置，烧结，优势，应用前景

Flexible Sealing Device in the Advantages and Application Prospects of Sintering

Tan Zhexiang, Zang Jiangwen

(Ironmaking Branch, Bayi Iron & Steel Co.Baowusteel Group, Urumchi 830022, China)

Abstract: Analysis of Bayi steel A265m^2 the cause of the high air leakage rate of sintering machine, using the flexible sealing function improved after Swing eddy current type flexible sealing device, effective area is expanding, sintering machine production increases, the power consumption is reduced, achieve the result of the high quality production and reducing consumption.

Key words: sealing device, sintering, advantages, application prospect

某高硅铁矿在湛江钢铁的烧结生产实践

李晓波，李孟土，章苇玲，金静忠，牛长胜

（宝钢湛江钢铁有限公司制造管理部，广东湛江 524072）

摘 要： 某高硅铁矿 TFe=59.84%，SiO$_2$=10.23%，二氧化硅主要以浸染状嵌布。烧结杯探索试验表明，在现有配矿结构中，以 10%左右的高硅铁矿替代巴西北部粉矿，可取消蛇纹石的配入，提高烧结矿铁品位。工业生产试验表明，在保证烧结矿 SiO$_2$=5.0%条件下，不配入蛇纹石，高硅铁矿最高可配入 12.22%，成品烧结矿 TFe=58.16%，烧结机利用系数达 1.274 t·m^2·h^{-1}，成品烧结矿转鼓强度达到 81.46%，产量和质量均满足湛江钢铁生产需要。通过烧结杯试验和工业生产试验结果的对比，建立起了湛江钢铁烧结杯试验指标与工业生产指标的对应关系，为后续烧结杯试验提供了参考依据。

关键词： 高硅铁矿，蛇纹石，烧结，烧结杯试验，工业生产试验

Sintering Production Practice of a High Silicon Iron Ore Fines in Zhanjiang Iron and Steel Co., Ltd.

Li Xiaobo, Li Mengtu, Zhang Weiling, Jin Jingzhong, Niu Changsheng

(Products & Technique Management Department, Baosteel Zhanjiang Iron and Steel Co., Ltd., Zhanjiang 524072, China)

Abstract: A high silicon iron ore fines is characterized by 59.84 percent total Fe and 10.33 percent SiO$_2$ which disseminated in patches. The sintering pot test results indicate that, when10 percent of Brazil northern iron ore was replaced by the high silicon iron ore in current blending structure, the serpentine can be canceled and the Fe content in sinter improved. The results of the industrial test demonstrate the above conclusions, under the condition of SiO$_2$=5.0% and without serpentine blended in, the proportion of high silicon iron ore can be as high as 12.22% with the total Fe of sinter equal to 58.16%, productivity of sintering machine is 1.274 t·m^2·h^{-1} and tumble index reach 81.46%. By comparison the sinter pot trail and industry test result, we can establish the correspondence between the two indicators, which can provides guidance for subsequent sintering pot tests.

Key words: high silicon iron ore, serpentine, sinter, sintering pot test, industry test

大型原料场智能化系统应用实践

魏玉林，宋宜富，杨广福，高　峰，邱懿飞

（宝钢湛江钢铁有限公司炼铁厂，广东湛江　524000）

摘　要： 大型钢铁企业原料场作为生产流程的起始环节，自动化水平总体偏低，不少环节仍然依靠人工操作，强度大且环境恶劣，劳动效率与作业效率都受到较大限制。宝钢湛江钢铁原料引入 3D 激光云图采集与处理、超声波与 RFID 定位等技术，通过集成与软件开发，率先在原料场实现了智能化系统，使得生产劳动效率与作业效率大幅提升，人力需求大幅下降，在行业内开创了智能化先河，提供了重要借鉴。

关键词： 原料场，智能化，堆取料，盘库

Application of Intelligent System in Large-scale Raw Material Field

Wei Yulin, Song Yifu, Yang Guangfu, Gao Feng, Qiu Yifei

(Ironmaking Plant, Baosteel Zhanjiang Iron & Steel Co., Ltd., Zhanjiang 524000, China)

Abstract: The raw material field of large iron and steel enterprises as the starting point of the production process, the overall level of automation is low, many links still rely on manual operation, strong and poor environment, labor efficiency and operating efficiency are subject to greater restrictions. Baosteel Zhanjiang Iron and Steel raw materials introduced 3D laser cloud image acquisition and processing, ultrasonic and RFID positioning technology, through the integration and software development, the first in the raw material field to achieve an intelligent system, making the production efficiency and operating efficiency increased significantly, manpower demand dropped significantly , In the industry to create a precedent for intelligent, providing an important reference.

Key words: raw material field, intelligent, stacking material, library

提高 2#干熄炉干熄率的生产实践

赵　华，周　鹏，马银华，李雪冬，魏世明，代　成

（鞍钢股份鲅鱼圈钢铁分公司炼焦部，辽宁营口　115007）

摘　要： 本文通过围绕制约 2#干熄焦的各种影响因素，现场标定 2#干熄焦干熄率情况，结合现场实际情况进行了多次技术改造，陆续改善 2#干熄焦排焦温度高、排焦不均匀以及排焦量小等问题，使 2#干熄焦干熄率有了较大提高，逐步提高干熄焦产量。

关键词： 干熄炉，干熄率

Improve of No.2 CDQ Dry Quenching Rate Production Practice

Zhao Hua, Zhou Peng, Ma Yinhua, Li Xuedong, Wei Shiming, Dai Cheng

(Ansteel Bayuquan Iron & Steel Subsidiary Department of Coking, Yingkou 115007, China)

Abstract: This paper describes the CDQ coke cans transmission, braking, rotation around the constraints CDQ Coke cans normal transmission, brake a variety of factors, by setting the transmission gear CDQ Coke board, turning the audience roller slot row gray drainage holes and parallel increase slowdown stop limit and other measures to effectively reduce the CDQ coke cans rotation frequency is not in place appear to meet the production needs of CDQ.

Key words: CDQ dry, quenching rate

含碳球团还原焙烧过程中的热态强度研究

孟庆民[1]，李家新[1,2]，春铁军[1]，魏汝飞[1]，王　平[1]，龙红明[1]

（1. 安徽工业大学冶金工程学院，安徽马鞍山　243032；

2. 东北大学冶金学院，辽宁沈阳　110819）

摘　要：利用含碳球团高温强度在线测试装置，研究了非等温加热条件下还原剂种类、配碳量等对铁精矿含碳球团热态强度的影响，结合含碳球团还原后试样的气孔率和微观结构分析，讨论了热态强度变化机理。结果表明：含碳球团还原焙烧过程中受到外力破坏作用时，存在压溃和塑性变形两种状态。球团热态强度在 200~800℃时较干燥后强度略有增加，超过 800℃后强度显著下降，在 1000℃左右达到强度最低值。含碳球团热态强度变化与球团气孔率变化具有一定关联性，200~1000℃时，球团热强度源自颗粒间的分子引力和粘结剂膨润土的粘滞力作用，高于 1000℃后，球团热强度靠金属铁连晶维系。

关键词：含碳球团，还原焙烧，热态强度，显气孔率，微观结构

Research on the Thermal Strength of Carbon-bearing Iron Ore Pellets During the Process of Reduction Roasting

Meng Qingmin[1], Li Jiaxin[1,2], Chun Tiejun[1], Wei Rufei[1], Wang Ping[1], Long Hongming[1]

(1. School of Metallurgical Engineering, Anhui University of Technology, Maanshan 243032, China;

2. School of Metallurgy, Northeastern University, Shenyang 110819, China)

Abstract: The thermal strength characteristics and mechanism of the carbon-bearing iron ore pellets in the non-isothermal heating process, including the effect of reducing agent and carbon content on the thermal strength, were studied by the on-line test device of thermal strength, combined with the samples microstructure and porosity after reduction. When the carbon-bearing iron ore pellets was damaged by external force at high temperatures, it existed in two states: crush and plastic deformation. The thermal strength of the pellets slightly increased at 200~800 ℃, significantly decreased after 800 ℃ and reached the minimum value at 1000 ℃. There is a correlation between the thermal intensity of the carbon-bearing iron ore pellets and its porosity, at 200~1000 ℃, the thermal strength of the pellets was derived from the molecular attraction between the particles and viscous force from bonding agent. But when the temperature higher than 1000 ℃, the thermal strength was served by aggregation of iron crystals.

Key words: carbon-bearing iron ore pellets, reduction roasting, thermal strength, porosity, microstructure

八钢烧结厂 265 烧结机降低煤气单耗生产实践

冯 齐

（宝武集团新疆八一钢铁有限公司，新疆乌鲁木齐　830022）

摘　要： 通过采取强化烧结点火操作，根据热值调整空燃比、投入煤气流量自动调节装置，八钢烧结厂煤气单耗逐步降低，265 烧结机由原来的 8.61Nm³/t 降低到现在的 5.50Nm³/t。有效地降低了烧结煤气消耗，改善了烧结生产过程中的燃耗。

关键词： 煤气单耗，热值，自动调节

八钢烧结降低烧结矿电耗的实践

石红兵

（八钢炼铁分公司烧结分厂，新疆乌鲁木齐　830022）

摘　要： 烧结工序在钢铁冶炼过程是仅次于高炉的用能大户，占钢铁生产用能 10%~20%，八钢烧结用电量占铁前系统总用电量的 50%，降低烧结工序电耗对降低铁水成本具有重要的意义，特别是钢铁市场低迷钢材价格下跌，降低制造成本更显得非常重要。2015 年下半年开始八钢烧结产能只有高峰时期产能的 40%，各种制造成本不断上升。八钢烧结分厂通过技术改造和在管理上下功夫，在产能降低的情况下保证电耗不上升。

关键词： 烧结机，电耗，改造

The Practice of Bagang Sintering Sinter Reducing Power Consumption

Shi Hongbing

(Bayi Iron Company Sintering Plant, Urumchi 830022, China)

Abstract: Sintering process in the iron and steel smelting process is second only to the blast furnace energy hungry, accounting for iron and steel production by 10%~20%, Bagang sintering electricity accounts for the iron before the system total energy consumption 50%, reduce power consumption in sintering process to reduce the cost of hot metal has important significance, especially iron and steel market downturn in steel prices fell, reduce the cost of manufacturing is very important. The second half of 2015 during the peak period of Bagang sintering production capacity is only 40% of capacity, various manufacturing costs continue to rise. Bayi Steel sintering branch through technical transformation and efforts in management, production capacity in lower power consumption that does not rise.

Key words: sintering machine, power consumption, reform

烧结混料机开式齿轮调整实践

李小治

（宝钢集团八钢公司炼铁分公司，新疆乌鲁木齐　830022）

摘　要：烧结混合就是为使烧结的物料物化性质充分均匀，使烧结料内微粒物料造成适宜的小球，而混料机就是用于烧结混合和造球的设备，随着混料机的大型化，其运行的稳定性对企业生产的连续性影响很大，本文针对混料机开式齿振动进行分析，通过齿轮顶隙、侧隙的测量、调整，旨在控制齿轮的振动，保证混料机稳定运行。

关键词：齿轮顶隙，齿轮侧隙，齿轮顶隙和侧隙测量，齿轮顶隙和侧隙调整

The Practice of Open Gear Adjustment for Sintering Mixing Machine

Li Xiaozhi

(Sintering Branch of Ironmaking Branch, Bayi Iron & Steel Co. Baosteel Group, Urumchi 830022, China)

Abstract: Mixed sintering is to fully and evenly, so that particles in the sinter material caused by suitable ball sintering material properties, and mixer is used for mixing and sintering and pelletizing equipment, with the mixed feeding machine of large, the running stability of enterprise production continuity effect greatly. In this paper, the mixed material for open gear vibration analysis, through the gear top clearance, clearance measurement and adjustment, gear to control the vibration to ensure mixed feeding machine and stable operation.

Key words: top clearance of gear, gear backlash, backlash and backlash, backlash and backlash adjustment

烧结机台车车轮轴承故障分析及对策措施

方　明，薛林涛

（宝钢集团八钢公司炼铁分公司烧结分厂，新疆乌鲁木齐　830022）

摘　要：针对430m² 烧结机台投产以来，每年烧结机台车因车轮轴承损坏，频繁下线，对台车车轮轴承损坏的原因进行分析，通过改进台车车轮密封形式以及轴承隔环结构、改善台车车轮润滑的方法，来进一步提高台车车轮的寿命，提高设备的稳定性，同时降低设备修复费用。本文主要对烧结机台车车轮轴承故障进行分析，采取相应对策改善轴承工况，提高车轮轴承使用寿命。

关键词：烧结机台车轮，轴承密封，轴承隔环，改进

Fault Analysis and Countermeasure of Trolley Wheel Bearing of Sintering Machine

Fang Ming, Xue Lintao

(Baosteel Group Bayi Corporation Ironmaking Branch Sintering Plant, Urumchi 830022, China)

Abstract: For 430m² sintering machine has been put into production, annual sintering machine trolley wheel bearings due to damage, frequent offline, of trolley wheel bearing failure reason analysis, by improving the trolley wheel seal and bearing spacer ring structure, improve the lubrication of trolley wheels, to further improve the service life of the trolley wheel, improve equipment stability, at the same time, reduce the cost of equipment repair. In this paper, the failure of the sintering machine trolley wheel bearing is analyzed, and the corresponding measures are taken to improve the working condition of the bearing, and the service life of the wheel bearing is improved.

Key words: sintering machine wheel, bearing seal, bearing spacer ring, improvement

膨润土质量指标对球团性能的影响

张晓萍，李小静，孙社生

（马鞍山钢铁股份有限公司技术中心，安徽马鞍山　243000）

摘　要： 以膨润土质量指标对球团用膨润土进行了质量评价，并提出现行质量指标与球团性能的关系，认为膨润土的大吸水率和强的膨胀容指标在一定程度上弥补了蒙脱石含量低带来的不足，明显改善了生球及成品球团的抗压强度，但单一的吸水率或蒙脱石指标与球团性能并没有一一对应的关系。三种膨润土中 ZYHD 膨润土的稳定性优于 FS 膨润土。

关键词： 膨润土，质量指标，球团性能

The Effect of Bentonite Quality Indexes on the Performance of Pellet

Zhang Xiaoping, Li Xiaojing, Sun Shesheng

(Maanshan Iron & Steel Co., Ltd., Maanshan 243000, China)

Abstract: In this paper, the quality evaluation and the performance analysis of three types of bentonite were carried out with existing technical indexes of bentonite. Also the relationship between the present technical indexes and the pellet performance was put forward. The large water absorption rate and expansion capacity of bentonite to some extent made up for the deficiency of low content of montmorillonite, which improved the compressive strength of green-pellet and finished product. However, there was no correspondence between the single water absorption rate or montmorillonite index with the pellet performance. In three types of bentonite the stability of ZYHD bentonite was superior to that of FS bentonite.

Key words: bentonite, quality indexes, pellet performance

不同堆料方式下的物料匀化效果比较

艾　宇，陈尚伦

（中冶赛迪工程技术股份有限公司，重庆　400013）

摘　要：本文通过对取料过程中的实际料堆进行合理简化，建立了用于匀化效果计算的人字形、倾斜层料堆数学模型，采用计算机编程方法对不同堆料方式下的物料匀化效果进行了计算和比较分析。结果表明采用刮板取料机侧面取料时，倾斜层堆料方法匀化效果优于人字形堆料方法。利用本文所编制的匀化效果计算程序，能够快速计算和比较不同堆料方式下的物料匀化效果，定量地分析堆料过程中物料的堆积层数、堆积顺序、物料成分等对匀化效果的影响，供人们选择较佳的堆料方式参考，指导生产操作，具有一定的理论意义和工程实用价值。

关键词：散装料，匀化效果，堆料方式

Blending Effectiveness Comparison of Different Material Stacking Method

Ai Yu, Chen Shanglun

(CISDI Co., Ltd., Chongqing 400013, China)

Abstract: In this paper, the stockpiles was reasonably simplified and the herringbone and the oblique stockpiles mathematical mode was established, Blending effectiveness was calculated and compared by using the computer programming methods. The results show that when reclaiming by side, the blending effectiveness of the oblique stockpiles is better than the herringbone stockpiles. Furthermore using the program can quickly compare blending effectiveness of different material stacking method, and quantitative analyze the effect of stacking layers, stacking sequence, material composition in the process of stacking. The program is effective to help people to choose the stock mode better, and also can guide the production operation. It has certain theoretical significance and practical value in engineering.

Key words: bulk material, blending effectiveness, stacking method

钒钛磁铁矿焙烧过程中物相转变规律研究

丁成义，吕学伟，李　刚，宣森炜，唐　凯，白晨光

（重庆大学材料科学与工程学院冶金工程专业，重庆　400044）

摘　要：钒钛磁铁矿焙烧过程中，研究碱度和焙烧时间对物相转变规律的影响，对钒钛磁铁矿烧结过程的优化控制尤为重要。通过 X 射线衍射对不同条件下钒钛磁铁矿焙烧样品的物相进行定性或半定量分析，结果表明：不同碱度下钒钛磁铁矿焙烧主要物相有铁酸一钙（CF）、钙钛矿（CT）、Fe_2O_3、Fe_3O_4 和钛榴石（$Ca_3Fe_2Si_{1.52}Ti_{1.48}O_{12}$）。CF 在高碱度条件下更容易生成且随着碱度提高含量逐渐增多。钛榴石含量随不同碱度变化很小。碱度为 1.7 时，平衡态下 TiO_2 在钒钛磁铁矿焙烧样品中主要以 CT 和钛榴石形式存在，并没有 CF 相生成。碱度为 2.5 时，开始出现复

合铁酸钙（SFCA）。在钒钛磁铁矿焙烧样品中，CF 只在焙烧时间很短时出现，在长时间保温或者低碱度条件下，随着 CT 和钛榴石的生成，CF 很难稳定存在。

关键词： 钒钛磁铁矿，烧结，物相转变，碱度，钙钛矿

Phases Transition of Vanadium Titanium Magnetite in Roasting Process

Ding Chengyi, Lv Xuewei, Li Gang, Xuan Senwei, Tang Kai, Bai Chenguang

(Metallurgical Engineering in School of Materials Science and
Engineering, Chongqing University, Chongqing 400044, China)

Abstract: In the roasting process of vanadium titanium magnetite (VTM), the influence of basicity and roasted time on the phases transition, is especially important for the optimization control of the industrial sintering process of VTM. This study mainly carried out qualitative or semi-quantitative analysis of phases content of roasted samples of VTM in different experimental conditions through X-ray diffraction (XRD). The results showed that the main phases of the roasted VTM samples in different basicity are calcium ferrite (CF), perovskite (CT), Fe_2O_3, Fe_3O_4, and titangarnet ($Ca_3Fe_2Si_{1.52}Ti_{1.48}O_{12}$). CF is easier to generate in high basicity conditions and increases with the increase of basicity. The content of titangarnet varies slightly with different basicity. When the basicity is 1.7, TiO_2 in the equilibrium state is mainly in the formation of CT and titangarnet in the roasted samples of VTM, and there is no CF phase generation. When the basicity reaches 2.5, TiO_2 is still in the form of CT and titangarnet, but at this time, the quaternary calcium ferrate (SFCA) phase appears. In the roasted samples of VTM, CF only temporarily presents in the short roasted time. In long time of heat preservation or low basicity, CF is difficult to stabilize with the formation of CT and titangarnet.

Key words: vanadium titanium magnetite, sintering, phase transition, basicity, perovskite

华菱湘钢降低膨润土实验研究

刘敏媛，汤　勇，吴　云，侯盼盼，严永红，刘守文

（华菱湘钢技术质量部，湖南湘潭　411101）

摘　要： 2017 年以来膨润土配比达 3.80%，很高；膨润土质量较差，并且各厂家之间波动较大。为了降低膨润土用量，为此开展精矿和膨润土的造球和焙烧实验，通过实验得出：适宜的膨润土配比，和兴精为 3.5%、巴西精为 7.5%、澳洲精为 2.0%；因和兴精成品球的强度比澳洲精高 1000N 以上，并且铁水成本较低，因此应采购一部分来降成本和膨润土配比；湘钢目前的膨润土供应商有 6 家，为保供应一般三家就可，因此至少应淘汰宏利和众帮；多数情况下膨润土的性能检测指标能反映膨润土的造球性能，也有例外，如飞来峰膨润土，因此通过造球和焙烧实验来选择更好。

关键词： 精矿，膨润土，造球，焙烧

方大特钢 245m² 烧结机提高混合料温度生产实践

吴小辉，王金生

（方大特钢科技股份有限公司炼铁厂，江西南昌　330012）

摘　要：对方大特钢 245m² 烧结机提高混合料温度生产实践进行了总结。通过改善生石灰质量，引用热水消化生石灰，混料筒添加热水及混合料仓蒸汽预热等方式，使烧结混合料温度达到 60℃以上，减少了烧结过程中过湿带的形成，改善了料层透气性，烧结矿产、质量得到了提升。

关键词：烧结，混合料温度，过湿带，透气性

Production Practice of the 245m² Sintering Machine under Increased Mixture Temperature in Fangda Special Steel

Wu Xiaohui, Wang Jinsheng

(Ironworks of Fangda Special Steel Technology Co., Ltd., Nanchang 330012, China)

Abstract: The paper summarizes the production practice of the 245m² sintering machine when the mixture temperature is increased in Fangda Special Steel. By improving the raw lime quality, using hot water to dissolve raw lime, adding hot water into mixing drum, steam preheating of mixing bin, and so forth, the temperature of sinter mixture can reach over 60℃, which is conducive to reducing the formation of moisture concentration zone in the sintering process and improving the permeability of bed. Thus, the output and quality of sinter can be improved.

Key words: sintering, mixture temperature, moisture concentration zone, permeability

大尺寸粉尘团块还原历程研究

郝云东，张建良，刘征建，王耀祖，李世钦

（北京科技大学冶金与生态工程学院，北京　100083）

摘　要：分析了布袋除尘灰和转炉灰两种粉尘的化学组成、粒度组成和物相分布等物化性质。对两种粉尘按照 C/O 比为 1.1 进行混合压制成直径 40mm 的圆柱形团块在 1473K 温度下进行了直接还原焙烧实验，团块中心温度上升的速率是先快后慢，最终团块中心温度达到与炉温平衡的时间大约为 30min。对不同焙烧时间的团块进行了对比分析，随着焙烧时间的增加，还原的 Fe 量在逐渐增加，在团块的边缘处有大量的单质 Fe 产生，并且组织比较松散，气孔率比较大；在团块的中心处基本没有 Fe 单质产生，Fe 大多以 FeO 形态存在并且组织比较紧密，气孔率较小。靠近还原交界面处横截面生成的铁单质大多呈现圆形，而团块边缘处生成的铁大多联结在一起形成铁连晶。

关键词：布袋除尘灰，转炉灰，团块，气孔率，铁连晶

Study on Reduction Process of Large Size Dust Briquette

Hao Yundong, Zhang Jianliang, Liu Zhengjian, Wang Yaozu, Li Shiqin

(School of Metallurgical and Ecological Engineering, University of Science and Technology Beijing, Beijing 100083, China)

Abstract: Physics-chemical properties including chemical composition，particle size distribution and phases of bag dust and converter sludge were analyzed. A cylindrical briquette with 40mm in diameter was prepared by two kinds of dust with the C/O ratio 1.1 at the temperature of 1473K. The results showed that center temperature of the briquette is increasing slowly, and the time for briquette to achieve balance of internal and external temperature is about 30min. Briquette with different roasting time was analyzed, with the increase of roasting time, the amount of reducing Fe is gradually increasing. There is a large number of elemental Fe at the edge of the briquette, and the comparison of the organization is loose, porosity is larger. There is basically no metal Fe in the center of the briquette. Fe is mostly in form of FeO orderly with a low porosity. The iron produced at the reduction interface is mostly round, and the iron formed at the edge of the briquette is mostly linked together to form an iron joined crystal.

Key words: bag dust, converter sludge, briquette, porosity, iron joined crystal

高比例褐铁矿烧结生产实践

朱 镇

（湖南华菱湘潭钢铁有限公司炼铁厂）

摘　要： 提高褐铁矿在烧结中的比例是企业降低生产成本的有效措施之一。本文通过查阅相关文献，结合生产实践，对设备进行不断改进，实施厚料层低水分、调整熟熔剂比例等技术措施改善褐铁矿烧结，以应对不断提高的褐铁矿比例，降低原料成本，同时确保烧结矿转鼓强度。

关键词： 褐铁矿，烧结，工艺参数

邯钢 435m^2 烧结机工艺配置特点

李凤民，李立新

（河钢股份有限公司邯郸分公司，河北邯郸　　056015）

摘　要： 简要介绍了邯钢 435m^2 烧结机在工艺、设备及环保节能方面的工艺配置特点，并对各个环节采用的国内外先进、成熟、实用的工艺和技术情况进行了阐述。

关键词： 烧结机，配置，特点

Hangang 435m² Sinter Machine Process Configuration Features

Li Fengmin, Li Lixin

(Handan Iron and Steel Co., Ltd., Handan 056015, China)

Abstract: Hangang 435m² is introduced in the process of sintering machine, equipment and technology in environmental protection and energy saving configuration feature, and all aspects of the use of advanced, mature and practical processes and technologies are described.

Key words: sintering circuit, configuration, feature

标准化引领烧结球团转型升级

刘文权¹，吴记全²

（1. 冶金工业规划研究院，北京　100711；2. 九江萍钢钢铁有限公司，江西九江　332500）

摘　要： 本文论述了开展标准化的必要性、标准化在钢铁转型中的作用和标准化引领烧结球团转型升级，并以烧结烟气循环技术为例，列举烧结烟气循环利用技术规范立项的主要内容，并对未来烧结球团开展标准化工作进行展望。

关键词： 烧结球团，转型升级，标准化

Standardization Leads Sinter and Pellet to Transform and Upgrade

Liu Wenquan¹, Wu Jiquan²

(1. China Metallurgical Industry Planning and Research Institute, Beijing 100711, China;
2. Jiujiang Pinggang Iron and Steel Co., Ltd., Jiujiang 332500, China)

Abstract: This article expounds the necessity of carry out standardized and standardized in the role of steel transformation and standardization leads the transformation and upgrading of sinter and pellet, and sintering flue gas recycling technology, for example, lists the main content of sintering flue gas recycling project technical specification, and the future of sinter and pellet to carry out the standardization work.

Key words: sinter and pellet, transform and upgrade, standardized

FEMA 工具在提高混匀矿质量生产实践的运用

侯　通，王　泽，汪建斌，孙钦伟，伍　钢

（武钢有限原料分厂，湖北武汉　430083）

摘　要： 运用失效模式 FEMA 分析工具找出影响建堆质量的关键因素，通过开展降低进料杂物，皮带运行跑故障，

降低堆料机运行故障以及优化生产建堆计划，加强过程控制管理等技术措施，2016 年混匀矿 TFe 与 SiO$_2$ 合格率分别为 90.56%和 92.86%，分别较目标值提高 1.19%与 1.31%，为稳定烧结矿质量奠定坚实的基础。

关键词：精益管理，混匀矿质量，过程控制，堆料机

Application Lean Management Tool of FEMA in Improving the Quality of Blending Ore

Hou Tong, Wang Ze, Wang Jianbin, Sun Qinwei, Wu Gang

(Raw Material Plant of Wusteel Co., Ltd., Wuhan 430083, China)

Abstract: The lean Management tool of FEMA was applied to improve the blended ore quality at raw material plant of Wu steel Co., Ltd., by reducing the feeding stuff, belt running fault operation, material piling machine malfunction and optimizing production build plan, and technical measures such as strengthening the process control management, the blended ore quality of TFe and SiO$_2$ stability factors above the target by 1.19% and 1.31% respectively.

Key words: lean Management, blended ore quality, process control, material piling machine

智利精粉、米纳斯精粉的烧结性能比较研究

邸兰欣，王纪元，阎丽娟

（宝钢股份，上海　201941）

摘　要： 精矿粉硅、铝含量较低、铁品位较高、理化指标较好。混匀矿中配入精矿粉，能够提高烧结矿铁品位，有利于控制烧结矿中的 SiO$_2$ 和 Al$_2$O$_3$ 含量。但是其粒度较细的特点会导致混匀矿粒度下降和粒度偏析，使用量因此受到限制。本文分析使用两种精矿粉烧结杯的技术指标与过程参数的变化，综合评价烧结性能。

关键词： 烧结，铁矿石，精矿粉

Technical Analysis on the Addition of Chilean and Minas Concentrates at Sintering Process

Di Lanxin, Wang Jiyuan, Yan Lijuan

(Baosteel, Shanghai 201941, China)

Abstract: The concentrates have low alumina and silica contents, high Fe grade and good chemical characteristics. Adding the concentrates to the blended ore used for sintering production can increase the content of total Fe and is beneficial to the control of SiO$_2$ and Al$_2$O$_3$ contents. However, the fine Chilean concentrates will cause the granularity of blended ore to decrease and segregate. Therefore, the use of the concentrates are limited. According to the variation of technology indicators and process parameters when chilean concentrates were used in sintering production, this paper made a comprehensive evaluation on the sintering characteristics of Chilean concentrates.

Key words: sintering process, iron ore, the concentrates

MgO 对烧结的影响研究

王天雄[1,2]，白晨光[1]，吕学伟[1]，丁成义[1]

（1. 重庆大学材料科学与工程学院，重庆　400044；2. 重庆钢铁股份公司，重庆　401258）

摘　要：MgO 含量对烧结影响很大，它不仅影响烧结矿质量、产量，还影响生产成本，本文对重钢配矿条件下不同的 MgO 和配碳量进行研究，通过研究表明合理控制烧结 MgO 含量，对改善生产指标，生产优质、低成本烧结矿有重要意义。

关键词：烧结，MgO，影响，研究

Study on MgO Influence of Sintering

Wang Tianxiong[1,2], Bai Chenguang[1], Lv Xuewei[1], Ding Chengyi[1]

(1. College of Materials Science and Engineering of Chongqing University, Chongqing 400044, China;
2. Sintering Plant of Chongqing Iron and Steel Co., Ltd., Chongqing 401258, China)

Abstract: MgO had a great influence on sintering production, it not only affects the sinter quality and yield, also affect the cost of production, in this paper, under the condition of Chongqing Iron and Steel ,different MgO and carbon content were studied, through studying fand proper MgO of sinter has important significance of improve production index and product high quality and low cost of sinter.

Key words: sintering, MgO, influence, study

炼铁烧结原料经营分析系统功能研发介绍

王家军[1]，张晓冬[1]，王　曦[2]，蔡宝旺[1]

（1. 河钢承钢公司，河北承德　067102；2. 昆明理工大学，云南昆明　650500）

摘　要：介绍了炼铁烧结原料经营分析系统功能的研究开发情况，对原料的质量综合评价、性价比分析、指标分析、原料配料优化、原料需求计划的自动生成等功能进行了较详细的描述。系统应用于生产质量监控及原料计划、原料到货过程的监控。在炼铁烧结生产中发挥了重要作用。

关键词：炼铁烧结原料，经营分析，系统功能

Study on the Comprehensive Evaluation Method of Raw Material Quality of Sintering

Wang Jiajun[1], Zhang Xiaodong[1], Wang Xi[2], Cai Baowang[1]

(1. HBIS Chengde Branch Company, Chengde 067102, China; 2. Kunming University of Science and Technology, Kunming 650500, China)

Abstract: Ironmaking sintering raw material management analysis is introduced in the research and development situation of system function, to comprehensive evaluation of the quality of raw materials, ratio analysis, index analysis, raw material ingredient optimization, material requirement planning of automatically generated, etc were described in detail. The system is used to monitor the production quality monitoring and raw material delivery process. It has played an important role in the production of iron sintering.

Key words: sintered raw materials, quality of raw materials, comprehensive evaluation

方大特钢炼铁原料系统创新管理实践

葛晓柱，钟国英

（方大特钢科技股份有限公司炼铁厂，江西南昌　330012）

摘　要： 方大特钢炼铁原料系统瞄准市场脉络，以经济效益为中心，通技术过创新管理，推进优化配矿及合理回收利用含铁资源，建立原料性能数据库，指导生产，提高指标，降低生产成本；同时围绕原料加工系统创新管理，各工序开展了工艺、设备、环保配套改造等技术工作。实现了成本最低化、质量最优化，产能最大化，提升市场竞争力。

关键词： 原料场，创新，成本，低耗

The Innovation and Management Practice of Fangda Special Steel Raw Materials System

Ge Xiaozhu, Zhong Guoying

(Fangda Special Steel Technology Co., Ltd., Nanchang 330012, China)

Abstract: The iron and steel raw materials system of fangda steel, through innovation management, is promoted to optimize distribution and reasonable recycling of iron resources. Through the innovation management of the raw material system, the process is carried out the technical work such as the process equipment environmental protection. We have realized the minimum cost, optimized quality, maximized capacity, and improved the competitiveness of the market.

Key words: raw materials field, innovation, cost, low consumption

精粉比例对烧结的影响机理与强化技术研究

贺淑珍[1]，李　强[1]，甘　敏[2]

（1. 山西太钢不锈钢股份有限公司，山西太原　030003；2. 中南大学，湖南长沙　410083）

摘　要：为了使 660m[2] 烧结机高比例配用微细精矿粉，以降低烧结原料成本，烧结矿满足 4350m[3] 高炉的生产需要，进行了精矿比例对烧结的影响机理研究与强化技术研究。研究结果表明：随着微细精矿配用比例的提高，烧结料层透气性恶化；烧结液相生成能力减弱，铁酸钙生成量减小。为此，在优化熔剂结构的基础上，进行了预成核技术开发应用，将部分精矿粉预先制成制粒核，减少物料中粘附粉的量，以强化物料制粒效果和成矿反应，烧结矿结构改善。

关键词：超细精矿粉，高比例，预成核技术，产量，质量

Research of the Effect of Ultra Fine Ore Sintering and Strengthening Measures

He Shuzhen[1], Li Qiang[1], Gan Min[2]

(1. Shanxi Taigang Stainless Steel Co., Ltd., Taiyuan, 030003, China;

2. Central South University, Changsha 410083, China)

Abstract: A series of tests have been taken to help the 660m[2] sinter production with high proportion of ultra fine ore, to lower the raw material cost of sinter production, to satisfy the outstanding sinter quality requirement of 4350m[3] blast furnace. Several conclusions have been made. The bed permeability is worsening with the increasing proportion of the ultra fine ore; Moreover, liquid phase formation ability is weakened and calcium ferrite is reduced. The pre-pelletizing technology has been developed with the optimization of flux additives combination. Part of the fine ore is pre-pelletized to "grain core", decreasing the conglutinated powder, thus to strengthen the granulation and the pellet formation reaction. The microstructure of sinter has been improved.

Key words: ultra fine ore, high proportion, pre-pelletizing technology, production, quality

模拟炉顶煤气循环-氧气高炉条件下烧结矿的矿相结构演变

闫昊天，王静松，刘锦周，孙德鹏，刘迎立

（北京科技大学钢铁冶金新技术国家重点实验室，北京　100083）

摘　要：炉顶煤气循环-氧气高炉条件下煤气的还原势显著提高，对烧结矿在高炉块状带的还原行为和软熔带的软熔行为产生重要影响。本文在气固换热—化学反应耦合数学模型的基础上，进行程序还原实验和单颗粒荷重软化实验，采用扫描电镜对试样进行显微结构观察，研究氧气高炉条件下烧结矿还原和软熔过程中矿相结构的演变规律。

结果表明：相比传统高炉条件，氧气高炉条件下烧结矿的还原有大幅度提高，边缘区域更为明显；由于还原势的提高以及 H_2 的增加，氧气高炉条件下烧结矿还原速率比传统高炉更快，生成金属铁的量更多，有利于改善炉料的软熔行为。为未来氧气高炉炼铁工艺的实际应用提供理论基础及数据支撑。

关键词：氧气高炉，烧结矿，还原，软熔，矿相结构

Study on Sinter Ore Phase Structure Evolution on the Condition of Oxygen Blast Furnace with Top Gas Recycling

Yan Haotian, Wang Jingsong, Liu Jinzhou, Sun Depeng, Liu Yingli

(State Key Laboratory of Advanced Metallurgy, University of Science and
Technology Beijing, Beijing 100083, China)

Abstract: The gas reduction potential increased on the condition of simulated roof gas circulation -oxygen blast furnace，which has an important effect on the reduction behavior of the sinter in the blast furnace block and the reflow behavior of the soft melt. In this paper, based on the mathematical model of gas-solidification heat-chemical reaction coupling,the experiment was carried out to carry out the program reduction experiment and the single particle load softening experiment. The microstructure of the specimen was observed by scanning electron microscope, The evolution of ore structure in the process of reduction and softening of sinter in oxygen blast furnace was studied. The results show: Compared with the traditional blast furnace conditions, the reduction of sinter in oxygen blast furnace has been greatly improved and the edge area is more obvious. Due to the improvement of the reduction potential and the increase of H_2, the rate of sintering ore reduction in oxygen blast furnace is faster than that of traditionalblast furnace The amount of metal iron is more conducive to improving the reflow behavior of the charge. And provide theoretical basis and data support for the future application of oxygen blast furnace ironmaking process.

Key words: oxygen blast furnace, sinter ore, reduction, soft melt, ore structure

超高品位铁精矿煤基直接还原实验研究

李　峰，储满生，唐　珏，冯　聪，汤雅婷，柳政根

（东北大学冶金学院，辽宁沈阳　110819）

摘　要：本研究以无烟煤为还原剂，对铁品位为 71.87% 的铁精矿煤基直接还原进行了研究，结果表明：在 1000~1100℃范围内，随还原温度升高，还原产物金属化率逐渐增大。但当温度高于 1075℃时，继续提高温度，产物金属化率增幅不大，适宜的还原温度应为 1075℃；在相同还原温度下，随还原时间延长，金属化率逐渐增大，适宜的还原时间应为 5h；在 1075℃条件下还原 5h，直接还原铁金属化率可达 95.7%。还原用煤中添加 CaO 可抑制煤中硫元素向直接还原铁中迁移，当其添加量为 2.5% 时，产物中硫含量由 0.023% 降低到 0.017%。实验条件下的直接还原铁金属化率高，杂质含量少，可作为冶炼高纯净钢的基料。

关键词：超高品位铁精矿，煤基直接还原，还原时间，还原温度，金属化率，硫含量

Experimental Study on Coal-based Direct Reduction of Ultra High Grade Iron Concentrate

Li Feng, Chu Mansheng, Tang Jue, Feng Cong, Tang Yating, Liu Zhenggen

(School of Metallurgy, Northeastern University, Shenyang 110819, China)

Abstract: The coal-based direct reduction of iron concentrate with an iron grade of 71.87% was investigated. The results showed that in the temperature range of 1000~1100℃, the metallization rate of direct reduction iron (DRI) increased with the reduction temperature. When the temperature was over 1075℃, the metallization rate increased little with the increasing temperature. The metallization rate of DRI also increased with the reduction time at the reduction temperature of 1100℃. The reasonable reduction temperature and time is 1075℃ and 5h respectively. Under this reduction condition, the metallization rate of DRI can reach 95.7% adding CaO can inhibit the migration of sulfur from coal into DRI. When the amount of CaO was 2.5%, the content of sulfur in the DRI decreased from 0.023% to 0.017%. The DRI with high metallization rate and low impurity content can be used as the base material for smelting high purity steel.

Key words: ultra-high grade iron concentrate, coal-based direct reduction, reduction temperature, reduction time, metallization rate, sulfur content

高磷鲕状赤铁矿深度还原分选试验研究

孙永升，韩跃新，栗艳锋，李艳军

（东北大学资源与土木工程学院，辽宁沈阳　110819）

摘　要： 采用深度还原高效分选新技术处理高磷鲕状赤铁矿，在深度还原热力学分析的基础上，主要考察了还原温度、还原时间和配碳系数对还原效果和分选指标的影响。结果表明：铁矿物的还原历程为 $Fe_2O_3 \rightarrow Fe_3O_4 \rightarrow FeO(Fe_2SiO_4，FeAl_2O_4) \rightarrow Fe$，并在还原过程中出现了 $2FeO \cdot SiO_2$、$FeSiO_3$、$Fe_2Al_2O_4$ 和 $CaFe_2O_4$ 等含铁复杂化合物。经过对还原温度、还原时间和配碳系数的考察，最终确定适宜的还原条件为：还原温度 1250℃，还原时间 50min，配碳系数 2.0。经两段磨矿两段磁选可得到金属化率 91.41%、铁品位 90.41%、回收率 91.27% 的磁选铁粉。

关键词： 高磷鲕状赤铁矿，深度还原技术，热力学分析，磁选，金属铁粉

Study on the Coal-based Reduction Followed by Magnetic Separation of High-phosphorus Oolitic Hematite Ore

Sun Yongsheng, Han Yuexin, Li Yanfeng, Li Yanjun

(School of Resources and Civil Engineering, Northeastern University, Shenyang 110819, China)

Abstract: The coal-based reduction followed by magnetic separation process adopting for high-phosphorus oolitic hematite ore was investigated. On the basis of coal-based reduction thermodynamic analysis, the effects of reduction temperature, reduction time and distribution coefficient of carbon on the reduction and separation indexes. The results showed that the iron oxides in the oolitic ore were reduced to metallic iron following the route of $Fe_2O_3 \rightarrow Fe_3O_4 \rightarrow FeO(Fe_2SiO_4, FeAl_2O_4) \rightarrow Fe$, as well as some iron complex compounds like $2FeO \cdot SiO_2$、$FeSiO_3$、$Fe_2Al_2O_4$ 和 $CaFe_2O_4$, etc, were generated in reduction process. The reduction temperature, reduction time and distribution coefficient of carbon were investigated to explore the appropriate reduction conditions, finally, reduction temperature 1250℃, reduction time 50 min and distribution coefficient of carbon 2.0 were intended to be the most suitable conditions. The reduced sample was treated via two stages grinding and magnetic separation, the magnetic iron powder which metallization ratio was 91.41%, iron grade was 90.41% and iron recovery was 91.27%, respectively, was obtained.

Key words: high-phosphorus oolitic hematite ore, coal-based reduction, thermodynamic analysis, magnetic separation, metallic iron powder

东鞍山铁矿石预富集精矿悬浮磁化焙烧试验

高　鹏，余建文，韩跃新，李艳军

（东北大学矿物工程系，辽宁沈阳　110819）

摘　要： 磁化焙烧—磁选是复杂难选铁矿石开发利用的重要方法。本文介绍了一种半工业悬浮焙烧系统，并考察了焙烧温度、还原气体 CO 及流化气体 N_2 用量对东鞍山含碳酸盐铁矿石预富集粗精矿悬浮焙烧效果的影响。试验结果表明，在焙烧温度 540℃、还原气体 CO 用量 $4Nm^3/h$ 及流化气体 N_2 用量 $2Nm^3/h$ 的条件下，焙烧矿经磁选后可获得 TFe66.06%、回收率 91.16%的铁精矿。铁的化学物相、穆斯堡尔谱及光学显微结构分析表明，经悬浮焙烧后弱磁性的菱铁矿和赤铁矿转变为强磁性的磁铁矿，部分粗粒（>100μm）赤铁矿仅表面转变为磁铁矿，内核为赤铁矿晶格。这种内核为赤铁矿的新生磁铁矿颗粒磁性较强，经磨矿与脉石矿物解离，在后续磁选过程中依然会进入铁精矿产品中，并不会影响铁的回收。

关键词： 含碳酸盐铁矿石，悬浮磁化焙烧，物相演变，磁选

Investigation on Fluidized Magnetization Roasting of Pre-concentrate for Donganshan Iron Ore

Gao Peng, Yu Jianwen, Han Yuexin, Li Yanjun

(Department of Mineral Engineering, Northeastern University, Shenyang 110819, China)

Abstract: Magnetization roasting-magnetic separation is an important method for the beneficiation of refractory iron ores. In this investigation, a pilot-scale fluidized magnetization roasting reactor was introduced and used to study the effects of roasting temperature, flow rates of reducing gas CO and fluidizing gas N_2 on magnetization roasting performance of the pre-concentrate for Donganshan carbonate-bearing iron ore. The results shown that a high-grade magnetic concentrate containing 66.06% iron with an iron recovery of 91.16% was achieved after roasting at 540℃ with a gas mixture of CO $4Nm^3/h$ and N_2 $2Nm^3/h$. The iron chemical phase, Mossbauer spectrum, and optical microscopy analyses revealed that weakly magnetic siderite and hematite could be converted to ferromagnetic magnetite successfully. Some coarse hematite particles were hard to be reduced to magnetite completely, while such magnetite particles with a hematite core could also be effectively recovered by subsequent magnetic separation after liberating from gangue minerals due to their strong magnetism.

Key words: carbonate-bearing iron ore, fluidized magnetization roasting, phase conversion, magnetic separation

富氧喷吹焦炉煤气工艺下烧结矿低温还原粉化研究

王彬旭，刘迎立，王　广，王静松，薛庆国

（北京科技大学钢铁冶金新技术国家重点实验室，北京　100083）

摘　要：为了减少碳排放，实现低碳炼铁，将采用高炉煤气循环及喷吹重整后的焦炉煤气技术，这将导致高炉内还原气氛和温度改变，影响烧结矿的低温还原粉化行为。根据喷吹焦炉煤气数学模型计算结果，按照设定程序升温并改变煤气成分来模拟高炉上部的还原过程，通过转鼓实验分析烧结矿的粉化程度，采用 XRD 确定烧结矿中的矿物组成，使用偏光显微镜观察还原后烧结矿的微观结构，研究温度和气氛（传统高炉、氧气高炉（鼓风氧体积分数为40%）喷吹焦炉煤气）对烧结矿低温还原粉化的影响。结果表明：烧结矿中的 Fe_2O_3 并未全部还原为 Fe_3O_4，$RDI_{+3.15}$ 与失重率以及 Fe_3O_4 含量呈负相关。500℃时，在氧气高炉喷吹焦炉煤气气氛下，烧结矿边缘的气孔—赤铁矿结构产生网状裂纹，导致粉化较为严重；600℃和 700℃时，传统高炉气氛中，烧结矿的内部结构发生断裂，粉化程度更为剧烈。

关键词：富氧，喷吹焦炉煤气，烧结矿，低温还原粉化

Study on Low Temperature Reduction Degradation of Sinter in Oxygen-enriched Coke Oven Gas Injection Process

Wang Binxu, Liu Yingli, Wang Guang, Wang Jingsong, Xue Qingguo

(State Key Laboratory of Advanced Metallurgy, University of Science and Technology Beijing,
Beijing 100083, China)

Abstract: In order to reduce carbon emissions and achieve low-carbon ironmaking, blast furnace top gas recycle and injection of coke oven gas after reforming will be used, and these technologies will change the reduction atmosphere and temperature within the blast furnace to affect the low temperature reduction degradation behavior of sinter. To study the influences of temperature and atmosphere(traditional blast furnace(TBF)、oxygen blast furnace with coke oven gas injection whose blast oxygen content is 40% (40%OBF-COGI)) on the low temperature reduction degradation of sinter, according to the calculation results of the mathematical model of the coke oven gas injection, the temperature is raised by setting procedure and the gas compositions are changed to simulate the reduction process in the upper blast furnace. The degree of degradation of the sinter was analyzed by the drum test, XRD was used to determine the mineral composition in sinter, and the microstructure of the sinter was observed by polarizing microscope. The results show that, the Fe_2O_3 in the sinter is not completely reduced to Fe_3O_4, and $RDI_{+3.15}$ is negatively correlated with weight loss rate and the contents of Fe_3O_4. At 500℃, the pore - hematite structures at the edge of the sinter produce mesh cracks that causes more serious degradation in the 40%OBF-COGI. The internal structure of sinter is broken that leads to a more intense degree of degradation at 600℃ and 700℃ in the TBF.

Key words: oxygen enrichment, coke oven gas injection, sinter, low temperature reduction degradation

低品位锰铁矿选择性还原降铁提质工艺探索研究

李梦伟，储满生，任非凡，杨　晶，周熙涵，张俊铧

（东北大学冶金学院，辽宁沈阳　110819）

摘　要： 我国锰矿资源匮乏，锰矿品位偏低，为满足国内发展需求，充分利用锰矿资源，大力发展低品位锰矿成为研究焦点。传统还原焙烧和湿法能耗较高，因此本文针对低品位锰铁矿进行还原焙烧磁选降铁提质制备富锰渣工艺探索性试验研究。主要研究还原时间、还原温度、配碳比等参数对低品质锰铁矿还原磁选制得富锰渣 Mn 品位、Mn 回收率和锰铁比 Mn/Fe 的影响规律。实验结果表明：还原时间 6 h，还原温度 1050℃，配碳比 2.5，锰矿粒度 8~13 mm。在此条件下，得到非磁性物中锰品位为 53.30%，锰回收率可达 81.98%，锰铁比可达 5.92。该研究为利用回转窑处理难选低品位锰矿资源提供新思路。

关键词： 低品位锰铁矿，选择性还原，还原焙烧，磁选分离

Exploratory Research to Decrease Iron and Upgrade Low-grade Ferromanganese Ore by Selective Reduction Process

Li Mengwei, Chu Mansheng, Ren Feifan, Yang Jing, Zhou Xihan, Zhang Junhua

(School of Metallurgy, Northeastern University, Shenyang 110819, China)

Abstract: It is difficult to meet the domestic development of manganese ore resources, due to manganese ore resource is scarce and manganese grade is low mainly. In order to make full use of manganese resources, the low-grade manganese ore has become the focus of research. Because of the energy consumption of traditional reduction roasting and wet process is higher, exploratory research to decrease iron and upgrade low-grade ferromanganese ore by selective reduction process in this paper. Through single factor experiment study on reduction process parameters of reduction temperature, reduction time, ferromanganese ores particles size and C/O ratio for influence law and action mechanism of manganese grade, recovery rate of manganese, Mn/Fe ratio in the non-magnetics, optimize the process parameters. The result shows that the following conclusions are obtained: The optimal reduction parameters include a reduction time of 6 hours, a reduction temperature of 1050℃, a manganese ores size of 8~13mm, a FC/O ratio of 2.5 and a magnetic field intensity of 50mT. Under this condition, in the non-magnetics manganese grade 53.30%, recovery rate of manganese 81.98%, Mn/Fe ratio 5.92.

Key words: low-grade ferromanganese ore, selective reduction, roasting reduction, magnetic separation

石钢高比例澳矿的烧结配矿研究与应用

李　杰，白雄飞

（河钢集团石钢公司技术中心，河北石家庄　050000）

摘　要： 本研究主要根据各种进口澳矿的烧结基础特性、资源及价格情况，优化烧结配矿结构，在合理控制烧结矿化学组分的前提下，得到高澳矿配比的配矿方案，在澳矿性价有优势时应用于烧结生产，在生产中通过调整工艺保

证烧结矿质量的稳定提高，实现合理应用低价格澳矿资源，降低烧结配矿成本的目的。

关键词：烧结，高比例澳矿，铁前成本

Study and Application of Sinter Batching with High Proportion Australian Ores in Shijiazhuang I & S Co., Ltd.

Li Jie, Bai Xiongfei

(Technical Center of Hebei I & S Group Shijiazhuang I & S Co., Ltd., Shijiazhuang 050000, China)

Abstract: The main purpose of the study is to optimize ore batching structure based on the basic sintering characteristics, resources and price situation of various kinds of imported Australian ores. With the precondition that the chemical composition of sinter is under reasonable control, to obtain the batching solution with higher proportion of Australia ores, and apply it to the sintering production when the characteristics and price of Australia ores have the advantages. The reliable improvement of sintered ore quality is ensured by adjusting the process during the production, in order to realize reasonable application of low price Australian ores, and reduce the cost of ore batching.

Key words: sintering, high proportion Australian ores, cost of sintering

石钢烧结优化原料结构的生产实践

张 纯，张海岸，李 斌

（河钢集团石家庄钢铁有限责任公司，河北石家庄 050000）

摘 要： 石钢烧结工序以精粉和澳矿为主要入烧原料的入烧结构入烧，通过降低生产成本，增加公司产品盈利水平，作为生铁主要原料，优化原料结构，降低烧结矿成本为最快最好的降本路径。2017 年扩大经济料使用比例，降低铁料成本，实现烧结原料结构优化，烧结工序提高混匀料成分稳定性，优化配料操作，实施厚料层操作等稳定烧结矿质量的措施，在烧结矿质量未受明显影响的前提下，降低铁料成本。2017 年一季度石钢实现了大比例澳矿粉和经济料的入烧，降低生产成本，提高企业竞争力。

关键词： 烧结，配料，配比，经济料

含碳酸盐难选铁矿石新型氨基羧酸类捕收剂的捕收机理

谷晓恬[1]，朱一民[1,2,3]，李艳军[1,2,3]，韩跃新[1,2,3]

（1. 东北大学资源与土木工程学院，辽宁沈阳 110819；

2. 东北大学 2011 钢铁共性技术协同创新中心，辽宁沈阳 110819；

3. 辽宁省难采选铁矿石高效开发利用技术工程实验室，辽宁沈阳 110819）

摘 要： 针对含少量碳酸盐的难选铁矿石高效选别，本文对赤铁矿、磁铁矿、菱铁矿及石英四种单矿物分别进行浮选试验，考察了新型高效氨基羧酸类捕收剂 α-EDA-LA 对四种矿物在不同 pH 下的捕收行为；通过动电位测试与红

外光谱分析，重点分析 α-EDA-LA 与石英及菱铁矿的吸附机理。浮选结果显示在 α-EDA-LA 200 mg/L 及浮选温度 18℃ 条件下，在 pH 6～12 范围内 α-EDA-LA 对石英回收率均在 90 %以上；在 pH 为 8 时，菱铁矿回收率达最高 83.4 %。动电位测试结果显示在浮选 pH 条件下，α-EDA-LA 吸附在荷负电的矿物表面，使其动电位降低，说明同荷负电的药剂与矿物表面能够克服同性电荷间的排斥作用而发生吸附。红外光谱分析显示 α-EDA-LA 对石英的捕收主要以氢键吸附为主；α-EDA-LA 对菱铁矿的捕收以化学键合吸附为主，氢键耦合吸附为辅。

关键词： 难选铁矿石，赤铁矿，菱铁矿，石英，浮选，捕收剂

Collecting Mechanism of a Novel Amino-acid Based Collector for Carbonate-containing Refractory Iron Ores

Gu Xiaotian[1], Zhu Yimin[1,2,3], Li Yanjun[1,2,3], Han Yuexin[1,2,3]

(1. College of Resource and Civil Engineering, Northeastern University, Shenyang 110819, China; 2. 2011 Collaborative Innovation Centre of Steel Technology, Northeastern University, Shenyang 110819, China; 3. Liaoning Technology and Engineering Laboratory of Effective Exploitation of Refractory Iron Ores, Shenyang 110819, China)

Abstract: A novel and highly-efficient collector was studied, aiming at effective beneficiation of carbonate-containing refractory hematite ores. The collector is amino-acid-based, of which the main composition is α-ethylenediamine lauric acid (α-EDA-LA). Single mineral flotation tests were carried out for hematite, magnetite, siderite and quartz, respectively. The objective was to investigate the floatability of the single minerals across the pH range from 6 to 12. All tests were conducted with 200mg/L of α-EDA-LA and 18℃ of flotation temperature. Zeta potential and Fourier transform infrared spectroscopy (FTIR) were used to identify the adsorption mechanisms of α-EDA-LA on quartz and siderite surfaces. The flotation results showed that the quartz recovery was over 90% at pH of 6～12. For siderite, the recovery peaked at 83.4% when pH was 8.0, where siderite presented different floatability from magnetite and hematite. Exploiting such difference, the separation of siderite could be achieved. Zeta-potential measurements showed that α-EDA-LA adsorption on the surfaces of siderite and quartz decreased the respective zeta potential at pH of 8～10 and 8～12, respectively. This means the adsorption overcome the electrostatic repulsion between similar charges of α-EDA-LA and the mineral surfaces. FTIR results suggest that α-EDA-LA was adsorbed on quartz via hydrogen bond formation. Between α-EDA-LA and siderite, the adsorption was dominated by chemisorption, while further enhanced by hydrogen bonding.

Key words: refractory iron ores, hematite, siderite, quartz, flotation, collector

超级铁精矿制备试验研究

刘　杰，李艳军，韩跃新，宫贵臣

（东北大学资源与土木工程学院，辽宁沈阳　110819）

摘　要： 本研究以辽宁某地出产的普通磁铁精矿为原料进行了超级铁精矿制备试验研究。实验室试验流程为预选抛尾-阶段磨矿-阶段磁选-反浮选，可获得 TFe 品位 71.88%、回收率为 64.01%、酸不溶物 0.19%的超纯铁精矿。在实验室研究的基础上进行了扩大连续试验，获得了 TFe 品位为 72.19%、回收率为 25.57%的超纯铁精矿和 TFe 品位为 71.82%、回收率为 68.65%的高纯铁精矿，二者总回收率达 94.22%。为我国发展高端钢材解决了优质原料来源问题。

关键词： 超级铁精矿，实验室研究，扩大连续试验，高纯铁精矿，超纯铁精矿

Experimental Study of the Preparation of Super Iron Concentrate

Liu Jie, Li Yanjun, Han Yuexin, Gong Guichen

(College of Resource and Civil Engineering, Northeastern University, Shenyang 110819, China)

Abstract: Experiments were carried out to investigate the production of super iron concentrate using a common magnetite iron concentrate as the raw material which was produced in Jianping, Liaoning province, China. The optimal laboratory experimental flowsheet was preconcentration-stage grinding-stage magnetic separation-reverse flotation, and a super iron concentrate containing 71.88% TFe, 0.19% acid insoluble matters was obtained at the recovery 64.01%. Continuous tests were performed based on the results of the laboratory experiments, a super pure iron concentrate with a grade 72.19%, a recovery 25.57% and a high pure iron concentrate with a grade 71.82%, a recovery 68.65% were obtained, the total recovery of these two products was 94.22%. The high-quality raw materials could be used to produce High quality steel.

Key words: super iron concentrate, laboratory research, continuous tests, high pure iron concentrate, super pure iron concentrate

烧结机提产保质的研究与实践

夏金生，吕定建

（日照钢铁有限公司，山东日照　276806）

摘　要： 为满足日钢年产 1700 万吨生铁烧结矿需求，充分挖潜烧结机设备能力，为高炉提供优质的烧结矿，通过采取了降低烧结机漏风率、提高混合料温、提高料层厚度、加强烧结过程控制等一系列措施，烧结利用系数较 2015 年提高了 0.018t/(m²·h)，烧结矿其他各项质量指标也得到了改善，达到了烧结提产保质的目的。

关键词： 能力，漏风率，料温，料层，过程控制

The Research and Practice of Sintering Machine to Produce with Good Quality

Xia Jinsheng, Lv Dingjian

(Rizhao Steel Co., Ltd., Rizhao 276806, China)

Abstract: To meet the annual output of 17 million tons of pig iron, steel sinter demand, fully tap potential sintering machine equipment capability, provide high quality sinter for blast furnace, by taking the lower sintering temperature, air leakage rate and improve the mixture increase the material thickness, a series of measures such as strengthening sintering process control, the utilization coefficient of sintering 0.018 t/(m²·h) increased from 2015 year, the sinter quality index of other items also have improved, achieve the goal of on sintering production with good quality.

Key words: ability, air leakage rate, metal temperature, bed of material, process control

烧结使用生白云石粉生产实践

夏金生，吕定建，朱　勇

（日照钢铁有限公司，山东日照　276806）

摘　要： 通过对烧结过程中白云石粉替代轻烧白云石粉理论测算降本，以及在生产实践中随着替代比例变化对烧结矿技术经济指标的变化影响，寻求最佳的白云石粉配加比例，实现在烧结过程中"烧结矿成本最低、产量最高、质量最好"的目的。

关键词： 白云石粉，轻烧白云石粉，理论测算，生产实践

Analysis and Production Practice of Raw Dolomite Powder for Sintering

Xia Jinsheng, Lv Dingjian, Zhu Yong

(Rizhao Iron and Steel Co., Ltd., Rizhao 276806, China)

Abstract: Dolomite fines in sintering process of substitute light burned dolomite fines theory calculate authors, as well as in the production practice with the change of replacement ratio of sinter with technical and economic indicators of change, to seek the best dolomite fines with addition proportion, realize in the process of sintering "sinter cost minimum, the highest yield, the best quality" purpose.

Key words: dolomite dust, the dolomite dust of calcination, theory of measuremen, the production practice

MgO 对低品位含铬型钒钛磁铁矿球团的冶金特性影响研究

程功金[1]，高子先[1]，薛向欣[1,2]

（1. 东北大学冶金学院，辽宁沈阳　110819；

2. 辽宁省冶金资源循环科学重点实验室，辽宁沈阳　110819）

摘　要： 低品位含铬型钒钛磁铁矿资源有望作为一种潜力的铁矿资源被大规模开采利用，并应用于实际生产。本文研究了 MgO 对低品位含铬型钒钛磁铁矿球团抗压强度、还原膨胀和高温软熔滴落特性的影响，考察了各种冶金特性的指标。研究表明：随 MgO 配入量从 0%增加到 3%时，低品位含铬型钒钛磁铁矿球团抗压强度均大于 2000N，但整体上降低；球团还原膨胀率均小于 20%，但先升高后降低，MgO 为 1%时还原膨胀率最高；软化开始温度逐渐升高，软化终了温度整体上升高，软化区间先降低后升高，MgO 为 2%时软化区间最低；熔化开始温度先降低后升高，MgO 为 1%时熔化开始温度最低，滴落温度和熔滴区间均先升高后降低，MgO 为 2%时滴落温度和熔滴区间最高；MgO 含量为 1%时，炉料透气性最好。

关键词： 低品位含铬型钒钛磁铁矿，球团，MgO，冶金特性

Effect of MgO on the Metallurgical Characteristics of Low-grade Chromium-bearing Vanadium-Titanium Magnetite Pellets

Cheng Gongjin[1], Gao Zixian[1], Xue Xiangxin[1,2]

(1. School of Metallurgy, Northeastern University, Shenyang 110819, China; 2. Liaoning Key Laboratory of Recycling Science for Metallurgical Resources, Shenyang 110819, China)

Abstract: Low-grade chromium-bearing vanadium-titanium magnetite is hopeful to be one kind of promising iron ores to be exploited on a large scale and applied to practical production. Effect of MgO on the metallurgical characteristics including crushing strength, reduction swelling and elevated softening-melting-dripping properties of low-grade chromium-bearing vanadium-titanium magnetite pellets (LCVTMP) was investigated in this paper. It is found that the crushing strength, all above 2000N, decreased wholly with increasing MgO addition from 0% to 3%. The reduction swelling degree, all lower than 20%, increased with increasing MgO addition from 0% to 1% and decreased with further increasing MgO addition. The softening start temperature increased and the softening end temperature wholly increased. The softening zone firstly decreased and then increased, and was minimum when MgO addition was 2%. The melting start temperature firstly decreased and then increased, and was minimum when MgO addition was 1%. The dripping temperature and melting-dripping zone increased with increasing MgO addition from 0% to 2% and decreased with increasing MgO addition further. It is also found that the gas permeability of furnace burden was optimized when MgO addition was 1%.

Key words: low-grade chromium-bearing vanadium-titanium magnetite, pellet, MgO, metallurgical characteristics

烧结半干法脱硫灰中亚硫酸钙浆液氧化试验研究

王毅璠[1]，龙红明[1,2]，魏汝飞[1]，周　笛[1]，钱立新[1]

（1. 安徽工业大学冶金工程学院，安徽马鞍山　243002；2. 冶金减排与资源利用教育部
重点实验室（安徽工业大学），安徽马鞍山　243002）

摘　要： 半干法脱硫灰中含有大量的 $CaSO_3$，限制了其资源化应用，对 $CaSO_3$ 进行氧化改性是其资源化应用的前提。本文在自制浆液氧化实验装置中，对不同氧化温度、氧化时间、固液比、鼓风流量、搅拌速度下脱硫灰中 $CaSO_3$ 的氧化进行了研究。研究结果表明，提高氧化温度、延长氧化时间、增大气流量和搅拌速度有利于 $CaSO_3$ 的氧化，降低浆液的固液比能够显著加快 $CaSO_3$ 的氧化。但总体而言，在低温下（<100℃）向脱硫灰浆液中通入空气，$CaSO_3$ 氧化非常缓慢，尚需寻找一条高效率低成本的氧化途径。

关键词： 烧结，半干法脱硫灰，亚硫酸钙，氧化

Experimental Study on Oxidation of Calcium Sulfite Slurry in Sintering Semi-Dry Desulfurization

Wang Yifan[1], Long Hongming[1,2], Wei Rufei[1], Zhou Di[1], Qian Lixin[1]

(1. School of Metallurgical Engineering, Anhui University of Technology, Maanshan, 243002, China; 2. Key Laboratory of Metallurgical Emission Reduction & Resources Recycling (Anhui University of Technology), Ministry of Education, Maanshan 243002, China)

Abstract: Half dry desulfurization ash contains large amounts of $CaSO_3$, limits its applications, $CaSO_3$ oxidation modification is the precondition of its resource application. Homemade slurry oxidation experiment device is presented in this paper, the different oxidation temperature, oxidation time, solid-liquid ratio, blast air flow rate, stirring speed of oxidation desulfurization ashes $CaSO_3$ were studied. The results show that the higher oxidation temperature, longer oxidation time, increased air flow rate and stirring speed can promote the oxidation of $CaSO_3$, and the lower solid-liquid ratio can accelerate the oxidation of $CaSO_3$. But in general, the desulfurization slurry is oxidized at low temperatures (<100 ℃), and $CaSO_3$ oxidation is very slow, so there is still a need to find efficient and low-cost oxidation pathways.

Key words: sintering, semi-dry desulfurization ash, calcium sulfite, oxidation

宣钢烧结工艺改进技术进步

韩　涛

（河钢集团宣钢公司技术中心，河北宣化　075100）

摘　要： 宣钢装备 3 台 360m² 烧结机，但投产以来主要技经指标在河钢集团内处于中下游水平，为此从设备设施功能完善、科技支撑体系建设、稳定原料质量、烧结矿成分优化控制、新技术应用等方面开展工作，烧结各项技经指标取得较大提升。

关键词： 烧结，工艺改进，技术进步，指标提升

Process Improvement and Technological Progress of Xuansteel Sintering

Han Tao

(Technique Center，Hesteel Group Xuansteel Company, Xuanhua 075100, China)

Abstract: There are three 360m² sintering machines in Xuansteel, but the main technical indexes in HBIS group in the middle and lower level, therefore, from the device function ware perfect, technology support system construction, stable raw material quality, sinter composition optimization control, application of new technology etc., the indicators of sintering have been greatly improved.

Key words: sintering, process improvement, technological progress, indicators promotion

利用悬磁干选机选别含碳酸铁磁铁矿试验研究

孙景新[1]，高志喆[1]，刘兴全[1]，孟　娜[2]，陈小艳[1]，于　凤[1]

（1. 鞍钢集团大孤山球团厂，辽宁鞍山　114000；2. 鞍钢集团矿业设计研究院，辽宁鞍山　114000）

摘　要： 悬磁式干选机是鞍山金裕丰选矿科技有限公司研制的一种新型磁铁矿石干式预选设备。大孤山球团厂三选作业区通过采用该设备对鞍山大孤山地区含碳酸铁磁铁矿石进行干选抛尾试验，结果表明，采用该设备抛尾率可达

20%，入磨矿石品位提高 0.2 个百分点，抛出尾矿的磁性铁含量降到 1% 以下的较好指标。为鞍钢开发小孤山矿区供矿问题及合理利用矿山资源提供可靠的技术依据。

关键词：含碳酸铁磁铁矿石，悬磁干选机，预选抛尾，试验研究

Experimental Study on Separating Magnetite Ore Containing Carbonate Iron by Suspension Magnetic Dry Separating Machine

Sun Jingxin[1], Gao Zhizhe[1], Liu Xingquan[1], Meng Na[2], Chen Xiaoyan[1], Yu Feng[1]

(1. Dagushan Pelletizing Plant of Ansteel Group Co., Anshan 114000, China;

2. Ansteel Mining Engineering Corporation, Anshan 114000, China)

Abstract: The suspension magnetic dry separating machine is a new type of magnetite dry preselection equipment developed by Anshan Jinyufeng Mineral Processing Technology Co., Ltd. The results show that by using of the equipment in the third separating workshop of the Dagushan Pellet Plant, the throwing tailing rate can reach 20%, grades of ore feeding into mill increase by 0.2 percentage points, and throwing tailings of the magnetic iron ontain magnetic iron below 1%, The thesis provide a reliable technical basis for solving the problem of ore supply in Xiaogushan of Ansteel and properly utilizing the mine resources in this mineral area.

Key words: magnetite ore containing carbonate iron, suspension magnetic dry separating machine, preselection throwing tail, experimentai research

高比例褐铁矿烧结生产实践

朱 镇

（湖南华菱湘潭钢铁有限公司炼铁厂，湖南湘潭 411100）

摘 要：提高褐铁矿在烧结中的比例是企业降低生产成本的有效措施之一。本文通过查阅相关文献，结合自身情况，对设备进行不断改进，实施厚料层低水分，调整熟熔剂比例等技术措施均有利于改善褐铁矿烧结，以应对不断提高的褐铁矿比例，降低原料成本，同时确保烧结矿转鼓强度。

关键词：褐铁矿，烧结，工艺参数

High Propotion Limonite in Sintering Production Practice

Zhu Zhen

(Ironmaking Plant of Hunan Valin Xiangtan Iron and Steel Co., Ltd., Xiangtan 411100, China)

Abstract: It is one of the effective measures to lower production cost that we increase the proportion of limonite in sintering. This discourse, by consulting relevant literature, combined with Xiangtan Steel's Sintering Plant practical situation, is to prove the feasibility to lower raw material cost and maintain the tumbler strength by continuously improve equipments, implementing thick layer of low moisture, adjusting the medium flux ratio and other technical measures, to deal with continuously increasing limonite proportion in sintering.

Key words: limonite, sintering, process parameters

铁矿石冶金性能对高炉的影响及应对措施

胡启晨

（河钢集团钢研总院）

摘　要： 文通过对国内现有高炉炉料结构对高炉指标的影响分析，针对铁矿石的低温还原粉化、还原性、高温冶金性能对高炉炉况的影响，针对各段影响规律进行研究，发现低温还原粉化对高炉上部透气性影响比较明显，$RDI_{+3.15}$ 每增加 1%，上部压差降低 0.7% 百分点，烧结矿还原性每提高 1%，燃料比降低 2.3 个百分点。随软化温度区间的增大，炉料压差增大，呈现正相关对应关系，每增加 1℃，压差增加 0.8kPa。不同品种天然块矿爆裂指数相差很大，每增加 1%，压差升高 0.1 个百分点。

关键词： 矿石，冶金性能，高炉，压差

新区烧结厂提高料层透气性的生产实践

桂林峰，李光军，赵江红，张桐亮，陈胜涛

（武钢集团昆明钢铁股份有限公司安宁分公司）

摘　要： 烧结料层的透气性是指固体散料层允许空气通过的难易度，也是衡量混合料空隙度的标志。它是影响烧结矿产、质量的一个重要因素，对烧结过程的强化有很大意义。昆钢新区 $300m^2$ 烧结机面对设计产能不足及混匀矿中进口矿配比持续下调的双重困境，通过一系列的工艺和设备技术改造，改善了烧结料层透气性，提高了烧结矿产能。

关键词： 技术改造，透气性，烧结性能

首秦 No.1 高炉末期 6.0 负荷的运行实践

徐　萌[1]，朱　利[2]，王　凯[2]，孙　健[1]，李荣升[1]

（1. 首钢集团有限公司技术研究院，北京　100041；2. 首秦公司，河北秦皇岛　066000）

摘　要： 在高炉适当降低冶炼强度前提下，调整原燃料结构改善烧结矿质量和焦炭整体性能、规整高炉炉型、提高送风风速和动能、调整高炉装料制度妥善引导中心和边缘两股煤气的合理分布、降低高炉碱负荷。通过以上多项措施协同，2016 年初综合入炉品位仅 54%~55% 时，1 号高炉实现 6.0 的高负荷水平并连续保持 6.0 负荷 42 天，2016 年 1 月、2 月月均燃料比约 497kg/t Fe，比 2015 年年均水平降低燃料消耗 30kg/t Fe。

关键词： 高炉，冶炼强度，焦炭性能，炉型，焦炭负荷

Production of No.1 BF at the Later Stage of Campaign up to 6.0 Coke Load at Shouqin Company

Xu Meng[1], Zhu Li[2], Wang Kai[2], Sun Jian[1], Li Rongsheng[1]

(1. Research Institute of Technology under Shougang Group Limited Company, Beijing 100041, China;
2. Shouqin Corporation under Shougang Group Limited Company, Qinhuangdao 066000, China)

Abstract: By decreasing the combustion intensity, improving the quality of sinter and the overall performance of coke with the adjustment of ore and fuel structure, putting in order of BF profile, increasing the blast velocity and kinetic energy, adjusting the charging scheme to rationally distribute the central and marginal gas flow and controlling the alkali load, the high level operation of No.1 BF under 54%~55% feed grade had been held 42 days in succession for the ore to coke ratio at 6.0 in January and February, 2016. The average monthly fuel ratio had been reduced to 497 kg/THM, about 30 kg/THM lower than last year.

Key words: blast furnace, BF profile, combustion intensity, coke performance, ore to coke ratio

风速对高炉煤气流分布的影响

徐 辉

（宝山钢铁股份有限公司炼铁厂高炉分厂，上海 200941）

摘 要： 风速是高炉下部送风制度的重要控制参数之一，基于高炉动力学模型和高炉生产实践研究了风速对高炉煤气流分布的影响。研究结果表明：通过风口衬套改变而改变风速时，高炉下部的软熔带略有变化，但在正常生产的范围内软熔带的变化很小；风速的改变对高炉上部的煤气流分布没有影响。

关键词： 高炉，风速，煤气流分布

1000m^{3}高炉在"经济产能"下指标的优化与提升

潘 林，孙连生

（山东钢铁股份有限公司莱芜分公司炼铁厂，山东莱芜 271104）

摘 要： 系统分析高炉生产关键控制点，分析炉料性价比，创建低成本炉料优化模型，为推行经济冶炼技术做好基础工作；为高炉提升产能创造条件，以风为纲、提高氧量、扩大矿批、提高高炉利用系数，在炉况稳定的基础上优化各项经济技术指标，使炉况保持了长时间的稳定顺行，实现了高炉的高产低耗。

关键词： 高炉，产能，炉况稳定，经济冶炼

Optimization and Promotion of 1000m³ Blast Furnace under "Economic Productivity"

Pan Lin, Sun Liansheng

(Shandong Iron and Steel Co., Ltd., Laiwu Branch Ironworks, Laiwu 271104, China)

Abstract: System analysis of critical control points of blast furnace production, analysis of charge price, creating a low cost burden for the implementation of the economic optimization model, smelting technology to do the basic work; to create conditions for the blast furnace to improve productivity, with the wind as the key link, improve oxygen, expand ore batch, improve blast furnace utilization coefficient, optimizing the economic and technical indicators based on the stability of furnace the furnace, to maintain a stable direct long time, achieve high yield and low consumption of blast furnace.

Key words: blast furnace, productivity, stable furnace condition, economical smelting

焦炭粉化对高炉透气性的影响研究

张文成[1,2]，张小勇[2]，郑明东[2]，任学延[3]

（1. 宝钢股份研究院梅钢技术中心，江苏南京　210039；2. 安徽工业大学
化学与化工学院，安徽马鞍山　243002；3. 梅钢制造部，江苏南京　210039）

摘　要： 本文为了探讨焦炭在高炉中的劣化对高炉压差的影响，对焦炭反应后的劣化进行了研究，提出了表征焦炭在二氧化碳反应后的粉化指数，表征焦炭在高炉内中下部粉化程度，更直接的解释了焦炭粉化对高炉透气性的影响。研究表明较低的配煤挥发分和适中的粘结性有利于降低焦炭 CPR；焦炭 CPR 与现有指标之间有一定的关系，随着灰分降低、焦炭 M_{10} 降低及焦炭 CRI 降低，有利于焦炭 CPR 降低。相同焦炭 CRI 下，焦炭的 CPR 呈现较宽的带宽，说明在相同溶损下，焦炭的反应后粉化程度不同，粉化越少则焦炭越好。从影响高炉压差和透气性的角度分析，对于 5BF 焦炭反应后粉化指数 CPR 小于 25% 为宜，相对于传统指标而言，要求其灰分小于 12.2%，焦炭 M_{10} 小于 6.5%，焦炭 M_{40} 大于 87%，焦炭 CRI 小于 26%，焦炭 CSR 大于 66%，为保证高炉压差的稳定降低创造条件。

关键词： 焦炭，反应性，粉化，压差

高铝型高炉渣物理化学性质

严志明，庞正德，白晨光，吕学伟

（重庆大学材料科学与工程学院，重庆　400044）

摘　要： 本文研究了 $CaO\text{-}SiO_2\text{-}Al_2O_3\text{-}MgO\text{-}TiO_2$ 五元渣系在不同 Al_2O_3 含量和不同 Al_2O_3/SiO_2 时炉渣的流动性能、表面性能和脱硫能力。研究结果表明：二元碱度为 1.20，随着 Al_2O_3 含量的增加，粘度先增加后减小，转折点为 24%。随着 Al_2O_3/SiO_2 增大炉渣粘度先减小后增大；表面张力随着 Al_2O_3 含量和 Al_2O_3/SiO_2 的增加而增加，渣系范围内的

表面张力介于 450~550mN/m 之间。二元碱度为 1.20，随着 Al_2O_3 含量的增加，炉渣脱硫能力降低；用 Al_2O_3 代替 SiO_2 使 Al_2O_3/SiO_2 含量从 0.47 变化到 1.50，随着 Al_2O_3/SiO_2 增大炉渣脱硫能力先升高后趋于平稳。

关键词：高铝高炉渣，粘度，表明张力，脱硫能力

Physicochemical Properties of Blast Furnace Slag with High Alumina

Yan Zhiming, Pang Zhengde, Bai Chenguang, Lv Xuewei

(School of Materials Science and Engineering, Chongqing University, Chongqing 400044, China)

Abstract: The effect of Al_2O_3 and the Al_2O_3/SiO_2 ratio on the viscosity, surface tension and sulfide capacity of the CaO-SiO_2-Al_2O_3-MgO-TiO_2 blast furnace slag system were studied in the present work. At a fixed CaO/SiO_2 ratio of 1.20, 9 mass pct MgO, and 1 mass pct TiO_2, the viscosity increases with an increase in Al_2O_3 content at a range of 16 to 24 mass pct, and decreases when the Al_2O_3 is higher than 24 mass pct. Increasing Al_2O_3/SiO_2 from 0.47 to 0.92 causes a slight decrease in viscosity of the slags and has an opposite effect when Al_2O_3/SiO_2 is more than 0.92. The surface tension increases with increasing Al_2O_3 content at a fixed Al_2O_3 content and Al_2O_3/SiO_2 at a fixed CaO content. The desulfurizing capacity decreases with increasing of Al_2O_3 content, while increases with increasing Al_2O_3/SiO_2 at a fixed CaO content.

Key words: high alumina BF slag, viscosity, surface tension, desulfurizing capacity

首钢京唐 1#高炉安全状态分析

马洪修[1]，宁晓钧[1]，张建良[1]，焦克新[1]，郑朋超[2]，陈艳波[2]

（1. 北京科技大学冶金与生态学院，北京　100083；

2. 首钢京唐钢铁联合有限责任公司，河北唐山　063200）

摘　要：本文重点对首钢京唐 1#高炉的基本情况、焦炭质量、炉缸残余厚度进行了整体的分析，统计了首钢京唐 1#高炉开炉以来所有的历史温度数据，根据其历史最高温度，计算出了炭砖的残余厚度。结合 1#高炉的实际风压、死铁层深度等，计算了其炉缸安全厚度，得出了炉缸侧壁炭砖的安全系数，为中国大型高炉提供参考。

关键词：高炉长寿，大型高炉，安全厚度

Analysis on Safety State of No.1 BF in Shougang Jingtang

Ma Hongxiu[1], Ning Xiaojun[1], Zhang Jianliang[1], Jiao Kexin[1],
Zheng Pengchao[2], Chen Yanbo[2]

(1. School of Metallurgical and Ecological Engineering，University of Science and Technology Beijing, Beijing 100083, China; 2. Shougang Jingtang United Iron and Steel Co., Ltd., Tangshan 063200, China)

Abstract: This paper focuses on the basic situation of Shougang Jingtang 1 # blast furnace, the quality of coke, the residual thickness of the hearth is analyzed. Statistics of the Shougang Jing Tang 1 #blast furnace since the opening of all the historical temperature data, according to its highest temperature, calculated the residual thickness of carbon brick. Combined with the actual wind pressure and the diameter of the hearth of 1 # blast furnace, the safety thickness of the hearth is calculated, and the safety factor of the carbon brick in the hearth of the hearth is obtained, which provides reference for the large blast furnace in China.

Key words: blast furnace longevity, large blast furnace, safety thickness

沙钢 5800m³ 高炉气流监控系统的开发和应用

赵华涛[1]，翟　明[1]，卢　瑜[1]，杜　屏[1]，朱　华[2]

(1. 江苏省（沙钢）钢铁研究院炼铁与环境研究室，江苏张家港　215625；

2. 江苏省沙钢集团有限公司炼铁厂，江苏张家港　215625)

摘　要： 开发了沙钢 5800m³ 高炉气流监控系统，包括对高炉高度方向气流分布、圆周方向气流分布以及径向气流分布进行全方位监控。从高度方向气流温度可以监控到高炉热量的上行还是下移以及软熔带根部位置的变化；通过对各层圆周方向气流分布可以监测到圆周方向的气流均匀性、边缘管道发生和演变过程；通过对十字测温枪和布料信号进行关联，提取布矿后，提尺时的中心、边缘和次中心的温度变化，排除了布料过程对气流变化的影响，气流分布更为准确。该系统为操作者及时准确提供了高炉内部气流的全部信息，帮助操作者及时进行炉况诊断和调剂，得到广泛的应用。

关键词： 高度方向，周向，径向，气流分布

Development and Application of Gas Flow Monitoring System in 5800m³ Blast Furnace of Shasteel

Zhao Huatao[1], Zhai Ming[1], Lu Yu[1], Du Ping[1], Zhu Hua[2]

(1. Ironmaking & Environment Research Group, Institute of Research of Iron & Steel, Shasteel, Zhangjiagang 215625, China; 2. Ironmaking Plant of Shasteel, Zhangjiagang 215625, China)

Abstract: The gas flow monitoring system was developed for 5800m³ shasteel to monitor the gas flow distribution in blast furnace axial direction, circumferential direction and radial direction. The thermal change in height direction and the cohesive zone level can be got from the axial temperature distribution graph; the uniformity of gas flow in circumferential direction and the peripheral channeling can be monitored timely from the circumferential temperature distribution graph; the gas temperature at center point, center-adjacent point and wall side of cross-sonde at winding up of mechanical probe was extracted to accurately reflect the radial distribution of top gas, eliminating the effect of burden charging process on gas flow distribution. This system provides the operator with comprehensive information of gas flow inside blast furnace, assisting the operator to diagnose and adjust the blast furnace in time. It was running well in the main control room of blast furnace, popular with each operator.

Key words: axial direction, circumferential direction, radial direction, gas flow distribution

武钢高炉炉料的中温还原过程分析

余珊珊，文志军，王素平

（武钢研究院铁前所，湖北武汉　430080）

摘　要： 文章对武钢高炉烧结矿的配矿结构调整后成品烧结矿的中温还原性能以及武钢现用球团矿以及块矿的中温还原性能进行了仔细的研究。经试验研究表明，烧结矿在 100~120min 时还原反应是个瓶颈，而在 160min 时，还原反应又再次得到突破，整个还原过程伴随有微观组织结构的变化；球团矿还原性能明显较烧结矿要差且结构改变不是很大；块矿的还原性能明显弱于球团矿以及烧结矿，且在 150min 之后还原反应几乎停滞不前。在目前高产低耗的生产需求下，对于高炉原料的选择，不应该仅仅从原料品位考虑，更应该考虑到能耗与高产之间的平衡。

关键词： 高炉，烧结矿，球团矿，块矿，中温还原度

Analysis of Reduction Process of Blast Furnace Burden in WISCO

Yu Shanshan, Wen Zhijun, Wang Suping

(Research and Development Center of Wuhan Iron & Steel Group, Wuhan 430080, China)

Abstract: In this paper, the middle temperature reduction performance of the sinter, the pellets and lump ores in the blast furnace of wisco are studied. The experimental results show that sinter in 100~120min reduction reaction is a bottleneck, and in the 160min, and once again break the reduction reaction, the reduction process is accompanied by changes in the microstructure; pellet performance is significantly worse than sinter and the structure change is not great; the reduction performance of lump ore was weaker than pellet and sinter, and after the 150min reduction reaction is almost stagnant. In the current high production and low production demand, the choice of raw materials for blast furnace, should not only consider the quality of raw materials, but also should take into account the balance between energy consumption and high yield.

Key words: blast furnace, sinter, pellet, ore lump, RI

鞍钢本部高炉系统降低生产成本实践

姜　喆，车玉满，郭天永，姚　硕，孙　鹏，费　静

（鞍钢股份有限公司技术中心，辽宁鞍山　114009）

摘　要： 本文对鞍钢降低高炉生产成本实践进行了总结。通过改变高炉造渣制度与炉料结构、调整喷吹煤配比和烧结配矿中白云石配比、加强生产与成本管理和优化高炉操作，达到了提高高炉技术指标降低生产成本的目的。

关键词： 高炉，成本，经济技术指标

Practice of Decreasing Production Cost in Angang Blast Furnace System

Jiang Zhe, Che Yuman, Guo Tianyong, Yao Shuo, Sun Peng, Fei Jing

(Technology Center of Angang Steel Co., Ltd., Anshan 114009, China)

Abstract: The paper summarizes the low-cost production experience in BF production in Ansteel. Economic and technical index was improved and production cost was reduced by changing slagging system, choosing rational burden composition, adjusting ratio of pulverized coal, decreasing dolomite usage in sinter, strengthening cost and production management and optimizing BF operation.

Key words: BF, cost, economic and technical index

马钢 3200m³ 高炉开炉工艺优化与提升

赵淑文，聂长果，王志堂，陈　军，张　群

（马钢股份公司第二炼铁总厂，安徽马鞍山　243000）

摘　要： 阐述了马钢 4 号 3200m³ 高炉根据采用的新工艺，新装备特点，对高炉顺利开炉、快速达产的技术进行了研究，采取了一系列针对性的工艺创新优化措施，实现了高炉快速开炉达产的目标。

关键词： 大型高炉，开炉，工艺优化

The Process Optimization and Improvement of Masteel 3200m³ BF Blowing-in

Zhao Shuwen, Nie Changguo, Wang Zhitang, Chen Jun, Zhang Qun

(No.2 Iron Making Plant of Masteel Co., Ltd., Maanshan 243000, China)

Abstract: Based on the characteristics of new process and equipment of No.4 3200m³ blast furnace in Masteel, the technology of the blast furnace's smooth blowing-in and rapid production was studied. A series of targeted optimization and innovation measures have been taken to achieve the goal of the blast furnace's smooth blowing-in and rapid production.

Key words: large blast furnace, blowing-in, optimization

八钢 C 高炉开炉实践

张文庆，田新龙

（新疆八钢炼铁分公司第二炼铁分厂，新疆乌鲁木齐　830022）

摘　要：对八钢 C 高炉开炉进行分析，主要介绍开炉料加入，点火送风，出铁及高炉降硅喷煤情况，总结开炉得与失，为下次开炉更快达产积累经验。
关键词：大型高炉，开炉，降硅，煤气利用率

八钢欧冶炉两次开炉实践对比分析

田宝山，陈若平

（宝钢集团新疆八一钢铁有限公司炼铁分公司，新疆乌鲁木齐　830022）

摘　要：经过一年半的优化改造，欧冶炉于 2017 年 3 月 25 日成功点火，26 日顺利出铁。对比两次开炉操作实践，2017 年开炉实现了安全稳定开炉、快速降硅的目标，标志着八钢欧冶炉开炉技术更加成熟。欧冶炉工艺技术的提升也进入了新的征程。
关键词：欧冶炉，开炉，烘炉，降硅

The Comparative Analysis between Two Times Blow-in Practice of Bayi OY

Tian Baoshan, Chen Ruoping

(Baosteel Group Xinjiang Bayi Iron and Steel Co., Urumchi 830022, China)

Abstract: After a year and a half of optimized retrofit, the European smelting furnace successfully ignited on March 25, 2017, and the iron was successfully released on June 26. Compared with the operation practice of the furnace, the target of a safe and stable open furnace and rapid drop of silicon in the furnace was realized in 2017, which marks the more mature technology of the furnace. The promotion of the process technology of the furnace has also entered a new journey.
Key words: european metallurgical furnace, furnace, the oven, silicon reduction

八钢炼铁低成本运行冶炼实践

王雪超，李胜强

（宝钢集团八钢公司炼铁分公司，新疆乌鲁木齐　830022）

摘　要：八钢公司炼铁分公司通过一系列的降低成本的措施：配煤结构优化，建立高炉成本模型，优化高炉用矿结构，改善工艺技术和管理人员操控水平，强化高炉冶炼，使炼铁成本得到大幅度降低。
关键词：低成本，配煤，冶炼实践

Low Cost Running Practice of Ironmaking in Bagang Group

Wang Xuechao, Li Shengqiang

(Iron-Smelting Branch, Bayi Iron & Steel Co., Ltd., Urumchi 830022, China)

Abstract: The ironmaking plant of Bagang has adopted a series of measures to reduce the cost.optimize the ratio of coal blending of coking, optimize the burden structure of the blast furnace, establishment of blast furnace cost model, to intensify the blast furnace smelting and improve the level of management, so that the cost of ironmaking has a significant reduction.

Key words: low cost, coal blending, smelting practice

本钢北营新 2#高炉紧急休复风操作实践

郑　文，王光亮，刘　森

（北营炼铁厂新 2 炉作业区，辽宁本溪　117000）

摘　要： 本钢北营新 2 号高炉成功处理风机断风导致的紧急休风：休风时处理好炉顶煤气确保煤气系统安全；快速恢复送水避免冷却系统烧坏；抢修送风系统及相关设备，争取及早送风；精心操作严格控制压差；及时排净渣铁等方面积累了经验，实现了炉况的快速恢复，避免了炉凉等恶性事故的发生，最大限度减少损失。

关键词： 大型高炉，停电休风，炉况恢复

Practice on Emergency Operation of New 2 # Blast Furnace in Beiying Company of Benxi Steel

Zheng Wen, Wang Guangliang, Liu Sen

(The New 2# Balst Furnace Ironmaking Plant of Beiying Operation Arer, Benxi 117000, China)

Abstract: Beiying Company of Benxi Steel No.2 blast furnace successfully deal with the wind off the wind caused by the emergency wind: the wind when dealing with a good roof gas to ensure the safety of gas systems; rapid recovery of water to avoid cooling system burned; repair air supply system and related equipment , for the early delivery of air; careful control of the pressure to strictly control; timely row of slag iron and other aspects of the accumulation of experience, to achieve the rapid recovery of the furnace conditions, to avoid the furnace cool and other vicious accidents occur, to minimize losses.

Key words: large blast furnace, power off the wind, furnace condition recovery

本钢北营新 2#高炉开炉装料实测

李　杰，孙世家，王光亮

（北营炼铁厂新 2#高炉作业区，辽宁本溪　117000）

摘　要：本钢北营新 2#高炉采用 PW 串罐无料钟炉顶装料设备，为探索北营新 2#高炉装料设备及在北营公司现有原燃料条件下的高炉布料规律，在投产前由北京神网对本高炉进行了开炉装料的实际测量，得出了料线 10m 内的料面形状、矿石和焦炭的料流轨迹及 FCG 曲线，为高炉开炉及日后高炉布料矩阵制定提供了有利依据。

关键词：高炉，装料设备，布料矩阵，料流轨迹

Measurement of Burden Profile of the New No.2 BF of Bengang Beiying

Li Jie, Sun Shijia, Wang Guangliang

(The New 2# Balst Furnace Ironmaking Plant of Beiying Operation Arer, Benxi 117000, China)

Abstract: The new 2# PW on the north of blast furnace top charging equipment tank without bell, In order to explore the new North 2# blast furnace charging equipment and blast furnace burden distribution in the north, the existing fuel conditions, Before the operation by the Beijing God Network for the blast furnace was measured in charge, The feed line 10 meters, surface shape of ore and coke material flow path and FCG curve, Provides a favorable basis for the blast furnace burden distribution matrix formulation and the future.

Key words: blast furnace, charging equipment, the fabric matrix, material flow track

莱钢提高 1000m³ 高炉炉顶压力操作实践

姬光刚，孙连生

（山东钢铁股份有限公司莱芜分公司炼铁厂，山东莱芜　271104）

摘　要：莱芜分公司炼铁厂有 3 座 1000m³ 高炉生产，在顶压的使用上与先进企业存在较大差距，通过采取加强原燃料管理、调整送风制度和优化炉前出铁组织等措施，逐步提高了顶压使用水平，从而提高了生铁的产量和质量，同时吨铁发电量有了明显提高，燃料比有了大幅度降低，取得了显著的经济效益。

关键词：高炉，顶压，送风制度，筛分速度

The Top Pressure of Blast Furnace Operation Practice to Improve the Laiwu Steel 1000m³

Ji Guanggang, Sun Liansheng

(Shandong Iron and Steel Co., Ltd., Laiwu Branch Ironworks, Laiwu 271104, China)

Abstract: Laiwu branch has 3 1000m³ blast furnace ironmaking plant production, and there is a big gap between the advanced enterprises in the use of the top pressure, by strengthening the management of raw materials, adjust the air supply system and the optimization of tapping organization measures, gradually improve the top pressure level, thereby increasing the yield and quality of pig iron, and T iron capacity has been significantly improved, the fuel ratio has been greatly reduced, and has achieved remarkable economic benefits.

Key words: blast furnace, top pressure, blast system, sieving speed

梅钢 4 号高炉中修开炉实践

董跃玲

（上海梅山钢铁股份有限公司炼铁厂，江苏南京　210039）

摘　要： 对梅钢 4 号高炉开炉前准备和开炉操作进行总结。系统梳理各项准备工作，主要包括清料、烘炉、试压捉漏以及开炉装料等，各自制定实施方案，严格有序落实，于 2016 年 11 月 3 日顺利开炉。开炉后通过合理控制送风制度、上部制度、热制度、造渣制度，达到高炉各项预期指标，燃料比大幅度下降，产能稳步提升，实现了开炉后 8 天顺利达产。

关键词： 高炉，烘炉，开炉，达产

Starting up Practice in Medium Maintenance of the No.4 BF in Meishan Iron & Steel

Dong Yueling

(Ironmaking Plant of Shanghai Meishan Iron & Steel Co., Ltd., Nanjing 210039, China)

Abstract: The paper summarized the preparation and operation of the starting up of the No.4 blast furnace in Meishan Iron&Steel. With the preparation of purging,heating-up, pressure and leak testing and charging, etc.,the blast furnace was opened smoothly on November 3rd, 2016.With the controlling of the blower system, charging system, heating system, slagging system,the expected targets of the furnace were achieved. The fuel ratio was reduced significantly ,the production capacity was improved steadily.The blast furnace reached the target capacity after 8 days of the starting-up.

Key words: blast furnace, heating-up, starting-up, reaching target capacity

大型高炉装料制度与炉况参数的数据挖掘

李壮年 [1,2]，阮根基 [2]，李宝峰 [2]，柳政根 [1]，储满生 [1]

（1. 东北大学冶金学院，辽宁沈阳　110819；2. 山西太钢不锈股份有限公司炼铁厂，山西太原　030003）

摘　要： 装料制度与炉况参数存在着内在的紧密联系，研究装料制度与炉况参数的规律对高炉操作意义重大。本文利用自主开发的高炉布料模型对高炉布料料面进行了模拟计算，对计算结果与炉况参数数据进行了挖掘、分析、总结，得出了大型高炉在不同装料制度下的一些冶炼规律，可以用于指导高炉生产。

关键词： 大型高炉，装料制度，炉况参数，数据挖掘，布料模型

Data Mining of Lager BF Charging System and Furnace Condition Parameters

Li Zhuangnian[1,2], Ruan Genji[2], Li Baofeng[2], Liu Zhenggen[1], Chu Mansheng[1]

(1. School of Metallurgy, Northeastern University, Shenyang 110819, China; 2. Ironmaking Plant, Shanxi Taigang Stainless Iron Shares Co., Ltd., Taiyuan 030003, China)

Abstract: The charging system and the furnace conditions have a close connection, and the study of the rules is of great significance to the operation of BF. By the self-developpe burden charging model , parameters under different charging system was simulated.The data of calculation results and furnace condition parameters are excavated, analyzed, and summarized, and some smelting rules under different charging system of large BF are obtained, which can be used to guide BF production.
Key words: large BF, charging system, furnace condition parameters, data mining, burden charging model

日照钢铁炼铁指标改善技术研究

范兴新，吕定建，董　岩

（日照钢铁有限公司，山东日照　276800）

摘　要：总结了日照钢铁高炉通过技术和管理的改善，通过上下部制度调剂、低硅冶炼、定期排碱等措施，高炉炉况达到长期稳定顺行，指标得到持续改善。
关键词：高炉，高炉上下部操作制度，低硅冶炼，锌碱金属富集

Technical Research on Improvement of Blast Furnace Index in Rizhao Steel

Fan Xingxin, Lv Dingjian, Dong Yan

(Rizhao Steel Co., Ltd., Rizhao 276800, China)

Abstract: Through the improvement of the technology and management of the blast furnace of Rizhaosteel Group, through the adjustment of the upper and lower parts of the system, the smelting of low silicon, and the regular discharge of alkali, the blast furnace condition has been stable, and the index has been continuously improved.
Key words: blast furnace, upper and lower parts of blast furnace system, low silicon smelting, zinc alkali concentration

提高炭素捣打料导热系数分析

邹祖桥，刘栋梁，胡正刚

（武汉钢铁有限公司研究院，湖北武汉　430080）

摘　要：针对高炉炭素捣打料实际工作温度低，而炭素捣打料导热系数检测制样高温焙烧情况，进行了炭素捣打料标准制定，该标准不仅要求导热系数高，而且导热系数测量制样只进行低温固化；为提高炭素捣打料导热系数，通过实验研究分析，提出了高导热炭素捣打料生产技术关键；并通过实验进行了不同结合剂炭素捣打料适用部位分析。

关键词：高炉，炭素捣打料，导热系数

Analysis of Improving Thermal Conductivity of Carbon Ramming Material

Zou Zuqiao, Liu Dongliang, Hu Zhenggang

(Research and Development Ceter of WISCO, Wuhan 430080, China)

Abstract: Because the working temperature of carbon ramming material in blast furnace was low and the thermal conductivity of carbon ramming material was detested after high-temperature roasting, the standard formulation of carbon ramming material was carried out, which requires high thermal conductivity and that the thermal conductivity measurement sample was prepared only by low-temperature curing; In order to improve the thermal conductivity of carbon ramming material, the key technology of high thermal conductive carbon ramming material was put forward through experiments research and analysis; The suitable parts for different anchoring agent of carbon ramming materials were analyzed through the experiments.

Key words: blast furnace, carbon ramming material, thermal conductivity

高炉炉底侵蚀模型技术研究

付　涛

（宝钢集团八钢公司炼铁分公司第二炼铁分厂，新疆乌鲁木齐　830022）

摘　要：通过对目前炉底侵蚀模型技术的发展动态和研究方向，提炼对目前大多数高炉可以使用的炉底侵蚀模型技术。

关键词：高炉，炉底侵蚀模型，有限元法

炉缸整体浇注技术的优势与应用

章荣会，邓乐锐，刘贯重，孙赛阳

（北京联合荣大工程材料股份有限公司，北京　101400）

摘　要：高炉炉缸陶瓷杯整体浇注技术，以其对炉缸原理透彻的领悟和行之有效的实践，逐渐被广大高炉炼铁技术所接受。相较于传统的砌砖陶瓷杯，浇注陶瓷杯，是真正意义上的"杯"结构，起到良好的隔离和隔热的作用，更有助于高炉长寿。另外，炉缸整体浇注技术，简单快捷，大大缩短工期；更有效利用残余碳砖，节省检修费用。结合不定形耐火材料的优点，炉缸整体浇注技术，必将在未来高炉检修中占据一席之地。

关键词：炉缸，陶瓷杯，不定形耐材，整体浇注

八钢 C 高炉铁口框架漏铁的原因分析及处理

陈佰军

（宝钢集团八钢制造管理部，新疆乌鲁木齐　830022）

摘　要：本文通过对八钢 C 高炉铁口框架漏铁事故原因的深入分析，提出了处理措施及后期的改善优化措施，对高炉铁口维护及事故预防具有借鉴作用。

关键词：铁口，漏铁，高炉

Analysis and Treatment of the Leakage Iron in the Frame of C Blast Furnace in Bayi Iron & Steel

Chen Baijun

(Manufacturing Department of Bayi Iron & Steel Co., Baosteel Group, Urumchi 830022, China)

Abstract: This article through to iron frame leakage of C blast furnace in Bayi Steel deep analysis the cause of the accident, put forward the treatment measures and improve the optimization measures, in the late of blast furnace iron mouth maintenance could be a reference and accident prevention.

Key words: iron mouth, leakage of iron, the blast furnace

八钢高炉配加高硅块矿的成本分析

吴瑞琴，陈佰军，王梅菊，马　雄

（宝钢集团八钢制造管理部，新疆乌鲁木齐　830022）

摘　要：配加块矿是降低炼铁成本的主要措施之一，本文通过对八钢炼铁使用量较多的三种块矿的成本分析，为八钢高炉高比例配加块矿以及采购高性价比块矿提供依据。

关键词：块矿，成本，采购

The Cost Analysis with High Silicon Ore in the Blast Furnace of Bayi Steel

Wu Ruiqin, Chen Baijun, Wang Meiju, Ma Xiong

(Manufactuing Department Bayi Iron & Steel Co., Baosteel Group, Urumchi 830022, China)

Abstract: With addition of lump ore is one of the main measures of lowering ironmaking cost,this article through to Bayi steel ironmaking usage more cost analysis of the three kinds of ore, to Bayi steel blast furnace high proportion with addition of lump ore and cost-effeceive ore procurement provides the basis.

Key words: ore, cost, purchase

八钢 A 高炉回收煤气降料面操作实践

李刚义，黄银春

（新疆八一钢铁有限公司炼铁分公司，新疆乌鲁木齐　830022）

摘　要： 对八钢 A 高炉(2500m³)干法除尘回收煤气降料面操作经验进行总结。在保证布袋入口温度≤280℃前提条件下，控制高炉炉顶温度≤400℃，采用大风量、高顶压，缩短降料面进程，安全将降料面降至风口。

关键词： 降料面，回收煤气，干除尘法

浅谈高炉热制度与炉缸活性的关系

代　兵[1]，姜　曦[2]，王运国[1]，张建良[3]，刘云彩[4]

（1. 本钢板材股份有限公司炼铁厂，辽宁本溪　117000；2. 中国钢铁工业协会，北京　100711；
3. 北京科技大学冶金与生态工程学院，北京　100083；4. 首钢总公司总工办，北京　100041）

摘　要： 通过分析高炉热制度和炉缸活性的意义，明确给出了热制度与炉缸活性的关系，即渣铁流动性能是连接炉缸活性和热制度的纽带，合理稳定的热制度是维护炉缸活性的前提条件。良好的炉缸活性状态需要有良好的渣铁流动性能，良好的渣铁流动性能需要合理稳定的热制度，而合理稳定的热制度则是高炉操作者日常工作的主要内容，本文明确给出了控制炉温的主要方法和思路。

关键词： 高炉，热制度，炉缸活性，关系

Elementary Introduction to the Relationship between Blast Furnace Thermal System and Hearth Activity

Dai Bing[1], Jiang Xi[2], Wang Yunguo[1], Zhang Jianliang[3], Liu Yuncai[4]

(1. Ironmaking Works of Benxi Steel Plate Co., Ltd., Benxi 117000, China; 2. China Iron and Steel Association, Beijing 100711, China; 3. School of Metallurgical and Ecological Engineering, University of Science and Technology Beijing, Beijing 100083, China; 4. General Engineer Office, Shougang Corp., Beijing 100041, China)

Abstract: By analyzing the significance of blast furnace thermal system and hearth activity, the relationship between the thermal system and hearth activity has been specifically put forward, which the slag and iron flow performance is the link between thermal system and hearth activity, and reasonable and stable thermal system is the precondition to maintain good hearth activity. Good hearth activity needs good slag and iron flow performance which needs reasonable and stable thermal system which is the main content of the daily work of blast furnace operators. The main methods and ideas of controlling thermal system also have been specifically put forward.

Key words: blast furnace, thermal system, hearth activity, relationship

铁水脱钛技术研究

张雪松[1]，梁海龙[1]，张殿伟[1]，周继良[1]，张　勇[1]，刘文运[1]，
赵满祥[2]，赵铁良[2]，张海滨[2]，贾国立[2]，罗德庆[2]，贾军民[2]

（1. 首钢技术研究院，北京　100043；2. 首钢京唐公司，河北唐山　063200）

摘　要： 实验室开展不同脱除剂、不同配比、不同铁中钛含量情况下的脱钛试验，同样配比情况下轧钢铁皮作为脱除剂比烧结矿脱钛率稍好，但差别不是很大；脱除剂配比 1% 与 2% 时脱除效果相差不大；钛脱除效率在 30%~35% 左右；加白灰的混合脱除剂未取得更好的钛脱除效果；铁中钛含量高（0.13%）和低（0.09%）时钛脱除效率变化不大。由于实验室铁水量 5kg 量小，搅拌 4min 左右基本已经粘稠，影响脱除效果，工业生产脱除效率会好于实验室结果。

关键词： 铁水，脱钛，脱除率

Abstract: Laboratory test of Ti removing is developed in different remover, different proportion and different Ti content of hot metal. Under the same conditions Ti removal rate using steel rolled iron scale is better than using sinter but the difference is not great. Ti removal rate is almost same in the ratio of the 1% and 2% remover. Ti removal rate is around 30%~35%. Ti removal rate has not obtained the better effect when adding lime. Ti removal rate is little changed under high titanium content in hot metal (0.13%) and low (0.09%). As the amount of hot metal only 5kg in the laboratory is small, the hot metal is viscous after stirring 4 minutes, Ti removal rate is affected. Ti removal rate of industrial production will be better than the result of the laboratory.

Key words: hot metal, removal of titanium, removal rate

宁钢高炉含 Zn 物料使用实践

刘维勤，丁德刚，廖洪亮

（宁波钢铁有限公司炼铁厂，浙江宁波　315800）

摘　要： 宁钢高炉含 Zn 物料的使用，曾经走过一段弯路，由于缺乏实际使用经验，入炉 Zn 负荷未能得到有效控制，对高炉炉况产生影响。通过对含 Zn 物料使用过程几起典型案例的经验教训总结，制定严格的控制标准，完善和改进含 Zn 物料使用原则，建立完善的使用测算系统，含 Zn 物料的使用有序可控，高炉炉况明显改善，各项技术经济指标有了较大的提升。

关键词： 含 Zn 物料，标准，使用，测算

Practice on Blast Furnace Operation Using Zinc-containing Materials at Ningbo Steel

Liu Weiqin, Ding Degang, Liao Hongliang

(Ironmaking Plant of Ningbo Iron & Steel Co., Ltd., Ningbo 315800, China)

Abstract: Due to insufficiency in using experience, Ningbo Steel blast furnace operators took lessons from consuming zinc-containing materials. They used to ignoring the control on zinc input which led to worsening working condition of blast furnace. By studying a few of typical cases related to use of zinc-containing materials, the plant stipulates stringent control standard, perfect and improve the principles in using zinc-containing materials and build up complete measurement system to make zinc-containing material usage under control which leads to obviously improved blast furnace working condition and in turn upgraded technical and economical indices substantially to extent.

Key words: zinc-containing materials, standard, use, measurement

本钢北营 3200m³ 高炉装料制度的优化

刘　森，王光亮，李　杰

（本钢集团北营炼铁厂，辽宁本溪　117017）

摘　要： 结合本钢北营新 2 号高炉降低燃料比的相关生产实践，基于理论分析和现场试验，依据对大量生产数据的对比分析，研究了降低北营 3200m³ 高炉燃料比的可行途径。通过采用优化装料制度等技术措施，提高了高炉煤气利用率，高炉燃料比大幅度降低。

关键词： 3200m³ 高炉，燃料比，装料制度

Analysis of Lower Fuel Ratio of 3200m³ Blast Furnace in Blant Optimization of 3200m³ Blast Furnace Charging System for Beiying Company of Benxi Steel

Liu Sen, Wang Guangliang, Li Jie

(The New 2# Balst Furnace Ironmaking Plant of Beiying Operation Arer, Benxi 117017, China)

Abstract: In this paper, the Beiying No.2 blast furnace production practices associated with reduced fuel ratio, based on theoretical analysis and field tests, according to the comparison of mass production data analysis, study possible ways of reducing North Camp 3200m³ blast furnace fuel ratio. By using technical measures to optimize the charging system, improve the utilization of blast furnace gas, blast furnace fuel ratio significantly reduced.

Key words: 3200m³ blast furnace, fuel ratio, charging system

莱钢 5#1080m³ 高炉护炉生产实践

张 明，张 振

（山东钢铁股份有限公司莱芜分公司炼铁厂，山东莱芜 271104）

摘 要： 本文主要介绍了莱钢 5#1080m³ 高炉通过采取降低冶炼强度、提高炉温、配加钛球及风口喂钛线护炉等措施，选择合理的操作制度，达到炉役后期生产安全和炉况顺行的效果。

关键词： 高炉，炉役后期，护炉

炉顶空间温度的降低与测量

汪 银，袁本雄

（中信泰富特钢集团铜陵泰富特种材料有限公司，安徽铜陵 244100）

摘 要： 焦炉加热水平高度设计为 1100mm 的 JNX3-70-1 型焦炉，投产后存在炉顶空间温度达过高的问题。本文主要介绍使用焦炉煤气加热的 7m 焦炉如何降低炉顶空间温度，减少炉顶空间温度过高造成的石墨等问题。另外对炉顶空间温度的测量进行不同方式的测量与比较，减少测量方式对测量数据造成的偏差。

关键词： 焦炉，炭化室，石墨

Reduction and Measurement of Roof Space Temperature

Wang Yin, Yuan Benxiong

(CITIC Pacific Special Steel Holdings, Tongling Pacific Special Materials Co., Ltd., Tongling 244100, China)

Abstract: The heating level of coke oven is designed to be a 1100mm jnx3-70-1 coke oven, and the temperature of the roof space after production is too high. This paper mainly introduces how to reduce the temperature of the top space of the furnace and reduce the graphite caused by excessive temperature in the roof space. In addition, the measurement and comparison of the temperature of the roof space of the furnace are measured, and the deviation of the measurement data is reduced.

Key words: coke oven, chamber, graphite

基于 StreamDEM 的高炉布料首层料面研究

张 瑶，钟星立，钟 渝，张 波

（中冶赛迪技术研究中心有限公司，重庆 401133）

摘　要： 本文在离散元仿真软件 StreamDEM 的基础上，二次开发了高炉布料模块。用该模块对高炉首层料面进行了离散元仿真研究。通过软件分析出首层料面形状，首层料面结构分布以及首层料的矿焦比例，为布料矩阵调整提供理论依据，为高炉仿真中的料流模型提供输入条件。

关键词： 离散元，StreamDEM，高炉布料，料面形状

Study on the First Layer of Blast Furnace Fabric Based on StreamDEM

Zhang Yao, Zhong Xingli, Zhong Yu, Zhang Bo

(CISDI Research & Development Co., Ltd., Chongqing 401133, China)

Abstract: In this paper, based on the discrete element simulation software StreamDEM, the blast furnace fabric module was developed twice. The module is used to study the discrete element simulation of the first layer of blast furnace. The distribution of the first layer surface, the distribution of the first layer of the surface structure and the ratio of the coke in the first layer were analyzed by software, which provided the theoretical basis for the adjustment of the fabric matrix and provided the input conditions for the flow model in the blast furnace simulation.

Key words: DEM, StreamDEM, blast furnace fabric, surface shape

方大特钢炼铁原料系统创新管理实践

葛晓柱，钟国英

（方大特钢科技股份有限公司炼铁厂，江西南昌　330012）

摘　要： 方大特钢炼铁原料系统瞄准市场脉络，以经济效益为中心，通技术过创新管理，推进优化配矿及合理回收利用含铁资源，建立原料性能数据库，指导生产，提高指标，降低生产成本；同时围绕原料加工系统创新管理，各工序开展了工艺、设备、环保配套改造等技术工作。实现了成本最低化、质量最优化，产能最大化，提升市场竞争力。

关键词： 原料场，创新，成本，低耗

The Innovation and Management Practice of Fangda Special Steel Raw Materials System

Ge Xiaozhu, Zhong Guoying

(Fangda Special Steel Technology Co., Ltd., Nanchang 330012, China)

Abstract: The iron and steel raw materials system of Fangda Steel, through innovation management, is promoted to optimize distribution and reasonable recycling of iron resources. Through the innovation management of the raw material system, the process is carried out the technical work such as the process equipment environmental protection. We have realized the minimum cost, optimized quality, maximized capacity, and improved the competitiveness of the market.

Key words: raw materials field, innovation, cost, low consumption

基于相控阵雷达数据的高炉三维料面可视化研究

张　森，魏凯东，张海刚，尹怡欣，陈先中

（北京科技大学自动化学院工业过程知识自动化教育部重点实验室，北京 100083）

摘　要： 在高炉工业现场高温高危环境中，由于工人不能直接观测高炉的实时炉况与料面，造成了很大程度上的资源浪费，对高炉的生产造成了很大的干扰。本文为解决上述问题，提出了高炉三维料面的重构算法，基于高炉现场使用的相控阵雷达采集的数据，首先对雷达测量获取的三维点云数据进行去噪，然后使用 Delaunay 算法对去噪后的点云三角化处理，最终形成真实的三维料面。为提高数据的利用率，运用 K 近邻算法进行去噪，并提出改进的 Delaunay 算法，利用该算法把散乱点云数据进行三维重构，逼近真实的三维高炉料面。根据形成的三维料面，将特征提取压缩至二维料线层面，并提取料线特征的做法，本文提出的算法具有数据利用率大，速度快，精度高的优点。

关键词： 相控阵雷达；三维点云，K 近邻算法，Delaunay 算法，三维料面

3D Surface Reconstruction of Blast Furnace Based on Phased Array Radar Data

Zhang Sen, Wei Kaidong, Zhang Haigang, Yin Yixin, Chen Xianzhong

(Key Laboratory of Knowledge Automation for Industrial Processes of Ministry of Education, School of Automation and Electrical Engineering, University of Science and Technology Beijing, Beijing 100083, China)

Abstract: In the industrial field of blast furnace high temperature high risk enviroment, the staff can not directly see that surface condition and real-time blast furnace, this situation caused largely a waste of resources, for the efficient production of the blast furnace caused great disturbance . In order to solve the above this problem, I proposed a 3D reconstruction algorithm of blast furnace burden. With the blast site using the phased array radar data acquisition, 3D point cloud data to obtain the first measurement of cleaning denoising, and then use the Delaunay triangulation algorithm for triangulation to take denoising processing of point cloud, the final form of three dimensional surface is relatively flat. In order to improve the utilization of data, in the process of processing the data using the KNN algorithm is applied to denoising, and then use the Delaunay triangulation algorithm improved the algorithm to get the scattered point cloud data form connected, continuous approximation to the true three-dimensional high furnace burden. Use with the existing blast furnace material extraction line algorithm are compared, the proposed algorithm has high data utilization rate, fast speed, high accuracy. The speed of the algorithm is also calculated, and the feasibility of the method is verified in the MATLAB simulation environment.

Key words: phased array radarc 3D point cloud, KNN algorithm, Delaunay triangulation algorithm, 3D surface

太钢 3 号高炉长期稳定顺行生产实践

汤海明，徐纪山，唐顺兵，曹建华，赵雪斌，师　青

（山西太钢不锈钢股份有限公司，山西太原　030003）

摘　要：对太钢 3 号高炉开炉 10 年来保持长期稳定顺行的实践进行了总结。从原燃料管理，装料制度，送风制度，造渣制度，铁口维护和设备管理这六个方面介绍了太钢 3 号高炉的典型做法。重点阐述了在大比例使用小粒级焦炭时如何排料，大比例使用球团时如何排料，如何实现高风温冶炼以及确定风速和鼓风动能的方法和原则。

关键词：风速和鼓风送能，大比例小粒级焦炭，大比例球团，铁口维护

邢钢 6 号高炉降低侧壁温度操作实践

郭　利，王一杰，谷　松

（邢台钢铁责任有限公司炼铁厂六高炉车间，河北邢台　054027）

摘　要：邢钢 6 号高炉炉缸侧壁温度最高 555℃，2016 年 10 月份开始常规护炉，效果不明显，侧壁升高趋势仍在继续。2017 年 3 月份开始，调整操作思路。以减少侧壁环流为主要目的。通过调整布料，提高鼓风动能，活跃炉缸，提高铁口深度和稳定性，等措施实现了侧壁温度降低的目标，为高炉提产、降耗，创造条件。

关键词：炉缸侧壁，最高 550℃，减少侧壁环流，温度降低

邯钢 1 号高炉长寿生产实践

邓　涛，陈　奎，王　云，夏万顺

（邯郸钢铁集团邯宝炼铁厂，河北邯郸　056000）

摘　要：邯钢 1 号高炉在日常的生产中高度重视高炉长寿问题,在技术人员的精心操作下使得自开炉以来持续安全、高效、稳定运行。经过生产实践的总结与分析，得出了 1 号高炉长寿的操作经验与规律，合理运用操作制度保持高炉长期稳定顺行，保证炉缸活跃、强化中心减少边缘侵蚀，加强原燃料质量管控及有害元素的入炉和富集，检修时候进行定点灌浆修补等措施确保了高炉的安全运行。

关键词：高炉，长寿，稳定顺行，定点灌浆

河钢石钢 2 号高炉炉缸侵蚀的控制

王贵宝，张余斌，习立阳，王彩永，任艳军，赵培义

（河钢集团石家庄钢铁有限责任公司，河北石家庄　050031）

摘　要：对石钢 2 号高炉（1080m³）炉缸局部热流强度升高进行了研究与分析，并通过采取炉皮灌浆、加装炉衬监测电偶、调整风口布局、改高压水加强冷却、局部堵风口、降低冶强等一系列可行性控制措施，炉缸热流强度降到了安全范围之内，保证了炉缸的安全运行。

关键词：高炉，炉缸侵蚀，控制

Control Practice of Hearth Erosion of No.2 BF at Shijiazhuang Iron and Steel Company

Wang Guibao, Zhang Yubin, Xi Liyang, Wang Caiyong,

Ren Yanjun, Zhao Peiyi

(Shijiazhuang Iron and Steel Company of Hebei Iron and Steel Group, Shijiazhuang 050031, China)

Abstract: This study analyzes the part thermal flow strength increasing in hearth of the No.2 blast furnace (1080m^3) in Shijiazhuang Iron & Steel. Co, Ltd. and adopts a series of reliable improvement including the furnace grouting, adding the lining thermal couple, adjusting the outlet, increasing the high pressure water to strengthen the cooling, plugging part tuyere, reduce the smelting strength, thus the hearth heat flow down to a sale range to ensure safe operation of the hearth.

Key words: blast furnace, hearth erosion, regulatory measure control measures

石钢 0 号高炉控制炉缸热流、高效低耗生产实践

袁兆锋，李 斌，姜勇硕，任艳军，刘国文

（河北钢铁集团石钢公司炼铁厂，河北石家庄 050031）

摘 要： 石钢 0 号高炉于 2012 年 11 月 18 日大修投产；由于受年修及环保减排影响，0 号炉经历多次长期焖炉操作，至 2015 年第四季度铁口区域冷却壁热流强度逐步升高到事故值，12 月被迫堵风口控产。2016 年 0 号高炉采取多种技术措施，开拓创新，通过多种手段确保炉缸热流稳定受控，实现公司的铁、钢平衡及煤气平衡；保持了炉况长期稳定、安全运行，实现稳产、安全、低耗生产。

关键词： 调剂原料结构，改造灌浆，安全受控，管理

Production Practice of Controlling Furnace Heat Flow and High Efficiency and Low Consumption in No. 0 BF of Shisteel

Yuan Zhaofeng, Li Bin, Jiang Yongshuo, Ren Yanjun, Liu Guowen

(Hebei Iron and Steel Company Ironmaking Plant, Shijiazhuang 050031, China)

Abstract: No.0 blast furnace of Shi Steel production in November 18, 2012 due to repair and overhaul; environmental protection and emission reduction effect, No.0 boiler experienced many long-term soaking operation, to the fourth quarter of 2015 Tiekou regional cooling wall heat flux intensity gradually increased to the value of December forced the accident, blocking air control production. No.0 BF 2016 adopting various technical measures, innovation, through a variety of means to ensure stable and controlled to achieve the balance of hearth heat, gas and iron and steel company balance; to maintain a long-term stable and safe operation of the furnace, to achieve stability, safety and low cost production.

Key words: adjustment of raw materials, structural modification, grouting safety, controlled management

薄壁炉型在中小高炉升级改造中的应用

吕定建，董　岩，范兴新

（日照钢铁控股集团有限公司，山东日照　276806）

摘　要： 为实现低成本炼铁以及实现节约改造投资目的，日钢高炉使用原设计的配套公辅设施原地进行扩容大修。该炉型采用更低的高径比 Hu/D 和较小的炉腹角和炉身角，更大的 Vu/A、D/d、d1/d，超过了高炉设计推荐的薄壁高炉内型尺寸范围，该炉型满足灵活选择炉料结构并能实现较差原燃料条件下的高产低耗，使薄壁"经济"炉型高炉能够取得传统炉型无法达到的指标。

关键词： 经济炉型，煤气炉内停留时间，经济炼铁

Application of Thin Walled Blast Furnace in Small and Medium Sized Blast Furnace

Lv Dingjian, Dong Yan, Fan Xingxin

(Rizhao Steel Co., Ltd., Rizhao 276806, China)

Abstract: In order to achieve low-cost iron and save investment purposes, Rizhao Steel blast furnace using the original design of the supporting and auxiliary facilities in situ super-popular overhaul. Of No.1 using lower high diameter smaller than Hu/D and bosh Angle and shaft Angle, the bigger Vu, d1/D/A, D/D, More than the recommended thin wall inside the blast furnace blast furnace design type size range, No.1 satisfy flexible choice burden structure and can realize the original fuel poor under the condition of high yield and low energy consumption, less the original fuel to make thin wall "economic" type furnace blast furnace can make the traditional type furnace cannot reach the target.

Key words: economical BF profile, gas retention period in BF, economical ironmaking

日钢高炉锌平衡研究及锌负荷对燃料比的影响

董　岩，吕定建，范兴新

（日照钢铁有限公司，山东日照　276806）

摘　要： 介绍了锌在高炉内的还原反应、锌的循环富集和对高炉的不利影响。分析了日钢一年来高炉锌负荷变化情况，日钢高炉锌负荷平衡状况及近期高炉锌负荷偏高的主要原因，锌负荷变化对日钢高炉燃料比影响的定量分析以及日钢高炉的适宜锌负荷。

关键词： 锌负荷，锌平衡，燃料比

Rizhao the Influence of High Furnace Zinc Balance Study and Zinc Load on Fuel Ratio

Dong Yan, Lv Dingjian, Fan Xingxin

(Rizhao Iron and Steel Co., Ltd., Rizhao 276806, China)

Abstract: This paper introduces the reduction reaction of zinc in blast furnace and the circulation of zinc，and the harmful effect of zinc overload on blast furnace. The change of zinc load in blast furnace for a year is analyzed. The main reasons of the load balance condition of steel blast furnace and high furnace zinc load，the change of zinc load is the quantitative analysis of daily steel blast furnace fuel ratio and the suitable zinc load of daily steel blast furnace.

Key words: zinc load, zinc balance, fuel ratio

炉缸侧壁温度升高原因分析及处理措施

王玉会

（日照钢铁集团有限公司炼铁制造部，山东日照 276806）

摘 要： 高炉长寿取决于炉缸寿命，炉缸侧壁温度升高是制约高炉长寿的重要因素。本文分析了炉缸侧壁温度升高的原因并介绍了主要处理措施，对相关高炉操作提供借鉴。

关键词： 炉缸，侧壁温度，长寿

Analysis and Processing of Abnormal Temperature Rise on Side Wall of Blast Furnace Hearth

Wang Yuhui

(Rizhao Steel Holding Group Co., Ltd., Ironmaking Department, Rizhao 276806, China)

Abstract: Long campaign life of blast furnace depends on blast furnace hearth.The reason for abnormal temperature rise was analyzed ,the processing was introduced. It is a very good reference for blast furnace presenting abnormal temperature rise on side wall.

Key words: blast furnace hearth, temperature on side wall of blast furnace, hearth long campaign

红钢 3#高炉提高煤气利用率生产实践

普 松，唐德元，钟 伟，赵树逵

（红河钢铁有限公司，云南蒙自 661100）

摘　要：对红钢 3#高炉提高煤气利用率进行了总结，通过从原燃料质量管理、装料制度探索、操作管理三方面入手，使煤气利用率从 2015 年的 40.57%提高到 44.52%，8 月份完成燃料比 545kg/t，突破了历史最好水平。
关键词：高炉，煤气利用率，原燃料，达产

Production Practice of Increasing Gas Utilization Rate in No.3 BF of Honghe Iron & Steel Co., Ltd.

Pu Song, Tang Deyuan, Zhong Wei, Zhao Shukui

(Kunming Iron & Steel Co., Ltd. Honghe Iron & Steel Co., Ltd., Mengzi 661100, China)

Abstract: This paper summarizes the improvement of gas utilization rate in No.3 blast furnace of Honggang Iron and Steel Company, and improves the utilization rate of gas from 40.57% in 2015 to 44.52% through the three aspects of fuel quality management, loading system exploration and operation management. August to complete the fuel ratio of 545kg / t, breaking the best level in history.
Key words: blast furnace, gas utilization, fuel, production

鞍钢 3 号高炉铜冷却壁破损维护实践

谢明辉，曾　宇，佟敏英，李晓春

（鞍钢股份有限公司，辽宁鞍山　114021）

摘　要：对鞍钢 3 号高炉炉体维护技术进行总结。针对鞍钢 3 号高炉炉腹、炉腰、炉身下部区域铜冷却壁大量破损、炉壳温度高等问题，通过采取破损冷却壁穿金属软管、安装微型冷却器、压浆造衬、炉壳洒水等一系列措施，有效控制了炉壳温度升高，保证了高炉安全生产及稳定顺行。
关键词：穿管，微型冷却器，压浆造衬

Wearing and Maintenance Practice of No.3 BF Copper Stave in Angang

Xie Minghui, Zeng Yu, Tong Minying, Li Xiaochun

(Angang Steel Company Limited, Anshan 114021, China)

Abstract: For years maintaenance practice in No.3 BF, a few experience has been attained. For the copper stave damage problem in bosh, belly and down shaft, and the shell temperature is higher, the compressive measurement, the likes of applying metal flexible hose, setting up micro-cooler, utilization of grouting Lining technology, using water-spray cooling method, has been conducted to control the shell temperature, and maintain the smooth operation for Angang BF.
Key words: copper stave, metal flexible hose, micro-cooler, grouting lining technology

日钢高炉提高利用系数生产实践

范兴新，吕定建，董 岩

（日照钢铁有限公司，山东日照 276806）

摘 要：为满足炼钢生产，提升公司效益，炼铁进行了高炉提产攻关，通过采取改善原燃料质量，高炉进行上下部调剂提高入炉风量，维持高富氧率等措施，在燃料比不升高的情况下，提升了炼铁产量。

关键词：高炉，富氧率，加风，煤气流分布

Practice of High Productivity for BF in Rizhaosteel

Fan Xingxin, Lv Dingjian, Dong Yan

(Rizhao Iron and Steel Company Limited, Rizhao 276806, China)

Abstract: In order to meet steel production, enhance the company's efficiency and ironmaking of blast furnace production research, through improved fuel quality and transfers up and down increasing the air flow into the furnace the blast furnace, high oxygen enrichment measures such as, in the case of fuel than increased, increased iron production.

Key words: blast furnace, oxygen enrichment, increase the air volume, gas distribution

不同炉料还原软化熔融及成渣过程

秦喧柯[1]，吴 铿[1]，陈小敏[1]，赵路朋[1]，赵 勇[1,2]，潘 文[2]

（1. 北京科技大学钢铁冶金新技术国家重点实验室，北京 100083；
2. 首钢技术研究院，北京 100041）

摘 要：为研究含铁炉料软化熔融及成渣过程，本试验综合利用了两种试验装置，熔滴炉和卧室高温可视反应炉，研究了 4 种单类含铁炉料和 2 种混合炉料的软化熔融特性，并观察到含铁炉料在高温下软熔及成渣行为，得到了直观图像，采用矿相解离分析仪分析生成物质的物相组成。结果表明：不同类型含铁炉料 C 还原形成的渣相中物相组成差异很大，高碱度烧结矿主要形成 $CaO\text{-}SiO_2\text{-}Al_2O_3\text{-}FeO\text{-}MgO$ 五元系渣，球团矿形成 $SiO_2\text{-}Al_2O_3\text{-}FeO\text{-}CaO$ 四元系渣，块矿形成 $FeO\text{-}SiO_2$ 二元系渣；试验中有气泡生成现象，该气泡生成现象反应了初渣的高温物理性能，渣相中 FeO 含量越高，其与内生气源生成气泡能力越强。

关键词：高炉，炉料，软化熔融，初渣，气泡，FeO

The Softening-melting and Slag Forming Process of the Different Ferrous Burden Materials

Qin Xuanke[1], Wu Keng[1], Chen Xiaomin[1], Zhao Lupeng[1], Zhao Yong[1,2], Pan Wen[2]

(1. State Key Laboratory of Advanced Metallurgy, University of Science and Technology Beijing, Beijing 100083, China; 2. Shougang Research Institute of Technology, Beijing 100041, China)

Abstract: In order to study the softening melting and slag forming process of the ferrous burden materials, two kinds of experimental devices, the softening-melting furnace and the high-temperature visualization furnace were used in this study. In this study, four kinds of single-type furnace burden and two kinds of mixed furnace burden were studied. The intuitive image of the test process was obtained. The phase composition of the material was analyzed by the dissociation analyzer. The results show that the phase composition of the slag phase formed by different furnace burden is very different. The high-alkali sintered ore mainly forms CaO-SiO_2-Al_2O_3-FeO-MgO five-element slag and the pellets form SiO_2-FeO-CaO quaternary slag and lumps to form FeO-SiO_2 binary slag. There is a bubble generation phenomenon in the experiment. The bubble generation phenomenon reflects the high temperature physical properties of the primary slag. And the higher FeO content in the slag phase, the stronger the ability to generate bubbles with endogenous gas.

Key words: blast furnace, burden, softening-melting, primary slag, bubble, FeO

水框架结构对风口小套传热的影响研究

郝良元[1]，吴振威[2]，张建良[2]，马富涛[2]，焦克新[2]

（1. 河钢集团钢研总院，河北石家庄　050000；
2. 北京科技大学冶金与生态工程学院，北京　100083）

摘　要：本文利用 SolidWorks 软件建立风口小套三维传热模型，利用计算流体力学的方法，计算了风口小套的温度场，并分析了进口水速以及水框架层数对风口小套温度场的影响，得到如下结论：在计算范围内，进口水速在12m/s 以下时对风口小套的传热影响较大，而进口水速在 12m/s 以上时对风口小套的传热影响较小。说明在实际高炉生产中，宜选取 12m/s 的进口水速；在计算范围内，水框架结构对风口小套前端面的温度场分布没有显著影响，而对风口小套侧壁的温度场有较大影响。水框架层数为 2 层时，侧壁温度最大，但是依然远远小于铜的许用温度，从温度梯度的角度来看，水框架层数为 2 时最佳。

关键词：风口小套，数值模拟，进口水速，水框架结构

Study on Influence of Water Frame Structure on Heat Transfer of Tuyere

Hao Liangyuan[1], Wu Zhenwei[2], Zhang Jianliang[2], Ma Futao[2], Jiao Kexin[2]

(1. HBIS Group Technology Research Institute, Shijiazhuang 050000, China; 2. School of Metallurgical and Ecological Engineering, University of Science and Technology Beijing, Beijing 100083, China)

Abstract: The software SolidWorks was used to build the three-dimensional heat transfer model of tuyere,the CFD was used to study the temperature field of tuyere,and discussed the influence of inlet hydraulic velocity,front wall thickness upon the temperature field of tuyere.The simulation result indicated that: In the calculation range,the inlet hydraulic velocity strongly effect the heat transfer on the tuyere when it is less than 12m/s, Conversely ,the effect of heat transfer on the tuyere is minimal when the inlet hydraulic velocity is more than 12m/s. Therefore, in the actual production process of the blast furnace, it is best to keep the water velocity at 12m/s; In the calculation range, the water frame structure has no significant influence on the temperature distribution of the front face of the tuyere, but yes to side wall of tuyere, that is, the temperature of the side wall of the tuyere is highest when water frame structure layer is 2, but still far less than the copper needed temperature, from the point of view of the temperature gradient, the optimum water frame structure layer is 2.

Key words: tuyere, numerical simulation, inlet hydraulic velocity, water frame structure

锌对高炉焦炭性质的影响研究

李 曼，薛庆国，佘雪峰，王 广，刘迎立，王静松

（北京科技大学钢铁冶金新技术国家重点实验室，北京 100083）

摘 要： 在管式炉中，在 1000℃下对取自莱钢的焦炭在不同的锌蒸汽浓度下进行吸附试验，结果表明随着锌蒸汽含量的增加，焦炭吸附锌的含量越高，对应的焦炭 CRI 性增大，反应后强度 CSR 降低。当锌蒸汽浓度较高时，焦炭会出现微观裂纹，但是却不会造成十分明显的粉化现象。通过 XRD、SEM、EPMA 观察发现经过锌蒸汽渗透后的焦炭中含有锌单质和 ZnO，但是却没有检测到含锌矿物，这说明在此温度下锌蒸汽并不会和焦炭中的矿物发生化学反应生成含锌矿物。锌对焦炭气化反应的催化作用主要是促进酮基的分解，加速 CO 的生成。由于焦炭中的各向异性组织表面较粗糙，相比表面平滑的各项同性组织更容易吸附锌，这就造成了吸附锌后的焦炭其各向异性的反应性要强于各项同性的反应性。

关键词： 锌，催化作用，矿物，反应性

Effect of Zinc on the Properties of Coke in Blast Furnace

Li Man, Xue Qingguo, She Xuefeng, Wang Guang, Liu Yingli, Wang Jingsong

(State Key Laboratory of Advanced Metallurgy, University of Science and
Technology Beijing, Beijing 100083, China)

Abstract: In the tubular furnace, the adsorption of coke from Laiwu Steel was carried out under different zinc vapor concentration at 1000℃. The results showed that with the increase of zinc vapor content, the higher the content of zinc in coke adsorption, the corresponding coke CRI increased and CSR decreased. By X-ray diffraction (XRD)、scanning electron microscope (SEM) and electron microprobe, it was found that the coke contained zinc and ZnO, but there was no zinc mineral，which shows that the zinc vapor does not react with the minerals in the coke to produce zinc minerals. The catalytic effect of zinc on coke gasification is mainly to promote the decomposition of ketone and accelerate the formation of CO. In coke, compared with the isotropic texture with smooth surface, the surface roughness of the anisotropic texture is more easily adsorbed zinc, which results in the reactivity of the anisotropic texture of the coke after zinc adsorption is stronger than that of the isotropic texture.

Key words: zinc, catalysis, mineral, reactivity

八钢喷煤制粉系统出粉低原因分析及对策

胡　滨

（新疆八钢炼铁分公司第二炼铁分厂，新疆乌鲁木齐　830022）

摘　要： 对原煤、制粉系统进行查找原因分析，查找对磨煤机出粉能力低于设计能力原因。对原煤、制粉系统本体进行跟踪检查分析，制定整改措施，提高制粉能力。制粉系统满负荷生产，喷煤三套制粉系统开启两套，满足三座 $2500m^3$ 高炉 $100kg/t$ 煤比。喷煤三套制粉系统循环开启，满足制粉系统的检修需求，同时降低电耗，降低煤粉制造成本。

关键词： 制粉量，磨煤机，煤粉收集器

高炉经济喷煤量的研究与实践

侯李平

（河南济源钢铁（集团）有限公司炼铁厂，河南济源　459000）

摘　要： 为了实现高炉的经济喷吹，本文研究了经济煤比的评价方法，定性地分析了影响高炉经济喷吹的因素；同时根据高炉的喷吹实践，利用数据统计和回归的方法，对不同煤比下的煤粉利用率、燃料比和吨铁燃料成本进行分析，得出了在一定的原燃料条件和操作水平下高炉的经济喷吹量；同时，结合目前高炉的外围生产条件，提出了进一步提高高炉经济喷煤量的途径。

关键词： 经济煤比，煤粉利用率，燃料比，燃料成本

高炉喷煤技术现状

郝良元[1]，王海洋[2]，张建良[2]，王广伟[2]

（1. 河北钢铁集团河钢研究院，河北石家庄　050000；
2. 北京科技大学冶金与生态工程学院，北京　100083）

摘　要： 高炉喷煤是降低焦比，节约铁水成本的重要措施之一，本文系统地介绍了我国高炉喷吹的煤种和高炉配煤方法的进展及存在的问题。有必要开展兰炭等新型喷吹燃料用于高炉喷吹的研究，利用其价格低、污染小的特点缓解成本和环保压力。传统的高炉喷吹配煤方法以按挥发分配煤为主，较为片面，需要综合考虑煤粉有效发热值、燃烧性、价格和相互间作用等因素。

关键词： 高炉，喷煤，配煤

本钢北营大高炉旋切顶燃热风炉优化控制系统的应用

程 英，刘仁军

（本钢北营炼铁厂，辽宁本溪 117017）

摘 要：热风炉自动控制包括完整的基础自动化和过程自动化，对提高热效率、节约能源、提高风温以及保护设备和延长炉子的寿命有重要作用。

关键词：热风温度，寿命，燃烧，控制

Blast Furnace Rotary Top Combustion Hot Blast Stove Optimization Control System Application

Cheng Ying, Liu Renjun

(Iron Mill, Benxi Steel North Camp, Benxi 117017, China)

Abstract: The hot blast stove automatic control contains a complete set of basic automation and process automation, to improve the thermal efficiency, save energy and improve blast temperature and protection devices, and prolongs the service life of the furnace play an important role.

Key words: hot blast temperature, life, burning, control

日照钢铁球式热风炉应用高辐射覆层技术的节能实践

吕定建，朱 勇

（日照钢铁有限公司，山东日照 276806）

摘 要：日钢在高炉热风炉上广泛使用高辐射覆层技术，耐火球体的蓄热效率和热风炉热效率明显提高，获得良好的节能效果。通过对热风炉高温区的 ϕ60mm、ϕ80mm 球覆层，改善蓄热球体的表面结构，提高热量吸收量值和速率并改善表面强度。在生产中测定在全部使用高炉煤气条件下可以节约煤气烧炉煤气 6.91%，节约高炉煤气消耗，增加煤气发电量。通过对所有高炉热风炉占全部蓄热球质量 47.5% 的球体进行高辐射覆层涂覆，近三年节约煤气平均年效益达到 6200 万元以上。

关键词：高炉，球式热风炉，高辐射覆层技术，节约煤气量

Energy Saving Practice of High Temperature Radiation Coating Technology for Ball Type Hot Blast Stove of Rizhao Iron and Steel Co., Ltd.

Lv Dingjian, Zhu Yong

(Rizhao Iron and Steel Co., Ltd., Rizhao 276806, China)

Abstract: Rizhao Iron and steel used in high radiation coating technology in blast furnace, thermal efficiency and thermal efficiency of hot blast stove refractory ball was improved to obtain good energy-saving effect. The high temperature zone of hot blast stove Phi 60mm, Phi 80mm ball layer, improve the surface structure of regenerative spheres, improve heat absorption quantity and rate to improve the surface strength and determination. In all of the use of blast furnace gas under the condition of gas burning furnace gas can save 6.91% in the production of blast furnace gas, save gas consumption, increase power generation. High radiation coating ball body through all of the hot blast furnace accounted for 47.5% of the quality of the regenerative ball, saving nearly three years the average annual benefit to gas more than 62 million yuan.

Key words: blast furnace, ball type hot blast stove, high radiation coating technology, gas saving

日钢高炉热风炉热效率提升实践

朱 勇

（日照钢铁有限公司钢铁研究发展处，山东日照　276806）

摘　要：日钢投产以来，对高炉热风炉能耗情况,热风炉热效率未进行过检测，本文通过对日钢高炉热风炉实际生产数据进行收集，并校对相关参数，在此基础上进行了热效率的计算，并对计算结果进行了分析。结果表明 1#～16#高炉整体热效率处于较低水平，主要原因是化学能和热能利用较低，热风炉燃烧不充分，烟气余热回收装置利用水平较低。提出提高热风炉热效率措施并实施，取得较好的效果，采取的措施主要通过生产组织方面稳定煤气压力,改进操作合理化空气过剩系数（降低空气过剩系数），富氧燃烧以及提高烟气余热回收装置效率的方式，提高热风炉热效率。

关键词：热风炉，热效率，空气过剩系数，富氧烧炉

Practice of Improving the Thermal Efficiency of Hot Blast Stove in Rizhao Iron and Steel Blast Furnace

Zhu Yong

(Company Research and Development in Rizhaosteel, Rizhao 276806, China)

Abstract: Rizhao Iron and Steel production, energy consumption of hot blast stove, the thermal efficiency of hot blast stove was not detected. In this paper, the actual production data of blast furnace hot blast furnace is collected, and check the relevant parameters. On the basis of the calculation, the thermal efficiency is calculated, and the results are analyzed. The results show that the overall thermal efficiency of 1#~16# blast furnace is at a low level, the main reason is that the chemical energy and heat energy utilization is low, the hot air stove burning is not sufficient, and the utilization level of the waste heat recovery device is lower. It is proposed to improve the thermal efficiency of hot blast stove and implement it, The main measures to take measures to stabilize the gas pressure through the production organization, improve the operation of reasonable air excess coefficient (Reduce air excess coefficient), Oxygen enriched combustion and the way to improve the efficiency of the flue gas waste heat recovery device, improve the thermal efficiency of the hot air stove.

Key words: hot air furnace, heat efficiency, air excess coefficient, oxygen enriched combustion furnace

干熄焦除尘灰在高炉中的喷吹应用探讨

姜　曦[1]，白兴全[2]，周东东[3]，王海洋[3]

（1. 中国钢铁工业协会，北京　100711；2. 酒钢集团宏兴股份公司炼铁厂，甘肃嘉峪关　735100；

3. 北京科技大学冶金与生态工程学院，北京　100083）

摘　要： 我国炼铁固体废弃物等资源循环利用率较低。干熄焦除尘灰是炼焦工艺的副产品之一，长期以来无法高效率的利用该资源。高炉中喷吹干熄焦除尘灰，不但可以降低生铁成本，在一定程度上也可以缓解适合高炉喷吹的无烟煤资源短缺的弊端，对环境保护也有积极的意义。本文主要通过对干熄焦除尘灰代替部分高价无烟煤的燃烧率、可磨性及着火点等技术可行性进行了论证，并在酒钢的高炉上进行了工业试验。通过分析试验结果提出了一些使用建议。本文的研究有利于钢铁企业降低成本、提高废弃资源的利用率，也有利于节能减排。

关键词： 高炉喷吹，干熄焦除尘灰，可行性分析，工业试验

The Injection of the Coke Dry Quenching Dust in Blast Furnace

Jiang Xi[1], Bai Xingquan[2], Zhou Dongdong[3], Wang Haiyang[3]

(1. China Iron and Steel Association, Beijing 100711, China; 2. The Ironmaking Plant, Gansu Jiusteel Group Hongxing Iron & Steel Co., Ltd., Jiayuguan 735100, China;

3. University of Science and Technology Beijing, Beijing 100083, China)

Abstract: The main characteristics of ironmaking in China are high costs of the hot metal and the low circulation utilization rate of the solid waste resources. For instance, the coke dry quenching dust, which was produced in the coking process, has not utilized effectively for a long period. Thus, the injection of coke dry quenching dust in blast furnace not only could reduce the smelting costs, but also could relief the short of the anthracite and benefits to the environment protection. In this paper, the technical feasibility to use the coke dry quenching dust replace some high cost anthracite was analyzed by analyzing the burnout rate, grinding performance and coal ignition point, while the industry experiments were conducted in the blast furnace in Jiusteel. In addition, some measurements for injecting the coke dry quenching dust were proposed. The study could contribute to reduce the costs of hot metal, improve the utilization of waste resources as well as the energy conservation and emission reduction.

Key words: blast furnace injection, coke dry quenching dust, technical feasibility analysis, industry experiment

昆钢 2500m³ 高炉风口小套长寿维护

麻德铭, 马杰全, 仇友金, 马晓峰

（武钢集团昆明钢铁股份有限公司安宁分公司）

摘　要　对昆钢新区 2500m³ 高炉风口频繁损坏的原因进行了分析，并且针对风口小套寿命较短的这一主要情况，针对性的采取了诸如风口校正、提高风口耐磨性、改进煤枪结构和插枪操作等措施，使风口小套寿命得到有效延长，为高炉稳定、低耗生产提供了有力保障。

关键词　风口磨损，风口上翘，操作制度

昆钢 2500m³ 高炉热风管道温度异常处理实践

麻德铭，张正峰，李　淼，李信平

（武钢集团昆明钢铁股份有限公司安宁分公司）

摘　要：本文介绍了热风管道各温度升高区域的详细处理过程，对灌浆与浇注的关键控制点进行了总结归纳，有效地抑制了温度的进一步升高，避免了热风管道烧穿等恶性事故的发生。

关键词：高炉，热风管道，温度异常，灌浆

昆钢 2500m³ 高炉铁口喷溅治理实践

李晓东，李　淼，李鸿雁

（武钢集团昆明钢铁股份有限公司安宁分公司）

摘　要：昆钢 2500m³ 高炉自开炉后铁口喷溅问题一直严重制约炉前生产，影响了高炉技术经济指标的提高。铁口喷溅严重时，在出铁过程中喷溅在大沟两侧的渣铁量较大，清理工作相当繁重，影响渣铁排放亏铁量，造成高炉憋风情况较多。高炉通过采取一系列行之有效的措施，有效治理了铁口喷溅，保证了高炉出铁顺畅，高炉技术经济指标得到了稳步提升。

关键词：铁口喷溅，憋风，治理

昆钢新区 2500m³ 高炉炉型维护探索

卢郑汀，张志明，李 淼

（武钢集团昆明钢铁股份有限公司安宁分公司）

摘 要：根据昆钢新区 2500m³ 高炉炉型设计特点和原燃料条件，分析了影响该高炉炉型维护的因素，采取高炉精料技术、调整煤气流分布、加强冷却系统的管理、加强高炉炉况调剂管理等措施，实现了对昆钢新区 2500m³ 高炉炉型的合理控制，保证了炉况稳定顺行。

关键词：高炉，炉型维护，布料，炉况调剂

首钢京唐烧结配加智利精粉的烧结杯试验研究

王同宾，赵景军，石江山，史凤奎，康海军，罗尧升

（首钢京唐钢铁联合有限责任公司，河北唐山 063200）

摘 要：通过智利精粉代替澳粉、罗伊山粉、巴卡和杨地粉，烧结矿的品位为上升趋势，当替代其它矿粉时，烧结矿的品位为下降趋势，MgO 含量为下降趋势；转鼓指数为上升趋势，成品率先下降后上升；垂直烧结速度和利用系数为先上升后下降，混匀矿有害元素 K_2O、Na_2O、Cl 和 S 均为增加趋势。综合考虑，使用新矿粉智利精粉替代其它矿粉时，应优先考虑替代澳粉、巴卡和杨地粉，对烧结矿指标影响不大。建议智利精粉替代澳粉配加比例控制在 3%左右，替代巴卡和杨地的基础上，将比例可提高至 6%，智利精粉替代比例增加时应随时关注有害元素的变化。

关键词：烧结，质量，智利精粉，有害元素

The Sintering Cup Test Study on Sintering with Chile Powder of Shougang Jingtang United Iron & Steel Co., Ltd.

Wang Tongbin, Zhao Jingjun, Shi Jiangshan, Shi Fengkui, Kang Haijun, Luo Yaosheng

(Shougang Jingtang United Iron & Steel Co., Ltd., Tangshan 063200, China)

Abstract: The grade of sinter is the upward trend through the substituting of Chilean fine powder for Australian powder, Roy mountain powder, barca and poplar powder.When replacing other mineral powder, the grade of sinter is decreasing and MgO content is decreasing.The drum index is on the upward trend, and the finished product is the first to rise.The vertical sintering rate and the utilization coefficient decrease after first rise.The harmful elements K_2O, Na_2O, Cl and S are all increasing trend.Consider using new mineral powder Chile powder instead of other mineral powder.Priority should be given to the substitution of Australian powder, barca and poplar powder, which has little impact on the sinter index.It is suggested that the proportion of Chilean fine powder substitute for Australian powder should be controlled at about 3%. On the basis of replacing barca and Yang ground, the proportion can be increased to 6%, and the change of harmful elements should be paid attention to when the replacement ratio of Chilean powder is increased.

Key words: sintering, quality, chile powder, the harmful elements

太钢 5 号炉缸堆积处理实践

张 智

（太钢炼铁厂，山西太原　030003）

摘　要：钢 5 号高炉进入炉役后期，护炉过程中长期堵 2~3 个风口，低冶强利用系数 1.9 组织生产运行，2017 年 4 月 16 日 8.68m 270°方向炉缸温度上升至 445℃，休风堵风口，送风炉况恢复困难，管道崩料频繁，高炉出现炉缸堆积，频繁损坏风口，4 月 21 日至 5 月 1 日处理炉缸堆积恢复炉况，主要采取措施：（1）调整下部送风制度，控制加风节奏；（2）全焦冶炼；（3）配加锰矿。

关键词：型高炉，堆积

八钢 2500m³ 高炉磨煤机磨辊检修技术

于 涛

（宝钢集团八钢公司炼铁分公司第二炼铁分厂，新疆乌鲁木齐　830022）

摘　要：磨煤机是一种宜于长期连续工作，其工作部件之间相互磨损很少的大型机械。然而，由于被碾磨的煤中混入有硫化铁类和其它硬质的物质。这些物质会降低磨煤机的零部件，诸如磨辊套、磨碗衬板、折向衬板、叶轮衬板、侧机体衬板、内锥体衬板等的使用寿命，喷煤 B 系列磨机磨辊、磨碗衬板、分离器体衬板及长短刮板装置都已磨损严重，需进行更换，本论文是对磨煤机检修的概述。

关键词：磨煤机，磨辊，检修

鞍钢鲅鱼圈真空碳酸钾脱硫贫富液换热器优化与改进

武 斌

（鞍钢股份鲅鱼圈钢铁分公司炼焦部，辽宁营口　115007）

摘　要：鞍钢鲅鱼圈焦化真空碳酸钾脱硫贫富液换热器原设计采用Ⅲ型可拆卸式螺旋板换热器，在 8 年多的使用过程中，其流道靠压盖的密封垫密封，使用一段时间后，密封垫老化，易出现短流现象，影响换热效果，而且压盖最外层垫圈长时间运行后易变性而泄漏，污染环境。通过对粗苯贫富油Ⅰ型不可拆卸螺旋板换热器应用研究，并对脱硫贫富液换热器换热介质和换热器内部研究，发现打开封头后流道没有任何堵塞甚至没有任何杂质，消除了贫富液换热器内部堵塞的担忧，2015 年开始对三台贫富液换热器进行改型，将三台贫富液Ⅲ型可拆卸式螺旋板换热器改为Ⅰ型不可拆卸螺旋板换热器，取得了良好的运行效果和显著的经济效益。

关键词：脱硫，换热器，改型

八钢2500m³高炉开口机液压故障诊断处理

张海涛

（宝钢集团八钢公司炼铁分公司第二炼铁分厂，新疆乌鲁木齐 830022）

摘 要：本文对八钢 2500m³ 高炉开口机小车回退慢的故障处理进行了总结。通过滤油、更换液压马达、更换液压阀组等措施，对故障进行了排查处理，为高炉后期的液压设备管理和故障诊断提供了参考。

关键词：开口机，液压，故障

打击机连接套与钻杆插销装配损坏分析及应对措施

李海云

（宝钢集团八钢公司炼铁分公司，新疆乌鲁木齐 830022）

摘 要：文章介绍了针对欧冶炉炉前区域开口机打击机连接套与钻杆的连接形式对开口影响的制约性，重点分析了连接插销的变形过程及通过新区 2500 高炉使用开口机拆装钻杆情况对比分析提出了相应合理的改进优化措施。

关键词：打击机连接套，钻杆，拆卸，改进优化

The Damage Analysis and the Measures for the Assembly of the Blow Machine's Connection Sleeve and Drill Pipe are Broken

Li Haiyun

(Ironmaking Branch, Bayi Iron & Steel Co., Baosteel Group, Urumchi 830022, China)

Abstract: The article introduces the against Ouye furnace area before opening machine blow machine connection set with the drill pipe connection form impact on the opening of conditionality, analyzed the deformation process of the connecting bolt and through the new 2500 blast furnace tapping machine dismounting drill pipe situation analysis puts forward the reasonable optimization measures for improvement.

Key words: blow machine connection set, drill pipe, remove, improve optimization

八钢欧冶炉双油缸驱动的布料器控制原理及故障分析

杨红松

（新疆八一钢铁集团有限责任公司炼铁分公司，新疆乌鲁木齐　830022）

摘　要： 本文对欧冶炉布料器控制系统原理进行了说明，并且对布料模式及控制器故障处理的方法进行描述分析。

关键词： 油缸，控制器，布料器

热矿振动筛故障原因分析与技术改造

曹宝库

（八钢公司炼铁分公司第二高炉分厂，新疆乌鲁木齐　830022）

摘　要： 对新区高炉配料振动筛投产以来出现的筛板易松动、盲板变形严重、轴承损坏频繁等故障原因进行分析，采取了盲板边缘加焊加强筋、改变落料点、降低激振器工作环境温度、减小激振力、改进轴承固定方式和润滑方式等技改措施，使故障停机率降低，年经济效益288.6万元。

关键词： 配料振动筛，激振力，轴承，盲板，筛板

降低工艺皮带秤计量误差率

王立峰，高振亮，王志刚

（首钢京唐公司供料作业部，河北唐山　063200）

摘　要： 本文首先对钢铁企业铁前工艺皮带秤的现状进行了调研，针对问题导向，通过鱼骨图、因果矩阵图等方法对影响皮带秤计量误差率的因素进行筛选、分析，进而有的放矢的提出了引进点检信息化系统、量化调平等对策，最后对实施效果进行了综合评定。

关键词： 皮带秤，误差率，稳定性，过程控制

Reduce the Error Rate of Process Belt Scale

Wang Lifeng, Gao Zhenliang, Wang Zhigang

(Shougang Jingtang United Iron & Steel Co., Ltd., Tangshan 063200, China)

Abstract: This article for the iron and steel enterprises have investigated the current situation of the iron before theprocess of belt scale, in view of the problem, through such means as fishbone diagram, causal matrix listed on influencing factors of belt scale measurement error rate, screening, and targeted puts forward introducing the tally information systems, quantitative equality countermeasure, finally made a comprehensive evaluation of the effect.

Key words: belt scale, error rate, stability, process control

阵列式皮带秤在原燃料管理中的应用

张仲元，丁国一，王立峰，鞠洪刚

（首钢京唐公司供料作业部，河北唐山　063200）

摘　要： 本文概述了"阵列式皮带秤"的原理、结构及特点，通过在京唐料场原燃料计量上的成功应用，阐述了"阵列式皮带秤"以较高的计量精度和计量效率，实现了以皮带秤方式进行长期稳定计量，为库存管理和成本核算提供了良好的依据。

关键词： 阵列式皮带秤，准确性，结算计量

Application of Array Belt Scale in Raw Fuel Management

Zhang Zhongyuan, Ding Guoyi, Wang Lifeng, Ju Honggang

(Shougang Jingtang Company Supplement Operation Department, Tangshan 063200, China)

Abstract: This paper summarizes the "array type belt weigher" principle, structure and characteristics, through the successful application of raw material in the field of fuel metering Jingtang that describes the array type belt weigher to measurement accuracy and measurement of high efficiency, to achieve long-term stability measurement by belt scale, provide a good basis for inventory management and cost accounting.

Key words: array belt scale, accuracy, settlement measurement

贾家堡铁矿尾矿长距离输送应用与实践

邓庆超，赵　尚

（本钢集团矿业辽阳贾家堡铁矿有限公司，辽宁辽阳　111219）

摘　要： 本文简要的介绍了本钢贾家堡铁矿处理富含部分粗颗粒的尾矿浆体，运用隔膜泵采用高浓度、长距离输送方式，在设计、运行中的一些关键措施的实践经验。介绍了长距离、高浓度、高扬程、高粘度尾矿浆体，运用国产往复式隔膜泵在输送运行过程中的应注意的问题和取得效果及经验。

关键词： 尾矿输送，隔膜泵，高效浓密机及内循环系统，隔渣筛，搅拌槽

The Application and Practice of Long Distance Pipage in Jiajiapu Dressing Plant

Deng Qingchao, Zhao Shang

(Jiajiapu Iron Mine, Liaoyang, Mining Company, Benxi Steel, Liaoyang 111219, Chian)

Abstract: This article briefly introduces the Benxi Steel Jiajiapu Ironmine ore processing is rich in coarse particle tail slurry body, using the diaphragm pump adopts the high concentration of long-distance conveying method, some key measures in design operation of practical experience for long and high concentration of high lift tail high viscosity pulp, using the domestic reciprocating diaphragm pump problems that should be paid attention to in the process of conveying operation and the good results and experience.

Key words: tailings transportation, diaphragm pump, efficiency thickener and inner circulation system, slag sifter, agitator bath

南芬选矿厂磨选系统旋流器参数优化试验研究

李　永，韩　爽，李庚辉

（本钢南芬选矿厂，辽宁本溪　117014）

摘　要： 介绍了本钢南芬选矿厂三段磨选工艺改造完成后，对旋流器主要参数进行优化研究，工业试验结果表明：旋流器参数优化后，三段旋流器溢流粒度平均为-0.074mm 占91.8%，分级效率达到47.27%的较好指标，为后续选别工序取得精矿质量 66.89%打下基础，可为其它类似改造提供借鉴和推广价值。

关键词： 旋流器，改造，优化，效果

Test of Optimizing Parameter Cyclone of Grinding and Separation in Nanfen Concentrator

Li Yong, Han Shuang, Li Genghui

(Bengang Nanfen Concentrator, Benxi 117014, China)

Abstract: The paper described the three-stage grinding and separation technology transformation in Nanfen concentrator and the main parameters of cyclone were optimized. Industrial test results show that after the cyclone parameter optimization system is stable, three-stage cyclone overflow granularity −0.074mm average 91.8% fractional efficiency reached 47.27% better for the subsequent sorting process to achieve the quality of concentrate 66.89%.The test of optimizing parameter cyclone can provide reference and promotional value of other similar transformation.

Key words: cyclone, reform, optimizing, effects

高炉 2500m³ 渣处理转鼓设备的自动控制升级改造

田新龙　刘忠明

（新疆八一钢铁公司炼铁分公司二炼铁分厂，新疆乌鲁木齐 830022）

摘　要： 根据 2500m³ 高炉渣处理工艺要求，分析了渣处理的关键设备——转鼓的控制思路及控制原理。详细介绍了转鼓的自动控制升级改造的全过程及重点注意事项，包括变频器系统的安装、调试、控制方式、参数设置等，以及调试中出现的问题及解决办法，对生产实际具有一定的指导意义。

关键词： 转鼓，变频器，参数设置，自动调速控制

从站模块故障报警监控在迁钢高炉工业网中的应用

袁　宇

（北京首钢自动化信息技术有限公司首迁运行事业部炼铁作业区，河北迁安　064400）

摘　要： 首钢迁钢高炉 PLC 的工业网是由多个岗位通过以太网通讯的工业环网，主要由昆腾系列 PLC、西门子 300 系列 PLC 组成，其中昆腾系列 PLC 的下位软件为 Concept 和 Unity，所有的扩展 I/O 从站都使用了 RIO 通讯。RIO 从站监控模块主要应用在首钢迁钢高炉工业网中，它能迅速判断各个岗位扩展 I/O 从站的模块状态，减少故障处理时间。

关键词： PLC，concept，unity，从站，RIO 通讯

The Application of the RIO Alarm Monitoring on Qiangang Blast Furnace PLC Industrial Network

Yuan Yu

(Beijing Shougang Automation & Information Technology Co. Shouqian Maintain Department, Ironmaking Department, Qian'an 064400, China)

Abstract: The PLC industrial network of Shougang Qiangang blast furnace is a ring network that contains multiple jobs through the industrial Ethernet communication, and is composed of Quantum series PLC, Siemens 300 series PLC. The Quantum series PLC uses Concept and Unity as its inferior software, and all of the extension I/O slave stations use RIO communications. The RIO slave station monitoring module is mainly used in the industrial network of Shougang Qiangang blast furnace. It can quickly judge the module state of extended I/O slave stations, and reduce the fault processing time.

Key words: PLC, concept, unity, slave station, RIO communication

智能 RFID 管理系统在冶金行业应用介绍

王 磊

（迁安首信自动化信息技术有限公司，河北迁安　064400）

摘　要： 企业产品具有分布广、数量大、单位价值高和调整频繁等特点，在目前的管理模式下，信息在传递过程中人为因素造成的信息失真和滞后引起滞后无法同步一致，网络部门无法及时进行设备优化调配，致使大量高价值网络类资产闲置浪费，严重影响信息的真实性。采用 RFID 射频识别技术可以发送准确、可靠的实时信息流，从而创造附加值，提高生产率和大幅度地节省投资。

关键词： 智能，RFID，射频技术，生产率

The Introduction of Intelligent RFID Management System Application in Metallurgical Industry

Wang Lei

(Qian'an Capitel Automation Information Technology Co., Ltd., Qian'an 064400, China)

Abstract: Enterprise products has a wide distribution, large quantity, high value per unit and frequent adjustment etc, in the current management mode, the information in the process of passing information distortion and lag caused by human factors cause lag could not synchronize, network department cannot promptly equipment optimization deployment, the high value of asset class Internet idle waste, seriously affect the authenticity of information.RFID technology can send accurate and reliable real-time information flow to create added value, increase productivity and save investment substantiall.

Key words: intelligent, RFID, RF technology, productivity

地坪材料在钢铁行业的应用和选择

姚俊海，朱　硕，张　迪

（北京联合荣大工程材料股份有限公司，北京　101400）

摘　要： 本文立足地坪材料在工业地坪中的应用，分析了几类传统地坪材料在钢铁行业高炉平台中的应用及存在的问题，并针对高炉平台地面施工或改造工程，选择工业地坪时，不仅要明确地坪的使用环境，而且对地坪材料的功能性、环保性、经济性需求提出更合理的选择。

关键词： 钢铁行业，工业地坪，环保性

The Best Choice and Application of Flooring Materials in Steel Industry

Yao Junhai, Zhu Shuo, Zhang Di

(Beijing Allied Rongda Engineering Material Co., Ltd., Beijing 101400, China)

Abstract: Based on the application of flooring materials in industrial floor, the application and problems of several kinds of traditional floor materials in the blast furnace platform of the steel industry are analyzed. Industrial floors were applicated in steel industry. Not only the application environment of the floor material is made clear, but also more reasonable solutions to the function, environmental protection and economic demand of the flooring material are proposed.
Key words: steel industry, industrial floor, environmental protection

高压变频器在高炉除尘风机上应用

洪 伟

（宝钢集团炼铁分公司二炼铁分厂，新疆乌鲁木齐 830022）

摘 要：本文通过对高炉铁口除尘风机改造方案的阐述以及节能效果对比，说明在满足高炉环境除尘的基础上，高压变频调速系统在除尘风机应用技术优点，以及实际运行效果。
关键词：高压变频器，风机

环冷机密封对烧结余热回收效率影响的实践与探讨

冶 飞

（宝钢集团八钢公司炼铁分公司，新疆乌鲁木齐 830022）

摘 要：本文介绍了八钢烧结环冷机余热回收系统采用新型环冷机密封技术，优化生产工艺控制参数，烧结余热回收系统的蒸汽回收效率和发电量大幅提升，取得了较好效果。并对环冷机密封改善前后对烧结余热回收效率的影响进行了研究和探讨。
关键词：环冷机，密封，烧结余热回收

The Practice and Discussed on Annular Cooler Sealing Influence to Sinter Waste Heat Recovery Efficiency

Ye Fei

(Sintering Branch of Ironmaking Branch, Bayi Iron & Steel Co., Baosteel Group, Urumchi 830022, China)

Abstract: The paper introduces the Bayi Iron & Steel sintering circular cooler waste heat utilization system using the new annular cooler sealing technology, optimize production process control parameters, the steam recovery efficiency and power generation of sintering waste heat recovery system are greatly improved, it has been a better effect.The effects of waste heat recovery efficiency on the sintering waste heat recovery system were studied and discussed in the before and after the annular cooler sealing.

Key words: annular cooler, sealing, sintering waste heat recovery

八钢欧冶炉处置消纳含油污泥生产实践

李维浩

（宝钢集团八钢公司炼铁分公司，新疆乌鲁木齐　830022）

摘　要：通过对欧冶炉热解含油污泥机理和优势分析，符合实现无害化、减量化和资源化处理废弃物原则，能有效克服焚烧处理含油污泥产生的二噁英污染问题。通过对欧冶炉三种热解处置含油沉淀污泥方式同传统焚烧炉处理工艺进行对比，利用欧冶炉热解处置含油沉淀污泥，具有前期工程费用投入少、二次污染小，无害化彻底，资源化利用程度高等优点。通过欧冶炉配加含油污泥球工业试验，其工艺流程能满足热解处置含油污泥生产组织和操作过程需要，不需要增加新的辅助处理设施，具有可处理多样形态含油污泥特点。

关键词：欧冶炉，含油污泥，热解处置

八钢欧冶炉煤气干法除尘技术的首次应用及实践

田　果，王　忠，陈若平，刘鹏南

（八钢公司炼铁分公司第一炼铁分厂，新疆乌鲁木齐　830022）

摘　要：煤气布袋除尘技术是现代炼铁系统实现节能减排、清洁生产的重要技术创新，可以显著的降低炼铁生产过程的新水消耗、减少环境污染，已成为炼铁技术煤气净化处理的方向。欧冶炉是非高炉炼铁的相对先进技术，八钢欧冶炉是在非高炉炼铁领域首次实践使用干法布袋除尘设备进行煤气净化的系统，本文就欧冶炉煤气干法除尘技术的应用实践及出现的新情况进行分析。

关键词：非高炉炼铁，煤气布袋除尘器，煤气净化

移动式皮带机除尘设施安装与应用研究

刘　鑫，张建伟，苏胜利，康志鹏

（首钢京唐公司供料作业部，河北唐山　063200）

摘　要：大宗铁矿石原料在转运站内运输过程中产生大量粉尘，在原设计中移动式皮带机机头没有除尘设施，通过对移动式皮带机机头自制安装除尘设施，并对安装后的现场粉尘浓度进行持续跟踪。结果显示，现场粉尘浓度明显降低并持续稳定处于低粉尘浓度状态，从而改善了作业现场的环境，提高了含铁粉尘的回收率。

关键词：除尘设施，结构原理，应用效果

浅谈贾家堡铁矿中水循环利用系统

宋万强，邓庆超，赵　尚

（本钢集团矿业辽阳贾家堡铁矿有限公司，辽宁辽阳　111219）

摘　要：贾家堡铁矿是采选一体的综合矿山企业，生产水是重要能耗之一，贾家堡铁矿利用现有工艺设备资源，构成中水循环利用系统，通过对中水系统的改造，降低中水内循环系统的能耗成本，并以贾矿为核心联合南芬选矿厂、北营过滤间建成中水循环利用区域模式，实现中水循环利用无外排，在节约资源、降低成本、保护环境方面发挥了重要作用，也为中水利用提供经验。

关键词：中水，循环利用，成本，环保

The Recycled Water System of Jiajiapu Iron Mine

Song Wanqiang, Deng Qingchao, Zhao Shang

(Jiajiapu Iron Mine, Liaoyang, Mining Company, Benxi Steel, Liaoyang 111219, China)

Abstract: The Jiajiapu Iron Mine is a comprehensive mining enterprise. The water used by production is one of the important energy consumptions. The recycled water system was built based on existing process equipment, and realized the water recycling without draining. The energy consumption costs of the internal recycle system of recycled water were reduced. A kind of regional patterns was built for recycled water utilization, which based on Jiajiapu iron mine and combined with Nanfen concentrator and Beiying filter plant. By this way, the water resource is saved, the environment is protected, and the production cost is reduced. Also the experiences for the of use recycled water are provided.

Key words: recycled water, cyclic utilization, cost, environmental protection

除尘器泄漏故障处理

宋万强，孟宪权，程凤利

（本钢集团矿业辽阳贾家堡铁矿有限公司，辽宁辽阳　111219）

摘　要：随着国家环境保护法规日益完善，企业对环保的重视程度越来越高，环保设备成为重点设备进行管理。本文通过对现场除尘系统故障的排查与分析，提出科学合理的维修方案，恢复设备使用性能，延长设备使用寿命，提高环保设备运行管理水平。

关键词：除尘器，过滤，布袋

The Leakage Fault Treatment of Dust Catcher

Song Wanqiang, Meng Xianquan, Cheng Fengli

(Jiajiapu Iron Mine, Liaoyang, Mining Company, Benxi Steel, Liaoyang 111219, China)

Abstract: With the development of legislation of environmental law, more attentions have been paid to the environmental protection by the enterprise. The environmental protection equipment becomes the key equipment. The scientific and reasonable maintenance scheme was studied based on the fault analysis of dust-cleaning apparatus. The equipment performance was restored and its service life was prolonged. The management level of environmental protection equipment is improved.

Key words: dust collector, filtration, bag

煤粉流量计在欧冶炉中的运用及准确性和可重复性验证

王金诚

（宝钢集团八钢公司炼铁分公司，新疆乌鲁木齐　830022）

摘　要： 在欧冶炉煤制气系统中四个喷煤支管上安装煤粉流量计及其控制机构的目的在于能够合理的控制每个支管喷入气化炉的煤粉量，使每个支管的喷煤量相同。在每个支管上实现煤粉的均匀吹喷有助于煤粉的充分燃烧，进而实现气化炉输出煤气含量均匀，输出高品质的煤气进入还原竖炉，以提高还原竖炉的金属化率，保障欧冶炉的稳定顺行。

关键词： 煤粉流量计，均匀吹喷，气化炉，还原竖炉

Bayisteel OY in Coal Powder Flowmeter and Verify the Accuracy and Repeatability

Wang Jincheng

(Ironmaking Branch, Bayi Iron & Steel Co., Baosteel Group, Urumchi 830022, China)

Abstract: In Bayisteel OY coal gas system in four coal injection pipe installed on the pulverized coal flow meter and control mechanism is designed for the purpose of rational control of each branch is injected into the gasification furnace of pulverized coal, pulverized coal injection to the same amount of each branch pipe. The pulverized coal in each branch on the uniform jet contributes to the full combustion of pulverized coal gasification furnace, so as to realize the output of the gas content is uniform, the output of high-quality gas into the reduction shaft, in order to improve the rate of metal reduction shaft, ensure stable OY of Bayisteel.

Key words: pulverized coal flow meter, uniform blowing, gasifier, reduction shaft furnace

液压油液的种类和选择污染及控制

王晓平

（宝钢集团新疆八一钢铁有限公司炼铁分公司焦化分厂，新疆乌鲁木齐 830022）

摘　要：文章分析了液压油的种类及性质，液压油污染的原因和防控方法，对提高液压设备经济、高效、环保、安全的运行提供借鉴。

关键词：液压油，危害，防控措施

Types of Hydraulic Fluid and Selection of Pollution and Control

Wang Xiaoping

(Baosteel Group Xinjiang Bayi Steel & Iron Company, Urumchi 830022, China)

Abstract: The paper analyzes the types and properties of hydraulic oil, causes and prevention and control methods of hydraulic oil pollution, and provides reference for improving the operation of hydraulic equipment economy; efficiency; environmental protection and safety.

Key words: hydraulic oil, harm, the prevention and control measures

焦炉高压氨水泵控制系统故障分析及改进

陈　刚

（攀钢集团西昌钢钒公司维修中心，四川攀枝花 617000）

摘　要：本文重点针对攀钢西昌钢钒公司煤化工厂高压氨水系统在焦炉无烟装煤和集气管清扫时存在的问题，从高压氨水控制系统角度来分析影响系统稳定运行的原因，运用变频器技术与 DCS 系统相结合对高压氨水系统进行改进，从而实现高压氨水系统长期稳定运行。

关键词：高压氨水，变频器，DCS 系统，改进

煤化电捕焦油器典型电气故障分析及解决方案

杨　伟，杜　刚

（攀钢集团西昌钢钒公司维修中心，四川攀枝花 617000）

摘　要：简要介绍了西昌钢钒炼铁总厂化产工序电捕焦油器内部构造特点及工作原理，并对电捕焦油器投产运行以来出现的较为典型的电气故障，进行分析和提出相应的解决方案，保证电捕焦油器安全、稳定的运行，提高电捕焦油器后的煤气质量，同时也降低职工作业安全风险，减少职工劳动强度和安全环保事故的发生。

关键词：电捕焦油器，故障分析，解决方案

高炉炼铁工艺上几个节能减排新技术的实践

汤清华，王宝海，张洪宇，王再义

（鞍钢股份公司，辽宁鞍山　114021）

摘　要：笔者根据炼铁生产多年的实践，开发了几项节能减排的工艺技术：炉顶装料过程中均压放散的煤气回收；加压热风炉烟道废气作为高炉煤粉喷吹用惰性气体；高炉煤气干法除尘系统煤气回收时的温度撑控；高炉干法除尘的防腐技术；新型旋风除尘器；高炉低噪音煤气压力调节阀；高炉大中修停炉降料面煤气全回收操作技术。试图从原理和实际应用上加以介绍，以引起同仁们的讨论和应用，想起到抛砖的作用。

关键词：均压放散煤气回收，烟道废气，新型旋风除，惰性气体，压力调节

2017 年上半年中钢协会员单位炼铁技术评述

王维兴

（中国金属学会，北京　100010）

摘　要：2017 年上半年比上年同期我国铁产量升高，产业集中度提高；中钢协单位炼铁技术下滑，高炉燃料比和工序能耗上升，入炉铁品位下降，热风温度无大改进；焦比下降和喷煤比升高，表明炼铁技术有进步。要加大化解铁能过剩的工作力度，努力提升我国炼铁技术水平。

关键词：17 年上半年，重点企业，炼铁技术，评述

Abstract: In the first half of 2017, China's iron output increased and industry concentration increased. The iron and steel technology of cisa is declining, the fuel ratio of the blast furnace is higher than that of the process, the steel grade of the furnace is reduced, and the temperature of the hot air is not improved greatly. The decrease in coke ratio and the increase of coal injection ratio indicate that the iron refining technology has improved. We should increase efforts to reduce excess iron capacity and improve the level of iron smelting in our country.

Key words: first half of 2017, key enterprises, ironmaking technology, review

高钛型高炉渣环境友好资源化高效综合利用研究

柯昌明，刘学新，韩兵强，张锦化，王景然

（武汉科技大学耐火材料与冶金国家重点实验室，湖北武汉 430081）

摘 要：以高钛型高炉渣环境友好、资源化高效综合利用为研究目标，采用等离子熔融还原等技术对高钛型高炉渣中 Ti，Si，V，Mn，Fe 等有价元素进行回收利用，同时对提取 Ti，Si，V，Mn，Fe 等有价元素后的尾渣进行了铝酸盐水泥、精炼渣等高附加值开发利用。研究开发了具有自主知识产权的高钛型高炉渣中钛等有价金属元素的环境友好，高效提取及应用、提钛后尾渣高附加值利用等系列技术，攻克了长期以来高钛型高炉渣中的钛资源提取、利用工作中普遍存在的"提钛成本高、钛回收率低、尾渣无法利用造成环境严重二次污染"等技术难题，在工业规模上实现了高钛型高炉渣中 Ti、V 等有价元素的高效回收（回收率＞90%）。研究开发了新型钛硅铁合金及尾渣铝酸盐水泥、钢水精炼渣等高附加值产品，并进行了工业生产及应用，使高钛型高炉渣的高附加值综合利用率达到 95%以上，实现了高钛型高炉渣的整体、无二次污染的规模化消纳及高附加值资源化综合利用。

关键词：环境友好，资源化高效利用，高钛型高炉渣，钛硅铁合金，铝酸盐水泥，精炼渣

Eco-efficient Titanium-bearing Blast Furnace Slag Recycling

Ke Changming, Liu Xuexin, Han Bingqiang, Zhang Jinhua, Wang Jingran

(The State Key Laboratory of Refractories and Metallurgy, Wuhan University of Science and Technology, Wuhan 430081, China)

Abstract: A new eco-efficient recycling route has been implemented to maximise resources from titanium -bearing blast furnace slag (TBBFS), through recovering valuable metal elements such as Ti, Si, V, Mn, Fe, by means of plasma or electric arc furnace smelting reduction process, and recycling the process slag as raw material for flux, aluminate cement. The industry trial results show that the average recovery of the Ti, V, Mn and Fe are over 90% and Ti-Si-V-Mn-Fe alloys were developed. Industry applications of as-prepared Ti-Si-V-Mn-Fe alloy, flux and aluminate cement were conducted in Zhujiangsteel, Baosteel and refractory manufacturer respectively. Integrated eco-efficient titanium-bearing blast furnace slag recycling has been achieved.

Key words: eco-efficiency, recycling, titanium-bearing blast furnace slag, Ti-Si-V-Mn-Fe alloy, aluminate cement, flux

高辐射覆层技术促进炼铁绿色发展

田凤军，孙传胜，刘常富，张绍强，杨秀青

（山东慧敏科技开发有限公司技术中心，山东济南 250100）

摘 要：高辐射覆层技术已在国内外 70 余家钢铁企业近 500 座高炉热风炉和焦炉上应用，获得了显著的节能效果；已获得国内外发明专利 10 项，形成了完整的知识产权保护体系；制修订相关国家标准、行业标准 14 项，形成了覆

盖高辐射覆层技术、产品生产、技术应用、效果检测等各环节的全方位标准化体系；按国家标准测定的蓄热量已成为高辐射覆层技术应用的考核指标。

关键词：高辐射覆层技术，节能涂料，专利，标准，蓄热量

The Applied Research and Progress of High Radiative Coating Technology

Tian Fengjun, Sun Chuansheng, Liu Changfu, Zhang Shaoqiang, Yang Xiuqing

(Shandong Huimin Science & Technology Co., Ltd., Technical Centre, Jinan 250100, China)

Abstract: High Radiative Coating Technology has been applied on nearly 500 BF hot stoves and coke ovens in more than 70 iron and steel enterprises, obtained obvious energy saving effect; HRC have already won 10 invention patents home and abroad, and formed a complete system of intellectual property rights protection. Drawing 14 items of related national standards and industry standards, we formed comprehensive standardization system, which covered HRC tech, production, application, inspection etc. Heat storage capacity inspection according to the national standard has became the evaluation index of HRC applied.

Key words: high radiative technology, energy saving coating, patent, standard, heat storage capacity

降低鞍钢西区烧结工序能耗的实践

夏铁玉，张飞宇

（鞍钢股份有限公司炼铁总厂，辽宁鞍山　114001）

摘　要：在对鞍钢西区烧结能耗现状进行分析的基础上，介绍了该厂降低烧结工序能耗的措施，包括：加强基础管理、优化配矿、强化操作及技术改造等方面。通过这些手段，该厂的节能工作取得了很大成效，2015年，工序能耗达到了42.37kg/t的较好水平。

关键词：烧结，工序能耗，节能措施

Practice of Reducing Sintering Process Energy Consumption in Ansteel No.1 Iron-making Plant

Xia Tieyu, Zhang Feiyu

(General Ironmaking Plant of Angang Steel Co., Ltd., Anshan 114001, China)

Abstract: Based on the analysis of current situation of sintering process energy consumption in Ansteel No.1 iron-making Plant, the measures taken for reducing the sintering process energy consumption in the plant were introduced, including: strengthening basic management, optimizing ore proportioning,strengthening operation and technical modification etc. Through combination of these measures, thenergy saving work of the plant is very effective, and the process energy consumption in year 2015 met the good level of 42.37kg/t.

Key words: sintering, process energy consumptions, energy saving measures

转底炉珠铁工艺的发展及建议

王　广，薛庆国，王静松

（北京科技大学钢铁冶金新技术国家重点实验室，北京　100083）

摘　要：资源短缺和环保压力是我国钢铁企业现阶段面临的严峻挑战，传统高炉炼铁工艺已接近完美，进一步发展乏力，开发新型炼铁工艺成为必然。转底炉煤基直接还原熔分珠铁工艺能实现渣铁快速分离，符合我国能源和资源结构，在成本和节能减排方面具有一定优势。我国应加强该技术的基础研究，促进钢铁工业的可持续发展。

关键词：转底炉，珠铁工艺，发展建议

Development of Rotary Hearth Furnace Iron Nugget Technology and the Suggestion

Wang Guang, Xue Qingguo, Wang Jingsong

(State Key Laboratory of Advanced Metallurgy, University of Science and Technology Beijing, Beijing 100083, China)

Abstract: Resources shortage and environmental pressure are severe challenges that the Chinese iron and steel industry faces. The traditional blast furnace ironmaking technology has become nearly perfect and the development of new ironmaking technology becomes inevitable. The coal-based rotary hearth furnace reduction and melting separation iron nugget technology can realize fast iron and slag melting separation and it fits China's energy and resource structure. It has a certain advantage in terms of cost and energy saving. China should strengthen the basic research of the technology to promote the sustainable development of the iron and steel industry.

Key words: rotary hearth furnace, iron nugget technology, development suggestion

COSRED 煤基竖炉直接还原工艺的特点和作用

周　强，李森蓉，李建涛，唐　恩，汪　朋，陈泉锋

（武汉科思瑞迪科技有限公司，湖北武汉　430223）

摘　要：本文简要说明了目前直接还原工艺的现状以及 COSRED 工艺的发展历程，详细说明了 COSRED 工艺的工艺流程、工艺参数、原燃料条件、关键技术、规模大小、工序能耗、环境保护、生产成本和操作维护等工艺特点，本文还说明了我国发展电炉炼钢短流程的重要性和紧迫性以及 COSRED 工艺对我国短流程炼钢发展的积极作用，最后本文介绍了 COSRED 工艺适合复合资源的综合利用以及开展钒钛磁铁矿和含铁赤泥的工业实验情况。

关键词：直接还原，COSRED 工艺，特点，作用

Characteristics and Function of COSRED Coal Based Direct Reduction Process

Zhou Qiang, Li Senrong, Li Jiantao, Tang En, Wang Peng, Chen Quanfeng

(Wuhan COSRED Science and Technology Co., Ltd., Wuhan 430223, China)

Abstract: This paper briefly illustrates the current situations of direct reduction process and the development history of COSRED, meanwhile, it minutely describes the process characteristics of COSRED from the following aspects, such as the process flow，the process parameters, raw materials and fuel, the key technology, production scale, process energy consumption, environmental protection, production cost, production operation and so on. Moreover, it has stated the importance and necessity of developing the short process of electric furnace steelmaking in our country, and the positive effects of COSRED process in these respects. Finally, it also gave an outline of the industrial experiments of COSRED process dealing with unmanageable iron material, such as the vanadium titano-magnetite and the red mud containing iron.

Key words: direct reduction, process of COSRED, characteristics, function

气基直接还原竖炉生产过程的数值模拟研究

吕建超，白明华，徐　宽，龙　鹄，任素波，何云华

（燕山大学国家冷轧板带装备及工艺工程技术研究中心，河北秦皇岛　066004）

摘　要： 为了模拟直接还原竖炉的生产过程，建立了气固化学反应模型，通过计算流体动力学对竖炉还原段进行模拟计算。含铁原料的还原反应采用三界面未反应核模型。结果表明：当还原气体为纯氢气时，含铁原料下降约 0.42m 即可完全变成 Fe_3O_4，下降约 0.82m 即可完全变成 FeO，含铁原料下降约 3m 时，已基本被还原成 Fe；当还原气体为纯一氧化碳时，含铁原料下降约 0.8m 即可完全变成，下降约 2m 即可完全变成 FeO，含铁原料下降约 3m 时，已基本被还原成 Fe；还原气的温度及颗粒直径对还原率及化学反应速率有较大影响，通过模拟可知：还原气温度越高，颗粒直径越小化学反应速率越快。通过对还原竖炉生产过程进行数值模拟，对实际生产中提高海绵铁的产量有指导意义。

关键词： 直接还原，竖炉，生产过程，数值模拟，计算流体动力学

Numerical Simulation of Gas-based Direct Reduction Furnace Production Process

Lv Jianchao, Bai Minghua, Xu Kuan, Long Hu, Ren Subo, He Yunhua

(Nation Engineering Research Center for Equipment and Technology of Cold Strip Rolling, Yanshan University, Qinhuangdao 066004, China)

Abstract: A gas-solid chemical reaction model is built to simulate the reduction section of shaft furnaceby Computational Fluid Dynamics. Three interface unreacted core model is adopted for reduction of iron raw materials. Results show that: when reducing gas is pure hydrogen,while the Fe_3O_4 and FeO reductions occur in the upper parts of 0.42 meter and 0.82

meter, Iron- containing materials has basically been reduced to Fe, which occur in the upper parts of 3 meters; When reducing gas is pure carbon monoxide, while the Fe_3O_4 and FeO reductions occur in the upper parts of 0.8 meter and 2 meters, Iron-containing materials has basically been reduced to Fe, which occur in the upper parts of 5 meters: Reducing gas temperature and particle diameter have an important influence on the reduction rate. Seen through simulation: The higher temperature of reduction gas and the smaller particle diameter make the rate of chemical reactions faster. Results also show that reduction rate grows fast with the increase of the temperature and decrease of pellets diameter. Through the simulate of reduction shaft furnace production process, which is instructive for increasing the production of sponge iron.

Key words: direct reduction, shaft furnace, production process, numerical simulation, computational fluid dynamics

钢铁厂含铁尘泥的利用途径分析

余正伟，穆固天，龙红明，春铁军，钱立新，李　宁

（安徽工业大学冶金工程学院，安徽马鞍山　243002）

摘　要： 含铁尘泥是钢铁工业生产过程中产生的主要固体废弃物之一，其产量巨大，富含 Fe、C、Pb、Zn 等有价元素，是重要的二次资源，但若处置不当也极易造成大气、土壤和水资源的污染。本文通过分析含铁粉尘的利用现状，阐明了各类尘泥利用方法的适应性和优缺点。通过分析可知，含铁尘泥直接堆存或填埋占用土地、浪费资源且造成污染，已不被业界接受。含铁尘泥分类返回烧结、炼铁和炼钢工序是比较有效、可行的方案。采用物理化学方法综合提取回收尘泥中有价组分的方法是尘泥高效回收的又一途径。上述方法中，铁矿复合造块法、含铁尘泥成金属化球团法、粉尘高炉喷吹法和熔融还原法应用前景广阔。

关键词： 钢铁生产，固体废弃物，尘泥，有价元素，环境污染

Utilization Method Analysis of Iron-bearing Dust and Sludge from Iron and Steel Plant

Yu Zhengwei, Mu Gutian, Long Hongming, Chun Tiejun, Qian Lixin, Li Ning

Abstract: Iron-bearing dust and sludge is one of the main solid wastes, which characterized by tremendous production and Fe, C, Pb, Zn-containing. The mismanagement of this important secondary resources easily leads to pollution of air, soil and water. In this paper, the utilization method analysis of dust and sludge was analyzed and the raw material adaptability, advantages and disadvantages of all kinds of utilization methods of iron-bearing dust and sludge were illuminated. Through the analysis, the methods of stockpiling or landfill has been unaccepted by ferrous metallurgy fields. Taken the classified dust and sludge back to sintering, ironmaking or steelmaking process is the more effective method of in-plant recycling. Also, the physical dressing and hydrometallurgical leaching can recycle the valuable elements from dust and sludge. Among them, composite agglomeration of iron ore fines, direct reduction for producing metalized pellets, injecting dusts into blast furnace and melting reduction process has broad prospects in application.

Key words: iron and steel production, solid waste, dust, valuable elements, environmental pollution

COSRED 煤基竖炉直接还原工艺的特点

周　强，李森蓉，李建涛，唐　恩，汪　朋，陈泉锋

（武汉科思瑞迪科技有限公司，湖北武汉　430223）

摘　要： 本文简要说明了目前直接还原工艺的现状以及 COSRED 工艺的发展历程，详细说明了 COSRED 工艺的工艺流程、工艺参数、原燃料条件、关键技术、规模大小、工序能耗、环境保护、生产成本和操作维护等工艺特点，本文还说明了我国发展电炉炼钢短流程的重要性和紧迫性以及 COSRED 工艺对我国短流程炼钢发展的积极作用。

关键词： 直接还原，COSRED 工艺，特点

Characteristics and Function of COSRED Coal Based Direct Reduction Process

Zhou Qiang, Li Senrong, Li Jiantao, Tang En, Wang Peng, Chen Quanfeng

(Wuhan COSRED Science and Technology Co.,Ltd., Wuhan 430223, China)

Abstract: This paper briefly illustrates the current situations of direct reduction process and the development history of COSRED, meanwhile, it minutely describes the process characteristics of COSRED from the following aspects, such as the process flow, the process parameters, raw materials and fuel, the key technology, production scale, process energy consumption, environmental protection, production cost, production operation and so on. Moreover, it has stated the importance and necessity of developing the short process of electric furnace steelmaking in our country, and the positive effects of COSRED process in these respects.

Key words: direct reduction, process of COSRED, characteristics

中温改质沥青产品的调试生产

刘海波，穆春丰，张大奎，贾楠楠

（鞍钢化工事业部，辽宁鞍山　114000）

摘　要： 以中温沥青（指标符合 GB/T 2290—2012）为原料，利用两釜串联的改质沥青生产装置，通过热缩聚反应，合理控制工艺参数，得到中温改质沥青产品，其指标参考中温沥青和改质沥青（YB/T 5194—2003 一级品）的相关标准，产品性能为软化点 90～96℃，甲苯不溶物 26%～33%，喹啉不溶物 7%～11%，结焦值≥54%。

关键词： 中温沥青，改质沥青，中温改质沥青

Study on the Production Adjustment of the Mid-temperature and Modified Quality Pitch

Liu Haibo, Mu Chunfeng, Zhang Dakui, Jia Nannan

(Chemical Segment Ansteel Group, Anshan 114000, China)

Abstract: Using mid-temperature pitch as raw material (product index: GB/T 2290—2012), the production device to two Kettle series is applied to produce mid-temperature and modified-pitch. The type of reaction is hot condensation along with controlling reasonable process parameter. Its index refers to the mid-temperature pitch and modified quality pitch (YB/T 5194—2003 first level product) related standard. The performance of products: softening point 90~96℃, toluene insolubles 26%~33%, quinoline insolubles 7%~11%, coking value ≥54%.

Key words: mid-temperature pitch, modified-pitch, mid-temperature and modified-pitch

干熄焦预存段压力波动问题分析与改进

朱庆庙[1]，庞克亮[1]，王　超[1]，谭　啸[2]，武　吉[1]

（1. 鞍钢股份有限公司技术中心，辽宁鞍山　114009；

2. 鞍钢股份有限公司炼焦总厂，辽宁鞍山　114031）

摘　要： 在干熄焦实际生产过程中频繁出现预存段压力波动的问题，分析循环气体系统各部位阻力结合气体循环系统的特点开展问题分析及现场标定。标定分析空气导入阀门及预存段压力调节阀门的关系；利用现场实际条件标定预存段四周各部位压力；标定预存段放散管处压力；发现预存段压力取出口内部的损害是预存段压力波动的根本原因，为干熄焦预存段压力波动问题处理提供科学依据，有效的保证了干熄焦生产的稳定运行，并为干熄焦故障处理提供科学的处理方法。

关键词： 干熄焦，预存段压力，波动原因，分析改进

Pressure Fluctuation Analysis and Improvement of Coke Dry Quenching in Pre-stored Segement

Zhu Qingmiao[1], Pang Keliang[1], Wang Chao[1], Tan Xiao[2], Wu Ji[1]

(1. Technology Center of Angang Steel Co., Ltd., Anshan 114009, China;

2. Coking Plant of Angang Steel Co., Ltd., Anshan 114031, China)

Abstract: Pressure frequent fluctuation problems of the pre-stored segement on coking dry quenching (CDQ) existed in actual production process, analysising of each part of the cycle gas system resistance combined with the characteristics of gas circulation system conduct problem analysis and field calibration. Analysising the relationship between air import valve calibration and the pressure regulating valve of the pre-stored segement; Using the actual conditions of the site to calibrate the pressure of each part of the pre-stored segement; calibration of the pressure in the discharge section. the calibration period of stored pressure at the exhaust pipe; that stored pressure taking out port internal damage is a fundamental cause of stored pressure fluctuation, pressure fluctuation of deposit and provide scientific basis for CDQ pre, effectively guarantee the stable operation of CDQ production, and provide the scientific method for CDQ fault handling.

Key words: coke dry quenching, pressure frequent fluctuation problems, fundamental cause, scientific method

混配生产中温改质沥青的研究

贾楠楠，穆春丰，张大奎，刘海波，张　展，王晓楠

（鞍钢集团化工事业部，辽宁鞍山　114000）

摘　要： 近年来，为了满足下游产品的日益发展，客户对于沥青产品的质量要求越来越高，普通的中温沥青、改质沥青已经不能满足客户需求，通过客户反馈和市场调研，沥青产品的质量指标已逐步趋向于高甲苯不溶物、高喹啉不溶物、高β树脂、低软化点、无中间相。而生产煤系针状焦时的副产品尾料沥青的指标又接近于中温沥青，因此选用尾料沥青和改质沥青为原料，通过混配实验生产中温改质沥青是最直接可行的方法。

关键词： 中温改质沥青，尾料沥青，改质沥青，混配生产

Study on Medium Temperature Modified Asphalt in Blending Production

Jia Nannan, Mu Chunfeng, Zhang Dakui, Liu Haibo, Zhang Zhan, Wang Xiaonan

(Angang Group Chemical Business Division, Anshan 114000, China)

Abstract: In recent years,in order to meet the development of downstream products,the higher the quality of asphalt products for customer demands,in ordinary temperature asphalt, modified asphalt has been unable to meet the needs of customers,through customer feedback and market research,the quality indexes of asphalt products have been gradually tend to be high,high toluene insoluble quinoline insolubles high beta resin,low softening point,no intermediate phase.A by-product of tailings and asphalt production when the index of coal based needle coke is close to the temperature asphalt,so the tailings asphalt and modified asphalt as raw material,through mixing temperature modified asphalt production experiment is the most direct way possible.

Key words: temperature modified asphalt, tailing asphalt, modified asphalt, mixed production

焦煤预处理工艺技术经济分析

胡德生[1]，Alex Wong[2]，牛　虎[2]

（1. 宝山钢铁股份有限公司研究院，上海　201900；

2. 济南戴尔塔干燥技术有限公司，山东济南　250101）

摘　要： 通过对不同的焦煤及煤焦油沥青价格计算出不同炼焦煤预处理工艺（DAPS，SCOPE21 及 DELTA 工艺）所产生的净收益对总投资的比率(α)并对其进行分析比较，得出如下主要结论：（1）DELTA 工艺的 α 值最高；SCOPE21 工艺的 α 值最低；（2）在任何工况下，SCOPE21 工艺的 α 值都比 DAPS 工艺的低；（3）在最不利工况

下，SCOPE21 工艺的 α 值为负；而 DELTA 工艺的 α 值仍为正。

关键词：DAPS 工艺，SCOPE21 工艺，DELTA 工艺，弱粘结煤，煤焦油沥青

Technical and Economic Analysis of Pretreatment Processes for Coking Coal

Hu Desheng[1], Alex Wong[2], Niu Hu[2]

(1. Research Institute of Baoshan Iron & Steel Co., Ltd., Shanghai 201900, China;
2. Delta Drying Technology Ltd., Jinan 250101, China)

Abstract: The ratio of net income generated by different coking coal pretreatment processes (DAPS, SCOPE21 and DELTA) to total capital investment (α) is calculated and analyzed under different price of coal and coal tar pitch, the main conclusions are drawn as follows: (1) DELTA process has the highest α value;　SCOPE21 process has the lowest α value; (2) SCOPE21 process' α value is lower than DAPS process' under any operating conditions; (3) SCOPE21 process's α value is negative while DELTA process' α value is still positive under the most unfavorable conditions.

Key words: DAPS process, SCOPE21 process, DELTA process, low grade coking coal, coal tar pitch

苯加氢气提塔操作参数优化

董　毅，陶帅江，杨　峰，穆春丰

（鞍钢集团化工事业部，辽宁鞍山　114200）

摘　要：鞍钢集团化工事业部苯加氢装置于 2009 年投产，现已连续运行 8 年。为增加规模效益，降低加工成本，在事业部组织下，苯加氢作业区进行了 110% 负荷试生产标定，试生产期间，出现气提塔塔底泵震动的现象，影响了进一步提高处理量。本文通过对苯加氢气提塔塔底泵震动原因进行分析，分析得出塔底泵因气蚀产生震动，通过对气提操作参数进行优化，有效解决了气提塔塔底泵气蚀的问题。

关键词：操作参数，气提塔，苯加氢，气蚀

Optimization of Operating Parameters for the Stripper Column in Crude Benzene Hydrogenation Process

Dong Yi, Tao Shuaijiang, Yang Feng, Mu Chunfeng

(Angang Group Chemical Business Division, Anshan 114200, China)

Abstract: The process of crude benzene hydrogenation of CIDA has run 8 years since putted into operation in 2009.Inorder to increase economies of scale, reduce the processing cost, the department of crude benzene hydrogenation made a 110% load trail production under the guidance of CIDA. But the bottom pump of the stripper column vibrated during 110% load trail production, which posed problems for further improving the capacity. To address this issue, the causes of the bottom pump vibration are resolved. Cavitation is the primary cause of the vibration. And it has been resolved efficiently by optimizing the operating parameters of the stripper column.

Key words: operating parameters, stripper column, process of crude benzene hydrogenation, cavitation

检修模式对焦炉生产操作的影响

赵恒波，周　鹏，党　平，马富刚，彭　磊

（鞍钢股份鲅鱼圈钢铁分公司炼焦部，辽宁营口　115007）

摘　要：本文通过对焦炉"定时"检修模式和"定炉号"检修模式进行定义、优缺点对比，结合鲅鱼圈炼焦部 7m 焦炉上的实际使用效果，论证了"定炉号"检修模式对焦炉生产操作的优势，阐述了检修模式对焦炉生产操作的影响。

关键词：检修，焦炉，操作，影响

干熄炉水封槽结构改进

刘占博，赵恒波，周　鹏，党　平

（鞍钢股份鲅鱼圈钢铁分公司炼焦部，辽宁营口　115007）

摘　要：本文主要针对干熄炉水封槽内易沉积焦粉，无法实现及时有效的清理，造成水封槽内局部热应力较高，水封槽底面与侧面焊口开裂，从而引起水漏入炉内，干熄炉 H_2 含量超标且砌体出现侵蚀，改进水封槽结构形式，实现焦粉的自动清除，延长水封槽使用寿命，稳定干熄焦生产操作。

关键词：干熄炉，水封槽，结构，改进

除尘灰回配炼焦生产实践

闪振国，李培铖

（宝钢集团新疆八钢炼铁分公司，新疆乌鲁木齐　830022）

摘　要：近几年，随着焦化产能的日益扩大，对炼焦煤的需求量也与日俱增，导致部分炼焦煤煤种供应紧张，炼焦煤价格持续走高，铁矿石原料的涨价以及国家宏观政策的调控也制约了钢铁企业在成本控制上对焦炭价格的要求，因此市场上炼焦煤价格与焦炭价格产生倒挂现象。在全球金融危机的背景下，全国的炼焦行业陷入前所未有的经营困境，我公司同样面临亏损大、亏损周期长的严峻局面。适当采用一些低成本原料，成为公司可持续发展的必由之路。本文论述了近年来八一钢铁炼铁分公司焦化分厂为了回收利用除尘焦粉，减少配合煤成本，通过小焦炉试验和大炉生产，在除尘焦粉回配炼焦的生产实践中取得良好效果。

关键词：配煤，除尘灰，焦炭质量

Coking Dust and Ashes Rematch Coking Production Practice

Shan Zhenguo, Li Peicheng

(Baosteel Group Xinjiang Bayi Steel & Iron Company, Urumchi 830022, China)

Abstract: In recent years, with the coking capacity increasing, the demand for coking coal is growing, the cause of coking coal supplies of coal, coking coal prices rising, prices of raw materials for iron ore, the national macro policy regulation is restricted the iron and steel enterprises in the cost control on the demand for coking coal prices, therefore coking coal prices in the market and the coke price inversion phenomenon. In the context of the global financial crisis, the national coking industry has been in an unprecedented situation of operation, and our company is also facing a serious situation of large losses and long loss cycle. It is the only way for sustainable development of the company to adopt some low-cost raw materials. This paper discusses the Bayi Iron and Steel iron branch coking plant in recent years in order to recycle shaker coke powder, reduce the cost of coal, and large furnace production, through small coke oven test in the production of dust removing coke powder back with coking achieve good effect in practice.

Key words: coal blending, coking dust, coke qualities

浅析市场经济下化工油库粗苯区域安全除锈刷漆维护管理

张　亮

（宝钢集团新疆八钢炼铁分公司，新疆乌鲁木齐　830022）

摘　要：在市场经济模式下，如何安全地高效地完成油库粗苯区域维护需求，在较低运行成本下保证维护作业的安全，是企业生产发展的基础，如何通过安全技术来保证化工油库粗苯区域的在线维护。

关键词：安全高效，安全技术，除锈刷漆，粗苯罐区在线维护

Discussion on the Maintenance and Management of the Safety Derusting Brush Paint in the Crude Benzol Area of Chemical Oil Depot under the Market Economy

Zhang Liang

(Ironmaking Branch, Bayi Iron & Steel Co., Baosteel Group, Urumchi 830022, China)

Abstract: In the market economy mode, how to ensure the safety of the crude benzene region maintenance in the oil depot, and the low operating cost, is the basis of the enterprise production development, how to ensure the online maintenance of the crude benzene region in the chemical industry through safety technology.

Key words: safety, high efficiency, safety technology, derusting brush paint, crude oil tank area online maintenance

型煤强度影响因素及配煤炼焦初步探讨

李培铖，韩　阁，何肖凯

（宝钢集团新疆八钢炼铁分公司，新疆乌鲁木齐　830022）

摘　要： 主要介绍了煤料水份、焦油渣配量、型煤生球干燥温度、干燥时间对不同煤化度煤所制型煤强度的影响相关因素，并对强度较高的气煤进行了配入焦化生产过程中所产生的除尘灰后型煤强度的变化，旨在将除尘灰融入型煤工艺中，解决生产中除尘灰配加的不稳定因素，同时对不同比例型煤的配入对焦炭质量的影响进行了试验研究。在八钢目前配煤结构下，型煤配入可明显增加入炉煤堆密度，增加焦炭产量，同时改善焦炭质量，也为型煤工艺的确立提供参考依据。

关键词： 型煤，强度，配煤炼焦

The Factors Influencing the Intensity of Coal Briquette and Coal Blending Coking Type

Li Peicheng, Han Ge, He Xiaokai

(Baosteel Group Xinjiang Bayi Steel & Iron Company, Urumchi 830022, China)

Abstract: The influence factors of coal material moisture, tar residue distribution, dry temperature and drying time on coal strength of different coal coal are introduced. In addition, the strength of coal dust in the coking process is carried out. In order to integrate the dust into the coal process. Solve the unstable factor of dust distribution in production. At the same time, the influence of different proportion coal on coke quality was studied. Under the current blending structure of the Bayi Steel & Iron. The density of coal pile can obviously be increased by the distribution of type coal. Increase coke production. while improving the coke quality. It also provides reference basis for the establishment of type coal technology.

Key words: coal briquette, intensity, coal blending coking

斜孔舌形塔板在蒸氨塔的应用分析

戴　杰

（宝钢集团新疆八钢炼铁分公司，新疆乌鲁木齐　830022）

摘　要： 介绍了斜孔舌形塔的结构、工作原理及特点，结合其在蒸氨系统应用中存在的问题对液体的流量、质量、温度及入塔蒸汽对蒸馏效果及生产状况的影响进行了讨论，提出了改进意见。

关键词： 斜孔舌形塔，喷射，蒸氨

Oblique Hole Tray Tongue Analysis Application in Ammonia Distillation Tower

Daiger

(Baosteel Group Xinjiang Bayi Iron Branch, Urumchi 830022, China)

Abstract: This paper introduces the structure, working principle and characteristics of inclined hole tongue tower, with its presence in the ammonia evaporation system in the application of liquid flow, quality, temperature and steam distillation effect of tower into effect and production conditions are discussed, and puts forward improvement suggestions.

Key words: inclined hole tower, ammonia, injection tongue

焦炉集气管压力波动的原因及处理方法

苗　钧，黄洁洁

（宝钢集团八钢公司炼铁分公司焦化分厂，新疆乌鲁木齐　830022）

摘　要：焦炉集气管压力是炼焦生产过程中的重要参数，它对化产品回收、炉体寿命及环境污染都会产生直接的影响。针对八钢焦化厂集气管压力波动大的情况，通过调节控制器参数、改变调节系统取压点等措施，稳定了集气管压力，取得了良好的效果。

关键词：集气管压力，波动，调节

The Cause and Measurement for the Pressure Fluctuation of Coke Oven Collecting Main

Miao Jun, Huang Jiejie

(Ironmaking Branch, Bayi Iron & Steel Co., Baosteel Group, Urumchi 830022, China)

Abstract: The coke oven collector main pressure is an important industrial parameter, it has direct influence to coke quality, air pollution and coke oven life. By adjusting the parameters of the controller, changing the pressure measuring point of the regulating system, we can solve the problem of pressure fluctuation on the coke oven of Bayi coking plant, and gain a good effect.

Key words: coke oven collecting main pressure, fluctuation, adjustment

八钢焦化厂降低电耗管理实践

苗　钧，李雪龙

（宝钢集团八钢公司炼铁分公司焦化分厂，新疆乌鲁木齐　830022）

摘　要：在当前钢铁产能过剩的大背景下，企业为了生存和发展，降本增效工作显得尤为重要。电耗作为体现企业管理水平的重要指标之一，本文结合现阶段生产状况，围绕降低电耗为主题，通过系统管理措施的实施，既锻炼了队伍又降低了电耗。

关键词：管理，电耗，降低

贵州 1/3 焦煤煤质与炼焦配煤研究

刘　全，张启锋

（宝山钢铁股份有限公司制造管理部，上海　201941）

摘　要：通过全面分析贵州 1/3 焦煤质，并与其他 5 种煤的性质进行比较，掌握了贵州 1/3 焦煤煤质特性，贵州 1/3 焦煤具有肥煤部分作用。合理利用贵州 1/3 焦性质和比较优势，通过系列工业配煤试验，分析和总结其在配煤中的应用效果，最大化的发挥其作用。

关键词：贵州 1/3 焦，煤质，炼焦配煤，研究

The Study on the Coal Quality and Coking Coal Blending of 1/3 Coking Coal in Guizhou

Liu Quan, Zhang Qifeng

(Manufacturing Management Department of Baosteel, Shanghai 201941, China)

Abstract: Through comprehensive analysis of 1/3 coking coal quality in Guizhou, and compared with the other five kinds of coal properties, master of Guizhou 1/3 coking coal quality characteristics, Guizhou 1/3 coking coal has the role of fat coal. Reasonable utilization of the nature and comparative advantage of 1/3 coke in Guizhou, through series of industrial coal blending test, analysis and summary of its application in coal blending effect, maximize its role.

Key words: Guizhou 1/3 coking coal, quality, the coal of blending, study

装煤除尘车爆鸣的原因分析及改进措施

孙红艳[1]，刘武镛[2]

（1. 本钢集团北营公司，辽宁本溪　117017；2. 本钢集团供应商管理部，辽宁本溪　117000）

摘　要：介绍了本钢集团北营焦化厂装煤除尘车除尘的工作原理及操作程序，分析了装煤末期产生爆鸣的原因是可燃气体达到了爆炸极限，并针对可燃气体达到爆炸极限的两大原因——吸入除尘干管的空气量少和烟气量多制定了相应的措施，从降低除尘系统阻力、保证高压氨水系统压力和规范装煤操作等方面进行改进，实施后效果很好，没有发生过装煤除尘车爆鸣的现象。

关键词：装煤除尘车，分析，措施，效果

Analysis and Improvement Measures for the Detonation of Coal Dust Removal Vehicle

Sun Hongyan[1], Liu Wuyong[2]

(1. Beiying Company of BX Steel, Benxi 117017, China;
2. Supplier Management Department of BX Steel, Benxi 117000, China)

Abstract: This paper introduces the working principle and operating procedures of coal dust removal vehicle in Beiying Coking plant of BX Steel.This paper analyses the cause of detonation at the end of coal loading,it's because that the flammable gas reaches the explosive limit.The two reasons for the explosion limit of flammable gases is that the amount of air is less and the amount of flute gas is more than normal amount that to be inhaled and removed in the dust removal pipe.We have improved the resistance of the dust removal system,the pressure of the high pressure ammonia system and the regulation of the coal operation.We have done very well after the implementation,there hasn't been the phenomenon of the coal dust removal removal detonation.

Key words: coal dust removal vehicle, analysis, measures, the effect

190t/h 干熄焦斜道损坏机理分析及处理

袁本雄，邵胜平

（中信泰富特钢集团铜陵泰富特种材料有限公司，安徽铜陵　244199）

摘　要：190t/h 干熄焦是目前国内投产较多的大型焦炭干熄装置，整体结构矮胖，冷却室高径比小于 0.6，采用双斜道技术。本文介绍了在实际生产中斜道部位易出现的问题和处理方法。

关键词：干熄焦，斜道修复，浇注料

Analysis on 190t/h CDQ Chamber Chute Damage and Treament

Yuan Benxiong, Shao Shengping

(Tongling Pacific Special Materials Co., Ltd., CITIC Pacific
Special Steel Holdings, TongLing 244199, China)

Abstract: 190t/h CDQ is promoting around China this years. The specific value of the cooling chanamber's height and diameter less than 0.6 make it look like dumpy, and Double Chute Process was applied in the chute area.This article introduced the problems in actual production and the maintenance.

Key words: CDQ, chute maintenance, castables

一种在复杂焦炉生产模型下焦罐车控制系统的设计及应用

苟晓峰

（山西太钢不锈钢股份有限公司焦化厂，山西太原　030003）

摘　要： 本文主要介绍了是在三座焦炉、三套车辆、四座干熄炉和三台干熄焦提升机的联合生产单元下，实现同一轨道上高速运行着两组焦罐车动态稳定可靠。此设计主要通过单台焦罐车单机控制系统、焦罐车同焦炉车辆地面协调控制系统的协调关系和三台焦罐车同南北区四座干熄炉控制系统的连锁来完成支持复杂焦炉生产模式下的焦炭全干熄车辆控制方案。此设计目前应用于太钢焦化厂焦炉三台焦罐车的控制系统，效果良好。

关键词： 复杂焦炉，生产模型，焦罐车，控制系统

太钢 7.63m 焦炉工艺技术创新与实践

贺　佳，梁　杰，马卫华

（山西太钢不锈钢股份有限公司焦化厂，山西太原　030003）

摘　要： 山西太钢不锈钢股份有限公司焦化厂（以下简称太钢焦化厂）先后于 2006 年 11 月、2007 年 8 月和 2013 年 11 月建成投产三座 7.63m 焦炉，焦炉由德国 Uhde 公司进行基础设计，中冶焦耐工程技术有限公司进行详细设计，移动车辆由德国 SCHALKE 公司设计。建成投产后太钢焦化厂根据焦炉实际生产运行情况，对部分工艺技术进行了改进，以满足对焦炉高产、优质、环保的要求。本文介绍了太钢 7.63m 焦炉生产组织模式 I、II 及两种生产模式的优势和弊端及对干熄焦工序的影响，同时通过 3 种方式对焦炉的单孔装煤量进行提高。

关键词： 7.63m 焦炉，工艺技术，改进，实践

太钢降低肥焦煤比例的研究

吴　波，屈秀珍，刘昔跃，梁　杰，殷龙腾

（山西太钢不锈钢股份有限公司焦化厂，山西太原　030003）

摘　要： 在保证焦炭质量的前提下，降低肥焦煤比例，从而降低配煤成本，是焦化企业的一项重要工作。本文以 300kg 试验焦炉为主要技术手段，结合常规指标（灰分、硫分、G 值、Y 值等）、300kg 试验焦炉炼焦指标（M40、M10、CRI、CSR 等）开展了一系列降低肥焦煤比例的实验，对太钢的用煤结构的优化方向进行了分析与探讨，在

降低肥焦煤比例上取得了较好的进展。

关键词：300kg 试验焦炉，降低肥焦煤比例

浅谈块煤防破碎技术的研究与实践

蒲继承，夏光勇

（中冶赛迪工程技术股份有限公司散料物流事业部，重庆 401122）

摘　要：文章从破碎机理出发，研究了目前几种常用的块煤防破碎方法。结合工程实例，探讨了以块煤为原料的某企业在原料受卸、贮存和转运环节中采取的块煤防破碎的措施，以降低块煤的粉粹率。

关键词：块煤，防破碎，贮存，转运，粉粹率

Research and Practice on the Anti-crushing Technology of Lump Coal

Pu Jicheng, Xia Guangyong

(Bulk Material Handling Business Section of CISDI Engineering Co., Ltd., Chongqing 401122, China)

Abstract: The paper studies the several common methods used for lump coal anti-crushing from the crushing mechanism. Combined with engineering projects, this paper discusses the anti-crushing measures in the unloading, storage and transport of raw materials to reduce the crushing rate of lump coal in some industry.

Key words: lump coal, anti-crushing, storage, transport, crushing rate

莱钢 1#1080m³ 高炉合理煤焦结构研究

张　明，王庆会

（山东莱芜钢铁股份有限公司炼铁厂，山东莱芜 271104）

摘　要：本文主要介绍了莱钢 1#1080m³ 高炉合理煤焦结构的研究，通过数据分析高炉生产历史数据，探索高炉焦比、煤比、燃料比控制关系，找出高炉最为经济合理的煤焦结构模型，降低高炉燃料消耗，为高炉绿色生产、低成本生产创造好条件。

关键词：高炉，煤焦结构，燃料比

Study on the Rational Coal Coke Structure of Laiwu Steel 1 # 1080m³ Blast Furnace

Zhang Ming, Wang Qinghui

(Laiwu Steel Corporation Ironworks, Laiwu 271104, China)

Abstract: 1 # 1080 m³ blast furnace in Laiwu Steel is mainly introduced in this paper the reasonable capital structure research, the blast furnace production history data through the data analysis, explore the coke rate of blast furnace, the relationship between the coal ratio, fuel ratio control, find out the most economic and reasonable blast furnace coal tar structure model, reduce the fuel consumption in blast furnace, for the green production, low cost of production to create good conditions for blast furnace.

Key words: the blast furnace, coke structure, fuel ratio

铁矿石对焦炭高温性能的影响研究

梁南山

（涟钢技术中心，湖南娄底　417009）

摘　要： 以热分析仪研究了焦炭在配加 10% 和 74% 两种比例不同铁矿石后在空气及 CO_2 环境中的高温反应性能。研究表明，铁矿石能显著降低焦炭燃烧的起始温度及燃烬温度，且高比例矿石比低比例矿石、熟矿比生矿对焦炭燃烧具更大的促进作用。与全焦对比，配入 10% 矿石的焦炭气化反应起始温度升高了 183~274℃，但高温气化反应大为加速，整体促进效果不大。而配入 74% 矿石的焦炭，气化反应起始温度升高 13~76℃，而结束温度要低于全焦 150~187℃，对焦炭低温与高温气化反应均具显著促进作用，其中以含较高碱金属的南非矿最为突出。

关键词： 高炉，矿石，焦炭，性能

Research on Effect about Iron Ore to Coke Performance at High Temperature

Liang Nanshan

(Technology Center of Lianyuan Iron & Steel Group Co., Ltd., Loudi 417009, China)

Abstract: The performances in air and CO_2 at high temperature of coke being mixed with 10% and 74% various iron ore are studied by thermogravimetric analysis. It indicated that, Comparing with 100% coke, the starting and ending temperature of coke burning has been cut down greatly after mixed with iron ore .On the promotion to coke burning, high proportion of iron ore is more than low proportion of iron ore, ripe ore is more than fresh ore. Comparing with 100% coke,the starting temperature of gasification in CO_2 of coke mixed with 10% iron ore rise up 183~274℃, but the reaction of gasification at high temperature are speeded up greatly. The total effect of promotion is not great. Nevertheless, the starting temperature of gasification in CO_2 of coke mixed with 74% iron ore only rise up 13~76℃, but the ending temperature cut down 150~187℃, it shows remarkable promotion to reaction of gasification at low temperature and high temperature, specially for South Africa ore that contains high alkali metal.

Key words: blast furnace, iron ore, coke, performance

稳定焦炭质量的技术管理方法及生产应用

彭永根，周云花

（华菱湘钢技术质量部，湖南湘潭　411101）

摘　要：本文介绍了焦炭质量稳定性对高炉的意义，分析了影响焦炭质量稳定的各种因素，重点对炼焦煤的质量评价技术、炼焦煤的采购管理、炼焦配煤结构的优化以及焦化原料及焦炉生产流程等技术管理方法进行了分析；结合湘钢生产实际，对 20 多个炼焦煤开展了多指标质量评价研究，提出了炼焦煤质量采购要求和配煤结构调整建议；从生产实际应用效果看，降低了配煤成本，稳定了焦炭质量，为高炉的高产顺行提供了焦炭质量保证。

关键词：焦炭质量，稳定质量，质量评价，配煤结构

Technical Management Method and Production Application for Stabilizing Coke Quality

Peng Yonggen, Zhou Yunhua

(Technical Quality Department of Valin Xiangtan Steel, Xiangtan 411101, China)

Abstract: This thesis introduces the significance of coke quality and stability to blast furnace, and analyzes various stabilizing factors that affect coke quality. The quality evaluation technology of coking coal ,the purchasing management of coking coal, the optimization of coking coal blending structure, and the technological management methods of coking raw material and coke oven production process are emphatically analyzed. According to the production practice, the quality evaluation technology of more than 20 coking coal was studied, the quality requirement of coking coal and the suggestion of coal blending structure adjustment are put forward. From the practical application effect, the coal blending cost is reduced, the quality of coke is stabilized, and the quality guarantee for the high production of the furnace is provided.

Key words: coke quality, steady quality, quality evaluation, coal blending structure

感应电使皮带机拉绳开关和跑偏开关不能动作的故障分析与处理

刘忠明

（新疆八钢炼铁分公司，新疆乌鲁木齐　830022）

摘　要：八钢南疆炼铁厂 1800m³ 高炉有许多皮带机运输系统，上料系统和渣处理系统都是高炉生产中的重要环节，

上料系统将球团矿和焦炭通过皮带机送到高炉，渣处理系统将炼铁产生的水渣，经皮带机运往水渣堆场，在皮带机运行和检修过程中，发现皮带机电机电气控制系统出现问题，由于感应电使皮带机拉绳开关和跑偏开关不能动作的故障，下面进行分析和探讨解决方法。

PLC 在大型筒仓配煤控制系统中的应用

袁文兵，张　车

（江苏沙钢集团有限公司，江苏张家港　215600）

摘　要：在焦炭生产工艺过程中，需要将气煤、肥煤、焦煤、瘦煤、1/3 焦煤五种煤按一定比例配成混合煤，然后经粉碎送入焦炉进行高温炼焦。配比的科学性以及配料系统的可靠性、稳定性将直接影响焦炭产品的质量。因此，通过提高配煤系统的可靠性、稳定性、准确性来提高焦炭的质量具有非常重要经济效益。

关键词：筒仓，圆盘，PLC 系统，原理，控制，申克秤

2　炼钢与连铸

炼铁与原料
★　炼钢与连铸
轧制与热处理
表面与涂镀
金属材料深加工
钢铁材料
汽车钢
海洋工程用钢
轴承钢
电工钢
粉末冶金
非晶合金
高温合金
耐火材料
能源与环保
分析检测
冶金设备与工程技术
冶金自动化与智能管控
冶金技术经济

连铸重压下技术研究及其应用

朱苗勇，祭　程

（东北大学冶金学院，辽宁沈阳　110819）

摘　要：连铸坯重压下技术是改善铸坯中心疏松、提升铸坯致密度的有效手段。本文简述了国内外在此方面的工程实践进展，介绍了东北大学自主开发的连铸坯凝固末端重压下新工艺，包括关键工艺控制技术与相应的重压下装备设计要求，以及在唐钢宽厚板坯连铸机和攀钢大方坯连铸机的工业应用效果。

关键词：连铸，重压下，宽厚板坯，大方坯

The Study and Application of Strand Heavy Reduction in Continuous Casting Process

Zhu Miaoyong, Ji Cheng

(School of Metallurgy, Northeastern University, Shenyang 110819, China)

Abstract: Heavy reduction is an effective method to improve center porosity of continuous casting strand. In this paper, the exploratory research and progress in heavy reduction area were presented, and the key technologies and industrial application results in Tanshan Steel and Panzhihua Steel of self-developed heavy reduction process by Northeastern University were introduced.

Key words: continuous casting, heavy reduction, wide-thick slab, bloom

红钢 3#高炉提高煤气利用率生产实践

普　松，唐德元，钟　伟，赵树逵

（红河钢铁有限公司，云南蒙自　661100）

摘　要：对红钢 3#高炉提高煤气利用率进行了总结，通过从原燃料质量管理、装料制度探索、操作管理三方面入手，使煤气利用率从 2015 年的 40.57%提高到 44.52%，8 月份完成燃料比 545kg/t，突破了历史最好水平。

关键词：高炉，煤气利用率，原燃料，达产

Production Practice of Increasing Gas Utilization Rate in No.3 BF of Honghe Iron & Steel Co., Ltd.

Pu Song, Tang Deyuan, Zhong Wei, Zhao Shukui

(Kunming Iron & Steel Co., Ltd. Honghe Iron & Steel Co., Ltd., Mengzi　661100, China)

Abstract: This paper summarizes the improvement of gas utilization rate in No.3 blast furnace of Honggang Iron and Steel Company, and improves the utilization rate of gas from 40.57% in 2015 to 44.52% through the three aspects of fuel quality management, loading system exploration and operation management. August to complete the fuel ratio of 545kg / t, breaking the best level in histor.

Key words: blast furnace, gas utilization, fuel, production

SrO 脱硫和脱磷问题初步探讨与钢厂索探性试验

颜根发[1]，黄志勇[2]，徐广治[1]，蔡文藻[1]，夏晶晶[2]，沙宣保[2]

（1.马鞍山钢铁股份有限公司，安徽马鞍山　243000；

2.马鞍山中冶高新技术公司，安徽马鞍山　243000；）

摘　要： 由于超低碳钢（[C] <0.03%）在炉外精炼时脱硫困难，另外钢液在炉外精炼时，CaO 系熔剂在钢包渣调质变性之后回磷严重，因此必须寻找比 CaO 碱性更强的氧化物，才可能达到对钢液进行钢包脱硫、脱磷，并且不回磷的要求,而 SrO 就是寻找的重点对象之一。本文对 SrO 在钢液炉外精炼过程中脱硫、脱磷和防回磷问题从理论上进行了初步探讨，并在钢厂的 300tRH 炉进行了探索性试验。试验表明在精炼剂平均耗量 6.27kg/t 情况下，平均脱硫率 28.6%，脱磷率达 9.2%。

关键词： 炉外精炼，SrO，脱硫，脱磷，试验

The Preliminary Discussion for SrO Desulfuration and Dephosphorization as Well as the Exploratory Experiment of the Steel Mill

Yan Genfa[1], Huang Zhiyong[2], Xu Guangzhi[1], Cai Wenzao[1],
Xia Jingjing[2], Sha Xuanbao[2]

(1. Magang (Group) Holding Co., Ltd., Maanshan 243000, China;

2. Maanshan MCC Hi-Tech Company, Maanshan 243000, China)

Abstract: Because of the difficulty of desulfuration for the external refining of ultra-low carbon steel ([C] <0.03%), in addition, when conducting the external refining on the liquid steel, there is severe rephosphorization on the CaO rholite after the steel ladle slag tempers and transforms, we must look for the oxide with stronger alkalinity than CaO, then we can reach the requirement of conducting the desulphurization and dephosphorization in steel ladle without rephosphorization, however, SrO is one of the key objects we are looking for. This paper has theoretically conducted the preliminary discussion for SrO desulfuration, dephosphorization and rephosphorization prevention in the process of the external refining of liquid steel and conducted the exploratory experiment in the 300 ton RH furnace of the steel mill. The experiment shows that in case of the average consumption of the refining agent being 6.27kg/t, the average desulfurization ratio is 28.6% and the dephosphorization ratio reaches 9.2%.

Key words: external refining, SrO, desulfurization, dephosphorization, experiment

钛微合金化钢的增钛新工艺的试验

黄志勇 [1]，徐广治 [2]，颜根发 [2]，蔡文藻 [2]，夏晶晶 [1]，沙宣保 [1]

（1. 马鞍山中冶高新技术公司，安徽马鞍山 243000；

2. 马鞍山钢铁股份有限公司，安徽马鞍山 243000）

摘　要：本文概述了钛对改善低合金高强度钢性能的作用，介绍了钛微合金化钢增钛新工艺的试验情况。试验表明，钢包喂钛铁合金包芯线的新工艺，钛收得率平均高达 97.8%；而钛铁合金投入的旧工艺，钛收得率平均仅为 68%。钢包喂钛铁合金包芯线新工艺生产低合金高强度的海洋平台用钢等钢种，钢材各项力学性能满足国家标准要求，并且降低了生产成本、减少环境污染、降低工人劳动强度。

关键词：低合金高强度钢，钛微合金化，增钛新工艺，钢包喂线

Experiment of New Technology on Increasing Titanium for the Titanium Microalloyed Steel

Huang Zhiyong[1], Xu Guangzhi[2], Yan Genfa[2], Cai Wenzao[2], Xia Jingjing[1], Sha Xuanbao[1]

(1. Maanshan MCC Hi-Tech Company, Maanshan 243000, China;

2. Magang (Group) Holding Co., Ltd., Maanshan 243000, China)

Abstract: This paper summarizes the functions of the titanium on improving the performance of low-alloy high-strength steel and introduces the experiment situation of increasing titanium new technology for the titanium microalloyed steel. The experiment shows that as for the new technology of steel ladle feeding ferro-titanium cored wires, the average yield of titanium is up to 97.8%; as for the old technology devoted by the ferro-titanium, the average yield of titanium is only 68%. The new technology of steel ladle feeding ferro-titanium cored wires produces low alloy and high strength sea platform with steel and other types of steel, the compositive mechanical properties of the steel materials meet the requirements of national standard, lower the cost of production, reduce the environmental pollution and decrease the labour strength of workers.

Key words: low alloy and high strength steel, titanium micro-alloying, increasing titanium new technology, wire feeding of steel ladle

抑制钢渣激烈飞溅可能性的探讨

颜根发 [1]，黄志勇 [2]，徐广治 [1]，蔡文藻 [1]，夏晶晶 [2]，沙宣保 [2]

（1. 马鞍山钢铁股份有限公司，安徽马鞍山 243000；

2. 马鞍山中冶高新技术公司，安徽马鞍山 243000）

摘　要：向钢液喂实芯纯钙包芯线的炉外精炼新技术，还存在喂线时，钢渣激烈飞溅的问题，这不仅会带来种种危

害，并且恶化了多项技术经济指标。本文从理论上分析了向钢液喂实芯纯钙包芯线时，钢渣激烈飞溅的原因；探讨了抑制钢渣激烈飞溅不同工艺的技术方针；研究了不同的实芯纯钙复合层包芯线，抑制钢渣激烈飞溅的功能与特点。研究表明：在向钢液喂实芯纯钙包芯线的过程中，采用中间层为 CaO 基材质的新型实芯纯钙复合层包芯线，有抑制钢渣激烈飞溅的可能性。

关键词：喂线，纯钙包芯线，钢渣，激烈飞溅，抑制

Discussion on the Possibility of Controlling the Fierce Splash of the Steel-slag

Yan Genfa [1], Huang Zhiyong [2], Xu Guangzhi [1], Cai Wenzao [1], Xia Jingjing [2], Sha Xuanbao [2]

(1. Magang (Group) Holding Co., Ltd., Maanshan 243000, China;

2. Maanshan MCC Hi-Tech Company, Maanshan 243000, China)

Abstract: As for the new technology of external refining feeding the cored wires of solid core and pure calcium to the liquid steel, when conducting the wire feeding, the problem of the fierce splash of the steel-slag has existed, which will not only bring all kinds of hazards but also deteriorate the various technical and economic indicators. This paper analyzes the cause of fierce splash of the steel-slag theoretically when feeding the cored wires of solid core and pure calcium to the liquid steel; discusses the technology policy of different technologies that controls the fierce splash of the steel-slag; study the functions and features of different composite bed cored wires of solid core and pure calcium that controls the fierce splash of the steel-slag. The research shows that: in the process of feeding the cored wires of solid core and pure calcium to the liquid steel, there is possibility of controlling the fierce splash of the steel-slag by using the new-type composite bed cored wires of solid core and pure calcium with the interlayer of CaO basic material.

Key words: wire feeding, cored wires of pure calcium, steel-slag, fierce splash, control

淮钢高废钢比转炉工艺优化

车从荣 [1]，王建宇 [1]，刘从德 [1]，蒋栋初 [1]，杨　跃 [1]，吴　伟 [2]

（1. 江苏沙钢集团淮钢特钢股份有限公司炼钢厂，江苏淮安　223002；

2. 钢铁研究总院炼钢室，北京　100081）

摘　要：本文通过介绍淮钢针对转炉高废钢比冶炼存在的问题，通过理论分析以及一系列试验研究等手段、方法等，以期解决高废钢比转炉冶炼技术难题，为淮钢转炉高废钢比炼钢实践提供技术依据及理论支持。

关键词：转炉，高废钢比，试验研究，终点控制

Optimization of Converter Process Parameter on High Scrap Ratio at Huaigang Steel

Che Congrong [1], Wang Jianyu [1], Liu Congde, Jiang Dongchu [1], Yang Yue [1], Wu Wei [2]

(1. Jiangsu Shagang Group Huaigang Special Steel Co., Ltd., Huaian 223002, China;

2. Central Iron Steel Research Institute Central, Beijing 100081, China)

Abstract: The problem of the high scrap smelting in Huaigang converter were introduced in this paper. The problem of high scrap smelting were solved by theoretical analysis, and a series of experimental research and methods, which is in order to provide the technical basis for theoretical support and practice for Huaigang high scrap steel.
Key words: converter, high scrap ratio, experimental study, end control

首钢京唐热态钢渣高效循环利用体系

董文亮[1,2]，田志红[1,2]，罗　磊[3]，季晨曦[1,2]，潘宏伟[1,2]

（1.首钢集团有限公司技术研究院，北京　100043；

2. 绿色可循环钢铁流程北京市重点实验室，北京　100043；

3. 首钢京唐钢铁联合有限责任公司，河北唐山　063200）

摘　要： 在"全三脱"工艺流程中，将KR脱硫渣、转炉渣、精炼渣及铸余在热态的条件下循环利用，循环工艺利用热态炉渣的热量，产生较好的冶金效果，并且节约石灰消耗，不污染环境。在KR脱硫预处理阶段，进行热态脱硫渣循环利用实验，在配合少量脱硫剂的情况下，终点硫含量控制在20ppm以内，温度下降减少10~15℃，石灰消耗控制在1.8~3.0 kg/t。在转炉双联脱磷炉阶段，成功开发了热态转炉渣循环利用工艺，将P$_2$O$_5$含量较低的液态脱碳炉终渣通过渣罐兑入脱磷炉继续发挥脱磷作用，与常规工艺相比，脱磷率提高6%，石灰、轻烧白云石和萤石分别节约9.37 kg/t、1.15 kg/t和2.45 kg/t，半钢温度提高约7℃。在转炉双联脱碳炉阶段，将精炼工序RH/LF/CAS产生的热态精炼渣及大包铸余兑入半钢包，连同半钢一起兑入脱碳炉中进行冶炼，解决了石灰消耗并且利用热态渣钢的热量。

关键词： KR脱硫渣，转炉渣，精炼渣，循环利用

System of Slag Hot Recycling in Shougang Jingtang

Dong Wenliang[1,2], Tian Zhihong[1,2], Luo Lei[3], Ji Chenxi[1,2], Pan Hongwei[1,2]

(1. Shougang Research Institute of Technology, Beijing 100043, China;

2. Beijing key Laboratory of Green Recyclable Process for Iron & Steel Production Technology, Beijing 100043, China;

3. Shougang Jingtang United Iron and Steel Co., Ltd., Tangshan 063200, China)

Abstract: In duplex process in converter, hot slag produced by KR desulphurization, converter and refining processes was used to save energy and reduce lime consumption. In KR process, hot desulphurization slag was recycled and the sulphur content was controlled below 20 ppm, the temperature was increased by 10~15 ℃, and the lime consumption was controlled between 1.8~3.0 kg/t. In converter, decarburization-slag of low P$_2$O$_5$ content was added into dephosphorization-furnace through slag-vessel to realize the recycle of slag. The recycle slag process was developed in the plant trials, the lime , light-burned dolomite and fluorite was saved by 9.37 kg/t, 1.15 kg/t and 2.45 kg/t respectively, and the temperature was increased by 7℃ around., the average dephosphorization rate was increased by 6%.
Key words: KR desulphurization slag, decarburization-slag, refining slag, recycle

300t 转炉长寿命复合吹炼工艺及实践

杨利彬，徐晓伟

（钢铁研究总院，北京　100081）

摘　要： 国内 300t 转炉炼钢起步较晚，冶炼过程存在喷溅、终点[O]含量高、铁损大、冶金效果不稳定等现象，本文针对其面临的实际问题，在实验研究及分析的基础上，结合现场实际，优化了马钢 300t 转炉喷头设计、对顶、底复合吹炼工艺进行了积极的探索，初步实现了高效吹氧，炉役内有效复吹大幅提高，取得了较好的冶金效果。在保证高附加值炉龄要求的前提下实现了长寿底吹，冶炼终点碳氧积和炉渣氧化性大幅降低。

关键词： 大型转炉，复吹，冷态模拟，碳氧积

Research and Application of Long Campaigns Combined Blowing Steelmaking Technique of 300t Converter

Yang Libin, Xu Xiaowei

(Central Iron and Steel Research Institute, Beijing 100081, China)

Abstract: 300t steelmaking converter put into production lately contrast to foreign. Based on the large combined-blowing converter low phosphorus steel steelmaking as the research object, through research of conbined blowing and dephosphorization anlysis, based on the water modelling and silulation experiments, high efficient conbined blowing technology has been eatablished , during the furnace campaign, bottom blowing getting good results and the metallurgical effect improved.

Key words: 300t converter, combined blowing, long campaign, $w(C) \cdot w(O)$ product

强底吹搅拌下的转炉终点碳氧反应平衡

舒宏富，张晓峰

（马鞍山钢铁股份有限公司技术中心，安徽马鞍山　243000）

摘　要： 研究了不同底吹强度对转炉终点[C]、[O]、[Mn]和炉渣 TFe%含量、脱碳速率、熔池搅拌能和 Pco 的影响关系，着重考察提高转炉底吹强度以后对改善熔池搅拌促进碳氧反应平衡、降低转炉终点氧含量的效果。结果表明，转炉底吹强度由 0.05 提高到 0.12 Nm³·t⁻¹·min⁻¹后，转炉终点氧含量由627ppm 降低到494ppm，下降率为 21.21%，并且转炉终点碳氧积由 0.0025 下降到 0.0019。

关键词： 顶底复吹转炉，脱碳反应，碳氧反应平衡，碳氧积

Balance of Carbon and Oxygen at the End of Converter under Strong Bottom Blowing

Shu Hongfu, Zhang Xiaofeng

(Technology Center of Maanshan Iron and Steel Co., Ltd., Maanshan 243000, China)

Abstract: The influence relationship between [C], [O], [Mn] of molten steel at the end point of converter and TFe% content of slag, decarburization rate with the effects of different strength of bottom blowing stirring was studied. The effect of improving the bottom blowing strength of converter on improving the balance of carbon and oxygen reaction and reducing the final oxygen content of converter is investigated.The results show that the end point oxygen content decreased from 627ppm to 494ppm with the converter bottom blowing intensity increased from 0.05 to 0.12 $Nm^3.T^{-1}.min^{-1}$.   The decline rate is 21.21%, and the converter end point carbon oxygen product decreased from 0.0025 to 0.0019.
Key words: BOF, decarburization reaction, carbon oxygen reaction equilibrium, carbon oxygen product

转炉高废钢比冶炼的技术进展

朱　荣，胡绍岩

（北京科技大学冶金与生态工程学院，北京　100083）

摘　要： 日益严苛的碳排放政策和逐渐增加的废钢积累量将推动转炉流程消耗更多的废钢资源。本文介绍并对比了国内外现有的转炉高废钢比冶炼工艺技术，并对废钢预热、二次燃烧、燃料添加和转炉底喷粉等关键单元技术进行了详细论述。结合国内外研究和应用现状指出，既能喷吹碳粉又能喷吹氧气和石灰粉的转炉底喷粉技术将成为转炉高废钢比冶炼和转炉降本增效的关键。
关键词： 转炉，高废钢比，喷粉，碳排放

Development of High Scrap Ratio Smelting Technology in Converter Steelmaking

Zhu Rong, Hu Shaoyan

(University of Science and Technology Beijing, Beijing 100083, China)

Abstract: Strict carbon-dioxide emission policy and increasing scrap accumulation will promote more scrap consumption in converter steelmaking process. This paper introduced the existing high scrap ratio smelting technology of converter, and discussed the key unit technologies such as scrap preheating, post combustion, fuel addition and converter bottom powder injection. Based on the research and application situation at home and abroad, it is indicated that the bottom powder injection will be the key technology to increase scrap ratio and reduce production cost in converter.
Key words: converter, high scrap ratio, powder injection, carbon-dioxide emission

尘泥复合球团爆裂反应诱发微小异相研究与实践

于淑娟[1]，侯洪宇[1]，钱　峰[1]，张新义[2]，王晓峰[3]

（1. 鞍钢股份有限公司技术中心，辽宁鞍山　114009；2. 鞍钢股份有限公司，辽宁鞍山　114009；

3. 鞍钢股份有限公司炼钢总厂，辽宁鞍山　114009）

摘　要： 在现有技术中，冶金尘泥循环利用的关键是环境保护因素和直接有效的利用。因此，本文研发了一种利用能够作为造渣剂的复合球团爆裂反应诱发微小异相再利用冶金尘泥的新工艺。通过圆盘造球机（12r/min）设计和制造出一种功能性复合球团，并且优化了球团的加入方式。结果表明：在转炉中加入复合球团是一种全新有效的尘泥处理新工艺。在转炉兑铁后，吹氧前一次性加入尘泥球团是最有效、可行的加入方式。复合球团技术不适合在冶炼低硫钢时使用。球团最大加入量为10t。铁的收得率大于95%。与传统的尘泥处理工艺相比，该技术更快捷、有效，每吨可降低成本309元。但我们还要对该尘泥处理技术进行更深入的研究。

关键词： 冶金尘泥，复合球团，转炉，爆裂，再利用

Study and Practice through Dispersed in-situ Phase Induced by Composite of Sludge and Dust Ball Explosive Reaction

Yu Shujuan[1], Hou Hongyu[1], Qian Feng[1], Zhang Xinyi[2], Wang Xiaofeng[3]

(1. Technical Center,Ansteel Co., Ltd., Anshan 114009, China; 2. Ansteel Co., Ltd., Anshan 114009, China;

3. General Steelmaking Plant,Ansteel Co., Ltd., Anshan 114009, China)

Abstract: Recycling of iron and steelmaking dusts is a key issue for environmental protection and efficient utilizations in the present investigation, a novel recovery process has been put forward using a dispersed in situ phase induced by an explosive reaction of a composite ball of slag forming materials. A composite ball with this function has been designed and prepared using a laboratory model batch type balling disc (at 12 rev min^{-1}) and the feeding modes has been optimized. The results indicate that feeding composite ball in BOF converter is a novel technology and the iron and steelmaking dusts can be recovered effectively. After hot metal charging, feeding all the composite ball is a most reasonable feeding before oxygen blowing. Composite ball treatment is not appropriate to the low sulfur steel production. The maximum composite ball-feeding amount is 10t. The iron yield rate is more than 95%. Compared with conventional recycling process of sludge and dust, this novel technology is more convenient and efficient, which save the cost of 309 RMB per ton steel. Further investigation on this novel inclusion removal technology should be carried out in future.

Key words: iron and steelmaking dusts, composite ball, recovery, convert, explosion

日照钢铁80吨钢包全程自动加盖装置应用及效益分析

贾敬伟，李宗一

（日照钢铁控股集团有限责任公司，山东日照　276806）

摘　要：根据日钢 80t 钢包全程自动加盖实践，介绍钢包全程自动加盖装置特点、操作工艺和取得效果，通过钢包全程自动加盖装置有效的降低了钢包周转过程中的热量损失，有效地节能降耗并降低生产成本。

关键词：钢包，自动加盖，效果，效益

The Analysis and Improvement of the 80 tons of Converter Gas Recovery

Jia Jingwei, Li Zongyi

(Rizhao Steel Holding Group Co., Ltd., Rizhao 276806, China)

Abstract: According to rizhao steel 80 tons ladle full automatic affixed with practice, this paper introduces ladle full automatic capping device features, operating process and achieved effect, by ladle full automatic capping device effectively reduces the heat loss in the process of ladle turnaround, effectively save energy and reduce production costs.

Key words: ladle, automatic capping device, effect, benefits

电弧炉炼钢绿色及智能化技术进展

朱　荣 [1,2]，魏光升 [1,2]，董　凯 [1,2]

（1. 北京科技大学冶金与生态工程学院，北京　100083；

2. 北京科技大学高端金属材料特种熔炼与制备北京市重点实验室，北京　100083）

摘　要：本文从绿色化和智能化技术在电弧炉炼钢领域的应用出发，介绍并分析了近年来电弧炉炼钢在高效洁净化冶炼、绿色清洁化生产和智能检测与控制等领域的技术发展状况。结合国内外研究现状，指出绿色化和智能化技术在电弧炉炼钢领域的作用将日益突出，更先进的绿色清洁化生产技术以及更可靠全面的流程智能化检测与控制将成为今后电弧炉炼钢技术的发展趋势。

关键词：电弧炉炼钢，绿色化，智能化

Development of Green and Intelligent Technologies in Electric Arc Furnace Steelmaking Processes

Zhu Rong[1,2], Wei Guangsheng[1,2], Dong Kai[1,2]

(1. Metallurgical and Ecological Engineering School, University of Science and Technology Beijing, Beijing 100083, China;

2. Beijing Key Laboratory of Special Melting and Preparation of High-End Metal Materials, Beijing 100083, China)

Abstract: Based on the application of greenization and intellectualization technology in electric arc furnace (EAF) steelmaking processes, the article analyzes the technology development in high efficiency and purification, green and clean production, and intelligent detection and control. According to the research status at home and abroad, it's indicated that purification and intellectualization will matter more in EAF steelmaking. More advanced green production technologies and more reliable detection and control will be the important development tendency of EAF steelmaking.

Key words: EAF steelmaking, greenization, intellectualization

高炉含钒铁水 CO_2 脱碳保钒的研究

刘卓林, 张廷安, 牛丽萍

（东北大学冶金学院, 辽宁沈阳　110819）

摘　要: 钒在钢中的作用是增强淬透性和碳化物, 并且能耐高温, 有强烈的二次硬化作用, 对提高硬度有显著作用。目前钒钢的生产主要通过钒铁合金化、钒渣合金化、钒氧化物直接合金化三种方式, 但其都存在冶炼流程长、钒利用率低、能量消耗大的问题。针对传统钒钢冶炼工艺不足, 提出了以高炉含钒铁水为原料, 直接喷吹 CO_2 进行铁水脱碳保钒的研究。单因素实验结果表明: CO_2 流量 0.6L/min, 实验温度 1550℃, 反应时间 60min, 碳含量降至 1.81%, 钒含量为 0.327%, 达到脱碳保钒目的。

关键词: 钒, CO_2, 单因素实验, 脱碳保钒

Study on Decarburization and Vanadium Protection by CO_2 from Vanadium Bearing Hot Metal in Blast Furnace

Liu Zhuolin, Zhang Tingan, Niu Liping

(School of Metallurgy of Northeastern University, Shenyang 110819, China)

Abstract: The role of vanadium in steel is enhancing the hardenability and carbides, high temperature resistance, a strong secondary hardening effect and improving hardness. At present, the production of vanadium steel is mainly in three ways, vanadium iron alloying, vanadium slag alloying and vanadium oxide direct alloying. All of them had the problems that long technical process, lower usage rate of vanadium, high energy consumption. In view of the high energy consumption, long technical process, low usage rate of vanadium during the traditional smelting vanadium steel process, a new method for decarburization by CO_2 to directly smelt vanadium-containing steel in blast furnace was proposed. Single factor experiment results showed gas flow 0.6L/min, reaction temperature 1550℃, reaction time 80min, carbon content could be decreased to 1.81%, vanadium content was 0.327%, which meets the request of decarburization and vanadium protection.

Key words: vanadium, CO_2, single factor experiment, decarburization and vanadium protection

350t 转炉留渣法冶炼技术研究及实践

兰　天, 潘永飞, 靳随城, 吴　政

（宝钢湛江钢铁有限公司炼钢厂, 广东湛江　524000）

摘　要: 本文主要对转炉留渣冶炼技术的发展和现状进行了回顾, 对留渣操作的安全性和可操作性两个方面进行探讨。结合湛钢 350t 转炉生产条件, 开展了一系列留渣法工业试验, 提出了转炉留渣操作时确保安全的技术要点和操作要点。研究表明: 与常规转炉炼钢工艺（单渣法）相比, 转炉留渣操作法降本增效效果显著, 吨钢钢铁料消耗降低 5.4kg, 吨钢造渣溶剂（石灰料和轻烧）耗降低 4.45kg。

关键词: 转炉炼钢, 留渣法, 节约成本

The Research and Practice of the Slag-remained Steelmaking Technology in 350t Converter

Lan Tian, Pan Yongfei, Jin Suicheng, Wu Zheng

(Steelmaking Plat of Zhanjiang Iron and Steel Co., Ltd., ZhanJiang 524000, China)

Abstract: The paper reviews the development and the present situation of the slag-remained technology of converter, discusses the safety and operative feasibility on the slag-remained process. Based on the production condition of the 350t converter of ZhanJiang Steel Co., Ltd., a series of industrial tests of the slag-remained process had been carried out, and the main points for technology and operation were put forward. The paper shown: Compared with the conventional converter steelmaking method(single slag steelmaking process), the slag-remained technology of converter bright a significant benefits, the ferrous charges consumption was reduced by 5.4 kg/t.s, the slag flux cost(both the lime and the magnesia)were saved 4.45 kg/t.s.

Key words: converter steelmaking, the slag-remained technology, cost reduction

大方坯连铸末端电磁搅拌交替搅拌工艺参数研究

孙海波[1]，李烈军[2]，刘成斌[3]，周成宏[3]

（1. 佛山科学技术学院，广东佛山　528000；

2. 华南理工大学，广东广州　510641；

3. 宝武集团广东韶钢松山股份有限公司，广东韶关　512123）

摘　要： 基于工业试验及射钉试验验证的二维凝固传热模型，分析了末端电搅（F-EMS）交替搅拌工艺参数对铸坯内部质量的影响规律。结果表明，F-EMS 采用连续搅拌方式时，易促使其工作区域内铸坯凝固前沿白亮带及中心缩孔的形成；采用交替搅拌模式工况下，顺（逆）时针搅拌时间分别为 15s、25s 和 35s 时，铸坯中心偏析度呈先减小后增加趋势，分别为 1.18、1.07 和 1.24，且交替搅拌时间为 35s 时，易引起白亮带现象的产生；与 F-EMS 交替搅拌为 15s（顺时针）-5s（停止）-15s（逆时针）工况相比，交替搅拌参数为 25s-5s-25s 下，汽车传动轴钢 XC45 轧材产品中心疏松等级≤1.0 的比例由 87.56%增加至 98.10%，利于产品致密度的改善。

关键词： 大方坯连铸，末端电磁搅拌，交替搅拌，中心偏析

Investigation on the Alternate Stirring Parameters of F-EMS for Bloom Continuous Casting

Sun Haibo[1], Li Liejun[2], Liu Chengbin[3], Zhou Chenghong[3]

(1. Foshan University, Foshan 528000, China;

2. South China University of Technology, Guangzhou 510641, China;

3. Baowu steel Group Guangdong Shaoguan Iron & Steel Co., Ltd., Shaoguan 512123, China)

Abstract: Based on plant trials and the 2D slice solidification and heat transfer model validated by pin shooting testes, the effect of the F-EMS alternate stirring parameters on the strand internal quality has been compared and analyzed. The white band at the solidification front in the working region of F-EMS and center shrinkage cavity will be presented in the strand by adopting the constant stirring mode for F-EMS. For the alternate stirring mode, the degrees of strand center segregation are 1.18, 1.07 and 1.24 when the alternate stirring durations within 15s, 25s and 35s in the clockwise or anti-clockwise direction. Moreover, the white band phenomena can be appeared when the duration alternate stirring is 35s. The ratio of center porosity degree less than 1.0 for the rolled product of XC45 can be increased from 87.56% to 98.10% when the alternate stirring parameters within 25s(anti-clockwise direction duration)-5s(stop duration)-25s(clockwise direction duration) as compared to that with 15s-5s-15s.

Key words: bloom continuous casting, final electromagnetic stirring, alternate stirring, center segregation

连铸长水口吹氩气泡大小及分布试验研究

李晓祥[1]，钟良才[1]，阳祥富[2]，庞立鹏[1]，季伟烨[1]，郝培锋[3]

（1. 东北大学冶金学院，辽宁沈阳　110004；　2. 上海宝山钢铁股份有限公司炼钢厂，
上海　201900；3. 东北大学软件学院，辽宁沈阳　110004）

摘　要： 通过连铸长水口吹氩试验，研究了不同吹氩条件下获得的小气泡的尺寸及分布。实验结果表明，0.25mm 单孔气嘴和 0.25mm×2 双孔气嘴在水量为 3.0m³/h 时，形成的微小气泡（0.1–0.5mm）比例均达到 80% 以上。在较低水量（≤2.4m³/h）下，这两种气嘴形成的微小气泡比例相近，但 0.25mm×2 气嘴比 0.25mm 气嘴产生的较大气泡（>1.0mm）比例要小且气泡数量多。低水流量（≤2.4m³/h）时，0.11mm 气嘴产生微小气泡（0.1mm–0.5mm）比例最大达到 60%；0.25mm 气嘴微小气泡比例达到 45%；0.58mm 气嘴在 35% 左右，且 0.58mm 气嘴产生较大气泡（>1.0mm）比例最多。

关键词： 连铸，长水口，吹氩，气嘴，气泡直径分布，流体流量

Experimental Study on Bubble Size and Its Distribution with Argon Blowing through Long Shroud in Continuous Casting

Li Xiaoxiang[1], Zhong Liangcai[1], Yang Xiangfu[2], Pang Lipeng[1],
Ji Weiye[1], Hao Peifeng[3]

(1. School of Metallurgy, Northeastern University, Shenyang 110004, China;
2. Steelmaking Plant, Shanghai Baoshan Iron and Steel Co., Ltd., Shanghai 201900, China;
3. School of Software, Northeastern University, Shenyang 110004, China)

Abstract: Gas bubble size and its distribution have been investigated at different conditions of argon blowing through long shroud in continuous casting with experiments. The experimental results show that the proportion of tiny bubbles (0.1–0.5mm) formed with a 0.25mm diameter single hole nozzle and a 0.25mm×2 double-hole nozzle is more than 80% when the water flowrate is 3.0m³/h. In the lower water flowrate (≤2.4m³/h), the proportion of tiny bubbles formed with the two nozzles is similar, but the proportion of larger bubbles (>1.0mm in diameter) produced with the 0.25mm×2 gas nozzle is smaller than that with the 0.25mm gas nozzle and the 0.25mm×2 gas nozzle produces larger number of bubbles. In the lower water flowrate (≤2.4m³/h), 0.11mm diameter gas nozzle produces 60% proportion of tiny bubbles (0.1mm–0.5mm),

0.25mm gas nozzle 45%, and 0.58mm nozzle about 35%. And the proportion of larger bubbles (>1.0mm) produced with the 0.58mm gas nozzle is maximal.

Key words: continuous casting, long shroud, argon blowing, gas nozzle, bubble diameter distribution, fluid flowrate

现代合金钢大方坯连铸机设计特点及生产实践

吴国庆[1]，邱明罡[1]，侯晓林[1]，周善红[1]，沈 克[2]，杨晓成[2]

（1. 中冶京诚工程技术有限公司，北京 100176；
2. 常熟市龙腾特种钢有限公司，江苏常熟 215511）

摘 要： 龙腾特种钢有限公司6机6流合金钢大方坯连铸机由中冶京诚工程技术有限公司自主设计，铸机成套设备立足国内制造。连铸机主要生产310mm×360mm断面高碳磨球钢，采用了多项提升铸坯质量的先进技术和关键装备，铸机投产以来运行稳定、铸坯质量好，为企业创造了良好的经济效益。本文简要介绍该连铸机的技术特点，同时详细描述了末端电磁搅拌以及组合压下技术有关的设计细节和实际的冶金效果。

关键词： 中冶京诚，大方坯连铸，磨球钢，电磁搅拌，轻压下，重压下

The Design Features and Production Practice of Modern Alloy Steel Bloom Caster

Wu Guoqing[1], Qiu Minggang[1], Hou Xiaolin[1], Zhou Shanhong[1], Shen Ke[2], Yang Xiaocheng[2]

(1. Capital Engineering & Research Incorporation Limited, Beijing 100176, China;
2. Changshu Longteng Special Steel Co., Ltd., Changshu 215511, China)

Abstract: Longteng Special Steel Co.,Ltd. built a 6 – strand bloom caster, which was primarily designed by Capital Engineering & Research Incorporation Limited and manufactured in China. The caster is mainly used to produce 310mm × 360mm grinding ball steel. A number of advanced technologies and critical equipment have been adopted in the caster, which made it possible to achieve top performances in both productivity and quality right from the beginning. This paper introduced the technical characteristics of the casting machine, at the same time described the details at final electromagnetic stirring and dynamic reduction and the results of application.

Key words: CERI, bloom caster, grinding ball steel, electromagnetic stirring, soft reduction, hard reduction

CEM® Process: the Most Efficient Process for Thin Gauge Products

KO Young-ju, CHO Myung-jong, CHUNG Jae-sook, HWANG Jong-yeon

(Global POIST Project Department CEM Project, POSCO, Korea)

Abstract: POSCO has developed CEM® (Compact Endless casting and rolling Mill) process, which is based on high casting speed with 6~8 m/min and endless rolling from mold meniscus to down coiler. Mechanical design improvement and high speed technology development have been achieved through various efforts such as the molten flow control in the mold, water cooling in the caster and the mold level hunting control. For stable change of coil thickness and width during endless rolling, the control software of casting and rolling have been reinforced continuously against a lot of abnormal situation.

Owing to the stable endless rolling technology, the production ratio of thin gauge HR coils was increased significantly to 70% but, on the contrary, the cobble ratio in the rolling region was greatly decreased. Because the temperature and the speed of the strip are constant during endless rolling in finishing mill region, and it makes easy to control the strip temperature to produce AHSS in the run out table.

Key words: CEM®, high speed casting, endless rolling, thin gauge, advanced high strength steel

连铸结晶器无氟保护渣熔化及传热性能的研究

张 磊[1,2]，王万林[1,2]，Shon I. L.[3]

（1. 中南大学冶金与环境学院，湖南长沙 410083；2. 中南大学清洁冶金国际联合研究中心，湖南长沙 410083；3. 延世大学材料科学与工程学院，韩国首尔 120-749）

摘 要： 本文利用单丝法（SHTT）和红外发射技术（IET）分别研究了碱度（CaO/SiO₂）和 B₂O₃ 对中碳钢无氟保护渣熔化和传热行为的影响。熔化研究结果表明碱度的增大会使初始熔化温度和最终熔化温度降低，而 B₂O₃ 则具有相反的作用；传热研究结果表明碱度的增加会抑制保护渣的传热，B₂O₃ 在 4%~6% 会促进保护渣的传热，而 6%~8% 会抑制传热。本文研究结果将为如何获得熔化行为和传热行为都适合的保护渣提供理论指导。

关键词： 无氟保护渣，碱度，B₂O₃，熔化行为，传热行为

Research on the Melting Behavior and Heat Transfer Properties of Fluoride-Free Mold Flux

Zhang Lei[1,2], Wang Wanlin[1,2], Shon I. L.[3]

(1. School of Metallurgy and Environment, Central South University, Changsha 410083, China;
2. National Center for International Research of Clean Metallurgy, Central South University, Changsha, 410083, China;
3. Department of Materials Science and Engineering, Yonsei University, Seoul, 120-749, South Korea)

Abstract: This article had conducted the influence of basicity and B_2O_3 on the melting behavior and heat transfer behavior of fluorine-free mold flux for casting medium carbon steels by using Single Hot Thermocouple Technique (SHTT) and Infrared Emitter Technique (IET), respectively. The results of melting test indicated initial melting temperature and the complete melting temperature increased with the basicity increased, while depressed with the augment of B_2O_3 contents and the amplitude of debasement become more obvious. Besides, the results of heat transfer test showed that the heat fluxes at steady state attenuate with the increase of the basicity. And B_2O_3 enhanced the heat flux at its contents of 4 mass% to 6 mass%, but decreased the heat flux at its contents of 6 mass% to 8 mass%. Those results will provide a theoretical reference for acquiring a suitable melting temperature and a benign radiative heat transfer by adjusting basicity and B_2O_3 contents of mold flux.

Key words: F-free mold flux, basicity, B_2O_3, melting behavior, heat transfer behavior

首钢京唐 FC mold 工艺技术研究

邓小旋 [1,2]，潘宏伟 [1,2]，季晨曦 [1,2]，田志红 [1,2]，崔　阳 [1,2]，何文远 [3]，杨晓山 [3]

（1. 首钢集团有限公司技术研究院薄板所，北京　100043；

2. 绿色可循环钢铁流程北京市重点实验室，北京　100043；

3. 首钢京唐钢铁联合有限责任公司，河北唐山　063200）

摘　要：采用插钉法对结晶器液面特征进行评价，并研究了 FC mold 对结晶器液面特征的影响。研究发现插钉法可有效检测结晶器液面波形、液渣层厚度和表面流速等特征参数。不使用 FC mold 时，在当前浇铸条件下液面波动与表面流速较大。随上线圈电流增大，液面波形最高点逐渐从窄面移到 1/4 处，结晶器内流场由"双股流"转变为"单股流"，不利于铸坯质量提高。研究了 FC mold 对钩状坯壳深度影响，发现使用 FC mold 后，铸坯钩状坯壳深度从 2.3mm 降低至 1.7mm。原因在于使用 FC mold 后，尽管上回流受到抑制，但下部磁场对主流股的刹车作用使钢液在结晶器内停留时间延长，提高了弯月面温度，减轻了钩状坯壳深度。

关键词：FC mold，板坯，连铸，结晶器，插钉法

The Application of FC Mold in Conventional Slab Casting Mold in SGJT

Deng Xiaoxuan[1,2], Pan Hongwei[1,2], Ji Chenxi[1,2], Tian Zhihong[1,2], Cui Yang[1,2],

He Wenyuan[3], Yang Xiaoshan[3]

(1.Strip Technology Department, Shougang Research Institute of Technology, Shougang Group, Beijing 100043, China; 2. Beijing Key Laboratory of Green Recyclable Process for Iron & Steel Production Technology, Beijing 100043, China; 3. Shougang Jingtang United Iron & Steel Co., Ltd., Tangshan 063200, China)

Abstract: The Effect of FC mold on the meniscus characteristics in the continuous casting mold was measured with nail-dipping methods. The results show that the level profile, slab pool thickness and surface velocity can be easily measured by nail-dipping methods. Without applying FC mold, the level is waved with high surface velocity at the present casting condition. The level become quieter using the combination of the current 0/c A。The highest point of the level profile transfer from the narrow face to the quarter width as the coil current increase from 0 to 0.8c A, indicating the flow pattern in the mold evolution from double roll flow to single roll flow. The hook depth decrease from 2.3mm to 1.7mm applying the FC mold, which is attributed to the braking effect of FC mold and elongation of the residence time in the mold.

Key words: FC mold, slab, continuous casting, mold, nail-dipping methods

连铸结晶器喂钢带工艺的数值研究

牛　冉，李宝宽，刘中秋，李向龙

（东北大学冶金学院，辽宁沈阳　110819）

摘　要： 中心疏松与中心偏析是大规格连铸坯中常见的缺陷，通过结晶器向连铸坯中连续喂入薄钢带，借助钢带熔化吸热降低铸坯中心过热度是解决这一问题的新方法。建立了连铸喂钢带过程的数学模型，耦合钢液内部流动与传热，同时考虑铸坯的表面凝固与中心处运动钢带的熔化，得到运动钢带的熔化规律以及铸坯内流场及温度场发生的变化。结果表明，钢带在浸入过热钢液后其厚度经历先增大后减小的过程，同时存在一个动态平衡状态即钢带端部到达最大插入深度。较大速度的钢带的喂入改变了铸坯内流场以及过热度分布。本研究为深入探究喂钢带工艺过程中的相变规律提供理论依据，为发展这一提高铸坯质量的技术奠定基础。

关键词： 喂钢带，连铸，过热度，凝固，熔化

Numerical Study on the Strip Feeding into Continuous Casting Mold Process

Niu Ran, Li Baokuan, Liu Zhongqiu, Li Xianglong

(School of Metallurgy, Northeastern University, Shenyang 110819, China)

Abstract: The major problem of large size continuous casting strand quality is centerline segregation and porosity, a new technology is proposed that feeding strip to the continuous casting mold to decrease the superheat of inner zone by the heat absorption during strip phase change process. A mathematical model is built to describe the adding strip process, the melt flow and heat transfer in the strand are coupled and the freezing strand and moving melting strip melting are taken into considered, simultaneously. The thickness variation of the strip, the flow and temperature field distribution are got. The results indicates that the thickness of the strip increases firstly and then decreases. A pseudo-steady state is found during the process that the maximum distance the strip could reach to a steady position. The feeding of the high speed strip changes the flow and temperature field distribution of the cast strand. The work provides theory basis for the phase change evolution in the process of feeding strip into continuous casting strand and establish theory basis for the development of technology.

Key words: steel strip feeding, continuous casting, superheat, solidification, melting

低碳钢连铸过程保护渣润滑行为的数学模拟研究

王强强，张少达，李玉刚，王　谦

（重庆大学材料科学与工程学院冶金系，重庆　400044）

摘　要： 本研究基于多相流和热焓-多孔介质模型，建立二维非稳态模型研究结晶器振动过程弯月面处液渣的流动行为及保护渣的润滑机理。计算结果表明，在整个振动周期内没有出现负的润滑现象，吨钢渣的消耗量为

0.39~0.51 kg/t，与现场统计及文献中报道一致。负滑脱期内保护渣的消耗量较正滑脱期间大，但是保护渣的消耗与结晶器的位置有明显的关系，当结晶器在初始位置上方运动时，随着振动的进行，保护渣耗量逐渐增大，而低于初始结晶器位置时，随着振动的进行，保护渣的耗量逐渐减小。

关键词：连铸，保护渣，润滑，数值模拟

Numerical Simulation on Slag Infiltration in Slab Continuous Casting of Low-carbon Steel

Wang Qiangqiang, Zhang Shaoda, Li Yugang, Wang Qian

(College of Materials Science and Engineering, Chongqing University, Chongqing 400044, China)

Abstract: In the current study, a two-dimensional model involving multi-phase fluid flow, heat transfer and solidification of molten steel, slag and air was established to study the infiltration of liquid slag into the mold-strand gap during continuous casting of low-carbon steel. The models were validated by comparing the slag consumption between the calculation results and the measurements from industrial practice. Main results show that the calculated slag consumption per ton steel at 0.9 m/min ranged from 0.39 kg to 0.51 kg, and was related to the profile of liquid slag channel. When the mold moved above the initial level, the flux channel was wide enough for the infiltration of liquid slag, while when the mold moved below the initial level, a little of liquid slag was even squeezed out into the slag pool in the negative strip time.

Key words: continuous casting, slag, infiltration, numerical simulation

现代连铸技术在武钢三炼钢 2 号连铸机的应用

闵浩然，宋泽啟，刘光明，杨新泉

（宝钢股份武汉钢铁有限公司炼钢厂，湖北武汉　430083）

摘　要：本文具体介绍了由宝钢股份武钢炼钢厂对现有先进连铸技术进行充分对比，根据实际需求，博采众长，形成武钢独有连铸技术，并运用自主技术集成对三炼钢 2# 连铸机进行改造，改造后铸机装备先进，满足品种质量需求。铸机改造后，采用直弧形铸机机型有效促进夹杂物的上浮；应用连续弯曲、连续矫直技术分散变形应力，从而减轻铸坯内部裂纹的产生；应用在线调宽结晶器大大减少了备件数量及再生产准备时间；应用动态二冷水及动态软压下技术则可大幅提高非稳态浇铸条件下的铸坯内外部质量；多模式板坯结晶器电磁搅拌技术减少了回流区和冲击深度，有利于钢液中夹杂物和气泡的上浮；钢包下渣检测技术减少钢包渣进入到中间包内污染钢水，提高了钢水的纯净度。

关键词：铸机改造，塞棒控流，在线调宽，动态二冷水，动态软压下，结晶器电磁搅拌，钢包下渣检测

The Application of Effective Technologies in 2# Caster at No.3 Steelmaking Plant

Min Haoran, Song Zeqi, Liu Guangming, Yang Xinquan

(Wuhan Iron & Steel Co., Ltd., Wuhan 430083, China)

Abstract: This article detailed introduces the Wuhan Iron & steel Co., Ltd., fully compared to the existing advanced continuous casting technology, according to the actual demand, learn widely from other's' strong points, formation of Wisco unique continuous casting technology, and using its own technology integration to modify the No. 2 caster of No.3 Steelmaking Plant, after transforming advanced slab caster equipment, meet the quality requirements. Slab caster modified, using direct arc slab caster effectively promote inclusion floatation; application of continuous bending and straightening technology disperse deformation stress to reduce the internal cracks of slab; application online adjustable wide crystallizer reduces the spare parts and production preparation time; application of Dynamic secondary cooling and soft reduction can greatly improve the unsteady casting slab under the condition of internal and external quality; FC-MEMS reduces the recirculation region and impact depth, conducive to inclusionsin molten steel and floating of bubbles; under the ladle slag detection technology to reduce the ladle slags into the molten steel in tundish pollution, to improve the purity of molten steel.

Key words: caster modification, flow controlling, online adjustment, dynamic secondary cooling, dynamic soft reduction, FC-MEMS, ladle slag detection

IF 钢铸坯表层大尺寸夹杂物分布的研究

卢　鹏，黄福祥，冀云卿，毛明坤，王新华，王万军

（北京科技大学冶金与生态工程学院，北京　100083）

摘　要：为减少 IF 钢产品表面质量缺陷，明确铸坯厚度方向洁净度的差异性，对立弯式板坯连铸机生产的 IF 钢连铸坯沿宽度方向不同位置截取 50mm×70mm 面积的金相试样，采用 Aspex explorer 自动扫描电镜，对每个试样距内弧表面 20mm 内每隔 1mm 的夹杂物的成分、形貌、数量、分布等特征进行检测统计。结果表明，铸坯表层夹杂物主要为簇群状 Al_2O_3，尺寸在 20μm 到 300μm 均有分布，其次还有"气泡+Al_2O_3"类夹杂物、Al-Ti 复合类夹杂物等，这些大尺寸夹杂物主要在距内弧表面 1-2mm 处、6mm 左右及 14-16mm 处数量密度较大，这跟结晶器内钢液流场密切相关。

关键词：IF 钢，夹杂物，连铸坯，三氧化二铝，气泡

Research on the Distribution of Large Inclusions in the Surface Layer of IF Steel Slab

Lu Peng, Huang Fuxiang, Ji Yunqing, Mao Mingkun, Wang Xinhua, Wang Wanjun

(School of Metallurgical and Ecological Engineering, University of Science and Technology Beijing, Beijing 100083, China)

Abstract: To decrease the surface defects on IF steel and clear differences of the purity in thickness direction of the slab, metallographic samples of 50mm×70mm were specially prepared from IF steel slab produced by vertical bending continuous casting, and the composition, morphology, quantity and distribution of inclusions were inspected by the automatic secondary electron microscopy ASPEX explorer from the inner arc within 20mm by every 1mm. It was found most majority of large inclusion were alumina clusters within 20~300μm, and the others were the "bubble + Al_2O_3" type, Al-Ti composite inclusions, etc. These large inclusions are mainly in the area of 1~2mm, 6mm and 14~16mm, which is closely related to the fluid flow field in the mold.

Key words: IF steel, inclusion, continuous cast slab, alumina, bubble

薄板坯连铸中碳钢角横裂缺陷成因及控制

张剑君[1,2]，张　慧[1]，席常锁[1]，谭　文[2]，王春峰[3]

（1. 钢铁研究总院，北京　100081；2. 武汉钢铁有限公司研究院，湖北武汉　430080；

3. 武汉钢铁有限公司条材厂，湖北武汉　430083）

摘　要： 针对薄板坯连铸中碳钢边角部出现的角横裂纹缺陷，通过典型中碳钢（50CrV4）缺陷试样的金相分析、热塑性分析和铸坯二冷模拟仿真计算等方法，确定了原工艺条件下典型中碳钢铸坯边角部在铸机弯曲和矫直处的温度为850℃，处于第Ⅲ脆性区650~945℃范围内，是造成铸坯边角部裂纹的主要原因。本文结合武钢CSP铸机工艺特点，提出了连铸高拉速4.0~4.2 m/min，二冷前段强冷、后段和边部弱冷，以及提高铸机设备精度等控制措施。各项措施实施后，中高碳钢铸坯边角部温度显著提升，热轧板表面横裂纹基本消失，各项性满足客户要求。

关键词： 薄板坯，中碳钢，角横裂纹，高温热塑性

Analysis and Control of Transverse Corner Cracking Formation on Middle Carbon Steel in Thin Slab Casting

Zhang Jianjun[1,2], Zhang Hui[1], Xi Changsuo[1], Tan Wen[2], Wang Chunfeng[3]

(1. Central Iron and Steel Research Institute, Beijing 100081, China; 2. Research and Development Center of WISCO, Wuhan 430080, China; 3.CSP Mill of WISCO, Wuhan 430083, China)

Abstract: Aim at transverse corner cracks on middle carbon steel in thin slab casting. The main reason of slab transverse corner cracks is that the temperature of slab corner is 850℃, which is in No.3 Brittleness Temperature area between 650~945℃. That was be found at the old continue cast parameter from methods of Metallographical observation, high temperature plasticity analysis and second cooling numerical simulation, which based on typical middle carbon steel (50CrV4). After several key measures was be carry out include high cast speed (4.0~4.2 m/min), increasing second cooling water of the front segment, decreasing second cooling water of the back segment and edge of slab, which base on the features of cast machine of WISCO CSP plant. The edge temperature of middle carbon steel slab is increased. Transverse corner cracking of hot-rolled sheets were be controlled. Each performance of hot-rolled product can meet customers' requirements.

Key words: thin slab, middle carbon steel, transverse corner cracks, high temperature plasticity

大涡模拟结晶器电磁搅拌下的三相流动及凝固

李向龙，刘中秋，牛　冉，李宝宽，杨　凯，王喜春

（东北大学冶金学院，辽宁沈阳　110819）

摘　要： 本文采用Euler-Euler方法，发展了新的LES模型，分析了在电磁搅拌作用下，结晶器内的空气-保护渣-

第十一届中国钢铁年会论文集

钢液的三相流动以及凝固过程。为了保证模型的准确性，将磁感应强度和圆坯表面温度与实验数据进行对比、校核，此外还对有无凝固坯壳时候的钢液的流场进行了对比。结果发现，不考虑凝固坯壳的情况下，计算结果会高估铸坯内的流场速度及渣金界面波动。在电磁搅拌作用下，水口射流会在竖直方向上有所偏移，围绕结晶器中心旋转运动，表现出了非稳态、周期性的特点。这个非对称流动会对在靠近偏流一侧的渣金界面会有升高，出现了两边高，中间低，非稳定，非对称的界面状态。研究结果对预测弯月面波动具有指导意义。

关键词： 大圆坯，电磁搅拌，大涡模拟，多相流，凝固

Large Eddy Simulation of Three-phase Flow and Solidification in a Large Round Bloom with Electromagnetic Stirring

Li Xianglong, Liu Zhongqiu, Niu Ran, Li Baokuan, Yang Kai, Wang Xichun

(School of Metallurgy, Northeastern University, Shenyang 110819, China)

Abstract: Based on Euler-Euler approach, a new LES model coupling with electromagnetic stirring (M-EMS) and multiphase is established to study the air-slag-steel three-phase flow and solidification in a jumbo bloom. In order to ensure accuracy of our mathematical model, the magnetic flux density and surface temperature is compared with local measurements, and good agreements are obtained. Then the flow field considering solidification is compared with that without solidification, showing that the ignorance of solidification would overestimate the velocity in bloom strand, and thus the slag-metal interface fluctuation. Under the effect of Lorentz force, a eccentric flow is observed due to the deflection jet of the nozzle, rotating with the center of the bloom, presenting a instantaneous and periodical behavior in mold. This eccentric flow may have impact on the slag-metal interface, which lifting the slag-metal interface on the side of the eccentric direction. Results are helpful for predicting slag-metal interface fluctuation.

Key words: large round bloom, electromagnetic stirring, large eddy simulation, VOF, solidification

熔融保护渣电导率研究

赵潞明，王　雨，赵　立

（重庆大学材料科学与工程学院，重庆　400044）

摘　要： 试验研究了温度、碱度及 Al_2O_3 等组分对连铸保护渣电导率的影响规律，通过 PARSTST 2273 电化学工作站来测定保护渣的电导率，温度从 1523~1573K，光学碱度在 0.707~0.751 之间，比较 Al_2O_3、MgO、CaF_2、Na_2O 组分对熔渣电导率的影响程度。结果表明：保护渣的电导率随着温度升高而升高，随着碱度升高而增大。各组分对保护渣电导率的影响程度排序为：$Na_2O>CaF_2>Al_2O_3>MgO$。

关键词： 保护渣，组分，温度，电导率

Research on Electrical Conductivity of Molten Slag

Zhao Luming, Wang Yu, Zhao Li

(College of Materials Science and Engineering, Chongqing University, Chongqing 400044, China)

Abstract: The experiment studied the influences of temperature, alkalinity and components such as Al_2O_3 on the conductivity of continuous casting mold powder. The conductivity of the mold powder was measured by the PARSTAT 2273 electrochemical workstation at a temperature from 1523 to 1573K, with optical basicity between 0.707 and 0.751. The influence of Al_2O_3, MgO, CaF_2 and Na_2O on the conductivity of slag was compared in present work.The results shown that the electrical conductivity of the mold powder increases with the increase of temperature and increases with the increase of optical basicity. The influence degree of each component on the conductivity of the mold powder is $Na_2O> CaF_2> Al_2O_3>$ MgO.

Key words: mold powder, components, temperature, electrical conductivity

连铸结晶器角部附近钢液初始凝固的热模拟研究

吕培生，王万林，钱海瑞

（中南大学冶金与环境学院，湖南长沙 410083）

摘 要： 本文使用带直角结晶器的热模拟装置研究了钢液在结晶器角部附近的初始凝固。许多铸坯表面缺陷来源于结晶器弯月面区域的初始凝固过程，所以准确了解钢液在结晶器角部附近的初始凝固行为对于减少甚至消除角部裂纹等其它角部缺陷至关重要。首先，基于热电偶测量的温度，使用 2D-IHCP 模型计算出结晶器的温度场以及结晶器热面的热流密度。 通过定量比较结晶器在角部和宽面的温度场和热面热流密度，可以发现结晶器在角部的冷却强度大于宽面的冷却强度。然后借助 FFT 分析和 PSD 分析，将热流密度分解为低频热流密度和高频热流密度，并且发现了四个特征热流信号 $f1, f2, f3$ 和 $f4$。接着用凝固平方根定律拟合了坯壳在宽面和角部附近的厚度与拉坯时间的关系，平均凝固系数分别为 $2.32mm/s^{1/2}$ 和 $2.77mm/s^{1/2}$。 对于同一根振痕，沿拉坯方向，其在角部的根部位置要低于其在宽面的根部位置，这是由于结晶器角部处冷却能力强，角部坯壳产生更大的冷却收缩，使得钢液在角部的溢流程度大于在宽面的溢流程度。

关键词： 热模拟，初始凝固，弯月面，热流，振痕

An Investigation on Initial Solidification of Molten Steel Near the Mold Corner Using Mold Simulator

Lv Peisheng, Wang Wanlin, Qian Hairui

(School of Metallurgy and Environment in Central South University, Changsha 410083, China)

Abstract: A study on initial solidification of molten steel near the corner of continuous casting mold have been conducted in this paper using mold simulator equipped with right-angle copper mold. As we all know, many surface defects originate from meniscus region during the early stage of molten steel solidification, so a clear understanding of molten steel initial solidification around mold corner would be of great significance for decreasing corner cracks and other corner surface defects. Based on measured temperatures, temperature field and heat fluxes were calculated both at mold corner and face through 2D-IHCP mathematical model. The cooling ability of mold corner is stronger than mold face during continuous casting because of the two-dimensional heat transfer of mold corner. With the help of FFT and PSD analysis, the heat fluxes near the meniscus were split into low- and high-frequency components and four characteristic heat flux signals ($f1, f2, f3$ and $f4$) were found. Next, the relation between thickness of solidified shell and solidification time was fitted with solidification square root law; as a result, the average solidification factor K of face shell is $2.32mm/s^{1/2}$ and near-corner shell is 2.77mm/s$^{1/2}$. For the same oscillation mark (OM), roots of OM at shell corner are lower than roots of OM at shell face in the casting direction because stronger shrinkage of shell corner allows the overflowing steel to penetrate deeper into larger gap between shell corner and mold.

Key words: thermal simulation, initial solidification, meniscus, heat flux, oscillation mark

线棒材免加热直接轧制技术的研究

佐祥均，袁钢锦，雷　松，阎建武

（中冶赛迪工程技术股份有限公司连铸事业部，重庆　400013）

摘　要： 连铸方坯免加热直接轧制工艺是一个关系连铸和轧钢两个工序的综合系统工程。本文着重研究和分析了实现连铸方坯免加热直接轧制所涉及到的高效连铸技术，连铸轧钢衔接技术和 DROF 免加热直接轧制工艺等三种关键技术。并对每一项关键技术所涉及到的工序，设备及改造措施进行了详细的讨论。从连铸到轧钢，整体对免加热直接轧制工艺流程的各个工序进行了整理和梳理，并给出了相关的解决方案和参考数据。

关键词： 连铸方坯，免加热，直接轧制

Investigation on Free-Heating Direct Rolling Wire and Rod

Zuo Xiangjun, Yuan Gangjin, Lei Song, Yan Jianwu

(Continuous Casting Department of MCC CISDI Engineering Co., Ltd., Chongqing 400013, China)

Abstract: Driect rolling of free-heating (DROF) technology is a systems engineering involved in continuous casting and rolling. In this paper, three key technologies of high-efficient continuous casting technology, connection technology of continuous casting and rolling, and DROF technology are investigated and analyzed. And each key technology about the process, equipment and renovation measures is also discussed. All procedures of DROF technological process from continuous casting to rolling are integrally sorted out, organized and analyzed, and the relevant solutions and reference data are proposed.

Key words: billet continuous casting, free-heating, direct rolling

引流砂对钢液洁净度的影响研究

杨　文[1]，李　超[2]，张立峰[1]

（1. 北京科技大学冶金与生态工程学院，北京　100083；

2. 河北钢铁集团河钢研究院，河北石家庄　050023）

摘　要： 针对钢液连浇过程中每炉钢包开浇时中间包钢液中 T.O.含量升高的现象，本文对钢包开浇用引流砂对铝脱氧钙处理钢液洁净度的影响开展了实验室研究。研究发现随着引流砂加入量即砂/钢比的增加，在钢液中[N]含量基本不变的情况下，钢液中 T.O.含量逐渐升高；夹杂物平均成分中 Al_2O_3 含量逐渐升高，而 CaO、MgO、CaS 的含量

逐渐降低，同时随着引流砂加入量的增加，夹杂物的数量密度显著增加。钢包开浇引流砂会对钢液造成二次氧化，恶化钢液的洁净度。

关键词：引流砂，铝脱氧钙处理钢，洁净度，夹杂物

Effect of Ladle Filler Sand on the Cleanliness of Molten Steels

Yang Wen[1], Li Chao[2], Zhang Lifeng[1]

(1. School of Metallurgical and Ecological Engineering, University of Science and Technology Beijing, Beijing 100083, China;
2. R&D Institute, HBIS Group Co., Ltd., Shijiazhuang 050023, China)

Abstract: In view of the phenomenon that the T.O. content of the molten steel in tundish would increase at the casting start of each heat, the effect of ladle filler sand on the cleanliness of Al-killed Ca-treated molten steels was studied in laboratory in the current work. It was found that with the increase of the adding amount of ladle filler sand, the T.O. content in molten steel increased gradually although the [N] content changed little. Meanwhile, the average Al_2O_3 content in inclusions increased, while the average contents of CaO, MgO, and CaS decreased. Moreover, the number density of inclusions increased significantly. It was indicated that the ladle filler sand would induce the reoxidation of molten steel, and deteriorate the steel cleanliness.

Key words: ladle filler sand, Al-killed Ca-treated steel, cleanliness, inclusion

立式电磁制动及其对结晶器内钢液流动的控制

王恩刚，李　壮，李　菲

（东北大学材料电磁过程教育部重点实验室，辽宁沈阳　110819）

摘　要：针对现行应用的水平式布置磁极的电磁制动存在的问题，提出了立式布置磁极的立式电磁制动（V-EMBr）和立式-水平组合电磁制动（VC-EMBr）新技术，并采用数值模拟和物理模拟实验的研究方法，研究了其电磁制动效果以及工艺参数和电磁参数对其影响规律；研究结果表明：施加立式电磁制动和立式-水平组合电磁制动能够有效抑制水口出流钢液对结晶器窄面的冲击，进而降低钢液表面流速，稳定结晶器液面的波动，而且立式制动能更好地适用于连铸工艺的动态变化，有利于连铸稳定操作和铸坯质量的提高。

关键词：立式电磁制动，钢液流动，结晶器，连铸

The New Technology of Vertical Electromagnetic Brake and Control Effect to Molten Steel Flow in Mold

Wang Engang, Li Zhuang, Li Fei

(Key Laboratory of Electromagnetic Processing of Materials of Education Ministry, Northeastern University, Shenyang 110819, China)

Abstract: In this paper, directing against the problems existing in the conventional electromagnetic brake with horizontal pole applied, the vertical electromagnetic brake (V-EMBr) and vertical–horizontal combination electromagnetic brake (VC-EMBr) were proposed. The brake effect and the influence of process and electromagnetic parameters on the brake were studied using numerical simulation and physical experiment. The results show that the application of V-EMBr and VC-EMBr can effectively depress the impact of jet flow to the narrow face of the mold and reduce the surface velocity of molten steel, and then stable the fluctuation of free surface in the mold. The application of V-EMBr and VC-EMBr are more suited for the dynamic changes of process parameters and more beneficial for the improvement of continuous casting slab quality.

Key words: vertical electromagnetic brake, molten steel flow, mold, continuous casting

板坯连铸设备与工艺参数监控系统的开发与应用

范　佳[1]，成旭东[2]，高福彬[1]，丁　剑[1]

（1. 河钢集团邯钢公司技术中心，河北邯郸 056015；

2. 河钢集团邯钢公司三炼钢厂，河北邯郸 056015）

摘　要： 本文从邯钢三炼钢厂连铸现场应用角度出发，开发了基于 PLC 的板坯连铸设备与工艺参数监控系统，详述了包括系统配置、数据库开发、数据通信在内的系统构成，介绍了系统的设计流程及主要功能。工艺人员利用该系统对连铸生产过程中的浇铸准备、正常浇铸、拉尾坯等各个生产环节的结晶器热流、二冷水流量、扇形段辊缝、扇形段电流、液压振动等关键工艺、设备参数的全自动监控，板坯温度预报、动态配水、动态轻压下、漏钢预报等参数智能预报及历史查询等功能，制定合理的工艺模式，并督导现场操作工按照工艺标准进行操作，从而实现生产现场的精细化控制，使板坯连铸现场的铸坯质量，生产运行的稳定性得到较大提升，取得了良好的经济效果。

关键词： 连铸，全自动监控，智能，精细化控制，质量

Development of Monitoring System for Slab Continuous Casting Equipment and Application

Fan Jia[1], Cheng Xudong[2], Gao Fubin[1], Ding Jian[1]

(1. HBIS Group HanSteel Company Technology Center, Handan 056015, China;

2. HBIS Group HanSteel Company Third Steel-making, Handan 056015, China)

Abstract: In this paper, from the third steel-making mill of HanDan Iron and steel fields application perspective, a system based on PLC for slab continuous casting equipment and process parameters is developed, and the system structure including system configuration, database development and data communication is described in detail, and the design process and main function of the system are introduced. Using by the function of full automatic monitoring of key process equipment parameters about heat flow, second cooling water flow rate, segment roll gap, segment current, hydraulic vibration etc and intelligent parameter prediction about slab temperature prediction, dynamic water distribution, dynamic soft reduction, breakout prediction etc and historical inquiry etc, technical personnel can make reasonable process model, and supervise the operator to operate according to technical standard. So it implements the fine controlling in the field and improves the slab quality and the stability of production and operation. And That obtains the good economic results.

Key words: continuous casting, full automatic monitoring, intelligence, fine control, quality

非对称 4 流中间包优化数值模拟及冶金效果

俞赛健，刘建华，苏晓峰，张游游，季益龙

（北京科技大学工程技术研究院，北京 100083）

摘 要：本文以澳森钢厂中间包的流场情况为研究对象，采用 ANSYS 数值模拟方法进行中间包流场研究。在湍流流动状态下，利用连续性方程和 k-ε 双方程模型，分析了添加控流装置前后中间包内流场和温度场。结果表明，原中间包内钢液流动不合理，各流差异性较大，温差大于 5℃，死区比例为 39.8%；在原中间包冲击区加入导流隔墙、原中间包 2、3 流之间加入 300mm 挡坝进行优化，优化后中间包流场趋于合理，明显消除了短路流，各流一致性变好，温差小于 3℃，死区比例下降了 15.2%。对优化后中间包进行工业实验，实验结果表明：相对于原中间包，优化包不同流次所含的大型夹杂物数量更为接近；优化包对应原中间包各流次单位重量钢水所含大型夹杂物数量有明显下降。

关键词：中间包，数值模拟，夹杂物去除，中间包流场

Numerical Simulation and Metallurgical Effect of Asymmetric 4 - flow Tundish's Optimization

Yu Saijian, Liu Jianhua, Su Xiaofeng, Zhang youyou, Ji Yilong

(University of Science and Technology Beijing, Institute of Engineering Technology, Beijing 100083, China)

Abstract: Based on the object research of flow field in the tundish of Aosen steel mills and the numerical simulation method was used to study the tundish flow field by ANSYS software. In the turbulent state of molten steel flow, continuity equation and k-ε equation model are used to analyses the flow and temperature field with and without the flow control device. The results in the original tundish show that the temperature of different outlet is different from each other and the difference is more than 5℃. The ratio of dead is 39.8%. Diversion walls and dams are added to the original tundish for optimization can significantly improve the flow characteristics within the tundish and eliminate the short circuit current. What's more, the temperature difference among different outlets is less than 3℃. The ratio of dead decreased by 15.2%. The industrial experiment results show that comparing with tundish, the amount of large inclusions are close in optimizational tundish with different flows and the amount of large inclusions contained in the tundish is significantly decreased.

Key words: tundish, numerical simulation, inclusion removal, tundish flow field

非等温条件下中间包夹杂物的运动行为

潘宏伟[1]，何文远[2]，关顺宽[2]，季晨曦[1]，邓小旋[1]，程树森[3]

（1.首钢技术研究院，北京 100043；2. 首钢京唐钢铁联合有限责任公司，河北唐山 063200；

3. 北京科技大学，北京 100083）

摘　要： 采用数值模拟与工业试验相结合的研究方法，将钢包和中间包视为一个整体，分析了在非等温条件下，中间包内夹杂物颗粒运动行为的规律。数值模拟结果表明，在浇注过程的前期，<30μm 夹杂物由于流体的离心运动，难以上浮和去除。在浇注中后期，跟随流体沿中间包底部流出中间包，去除效果进一步变差。80t 中间包的工业实验结果表明，通过在中间包底部吹入氩气的方式，改善了<30μm 夹杂物去除条件，中间包底吹氩气后，T.[O]平均值降低了 1.5ppm，中间包 N 平均值降低了 3ppm。分别对采用底吹氩气、未采用底吹氩气铸坯试样进行了 1607mm²、1599mm² 电镜扫描，中间包底吹氩气前，铸坯中 10-20μm、20-50μm、>50μm 夹杂物的数量密度分别为 42.46、6.44、0.88 个/100mm²；中间包底吹氩气后，铸坯中 10-20μm、20-50μm、>50μm 夹杂物的数量密度分别为 33.66、6.35、0.25 个/100mm²。

关键词： 夹杂物，中间包，颗粒运动，非等温，连铸

Inclusions Movement Behavior in Tundish on Non-isothermal Condition

Pan Hongwei[1], He Wenyuan[2], Guan Shunkuan[2], Ji Chenxi[1],
Deng Xiaoxuan[1], Cheng Shusen[3]

(1. Shougang Research Institute of Technology, Beijing 100043, China;
2. Shougang Jingtang United Iron & Steel Co., Ltd., Tangshan 063200, China;
3. University of Science and Technology Beijing, Beijing 100083, China)

Abstract: On non-isothermal condition, inclusions movement behavior in tundish is studied, by means of Mathematical simulation and industrial experiment. Simulation results shows, during the earlier stage of continuous casting period, inclusions with different sizes tend to separate, and removing rate of inclusions with size smaller than 30 μm get reduced, because these small inclusions get into the center of vertex region caused by liquid flow. While during the later stage, ,inclusions smaller than 30 μm get worse and worse, some even gets out of tundish. Industrial experiment carried out in 80 tons tundish results show that, with bottom-blowing in tundish, it changes the inclusions movement behavior in tundish, inclusions removing effect in tundish get improved. After bottom-blowing in tundish, T.[O] in tundish is reduced 1.5ppm, and N is reduced 3 ppm. Before bottom-blowing in tundish, the inclusions quantity of 10-20μm, 20-50μm, >50μm is 42.46, 6.44, 0.88-/100mm²,while after bottom-blowing, it is 33.66, 6.35, 0.25-/100mm², which is reduced by 20.7%, 1.4%, 71.6%-/100mm² respectively.

Key words: inclusion, tundish, particle movement, non-isothermal, continuous casting

钢液凝固、冷却和轧制过程夹杂物的演变

张立峰，杨　文，任　英

（北京科技大学冶金与生态工程学院，北京　100083）

摘　要： 夹杂物的控制已经成为钢生产关键任务之一。冶炼过程钢液中各类氧化物夹杂和控制技术已日臻成熟，已经形成了一系列脱氧、渣改性、钙处理改性等成熟的夹杂物控制方法。然而，在凝固过程中，随着温度的降低，钢和夹杂物之间的平衡也会随之变化，从而导致凝固过程中夹杂物的成分转变。因此，研究钢液凝固、冷却和轧制过程夹杂物的演变非常重要。本文对以管线钢、帘线钢和不锈钢为研究对象，分别研究了钢液凝固、冷却和轧制过程

夹杂物的演变。研究发现，管线钢钢液凝固和冷却过程中，夹杂物中的 CaO 发生了向 CaS 的转变。在帘线钢热轧过程中，由于夹杂物的变形，Al$_2$O$_3$-SiO$_2$-CaO 类夹杂物的尺寸变化导致了夹杂物成分的变化。在 18Cr-8Ni 不锈钢热处理过程中，304 不锈钢氧化物由球形 MnO-SiO$_2$ 夹杂物转变为纯 MnO·Cr$_2$O$_3$ 尖晶石夹杂物。这些低温下夹杂物的转变行为需要我们更多的研究和关注。

关键词：夹杂物，凝固，冷却，轧制，热处理

Evolution of Inclusions in Steel during Solidification, Cooling, and Rolling Process

Zhang Lifeng, Yang Wen, Ren Ying

(School of Metallurgical and Ecological Engineering, University of Science and Technology Beijing, Beijing 100083, China)

Abstract: Control of inclusions in steel is one of the main task of steelmakers. Many technologies have been developed to achieve to the precise control of inclusions in the liquid steel, such as deoxidation, slag modification, calcium treatment, etc. However, during the solidification and cooling process, the equilibrium between steel and inclusions changed with the decrease of temperature. Thus, it is of great importance to investigate the evolution of inclusions in steel during solidification, cooling, and rolling process. It was found that the CaO changed to CaS in inclusions in linepipe steels during the solidification and cooling process. During the hot-rolling process of tire cord steels, there was an composition evolution of Al$_2$O$_3$-SiO$_2$-CaO inclusions during to the deformation of inclusons. In 18Cr-8Ni stainless steel, the MnO-SiO$_2$ inclusions tranformed to MnO·Cr$_2$O$_3$ spinal inclusions during the heat treatment process. It is significant to investigate the evolution of inclusions in steel under low temperature.

Key words: inclusions, solidificaiton, cooling, rolling, heat treatment

控制浸入式水口带电行为改善水口堵塞的研究

杨　鑫[1]，周秀丽[2]，于景坤[1]

（1. 东北大学 材料与冶金学院，辽宁沈阳　110004；

2. 梅山钢铁股份有限公司，江苏南京　210039）

摘　要： 在连铸生产过程中，水口内壁因钢液流经产生的摩擦而带有电荷，水口内壁因带电产生的静电力会对钢液进行吸附，并使得钢液流动状态不稳定，另外由于水口内壁与钢液固液两相间产生的电势差引起的电润湿也促进了钢液润湿，这些都是导致氧化铝夹杂附着水口的原因。通过对浸入式水口通电处理，使得水口内壁多余的静电荷被中和，改变了因水口内壁带电后与钢液间的界面相互作用，从而改善水口堵塞，提高钢坯质量。

关键词：连铸，浸入式水口，界面作用，静电力吸附，电润湿

Research on Reducing Nozzle Clogging by Controlling the Electrical Characteristics of SEN

Abstract: In the process of continuous casting(CC), the inner surface of submerged entry nozzle(SEN) is charged by the friction, which is generated by the molten steel flowing through SEN. Electrostatic force is generated by the accumulation of charge in the inner surface of SEN will adsorb molten steel, and then it will make the flow of the molten steel unstable. In addition, electrowetting generated by the potential difference between the inner surface of the nozzle and molten steel also promotes the wetting of molten steel. All these lead to the alumina inclusions attaching to nozzle. The excess charge in the inner surface of nozzle is neutralized through energizing or using static eliminator. This change the interaction between the charged inner surface of nozzle and molten steel, reduce nozzle clogging and then improve the quality of billet.

Key words: continue casting, submerged entry nozzle, interfacial interaction, electrostatic adsorption, electrostatic adsorption

基于 ANSYS Workbench 的钢包回转台叉形臂的有限元分析

孙晓娜[1]，郑 浩[2]，孟 岩[1]

（1. 鞍钢重型机械设计研究院有限公司炼钢室，辽宁鞍山 114031；

2. 鞍钢汽车运输有限责任公司设备部，辽宁鞍山 114110）

摘 要： 为了延长钢包回转台的使用寿命，有必要探究其关键零件的可靠性。本文在对叉形臂进行受力分析的基础上，分别利用三维建模软件 Inventor 和有限元分析软件 ANSYS Workbench 构建叉形臂实体模型和有限元模型，并利用有限元分析软件分别对处于静载荷和冲击载荷工况下的叉形臂进行有限元分析，找到零件的薄弱环节。结果表明：叉形臂结构基本满足强度和刚度要求，但是要特别注意冲击载荷对其疲劳寿命的影响。本方法为类似零件提供量化的设计依据，具有较高的实用价值。

关键词： 钢包回转台，叉形臂，有限元分析，ANSYS Workbench

Finite Element Analysis for the Fork Arm of Ladle Turret Based on ANSYS Workbench

Sun Xiaona[1], Zheng Hao[2], Meng Yan[1]

(1. Angang Heavy Machine Design and Research Institute Co., Ltd. Steelmaking Branch, Anshan 114031, China;

2. Angang Automobile transportation Co., Ltd. Equipment Department, Anshan 114110, China)

Abstract: To extend the service life of ladle turret, it is necessary to study the reliability of its critical parts. On the basis of stress analysis on the fork arm, this paper makes use of three-dimensional modeling software Inventor and finite element analysis software ANSYS Workbench to create the fork arm entity and finite element models respectively, and does finite element analysis for the fork arm under static load and impact load conditions and the weak position is found. The results show that the fork arm's structure meets mainly the strength and the rigidity requirements, but it is important to pay special attention to the impact load that can affect it's fatigue life. The method can provide quantitative basis for designing similar parts and has high practical value.

Key words: ladle turret, the fork arm, finite element analysis, ANSYS Workbench

韶钢 7 号大方坯连铸机工艺设备的优化与改进

付谦惠，郭峻宇

（宝钢特钢韶关有限公司生产技术室，广东韶关　512123）

摘　要：韶钢 7 号大方坯连铸机于 2012 年 10 月 18 日顺利投产，7 号大方坯连铸机由 Danieli 设计。生产的钢种主要有优质碳素结构钢、合金钢、齿轮钢、轴承钢等。投产以来，设备故障多、工艺不合理等因素严重制约了特钢的生产，产量与质量远远达不到要求。为了满足特钢生产的要求，进一步释放了铸机的产能，改善钢坯的质量。2012 年 11 月至 2013 年 7 月，对现有工艺设备 12 个方面进行了优化与改进。结果表明，铸机的工艺制度合理，铸机的设备稳定顺行，产能大幅提高，钢坯的质量得到明显改善，生产事故大幅减少，为开发更高等级的钢种创造了条件。

关键词：大方坯连铸机，特钢，设备，工艺，优化与改进

Optimization and Improvement of Technology Equipment of No.7 Bloom Caster in Shaogang Iron and Steel

Fu Qianhui，Guo Junyu

(BAO Steel Special Steel Shaoguan Co., Ltd., Production Engineering Office，Shaoguan 512123, China)

Abstract: Shaosteel No.7 bloom caster was successfully put into production on 18 October 2012, it is designed by Danieli. The steel of production is high quality carbon structural steel, alloy steel, gear steel, bearing steel etc. Since putting into production, equipment, process more unreasonable factors such as serious restricted the production of special steel, yield and quality far can not meet the requirement of steel grade mainly .In order to meet the requirement of special steel production, the production capacity of casting machine was further released, and the quality of billet was improved from 2012.11 to 2013.7, the existing process equipment is optimized and improved by 12 aspects. The results show that the casting machine system is reasonable, the technology of casting machine equipment stability along the line, a substantial increase in capacity, dramatically improve the quality of the bloom, sharply reduce production accident, create conditions for the development of higher grades of steel.

Key words: bloom caste, special steel, equipment, technology, optimization and improvement

钢包浇注过程环出钢口吹氩控制下渣新工艺的数学物理模拟

郑淑国，朱苗勇

（东北大学冶金学院，辽宁沈阳　110819）

摘　要： 洁净钢连铸生产中的钢包下渣难题一直缺乏经济有效的控制技术，本文提出了一种钢包浇注过程环出钢口吹氩控制下渣新工艺，并利用物理和数学模拟研究了该新工艺的控制下渣行为机理及可行性。结果表明：钢包浇注过程中汇流漩涡先于排流沉坑形成，且存在从前者向后者的过渡；新工艺可消除汇流漩涡下渣，并能抑制排流沉坑下渣，显著降低下渣临界高度；随环出钢口吹氩量增加，钢包浇注过程下渣临界高度呈减小趋势，且存在最佳控制下渣气量；水模和数模结果有很好的一致性，且验证了该控制下渣新工艺的可行性，为其工业应用奠定了基础。

关键词： 连铸，浇注钢包，环出钢口吹氩，控制下渣新工艺，物理模拟，数学模拟

Physical and Mathematical Simulation of the New Process with Argon Injected into Ladle Around the Tapping Hole for Controlling Slag Carry-over During Ladle Teeming Process

Zheng Shuguo, Zhu Miaoyong

(School of Metallurgy, Northeastern University, Shenyang　110819, China)

Abstract: There are no economical and effective measures to control the slag carry-over during continuous casting ladle teeming process of clean steel. A new process with argon injected into ladle around the tapping hole for controlling slag carry-over in a teeming ladle was presented. And physical and mathematical modelling were used to study the mechanism of controlling slag carry-over and the feasibility of the new process. The results show that vortex forms firstly and then converts to drain sink, and there exits the transition from vortex to drain sink. The new controlling slag carry-over process can eliminate the slag carry-over caused by vortex, and it can obviously decrease the critical height of slag carry-over caused by drain sink. With the increase of the flow rate of argon injected into ladle around the tapping hole, the critical bath height for slag carry-over during the ladle teeming process decreases. And there is an optimal one to control the slag carry-over. The results of physical and mathematical simulation show good coherence, and by them the application feasibility of the new controlling slag carry-over process is verified, which lay foundation for its industrialization.

Key words: continuous casting, teeming ladle, argon injected into ladle around the tapping hole, new controlling slag carry-over process, physical modeling, mathematical simulation

引流砂：钢中大型夹杂物的重要来源

邓志银，朱苗勇

（东北大学冶金学院，辽宁沈阳　110819）

摘　要： 基于工业实验和大样电解分析，通过对比钢中的显微夹杂物和引流砂的烧结行为，研究了钢中部分大型夹杂物的来源。研究结果表明：钢中的许多大型夹杂物与细小显微夹杂物没有明显关联，这些大型夹杂物并不是源自钢中的细小夹杂物；而基于铬质引流砂烧结机理和大型夹杂物的化学成分，可以确定这些大型夹杂物为引流砂及其烧结产物。当引流砂进入钢液后，要在中间包内完全去除是十分困难的。在钢包开浇时，应尽可能移除引流砂，这对提升钢的质量具有重要的意义。

关键词： 引流砂，大型夹杂物，大样电解，铬铁矿，二氧化硅

Ladle Filler Sand: An Important Source of Macro-Inclusions in Steel

Deng Zhiyin, Zhu Miaoyong

(School of Metallurgy, Northeastern University, Shenyang 110819, China)

Abstract: The aim of the present study is to find the source of macro-inclusions on the basis of the results of Slime Method and industrial trails. According to the comparison with micro-inclusions in steel, it is found that there is no obvious relationship between macro-inclusions and micro-inclusions in steel, while it is confirmed that most of the macro-inclusions are the sintering products of ladle filler sands based on their compositions and the sintering behaviors of ladle filler sand. When ladle filler sands fall into liquid steel, it is very difficult to remove in tundish, even after the middle casting stage of each heat. At the beginning of the teeming stage of each heat, the removal of ladle filler sand therefore becomes a very important step to control macro-inclusions during industrial process.

Key words: ladle filler sand, macro-inclusions, slime method, chromite, silica

水平和垂直电极振动对电渣重熔过程温度场分布的影响

王　芳，李宝宽

（东北大学冶金学院，辽宁沈阳　110819）

摘　要：振动电极既可以独立使用，又可与"旋转结晶器和旋转电极"或"电磁搅拌"等其他电渣重熔领域新技术组合使用，不受电渣重熔其他技术的限制，具有很好的实用性和适应性。本文探究了振动方式对电渣重熔系统温度场、速度场、液滴和熔池形状的变化规律。结果表明：竖直振动渣池内的最高温度最高，无振动渣池内的最高温度次之，水平振动渣池内的最高温度最低。振动电极使得渣池中的高温区范围扩大，温度分布更加均匀。水平振动使得高温区向电极两侧转移，而竖直振动使得高温区集中在电极正下方及其附近。

关键词：水平振动，垂直振动，电渣重熔，数值模拟

The Effect of Horizontal and Vertical Vibration of Electrode on Temperature Distribution in the ESR Process

Wang Fang, Li Baokuan

(Department of Metallurgy, Northeastern University, Shenyang 110819, China)

Abstract: The effect of vibration mode on temperature field, velocity field, droplets and molten pool shape in the vibating electrode electroslag remelting process is investigated. She results show that the highest temperature of slag pool in ESR under vertical vibration is highest. The highest temperature of slag pool in ESR under no vibration takes the second place. The highest temperature of slag pool in ESR under horizontal vibration is lowest. Vibrating electrode makes high temperature zone in the slag pool more largely and temperature distribution more uniformly. Horizontal vibration makes high temperature zone transfer to both sides of the slag pool, while vertical vibration makes high temperature zone focused on the below and nearby of the electrode.

Key words: horizontal vibration, vertical vibration, remelting electroslag, numerical simulation

帘线钢中钛夹杂析出机理的研究

蒋跃东[1]，薛正良[2]，吴　杰[1]，齐江华[1]，张　帆[1]

（1. 武汉钢铁有限公司研究院条材所，湖北武汉　430080；
2. 武汉科技大学材料与冶金学院，湖北武汉　430081）

摘　要：热力学分析不同碳含量的帘线钢（72A、82A）中钛夹杂形成，在连铸坯上取样，在实验室热处理后，通过光学显微镜、扫描电镜检测钛夹杂的成分和形貌，探索钛夹杂析出的主要影响因素及控制措施。
关键词：帘线钢，钛夹杂，碳含量，热力学

The Research of Ti-inclusion Precipitation in Tire Cord Steel

Jiang Yuedong[1], Xue Zhengliang[2], Wu Jie[1], Qi Jianghua[1], Zhang Fan[1]

(1. Research and Development Center, Wuhan Iron and Steel Corporation, Wuhan 430080, China;
2. College of Materials and Metallurgical Engineering, Wuhan University of Science and Technology,
Wuhan 430081, China)

Abstract: Thermodynamic analysis of cord steel of different carbon contents (72A, 82A) of titanium in mixed form, sampling in continuous casting, in the laboratory after heat treatment by optical microscope and scanning electron microscope titanium inclusion composition and morphology, to explore the main influence factors and the inclusion of titanium precipitation control measures.
Key words: tire cord steel, Ti inclusion, carbon content, thermodynamics

钢水温度对铁素体不锈钢铸坯宽度的影响

王　伟，孙仁宝，陈法涛

（山西太钢不锈钢股份有限公司炼钢二厂，山西太原　030003）

摘　要：本文结合铁素体不锈钢的凝固特点和高温力学性能，分析了铁素体不锈钢宽度变化的原因。对钢水温度和铸坯宽度数据进行了统计分析，认为钢水温度的变化对铸坯宽度有显著影响。随着温度变化量的增大，铸坯宽度也相应增大，当温度变化量为12~15℃，铸坯宽度变化量约为14~20mm。根据钢包状况、LF到站温度、生产节奏，采用合适的软搅拌参数，控制合理的精炼结束温度，同时，适当延长中间包的烘烤，增加冲击区的覆盖剂加入量，可以有效的控制钢水温度的稳定。通过上述措施，中包温度合格率有所升高，铁素体不锈钢铸坯宽度提高了7.6个百分点。
关键词：铁素体不锈钢，钢水温度，板坯宽度，凝固

Effect of Molten Steel Temperature on Width of Ferrite Stainless Steel Slab

Wang Wei, Sun Renbao, Chen Fatao

(No.2 Steelmaking Plant of Shanxi Taigang Stainless Steel Co., Ltd., Taiyuan 030003, China)

Abstract: Combined with ferrite stainless steel solidification characteristics and high temperature mechanical properties, the reason of the change of the width of ferritic stainless steel was analyzed in this paper. The data of molten steel temperature and slab width were statistically analyzed. It was found that the change of molten steel temperature had a significant effect on the slab width. With the increase of the temperature change, the slab width increases correspondingly. When the temperature change from 12℃ to 15℃, and the width of the slab varies from 14mm To 20mm. According to the ladle situation, LF station temperature, production rhythm and the use of appropriate soft mixing parameters, the molten temperature can be controlled reasonably. At the same time, the molten temperature would be controlled by the appropriate extension of the tundish baking and increasing the cover flux. Through the above measures, the rate of the molten temperature hit has increased and the width of slab was increased by 7.6 percentage points.

Key words: ferritic stainless steel, molten steel temperature, slab width, solidification

薄板坯连铸 SPHC 钢硅质量分数控制研究

孙　波，张良明，吴耀光，解养国，刘前芝，万　栋

（马鞍山钢铁有限公司第一钢轧总厂，安徽马鞍山　243000）

摘　要： 为了减少低碳低硅铝镇静钢精炼过程中的增硅问题，本文结合马鞍山钢铁股份有限公司 CSP 流程 SPHC 钢生产过程，分析了转炉下渣量、连铸热态铸余回渣量、钢中 Als 含量、精炼炉渣成分和精炼处理时间对钢水中硅质量分数增加的影响。严格控制转炉下渣量≤3.0kg/t、减少热态铸余回渣量和钢中 Als 含量、调整精炼渣系成分、提高炉渣碱度和合理缩短 LF 精炼处理时间，有利于控制 LF 出站钢水硅的质量分数 ω([Si])≤0.03%，满足后续加工需求。

关键词： CSP 流程，SPHC 钢，低碳低硅，增硅，LF 精炼，炉渣

Study on Silicon Content Control of SPHC Steel in Thin Slab Continuous Casting

Sun Bo, Zhang Liangming, Wu Yaoguang, Xie Yangguo, Liu Qianzhi, Wan Dong

(Steel Making and Rolling General Plant of Maanshan Iron and Steel Co., Ltd., Maanshan 243000, China)

Abstract: In order to reduce the increase of silicon in refine process of low carbon and low silicon Al-Killed steel. This paper combines the production of SPHC steel in CSP process in Ma An Shan Iron & Steel Company, The influence of the quantity of slag in tapping, the hot steel & slag recovery, the content of Als, the composition of refining slag and the time of refining treatment on the increase of silicon in the molten steel was analyzed. Controlling the quantity of slag in tapping less than 3.0kg/t, reducing the amount of hot steel & slag recovery and the content of Als in steel, Adjusting the composition &

alkalinity of the slag and shortening the time of LF refine process is beneficial to control silicon content of LF outlet steel less than 0.03%, to meet the subsequent processing demand.

Key words: CSP process, SPHC steel, low carbon and low silicon, silicon increasing, LF refining, slag

转炉连铸工艺生产弹簧钢 55SiCrA 的夹杂物控制

阎丽珍，孟耀青，姬旦旦，王秋坤，和红杰，苏庆林

（邢台钢铁有限责任公司，河北邢台　054027）

摘　要： 本文对转炉连铸工艺生产弹簧钢 55SiCrA 的夹杂物控制进行了分析，通过合金选择、渣系设计、耐材选择、RH 真空处理和浇铸过程结晶器液面波动控制等关键工艺措施实施，实现了弹簧钢 55SiCrA 的非金属夹杂物塑性化和小径化目标，满足了高端弹簧钢对夹杂物的质量要求。

关键词： 弹簧钢，夹杂物，塑性化

Inclusion Control for Spring Steel 55SiCrA of Converter Steelmaking and Continuous Casting Process

Yan Lizhen, Meng Yaoqing, Ji Dandan, Wang Qiukun, He Hongjie, Su Qinglin

(Xingtai Iron & Steel Co., Ltd., Xingtai 054027, China)

Abstract: This paper carried on analysis to the inclusion control of spring steel 55SiCrA，adopt the key process craft measures as selection of alloy、slag system design、selection of refractory material、RH vacuum treated、control of mold level fluctuation during casting process, and so on, the inclusions in spring steel 55SiCrA could be plasticized and smaller, meets the inclusions quality requirement of spring steel.

Key words: spring steel, inclusion, plasticization

时间扰动下的炼钢-连铸调度与铸机拉速协同优化设定方法

宋雨桥，罗小川

（东北大学流程工业综合自动化国家重点实验室，辽宁沈阳　110004）

摘　要： 炼钢-连铸生产工况复杂，生产过程中会因各种因素出现时间扰动使生产调度发生变化，导致连铸出现"断浇"现象，这就需要对生产调度和连铸拉速进行协同优化。本文提出了炼钢调度与连铸拉速协同优化设定方法，以冗余等待时间最小，"断浇"时间最小以及拉速调整尽量平缓为目标，建立数学模型。将模型分为炼钢-连铸调度调整模型和连铸拉速优化模型，分别采用字典序方法和控制变量参数化方法进行求解。仿真实验结果表明，本文提出的方法与人工操作方式相比更具准确性和优越性。

关键词： 炼钢-连铸，连铸拉速，协同优化，控制变量参数化

Coordination Optimization Setting Method for Steelmaking Scheduling and Casting Speed under Time Disturbance

Song Yuqiao, Luo Xiaochuan

(State Key Laboratory of Synthetical Automation for Process Industries, Northeastern University, Shenyang 110004, China)

Abstract: The condition of continuous casting production process is complex, and it will be affected by various disturbance time factors in the process of production, which lead to the broken problem. This requires coordination optimization of production scheduling and continuous casting speed. A coordinated optimization method is proposed in this paper. The mathematical model is set up with the minimum waiting time, the minimum casting time and the smooth adjustment of the speed. The model is divided into steelmaking continuous casting scheduling adjustment model and continuous casting speed optimization model, which are solved by lexicographic order method and control variable parameterization. Simulation results show that the method proposed in this paper is more accurate and superior than manual operation.

Key words: steelmaking continuous casting, continuous casting speed, coordinated optimization, control variable parameterization

抗震钢筋钢 LF 吹氮气合金化工艺研究与应用

刘晓峰[1]，程　殿[1]，杜亚伟[2]

（1. 重庆钢铁股份有限公司炼钢厂；2. 安阳钢铁集团有限责任公司）

摘　要： 吹氮气合金化是一种新型氮化合金技术。通过理论分析抗震钢筋钢钢水吹氮气合金化增氮的热力学和动力学影响因素，在重钢 80 t LF 进行吹氮气合金化工艺现场试验。提出了合理控制 LF 大流量吹氮气前的钢水温度、氮气流量、吹氮气时间等工艺控制措施。实践结果表明：抗震钢筋钢未加入任何含氮合金，经 LF 吹氮气合金化工艺后钢水增氮效果良好，氮含量稳定，钢材力学性能够满足国家标准。

关键词： 抗震钢筋钢，LF，氮气，合金化，实践

Study and Application of LF Blowing Nitrogen Alloying Process for Earthquake Resistant Steel

Liu Xiaofeng[1], Cheng Dian[1], Du Yawei[2]

(1. Steelmaking Plant of Chongqing Iron and Steel Company Limited; 2. Anyang Iron and Steel Group Co., Ltd.)

Abstract: Blowing nitrogen alloying is a new kind of nitriding alloy technology. Through theoretical analysis of influencing factors of thermodynamics and kinetics of aseismic reinforcement steel grade blowing nitrogen alloying nitriding, by blowing nitrogen alloying process in CISC 80t LF field test. The measures to control the temperature of molten steel, nitrogen flow rate and blowing time before controlling LF large flow rate blowing nitrogen are put forward. The practice results show that the nitrogen free alloy is not added to the anti-seismic steel bar, the nitrogen adding effect is good after the

LF blowing nitrogen alloying process, and the nitrogen content is stable, the mechanical properties of the steel could meets the national standard as well.

Key words: earthquake resistant steel, LF, nitrogen, alloying, practice

国际热核聚变实验堆磁体支撑用 316LN 板材制造技术开发

刘承志 [1,2]，姜周华 [1]，李花兵 [1]，舒　玮 [2]，张文茹 [3]，李志斌 [3]

（1. 东北大学冶金材料学院，辽宁沈阳　110004；2. 山西太钢不锈钢股份有限公司技术中心，
山西太原　030003；3. 山西太钢不锈钢股份有限公司军工与核电产品开发业务部，
山西太原　030003）

摘　要：本文针对国际热核聚变实验堆（简称 ITER 计划）磁体支撑所用 316LN 钢板关键性能要求，研究了不锈钢冶炼过程中 Co、P、Nb、Ti 杂质元素的来源，以及 316LN 不锈钢材料成分与 Fe-δ 含量、室温相对磁导率性能的相关性，开发了 Co、P、Nb、Ti 杂质元素及夹杂物控制技术，所开发的 316LN 中厚板产品 Co 含量在 0.08% 以下、P 在 0.022% 以下、Nb 在 0.015% 以下、Ti 在 0.020% 以下，全厚度组织为完全奥氏体组织，无 Fe-δ 痕迹，且室温磁导率性能达到 1.010 以下，钢板平均夹杂物级别小于 2.0 级，4.2K 温度下钢板屈服强度（$R_{P0.2}$）、抗拉强度（R_M）、延伸率（A）、断裂韧性（K_{1c}）平均值分别为 1023MPa、1608MPa、53%、249MPa·m$^{1/2}$，全部指标完全满足 ITER 计划要求，且钢板具有更加优异的低温强韧性。

关键词：国际热核聚变实验堆，磁体支撑，316LN 钢板，制造技术，开发

Development of 316LN Plates Critical Manufacturing Process for Magnet Supports of ITER

Liu Chengzhi[1,2], Jiang Zhouhua[1], Li Huabin[1], Shu Wei[2], Zhang Wengru[3], Li Zhibin[3]

(1. Metallurgical Material College of Northeastern University, Shenyang 110004, China; 2. Technology Center of Shanxi Taigang Stainless Steel Co., Ltd., Taiyuan 030003, China; 3. Products Developing & Marketing Department of Military& Nuclear Power of Shanxi Taigang Stainless Steel Co., Ltd., Taiyuan 030003, China)

Abstract: Aiming at critical properties of 316LN plates for magnet supports of International Thermonuclear Experimental Reactor(ITER), source of the impurity elements such as Co, P, Nb and Ti during stainless steel smelting, and the relativity of chemical composition of the steel grade and Fe-δ content or relative magnetic permeability at room temperature were studied. And control technique of impurity elements of Co, P, Nb or Ti and inclusion content have been developed. Co content is less than 0.08%, Phosphors less than 0.022%, Niobium less than 0.05% and Titanium less than 0.020% in the 316LN plates products. Microstructure of full thickness plates is complete austenite and no trace of Fe-δ content. The relative magnetic permeability at room temperature is lower than 1.010. Additionally, average inclusion level of plates is lower than 2.0. Average value of yield strength, tensile strength, elongation and fracture toughness at 4.2K temperature is 1023MPa, 1608MPa, 53% and 249MPa·m$^{1/2}$, which meet the technical requirement of ITER so products are more excellent in cryogenic toughness properties.

Key words: ITER, magnet supports, 316LN plates, manufacturing process, development

夹杂物塑性化处理技术在酒钢 82B 中的应用

常全举，任文卓，晁增武

（酒泉钢铁集团公司宏兴股份公司）

摘　要：酒钢通过在 J82B 精炼处理过程中加入石英砂降低精炼渣碱度，将精炼渣碱度控制在 0.72-1.0，使得精炼渣熔点处于 1300~1500℃，钢中夹杂物得到塑性化处理。在拉拔过程中塑性夹杂物变形为长条状，提高钢材拉拔性能。

关键词：夹杂物，塑性化，碱度，变形

The Plastic Processing Technology of Inclusions was Applicated in Jisco 82B

Chang Quanju, Ren Wenzhuo, Chao Zengwu

(Jiuquan Iron & Steel (Group) Co., Ltd.)

Abstract: While the refining slag basicity was controlled between 0.72 and 0.80 by adding quartz sand during the processing of J82B refining treatment, the melting point of refining slag is between 1300 ℃ and 1400 ℃.The drawing pertormance was improved while the plastic inclusions are longer in the process of drawing.

Key words: inclusions, plasticizing, basicity, deformation

电渣重熔钢液滴落过程中夹杂物运动行为的模拟研究

王　强[1,2]，汪瑞婷[1,2]，刘　昱[1,2]，张　钊[1,2]，李光强[1,2]，李宝宽[3]

（1．武汉科技大学耐火材料与冶金省部共建国家重点实验室，湖北武汉　430081；
2．武汉科技大学钢铁冶金及资源利用省部共建教育部重点实验室，湖北武汉　430081；
3．东北大学冶金学院，辽宁沈阳　110819）

摘　要：采用双向耦合的欧拉—拉格朗日方法追踪电渣重熔钢液滴落过程中非金属夹杂物的运动轨迹，考虑了夹杂物受到的重力、浮力、阻力、附加质量力、升力、独特的电磁压力和渣钢界面反弹力。为了体现湍流对夹杂物运动轨迹的影响，本文使用了随机游走模型。使用数学模型计算了电流、夹杂物直径和密度对去除率的影响。结果表明当液滴内的夹杂物运动至渣钢界面时，大部分夹杂物能够穿过渣金界面进入到熔渣中，1μm 和 3μm 等小粒径夹杂物则会部分残留在钢液滴中，然后跟随着液滴进入到钢液熔池中。

关键词：电渣重熔，夹杂物，运动轨迹，电流，粒径，密度，数值模拟

Numerical Analysis of Inclusion Motion Behavior in Electroslag Remelting Process

Wang Qiang[1,2], Wang Ruiting[1,2], Liu Yu[1,2], Zhang Zhao[1,2], Li Guangqiang[1,2], Li Baokuan[3]

(1. The State Key Laboratory of Refractories and Metallurgy, Wuhan University of Science and Technology, Wuhan 430081, China; 2. Key Laboratory for Ferrous Metallurgy and Resources Utilization of Ministry of Education, Wuhan University of Science and Technology, Wuhan 430081, China; 3. School of Metallurgy, Northeastern University, Shenyang 110819, China)

Abstract: In order to understand the movement of the inclusion in the electroslag remelting (ESR) process, a transient three-dimensional (3D) comprehensive mathematical model has been established. The finite volume method was employed to simultaneously solve the mass, momentum and energy conservation equations as well as the Maxwell's equations. The volume of fluid (VOF) approach was used to define the redistribution of the metal and the slag. Moreover, the inclusion trajectory was described through the application of the two-way coupled Euler-Lagrange approach. The gravity, buoyancy, drag, added mass, lift, electromagnetic pressure and rebound forces were taken into account. The random walk module was invoked to examine the chaotic effect of the turbulence. Experiments were conducted to verify the proposed model. The influences of the current, inclusion diameter and density on the removal ratio were clarified. The results indicate that the slag-metal droplet interface is constantly renewed with the growing of the droplet. Some inclusions could pass through the interface when they move to the interface, while other inclusions are bounced back to the inside of the droplet. Most inclusions are removed during the formation and the growing of the droplet.

Key words: electroslag remelting, inclusion, movement trajectory, numerical simulation

高效率低成本 RH 生产技术

付中华[1]，刘向东[2]，艾　磊[1]，吴　令[1]，行开新[3]

（1.中冶赛迪工程技术股份有限公司，重庆　400013；2.中冶赛迪技术研究中心有限公司，重庆　401122；3. 中冶赛迪装备有限公司，重庆　400013）

摘　要： 本文主要介绍了三种高效、低成本 RH 生产技术。在吹氧脱碳及升温阶段，应用氩气、O_2 共同吹入钢液代替纯氩气驱动，提高循环效果，提高氧气利用率并减少氧气、氩气消耗。一体式浸渍管技术充分利用上升管、下降管之间的缝隙，增加浸渍管直径，从而提高钢液循环效果，加快脱碳、脱气反应速率。RH 冶金生产模型的应用有利于合金成本优化，减少定氧、定温次数，减少夹杂，稳定钢液终点成分与温度。

关键词： 氩氧共吹，一体式浸渍管，RH 冶金模型，高效制造技术，低成本生产技术

The Effective RH Producing Technologies with Low Cost

Fu Zhonghua[1], Liu Xiangdong[2], Ai Lei[1], Wu Ling[1], Xing Kaixin[3]

(1. CISDI Engineering Co., Ltd., Chongqing 400013, China; 2.CISDI Research & Development Co., Ltd., Chongqing 401122, China; 3. CISDI Equipment Co., Ltd., Chongqing 400013, China)

Abstract: Three kinds of RH technologies with great efficiency and low cost are mainly introduced in this article. In the phases of decarburization and temperature elevation, the way of blowing Argon is replaced by the way of blowing both Argon and Oxygen to drive the liquid steel, which can improve circulation efficiency of liquid steel, increase use ratio of O_2 and decrease the consumption of Ar and O_2. In the technology of integrated immersion tube, the crevice between ascension and descent pipes is utilized effectively, which increases the pipe diameter. Therefore, the circulation efficiency of liquid steel is improved and decarburization rate is enhanced as well as degasification rate. The RH metallurgical producing model is useful for cost optimization of alloys, the reduction of measuring oxygen and temperature, stabilization of ultimate contents and temperature.

Key words: argon and oxygen blowing, integrated immersion tube, RH metallurgical model, efficient manufacturing technology, producing technology with low cost

含镁 Fe-1.2Mn-0.2Si-0.2S 钢液体系中夹杂物的演变行为

张庆松[1,2]，闵 义[1,2]，华 瑶[1,2]，许海生[1,2]，刘承军[1,2]

（1. 东北大学多金属共生矿生态化冶金教育部重点实验室，辽宁沈阳 110819；

2. 东北大学冶金学院，辽宁沈阳 110819）

摘 要： 采用高温模拟实验与热力学分析相结合的方法，考察了含镁 Fe-1.2Mn-0.2Si-0.2S 钢液体系中夹杂物的演变行为。研究结果表明，当钢液镁含量为 0.0005% 时，加硫前夹杂物平衡相由 SiO_2-MnO 液相氧化物变质为 MgO-SiO_2-MnO 液相镁系氧化物。加硫后钢液中 S 含量达到 0.2%，夹杂物平衡相为 MgO-SiO_2-MnO-MnS 液相镁系氧硫化物。凝固初期随着钢液中 S 含量的提高，S 元素不断由钢液向液相氧硫化物内部扩散。然而在淬冷过程中，由于 S 元素在夹杂物内部扩散不充分而导致 S 元素在夹杂物边缘富集。

关键词： 硫系易切削钢，硅锰脱氧，镁处理，富 S 层

Effect of Magnesium Addition on Formation and Evolution of Inclusions in Fe-1.2Mn-0.2Si-0.2S Molten Steel

Zhang Qingsong[1,2], Min Yi[1,2], Hua Yao[1,2], Xu Haisheng[1,2], Liu Chengjun[1,2]

(1. Key Laboratory for Ecological Metallurgy of Multimetallic Ores (Ministry of Education), Northeastern University, Shenyang 110819, China;

2. School of Metallurgy, Northeastern University, Shenyang 110819, China)

Abstract: To reveal the effects of magnesium on the formation and evolution of inclusions in Fe-1.2Mn-0.2Si-0.2S molten steel, both thermodynamic calculation and deoxidization experiments were carried out in the present work. The results showed that SiO_2-MnO was modified into MgO-SiO_2-MnO liquid oxide inclusions when the content of magnesium in molten steel was 0.0005%. When the content of sulfur in molten steel is 0.2%, the equilibrium phase of inclusions were MgO-SiO_2-MnO-MnS liquid oxysulfide. During the initial stage of solidification process, S element was continuously diffused into liquid oxysulfide inclusions from molten steel with the increase of S content in steel. However, S element was enriched at the edges of inclusions due to the insufficient diffusion in the inclusions during the quenching process.

Key words: sulfur-containing free-cutting steel, Si-Mn deoxidization, magnesium addition, S enrichment layer

精炼渣成分对硅锰脱氧弹簧钢夹杂物影响的实验室研究

殷　雪[1]，孙彦辉[2]，焦　帅[1]，牛阿朋[2]

（1. 冶金与生态工程学院，北京科技大学，北京　100083；

2. 钢铁共性技术协同创新中心，北京科技大学，北京　100083）

摘　要：本文通过实验室钢渣反应实验研究了不同组分的精炼渣系($CaO-SiO_2-Al_2O_3$)对硅锰脱氧弹簧钢中夹杂物特征的影响。研究表明：反应结束后钢中氧化物夹杂主要为 $SiO_2-CaO-Al_2O_3$ 系，夹杂物呈球形、椭球形和不规则形貌，尺寸均在 5μm 以下。顶渣碱度及渣中 Al_2O_3 含量的增加均可导致钢中 Als 含量的增加，但与碱度相比，渣中 Al_2O_3 含量的增加对钢中 Als 影响更大。夹杂物中 Al_2O_3 含量随着渣中 Al_2O_3 含量的增加而升高，当精炼渣碱度控制在 1.5，渣中 Al_2O_3 含量控制在 3% 时，可将钢中的夹杂物组分控制在 $CaO-SiO_2-Al_2O_3$ 相图中 1400℃ 低熔点区域。

关键词：硅锰脱氧，弹簧钢，夹杂物，精炼渣

Laboratory Study on the Effect of Slag Composition on Inclusion Characteristics in Si/Mn- Deoxidized Spring Steel

Yin Xue[1], Sun Yanhui[2], Jiao Shuai[1], Niu Apeng[2]

(1. Department of Metallurgical and Ecological Engineering，University of Science and Technology Beijing，Beijing 100083，China; 2. Collaborative Innovation Center of Steel Technology, University of Science and Technology Beijing，Beijing 100083, China)

Abstract: In order to investigate the effect of top slag on inclusion characteristics in Si/Mn-Deoxidized spring steel, laboratory experiments were performed to optimize the composition of $CaO-SiO_2-Al_2O_3$ slag system. It is proved that inclusions after steel-slag reaction mainly contain $CaO-SiO_2-Al_2O_3$ shown as spherical, elliptic and irregular morphology with the size less than 5μm. Both higher slag basicity and Al_2O_3 in slag could increase [Al]s contents in steel. The slag basicity should be controlled at 1.5 to increase the CaO content in inclusions and the Al_2O_3 in slag should be lowered to 3% to reduce the Al_2O_3 content in inclusions for distributing the inclusion compositions in the low melting zone (1400℃) and improving the deformability of inclusions.

Key words: Si/Mn-Deoxidized, spring steel, inclusion, slag

高强度耐磨钢中非金属夹杂物控制研究

初仁生，李战军，刘金刚，陈　霞，王卫华，郝　宁

（首钢技术研究院宽厚板所，北京　100043）

摘　要：磨损失效是耐磨钢失效方式，提高耐磨钢的洁净度，控制钢中非金属夹杂物是保证上述性能和延缓磨损失

效的关键技术之一。本文通过分析 SSAB 耐磨钢和某厂生产的高强度中厚板耐磨钢中的非金属夹杂物的演变规律进行分析，确定耐磨钢中洁净度和非金属夹杂物的控制策略。通过扫描电镜对耐磨钢中非金属夹杂物的尺寸、组成和形状进行分析。研究发现生产的耐磨钢中非金属夹杂物的类型不单一而是由钙铝酸盐和硫化锰夹杂物共同组成，夹杂物的尺寸控制为：Ds 类控制小于 20μm，A 类 MnS 夹杂物和 B 类钙铝酸盐夹杂物控制为 1.5 级以下，可以较好的满足耐磨钢性能要求。

关键词：耐磨钢，非金属夹杂物，中厚板，控制

Study on the Control of Cleanliness for High-strength Wear-resistance Steel

Chu Rensheng, Li Zhanjun, Liu Jingang, Chen Xia, Wang Weihua, Hao Ning

(Shougang Research Institute of Technology, Beijing 100043, China)

Abstract: High-strength wear-resistance steel has been widely used in various wear conditions of wear-resistant materials which has to meet the high strength with good ductility, toughness, bending and welding performance. In the current paper, the cleanliness control method was analyzed for the evolution of non-metallic inclusions produced by SSAB and some company. The size, composition and the type of the non-metallic inclusions were analyzed by SEM and EDS. The results show that the type of non-metallic inclusions is not single, it consists of calcium aluminate and MnS for the non-metallic inclusions. To meet the performance, the rating of the non-metallic inclusions is 1.5 or less for MnS for class A and calcium aluminate for class B. Also the size control for class Ds is less than 20μm. The control strategy for the inclusions is small size, diffuse distribution and little amount of deformation after rolling.

Key words: high-strength wear-resistance steel, heavy and medium plate, nonmetallic inclusion

精炼渣中 Al$_2$O$_3$ 含量对钢洁净度的影响

于会香，王新华，姜 敏

（北京科技大学冶金与生态工程学院，北京 100083）

摘 要：为了研究精炼渣中 Al$_2$O$_3$ 含量对钢洁净度，尤其是非金属夹杂物的影响，选用合结钢和高强度低合金钢两类钢种，分别开展了不同 Al$_2$O$_3$ 含量的高碱度（B=CaO/SiO$_2$，B 为 7 左右）和中等碱度（B 为 3.5 左右）CaO-SiO$_2$-Al$_2$O$_3$-MgO 系炉渣与实验钢的平衡实验，并采用 Thermo-Calc 热力学软件计算了该渣系组元的活度。结果发现，两种炉渣条件下，当碱度一定时，随着渣中 Al$_2$O$_3$ 含量的降低，与之平衡的钢液总氧和硫含量降低；渣中 $a_{Al_2O_3}$ 减小，a_{MgO} 显著增大；夹杂物中 MgO 含量增加，Al$_2$O$_3$ 含量降低；夹杂物总数量和大尺寸夹杂物数量减少。

关键词：精炼渣，Al$_2$O$_3$ 含量，洁净度，夹杂物，渣-钢反应

Effect of Al$_2$O$_3$ Content of Refining Slag on Steel Cleanliness

Yu Huixiang, Wang Xinhua, Jiang Min

(School of Metallurgical and Ecological Engineering, University of Science and Technology Beijing, Beijing 100083, China)

Abstract: To investigate the effect of Al_2O_3 content of refining slag on steel cleanliness, especially on non-metallic inclusions, slag-metal equilibrium experiments were carried out between $CaO-SiO_2-Al_2O_3-MgO$ system containing different Al_2O_3 content with high basicity (B=CaO/SiO$_2$, B is around 7) and alloyed structural steel, as well as with medium basicity (B is around 3.5) and high strength low alloyed steel. The activity of slag components was also calculated by Thermo-Calc software. The results show that, For the two types of slag, with Al_2O_3 content decreasing when slag basicity is fixed, the total oxygen content and Sulphur content in steel in equilibrium decrease, the calculated activity of Al_2O_3 decreases and activity of MgO increases greatly, the observed MgO content increases and Al_2O_3 content decreases in inclusions in equilibrium with top slag, and the amount of total inclusions, particularly, large sized inclusions decrease.

Key words: refining slag, Al_2O_3 contentl, cleanliness, inclusions, slag-metal equilibrium

精炼过程 304 不锈钢夹杂物的演变

张井伟[1]，张立峰[1]，任　英[1]，任　磊[2]，杨　文[1]，翟　俊[3]

（1. 北京科技大学冶金与生态工程学院，北京　100083；

2. 内蒙古科技大学材料与冶金学院，内蒙古包头　014000；

3. 太原钢铁（集团）有限公司，山西太原　030003）

摘　要： 本文通过 ASPEX 自动扫描电镜分析研究了 AOD、LF 冶炼过程中 304 不锈钢夹杂物的类型、形态、尺寸和成分，并通过 Factsage7.0 软件对夹杂物中 MgO 的演变进行热力学分析。结果表明 AOD 氧化期夹杂物主要为 CrS、MnS 及 CaO、SiO_2、MgO、Al_2O_3、Cr_2O_3、TiO_2 的复合夹杂物，AOD 还原期及 LF 过程夹杂物主要为 CaO、SiO_2、MgO、Al_2O_3、MnO、TiO_2 复合夹杂物，夹杂物尺寸集中在<10 μm。本文重点结合热力学计算分析了夹杂物中 MgO 和 CaO 的含量变化。

关键词： 304 不锈钢，AOD，LF，夹杂物

Evolution of Inclusions in 304 Stainless Steels during Refining Process

Zhang Jingwei[1], Zhang Lifeng[1], Ren Ying[1], Ren Lei[2], Yang Wen[1], Zhai Jun[3]

(1. School of Metallurgical and Ecological Engineering, University of Science and Technology Beijing, Beijing 100083, China;

2. School of Material and Metallurgical Engineering, Inner Mongolia University of Science and Technology, Baotou 014000, China;

3. Taiyuan Iron & Steel (Group) Co., Ltd.,Taiyuan 030003,China)

Abstract: In this study, the type, morphology, size, and composition of 304 stainless steel inclusions in AOD and LF refining process were analyzed using ASPEX. The evolution of MgO in inclusions was calculated using Factsage7.0 software. The results show that inclusions during AOD oxygen blowing process are mainly composed of CrS, MnS. During AOD reduction period, inclusions change to CaO, SiO_2, MgO, Al_2O_3, Cr_2O_3, TiO_2. The inclusions during LF process are mainly CaO,SiO_2,MgO,MnO,TiO_2. The inclusions size were less than 10μm. The changes of MgO and CaO in inclusions are thermodynamically discussed.

Key words: 304 stainless steel, AOD, LF, inclusion

铁水进 LF 炉造渣深脱硫工艺实践

廖扬标，孟　磊，耿恒亮，吴义强，朱志鹏

（武钢条材厂一炼钢分厂，湖北武汉　430083）

摘　要： 分析了转炉冶炼低硫钢回硫的原因，促使喷镁脱硫产物的上浮和排除是最主要的措施。在实践阶段，利用 LF 炉电极加热提供高温条件，对喷镁脱硫后的铁水进行造高碱度还原渣，稀释并吸附镁脱硫产物，再返脱硫站进行扒除。结果表明，起到了较好的效果，出钢硫小于 0.004%的比例达到 88.3%，解决了喷吹镁粉脱硫易造成低硫钢回硫的问题。

关键词： 回硫，铁水造渣，深脱硫，吸附

Process Practice of Hot Metal Deep Desulphurization by Making Slag in Ladle Furnace

Liao Yangbiao, Meng Lei, Geng Hengliang, Wu Yiqiang, Zhu Zhipeng

(No.1 Steel Making Branch of General Wire Rod Mill of WISCO, Wuhan 430083, China)

Abstract: The proper measure are promoting the desulphurization product float and rule out by analyzing the reason to the resulfurization in refining low sulfur steel. In practice stage, using Ladle Furnace electrode heating to provide high temperature conditions, slagged on deep desulphurization hot metal, made of high basicity slag, diluted and adsorbed of desulphurization product, then skimmed the slag in the desulphurization station. The results show that the effect is good, the proportion of the tapping sulfur less than 0.004% reached 88.3%, solved the problem of low sulfur steel resulfurization by Mg-injection process.

Key words: resulfurization, slagged on hot metal, deep desulphurization, adsorbed

20Cr13 不锈钢脱氧夹杂物研究

屈志东[1]，成国光[2]，万文华[1]，滕力宏[1]，黄永生[1]，王日红[1]

（1. 中天钢铁集团技术中心，江苏　213011；2. 北京科技大学国家重点实验室，北京　100083）

摘　要： 为了研究 20Cr13 不锈钢在不同脱氧制度下的夹杂物，本文结合实际生产以及 FactSage 热力学软件计算，分析了 20Cr13 不锈钢单独 Si 脱氧，单独 Al 脱氧，Si-Al 复合脱氧条件下夹杂物的生成情况。得出单独 Si 脱氧条件下体系随 Si 含量增加夹杂物由 MnO•Cr$_2$O$_3$ 转变为 MnO-SiO$_2$-CrO$_x$ 再变为 SiO$_2$；单独 Al 脱氧条件下体系随 Al 含量增加夹杂物由 MnO•Cr$_2$O$_3$ 直接转变为 Al$_2$O$_3$；Si-Al 复合脱氧条件下在高 Si 低 Al 时产物是 SiO$_2$，高 Al 条件下是 Al$_2$O$_3$ 类夹杂物，合适的 Si 及 Al 条件下是 Al$_2$O$_3$-SiO$_2$-MnO-CrO$_x$ 类夹杂物。

关键词： 20Cr13 不锈钢，脱氧，夹杂物，MnO•Cr$_2$O$_3$，MnO-SiO$_2$-CrO$_x$，Al$_2$O$_3$-SiO$_2$-MnO-CrO$_x$

Study on the Inclusion of 20Cr13 Stainless Steel Deoxidization Process

Qu Zhidong[1], Cheng Guoguang[2], Wan Wenhua[1], Teng Lihong[1],
Huang Yongsheng[1], Wang Rihong[1]

(1. Technical Center, Zenith Steel Group Co., Ltd,. Jiangsu 213011,China;
2. State Key Laboratory,University of Science and Technology Beijing,Beijing 100083,China)

Abstract: To study the inclusion of 20Cr13 stainless steel with different deoxidization processes, thermodynamic calculation software FactSage was used based on the manufactured production，and the formation of inclusions of 20Cr13 stainless steel deoxidized with Si, Al and Si-Al compound deoxidization was discussed in this paper. It was concluded that with the increasing of Si content，the inclusion will transform from $MnO \cdot Cr_2O_3$ to $MnO\text{-}SiO_2\text{-}CrO_x$ and then to SiO_2. Under the condition of Al deoxidization, the inclusion system will directly transform from $MnO \cdot Cr_2O_3$ to Al_2O_3 with increasing of Al content. Under the condition of Si - Al compound deoxidization, the inclusion is SiO_2 with high Si content and low Al content, Al_2O_3 inclusions will be formed with high Al content, and $Al_2O_3\text{-}SiO_2\text{-}MnO\text{-}CrO_x$ inclusions is produced with suitable Si and Al contents.

Key words: 20Cr13 stainless steel, deoxidization, inclusion, $MnO \bullet Cr_2O_3$, $MnO\text{-}SiO_2\text{-}CrO_x$, $Al_2O_3\text{-}SiO_2\text{-}MnO\text{-}CrO_x$

RH 真空室烘烤温度场的数值模拟仿真与优化

孙　晓[1]，徐安军[1]，贺东风[1]，汪红兵[2]，袁　飞[1]

（1. 北京科技大学冶金与生态工程学院，北京　100083；
2. 北京科技大学计算机与通信工程学院，北京　100083）

摘　要： 为了优化某厂的 RH 真空室烘烤工艺制度，将烘烤过程中的流体流动、燃烧和换热过程进行了耦合，建立了用于数值模拟计算的三维 RH 真空室数学模型，计算了 RH 真空室内烘烤过程的燃烧现象，重点分析了 RH 烘烤过程不同顶枪位置下 RH 真空室内煤气燃烧温度场的变化规律。结果表明，烘烤温度场随着顶枪枪位的改变在垂直方向上改变明显，烘烤枪位 4.5m 时，烘烤温度场较为理想，烘烤均匀性高，对去除真空室内衬凝结的钢水有积极的作用。烘烤枪位 5m 时，烘烤高温区处于真空室底部与浸渍管部分，对真空室底部与浸渍管耐材烘烤效果较好，生产过程中应该将两种烘烤枪位结合使用以达到提高烘烤效果的目的。计算结果与实验结果基本吻合，提出可供现场参考的烘烤工艺参数。

关键词： RH 真空室，烘烤，数值模拟

Numerical Simulation and Optimization of Temperature Field in the Baking Vacuum Space of RH

Sun Xiao[1], Xu Anjun[1], He Dongfeng[1], Wang Hongbing[2], Yuan Fei[1]

(1. School of Metallurgical and Engineering, University of Science and Technology Beijing, Beijing 100083, China;
2. School of Information Engineering, University of Science and Technology Beijing, Beijing 100083, China)

Abstract: In order to optimize the vacuuun space of RH baking system of a factory,numerical simulation of gas flow,combustion and heat transfer in vacuum space of RH during the baking process was carried out. The temperature distribution of vacuum space of RH baking at different position of RH top lance was investigature numerically and experimentally.Results show that the different position of RH top lance can lead to the temperature field change dramatically in vertical direction.4.5m is suitable position of RH top lance,in this condition,there is a ideal and uniformity temperature field,it's benefited for removing the accumulation of steel in cacuum space of RH. When the RH top lance is in 5m,the bottom of the vacuum space and the immerge tube are in hige baking temperature, two baking technology should combine in the production process to achieve the purpose of improving the baking effect. The caculated results agree with the experimental observation. In addition,the optimal baking parameters for the actual performance were presented.

Key words: vacuum space of RH, baking, numerical simulation

电渣重熔过程中渣壳动态形成的数值模拟

余　嘉，刘福斌，姜周华，陈　奎，钱瑞清

（东北大学冶金学院，辽宁沈阳　110819）

摘　要： 电渣重熔过程中渣壳的动态形成过程决定着铸锭最终的表面质量。本文基于 VOF 模型和动网格技术，考虑固态渣壳和液态熔渣电导率的差异，建立一个二维轴对称的瞬态模型，研究了电渣重熔过程中渣壳的动态形成过程。研究结果表明：渣池底部形成了较厚的渣壳，其产生的焦耳热大于周围渣池区域，这与不考虑渣壳形成时的焦耳热分布明显不同；根据电极端部的热平衡计算的实验室规模的电渣炉的熔速约为 83kg/h；金属熔池中的液态金属与渣壳接触并使其局部熔化，形成弯月面形状的渣壳；随着铸锭的连续生长，弯月面型的渣壳将随着铸锭一起推进，最终在铸锭表面形成均匀的渣壳。

关键词： 电渣重熔，渣壳，动态形成，数值模拟

Numerical Investigation on the Dynamic Formation of Slag Skin in Electroslag Remelting Process

Yu Jia, Liu Fubin, Jiang Zhouhua, Chen Kui, Qian Ruiqing

(School of Metallurgy, Northeastern University, Shenyang 110819, China)

Abstract: The surface quality of ingot is determined by the dynamic formation of slag skin in electroslag remelting process. In this paper, a 2D transient mathematical model is established to investigate the dynamic formation of slag skin, based on the volume of fluid and dynamic mesh technique. The results indicate that the Joule heat induced from the slag skin is larger than the surrounding zones at the bottom of slag bath, and the distribution of which is obviously different from that in the case ignoring the formation of slag skin; According to the heat balance at the electrode tip, the calculated melt rate in the laboratory scale electroslag remelting process is approximately 83 kg/h. The slag skin contacted with the liquid metal is partially remelted, and forms a meniscus surface. It moves upward accompanied by the rise of ingot, which results in the generation of a layer of uniform slag skin around the ingot surface.

Key words: electroslag remelting, slag skin, dynamic formation, numerical simulation

提高复吹转炉透气砖寿命和冶金效果的新技术

杨文远[1]，李　林[2]，彭小艳[2]，王明林[3]，高　飞[2]，史晓强[2]

（1. 钢铁研究总院冶金工艺研究所，北京　100081；

2. 新冶高科技集团有限公司热工装备与材料事业部，北京　100081；

3. 钢铁研究总院连铸中心，北京　100081）

摘　要： 为了提高复吹转炉透气砖寿命和冶金效果，研究了大流量透气砖底吹不对称供气技术。采用水模试验方法，底吹气量按 3.3:1 分两路供气，每隔 1 炉交换一次。试验结果表明，供气强度在 0.2 Nm³/t.min 时，透气砖侵蚀速度与单根毛细管的气体流量成正比。毛细管根数增加 1 倍，透气砖的供气能力提高 1 倍。采用大流量大尺寸透气砖不对称交错供气的技术，可使透气砖的侵蚀速度减少 50%，寿命提高 1 倍。转炉炼钢的熔池混匀时间缩短 19.2%，铁水脱磷预处理熔池混匀时间缩短 63%，复吹转炉的冶金效果得到明显改善。与复吹转炉预埋透气砖和更换透气砖的方法相比，可以更有效的提高转炉的复吹炉龄和冶金效果。

关键词： 转炉，复合吹炼，炉龄，大流量，不对称供气

New Technology for Improving Service Life of Multiple Hole Plug and Metallurgical Effect for Combined Blowing Converter

Yang Wenyuan[1], Li Lin[2], Peng Xiaoyan[2], Wang Minglin[3], Gao Fei[2], Shi Xiaoqiang[2]

(1.Metallurgical Technology Institute, Central Iron and Steel Research Institute, Beijing 100081, China;

2.Thermal Equipment and Material Department, New Metallurgy Hi-Tech Group Co., Ltd., Beijing 100081, China;

3. National Engineering Research Center of continuous Casting Technology, Central Iron and Steel Research Institute, Beijing 100081, China)

Abstract: In order to improve service life of Multiple Hole Plug（MHP）and metallurgical effect for combined blowing converter, new asymmetric alternating gas supplying technology for large flow capillary MHP was investigated. Use water model experiment, as well as the gas flow ratio of the two branches is 3.3:1 and alternates each heat. The results illustrated that the corrosion rate of the MHP is proportional to the gas flow of each capillary when the bottom gas supply intensity is 0.2 Nm³/t. min. The quantity of capillary increases 1 time, and then the gas supply ability of MHP rise 1 time. The corrosion rate of MHP can be reduced 50% and the life of MHP can be increased 1 time through the application of large flow capillary MHP with big size and asymmetric alternating gas supplying technology at bottom. The mixing time of converter molten steel bath and hot metal dephosphorization pretreatment bath can be reduced 19.2 and 63% respectively by using asymmetric alternating gas supplying technology, therefore the metallurgical effect of the converter are improved remarkably. The asymmetric alternating gas supplying technology of large flow capillary MHP can improve the combined blowing furnace campaign and metallurgical effect significantly compared with the method of pre-embedded MHP and replacing the MHP in combined blowing converter.

Key words: converter, combined blowing, furnace campaign, large flow, asymmetric gas supplying technology

粉煤灰碳热还原制备硅铁合金的研究

余文轴，李　杰，游志雄，党　杰，吕学伟

（重庆大学材料科学与工程学院，重庆　400044）

摘　要： 粉煤灰综合利用是一个技术含量高、应用潜力大、具有广阔市场前景、集环保与资源再生利用为一体的极具发展前途的新兴产业。为了实现粉煤灰的高附加值、全资源化利用，在实验室条件下进行了粉煤灰碳热还原制备硅铁合金并富集氧化铝的探索研究。研究表明，碳热还原过程中，随着碳配比的增加，粉煤灰中的莫来石相可以更充分地分解，转变为氧化铝和二氧化硅，而二氧化硅则在碳热还原过程中最终生成硅铁合金和碳化硅；在配加 Fe_2O_3 进行碳热还原的条件下，莫来石相中的二氧化硅更易被还原成硅，并与金属铁结合生成硅铁合金粒，这为后续硅铁合金和氧化铝的分离创造了条件。此外，粉煤灰中各氧化物的还原热力学也表明，在有 Fe_2O_3 存在的条件下，SiO_2 可以在更低温度下被碳还原，并生成硅铁合金，这与实验结果一致。

关键词： 粉煤灰，碳热还原，氧化铝，硅铁合金

Study on Preparation of Ferrosilicon Alloy by Carbothermal Reduction of Coal Fly Ash

Yu Wenzhou, Li Jie, You Zhixiong, Dang Jie, Lv Xuewei

(College of Materials Science and Engineering, Chongqing University, Chongqing 400044, China)

Abstract: Comprehensive utilization of coal fly ash is a promising industry with high technology, great potential application, broad market prospect, environmental protection and resource recycling. In order to realize the high added value and full utilization of fly ash, the research on the preparation of ferrosilicon alloy and the enrichment of alumina by carbothermal reduction of fly ash under laboratory conditions has been carried out. Research shows that in the carbothermal reduction process with the increase of carbon ratio, mullite in fly ash was more fully decomposed into alumina and silica, and the silica was finally reduced to ferrosilicon alloy and silicon carbide in the carbothermal reduction process. Under the condition of carbothermal reduction by adding Fe_2O_3, the silicon dioxide in the mullite phase was more easily reduced to silicon and was synthesized from iron alloy particles with metal iron. This provides the conditions for the subsequent separation of ferrosilicon and alumina. In addition, the reduction thermodynamics of the oxides in the fly ash also showed that under the presence of Fe_2O_3, SiO_2 could be reduced to carbon at lower temperatures and generation of ferrosilicon alloys, which was consistent with the experimental results.

Key words: coal fly ash, carbothermal reduction, alumina, ferrosilicon alloys

高品质冷轧板精炼渣优化控制技术与应用

季晨曦[1]，董文亮[1]，崔　阳[1]，田志红[1]，单　伟[2]，袁天祥[2]，朱国森[2]

（1.首钢集团有限公司技术研究院，北京　100043；

2. 首钢京唐钢铁联合有限责任公司，河北唐山　063200）

摘　要： 钢包精炼渣对夹杂物溶解与吸收起决定性影响，其熔点、组元活度、粘度、界面张力等理化性能起到了关键作用。本文基于首钢京唐炼钢厂精炼渣成分，采用 Factsage 软件分析了渣中 CaO/Al$_2$O$_3$ 比对 FeO、CaO 和 Al$_2$O$_3$ 三个组元活度和粘度及熔点的影响，并分析了精炼渣粘度随温度的变化规律。结合钢包加盖技术对精炼渣温度的提升，分析了钢包加盖后对渣–钢反应的影响。分析认为精炼渣温度提高后液态化明显，从而提高了渣中(FeO)活度对渣钢反应的敏感度，从而采用了新的精炼渣控制思路。结果表明，FeO 平均含量降低至 4.6%、平均 CaO/Al$_2$O$_3$ 提高至 1.7 后，钢水洁净度改善明显。

关键词： 精炼渣控制，钢渣反应，钢包加盖技术，CaO/Al$_2$O$_3$ 比

Application of Refining Slag Optimum Control Process for High Quality Cold-Rolled Sheet

Ji Chenxi[1], Dong Wenliang[1], Cui Yang[1], Tian Zhihong[1],
Shan Wei[2], Yuan Tianxiang[2], Zhu Guosen[2]

(1. Research Institute of Technology, Shougang Group Co., Ltd., Beijing 100043, China;
2. Shougang Jingtang Iron and Steel United Co., Ltd., Tangshan 063200, China)

Abstract: Refining slag plays an important role in the inclusion dissolution and absorption. Its physical and chemical properties like melting point, activity of components, viscosity, interfacial tension is crucial for inclusion removal. Based on the refining slag compositions of Jingtang Steelworks, Factsage was used to clarify the effect of CaO/Al$_2$O$_3$ ratio on melting temperature, viscosity, activities of FeO, CaO and Al$_2$O$_3$. Furthermore, the relationship between slag viscosity and temperature was investigated by calculation. For ladle top covering technique, slag temperature increased obviously. The effect of ladle top covering on slag-metal reaction was analyzed and found this reaction could be sensitively affected by the FeO activity. New tactic of refining slag control was applied for high quality cold-rolled sheet steel production to improve the ability of inclusion removal. The result showed that average FeO content decreased to 4.6%, average CaO/Al$_2$O$_3$ ratio increased to 1.7 could improve the steel cleanliness.

Key words: refining slag control, slag-metal reaction, ladle top covering, CaO/Al$_2$O$_3$ ratio

镁铬以及镁碳质耐材向铝脱氧钢中溶解 Mg 能力的比较

刘春阳[1]，八木元己[2]，高　旭[3]，金宣中[4]，植田滋[3]，北村信也[3]

（1. 工学研究科，东北大学，仙台市，日本；2. 原工学研究科，东北大学，现 Japan Railways，名古屋市，日本；3. 多元物质科学研究所，东北大学，仙台市，日本；
4. Department of Materials Science & Engineering, Chosun Unversity, Gwangju, Korea）

摘　要： 镁铬和镁碳耐材在炼钢中应用广泛，同时也是钢中 MgO•Al$_2$O$_3$ 尖晶石夹杂物（MA）形成过程中 MgO 的供给源。本文将镁铬和镁碳耐材棒插入铝脱氧钢液中反应，通过研究钢液中溶解 Mg 浓度以及夹杂物成分的变化，来比较两种耐材向钢液中溶解 Mg 能力的差异。研究发现：镁铬耐材与钢液反应 60min 后，钢中溶解 Mg 含量仅为 0.7ppm，钢中夹杂物仍然为 Al$_2$O$_3$；而镁碳耐材反应 60min 后，钢中溶解 Mg 含量增至 3.5ppm，钢中初始的 Al$_2$O$_3$ 夹杂物转变为 MA 尖晶石夹杂物。对耐材/钢液反应界面分析发现：镁铬耐材棒表面形成了 MA 尖晶石反应层。由于该尖晶石层完全包裹住耐火材料与钢液的接触面，从而抑制了 Mg 的溶出。而在镁碳耐材实验中，耐材表面没有

反应层产生从而钢水与耐材中的 MgO 相始终保持直接接触，因此耐材中的 Mg 较易溶解进入钢液。

关键词：镁溶解，镁铬耐材，镁碳耐材，铝脱氧钢，耐材/钢液界面

钢包底喷粉过程多相流传输及脱硫动力学的数值模拟

娄文涛，朱苗勇，王晓雨

（东北大学冶金学院，辽宁沈阳 110819）

摘　要：建立 CFD-PBM-SRM 耦合模型描述 80t 钢包底喷粉过程中多相流传输行为及精炼反应动力学。提出了底喷粉过程中顶渣-钢液、空气-钢液、粉剂-钢液、气泡-钢液多界面-多组分同时反应模型，并考虑了气液两相流、粉剂传输及脱硫产物饱和对精炼反应动力学的影响，考察了喷粉参数对粉剂传输和脱硫效率的影响规律。结果表明：在底喷粉过程中，粉剂颗粒尺寸逐渐增大到一个稳定尺寸，在低喷粉量时，钢液脱硫主要依赖于粉剂－钢液和顶渣－钢液反应共同作用，其中顶渣－钢液反应为主导机制。当喷粉量超过 0.75kg/t 时，粉剂－钢液反应成为主导机制，当喷粉量大于 2.25kg/t 时，气泡－钢液反应脱硫作用增强。随着喷气量的增加，钢液脱硫效率降低；随着喷粉量的增大，钢液脱硫效率增大。

关键词：底喷粉，脱硫，多相流，数值模拟，钢包

Numerical Simulation of Multiphase Flow Transport Behavior and Desulfurization Kinetics in Ladle with Bottom Powder Injection

Lou Wentao, Zhu Miaoyong, Wang Xiaoyu

(School of Metallurgy, Northeastern University, Shenyang 110819, China)

Abstract: A computation fluid dynamics–population balance model–simultaneous reaction model (CFD–PBM–SRM) coupled model was used to predict the multiphase flow behavior and reaction kinetic in 80 ton ladle with bottom powder injection. The multiple interface reactions models including top slag-liquid steel reaction, air-liquid steel reaction, powder-liquid steel reaction and bubble-liquid steel reaction were proposed in present work, and the effects of powder transport and desulfurization products saturation on the desulfurization behavior were considered as well. The influence of different powder injection parameters on the desulfurization and powder removal efficiency were investigated, and the importance of different transport and reaction mechanisms were discussed and clarified. The results are shown as follows: At the low powder injection rate less than 0.75kg/t, the desulfurization is mainly attributed to the joint effort of both powder-liquid steel reaction and top slag-liquid steel reaction which is the prevailing mechanism. With the increase of the powder injection rate, the powder-liquid steel interface reaction becomes more important and then predominates the desulfurization behavior, and when the powder injection rate exceeds 2.25kg/t, the role of bubble-liquid steel interface reaction becomes important and exceeds the top slag-liquid steel interface reactions. With the increase of gas flow rate, the desulfurization ratio gradually decrease. With the increase of powder injection rate, the desulfurization ratio increase.

Key words: bottom powder injection, desulfurization, multiphase flow, numerical simulation, ladle

铁水喷吹脱硫过程的模拟和分析

马文俊，刘国梁，陈　斌

（首钢技术研究院，北京　100043）

摘　要： 通过对喷吹复合脱硫剂（60%Mg+40%CaO）铁水脱硫过程动力学的分析，建立了复合脱硫剂喷吹脱硫的动力学模型。通过模型分析了脱硫剂粒径对脱硫效果的影响，在现有的喷吹条件下，粉剂粒度控制在0.9mm到1.5mm时，粉剂能有效突破气液界面进入铁水，发生脱硫反应。通过现场取样分析，验证了模型的模拟结果与现场实际结果具有良好的一致性。硫含量的≥0.035%高硫铁水在脱硫过程存在不同的三个阶段：孕育阶段、快速脱硫阶段和缓慢脱硫阶段。针对高硫铁水，本研究开发了两阶段喷吹脱硫工艺，根据现场试验结果与传统工艺相比，采用两段喷吹法可使脱硫剂用量减少10%~20%。

关键词： 铁水预处理，脱硫，喷粉，动力学，两阶段

Simulation and Analysis of Injecting Desulfurization Process in Hot Metal

Ma Wenjun, Liu Guoliang, Chen Bin

(Shougang Research Institute of Technology, Beijing 100043, China)

Abstract: Based on the analysis of the kinetics of hot metal desulfurization by injecting the desulfurizer (60%Mg+40%CaO), a kinetic model of desulfurization with compound desulfurizer was established. The influence of particle size of desulfurization agent on desulfurization efficiency was analyzed. Under the existing injection conditions, when the powder size is controlled from 0.9mm to 1.5mm, the powder can effectively break through the gas-liquid interface, enter the molten iron, and then desulfurization. Field sampling analysis shows that the simulation results of the model are in good agreement with the actual results. The sulfur content of more than 0.035% high sulfur iron has three different stages in the desulfurization process: embryonic stage, fast and slow stage desulfurization. In view of the high sulfur hot metal, the two stage injection desulfurization method has been developed. According to the field test results, compared with the traditional process, the two stage injection method can reduce the dosage of desulfurizer by 10%~20%.

Key words: hot metal pretreatment, desulphurization, powder injection, dynamics, two stage

湛钢钢包粘渣增重原因分析及对策

吴　政[1]，李洪涛[2]，潘永飞[2]，王洪亮[2]，兰　天[1]

（1. 宝钢湛江钢铁有限公司炼钢厂，广东湛江　524000；
2. 宝山钢铁股份有限公司炼钢厂，上海　201900）

摘　要： 介绍了宝钢湛江钢铁有限公司炼钢厂350t钢包在使用过程中粘渣情况，为了解决钢包粘渣问题，本文主要从钢包渣熔点进行分析，通过分析钢包渣成分，确定影响钢包渣熔点的因素，并且使用Factsage 6.4热力学计算软件绘出

SiO$_2$-CaO-Al$_2$O$_3$-7%MgO 四元相图，根据 SiO$_2$-CaO-Al$_2$O$_3$-7%MgO 四元相图分析钢包渣熔点。研究结果表明，湛江钢铁钢包粘渣主要原因为钢包渣熔点高，主要分布在 1530～1750℃之间，为了使钢包渣熔点降低到 1500℃以下，钢包渣钙铝比（CaO/Al$_2$O$_3$）需要控制在 0.7~2 范围之间，碱度（CaO/SiO$_2$）需要达到 2.3 以上，经过理论计算，要控制碱度与钙铝比在目标范围内，需要在出钢过程中加入 400～800kg 石灰。在现场试验时，转炉出钢时加 400～800kg 石灰后，钢包渣熔点在四元相图中主要分布在 1500℃以下区域内，并且钢包重量明显降低，最终控制在 142t 左右。

关键词：钢包，粘渣，熔点，石灰

Causes of Analysis and Countermeasure on Slag Adhesion and Over Weight of Ladle at Zhangang

Wu Zheng[1], Li Hongtao[2], Pan Yongfei[2], Wang Hongliang[2], Lan Tian[1]

(1. Steelmaking Plant of Baosteel Zhanjiang Iron & Steel Co., Zhanjiang 524000, China;

2. Steelmaking Plant of Baoshan Iron & Steel Co., Shanghai 201900,China)

Abstract: This Paper Introduces slag adhesion and over wight of the 350t ladle of steelmaking plant of Baosteel Zhanjiang Iron & Steel Co. In order to solve the problem of slag adhesion, this paper mainly analyses the melting point of the ladle slag. By analyzing the composition of the ladle slag, the factors affecting the melting point of ladle slag are determined. Using the Factsage 6.4 thermodynamics calculation software draws SiO$_2$-CaO-Al$_2$O$_3$-7%MgO quaternary phase diagram and analyzed the he melting point of the ladle slag. The results show that the reason of slag adhesion and over weight of ladle at Zhangang is the high melting point of ladle slag, mainly distributed between 1530℃ to 1750℃. In order to reduce the melting point of ladle slag to 1500℃ below, the ratio (CaO / Al$_2$O$_3$) of ladle slag needs to be controlled in the range of 0.7~2, and alkalinity (CaO/SiO$_2$) needs to reach 2.3 or more. To control the alkalinity and the ratio (CaO/Al$_2$O$_3$) in the target range, the weight of lime needing to be added into the ladle in the tapping process is 400 kg to 1000 kg. In the field test, when the weight of lime added into the ladle in the tapping process is 400 kg to 800 kg. The melting point of ladle slag is mainly distributed in the quaternary phase diagram in the area where the melting point is below 1500 ℃, and ladle weight significantly reduces to 142t or so.

Key words: ladle, slag adhesion, melting point, lime

传动壳体铸铁件的数值模拟和铸造工艺优化

范　涛[1,2]，张巨成[2]，丁翔祺[2]，李红波[2]，唐伟忠[1]，贾成厂[1]

（1. 北京科技大学粉末冶金研究所，北京　100083；

2. 北华航天工业学院材料工程学院，河北廊坊　065000）

摘　要：根据传动壳体铸铁件的结构特点、工作要求及生产性质等，对传动壳体进行铸造工艺设计。对零件设计了两个铸造工艺方案，采用金属液由铸件上下端面处引入的封闭式浇注系统，用 SolidWorks 软件建立铸件的三维模型，运用华铸 CAE8.0 软件对传动壳体凝固过程进行数值模拟。通过模拟结果和分析选择较好的方案，通过改进远离直浇道处设置一个搭边暗冒口的方法优化铸造工艺。结果表明，铸造工艺出品率提高 20%达到了 88.92%，铸件部分已无明显缩孔缩松缺陷，铸件符合要求。

关键词：传动壳体，数值模拟，凝固过程，工艺优化

Numerical Simulation of Cast Iron Transmission Shell and Its Casting Process Optimization

Fan Tao[1,2], Zhang Jucheng[2], Ding Xiangqi[2], Li Hongbo[2], Tang Weizhong[1], Jia Chengchang[1]

(1. Institute for Advanced Materials and Technology, University of Science and Technology Beijing, Beijing 100083, China; 2. School of Materials Engineering, North China Institute of Aerospace Engineering, Langfang 065000, China)

Abstract: The casting process of cast iron transmission shell body was designed according to the structrue characteristics requirements and productive characteristic of the castings. Two casting process were designed for the parts, and the enclosed pouring system was introduced by the metal liquid from the upper and lower end of the casting. The three-dimensional model of the casting was built by SolidWorks software. The casting simulation software Huazhu CAE 8.0 was employed to simulate the solidification process of the casting. According to the results of numerical simulation and analysis, the casting process was optimized. A side-blind riser which kept away from sprue was designed to feed the casting. The results show that the process production rate reach to 88.92% increasing by 20%, while the shrinkage defects are basically eliminated and the cast meets product requirements.

Key words: transmission shell, numerical simulation, solidification process, process optimization

转炉双联法开发超低磷钢实践研究

李福高，袁理想，杨　波，贺旭阳，马少青，王云波

（青岛特殊钢铁集团公司，山东青岛　266000）

摘　要： 伴随着我国钢铁企业产能过剩的现状和国家产业结构调整的契机，青岛特殊钢铁集团炼钢厂于 2015 年 11 月份正式投产。随着新厂区投产、达产阶段的顺利完成，为适应市场需求和完成公司领导的前瞻性部署，开发低磷钢和超低磷钢，同时完成对现有钢种质量升级已是当前需迫在眉睫完成的目标，而钢种质量的进一步提升则要求转炉脱磷必须提高到一个新的台阶。常规转炉冶炼工艺难以完成低成本低磷钢品种的冶炼，为此我厂对铁水进行了转炉脱磷的试验研究，即转炉双联法炼钢工艺[1]，并取得了优良的应用效果。

关键词： 转炉，双联法，脱磷转炉，脱碳转炉，超低磷钢

Practice Research on Development of Super Low Phosphorus Steel by Converter Duplex Method

Li Fugao, Yuan Lixiang, Yang Bo, He Xuyang, Ma Shaoqing, Wang Yunbo

(Qingdao Special Steel Group Company, Qingdao 266000, China)

Abstract: With the status of overcapacity in China's iron and steel enterprises and the opportunity of adjustment of national industrial structure, The steel-making plant of Qingdao special steel group was put into production in November 2015.With the successful completion of the new plant production and production phase, to meet the market demand and the prospective deployment of the company leaders, to develop the low phosphorus steel and ultra-low phosphorus steel, at the same time to complete the existing quality upgrading of steel has become an imminent goal, and the further improvement of steel quality requires that the dephosphorization of the converter must be improved to a new level.The conventional converter smelting process is difficult to complete the low cost and low phosphorus steel smelting. in this paper, the experimental study on dephosphorization of hot metal in converter is carried out, that is, the duplex steelmaking process, and the excellent application effect has been obtained.

Key words: converter, double-line method, dephosphorization converter, decarburization converter, ultra-low phosphorus steel

滚筒钢渣用于建筑集料的性能研究

朱 珉[1,2]

（1. 北京科技大学，北京 100083；2. 青岛特殊钢铁有限公司，山东青岛 266409）

摘 要：滚筒钢渣由于具有游离氧化钙含量低、粒形好、颗粒级配好等优点，可在一定程度下取代天然砂配制砂浆，这既可降低钢铁行业钢渣对大气、水资源及土资源等环境的影响，而且可以提高处理钢渣的经济效益和附加值。

本研究在对钢渣砂进行基本性质测试、岩相分析、成分的基础上，充分论证滚筒钢渣代替天然河砂的可行性。

关键词：钢渣砂，滚筒渣，砂浆，和易性，安定性，强度

Study on Performance of Roller Slag Used in Building Aggregate

Zhu Min[1,2]

(1. University of Science and Technology Beijing, Beijing 100083, China;

2. Qingdao Special Steel Co., Ltd., Qingdao 266409, China)

Abstract: The roller steel slag has the advantages of low content of free calcium oxide, grain shape, particle size distribution and so on, can to some extent replace natural sand mixing mortar, it can reduce the influence of iron and steel industry slag on atmosphere, water resources and soil resources and environment, but also can improve the slag treatment economic benefits and added value.

Based on the test of basic properties, petrographic analysis and composition of steel slag sand, the feasibility of roller slag instead of natural river sand is fully demonstrated.

Key words: steel slag sand, roller slag, mortar, workability, stability, strength

真空感应炉冶炼含氮高合金钢控氮工艺

彭雷朕，姜周华，耿 鑫，李 星，师 帅

（东北大学材料与冶金学院，辽宁沈阳 110819）

摘 要：实验研究了30kg真空感应炉冶炼含氮合金钢9Cr（0.01%~0.03%N），采用充入不同压力的氮气进行保护

的渗氮工艺。结果表明：真空感应炉采用氮气渗氮，渗氮时间为15min，能够达到该氮分压下的平衡氮含量；使用Factsage软件计算的氮分压与氮含量之间的关系与经验公式计算的值相吻合，该经验公式适用于中低氮高合金钢的控氮工艺研究；采用6kPa的氮分压渗氮时间约为15min，能够保证9Cr钢中氮含量在目标范围内。

关键词：真空感应炉，含氮9Cr钢，氮的合金化，氮气分压

The Process of Controlling Nitrogen in the High Alloy Steel Melting in the Vacuum Induction Furnance

Peng Leizhen, Jiang Zhouhua, Geng Xin, Li Xing, Shi Shuai

(School of Metallurgy, Northeastern University, Shenyang 110819, China)

Abstract: This paper is mainly focused on the nitrogen alloying by using the nitrogen atmosphere only under different pressure in the 30kg vacuum induction furnace (VIM) for remelting the Cr9 (0.01%~0.03%N). The conclusions can be drawn through analyzing the results: the content of nitrogen in the steel under different partial pressure of N_2 after 15 minutes will be reached to the equilibrium content; The relationship between the nitrogen partial pressure and the nitrogen content calculated by the Factsage software is consistent with the value calculated from the empirical formula. The empirical formula is applicable to the nitrogen control process of medium and low nitrogen high alloy steels; When the pressure of the nitrogen is 6kPa and the nitriding time is 15min, the content of nitrogen in the molten steel is just aimed in the nitrogen range of the 9Cr.

Key words: VIM, nitrogenous 9Cr steel, nitrogen alloying, partial pressure of N_2

无磁钢 20Mn23AlV 表面纵裂及轧材表面皮下裂纹研究

李晓军，张润平，李忠利，李　振，张　彬，师国平

（山西太钢不锈钢股份有限公司炼钢二厂，山西太原　030003）

摘　要：本文结合无磁钢 20Mn23AlV 钢种成分特点及连铸浇注特性，研究了其表面纵裂、皮下裂纹产生机理。研究表明：该钢种表面纵裂产生与结晶器保护渣中保护渣变性有关，随浇注过程进行，结晶器保护渣中 Al_2O_3 含量逐步升高，导致保护渣熔点升高、黏度增大，保护渣渣条增多增大，导致连铸坯因传热不均而产生纵裂；连铸坯皮下裂纹与连铸工序吹氩有关，关闭塞棒及滑板吹氩系统后，连铸坯皮下裂纹消失。

关键词：表面纵裂，PT检验，皮下裂纹，保护渣变性，高铝钢

新型环保钢水精炼剂在炼钢过程中的应用

李志广，曹树卫，郭永谦

（安阳钢铁股份有限公司第二炼轧厂，河南安阳　455004）

摘　要：从解决烟尘排放问题和降低炼钢生产成本角度出发，对原有 LF 炉用钢水改质剂进行改良优化，替换为新

型环保钢水精炼剂，同时优化工艺操作，外排烟气粉尘含量明显降低，满足了日益增长的环保要求，改善了岗位工人的作业环境，减少了恶劣的作业环境对工人身体的伤害。

关键词：优化工艺，降低粉尘排放，改善作业环境

The Application of New-type Environmental Refining Flux in Steelmaking Process

Li Zhiguang, Cao Shuwei, Guo Yongqian

(Anyang Iron and Steel Co., Ltd., No.2 Steel Making & Rolling Plant, AnYang　455004, China)

Abstract: In order to solvedust emission problem and reduce the steelmaking production cost, the original conditioning flux had beenreplaced by with anew-type environmental refining flux，as well as optimizing process operation，the dust content in discharged off—gas is obviously reduced, the increasingly demand of environmental requirements is met，working environment of post workers is improved，harms of bad working environment on workers health is reduced.

Key words: optimization process, reduce dust emission, improve the working environment

IF 钢在多晶氧化铝基底上异相行核行为的研究

路　程，王万林，邹　格，高尔卓

（中南大学冶金与环境学院，湖南长沙　410083）

摘　要： 本文通过卧滴法研究了在不同氧分压下钢液与多晶氧化铝之间的过冷度行为，结果显示氧分压对 IF 钢/多晶氧化铝基底间过冷度的影响分析中出现了临界点 10^{-23} atm，在低于及高于此氧分压的条件下，过冷度均有不同程度地减少。扫描电镜分析显示，临界氧分压条件下无界面层存在；在氧分压偏高时，有连续的界面层存在，其化学成分接近铁铝尖晶石；氧分压偏低时界面层呈现出裂解状态，分析其成因在于基底发生了分解。经典异相形核模型计算的结果显示，δ 铁素体晶核与基底间的临界接触角是控制过冷度的决定性因素，过冷度随着临界接触角的增加而呈线性增加的趋势，此外氧分压会影响基底的稳定性以间接地改变临界接触角，从而影响异相行核效果。

关键词：氧分压，卧滴法，过冷度，行核效率

Heterogeneous Nucleation Behavior of Interstitial-Free Steel on Poly-crystal Al₂O₃ Substrate

Lu Cheng, Wang Wanlin, Zou Ge, Gao Erzhuo

(School of Metallurgy and Environment in Central South University, Changsha 410083, China)

Abstract: The undercooling degreeof molten steel on poly-crystal Al_2O_3 substrate under different oxygen partial pressure were studied by using sessile drop method. The results showed that there existed a critical oxygen partial pressure on the curve of undercooling degree versus oxygen partial pressure, the undercooling degree of liquid steel on poly-crystal Al_2O_3 substrate decreased whenever the oxygen partial pressure was lower or higher than the critical one. The SEM analysis indicated that there was no interfacial layer between the droplet and the substrate at the critical point of 10^{-23}atm, while oxygen partial pressure above 10^{-23}atm, there was a continuous interfacial layer, its chemical composition was analogous to that of hercynite; when oxygen partial pressure below 10^{-23}atm, there was a splitting layer due to thesubstrate decomposition. The calculation results of the classical heterogeneous nucleation model suggested that the critical contact angle between the crystal nucleus of δ-Fe and the nucleant substratewas the decisive factor to control the undercoolingdegree, and the undercooling degree increased linearly with the increase of critical contact angle, besides, the oxygen partial pressure wouldaffect the stability of the substrate to change the critical contact angleindirectly, thus affecting the heterogeneous nucleating effect.

Key words: oxygen partial pressure, sessile drop method, undercooling degree, nucleation efficiency

首钢京唐液态渣、钢高效循环利用工艺开发

刘延强，罗　磊，张　鹏，赵长亮，袁天祥

（首钢京唐钢铁联合有限责任公司，河北唐山　063200）

摘　要： 为实现"全三脱"工艺少渣冶炼，进一步降低辅料消耗，首钢京唐开发了热态脱硫渣、热态脱碳渣和铸余渣钢直接返回利用工艺。本文对热态渣、钢的可回收性进行了分析，并通过工业试验验证了工艺的应用效果。结果表明，回收利用 5t 的脱硫渣，脱硫剂消耗可降低 30%～40%，铁水温降相对减少 10～15℃，总渣量减少 30%～40%，同时可降低铁损，减少对环境的污染；每炉回收热态渣 10～15t，可节约石灰 1.6～2.4t，若铁水 Si 含量小于 0.15%，脱磷炉可不加石灰，钢铁料消耗相应减少 2.4kg/t 钢，并且可取消萤石及轻烧的使用；通过控制高炉出铁量，铸余钢回包次数可达到 6～8 次，实现液态铸余直接回收。

关键词： 铁水预处理，热态渣，脱硫渣，脱磷渣，铸余，资源化利用

Process Development of Efficient Recycling of Molten Steel and Slag at Shougang Jingtang

Liu Yanqiang, Luo Lei, Zhang Peng, Zhao Changliang, Yuan Tianxiang

(Shougang Jingtang Iron and Steel Co., Ltd., Tangshan 063200, China)

Abstract: In order to achieve less slag of "Full-Tri-de" smelting process, new technology which was study on KR using recycling of hot desulfurization slag, dephosphorization converter using recycling of hot steel slag of decarburization converter and casting residual direct return utilization was employed to further reduce material consumption of semi steel smelting process by Shougang Jingtang. The recyclability of hot steel and slag of "Full-Tri-de" smelting process. Simultaneously, the effect of hot steel and slag on application effects were investigated. The results show that desulfurization agent consumption can be reduced by 30% ~ 40% through recycling of 5t desulfurization slag, and the temperature drop of hot metal is reduced by 10 to 15℃, the total amount of slag is reduced from 30% to 40%; average consumption of 1.6~2.4t lime was saved through recycling hot steel slag 10~15t per ladle. If silicon content in molten iron is lower than 0.15%, lime could not be added to dephosphorization furnace. The consumption of 2.4kg/t iron and steel raw material was correspondingly cut. Meanwhile, light-burned dolomite and fluorite as burden materials were eliminated in dephosphorization furnace. through the control of blast furnace, the time of recycling casting residual slag and steel can reach 6 ~ 8 times and may achieve direct recovery of liquid casting.

Key words: hot metal pretreatment, molten slag, desulfurization slag, dephosphorization slag, cast slag, resource utilization

高阻抗电弧炉短网系统三相不平衡度的研究

吕　明[1]，李小明[1]，朱　荣[2]

（1. 西安建筑科技大学冶金工程学院，陕西西安　710055；

2. 北京科技大学冶金与生态工程学院，北京　100083）

摘　要：本文详细分析了 100t 高阻抗电弧炉短网系统各部分的电阻和电抗，发现电极电阻为短网系统电阻的主要组成部分，中相和边相电极电阻分别占各相总电阻的 52.72% 和 54.32%；导电横臂、水冷电缆、导电铜管的电抗为短网系统电抗的主要组成部分，中相和边相的三部分电抗分别占各相总电抗的 82.49% 和 82.24%。电弧炉的三相阻抗不平衡度为 1.50%，满足设计要求。本文的计算结果和某厂 100t 高阻抗电弧炉的实际测量结果吻合，表明该计算方法在工程中应用是可行的，为炼钢电弧炉短网系统的优化设计提供了参考。

关键词：炼钢，高阻抗电弧炉，短网，阻抗，不平衡度

Research on the Three-Phase Unbalance Degree of Large Current System for High-impedance EAF

Lv Ming[1], Li Xiaoming[1], Zhu Rong[2]

(1. School of Metallurgical Engineering, Xi'an University of Architecture and Technology, Xi'an 710055, China;

2. School of Metallurgical and Ecological Engineering, University of Science and Technology Beijing, Beijing 100083, China)

Abstract: Resistance and reactance of large current system in 100t high-impedance EAF were analysed in this paper, which found that the electrode was the main part of system resistance, and electrode resistance in medium phase and lateral phase were respectively 52.72% and 54.32% of the total resistance. The reactance of current conducting arm, water-cooled cable, conductive copper tube were major components of large current system, and their reactance in medium phase and lateral phase were respectively 82.49% and 82.24% of the total reactance. The three-phase unbalance degree of impedance was 1.50%, which meets the design requirement. The calculation results agreed well with actual measured results in 100 t high-impedance EAF. The calculation method is feasible in engineering application, in favor of optimization design of large current system.

Key words: steelmaking, high-impedance EAF, large current system, impedance, unbalance degree

八钢风电钢钢中夹杂物探讨

李立民

（新疆八一钢铁股份有限公司，新疆乌鲁木齐　830022）

摘　要：通过八钢风电钢生产工艺优化，系统性分析了 LF 精炼、RH 真空处理、连铸浇注过程钢水中氧氮变化、

夹杂物的组成变化、夹杂物的数密度和面密度变化及不同夹杂物百分数变化，得出八钢风电钢洁净化水平为钢中氧含量≤12ppm、氮含量≤50ppm，夹杂物数量≤20 个/mm²，10μm 以上夹杂物约占 2%。

关键词：工艺优化，夹杂物，炉外精炼，保护浇注

Eight Steel Wind-electric Steel Inclusion in Steel Discuss

Li Limin

(Xinjiang Bayi Iron & Steel Stock Co., Ltd., Urumchi 830022, China)

Abstract: Through eight steel wind steel production process optimization, systemic analyzed the LF refining, RH vacuum treatment, oxygen and nitrogen in continuous casting steel casting process change, change the composition of inclusions, inclusion of number density and density changes and different percentage of inclusions, it is concluded that the eight wind steel and clean steel level for 12ppm or less oxygen content and nitrogen content in steel 50ppm or less, the number of inclusions acuities were/was 20, more than 10 microns inclusions accounted for about 2%.

Key words: process optimization, inclusions, refining outside the furnace, protection of pouring

30CrMoA 抽油杆用圆钢热顶锻裂纹原因分析

孙学刚，苏　磊，张怀忠

（新疆八一钢铁股份有限公司第一炼钢厂，新疆乌鲁木齐　830022）

摘　要：针对 30CrMoA 抽油杆用圆钢出现的热顶锻裂纹的原因进行了分析，通过宏观和微观分析，找到了影响热顶锻不合格的主要铸坯缺陷，并提出了改进措施，实施后达到了预期的效果。

关键词：抽油杆用圆钢，热顶锻，裂纹，铸坯质量

Analysis on the Cracks of 30CrMoA Steel Bars for Sucker Rods on Hot Forming Test

Sun Xuegang, Su Lei, Zhang Huaizhong

(No. 1 Steel-making Plant, Xinjiang Bayi Iron & Steel Co., Ltd., Urumchi 830022,China)

Abstract: 30CrMoA steel bars for sucker rods of hot forming the cause of the cracks are analyzed, through the macroscopic and microscopic analysis, found the main casting billet defect affecting the quality of hot forming test, and put forward the improvement measures, to achieve the desired effect after implementation.

Key words: steel bars for sucker rods, hot forming, the crack, casting billet qualities

智能工厂设计及智能制造关键技术

田　陆[1]，胡志刚[1,2]，徐祖宏[1]

（1. 湖南镭目科技有限公司，湖南长沙　410000；2. 湖北理工学院材料与科学学院，
湖北黄石　435000）

摘　要：本文分析了工业 4.0 和智能工厂设计的新特点，依据目前镭目公司在沙钢等钢厂生产线上实施的智能工厂设计方案和实施情况，重点介绍了智能制造关键技术和实施效果。

关键词：智能工厂，自动化炼钢，炉气分析，自动化出钢，精炼自动化

Smart Plant Design and the Key Technologies of Intelligent Manufacturing

Tian Lu[1], Hu Zhigang[1,2], Xu Zuhong[1]

(1. Ramon Technology Co., Ltd., Changsha 410000, China; 2. Hubei Polytechnic University,
Materials and Sciences Institute, Huangshi 435000, China)

Abstract: The new characteristics of industrial 4.0 and smart plant designing were analyzed. According to the application of smart plant designing in Shagang by Ramon, the key technologies of intelligent manufacturing and implementation effects were mainly introduced.

Key words: smart factory, automation of steelmaking, gas analysis, automation of tapping, refining automation

Al 脱氧 961 不锈钢夹杂物研究

屈志东[1]，成国光[2]，万文华[1]，滕力宏[1]，黄永生[1]，王日红[1]

（1. 中天钢铁集团技术中心，江苏　213011；2. 北京科技大学国家重点实验室，北京　100083）

摘　要：根据某钢厂生产的 961（1Cr11Ni2W2MoV）不锈钢，对其进行了夹杂物的检测以及分析计算。检测结果表明夹杂物主要为 Al_2O_3、Al_2O_3-MnO-CrO_x 以及内层为 Al_2O_3 外层为 Al_2O_3-MnO-CrO_x 的复合夹杂物。利用热力学软件 FactSage 对夹杂物进行了相关的计算，结果表明高温、高 O、高 Mn、低 Cr 以及合适的 Al 含量才会生成 Al_2O_3-MnO-CrO_x 类夹杂物。钢液中 Al、Mn 及 Cr 含量的增加会使得各自氧化物在 Al_2O_3-MnO-CrO_x 类夹杂物中比例升高；温度以及 O 含量的升高均会使得该类夹杂物中 Al_2O_3 的含量降低。温度升高、Al 含量降低以及 O 含量升高均会促进体系 Al_2O_3-MnO-CrO_x 类夹杂物区域的扩大，但温度升高到一定程度便没有很大的变化。相较于单纯的 Al 脱氧，改用 Si 或者 Si-Al 脱氧能很好的改善不锈钢体系夹杂物。

关键词：1Cr11Ni2W2MoV, Al_2O_3-MnO-CrO_x, 夹杂物，热力学计算

Study on the Inclusion of 961 Stainless Steel with Al Deoxidization

Qu Zhidong[1], Cheng Guoguang[2], Wan Wenhua[1], Teng Lihong[1],
Huang Yongsheng[1], Wang Rihong[1]

(1. Technical Center, ZENITH Steel Group Co., Ltd., Jiangsu 213011,China;
2. State Key Laboratory, University of Science and Technology Beijing, Beijing 100083, China)

Abstract: Depending on the 961 (1Cr11Ni2W2MoV) stainless steel produced by a steel plant, the inclusions were tested and calculated. The results showed that the inclusions were mainly composed of Al_2O_3, Al_2O_3 -MnO-CrO_x and the composite inclusions of Al_2O_3 -MnO-CrO_x. With the thermodynamic software FactSage, the calculation results indicated that at the condition of high temperature, high oxygen content, high Mn content, low Cr content, and proper Al content, Al_2O_3-MnO-CrO_x inclusions could be formed. Increasing of the content of Al, Mn and Cr in molten steel will lead to the increasing of the proportion of the oxide in the Al_2O_3-MnO-CrO_x inclusions. The increasing of temperature and Oxygen content will decrease the Al_2O_3 content of the inclusions.The increasing of temperature and oxygen content, the decreasing of Al content will lead to the expansion of the Al_2O_3-MnO-CrO_x inclusion area, as the temperature rise to a certain degree,there will be will no great change.Compared with simple Al deoxygenation, switching to Si or Si-Al deoxidization can improve the inclusion system of stainless steel.

Key words: 1Cr11Ni2W2MoV, Al_2O_3-MnO-CrO_x, inclusion, thermodynamic calculation

含铜铬铁水喷吹 CO_2 脱碳的探索研究

张保敬，张廷安，牛丽萍，李志强，郑　超，张东亮

（东北大学冶金学院，辽宁沈阳　　110819）

摘　要： 为了使铜渣得到最大价值化的利用，铜渣彻底还原冶炼含铜抗菌不锈钢的新工艺被提出。本文在还原得到的含铜铁水中加入铬铁后，进行了 CO_2 脱碳的工艺研究，对 CO_2 与铁水中主要元素的反应进行热力学计算，并从反应温度、通气时间、CO_2 流量及 CO_2 分压四个因素来考察所得金属中铜铬碳的含量变化规律。结果表明，反应温度1650℃，反应时间60min，CO_2 流量590mL/min，Ar 占比20%时，得到的金属成分最为符合马氏体含铜抗菌不锈钢的要求。

关键词： 含铜铬铁水，CO_2，脱碳，含铜抗菌不锈钢

Decarburization of Iron Containing Copper and Chrome with CO_2

Zhang Baojing, Zhang Tingan, Niu Liping, Li Zhiqiang, Zheng Chao, Zhang Dongliang

(School of Metallurgy, Northeastern University, Shenyang 110819, China)

Abstract: In order to make copper slag with maximum value, a new technology that copper slag reduced to smelt copper-containing antimicrobial stainless steel is proposed. In this paper, CO_2 decarburization was studied after ferrochromium being added in copper-containing molten iron reduced from copper slag. Thermodynamic of the reactions between CO_2 and the main elements in molten iron were calculated. Four factors that reaction temperature, ventilation time, CO_2 flow rate and CO_2 partial pressure were studied to investigate changing law of Cu, C and Cr in iron. The results show that reaction temperature 1650℃, reaction time 60min, CO_2 flow rate 590mL/min and Ar proportion 20%, composition of metal after decarburization are suitable conditions for martensite copper-containing antimicrobial stainless steel.

Key words: iron containing Copper and chrome, CO_2, decarburization, copper-containing antimicrobial stainless steel

提高 CrMnTi 系齿轮钢窄成分命中率工艺研究与实践

谷志敏，杨锋功，杨华峰

（河钢集团石家庄钢铁有限责任公司，河北石家庄　050031）

摘　要： 分析了 CrMnTi 系齿轮钢整个生产过程，从中找出了该钢种化学成分稳定的影响因素，并制定技术工艺改进优化措施，实施效果良好，窄成分合格率由 41.1%提高到 80.15%，提高了齿轮钢的端淬稳定性。

关键词： 齿轮钢，窄成分控制，端淬

Research and Practice on Improving Narrow Component Percentage Hit Ratio of CrMnTi Gear Steel

Gu Zhimin, Yang Fenggong, Yang Huafeng

(Hesteel Group Shisteel Company, Shijiazhuang 050031, China)

Abstract: Analyses entire production process of CrMnTi gear steel, found out the influence factor of steady chemical composition, and institute measures to improve technical methods, the effectiveness is good, the narrow component percentage of CrMnTi gear steel to 80.15 percent from 41.1 percent, improved stability of gear steel jominy.

Key words: gear steel, narrow component control, jominy

国内转炉煤气回收概况与研究展望

于鹏飞，曾加庆，林腾昌

（钢铁研究总院冶金工艺研究所，北京　100081）

摘　要： 本文概括了 LT 法与 OG 法的优缺点，论述了国内主要钢铁企业转炉煤气回收现状。调研发现国内钢厂主要采用煤气系统设备改进、炉口微压差控制、调整煤气回收条件和优化供氧制度等方法提高煤气回收量，但煤气热值比日本转炉低约 15%~20%。提出从转炉熔池反应原理与烟气回收设备匹配的角度开展深入研究提高转炉煤气回收量和热值的新思路。

关键词： 转炉，煤气回收，回收量，热值

General Situation and Research Prospect of Converter Gas Recovery in China

Yu Pengfei, Zeng Jiaqing, Lin Tengchang

(Institute for Metallurgical Process, Central Iron and Steel Research Institute，Beijing 100081, China)

Abstract: The advantage and weakness of the Oxygen Gas Recovery System and Lurgi-Thyssen System were generalized. The status of converter-gas-recovery in major steel enterprises in China was discussed in this paper. By investigating steel enterprises in China, we find that steel enterprises mainly use these methods such as gas system equipment improvement, micro pressure control, adjustment of gas recovery conditions and optimization of oxygen supply system to increase the amount of gas recovery in domestic steel mill. And the calorific value of gas is about 15%~20% lower than that in Japan. A new idea based on the matching of Converter-Bath-reaction-principle and flue gas recovery equipment to improve recovery ratio and calorific value of converter gas put forward in this paper.

Key words: converter, gas-recovery, recovery, calorific value

桥梁钢探伤不合原因分析

于赋志[1]，史页殊[2]，许孟春[1]，王鲁毅[2]，李德军[1]，李晓伟[1]

(1. 海洋装备用金属材料及其应用国家重点实验室，辽宁鞍山　114009；
2. 鞍钢股份鲅鱼圈钢铁分公司，辽宁营口　115007)

摘　要： 针对某厂 Q370qE 等桥梁钢钢板超声波探伤不合的问题，采用金相检验、扫描电镜等分析手段，对钢板探伤不合部位取样进行检测与分析，发现中心偏析组织中有微裂纹缺陷。分析结果表明，钢板中心存在金属铌块与 Nb、Ti 碳氮化物，在受热应力、机械应力作用时产生微裂纹是造成 Q370qE 等桥梁钢探伤不合格的主要原因。通过采取相应措施，延长合金溶解时间与降低铌铁粒度，适当降低 [Nb]、[Ti] 含量、降低降低 [S]、[P] 含量到较低水平，优化轻压下工艺，减少凝固前沿析出的 Nb、Ti 碳氮化物的数量，有效提高了 Q370qE 等桥梁钢的探伤合格率。

关键词： 桥梁钢，超声波探伤，微裂纹，碳氮化物

An Analysis on Cause of Disqualification for Flaw Detection of Bridge Steel Plates

Yu Fuzhi[1], Shi Yeshu[2], Xu Mengchun[1], Wang Luyi[2], Li Dejun[1], Li Xiaowei[1]

(1. State Key Laboratory of Metal Material for Marine Equipment and Application，Anshan 114009, China
2. Bayuquan Iron & Steel Subsidiary Company of Angang Steel Co., Ltd., Yingkou 115007, China)

Abstract: In order to investigate the causes of disqualification plates in ultrasonic flaw detection on bridge steel such as Q370qE, metallographic examination and SEM analysis were carried out on the sampling from the disqualification plate. It was found that there were micro-cracks in the center segregation microstructure. Further analysis met there were metal niobium blocks and Nb, Ti carbonitrides in the center of the steel plate. Then micro-cracks were generated under the heat stress and mechanical stress, as is the cause of the disqualification for bridge steel plates of Q370qE in ultrasonic flaw detection.The qualified rate of ultrasonic flaw detection was increased on bridge steel plates by taking appropriate measures to decrease the content of [Nb] and [Ti], reduce the content of [S] and [P] to a low level, optimize the soft reduction process to reduce the amount of Nb, Ti carbonitrides precipitated at the solidification front.

Key words: bridge steel, ultrasonic flaw detection, micro-crack, carbonitride

CSP 低碳低硅铝镇静钢夹杂物控制实践

吴维轩

（武汉钢铁有限公司条材厂 CSP 分厂，湖北武汉　430083）

摘　要：通过优化转炉、精炼工序过程控制，降低出钢氧含量，采用下渣检测，降低熔渣氧化性，稳定钙处理效果，提高钢水纯净度，细化连铸浇钢工序标准，做好保护浇铸和稳态浇铸，稳定中间包液位等措施，有效减少了带钢表面夹渣缺陷的发生，铝镇静钢夹杂缺陷改判率降至 1.56%以下。

关键词：CSP，夹杂物，钙处理，保护浇铸，稳态浇铸

邯钢连铸浸入式水口堵塞和熔损问题分析及改善

徐　晓，朱文玲

（河钢集团邯钢公司技术中心，河北邯郸　056015）

摘　要：连铸采用浸入式水口浇注钢水的过程中，尤其在浇注超低碳钢时，经常因为 Al_2O_3 附着在水口内壁发生结瘤，或者钢水温度过高等造成水口熔损的现象发生，不仅降低连铸机的生产效率，而且也是引起钢铁产品产生缺陷的主要原因之一。从堵塞机理入手，提出钢液品质控制和浸入式水口形状、材质等改善措施。

关键词：浸入式水口，Al_2O_3，堵塞机理，防止措施

Analysis and Improvement of Clogging and Melting Loss of Submerged Entry Nozzle for Continuous Casting in Handan Iron and Steel Company

Xu Xiao, Zhu Wenling

(Technology Center, HBIS Group Hansteel Company, Handan 056015, China)

Abstract: During the continuous casting process with submerged entry nozzles for molten steels, especially for the casting of ultra-low carbon steel, the phenomenon of clogging and melting loss of nozzles would always be encountered. Because of the attachment of alumina to the inner nozzle wall, the nodulation of submerged entry nozzles would often happen. Besides, factors such as too high temperature of molten steel usually cause the serious melting loss of nozzles. These phenomena would not only decrease production efficiency of the caster, but also be the main reasons causing the defects of steel products. Beginning from the analysis on clogging mechanism, some improvement measures such as the quality control of the liquid steel and the adjustments on shape and material of submerged entry nozzles were proposed.

Key words: submerged entry nozzle, Al₂O₃, pluggingmechanism, prevention measures

中间包方形稳流器的数值模拟与工业实验

罗衍昭[1]，潘宏伟[1]，季晨曦[1]，邓小旋[1]，曾　智[1]，安泽秋[1]，

李　峰[2]，崔　阳[1]

（1.首钢集团有限公司技术研究院，北京　100043；

2. 首钢京唐钢铁联合有限责任公司制造部，河北唐山　063200）

摘　要： 本文通过 ANSYS Fluent 软件对 80 吨中间包浇铸过程流场、温度场进行数值模拟计算。结果表明：对圆形与方形稳流器，滞止时间由 136s 提高至 147s，平均停留时间由 655s 提高至 663s，死区 5.65%降低至 4.26%，活塞区由 68.4%提高至 74.89%，均可显示有利于夹杂物碰撞上浮去除，提高钢水洁净度。圆形与方形稳流器中间包氧含量分别为 22.8ppm、21.3ppm，Al 损分别为 37.9ppm、41.9ppm，Al 损变化不大。对比方形稳流器与原工艺热轧冷轧轧制情况，热轧带出品率下降为 0.6%，冷轧带出品率下降为 0.55%，可见方形稳流器效果良好。

关键词： 方形稳流器，数值模拟，工业实验，中间包

Mathematical Simulation and Industrial Experiment of Square Turbulence Inhibitor in Tundish

Luo Yanzhao[1], Pan Hongwei[1], Ji Chenxi[1], Deng Xiaoxuan[1], Zeng Zhi[1],

An Zeqiu[1], Li Feng[2], Cui Yang[1]

(1. Shougang Research Institute of Technology, Shougang Group Corporation, Beijing　100043, China;

2. Steel Making Department，Shougang Jingtang United Iron and Steel Co., Ltd., Tangshan　063200,China)

Abstract: In this paper, the flow field and temperature field in the casting process of 80 ton tundish were numerically simulated by ANSYS Fluent software. The results showed that: the round and square turbulence inhibitor, the stagnation time increased from 136s to 147s, the average residence time increased from 655s to 663s, the dead zone percent is from 5.65% to 4.26%, the flow area increased from 68.4% to 74.89%, which are good for the inclusion collision and floating to remove, improve the cleanliness of molten steel. The oxygen content in the tundish of the round and square turbulence inhibitor is 22.8ppm, 21.3ppm, respectively, the Al loss is 37.9ppm, 41.9ppm, and the Al loss varies little. Compared with the hot rolling and cold rolling results with different turbulence inhibitor, the rates reduced 0.6% and 0.55%. The square turbulence inhibitor shows the good results.

Key words: square steady flow device, numerical simulation, industrial experiment, tundish

八钢 Q345D 铸坯中心偏析的研究

陈　刚[1]，丁　寅[1]，郭　鹏[1]，卜志胜[1]，刘　青[2]

（1. 宝钢集团新疆八一钢铁股份有限公司，新疆乌鲁木齐　830022;
2. 北京科技大学钢铁冶金新技术国家重点实验室，北京　100083）

摘　要：本研究针对 Q345D 偏析缺陷严重的连铸坯，通过钻孔取样加成分分析的方法，分析了 Q345D 铸坯中心偏析缺陷。结果表明，Q345D 中心偏析缺陷具体表现为中心负偏析，负偏析元素包括 C、P、S、V、Nb 元素，枝晶间液相流动与凝固速度相反是负偏析产生的必要而非充分条件。

关键词：Q345D，中心偏析，凝固速度，负偏析元素

关于板坯连铸结晶器间隙与锥度偏移的分析处理

吴　伟

（宝武集团新疆八一钢铁股份有限公司，新疆乌鲁木齐　830022）

摘　要：本文结合八一钢铁股份公司板坯连铸机在生产过程中产生的结晶器锥度偏移与结晶器在制造过程中，转配过程中产生的间隙进行关联分析，针对结晶器夹紧装置、驱动系统、调宽装置等进行了深层次的分析。以对制造过程、装配过程中产生的间隙进行处理，解决结晶器生产过程中产生的锥度偏移问题。

关键词：结晶器，锥度偏移，夹紧装置，制造间隙，安装间隙

Analysis and Treatment of Mould Clearance and Taper Offset in Slab Continuous Casting

Wu Wei

(Xinjiang Bayi Iron & Steel of Bao Wu group, Urumchi　830022, China)

Abstract: This combination of Bayi Iron and Steel Co., slab continuous casting machine produced in the production process of mould taper offset and mold in the manufacturing process, correlation analysis creates a gap in the process of assembly, the mold clamping device, driving system, width adjustment device are analyzed deeply. In order to solve the problem of taper offset in the process of mould production, the gap between the manufacturing process and the assembly process is dealt with.

Key words: mould, taper offset, clamping device, manufacturing clearance, installation gap

Q345E+Z35 大圆坯成分偏析及内部质量控制的研究

李 颇，张立明，申祖锋

（东北特钢集团北满特殊钢有限责任公司技术中心，黑龙江齐齐哈尔 161041）

摘 要： 为了改善 Q345E+Z35（ϕ650mm）大圆坯低倍表面质量及了解大圆坯成分的偏析程度对低倍表面质量有何影响，特制定试验计划，通过改变电磁搅拌电流频率和拉速等连铸工艺参数，来探索合理的工艺。本文利用碳偏析检测法，在电磁搅拌电流、拉速和比水量等连铸工艺参数的不同匹配下，研究了不同连铸工艺对 Q345E+Z35 圆坯成分偏析、宏观形貌等方面的影响，理论和实验表明：在拉速、电磁搅拌电流增大的情况下，合理的匹配连铸工艺参数，有助于改善 Q345E+Z35 圆坯成分偏析、减少铸坯质量缺陷以及提高成品锻件质量。

关键词： 电磁搅拌，成分偏析，Q345E+Z35，圆坯

Research on Segregation and Internal Quality Control of Q345E+Z35 Round Billet

Li Po, Zhang Liming, Shen Zufeng

(Dongbei Special Steel Group Beiman Special Steel Co., Ltd., Qigihar 161041, China)

Abstract: In order to improve the Q345E+Z35 (650mm) bloom low times the surface quality and understanding of bloom composition segregation degree of impact on the low surface quality, special test plan by changing the mold cooling water flow and the end of mixing frequency, to explore a reasonable process. In this paper, through the use of carbon segregation detection method in the electromagnetic stirring current, casting speed and casting process parameters than water under different matching, to study the effects of continuous casting process for Q345E+Z35 round billet segregation, macro morphology and other aspects of the theory and experimental results show that the casting speed, the electric magnetic stirring current increases, by reasonable matching parameters of continuous casting process, contribute to the improvement of Q345E+Z35 round billet segregation, reduce quality defects and improve the quality of the finished forging.

Key words: electromagnetic stirring, composition segregation, Q345E+Z35, round billet

改善 55SiMnMo 钢表面与内部质量工艺的研究与应用

李反辉

（河北钢铁集团石钢公司，河北石家庄 050031）

摘 要： 本文针对 55SiMnMo 铸坯因表面存在一些缺陷而被大量判废的情况，本厂技术人员通过改善钢水质量、优

化连铸工艺参数、优化结晶器振动系统，逐步解决了铸坯表面振痕深、中心偏析等缺陷，大大提高了钢的内部质量。

关键词：55SiMnMo，表面，中心偏析，质量

Abstract: This paper introduces the 55 simnmo slab due to there are some surface defects was sentenced to waste a lot of situations.Our factory technical staff by improving the quality of molten steel, optimization, optimization of process parameters of continuous casting crystallizer vibration system, and gradually solve the casting defects such as surface vibration mark depth, center segregation, greatly improving the internal quality of steel.

Key words: 55 SiMnMo, surface, center segregation, the quality

连铸结晶器保护渣非等温结晶动力学的研究

李　欢，王万林，周乐君

（中南大学冶金与环境学院，湖南长沙　410083）

摘　要： 本文采用单丝热电偶技术（SHTT），扫描电镜（SEM）等分析手段对保护渣的结晶机理进行了详细研究。实验结果表明，随着冷却速率的增大，保护渣开始结晶的温度随之降低，其主要原因是由于在较高冷却速率下，保护渣的粘度会随着温度骤降而迅速增大，保护渣形核需要更大的驱动力。通过 Avrami-Ozawa 理论、SHTT 图像分析和对单丝样品进行扫描电镜的分析结果可以准确地判定其结晶机理为常形核速率、三维生长、界面反应控制。同时通过 SEM 和 EDS 图可以看出，随着冷却速率升高，保护渣晶体的粒度越来越小，且可以初步判断其形成的晶相为硅酸钙（Ca_2SiO_4）。

关键词：非等温，动力学，结晶，结晶器保护渣

Non-Isothermal Crystallization Kinetics Study of Mold Flux for Continuous Casting

Li Huan, Wang Wanlin, Zhou Lejun

(School of Metallurgy and Environment in Central South University, Changsha 410083, China)

Abstract: In this paper, the crystallization mechanism of mold flux was investigated detailed by Single Hot Thermocouple Technology (SHTT) and Scanning Electron Microscope (SEM). The results showed that the initial crystallization temperature of mold flux decreased with the increase of cooling rate. The reason for that is because the viscosity of molten slag increases rapidly with the temperature drop, which requires a larger driving force for the nucleation of the mold flux. And the crystallization mechanism of mold flux was accurately determined constant nucleation rate, 3-dimensional growth, and interface reaction control through the Avrami-Ozawa theory, SHTT image and scanning electron microscopy (SEM) of SHTT samples analysis. At the same time, it can be seen from the SEM and EDS diagram that with the increase of the cooling rate, the crystal phase of the mold flux becomes smaller, and it can be judged that the crystalline phase is calcium silicate (Ca_2SiO_4).

Key words: non-isothermal, kinetics, crystallization, mold flux

IF 钢热轧线性缺陷分析及控制实践

秦 伟

（华菱涟源钢铁公司 210 转炉厂，湖南娄底　417000）

摘　要： 本文根据热轧板表面线性缺陷的形貌特征，通过对线性缺陷试样应用扫描电镜分析、能谱分析和金相分析等手段，得出了 IF 钢线性缺陷的产生原因，主要是由连铸坯表面皮下夹渣引起。通过改善钢水洁净度、改进浸入式水口插入深度，规范氩气流量，优化保护渣粘度，IF 钢板坯表层夹渣缺陷显著减少，热轧线性缺陷发生率明显下降。

关键词： IF 钢，线性缺陷，皮下夹渣，连铸

Analysis and Control Practice of Linear Defect in Hot Rolling of IF Steel

Qin Wei

(210 Converter Plant, Lianyuan Steel Corp, Loudi 417000, China)

Abstract: This paper according to the morphological features of linear surface defect of hot rolled plate through the analysis of linear defects samples using scanning electron microscope (SEM), energy spectrum analysis and metallographic analysis, it is concluded that the IF steel of linear defects and causes, mainly by the continuous casting billet surface subcutaneous slag caused. By Improving cleanliness of molten steel，Sen insertion depth, argon flow regulation, optimization of protecting slag, IF steel casting billet surface inclusions significantly reduced, the hot rolling linear defects occur rate decreased significantly.

Key words: IF steel, linear defect, mold casting, continuous casting

异钢种连浇过程生产实践

吕志勇，周刘建，殷东明，于海岐，苏小利

（鞍钢股份有限公司鲅鱼圈钢铁分公司，辽宁营口　115007）

摘　要： 随着市场竞争加剧，客户订单量小的生产计划逐渐增加，异钢种连浇生产越来越多，必须进行专门管理。分析了异钢种连浇时的钢水成分变化规律，确定交接坯长度划定规则。优化异钢种连浇的中间包钢水量、中间包钢水温度、铸机拉速、结晶器保护渣等工艺控制参数，实现异钢种连浇的稳定生产，连铸机单中包连浇罐数由 9.3 罐提高至 9.6 罐。

关键词： 异钢种，连浇，成分，保护渣

Practice of Different Steel Grade Continuous Casting Process

Lv Zhiyong, Zhou Liujian, Yin Dongming, Yu Haiqi, Su Xiaoli

(Bayuquan Iron & Steel Subsidiary Co., of Angang Steel Co., Ltd., Yingkou 115007, China)

Abstract: As market competition intensifies, customer orders for a small amount of production plans gradually increased. Different steel grade casting production is becoming more and more, specialized management must be conducted. Analyzed the different steel grade when casting steel composition change rule, to determine the transition slab length defined rules. Optimization the different steel grade casting tundish weight, the tundish steel temperature, the casting speed, the mold powder, and other process control parameters, to achieve the stable production of different steel grade casting . The number of single tundish casting ladles increased from 9.3 to 9.6.

Key words: different steel grade, continuous casting, component, casting powder

稀土在 30CrMnMo 钢中的应用研究

陈本文 [1,2]，苏春霞 [1,2]，赵 刚 [1,2]，付 超 [1,2]，杨 晰 [1,2]，杨 成 [1,2]

（1. 海洋装备用金属材料及其应用国家重点实验室，辽宁鞍山 114009；
2. 鞍钢股份技术中心，辽宁鞍山 114009）

摘 要： 本文以中碳低合金 30CrMnMo 钢为研究对象，在连铸结晶器中喂入稀土丝，以期最终产品获得良好的力学性能。试验结果表明，在结晶器两侧喂入稀土丝，可使稀土元素分布均匀，且收得率可达 95%以上，同时稀土加入后改变了钢中夹杂物形态，使钢板的横向冲击韧性得到提高，而钢板的拉伸、硬度未得到提高。

关键词： 稀土，夹杂物，稀土丝

Research of Rare-earth Element in 30CrMnMo Steel

Chen Benwen[1,2], Su Chunxia[1,2], Zhao Gang[1,2], Fu Chao[1,2], Yang Xi[1,2], Yang Cheng[1,2]

(1. State Key Laboratory of Metal Material for Marine Equipment and Application, Anshan 114009, China;
2. Angang Steel Company Limited Technology Centre Liaoning, Anshan 114009, China)

Abstract: This article introduce an experimentation that rare-earth thread is put in cast crystallizer liquid steel of 30CrMnMo steel. Final produts expect to gain favorable mechanical property. The results of experimentation proved that the distribute of rare-earth element is well-proportioned in steel, and the percent of rare-earth element is up 95%, and the inclusions of shape are changed in steel. It will improve impact ductility of steel, but it can not improve intensity and hardness of steel.

Key words: rare-earth element, inclusions, rare-earth thread

常见浸入式水口质量缺陷对铸坯质量的影响及应对措施的研究

赵晨光，孙振宇，李叶忠，王华东，王成青

（鞍钢股份炼钢总厂，辽宁鞍山　114000）

摘　要： 浸入式水口是与钢水直接接触的耐材之一，其浇铸状态的稳定与否对钢水的纯净度有着直接影响。通过对生产的跟踪发现，浸入式水口的缺陷种类多样化，随之，不同的缺陷种类对铸坯夹杂的影响也不相同。为了避免造成轧后废品，应该根据浸入式水口的缺陷种类，有针对性的对铸坯制定相应的处理措施。

关键词： 浸入式水口，缺陷，夹杂

The Common Infuence of SEN Defect on the Slab and the Research on Countermeasures

Zhao Chenguang, Sun Zhenyu, Li Yezhong, Wang Huadong, Wang Chengqing

(Ansteel factory, Anshan 114000, China)

Abstract: SEN is one of the refractory contacting with the molten steel directly. The casting state stability of SEN influences molten steel cleanliness directly. By tracking the production process, the diversification defect types of the SEN have been discovered, which lead to different defect types of the casting slabs. In order to avoid rolling waste, treatment measures should be carried out to casting slabs according to different defect types of the SEN.

Key words: submerged nozzle, defect, inclusion

连铸坯新矫直方法探索与研究

张兴中，郭　龙

（燕山大学国家冷轧板带装备及工艺工程技术研究中心，河北秦皇岛　066004）

摘　要： 本文综合考虑了几种连铸坯矫直理论，并利用材料的高温力学性能，提出对连铸坯矫直段进行蠕变矫直的新方法。在 Gleeble-3800 型热/力模拟试验机上对 Q460E 钢进行了热塑性试验和蠕变拉伸试验，确定了该钢种的相关热物性参数，得出蠕变本构方程，确定了该钢种的蠕变应变速率。针对某钢厂 R9300 连铸机提出了一条新型的拥有较低曲率变化率的矫直曲线，通过理论计算应变速率对比蠕变应变速率，新曲线的应变速率比蠕变应变速率低，连铸坯在矫直变形过程中更多依靠蠕变变形，确定了充分利用材料蠕变变形进行弯曲矫直的铸机曲线设计方法，该方法大大降低了铸坯的应变速率，避免矫直裂纹。

关键词： 蠕变矫直，蠕变占比，蠕变拉伸试验，矫直曲线

Research and Discussion of the New Straightening Theory in Continuous Casting of Steel Slabs

Zhang Xingzhong, Guo Long

(National Engineering Research Center for Equipment and Technology of Cold Strip Rolling, Yanshan University, Qinhuangdao 066004, China)

Abstract: A new straightening method called creep straightening by the full use of the high temperature mechanical property of the continuous casting slabs was proposed in this paper. Thermophysical properties parameters of Q460E steel and the creep constitutive equation were determined through the uniaxial tensile test on the Gleeble-3800. The new straightening curve which had lower curvature variation was designed. Based on the new method, high temperature creep property could be used fully in the straightening process of slabs. The straightening strain rate of slabs can be less than creep strain rate so that the slabs can be straightened by creep deformation. It is helpful to avoid the straightening cracks by means of the new straightening method.

Key words: creep straightening, the percentage of creep strain in the total plastic strain, creep tensile test, straightening curve

连铸板坯角横裂的成因及改善

丁占元，冯长宝，胡　娇

（宝钢湛江钢铁有限公司炼钢厂，广东湛江　524000）

摘　要： 本文介绍了宝钢湛江钢铁 2300mm 连铸机主要生产工艺，分析了开工以来该连铸机板坯发生角横裂缺陷的原因，并结合现场生产实践，通过优化结晶器锥度、结晶器振动、二次冷却、弯曲段辊缝参数等一系列措施，使角横裂缺陷的发生率明显减少。

关键词： 角横裂，二次冷却，辊缝收缩，工艺控制

Formation and Improvement of Transverse Corner Cracks in Continuous Cast Slab

Ding Zhanyuan, Feng Changbao, Hu Jiao

(Steel-Making Plant of Baosteel Zhanjiang Iron and Steel Co., Ltd., Zhanjiang 524000, China)

Abstract: This paper introduces Baosteel Zhanjiang Iron and Steel Co.,Ltd 2300mm Continuous Caster main processing and analyses the cause of transverse corner crack from the beginning of the production. A series of measures including optimizing taper of mold, oscillation of mold, secondary cooling water, bending segment roll gap and so on have been taken based on the production practice. Finally, the rate of transverse corner crack of slab has been reduced efficiently.

Key words: transverse corner crack, secondary cooling, roll gap shrinkage, process control

长水口吹氩小气泡在中间包内流动数学模拟研究

钟良才[1]，李晓祥[1]，阳祥富[2]，周小宾[3]，郝培锋[1]

（1. 东北大学冶金学院，辽宁沈阳　110004；2. 上海宝山钢铁股份有限公司炼钢厂，上海　201900；
3. 安徽工业大学冶金工程学院，安徽马鞍山　243032）

摘　要：建立了长水口吹氩形成的小气泡随钢流进入中间包的流动数学模型，研究了不同直径小气泡在不同中间包结构的流动。结果表明，数学模拟得到的小气泡在中间包中的流动行为与实验室观察小气泡在不同中间包结构的流动吻合。在长水口吹氩时，中间包安装湍控器不利于气泡与钢水的相互作用，气泡的作用区域变小，气泡在中间包的停留时间短。在只有堰坝的两流中间包中，直径 0.1mm 的气泡主要分布在两堰之间的区域内，气泡的停留时间长，0.3~1.0mm 直径的气泡在中间包内的分布区域和停留时间比有湍控器的中间包有所改善。

关键词：钢包长水口吹氩，中间包，小气泡流动，气泡作用区域，气泡停留时间，数学模拟

Mathematical Simulation of Flow of Fine Bubbles in a Tundish with Ar Injection through a Shroud

Zhong Liangcai[1], Li Xiaoxiang[1], Yang Xiangfu[2], Zhou Xiaobin[3],
Hao Peifeng[1]

(1. School of Metallurgy, Northeastern University, Shenyang 110004, China; 2. Steelmaking Works,
Baoshan Iron & Steel Co., Shanghai 201900, China; 3. College of Metallurgical Engineering,
Anhui University of Technology, Maanshan 243032, China)

Abstract: A mathematical model was set up for the flow of fine Ar bubbles in a tunsidh with Ar injection through a shroud and the flowing behavior of fine bubbles with different diameters in different configurations of the tundish was investigated. It was shown from the results that the flowing behavior of fine bubbles in the tundish from the mathematical simulation was agreed with that observed in different tundish configurations in laboratory. It was found that a turbulent inhibitor in the tundish with Ar injection through a shroud was harmful to the interaction between bubbles and the liquid steel, the function region of the bubbles became smaller, and the residence time of the bubbles in the tundish was short. In the tundish with two weirs and two dams, the fine bubbles with 0.1 mm diameter mainly distributed in the zone between the two weirs, the residence time of bubbles was long, and the distribution region and residence time of fine bubbles with 0.3~1.0mm diameter in such tundish configuration were improved in comparison with those in the tundish with a turbulence inhibitor.

Key words: Ar injection through ladle shroud, tundish, fine bubble flow, interaction region of bubble, residence time of bubble, mathematical simulation

微合金钢连铸坯表面裂纹控制新工艺开发及应用

蔡兆镇[1]，牛振宇[1]，安家志[1]，朱苗勇[1]，张洪波[2]，刘志远[2]，王重君[2]

（1. 东北大学冶金学院，辽宁沈阳　110819；2. 唐山中厚板材有限公司，河北唐山　063000）

摘　要：弥散化微合金碳氮化物析出并细化连铸坯角部组织晶粒，可提高微合金钢高温热塑性，防止铸坯角部裂纹产生。本文基于含铌钢板坯连铸生产实际，采用数值模拟方法，设计开发了新型内凸型曲面结晶器及 Q345B-Nb 含铌钢板坯连铸二冷高温区角部晶粒超细化控冷新工艺。工业应用表明，实施内凸型曲面结晶器及铸坯二冷高温区多相变控冷新工艺后，铸坯角部距表面 0~20mm 范围内的组织晶粒均可由传统工艺下"奥氏体+先共析铁素体膜"结构转变成"铁素体+珠光体"结构，且晶粒细化至≤20μm，铸坯抗裂纹能力大幅提高，稳定控制含铌钢连铸坯角部裂纹发生。

关键词：含铌钢，连铸板坯，角部裂纹，结晶器，晶粒细化

Development and Application of New Process for Micro-alloyed Steel Slab Surface Crack Control during Continuous Casting

Cai Zhaozhen[1], Niu Zhenyu[1], An Jiazhi[1], Zhu Miaoyong[1], Zhang Hongbo[2], Liu Zhiyuan[2], Wang Chongjun[2]

(1. School of Metallurgy, Northeastern University, Shengyang 110819, China;

2. Tangshan Heavy Plate Co., Ltd., Tangshan 063000, China)

Abstract: Dispersing the precipitation of carbonitride and refining the grains size of slab corner of micro-alloyed steel would improve its hot ductility, and the transverse corner cracks would be prevented. In the present work, a new ICS-Mold and a process of Q345B-Nb steel slab corners grains refinement at high temperature were proposed by mathematical simulation according to the practical Nb micro-alloyed steel continuous casting. The application results show that the slab corners microstructure transfer from the austenite + ferrite film structure to the ferrite + pearlite structure, and the grains were greatly refined to ≤20μm by using the new technology in the range of 0~20mm under the slab corner surface, and the cracking rate of the Nb micro-alloyed steel slab corner could be greatly reduced.

Key words: Nb micro-alloyed steel, slab, transverse corner crack, mold, grain refinement

过热度对 82B 帘线钢连铸方坯碳偏析程度的影响及机理研究

侯自兵，郭东伟，曹江海，常　毅，唐　萍，文光华

（重庆大学材料学院冶金系，重庆　400044）

摘　要：高碳钢连铸坯内碳元素的宏观/半宏观偏析对产品质量有重要影响，但过热度对该类缺陷的影响规律一直存在争议。本论文以某厂 82B 帘线钢连铸方坯（150mm×150mm）为例，通过横纵断面低倍组织、碳元素含量（偏析指数）、偏析率以及枝晶间距等分析过热度对碳偏析程度的影响规律及机理。研究发现：低过热度相比于高过热度，等轴晶率更高，但 V 形偏析夹角更大，中心等轴晶区偏析更大，中心偏析指数更高。因此，虽然低过热度条件下等轴晶率增加，其对中心区域液相的宏观均化效果可能提高，但凝固后期由于抽吸作用流到中心位置的高溶质液相量增加；并且此时由于中心凝固时间增加使得等轴晶区枝晶间距变大，枝晶间液相的流动阻力将减小，二者共同作用最终导致低过热度下铸坯的中心偏析与 V 形偏析变得更为严重。

关键词：偏析，过热度，高碳钢，连铸方坯，帘线钢

Influence Mechanism of Superheat on Carbon Element Segregation in Continuous Casting Billet of 82B Cord Steel

Hou Zibing, Guo Dongwei, Cao Jianghai, Chang Yi,

Tang Ping, Wen Guanghua

(College of Materials Science and Engineering, Chongqing University, Chongqing 400044, China)

Abstract: Macro/Semi-Macro segregation in high carbon steel play an important role in the quality of the product, but the influence mechanism of superheat on this kind of segregation is still a debatable point. In this article, on the basis of transverse and longitudinal macrostructure, carbon element content (segregation index), segregation ratio and grain spacings, the continuous casting billet (150mm×150mm) of 82B Cord Steel in some factory is investigated. It is found that, comparing the high superheat, the higher equiaxed grain ratio, the larger angle of the V-shape segregation, the higher segregation ratio and index are all gained in the billet under low superheat. The equiaxed grain ratio is increased with the decrease of superheat which may induced the macro homogenizing effect more powerfully, whereas the liquid amount of high solute to the centerline is increased in the later period of solidification and the flow resistance in the interdendritic area is decreased for the reason that the grain spacing in the equiaxed grain zone becomes large when increasing the centre solidification time under low superheat. Consequently, as to the studied continuous casting billet of 82B Cord Steel, the centre segregation and the V-shape segregation become severe when decreasing the superheat.

Key words: segregation, superheat, high-carbon steel, continuous casting billet, cord steel

180×180mm 方坯连铸结晶器内流场及温度场模拟研究

汪成义，崔怀周，赵　斌

（钢铁研究总院冶金工艺研究所，北京　100081）

摘　要：进行 180×180mm 方坯连铸结晶器内的流场和温度场数值模拟，计算了铸坯拉速变化、水口插入深度、水口直径大小、浇铸过热度对结晶器内钢液运动和传热的作用情况。为得出合理的结晶器工艺参数匹配关系，分析了各个影响因素对结晶器流场和温度场的显著性影响关系，并最终提出以下参数建议：在铸坯拉速保持不变的情况下，水口插入深度控制在 50~80mm 之间，并按下限进行控制，水口直径选择在 30~35mm 较合适，浇铸过热度应控制在 15~30℃之间。

关键词：180×180mm 方坯，结晶器，流场，温度场，数值模拟

Simulation Study of Flow & Temperature Field in Continuous Casting Mold of 180×180mm Billet

Wang Chengyi, Cui Huaizhou, Zhao bin

(Iron & Steel Research Institute, Beijing 100081, China)

Abstract: In this paper the numerical simulation of flow & temperature field in continuous casting mold of 180×180mm billet was carried out， and the change of casting speed, the depth of nozzle inlet, the diameter of nozzle, the superheat of casting were calculated, through which the movement and heat transfer of molten steel could be understand. In order to obtain a reasonable relationship among various factors of the mold parameters, the influence factors of the mold flow & temperature field were analyzed. Finally, the following parameters were put forward: Under the condition that the casting speed remained unchanged, the depth of the nozzle is controlled between 50mm and 80mm and controlled by the lower limit. The diameter of the nozzle is 30~35mm. The superheat of casting should be controlled between 15℃ and 30℃.

Key words: 180×180mm billet, mold, flow field, temperature field, numerical simulation

双伺服电机同步驱动连铸结晶器非正弦振动装置

张兴中[1]，周 超[1,2]

（1. 燕山大学国家冷轧板带装备及工艺工程技术研究中心，河北秦皇岛 066004；

2. 河北农业大学，河北秦皇岛 066003）

摘 要： 通过改变伺服电机转速实现连铸结晶器非正弦振动波形偏斜率的在线调节，提出一种新型双伺服电机空间双侧同步驱动结晶器的振动装置。首先阐述装置的工作原理,并建立装置的三维模型；其次，分析了正弦与非正弦振动波形的特点，并绘制德马克，整体函数，复合函数非正弦振动的波形曲线，给出实现不同波形偏心轴的角速度的计算方法。该装置的研究对于进一步发挥非正弦振动的特性，提高铸坯质量有一定的参考意义。

关键词： 连铸，结晶器，振动装置，工艺参数

Non-sinusoidal Oscillator of Continuous Casting Mold Driven by Double Servomotors

Zhang Xingzhong[1], Zhou Chao[1,2]

(1. National Engineering Research Center for Equipment and Technology of Cold Strip Rolling, Yanshan University, Qinhuangdao 066004, China;

2. Hebei Agricultural University, Qinhuangdao 066003, China)

Abstract: By control angular speed of servomotor to realize modification ratio of continuous casting mold non-sinusoidal oscillation adjusted online, a new oscillator synchronously driven by two spatial servomotors was proposed. Firstly, the working principle of the oscillator was described. And the three-dimensional model of the system was established. Secondly, the characteristic of sinusoidal and non-sinusoidal oscillation waveform was analyzed. Meanwhile, non-sinusoidal oscillation waveform of DEMAG, entire and composite function were given. And the angular speed calculation method of servomotor was presented. The investigation of oscillator will provide reference for realizing the characteristic of non-sinusoidal oscillation further and improving the quality of slab.

Key words: continuous casting, mold, oscillator, technological parameters

CSP 含铌钢边裂缺陷产生原因分析与控制

巩彦坤，吕德文，席江涛，靖振权，张志克

（河钢集团邯钢公司，河北邯郸　056015）

摘　要：本文通过金相组织检验、扫描电镜能谱分析和 Geleeble3500 高温热塑性研究，结合生产工艺实践，确定了 CSP 含铌钢边裂产生原因是由于钢水成分 N 不稳定，在二冷强冷工艺特点下，铸坯矫直时处于高温脆性区。通过在生产上严格控制含铌钢成分 N 含量，提高拉速，优化连铸二冷工艺与设备，保证铸坯边角部温度矫直时避开高温脆性区间，有效控制了边裂缺陷产生。生产实践表明：CSP 含铌钢边裂由原来的 0.75% 降低到了 0.21% 以下。

关键词：CSP，含铌钢，边裂缺陷，高温热塑性

The Cause Analysis and Control of Edge Crack Defects in CSP Nb Containing Steel

Gong Yankun, Lv Dewen, Xi Jiangtao, Jing Zhenquan, Zhang Zhike

（Handan Iron and Steel Co., Ltd. Corporation, Hebei steel and iron group, Handan 056015, China）

Abstract: In this paper，the reason of the edge crack defects in CSP Nb containing steel is determined through microstructure examination SEM-EDS and high temperature thermoplastic study of Geleeble3500. It is because that the steel composition N is unstable，Strong cold process characteristics，and slab straightening in the high temperature brittle area. It is necessary to strictly control the N content of Nb-containing steel components，improve the drawing speed，optimize the continuous cooling process and equipment to ensure that the billet corner temperature straightening to avoid high temperature brittle interval，effectively control the edge crack defects. Production practice shows that：CSP containing niobium steel edge crack from the original 0.75% to 0.21% below.

Key words: CSP, niobium-containing steel, edge crack defects, high temperature thermoplastic

八钢高建钢连铸冷却配水优化的数值模拟

卜志胜[1]，郭　鹏[1]，丁　寅[1]，陈　刚[1]，刘　青[2]

（1. 宝武集团新疆八一钢铁股份有限公司，新疆乌鲁木齐　830022;

2. 北京科技大学钢铁冶金新技术国家重点实验室，北京　100083）

摘 要：本研究以八钢二炼钢 4 号连铸机为研究对象，针对高建钢连铸生产时出现的中心偏析等缺陷，利用 PROCAST 计算软件建立板坯凝固传热数学模型，模拟其板坯的凝固传热过程，分析板坯连铸过程中铸坯温度场随时间变化的规律，为实际生产工艺控制提供参考，进而对高建钢二冷配水进行优化，确保二冷系统水量的合理化，实现板坯的纵向均匀冷却，提高铸坯质量。

关键词：高建钢，冷却配水，连铸，数值模拟，中心偏析

冷却条件对铸坯三角区及中心线裂纹的影响

吴 军

（宝武集团八钢公司第二炼钢分厂，新疆乌鲁木齐 830022）

摘 要：通过对八钢四台板坯连铸机二次冷却系统的参数调整及技术改造，消除铸坯宽面表面亮度不均匀，中间铸坯亮度低,而造成的中心冷却过强及喷水覆盖面不足以覆盖铸坯的问题。通过加强了辊列精度的控制、改进喷嘴布置改善铸坯宽度方向冷却均匀性、优化二冷温度控制模式及相应的各冷却区水量优化的方法，严格控制凝固终点附近扇形段辊缝和加强二冷区后程冷却并改善沿宽度方向铸坯冷却不均匀程度，使铸坯的裂纹发生率和中心偏析程度得到显著改善。

关键词：连铸板坯，三角区裂纹，控制

淮钢大断面高碳钢圆坯保护渣的优化及应用

车从荣，戈强旺，许光乐，王建宇

（江苏沙钢集团淮钢特钢股份有限公司，江苏淮安 223002）

摘 要：分析了淮钢炼钢厂 3#连铸机大断面高碳钢圆坯表面缺陷产生的原因，通过优化保护渣，改善浇注性能，铸坯表面缺陷得到有效控制。

关键词：渣条，渣沟，裂纹，高黏度，保护渣

Optimization and Application of Mold Powder for High Carbon Steel Big Round Billet at Huaigang

Che Congrong, Ge Qiangwang, Xu Guangle, Wang Jianyu

(Jiangsu Shasteel Group Huaigang Special Steel Co., Ltd., Huaian 223002, China)

Abstract: It analyzes the causes of the surface defects of the high carbon steel round billets in the 3 # round caster of Huaigang steelmaking plant .By optimizing the mold powder and improving the casting performance, the surface defects of the round billets was effectively controlled.

Key words: slag strip, slag runner , crack, high viscosity, mold powder

连铸火焰切割用焦炉煤气管道积渣快速清管装置

曹　军，柴晓慧，刘建民

（宝钢集团新疆八一钢铁有限公司能源中心，新疆乌鲁木齐　830022）

摘　要： 八钢 120t 转炉炼钢和 40t 转炉炼钢火焰切割用的是焦炉煤气，压力 0.4~0.6MPa。120t 用量~1200m³/h，管径 DN250。40t 用量~300m³/h，管径 DN150。切割煤气的正常供应直接关系到炼钢的正常生产，若炼钢生产中断，上游将影响高炉的正常生产，下游将影响轧用户，固切割煤气的正常供应关系到八钢三大主工序炼铁、炼钢、轧钢工序的正常生产。

关键词： 焦炉煤气，火焰切割，管道

Coke Oven Gas Pipeline Slag Accumulation Quick Cleaning Device for Flame Cutting of Continuous Casting

Cao Jun, Chai Xiaohui, Liu Jianmin

(Energy Center of Baosteel Group Xinjiang Bayiiron & Steel Co., Ltd., Urumchi 830022, China)

Abstract: Eight steel 120t converter steelmaking and 40t converter steel casting with coke oven gas, pressure 0.4-0.6MPa. 120t dosage ~ 1200m³ / h, diameter DN250.40t dosage ~ 300m³ / h, diameter DN150. The normal supply of cutting gas is directly related to the normal production of steelmaking. If the steelmaking production is interrupted, the upstream will affect the normal production of the blast furnace, and the downstream will affect the rolling users. The normal supply of solid cutting gas is related to the three main ironworks , Steel, steel rolling process of the normal production.

Key words: coke oven gas, flame cutting, pipeline

45#板的内部质量优化

蔡常青

（福建三钢闽光股份有限责任公司炼钢厂，福建三明　365000）

摘　要： 福建三钢闽光股份有限责任公司采取通过低拉速、低过热度浇铸，恒拉速操作，辅以提高板坯连铸机辊缝精度、优化二冷水强度、调整动态轻压下区间和压下力、中板轧制工艺优化及铸坯与板材堆垛缓冷等手段，提高了45#板坯的内部质量，C 级比例提高到 83.8%，市场上构成质量异议下降到 0.19 次/万吨钢，用户满意度提高。

关键词： 45#板，低拉速，恒拉速，C 级比例，质量异议

Internal Quality Optimization of 45# Plate

Cai Changqing

(Steelmaking plant, Fujian steel Minguang Co., Ltd., Sanming 365000, China)

Abstract: Sansteel MinGuang Co.,Ltd., Fujian applies casting of low pulling rate and low degree of superheat, as well as constant pulling rate operation, which are helpful to the improvement of the accuracy of roll gap of slab caster, the optimization of intensity of second cooling water, the adjustment of dynamic soft reduction interval and reduction force, the optimization of medium plate rolling technology and the implementation of stacking for slow-cooling of casting blank and plate. By doing so, it can achieve improvement in the internal quality of 45# plate with the proportion of products of C-class up to 83.8%. Furthermore, times of objection about quality of steel on the market were declined to 0.19 per 10 thousand metric tons, Increased customer satisfaction.

Key words: 45# Plate, low speed, constant pulling speed, class c scale, quality objection

国内外铸坯表面淬火技术的发展及应用

李永超 [1,2]，王成杰 [1,2]，卢立新 [1,2]

（1. 邢台钢铁有限责任公司技术中心，河北邢台　054027；
2. 河北省线材工程技术研究中心，河北邢台　054027）

摘　要：本文概述了铸坯表面淬火技术的基本工作原理，重点介绍了意大利 Daniel、新日铁、韩国世亚、济钢、邢钢等国内外钢铁企业在铸坯淬火技术方面的发展及生产应用。理论和实践表明，表面淬火技术在减低热装铸坯表面裂纹和降本增效方面起到了显著效果。

关键词：表面淬火技术，热装热送，AlN 析出，发展应用

The Development and Application of Billet Surface Quenching Technology at Home and Abroad

Li Yongchao[1,2], Wang Chengjie[1,2], Lu Lixin[1,2]

(1. Technical Center Xingtai Iron and Steel Co., Ltd., Xingtai 054027, China;
2. Hebei Engineering Research Center for Wire Rod, Xingtai 054027, China)

Abstract: The paper briefly describes the basic principle of surface quenching technology, and the development and application at home and abroad such as Daniel, Nippon Steel & Sumitomo Metal, SeAH Besteel, Jigang and Xingtai steel are introduced. Theories and practice show that the surface quenching technology has played a significant effect in reducing surface crack and cost-saving and profit-increasing.

Key words: surface quenching technology, hot charging hot delivery, AlN precipitation, development and application

钢的初始凝固特性与连铸工艺

汪洪峰，王　勇

（宝钢股份梅山钢铁公司，江苏南京　210039）

摘　要：初始凝固行为和铸坯表面质量有明显的相关性，不规则的初始凝固引起近表面凝固组织粗化，这是连铸坯裂纹产生的主要原因。优化连铸工艺，预先改善钢的初始凝固，形成均匀的初始凝壳，降低钢的裂纹敏感性，才能稳定地改善连铸坯质量。

关键词：凝固，包晶，连铸工艺

Initial Solidification Characteristics and Continuous Casting Process of Steel

Wang Hongfeng, Wang Yong

(Baosteel Meishan Iron & Steel Co., Ltd., Nanjing 210039, China)

Abstract: The initial solidification behavior of steel has a significant correlation with the surface quality of slab. Irregular initial solidification causes near-surface grains to become rough, which is the main reason slab cracks. Optimize the continuous casting process, pre-improve the initial solidification of steel to form a uniform initial shell, reducing the crack sensitivity of steel in order to steadily improve the quality of continuous casting billet.

Key words: initial solidification, peritectic, continuous casting process

齿轮钢连铸坯宏观偏析的研究与控制

高　鹏，陈良勇，高　晗，秦　影

（河钢石钢炼钢厂，河北石家庄　050031）

摘　要：本文通过铸坯钻孔分析实验、射钉试验、原位分析及铸坯低倍酸洗组织统计，对 8620 齿轮钢的电磁搅拌和二冷水进行优化调整后，铸坯 1/4 和中心处的元素偏析均有所改善，铸坯质量得到提升。

关键词：齿轮钢，宏观偏析，电磁搅拌，二冷水

Research and Control of Macrosegregation of Gear Steel Continous Casting Billets

Gao Peng, Chen Liangyong, Gao Han, Qin Ying

(HBIS Group SHI Steel Company, Shijiazhuang 050031, China)

Abstract: In the current paper, the pin-shooting and drilling experiments were done.The original position analysis and the billet macrstructure statistics has been completed.The macroseregation at quarter and central position of the billet reduced after the distribution of water and EMS was adjusted,the inner quality of gear 8620 billets were improved.
Key words: gear steel, macrosegregation, EMS, second cooling water

结晶器电磁搅拌对连铸 GCr15 钢温度场分布的影响

张 静，赵登飞，吴会平

（燕山大学车辆与能源学院能源与动力工程系，河北秦皇岛 066004）

摘 要： 以某钢厂结晶器电磁搅拌作用下连铸 GCr15 钢为研究对象，基于有限体积法建立铸坯三维温度场模型，分析结晶器电磁搅拌对铸坯温度场分布的影响。研究表明，电磁搅拌产生的电磁力使结晶器内钢液传热速度加快，热区位置提高，钢液向壁面传热加快，铸坯内部温度降低，坯壳生长加快，二冷区出口凝固率增加。随着电流强度增大，二冷区铸坯温度降低，坯壳增长速度加快。
关键词： 圆坯，结晶器电磁搅拌，温度场，二冷区，数值模拟

Effect of Mold Electromagnetic Stirring on Temperature Field of GCr15 Casting Steel

Abstract: Based on GCr15 steel billet continuous casting with mold electromagnetic stirring(M-EMS)，the temperature field distribution was studied. The method of finite volume method was used to establish three-dimensional temperature field model. The results shows that the heat transfer rate of molten steel accelerated in the mold by the electromagnetic force, which generated by electromagnetic stirring. The hot spot position improved. The heat transfer from molten steel to wall accelerated. The inner temperature of billet decreased, the growth of shell accelerated, the solidification rate of secondary cold zone outlet increased.With the current intensity increase, the temperature of secondary cold zone decreases, and the shell growth rate accelerates.
Key words: round billet, mold electromagnetic stirring, temperature field, secondary cold zone, numerical simulation

板坯亚包晶钢结晶器保护渣的优化

尹 娜 [1,2]，景财良 [3]，崔园园 [1,2]，李海波 [1,2]，曹 勇 [1,2]

（1. 首钢集团有限公司技术研究院，北京 100043；
2. 北京市能源用钢工程技术研究中心，北京 100043；
3. 中国首钢国际贸易工程公司，北京 100082）

摘 要： 通过实验研究，对某厂现用板坯亚包晶钢结晶器保护渣进行了优化，开发出了一种低粘度、低熔化温度、高熔化速度和高结晶温度的保护渣，并在现场试验。试验表明，在拉速 1.2m/min 下，使用优化渣浇铸，保护渣液渣层厚度增加约 2mm，单耗增加 0.08kg/m²，结晶器内弧面平均热流降低了 0.2MW/m² 左右，外弧面平均热流降低

了 0.15MW/m² 左右；在拉速从 1.2m/min 提高到 1.4m/min 后，纵裂发生率明显降低，铸坯质量保持良好，生产过程稳定，工业试验很成功。

关键词：亚包晶钢，高拉速，优化保护渣，工业试验

Development on Mold Flux for Hypo-peritectic Steel Slab

Yin Na[1,2], Jing Cailiang[3], Cui Yuanyuan[1,2], Li Haibo[1,2], Cao Yong[1,2]

(1. Shougang Research Institute of Technology, Beijing 100043, China;
2. Beijing Engineering Research Center of Energy Steel , Beijing 100043, China;
3. China Shougang International Trade and Engineering Corporation, Beijing 100082, China)

Abstract: A kind of mold flux with low viscosity, low melting temperature, high melting speed and high crystallization temperature was developed, which was used in the plant trial for increasing casting speed of Hypo-peritectic steel slab. Plant test results showed that: when casting speed kept 1.2m/min, the liquid layer was about 2mm thicker, and the consumption of powder increased 0.08kg/m² or so, moreover, the average heat flux at inner and outer side of mold were reduced by 0.2MW/m² and 0.15MW/m² respectively, comparing optimized mold powder with original mold powder. When the casting speed increased from 1.2m/min to 1.4m/min, the slab had its quality and the production was stable. In conclusion, the industrial test for increasing casting speed of Hypo-peritectic steel slab was successfully achieved.

Key words: hypo-peritectic steel, high casting speed, optimized mold flux, industrial trial

ER70S-6-RG 中间包水口结瘤分析

田维波，张新江，宋建国

（日照钢铁控股集团有限公司，山东日照　276806）

摘　要： 日照钢铁生产 ER70S-6-RG 时出现中间包水口结瘤的现象，影响产量，通过荧光分析水口结瘤物的组成成分，结瘤物中的主要是大量熔点相对较高的 MgO·Al₂O₃，少量的 Al₂O₃ 及 MgO 等高熔点夹杂物在中间包上水口表面附着和聚集是导致水口结瘤的主要原因。采取降低钢中 Al 质量分数，加强钢水保护浇注，减少二次氧化，优化脱氧工艺及 LF 精炼工艺，可有效控制产生的 MgO-Al₂O₃ 夹杂物。

关键词：ER70S-6-RG，结瘤，非金属夹杂

Cause Analysis and Countermeasure of Nozzle Clogging Inalloy Welding Wire Steel Casting

Tian Weibo, Zhang Xinjiang, Song Jianguo

(Rizhao Iron ＆ Steel (Group) Co., Ltd., Rizhao 276806, China)

Abstract: The clogging occurs in nozzle in the process of ER70S-6 alloy welding wire steel casting by the Rizhao steel Limited Liability Company, which decrease production efficiency. The clogging sample was analyzed by fluorescence analysis, the calcium aluminates in clogging were mainly relative high melting point MgO·Al₂O₃. The high melting point

inclusions such as large quantity not floating magnesium aluminates, Al_2O_3 spinel attached and gathered on the surface of tundish upper nozzle were the main reason of nozzle clogging.The MgO- Al_2O_3 inclusions produced in deoxidization can be removed by adopting the following measures such as to decrease aluminum mass fraction in steel, to strengthen molten steel protection casting, to decrease secondary oxidation, to optimize deoxidization process and Ladle Furnace refine process.
Key words: ER70S-6RG, gas shielded welding wire, clogging, non-metallic inclusion

异型坯表面裂纹成因分析及控制措施

贾敬伟，李宗一，杨鹏辉

（日照钢铁控股集团有限公司，山东日照　276806）

摘　要：根据连铸坯凝固特性、异型坯应力分析以及铸坯裂纹产生的机理，结合异型坯断面特性分析其表面裂纹形成原因，通过生产工艺试验，制定相关措施控制异型坯表面裂纹。
关键词：异型坯，裂纹，控制

Cause Analysis and Control Measures of Surface Crack of Beam Blank

Jia Jingwei, Li Zongyi, Yang Penghui

(Rizhao Steel Holding Group Co., Ltd., Rizhao 276806, China)

Abstract: According to the continuous casting billet solidification characteristics,beam blank stress analysis and casting billet crack mechanism,combined with beam blank characteristics and the analysis of the forming reason of surface crack,through production trials, formulate relevant measures to control the beam lank surface crack.
Key words: beam blank, cracking, control

管线钢钢板探伤不合原因分析

于赋志[1]，许孟春[1]，王鲁毅[2]，李海峰[2]，方恩俊[2]，李德军[1]

（1. 海洋装备用金属材料及其应用国家重点实验室，辽宁鞍山　114009；
2. 鞍钢股份鲅鱼圈钢铁分公司，辽宁营口　115007）

摘　要：针对某厂管线钢钢板超声波探伤不合的问题，采用金相检验、扫描电镜等分析手段，对钢板探伤不合部位取样进行检测与分析，发现组织中有中心裂纹和夹杂物缺陷。分析结果表明，连铸时尾坯中间包覆盖剂卷入是造成管线钢板探伤不合格的主要原因。通过采取必要措施稳定中间包液面，避免中间包覆盖剂卷入，优化尾坯切除长度，可有效提高管线钢板探伤合格率。
关键词：管线钢，超声波探伤，夹杂物，裂纹，保护渣

An Analysis on Cause of Disqualification for Flaw Detection of Pipeline Steel Plates

Yu Fuzhi[1], Xu Mengchun[1], Wang Luyi[2], Li Haifeng[2], Fang Enjun[2], Li Dejun[1]

(1. State Key Laboratory of Metal Material for Marine Equipment and Application, Anshan 114009, China;
2. Bayuquan Iron & Steel Subsidiary Company of Angang Steel Co., Ltd., Yingkou 115007, China)

Abstract: In order to investigate the causes of disqualification plates in ultrasonic flaw detection on pipeline steel plates, metallographic examination and SEM analysis were carried out on the samples from the disqualification plates by ultrasonic flaw detection. There were center cracks and inclusions in the microstructure of disqualification plates. The results showed that reason of the disqualification plates in ultrasonic flaw detection was that the covering agent involved in tail slabs. The pass rate of flaw detection for pipeline steel plates can be effectively improved by taking the necessary measures to stabilize the liquid level of tundish, avoid the covering agent involved in tail slabs and optimize the cutting length of tail slabs.
Key words: pipeline steel, ultrasonic flaw detection, inclusion, crack, covering agent

表面控制冷却改善特厚板坯表面组织研究

关春阳[1]，赵新宇[2,3]，孙　宇[1]

（1. 首秦金属材料有限公司制造部，河北秦皇岛　066326；2. 首钢技术研究院宽厚板所，
北京　100043；3. 北京市能源用钢工程技术研究中心，北京　100043）

摘　要： 表面控制冷却是一种新型的铸坯二冷控制技术，其是通过快速冷却-内部返温-正常冷却三步来实现的，通过再回温的过程控制析出物的弥散，从而通过细化晶粒、消除脆性组织来提高铸坯表面质量的效果，本文依托400mm 特厚铸坯进行表面控制冷却的工业试验，研究表面控制冷却的效果。试验得到如下结果：表面控制冷却通过控制铁素体和析出物在晶内的随机析出可以有效地消除膜状先共析铁素体在晶界位置的过量析出，细化晶粒，均匀组织，从而提高铸坯表面的高温热塑性，减少表面横裂纹的发生率。依托400mm 特厚板坯进行表面控制冷却试验，结果表明铸坯内弧中心的膜状先共析铁素体得到了有效的消除。
关键词： 特厚板坯，表面组织控制冷却，先共析铁素体，表面质量

Research on Improvement of Surface Quality of Ultra-thick Slab through Surface Structure Controlling

Guan Chunyang[1], Zhao Xinyu[2,3], SunYu[1]

(1. Shouqin Metal Material Co., Ltd., Qinhuangdao 066326, China; 2. Department of Plate technology, Shougang Research Institute of Technology, Beijing 100043, China;
3. Beijing Engineering Research Center of Energy Steel, Beijing 100043, China)

Abstract: Surface Structure Controlling (SSC) cooling is a novel pattern of second cooling, which is achieved through three steps of quick cooling-reheating-normal cooling. Then the surface quality of slab will be improved by grains refined and brittle structure decreased. A 400mm slab was tested by SSC cooling to research on improvement of surface quality of ultra-thick slab. The following results were drawn: pro-eutectoid ferrite was controlled and precipitation was precipitated randomly by SSC cooling, which will decrease the over-precipitating along grain boundary of film-like pro-eutectoid ferrite. The hot ductility of surface of slab was improved through grain refining and structure homogenizing, which decreased the rate of transverse cracking. The SSC cooling test was conducted to 400mm ultra-thick slab. And the results show that the pro-eutectoid ferrite on inner surface of slab was restrained through SSC cooling technology.

Key words: ultra-thick slab, SSC, pro-eutectoid ferrite, surface quality

高碳钢小方坯碳偏析优化研究

崔怀周[1]，赵　斌[1]，汪成义[1]，丁秀中[2]

（1. 钢铁研究总院冶金工艺研究所，北京　100081；

2. 山东寿光巨能特钢有限公司，山东寿光　262711）

摘　要：本文通过连铸机拉速、比水量以及末端电磁搅拌参数的优化，对高碳钢小方坯碳偏析进行了研究。结果表明：相同二冷水比水量下，拉速 2.3m/min 的铸坯碳偏析指数均高于拉速 2.0m/min。对于 150×150mm 小方坯，在 0.8~1.8L/kg 范围内，随着二冷水比水量的增加，铸坯的碳偏析指数呈下降趋势。冷却强度过大，中心缩孔等级升高，且会出现中心裂纹。末端电磁搅拌能够显著降低高碳钢小方坯的碳中心偏析指数。

关键词：高碳钢，小方坯，碳偏析，强冷，末端电磁搅拌

Study on Carbon Segregation Optimization of High Carbon Steel Billets

Cui Huaizhou[1], Zhao Bin[1], Wang Chengyi[1], Ding Xiuzhong[2]

(1. Central Iron & Steel Research Institute, Beijing 100081, China;

2. Shandong Shouguang Juneng Special Steel Co., Ltd., Shouguang 262711, China)

Abstract: In this paper, the carbon segregation of high carbon steel billets was studied by optimizing the casting speed, specific water quantity and electromagnetic stirring parameters at the end of the continuous casting machine. The results show that the carbon segregation index of the billet under the same cold water ratio of two and the pulling speed 2.3m/min is higher than that of the casting speed 2.0m/min. For the 150×150mm billets, in the range of 0.8~1.8L/kg, the carbon segregation index of the billet decreases with the increase of secondary cooling water ratio. When the billets excessive cooling strength，the central shrinkage grade increases, and the center cracks appear. The end electromagnetic stirring can significantly reduce the carbon center segregation index of high carbon steel billets.

Key words: high carbon steel, small billet, carbon segregation, strong cold, end electromagnetic stirring

L245 钢连铸板坯角部横裂纹的成因与控制措施

王亚栋[1]，张立峰[1]，张海杰[1]，李源源[2]，杨剑洪[2]，司海逢[2]

（1. 北京科技大学冶金与生态工程学院，北京　100083；

2. 柳州钢铁股份有限公司转炉炼钢厂，广西柳州　545002）

摘　要：针对 L245 管线钢连铸生产存在的角部横裂纹缺陷，本文进行了裂纹缺陷金相分析与 SEM 观察、铸坯高温力学性能测试与二冷模拟和现场优化试验等研究。金相检测发现裂纹沿振痕波谷开裂，未发现脱碳层；SEM 观察到裂纹内的钢基体被氧化，未发现保护渣成分和可能来自结晶器的磨损元素。模拟结果显示，原二冷配水条件下，弯曲段和矫直段铸坯角部温度均处于脆性温度区间，是铸坯角部横裂纹形成的主要原因；通过模拟计算优化了连铸坯的二冷配水，在新的配水条件下，铸坯弯曲段和矫直段角部温度均避开脆性温度区间。现场优化二冷配水后，铸坯角部横裂纹明显改善。

关键词：角部横裂纹，高温力学性能，二冷模拟，连铸板坯

Study on Forming and Controlling Transverse Corner Cracks for L245 Continuous Casting Slabs

Wang Yadong[1], Zhang Lifeng[1], Zhang Haijie[1], Li Yuanyuan[2], Yang Jianhong[2], Si Haifeng[2]

(1. School of Metallurgical and Ecological Engineering, University of Science and Technology Beijing, Beijing 100083, China; 2. Converter Steelmaking Plant, Liuzhou Iron & Steel Company, Ltd., Liuzhou 545002, China)

Abstract: To investigate the transverse corner cracks for L245 continuous casting slabs, the samples were studied by metallography and the cracks were analyzed by SEM. The high temperature mechanical properties were investigated. Besides, the simulation of second cooling and optimation experiments were conducted. It was found that cracks were generated on the valley of scillation marks without the decarburized layer by metallography. It was oxidized inside of cracks. There were no mold powder and wear elements of the mold detected. The slab corner temperature in the bending and the straightening zone was falling into the brittle zone in the original water cooling condition, leading to the formation of cracks. The secondary cooling scheme was optimized to prevent the corner temperature in the bending and straightening zone from entering the brittle zone, significantly reducing the occurrence of transverse corner cracks.

Key words: transverse corner cracks, high temperature mechanical properties, secondary cooling simulation, continuous casting slabs

韶钢连铸方坯脱方缺陷的成因及其改善研究

付谦惠，郭峻宇

（宝钢特钢韶关有限公司生产技术室，广东韶关　512123）

摘　要： 韶钢8号方坯连铸机自2011年投产以来，铸坯质量一直受脱方困扰，供轧钢厂连铸方坯脱方超标率最高达到1.74%。分析表明，浇铸过程中结晶器冷却的不均匀是连铸方坯产生脱方缺陷的直接原因。通过分析并调整转炉冶炼和中间包冶金工艺、结晶器参数、浇注温度、二冷配水、铸坯拉速等，强化铸坯对弧和喷水的精确度，使韶钢铸坯脱方率小于0.01%，整体铸坯合格率达到99.99%以上。

关键词： 连铸坯，结晶器，脱方，缺陷

Research on Cause Analysis of Off-square Defection and Improvement Method of Continuous Casting Billet in Shaogang Iron and Steel

Fu Qianhui，Guo Junyu

(Bao Steel Special Steel Shaoguan Co., Ltd. Production Engineering Office，Shaoguan 512123, China)

Abstract: Off-square Defection of Continuous Casting Billet has been a big problem since using the No.8 caster in 2011. The deformation rate sometimes will reach to 1.74%. Analysis results, the non-uniform cooling of mould during casting process is the direct cause of the off-square defect of continuous casting billet. According to the analysis results, off-square Defection rate will reach to less than 0.01% and the pass rate of Continuous Casting Billet will reach to more than 99.99% after adjusting the device configuration and optimizing the process of the tundish metallurgy, mould parameters, liquid steel temperature, secondary cooling water, casting speed, strengthen the billet on the arc and the accuracy of water.

Key words: continuous casting billet, mould, off-square, defection

结晶器振动频率对铜模与坯壳之间传热现象的影响

龙旭凯，　王万林，　张海辉

（中南大学冶金与环境学院，湖南长沙　410083）

摘　要： 本文中，采用mold simulator研究结晶器振动频率对铜模与坯壳之间传热现象的影响。首先，进行四组不同结晶器振动频率下的高温连铸模拟拉坯对比实验，再结合2DIHCP，1DITPS and heat transfer model这三个数学模型详细分析铜模结晶器表面温度及热流、初始凝固坯壳表面温度和铜模与坯壳间热阻。实验结果表明：随着结晶器振动频率的增高，结晶器表面温度和热流升高，坯壳表面温度降低；渗入铜模与坯壳间缝隙的渣膜厚度由于负滑脱时间的降低而变薄；振动频率的增高，也导致了铜模与坯壳间总热阻R_{tot}、渣膜热阻R_{slag}和铜模与渣膜间界面热阻R_{int}的减小。

关键词： mold simulator，振动频率，传热，渣膜，热阻

Influence of Oscillation Frequency on Heat Transfer Phenomenon during the Initial Solidification in Continuous Casting Mold

Long Xukai, Wang Wanlin, Zhang Haihui

(School of Metallurgy and Environment in Central South University, Changsha　410083, China)

Abstract: Heat transfer in continuous casting mold is important to decide the surface quality of the slab. In this paper, mold simulator was used to study the influence of oscillation frequency on heat transfer phenomenon between mold and shell during continuous casting. After four group of different parameters experiment was carried out, and through2DIHCP, 1DITPS and heat transfer model, the detailed information about temperatures and heat flux of mold surface, temperatures of solidified shell surface and thermal resistance between mold and shell were obtained. The experiment result shows that: the mold surface temperatures and heat flux increase with oscillation frequency. And the average thickness of infiltrated slag film increases with the NST increasing. When the mold oscillation frequency increases, the average of total mold/shell thermal resistance Rtot, slag thermal resistance Rslag and interfacial thermal resistance Rint all decrease. Moreover, the increase of oscillation frequency results in the thinning of air gap between mold and slag film.

Key words: mold simulator, oscillation frequency, heat transfer, mold flux film, thermal resistance

新型熔滴凝固技术对薄带连铸中界面传热及沉积膜的研究

朱晨阳，王万林，路　程

（中南大学冶金与环境学院，湖南长沙　410083）

摘　要： 在薄带连铸生产过程中，钢液中部分合金元素容易挥发，导致钢液在水冷铜辊上凝固过程伴随着膜的自然沉积现象。在凝固坯壳与结晶辊之间的沉积膜显著影响界面传热。本研究中，采用新型熔滴凝固技术，研究薄带连铸中钢液初始凝固、传热和自然沉积膜。通过热传导反问题算法对高灵敏度热电偶采集的温度数据进行计算，得到界面热流密度。研究结果表明：峰值热流密度随着实验次数的增加先增加后减小。这种自然沉积膜主要由多种氧化物构成。该氧化膜改善了润湿条件，增加了实际接触面积，导致热流密度增加。但随着沉积膜达到一定厚度，氧化物本身具有较大热阻抑制了传热。说明薄带连铸中自然沉积膜对界面传热有很大影响。

关键词： 熔滴凝固技术，界面传热，自然沉积膜，薄带连铸

An Investigation on the Interfacial Heat Transfer and Deposited Films in Strip Casting through Droplet Solidification Technique

Zhu Chenyang, Wang Wanlin, Lu Cheng

(School of Metallurgy and Environment, Central South University, Changsha 410083, China)

Abstract: In the process of strip casting, liquid steel solidified after being sprayed directly onto the mold, which tends to form the oxides films that naturally deposited on the surface of the mold due to the evaporation of the alloying elements. The films between solidified steel and mold would greatly influence the heat transfer from molten steel pool to the mold wall. In this study, a novel droplet solidification technique has been developed to study the process of solidification and films deposition in strip casting. The responding temperatures and heat fluxes across the mold surface were calculated by the Inverse Heat Conduction Program (IHCP) mathematical model. The results suggested that the maximum value of heat flux increased first and then decreased with the numbers of experiments being conducted. This particular film composition was detected as a variety of oxides. The oxides film facilitates the micro wetting phenomenon, increases the true surface area, which leads to the enhancement of the heat flux. When the thickness of films increased a lot, the thermal resistance would get higher, and inhibit the heat flux. This demonstrates that the formation of naturally deposited films contributes to the heat transfer behavior in strip casting.

Key words: droplet solidification technique, interfacial heat transfer, naturally deposited films, strip casting

TiN 对 430 铁素体不锈钢微观组织影响研究

秦国清，李　阳，姜周华，杜鹏飞，刘　浩，胡　浩

（东北大学冶金学院，辽宁沈阳　110819）

摘　要： 通过热力学计算确定了 TiN 在 430 铁素体不锈钢中的析出行为，计算得出本实验条件下 430 铁素体不锈钢凝固过程中生成 TiN 颗粒都是由于外加 TiN 微米级粉末。在实验室条件下，通过热态模拟实验冶炼 5 炉次不同 TiN 加入量的 430 铁素体不锈钢，TiN 的加入量分别为 0，0.01%，0.02%，0.03%，0.05%。通过金相显微镜观测发现，随晶粒细化剂加入量的增加，淬火状态下，钢中针状铁素体的数量增加，当加入量增至 0.03% 时，针状铁素体数量最多，继续增加加入量时，无明显变化。通过 SEM 观测发现 TiN 的形状是不规则带有棱角的，尺寸在 6~8μm。

关键词： TiN，铁素体不锈钢，微观组织，晶粒

The Study on the Influence of TiN to Ferrite Stainless Steel Microstructure

Qin Guoqing, Li Yang, Jiang Zhouhua, Du Pengfei,
Liu Hao, Hu Hao

（School of metallurgy, Northeastern University, Shenyang 110819, China）

Abstract: The sedimentary condition of TiN was determined by the thermodynamic calculation in the first stage of solidification of 430 ferrite stainless steel. The micron grade powder of TiN particles in the ferrite stainless steel are generated as a result of plus. In the laboratory condition, the TiN were added in five furnaces respectively, which were 0, 0.01%, 0.02%, 0.03%, 0.05%. Through metallographic microscope observation, it found the number of acicular ferrite in steel increased with the increase of the amount of grain refining agent in the quenching condition. When the addition amounted to 0.03%, the quantity of acicular ferrite was most, and the TiN were continued to increase, there was no obvious change. SEM observed that the TiN is a shape of irregular and angularity with a size of approximately 6~8μm.

Key words: TiN, ferrite stainless steel, microstructure, grain

经济视角下的转炉炼钢技术探讨

何宇明

(重庆钢铁股份有限公司一炼钢厂，重庆　401258)

摘　要： 在 2015 年全行业亏损的情况下生产，将成本控制作为第一要务，强化成本管理，2016 年行业实现扭亏，取得了一定的效果，但国内同类厂家之间相比，存在诸多差距，本文以经济的角度对转炉炼钢流程所采用的技术进行系统解析和讨论，提出一些目前可以实施的降本方法和今后成本控制的改进方向。

关键词： 经济，低成本，炼钢技术，铁素流，能量流

Discussion on Converter Steelmaking Technology from Economic Perspective

He Yuming

(Chongqing Iron and Steel Co., Ltd., The First Steelmaking Plant, Chongqing 401258, China)

Abstract: Production in 2015 industry wide loss situation, will cost control as the first priority, strengthen cost management, in 2016 the industry realized losses, and achieved certain results, but there are many domestic manufacturers compared gap, based on the economic perspective on the process of steelmaking techniques used in the system analysis and discussion, put forward some the implementation of this method can be reduced and the cost control in the future improvement direction.

Key words: economy, low cost, steelmaking technology, iron element flow, energy flow

高级别管线钢探伤合格率低原因分析及措施

苏小利，于海岐，王金辉，邢维义，吕志勇，方恩俊，殷东明

（鞍钢股份鲅鱼圈钢铁分公司炼钢部，辽宁营口　115007）

摘　要： 本文对探伤不合格的高级别管线钢板进行检验，认为高级别管线钢探伤合格率低的主要原因为钢板内含有 Al_2O_3 内生夹杂物和 MgO、Na_2O、K_2O 等中间包内衬和保护渣等外来夹杂物。通过分析，认为 LF 炉脱硫率（即顶渣改质程度），钢水镇静时间，喂线工位，铸机的中间包钢水过热度，不自浇率，保护渣性能波动，连浇罐数对探伤合格率有影响。并通过对以上工艺参数进行适当调整，探伤合格率由 70%～80%提高到了 97.36%。

关键词： 高级别管线钢，探伤，夹杂物，喂线

Analysis and Improvement Measures of the Ultrasonic Flaw Detection Low Qualification Rate for High Grade Pipeline Steel

Su Xiaoli, Yu Haiqi, Wang Jinhui, Xing Weiyi, Lv Zhiyong,

Fang Enjun, Yin Dongming

(Ansteel Bayuquan Iron and Steel Subsidiary Steelmaking Department, Yingkou 115007, China)

Abstract: In this paper, the high grade pipeline steel plates which be inspected having flaw by ultrasonic detection are observed under scanning electron microscopy (sem) and the OXFORD spectrometer component analysis. The result of analysis thinks that the main reasons of high grade pipeline steel ultrasonic flaw detection low qualified rate are containing Al_2O_3 endogenous inclusions in steel and MgO, Na_2O and K_2O tundish lining and powder slag inclusions from the outside world. Through the analysis, the paper thinks the LF furnace desulfurization rate (i.e., the top slag modification degree), the motel steel calm time, feed CaSi station, the motel steel overheat, ladle casting failing rate, the performance volatility of powder slag, continue casting ladles quantity have influence on the ultrasonic flaw detection qualified rate. At the same time, by means of improving the above process parameters, the ultrasonic flaw detection qualified rate increased to 97.36% by 70%～80%.

Key words: high grade pipeline steel, ultrasonic flaw detection, inclusion, feed SiCa wire

J55 石油套管钢生产过程纯净度控制

吾　塔[1]，张爱梅[1]，李立民[2]，孙学刚[2]，卜志胜[2]

（1. 宝钢集团新疆八一钢铁有限公司制造管理部，新疆乌鲁木齐　830022；

2. 八钢股份公司炼钢厂，新疆乌鲁木齐　830022）

摘　要：对 J55 石油套管钢从精炼开始到轧材整个生产过程中夹杂物的行为进行了研究。采用电镜分析手段，分析了精炼过程、连铸坯及热轧卷的洁净度的变化。试验结果表明：现有的生产工艺在稳态浇注时钢水的洁净度满足产品质量要求，LF 精炼后钢中夹杂物的含量明显降低，钢中没有发现大于 20μm 的夹杂。

关键词：J55 石油套管用钢，纯净钢，夹杂物

Production Process Control of Cleanliness in J55 for Oil Casing Steel

Wu Ta[1], Zhang Aimei[1], Li Limin[2], Sun Xuegang[1], Bu Zhisheng[2]

(1. Manufacturing Management Department, Bayi Iron & Steel Co., Baosteel Group, Urumchi 830022, China;

2. Xinjiang Bayi Iron & Steel Stock Co., Ltd., Urumchi 830022, China)

Abstract: The J55 steel casing to the inclusion of rolling material in the whole production process were investigated by electron microscopy. The refining analysis method, analysis of the refining process, continuous casting and hot rolling coil cleanliness changes. The experimental results show that the existing production process in the steady state when pouring the cleanliness of molten steel to meet the requirements of product quality, content after LF refining of inclusion in steel decreased obviously, the inclusion of more than 20μm was not found in steel.

Key words: J55 for oil casing steel, clean steel, inclusion

X90 管线钢冶炼工艺研究

李战军[1,2,3]，刘金刚[1]，初仁生[1]，郝　宁[1]，王东柱[4]

（1. 首钢技术研究院，北京　100043；2. 绿色可循环钢铁流程北京市重点实验室，北京　100043；

3. 北京市能源用钢工程技术研究中心，北京　100043；

4. 首秦金属材料有限公司，河北秦皇岛　066326）

摘　要：本文研究了采用"铁水脱硫预处理—转炉—LF 炉精炼—RH 炉精炼—连铸"生产 X90 管线钢的冶炼工艺，实现了高洁净度和良好内部质量的 X90 管线钢连铸坯的工业化生产。采用此工艺生产的 X90 管线钢成品成分控制水平达到[C]≤0.070%，[P]≤100ppm，[S]≤20ppm，[N]≤40ppm，[H]≤2.0ppm；钢中 T[O]≤13ppm，铸坯中的非金属夹杂物以球形或类球形的高熔点钙铝酸盐类夹杂物为主，尺寸控制≤15μm；连铸坯中心偏析≤C 类 1.0 级。

关键词：X90 管线钢，冶炼工艺，成分控制，非金属夹杂物，中心偏析

Research on Smelting Process in X90 Pipeline Steel

Li Zhanjun[1,2,3], Liu Jingang[1], Chu Rensheng[1], Hao Ning[1], Wang Dongzhu[1]

(1. Shougang Research Institute of Technical, Beijing 100043, China;

2. Beijing Key Laboratory of Green Recyclable Process for Iron & Steel Production Technology, Beijing 100043, China;

3. Beijing Engineering Research Center of Energy Steel, Beijing 100043, China;

4. Shouqin Metal Material Company Ltd., Qinhuangdao 066326, China)

Abstract: In this paper, the smelting process of "Hot metal desulphurization pretreatment-Basic oxygen furnace-LF refining-RH refining-continuous casting slab" is studied to produce X90 pipeline steel, meet the needs of X90 pipeline steel mass production. By this process of X90 pipeline steel finished product component control is [C]$\leqslant$0.070%, [P]$\leqslant$100ppm, [S]$\leqslant$20ppm, [N]$\leqslant$40ppm, [H]$\leqslant$2.0ppm; T[O]$\leqslant$13ppm; non-metallic inclusions in slab is the high melting point of spherical or similar spherical calcium aluminate salts inclusions is given priority to, size control $\leqslant$15μm; The center segregation of continuous casting billet is controlled blow C1.0.

Key words: X90 pipeline steel, smelting process, composition control, non-metallic inclusions, center segregation

底吹氩冶金系统中气泡形成过程数值模拟

黎　俊，张美杰，顾华志，黄　奥

（武汉科技大学耐火材料与冶金省部共建国家重点实验室，湖北武汉　430081）

摘　要： 微小气泡去除钢液中夹杂物技术在气泡冶金技术中占重要地位，为了在底吹氩冶金系统中形成小气泡群，本文针对底吹氩冶金过程中气泡所处环境，对气泡进行受力分析，并利用界面追踪模拟方法研究气泡形成过程的影响因素。结果表明：在中间包底吹氩气气幕挡墙中，气泡形成大小主要受到气流速率、气孔孔径及间距、钢液速率和接触角的影响。气流速率较小时，气泡开始沿水平方向拉长分离，随气流速率增大，气泡开始沿竖直方向拉长分离，气流速率越大气泡生成频率越大，气泡当量直径增大。气泡随气孔孔径增大而增大，气孔间距对气泡合并的影响较大，间距越小越容易合并。气泡随着钢液对透气耐火材料润湿角增大而增大，润湿角接近 90 度气泡形成过程中容易分裂。

关键词： 底吹氩，气泡形成，数值模拟，气泡大小

Numerical Simulation of Bubble Formation on Bottom with Argon Blowing

Li Jun, Zhang Meijie, Gu Huazhi, Huang Ao

(The State Key Laboratory of Refractories and Metallurgy, Wuhan University of Science and Technology, Wuhan 430081, China)

Abstract: Using micro-bubble to remove inclusions in molten steel is playing an important role in the bubble metallurgy technology. In order to form a small bubble group in the bottom argon argon metallurgical system, the force of bubbles is analyzed , and the influencing factors of bubble formation process are studied by interface tracking simulation method . The results show: the size of the bubble formation is mainly affected by the airflow rate, pore size and spacing, the molten steel rate and the contact angle in the argon gas curtain wall.When the airflow rate is small, the bubbles begin to elongate along the horizontal direction. As the air velocity increases, the bubbles begin to be separated in the vertical direction. The larger the airflow rate, the larger the bubble generation frequency and the bubble radius increases.Bubble increases with the increase of pore diameter, and the effect of stomatal spacing on bubble merging is larger, and the smaller the spacing is, the easier it is to merge.The bubbles increase as the wetting angle increases, and the wetting angle is closer to 90 degrees the bubble is easier to divide.

Key words: bottom blowing argon, bubble formation, numerical simulation, bubble size.

低碳汽车钢中非金属夹杂物的研究

龚 伟，万 万，庞 昇，郎凯旋，李 涵，张 楠

（东北大学冶金学院，辽宁沈阳 110819）

摘 要：本文采用大样电解与金相法相结合的方式，利用金相显微镜、扫描电镜分析低碳汽车钢中的大型夹杂物与显微夹杂物的尺寸、形貌、分布以及成分等。结果表明：铸坯宽度 1/4 处的非金属夹杂物总量高于中心和边部；铸坯中的大型夹杂物主要有 Al_2O_3 夹杂、SiO_2 夹杂、硅铝酸盐、SiO_2-TiO_2 夹杂、CaO-MgO 夹杂，其主要来源于脱氧产物、结晶器保护渣和耐火材料的侵蚀等；铸坯中的显微夹杂显微夹杂物尺寸主要为 0~3μm，随着夹杂物粒径的增加，其所占比重逐渐减小；显微夹杂物主要有单一的 Al_2O_3 夹杂、TiN 夹杂、Al_2O_3-TiN 复合夹杂以及少量的 Al_2O_3-TiO_x 复合夹杂。

关键词：低碳汽车钢，大型夹杂物，显微夹杂物，大样电解

Study on Non-metallic Inclusions in Low Carbon Automobile Steel

Gong Wei, Wan Wan, Pang Sheng, Lang Kaixuan, Li Han, Zhang Nan

(School of Metallurgy, Northeastern University, Shenyang 110819, China)

Abstract: In this paper, the size, morphology, distribution and composition of large inclusions and micro inclusions in low carbon automobile steels were analyzed by metallographic microscope and scanning electron microscopy using the method of large sample electrolysis and metallographic method. The results show that the non-metallic inclusions at the width of the slab are higher than the center and the edge. The large inclusions in the slab are mainly Al_2O_3 inclusions, SiO_2 inclusions, aluminosilicate, SiO_2-TiO_2 inclusions, CaO-MgO mixed, which mainly comes from deoxidation products, mold powder, refractory erosion. The size of the micro-inclusions in the slab is $0 \sim 3\mu m$, and the proportion of the inclusions is decreasing with the increase of the particle size. The micro-inclusions mainly have a single Al_2O_3 inclusions, TiN inclusions, Al_2O_3-TiN composite inclusions and a small amount of Al_2O_3-TiO_x composite inclusions.

Key words: low carbon automobile steel, large inclusions, microscopic inclusions, large sample electrolysis

添加稀土处理钢的炼钢生产工艺研究

智建国 [1]，吴章忠 [1]，栾义坤 [2]，陆　斌 [1]，宋　海 [1]，刘宏伟 [2]

（1. 包钢（集团）公司，内蒙古包头　014010；2. 中国科学院金属研究所，辽宁沈阳　110016）

摘　要： 在 LD-LF-VD-CC 和 LD-LF-RH-CC 生产工艺流程,采用纯稀土合金投入法进行稀土钢的炼钢生产技术研究。研究结果表明：采用投入纯稀土合金方法，能够生产低氧高洁净度的稀土重轨钢、稀土 Q345E、稀土 BT700。与 LD-LF-VD-CC 生产流程比，LD-LF-RH-CC 生产流程更易获得较高稀土含量的稀土钢；与常规钢相比，稀土重轨钢中的夹杂物分布具有如下特点，<5μm 的小尺寸夹杂略多，而>10μm 的大尺寸夹杂将变少，有助于提升材料性能。风电板中加入稀土所有厚度规格产品的夹杂物总数量均减少。

关键词： 稀土钢，炼钢–连铸工艺，纯稀土合金，投入法

Study on Steelmaking Process of Adding RE Treated Steel

Zhi Jianguo[1], Wu Zhangzhong[1], Luan Yikun[2], Lu Bin[1], Song Hai[1], Liu Hongwei[2]

(1. Baotou Iron & Steel (Group) Co., Ltd., Baotou 014010, China;

2. Institute of Metal Research, CAS, Shenyang 110016, China)

Abstract: In the production process of LD-LF-VD-CC and LD-LF-RH-CC, the steelmaking process of RE(rare earth) steel was studied by pure RE alloy putting in method. The results show that RE heavy rail steel with low oxygen content and high clean can be produced by using pure RE alloy, similarly, RE Q345E and RE BT700. Compared with the LD-LF-VD-CC production process, the LD-LF-RH-CC production process is more likely to obtain the more RE content of RE steel. Compared with the conventional steel, the inclusion distribution in RE railsteel has the following characteristics: less than 5 micron, small size inclusions, slightly larger inclusions than 10 micron will be helpful to improve material properties. The total number of inclusions of all the thickness specifications of the products in the wind plate has been reduced.

Key words: RE（rare earth）steel, steelmaking and continuous casting process, pure RE alloy, putting in method

低合金高强钢生产过程氮含量的控制

张　东，薛文辉

（本钢板材股份有限公司炼钢厂，辽宁本溪　117000）

摘　要：本文介绍了本钢炼钢厂在开发低合金钢高强的生产过程中，为降低生产成本并减少对 RH 的设备资源的占用，采用 LD—LF—CC 工艺路径替代 LD—LF—RH—CC 工艺路径。通过对转炉冶炼及 LF 处理过程的工艺改进，在没有降氮措施的情况下有效控制各环节的增氮量.生产实践表明，可以满足低合金钢高强钢的成品氮含量要求。

关键词：低合金高强钢，工艺改进，氮含量

Control on Nitrogen Content during Production of HSLA Steel

Zhang Dong, Xue Wenhui

(Steel-making Plant of Bengang Steel Plates Co., Ltd., Benxi 117000, China)

Abstract: This paper introduces that Bengang steel plant adopts the LD—LF—CC process path instead of LD—LF—RH—CC process path during the production of HSLC steel in order to reduce production cost and occupancy of RH equipment resources. By process improving for BOF melting and LF treatment, the N-pick up of every procedure can be controlled effectively in the absence of nitrogen reduction measures. The production practise shows that the Ni-content requirement of final HSLC steel can be met.

Key words: HSLC steel, process improvement, Ni-content

直接还原铁与电弧炉炼钢的关联性综述

唐　恩, 李森蓉, 李建涛, 周　强, 汪　朋, 陈泉锋

（武汉科思瑞迪科技有限公司，湖北武汉　430223）

摘　要：本文主要基于对直接还原铁(DRI)的品质及电弧炉炼钢生产操作的工艺制度出发，较为全面的分析探讨了电弧炉采用直接还原铁炼钢的影响，指出了全铁含量高、金属化率高、有害元素少、粒度和密度适中的直接还原铁是电弧炉炼钢的优质原料，将对电弧炉炼钢的各项指标有积极的作用，并且，在电弧炉炼钢生产过程中大比例使用直接还原铁已较为成熟。

关键词：直接还原铁，海绵铁，电弧炉，炼钢

Relevance Study between Direct Reduction Iron and EAF Steelmaking

Tang En, Li Senrong, Li Jiantao, Zhou Qiang, Wang Peng, Chen Quanfeng

(Wuhan COSRED Science and Technology Co.,Ltd., Wuhan 430223, China)

Abstract: It was fully analyzed and discussed the impacts of direct reduction iron to electric arc furnace(EAF) steelmaking considering the quality of DRI as well as the process operation of EAF. It was pointed the good quality material of DRI should have higher total ferrous, higher metallization rate, lower other metal elements, the right particle and reasonable density which will positively push the steelmaking process of EAF. Moreover, currently it indicated it is common and mature to use large percentage of DRI in EAF.

Key words: direct reduction iron, sponge iron, electric arc furnace, steel making

精炼双渣冶炼高品质弹簧钢 55SiCrA 盘条

陈志亮，黄永生，孙光涛，仝太钦

（中天钢铁集团公司技术中心，江苏常州　213000）

摘　要： 精炼 LF 炉采用双渣法冶炼高品质汽车用弹簧钢 55SiCrA 盘条，使其夹杂物大小和数量及塑性均达到较高控制水平。精炼前期和中期使用高活性石灰造高碱度还原白渣，精炼白渣碱度为 2.0~2.4，精炼成分温度控制合适后精炼后期使用高纯石英砂造低碱度终渣，精炼终渣碱度为 0.75~1.0。采用电子扫描电镜检测分析，经精炼 LF 炉双渣冶炼后，非金属夹杂位于 CaO-SiO$_2$-Al$_2$O$_3$ 相图低熔点理想区域。对轧材非金属夹杂评级，各类夹杂评级基本位于 0.5 级和 0 级，夹杂物大小在 0~5μm 范围内，轧材氧含量控制在 12ppm 以内，轧材力学性能抗拉强度为 1000~1080MPa，平均面缩可以达到 50 以上，对其进行疲劳寿命检测，试验 20 万次未断裂。该精炼 LF 炉双渣法使得弹簧钢 55SiCrA 纯净度和夹杂物塑性大幅提高，为其适应更严苛环境奠定良好基础。

关键词： 55SiCrA，双渣，非金属夹杂，力学性能，疲劳寿命

Refining Double Slag Smelting High Quality Spring Steel 55SiCrA Wire Rod

Chen Zhiliang, Huang Yongsheng, Sun Guangtao, Tong Taiqin

(Technical Center Department of ZENITH Iron and Steel Group, Changzhou 213000, China)

Abstract: Refining LF furnace adopts double slag method to smelt high quality spring steel 55SiCrA wire rod for automobile. The size, quantity and plasticity of the inclusions reached a high level of control. In the early and middle stage of refining, high active lime is used to produce high basicity and reduce white slag. The basicity of refined white slag is 2.0~2.4. After refining in the late refining stage, the high temperature quartz sand is used to produce low basicity slag. The basicity of refining final slag is 0.75~1.0. By means of SEM analysis, the non-metallic inclusion is located in the ideal zone of low melting point of CaO-SiO$_2$-Al$_2$O$_3$ phase in refining LF furnace. Rating of rolled materials, non-metallic inclusions, all kinds of inclusion ratings are basically at grade 0.5 and 0. The inclusion size in 0~5 μm range. The oxygen content of rolled products is less than 12ppm. The mechanical properties and tensile strength of the rolled material are 1000~1080MPa, and the average surface shrinkage can reach more than 50. The fatigue life test was carried out, and the test was not broken for 200 thousand times. The double slag method of the refining LF furnace makes the 55SiCrA purity and inclusion plasticity of the spring steel substantially increase, and lays a good foundation for adapting to more stringent environment.

Key words: 55SiCrA, double slag, nonmetallic inclusion, mechanical property, fatigue life

KR 脱硫法脱硫剂经济用量计算模型研究

闫小柏[1,2]，赵晓东[3]，张立国[3]，刘丹妹[1,2]，杨伟强[1,2]，

李亚宁[1,2]，孙　丹[1,2]，廖　慧[1,2]

（1. 北京首钢自动化信息技术有限公司自动化研究所，北京　100041；

2. 混合流程工业自动化系统及装备技术国家重点实验室首钢分实验室，北京　100041；

3. 首钢股份公司迁安钢铁公司炼钢部，河北迁安　064404）

摘　要：KR 脱硫工艺生产的主要成本包括脱硫剂消耗、电耗、设备损耗及人工成本。在不对设备进行改造和操作员人员的大规模调整的情况下，利用现有的技术手段，采用经济的脱硫剂用量是降低 KR 脱硫生产成本的重要手段。本研究在总结分析现有算法利弊的基础上，开发了一种基于参考炉次动态库的 KR 脱硫法脱硫剂经济用量计算模型。该计算模型通过对大量实际历史炉次分析，获得满足当前铁水条件冶炼的最经济脱硫剂用量。通过该模型，可以为降低脱硫剂消耗提供有效参考。

关键词：脱硫剂，历史炉次，经济用量，模型

The Study of Economical Desulphurizer Consumption Computational Model Utilized on KR Desulfurization Process

Yan Xiaobai[1,2], Zhao Xiaodong[3], Zhang Liguo[3], Liu Danmei[1,2],

Yang Weiqiang[1,2], Li Yaning[1,2], Sun Dan[1,2], Liao Hui[1,2]

(1. Institute of Automation, Beijing Shougang Automation & Information Technology Co., Ltd.,
Beijing 100041, China;

2. Shougang Branch of State Key Laboratory of Hybrid Process Industry Automation System
and Equipment Technology, Beijing 100041, China;

3. Steelmaking Department, Shougang Qian'an Steel Company, Qian'an 064404, China)

Abstract: The main cost of KR desulfurization process includes desulfurizer consumption, power consumption, equipment depreciation and labor cost. In the case of no equipment alternation and large scale of labor adjustment, the use of economic desulfurizer is an important means to reduce the production cost of KR desulfurization process. In this paper, the advantages and disadvantages of the existing algorithms are analyzed. An economical desulphurizer consumption computational model utilized on KR desulfurization process is developed。The model is based on a large number of dynamically historical production data. Using this model, a most economical desulphurizer value is obtained which satisfying the demand of desulfurization of current hot metal. The model can provide an effective reference for reducing the consumption of KR desulfurization process.

Key words: desulphurizer, historical production data, economical consumption, model

复合旋转喷吹铁水脱硫预处理工艺研究与应用

徐延浩，徐向阳，马　勇，高学中，潘瑞宝

（鞍钢股份有限公司炼钢总厂，辽宁鞍山　114021）

摘　要： 本文对原有复合喷吹铁水预处理脱硫喷枪进行了工艺研究改进，并实现了复合旋转喷吹。摸索出了复合旋转喷吹铁水脱硫预处理工艺制度。该应用实践表明，采用复合旋转喷吹铁水脱硫预处理工艺，与原有复合喷吹铁水预处理脱硫工艺相比，可以进一步降低脱硫粉剂吨铁消耗约 30%。

关键词： 铁水脱硫预处理，复合喷吹，旋转喷枪

Research and Practice on Rotary Co-injection in Hot Metal Desulphurization

Xu Yanhao, Xu Xiangyang, Ma Yong, Gao Xuezhong, Pan Ruibao

(General Steelmaking Plant of Angang Steel Co., Ltd., Anshan 114021, China)

Abstract: Research and improvement for lance of co-injection in hot metal desulphurization are made. A new rotary co-injection method of hot metal desulphurization becomes true. The technological regulation of rotary co-injection method of hot metal desulphurization has been studied. Practice showed that desulphurization powder consumption had been reduced about 30% per ton of iron compared with that of co-injection in hot metal desulphurization

Key words: hot metal desulphurization, co-injection, rotary lance

鞍钢 KR 高效脱硫工艺技术

乔冠男，何海龙，李伟东，刘鹏飞，方　敏

（鞍钢股份有限公司炼钢总厂，辽宁鞍山　114021）

摘　要： 鞍钢股份炼钢总厂三分厂 5# 线于 2015 年 1 月开工，其中铁水预处理采取 KR 脱硫工艺。新工艺上线，在生产过程中存在脱硫周期长、粉剂耗量高、扒渣大等问题。本文介绍了围绕提高 KR 脱硫效率展开的各项工艺实践，包括搅拌桨及扒渣板改型、搅拌桨浸深调整、吹气赶渣工艺优化等相关工艺技术。

关键词： 搅拌桨，扒渣，吹气赶渣，脱硫周期

Qiao Guannan, He Hailong, Li Weidong, Liu Pengfei, Fang Min

(General Steelmaking Plant of Ansteel Co., Ltd., Anshan 114021, China)

Abstract: The fifth production line of General Steelmaking Plant of Ansteel Co., Ltd., have start working on Jan. 2015, KR desulfurization's technology and equipment were used.But some problems like longer desulfurization cycle,higher desulfurizer consumption and loss of slagging-off were found in practice. This paper introduces some optimization measure in practice,include improvement of stirring paddle,adjustment of paddle immersion depth and optimization of air blowing.

Key words: stirring paddle, slagging-off, air blowing, desulfurization cycle

KR 法与钙镁复合喷吹法在铁水脱硫中应用对比研究

何海龙，方　敏，刘鹏飞，曹　祥，宋吉锁，吴跃鹏，曹　琳，姜　丰

（鞍钢股份炼钢总厂，辽宁鞍山　114021）

摘　要：通过对设备、脱硫效果、温降、铁损、脱硫剂、运行成本等因素及对脱硫流程的影响分析，对喷吹法和 KR 机械搅拌法两种铁水脱硫方法进行了全面的比较，结果表明，KR 机械搅拌法脱硫消耗成本、扒渣铁损、冶炼回流率等明显低于钙镁复合喷吹脱硫。对于大中型钢铁企业，KR 法脱硫预处理的总体优势较为突出。

关键词：铁水预处理，KR 法，喷吹法

Comparison of Application of KR Method with That of Injection Method in Hot Metal Desulphurization

He Hailong, Fang Min, Liu Pengfei, Cao Xiang, Song Jisuo,

Wu Yuepeng, Cao Lin, Jiang Feng

(General SteelmakingPlant of Ansteel Co., Ltd., Anshan 114021, China)

Abstract: Two different hot metal desulphurization methods, namely the injection method and KR mechanical stirring method have been compared comprehensively with each other in the aspects of technical equipments, desulphurization effect, temperature drop, iron loss, desulphurization agent, running cost and effect on the operating flow. Results show that for large and medium iron and steel enterprises the KR hot metal desulphurization method is generally superior over the other method.

Key words: hot metal treatment, KR, powder injection

100t 转炉氧枪喷头参数优化实践

梁祥远，王　兴

（鞍钢股份有限公司炼钢总厂，辽宁鞍山　114021）

摘　要：针对鞍钢股份有限公司炼钢总厂二分厂百吨转炉冶炼供氧时间长、生产每吨钢消耗氧量过大、钢铁料消耗高等问题，对原氧枪喷头参数进行优化设计，通过比较原氧枪喷头参数和优化后氧枪喷头参数 400 炉的转炉冶炼数据，转炉冶炼供氧时间缩短 100.8s，氧气的消耗量减少 1.81 m^3/t，终点磷控制能力增强，氧枪使用寿命提高。

关键词：氧枪喷头，供氧时间，脱磷

Parameter Optimization Design of Lance Head for 100t Converter

Liang Xiangyuan, Wang Xing

(General Steelmaking Plant of Angang Steel Co., Ltd., Anshan 114021, China)

Abstract: In order to optimize the design of the original oxygen lance nozzle parameters and compare the parameters of the original oxygen lance nozzle and optimize the oxygen after a long time, the production of oxygen per ton of steel is too high and the consumption of iron and steel is high in No.2 Subsidiary Plant of General Steelmaking Plant of Angang Steel Co., Ltd. Gun nozzle parameters of 400 furnace converter smelting data. The oxygen consumption of converter smelting is reduced by 100.8 s, the consumption of oxygen is reduced by 1.81 m³/t, the end point phosphorus control ability is enhanced, and the service life of oxygen lance is improved.

Key words: oxygen lance nozzle, oxygen time, dephosphorization

铝镇静中锰钢非金属夹杂物的演变行为

孔令种，邓志银，朱苗勇

(东北大学冶金学院，辽宁沈阳　110819)

摘　要： 通过在中锰钢精炼过程的不同工位取样分析，研究了中锰钢中非金属夹杂物的演变行为。研究结果表明，精炼过程中，钢中共发现七种不同类型的夹杂物，其中(Mn, Mg)O·Al₂O₃夹杂物是 MgO·Al₂O₃夹杂物与钢液中 Mn反应的产物；当钢液中有 Ca 存在的情况下，(Mn, Mg)O·Al₂O₃夹杂物并不不稳定，会向铝酸钙夹杂物转变。中锰钢中非金属夹杂物的转变路径为：Al₂O₃→MgO·Al₂O₃→(Mn, Mg)O·Al₂O₃→CaO-MnO-MgO-Al₂O₃。

关键词： 精炼，中锰钢，非金属夹杂物，MgO·Al₂O₃夹杂物，(Mn, Mg)O·Al₂O₃夹杂物

Evolution Behaviors of Non-Metallic Inclusions in Al-Killed Medium Manganese Steel

Kong Lingzhong, Deng Zhiyin, Zhu Miaoyong

(School of Metallurgy, Northeastern University, Shenyang 110819, China)

Abstract: The evolution behaviors of non-metallic inclusions in medium manganese steel are investigated by the means of sampling at different stages during secondary refining process. Seven types of inclusions are found in the steel samples, in which (Mn, Mg)O·Al₂O₃ inclusions are formed by the reaction between Mn in steel and MgO·Al₂O₃ inclusions. When a trace of Ca is presented in steel, (Mn, Mg)O·Al₂O₃ inclusions are not stable, and would transform into calcium aluminates. The evolution of inclusions in medium steel follows along the route of: "Al₂O₃→MgO·Al₂O₃→(Mn, Mg)O·Al₂O₃→CaO-MnO-MgO-Al₂O₃".

Key words: secondary refining, medium manganese steel, non-metallic inclusions, MgO·Al₂O₃ inclusions, (Mn, Mg)O·Al₂O₃ inclusions

一体式浸渍管深熔池 RH 新型高效真空槽

刘向东[1]，行开新[2]，王　翔[2]

（1. 中冶赛迪技术研究中心有限公司，重庆　401122；2. 中冶赛迪装备有限公司，重庆　400013）

摘　要：一体式浸渍管深熔池 RH 高效真空槽在扩大环流管内径的同时，增加了真空槽底部熔池的有效容积，通过提高钢液循环流量同时延长钢液在真空室内停留时间来提高 RH 真空槽的处理效率。工程应用证明，一体式浸渍管深熔池新型真空槽能够满足 RH 真空精炼处理的要求，对于 RH 高效生产具有普遍意义。

关键词：一体式浸渍管，深熔池真空槽，高效制造技术

Vacuum Tank of Integrated Immersion Tube

Liu Xiangdong[1], Xing Kaixin[2], Wang Xiang[2]

(1. CISDI R&D Co., Ltd., Chongqing 401122, China；

2. CISDI Engineering Co., Ltd., Chongqing 400013, China)

Abstract: A high-efficiency RH vacuum tank with intergrated immersion tube and deep molten bath, which increases the inner diameter of immersion tube and the effective volume of molten bath, can improve the efficiency of RH vacuum treatment by increasing the circulation rate of liquid steel and extending the stay time of liquid steel in the tank during treatment. It has been proved by engineering application that high-efficiency RH vacuum tank with intergrated immersion tube and deep molten bath meets the requirements of RH vacuum treatment well and has universal significance for efficient RH treatment.

Key words: integrated immersion tube, vacuum tank with deep molten bath, efficient manufacturing technology

不同形态的夹杂物在钢–渣界面处的运动行为研究

周业连，邓志银，朱苗勇

（东北大学冶金学院，辽宁沈阳　110819）

摘　要：为了加深理解固态夹杂物更易被顶渣吸收的现象，本文分别对固态夹杂物和液态夹杂物在钢渣界面处的运动行为进行了物理模拟。结果表明：物理模拟实验选择的夹杂物模拟物的尺寸远大于实际夹杂物的尺寸。这导致了本实验仅能定性描述非金属夹杂物穿过钢–渣界面的运动过程。钢–夹杂物接触角是液膜形成的原因。由于固态夹杂物–钢液的接触角大于 90° 不被钢液润湿，其穿过钢–渣界面过程时无液膜形成。由于液态夹杂物–钢液的接触角小于 90° 被钢液润湿，在分离过程中，液态夹杂物和界面之间形成一液膜。固态夹杂物一接触钢–渣界面就被吸收去除。然而，液态夹杂物将停留在钢–渣界面处。这是固态夹杂物更易去除的原因。

关键词：非金属夹杂物，形态，钢–渣界面，分离过程，液膜

Study on Motion Behavior of Inclusions with Different State at the Steel-slag Interface

Zhou Yelian, Deng Zhiyin, Zhu Miaoyong

(School of Metallurgy, Northeastern University, Shenyang 110819, China)

Key words: non-metallic inclusion, state, steel-slag interface, separation process, liquid film

超低碳冷轧搪瓷钢化学成分对第二相影响的热力学分析

邵肖静，刘再旺，黄学启，邓小旋，吕利鸽，季晨曦，张志敏，崔　阳

（首钢集团有限公司技术研究院，绿色可循环钢铁流程北京市重点实验室，北京　100043）

摘　要： 目前超低碳冷轧搪瓷钢大多采用在钢中加入 Ti，形成大量的 TiN、$Ti_4C_2S_2$、TiS 和 TiC 第二相作为不可逆氢陷阱来提高超低碳搪瓷钢的抗鳞爆性能。本文采用 thermo-calc 软件计算了 Ti、C、N 含量变化对第二相生成温度和生成量的影响规律。在热力学计算基础上开发了首钢超低碳冷轧搪瓷钢，并对钢中夹杂物和析出相进行的观察和抗鳞爆性能检测。

关键词： 搪瓷钢，鳞爆，第二相

Thermodynamic Analysis of the Influence of Chemical Composition on the Second Phase of Ultra-low Carbon Cold-rolled Enameled Steel

Shao Xiaojing, Liu Zaiwang, Huang Xueqi, Deng Xiaoxuan, Lv Lige, Ji Chenxi, Zhang Zhimin, Cui Yang

(Shougang Research Institute of Technology, Beijing Key Laboratory of Green Recyclable Process for Iron & Steel Production Technology, Beijing 100043, China)

Abstract: At present, ultra-low carbon cold-rolled enameled steel is mainly used in the steel by adding Ti, forming a large number of TiN, $Ti_4C_2S_2$ and TiC second phase as irreversible hydrogen trap to improve ultra-low carbon enamel steel anti-scouring performance. In this paper, the influence of Ti, C, N content on the formation temperature and formation amount of the second phase was calculated by thermo-calc software. On the basis of thermodynamic calculation, Shougang ultra low carbon cold rolled enamel steel was developed, and the observation of inclusions and precipitates in steel was carried out.

Key words: enameled steel, fish scale, second phase

C70S6 开发生产实践

车从荣，蒋栋初

（江苏沙钢集团淮钢特钢股份有限公司，江苏淮安 223002）

摘　要：C70S6 作为汽车胀断连杆用非调质钢，其机械性能和稳定性要求非常高。淮钢通过采用初炼炉合理配渣，特殊的脱氧制度，合适的合金化、脱硫、增硫、增氮与连铸工艺，保证了其合金成分、连铸坯偏析满足控制要求，开发出力学性能满足使用要求的 C70S6 涨断连杆用钢。

关键词：C70S6，造渣，合金化，弱冷

C70S6 Development and Production Practice

Che Congrong, Jiang Dongchu

(Jiangsu Shasteel Group Huaigang Special Steel Co., Ltd., Huaian 223002, China)

Abstract: As hot rolled high strength steel for autocar splitting connecting rod, it is very important to the mechanical property. Stable alloying component could be gotten by technology for building of slag、deoxidation、alloying、desulphurizing. How to increase sulful content and nitrogen content were studied in the paper. It reduced the cosegregation degree of sulful and nitrogen to reduce intensity of cooling and use suitable electromagnetic stirring. So, the mechanical property of C70S6 met the challenge.

Key words: C70S6，building of slag，alloying，reduce intensity of cooling

钢液内液滴的聚合现象

倪　冰[1]，张　涛[2]，罗志国[2]

（1. 钢铁研究总院冶金工艺研究所，北京　100081；2. 东北大学冶金学院，辽宁沈阳　110004）

摘　要：通过观察凝固后的试验钢样中夹杂物的分布形态，推测了钢液中液滴聚合行为。利用水模实验分析了液滴夹杂物聚合的机理，液滴聚合的机理是缓慢聚合。提出了韦伯数 We 与碰撞参数 B 之间的函数关系式作为液滴缓慢聚合与分离的标准。以钢液中喂入碳酸盐线的操作为例，采用计算流体力学（CFD）和离散单元法（DEM）耦合的数学模型，对底吹氩钢包内不同初始直径的液滴在连续相中的运动行为和碰撞聚合行为进行了模拟，数值模拟结果和试验钢样中夹杂物的聚合现象是一致的。

关键词：计算流体力学，多相流，钢液，液滴，碰撞，聚合

Phenomenon of Droplets Coalescence in Molten Steel

Ni Bing[1], Zhang Tao[2], Luo Zhiguo[2]

(1. Central Iron and Steel Research Institute, Department of Metallurgical Technology, Beijing 100081, China;
2. School of Materials Science and Metallurgy, Northeastern University, Shenyang 110004, China)

Abstract: The behavior of droplets coalescence in the liquid steel was investigated carefully by observing the distribution patterns of the inclusions in the solidified steel sample. This article analyzed the mechanism of droplets coalescence that is slow coalescence in the water model. The function relationship between Weber number (*We*) and space collision parameter (*B*) is recommended as a standard to judge the process coalescence or separation of droplet collision. In carbonate wire operation fed into the liquid steel, for example, a mathematical model using CFD and DEM method was developed to simulate the movement behavior and the coalescence process of different size droplets in continuous phase in the bottom blowing argon ladle. The numerical simulation results and the coalescence phenomenon of inclusions in steel samples are consistent.

Key words: CFD, multiphase flow, liquid steel, droplet, collision, coalescence

铁酸钙球团在转炉铁水脱磷预处理中的应用

吴　伟[1]，王　鹏[1]，刘　跃[2]

（1. 钢铁研究总院工艺所，北京　100081；2. 武汉市海易通特种材料有限公司，湖北武汉　430082）

摘　要：为了满足转炉炼钢化渣和铁水脱磷的需求，在实验室条件下研究开发了铁酸钙球团。采用的方法是以不同配比的石灰粉（筛下物）和铁的氧化物为原料试制铁酸钙球团，在电阻炉测定铁酸钙球团的熔化性能以此作为配方的依据。测试结果表明，配方 C 具有较好的熔化性，由此制作铁酸钙球团，铁酸钙与石灰的配料比例为(1.5~2):1。在 120t 转炉中把铁酸钙球团作为脱磷剂用于双渣冶炼中的转炉前期脱磷工艺并与原工艺相比较，结果表明，与原工艺相比，加铁酸钙球团的炉次转炉前期脱磷率稳定在 50~60%，磷的分配比平均为 55.9，均好于原工艺的脱磷效果；炉渣中 T.Fe 含量和炉渣碱度较稳定，炉渣岩相中低熔点物质较多，流动性好。同时加铁酸钙球团的炉次石灰的加入量也有所降低。

关键词：铁酸钙球团，脱磷剂，铁水脱磷预处理，转炉前期

The Application of Calcium Ferrite Pellet in the Hot Metal Dephosphorization Pretreatment of BOF

Wu Wei[1], Wang Peng[1], Liu Yue[2]

(1. Metallurgical Technology Research Department of Central Iron & Steel Research Institute, Beijing 100081, China;
2. Hai Yi Tong Special Material Co., Ltd., Wuhan 430082, China)

Abstract: In order to meet the demands of slag formation without fluorite and hot metal dephosphorization pretreatment, the calcium ferrite pellet is developed in laboratory. The lime and iron oxides combining in different proportion are used as raw material of the calcium ferrite pellet. The melting properties of the calcium ferrite pellet is determined in resistance furnace as basis for the optimum recipe. The results of test show that the recipe C has better melting, which is used as composition of the calcium ferrite pellet. The melting characteristic of the proper proportion for the calcium ferrite and lime being (1.5~2):1 is the best. The calcium ferrite pellet is used as the dephosphorizing agent in the hot metal dephosphorization pretreatment during the earlier stage of smelting process of 120t converter. The experimental results of the hot metal dephosphorization pretreatment between two processes are compared. The results show that the dephosphorization rate reaches to 50 ~60% steadily and the phosphorus distribution ratio is averagely 55.9, which is both higher than that without the calcium ferrite pellet. The slag basicity and total iron content in slag are more stable and have little fluctuation by the slag formation of the calcium ferrite pellet than that without the calcium ferrite pellet. According to the analysis results of slag petrographic there are many the materials of low melting point and better fluidity in slag by the slag formation of calcium ferrite pellet than that without the calcium ferrite pellet.

Key words: the calcium ferrite, the dephosphorizing agents, hot metal dephosphorization pretreatment, the earlier stage of converter process

转炉冶炼含钛铁水成渣机理研究

王海宝[1,2,3]，吕延春[1]，秦丽晔[1]，刘　洋[1]，马长文[1]，孔祥涛[1]

（1. 首钢集团有限公司技术研究院，北京　100043；2. 绿色可循环钢铁流程北京市重点实验室，
北京　100043；3. 北京市能源用钢工程技术研究中心，北京　100043）

摘　要：含钛铁水的使用可以扩大含铁原料来源，降低成本，但是冶炼过程会出现炉渣泡沫化严重、脱磷效果差以及石灰消耗高等问题，技术人员通过试验，改进了操作工艺，解决了高效、平稳冶炼含钛铁水的难题。利用扫描电镜和能谱仪对冶炼过程炉渣进行矿相组成分析，发现了钛元素在炉渣中的存赋状态，磷元素的存赋规律，并提出了防止转炉后期产生回磷的措施。

关键词：复吹转炉，含钛铁水，炉渣，矿相组成

Study on Slag Forming Mechanism of Hot Metal Containing Titanium in Converter

Wang Haibao[1,2,3], Lv Yanchun[1], Qin Liye[1], Liu Yang[1], Ma Changwen[1],
Kong Xiangtao[1]

(1. Shougang Research Institute of Technology, Beijing 100043, China; 2. Beijing Key Laboratory of Green Recyclable Process for Iron & Steel Production Technology, Beijing 100043, China;
3. Beijing Engineering Research Center of Energy Steel, Beijing 100043, China)

Abstract: The use of titanium containing molten iron can expand the sources of raw materials, reduce the cost, but the process produce serious foaming slag, low dephosphorization proportion and the consumption of lime is high. The technicians have improved the operation process through experiments and solved the problem of efficient and smooth blowing of hot metal containing titanium. Through mine phase composition analysis of slag using SEM and EDS, the distribution of titanium and the regularity of phosphorus retention are found, and measures to prevent phosphorus recovery in the later stage of converter are put forward.

Key words: BOF, molten iron containing titanium, slag, mine phase composition

北营炼钢厂 120t LF 炉造渣工艺研究及应用

富　强，郭晓春，姚志龙

（本钢集团北营公司炼钢厂，辽宁本溪　117017）

摘　要： 为解决 LF 冶炼钢水可浇性差，浇注水口结瘤以及连浇炉次少等问题，北营炼钢厂通过优化 LF 炉渣组分、降低熔点、提高流动性等方式，降低了钢水夹杂物含量，铸坯质量明显提高，提高了钢水纯净度以及连浇炉数，有效的降低了生产成本。

关键词： 精炼渣，低熔点，流动性，夹杂物

North Camp 120 tons of Steel Slag Formation of LF Furnace Research and Application

Fu Qiang, Guo Xiaochun, Yao Zhilong

(Benxi Steel Group North Camp Company Steel Mills, Benxi 117017, China)

Abstract: To solve poor LF refining liquid steel can be poured, pouring nozzle nodulation and even less casting furnace time, north camp steel mills by optimizing LF slag composition, lower melting point and improve liquidity, reduce the content of molten steel inclusion, obviously improve the slab quality, improve the molten steel purity, and even the number of casting furnace, effectively reduce the production cost.

Key words: refining slag, low melting point, liquidity, inclusions

120t 转炉少渣冶炼工艺研究和实践

富　强，郭晓春，王志强，王　健

（本钢集团北营炼钢厂，辽宁本溪　117017）

摘　要： 通过渣系理论摸索出转炉前期渣最适合组分为碱度 R 控制在 1.5-1.7 之间；前期 FeO≥20%；易造低熔点渣。因钢渣反应界面增大，从而使少渣冶炼实施后半钢脱磷率由 32.02% 增大到 59.75%。并通过对少渣冶炼前后终点脱磷率比较发现，少渣冶炼工艺的实施，对质量控制也没有太大影响，钢水洁净度未因工艺发生变化而恶化；转炉灰耗由 42kg/t 降低到 32kg/t，显著降低成本。

关键词： 低熔点渣，少渣冶炼，半钢脱磷率

Study and Practice of Less Slag Smelting Technology for 120 ton Converter

Fu Qiang, Guo Xiaochun, Wang Zhiqiang, Wang Jian

(Benxi Steel Group Beiying Steel Mill, Benxi 117017, China)

Abstract: The most suitable component of prior converter slag was designed by slag system theory, the alkalinity was in 1.5~1.7; the content of FeO was more than 20%, it was easy to slag-making because of low melting point. The reason of dephosphorization rate of semisteel increased form 32.02% to 59.75% after using less slag melting was the steel-slag interface increasing. With the slagless steelmaking operating: through the comparison of the initial stage of smelting's dephosphorization rate with the final, the quality of steel change little and the cleanliness of steel was no worse; the cement consumption of converter decreased form 42kg/t to 32kg/t, there was a significant reduction in operating costs.

Key words: low melting point, less slag melting, dephosphorization rate of semisteel

电炉渣常压下碳酸化反应过程表面覆盖模型构建

张慧宁[1]，李　辉[2]，董建宏[1]，熊辉辉[1]，徐安军[3]

（1. 冶金与化学工程学院，江西理工大学，江西赣州　341000；2. 冶金工业规划研究院，北京　100711；3. 冶金与生态工程学院，北京科技大学，北京　100083）

摘　要： 为了研究电炉渣在水中碳酸化过程的包裹行为，本文构建了基于反应速率随电炉渣表面活性含钙相微粒占位百分数变化的表面覆盖模型，并通过电炉渣碳酸化动力学实验验证了表面覆盖模型预测碳酸钙转化率的准确性。研究结果表明电炉渣碳酸化钙转化的最佳反应温度为60℃，这是受到 CO_2 气泡在水中溶解度动力学及反应速率控制动力学综合决定所致。实验条件下电炉渣碳酸化反应过程的反应速率在 0.0083-0.0114mol/(l*h) 之间波动，自阻系数是 0.901-1.036。通过实验验证发现基于反应速率随电炉渣表面活性含钙相微粒占位百分数变化的表面覆盖模型预测碳酸钙转化率精度能控制在[-2%, 2%]。

关键词： 电炉渣，碳酸化，表面覆盖模型，动力学

Application of Surface Coverage Modeling on EAF Slag Aqueous Carbonation Process at Environmental Pressure

Zhang Huining[1], Li Hui[2], Dong Jianhong[1], Xiong Huihui[1], Xu Anjun[3]

(1. School of Metallurgy and Chemical Engineering, Jiangxi University of Science and Technology, Ganzhou 341000, China; 2. China Metallurgical Planning Net, Beijing 100711, China; 3. School of Metallurgical and Ecological Engineering, University of Science and Technology, Beijing, Beijing 100083,China)

Abstract: In order to investigate calcite coating behavior for electric furnace slag aqueous carbonation process, this paper establishes surface coverage model in order to analysis the carbonation kinetics on the conditions that the reaction rate is dependence with active calcium particles site percentage in EAF surface, which is verified by slag carbonation experiments at 25-80℃. The results show that the optimal reaction temperature for calcium carbonate conversion is 60℃, which is significantly influenced by CO_2 solubility kinetics and carbonation reaction rate. Under the experimental conditions, the reaction rate of EAF aqueous carbonation reaction is fluctuated between 0.008~0.0114 mol/(l*h) and the resistance coefficient is 0.901-1.036. Established surface coverage model prediction results coincides with experimental results better by comparison, the accuracy of calcite conversion rate can be control between −2% and 2%.

Key words: slag, aqueous carbonation, surface coverage model, reaction kinetics

唐山中厚板公司 120t 转炉自动炼钢的实践

刘志远，栾文林，王重君，刘晓娟，朱斐斐

（唐山中厚板材有限公司，河北唐山　063000）

摘　要： 唐山中厚板材有限公司采用了国内先进的副枪炼钢技术，通过副枪的成功应用，有效地减少倒炉时间，提高了生产节奏，而且优化了降本增效目的，最终钢水质量得以大幅提高。

关键词： 副枪，自动炼钢，转炉

Application of Sublance Technology for the 120t-Converter

Liu Zhiyuan, Luan Wenlin, Wang Chongjun, Liu Xiaojuan, Zhu Feifei

(Tang Shan Heavy Plate Co., Ltd., Tangshan 063000, China)

Abstract: An advanced sublance technology is applied in 120t-converter in heavy plate of tangsteel, which could improve production efficiency based on the time of steel liquid transferring, and the quality of product is improved also.

Key words: sublance technology, auto steelmaking, converter

碳、硅热还原转炉渣脱磷的试验研究

杨　建，盛鹏飞，王海军，凌海涛，彭世恒

（安徽工业大学 冶金工程学院，安徽马鞍山　243032）

摘　要： 分别对碳、硅还原转炉渣脱磷进行热力学分析，并在真空电阻炉中用碳粉、硅粉进行还原转炉渣的脱磷试验。计算、试验结果表明，在1300~1500℃时，碳和硅还原转炉渣脱磷反应均可以发生，且反应温度和还原剂种类对脱磷率有较大影响；在温度1573~1773K范围内，硅气化脱磷率随温度升高而增大，硅在1500℃左右时还原脱磷效果较好；碳气化脱磷率随温度变化是先增加后减少，1400℃左右时脱磷率最高，在相同试验条件下，硅还原转炉渣的脱磷效果优于碳还原脱磷效果。

关键词： 转炉渣，碳热还原，硅热还原，脱磷

Experiments on Dephosphorization of BOF Slag by Carbothermal Reduction and Silicon Thermal Reduction

Yang Jian, Sheng Pengfei, Wang Haijun, Ling Haitao, Peng Shiheng

(School of Metallurgical Engineering, Anhui University of Technology, Maanshan 243032, China)

Abstract: Thermodynamic analysis of dephosphorization of BOF slag by carbothermal reduction and silicon thermal reduction were carried out, and the experiment of dephosphorization by carbothermal reduction and silicon thermal reduction were carried out in a vacuum resistance furnace. The results show that, the dephosphorization reaction of BOF slag with carbon and silicon can be carried out at 1300-1500℃, and the reaction temperature and the type of reductant hava a great influence on the dephosphorization rate of the reduction. In the temperature range of 1573K~1773K dephosphorization rate of silicon increases with the increase of temperature, and the dephosphorization effect is better at 1500℃. The dephosphorization rate of carbon varing with temperature increases first and then decreases, and the dephosphorization rate is the highest at 1400℃. At the same experiment conditions, the dephosphorization effect of silicon thermal reduction is better than carbotthermal reduction.

Key words: BOF slag, carbothermal reduction, silicon thermal reduction, dephosphorization

武钢炼钢厂三炼钢分厂铁水脱硫改造实践

张利锋[1]，肖邦志[1]，李明晖[2]

（1. 武钢有限公司炼钢厂三炼钢分厂，湖北武汉 430080；

2. 武钢有限公司研究院，湖北武汉 430080）

摘　要： 武钢有限公司炼钢厂三炼钢厂，装备一座厂外4工位鱼雷罐喷吹脱硫站，厂内一座单工位双扒渣喷镁剂铁水脱硫站和一座独立扒渣站，具备铁水全脱、全扒能力。因鱼雷罐脱硫效率低下，成本高，返硫严重，且不利于鱼雷罐运行。为此进行了脱硫改造，在厂内新建两座KR脱硫站，关停厂外鱼雷罐脱硫。新建KR脱硫站采用一键脱硫、远程扒渣等技术手段，降低了劳动强度，减少了铁水返硫，大幅降低了脱硫成本。

关键词： 脱硫改造，KR脱硫，一键脱硫，远程扒渣

Practice of Hot Metal Desulphurization in No. three Steel Making Mill of WISCO

Zhang Lifeng[1], Xiao Bangzhi[1], Li Minghui[2]

(1. No.3 Steelmaking Plant of WISCO, Wuhan 430080, China;

2. Research and Development Center of WISCO, Wuhan 430080, China)

Abstract: Steelmaking plant of three steelworks of WISCO Equipment Co. Ltd., a factory outside the 4 station torpedo blowing desulfurization station in a single station, double slag spray magnesium desulfurization station and a slag independent station, with hot metal removal, grilled ability. Because of the low efficiency of the torpedo tank, the cost is high, the return sulfur is serious, and it is not conducive to torpedo tank operation. To this end, desulfurization has been carried out, two new KR desulfurization stations have been built in the factory, and the torpedo tanks are shut down. The new KR desulfurization station adopts automatic desulfurization and remote skimming technology to reduce the labor intensity, reduce the sulfur return of the hot metal and greatly reduce the desulfurization cost.

Key words: desulfurization transformation, KR desulfurization, automatic desulfurization ,emote skimming

20CrMnTi 齿轮钢中氮含量控制的研究

赵华森，华祺年，王恭亮，谷志敏，杨锋功

（石家庄钢铁有限责任公司，河北石家庄　050000）

摘　要： 对转炉流程 20CrMnTi 齿轮钢生产工艺进行了跟踪，通过 BOF-LF-VD-CC 各个环节中氮含量的变化分析，研究了影响钢水中氮含量的因素及控制措施。

关键词： 20CrMnTi，齿轮钢，氮含量

Study on Nitrogen Content of Gear Steel 20CrMnTi

Zhao Huasen, Hua Qinian, Wang Gongliang, Gu Zhimin, Yang Fenggong

(HBIS Group Shisteel Company, Shijiazhuang 050000, China)

Abstract: The factors and control measures affecting the nitrogen content of liquid steel are studied through the tracking to production process of BOF steelmaking for gear steel 20CrMnTi and analysis to the change of nitrogen content along the process of BOF- LF - VD - CC.

Key words: 20CrMnTi, gear steel, nitrogen content

LF 生产 SPHC 钢工艺实践

王重君，幺敬文，李　雷，孔明姣，朱斐斐，安海瑞

（唐山中厚板材有限公司，河北唐山　063000）

摘　要： 目前钢铁市场对钢材纯净度要求越来越高，促使全国各大钢厂的钢铁生产工艺愈发成熟。在生产 SPHC 钢的过程中，根据唐钢中厚板厂精炼作业区的生产环境和工艺，分析了影响 SPHC 钢脱硫率的主要因素。通过分批次的配加渣料，提高炉渣碱度，合理控制炉渣成分等措施，快速有效地形成精炼白渣，SPHC 钢的脱硫率达到了 87% 以上。为了净化钢液去除夹杂，生产中氩气流量的控制要在 60Nl/min，软吹时间控制在 8～10min。目前该厂具备了批量生产优质 SPHC 钢的能力，并且大大缩短了冶炼周期，提升了生产效率。

关键词： SPHC 钢，LF，脱硫率，夹杂物

The Technology Practice of SPHC Steel in LF Production

Wang Chongjun, Yao Jingwen , Li Lei, Kong Mingjiao, Zhu Feifei, An Hairui

(Tang Shan Heavy Plate Co., Ltd., Tangshan 063000, China)

Abstract: At present, the steel market is more and more demanding for steel purity, and the steel production process is getting more and more mature. In the process of producing SPHC steel, the main factors affecting the desulfurization rate of SPHC steel are analyzed according to the production environment and process of the refining operation area of the thick plate mill in tangshan steel.Through batch process of adding slag, improving the alkalinity of slag, controlling the composition of slag and so on, quickly and effectively forming the refined white slag, the desulfurization rate of SPHC steel reached 87%. In order to purify the inclusion of molten steel, the control of argon flow in the production is controlled at 60Nl/min, and the soft blowing time is controlled at 8 ~ 10min. At present, the plant has the ability to mass-produce high quality SPHC steel, which greatly reduces the smelting cycle and improves production efficiency.

Key words: SPHC steel, LF, the desulfurization rate, inclusions

转炉吹炼智能控制系统的开发与实践

刘书超

（本溪钢铁集团公司炼钢厂，辽宁本溪 117000）

摘 要： 转炉炼钢冶炼周期短、人工操作为主，冶炼过程中的化渣效果和冶炼终点的碳温命中情况，直接影响到钢水的质量与炼钢效率，传统的转炉自动化炼钢系统大多不能满足实际转炉生产的实际需要。本文着重介绍了基于转炉声呐信号和烟气 CO 信号辅助下的转炉吹炼智能控制系统的开发和实践，从系统构成、系统功能及实际效果展示了该系统在转炉炼钢过程的应用情况，事实数据表明该系统的多个创新点为国内外同类项目首创，具有较高实用价值。

关键词： 转炉吹炼，化渣，碳温命中，智能控制

The Development and Practice of Converter Intelligent Controlling System

Liu Shuchao

(Steel-making Plant of Benxi Iron and Steel (Group) Co., Ltd., Benxi 117000, China)

Abstract: The main smelting cycle of BOF is short, the manual operation of converter steelmaking slag and the effect of temperature, carbon smelting end point in the smelting process of the hit, directly affects the quality and efficiency of the converter steelmaking molten steel, the automation steelmaking system usually can not meet the actual needs of the actual production of converter. This paper introduces the development and practice of BOF intelligent control system of converter sonar signal and flue gas assisted CO signal based on the system structure, system function and practical results demonstrate the system application in converter steelmaking process, data show that the first fact several innovations of the system for domestic and foreign similar projects, has high practical value.

Key words: blowing and smelting, slagging, carbon-temperature hitting rate, intelligent control

含碳球团在转底炉工艺中的应用

赵 刚

（日照钢铁循环经济部，山东日照 276806）

摘 要： 对含碳球团固结机理进行分析、理论计算并结合实际生产得出最佳的配碳量为9.54%。另外，通过对含碳球团在转底炉自还原机理分析，得出日钢转底炉最佳的工艺运行参数，为转底炉生产提供理论性依据。并指出日钢转底炉存在产量低、料耗高等生产上的不足。

关键词： 含碳球团，转底炉，配碳量，自还原机理，工艺运行参数

The Applicetion of Carbon Pellet in the Process of Rotary Hearth Furnace

Zhao Gang

(Rizhao Steel Plate Manufacturing Department, Rizhao 276806, China)

Abstract: Through study of the consolidation of carbon pellet and theoretical calculation，it is getting the best carbon content of carbon pellet is 9.54% in the actual production process. Additional, through study of the self-reduction mechanism of pellet in the rotary hearth furnace，it is getting the best process parameter in the rotary hearth furnace of Rizhao steel. It provides theoretical foundation in production process of rotary hearth furnace. Besides, it indicates the shortcoming of low yield and high material consumption in the actual production process.

Key words: carbon pellet, rotary hearth furnace, carbon content of pellet, mechanism of self-reduction, parameters of process

攀钢钒转炉复吹技术实践

施明川，李扬洲

（攀钢集团攀枝花钢钒有限公司提钒炼钢厂，四川攀枝花 617000）

摘 要： 本文简要介绍了转炉复吹工艺，对比分析了没有复吹的 1#炉、低强度复吹和高强度复吹的 3#炉的冶金效果，强搅复吹模式能将钢水中的氧含量降至 579.1ppm，转炉终碳氧积降至 0.00257，渣中 TFe 降至 16.6%，低碳硅镇静钢的 Si、Mn 收得率提升至 88.21%、95.45%，低碳铝镇静钢的 Al、Mn 收得率提升至 31.12%、90.12%。结果表明有复吹的转炉冶金效果优于没有复吹，强复吹的转炉冶金效果优于弱复吹。

关键词： 转炉复吹，脱磷，冶金效果

Converter Blowing Technology Research in PZH Steel and Vanadium

Abstract: This paper briefly introduces the blowing converter technology, comparative analysis metallurgical effect of the $1^#$ converter of no-blowing with the $3^#$ converter of blowing. The strong blowing mode can reduce oxygen content of molten steel to 579.1ppm, [%C][%O] decreased to 0.00257，the TFe in the slag decreased to 16.6%，the yield of Si and Mn in low carbon killed steel was increased to 88.21%, 95.45%，the yield of Al and Mn in low carbon Al-killed steel was increased to 31.12%, 90.12%. Found that the blowing converter metallurgical effect is better than no-blowing, the strong-blowing converter metallurgical effect is better than the weak-blowing.

Key words: blowing converter, dephosphorization, metallurgical effect

关于提高挡渣出钢效果问题的探讨

颜根发，金永明

（马鞍山劲鹰科技公司，安徽马鞍山 243000）

摘　要：简述了转炉出钢过程中钢包下渣的情况下，分析了转炉出钢过程中钢、渣的流动情况与"汇流漩涡"形成的原因；探讨了在不影响减少一般性下渣和"漩涡卷渣"的前提下，尽可能将钢水出完的挡渣块密度；研究了钢、渣的径向分流与挡渣器命中率的关系和漩涡强度与下渣量的关系。得出新型的横剖面为齿轮型的挡渣块挡渣效果最好。新型的横剖面为齿轮型的挡渣块，优于相对带浅凹槽的挡渣块；带浅凹槽的挡渣块，优于表面基本光滑的挡渣球。

关键词：挡渣块，密度，汇流漩涡，命中率，下渣厚度

Discussion on Improving the Effect of Slag - tapping

Yan Genfa, Jin Yongming

(Maanshan JinYing Science and Technology Ltd., Maanshan 243000, China)

Abstract: In this paper, the flow of steel and slag during the process of converter tapping and the causes of the formation of "confluence whirlpool" during the process of converter tapping are briefly introduced. Explored in the absence of the general reduction of slag and "whirlpool slag" under the premise, as far as possible the finished steel slag density. Study the relationship between the radial shunt of steel and slag and the hit rate of the sluice and the relationship between the vortex intensity and the slag.It is concluded that the new cross section is the best for the gear block. The new cross section is gear type block block, which is superior to the block with relatively shallow groove. The block with shallow groove is better than the basic slag ball.

Key words: block block, density, confluence whirlpool, hit rate, slag thickness

鞍钢 100t 顶吹转炉双联法脱磷工艺的研究

朱晓雷[1]，栾花冰[2]，吴世龙[2]，安晓光[3]，廖相巍[1]，李叶忠[2]

（1. 鞍钢股份技术中心，辽宁鞍山　114021；2. 鞍钢股份炼钢总厂，辽宁鞍山　114021；

3. 鞍钢股份产品发展部，辽宁鞍山　114021）

摘　要： 对转炉脱磷的原理进行分析，找出顶吹转炉进行双联法脱磷的影响因素，并在 100t 转炉进行了相应工业试验。结果表明，炉渣碱度的提高能明显提高前半钢的脱磷率，同时温度控制不要过高；炉渣氧势、炉渣流动性和温度是后半钢脱磷反应的前提条件。在不影响钢质的前提下，适当延长吹炼时间更利于后半钢的进一步脱磷。

关键词： 转炉，双联法，脱磷，低磷钢

Research on Duplex Dephosphorization Process in 100t Top-blowing Converter of Ansteel

Zhu Xiaolei[1], Luan Huabing[2], Wu Shilong[2], An Xiaoguang[3], Liao Xiangwei[1], Li Yezhong[2]

(1. Technology Center of Angang Steel Co., Ltd., Anshan 114021, China；

2. General Steelmaking Plant of Angang Steel Co., Ltd., Anshan 114021, China；

3. Product Development Department of Angang Steel Co., Ltd., Anshan 114021, China)

Abstract: This paper analyses the principle of dephosphorization process in converter and the effect factors of duplex dephosphorization process in top-blowing converter are found out. The corresponding industrial tests were performed in the 100t converter. The results showed that the increasing of slag basicity could obviously improve the dephosphorization rate of the first half steel, the temperature controlled to avoid overheating. Oxygen potential slag, slag fluidity and temperature are the precondition for dephosphorization reaction of the second half steel. Appropriate extend blowing time was benefic to dephosphorization of the second half steel while not effecting the quality of steel.

Key words: converter, duplex process, dephosphorization, low phosphor steel

喷枪喷口数量对铁水喷吹脱硫动力学条件影响研究

李明晖，欧阳德刚，饶江平，朱善合，邓志方，罗　巍

（1. 武汉钢铁有限公司研究院，湖北武汉　430080；

2. 武汉钢铁有限公司炼钢总厂，湖北武汉　430083）

摘　要： 通过水力学模型试验，分析了铁水喷吹脱硫工艺中不同喷口数量喷枪工艺的熔池流动状态，研究了喷枪喷口数量对铁水喷吹脱硫混匀时间、液面波动高度和喷吹气体压力的影响规律。

关键词： 铁水脱硫，喷枪，喷口数量，水模试验，动力学条件

Hydraulics Simulating Experiment Study on Dynamic Condition of Hot Metal Desulphurization by Injection Lance with Different Nozzle Number

Li Minghui, Ouyang Degang, Rao Jiangping, Zhu Shanhe, Deng Zhifang, Luo Wei

(1. Research and Development Center of Wuhan Iron and Steel Co., Ltd., Wuhan 430080, China；

2. Steelmaking General Plant of Wuhan Iron and Steel Co., Ltd., Wuhan 430083, China)

Abstract: Through hydraulic model experiments, the flow state of hot metal desulphurization by injection lance with different nozzle number was studied. Effects of process parameters on mixing time, surface wave and blowing gas pressure was also been discussed in this paper.

Key words: hot metal desulphurization, injection lance, nozzle number, hydraulic model experiment, dynamic conditions

汽车用弹簧钢中氮含量的控制

王旭冀

（湖南华菱湘潭钢铁有限公司技术质量部）

摘　要：钢中氮含量为汽车用弹簧钢的一个重要指标，湘钢弹簧钢中 X55SiCrA 的氮含量为 0.0060%左右，较国内同类钢厂偏高。通过对转炉出钢后各个时期增氮因素进行分析：原料气体来源为主要原因，其次为二次氧化。通过更换炼钢增碳剂、减少出钢口出钢时间、吹氩等控制二次氧化措施，最终将弹簧钢中的氮含量控制在 0.0040%以内。

关键词：弹簧钢，氮，分析

Control of Nitrogen Content in Automobile Spring Steel

Wang Xuji

(Hunan Valin Xiangtan Iron and Steel Co., Ltd., Technical Quality Department)

Abstract: The nitrogen content in steel for automobile spring steel is an important indicator, Nitrogen content in Xiangtan Iron spring steel X55SiCrA is approximately 0.0060% moer than like domestic steel mills .Through to the various periods after the converter steel factors are analyzed, the gas sources of raw materials as the main reason ,followed by secondy oxidation.By chananging the steel recarburezer,reduce the tapping hole tapping time, argon blowing and molten steel secondary oxidation,made of nitrogen content in molten steel within 0.0040%.

Key words: spring steel, nitrogen content, analysis

铝镇静钢中铝酸钙夹杂物的来源

迟云广，邓志银，成　刘，朱苗勇

（东北大学冶金学院，辽宁沈阳　110819）

摘　要：本文通过实验室实验研究了转炉钢液中的夹杂物在 Al 脱氧过程中的转变行为，同时分析了钢包釉对夹杂物转变的影响，重点考察了铝镇静钢中铝酸钙夹杂物的来源。研究结果显示：转炉钢液中存在三种夹杂物类型，分别是硅酸钙夹杂物、含有固体颗粒的硅酸钙夹杂物和(Fe, Mn)O 类型夹杂物。硅酸钙类型夹杂物在 Al 脱氧过程中可以转变成为铝酸钙类型夹杂物，(Fe, Mn)O 类型夹杂物在 Al 的作用下生成 Al_2O_3 夹杂物。在钢包釉的作用下，钢液中的氧化铝夹杂物首先转变成为尖晶石夹杂物，再转变成为铝酸钙夹杂物。钢包釉脱落也是铝酸钙夹杂物的一个重要来源。

关键词：转炉钢液，精炼渣，钢包釉，铝酸钙夹杂物

Study on the Source of Calcium Aluminate Inclusions in Al-Killed Steel

Chi Yunguang, Deng Zhiyin, Cheng Liu, Zhu Miaoyong

(School of Metallurgy, Northeastern University, Shenyang 110819, China)

Abstract: The evolution behavior of inclusions in BOF crude steel during Al deoxidation was studied by laboratory experiments, while the effect of ladle glaze on the evolution of inclusions in steel was also investigated. The results show that three types of inclusions were found in BOF crude steel, and they are calcium silicate, calcium silicate with some solid particles and (Fe, Mn)O inclusions respectively. During Al deoxidation, calcium silicate system inclusions would transform into calcium aluminate system inclusions, and (Fe, Mn)O inclusions will be reduced by Al to form Al_2O_3 inclusions. With the help of ladle glaze, calcium aluminates could also be generated from Al_2O_3 inclusions *via* spinel inclusions. The peel-off of ladle glaze could result in calcium aluminate inclusions as well.

Key words: BOF crude steel, Al deoxidation, ladle glaze, inclusions

干法除尘在 180t 转炉的应用

王小善，李　冰，李　泊，曹　琳，宋吉锁

（鞍钢股份有限公司炼钢总厂，辽宁鞍山　114021）

摘　要：鞍钢 180t 转炉在干法除尘投入使用初期卸爆炉次较多，对转炉生产干扰较大，同时影响静电除尘器的稳定运行。通过对静电除尘器卸爆产生的原因进行详细分析，不断优化转炉供氧制度、氧枪枪位控制、废钢铁水条件、物料加入等，并采用转炉自动化炼钢，大大降低了转炉干法除尘的卸爆率。卸爆比例从生产初期的 4.86%，降低到了目前 0.19‰。同时实现了转炉干法除尘在留渣条件下的稳定运行。

关键词：干法除尘，卸爆，静电除尘器

Application of Dry Dedusting System in 180t Converter

Wang Xiaoshan, Li Bing, Li Bo, Cao Lin, Song Jisuo

(General Steelmaking Plant of Angang Steel Co., Ltd., Anshan 114000, China)

Abstract: The 180 tons converter of Anshan Iron and Steel Co., Ltd has more blast furnace times in the early stage of dry dedusting, which has great interference to converter production and affects the stable operation of electrostatic precipitator. Through detailed analysis of the reason of explosion of electrostatic precipitator discharge, constantly optimize the converter oxygen lance position control system, and scrap iron, material added, and the use of automatic steelmaking, greatly reducing the discharge rate of converter dry dedusting. The proportion of explosive decompression from the initial production of 4.86%, reduced to the present 0.19 per thousand. At the same time, the stable operation of converter dry dedusting under residue condition was realized.

Key words: dry dedusting, explosion venting, electrostatic precipitator

鞍钢大型转炉高效生产技术

朱国强，王　鹏，张志文，毛志勇，李　冰，宋　宇

（鞍钢股份公司炼钢总厂，辽宁鞍山　114021）

摘　要： 随着炼钢四分厂 3#铸机的正式投产，形成三座转炉兑三台铸机的生产局面，产能限制环节由连铸转变到炼钢。为了解决炼钢四分厂炼钢工序严重限制产能释放的问题，对氧枪的工艺参数和出钢口尺寸进行了优化并改善了冶炼操作制度。实践表明，优化后转炉的冶炼周期大幅度缩短，改善了转炉吹炼效果，脱磷率和金属收得率有了不同程度的提高，实现了转炉的高效冶炼，满足了公司对四分厂产量任务的要求。

关键词： 转炉，冶炼周期，氧枪，出钢口

High Efficient Production Technology of Large Converter in Ansteel

Zhu Guoqiang, Wang Peng, Zhang Zhiwen, Mao Zhiyong, Li Bing, Song Yu

(Steelmaking Plant of Angang Steel Co., Ltd., Anshan 114021, China)

Abstract: With the production of No.3 caster in the No.4 branch of steelmaking plant.Production situation of three converters with three converters. Production capacity limit change from continuous casting to steelmaking. In order to solve the problem of severe restriction on the production of steelmaking process in the four branch factory of steelmaking.The process parameters and outlet size of oxygen-lance were optimized and the smelting operation system was improved. Practice shows.After optimization, the smelting cycle of the effect of converter blowing is improved. The dephosphorization rate and the yield of the metal have been improved, and the efficient smelting of the converter has been realized.Meet the company's requiremaents for the four branch factory output.

Key words: converter, residence time in blast furnace, oxygen-lance, tap hole

海工钢 EQ460 冶炼工艺研究

刘金刚 [1,2,3]，李战军 [1]，史志强 [4]，王东柱 [4]，初仁生 [1]

（1. 首钢集团有限公司技术研究院，北京　100043；2. 绿色可循环钢铁流程北京市重点实验室，北京　100043；3. 北京市能源用钢工程技术研究中心，北京　100043；4. 秦皇岛首秦金属材料有限公司，河北秦皇岛　066326）

摘　要： 通过对某 100t 转炉的中厚板炼钢厂采用"颗粒镁铁水脱硫－转炉炼钢－LF 精炼－RH 精炼－板坯连铸"工艺生产海工钢 EQ460 的冶炼过程分析，得出该生产工艺可以满足海工钢 EQ460 低碳、低磷、超低硫、低氢氮含量的要求，其中 C%、P%、S%|、[N]、[H] 分别可以控制在 0.058%、0.006%、0.0018%、0.0034%、0.00007%；铸坯质量良好，低倍检验中心偏析达到 C 类 1.0 级，中间疏松为 0.5 级，夹杂物满足≤2.0 级的要求，其中最大级别为 B 类粗系 1.5 级和 DS 类 1.5 级，其他类别平均≤0.69 级，均小于 1.0 级。

关键词： 海工钢，EQ460，成分控制，铸坯内部质量，夹杂物

Research of Marine Steel EQ460 Steelmaking Process

Liu Jingang[1,2,3], Li Zhanjun[1], Shi Zhiqiang[4], Wang Dongzhu[4], Chu Rensheng[1]

(1. Shougang Research Institute of Technical, Beijing 100043, China; 2. Beijing Key Laboratory of Green Recyclable Process for Iron & Steel Production Technology, Beijing 100043, China; 3. Beijing Engineering Research Center of Energy Steel, Beijing 100043, China; 4. Shouqin Metal Material Company Ltd., Qinhuangdao 066326, China)

Abstract: This paper analyzes the steelmaking process of marine steel EQ460 by " Magnesium desulfurization - a 100 ton converter blowing-LF refining-RH refining-slab continuous casting". This process can meet the requirements of low-carbon, low-phosphorus, ultra-low sulfur, low hydrogen, low nitrogen and high quality of slab in EQ460, among which C%, P%, S% , [N] and [H] can be controlled at 0.058%, 0.006%, 0.0018%,0.0034% , 0.00007%, the center segregation reaches C 1.0, the center porosity is 0.5, the inclusions level meet the requirement of ≤ 2.0, and the maximum level is B 1.5 and DS 1.5, other inclusions average ≤ 0.69 , all less than 1.0.

Key words: marine steel, EQ460, composition control, slab internal quality, inclusion

150t 复吹转炉双渣生产低磷钢的工艺实践

苏　磊，孙学刚，韩雨亮

（新疆八一钢铁股份有限公司，新疆乌鲁木齐　830022）

摘　要： 文章介绍了 150t 复吹转炉在干法除尘的条件下采用双渣法进行了低磷钢的冶炼试验，通过优化转炉造渣、

供氧、温度等工艺制度，采用造双渣的方法，一次倒渣脱磷率达到 50%，终点磷含量控制在 0.010%以下，满足了低磷钢的生产要求。

关键词： 双渣操作，低磷钢，复吹转炉

Practices of Produce Low Phosphorus Steel by Double Slag Process in 150t BOF

Su Lei, Sun Xuegang, Han Yuliang

(Xinjiang Bayi Iron & Steel Stock Co., Ltd., Urumchi 830022, China)

Abstract: This article introduces the 150t BOF under the condition of dry dust removal method adopts double slag has carried on the experiment of smelting low phosphorus steel, by optimizing the converter slag, oxygen, temperature and other process system, adopt the method of making double slag, pour a slag dephosphorization rate of 50%, the end phosphorus content is controlled under 0.010%, satisfying the requirements of low phosphorus steel production.

Key words: double slag process, low phosphorus steel, BOF

50t 转炉炼钢过程钢水及炉渣成分分析

赵　斌，吴　伟，吴　巍，崔怀周，汪成义，王天明

（钢铁研究总院冶金工艺研究所，北京　100081）

摘　要： 研究了某钢铁厂 50t 转炉炼钢过程中熔池金属成分、炉渣成分、温度的变化以及熔池脱碳、脱磷、脱硫的情况，检测了炉渣的成分变化和岩相结构。试验分析表明，吹炼终点时脱磷、脱硫反应偏离平衡值较远，转炉炼钢平均脱磷率为 87%，平均脱硫率为 30%。[C][O]积为 0.0045，降碳速度为 0.429%/min，熔池平均升温速度为 33.46℃/min，每增加 1%的碳，钢水温度增加 76.87℃。该厂炉龄大于 5000 导致碳氧积升高，从而影响了碳氧反应的动力学条件，炉渣碱度的变化对转炉脱磷率没有明显影响，增加初期烧结矿用量，提高前期化渣速度，避免后期炉渣返干。

关键词： 转炉，钢水，炉渣，岩相分析

Analysis on the Variation Law of Mineral Phase and Sulfur-phosphorus in the Process of 100t BOF

Zhao Bin, Wu Wei, Wu Wei, Cui Huaizhou, Wang Chengyi, Wang Tianming

(Central Iron & Steel Research Institute，Beijing 100081, China)

Abstract: The change of liquid steel composition, slag composition, and temperature and the decarburization, dephosphorization and desulfurization of molten pool in a 100t converter in a steel plant were studied. The composition change and petrographic structure of the slag were examined. The experimental analysis shows that the dephosphorization and desulfurization reaction is far from the equilibrium value at the end of the blowing finish. The average dephosphorization rate is 87%, and the average desulfurization rate is 30%.The product of [C][O] is 0.0045, the rate of carbon reduction is 0.429%/min, the average heating rate of molten pool is 33.46 /min, and the temperature of molten steel increases by 76.87℃ when 1% carbon is added. The plant life of more than 5000 in carbon oxygen product increased, thereby affecting the dynamic conditions of carbon oxygen reaction. There is not effect on dephosphorization rate by changing of slag basicity. Increased in the average amount of pre sinter early, improving the slagging speed, to avoid the late slag getting dry.

Key words: BOF, liquid steel, slag, petrographic analysis

一种提升常温氧化铁脱硫剂使用效果的方法

刘芳荣，褚庆枢

（宝钢集团新疆八一钢铁有限公司能源中心，新疆乌鲁木齐　830022）

摘　要：煤气中硫化氢气体的脱除方法较多，其中常温氧化铁脱硫法是一种经典而有效的脱硫方法，其优点是工艺简单、操作容易、能耗低。但在使用中后也存在脱硫剂粉尘含量增大、脱硫效果下降等现象。目前使用的常温氧化铁脱硫剂外部再生环境污染严重且劳动强度大已被许多使用单位淘汰，通入氧气在线再生对装置密封性及装备水平要求较高且再生过程风险大很少有使用单位选择使用。而此方法在一般装备水平的装置中均可使用，使用风险率几乎为零，可推广性强。

关键词：常温氧化铁脱硫剂，加湿，使用寿命，硫容

Method for Promoting the Use Effect of Iron Oxide Desulfurizer at Normal Temperature

Liu Fangrong, Chu Qingshu

(Energy Center of Baosteel Group Xinjiang Bayiiron & Steel Co., Ltd., Urumchi 830022, China)

Abstract: There are many methods to remove hydrogen sulfide gas in gas. At room temperature, iron oxide desulfurization is a classic and effective desulfurization method. The advantage is simple process, easy operation and low energy consumption. But also in use after the existence of desulfurization agent dust content increased desulfurization effect and so on. At present, the use of room temperature iron oxide desulfurization agent external regeneration environment is serious and the labor intensity has been eliminated by many units of use, access to oxygen on-line regeneration of the device sealing and equipment level requirements are higher and the regeneration process risk is very little use of the unit selection use. And this method can be used in general equipment level devices, the use of the risk rate is almost zero, can promote the strong.

Key words: iron oxide desulfurizer at normal temperature, damping, application life, sulfur capacity

含铜铁水喷吹 CO_2 脱碳动力学研究

李志强，牛丽萍，张廷安，张保敬，郑　超，张东亮

（东北大学冶金学院，辽宁沈阳　110819）

摘　要： 研究了含铜铁水喷吹 CO_2 脱碳过程的动力学条件，发现在较低温度条件下，喷吹 CO_2 流量虽然加大了反应面积，由于 CO_2 与 C 的反应是吸热反应，因此并不是流量越大反应越好，这可能是因为大流量带走了热量，导致反应速率减小，即流量导致的反应面积的变化与流量导致的反应温度的变化产生了交互作用。由计算所得的活化能值在 14~35kJ/mol 可知 CO_2 气相传质是 CO_2 脱碳反应的动力学限制条件。此外喷吹 CO_2 脱碳过程中对 Cu 含量基本没有影响。

关键词： 含铜铁水，CO_2，脱碳，动力学

Kinetics Research on Decarburization of Hot Metal Contains Copper Reaction with Carbon Dioxide

Li Zhiqiang, Niu Liping, Zhang Tingan, Zhang Baojing, Zheng Chao, Zhang Dongliang

(School of Metallurgy, Northeastern University, Shenyang 110819, China)

Abstract: This paper analysis the dynamic law of CO_2 injected into hot metal contains copper, using the CO_2 as an oxidant removing the carbon contained in hot metal. According to the experiment, the middle flow got a good result in lower temperature because of the interaction between the change of reaction surface and the change of temperature caused by flow of CO_2. The CO_2 decarburization reaction kinetics constraints is the mass transfer of CO_2 in gas phase because of the activation energy was between 14 and 35 kJ/mol obtained by calculation. In addition, the decarburization process by CO_2 has no effect on the content of Cu in the molten iron.

Key words: hot metal contains copper, CO_2, decarburization, kinetics

精炼渣碱度及 Al_2O_3 含量对 55SiCrA 弹簧钢中夹杂物的影响

林　路[1]，顾　超[2]，曾加庆[1]

（1. 钢铁研究总院冶金工艺研究所，北京　100081；
2. 北京科技大学钢铁冶金新技术国家重点实验室，北京　100083）

摘　要： 采用 FactSage 热力学软件计算，并通过实验室热态试验，研究了精炼渣碱度及 Al_2O_3 含量对 55SiCrA 弹簧钢中夹杂物的影响规律，得到合理的精炼渣控制条件。研究结果表明：精炼渣成分对 55SiCrA 弹簧钢中夹杂物行为有着重要影响；在试验渣系范围内，对夹杂物塑性的控制效果为 C2 方案（碱度 0.9、Al_2O_3 含量 3%）＞C3 方案（碱度 1.2、Al_2O_3 含量 3%）＞C4 方案（碱度 0.9、Al_2O_3 含量 8%），C2 方案中最终夹杂物熔点均在 1600℃ 以下。渣中 Al_2O_3 含量的降低可以降低钢中夹杂物的尺寸，并降低夹杂物中 Al_2O_3 含量，从而降低夹杂物的熔点，提高夹杂物塑性化程度；渣碱度的降低会引起夹杂物熔点的降低，塑性化程度增加，因此在实际生产选择精炼渣系时，尽量降低精炼渣中 Al_2O_3 含量及碱度（0.9-1.2）。

关键词： 精炼渣，夹杂物塑性化，FactSage，55SiCrA 弹簧钢

Effect of Basicity and Al₂O₃ Content of Refining Slag on Inclusions in 55SiCrA Spring Steel

Lin Lu[1], Gu Chao[2], Zeng Jiaqing[1]

(1. Metallurgical Technology Institute, Central Iron and Steel Research Institute, Beijing 100081, China;
2. State Key Laboratory of Advanced Metallurgy, University of Science and Technology Beijing, Beijing 100083, China)

Abstract: The influence of the basicity and Al₂O₃ content of refining slag on inclusion in 55SiCrA spring steel was studied by FactSage thermodynamic software and laboratory hot test. Reasonable control condition of refining slag was obtained. The results show that the composition of refining slag has an important influence on the inclusion behavior of 55SiCrA spring steel slag. In the range of test slag, the decreasing of Al₂O₃ content in slag can decrease the size of inclusions and the content of Al₂O₃ in inclusions, thus reducing the melting point of inclusions and improving the plasticity of inclusions. The decreasing of slag basicity will lead to lower melting point of inclusion and improve the plasticity of inclusions. Therefore, the Al₂O₃ content and basicity (0.9-1.2) of the refining slag can be decreased as much as possible in selecting the refining slag system.

Key words: refining slag, plasticity of inclusion, FactSage, 55SiCrA spring steel

电梯曳引机用 C45E+N 钢的生产工艺研究

齐　峰，卢秉军，阚　开

（本钢板材股份有限公司产品研究院，辽宁本溪　117000）

摘　要： 介绍了采用转炉+矩形坯连铸工艺生产电梯曳引轮用钢 C45E+N 的生产过程。电梯曳引轮涉及电梯安全，钢的纯净度要求高，综合力学性能要求高。本钢根据产品质量要求设计了 C45E+N 的化学成分及生产工艺并进行了工业试验，结果表明，C45E+N 各项性能满足电梯曳引轮质量要求。

关键词： 电梯曳引机，C45E+N，力学性能，正火

Process Study of the Steel C45E+N for Elevator Traction Machine

Qi Feng, Lu Bingjun, Kan Kai

(Products Research Institute of Benxi Steel Plates Co., Ltd., Benxi 117000, China)

Abstract: Introduces the production process of Benxi steel converter plus rectangular billet continuous casting process to produce elevator traction wheel steel C45E + N .Elevator traction wheel relates to elevator safety, it require high purity, high comprehensive mechanical performance. Benxi steel according to the product quality requirements designed C45E + N chemical components and production technology and carry on in the industrial test, the results show that the design of Benxi steel C45E + N all performance meet user requirements.

Key words: elevator traction machine, C45E+N, mechanical property, normalizing

热作模具钢 4Cr5MoSiV1 电炉模铸生产工艺研究

王德勇 [1]，李红梅 [2]

（1. 本钢板材股份有限公司产品研究院，辽宁本溪 117000；
2. 本钢板材板材股份有限公司质量管理中心，辽宁本溪 117000）

摘 要：4Cr5MoSiV1 作为模具用钢的典型钢种，在模具加工制造业中得到了长足的发展和广泛的应用。通过采用电炉模铸和 800 棒线材轧机进行生产试制，试制结果表明此工艺可行。产品磷含量≤0.010%，硫含量≤0.004%，晶粒度达到 10.5 级，横向冲击≥11J，夹杂物 A 类≤1 级，B 类≤0.5 级，C、D 类均为 0 级。

关键词：模具钢，模铸，夹杂物，横向冲击

Production Technology Study of Hot-work Die Steel 4Cr5MoSiV1 by EF Plus Mold Casting

Wang Deyong[1]，Li Hongmei[2]

(1. Products Research Institute of Benxi Steel Plates Co., Ltd., Benxi 117000, China；
2. Quality Control Center of Benxi Steel Plates Co., Ltd., Benxi 117000, China)

Abstract: 4Cr5MoSiV1 is typical die casting steel, it is widely used and rapidly developmented in the mould manufacturing industry. By using electric furnace and mold casting and 800 bar rolling mill in trial production,the trial production results show that the technology is feasible.Phosphorus content is 0.010% or less, the sulfur content is 0.004% or less, the grain size is 10.5 level, transverse impact is more than 11 J, inclusions grade A≤1, B≤0.5, C and D are 0.

Key words: die casting steel, mold casting, inclusion，lateral impact

镁蒸气铁水脱硫的气液吸收过程的水模型实验研究

王 坤，刘 燕，侯君洋，杨永坤，张廷安

（东北大学多金属共生矿生态化冶金教育部重点实验室，特殊冶金与过程工程研究所，
辽宁沈阳 110819）

摘 要：本文针对镁蒸气铁水预处理脱硫的气泡细化和分散过程，利用物理模拟的方法，针对某钢厂 120t 底吹钢包脱硫过程进行实验研究。基于相似原理的水模型实验，采用高速照相机来获得气泡在不同搅拌桨桨型、搅拌转速和搅拌桨偏心度等条件下分布的瞬态图像。采用 NaOH 溶液吸收 CO_2 的方法模拟机械搅拌镁脱硫工艺中镁蒸气的吸收速率和利用率。结果表明：使用 SSB-D 桨，搅拌转速 200rpm，通气流量为 1.0m³/h，偏心度 0.4 时，熔池内的气泡细化和分散得更均匀，增加了气液接触面积，提高了气体的吸收速率和气泡的利用率。

关键词：气泡分散细化，脱硫，物理模拟，吸收速率，气泡利用率

Water Model Study of Desulfurization of Molten Iron with Magnesium Vapor

Wang Kun, Liu Yan, Hou Junyang, Yang Yongkun, Zhang Tingan

(Northeastern University, Key Laboratory of Ecological Metallurgy of Multi-metal Intergrown Ores of Ministry of Education, Special Metallurgy and Process Engineering Institute, Shenyang 110819, China)

Abstract: In this paper, for the process of bubble refinement and dispersion in desulfurization of hot metal pretreatment with magnesium vapor, the physical simulation method was used to study on desulfurization process of 120 ton bottom blown ladle. On the base of cold water model experimental result basing on principle of similitude, high speed camera was used to obtain bubble dispersion phenomena under the conditions of different impeller structures, rotation speeds and eccentricity. Absorption rate and utilization rate of magnesium vapor were simulated by the method of CO_2 absorption by NaOH solution. Results show that: under the conditions of using the SSB-D impeller, rotation speed 200rpm, gas flow rate $1.0m^3/h$, the eccentricity 0.4, the bubble dispersion and disintegration in the molten iron were more uniform. The gas-liquid contact area was increased to improve the gas absorption rate and bubble utilization.

Key words: bubble dispersion and disintegration, desulfurization, physical simulation, absorption rate, bubble utilization

高品质活塞杆电镀不良的原因分析及改善

丁 辉

（石家庄钢铁有限责任公司技术中心，河北石家庄　050031）

摘　要：活塞杆是工程机械液压油缸的关键部件，对钢材质量要求极高，因国内钢材较进口钢材电镀不良率高，知名外资企业全部从国外进口钢材。重点介绍了活塞杆电镀不良的原因、改善措施和改善效果，最终实现了不良率的降低并成功替代进口。

关键词：活塞杆，电镀不良的原因，改善措施，改善效果

Reasons Analysis and Improvement on Poor Plating of High Quality Piston Rod

Ding Hui

(Shijiazhuang Iron & Steel Group Company Technology Center, Shijiazhuang 050031, China)

Abstract: Piston rod is the key component of hydraulic cylinder of construction machinery, Very high quality steel requirements. Because the domestic steel is more poor than the imported steel. Well-known foreign enterprises all import steel from abroad. This paper discussed mainly the reasons for the poor electroplating of piston rod, improvement measures and improvement effect. Finally, the rate of bad was reduced and the import was successfully replaced.

Key words: piston rods, the reasons of poor plating, improving measures, improvement effect

直筒真空精炼装置(MSR)的真空脱碳脱硫试验研究

沈　昶，乌力平，潘远望，张良明

（马钢（集团）控股有限公司，安徽马鞍山　243000）

摘　要：本文基于 120t RH 真空装置开发了直筒型真空精炼装置(MSR)，该装置特点为：（1）与真空室下部槽等内径的直筒型浸渍管；（2）在浸渍管下沿周向分段布置了独立控制流量的侧吹孔控制钢水和顶渣行为；（3）真空精炼过程采用顶渣还原脱硫工艺。物理模拟研究了 MSR 浸渍管侧吹及辅助钢包底吹工艺对混匀时间的影响。工业试验研究了真空深脱碳、顶渣还原脱硫工艺，真空精炼终点碳平均为 14×10^{-6}，最低为 11×10^{-6}，顶渣还原脱硫的平均脱硫率为 73%，最高达到 81%。

关键词：直筒型真空精炼装置（MSR），真空精炼，脱硫，脱碳

Research of Vacuum Decarbonization and Desulfurization by Cylinder Snorkel Refining Furnace(MSR)

Shen Chang, Wu Liping, Pan Yuanwang, Zhang Liangming

(Masteel, Maanshan 243000, China)

Abstract: Based on the 120 tons RH degasser, a Cylinder Snorkel Refining furnace(MSR-Multipurpose Secondary Refining Degasser) is developed in this paper, the characteristics are (1) the cylinder snorkel with the same inner diameter as lower vessel, (2) several segment independently flow control side blow gas tubules which circumference arrange in the lower edge of snorkel to control the behaviors of liquid steel and top slag, (3) during vacuum process the desulfurize is achieved by mixing of slag and metal. Using physical simulation, the effects of snorkel side blow and bottom blow on mixing time are studied. The vacuum deep decarbonization and slag-metal mixing desulfurization are studied in industry research, the average end carbon of vacuum decarbonization is 14×10^{-6}, the min is 11×10^{-6}, the average slag-metal mixing desulfurization rate is 73%, the max is 81%.

Key words: cylinder snorkel refining furnace-MSR, vacuum refining, desulfurization, decarbonization

大型转炉高供氧强度吹炼的水模实验

杨文远[1]，冯　超[2]，王明林[1]，吕英华[2]，胡砚斌[1]，彭小艳[3]

（1. 钢铁研究总院工艺所，北京　100081；2. 中钢集团鞍山热能研究院有限公司设备研制厂，辽宁鞍山　114044；3. 新冶高科技集团有限公司热工装备与材料事业部，北京　100081）

摘　要：为使我国大型转炉的生产率达到国际先进水平，在 300~350t 大型氧气转炉上实现 4.0~5.0Nm³/（t·min）的高供氧强度吹炼，设计了新的大流量氧枪喷头，并在 1:10 的有机玻璃模型上进行氧射流与熔池作用的水模实验。大流量氧枪喷头的主要参数为：喷头孔数 6~8 个，采用双角度交错布置；喷孔倾角 10~17°，喷孔出口马赫数 2.0~2.2。分别测定了枪位高度 1.8~2.6m 时的喷溅率、熔池混匀时间和穿透深度。研究结果表明，大流量新喷头的喷溅量和

射流对熔池的穿透深度都在转炉正常吹炼范围内，熔池混匀时间平均缩短 6s，泡沫渣可降低喷溅率 50%。大流量新喷头良好的吹炼性能，为大型转炉高供氧强度吹炼的氧枪喷头设计提供了可靠的数据。

关键词：顶吹转炉，供氧强度，喷溅，穿透深度，泡沫渣

Water Model Experiment of High Supplying Oxygen Blowing in Large Converter

Yang Wenyuan[1]，Feng Chao[2]，Wang Minglin[1]，Lv Yinghua[2]，
Hu Yanbin[1]，Peng Xiaoyan[3]

(1. Metalllurgical Technology Institute, Central Iron and Steel Research Institute, Beijing 100081, China;
2. Equipment Factory, Sinosteel Anshan Research Institute of Thermo-energy Co., Ltd., Anshan, 114044, China;
3. Thermal Equipment and Material Department, New Metallurgy Hi-Tech Group Co., Ltd., Beijing 100081, China)

Abstract: In order to make the productivity of large converter in our country reach to the international advanced level, a new oxygen lance nozzle was designed to carry out high oxygen supply intensity blowing in the range of 4.0~5.0 $Nm^3/(t\cdot min)$ in 300~350 tons of large oxygen converter, and the water model of oxygen jet and molten bath was carried out in the 1:10 organic model. The parameters of the nozzles are as follows: 6~8 holes, two kinds of angles staggered arrangement, 10 °~17 ° inclination angle and mach number is 2.0~2.2. Splash rate, mixing time and penetration depth were measured when the lance level was between 1.8 m and 2.6 m. The results indicate that the splash rate of the new large flow oxygen lance nozzle and penetration depth of oxygen jet on the liquid bath are in the normal blowing range of large-scale converter. The mix time of liquid bath decreases 6 s at the average level, as well as foaming slag can reduce splash rate by 50% in water model test. The exploratory test provides reliable data for nozzle design in large converter with high intensity of oxygen blowing as the new large flow oxygen lance nozzle has the excellent blowing performance.

Key words: top blowing converter, oxygen supply intensity, splash, penetration depth, foaming slag

高品质切割钢丝中夹杂物控制技术研究

姜　敏[1]，王新华[1]，王　郢[2]，王昆鹏[2]，赵昊乾[2]

（1. 北京科技大学冶金与生态工程学院，北京　100083；
2. 邢台钢铁有限责任公司，河北邢台　054027）

摘　要： 本研究针对切割丝生产中夹杂物控制这一核心技术，原创性提出应将夹杂物控制为 $MnO\text{-}Al_2O_3\text{-}SiO_2/SiO_2$ 复相体系，并深入地揭示了夹杂物的控制机理。通过控制 $MnO\text{-}Al_2O_3\text{-}SiO_2$ 夹杂物中析出 SiO_2 而形成 $MnO\text{-}Al_2O_3\text{-}SiO_2/SiO_2$ 复相结构，利用不变形的 SiO_2 在多道次热轧与冷拉拔过程中分割塑性变形的 $MnO\text{-}Al_2O_3\text{-}SiO_2$ 母体相，实现夹杂物最终尺寸的极细小化。基于这一夹杂物控制目标，本研究首次在国内开发成功高品质切割丝（0.11~0.12mm 直径规格），连续拉拔超过 29000km 以上不断丝。

关键词： 夹杂物，复相，炉渣，硅锰脱氧，变形性能，切割丝

Study on Control of Inclusions in High Quality Saw Wires

Jiang Min[1], Wang Xinhua[1], Wang Ying[2], Wang Kunpeng[2], Zhao Haoqian[2]

(1. School of Metallurgical and Ecological Engineering, University of Science & Technology Beijing, Beijing 100083, China;

2. Xingtai Iron & Steel Company, Xingtai 054027, China)

Abstract: A innovative technical concept was proposed on the control of inclusions in saw wires in this paper, which targets the formation of multi-phased MnO-Al_2O_3-SiO_2/SiO_2 inclusions. Related mechanisms on control of inclusions can be described as following: multi-phased MnO-Al_2O_3-SiO_2/SiO_2 inclusions were featured by un-deformable SiO_2 precipitating on the plastic MnO-SiO_2-Al_2O_3 inclusion matrix; (2) Because of distinctive deformability, the un-deformable SiO_2 would help to "cut" the ever-deforming MnO-SiO_2-Al_2O_3 matrix into tiny strips with the proceeding of multi-pass hot rolling and cold drawing; (3) in the end, extremely minized sizes of inclusions would be obtained. Based on this method, domestic production of saw wire (with diameter about 0.11~0.12mm) in China was successfully realized for the first time, which can be continuously cold drawn to a length of more than 29000km without breakages.

Key words: inclusion, multi-phase, refining slag, Si-Mn deoxidization, deformability, saw wire

S355NL 风电法兰钢精炼渣系优化与应用

欧西达，杨永超

（山西太钢不锈股份有限公司技术中心，山西太原　030003）

摘　要： S355NL 风电法兰钢的精炼渣系使用的是 CaO-Al_2O_3-SiO_2 渣系，在生产过程中发现，该渣系埋弧效果不好，精炼结束后结壳严重，总氧含量较高。针对以上问题，通过热力学和动力学分析，设计了目标精炼渣系，在电炉炉后使用复合脱氧方式进行精炼渣系优化后，降低了炉渣碱度，总氧含量由 13-16ppm 降低到 6-8ppm；炉渣发泡性能和流动性良好，解决了炉渣埋弧效果不好和精炼结束后炉渣结壳的问题，并通过减少原辅料加入量降低了成本 10.20 元/t，取得了良好的效果和经济效益。

关键词： LF 炉，精炼渣，优化，脱氧

Optimization and Application of Refining Slag in Ladle Furnace

Ou Xida, Yang Yongchao

(Technology Centerof Shanxi Taigang Stainless Steel Co., Ltd., Taiyuan 030003, China)

Abstract: S355NL wind power flange steel refining slag system was CaO - Al_2O_3 - SiO_2 slag series.In the process of production,this LF slag was found to be poor effect ofsubmerged arc. After therefining, the crusts were serious and the total oxygen content was high. To solve above problems, through the analysis of thermodynamics and kinetics, a new refining slag system was designed, and it was added after the electric furnace using compound deoxidizing manner. Through the optimization of refining slag, the basicity of slag reduced and the total oxygen content decreased from 13-16ppm to 6-8ppm; the slag foaming performance and fluidity were well, which solved the problem of the submerged arc and the slag crust after refining. By reducing the amount of raw materials,this new process reduced the cost of10.20yuan/ton, and achieved good results and economic benefits.

Key words: LF(ladle furnace), refining slag, optimization, deoxidation

首钢转炉炼钢高效复吹技术开发与应用

高　攀[1,3]，李海波[1,3]，郭玉明[1]，张　勇[1]，赵晓东[1]，
于会香[2]，王新华[1,3]

（1. 首钢集团有限公司，北京　100043；2. 北京科技大学冶金与生态工程学院，北京　100083；
3. 绿色可循环钢铁流程北京市重点实验室，北京　100043）

摘　要： 为解决转炉高炉龄条件下，由于采用低底吹搅拌强度和溅渣护炉工艺导致的转炉实际底吹效果难以保证的问题，首钢股份公司迁安钢铁公司（首钢迁钢）进行了转炉炼钢高效复吹技术开发。开发过程中，形成了一系列关键工艺技术，主要包括：底吹风口布置与熔池流场优化、溅渣护炉条件下的底吹风口裸露技术、底吹风口长寿命技术等，并开展了底吹模式优化，实现了转炉全炉役底吹效果稳定控制，在炉龄 6500 炉左右时，炉役后期终点钢水碳氧积依旧能稳定控制在 0.0020 以下。本工艺技术应用后，生产低碳、超低碳钢（终点碳为 0.025~0.05%），吹炼终点钢水氧含量较前工艺平均降低 0.025% 以上，终渣 T.Fe 含量降低 4% 以上，耐火材料侵蚀明显降低，转炉脱磷效率明显提高。

关键词： 转炉，复合吹炼，底吹搅拌强度，碳氧积，钢水氧含量，脱磷

Development and Application of High Efficiency Blowing Technology of Converters in Shougang

Gao Pan[1,3], Li Haibo[1,3], Guo Yuming[1], Zhang Yong[1], Zhao Xiaodong[1],
Yu Huixiang[2], Wang Xinhua[1,3]

(1. Shougang Group Co., Ltd., Beijing 100043, China; 2. University of Science and Technology Beijing,
Beijing 100083, China; 3. Beijing Key Laboratory of Green Recyclable Process for Iron &
Steel Production Technology, Beijing 100043, China)

Abstract: In order to solve the problem that the actual converter bottom blowing effect was difficult to guarantee due to low bottom blowing intensity and the slag splashing technology under high converter campaign life, high efficiency bottom blowing technology for converter steelmaking was developed in Shougang Co., Ltd. (SGQG). During the development process, a series of key technologies such as: optimization of bottom blowing tuyeres arrangement and pool mixing flow field, exposure technology of bottom blowing tuyeres under the conditions of slag splashing, bottom blowing tuyeres long life technology, and optimization of bottom blowing mode, the stable control of converter bottom blowing effect was achieved, even at the late age of the campaign, the carbon oxygen product can still be less than 0.0020 under the converter campaign life of about 6500. By applying the above technologies, in the production of ULC and IF steel (converter end point carbon content is 0.025~0.05%), the average reduction of converter end oxygen content is more than 0.025%, T.Fe content in end slag is more than 4%, and the erosion of refractory materials was significantly reduced, converter dephosphorization efficiency was significantly improved.

Key words: converter, combined blowing, bottom blowing intensity, carbon oxygen product, oxygen content of molten steel, dephosphorization

基于 BOF-RH-LF-ESP 生产 SPHE 炼钢工艺的优化与实践

谷庆龙，陆显然，杨兆成，李玉春

（日照钢铁控股集团有限公司，山东日照　276806）

摘　要： 深冲用 SPHE 产品具有强度高、韧性好，易于加工成型及良好的可焊接性等优良性能，被广泛用于船舶、汽车、压力容器等制造行业。ESP 作为短流程连铸连轧生产技术的代表作，以 ESP 为依托生产深冲级别 SPHE 冷轧基板具有很高的经济优势，同时也对常规的炼钢工艺提出了更深层次的要求、更大的挑战。本文首先对冷轧 E 料生产流程做简单介绍，其次重点分析冷轧 E 料在 BOF、RH、LF 工序上的成分和工艺控制，最后对 ESP 上生产冷轧 E 料的成本优势进行阐述。

关键词： SPHE BOF，RH，LF，工艺

BOF-RH-LF-ESP Process Production of SPHE Steelmaking Process Optimization and Practice

Gu Qinglong, Lu Xianran, Yang Zhaocheng, Li Yuchun

(Rizhao Steel Holding Group Co., Ltd., Rizhao 276806, China)

Abstract: SPHE production has high strength, good toughness, easy to processing molding, good weldability and other excellent properties. They are widely used in shipbuilding, automobile, pressure vessel and other manufacturing industries. ESP as the representative work of continuous casting and rolling production technology, short process, which is based on the ESP production level of deep drawing SPHE cold-rolled base board has the very high economic advantage, and also puts forward deeper on conventional steelmaking process requirements, and even more of a challenge. This article first cold-rolled SPHE production process to do a simple introduction, secondly analyzed cold-rolled SPHE in BOF, RH, LF process on the composition and process control, the last of the production of cold rolled on ESP E cost advantage.

Key words: SPHE BOF, RH, LF, process

大型转炉复吹与炉龄同步技术在宣钢的研究与应用

郑抗战，于春强

（宣化钢铁集团有限责任公司，河北宣化　075100）

摘　要： 复吹工艺是从转炉底部吹入少量惰性气体，改善顶吹转炉搅拌力不足的冶炼工艺，在生产实践中为提高复吹效果，宣钢二钢轧厂炼钢作业区在转炉冶炼工艺上积极创新，并采取多项措施加强对复吹系统的设备维护和工艺优化。采用优质镁碳砖及高寿命底吹供气元件材质、合理炉底砌筑工艺、优化转炉冶炼工艺，加强复吹设备及维护

制度方面的改进，使复吹供气元件的寿命大大提高。

关键词：复吹，炉龄，同步，底吹元件

Combined Blown and Life Synchronization Technology in the Research and Application of Xuanhua Steel

Zheng kangzhan, Yu Chunqiang

(Xuanhua Steel Refco Group Ltd., Xuanhua 075100, China)

Abstract: The combined blowing process from converter bottom blowing a small amount of inert gas, improve the smelting process of BOF stirring force is insufficient, in the production practice to improve the blowing effect of Xuanhua Steel, two steel rolling plant of steelmaking operation area and positive innovation in the smelting process, and to take various measures to strengthen the blowing system of equipment maintenance and process optimization. The quality and service life of MgO-C bottom blowing elements material, reasonable bottom laying process, optimization of converter smelting process, improving blowing equipment and maintenance system, the complex blowing elements in life is greatly improved.

Key words: blowing, furnace synchronous, bottom blowing elements

宣钢 150t 转炉氧枪粘枪原因及处理方法

刘　琛，陈东辉，吕艳伟

（宣化钢铁集团有限责任公司，河北宣化　075100）

摘　要：氧枪系统是转炉生产冶炼过程中的重要环节，氧枪系统的正常与否直接关系到转炉以及整个生产系统的运行。本文分析了宣钢 150t 转炉氧枪粘枪的原因并且有针对性的提出合理的操作方法、控制措施和解决办法，为保证氧枪减少粘枪，转炉平稳生产提供了实践指导。

关键词：喷溅，氧枪，粘枪

Onverter Oxygen Lance Causes and Treatment Methods in Xuanhua Steel

Liu Chen, Chen Donghui, Lv Yanwei

(No.2 Sreel Making and Rolling Plant of Xuanhua Steel, Xuanhua 075100, China)

Abstract: Oxygen lance system is an important part in the process of converter production and smelting. The normal operation of oxygen lance system is directly related to the operation of converter and the whole production system. This paper analyzes the causes of gun sticking in 150t converter oxygen lance and put forward the operation method, reasonable control measures and solutions, to ensure the reduction of oxygen lance sticking, and provides practical guidance and smooth production of converter.

Key words: splashing, lance, sticking gun

氮的行为及控制的一些看法

车从荣，王建宇，张益民，蒋栋初，韦　波

摘　要：氮在大多数特殊钢产品中是一种有害的元素，如在钢液凝固过程中析出的 TiN 夹杂尺寸与钢液中氮含量及其凝固偏析有重要关系，TiN 对高强度材料疲劳性能的影响远比氧化物夹杂物大。

关键词：钢液凝固，TiN，凝固偏析，夹杂物

Abstract: Nitrogen is harmful for most special steel products. For example, the sized of TiN inclusions precipitated during solidification is related to the nitrogen content of molten steel and segregation. TiN inclusion has bad effect on the fatigue life of high strengh steel, and the effect is much worse than oxide inclusions.

Key words: solidification of molten steel, TiN inclusion, segregation, inclusions

KR 铁水脱硫渣再循环利用的可行性研究

赵中福

（宝钢股份武汉钢铁有限公司条材厂，湖北武汉　430083）

摘　要：本文对 KR 脱硫渣回收再利用于 KR 脱硫的必要性和 KR 脱硫渣的特性进行了分析，通过实验室铁水脱硫试验，对影响铁水脱硫的热力学因素如不同 KR 脱硫渣的比例、CaF_2 含量、脱硫渣碱度、脱硫渣的熔点、铁-渣硫分配比和脱硫渣的硫容量进行了全面的分析讨论，通过实验室大量的试验，论证了 KR 脱硫渣应用于 KR 铁水脱硫的可能性和实用性。

关键词：KR 脱硫渣，铁水脱硫，可行性

Feasibility Study on Hot Metal Desulfurizing Flux with KR Recycling Slag

Zhao Zhongfu

(Long Products Plant Wuhan Iron & Steel Co., Ltd. of Baoshan Steel, Wuhan 430083, China)

Abstract: The characteristics of KR recycling slag was studied, it was reused for hot metal desulfurization by adding some other new fresh reagents in laboratory. The kinetic and thermodynamic factors, which influenced the effect of hot metal desulfurization, such as KR slag weight proportion, CaF_2 content, basicity of desulfurizing slag, stirring power, fusion temperature of desulfurizing flux, sulfur distribution ratio (Ls), and sulfide capacity (C_S) were investigated comprehensively. The practicality and of desulfurizing flux containing KR slag was illustrated by experiments in laboratory.

Key words: KR recycling slag, hot metal desulfurizing, feasibility

顶底复吹转炉底吹工艺优化

韩鹏龙，王建林，赵彦岭，张进红

（邢台钢铁有限责任公司，河北邢台　054027）

摘　要： 介绍了转炉底吹系统现状，并对其进行工艺优化，实现了由静态控制向动态控制的转变。优化后降低了渣中的全铁含量，提高了转炉的整体脱磷率及终点的余锰含量，降低了钢铁料消耗及合金消耗成本；并使平均枪龄由27.18炉提高到150.4炉，杜绝了环保事故的发生，降低了能源介质成本约4.63万元，使透气砖的使用寿命基本达到与炉龄同步。

关键词： 转炉，底吹，工艺优化

Process Optimization of Bottom Blowing Converter

Han Penglong, Wang Jianlin, Zhao Yanling, Zhang Jinhong

(Xingtai Iron and Steel Compangy, Xingtai 054027, China)

Abstract: The present situation of converter bottom blowing system is introduced, and its process optimization is carried out to realize the transition from static control to dynamic control. After optimization, the total iron content in the slag is reduced, the dephosphorization rate of the converter and the residual manganese content at the end point are improved, and the consumption of steel and the cost of the alloy are reduced. And the average gun age increased from 27.18 furnace to 150.4 furnace, put an end to the occurrence of environmental accidents, reducing the cost of energy media 46,300 yuan, so that the basic life of the ventilation brick to achieve synchronization with the furnace age.

Key words: converter, bottom blowing, process optimization

Q235B 板不进精炼生产工艺的实践

蔡常青

（福建三钢闽光股份有限公司炼钢厂，福建三明　365000）

摘　要： 福建三钢闽光股份有限公司自2012年4月份开始了Q235B板坯不进精炼生产工艺试验，通过转炉加强脱氧及定氧措施解决板边部气孔问题；通过钢包顶渣改质和保证吹氩时间，解决大型夹杂物导致Q235B冷弯开裂；通过选择合适的中间包塞棒和降低钢中氧含量解决中间包寿命低的难题，实现了规格小于50mm Q235B板坯不进精炼直上连铸浇铸的常态化生产，质量稳定，取得可观效益。

关键词： Q235B，不精炼，气孔，冷弯开裂，塞棒侵蚀

Practice of Q235B Steel without Refining Process

Cai Changqing

(Steelmaking Plant, Fujian Steel Minguang Co., Ltd., Sanming 365000, China)

Abstract: The production test of Q235B steel without refining at Fujian Sanming Iron and Steel Company in April 2012. Through strengthening deoxidation capacity and equipping with oxygen probes in converter steelmaking process, the edge gassiness problem was solved. By altering the slag modifier in ladel and guarantying the blowing argon time, the cold cracking of Q235B caused by large-scale inclusions was solved. By choosing reasonable tundish stopper and reducing the oxygen content, the problem of low service life of tundish was solved. Now, Sanming Steel has achieved a normal production of Q235B (the size is smaller than 50mm) without refining process, quality is stable and considerable benefits will be obtained.

Key words: Q235B，not refining，gassines，cold-bending cracking，stopper erosion

钙基氧化物冶金生产技术的发展与展望

赵福才，刘 创，孔维明

（西安建筑科技大学冶金工程学院，陕西西安 710055）

摘 要： 氧化物冶金技术解决了大线能量焊接粗大的 HAZ 组织缺陷问题，成为研究的重点。本文简单介绍了氧化物冶金技术的产生、发展，目前研究的重点和存在的问题以及国内外钙基氧化物冶金技术研究的概况。借助氧化物冶金思想，提出了钙基氧化物冶金生产技术，分析了氧化物冶金生产技术与氧化物冶金的区别，简述了钙基氧化冶金生产技术在普通建筑钢材和特殊船板钢的应用，并对此技术进行了展望。

关键词： 钙基氧化物冶金生产技术，有效夹杂物，形核，针状铁素体

Development and Prospect of Calcium Oxide Metallurgical Production Technology

Zhao Fucai, Liu Chuang, Kong Weiming

(College of Metallurgical Engineering, Xi'an University of Architecture and Technology, Xi'an 710055, China)

Abstract: Oxides metallurgy technology has solved the big line energy welding HAZ tissue defects,for which becomes the focus of research. This article simply introduces the generation, development, current research focus,problems and the general situation of the study of calcium oxide metallurgy technology at home and abroad of oxide metallurgy technology. With the aid of oxides metallurgy thought, calcium oxide metallurgy technology is put forward, the difference between the oxide metallurgy technology and oxides metallurgy is analyzed, Calcium oxide metallurgy technology applications are briefly discussed in the General construction steel and special ship plate steel, and this technology is prospected.

Key words: calcium oxide metallurgy technology, effective inclusion, nucleation, acicular ferrite

提高废钢比降低铁水消耗的可行性分析

景琳琳，周　详

（陕西汉中钢铁有限责任公司，陕西汉中　724200）

摘　要： 本文通过热平衡理论计算，对比不同铁水温度，不同铁水硅成分下的废钢加入量及热量消耗。得出将铁水温度提高稳定到 1350℃，铁水硅提高到 0.5%，采取提高热收入，降低热支出的相关技术措施，可将废钢比由 12.95% 提高到 14.54%，降低铁水消耗，从而降低生产成本。

关键词： 热平衡理论计算，铁水温度，铁水硅，废钢比，铁水消耗

Feasible Analysis of Reducing Metal Consumption Improving Scrap Ration

Jing Linlin, Zhou Xiang

(Hanzhong Iron & Steel Co., Ltd. of Shanxi Iron & Steel Group Co., Ltd., Hanzhong 724200, China)

Abstract: Based on the heat balance calculation, comparing different hot metal temperature, different components of hot metal silicon and heat for the amount of scrap consumption. It is concluded that the molten iron temperature stable increase to 1330℃, hot metal silicon increased to 0.5%, raisingheatincome, heat reduce spending related technical measures, scrap ration can be increased from 13.24% to 14.54%, reducing consumption of molten iron, therby reducing the cost of production.

Key words: heat balance calculation, hot metal temperature, hot metal silicon, scrap ratio, consumption of hot metal

特厚超宽钢板冶炼工艺生产实践

赵红康，赵向政

（舞阳钢铁有限责任公司，河南舞阳　462500）

摘　要： 通过采用高铁水比例操作技术、电炉终点碳控制、电炉出钢过程不同预脱氧制度、炉渣碱度的控制以及采用新型保温剂等技术，大幅度提高了钢水纯净度，改善了钢锭内部凝固质量，保证了采用扁钢锭生产要求 JB/T4730.3—2005 标准 150~210mm 特厚钢板，要求 EN10160 S2E2 标准 260mm 以上特厚钢板，要求 EN10160 标准 S1E1 级或 SEL072 标准Ⅲ级 350mm 以上特厚钢板的探伤合格率，从而大幅度的降低生产成本。

关键词： 特厚钢板，无损探伤，新型保温剂，高碱度炉渣

Production Practice of Special Thick Ultra Wide Steel Plate Smelting Process

Zhao Hongkang, Zhao Xiangzheng

(Wugang Iron and Steel Co., Ltd., Wuyang 462500, China)

Abstract: By using a high proportion of iron water operation technology, electric furnace end-point carbon control, different pre deoxidation system in the process of electric furnace tapping, control of slag basicity and adopt new insulation agent and other technologies, the purity of molten steel was greatly improved, and the solidification quality of ingot was improved. Thus ensure the flat steel ingot production requirements of JB/T4730.3—2005 standard 150~210mm thick steel plate, EN10160 S2E2 requires more than standard 260mm thick steel plate, inspection requirements above EN10160 standard S1E1 or SEL072 standard grade 350mm thick plate pass rate, thereby greatly reducing the cost of production.

Key words: extra heavy steel plate, nondestructive inspection, new insulation agent, high basicity slag

舞钢电炉热装铁水冶炼工艺实践与改进

贺春阳

（舞阳钢铁有限责任公司，河南舞阳 462500）

摘　要： 通过对电炉热装铁水工艺实践的研究，找出舞钢 90t 电炉目前设备下的最佳热装铁水比例为 40%左右，但随着供氧强度的增加，能有效提高热装铁水比例。热装铁水模式下冶炼前期炉渣碱度在 1.0~2.0 之间时，易出现乳化跑钢现象，前期加灰提高炉渣碱度至 2.5 左右，能有效防止炉渣乳化喷溅跑钢。优化改进后的热装铁水冶炼工艺，能较好的适应热装铁水比例 30%~80%的变化，大幅缩短冶炼周期，降低冶炼成本，满足电炉热装高磷铁水情况下出钢磷含量的要求，冶炼低磷钢的出钢磷含量稳定控制在 0.008%以下。

关键词： 电炉，热装铁水，碱度，造渣制度

Practice and Improvement of Hot Metal Charging Smelting Process in Wusteel Electric Furnace

He Chunyang

(HBIS Group Co,. Ltd., Wusteel Company, Wuyang 462500, China)

Abstract: By studying the practice of hot metal charging in electric furnace, it is found that the optimum hot metal ratio is about 40% in the 90t electric furnace in Wusteel. But with the increase of oxygen supply intensity, the proportion of hot metal can be effectively raised. When the slag basicity is between 1.0~2.0 in hot charging mode, the phenomenon of emulsified running steel is easy to occur. The addition of ash in the early stage can increase the basicity of slag to about 2.5, which can effectively prevent the slag ran out of the furnace door. With the hot metal charging smelting process optimization improved heat, can better adapt to changes in the proportion of hot metal charging 30%~80%, greatly shorten the smelting cycle, reduce smelting costs, meet the requirement of the target phosphorus content in the case of high phosphor, control of smelting low phosphorus steel tapping stable phosphorus content below 0.008%.

Key words: electric furnace, hot metal charging, basicity, slag-making

3　轧制与热处理

炼铁与原料
炼钢与连铸
★ 轧制与热处理
表面与涂镀
金属材料深加工
钢铁材料
汽车钢
海洋工程用钢
轴承钢
电工钢
粉末冶金
非晶合金
高温合金
耐火材料
能源与环保
分析检测
冶金设备与工程技术
冶金自动化与智能管控
冶金技术经济

轧机柔性辊型调控技术实验与仿真研究

杜凤山，刘文文，冯岩峰，孙静娜

（燕山大学国家冷轧板带装备及工艺工程技术研究中心，河北秦皇岛 066004）

摘　要：柔性辊型调控技术在板形控制方面具有巨大的优势。为了使轧辊具备柔性辊型调控能力，本文提出了辊型电磁调控技术，设计制造了单一电磁棒的辊型电磁调控实验平台，建立了电磁-热-力耦合仿真模型。通过实验对所建仿真模型进行校核，发现模型具有较高精度。依托校核后模型，分析了单一电磁棒和多组电磁棒时电磁调控轧辊辊型曲线及特征。结果表明，辊型电磁调控技术可实现柔性辊型调控，而采用单一电磁棒的电磁调控轧辊可形成高阶凸函数辊型曲线，采用多组电磁棒的电磁调控轧辊可形成由多段高阶凸函数组合而成的高阶辊型曲线。

关键词：电磁调控轧辊，电磁棒，柔性调控，高阶辊型曲线

Experimental and Simulation Study on Flexible Roll Profile Control Technology in Rolling Mill

Du Fengshan, Liu Wenwen, Feng Yanfeng, Sun Jingna

(National Engineering Research Center for Equipment and Technology of Cold Strip Rolling, Yanshan University, Qinhuangdao 066004, China)

Abstract: Flexible roll profile control technology has a huge advantage in flatness control. In this paper, in order to make the roll with flexible roll profile control capability, the roll profile electromagnetic control technology was proposed, a roll profile electromagnetic control experimental platform with single electromagnetic stick was designed and built, and electromagnetic-thermal-mechanical coupled simulation model was established. The simulation model was checked through the experiment, it found that the model had high accuracy. Based on the model, roll profile and characteristics of electromagnetic control roll which have single electromagnetic stick or multiple sets of electromagnetic sticks were analyzed. The results showed that the roll profile electromagnetic control technology can achieve flexible roll profile control, and electromagnetic control roll with a single electromagnetic stick can form a high-order convex function roll profile curve, and electromagnetic control roll with multiple sets of electromagnetic sticks can form a high-order roll profile curve which was made up of multi-stage high-order convex functions.

Key words: electromagnetic control roll, electromagnetic stick, flexible control, high-order roll profile curve

大型镶套式支撑辊过盈量调控技术的研究

杜凤山，冯岩峰，刘文文，孙静娜

（燕山大学国家冷轧板带装备及工艺工程技术研究中心，河北秦皇岛 066004）

摘　要：大型镶套式支撑辊由于刚度大、韧性好、硬度高、耐磨性良好等优点，在实际生产中获得了大量的应用。由于受到周期性作用力，其辊套与芯辊之间的过盈量逐渐降低导致轧辊失效，严重影响了轧辊使用寿命。为了解决

该问题，本文提出了大型镶套式支撑辊过盈量调控技术，并以 $\phi1400\times1420$ 镶套式支撑辊作为研究对象，建立了电磁-热-力耦合数学模型，并依托物理模拟平台对模型进行验证。利用验证后的模型，分析了本技术对辊套和芯辊间过盈量的调节作用。结果表明：该技术可以有效的增加补偿辊套和芯辊接触面之间的过盈量，补偿界面磨损导致的孔隙；同时增加支撑辊凸度，补偿支撑辊挠曲量，也可以使支撑辊出现特定的辊型，从而改善板形。

关键词：镶套支撑辊，过盈量，接触压力，辊型曲线

Study on the Control Technique for the Interference of Large-size Sleeved Backup Roller

Du Fengshan, Feng Yanfeng, Liu Wenwen, Sun Jingna

(National Engineering Research Center for Equipment and Technology of Cold Strip Rolling,
Yanshan University, Qinhuangdao 066004, China)

Abstract: Large-size sleeved backup roller was becoming widely used because of its high stiffness, ductile, hardness and wear resistance. Long-time cyclic heavy-load could decline magnitude of interference of roll shell and roll core, which would shorten the service life of roller. Control technique for magnitude of interference of large-size sleeved backup roller was proposed to prevent the decline of interference, and the finite element model of a $\phi1400\times1420$ size sleeved backup roller was created and verified. The regulating effect on magnitude of interference of the technique was analyzed with the finite element calculation. The analysis shows that the technique can successfully compensate interference of large-size sleeved backup roller, and also can produce specific roll shape to obtain good profiles of strip.

Key words: sleeved backup roller, magnitude of interference, contact pressure, roll shape curve

轧钢加热炉分段燃烧优化控制系统实践

丁　毅，史德明，周劲军，翟　炜，刘自民，刘其明

（马钢（集团）控股有限公司，安徽马鞍山　243000）

摘　要： 加热炉是轧钢工序的主要耗能设备。在满足轧制工艺要求的同时，降低氧化烧损和能耗是轧钢加热炉生产工艺追求的目标。轧钢加热炉氧化烧损偏高，影响轧钢成材率，增加生产成本。针对炉内气氛、加热温度等影响氧化烧损及能耗的主要因素，本文采用分段燃烧控制技术及炉内钢坯在线红外精确测温技术对其进行了燃烧优化，旨在对加热炉控制最终目标值（各加热段烟气中 O_2 和 CO 及钢坯炉内温度）进行在线监测，指导加热炉操作。通过在 1580 加热炉上的成功实践，相对单位面积烧损率减少 15% 以上。以加热炉年产量 250 万吨测算，年综合经济效益 900 万元以上。

关键词：轧钢加热炉，分段燃烧，控制系统

Practice of Optimized Control System for Sectional Combustion of Steel Rolling Heating Furnace

Ding Yi, Shi Deming, Zhou Jinjun, Zhai Wei, Liu Zimin, Liu Qiming

(Masteel (group) Holding Co., Ltd., Maanshan, 243000, China)

Abstract: Heating furnace is the main energy dissipation equipment in rolling process. While satisfying the requirements of rolling process, reducing oxidation loss and energy consumption is the goal of production technology of steel rolling reheating furnace. The oxidation burn loss is high in steel rolling heating furnace, which affects the yield of rolling mill and increases production cost. The main factors influencing the oxidation loss and energy consumption of the furnace atmosphere, heating temperature, etc. In this paper, the combustion optimization is carried out by using segmented combustion control technology and on-line infrared precise temperature measurement technology of billet in furnace, aiming to monitor the final target value (O_2 and CO in the flue gas of each heating section and temperature of billet furnace), and instruct the heater operation. By the successful practice of 1580 reheating furnace, the rate of burn loss is decreased by 15%. With the annual output of 2.5 million tons of heating furnace calculation, annual comprehensive economic benefits of more than 9 million yuan.

Key words: steel rolling heating furnace, segmented combustion, control system

纵向变厚度（LP）钢板轧制技术研究

丛津功，李新玲，李靖年，姚　震，隋广雨

（鞍钢股份有限公司鲅鱼圈钢铁分公司厚板部，辽宁营口　115007）

摘　要： 国内对纵向变厚度钢板的研究尚在起步阶段，且进展非常缓慢，主要在于钢板在楔形段轧制过程中，金属沿长度与宽度的延伸量无法精确计算，导致前期坯料设计精确度不够、轧制过程中变厚度阶段尺寸规格误差较大以及没有稳定轧制工艺等诸多原因。本文主要利用原有的理论知识，研究纵向变厚度轧制过程中金属流动规律，制定出合理的坯料设计、加热工艺以及轧制过程中金属沿长度、宽度方向的流动规律。

关键词： 纵向变厚度，坯料设计，轧制技术，形状控制

The Research of Longitudinally Profiled (LP) Plate Rolling Technology

Cong Jingong, Li Xinling, Li Jingnian, Yao Zhen, Sui Guangyu

(Bayuquan Iron & Steel Subsidiary Company of Angang Steel Co., Ltd., Yingkou 115007, China)

Abstract: Domestic study of longitudinally profiled steel plate is still in its infancy, and the progress is very slow, mainly lies in the steel plate in the wedge rolling process, the metal along the length and the width elongation can't accurate calculation, lead to early in the process of rolling billet design accuracy is not enough, variable thickness phase size error is bigger and the lack of stable rolling technology, and many other reasons. In this paper, we use the original theory knowledge, the vertical variable thickness metal flow law in the process of rolling, formulate reasonable blank design in the process of rolling, heating process and flow law of metal along the length, width, in combination with the practical situation of angang

5500 production line, summarized a set of longitudinal variable thickness steel rolling technology.

Key words: longitudinal variable thickness, slab design, rolling, shape control

板带轧制过程动态理论的建立及应用发展过程

张进之, 吴增强

（中国钢研科技集团有限公司, 北京　100081）

摘　要： 连轧的发明是金属塑性加工工艺最重要的进步。连轧技术发明以来，发生了两次革命性进步。第一次连轧技术革命发生在西方工业发达国家，首先是计算机在连轧机上应用，其实现的理论基础是英国人开创的以秒流量相等为基本方程的影响系数仿真计算，弄清了控制量（辊缝和速度）与轧件厚度、张力、压力、力矩和功率的定量关系；其次是美国人的以张力微分方程和厚度延时方程为动态的过程方程的定量分析。之后日本人进一步发展了英、美的技术，使连轧技术发展到了一个新的高度。目前国内外应用的还是日本、德国为代表的连轧技术。第一次连轧技术革命是在轧制工艺传统理论基础上，加上计算机和控制理论发展的应用，而没有轧制过程的动态解析理论（国外由计算机仿真实验方法可代替轧制过程动态理论）。第二次连轧技术革命是在我国发生的，其理论基础为连轧张力公式、动态设定型变刚度厚控方法（DAGC）、解析板形刚度理论和Φ函数及 dΦ/dh。这些动态理论创建的应用花了近 60 年的时间。特点是建立轧制过程的广义空间（辊缝、轧辊速度）基础上，由数学分析方法建立起来的新型轧制理论。它已在生产上取得明显效果，可以在装备落后的连轧机上使轧件尺寸达到从国外引进的轧机的水平，在从德、日引进的热连轧机上使产品精度大幅度提高。第二次技术革命的特点是在简化装备的条件下，大幅度提高产品质量。最终将其推广应用会改变目前轧制装备极端复杂的状况。

关键词： 厚度，张力，板形，连轧张力理论，DAGC，解析板形刚度理论，Φ函数 dΦ/dh

Establishment and Application Development of Dynamic Theory in Plate Rolling Process

Zhang Jinzhi, Wu Zengqiang

(China Iron & Steel Research Institute Group Co., Ltd ., Beijing 100081, China)

Abstract: The invention of continuous rolling is the most important process for metal plastic working. Since then, two revolutionary processes have occurred.

The first technical revolution of continuous rolling took place in western developed countries. Above all is the application of computer science on continuous mills. The technical base for that was influence coefficient calculation of basic function of equal second flow, which was proposed by British engineers.Then the relationships between controlled variable (roll gap and speed) and the thickness of rolled piece, tension, pressure, torque and power were understood. Engineers in United states then reported the quantitative analysis of process formula, and the dynamic of that is tension and thickness delay differential equations. This was followed by the development of technology from Britain and Unites States in Japan, reaching new levels of continuous rolling. Nowadays, the dominant applications of continuous rolling in China and overseas are imported from Japan and Germany. Based from classical theory of rolling process, the first technical revolution of continuous rolling combined with the application and development of computer science and controlling theories, rather than dynamic analytic theory of rolling process. In foreign countries, the dynamic theory of rolling could be substituted by computational simulation methods.

The second technology revolution of continuous rolling occurred in China, and the theoretical basis for it are continuous rolling tension formula, dynamic automatic gauge control (DAGC) set of variable stiffness thick control methods, the analytic shape stiffness theory, Φ formula and dΦ/dh. It took nearly 60 years for these dynamic theories to be created and applied. The main specialty is the new rolling theory by mathematic analysis methods, which was based on the establishment of generalized space in rolling process. An obvious effect in the production has been obtained, making rolled piece produced by continuous mills with backward equipment reach size of those pieces by imported mills, resulting in higher accuracy of products fabricated by hot continuous rolling mills imported from Germany and Japan. The feature of the second technical revolution is the substantial improvement of products quality under the condition of simplified equipment. The final promotion and application would change the status of extremely complicated rolling equipment.

This paper would give a detailed introduction of the application process of dynamic theory.

Key words: thickness, tension, strip shape, and rolling tension theory, DAGC, the analytic shape stiffness theory, Φ function, and dΦ/dh

热轧带钢超快冷物理模拟及 TRIP 钢相变行为研究

赵日东[1]，黄华贵[1]，陈　雷[1]，高　林[2]，张尚斌[2]

（1. 燕山大学国家冷轧板带装备及工艺工程技术研究中心，河北秦皇岛　066004；

2. 二重集团（德阳）重型装备股份有限公司，四川成都　610052）

摘　要： 现代热轧超快冷装备冷却能力可达 100~200℃/s。为满足超快速冷却条件下材料连续冷却相变行为物理模拟要求，本文通过对传统膨胀法测 CCT 曲线的实验方法进行改进，采用空心圆柱试件和喷气冷却方式，实现在超快速冷却条件下（大于 100℃/s）较稳定的温度控制，提高了实际冷速与设定冷速的一致性。基于新方法，获得了 TRIP780 钢在大冷速范围（1~200℃/s）的 CCT 曲线。同时，对不同冷速下的显微组织与硬度进行了分析测定。结果表明，随冷速的增加，TRIP780 钢硬度逐渐增大，这主要是由于马氏体含量随冷速增加而增大导致的；当冷速超过 100℃/s 时，随冷速增加，试验钢均可获得均匀的全马氏体组织，但硬度基本不再增加，保持在 527HV5。

关键词： 超快冷，物理模拟，TRIP 钢，CCT 曲线，显微组织，硬度

Research on Physical Simulation of Ultrafast Cooling of Hot Rolling Steel Strip and Phase Transition Behavior of TRIP Steel

Zhao Ridong[1], Huang Huagui[1], Chen Lei[1], Gao Lin[2], Zhang Shangbin[2]

(1. National Engineering Research Center for Equipment and Technology of Cold Strip Rolling, Yanshan University, Qinhuangdao 066004, China;

2. Heavy Machinery Research Institute, China National Erzhong Group Company, Chengdu 610052, China)

Abstract: The cooling ability of the modern ultrafast cooling device is up to 100~200 ℃/s. To meet the requirement of the physical simulation for continuous cooling phase transition under ultrafast cooling condition, a new approach was proposed to achieve high cooling rate in traditional expansion method. A hollow cylinder was adopted as test specimen in cooling process, accompanied with air blowing, the temperature and cooling rate can be controlled stably in accordance with the manual setting even under ultrafast cooling rate higher than 100 ℃/s. With this new approach, the CCT curve of TRIP 780 was achieved and the microstructure and hardness under the cooling rate from 1 ℃/s to 200 ℃/s was analyzed. The results show that the hardness of TRIP 780 steel increases with the increase of cooling rate, which is mainly due to the increase of martensite content with the increase of cooling rate. The final microstructure is all of homogeneous martensite when the cooling rate higher than 100 ℃/s, and the hardness is no longer increased and remains at 527HV5.

Key words: ultrafast cooling, physical simulation, TRIP steel, CCT curve, microstructure, hardness

武钢CSP极薄规格轧制中双侧刚度技术集成及实践应用

赵　敏，宋　波，高　智，王　坤，杨　洲，

杨　光，李　波，王　青，赵　强

（武钢有限公司条材厂，湖北武汉　430080）

摘　要： 本文针对CSP极限规格薄材单块轧制时，总结PDA数据、设备精度、板型控制、轧机刚度等相关专业集成技术并实践，形成了一套完整的、具有独立知识产权的短流程极限规格薄材单块轧制技术。目前武钢CSP批量供货最薄规格为1.0mm，并成功批量生产1.1mm厚度出口泰国以及中东地区国家，由于极限规格薄材完全超出了常规热轧产线的生产能力，对CSP产线板型控制能力提出了更高的技术要求。武钢CSP成功以单块轧制的方式连续生产了5卷规格为0.8×1250mm的SPHC-B极薄材，为拓展武钢薄规格热轧产品市场奠定了坚实基础。此前，国内仅有涟钢用半无头、日照用全无头（ESP）轧制方式试制，韩国浦项采用全无头（CEM）轧制方式试制，而武钢CSP产线单块生产0.8mm热轧卷在国内外同类产线尚属首次，标志着武钢CSP产线极薄材生产工艺技术和批量生产能力达到国际领先水平。

关键词： CSP，极薄规格，PDA数据，设备精度，板型控制，轧机刚度，技术集成

Technical Integration and Practical Application of Bilateral Rigidity in CSP Ultra Thin Gauge Rolling of WISCO

Zhao Min, Song Bo, Gao Zhi, Wang Kun, Yang Zhou,

Yang Guang, Li Bo, Wang Qing, Zhao Qiang

(Wuhan Iron & Steel Co., Ltd., Plant of Long Product, Wuhan 430080, China)

Abstract: According to the CSP specification limit single rolling thin material, summarize the PDA data, equipment precision, shape control, mill stiffness and other related professional integration technology and practice, form a complete set, with independent intellectual property rights of short process limit specifications sheet single piece rolling technology. At present, Wuhan CSP batch supply the thinnest specifications for the 1.0mm, and the successful state of mass production of 1.1mm thickness exported to Thailand and the Middle East, due to the limit of thin material specifications completely beyond the production capacity of conventional hot rolling production line, CSP production line shape control ability put forward higher technical requirements. Wuhan Iron and steel company CSP successfully produced 5 coils of SPHC-B extremely thin material of 0.8×1250mm with single block rolling method, which laid a solid foundation for expanding the market of thin gauge hot rolled products of WISCO. Previously, domestic only half Lianyuan Steel without head, with no sunshine head (ESP) rolling trial, South Korea Pohang with no head (CEM) rolling trial, and Wuhan CSP production line of single block production of 0.8mm hot rolled coil production line home and abroad for the first time, marking the Wuhan CSP production line very thin material production technology and production capacity has reached the international advanced level.

Key words: CSP, very thin specification, PDA data, equipment accuracy, plate control, rolling mill rigidity, technology integration

宽厚板轧机功能精度对轧机状态的影响

王新科

（宝钢湛江钢铁厚板厂，广东湛江 524072）

摘　要：介绍了轧机轧辊交叉的关键原因。对轧机辊系交叉带来的轴向力进行深入的研究，分析工作辊与支撑辊交叉带来的轴向力、工作辊与工作辊交叉带来的轴向力，得到轴向力的计算公式并进行计算。然后针对轴向力过大提出了相应的控制措施，以保证轧机的功能精度。

关键词：精度控制，轴向力，精轧机

Precision Influence of Complex Thick Plate Mill

Wang Xinke

(Heavy Plate Plant, Zhanjiang Iron & Steel, Zhanjiang 524072, China)

Abstract: This paper introduces in-depth axial force for roll crossing, and reveals the influence of key process parameters on roll crossing, with analyzing axial force between work roller and back-up roller and between work roll and work roll，it establishes a calculation formula of axial force. To ensure the precision of the finish mill, we put forward some measures.

Key words: precision control, axial force, finish mill

冷却速率对高铝低硅 TRIP 钢组织性能及织构组成的影响

黄慧强，邸洪双，闫　宁，李云龙，邓永刚

（东北大学轧制技术及连轧自动化国家重点实验室，辽宁沈阳 110819）

摘　要：通过对冷轧 TRIP 钢进行连续退火处理，最后冷却阶段采用不同冷却方式进行处理，获得 1 mm 厚的热处理薄板。采用扫描电镜对试样不同组成相进行金相、EBSD 形貌和织构分析，并对不同工艺试样进行力学性能测试。研究表明，连续退火处理后，钢中组织主要由铁素体、贝氏体和残余奥氏体组成。空冷处理后，抗拉强度为 690 MPa，延伸率为 32%。最终冷却方式改为水淬后，实验钢中贝氏体增多，残余奥氏体稳定性增大，抗拉强度增大，延伸率下降。钢中存在 {113} <110>、{111} <112>、{011} <100> 和 {331} <136> 四种主要织构，两种工艺钢中 {113} <110>织构变化不大，空冷改为水淬之后，{011} <100> 和 {331} <136>织构减少，{111} <112>织构增多。

关键词：TRIP 钢，残余奥氏体，力学性能，织构

Effect of Cooling Rate on Microstructure, Mechanical Properties and Texture Evolution of TRIP Steel

Huang Huiqiang, Di Hongshuang, Yan Ning, Li Yunlong, Deng Yonggang

(State Key Laboratory of Rolling and Automation, Northeastern University, Shenyang 110819, China)

Abstract: After continuous annealing the cold rolled TRIP steel sheets were cooled with two different cooling modes and sheets of 1 mm thick were obtained. The phase components, EBSD morphology and texture distribution of the specimen were observed and analyzed with field emission SEM. The retained austenite content was measured by EBSD analysis. Mechanical properties were tested by universal testing machine. The results indicate that the microstructures of specimens were both mainly composed by ferrite, bainite and retained austenite. When cooled in the air, the tensile strength of experimental TRIP steel was 690 MPa and the percentage of elongation was 32 %. After cooled by water quenching, in the microstructure of the steel more bainite was observed. The tensile strength increased and the percentage of elongation decreased. Four textures, {113} <110>、{111} <112>、{011} <100>和{331} <136>, existed in the specimens of two annealing processes. The texture {113} <110> nearly performed the same morphology. The volume fraction of {111} <112>was obvious increased in the steel sheet which was cooled by water quenching than by air cooling.

Key words: TRIP steel, retained austenite, mechanical properties, texture

新型轧制耐磨复合板的试验研究

田志强[1]，吝章国[1]，刘建磊[2]，孙　力[1]，张雲飞[1]，陈振业[1]，赵燕青[1]

（1. 河钢集团钢研总院，河北石家庄　050023；2. 河钢集团舞钢公司二轧厂，河南舞阳　462400）

摘　要： 研究了一种采用直接轧制复合并且免热处理的耐磨复合板，新型耐磨复合板具有良好的结合强度和高的表面硬度，抗剪强度达到420MPa，耐磨层硬度达到HRC55，新型耐磨复合板解决了堆焊耐磨复合板生产效率低，作业环境差的问题，也解决了其他轧制复合耐磨板由于后续调质处理造成的工艺成本增加、板形不良的问题，具有很大的经济效益和社会效益。

关键词： 轧制复合，免热处理，耐磨复合板

Test Research of New Wear-resistant Composite Plate by Rolling

Tian Zhiqiang[1], Lin Zhangguo[1], Liu Jianlei[2], Sun Li[1], Zhang Yunfei[1],
Chen Zhenye[1], Zhao Yanqing[1]

(1. HBIS Group Technology Research Institute, Shijiazhuang 050023, China;
2. HBIS Group Wusteel Company, Wuyang 462400, China)

Abstract: Study on the wear-resistant composite plate, by direct rolling compound and free of heat treatment for the production. The new wear-resistant composite plate with good bonding strength and high surface hardness, shear strength reached 420MPa, wear-resisting layer hardness reached HRC55. New wear-resistant composite plate solves the problems of traditional surfacing production with low efficiency and bad environment, but also solves the other rolling composite wear-resistant plate due to heat treatment causing process cost increase and destroys the final strip shape, which has great economic and social benefits.

Key words: composite rolling, free of heat treatment, wear-resistant composite plate

半连续轧机上实现铁素体轧制关键技术研究

周　旬[1]，王晓东[1]，王建功[1]，夏银锋[1]，赵　虎[2]，艾矫健[1]，
邱晨阳[3]，李　浪[3]，郝磊磊[3]

（1. 首钢京唐钢铁联合有限责任公司，河北唐山　063000；2. 首钢技术研究院薄板所，北京
100043；3. 北京科技大学材料科学与工程学院，北京　100083）

摘　要： 本文研究了半连续轧机上实现铁素体区轧制批量生产的技术。在首钢京唐公司万吨级工业试验及生产的基础上，探讨了铁素体轧制和常规轧制的力学性能、微观组织和表面质量的区别。分析了现场批量生产中粗轧温度、终轧温度、卷取温度、退火温度及轧制润滑控制等典型工艺要点对产品的微观组织和性能的影响。提出采用合理工艺进行铁素体区轧制，退火产品可获得更有利的{111}织构，进而获得更优异的屈强比和塑性应变比，最后探讨了铁素体轧制下一步的发展趋势。

关键词： 半连续轧机，铁素体轧制，力学性能，织构

Research on Key Technologies of Ferrite Rolling on Semi Continuous Rolling Mill

Zhou Xun[1], Wang Xiaodong[1], Wang Jiangong[1], Xia Yinfeng[1], Zhao Hu[2],
Ai Jiaojian[1], Qiu Chenyang[3], Li Lang[3], Hao Leilei[3]

(1. Hot Rolling Department, Shougang Jingtang Iron & Steel Co., Ltd., Tangshan 063000, China;　2. Shougang Research Institute of Technology, Beijing 100043, China;　3. University of Science and Technology Beijing, School of Materials Science and Engineering, Beijing 100086, China)

Abstract: This paper study on technology of ferrite rolling on semi continuous rolling mill. Based on the industrial test in Shougang Jingtang Iron & Steel Company, the differences of the mechanical property and microstructure between the ferrite rolling technology and normal austenite rolling technology were discussed. The effect of rough rolling temperature, finishing temperature, coiling temperature, annealing temperature and rolling lubrication on the mechanical property and microstructure were analyzed. the more favorable {111} crystal texture could formed with reasonable ferrite rolling technology, and then, the better yield ratio and plastic strain ratio could be got. the trend of ferrite rolling technology were discussed at last.

Key words: semi continuous rolling mill, ferrite rolling technology, mechanical property, crystal texture

卷取温度对 Nb-Ti 微合金钢组织与析出行为的影响

许立雄[1]，武会宾[1,2]，汤启波[1]，王鑫田[1]，牛　刚[1]，顾　洋[1]，于新攀[1]

（1. 北京科技大学钢铁共性技术协同创新中心，北京　100083；

2. 北京科技大学高效轧制国家工程研究中心，北京　100083）

摘　要： 本文基于日钢的 ESP(endless strip processing)生产线，介绍了无头轧制过程中卷取温度对 Nb-Ti 微合金钢的显微组织、析出行为及力学性能的影响规律。当卷取温度为 560℃时，卷取前 63.4%的 Nb、44.0%的 Ti 和 74.8%的 C 仍处于固溶状态；当卷取温度从 600℃降低到 560℃，室温组织中针状铁素体明显增多且平均晶粒直径由 2.01 nm 减小到 1.29 nm，组织强化使钢的屈服强度和抗拉强度分别增加了 22 MPa 和 20 MPa。当卷取温度由 600℃升高到 640℃，析出物质量分数由 0.083%增加到 0.110%，且 18 nm 以下的小尺寸析出物的质量百分比由 12.2%增加到 14.7%，此时析出强化使屈服强度和抗拉强度分别增加了 35 MPa 和 42 MPa。因此，当卷取温度由 640℃降低到 560℃时，试验钢的强度先降低后升高，而低温转变组织的增多使钢的延伸率由 18.9%降低到 14.1%。

关键词： Nb-Ti 微合金钢，无头轧制，卷取温度，显微组织，析出行为，力学性能

The Influence of Coiling Temperature on Microstructure and Precipitation Behavior in Nb-Ti Microalloyed Steels

Xu Lixiong[1], Wu Huibin[1,2], Tang Qibo[1], Wang Xintian[1], Niu Gang[1], Gu Yang[1], Yu Xinpan[1]

(1. Collaborative Innovation Center of Steel Technology of USTB, Beijing 100083, China;

2. National Engineering Research Center of Advanced Rolling of USTB, Beijing 100083, China)

Abstract: This paper presents our latest results on how coiling temperature affects the microstructure, precipitation behavior, and mechanical properties of a Nb-Ti microalloyed steel, which was produced on endless strip processing (ESP) line in Rizhao steel and coiled at different temperatures. The results reveal that 63.4 % of Nb, 44.0 % of Ti, and 74.8 % of C were still in solution before coiling at the coiling temperature of 560℃. As the coiling temperatures decreased from 600℃ to 560 ℃, the proportion of acicular ferrites (AF) increased and the average grain size was refined from 2.01 nm to 1.29 nm, then the yield strength and tensile strength of the tested steel increased by 22 MPa and 20 MPa under the effect of microstructural strengthening. As the coiling temperatures increased from 600℃ to 640℃, the mass fraction of the precipitates increased from 0.083 % to 0.110 % and the percentage of fine precipitates (smaller than 18 nm) increased from 12.2 % to 14.7 %, the intense precipitation strengthening increased the yield strength and tensile strength by 35 MPa and 42 MPa respectively. In consequence, as the coiling temperatures decreased from 640℃ to 560℃, the strength of the tested steel decreased first and then increased, while the elongation of steel decreased monotonously from 18.9 % to 14.1 % due to the increasing proportion of AF.

Key words: Nb-Ti microalloyed steels, ESP endless rolling, coiling temperature, microstructure, precipitation behavior, mechanical properties

基于容错服务器实现鞍钢热轧钢卷库管理系统

高 松，车志良，赵 勐，张 喆

（鞍钢股份有限公司热轧带钢厂，辽宁鞍山 114000）

摘 要： 1780 钢卷库管理系统是热轧 MES 系统的重要组成部分，主要负责 1780 线钢卷库物料入库管理、库内管理、包装、分卷、发货管理，同时与公司企业资源计划系统（ERP）、二级过程控制系统进行通讯，收集和传递生产数据，实现信息共享，保证各级系统信息的数据一致性。该系统建立在 Stratus 容错服务器平台上，相比传统服务器集群，具有更高的安全性和可靠性。该系统投入运行以来，充分发挥出钢卷库管理的指导性作用，将鞍钢 1780 生产线的钢卷库管理和 L2 的过程控制、ERP 管理等功能紧密地连接在一起，大大地提高了钢卷库生产管理的自动化水平，优化了生产线人力资源配备，提高了企业的劳动生产力，对促进企业的信息化进程具有重要的理论和实际意义。

关键词： 鞍钢，钢卷库管理系统，MES，Stratus

Implementation of the Management System of Hot-rolling Coil-yard in Ansteel Corp. Based on FT Server

Gao Song, Che Zhiliang, Zhao Meng, Zhang Zhe

(Hot Strip Rolling Mill, Ansteel Corp., Ltd., Anshan 114000, China)

Abstract: The management system of 1780 coil-yard is the key composition part of MES of Hot Strip Rolling Mill. It mainly takes charge of the management of coil in-yard,management of coil-yard,package,HDL,shipment and crane guidance of 1780mm product line. Meanwhile,it communicates with enterprise resource planning system(ERP) of company and L2 process control system,to transfers and gathers various production data,making information sharing possible,and keep the information data consistency at all levels system. This system is built on the platform of stratus fault-tolerant server,compared with traditional cluster server,it will takes us higher safety and reliability.Since the system was brought into operation，it takes guidance for the management of coil-yard,and realizes the high-class technology by firmly connecting the production planning management in 1780mm product line of Ansteel,L2 process control and the production management of ERP.The system greatly promotes the automation level of production.It has an important theoretical as well as actual meaning by accelerating the informatization process of Ansteel group.

Key words: Ansteel, management system of coil-yard, MES, Stratus

热轧高强双面搪瓷钢开发

黄学启，张志敏，李晓林，刘再旺

（首钢集团有限公司技术研究院，绿色可循环钢铁流程北京市重点实验室，北京 100043）

摘　　要： 热轧高强双面搪瓷钢在环保大型工程上应用日益普及，要满足钢板高温搪烧后高强度及双面涂搪条件下的抗鳞爆性能要求。研究表明 Ti 既作为钢板的析出强化元素，同时 TiC 的析出提供了不可逆的氢陷阱。通过实验钢成分设计，在 600℃、650℃卷取温度获得的实验钢在搪烧后其强度均下降显著，但 TH 值的实验结果相近。在抗鳞爆性能相近条件下，低温卷取可获得好的抗高温搪烧软化效果。

关键词： 搪瓷钢，析出物，鳞爆，卷取温度

Development of Hot Rolled High Strength Enameled Steel

Huang Xueqi, Zhang Zhimin, Li Xiaolin, Liu Zaiwang

(Shougang Research Institute of Technology, Beijing Key Laboratory of Green Recyclable Process for
Iron & Steel Production Technology, Beijing 100043, China)

Abstract: Nowadays hot rolled double-side high strength enamel steel is adopted in large environment project, and it is urgent to improve the fish-scaling resistance and insure proper strength after enameling. The result shows that as the main strengthening element, the process of TiC precipitation provides irreversible hydrogen traps. Based on the chemical composition of test samples, the yield strength decrease significantly under 600℃, 650℃ coiling temperature, with similar TH values, which suggests that lower coiling temperature could be adopted for higher TH value and lower yield strength.

Key words: enameled steel, properties, fish scale, coiling temperature

热轧 CVC 轧辊磨损曲线的试验分析

马占福

（新疆八一钢铁股份有限公司，新疆乌鲁木齐　830022）

摘　　要： 用 Matlab 软件进行 CVC 轧辊辊型曲线数学模型的建模分析，拟合求解得到辊型曲线方程，跟踪在一个轧制周期后轧辊的磨损曲线，分析了轧辊磨损的原因和规律，试验结果得到工作辊的实际磨损曲线与理论磨损的曲线相似，从获得的实测数据来看，与轧辊磨损的规律趋于一致，即上工作辊磨损量大于相应下工作辊磨损量，轧辊的磨损随着轧制单位压力增加而增加等，反映出轧辊磨损模型计算精度较高，有利于轧辊磨损的在线预报。

关键词： 热连轧机，CVC 轧辊，轧辊磨损

Experimental Analysis of Wear Curve of Hot Rolled CVC Roll

Ma Zhanfu

(Xinjiang Bayi Iron & Steel Co., Ltd., Urumchi 830022, China)

Abstract: The mathematical modeling and analysis of CVC roll profile curve model with Matlab software, and get a curve fitting equation solving, tracking the wear curve of roll in a rolling cycle, analyzes the reasons and rules of roll wear and test results of the actual work roll wear curve similar to the theory of wear curve from measured the data obtained, consistent with the rules of the roll wear, the wear of up work roll is larger than the corresponding lower work roll wear, wear of the roller increases　with the rolling unit pressure increases, reflecting that the roll wear model has higher calculation precision,

is conducive to the on-line prediction of roll wear.
Key words: hot strip mill, CVC roll, roll wear

热轧平整轧辊磨损分析和辊形优化

王立旗，田吉祥，樊国齐

（河钢集团邯钢公司连铸连轧厂，河北邯郸　056015）

摘　要： 为了减少板形缺陷数量，使得平整机发挥更大的效益，提高平整机的板形调控能力，通过跟踪分析轧辊磨损情况，对原有工作辊和支持辊辊形曲线进行优化，提高了轧辊的平整量，降低辊耗，板形缺陷数量也降低，取得较好效果。

关键词： 平整，辊形，板形控制

Analysis for of Roll Wear and Roll Contour Optimization on Hot-rolled Skin Pass

Wang Liqi, Tian Jixiang, Fan Guoqi

(Hesteel Group Hansteel Company，CSPMill，Handan 056015, China)

Abstract: In order to reduce the quantity of flatness defect, make temper mill to play a greater benefit，increase the ability of flatness control. Through analysis of roll wear is tracked, the original roll contour of work and back roll optimization, not only increase the amount of smooth roll, reduce roll consumption, but also reduce the quantity of flatness defect, obtain the good effect.

Key words: skin pass, roll contour, flatness

热轧普碳钢厚板冷弯断裂失效分析

叶其斌，李　勇

（东北大学轧制技术及连轧自动化国家重点实验室，辽宁沈阳　110819）

摘　要： 采用断口分析、显微观察和硬度、冷弯、拉伸、冲击实验方法，分析了 50 mm 厚 235MPa 级普碳钢板在冷弯成形过程中的贯穿断裂失效原因。钢板失效微观断裂模式为穿晶解理断裂，位于钢板切边截面的裂纹源在冷弯时承受拉应力载荷。钢板基体试样的冷弯、拉伸、冲击性能均满足技术指标要求，因此可以排除钢板冶金质量原因。显微分析显示，切边截面有大量毫米级穿晶裂纹，且存在约 4 mm 深度的严重加工硬化层，局部应变引起脆性相珠光体的断裂。钢板切边截面裂纹及其附近加工硬化降低了解理裂纹扩展所需临界应力和应变，使钢板冷弯变形时的断裂不可避免。

关键词： 热轧厚板，冷弯，解理断裂，切边裂纹，珠光体断裂

Failure Analysis on Fracture of Hot-rolled Mild Steel Plate during Cold Bending

Ye Qibin, Li Yong

(The State Key Laboratory of Rolling and Automation, Northeastern University, Shenyang 110819, China)

Abstract: A catastrophic fracture of 50 mm-thickness 235 MPa grade plate during cold-bending was investigated using fractography analysis, microstructure characterization, testing of hardness, bending, tension and impact. Transgranular cleavage fracture was found to be the major micromechanism of the fracture, which was initiated from the one side which was shear cut for delivery. The fracture initiation position was undergone tension stress during bending. The testing of bending, tension and impact shows the basic mechanical properties fulfill the requirements for Q235B steel, which can exclude the factor of metallurgical quality causing the fracture. Microscopic analysis reveals a large amount of millimeter-scale brittle cracks in the shear-cut side section, and the work-hardening depth was about 4 mm with severe plastic deformation, resulting in cracking of pearlite. The existence of large-size cracks and severe work-hardening zone decrease both the critical stress and strain for propagation of cleavage fracture, and therefore, the disastrous failure cannot be prevented.

Key words: hot-rolled heavy plate, cold-bending, cleavage fracture, shear-cutting crack, work-hardening

生产工艺对 65Mn 带钢奥氏体化行为的影响

张　星，李秀景，梅淑文

（河钢集团股份有限公司唐山分公司技术中心，河北唐山　063016）

摘　要： 通过奥氏体化实验对比分析了薄板坯连铸连轧工艺（FTSR）和常规中板坯连铸连轧工艺（CCDR）生产的同规格 65Mn 带钢在奥氏体化过程中晶粒长大行为，结果表明：相比较于 FTSR，CCDR 生产的带钢珠光体组织更细小、均匀，层片间距更小；在较低奥氏体化温度下，晶粒正常形核、长大，CCDR 生产的带钢可以获得更加细小、均匀的晶粒；而在较高奥氏体温度下，晶粒有异常长大的趋势，两种工艺生产的带钢获得了相近尺寸的奥氏体晶粒。

关键词： 65Mn 钢，连铸连轧，奥氏体化，晶粒长大

The Effect of Austenitization Behavior in 65Mn Strip Based on Different Processes

Zhang Xing, Li Xiujing, Mei Shuwen

(HBIS Group Tangsteel Company Technical Center, Tangshan 063016, China)

Abstract: Austenitization experiment is carried on 65Mn steel strips with the same thickness that produced by thin slab continuous casting and hot rolling (FTSR) and conventional continuous casting and hot rolling processes (CCDR); the austenite grains growth behavior is analyzed at the same time. The result shows that: compared with FTSR, pearlite in the strip based on CCDR is more fine and uniform, with the spacing of pearlite layer closer; when austeniting temperature is lower, nucleation and growth of the grains is normal and more fine and uniform grains are gained in CCDR steel; conversely, at higher temperature, austenite grain is more likely to grow abnormally, and the size of grains in FTSR and CCDR is approximate.

Key words: 65Mn steel, continuous casting and hot rolling, austenitization, grain growth

热连轧薄带板形板厚控制系统多时标建模与稳定性分析

陈金香[1]，陈其安[2]

（1. 中国钢研冶金自动化研究设计院混合流程工业自动化系统及装备技术国家重点实验室，
北京　100081；

2. 中国钢研钢铁研究总院，北京　100081）

摘　要： 针对热连轧薄或超薄带钢板形板厚综合控制系统，提出多时标建模与稳定性分析方法，解决现有理论与方法较难获得高控制性能的问题。多时标建模采用奇异摄动技术，将系统状态变量分解为慢、快两类，建立热连轧薄或超薄带钢板形板厚控制系统的离散时间线性奇异摄动模型。基于此模型，设计慢状态反馈与输出积分器组合控制器，并采用线性矩阵不等式方法推导出了求解控制器增益的定理。该方法无需对板形板厚进行解耦，且能够较准确的表征被控系统的动力学，实现板形板厚的高精度控制性能指标，仿真结果表明了该方法的有效性。

关键词： 多时标，奇异摄动，薄带板形板厚，热连轧

Multiple Time-scale Modeling and Stability Analysis for Hot Continuous Rolling Thin Strip Shape and Gauge Control Systems

Chen Jinxiang[1], Chen Qi'an[2]

(1. State Key Laboratory of Hybrid Process Industry Automation Systems and Equipment Technology,
Automation Research and Design Institute of Metallurgical Industry, China Iron & Steel
Research Institute Group, Beijing 100081, China；

2. Central Iron and Steel Research Institute, China Iron & Steel Research Institute Group,
Beijing 100081, China)

Abstract: Multiple time-scale modeling and stability analysis approaches are presented for hot continuous rolling thin steel shape and gauge control systems to solve the problem of low control performance obtained by the existing control methods. By using singular perturbation technology, the system state variables are firstly classified as slow and fast state variables, then a discrete-time linear singular perturbation model is build for hot continuous rolling thin steel shape and gauge control systems. Based on the model, a composed controller with the slow state feedback and output integrator is designed. The controller gains can be obtained by Theorem 1, which is derived by applying a linear matrix inequality approach. The dynamic characteristics of the controlled systems can be described accurately by the proposed methods. The high control performances for thin steel shape and gauge can be obtained, which is verified by the simulation results.

Key words: multiple time-scale, singular perturbation, shape and gauge of thin-strip, hot continuous rolling

基于热轧轧辊磨损模型优化和带钢轮廓质量的
窜辊工艺研究

张长利[1,2]，王　策[1]，刘子英[1]

（1. 首钢技术研究院，北京　100064;

2. 绿色可循环钢铁流程北京市重点实验室，北京　100064）

摘　要： 根据某厂热轧轧辊磨损数据的分析，考虑轧制钢种以及轧辊非均匀磨损的影响因素，对原有轧辊磨损预报模型进行改进。基于磨损预报模型，研究了热轧下游机架周期窜辊的不同方式对带钢厚度轮廓的影响，变行程窜辊对轧辊的"猫耳"非均匀磨损、带钢边部反翘以及边部减薄区也有较大影响，实际窜辊工艺优化时应综合考虑窜辊行程及周期的影响，达到最优带钢轮廓、板形质量及轧辊轧制公里数的要求。

关键词： 热轧，轧辊磨损，窜辊，轮廓

Impact of Cyclic Roll Shifting on Thickness Profile Basing Optimization of Hot Strip Work Roll Wear Model

Zhang Changli[1,2], Wang Ce[1], Liu Ziying[1]

(1. Shougang Research Institute of Technology, Beijing 100064, China;

2. Beijing Key Laboratory of Green Recyclable Process for Iron & Steel Production

Technology, Beijing 100064, China)

Abstract: Based on the analysis of roll wear for a hot strip mill(HSM), considering of the influence of grades and non-uniform wear, the mathematical roll wear model was improved and parameters were identified. With the optimized mathematical model, the Impact of cyclic roll shifting on strip thickness profile was studied, which includes strip edge ridge and edge drop zone. Actual roll shifting strategy should be designed and optimized with consideration of the combined acquirements of strip profile, shape and work roll rolling kilometer.

Key words: HSM, roll wear, roll shifting, thickness profile

免预处理集装箱用钢的开发与应用

杨　奕[1,2]，刘　洋[2]，韩　斌[2]，魏　兵[2]，汪水泽[2]

（1. 东北大学轧制技术及连轧自动化国家重点实验室，辽宁沈阳　110004;

2. 武汉钢铁有限公司研究院，湖北武汉　430080）

摘　要： 免预处理集装箱用钢能够取消喷砂工序，减少噪声和粉尘污染，在经济性及环保性方面具有较大优势，具有非常大的发展前景。通过分析集装箱对表面质量的要求，免预处理集装箱用钢确定了氧化铁皮结构控制目标。通过开展工艺试制，试制的免预处理集装箱钢的氧化铁皮由外层较薄的 Fe_3O_4 层和内部 $\alpha\text{-}Fe+Fe_3O_4$ 层组成，氧化铁皮

结构基本达到预期目标，涂装后的耐盐雾试验、二次附着力和成型性能基本满足集装箱用钢的要求。

关键词：免预处理，喷砂，氧化铁皮，热轧，集装箱钢

Development and Application of Free Pretreatment Container Steel

Yang Yi[1,2], Liu Yang[2], Han Bin[2], Wei Bing[2], Wang Shuize[2]

(1. The State Key Laboratory of Rolling and Automation, Northeastern University, Shenyang 110004, China；
2. Research and Development Institute of Wuhan Iron & Steel Co., Ltd., Wuhan 430080, China)

Abstract: It is a very big development prospects for free pretreatment container steel in terms of its economic and environmental advantages, which can cancel shot blasting process, decrease the noise and dust pollution. By analyzing requirements of the container steel surface quality, target oxide scale structure of free pretreatment container steel has been determined. Trial process was carried out and test results showed outer oxide scale was thin Fe_3O_4 layer and inner eutectoid α-Fe+Fe_3O_4 layer, the oxide scale achieved the desired target. Salt spray test, second adhesion test and modeling performance were basically satisfied with container needs.

Key words: free pretreatment, shot blasting, oxide scale, hot rolling, container steel

AlN 析出对 DC01 搪瓷用钢鳞爆性能的影响

李 黎 [1,3]，肖宗益 [2,3]，周焕能 [1,3]，贾冬梅 [1,3]

（1. 产品技术部华菱安赛乐米塔尔汽车板有限公司，湖南娄底 417000；
2. 生产工艺部华菱安赛乐米塔尔汽车板有限公司，湖南娄底 417000；
3. 技术中心华菱安赛乐米塔尔汽车板有限公司，湖南娄底 417000）

摘 要： 为了研究 AlN 颗粒析出对 DC01 搪瓷用钢鳞爆性能的影响，通过对 DC01 搪瓷用钢采用不同卷取温度（580℃、610℃及 640℃）控制退火成品第二相粒子 AlN 颗粒析出并采用模拟搪瓷试验对退火成品进行涂搪对比。结果表明，当卷取温度为 580℃时，退火后析出相以 Fe_3C 颗粒为主，平均尺寸约为 500nm；随着卷取温度升高到 640℃，在钢板内形成大量、弥散分布的第二相粒子 AlN 颗粒，其平均尺寸为 50nm 左右，从而形成了大量的贮氢陷阱，能有效提高钢板的贮氢性能，避免鳞爆缺陷产生。

关键词： 卷取温度，搪瓷钢，鳞爆，贮氢性能

Effects of Precipitation AlN on Fish-scaling of Cold-rolled Enamel Steel DC01

Li Li [1,3], Xiao Zongyi [2,3], Zhou Huanneng [1,3], Jia Dongmei [1,3]

(1. Valin ArcelorMittal Automotive Steel Co., Ltd., Product Technology Department, Loudi 417000, China;
2. Valin ArcelorMittal Automotive Steel Co., Ltd., Product Process Department, Loudi 417000, China;
3. Valin ArcelorMittal Automotive Steel Co., Ltd., Technology Center, Loudi 417000, China)

Abstract: In order to study the effects of precipitation AlN on fish-scaling of Cold-rolled Enamel Steel DC01, using the different coiling temperature, 580℃, 610℃ and 640℃, to control the precipitation of AlN after annealing and the simulation enamel test have been done to analyses the effects for the different coiling samples after annealing. The result shows that when the coiling temperature is 580℃, the precipitation is Fe_3C particles mainly after annealing and the size of precipitation is about 500nm; there are a large dispersion of precipitation of AlN in the sheet when the coiling temperature increased to 640℃, the size of precipitation is about 50nm.the precipitation of AlN will improve the hydrogen storage performance of steel and avoid fish-scale defects.

Key words: coiling temperature, enamel steel, fish-scaling, hydrogen storage performance

高级别管线钢快冷工艺内应力控制及板形平直度优化

刘红艳[1]，杜琦铭[1]，左丽峰[1]，管连生[2]

（1. 河钢集团邯钢公司技术中心，河北邯郸　056015；

2. 河钢集团邯钢公司连轧厂，河北邯郸　056015）

摘　要： 以中厚板高级别管线钢 X70M 为目标钢种，研究钢板在快冷过程中厚度和宽度方向上温度和相变的变化和分布规律、以及终冷后钢板厚度方向上温度和相变规律，终冷温度越低，相变越充分组织分布越均匀，最终钢板残余应力越小。为了提高高级别管线钢快冷工艺下板形质量，通过增加冷却水流量、增加冷却水流密度，钢板在宽度方向上应变量的差异变小，从原来的 $7.69×10^{-6}$ 减少到 $3.71×10^{-6}$，同时 DQ+ACC 上下水比由原来 1：2.0 优化为 1：2.6，薄规格产品终冷温度降低 40~70℃，提高厚度方向上的冷却均匀性，降低厚度方向上内应力。工艺优化后，X70M 快冷过程中板形质量、尤其是厚度小于等于 14mm 的规格产品，板形平直度控制良好，对提高快速冷却过程钢板质量等级具有重要的意义。

关键词： 中厚板管线钢，快冷工艺，内应力，板形平直度

Stress Control and Shape Flatness Optimization of High-grade Pipeline Steel in Rapid-cooling Process

Liu Hongyan[1], Du Qiming[1], Zuo Lifeng[1], Guan Liansheng[2]

(1. HBIS Group Hansteel Company Technology Center, Handan 056015, China;

2. HBIS Group Hansteel Company Hot-rolling, Handan 056015, China)

Abstract: The plate of high grade pipeline steel X70M as the goal, study on the change and distribution of temperature and phase transition in the thickness and width of steel plate during rapid cooling, and the temperature and phase transition in the thickness direction of the steel plate after the final cooling, the more uniform the distribution of the structure, the more the residual stress of the steel plate is. In order to improve the shape quality of fast cooling, increasing the cooling water flow rate, increase cooling water flow density, plate strain variation in the width direction, reduced from the original $7.69×10^{-6}$ to $3.71×10^{-6}$, at the same time, the ratio of DQ+ACC is optimized from 1：2.0 to 1：2.6, and the final cooling temperature of thin gauge product is reduced by 40~70 degrees, the uniformity of cooling in the thickness direction is increased, and the internal stress in the thickness direction is decreased. Process optimization, X70M fast cooling process of plate shape quality, especially the thickness of less than 14mm is equal to the product specifications, profile and flatness control is good, has important significance to improve the quality level of the rapid cooling process of steel plate.

Key words: pipeline plate, rapid cooling process, internal stress, plate flatness

热轧温度及压下量对带钢表面氧化铁皮的影响

韩会全[1]，王万慧[1]，周天鹏[2]，陈泽军[2]，杨春楣[1]

（1. 中冶赛迪技术研究中心有限公司，重庆　401122；
2. 重庆大学材料科学与工程学院，重庆　400044）

摘　要：采用高温热轧实验，通过分光光度计，SEM 和酸洗实验检测，研究了轧制温度和轧制变形量对普碳钢氧化铁皮表面形貌，厚度及红锈程度的影响。结果表明：同样轧前和轧制温度下，轧制变形量越大，氧化铁皮延展的越薄；随轧制温度的升高，氧化铁皮塑性增强，在较大轧制变形量下，能够将氧化铁皮压实，增加与基体的粘附性，进而减少了氧化铁皮的脱落；轧制温度越高，其产生红锈的临界变形量越大，能容忍的轧前铁皮厚度越厚；热连轧过程中，提高前机架的轧制负荷分配，减少尾机架的轧制变形量，有利于控制带钢红锈倾向和氧化铁皮的均匀变形。

关键词：氧化铁皮，热轧温度，压下量，分光光度计，低碳钢

Effect of Hot Rolling Temperature and Reduce on Surface Oxide Scale of Strip

Han Huiquan[1], Wang Wanhui[1], Zhou Tianpeng[2], Chen Zejun[2], Yang Chunmei[1]

(1. CISDI Research & Development Co., Ltd., Chongqing 401122, China; 2. College of Materials Science and Engineering, Chongqing University, Chongqing 400044, China)

Abstract: Through hot rolling experiments at high temperature, the effect of rolling temperature and rolling deformation on carbon-steel oxide scale microstructure, thickness and surface red rust were investigated by spectrometer, SEM and pickling experiment. The results show that the same other experiment conditions, the greater the rolling deformation, the thinner the oxide scale extend; with the increase of rolling temperature, iron oxide scale plasticity enhanced, the oxide scale can be compacted under a larger rolling deformation, increase the adhesion between oxide scale and substrate, reduce the oxidation of iron loss; the higher the rolling temperature, the greater critical deformation of the red rust, and can tolerate a thicker oxide scale thickness before hot rolling; in the process of hot continuous rolling, improve the front frame of rolling load distribution, reduce the rolling deformation tail frame, is helpful to control strip red rust tendency and oxide scale uniform deformation.

Key words: oxide scale, hot rolling temperature, hot rolling reduce, spectrophotometer, mild steel

韶钢 3450mm 中厚板板形质量改善研究

戴杰涛[1]，李烈军[2]，张祖江[3]，周　峰[2]

（1. 广州大学机械与电气工程学院，广东广州　510006；

2. 华南理工大学机械与汽车工程学院，广东广州　510641；

3. 宝武集团韶关钢铁股份有限公司，广东韶关　512223）

摘　要： 针对韶钢 3450mm 中厚板生产线存在的凸度过大、边浪严重和楔形大的板形问题进行了分析，通过大量的现场跟踪数据和理论分析发现造成凸度和边浪严重的主要原因在轧制过程中钢板宽度方向上的压下量中间小两边大，而造成楔形的主要原因是由于轧机操作侧和传动侧的牌坊刚度不一致。针对这些原因对韶钢 3450mm 生产线的工艺进行了优化改进：（1）设计了凸度更大的工作辊辊形曲线；（2）优化了机组的弯辊力投入制度；（3）减少了末3 道次的压下量；（4）对换辊和配辊制度进行了优化；（5）通过机组的垫片调整了轧机的辊缝值。通过上述工艺措施，机组的凸度值由 0.2mm 以上降低到了 0.07mm 以内，楔形值由 0.17mm 降低到了 0.06mm 左右，同时有效地控制了机组薄规格板材轧制过程中的边浪问题，将机组的轧制厚度由 9mm 拓展到了 8mm，并实现了稳定轧制。

关键词： 中厚板，楔形，辊形，凸度

Improve the Peofile and Flatness Quality of Medium Plate on Shaoguan Steel 3450 Mill

Dai Jietao[1], Li Liejun[2], Zhang Zujiang[3], Zhou Feng[2]

(1. School of Mechanical & Electric Engineering, Guangzhou University, Guangzhou 510006, China；

2. School of Mechanical & Automotive Engineering, SCUT, Guangzhou 510641, China；

3. Shaoguan Iron & Steel Co., Baowu Steel Group, Shaoguan 512223，China)

Abstract: Analysis of the problem that the crown too large, wave and wedge serious on shaoguan steel 3450 mill. Through a large number of on-site tracking data and theoretical analysis found that the main reason for serious crown and waves is that the reduction inconsistent during the rolling process on the width of the steel sheet, and the main reason foe wedge is rolling mill stiffness difference on the drive side and operating side. According to these main reasons, some process measures were proposed: (1) Designed a new work roll shape curve; (2) Improved bending force using technology; (3) Reduced the amount of the last 3 passes reduction; (4) Optimized with roller and roller system; (5) Adjust the rolling mill roll value.Through the above process measures, the unit's convexity value is reduced from 0.2mm to 0.07mm, the wedge value is reduced from 0.17 mm to about 0.06 mm, effectively control the problem of the edge wave in thin plate rolling, created significant economic benefits.

Key words: medium plate, wedge, roll curve, crown

楔形板矫直过程建模和仿真研究

杜兴明，孙建亮，李　凯，李　硕

（燕山大学国家冷轧板带装备及工艺工程技术研究中心，河北秦皇岛　066004）

摘　要： 楔形板为纵向变截面钢板，矫直是楔形板生产过程的关键工序，本文针对楔形板矫直过程进行了仿真研究。首先，设计了针对楔形板矫直的在线动态调整辊系和矫直工艺；其次，考虑矫直时板厚变化的特点，建立了楔形板矫直过程以矫直时间为参数的数学模型，主要包括压下量模型和矫直力模型；最后，基于有限元软件 ABAQUS 建立了楔形板矫直有限元模型，对矫直过程进行了仿真模拟，并与 MATLAB 计算得到的矫直力进行了对比分析。仿真计算结果表明，压下量随着板厚增加逐渐减小，且减小量越来越小，矫直力随着板厚增加逐渐增大，且增加量越

来越大。

关键词：楔形板，矫直，动态调整，压下量，矫直力

Modeling and Simulation Research on Longitudinal Profiled Plate Leveling

Du Xingming, Sun Jianliang, Li Kai, Li Shuo

(National Engineering Research Center for Equipment and Technology of Cold Strip Rolling,
Yanshan University, Qinhuangdao 066004, China)

Abstract: Longitudinal profiled plate is a plate with variable thicknesses along longitudinal direction. Leveling is the key process of LP plate production. In this paper, the LP plate leveling process is researched by simulation. First, the roll system of on-line dynamic adjustment and the leveling process for LP plate leveling are designed. Secondly, considering the variation of plate thickness during leveling, the mathematical models taking leveling time as a parameter of LP plate leveling process is built. Models mainly include reduction model and leveling force model. Last, the finite element model of LP plate leveling is built by ABAQUS, the leveling process is simulated and the leveling force by ABAQUS simulation is compared with the leveling force by MATLAB. The simulation and calculation results show that, the reduction decreases gradually with the increase of plate thickness with decrease number being smaller and smaller and the straightening force increases with the increase of plate thickness with increase number being bigger and bigger.

Key words: longitudinal profiled plate, leveling, dynamic adjustment, reduction, leveling force

板带轧制实现板厚和板形向量闭环控制的效果

张进之，张　宇，许庭洲

（中国钢研科技集团有限公司，北京　100081）

摘　要： 板带轧制动态理论适用于冷、热连轧和中厚板轧机。动态轧制理论的基本内容有四项：连轧张力公式；动态设定模型厚度控制方法（DAGC）；解析板形刚度理论；Φ函数及 dΦ/dh。该理论已成功应用在新建和改造的冷轧、热轧可逆式板带轧机，应用于新建四套热连轧和改进引进的热连轧机和宽厚板轧机。下面简述具体技术内容和主要的控制数学模型。

关键词：DAGC，流量 AGC，张力（或活套角）间接测厚，厚控精度，Φ函数，dΦ/dh，解析板形刚度理论

Control Mathematical Mode of High-precision Thickness

Zhang Jinzhi, Zhang Yu, Xu Tingzhou

(China Iron & Steel Research Institute Group, Beijing 100081, China)

Abstract: The dynamic theory of plate and strip rolling is suitable for cold and hot rolling and plate mill. The basic content of the theory of dynamic rolling has four items, the first one is continuous rolling tension formula, the second is model of dynamic setting control method (DAGC), the third is plate analytical theory for shape stiffness; the fourth is function Φ and dΦ/dh. The theory has been successfully applied in the newly built and modified cold rolling and hot rolling reversible strip rolling mill, which is applied to the newly built four sets of hot continuous rolling mill and the introduction of the hot rolling mill and wide plate mill. The following brief description of the specific technical content and the main control mathematical model.

Key words: DAGC, flow AGC, tension indirect thickness measurement, thickness control precision, function ϕ, dϕ/dh, analytical strip shape theory

热连轧中间坯凸度对比例凸度分配的影响研究

郭　薇[1], 王秋娜[2], 王凤琴[1], 李　飞[1]

（1. 首钢集团有限公司技术研究院冶金过程研究所，北京　100041；

2. 首钢股份公司迁安钢铁公司热轧作业部，河北迁安　064404）

摘　要： 本文根据某热连轧产线二级板形模型精轧来料凸度的控制现状，对中间坯来料凸度对机架间比例凸度的分配影响进行了大量的模拟研究，摸索出一种适用于不同钢种的来料凸度设定方法，将模型计算的来料凸度考虑钢种特性与规格变化，以减小与实际来料凸度的偏差，该方法在现场进行了优化实验。优化实验表明，精准的中间坯来料凸度能够有效的进行精轧机架间的比例凸度分配，对上下游窜辊量及弯辊力进行合理配置，能够有效的解决机架间和精轧出口的浪形问题，大幅提高精轧出口的凸度控制精度。

关键词： 热连轧，中间坯凸度，比例凸度分配，凸度控制

Research on the Influence of the Transfer Bar Profile on the Per Unit Profile Distribution of the Hot Strip Mill

Guo Wei[1], Wang Qiuna[2], Wang Fengqin[1], Li Fei[1]

(1. Shougang Research & Development Institute, Beijing 100041, China;

2. Shougang Co.. Ltd., Qian'an Steel Corp., Qian'an 064404, China)

Abstract: Aiming at the controlling actualities of the transfer bar profile on hot strip mill, an effective optimized setup method of various steel grades' transfer bar profile, which has considered characteristics of steel and specifications change, was proposed based on the amount simulation research of the influence of the transfer bar profile on the per unit profile distribution, in order to minimize the deviation between actual and model calculated transfer bar profile. This method implemented on the hot strip mill, and effectively has achieved the reasonable per unit profile distribution and roll shift and roll bending force configuration, resolving the wave shape problems and increasing the strip crown control precision.

Key words: hot rolling, transfer bar profile, per unit profile distribution, profile control

粗轧机立辊辊形技术对带钢边部翘皮缺陷控制的研究与应用

王晓东[1], 杨孝鹤[2], 刘靖群[1], 苏长水[1], 艾矫健[1]

（1. 首钢京唐钢铁联合有限责任公司热轧作业部，河北唐山　063200；

2. 首钢京唐钢铁联合有限责任公司制造部，河北唐山　063200）

摘 要：对带钢边部翘皮缺陷进行取样，采用扫描电镜观察缺陷的微观组织，通过试验分析结果确定翘皮发生的原因为两类。除了板坯气泡、夹杂导致的翘皮缺陷外，在板坯轧制过程中，带钢也会产生边部翘皮缺陷。根据板坯横断面温度分布测试、板坯立-平辊轧制数值模拟、板坯边角预制缺陷轧制试验等工作，验证轧制原因导致边部翘皮形成的机理是粗轧过程中由于板坯角部位置金属转移到带钢上、下表面距离边部为25mm左右的位置，并且该距离不受成品带钢厚度即整个热轧过程变形量的影响。为了降低板坯边角温降并改善金属流动，本文设计了一种新型的立辊辊形来对板坯边部进行倒角并对板坯侧面中部金属进行预压下，改变板坯边部形状，抑制平辊轧制过程中板坯角部金属向上下表面移动，进而减少边部翘皮缺陷的发生。通过粗轧立辊辊形技术的应用，对带钢边部翘皮缺陷改善效果显著，月均边部翘皮协议品钢卷数由78降低到17，超低碳钢边部翘皮缺陷发生率由应用前的22.2%降低至4.6%。

关键词：热轧带钢，粗轧机，立辊辊形，边部翘皮，金属流动

Research and Application of Contour Technology on Edge Rolls of Roughing Mill for Controlling Shell Defects on the Edges of Steel Strip

Wang Xiaodong[1], Yang Xiaohe[2], Liu Jingqun[1], Su Changshui[1], Ai Jiaojian[1]

(1. Hot Rolling Department of Shougang Jingtang United Iron and Steel Co., Ltd. Caofeidian Industrial Zone, Tangshan 063200, China; 2. Manufacturing Division of Shougang Jingtang United Iron and Steel Co., Ltd. Caofeidian Industrial Zone, Tangshan 063200, China)

Abstract: Through microstructure analysis works on the shell defect samples taken from in steel strip edge by Scanning Electron Microscope(SEM), it is determined that shell defects are produced during the hot rolling processes except the reasons of blister and inclusions in slab. Based on the conclusions of the works such as the cross section temperature profile of slab, numerical simulation of vertical – flat slab rolling, and rolling experiments of slabs with prefabricated defects, it is verified that the mechanism of shell defects induced by the rolling processes is that the corner parts of slab will move to the top and bottom surfaces with about 25mm distance to the slab edge during the roughing rolling process and this distance is not influenced by the total reduction of the slab. In order to decrease the temperature drop in slab edge and improve the metal flow, a new kind of edge roll contour was designed to make chamfer on slab edge and pre-reduction on the central part of the vertical edge of slab. The purpose is to restrain the transfer of the corner parts of slab to the top and bottom surface during the slab rolling and in the end to reduce occurrence of the shell defects. After the application of the edge roll contour on roughing mill, the improving effect on the edge shell defect of strip is remarkable. The per month amount of concession and degrade strip coils with edge shell defect is decreased from 78 to 17, and the incidence rate of edge shell defect of ultra-low carbon steel is reduced from 22.2% to 4.6%.

Key words: hot rolled strip, roughing mill, edge roll contour, edge shell, metal flow

热轧带钢表面折印原因分析及解决措施

张红斌，陈建泰，王卫东，杨春雷，张　瑜，刘红兵，陈　轩

（武钢集团昆明钢铁股份有限公司，云南昆明　650302）

摘 要：热轧带钢主要用于机械设备和桥梁、建筑物的结构件等，对带钢的表面质量要求较高。折印是热轧带钢常见的主要缺陷之一，出现折印的带钢一般在带钢正反表面对称出现，严重影响带钢的美观，导致用户质量异议或改

判。本文针对昆钢热轧板带生产线和热轧中宽带生产线生产中出现的折印缺陷，对其产生原因进行分析并提出解决措施。

关键词：热轧带钢，折印

The Reason Analysis and Solving Measure of Surface Cross Breaks of Hot Rolled Strip

Zhang Hongbin, Chen Jiantai, Wang Weidong, Yang Chunlei,

Zhang Yu, Liu Hongbing, Chen Xuan

(Technology Center of Wukun Steel Co., Ltd., Kunming 650302, China)

Abstract: Hot rolled steel strip is wide used for mechanical equipment, bridge engineering and structural parts of buildings, in that its surface is requested a highly quality. One of mainly disfiguenment is cross breaks in hot rolled steel strip, both sides of steel strip are symmetry arise cross breaks, it affected the artistic of the steel strip, and caused any doubt to the quality of our products or sentence for customer. In this paper, cross breaks in hot rolled steel strip produce line and hot rolled broadband production line at KISC was studied, an attempt to analyze the cause and solving measure were proposed.

Key words: hot rolled steel strip, cross breaks

解决钢板表面丸料压入问题的实践

王亮亮，周　强，罗　军，高　峰，高　强

（鞍钢股份有限公司鲅鱼圈钢铁分公司厚板部，辽宁营口　115007）

摘　要： 针对5500生产线2#抛丸机生产钢板表面残留丸料问题比较严重（30余个/m²）导致丸料压入的问题，通过研究压入机理，从设备功能完善和工艺技术参数优化等方面入手，采取对辊刷和挂蜡辊道进行更换，制作下刮板，优化辊刷高度等措施，解决了钢板表面丸料压入问题。

关键词：抛丸机，丸料，残留，压痕

To Solve the Problem of Pressure of Steel Plate Surface

Wang Liangliang, Zhou Qiang, Luo Jun, Gao Feng, Gao Qiang

(Bayuquan Iron & Steel Subsidiary Company of Angang Steel Co., Ltd., Yingkou 115007, China)

Abstract: Contraposed 2# shot blasting machine of 5500 production line about residual of pellet is serious(30/m² remain), cause press-in of the surface. By analyzing the causes, start from perfect machine function and optimized parameter, take replace rolling brush ande roll, make lower scraper, optimized the height of rolling brush ,solve the problem of pellet press-in.

Key words: shot blasting machine, pellet, residual, press-in

邯钢热轧带钢表面凹坑锈蚀缺陷成因分析及对策

李冠楠，赵林林，李红俊

（河钢集团邯钢公司，河北邯郸　056015）

摘　要：针对邯钢 2015 年出现的厚规格热轧带钢表面凹坑锈蚀质量问题，通过分析，确定邯钢出现的表面锈蚀凹坑质量问题是由于氧化铁皮结构过于疏松导致在加工过程中破裂，后续产生碾压或二次锈蚀所致。进而通过采取提高终轧温度，降低精轧阶段冷速，提高层冷段冷速和降低卷取温度等方式有效的改善氧化铁皮结构，其内部疏松的共析组织减少、距表面较远，结构趋于致密，同时带钢氧化铁皮平均厚度降低约 5μm，用户使用效果良好，没有再次出现类似质量问题。

关键词：凹坑，疏松，共析组织，冷速

Analysis and Countermeasures of Hot Strip Corrosion Pits on Surface Defects of Hansteel

Li Guannan, Zhao Linlin, Li Hongjun

(Hesteel Group Hansteel Company, Technology Center, Handan 056015, China)

Abstract: Through the analysis of the Handan 2015 appeared thick hot strip corrosion pits on surface quality problems，determine the quality problems of surface corrosion pits occur because Handan Iron oxide structure is too loose to rupture in the machining process. Subsequent caused by rolling or second corrosion. Then we raise the rolling temperature，reduce the cold finishing stage speed. Increase the cooling rate of the laminar cooling section and reduce the cooling temperature to improve the iron oxide structure effectively，reduce the internal defect of eutectoid， and it is far from its internal loose eutectoid structure tends to be dense，at the same time, the average thickness of strip steel scale is reduced to 5μm, the users effect is good，no similar quality problems appear again.

Key words: depression, osteoporosis, eutectoid structure, cooling rate

热处理工艺对钛钼化马氏体钢组织性能的影响

杨　晰[1,2]，苏春霞[1,2]，陈本文[1,2]，付　超[1,2]，杨　成[1,2]，易东升[3]

（1. 海洋装备用金属材料及其应用国家重点实验室，辽宁鞍山　114009；2. 鞍钢股份公司技术中，心辽宁鞍山　114009；3. 鞍钢股份公司产品制造部，辽宁鞍山　114009）

摘　要：本文利用拉伸试验机、冲击试验机、硬度计、金相显微镜、扫描电子显微镜、透射电子显微镜等设备分析了不同热处理工艺下钛钼微合金化马氏体钢的组织性能，结果表明：随回火温度升高，试验钢的硬度、强度呈下降趋势，延伸率、冲击功先略升高，随后出现平台，最后呈线性升高；经淬火，低温回火组织为碳化物细小、板条边

界清晰的回火马氏体，中温回火组织为碳化物呈棒状、边界较模糊的回火屈氏体；高温回火组织为碳化物球化、无板条状特征的回火索氏体；基体中存在钛钼复合析出物，其尺寸较大的为近似方形的、富钛低钼的碳氮化物，数量较少，较小的析出物主要为近似球形的、钛钼碳化物，数量较多，且高温回火可以促进细小钛钼复合相的析出。

关键词：钛钼，热处理，组织，力学性能，析出物

Effect of Heat Treatment Process on Microstructures and Mechanical Properties of Ti-Mo-bearing Martensitic Steel

Yang Xi[1,2], Su Chunxia[1,2], Chen Benwen[1,2], Fu Chao[1,2], Yang Cheng[1,2], Yi Dongsheng[3]

(1. State Key Laboratory of Mental Material for Marine Equipment and Application, Anshan 114009, China;

2. Technology Center of Angang Steel Co., Ltd., Anshan 114009, China;

3. Production Department of Angang Steel Co., Ltd., Anshan 114009, China)

Abstract: Microstructures and mechanical properties of Ti-Mo-bearing martensitic steel under different heat treatment process was studied in this paper, by Tensile tester, impact tester, hardness tester, metallographic microscope, scanning electron microscope, transmission electron microscopy and other equipments. The result shows that the hardness and strength of the test steel declines, elongation and impact energy slightly increased at first, then shows a curve platform and increases linearly at last with the increasing tempering temperature. After quenching, the microstructure under low temperature tempering is tempered martensite with small carbide and clear lath boundaries, the microstructure under medium temperature tempering is tempered troostite with rod Carbide and fuzzy boundaries, the microstructure under high temperature tempering is tempered sorbite with globular carbide and no lath boundaries. There are Ti-Mo composite precipitates in matrix organization. The larger size is approximate square, rich Ti and low Mo carbon nitride that quantity is small. The smaller size is approximate globular, low Ti and rich Mo carbon nitride that quantity is large. And high temperature tempering can promote small Ti-Mo compound phase precipitation.

Key words: Ti-Mo, heat treatment, microstructure, mechanical property, precipitate

热处理对异质复合钢板覆材晶间腐蚀和力学性能的影响

颜秉宇[1,2]，王永才[2]，孙殿东[1,2]，王　爽[1,2]，胡海洋[1,2]，王　勇[2]

（1. 鞍钢集团海洋装备用金属材料及其应用国家重点实验室，辽宁鞍山　114009；

2. 鞍钢股份有限公司，辽宁鞍山　114009）

摘　要： 异质复合钢板中的覆材需连同整张异质复合钢板进行热处理，本文中异质复合钢板的覆材选用 304L 奥氏体不锈钢，奥氏体不锈钢在 600℃ 开始析出碳化物，碳化物的析出量随温度的升高而增多，当碳化物析出达到一定程度会影响奥氏体不锈钢的耐晶间腐蚀能力和力学性能，所以本文对覆材在不同热处理状态下的耐晶间腐蚀能力和力学性能进行研究，研究结果如下：在固溶热处理、调质热处理、敏化热处理和模拟消应力热处理四种热处理状态下，覆材 304L 的力学性能和晶间腐蚀性能均能满足要求。

关键词：异质复合，覆材，热处理，晶间腐蚀

Effect of Heat Treatment on Intercrystalline Corrosion and Mechanical Properties of Stainless Steel Composite Plate

Yan Bingyu[1,2], Wang Yongcai[2], Sun Diandong[1,2], Wang Shuang[1,2], Hu Haiyang[1,2], Wang Yong[2]

(1. State Key Laboratory of Metal Material for Marine Equipment and Application, Ansteel Group Corporation,Anshan 114009, China;2.Angang Steel Company Limited, Anshan 114009, China)

Abstract: The clad material of the stainless steel composite plate must be attached with the heat treatment,this stainless composite steel cladding material selection of 304L austenitic stainless steel,the carbide precipitation of austenitic stainless steel begin at 600℃,the mount of carbide precipitation increased with the rise of temperature, when the carbide precipitation reaches to a certain extent affect the austenitic stainless steel intercrystalline corrosion resistant ability and mechanical properties,so the clad material were studied in different heat treatment, The results show that clad material can meet the requirements in the solution treatment, quenching and tempering treatment,sensitization treatment and simulated stress-relief heat treatment.

Key words: stainless steel composite plate, clad materials, heat treatment, intercrystalline corrosion

QLT 热处理工艺对低碳高强韧钢组织和性能的影响研究

唐文川，郭呈宇，康永林

（北京科技大学材料科学与工程学院，北京 100083）

摘 要： 对热轧高强韧中厚板进行 860℃淬火+临界区二次淬火+不同温度回火（QLT）的热处理，通过调整二次淬火温度及回火温度得到强度和韧性俱佳的热轧高强韧钢。探究了不同二次淬火热处理工艺下的力学性能趋势与显微组织变化，并分析二者之间的关系，总结强韧化机理，对后续更高性能产品的研发起着重要的指导作用。

关键词： 热处理，中厚板，二次淬火，贝氏体

Effect of QLT Heat Treatment Process on Microstructure and Properties of Low Carbon High Strength Steel

Tang Wenchuan, Guo Chengyu, Kang Yonglin

(School of Materials Science and Engineering, University of Science and Technology Beijing, Beijing 100083, China)

Abstract: Hot rolled high strength and toughness plate for direct quenching at 860℃, intercritical secondary quenching and different temperature tempering(QLT) heat treatment. The hot rolled high strength steel with good strength and toughness was obtained by adjusting the two quenching temperature and tempering temperature. Explored the different two quenching of the mechanical properties and microstructure change trend, and analyze the relationship between the two, summarized the toughening mechanism, research on subsequent higher performance products plays an important role in guiding.

Key words: heat treatment, medium and heavy plate, secondary quenching, bainite

8CrV 锯片钢奥氏体连续冷却相变规律研究

惠亚军[1,2]，刘阳春[1,2]，张云鹤[3]，徐　斌[3]，吴科敏[3]，周　娜[3]

（1. 首钢集团有限公司技术研究院薄板研究所，北京　100043；2. 绿色可循环钢铁流程北京市重点实验室，北京　100043；3. 首钢股份公司迁安钢铁公司制造部，河北迁安　064404）

摘　要： 在 Gleeble3500 热模拟试验机研究了 8CrV 锯片钢的连续冷却相变规律并绘制了 CCT 曲线。研究结果表明：8CrV 锯片钢的临界转变温度 Ac1、Accm、Ar1、Arcm 与 Ms 分别为 755、779、685、705 与 177℃。CCT 曲线分为 2 个区，高温铁素体与珠光体转变区，低温马氏体转变区，而无中温上贝氏体转变区。冷却速度对奥氏体相变过程及组织有较大影响，随着冷速增加，先共析渗碳体与珠光体的含量逐渐减少，马氏体含量逐渐增加；8CrV 锯片钢淬透性优异，在 3℃/s 时即出现马氏体，其临界淬火速度为 20℃/s。8CrV 锯片钢的 CCT 曲线的测定可为控冷工艺与热处理工艺的制定提供理论参考。

关键词： 8CrV 锯片钢，奥氏体，连续冷却相变，CCT 曲线，热模拟

Study on Continuous Cooling Transformation of Austenite for 8CrV Saw Blade Steel

Hui Yajun[1,2], Liu Yangchun[1,2], Zhang Yunhe[3], Xu Bin[3], Wu Kemin[3], Zhou Na[3]

(1. Technology Institute of Shougang Group Co., Ltd., Sheet Metal Research Institute, Beijing 100043, China;
2. Beijing Key Laboratory of Green Recyclable Process for Iron & steel Production, Shougang Group Co., Ltd., Beijing 100043, China; 3. Qian'an Iron and Steel Company of Shougang Group Co., Ltd., Manufacturing Department, Qian'an 064404, China)

Abstract: The continuous cooling phase transformation law of 8CrV saw blade steel was studied in the Gleeble3500 thermal simulation test machine, and the CCT curve was drawn. The results show that the critical transition temperatures of Ac1, Accm, Ar1, Arcm and Ms are 755,779,685,705 and 177℃. The CCT curves were divided into two zones: the high temperature of cementite and pearlite transition zone and the low temperature of martensitic transition region, while there is no the medium temperature of bainite transition zone. The cooling rate has a great effect on the austenite transformation process and microstructure, with the increase of the cooling rate, the content of the eutectoid cementite and pearlite decreases and the martensite increases. The 8CrV saw blade steel shows excellent hardenability, The martensite appears when cooling rate reached 3℃/s and the critical quenching speed is 20℃/s. The CCT curve of 8CrV saw blade steel could provide theoretical reference for the controlled cooling process and the heat treatment process.

Key words: 8CrV saw blade steel, austenite, continuous cooling transformation, CCT curve, thermal simulation

中厚板热处理淬火机国内外研制现状与发展

韩　钧，王昭东，付天亮，王国栋

（东北大学轧制技术及连轧自动化国家重点实验室，辽宁沈阳　110004）

摘　要： 中厚板热处理淬火机的研制近年来取得了较快发展，通过对国内外淬火机的简单对比介绍，从 20 世纪末的引进消化，到现在的国产化做优增量、原始创新、从跟跑者变为技术领跑者，为我国装备制造业的大型化、轻量化和高效化做出了贡献，为国民经济及国防建设所需的热处理产品研发生产奠定了坚实基础。

关键词： 热处理，淬火机，装备制造，国产化

Current Status and Development of Plate Heat Treatment Quenching Machine in China and Abroad

Han Jun, Wang Zhaodong, Fu Tianliang, Wang Guodong

(State Key Laboratory of Rolling Technology and Automation, Northeastern University, Shenyang 110004, China)

Abstract: Heat treatment quenching machine has been developed rapidly in recent years, through the simple comparison of domestic and foreign quenching machine, from the introduction of digestion at the end of the 20th century, to the present localization, the original innovation, to become a technology leader for China's equipment manufacturing industry, large-scale, lightweight and efficient to make a contribution to the national economy and national defense construction required for heat treatment product development and lay a solid foundation.

Key words: heat treatment, quenching machine, equipment manufacturing, localization

轧后回火工艺对厚规格 X80 钢组织性能的影响

徐　锋，徐进桥，李利巍，崔　雷，彭　周，邹　航

（宝武集团武钢研究院，湖北武汉　430083）

摘　要： 对热轧态 X80 钢板进行了 300~600℃回火热处理，分析了其组织、性能的变化规律及原因。研究结果表明：随回火温度的升高，屈服强度不断上升，在 600℃达到最大，与轧态相比增加约 100MPa。而抗拉强度则是在 400℃回火时达到最大，与轧态相比增加约 53MPa。当回火温度进一步升高至 500℃时，抗拉强度降至轧态水平。对回火前后钢板的析出物、金相组织及精细结构分析发现：在回火过程中 Nb 的补充析出是屈服强度不断增加的主要原因，在 600℃回火时，Nb 的补充析出增幅达到 16.28%。在 400℃时由于 MA 回火强化、位错密度上升导致抗拉强度不断升高，延伸率降至最低。而 500℃回火时，由于 MA 分解及位错的消失导致抗拉强度降低。屈服强度的不断升高和抗拉强度先小幅升高后降低的现象，导致屈强比随回火温度的升高不断增加，与轧态相比，在 600℃回火时，屈强比增加 0.15。

关键词： X80 管线钢，回火热处理，析出物，MA，位错密度

Effect of Temper After Rolling on Microstructure and Mechanical Properties of Thick Specification X80 Steel

Xu Feng, Xu Jinqiao, Li Liwei, Cui Lei, Peng Zhou, Zou Hang

(Research and Development Center of WISCO, Baowu Iron and Steel Co., Ltd., Wuhan 430083, China)

Abstract: Hot rolling X80 steel was tempered heat treatment from 300℃ to 600℃ for analysis the change regularity and reason on microstructure and mechanical properties. The results show that with the increase of tempering temperature the yield strength increase gradually. It reaches maximum at 600℃, compared with the rolling state increased about 100 MPa. The tensile strength is reached the maximum at 400℃, compared with the rolling state increased about 53 MPa. When tempering temperature raise to 500℃ in further, the tensile strength decreased to rolling state level. After analysis the precipitates and metallographic phase and fine structure for hot rolling and temper samples，Nb supplement precipitation is the main factor, which lead to yield strength increases. Nb precipitation add percentage rate up to 16.28% at 600℃ tempering. Because of MA tempering strengthening, dislocation density increases, the tensile strength rises up to maximum，meanwhile elongation decrease to minimum at 400℃ tempering. Due to the decomposition of MA and the disappearance of the dislocation results in the decrease of tensile strength at 500℃ tempering. Yield strength unceasing increase and tensile strength after rising first down lead to yield ratio gradually increase with the increase of tempering temperature. At 600℃ tempering, compared with the rolling state, yield ratio increased by 0.15.

Key words: X80 pipeline steel, Temper heat treatment, precipitation, MA , dislocation density

轧制工艺对 1300MPa 级低合金超高强钢组织性能的影响

温长飞，邓想涛，王昭东，王国栋

（东北大学轧制技术及连轧自动化国家重点实验室，辽宁沈阳　110819）

摘　要： 采用不同的轧制冷却工艺，包括轧制温度、道次压下率及轧后冷却路径，得到不同工艺下的轧态初始组织，对轧态钢板进行统一的再加热淬火和回火处理，研究了不同的轧态组织对 1300MPa 级超高强钢热处理后组织和性能的影响。结果表明：高温连续热轧工艺得到的组织为粒状贝氏体+少量板条贝氏体；两阶段控轧控冷的轧态组织为粒状贝氏体，并且能够增加轧态组织原始奥氏体晶界面积，有利于得到细化的再加热奥氏体晶粒；采用两阶段控轧控冷后超快速冷却到 600℃再空冷至室温的中断冷却工艺，可得到更细小的粒状贝氏体组织，并且析出的碳化物粒子尺寸也更小，具有最佳的热处理态性能，屈服强度为 1345MPa，抗拉强度为 1590MPa，-40℃冲击功为 44J。

关键词： 低合金超高强钢，控轧控冷，微观组织，力学性能

Effect of Rolling Process on Microstructure and Properties of Low Alloy Ultra-strength Steel Q1300

Wen Changfei, Deng Xiangtao, Wang Zhaodong, Wang Guodong

(The State Key Laboratory of Rolling and Automation, Northeastern University, Shenyang 110819, China)

Abstract: The initial microstructure was obtained by different rolling cooling technology which mainly contains rolling temperature, reduction and cooling process. The effects of different rolling microstructures on the microstructure and properties of Q1300 ultra-high strength steel after heat treatment were studied by reheating quenching and tempering. The results showed that the microstructure of the continuous hot rolling process is granular bainite and a small amount of lath bainite. Compared to hot rolling process, TMCP process is able to significantly increase the prior austenite grain boundary area is benefical to acquire more smaller reheating austenite grain. Finer granular bainite structure and carbide particle was obtained after control rolling when water cooling to 600 degree before air cooling. The experimental steel has a good combination of strength and toughness after the TMCP and quenching and tempering heat treatment, it's yield strength of 1345MPa, tensile strength of 1590MPa, and the Charpy impact energy at −40 ℃ could reach 44J.

Key words: low alloy ultra-strength steel, TMCP, microstructure, mechanical properties

宝钢湛江钢铁厚板厂热处理线工艺及设备特点

赵文涛，李 伟

（宝钢湛江钢铁厚板厂，广东湛江 524072）

摘 要：热处理工艺是提高材料性能的重要措施之一，热处理产线是厚板高附加值产品生产的重要产线之一，本文介绍了宝钢湛江钢铁厚板厂热处理线所采用的工艺及设备特点，主要阐述了湛江热处理线的抛丸机、无氧化辊底式热处理炉、淬火机、明火辊底式热处理炉、滚盘式冷床机组设备特点及工艺控制特点。

关键词：热处理炉，淬火机，燃烧系统，加热模型

The Technology and Equipment of Heat Treatment Line in Zhanjiang Department of Baosteel

Zhao Wentao, Li Wei

(Zhanjiang Department of Baosteel, Zhanjiang 524072, China)

Abstract: The heat treatment process is one of the important measures to improve the performance of materials, and it is also one of the important line for high value-added plates.This disquisition introduces the technology and equipment of heat treatment line in Zhanjiang department of Baosteel .It mainly introduces the equipment characteristics and process control characteristics of shot-blasting machine, non-oxidation roller hearth normalizing furnace ,quenching machine and open fire roller hearth tempering furnace，disc type cooling bed in Zhanjiang heat treatment line.

Key words: heat treatment furnace, quenching machine, burning system, mathematics model for heating

冲压用钢的耐时效技术开发

杨士弘，张　鹏，杨丽芳，梁媛媛

（河钢集团钢研总院工艺技术研究所，河北石家庄　050000）

摘　要：通过调整材料化学成分、热轧工艺、热轧卷取温度、冷轧退火工艺等，探索出一种通过控制板带的微观碳不均匀分布，形成高含碳相与低碳固溶度的铁素体组织，从而降低间隙原子对时效性的影响。材料经自然时效或烘烤，屈服平台长度明显减小，上屈服强度明显降低，最终提高了材料的抗时效性，满足了烘烤硬化效应的彩涂板在成型中，不发生吕德斯带的需求。

关键词：抗时效性，柯氏气团，平整，烘烤硬化

The Aging Stability Research of Deep Drawing Steel

Yang Shihong, Zhang Peng, Yang Lifang, Liang Yuanyuan

(HBIS Group Technology Research Institute, Shijiazhuang 050000, China)

Abstract: By means of controlling the chemical property of the steel, hot rolling temperature and winding temperature, as well as the annealing parameters. A way of getting an unbalanced microcosmic interstitial atom potency was found. Following the technology, the microcosmic organization is consisted of the high interstitial atom potency organization and the low interstitial atom potency ferritic solid solution. As well the influence of the interstitial atoms are controlled. After aging of storing or baking, the yielding phenomenon is less obvious than it was. Finally the aging is controlled. the DQ galvanizing strip can meet the requirement of non-Lüders bands during forming with the color coating sheets.

Key words: low carbon aluminum killing steel, aging resistance, skin-pass, hard-baking

不同退火温度下铁素体和珠光体组织的演变以及对力学性能的影响

陈　宇，胡　洋，关　琳，焦　坤

（本钢产品研究院，辽宁本溪　117000）

摘　要：本文通过光学显微镜观察不同退火温度下铁素体和珠光体组织的变化，通过 ZWICK 万能拉伸机测试不同退火温度下试样的力学性能，研究组织变化与力学性能的关系。结果显示：经退火后强度大幅度下降，珠光体未发生球化时，由于在加热过程中，金属原子活动能力增强，自发的向外扩散，大晶粒吞食小晶粒而重新长大引起强度下降；强度随退火温度的升高而降低的主要原因是球化珠光体的强度要比片层状珠光体的强度低，珠光体的球化使铁素体基体中的固溶原子扩散并在晶界附近形成碳化物，聚集的碳化物使得固溶原子越来越少，固溶强化作用越来越弱，引起材料整体的强度下降。

关键词：退火温度，组织变化，球化珠光体，固溶强化

Effects of Different Annealing Temperatures the Evolution of Ferrite and Pearlite and Mechanical Property

Chen Yu, Hu Yang, Guan Lin, Jiao Kun

(Product Research Institute of Benxi Iron & Steel Co., Ltd., Benxi 117000, China)

Abstract: In this paper, by means of optical microscope the change of ferrite and pearlite under different annealing temperature, the mechanical properties of the specimens under different annealing temperatures were tested by ZWICK universal tensile machine, Research the relationship between change of microstructure and mechanical properties. The results showed that the strength decreased greatly after annealing, when Pearlite spheroidization not occur, in the heating process, metal atoms activity ability enhancement, spontaneous outward diffusion, the decrease of strength because the grain growth by large-grained devour small-grained; The main reason for the decrease of strength with the increase of annealing temperature is that the strength of spheroidal pearlite is lower than lamellar pearlite, the spheroidization of pearlite results in solid solution of atomic diffusion and carbides are formed near the grain boundaries, because Aggregated carbon atoms lead to the solid solution atoms is less and less and the solid solution strengthening effect is getting weaker and weaker, the overall strength of the material decreased.

Key words: annealing temperature, the change of organization, spheroidal pearlite, solid solution strengthening

亚温淬火对 960MPa 级高强钢组织性能影响

卢茜倩，谷海容，汪永国，晋家春，崔　磊，詹　华

（马鞍山钢铁股份有限公司，安徽省马鞍山　243000）

摘　要： 借助热膨胀仪测定 Q960 相变点温度 Ac1、Ac3 以及静态 CCT 曲线，通过对 960MPa 级高强钢进行奥氏体化+两相区淬火+回火处理，探讨亚温淬火的奥氏体化温度、两相区淬火、回火温度对其组织和性能的影响。结果表明，Q960 的相变点 Ac1、Ac3 分别为 730℃、866℃；当奥氏体化温度达到 900℃时，在完全淬火状态下可以得到均匀的马氏体组织；采用在 900℃奥氏体化+810℃淬火+450℃回火处理，实验钢综合性能优良，具有更加良好的强韧性匹配，尤其具有较高的低温冲击韧性。

关键词： Q960，CCT 曲线，亚温淬火，热处理，组织与性能

Influence of Subcritical Quenching Process on Microstructure and Mechanical Properties of 960MPa Grade High-strength Steel

Lu Qianqian, Gu Hairong, Wang Yongguo, Jin Jiachun, Cui Lei, Zhan Hua

(Maanshan Iron and Steel Co., Ltd., Maanshan 243000, China)

Abstract: The phase change temperature Ac1 and Ac3 and static CCT curves of Q960 had been measured through the thermal expansion experiment. The influence of subcritical quenching process on microstructure and properties was studied through the process of austenitizing + two-phase zone quenching + tempering treatment on 960MPa Grade High-strength

Steel. The results showed that the transformation temperature Ac1 is 730℃ and Ac3 is 866℃. When the austenitizing temperature reach 900℃, the tissue of steel almost all of well-distributed martensite on the condition of complete quenching. With the 900℃ austenitizing + 810℃ quenching + 450℃ tempering process, the experimental steel showed a good comprehensive performance and strong toughness match especially a higher impact toughness at low temperature.

Key words: Q960, CCT curve, subcritical quenching process, heat treatment, microstructure and mechanical properties

Q960E 高强钢新品种开发及热处理工艺研究

于　浩[1]，王少阳[1]，谷海荣[2]，卢茜倩[2]

（1. 北京科技大学材料科学与工程学院，北京　100083；

2. 马鞍山钢铁股份有限公司技术中心汽车板研究所，安徽马鞍山　243000）

摘　要： 本文设计了一种低成本薄规格 Q960E 高强钢，在马钢 2250 生产线轧制出 8mm、10mm 和 12mm 三种厚度规格的热轧板，热轧板组织为铁素体和珠光体。通过扫描电镜、透射电镜和力学性能分析研究了热处理工艺对组织性能的影响规律。通过提高回火温度抗拉强度随之增加，屈服强度先增加后减小，并且在回火温度小于 250℃ 时表现为连续屈服，在回火温度大于等于 350℃ 时为不连续屈服，塑性和低温冲击韧性均随回火温度的增加先降低后增加，在 250～350℃ 回火时出现低温回火脆性。现场经过 900℃ 淬火和 450℃ 回火热处理后，三个厚度规格的钢板板形良好，屈服强度大于 1017MPa，抗拉强度大于 1051MPa，断后伸长率大于 14.2%，–40℃ 冲击功大于 105J。

关键词： Q960E 高强钢，热处理，组织结构，力学性能

Development and Heat Treatment Process of a New Q960E High Strength Steel

Yu Hao[1], Wang Shaoyang[1], Gu Hairong[2], Lu Qianqian[2]

(1. University of Science and Technology Beijing, School of Materials Science and Engineering, Beijing 100083, China

2. Magang (Group) Holding Co., Ltd., Technology Center, Automobile Plate Research Institute, Maanshan 243000 Chnia)

Abstract: A new thin plate of Q960E with low cost has been designed. The steel is hot rolled to 8mm, 10mm, and 12mm thickness with microstructure of ferrite and pearlite. The effect of heat treatment on its microstructure and properties is studied by SEM, TEM and mechanical testing. The tensile strength increases with the increase of tempering temperature while the yield strength goes up first and then goes down. When the tempering temperature is below 250℃ there is no obvious yield point but the yield platform appears when tempered over 350℃. The plastic and low-temperature impact toughness decrease first and then increase with the increase of temping temperature and the temper brittleness appears between 250℃ to 350℃. After industrial heat treatment of quenching at 900℃ and tempering at 450℃, all of the plates with different thickness have good plate shape, with yield strength over 1017MPa, tensile strength over 1051MPa, total elongation over 14.2% and impact energy at –40℃ over 105J.

Key words: Q960E high strength steel, heat treatment, microstructure, mechanical property

两种不同类型冷轧连续退火炉简介

王　静，孙荣生，辛利峰，林森木，张福义

（鞍钢股份冷轧厂，辽宁鞍山　114003）

摘　要：连续退火线是目前新建冷轧厂的必备机组之一，不同的立式退火炉组成部分相同，但由于加热段、冷却段及时效段等的不同设计而存在一定区别，本文简要介绍了比利时 DREVER 公司和法国 SELAS 公司设计的两条立式连续退火炉的差异。

关键词：连续退火炉，DREVER，SELAS

The Introduction of Two Different Continuous Annealing Lines

Wang Jing, Sun Rongsheng, Xin Lifeng, Lin Senmu, Zhang Fuyi

(The Cold-Rolling Plant of Ansteel Corporation, Anshan 114003, China)

Abstract: Nowadays, the continuous annealing line (CAL) is one of the necessary production line when people construct a new cold plant. All the CAL consist of the same parts, but because of the different designation of the heating-section、cooling-section and the overaging-section there are some differences between some CAL. This article mainly introduced the differences between the CAL designed by Belgium DREVER Corporation and France SELAS Corporation.

Key words: continuous annealing line, DREVER, SELAS

轧制工艺和冷却工艺对管线钢正火组织与性能的影响

张清清，章传国，郑　磊

（宝钢股份中央研究院，上海　201900）

摘　要：通过金相、拉伸、冲击及抗氢致裂纹试验等显微组织和性能分析测试手段，研究了轧制工艺和轧态组织对 C-Mn-Nb-V 管线钢正火组织与性能的影响规律。结果表明，经 880~900℃正火后试验钢的组织为铁素体+片状珠光体+粒状珠光体，控轧后低温停冷或空冷有利于 880℃正火组织的细化，随着正火温度提高，轧制工艺引起的组织差异弱化。控轧钢对正火温度较敏感，控轧+350℃停冷钢经 880℃正火后的强度最高，900℃正火条件下控轧空冷和普通热轧钢的强度较控轧水冷钢高。控轧钢经 880~900℃正火后的-40℃夏比冲击韧性优于普通热轧钢，轧制工艺和冷却工艺对管线钢正火后的-15℃ DWTT 性能无显著影响。控轧空冷钢经 880~900℃正火后片状珠光体部分转变为粒状珠光体，带状减弱，抗 HIC 性能提升，氢致裂纹在控轧空冷组织中沿片状珠光体带萌生和扩展，在正火组织中沿残留的片状珠光体扩展。

关键词：轧制工艺，冷却工艺，正火，管线钢，显微组织，性能

Effect of Rolling and Cooling Process on Microstructure and Mechanical Properties for the Normalized Pipeline Steel

Zhang Qingqing, Zhang Chuanguo, Zheng Lei

(Center Institute of Baosteel Group, Shanghai 201900, China)

Abstract: The effects of rolling and cooling process on microstructure and mechanical properties for the normalized C-Mn-Nb-V pipeline steel were studied by optical scope, tensile test, impact and hydrogen induced crack test. The results show that, the microstructure of 880~900℃ normalized steel is composed of ferrite, lamellar and granular pearlite. Stop cooling at low temperature or air cooling after control rolling is beneficial to refining the normalized grain at 880℃. Since the normalizing temperature increases to 900℃, the difference among the microstructure is weakening. The strength for controlled rolling steel is sensitive to the normalizing temperature. After 880℃ normalizing the strength for steel with stopping cooling at 350℃ followed after control rolling is optimal, while after 900℃ normalizing the strength of traditional hot rolled steel and steel with air cooling after control rolling are higher than steel with water cooling followed with control rolling. Followed with 880~900℃ normalizing, the −40℃ impact toughness of control rolled steels are better than traditional hot rolling. Little effect on the performance of −15℃ drop weight tear test can be showed for normalized steel with different rolling and cooling process. Some lamellar pearlites are replaced by dispersion granular pearlites after 880~900℃ normalizing. The number of hydrogen induced crack decreases, which is initiated and propagated along with the lamellar pearlitic band in controlled rolling steel, while at remained lamellar pearlite in normalized steel.

Key words: rolling porcess, cooling process, normalizing, pipeline steel, microstructure, mechanical properties

宝钢变厚板（VRB）生产情况介绍

熊　斐[1]，姜正连[2]，李山青[1]

（1. 宝山钢铁股份有限公司研究院；2. 宝山钢铁股份有限公司冷轧厂）

摘　要： 变厚板（Variable Rolled Blanks）是通过在轧制的时候周期地改变工作辊辊缝而获得的一种在纵向具有不同厚度的带钢。该产品与激光拼焊板（TWB）一样，可以在汽车减重方面发挥作用。本文将介绍宝钢 VRB 的生产情况及后续的发展。

关键词： 变厚板轧制，控制，辊缝

The Production Status of the Variable Rolled Blanks (VRB) in Baosteel

Xiong Fei[1], Jiang Zhenglian[2], Li Shanqing[1]

(1. Baoshan Iron & Steel Co., Ltd. Research Institute; 2. Baoshan Iron & Steel Co., Ltd. Cold Rolling Mill)

Abstract: VRB (Variable Rolled Blanks) is a kind of strip that obtained from periodical variation of work rolls' gap in rolling process. The strip can play a role in Lightweight Automobile like TWB. In this article, we will introduce the current production and subsequent development of VRB in Baosteel.

Key words: variable rolled blanks, rolling, control, roll gap

冷轧镀锌汽车板隆起缺陷研究

于　孟，林海海，常　安，王永强，靳振伟，陈　飞

（首钢技术研究院，北京　100043）

摘　要：结合对现场大量隆起数据的统计分析，从热轧、冷轧上下游工艺对几种典型带材隆起缺陷的产生机理进行研究；有限元仿真分析了轧制过程中热轧带材断面形状与冷轧承载辊缝的匹配关系，分析了不同弯辊力对轧制后带材横断面形状的影响；通过控制热轧带钢横断面形状、调整冷轧轧制工艺参数等一系列优化措施，大大降低了镀锌产线带材隆起缺陷的发生，实现镀锌汽车板下线后板形质量的大幅提升。

关键词：镀锌，隆起缺陷，板形，工艺参数

Research on Ridge Defect of Cold Rolled Galvanized Automobile Sheets

Yu Meng, Lin Haihai, Chang An, Wang Yongqiang,
Jin Zhenwei, Chen Fei

(Shougang Research Institute of Technology, Beijing 100043, China)

Abstract: Combined with the data analysis of ridge defect in the field, the mechanism of several typical ridge defects is studied in this paper in terms of hot and cold rolling's process. Using FEM simulation, the relation between shape of hot rolled strip section and roll gap of cold rolling in the rolling process, as well as the effects of bending force on shape of strip section after rolling, is analyzed. By a series of optimization measures, such as controlling the shape of hot rolled strip section and adjusting the cold rolling process parameters, the ridge defect of cold rolled galvanized sheet strip is greatly reduced and the quality of the galvanizing automobile strip shape off the production line is improved.

Key words: galvanizing, ridge defect, strip shape, technological parameter

四辊铝带冷轧机工作辊辊型对板形影响分析

王　伟，张　波，郑振环

（福州大学机械工程及自动化学院，福建福州　350116）

摘　要：针对四辊铝带冷轧机工作辊三角函数辊型曲线设计问题，分析了轧辊凸度和三角函数包角参数对轧辊辊型曲线形状的影响规律，并利用金属横向流动的变分法与辊系弹性变形的影响函数法耦合的板形模拟模型，研究了辊型曲线参数对带材前张应力横向分布和出口横断面厚度分布的影响。分析结果表明，轧辊凸度主要影响带材二次板形，曲线包角对1/4板宽处的板形有明显的影响。这些结果有助于三角函数工作辊辊型曲线在工业中的应用。

关键词：铝带冷轧机，三角函数曲线，工作辊辊型，板形

Effect of Work Roll Profile on Strip Flatness of Four-high Cold Aluminum Strip Rolling Mill

Wang Wei, Zhang Bo, Zheng Zhenhuan

(College of Mechanical Engineering and Automation, Fuzhou University, Fuzhou 350116, China)

Abstract: In order to investigate the design problem of the triangular function profile of work rolls for four - high cold aluminum strip rolling mill. The effect of roll crown and wrap angle on the work roll profile is analyzed. In addition, the transverse distribution of the front tension stress and the exit thickness of the cold rolled Aluminum strip are analyzed by using the shape simulation model, which couples the variation method of the cold rolled metal transverse flow and the influence function method of elastic deformation of roll stack. The analysis results show that the work roll crown mainly affects the secondary strip shape, the work roll wrap angle has a significant effect on the shape of the strip located at one-fourth width, and the analysis results facilitates the industrial application of the triangular function profile of work rolls.

Key words: cold aluminum strip rolling mill, triangular function, work roll profile, strip flatness

冷轧 ND 钢的耐硫酸腐蚀性能研究

李永灿，梁高飞，阎元媛

（宝钢股份中央研究院，上海　201900）

摘　要： 本文开发冷轧耐硫酸露点腐蚀钢 ND 钢薄板，应用于烟气处理系统。取耐候钢 SPA-C 作为对比材料，采用极化曲线和全浸试验评估冷轧耐硫酸钢 ND 钢和耐候钢 SPA-C 在 50vol.% H_2SO_4 腐蚀环境下的腐蚀行为。极化曲线和全浸试验表明，在 50vol.% H_2SO_4 腐蚀环境下，ND 钢腐蚀速率较 SPA-C 低。采用 XRD、SEM 和 EDS 分析腐蚀产物表明，Cr 和 Sb 的添加改变表层腐蚀产物构成，形成连续致密，与基体结合良好的腐蚀层，提高了 ND 钢的耐硫酸腐蚀性能。

关键词： 耐硫酸钢，耐候钢，冷轧，酸腐蚀

Research on the Sulfuric Acid Corrosion Resistance of Cold Rolled ND Steel

Li Yongcan, Liang Gaofei, Yan Yuanyuan

(Research Institute, Baoshan Iron & Steel Co., Ltd., Shanghai 201900, China)

Abstract: Cold-rolled sheet with good sulphuric acid dew point corrosion resistance steel ND was developed in this study, which is suitable for the using of flue gas treatment system. The corrosion behavior of cold-rolled ND steel was investigated using potentiodynamic polarization test and weight loss measurement in 50vol.% H_2SO_4, comparing with weathering steel SPA-C. Both potentiodynamic polarization test and weight loss measurement indicated that the corrosion resistance of ND steel was better than SPA-C. Surface analyses of the corroded surfaces conducted after the immersion test indicated that two layers were formed on the substrate of ND steel, with the outer layer composed of $FeSO_4·H_2O$ and the inner layer mainly composed of iron oxides. The additions of Cr and Sb refined the corrosion products, and improved the adhesion ability between the rust layer and the substrate.

Key words: sulfuric acid resistant steel, weathering steel, cold-rolled, acid corrosion

高强 IF 钢冷轧过程轧裂原因及机理研究

王　畅[1,3]，于　洋[1,3]，王　林[1,3]，刘国梁[1,3]，陈　瑾[2]，王明哲[2]

（1. 首钢技术研究院薄板研究所，北京　100043；2. 首钢股份公司制造部，河北迁安　064404；
3. 绿色可循环钢铁流程北京市重点实验室，北京　100043）

摘　要：本文观察了含磷高强 IF 钢冷轧轧裂缺陷的微观形貌及热轧卷拉伸试样断口情况，分析了轧裂形成原因主要始于连铸坯的中心组织及元素异常。采用热酸浸实验、铸坯刨层化学分析实验等方法发现铸坯心部存在偏析带和缩孔，中部磷偏析度为 1.15。根据现有连铸工艺条件提出了控制过热度低于 30℃以及加大轻压下量至 4 mm 等改善措施，可有效改善连铸坯内部质量，减轻中心裂纹及缩孔的产生。

关键词：冷轧，高强 IF 钢，中心偏析，过热度

The Causes and Mechanism of the Cracking of High Strength IF Steel during Cold Rolling Process

Wang Chang[1,3], Yu Yang[1,3], Wang Lin[1,3], Liu Guoliang[1,3], Chen Jin[2], Wang Mingzhe[2]

(1. Shougang Research Institute of Technology, Beijing 100043, China; 2. Shougang Qian'an Steel Company, Qian'an 064404, China; 3. Beijing Key Laboratory of Green Recyclable Process for Iron & Steel Production Technology, Beijing 100043, China)

Abstract: Fracture defect microstructure and hot-strip tensile sample of the high strength IF steel containing phosphorus was observed, the main cause was that the center of continuous casting billet is abnormal. In this paper, the determination of segregation and shrinkage in the center of casting billet was found in the experiment of hot acid leaching experiment and the chemical analysis of the casting blank plane, and the center phosphorus segregation was 1.15.Suggested improvement measures according to current continuous casting process is that the degree of superheat should be lower than 30℃and soft reduction amount should be improved to 4mm, which can effectively improve inner quality of the slab and alleviate the middle fissure and shrinkage defects.

Key words: cold-rolling, high strength IF steel, center segregation, degree of superheat

冷轧高强钢厚规格产品焊接工艺优化研究

王　焕，陈明希

（本钢板材股份有限公司第三冷轧厂，辽宁本溪　117000）

摘　要： 为满足汽车发展的要求，应对汽车"轻量化"和"提高安全性"，具有较低的屈强比、优良的成型性能的高强度双相钢得到了越来越广泛的应用。但随着高强钢的强度提升，合金元素相应增加，对焊接工艺的要求也越来越高，特别是对于厚度 2.0mm 以上的冷轧带钢，焊缝质量更是难以保证。本文分析了 2.0mm 厚度以上的高强钢在某钢厂连退线生产过程中焊缝质量不稳定的原因，以及对焊机设备本体及焊接工艺的优化，得到良好的焊缝质量，保证稳定生产。

关键词： 高强钢，窄搭接焊机，焊缝质量

Analysis and Optimization on Thick High Strength Steel Welding Process

Wang Huan, Chen Mingxi

(The Third Cold Rolling Mill, Bengang Steel Plates Co., Ltd., Benxi 117000, China)

Abstract: To meet the requirements of the automotive development and "lightweight" and "increased safety", high strength dual phase steel with low yield ratio and excellent formability has been more and more widely applied. However, as the strength increased, the more alloy elements increased, the high requirements of the welding process for high strength steel are needed, especially for the thickness up to 2.5mm cold rolled strip, which is difficult to guarantee the welding seam quality. Here analyzes the reason that welding seam quality is not stable on continuous annealing line 2150 of a steel plant, and optimizes the welder equipment and welding process to get good welding seam quality, ensure the stable production.

Key words: high strength steel, mash seam welder, welding seam quality

明暗场成像对冷轧表面缺陷检测仪的影响分析

蒋　渝

（宝山钢铁股份有限公司，上海　201900）

摘　要： 分析了冷轧表面缺陷检测仪采用的明暗双场照明设计的由来，结合缺陷样板的测试效果，比较了明场和暗场在各类缺陷检测中的差异，明确了相机成像的合理角度。并通过在生产现场的应用，证实了明暗双场可以提升缺陷的检出效果。

关键词： 表面缺陷，检测仪，明场，暗场

Analysis of the Influence of Bright and Dark Field Image on the Surface Defects Detector for Cold Rolled Strips

Jiang Yu

(Baoshan Iron & Steel Co., Ltd., Shanghai 201900, China)

Abstract: The origin of the surface defects detector for cold rolled strips adopts the design of bright and dark field are analyzed, Compares the difference of defect detection in bright-field and dark-field by defect sample test, And summarized the reasonable angle of image. Proved the bright and dark field can improve the detection effect by the application of surface defect detector on the production site.

Key words: surface defect, detector, bright field, dark field

1700 冷连轧机 FGC 断带原因分析与控制

张　良，杨洪凯，王少飞，万　军

（首钢京唐钢铁联合有限责任公司，河北唐山　063200）

摘　要： 冷连轧机动态变规格（Flying Gauge Change）是实现酸轧机组全连续轧制的关键技术，如何实现稳定动态变规格，防止在变规格过程中发生断带事故是保证轧制稳定，提高机组产能的关键所在。本文主要针对京唐 1700 酸轧动态变规格过程断带问题，介绍了其动态变规格参数的基本设定和变化时序，重点分析了影响动态变规格过程稳定控制的主要因素，以及动态变规格过程中辊缝和辊速的计算与控制，并通过对各种影响因子的分析，最终将复杂的各种因子利用二进制逻辑回归的方式，简化为几个主要因子建立起来的可控数学模型，通过控制相应因子的数值，从而达到解决动态变规格过程断带问题的目的，具有较强的现场指导意义。

关键词： 冷轧连轧机，动态变规格（FGC），断带，辊缝变化量，逻辑回归，数学模型，稳定轧制

Reason Analyzing and Control of 1700 Cold Continue Rolling Mill FGC Strip Break

Zhang Liang, Yang Hongkai, Wang Shaofei, Wan Jun

(SGJT Iron & Steel United Co., Ltd., Tangshan 063200, China)

Abstract: Tolerance and uniformity of cold rolling strip thickness is a very important standard to evaluate a line products quality and control level. Strip head and tail cold rolling under more parameter fluctuation and external factor influence, it is difficult to build up a precise math model, that leads to strip head and tail thickness over proof problem, the products yield is becoming lower. Controlling of cold rolling strip head and tail thickness over proof length is a challenge needs to be conquered urgently. This thesis mainly introduced the type、reason and control measures of strip head and tail thickness over proof problem on 2230 PLTCM, it is the first time to introduce a concept of strip head and tail thickness over proof rate, made a lot of technical improvement by long time data analyzing and tracking, reduced the length and possibility of strip head and tail thickness over proof, improved the strip head and tail thickness control accuracy, increased products yield.

Key words: cold continue rolling mill, FGC, strip break, roll gap changing volume, logistic regression, math model, stable rolling

高强汽车板生产连退线常见质量问题及解决方法

关淑巧，李文田，李建军

（唐钢高强汽车板有限公司，河北唐山　063000）

摘　要：连续退火机组高强钢的生产具有其自身的设备及工艺特点，在普材过渡至双相钢期间经常出现跑偏、瓢曲的问题，板面易出现斑迹、麻点等质量缺陷。通过系统分析研究，通过对温度调整、张力控制、排产等方面进行优化，解决了上述质量及影响生产稳定性问题，实现了高强钢在连续退火机组的批量稳定生产。

关键词：高强钢，跑偏，斑迹，麻点，稳定生产

浅谈宝钢湛江冷轧工程自主集成设计与创新实践

王海东，严江生

（宝钢工程技术集团有限公司冷轧事业部，上海　201900）

摘　要：本文介绍了宝钢湛江冷轧工程的产品结构，并论述了酸轧机组、连退机组、热镀锌机组工艺流程设计和主要工艺设备选型，并论述了自主集成设计与创新实践。本论文总结的湛江冷轧工程机组设计经验亮点和采取的引进少量关键设备基础上的自主集成及创新模式和优化设计，对于建设同类生产线及相关产线，具有很好的借鉴价值。

关键词：冷轧，自主集成，优化设计，创新，实践

Independent Integration Design & Innovation Practice of Baosteel Zhanjiang Cold Rolling Project

Wang Haidong，Yan Jiangsheng

(Cold Rolling Department of Baosteel Engineering & Technology Group Co., Ltd., Shanghai 201900, China)

Abstract: The product mix of Baosteel Zhanjiang cold rolling project is introduced. The production process design, main process equipment selection of PL-TCM, CAL, CGL and independent integration design & innovation practice are described. Based on introduction of a small number of key equipment, the independent integration, innovation, optimal design and design experience of Baosteel Zhanjiang Cold Rolling Project will provide the reference value for the similar production line.

Key words: cold rolling, independent integration, optimal design, innovation, practice

宝钢湛江一冷轧工程"降本增效"设计

李　佳

（宝钢工程技术集团有限公司，上海　201900）

摘　要：本文介绍了宝钢湛江一冷轧工程设计过程中，是如何考虑"降本增效"的。

关键词：宝钢，湛江，冷轧工程，降本增效，设计

"Cost Decreasing and Benefit Increasing" Design of Baosteel Zhanjiang Cold Rolling Project

Li Jia

(Baosteel Engineering & Technology Group Co., Ltd., Shanghai 201900, China)

Abstract: This article introduced the design of Baosteel Zhanjiang Cold Rolling Project, How to consider "Cost decreasing and benefit increasing".

Key words: Baosteel, Zhanjiang, cold rolling project, cost decreasing and benefit increasing, design

汽车用镀锌基板表面微腐蚀的研究

阮国庆，辛利峰，张学伟，孙晓宇，解志彪

（鞍钢股份冷轧厂，辽宁鞍山 114003）

摘　要： 随着汽车工业的发展，用户对汽车用镀锌板外观质量提出了更高的要求，镀锌板出现表面"粗糙"的问题，影响涂装后质量。本文主要对汽车用镀锌板表面"粗糙"问题进行分析，通过对冷轧酸工艺和轧制工艺的相关实验分析，确定镀锌板表面"粗糙"问题与酸洗微腐蚀有关，最终提出了改善镀锌板表面质量的有效措施。

关键词： 镀锌板，粗糙，鳞片，酸洗，微腐蚀

Research on the Surface Micro Corrosion on the Cold Rolled Strip for Car Use Galvanized

Ruan Guoqing, Xin Lifeng, Zhang Xuewei, Sun Xiaoyu, Xie Zhibiao

(The Cold-Rolling Plant of Ansteel Corporation, Anshan 114003, China)

Abstract: With the development of the auto industry, consumer sets higher demands for galvanized strip for automobile. galvanized sheet discovered surface "rough" problems, impacts on the quality of coating. We analyzed surface "rough" problems for galvanized strip for automobile in this paper, by means of cold rolling acid technology and rolling technology related experimental analysis, determine the surface of galvanized sheet "rough" issues related to micro pickling corrosion, Finally puts forward the effective measures to improve the surface quality of the galvanized sheet.

Key words: galvanized sheet, rough, flake, acid pickling, micro corrosion

对于薄带轧制的一种新型条元法

冯夏维[1]，杨荃[1]，Pierre Montmitonnet[2]

（1. 北京科技大学，北京 100083；2. 法国国立巴黎高等矿业学校，法国 06904）

摘　要：本文描述了一种针对薄带轧制过程中轧制压应力的计算方法，这种方法属于条元法的一个变形，既可以计算由于带钢塑性变形产生的轧制力，也可以计算弹性变形所产生的轧制压应力。随后，本文将所开发的方法同有限元方法进行了比较论证，结果显示，利用本方法计算所得结果同有限元方法所得结果相近，但本方法具有计算速度快的优势。本方法将同轧辊弹性变形模型进行耦合，求解轧制过程中的薄带断面。

关键词：薄带轧制，轧制力，条元法

An Advanced Slab Method for Rolling

Feng Xiawei[1], Yang Quan[1], Pierre Montmitonnet[2]

(1. University of Science and Technology Beijing, Beijing 100083, China;

2. Mines ParisTech, France 06904)

Abstract: This paper presents a mathematical method for the calculation of pressure during strip rolling process. It is a derivative form of Slab Method, and it is capable of calculating the rolling pressure in both the elastic and plastic deformation region. The present paper also provides a comparison study of our model with Finite Element Method, the results show that the difference between these two methods is small, but our model has the advantage of faster speed. This method will be coupled with the elastic deformation model of roller stack, in order to calculate the strip profile during rolling.

Key words: strip rolling, rolling pressure, slab method

基于多元线性回归的板形平直度预测控制研究

赵　强[2]，汪晋宽[1,2]，韩英华[2]

（1. 东北大学信息科学与工程学院，辽宁沈阳；

2. 东北大学秦皇岛分校工程优化与智能天线研究所，河北秦皇岛）

摘　要：在板形闭环控制系统中，板形检测仪一般安装在距末机架出口 3~5m 处，在低速过渡轧制过程中板形检测信号明显滞后后，导致带钢头尾很长一段（有时长达数百米）板形较差。本文系统分析了冷连轧机板形闭环控制系统的构成及其控制规律，深入研究板形表示方法、常见的板形缺陷及将其消除的控制策略。提出了基于小波分析的板形模式识别方法，并采用最小二乘法确定一次板形分量和二次板形分量。建立了板形分量的多元线性回归模型，采用带有遗忘因子的递推最小二乘法估计出模型系数，给出了板形滚动优化控制算法，得到局部最优控制量，有效的消除了板形缺陷。利用 Matlab 软件平台编写了板形预测控制程序，采用实际轧制数据进行仿真实验。仿真结果表明，本文所建立的板形预测控制系统具有良好的控制效果，改善了低速过渡轧制过程中的板形质量，具有一定的工程指导意义。

关键词：板形控制，小波分解，多元线性回归，预测控制

Flatness Prediction Control of Cold Rolling Using Multivariate Linear Regression (MLR) Model

Zhao Qiang[2], Wang Jinkuan[1,2], Han Yinghua[2]

(1. College of Information Science and Engineering, Northeastern university, Shenyang, China;
2. Engineering Optimization & Smart Antenna Institute, Northeastern University at Qinhuangdao, Qinhuangdao, China)

Abstract: The conventional method uses closed loop feedback controller which is located at the 5th housing to adjust the flatness. The flatness measurement sensor placed in the mill behind the last housing. Due to the distance between flatness measurement roll and the last housing is about five meters, it causes a delay of flatness measurement signal. So it is difficult to optimize the closed loop response and reduce the influence of load fluctuation for the present control system, especially in the low-speed cold rolling process. The research on flatness prediction control and automatic flatness control is of important practical value. The flatness pattern recognition based on wavelet analysis is proposed. The first-order coefficient and second-order coefficient of flatness is determined by Least Square Method. The multivariate linear regress equation of flatness coefficient is also established. The algorithm of Recursive Least Square with forgetting factor is used for parameter estimation of the prediction model of flatness coefficient. Receding horizon optimization is applied to obtain the control signal and online identification to ensure the accuracy of prediction model. In the end, the flatness prediction control program is compiled using Matlab software. Through the simulation with field-measured data, its results show that flatness prediction control system has sound effects. It also improves the quality of strip shape in the low-speed cold rolling process and has engineering guidance value.

Key words: flatness control, wavelet analysis, multivariate linear regression

不锈钢带材冷轧工艺控制模型及轧辊弹性变形

张宏亮，王宝山，靳书言，刘　鑫，冯光宏

（钢铁研究总院冶金工艺研究所，北京　100081）

摘　要： 通过对不锈钢精密带材冷轧过程进行有限元的模拟计算，在弹性辊的条件下，研究压下率、摩擦系数、变形抗力、前后张力等工艺参数的控制模型，以及弹性轧辊的应力应变场、空转应力场等，结果表明：（1）随着冷轧压下率的增加，沿接触弧长方向上的轧制压力不均匀程度提高，尤其在接近出口处接触弧长上的轧制压力峰值显著增加。（2）冷轧压下率与沿接触弧最大轧制压力呈线性关系，为 $y=2\times10^9x+4\times10^8$；冷轧压下率与轧辊的最大压扁变形呈线性关系，为 $y=0.3971x+0.0138$；冷轧前后单位张力与轧制力峰值呈线性关系，为 $y=-7\times10^7x+10^9$。（3）轧辊变形区主要集中于接触弧长上方的轧辊的浅表层，应力主要集中在接触弧区上方的区域，渗入深度为 1/4 半径。弹性轧辊空转过程存在环状的应力场分布，呈周期性变化。这与现场轧辊主要失效形式轧辊剥落相一致，可为现场提供技术支持。

关键词： 不锈钢带材，工艺控制，弹性辊

Process Control Model of Stainless Steel Strip Cold Rolling and Elastic Roll Deformation

Zhang Hongliang, Wang Baoshan, Jin Shuyan, Liu Xin, Feng Guanghong

(Metallurgy Department, Central Iron and Steel Research Institute, Beijing 100081, China)

Abstract: In this paper, the finite element calculation of the cold rolling process of stainless steel precision strip is carried out. Under the condition of the elastic roll, process control model for reduction rate, coefficient of friction, deformation resistance, front and rear tension is studied, including the elastic deformation of the roll. The results show that: (1) With the increase of the cold rolling reduction rate, the rolling force nonuniformity in the direction of the contact arc is increased, especially the peak value of the rolling pressure at the contact arc length near the outlet.(2) The relationship between the cold rolling reduction rate and the maximum rolling force along contact arc is basically linear. The relationship is $y=2\times10^9x+4\times10^8$; The cold rolling reduction rate and the maximum flattening deformation of the roll Of the relationship is a linear relationship, the relationship is $y=0.3971x+0.0138$; Cold rolling unit tension and rolling force peak linear relationship, $y=-7\times10^7x+10^9$. (3) The deformation area of the roll is mainly concentrated in the shallow surface of the roll above the arc length. The stress concentration area is mainly concentrated in the area directly above the contact arc area and the infiltration depth is 1/4 radius.The stress field of the elastic roll is cyclic and there is a periodic change. This is consistent with the roll off of the main failed form roll, providing technical support for the scene.

Key words: stainless steel strip, process control, elastic roll

冷轧复合钢板生产工艺及产品开发

原思宇[1]，杨建峰[2]，张万山[1]，胡志勇[2]，马普生[1]

（1. 鞍钢未来钢铁研究院，北京；2. 北钢联（北京）重工科技有限公司，北京）

摘　要：本文介绍了冷轧复合钢板生产工艺及装备，以及利用该技术生产的冷轧不锈钢复合钢板的检验结果，其力学性能满足国家标准要求。对比和分析了冷轧复合技术的先进性和优势，冷轧复合技术突破了传统爆炸复合和热轧复合技术的局限性，减少了坯料加热环节，能够实现钛、铜、不锈钢、铝等多品种复合钢板生产的需求，产品附加值大、应用广泛、市场竞争力强。

关键词：冷轧复合，冷轧不锈钢复合钢板，生产工艺，开发

Technology and Production of Cold Rolled Clad Steel Plate

Yuan Siyu[1], Yang Jianfeng[2], Zhang Wanshan[1], Hu Zhiyong[2], Ma Pusheng[1]

(1. Ansteel Beijing Research Institute, Beijing, China; 2. North Steel Union (Beijing) Heavy Industry Science & Technology Co., Ltd., Beijing, China)

Abstract: This paper introduces the production technology and equipment of cold rolled composite plate, and the test results of cold-rolled stainless steel clad plate produced by this technology. The mechanical properties meet the requirements of the national standard. Comparison and analysis of the advanced composite technology and advantages of cold rolling, cold rolling composite technology broke through the limitation of traditional explosive composite and hot rolling composite technology, with reducing the heating process. Cold rolling composite technology could realize the production demand of the titanium, copper and stainless steel, aluminum and other varieties composite steel plate which productions have added value, wide application and strong market competitiveness.

Key words: cold rolling compound, cold rolling stainless steel clad plate, production process, development

Nano-MoS$_2$ 在水基轧制液中的润滑作用机理研究

李 岩[1,2]，陈义庆[1,2]，高 鹏[1,2]，钟 彬[1,2]，

李 琳[1,2]，艾芳芳[1,2]，肖 宇[1,2]

（1. 海洋装备用金属材料及其应用国家重点实验室，辽宁鞍山 114009；

2. 鞍钢集团钢铁研究院，辽宁鞍山 114009）

摘 要：采用 MRS-10A 四球摩擦磨损试验机考察不同纳米 MoS$_2$ 添加量对水基轧制液承载能力的影响，确定其最佳含量为 0.4wt.%。通过研究不同载荷条件下 MoS$_2$ 水基纳米轧制液的润滑特性，利用 LEXT OLS4000 激光共聚焦显微镜分析磨斑表面纳米粒子的残留形态。结果表明纳米 MoS$_2$ 是一种理想的水基纳米添加剂，可有效降低金属变形过程中的摩擦阻力，提高轧后表面质量。根据其特殊的层状分子结构及不同压力条件下表现出的润滑特性，提出了纳米 MoS$_2$ 在水基轧制液中的润滑作用机理。

关键词：Nano-MoS$_2$，水基轧制液，摩擦学性能，润滑机理

Study on Lubrication Mechanism of Nano-MoS$_2$ in Water-based Rolling Fluid

Li Yan[1,2], Chen Yiqing[1,2], Gao Peng[1,2], Zhong Bin[1,2],
Li Lin[1,2], Ai Fangfang[1,2], Xiao Yu[1,2]

(1. State Key Laboratory of Metal Material for Marine Equipment and Application, Anshan 114009, China;

2. Iron & Steel Research Institutes of Ansteel Group Corporation, Anshan 114009, China)

Abstract: The goal of the work was to investigate the lubrication mechanism of nano-MoS$_2$ in water-based rolling fluid. The tribological performance of different content of nano-MoS$_2$ in water-based rolling fluid was tested by MRS-10A four-ball friction and wear testing machine. Lubrication properties of nano-MoS$_2$ water-based rolling fluid were studied under different loading conditions. The LEXT OLS4000 laser scanning confocal microscope was applied in analyzing different worn spots. The experiments results show that nano-MoS$_2$ was an ideal water-based nano additive, which can effectively reduce the friction resistance in the metal deformation process and improve the surface quality of rolled strip. According to its special layered molecular structure and the lubrication characteristics exhibited under different pressure conditions, the lubrication mechanism of nano-MoS$_2$ in water-based rolling liquid was put forward.

Key words: nano-MoS$_2$, water-based rolling fluid, tribology performance, lubrication mechanism

轧制及热处理对带状组织的影响

秦丽晔[1,2]，邹　扬[1]，赵新宇[1]，樊艳秋[1]

（1. 首钢集团有限公司技术研究院，　北京　100041；

2. 绿色可循环钢铁流程北京市重点实验室，北京　100041）

摘　要： 带状组织对材料的冲击韧性、抗氢致开裂性能、Z 向性能均有重要影响。本文研究了不同轧制工艺以及不同正火热处理工艺对碳锰钢带状组织的影响作用，分析带状组织产生机理以及减轻带状组织的条件，研究结果显示，随着终轧温度的降低，带状组织等级呈现先降低后升高的趋势，正火后采用加速冷却方法，增加冷却速度，抑制合金元素在相变过程中的扩散作用，有利于相变在不同区域内同时发生，可以减轻带状组织。

关键词： 带状组织，轧制工艺，正火热处理，成分偏析

The Influence of Rolling and Heat Treatment for Banding Structure

Qin Liye[1,2], Zou Yang[1], Zhao Xinyu[1], Fan Yanqiu[1]

(1. Shougang Research Institute of Technology, Beijing 100041, China

2. Beijing Key Laboratory of Green Recyclable Process for Iron & Steel Production
Technology, Beijing 100041, China)

Abstract: The banding structure can influence metal property, such as impact toughness, resistance hydrogen-induced cracking, Z-direction property. So the research on the causes and control methods for band structure is significant. The effects of rolling and normalization process factors on banding structure have been studied and the mechanism have been analyzed to decrease the banding pearlite. The results that: with the decrease of finishing temperature, the banding pearlite was decreased at first and then increased. The increased cooling rate by soft water after normalization, was beneficial to curb the diffusion of alloy elements. So the phase transformation at the different zones was likely happened at the same time, and the banding pearlite could be reduced.

热轧桥壳用钢 BQK580 钢板的研制开发

张爱梅，赵　亮，吾　塔

（宝钢集团八钢公司制造管理部,新疆乌鲁木齐　830022）

摘　要： 根据汽车桥壳用钢使用特点进行了成分和工艺设计。结合桥壳钢技术要求，分析了化学成分、工艺参数、金相组织对桥壳钢性能的影响，在热轧 1750 生产线研制开发了桥壳用钢 BQK580。

关键词： 桥壳用钢，复合微合金化，热轧工艺

Research and Development on Hot-rolled of BQK580 Automobile Axle Housings

Zhang Aimei, Zhao Liang, Wu Ta

(Manufacturing Management Department, Bayi Iron & Steel Co.,Baosteel Group, Urumchi 830022, China)

Abstract: On the basis of composition and process design according to the application features of automobile axle housing steel. Combined with technical requirements of the axle housings steel, the influences of chemical composition,process parameters and microstructure to axle housing steel's properties were analyzed. And BQK580 axle housing of hot-rolled stamping special steel was researched and developed in 1750 hot-rolled mill.

Key words: axle housing steel, compound micro-alloying, hot-rolled process

1450 连退机组连退线快冷段风机平衡控制实现

李 义

（鞍钢股份冷轧厂四分厂，辽宁鞍山　114009）

摘　要： 为提高连退线冷轧钢板的表面质量和安全生产对快冷段的冷却风机控制进行改进，以实现冷轧钢板在连退退火处理过程中，上下表面冷却压力平衡，避免钢板擦划伤。

关键词： 变频风机，可调速风机，快冷段，HMI 人机画面接口

基于 B 样条有线条法的带钢矫直过程建模与仿真

李 凯，孙建亮，高亚南，彭 艳

（燕山大学国家冷轧板带装备及工艺工程技术研究中心，河北秦皇岛　066004）

摘　要： 本文满足条单元简支端和自由端边界条件的 B 样条位移函数，在位移函数的基础上建立应力应变矩阵、条单元刚度矩阵和结线等效载荷列阵，揭示了 B 样条位移函数的局部便捷性和灵活性，且可以通过调节结线样条函数控制点数量来控制计算规模和精度；进而建立了基于弹塑性 B 样条位移函数的钢板连续矫直过程的有限条法数学模型,模拟计算了中厚板 9 辊矫直过程，计算出的矫直效果与现场实际矫直质量吻合较好，证明所建弹塑性 B 样条有限条法用于矫直过程的合理性。

关键词： 弹塑性 B 样条，钢板矫直，有限条法，位移函数

Application of Elastoplastic B Spline Strip Line Method in Plate Leveling

Li Kai, Sun Jianliang, Gao Ya'nan, Peng Yan

Abstract: This paper used unit simply supported end and free end boundary condition of B spline displacement function, establish strain matrix, stiffness matrix and equivalent load line array based on displacement function, reveals B spline displacement function of local convenience and flexibility, and can adjust the junction line like the function of control points to control the computational scale and the precision; and to establish the mathematical model of the finite strip method of elastic-plastic B spline displacement function of steel plate based on continuous straightening process, the simulation of plate 9 roller straightening process, calculate straightening effect and actual straightening quality is consistent, that the elastic-plastic B spline finite strip method for reasonable straightening process.

Key words: elastioplastic B spline, plate straightening, finite strip method, displacement function

热轧原料对冷连轧机板形控制的影响

王少飞，孙光中，窦爱民，时海涛，张　良

（首钢京唐钢铁联合有限责任公司，河北唐山　063200）

摘　要： 热轧带钢的截面形状和硬度差异对冷轧带钢板厚和板形具有重要影响。本文通过现场生产实际数据以及热轧板硬度测量，定量研究热轧来料对冷轧板形控制的影响。通过研究可以表明，来料的断面形状、硬度对冷轧的板形控制具有很大的影响。

关键词： 板形，控制，原料

Hot Rolling Material Influence on Cold Tandem Mill Flatness Control

Wang Shaofei, Sun Guangzhong, Dou Aimin, Shi Haitao, Zhang Liang

(Shougang Jingtang Iron & Steel Co., Ltd., Tangshan 063200, China)

Abstract: The cross section shape of hot rolling strip steel and hardness difference of cold rolled steel strip thickness and shape has an important influence. In this article, through the actual production data and hot rolling strip hardness measurement, quantitative research hot rolling material to manufacture the influence of shape control. Through research can show that the cross section shape, the hardness of the material has a great influence on cold rolled strip shape control.

Key words: shape，control，material

邯钢 CSP 轧制工艺设备数字化管控与生产应用

邓　科，王春生，伊晓亮，田吉祥

（河钢集团邯钢公司连铸连轧厂，河北邯郸　056015）

摘　要：本文根据邯钢 CSP 短流程薄规格轧制工艺特点，从剖析影响薄规格轧制稳定性的关键因素入手，通过提升相关轧制工艺设备功能数字化管控和工艺改进方面，有效解决了薄规格轧制时面临的生产、工艺及质量问题，实现了轧制技术科学化管理，显著提升了邯钢 CSP 薄规格轧制能力。

关键词：数字化，薄规格，精度

The Digital Evaluation and Practice of the Rolling Equipment in Hansteel CSP Mill

Deng Ke, Wang Chunsheng, Yi Xiaoliang, Tian Jixiang

(Hesteel Group Hansteel Company，CSPMill，Handan 056015, China)

Abstract: According to the CSP short process characteristics，face from the point of view of rolling process of thin gauge rolling stability of key factors are analyzed, through rolling equipment accuracy digital evaluation and process optimization, an effective solution to the thin gauge rolling production，technology and quality problems, realizes the scientific research of the rolling technology, significantly enhance the Hansteel CSP thin gauge rolling ability.

Key words: digital, thin gauge, accuracy

冷连轧机组轧后钢板表面形貌的控制

孙荣生，王　静，刘英明，辛利峰

（鞍钢股份冷轧厂，辽宁鞍山　114003）

摘　要：随着经济的发展，冷轧带钢使用范围日益广泛，用户对冷轧带钢的表面形貌也提出了越来越高的要求。本文主要对影响冷连轧机组轧后钢板表面形貌的重要因素进行了详细分析，主要包括轧辊原始表面形貌、轧制周期及第 5 架轧制力等几个主要方面，最终提出了控制冷连轧后钢板表面形貌的措施。

关键词：冷连轧机组，表面形貌，轧辊，粗糙度，PC 值

The Control Measure of the Surface Topography on the Cold Rolling Strip

Sun Rongsheng, Wang Jing, Liu Yingming, Xin Lifeng

(The Cold-Rolling Plant of Ansteel Corporation, Anshan 114003, China)

Abstract: With the development of the economy, the cold rolling strip is used widely and the surface topography of the strip was demanded higher and higher by the customers. The main factors that effect the surface topography on the clod rolling strip in the TCM are analyzed briefly in this paper, the main contents are as following: the original surface topography of the roll, the rolling circle and the roll force of the fifth stand, the control measures of the surface topography on the cold rolling strip are brought at last.

Key words: tandem cold mill, surface topography, roll, roughness average, PC value

生产工艺对高强低合金钢屈服点伸长率的影响

贾　岳[1,2]，黄学启[1,2]，刘李斌[1,2]，李　维[1,2]，黄　俊[3]，巫雪松[3]

（1. 首钢集团有限公司技术研究院，北京　100043；2. 绿色可循环钢铁流程北京市重点实验室，
北京　100043；3. 首钢京唐钢铁联合有限责任公司，河北唐山　063200）

摘　要： 低合金高强钢在屈服点伸长率较高时，其冲压汽车零件容易产生拉伸应变痕缺陷。以 H340LA 板材进行实验，考查了热轧卷取温度和冷轧平整延伸率生产工艺对屈服点伸长率的影响。实验结果表明，升高热轧卷取温度可以增加组织中碳化物含量；提高冷轧平整延伸率有利于增大位错密度。通过这两种途径，可以有效减小低合金高强钢的屈服点伸长率。

关键词： 低合金高强钢，屈服点伸长率，卷取温度，平整延伸率

Effect of Processing Parameters on Yield Point Elongation of High Strength Low Alloy Steel

Jia Yue[1,2], Huang Xueqi[1,2], Liu Libin[1,2], Li Wei[1,2], Huang Jun[3], Wu Xuesong[3]

(1. Shougang Research Institute of Technology, Beijing 100043, China;　2. Beijing Key Laboratory of Green
Recyclable Process for Iron and Steel Production Technology, Beijing 100043, China;
3. Shougang Jingtang United Iron & Steel Co., Ltd., Tangshan 063200, China)

Abstract: The large yield point elongation of high strength low alloy steel, readily gives rise to slip-line defects of the car parts. By the experiments on H340LA, the influences of coiling temperature and temper-rolling elongation on yield point elongation were investigated. It showed that the rising temperature of coiling increase the content of carbides, and the higher temper-rolling elongation contributed to increase the dislocation density. These two approaches can decrease the yield point elongation of high strength low alloy steel.

Key words: high strength low alloy steels, yield point elongation, coiling temperature, temper-rolling elongation

低成本调质型 Q690E 钢板的开发

陈振业[1]，陈　起[2]，杨现亮[1]，赵燕青[1]，孙晓冉[1]，杨　浩[1]

（1. 河钢集团钢研总院工艺研究所，河北石家庄　050023；
2. 河钢集团舞阳钢铁有限责任公司科技部，河南平顶山　462500）

摘　要： 通过低碳及 "Mn-Cr-Mo-B-V-Ti" 微合金化成分设计及轧制、热处理工艺选择，成功开发出了低成本调质型 Q690E 钢板。Q690E 钢板的微观组织为回火索氏体，屈服强度 811~891MPa，抗拉强度 852~938MPa，-20℃下冲击性能在 152~187J 之间，满足 GB/T 16270—2009 对 Q690E 工程机械用钢力学性能的要求。在满足用户越来

越严格的使用要求的同时降低了生产成本, 扩大了调质型高强钢产品的市场份额。

关键词: 低成本, 调质, 热处理, 力学性能

Research and Development of Low-cost Quenching and Tempering Q690E Steel

Chen Zhenye[1], Chen Qi[2], Yang Xianliang[1], Zhao Yanqing[1], Sun Xiaoran[1], Yang Hao[1]

(1. The Centre Iron and Steel Technology Research Institute of HBIS, Shijiazhuang 050023, China;

2. Hebei Iron and Steel Group, Wuyang Iron and Steel Co., Ltd., Pingdingshan 462500, China)

Abstract: The Q690E steel for engineering machinery has been developed through low carbon and "Mn-Cr-Mo-B-V-Ti" microalloyed components design, rolling and heat treatment technology. The microstructure of Q690E steel plate is tempered sorbite. Besides, the yield strength is from 811MPa to 891MPa, and the tensile strength is from 852MPa to 938MPa. The impact property below 20℃ keeps between 152J and 187J. It can meet the requirements of GB/ T16270—2009 on mechanical properties for Q690E Steel. The result is that, while meeting the increasingly stringent requirements of users, we have reduced the production costs and expanded the market share of quenching and tempering steel.

Key words: low-cost, quenching and tempering, heat treatment process, mechanical properties

四辊铝带冷轧机支撑辊辊型曲线优化

王 伟[1,2], 张 波[1], 郑振环[1]

(1. 福州大学机械工程及自动化学院, 福建福州 350100;

2. 福建省高端装备制造协同创新中心, 福建福州 350100)

摘 要: 支撑辊辊型曲线对于辊间压力分布, 减少有害接触有着重要的的影响。本文针对四辊铝带冷轧机, 分析了各种轧机辊型凸度配置方式, 比较了直倒角、圆弧倒角、幂函数倒角、无倒角辊端曲线对降低辊间压力峰值和均匀辊间接触压力分布的效果, 同时对支撑辊辊端幂函数倒角曲线的参数进行了优化设计。分析结果表明, 幂函数倒角曲线的辊间压力峰值有明显的降低, 优化后的辊间接触压力分布和弯辊控制效果得到了较大的改善, 有利于板形和板凸度的控制。这些工作对于四辊铝带冷轧机的使用寿命, 改善支撑辊边部磨损, 以及工业生产有重要的指导作用。

关键词: 铝带冷轧机, 支撑辊, 辊端曲线, 辊间压力

Backup Roll Profile Optimization of Four-high Cold Aluminum Strip Rolling Mill

Wang Wei[1,2], Zhang Bo[1], Zheng Zhenhuan[1]

(1. College of Mechanical Engineering and Automation, Fuzhou University, Fuzhou 350100, China;

2. Collaborative Innovation Center of High-End Equipment Manufacturing in Fujian, Fuzhou 350100, China)

Abstract: The backup roll has an important effect on the pressure distribution between the rollers and the reduction of harmful contact. In this paper, for four - high cold aluminum strip rolling mill, analyzes various roll crown configuration, compared with straight chamfer, arc chamfer, power function chamfer, unchamfered of the roll end curve to reduce the effect of peak pressure and uniform the contact pressure distribution between rolls, and the parameters of the chamfer curve of the power function of the backup roller are optimized. The results show that the peak pressure between the rolls of the power function is significantly reduced, and the effect of the contact pressure distribution, the bending roll control is improved greatly, which is beneficial to the shape of the plate and the crown control. These works have important guiding effect on the service life of four - high cold aluminum strip rolling mill, improving the wear of the backup roll, and the industrial production.

Key words: cold aluminum strip rolling mill, backup roll, roll end curve, pressure distribution between the rollers

低屈强比控制技术的开发与应用

黄乐庆[1], 杨永达[1], 王彦锋[1], 狄国标[1], 马长文[1], 韩承良[2]

（1. 首钢集团有限公司技术研究院宽厚板所，北京　100043;

2. 秦皇岛首秦金属材料有限公司，河北秦皇岛　066326）

摘　要： 本文以低屈强比的控制技术为研究对象，通过数据统计发现以 AR、CR 及 N 状态交货的钢板其屈强比较低，其次为 TMCP 态钢板，QT 态钢板的屈强比最高，并选择上述有代表性钢板的组织及应力应变曲线进行分析。后对以控轧-弛豫-控冷工艺生产 TMCP 交货状态的钢板其屈强比控制原理进行分析，并将上述工艺在 Q420qE 成功的应用，得到了较低屈强比和良好的综合力学性能。

关键词： 屈强比，控轧-弛豫-控冷，F/B 钢

Development and Application of Low Y/T Ratio Control Technology

Huang Leqing[1], Yang Yongda[1], Wang Yanfeng[1], Di Guobiao[1],

Ma Changwen[1], Han Chengliang[2]

(1. Plate Technology Department, Shougang Research Institute of Technology, Beijing 100043, China;

2. Qinghuangdao Shouqin Metal Materials Co., Ltd., Qinghuangdao 066326, China)

Abstract: Low Y/T ratio control technology was studied through data statistics and theoretical analysis. The results show that the steel with AR, CR and N delivery status has the lowest Y/T ratio, followed by TMCP status, and QT steel has the highest Y/T ratio. Then stress-strain curves and microstructure of the typical steel were chosen to analyze the law of its Y/T ratio. After that, the controlled rolling-relaxation-controlled cooling process were accepted to control Y/T ratio, good practice of Q420qE dictates controlled rolling-relaxation-controlled cooling process can decrease Y/T ratio and ensure good comprehensive mechanics performance.

Key words: Y/T ratio, controlled rolling-relaxation-controlled cooling process, F/B steel

重型矿山机械轻量化用高强结构钢的研制开发

顾林豪 [1,2,3]，路士平 [1]，初仁生 [1]，刘金刚 [1]，刘春明 [1]，张苏渊 [1]

（1. 首钢集团有限公司技术研究院，北京 100043；

2. 绿色可循环钢铁流程北京市重点实验室，北京 100043；

3. 北京市能源用钢工程技术研究中心，北京 100043）

摘 要： 首钢公司采用低碳低合金的化学成分设计，运用控制轧制和控制冷却技术，通过淬火+回火的热处理工艺，开发的重型矿山机械轻量化用 KQ690D 高强结构钢，具有较高的强度、优异的成型性能、良好的低温冲击韧性和较好的焊接性能。KQ690D 高强结构钢屈服强度大于 750MPa，抗拉强度在 800MPa 以上，延伸率大于 20%，−40℃低温冲击功均值都在 200J 以上。KQ690D 成功应用于首钢 SGE150 重型自卸矿车减量化车身的制造，较好的满足了重型矿山机械轻量化对钢板各项性能的要求。

关键词： 轻量化，高强结构钢，低碳当量，焊接，Nb 微合金化

Research and Development of the High Strength Structural for the Lightweight Heavy Mining Machinery

Gu Linhao[1,2,3], Lu Shiping[1], Chu Rensheng[1], Liu Jingang[1], Liu Chunming[1], Zhang Suyuan[1]

(1. Shougang Research Institute of Technical, Beijing 100043, China;

2. Beijing Key Laboratory of Green Recyclable Process for Iron & Steel Production Technology, Beijing 100043, China;

3. Beijing Engineering Research Center of Energy Steel, Beijing 100043, China)

Abstract: Shougang Group has developed high strength structural steel of KQ690D for the lightweight heavy mining machinery by using low carbon and low micro-alloyed design and two-stage rolling and quenching and tempering process. The quenched and tempered steel of 690MPa offers high strength, good formability, excellent low temperature impact toughness, and good weldability. For the high strength structural steel of KQ690D, the yield strength is greater than 750MPa, the tensile strength is above 800MPa, the elongation is higher than 20%, and the low temperature (−40℃) impact energy value is more than 200J. The steel plate is successfully applied in Shougang SGE150® heavy dump body tub, and it has better satisfied the requirements of the performance of the steel plate for the lightweight heavy mining machinery.

Key words: lightweight, high strength structural steel, low carbon equivalent, welding, Nb microalloying

CSP 板形控制模型优化提升

陈剑飞，吴利娟，杨 光

（武钢有限条材厂 CSP 分厂，湖北武汉 430083）

摘　要： 武钢 CSP 是第二代 CSP 短流程生产线，相对于常规热轧来说，其机型分为上中下三组，不同控制模式的精轧机组辊系配置，更强化了七个机架各自的板形控制重点。随着市场需求，极薄材、高强度结构钢及电工钢对于板形尺寸精度要求越来越高，已不满足于指标数据更看中"断面形状质量评级"，即在满足整体测量数据精度达标的同时断面形状也需要满足加工要求。

为紧跟市场需求，最大可能性的满足后工序客户加工要求，降低 CSP 生产成本提高产品质量，我们对 CSP 现有的 SMS 控制模型开展了深入的研究工作，以求提升 CSP 厂板形控制能力。

关键词： 板形控制，断面质量评定系统

The Model of Strip Profile Control for CSP

Chen Jianfei, Wu Lijuan, Yang Guang

(WISCO CSP, Wuhan 430083,China)

Abstract: WISCO CSP is the second generation CSP short process production line, ompared to conventional hot rolling, 7 mills were made using 3 different sets of CVC roll shapes, To strengthen the focus on their shape control for rolling strips. With the market demand, Extremely thin, high strength structural steel and electrical steel have higher requirements for shape dimension accuracy, Pay attention to "Section shape quality rating" is a new demand for raw materials of high value-added products.

Key words: PCFC, section shape quality rating

影响测速仪测量精度因素的分析

全利军，宋佳帅，尹啸峰

（北京首钢自动化信息技术有限公司首迁运行事业部，河北迁安　064400）

摘　要： 首钢迁钢酸连轧机组是非常重要的产线，酸连轧机组的激光测速仪对于机组的的 AGC 厚度控制起着非常重要的作用，测速仪测量精度直接影响带钢厚度的秒流量控制，秒流量的精度影响带钢的厚度控制，尤其对薄带钢得到厚度控制影响更大，厚度是带钢的质量判定的重要依据。因此轧机测速仪运行的精度与稳定性直接影响带钢质量。

关键词： 测速仪，秒流量控制，激光探测器，多普勒

首钢迁钢公司热轧粗轧 R2 轧机压下系统研究

康新勇，张　强，卢建军

（北京首钢自动化信息技术有限公司首迁运行事业部，河北迁安　064400）

摘　要： 轧机电动压下系统是一个典型的快速定位系统，对传动系统的启动和制动快速性都有很高的要求。首钢迁钢热轧的粗轧 R2 轧机压下系统采用 SIEMENS 公司的 6SE70 装机柜型变频器控制系统，电机通过电磁离合器同轴连接。为了提高传动性能，轧钢过程中当调整辊缝时离合器断电闭合，压下系统传动侧与操作侧联动，此时电动

压下系统采用主从控制方式；当两侧辊缝值出现偏差时，选择单动控制方式，此时电动离合器通电打开，两台电动机可任选其中一台单动，此时传动侧与操作侧电机都采用主方式控制。本文分析了热连轧 2160 生产线中粗轧机 R2 电动压下系统的结构、系统参数并分析系统运行中存在的问题，给出了粗轧 R2 压下系统两台交流电动机同轴传动时主从控制方式的实现方法，并对压下系统的转矩控制和转速控制这两种主从控制方式进行了试验和对比分析。试验结果表明，采用转矩输出作为从动装置的给定能够实现轧机压下系统的快速定位和主从控制，在稳定性和精确性上都完全能够满足工艺控制要求。

关键词：电动压下，主从控制，转矩

Study on the Reduction System of Hot Rolling R2 Mill in Shougang Group

Kang Xinyong, Zhang Qiang, Lu Jianjun

(Shougang Automatic Information Technology Co., Ltd., Qian'an 064400, China)

Abstract: The electric system of the rolling mill is a typical fast positioning system, which has a high demand for the starting and braking of the drive system. The 6SE70 hot rolling roughing mill R2 rolling down system of Shougang Group adopts SIEMENS installed cabinet type inverter control system of the company. In order to improve the transmission performance of the rolling process when the adjustment of the roll gap when the clutch power closed, pressure system drive side and operation side linkage, this electric screwdown system uses master-slave control mode; when the value deviation appears on both sides of the joint roller, select single control mode, the electric clutch open, two motors can choose one single action, the transmission side and the operating side are the main way to control the motor. This paper analyzes the problems in the production line of 2160 hot strip roughing machine R2 electric screwdown system structure, system parameters and analysis of the operation of the system, gives the realization method of R2 system under pressure roughing two AC motor coaxial transmission master-slave control mode, and torque control and speed control system under the pressure of the two a master-slave control method is analyzed to test and compare. The test results show that the torque output can be used as the driven device to realize the rapid positioning and master-slave control of the rolling down system.

Key words: electric power down, master-slave control, torque

热轧钢卷喷号机器人控制系统的研究与应用

张文宝

（迁安首信自动化信息技术有限公司技术中心，河北迁安 064400）

摘 要： 文章深入剖析热轧喷号系统的组成与工作原理，结合应用场景对机器人的控制系统进行了阐述，针对现场喷号系统应用存在的套接字故障，从硬件和软件两个方面进行了详细的分析，找到了故障发生的根本原因，提出了解决方案。

关键词：喷号系统，工作原理，机器人，控制系统，套接字

Research and Application on Control System of Steel Coil Spraying Robot

Zhang Wenbao

(Technology Department, Qian'an Shouxin Automatic Information Technology Co., Ltd., Qian'an 064400, China)

Abstract: In this paper, the composition and working principle of the hot spray system are analyzed, and the control system of the robot is expounded with the application scene, at last presents a solution to the socket failure of the spray system.

Key words: spraying system, operational principle, robot, control system, sockets

T2.5 调质度镀锡基板罩退工艺优化的试验研究

张 鹏[1]，孙 力[1]，李杰义[2]，郑传宝[2]，尹丽改[2]，王文宝[2]

（1. 河钢集团钢研总院工艺所，河北石家庄 050000；

2. 河钢集团衡水薄板有限责任公司，河北衡水 053000）

摘 要： 为降低燃料消耗，提高生产效率，对 T2.5 调质度的镀锡基板进行退火工艺优化试验，在实验室研究了退火温度、保温时间对镀锡基板组织与性能的影响，经现场工艺优化后，产品的保温时间降低了 1h，组织和性能满足要求，达到了预期目标。

关键词： 镀锡基板，T2.5，退火工艺，优化

Study on Annealing Technology Optimization of T2.5 Tinplate Substrate

Zhang Peng[1], Sun Li[1], Li Jieyi[2], Zheng Chuanbao[2], Yin Ligai[2], Wang Wenbao[2]

(1. Hesteel Technology Research Institute, Shijiazhuang 050000, China;

2. Hengshui Sheet Co., Ltd., Hesteel Group, Hengshui 053000, China)

Abstract: In order to reduce the fuel consumption and improve the production efficiency, the annealing process of T2.5 tinplate was optimized. The influence of annealing temperature and time on the microstructure and mechanical character of tin-plated substrate was researched. After the process is optimized, the annealing time of the product is reduced by 1 hour, and the organization and mechanical character meet the requirements and meet the expected target.

Key words: tinplate substrate, T2.5, annealing process, optimization

厚规格 Q460C 热轧卷板的研制

闵洪刚

（本钢产品研究院热轧高强钢研究所，辽宁本溪 117000）

摘 要：介绍了本钢 2300 热连轧机组厚规格 Q460C 热轧高强度卷板的产品设计、冶炼、连铸、轧制工艺和性能情况。结果表明在低碳、适量锰含量的基础上，采用 Nb、Ti 微合金化，合理设计生产工艺，成功开发了无拉伸试验试样断口分层的 25mm 厚的 Q460C 热轧卷板。所试制的厚规格 Q460C 卷板的强韧性匹配良好，各项性能指标全部满足了用户的要求，可替代中厚板轧机生产的厚板，在煤矿液压支架等结构上进行应用。

关键词：Q460C，厚规格，热轧卷板

Development of Thick Q460C Hot Rolled Strip

Min Honggang

(Hot Rolled High Strengh Steel Development Department of Bengang Product Research Institute, Benxi 117000, China)

Abstract: This paper introduces product design,smelting technics,continuous casting technics,rolling technics and performance of thick Q460C hot rolled strip produced on Bengang 2300 continuously hot-rolled production line. The results indicated that useing the lower carbon content, appropriate manganese content, adding a small amount of Nb and Ti elements and appropriate production process, can succeed develops 25mm thick Q460C hot rolled strip with tensile test specimen without layered crack. The thick Q460C hot rolled strip has excellent strength and toughness balance, All of its performance parameters meet the user's requirements, It can be used instead of thick plate that producted by the medium and heavy plate mill, It not only can be used for making the coal mine hydraulic support machinery but also be used for making other structure.

Key words: Q460C, thick, hot rolled strip

冷轧半自动包装技术优化与实践

孙 力[1]，伍昕忠[2]

（1. 河钢集团钢研总院，河北石家庄 050023；2. 机科发展科技股份有限公司，北京 100044）

摘 要：通过模拟运输条件进行冷轧包装卷浸水、浇水试验，对包装材料和方案进行改善，逐渐形成与不同用户不同产品定位相符合的简包、普包以及各种精包装方案体系，满足了不同用户和目标市场对不同冷轧产品差异化包装的要求。对半自动包装机组相关技术进行持续优化，形成了包装系统化操作规程和技术规范，做到包装机组操作标准化、定制化、规范化和目视化，确保冷轧产品包装质量的高效稳定，满足主线高速稳定和高产对包装高质量需求。

关键词：冷轧，钢卷，包装，半自动，优化

Optimization and Practice of Semi-automatic Packaging Technology for Cold Rolling

Sun Li[1], Wu Xinzhong[2]

(1. HBIS Group Technology Research Institute, Shijiazhuang 050023, China;
2. Machinery Technology Development Co., Ltd., Beijing 100044, China)

Abstract: By the water immersion test and watering test of cold rolled packaging under simulating transport conditions, the materials and methods of cold rolling packaging were improved. The packaging solution system has been formed gradually, in which simple packaging or normal packaging or elaborative packaging can be chosen according to different users and different product positioning. It is satisfied the requirements that different users and different target markets need different packaging of different cold rolled product. Technologies of semi-automatic packaging units have been optimized continually. The packaging systematic operation procedures and technical specifications have been formed. The operation of packaging units has been of standardization, customization, normalization and visualization. The packaging quality of cold rolled product has been ensured of high stability. It has been met the high quality requirements of packaging in mainline of high speed stability and high yield.

Key words: cold rolling, steel coil, packaging, semi-automatic, optimization

亚温淬火工艺对低合金高强钢连续冷却转变及其相变动力学的影响

童　志[1]，王红鸿[1]，覃展鹏[1]，胡　锋[2]，李　丽[2]，吴开明[1]

（1. 省部共建耐火材料与冶金国家重点实验室，高性能钢铁材料及其应用湖北省协同创新中心，
武汉科技大学，湖北武汉　430081；
2. 南京钢铁集团有限公司研究院，江苏南京　210035）

摘　要： 采用相同化学成分不同热处理工艺(淬火+回火，淬火+亚温淬火+回火)的两组钢，通过模拟峰值温度为1320℃不同 $t_{8/5}$ 下的焊接热循环，采用光学显微镜（OM）观察显微组织，显微硬度仪测定显微硬度，计算分析膨胀量-时间-温度曲线得到不同 $t_{8/5}$ 下相转变开始结束温度、SH-CCT 曲线、相变动力学曲线。通过亚温淬火的低合金高强钢在焊接热循环中，使贝氏体转变开始温度（B_s）和结束温度（B_f）得到提升，并且推迟了马氏体组织的产生；增大了在焊接热循环慢速冷却过程中的相转变温度区间；提升了焊接热循环快速冷却的相转变速率，但降低了在中慢冷速下的相转变速率。

关键词： 亚温淬火，低合金高强钢，焊接热循环，SH-CCT，相变动力学

Influence of Lamellarizing Process on Continuous Cooling Transformation and Phase Transformation Kinetics of High-strength Low-alloy Steels

Tong Zhi[1], Wang Honghong[1], Qin Zhanpeng[1], Hu Feng[2], Li Li[2], Wu Kaiming[1]

(1. International Research Institute For Steel Technology, Hubei Collaborative Innovation Center For Advanced Steels,
Wuhan University of Science and Technology, Wuhan 430081, China;
2. Research Institute Nanjing Iron And Steel Group Co., Ltd., Nanjing 210035, China)

Abstract: The two groups steel that have the same chemical composition and different heat treatment (Quench+ Tempering, Quench + Lamellarizing + Tempering) were simulated by the peak temperature 1320 °C at different $t_{8/5}$. Microstructure and morphology were observed by using Optical Microscope (OM). The start and finish temperature of phase transformation, SH-CCT curve and transformation kinetics curve were analyzed by CGauge-Time-Temperature curve at different $t_{8/5}$. Lamellarizing process raised the start temperature B_s and finish temperature B_f of bainite for high-strength low-ally steel during continuous cooling; martensite was delayed to bring up; the temperature range was broadened in slow cooling at welding thermal cycle; the phase transformation rate was improved in fast cooling rate and decreased medium or slow cooling rate.

Key words: lamellarizing, low-alloy and high strength steels, welding thermal cycle, SH-CCT, phase transformation kinetic

1.5mm 花纹板的研制开发

王茂伟，胡书茂

（河北钢铁集团承钢公司热轧卷板事业部，河北承德　067002）

摘　要：介绍了 1780 热连轧生产线 1.5mm 规格花纹板的研发情况。通过开发花纹辊标定程序和卷取机卷径计算输入模式程序、设备精度维护、优化负荷分配、优化卷取参数、提高卷形质量等方法，成功轧制 1.5mm 花纹板，并已批量生产，为我厂带来了可观的经济效益。

关键词：花纹板，负荷分配，豆高

Research and Development of the Pattern Plate of 1.5mm

Wang Maowei, Hu Shumao

(Hot Rolling Steel Sheet Coil Business, Chengde Iron and Steel Company,
Hebei Iron and Steel Group, Chengde 067002, China)

Abstract: It is introduced the development of very thin checkered steel plate of 1.5 mm in 1780 hot continuous rolling line. By developing pattern roller calibration procedure and coiling diameter calculating -inputting model，Precision of equipment maintenance，load distribution optimization and optimizing coiling parameters to improve coil shape. Now we have already have power to produce thin checkered steel plate in batch，and earn considerable economic profit.

Key words: checked plate, load distribution, beans height

沙钢 1450 热连轧加热炉加热质量及技术的研究

孙　林[1]，杨丽琴[1]，丁美良[1]，杨路元[2]，茅永明[2]

（1. 江苏省（沙钢）钢铁研究院，江苏张家港　215625；
2. 江苏沙钢集团有限公司，江苏张家港　215625）

摘　要：本文借助 CFD 建立模拟计算模型，分析沙钢 1450 热连轧加热炉炉内温度场、速度场、压力场，应用模型结合现场数据进行了加热炉煤气量、空燃比等加热工艺的优化研究。并且借助"黑匣子"检测板坯加热过程的温度数据，分析温度均匀性及其影响因素，找出提高加热质量的措施和方法。理论计算和检测数据分析结果，指导现场解决了炉门冒火、装坯方式改进等一系列技术难题。虽然研究成果应用后提升了加热炉的加热质量，但是与国内先进水平仍然差距较大，解决此问题的关键是改目前的水冷却为汽化冷却方式，改燃烧手动操作方式为自动化控制。

关键词：加热炉，热连轧，CFD，黑匣子

Research on Technology and Quality of Billet Heating in Heating Furnace of 1450 Hot Strip Mill in Shagang

Sun Lin[1], Yang Liqin[1], Ding Meiliang[1], Yang Luyuan[2], Mao Yongming[2]

(1. Institute of Research of Iron and Steel, Shasteel, Zhangjiagang 215625, China;

2. Jiangsu Shasteel Group Co., Ltd., Zhangjiagang 215625, China)

Abstract: With the help of CFD (short for Computational Fluid Dynamics) simulation model， those field were analyzed, such as temperature , velocity and pressure inside the heating furnace of 1450 hot strip mill in Shagang. With the aid of the simulation model combining with the real-time data, some parameters in heating process were optimized, such as gas volume, air-fuel ratio and so on. By acquiring the temperature of the gas inside furnace and the measuring point of billet in each direction from Black box system in the heating process, temperature uniformity and its influencing factors were comprehended. And the measures to improve the quality of billet heating were found out. Based on the above results both theoretical calculation and testing data analysis, A series of technical problems, such as the phenomena of severe furnace door fire, the positioning way of billet charging, etc were solved. Application of research result in the field had improved the quality of the heating furnace more than before, but there still was a big gap compared with the domestic advanced quality index. The key to closing the gap problem is to change the cooling system of heating furnace from water cooling to evaporative cooling, and change the combustion mode of operation from manual operation to automation control.

Key words: heating furnace, hot strip mill, CFD, black box

高硅双相钢低温卷取工艺卷取温度控制研究

夏银锋，徐　芳，王晓东，王建功，刘鸿明，周　旬，

杨孝鹤，齐　达，王智锋

（首钢京唐钢铁联合有限责任公司，河北唐山　063000）

摘　要: 本文研究了高硅双相钢在低温卷取工艺卷取温度的控制。生产初期低温卷取的温度波动较大，卷取温度命中率低。本文对轧制过程中工艺参数、带钢表面质量和板形质量对卷取温度命中率的相关性进行分析研究，最终得出表面红锈、板形翘曲等是影响卷取温度的关键。进一步制定了解决表面红锈和板形翘曲的措施，最终实现了低温卷取的卷取温度稳定控制。

关键词: 高硅双相钢，红锈，翘曲，卷取温度命中率

Research on Coiling Temperature Control of High Silicon Dual Phase Steel at Low Temperature Coiling

Xia Yinfeng, Xu Fang, Wang Xiaodong, Wang Jiangong, Liu Hongming,
Zhou Xun, Yang Xiaohe, Qi Da, Wang Zhifeng

(Hot Rolling Department, Shougang Jingtang Iron & Steel Co., Ltd., Tangshan 063000, China)

Abstract: This paper studies the control of the coiling temperature of high silicon duplex steel during low temperature coiling. Production of low temperature coiling temperature fluctuations in the early, coiling temperature hit rate is low. In this paper, the correlation between process parameters, strip surface quality and plate quality during the rolling process is analyzed. Finally, the surface red scale and the plate warping defects are the key factors affecting the coiling temperature. And then develop measures to solve the surface red scale and the plate warping defects，and finally achieved a low temperature coiling of the coiling temperature stability control.

Key words: high silicon duplex steel, red scale, warping, coiling temperature hit rate

冷轧低合金高强钢再结晶和析出行为研究

李志红，蓝慧芳，王明阳，王文博，李建平

（东北大学材料与冶金学院，辽宁沈阳　110819）

摘　要： 利用热模拟试验机详细研究了在新工艺控制下的冷轧低合金高强钢在退火过程中铁素体再结晶动力学规律，并基于 JMAK 方程建立了铁素体再结晶动力学模型。结果表明，温度越高，铁素体再结晶发生时间越早，达到完全再结晶时所用时间越短，同时通过数据拟合得到常规工艺和新工艺下两种实验钢的再结晶激活能分别为308.83kJ/mol 与 261.87kJ/mol；与初始组织为铁素体-珠光体时相比，贝氏体组织在退火的过程中再结晶激活能更小，完成再结晶所需时间更短，并且退火后显微组织中的渗碳体颗粒分布更加均匀、弥散；与5℃/s 加热时相比，80℃/s 加热条件下的回复过程被推迟发生，铁素体再结晶快速完成。同时对比研究了不同初始组织的实验钢在退火过程中的析出规律，新工艺下实验钢析出强化贡献更明显。

关键词： 冷轧，低合金钢，再结晶，退火，析出

Study on Recrystallization and Precipitation Behavior of Cold Rolled High Strength Low Alloy Steel

Li Zhihong, Lan Huifang, Wang Mingyang, Wang Wenbo, Li Jianping

(School of Materials & Metallurgy, Northeastern University, Shenyang 110819, China)

Abstract: The kinetics of ferrite recrystallization in cold-rolled low-alloy high-strength steel under the control of new technology was studied by thermal simulation test. The kinetic model of ferrite recrystallization was obtained based on JMAK equation, which provides a theoretical basis for the control of ferrite recrystallization. The results showed that the time to complete the recrystallization is shorter and the recurrence time of ferrite is earlier at the higher temperature, and obtain the recrystallization activation energy of the experimental steel under the conventional process and new process by 308.83kJ/mol and 261.87kJ/mol. The recrystallization activation energy of the bainite structure during the annealing process is shorter than that of the initial structure of the ferrite-pearlite, and the time required to complete the recrystallization is shorter, and the distribution of cementite particles in the microstructure is more uniform and dispersed after annealing. Compared with the heating rate at 5℃/s, the recovery process under 80℃/s heating rate is delayed, so that the ferrite recrystallization quickly completed. At the same time, the precipitation characteristics of experimental steel with different initial structures were studied. The results showed that the corresponding precipitation strengthening contribution is higher in the new process.

Key words: cold rolling, low alloy steel, recrystallization, annealing, precipitation

980MPa 级高强钢冷轧厚度波动的原因及解决方案

苏振军，杨建宽，李晓广

（河钢集团邯钢公司技术中心，河北邯郸　056015）

摘　要： 邯钢邯宝热轧厂生产 DP980 高强双相钢时，在冷轧环节中带钢头部和尾部会出现较大的厚度波动问题。对热轧及冷轧工艺参数与设备情况进行分析检查，认为带钢头部与尾部在冷轧过程中发生厚度波动的原因是热轧带钢头尾在冷却过程中存在差异，引起带钢通卷的组织及性能上有较大差别，在后序酸轧过程中头尾发生厚度及轧制力波动，影响酸轧生产稳定性，厚度波动长度 300~500m，波动范围 ±60μm 左右。通过控制层流冷却工艺，使带钢在热轧后的组织和性能更加均匀，避免了带钢长度方向的硬度强度差异造成的厚度波动。

关键词： 轧钢，高强钢，厚度波动，层流冷却

Causes and Solutions for Thickness Fluctuation of DP980 High Strength Steel at Cold Rolling Process

Su Zhenjun, Yang Jiankuan, Li Xiaoguang

(HBIS Hansteel Technology Center, Handan 056015, China)

Abstract: Thickness fluctuation at the head and tail of the strip usually happened when DP980 dual phase steel produced at Hanbao rolling mill. After analyze the rolling parameters and check the rolling facilities situation, it is thought that for local fast laminar cooling, microstructure、tensile and hardness of the strip at the head、tail and main body became differently, which induce rolling force fluctuate so as the thickness of the strip. At the pickling and cold rolling line, thickness and rolling force fluctuate largely at the head and tail of the strip, which impact the stability of the production. The length of the thickness fluctuation reaches 300~500m, thickness fluctuation ranges ±60μm. By controlling the laminar cooling process, the microstructure and mechanical properties are evenly distributed, so the tensile and hardness differential at various part is avoided so as the thickness fluctuation of the strip.

Key words: rolling, high strength steel, thickness fluctuation, laminar cooling

双相区轧制抗大变形管线钢 X70HD 的开发

李玉谦，刘红艳，杜琦铭，徐桂喜

（河钢集团邯钢公司技术中心，河北邯郸　056015）

摘　要： 采用 Gleeble 3500 得出 X70HD 抗大变形管线钢发生动态再结晶的临界压下率、Ar3 和 Ar1 温度等关键参数，实际生产中为了得到合理比例的细小铁素体和贝氏体双相组织，奥氏体再结晶区末道次压下率≥17%，未再结晶区采用"奥氏体+铁素体"双相区轧制，开轧温度低于 Ar3 温度 20~30℃，道次压下率不高于 18%，避免产生先共析铁素体；控冷模式采用 DQ+ACC 两段式冷却，开始冷却温度低于 Ar1 温度，DQ 控冷段以 20~30℃/s 的冷速快速通过贝氏体转变温度区，得到相应比例的贝氏体，ACC 控冷段以 10~20℃/s 的冷速冷却到终冷温度。经过检测，微

观组织结构为：30%~45%铁素体+40%~65%贝氏体+0~5% MA 岛，该方法生产的抗大变形管线钢 X70HD 具有良好的综合性能，形变强化指数＞0.12，均匀伸长率＞13%，较低的屈强比＜0.80，达到同行业领先水平。

关键词：双相区轧制，抗大变形，管线钢 X70HD，均匀延伸率

Development of Anti - large Defect Pipeline Steel X70HD for Double - zone Rolling

Li Yuqian, Liu Hongyan, Du Qiming, Xu Guixi

(HBIS Group Hansteel Company Technology Center, Handan 056015, China)

Abstract: Using Gleeble 3500 dynamic recrystallization of the critical reduction rate, Ar3 and Ar1 temperature and other key parameters of X70HD pipeline steel, in order to get a reasonable proportion of small ferrite and bainite biphase structure actual production, recrystallization end more than 17% rate reduction, in the recrystallization zone, the austenite or ferrite dual phase rolling is adopted. The rolling temperature is below Ar3, the temperature is 20~30℃, and the rate of pass reduction is not higher than 18%, so as to avoid the eutectoid ferrite; Controlled cooling mode using DQ+ACC two stage cooling, start cooling temperature is lower than Ar1, DQ during controlled cooling cooling rate to 20~30℃/s fast through the bainite transformation temperature of bainite zone, get the corresponding proportion, ACC controlled cooling cooling section to 10~20℃/s to the final cooling temperature. After testing, microstructure: 30~45% ferrite bainite+ 40~65%+0~5% MA, the production method of the high deformability pipeline steel X70HD has excellent comprehensive properties, the strain hardening index is more than 0.12, the elongation of more than 13%, the low yield ratio is less than 0.80, reached the leading level in the same industry.

Key words: dual phase rolling, anti-large deformation, pipeline steel X70HD, uniform elongation

薄带材高速连续退火机组调整材的使用分析

任予昌

（宝武钢铁武汉有限公司冷轧厂，湖北武汉　430080）

摘　要：连续退火机组生产具有品种多样化，带钢规格较薄，运行速度快等特点。其合理的计划排程不仅是机组稳定运行、过程控制的基础，而且是机组质量控制、消耗控制、成本管理的关键。本文以镀锡基板的生产排程为核心，对硬质材（T4、T5）到软质材（T3、T2.5）订单材过渡和规格过渡进行了优化分析。

关键词：连续退火，调整材，计划排程

The Optimization of Cycle Coils for Continuous Annealing Line

Ren Yuchang

Abstract: The continuous annealing process line have the character of complex varieties, thin specifications and high speed. The reasonable production schedule is not only the basis of stable operation and process control, but also is the key of quality control, cost control, consumption management. The article optimize the transition of T4 、T5↔T3 、T2.5 order

coils and specification.
Key words: continuous annealing, cycle coils, production schedule

冷连轧机薄料轧制出现的泡泡浪板形缺陷问题研究

戴竞舸[1]，朱　涛[2]，陈　光[2]，樊志强[2]，何小丽[2]，张清东[3]，刘巨双[3]

（1. 宝山钢铁股份有限公司研究院冷轧所，上海　201900；2. 宝钢集团上海梅山钢铁股份有限公司冷轧厂，江苏南京　210039；3. 北京科技大学机械工程学院，北京　100083）

摘　要： 针对某冷轧厂 1420UCM 冷连轧机组在更换五机架工作辊后，轧制硬质薄规格镀锡板时出现的板形缺陷问题，通过观察该实物带钢浪形分布以及根据浪形产生的机理定义了一种新的浪形——交替不对称泡泡浪。研究了该类板形缺陷产生时伴随的条件与现象，分析了其产生的原因，并提出了具体的应对措施。部分措施上机应用后，取得了良好的使用效果，板形缺陷出现的概率由原来的 14.8% 下降至 8.8%，为现场创造了较大的经济效益。

关键词： 换辊，冷轧，浪形，薄规格

Study on the Defect Problem of the Bubble Wave Plate in the Thin Rolling Mill of Cold Continuous Rolling Mill

Dai Jingge[1], Zhu Tao[2], Chen Guang[2], Fan Zhiqiang[2], He Xiaoli[2],
Zhang Qingdong[3], Liu Jushuang[3]

(1. Institute of Cold Rolling, Research Institute, Baoshan Iron & Steel Co., Ltd., Shanghai 201900, China;
2. Cold Rolling Plant of Meishan Iron & Steel Co., Ltd., of Baosteel Group, Nanjing 210039, China;
3. School of Mechanical Engineering, University of Science and Technology Beijing, Beijing 100083, China)

Abstract: After replacing the five stand work roll of UCM Cold Tandem Mill, flatness defect will happen when the hard and thin tinning plate is rolled. According to the distribution of the wave-hape and the induced mechanism, we define a new wave-shape——alternating asymmetrical bubble waves. We study the induced conditions of the wave-shape and analyze the causes of the defect and put forward corresponding measures. Some of the measures have been applied, which obtains good applicative effectiveness. The probability of plates defect occurs from 14.8% originally down to 8.8%, creating large economic benefits for the field.

Key words: roll changing, cold rolling, shape wave, thin gauge

滚切式定尺剪剪切机理与剪刃断裂分析

郝建伟，胡典章，陈玉柏

（中冶京诚工程技术有限公司，北京　100176 ）

摘　要：研究了滚切剪剪切钢板的过程，分析了影响剪切质量的主要因素，对剪刃使用寿命进行了详细分析。针对某现场 5m 滚切剪剪刃断裂事故进行了深入分析，避免再次发生类似事故。

关键词：滚切剪，剪刃，剪刃间隙，磨损，寿命，断裂

Fracture Analysis and Shearing Mechanism for Dividing Shear

Hao Jianwei, Hu Dianzhang, Chen Yubai

(Capital Engineering & Research Incorporation Limited, Beijing 100176, China)

Abstract: The main factors influencing the shear quality were analyzed in cutting process, and the service life of the cutting edge was analyzed in detail. A detailed investigation was carried out to investigate the accident of a 5m roll cutting shear fracture in this paper.

Key words: rolling shear, shear, knife gap, wear, life, fracture

加热炉煤气热值优化控制技术研究

胡凤洋

（宝钢股份有限公司热轧厂，上海　200941）

摘　要：煤气热值是工业炉窑设计过程中的重要参数，它影响到工业炉窑的炉型结构、管道设计以及工业炉窑的热效率。但在工业炉窑的使用过程中，对于混合煤气热值对加热炉燃烧效率的影响研究很少，一方面受混合煤气的来源问题限制，影响了煤气热值的灵活调整，另一方面煤气控制系统变化对工业炉窑的控制稳定性有一定的影响。通过煤气热值优化调整技术研究摸索出提升加热炉热效率的热值模式，进而降低加热炉燃耗。

关键词：煤气热值，加热炉热平衡，热效率

Research on Optimized Control Technology of the Furnace Calorific Value of Gas

Hu Fengyang

(Hot Rolling Mill Plant, Baoshan Iron & Steel Co., Ltd., Shanghai 200941, China)

Abstract: Calorific value of gas is an important parameter in process furnace design industry. It will affect the industrial kiln furnace structure, pipeline design and the thermal efficiency of industrial furnaces. But the heat value of the mixed gas of combustion efficiency of the heating furnace in the process of industrial furnace is few. It limits the source of mixed gas, effects the flexible of adjustment of gas calorific value, on the other hands it has certain influence gas control system change control stability of industrial furnaces. Through the study on the dynamic adjustment of the gas calorific value calorific value model to improve exploration efficiency of the heating furnace, and then reduce the heating furnace fuel consumption.

Key words: thermal efficiency, heating furnace

退火温度对 21Cr-0.3Cu 铁素体不锈钢组织性能的影响

尹鸿祥[1]，吴　毅[1]，赵爱民[2]，赵征志[2]，付秀琴[1]，张　弘[1]

（1. 中国铁道科学研究院金属及化学研究所，北京　100081；
2. 北京科技大学钢铁共性技术协同创新中心，北京　100083）

摘　要： 本文研究了退火温度对 21Cr-0.3Cu 超纯铁素体不锈钢显微组织、力学性能、成形性能和微观织构的影响规律。利用光学显微镜观察不同退火温度试验钢微观组织，通过拉伸试验测定试验钢力学性能和成形性能，通过背散射电子衍射测定试验钢微观织构。结果表明，试验钢经 970℃ 退火时，晶粒细小且均匀，组织处于完全再结晶状态。温度较低时，再结晶不完全；温度过高时，再结晶晶粒异常长大。均出现混晶组织。试验钢 970℃ 退火时，综合力学性能最佳，抗拉强度为 473MPa，屈服强度为 315MPa，延伸率 35.7%。随退火温度的升高，试验钢钢的平均塑性应变比 r_m 值先增加后减少。当退火温度达到 970℃ 时，r_m 值可以达到最大值 1.82。γ 纤维织构密度变化趋势与 r_m 值一致，退火温度为 970℃ 时，γ 纤维增强明显，此时其取向密度达到最大值 f（g）=20.56，成形性能最佳。

关键词： 铁素体不锈钢，平均塑性应变比，退火温度，织构，成形性能

Effect of Annealing Temperature on Microstructure and Properties of 21Cr-0.3Cu Ferritic Stainless Steel

Yin Hongxiang[1], Wu Yi[1], Zhao Aimin[2], Zhao Zhengzhi[2], Fu Xiuqin[1], Zhang Hong[1]

(1. Metal and Chemistry Research Institute, China Academy of Railway Sciences, Beijing 100081, China;
2. Collaborative Innovation Center of steel Technology, University of Science and Technology Beijing, Beijing 100083, China)

Abstract: The effects of annealing temperature on microstructure, mechanical properties, formability, and texture evolution of 21Cr-0.3Cu ultra-pure ferrite stainless steel was analyzed in the article. The microstructure of the steel obtained through different annealing temperature were investigated by means of optical microscope; The mechanical properties and forming properties were measured by tensile testing. The micro texture of steel was measured by using electron back scattering diffraction analysis. It shows that, it obtained small, uniform and completely recrystallized grains at 970℃. When the temperature is low, the recrystallization is not complete; when the temperature is too high, abnormal grain growth occurs. Both have mixed grain size structure. When it is annealed at 970 ℃, it obtained the best mechanical performance, with the tensile strength reaching 473 MPa, yield strength reaching 315 MPa and elongation reaching 35.7%. As the annealing temperature rises, average plastic strain ratio r_m first increases then decreases, getting the maximum value of 1.82 at 800 ℃. The sum of γ texture is consistent with the variation trend of average plastic strain ratio r_m. When the annealing temperature is 970 ℃, γ fiber texture was reinforced obviously. It gets the maximum orientation density f (g) = 20.56 and obtains the best formability.

Key words: ferritic stainless steel, average plastic strain ratio, annealing temperature, texture, formability

中锰TRIP钢在临界退火中的组织演变与数值分析

闫　宁，邸洪双，李　洋，刘剑锋

（东北大学轧制技术及自动化国家重点实验室，辽宁沈阳　110819）

摘　要：本文研究了逆相变退火过程中微观组织的演变，利用DICTRA模拟计算与实验，分析了奥氏体变退火过程中组织结构的演变规律，以及元素扩散行为。实验结果表明：逆相变退火后，热轧中锰TRIP钢的微观组织主要由铁素体、板条状逆相变奥氏体以及少量的HCP马氏体组成；数值模拟和EPMA分析得出，奥氏体逆相变由NPLE和PLE两种模式控制，相变过程中存在元素的扩散，Mn和Si元素向奥氏体内配分，奥氏体中存在明显地Mn富集。

关键词：逆相变退火，奥氏体，NPLE，PLE，元素扩散

Microstructure Evolution and Numeric Alanalysis during Intercritical Annealing in a Medium Mn TRIP Steel

Yan Ning, Di Hongshuang, Li Yang, Liu Jianfeng

(State Key Laboratory of Rolling and Automation, Northeastern University, Shenyang 110819, China)

Abstract: In current paper, microstructure evolution and elements diffusion during different intercritical annealing in 9 wt% Mn-containing steel was analyzed numerically and examined experimentally. The analysis of microstructure has shown that the microstructure of tested steel treated by intercritical annealing mainly consists of ferrite, lath-austenite and few HCP-martensite. The results of DICTRA simulation and EPMA show that the growth of austenite may be controlled by elements diffusion under NPLE and PLE. The concentrations of both Mn and Si are non-uniform in austenite and enrich in the austenite/ferrite boundary.

Key words: intercritical annealing, austenite, NPLE, PLE, element diffusion

Cr含量对钒氮微合金钢动态再结晶的影响

梁新增，衣海龙，朱伏先

（东北大学轧制技术及连轧自动化国家重点实验室，辽宁沈阳　110819）

摘　要：采用Gleeble-1500热模拟实验机对Cr含量不同的钒氮微合金钢进行了单道次压缩实验，研究了变形温度为850～1100℃、应变速率为0.01～1s^{-1}的动态再结晶行为。研究表明：实验钢的动态再结晶行为在较高温度和较低应变速率下更容易发生；且Cr含量越高，越易发生动态再结晶；对应力应变曲线处理得到的$\theta-\sigma$曲线都出现拐点特征，且$\partial\theta/\partial\sigma-\sigma$曲线都相应出现最值点，该最值点即为临界应变；建立动态再结晶数学模型。

关键词：Cr含量，钒氮微合金钢，动态再结晶，临界应变，数学模型

Effect of Cr Content on Dynamic Recrystallization of V-N Microalloyed Steel

Liang Xinzeng, Yi Hailong, Zhu Fuxian

(The State Key Laboratory of Rolling & Automation, Northeastern University, Shenyang 100819, China)

Abstract: The Gleeble-1500 thermal simulation experiment was used to study the dynamic recrystallization behavior of the vanadium-nitrogen microalloyed steel with different Cr content. The dynamic recrystallization behavior was studied at the deformation temperature of 850~1100 ℃ and the strain rate of 0.01~1s^{-1}. The results show that the dynamic recrystallization behavior of experimental steels is more likely to occur at higher temperatures and lower strain rates. The higher the Cr content is, the more likely the dynamic recrystallization occurs. The θ-σ curves obtained by the stress-strain curve Inflection point characteristics, and ∂θ/∂σ-σ curves corresponding to the most significant point, the most point is the critical strain; the establishment of dynamic recrystallization mathematical model.

Key words: Cr content, vanadium-nitrogen microalloyed steel, dynamic recrystallization, critical strain, mathematical model

热静压复合工艺条件对不锈钢/碳钢界面复合率的影响

徐 泷[1]，刘吉阳[2]，李 硕[2]，张清东[2]

（1. 安徽博特金属复合材料制造有限公司，安徽滁州 239000；

2. 北京科技大学机械工程学院，北京 100083）

摘 要： 在不锈钢/碳钢热静压复合过程中，恒温保温处理和压下形变都对界面复合率有显著的影响，且两者的影响作用存在相互耦合关系。为了研究该两个复合工艺环节在复合过程中的作用，利用 GLEEBLE 热模拟机进行热静压复合实验，采用超声波 C 扫描设备对试样进行复合率检测，以复合率表征复合界面的结合状态，采取相同的恒温保温处理时间和不同的恒温温度以及压下变形量与变形速率下，并分别进行先恒温保温处理后压下形变和先压下形变后恒温保温处理两种不同工艺组合，研究不同热静压复合工艺路线及工艺参数对不锈钢-碳钢复合行为及效果的影响。

关键词： 不锈钢，碳钢，保温处理，压下形变，复合率

Effect of Hot Hydrostatic Composite Process on Interfacial Recombination Rate of Stainless Steel / Carbon Steel

Xu Long[1], Liu Jiyang[2], Li Shuo[2], Zhang Qingdong[2]

(1. Anhui Bote Metal Composite Material Co., Ltd., Chuzhou 239000, China;

2. School of Mechanical Engineering, University of Science and Technology Beijing, Beijing 100083, China)

Abstract: In the process of hot hydrostatic composite of stainless steel / carbon steel, both the constant temperature heat treatment and the compressive deformation have significant influence on the interface recombination rate, and the influence of them has a mutual coupling relationship.In order to study the role of the two composite processes in the composite process, GLEEBLE thermal simulator was used to perform compound test of hot hydrostatic pressure, the composite rate was detected by ultrasonic C scanning equipment,characterization of bonding state of composite interface by composite rate, the same constant temperature holding time and different constant temperature, as well as the deformation and deformation rate are adopted, two kinds of different process combinations, which are treated by constant temperature heat preservation, press and deform first, and then be treated with constant temperature and thermal insulation, are respectively carried out, the effects of different hot and hydrostatic composite processing routes and technological parameters on the composite behavior and effect of stainless steel and carbon steel were investigated.

Key words: stainless steel, carbon steel, depressed deformation, insulation treatment, recombination rate

冷连轧机组的垂振与对策研究

魏立群[1], 董亚军[1], 瞿志豪[2]

（1. 上海应用技术大学材料科学与工程学院，上海 201418；

2. 上海第二工业大学，上海 201206）

摘 要： 本文针对我国某钢厂高速冷连轧机组轧制 T5 极薄带材出现的振动现象，从现场实测入手分析了它的振动源和振动频率，并结合有限元分析方法进一步验证了轧机振动的实测分析结果和振动形态。最后优化轧制润滑工艺，有效地抑制了该机组轧制 T5 极薄带材时的振动的瓶颈问题，使机组的轧制速度从 650 m/min 左右提高到 900 m/min 以上。

关键词： 冷连轧机，轧制，极薄带材，垂直振动，润滑

Research of Vertical Vibration and Strategy in Cold-rolling Mill

Wei Liqun[1]，Dong Yajun[1]，Qu Zhihao[2]

(1. Sechool of Materials Science and Engineering，Shanghai Institute of Technology，Shanghai 201418，China;

2. Shanghai Second Polytechnic University，Shanghai 201206，China)

Abstract: On account of the vibration problem in high speed cold-rolling mill in rolling T5 thin strip steel,the vibration resource and frequency of rolling mill was studied by measurement.The results of measurement and the mode of vibration were proved by using finite element methods. The vibration in high speed cold-rolling mill was restrained effectually by optimizing the lubrication technical parameters, and the rolling speed was increased from 650 m/min to above 900m/min.

Key words: cold rolling mill, rolling, thin strip steel, vertical vibration, lubrication

含磷高强 IF 钢连退板表面条纹色差缺陷成因分析及控制

张亮亮[1,2]，于　洋[1,2]，李晓军[3]，高小丽[1,2]，王　畅[1,2]，王泽鹏[1,2]

（1. 首钢技术研究院；2. 绿色可循环钢铁流程北京市重点实验室；

3. 首钢京唐钢铁联合有限责任公司）

摘　要： 针对某冷轧钢生产的含磷高强 IF 钢连退板表面条纹色差缺陷问题，从宏观表现和微观特征方面进行系统研究，并结合现场设备与该钢种工艺特点，分析该缺陷产生的机制。结果表明，条纹色差缺陷产生一方面主要是该钢种特有的铁皮界面特征导致冷轧后带钢表层出现不同程度破碎形貌，另一方面连退炉内气氛也会使合金元素在破碎处加重氧化富集，造成成品板表面色差更严重。通过优化精轧轧制规程可显著减少铁皮的不规整嵌入，同时控制连退炉内气氛、露点等参数，最终有效提高带材的表面质量，为现场创造了较大的经济效益。

关键词： 含磷高强 IF 钢，条状色差缺陷，成因分析，控制

Origin Analysis and Control for Anneal Striped Color Defects of the High Strength IF Steel Containing Phosphorus

Zhang Liangliang[1,2], Yu Yang[1,2], Li Xiaojun[3], Gao Xiaoli[1,2],
Wang Chang[1,2], Wang Zepeng[1,2]

(1. Shougang Research Institute of Technology; 2. Beijing Key Laboratory of Green Recyclable Process for Iron and Steel Production Technology; 3. Shougang Jingtang Iron and Steel Company)

Abstract: For solving the problem of the striped color defects on the anneal surface of the high strength IF steel, the macroscopic and microscopic characteristics were systematically investigated. Combined with the equipment of the production line and the characteristics of the the high strength IF steel technology, its mechanism of the color defects was analyzed simultaneously. The results showed that there are two main causes of striped color defects. On the one hand, the unique interface characteristics of hot rolled plate lead to the broken surface of varying degrees after cold rolling, on the other hand, alloying elements will be oxidized and enriched easier in the crushing area on the surface when annealing, which resulting more serious striped color defects. By optimizing the finishing rolling schedule, the irregular embedded scale can significantly be reduced, in addition controlling the furnace with the atmosphere, dew point and other parameters is necessary. thus, the strip surface quality is improved effectively and larger economic benefits are obtained.

Key words: high strength IF steel containing phosphorus, striped color defects, origin analysis, control

压力容器制管用 P235GH 钢板的研制

欧阳鑫，胡昕明，王　储，胡海洋，颜秉宇，金耀辉

（鞍钢集团钢铁研究院，辽宁鞍山　114009）

摘　要：鞍钢通过对 P235GH 钢板的成分设计、生产工艺路线设计、生产工艺的分析，研究了 P235GH 钢板的力学性能，观察了钢板在实际工业生产中轧态、正火态的金相组织。分析研究结果表明：钢板成分设计、生产工艺路线设计合理；钢板生产工艺制定准确、易于实现；钢板成分均匀、性能稳定。所开发的 P235GH 钢板满足交货技术条件要求，产品实物质量优于欧标标准要求，达到产品开发的目的。

关键词：钢板试制，成分，工艺，力学性能，金相

The Development and Study on Performance of Pressure Vessel Tube Steel P235GH Steel Plate

Ouyang Xin, Hu Xinming, Wang Chu, Hu Haiyang, Yan Bingyu, Jin Yaohui

(Anshan Iron and Steel Group Steel Research Institute，Anshan 114009, China)

Abstract: By means of chemical composition design，process route design and production process analysis of P235GH steel plate，An Steel Co.,Ltd. studies the mechanical properties，examines the mierostrueture of steel plate as—rolled and as—nor—malized during industrial production. The results of analysis and research show that the chemical composition design and process route design for the steel plate production are rational，the steel plate is of uniform chemical composition and stable mechanical properties. The developed P235GH steel plate satisfies the requirements of technical specification for delivery with more superior actual product quality than the requirements in European standard，achieving the aim of product development.

Key words: trial development of steel plate, chemical composition, process, mechanical properties, mierostrueture

板形功效系数在线计算模型的研究及应用

梁勋国

（中冶赛迪技术研究中心有限公司，重庆　401122）

摘　要：基于修正的影响函数法建立了板形功效系数理论计算模型，通过分析各种参数对板形功效系数的影响规律确定了关键的影响因素，根据实测值建立了包含关键影响因素的板形功效系数在线计算模型。实际应用效果表明，板形控制稳定，平直度偏差小于 5IU。

关键词：功效系数，板形控制，影响函数法，冷轧

A Study and Application on Online Calculation Model of Actuator Efficiency Factor

Liang Xunguo

(CISDI Research & Development Co., Ltd., Chongqing　401122, China)

Abstract: Theoretical model of actuator efficiency factor was built base on modified influence function method. Key influencing factors were determined by analyzing the relationship between actuator efficiency factor and the main parameters, and actuator efficiency factor online calculation model was developed according the measured value. Practical application results show that flatness deviation is less than 5 I-Unit and the distribution is very stable.

Key words: actuator efficiency factor, flatness control, influence function method, cold rolling

极薄带最小可轧厚度新公式

肖　宏，任忠凯，刘　晓

（燕山大学国家冷轧板带装备及工艺工程技术研究中心，河北秦皇岛　066004）

摘　要： 许多学者在实验中发现，实验得到的最小可轧厚度远小于常用最小可轧厚度公式计算得到的结果。针对上述问题，本文进行了最小可轧厚度理论分析，推导出新的考虑轧制力限制的条件最小可轧厚度和不考虑轧制力限制的理论最小可轧厚度公式，理论最小可轧厚度约为 Stone 公式计算结果的 22%。通过实验室二辊轧机进行 304 不锈钢、纯铝、纯铜极薄带轧制实验，实测了不同材质在不同轧制力下的实验最小可轧厚度，发现本文公式计算结果实测结果非常吻合，验证了本文最小可轧厚度理论的正确性，从而解决了存在半个多世纪的最小可轧厚度计算不准确问题。

关键词： 极薄带轧制，最小可轧厚度，理论及实验研究，Stone 公式

A New Formula of the Minimum Thickness for Ultra-thin Strip Rolling

Xiao Hong, Ren Zhongkai, Liu Xiao

(National Engineering Research Center for Equipment and Technology of Cold Strip Rolling,
Yanshan University, Qinhuangdao 066004, China)

Abstract: Most scholars have found that the minimum thickness obtained by the experiment is much smaller than that calculated by the common formula of the minimum thickness. In view of the above problem, the theory of the minimum thickness was analyzed, and then a new formula of the minimum thickness was deduced. The minimum thickness calculated by the new model is about 22% of the Stone formula. Finally, the minimum thickness of 304 stainless steel, pure aluminum and pure copper was obtained by a two-high mill. By comparing the measured results with the calculated results of the new model and the Stone formula, it is found that the calculations of the new model are closer to the measurement, which verifies the accuracy of the new model.

Key words: ultra-thin strip rolling, the minimum thickness, experimental and theoretical analysis, Stone formula

温度控制对薄规格钢带炉内跑偏的影响

黄海生

（新余钢铁集团有限公司，江西新余　338001）

摘　要： 厚度 0.2~2.0mm 连退炉生产厚度 0.5mm 以下钢带，炉内钢带跑偏除考虑来料板形、炉辊辊形及粗糙度、

炉内钢带张力等因素外，加热段钢带升温速率也是不可忽视的因素。原因是炉子热容量大，热惯性大，温度检测和控制固有的滞后性，钢带升温速率未得到有效控制，钢带升温速率过快，加热段钢带出现微瓢曲，在低张力下，钢带跑偏。本文简单介绍连退炉温控系统，并采用较为精准控制钢带温度 PID 调节器输出限幅方案，提高了钢带温控精度，降低了钢带升温速率，薄规格钢带炉内跑偏明显改善。

关键词：加热速率，钢带跑偏，温度控制，连退炉

连退平整机组带钢过焊缝过程中延伸率补偿技术的研究

沈青福[1]，袁文振[1]，陈双玉[2]，常金梁[2]，周莲莲[3]

（1. 宝山钢铁股份有限公司冷轧厂，上海　201900；

2. 燕山大学国家冷轧板带装备及工艺工程技术研究中心、燕山大学亚稳材料制备技术与科学国家重点实验室，河北秦皇岛　066004；

3. 燕山大学电子实验中心，河北秦皇岛　066004）

摘　要：针对连退平整机组带钢过焊缝过程中延伸率波动的问题，充分考虑连退平整机组的设备与工艺特点，以延伸率波动幅度最小为目标，建立了一套适合于连退平整机组过焊缝过程中的延伸率补偿技术，通过理论模型与现场试验相配合，实现了前后张力及轧制压力的动态补偿，最大程度地减少了延伸率的波动。

关键词：连退平整机组，过焊缝，延伸率，补偿技术

Research on Strip Steel Elongation Compensation Technique in the Process of CAPL Over Weld

Shen Qingfu[1], Yuan Wenzhen[1], Chen Shuangyu[2], Chang Jinliang[2], Zhou Lianlian[3]

(1. Baoshan Iron and Steel Co., Ltd., Shanghai 201900, China;

2. Yanshan University of National Engineering Research Center for Equipment and Technology of Cold Strip Rolling、State Key Laboratory of Metastable Materials Science and Technology, Qinhuangdao 066004, China;

3. Department of Electrical Engineering, Yanshan University, Qinhuangdao 066004, China)

Abstract: Aiming at the problems of strip steel elongation fluctuation in the process of CAPL over weld, fully combined with the equipment and process characteristics of CAPL, as the target of minimum elongation fluctuation range, the elongation compensation technique is established, combining the theoretical model and field test, the dynamic compensation of tension and rolling pressure is realized, and the fluctuation of elongation is reduced to the maximum extent.

Key words: CAPL, over weld, elongation, compensation technique

连退平整机组恒轧制力过焊缝强度分析研究

袁文振

（宝山钢铁股份有限公司，上海　200431）

摘　要：连退平整机组在带材过焊缝过程中，带头带尾存在很长的延伸率波动超差长度。结合连退平整机组的设备与工艺特点，提出了连退平整机组恒轧制力过焊缝的轧制方法，为保证包括焊接接头在内的带材各处强度均在允许范围之内，建立了连退平整机组带材非焊缝处的应力计算模型和带材焊缝处的应力计算模型，并将该模型应用到宝钢某连退平整机组，实现了恒轧制力过焊缝的轧制模式，有效地减少了带头带尾延伸率波动超差的长度，提高了带头带尾的质量，具有进一步推广应用的价值。

关键词：连退，平整，过焊缝，强度

Research on the Strength of the Constant Rolling Force Over Weld in the Continuous Annealing Line

Yuan Wenzhen

(Baoshan Iron Steel Co., Ltd., Shanghai 200431, China)

Abstract: The head and tail of strip exist extremely long oversize length of elongation fluctuation, during the process of welding in the continuous annealing line. Combine the equipment and technological features of the continuous annealing line, put forward a rolling method of the constant rolling force over weld in the continuous annealing line, in order to ensure that strength is within the allowable range throughout the strip including welding joints, establish the calculation model of stress in the welding seam and out the welding seam for the continuous annealing line, and the model is applied to the continuous annealing line in Baosteel, realize the rolling model of the welding seam with constant rolling force, reduce the head and tail of strip's oversize length of elongation fluctuation, improve the quality of the head and tail of strip, and have further promoted application value.

Key words: continuous annealing, leveling, over weld, strength

SPCC 表层粗晶缺陷原因分析和改进

程鹏飞

（首钢京唐钢铁联合有限责任公司，河北唐山　063200）

摘　要：有客户抱怨冷轧低碳铝镇静钢冲压后出现表层粗晶缺陷，通过该缺陷的形貌和形成机理的分析研究，提出改进思路，减少粗晶缺陷的发生。

关键词：冷轧，低碳铝镇静钢，粗晶，分析，改进

Analysis on Causes of Surface Coarse Grain of Steel Plate of Cold Coil and Improvement

Cheng Pengfei

(Shougang Jingtang United Iron & Steel Co., Ltd. Tangshan 063200, China)

Abstract: Customer complaints when stamping the low carbon aluminum killed steel,coarse grain appears on the surface. Through the analysis of the morphology and formation mechanism of the defects, and puts forward some ideas of improving and reducing the occurrence of coarse grain.

Key words: cold rolling, low carbon aluminium killed steel, coarse grain, analyses, improvement

Q295NHA 耐候钢板表面裂纹原因分析

肖　亚，黄微涛，唐志刚，戴　林，向浪涛，王　铎

（重庆钢铁股份有限公司技术中心，重庆　400084）

摘　要： 针对某公司生产耐候结构钢板 Q295NHA 时出现批量性表面裂纹的问题，通过分析，找出产生钢板表面裂纹的原因，对其微合金化方式进行优化，最终取得较好效果，基本解决 Q295NHA 钢板生产过程中的表面质量问题。

关键词： Q295NHA，耐候钢板，表面裂纹，原因分析

热轧板氧化皮结构及其对酸洗效果影响的研究

贾幼庆

（马鞍山钢铁股份有限公司技术中心，安徽马鞍山　243000）

摘　要： 采用金相显微镜、扫描电镜和色差仪（HUNTER LABSCAN XE），从氧化皮结构出发，研究某钢厂 CSP 和普通热连轧生产的热轧板表面氧化皮结构及其对酸洗效果的影响。结果表明：CSP 生产的 SPCC 热轧板表面氧化皮厚度均值为 13.70μm，由两层构成，表层为 Fe_2O_3，约占总厚度的 2%，分布不连续，内层为 Fe_3O_4，接近基体处存在大量块状 FeO；普通热连轧生产的 SPCC 热轧板表面氧化皮厚度为 13.70μm，单层结构，由 Fe_3O_4 构成，在接近基体处存在极少量 FeO，呈块状分布。普通热连轧生产的 SPCC 热轧板比 CSP 易酸洗。

关键词： CSP，普通热连轧，氧化皮，酸洗

Study on the Structure of Oxide Skin of Hot Rolled Plate and Its Influence on Pickling Effect

Jia Youqing

(Technology Center, Maanshan Iron & Steel Co., Ltd., Maanshan 243000, China)

Abstract: Using Metallurgical microscope, scanning electron microscopy and colorimeter (HUNTER LABSCAN XE), based on the structure of oxide skin, the structure of surface oxide scale of hot rolled plate produced by CSP and common hot strip mill in a steel mill and its influence on acid washing efficiency were studied. The result shows that the average thickness of oxide surface of CSP hot rolled plate is 13.70μm, which is composed of two layers, the surface layer is Fe_2O_3, about 2% of the total thickness, and this layer is discontinuous distribution, the inner layer is Fe_3O_4, there is a lot of massive FeO near the substrate. The surface oxide thickness of common hot rolled plate is 13.70μm, the single layer structure is composed of Fe_3O_4, and there is a small amount of FeO near the substrate, which is a block distribution. Ordinary hot rolled plate is easier to be pickled than CSP hot rolled plate.

Key words: CSP, hot strip mill, oxide skin, pickling

CSP 轧辊温差对薄材轧制的影响及改进方法

陈剑飞，吴利娟，杨　光，杨　洲

（武汉钢铁股份有限公司条材总厂 CSP 分厂，湖北武汉　430083）

摘　要： 本文主要以武钢 CSP 在轧制极薄材过程中的稳定性做为研究对象，主要分析了以下问题：上下工作辊辊温对极薄材轧制过程中的板形控制能力及稳定性的影响；基于该轧制工艺及 PCFC 模型计算原理提出了相应的控制手段和改进措施。

关键词： 工作辊，温度场，冷却水，板形控制，切水板

Work Roll Temperature Effect on Thin Strip Rolling and Improvement Method

Chen Jianfei, Wu Lijuan, Yang Guang, Yang Zhou

(WISCO CSP, Wuhan 430083, China)

Abstract: This paper mainly to the stability of WISCO CSP in very thin material rolling process as the research object, mainly analyses the following problems: upper and lower work roll temperature on the shape and very thin strip rolling process control ability and stability of the proposed control method; and the corresponding improvement measures of the rolling process and PCFC model based on the principle of.

Key words: work roll, temperature field, cooling water, PCFC, water cutting plate

冷轧普板连退产品性能调试改进工艺

赵艳涛，吴卫军，杨树鹏

（河北中重冷轧材料有限公司工艺技术科，河北沧州　061113）

摘　要： 本文主要介绍了沧州中铁装备制造材料有限公司冷轧厂普板连退冷轧产品性能工艺的改进过程，通过新生产线的调试、分析研究，并对影响产品力学性能的因素进行总结，并提出力学性能的改进措施，确保冷轧普板连退机组在实际生产过程中表面质量及力学性能指标达到客户使用要求，目前河北中重冷轧材料有限公司的普板连退机组已正常生产。

关键词： 冷轧带钢，力学性能，屈服强度，连续退火

Production Practice of Continuous Annealing Process Improvement of Cold Rolled Strip

Zhao Yantao, Wu Weijun, Yang Shupeng

(Hebei Zhongzhong Cold-rolled Sheet Co., Ltd., Cangzhou 061113, China)

Abstract: In this paper, the main content is about: Cangzhou China railway equipment manufacturing materials Co., Ltd's plate abrasion even retreat cold rolled product performance process improvement process, as well as the factors influencing the mechanical properties of products, and put forward the improvement measures of mechanical performance, to ensure that cold rolled at plate even return units in the actual production process of surface quality and mechanical performance indicators meet the requirements of customers to use.

Key words: cold rolled strip, mechanical property, yieldstrength, continuous annealing

武钢 CSP 热轧薄规格带钢亮带缺陷的原因分析与控制实践

姜　南，李国全，王　宝，高　智

（武钢有限公司条材厂 CSP 分厂，湖北武汉　430085）

摘　要： 武钢 CSP 作为具有国际先进水平的薄板坯连铸连轧生产线，其特点为：短流程、生产节奏紧。产品主要定位在薄材、硅钢、高强度钢等高端品种上，广泛应用于电力、机械、化工、造船、汽车结构、轻工业等领域。客户对板形的要求非常高，与同规格的冷轧材一致。这对于热轧产品而言提出很高的要求。而实际生产中薄规格带钢的亮带问题非常突出，对板形和表面的影响特别大。本文从亮带区域的组织差异，造成原因入手，开展实验并得到控制措施几个方面介绍亮带缺陷。措施采取后取得较好的实际效果。

关键词： 亮带缺陷，卷取，热轧薄材

The Cause Analysis and Control Practice of Bright of WISCO CSP Hot-rolled Thin Gauge Strip

Jiang Nan, Li Guoquan, Wang Bao, Gao Zhi

(CSP Plant of Wuhan Iron and Steel Co., Wuhan 430085, China)

Abstract: WISCO CSP as a thin slab casting and rolling production line with the international advanced level is characterized by: a short process, production rhythm tight. Products are mainly located on the thin material, silicon steel, high-strength steel and other high-end variety, widely used in electrical, mechanical, chemical, shipbuilding, automobile construction, light industry and so on. Customer demand for strip shape is very high, In line with with the specifications of the cold rolled material. It puts forward high requirements for hot-rolled products. And thin strip light with problems in the process of production is very prominent, particularly big to the influence of the flatness and surface. Group differences from the bright area, killing reason, this paper carried out the experiment and control measures from several aspects to introduce light with defects.After measures taken to obtain the good actual effect.

Key words: bright, coil, hot-rolled thin gauge strip

关于热轧粗轧机保护功能的研究与应用

高颖男，殷程飞，王文杰，张红军，殷　敏，陈百红

（鞍钢股份热轧带钢厂，辽宁鞍山　114014）

摘　要： 粗轧机做为热轧生产的开坯机,在生产中最大的设备安全隐患就是板坯来料错误，也就是混钢。板坯来料混钢后会造成粗轧机压下量过大，进而造成粗轧机损坏。第二个设备安全隐患就是板坯头部的弯曲变形。对于上述设备安全隐患各大热轧厂几乎都受到过危害，但由于产生原因很多，所以一直没有找到简便直接的消除设备安全隐患的有效方法。本文是针对上述问题，研究的一些解决对策。

关键词： 粗轧机，压下量，混钢，联锁保护

The Research and Application about the Protection Function of Rough Mill in Hot Rolling Plant

Gao Yingnan, Yin Chengfei, Wang Wenjie, Zhang Hongjun, Yin Min, Chen Baihong

(Hot Strip Mill of Angang Stock Co., Ltd., Anshan 114014, China)

Abstract: As the blooming machine of the hot rolling production line, the first biggest equipment security risk is the slap number error that is called the slap in disorder. The slap in disorder may cause the reduction oversize, and then cause the damage of rolling mill. The second biggest equipment security risk is that the middle of the slab usually bends up or down. Almost all the hot rolling plant has the same problem, there are a lot of reasons that can cause it, and there is no sample and effective method to solve the problem for a long time. This article is about the research and application to solve the problem above.

Key words: rough mill, reduction, slap in disorder, interlock protection

含 Ti 微合金低碳钢的力学性能及预测

任家宽，吴思炜，陈其源，周晓光，刘振宇

（东北大学材料科学与工程学院，辽宁沈阳　110819）

摘　要： 研究了 Ti 含量对低碳钢组织及力学性能的影响，并通过 Hall-Petch 细晶强化公式和 Ashby-Orowan 沉淀强化模型分析其变化规律的原因，在传统 C-Mn 钢力学性能预测模型的基础上，开发 Ti 微合金钢的力学性能预测模型，并对含 Ti 钢 Q345B 力学性能进行预测。结果表明，随 Ti 含量的增加，低碳钢的屈服强度、抗拉强度以及屈强比都有不同程度的增加，延伸率降低。细晶强化对强度贡献最大，而铁素体中析出 TiC 粒子的沉淀强化作用是导致实验钢强度以 Ti=0.042% 为界呈现显著差异的主要原因。力学性能预测模型对含 Ti 钢 Q345B 的屈服强度和抗拉强

度预测相对误差分别达到±8%和±6%，准确度高，稳定性可靠。

关键词：Ti 微合金钢，Ti 含量，沉淀强化，细晶强化，力学性能预测

Study on Mechanical Properties and Prediction of Ti Micro-alloyed Low Carbon Steel

Ren Jiakuan, Wu Siwei, Chen Qiyuan, Zhou Xiaoguang, Liu Zhenyu

(School of Materials Science & Engineering, Northeastern University, Shenyang 110819, China)

Abstract: Studying on the influence of different Ti content on mechanical properties of Ti microalloyed low carbon steels, Hall-Petch grain refinement formula and Ashby-Orowan precipitation strengthening model are carried out to analyse the reason for the variation. Based on the prediction model of mechanical properties of traditional C-Mn steel, the mechanical properties prediction model of Ti microalloyed steel is developed, and the mechanical properties of Ti steel Q345B are predicted. The results show, with the increase of Ti content, the yield strength, tensile strength and yield ratio of low carbon steel all increase to different degrees, and the elongation decreases. The strengthening effect of fine grain has the greatest contribution to the strength, and the precipitation strengthening effect of TiC particles is the main reason for the significant difference of Ti = 0.042%. The relative error of YS and TS for the prediction of Q345B steel is within ± 8% and ± 6%, respectively, and the accuracy is high and the stability is reliable.

Key words: Ti micro-alloyed steel, Ti content, precipitation strengthening, fine grain strengthening, mechanical properties

轧钢工艺对高韧性桥梁钢 Q370qE 力学性能的影响

黄微涛，唐志刚，戴 林，李宏伟，赵永峰，杜大松

（重庆钢铁股份有限公司技术中心，重庆 400084）

摘 要： 新国标执行后，高强桥梁钢 Q370qE 低温冲击功由 47J 提高至 120J。沿用原有轧钢工艺生产厚规格 Q370qE 钢板时，出现冲击性能不合的情况，对此进行研究和工业试验。通过对不同轧钢工艺方案的比较，得出采用 300mm 断面铸坯，并利用两阶段控轧+轧后控冷工艺，可生产出合格的高韧性厚规格 Q370qE 钢板。

关键词：高韧性，桥梁钢，Q370qE，轧钢工艺，优化，力学性能

热金属探测仪对飞剪系统影响及优化研究

何 苗，曹 凯，胡 杰，付守顺

（武汉钢铁有限公司计控厂，湖北武汉 430080）

摘 要： 热金属探测仪（HMD）作为飞剪系统的重用环节，用于带钢位置跟踪，其检测精确度直接影响带钢剪切质量，HMD 对飞剪系统的影响因素主要包括系统参数、带钢温度和现场环境。扫描时间和亮度阈值是 HMD 的重要参数，是 HMD 准确测量的前提。针对带钢温度影响，采用最小二乘法，研究了带钢温度与扫描时间的对应关系，

提出了基于带钢温度变化的扫描时间自调整方法。针对现场环境影响，提出了相对阈值的概念，可有效减小亮度迁移对检测的干扰。

关键词：热金属探测仪，飞剪系统，扫描时间，带钢温度，相对阈值

Studies for Influence and Optimization of Hot Metal Detector on Shear System

He Miao, Cao Kai, Hu Jie, Fu Shoushun

(Instrumentation and Control Company of Wuhan Iron and Steel Co., Ltd., Wuhan 430080, China)

Abstract: As an important part of shear system, Hot Metal Detector (HMD) is used to track the position of the strip. The accuracy of inspection directly affects the shear quality. The influence factors of HMD on shear system mainly include system parameters, strip temperature and site environment. Scanning time and lightness threshold are important parameters. It is the premise of accurate measurement of HMD. According to strip temperature, the relationship between strip temperature and scanning time is studied with least square method. It presents a method of scanning time self-adjustment based on temperature variation of strip. For the influence of site environment, the concept of relative threshold is puts forward, which can effectively reduce the interference of light migration on detection.

Key words: hot metal detector, shear system, scanning time, strip temperature, relative threshold

鞍钢朝阳钢铁 ASP1700 高强汽车大梁钢生产实践综述

郭洪河 [1,2]，周晓光 [1]，刘振宇 [1]，徐小科 [2]，乔立峰 [2]

（1. 东北大学轧制技术及连轧自动化国家重点试验室，辽宁沈阳　110819；

2. 鞍钢集团朝阳钢铁有限公司热轧厂，辽宁朝阳　122000）

摘　要：在 ASP1700 中薄板坯连铸连轧生产线上，在 C-Mn 普碳钢成分基础上调整 Ti、Nb 微合金含量进行成分设计，优化各工序关键工艺控制点，结合横切机组的高强矫直能力，实现了系列高强汽车大梁钢的柔性化制造，保证了高强汽车大梁钢性能合格、表面光洁和板形良好，满足了用户的需求。

关键词：汽车大梁钢，Ti、Nb 微合金化，成分设计，工艺控制要点，柔性化生产

Summary of Production Practice of ASP1700 High Strength Automobile Beam Steel in Chaoyang Iron and Steel Company of Anshan Iron and Steel Group

Guo Honghe,[1,2] Zhou Xiaoguang,[1] Liu Zhenyu[1], Xu Xiaoke[2], Qiao Lifeng[2]

(1. State Key Lab Rolling and Automation, Northeastern University, Shenyang 110819, China;

2. Hot Strip Plant Anshan Steel Group Chaoyang Iron & Steel Co., Ltd., Chaoyang 122000, China)

Abstract: thin slab continuous casting and rolling production line in ASP1700, Ti, C-Mn in the adjustment of steel components based on Nb micro alloy content composition design, optimization of the key technology of each process control points, combined with the strength of crosscut set straight, to achieve a flexible series of high strength automobile beam steel manufacturing, to ensure the performance of high strength steel automobile beam qualified, smooth surface and good shape, to meet the needs of users.

Key words: automobile beam steel, Ti and Nb microalloying, component design, process control compoints, flexible production

冷连轧机打滑机架判别与打滑防控手段研究

宋浩源[1]，陈甚超[1]，张　生[1]，曾卫仔[1]，齐海英[1]，郭立伟[2]

（1. 北京首钢冷轧薄板有限公司，北京　101304；

2. 北京首钢自动化信息技术有限公司，北京　100041）

摘　要：冷连轧生产中的打滑是轧制稳定性破坏的主要表现之一。本文根据轧制参数的变化，直观判断打滑发生机架。通过提高轧辊粗糙度，对轧制规程的设定调整，有效防治打滑发生。

关键词：打滑，机架，粗糙度，轧制规程

Distinguishing Slip Stands and Research on Preventing Slip on Cold Tandem Mill

Song Haoyuan[1], Chen Shenchao[1], Zhang Sheng[1], Zeng Weizai[1], Qi Haiying[1], Guo Liwei[2],

(1. Beijing Shougang Cold Rolling Co.,Ltd., Beijing 101304, China；

2. Beijing Shougang Automation Information Technology Co., Ltd., Beijing 100041, China)

Abstract: In cold tandem rolling, slip is one of the main feature of roling stability failuer. The slip stands are distinguished intuitively according to the change of rolling parameters. Through increasing roughness of rolls, adjusting the rolling regulation, the slip is prevented effectively.

Key words: slip, stand, roughness, rolling regulation

IF 钢连退板表面典型翘皮缺陷分析

王　林[1]，于　洋[1]，王　畅[1]，张　栋[1]，王明哲[2]，焦会立[2]

（1. 首钢技术研究院，北京　100043；2. 首钢股份公司迁安钢铁公司，河北迁安　064404）

摘　要：翘皮是 IF 钢产品最常见的表面缺陷之一，由于产品经历加工工序较多，缺陷成因复杂多样，对应缺陷形貌也有一定区别。结合实际生产情况，总结了 IF 钢连退板三种典型的表面翘皮缺陷的宏观和微观特征，分析了各类翘皮缺陷的形成原因以及对应的形貌特征，并给出了对应缺陷的预防措施，有利于进一步认识和区分不同的翘皮缺陷从而更好的指导生产。

关键词：翘皮，卷渣，热轧，冷轧

Analysis on the Typical Shell Defects on the Surface of Annealing IF Steel

Wang Lin[1], Yu Yang[1], Wang Chang[1], Zhang Dong[1], Wang Mingzhe[2], Jiao Huili[2]

(1. Shougang Research Institute of Technology, Sheet Metal Research Institute, Beijing 100043, China;
2. Shougang Qian'an Steel Company, Qian'an 064404, China)

Abstract: Shell is one of the common surface defects in IF steel, which has complicated reasons and different morphology, because of the long processing procedure. Three kinds of typical shell defects in the annealing IF steel was summarized, causes and corresponding morphology was researched and corresponding preventive measures was proposed, which is helpful to understand and distinguish different shell defects and thereby guide the production.

Key words: shell, entrapped slag, hot-rolling, cold-rolling

汽车板点/斑状缺陷分类解析

于　洋[1,3]，王　畅[1,3]，王　林[1,3]，刘文鑫[1,3]，陈　瑾[2]，王明哲[2]

（1. 首钢技术研究院薄板研究所，北京　100043；2. 首钢股份公司制造部，河北迁安　064404；
3. 绿色可循环钢铁流程北京市重点实验室，北京　100043）

摘　要：本文采用扫描电镜、能谱分析手段研究了汽车板表面频繁发生的点\斑状缺陷。结果表明，采用扫描电镜获得缺陷形貌特征，结合能谱分析可很好的分辨缺陷产生原因。目前产线点\斑状缺陷主要类型为表面小锈点，上游小结疤，镀锌过程锌灰三类缺陷。提出控制措施包括：锈蚀缺陷主要从平整液，防锈油，防锈包装入手；结疤缺陷主要控制连铸过程中气泡，切割瘤及毛刺等控制；锌灰缺陷主要控制锌锅相应参数。

关键词：汽车板，点状缺陷，斑状缺陷，锈蚀，锌灰，结疤

Analysis of the Spot/Porphyritic Defects of Automobile Steel

Yu Yang[1,3], Wang Chang[1,3], Wang Lin[1,3], Liu Wenxin[1,3], Chen Jin[2], Wang Mingzhe[2]

(1. Shougang Research Institute of Technology, Beijing 100043, China; 2. Shougang Qian'an Steel Company, Qian'an 064404, China; 3. Beijing Key Laboratory of Green Recyclable Process for Iron & Steel Production Technology, Beijing 100043, China)

Abstract: This paper use scanning electron microscopy and energy spectrum analysis to study the frequent occurrence of spot and porphyritic defects on the surface of the automobile steel. The results show that the characteristics of the defects are obtained by scanning electron microscopy, and the causes of defects can be obtained by combining the spectral analysis. At present, the main types of production line spot and porphyritic defects are surface small rust spots, small scarring in the continuous casting, and zinc ash in zinc process. The control measures include: The rust defects mainly start control the formation fluid, anti-rust oil and anti-rust packaging. Scar defects mainly control the air bubbles, cutting tumors and burrs in the continuous casting process. Zinc - ash defects mainly control the corresponding parameters of zinc pot.

Key words: automobile steel, spot defects, porphyritic defects, rust, zinc ash, scar

单机架冷轧机乳化液系统设计思路的改进

黄爱军，潘小成

（浙江龙盛薄板有限公司，浙江绍兴　312369）

摘　要： 乳化液在冷轧过程中起冷却和润滑作用，对轧制过程稳定和产品质量起关键作用，有效控制乳化液中铁粉和灰分含量是乳化液系统设计的关键。本文介绍了乳化液系统设计思路的改进。

关键词： 乳化液，冷轧，质量，改进

Improved Design of Emulsion System in Single-stand Cold Mills

Huang Aijun, Pan Xiaocheng

(Longsheng Steel Strip Co., Ltd., Shaoxing 312369, China)

Abstract: The cooling and lubrication role of emulsion during the cold rolling process of steel plate plays a key role in both stability of the rolling process and quality of the products. The effective control of iron powder and ash constituentis the key to design of emulsion system. In this paper, the improved design of emulsion system is introduced.

Key words: emulsion, cold roll, quality, improve

整流器道次管理及维护对锡层质量影响的探究

张超阳

（首钢京唐钢铁联合有限责任公司冷轧部点检作业区，河北唐山　063200）

摘　要： 电镀锡生产线中工艺段主要依赖于整流器输出电流，来依次实现电解清洗、电解酸洗、电镀段、钝化段各段的基本功能，其中整流器的规格清洗段为36V、14000A，酸洗段为36V、8000A，电镀段为36V、6000A，后处理为36V、4000A。电镀段部分阳极条因为跑线时长期小于500A运行，会导致电镀液中的锡吸附到阳极条上，当锡花掉落时会导致带钢与辊接触时产生碴坑，通过对这种运行电流小于500A的阳极进行1h计时报警，提示操作工及时更换整流器的优先级，可以极大避免阳极条上锡的吸附，对于改善产品质量具有重大作用。

关键词： 整流器，优先级，计时，锡花

The Study of Influence of Rectifier Management and Maintenance on Quality of Tin Layer

Zhang Chaoyang

(SGJT United Iron & Steel Co., Ltd., Tangshan 063200, China)

Abstract: Tin plating production line process section mainly depends on the rectifier output current, in order to realize the basic function of electrolytic cleaning, electrolytic pickling, electroplating, post paragraphs, the rectifier specifications of the cleaning section for 36 V, 14000 A, pickling section to 36 V, 8000 A, electroplating to 36 V, 6000 A, post is 36 V, 4000 A. Parts anode of Electroplating running for a long time　less than 500 A when line is running, can lead to the tin plating solution adsorption to the anode bar, strip, as a result of tin crystal flower drops from the bar when in contact with the rolls may has some pits, through the alarm of the anode running more than 1 hour When current is less than 500 A, prompting operator timely change the priority of the rectifier, can greatly avoid tin anode bar adsorption, plays an important role to improve the quality of the product.

Key words: rectifier, priority, yiming, yin crystal flower

数据回归优化模型在冷轧自动轧制控制中的实践

任延庆，张　巍

（本钢板材股份有限公司第三冷轧厂，辽宁本溪　117000）

摘　要： 冷连轧带钢轧制控制过程复杂，产品要求的精度高，轧制力计算是过程控制的核心，而轧制力的计算基础为变形抗力模型，因此提高变形抗力计算精度是提高轧制力计算精度的一条有效途径，为此首先通过实际轧制力数据反算变形抗力，然后使用分析软件对变形抗力进行曲线回归拟合，因此提高了轧制力的设定准确性，现场应用证明，这种方法能有效提高轧制力设定精度。

关键词： 轧制力，变形抗力，回归

Application of Data Regression Optimization Model in Cold Rolling Automatic Rolling Control

Ren Yanqing, Zhang Wei

(Bengang Steel Plates Co., Ltd., The Third Cold Rolling Mill, Benxi 117000, China)

Abstract: The cold mill control process is complex, product's precision is very high, Rolling force calculation model is the core and foundation of process control in the cold rolling ,The accuracy of rolling force calculation is based on the deformation resistance model ,so the precision of deformation resistance parameters is very important First based on actual rolling force calculate deformation resistance ,and compared with the actual rolling force then the software auto calculated the right parameters ,so it improve the accuracy of rolling force setuo value. Field application result shows that the accuracy of rolling force setpoint is improved effectively.

Key words: rolling force, deformation resistance, regression

二级过程控制系统在质量管理方面的应用

吴志明，姜 赫

（本钢板材股份有限公司第三冷轧厂， 辽宁本溪 117000）

摘 要：二级过程控制系统是整个机组自动化控制的重要环节，承担着数据处理的主要工作．传统的二级控制系统仅仅是对主要生产过程数据的数学特征值进行存储，无法给出给予钢卷长度上的数据统计和趋势图表，并且更多的过程数据沉积在一级，这些都给生产工艺人员对产品质量进行分析和管理带来了不便。QDMS 系统很好的实现了质量管理方面的功能，使镀锌机组二级过程控制系统不仅具有自动化控制的功能，同时也具备了质量管理方面的功能。

关键词：二级过程控制系统，质量管理，QDMS

Application of Level 2 Process Control System in Quality Management

Wu Zhiming, Jiang He

(Bengang Steel Plates Co., Ltd., The Third Cold Rolling Mill, Benxi 117000, China)

Abstract: The level 2 process control system is an important part of the automation control, bear the main work of data processing. The traditional level 2 control system is only for the mathematical characteristics of the main production process data storage, are unable to give the steel coil on the length of the data statistics and trend chart, and more process data deposition in level 1, All of these have caused inconvenience to the process personnel to analyze and manage the quality of the products. The QDMS system has realized the function of quality management well, so that the level 2 process control system of CGL not only has the function of automatic control, but also has the function of quality management.

Key words: level 2 process control system, quality management, QDMS

冷轧退火产品边部褶皱的原因分析与控制优化

尤 龙，冯 岗

（本钢板材股份有限公司第三冷轧厂， 辽宁本溪 117000）

摘 要：本文主要针对对冷轧退火板边部褶皱的特征、产生原因进行了分析，提出了控制边部褶皱的措施。

关键词：边部褶皱，屈服平台，平整

Analysis and Optimization of Edge Wrinkling in Cold-rolled Annealed Sheet

You Long, Feng Gang

(Bengang Steel Plates Co., Ltd., The Third Cold Rolling Mill, Benxi 117000, China)

Abstract: This article discusses feature，mechanism and causes of edge wrinkling are analyzed in cold—rolled annealed sheet and several control measures are proposed.

Key words: edge wrinkling, collapse region, levelling

高档汽车面板切边边浪问题机理研究与控制

计 琳，赵 琦，黄 涛

（本钢板材股份有限公司第三冷轧厂， 辽宁本溪　117000）

摘　要： 某冷轧厂重卷机组经过一段时间的生产，由于设备出现劣化问题，导致当进行对汽车板外板尤其为软钢切边时，出现单边浪情况。本文通过对圆盘剪的生产过程分析，产生原因分析及通过分析后几种调整方式，最终取得了比较理想的效果消除了切边产生单边浪问题。

关键词： 切边，单边浪，原因分析，改进措施

Mechanism Study and Control for Edge Wave Problem during Trimming High-level Automotive Sheet

Ji Lin, Zhao Qi, Huang Tao

(Bengang Steel Plates Co., Ltd., The Third Cold Rolling Mill, Benxi 117000, China)

Abstract: After production for several years, we find the equipment aging problem in a Recoiling Line of Bengang. There will be edge wave phenomenon when trimming soft steel material used for automotive industry. In this paper, the author will analyze the working principle of side trimmer and try several adjusting ways to get ideal result and solve the problem.

Key words: edge trimming, edge wave, cause analysis , improvement method

CPC 系统在酸轧机组的应用及常见故障分析

黄海鹏

（本钢板材股份有限公司第三冷轧厂，辽宁本溪　117000）

摘　要： 在现代带钢轧制过程中，仪表起着非常重要的作用。本文针对轧制生产中常用的一种仪表CPC，基于北方某冷轧厂的酸轧产线的实际情况，分别从系统本身的工作原理，结构组成，功能分析，工作过程描述，以及日常维护方面进行了详细的介绍。

关键词： 仪表，CPC 系统，测量原理

The Application of CPC System in PLTCM and Common Fault Analysis

Huang Haipeng

(Bengang Steel Plates Co., Ltd, The Third Cold Rolling Mill, Benxi 117000, China)

Abstract: The instrumentation plays a very important role in the strip rolling at present. This paper in views of CPC system which is common used in strip rolling, based on the actual situation of PLTCM in a cold rolling mill in the north of China, gives a introduce of measurement principle, internal composition, functional analysis, description of working process and maintenance.

Key words: instrumentation, CPC system, Measurement principle

轴承润滑系统温度控制的改进

张耀明

（本钢板材股份有限公司第三冷轧厂， 辽宁本溪 117000）

摘　要： 本文以北方某冷轧厂酸轧机组支承辊轴承润滑系统为案例，分析了润滑油温度的控制方法，结合现场实际情况，优化了温度控制方案，提高了系统运行的稳定性。

关键词： 冷轧，轴承润滑，温度控制

Improvement of Temperature Control for Bearing Lubrication System

Zhang Yaoming

(Bengang Steel Plates Co., Ltd., The Third Cold Rolling Mill, Benxi 117000, China)

Abstract: This paper takes a cold rolling mill in the North of china bearing lubrication system as a case, and analyzes the temperature control method of lubrication oil. And combined with the actual problems, gives the optimal temperature control scheme, improved system operation stability.

Key words: cold mill, bearing lubrication, temperature control

连退线卷取机拉杆断裂原因浅析

李子俊，刘　森，张森建

（首钢京唐钢铁联合有限责任公司冷轧作业部，河北唐山 063200）

摘　要： 介绍了连退线卷取机卷筒结构及工作原理，并从拉杆选材、机加工和结构设计等方面对卷取机在使用过程

中卷筒拉杆断裂问题进行研究分析，并提出了解决方案。

关键词：卷取机，拉杆，断裂

Analysis on the Fracture Reason of the Tensile Rod of CAL

Li Zijun, Liu Sen, Zhang Senjian

(Shougang Jingtang United Iron & Steel Co., Ltd., Cold Rolling Department, Tangshan 063200, China)

Abstract: Introduces the structure and working principle of the coiler mandrel, and analyzes the fracture problem of the tensile rod from rod material, machining and structural design, and puts forward the solution.

Key words: tension reel, tensile rod, fracture

S6 轧机工作辊端面接触行为研究

葛红洲，陈　军，战　波

（宝山钢铁股份有限公司，上海　201900）

摘　要： S6 轧机为国内外较为先进的一种改进型 Z-Hi 轧机机型。因辊径小，工作辊轴颈无轴承套结构，所以，在工作辊轴两端面各增加了轴向止推机构负责工作辊的轴向定位。本文以 S6 轧机轴向止推轴承为研究对象，对工作辊在不同工况下的轴向止推轴承转速进行了分析，揭示了工作辊不同工况对 S6 轧机工作辊端面接触行为的影响。研究表明：工作辊挠曲、上下平移以及水平平移，都会改变轴向止推轴承的转速以及转速波动规律，并且导致轴向止推轴承和工作辊轴端的磨损加剧。

关键词： S6 轧机，轴向力，轴向止推轴承

Study on Contact Behaviors of Work Roll End Face in S6 Rolling Mill

Ge Hongzhou, Chen Jun, Zhan Bo

(Baoshan Iron & Steel Co., Ltd., Shanghai 201900, China)

Abstract: S6 rolling mill is an advanced Z-Hi rolling mill type. Because the diameter of work roll is small, work roll has no bearing chock. There is a axial trust bearing contacting on both end faces of work roll in order to keep the roll axis positio. Focusing on the axial trust bearing, this paper mainly analyzes the rotate speed of axial trust bearing in different working conditions and reveals the influence of different work roll conditions on contact behaviors of work roll end face in S6 rolling mill. The study shows that the work roll deflection, vertical translation and horizontal translation can change the speed of axial trust bearing and its speed fluctuation pattern, which aggravates the wear between axial trust bearing and work roll end face.

Key words: S6 rolling mill, axial force, axial thrust bearing

2230 酸轧机组头尾厚度超差控制与改进

张　良，杨洪凯，王少飞，齐海峰

（首钢京唐钢铁联合有限责任公司，河北唐山　063200）

摘　要： 冷轧带钢厚度公差及均匀性是衡量一条产线产品质量好坏及其控制水平高低的重要指标。冷轧带钢头尾段轧制具有较大的参数波动和外界因素干扰，很难建立精确的数学模型，因此造成带钢头尾段厚度超差，产品成材率降低。控制冷轧带钢头尾超差段长度是冷带生产中亟待解决的难题。本文重点介绍了 2230 酸轧机组头尾厚度超差的类型、成因及控制措施，首次引入了头尾厚度超差率的概念，通过长期的数据分析与跟踪，进行了一系列技术改进，有效减少了头尾段厚度超差长度和发生概率，改善了头尾段厚度控制精度，提高了产品成材率。

关键词： 头尾厚度超差，厚度控制，张力，FGC，AGC

Control and Improvement for Strip Head and Tailthickness Over Proof on 2230 PLTCM Cold Rolling Department

Zhang Liang, Yang Hongkai, Wang Shaofei, Qi Haifeng

(SGJT Iron & Steel United Co. Ltd., Tangshan 063200, China)

Abstract: Tolerance and uniformity of cold rolling strip thickness is a very important standard to evaluate a line products quality and control level. Strip head and tail cold rolling under more parameter fluctuation and external factor influence, it is difficult to build up a precise math model, that leads to strip head and tail thickness over proof problem, the products yield is becoming lower. Controlling of cold rolling strip head and tail thickness over proof length is a challenge needs to be conquered urgently. This thesis mainly introduced the type、reason and control measures of strip head and tail thickness over proof problem on 2230 PLTCM, it is the first time to introduce a concept of strip head and tail thickness over proof rate, made a lot of technical improvement by long time data analyzing and tracking, reduced the length and possibility of strip head and tail thickness over proof, improved the strip head and tail thickness control accuracy, increased products yield.

Key words: strip head and tail thickness over proof, thickness control, tension, FGC, AGC

TMEIC 冷连轧机动态变规格控制方法与断带治理

王少飞，窦爱民，齐海峰，张　良

（首钢京唐冷轧作业部，河北唐山　063200）

摘　要： 冷连轧机组在轧制过程中发生的断带故障是冷轧工序的主要生产故障之一。首钢京唐 2230 酸轧从试生产到正式投产，轧机的高强接普碳钢动态变规格断带一直是生产的难题，很大程度上影响了机组的产能释放和产品质量。因此首钢京唐冷轧厂为降低轧机断带率，针对轧制过程中发生断带的原因和规律进行了大量统计和分析，最终通过调整动态变规格模式以及二级模型设定，成功解决了动态变规格断带问题。目前该机组实现连续多月动态变规格"零"断带。

关键词： 高强钢，断带，动态变规格

Control Method and Strip Break Control of TMEIC Tandem Cold Rolling Mill

Wang Shaofei，Dou Aimin，Qi Haifeng，Zhang Liang

(Iron and Steel Company SGJT, Tangshan 063200, China)

Abstract: It is one of the main production failures in cold rolling process that the strip breaking occurs in cold rolling mill. Shougang Jingtang 2230 acid rolling from pre production to formal production, high strength steel mill with FGC broken belt has been the problem of production, has great influence on the release of production capacity and product quality unit. Therefore, in order to reduce the rolling mill in cold rolling mill of Shougang Jingtang belt breaking rate, breaking for the reasons and rules with by a large amount of statistics and analysis in the process of rolling, final specifications electrical control program and operation method improved by adjusting the dynamic, has successfully solved the problem with broken fgc. At present, the unit has been changing the specifications of "zero" tape for more than a month.

Key words: HSS, strip break, FGC

鞍钢冷轧厂联合机组系统升级关键技术集成与应用

柳 军，张 哲

（鞍钢股份有限公司冷轧厂，辽宁鞍山　114021）

摘　要: 为满足市场的需求，鞍钢冷轧厂通过对工艺及设备都比较陈旧的联合机组进行改造，升级为一条装机水平先进，能够生产高档汽车及家电板的联合机组。结合产品定位，考虑生产能力和设备配置特点，合理整合资源，通过借鉴国内外先进工艺技术，利用自身技术优势，鞍钢研发并集成系统升级的关键技术，达到改善设备装机水平，有效规避设备运行风险，提高机组作业率，提高成品轧制厚度精度和板形精度，增加产能，节约投资的改造目的，另外电控系统的改造也为产品开发和产品结构升级提供有效的功能支持。

　　本文重点介绍了机组系统升级所应用的关键技术，对于建设同类生产线，尤其是对老机组实施新技术改造，并快速完成调试工作，具有很好的参考价值。

关键词: 联合机组，系统升级，关键技术，优化配置，集成

System Upgrade Key Technology Integration and Appliance for PLTCM in Cold Rolling Plant

Liu Jun, Zhang Zhe

(Cold Rolling Plant , Ansteel Limited Company, Anshan 114021, China)

Abstract: In order to meet the needs of the market, The old process equipment was revamped for PLTCM of cold rolling mill in Ansteel. After revamped, the mill became an advanced line that can produce high-level cars and appliances. Combined with product positioning, and considering the production capacity and equipment configuration characteristics, we integrated the resources rationally. Through the usage of advanced technology in the world and our technological advantages, we researched, developed and integrated the key technologies to upgrade the system. The revamping target is include improving the level of the equipment and the operating rate, the accuracy of thickness and shape, avoiding equipment operating risk effectively , increasing the production capacity, saving investment. In the other hand, the electronic control system revamping   provide effective functional support for product development and product structure upgrades.

This paper focuses on the key technologies that are used in the system upgrading. It has great reference value for the construction of similar production lines, especially for the old revamping line that applied the new technology and completed the commissioning wok quickly.

Key words: PLTCM, control system upgrade, key technology, optimization and configuration, integration

宝钢 DI 材边缘降控制及其优化

叶学卫，沈青福

（宝山钢铁股份有限公司冷轧厂，上海　200941）

摘　要： DI 材主要用于碳酸饮料和啤酒的包装，用户对厚度精度有十分苛刻的要求，成品规格 DI 材距边部 5mm 外所有点厚度都必须保证在±2%的范围内。同时，随着 DI 材厚度的持续减薄，规格不断拓展，DI 材的横向厚度精度控制（边缘降控制）显得越来越重要，越来越困难。因此，为了减少 DI 材边部减薄，宝钢在某冷连轧机组新增了边缘降仪、1 机架端部辊形设计和优化，结合连轧机设备特点，开发了边缘降自动控制系统，大幅度提高了 DI 材边缘降控制水平，满足了用户需求。本文就宝钢 DI 材几年来在边缘降控制方面所做的工作进行介绍和分析。

关键词： DI 材，横向厚差，边缘降控制

Edge Drop Control Application and Optimization of DI Tinplate at Baosteel

Ye Xuewei, Shen Qingfu

(Cold Rolling Plant, Baoshan Iron & Steel Co., Ltd., Shanghai 200941, China)

Abstract: DI tinplate are mainly used for drink packing of carbonic acid beverage and beer. Customers are very strict with the gauge precision of tinplate. Finished product's thickness deviation should be within ±2% except for the 5mm strip edge. It are important and difficult to decrease the transverse thickness deviation(it is mainly meant to thin at the edge in the rolling process) of DI tinplate Meanwhile the thinner thickness and specification development. In order to decrease the transverse thickness deviation of DI tinplate, A new edge drop meters was allocation, using circular work rolls of stand 1, in combination with characteristic of the tandem cold mill, an edge drop control system was developed in a certain tandem cold mill, the edge drop control level was greatly improved, which satisfied customers' demands. In this paper, the works of improve the edge Drop control of DI tinplate at Baosteel in several years were introduced and analyzed.

Key words: DI tinplate, transverse thickness deviation, edge drop control

连续酸洗机组带钢张力和设备能力分析

刘显军，王业科，辜蕾钢

（中冶赛迪工程技术股份有限公司，重庆　401122）

摘　要：本文根据连续酸洗机组的生产工艺特点，将机组按工序进行张力分区，各区采用不同的控制策略，保证各段张力能够独立控制。带钢在运行过程中会在辊子上缠绕、接触和加减速，会产生带钢弹塑性变形、摩擦、惯性等而导致张力损失，结合机组具体设备配置，可依次计算出机组带钢上各个区的张力分布。在设计时考虑设备自身特点（以活套摆臂距离为例）和各种工序要求（以入口 1#张力辊前进和倒带为例），结合投资成本，综合确定酸洗机组带钢张力和设备能力。

关键词：控制策略，张力损失，基准张力，工序要求

Analysis of Strip Tension and Equipment Capacity for the Continuous Pickling Line

Liu Xianjun, Wang Yeke, Gu Leigang

(CISDI Engineering Co., Ltd., Chongqing 401122, China)

Abstract: Based on the process characteristics of continuous pickling line, This thesis classifies the line into different tension zones ,and different control strategies are adopted for ensuring independent control of each tension zone. The strip is winding around and contacting with the bride rolls and accelerating/decelerating during the line running, as a result, the strip may lose a lot of tension because of the elastic-plastic deformation, friction and inertia. According to the process-required basic tension, the tension distribution on various points of the running strip along the strip flow direction can be calculated in turn. It can be used for defining the strip tension and equipment capacity of the pickling line by comprehensively considering the equipment (such as the looper swing arm distance) and process requirements such as the entry #1 bridle roll forward and backward moving) as well as the investment costs.

Key words: control strategies, tension loss, basic tension, process requirements

浅谈改善 MCJD5 家电板板型控制实践

胡　平

(马钢(合肥)板材有限责任公司，安徽合肥　230011)

摘　要：本文介绍了马钢（合肥）板材有限责任公司酸轧联合机组（PL-TCM）生产 MCJD5 高级家电板过程中出现的板型质量问题，并针对板型问题采取改进的板型控制技术使问题得以解决。

关键词：酸洗连轧，板型拉制，斜浪，单边浪

Practice with Control Technology of Improving MCJD5 Home Appliance Shape

Hu Ping

(Masteel (Hefei) Cold Rolled Sheet Co., Ltd., Hefei 230011, China)

Abstract: The article introduced the sheet shape problem of production MCJD5, and found advance shape control technology to solve the problem on PL-TCM of Masteel (Hefei) Cold Rolled Sheet Co.Ltd.
Key words: PL-TCM, shape cntrol, oblique wave, single edge wave

提高酸轧机组机架内激光测速仪稳定性的方法

王玉锋，张　伟，李　超

（宝钢新日铁汽车板有限公司，上海　201900）

摘　要：本文基于酸洗轧机联合机组激光测速仪进行优化，主要针对了测速仪外部固定架构、压缩空气喷嘴以及乳化液的吹扫方式进行了改造。其中，优化工作包括了测速仪保护结构的设计、吹扫方式的变更以及喷嘴小孔效应的消除。最终，通过流体力学、空气动力学进行分析计算，得出最优化的乳化液清除方式。经过整体改良后，测速信号得到稳定控制，并且反馈功率得到显著的提高，改变了在轧制过程中乳化液的大量喷洒使得激光测速仪信号不稳定的状况。
关键词：激光测速仪，乳化液，拉瓦尔喷嘴，小孔效应，屏蔽型吹扫

A New Method to Modify the Stablish of the Laser Speedometers Which are Used in the PL-TCM

Wang Yufeng, Zhang Wei, Li Chao

(Baosteel-NSC Automotive Steel Sheets Co., Ltd., Shanghai 201900, China)

Abstract: In this paper, the optimization is based on the laser speedometer which is used on the PL-TCM. The optimize work contains the external fixed structure, the compressed air nozzle and the emulsion purging mode of the tachometer. In detail, redesigning the protective structure of the speedometer, changing the purge mode, and eliminating the pinhole effect are all included in this work. Finally, through the analysis and calculation which rely on the fluid mechanics and aerodynamics, The optimized emulsion removal method is derived. After all of the improvements, the laser signal is able to be stably controlled, and the feedback power is significantly improved. These optimizations have changed the situation of the instability of the laser speedometer signal in which the emulsion is heavily sprayed during the rolling process.
Key words: laser speedometer, emulsion, laval nozzle, pinhole effect, shielding type purging

我国冷轧工程技术的发展与进步

王业科，辜蕾钢，刘显军，郑建华，沈继刚，王晓晶，李瑞华

（中冶赛迪工程技术股份有限公司，重庆　401122）

摘　要： 本文综述了国内冷轧工程的现况与装备配套能力、冷轧工程的实践、国内工程公司与世界标杆企业的差距、冷轧工程的建设模式与市场环境、冷轧技术的发展、冷轧工程的发展方向。经过多年的努力，冷轧设备及备件国产化取得显著成效，但与国际标杆企业还有一定的差距。提高自主设计和制造能力，提高装备质量才是提高我国冷轧设备工程水平及技术出口的根本途径。

关键词： 冷轧工程，冷轧技术，发展与进步

Development and Progress of Cold Rolling Engineering Technology in China

Wang Yeke, Gu Leigang, Liu Xianjun, Zheng Jianhua, Shen Jigang, Wang Xiaojing, Li Ruihua

(CISDI Engineering Co., Ltd., Chongqing 401122, China)

Abstract: This paper summarized the domestic cold rolling project status and equipment supporting capacity, cold rolling engineering practice, domestic engineering company and the world benchmark enterprise gap, cold rolling project construction mode and market environment, the development of the technology and project development direction of cold rolling. After years of efforts, the localization of cold rolling equipment and spare parts has achieved remarkable results, but there is still a gap between them and international benchmarking enterprises. Improving the capability of independent design and manufacturing and improving the quality of equipment are the basic ways to improve the level of engineering and technology export of cold rolling equipment in china.

Key words: cold rolling engineering, cold rolling technology, development and progress

本钢硅钢表面清洗脱脂工艺及生产实践探讨

李邦波

（本溪钢铁(集团)有限责任公司冷轧薄板厂，辽宁本溪　117000）

摘　要： 本文简述了冷轧硅钢表面清洗脱脂的原理，分析了各种清洗工艺参数对清洗脱脂效果的影响，并根据本钢特有设备条件和生产实际对设计工艺参数进行优化，取得较好效果。

关键词： 冷轧硅钢，清洗脱脂，工艺参数

Discussion on the Surface Cleaning Technology of Cold Rolled Silicon Steel and Production Practice

Li Bangbo

(Benxi Steel (Group) Co., Ltd., Cold Rolled Sheet Factory, Benxi 117000, China)

Abstract: This article described the principle of cold-rolled silicon steel surface cleaning, analysis of various cleaning effect of process parameters on the cleaning effect, and to analyze and solve problems in the production process.

Key words: cold-rolled silicon steel, surface cleaning, process parameters

罩式退火炉底部对流盘设计优化

张 勇[1]，张晓岳[2]，刘昌伟[1]

（1.本钢板材股份有限公司冷轧薄板厂，辽宁本溪 117021；2. 北京航空航天大学，北京 100083）

摘 要：根据冷轧罩式退火炉用底部对流盘使用现状、制造成本、轧制后钢卷端面带钢溢出尺寸，将底部对流盘的上下端面在适当位置进行开沟槽处理，并在满足屈服强度的前提下选择合适的制造材料。使退火堆垛钢卷的溢出伤卷率降低 15.18%、焊缝伤卷率下降 31.55%，单台对流盘材料成本下降 82.22%，对流盘重量减小使吊运能源消耗降低 9%，对流盘实现规格尺寸统一、不同炉型通用，提高了热处理机组作业率。

关键词：罩式退火炉，底部对流盘，屈服强度，伤卷

Design Optimization of Convection Plate at the Bottom of the Bell Type Annealing Furnace

Zhang Yong[1]，Zhang Xiaoyue[2]，Liu Changwei[1]

(1. The Cold Rolling Mill of Bengang Steel Plates Co., Ltd., Benxi 117021，China；

2. Beijing University of Aeronautics and Astronautics，Beijing100083，China)

Abstract: On the basis of cold-rolled bell type annealing furnace with convection at the bottom of the plate after use present situation, the manufacturing cost, rolling steel coil end strip overflow size, will be opened groove processing in proper place at the bottom of the convection of upper and lower end face, and on the premise of meet the yield strength to choose the appropriate manufacturing materials.Make annealing coil stacking overflow wounded volume ratio decreased 15.18%, weld volume rate fell by 31.55%, single convection plate material costs fell by 82.22%, convection disc weight decrease make lifting energy consumption reduced by 9%, convection plate realize size unification , different furnace convection type universal, which improves the operability of the heat treatment unit.

Key words: bell type annealing furnace, convection plate at the bottom, yield strength, wound roll

冷轧检查台钢卷溢出边的分析与对策

王　杨，李　超，张　旭

（鞍钢股份有限公司冷轧厂，辽宁鞍山　114000）

摘　要：冷轧轧后成品卷在检查过程中易发生跑偏而导致钢卷带尾产生溢出边问题，通过理论分析及实际情况的观察找出带钢跑偏的具体原因，反复摸索制定有效解决方案。最终研究决定提高卷筒与各辊系之间的水平及正交精度，并在原检查台压辊上新增对中装置来控制带钢跑偏，从而彻底解决溢出边的产生，本文着重从机械设备方面介绍分析了此改进项目的可行性和实际效果。

关键词：带钢，跑偏，溢出边，对中装置

The Analysis and Countermeasure of Steel Coil Overflow on Cold Rolling Inspection Table

Wang Yang, Li Chao , Zhang Xu

(Cold Strip Works of ansteel， Anshan 114000, China)

Abstract: After cold rolling, the finished product is prone to running deviation in the inspection process, which leads to the overflow side problem of steel coil tail， through theoretical analysis and actual situation, the paper finds out the specific reasons for the deviation of the strip， to develop effective solutions by trial and error. Finally， the horizontal and orthogonal accuracy of the drum and the roller system are improved， and in the original inspection table press roller to add the new to control the strip running deviation， the production of overflow side can be completely solved， this paper mainly introduces the feasibility and practical effect of this improvement project from mechanical equipment.

Key words: the strip, wandering, the overflow side, centring device

平整开卷机带钢擦划伤问题解决

曹　威，李　超

（鞍钢股份公司冷轧厂，辽宁鞍山　114021）

摘　要：针对双柱头开卷机高速生产时产生的带钢抖动、带尾卷筒涨径力过大和带钢间相对滑动等原因造成钢板表面擦伤、划伤的问题，本文提出了一种方法，即改双柱头开卷机为单柱头开卷机，并计算改进后液压系统压力，使得液压系统实现开卷涨径压力两机控制。该方法有效地解决了双柱头开卷机现存的问题。

关键词：开卷机，涨径压力，带钢，擦伤

Solution to the Problem of Steel Scratches on Level Uncoiling Machine

Cao Wei, Li Chao

(Cold Strip Work Angang Steel Company Limited, Anshan 114021, China)

Abstract: This paper presents a method to improve the surface abrasion and scratches caused by the high-speed production of the double-stigma uncoiler, the relative slippage of the tail drum and the relative sliding of the strip. Change the double stigma uncoiler into the single stigma uncoiler and then calculate the pressure of the improved hydraulic system, which makes the hydraulic system realize the control of the opening and closing pressure. This method effectively solves the existing problems of the double stigma unrollers.

Key words: uncoiling machine, diameter pressure, strip steel, scratch problem

在线油膜测厚仪工作原理及其发展现状

袁文振

（宝山钢铁股份有限公司，上海 201900）

摘 要： 由于在线油膜测厚仪具有实时监测、实时调整油膜状态的功能，可以解决带钢表面油膜所存在的漏涂、涂油不均等缺陷，因此得到钢铁行业的青睐。本文从在线油膜测厚仪的工作原理入手，对激光荧光法、红外光谱吸收法、红外光谱干涉法等三种类型的在线油膜测厚仪的工作原理进行了分析。在此基础上，比较了上述三种类型在线油膜测厚仪的优缺点，方便现场对在线油膜测厚仪的选择。最后，针对在线油膜测厚仪在钢铁行业的应用，简述了在线油膜测厚仪的发展现状。

关键词： 在线油膜测厚仪，工作机理，优缺点，钢铁行业，发展现状

Development Status of Oil Film Thickness Gauge and Its Application in Iron and Steel Industry

Yuan Wenzhen

(Baoshan Iron Steel Co., Ltd., Shanghai 201900, China)

Abstract: Because the on-line oil film thickness gauge has the function of real-time monitoring and adjusting the oil film state in real time, it can solve the uneven coating and oil coating defects in the oil film on the strip. Therefore, it has been favored by the steel industry. In this paper, the principle of on-line oil film thickness gauge is introduced, and the principle of three kinds of on-line oil film thickness measuring instrument, laser fluorescence method, infrared spectrum absorption method and infrared spectrum interferometry method, are analyzed. On this basis, the advantages and disadvantages of the above three types of on-line oil film thickness gauge are compared, which is convenient for the selection of on-line oil film thickness gauge. Finally, according to the application of on-line oil film thickness gauge in iron and steel industry, the development status of on-line oil film thickness gauge is introduced.

Key words: on-line oil film thickness gauge, working principle, advantages and disadvantages, iron and steel industry, development status

浅析热送热装连铸坯的加热方式

钱红兵

（无锡应达工业有限公司，江苏无锡　214028）

摘　要： 按工艺曲线和冶金学特点，同时考虑工艺流程，热送热装技术被分为 5 种类型，每种类型钢坯初始温度不同，加热要求也不同。采用火焰炉加热：热装钢坯可以缩短总的在炉时间，钢坯热送温度越高，加热时间越短；连铸坯热装温度越高，加热炉的排烟温度越高，虽然能耗降低但热效率也降低；钢坯在高温区加热时间减少，氧化烧损也减少。感应加热显著减少钢坯在高温区的加热时间，从而减少钢坯的氧化烧损量；感应加热可以对每个钢坯的温度精确控制，提高产品质量的同时减少能源消耗和氧化烧损。

关键词： 热送热装，火焰加热炉，感应加热，节能，氧化烧损

The Simple Analysis of Hot Delivery and Hot Charging of Continuous Casting Billet

Qian Hongbing

（Wuxi Inductotherm Industrial Company, Wuxi 214028, China）

Abstract: According to the process curve, the characteristics of the metallurgy and process flow, hot delivery and hot charging technology is divided into five types. Billet initial temperature is different, heating requirements is different. The characteristics of flame reheating furnace is that the total time of hot charging billet in furnace is less. The higher the temperature of billet is, the shorter the heating time. The higher the temperature of hot delivery billet, the higher the exhaust temperature of the heating furnace. Although the energy consumption decreases, the thermal efficiency is decreased. Billet heating time decreases in high temperature zone and oxidation loss decreases. Induction heating significantly reduces the heating time of the billet in the high temperature zone, thus the amount of oxidation of billet is reduced. Induction heating can control the temperature of each billet, and improve the quality of the products, and reduce energy consumption and oxidation loss.

Key words: hot delivery and hot charging, flame reheating furnace, induction heating, energy saving, oxidation loss

基于 V-N 微合金化的全厚度厚钢板的生产实践

陈定乾，赵　军

（湖南华菱湘潭钢铁有限公司宽厚板厂，湖南湘潭　411101）

摘　要： 我国是 V 资源较丰富的国家，以 V 代 Nb 可以减轻我国长久以来对 Nb 合金依赖。VN 促进针状铁素体形核以及针状铁素体组织的强韧性能，通过在冶炼过程中添加 V-N 合金，并控制轧制温度与道次间隙时间，中间坯

采用水冷，在奥氏体中获得大量的纳米级的 VN 质点，终冷后钢板全厚度形成针状铁素体组织。

关键词：V-N，针状铁素体，全厚度，厚钢板

Production Practice of the Whole Thickness Steel Plate Based on V-N Microalloy

Chen Dingqian, Zhao Jun

(Heavy Plate Factory, Xiangtan Iron and Steel (XISC) Co., Ltd., Xiangtan 411101, China)

Abstract: V resource is abundant in our country. It can reduce our country's dependence on the Nb alloy for a long time. VN promotes the nucleus of acicular ferrite and the strength of the acicular ferrite. By adding the V-N alloy in the process of smelting, and control the rolling temperature and the interval time, middle slab using water-cooling, as a result, A large number of nano-scale VN particle is obtained in the austenite. And the whole thickness of the steel plate is formed by acicular ferrite.

Key words: V-N, acicular ferrite, whole thickness, thick plate

河钢塞尔维亚钢厂高炉煤气双蓄热加热炉的技术特点

丁国伟，马中杰，王禹尧

（河钢唐钢公司，河北唐山 063016）

摘　要：介绍了河钢塞尔维亚斯梅带雷沃钢厂新建 3#加热炉的技术特点，采用高炉煤气替代天然气、烟气反吹燃烧等技术的应用，解决蓄热式 NOx 和 CO 超标排放的问题；同时，3#加热炉还采用汽化冷却、均热段倾斜式炉底和排渣孔、一级和二级控制技术等技术，提高加热炉的经济效益和操作便利性。

关键词：高炉煤气，双蓄热，供热分配，自动控制

The Technical Characteristics of the Dual Regenerative Reheating Furnace for the BF Gas in HBIS Serbian

Ding Guowei, Ma Zhongjie, Wang Yuyao

(No.1 Steel Making and Rolling Mill, Tangshan 063016, China)

Abstract: Introduces the technical characteristics of the No.3 new reheating furnace in HBIS Serbia, adopted the blast furnace gas instead of the natural gas, and the application of the technology such as flue gas blow burning, solve excess emissions of NOx and CO in regenerative burners. At the same time, the No.3 reheating furnace also adopts the vaporizing cooling technology, the slanting furnace bottom and set slag holes in the furnace, the L1 and L2 control technology, etc, to improve the economic efficiency and operation convenience of the reheating furnace.

Key words: blast furnace gas, dual regenerative, heating distribution, auto control

双蓄热式加热炉运行成本以及问题探讨

牛　琢

（太原钢铁公司不锈热轧厂设备能源科，山西太原　030003）

摘　要： 太原钢铁公司不锈热轧厂 2#加热炉为空、煤气双蓄热式加热炉，从 2006 年新建至 2017 年已投入运行 11 个年头。1#加热炉于 2016 年改造，为常规步进式加热炉。笔者以两种加热炉为对比，通过相关数据从能源消耗，能源成本，检修维护成本等方面对比了两种加热炉的优缺点。得出了目前虽然蓄热式加热炉仍有较大的问题，但仍具有较大的节能和运行成本低等优势，因而可选择性、有条件的推广使用。

关键词： 空、煤气双蓄热式加热炉，能耗，能源成本，运行成本

Double Regenerative Heating Furnace Operation Cost as well as the Problems Discussed

Niu Zhuo

(Taiyuan Iron and Steel Co., Ltd., Taiyuan 030003, China)

Abstract: Taiyuan Iron and Steel Co., Ltd., a non-rust hot rolling mill 2 # heating furnace for air and gas, has been in operation for 11 years since its construction in 2006. The 1 # heating furnace was reformed in 2016, which is a regular step type heating furnace. Compared with two heating furnaces, the author compared the advantages and disadvantages of two heating furnaces through related data from energy consumption, energy cost and maintenance cost. It is concluded that although the regenerative heating furnace still has a big problem, it still has the advantages of high energy saving and low running cost, so it can be used selectively and conditionally.

Key words: air and gas dual regenerative furnace, energy consumption, energy costs, operating costs

一种改型的辐射管换热器

符新涛，刘佑爽

（宝钢集团八一钢铁股份有限公司冷轧薄板厂，新疆乌鲁木齐　830022）

摘　要： 八钢热镀锌连续退火炉 U 型辐射管排烟系统设计采用翅片式换热器、空气进出口壳、排烟过渡管、热空气波纹管共四部分组成，在使用过程中存在翅片式换热器焊缝处断裂，导致烧损排烟过渡管、辐射管法兰及辐射管更换返修。经过将排烟系统改为一种以列管式换热器为原理制作的新型辐射管换热器，很好的避免了以上缺陷。

关键词： 排烟系统，翅片式换热器，空气进出口壳，排烟过渡管，热空气波纹管，辐射管换热器

Radiation Modified Tube Heat Exchanger

Fu Xintao, Liu Youshuang

(Baosteel Group Xinjiang Bayi Iron & Steel Co., Ltd., Urumqi 830022, China)

Abstract: U type radiant tube of Bayi Iron & Steel Continuous galvanizing furnace smoke extraction system is alice type heat exchanger, air import and export housing, exhaust transition pipe、hot air bellows a total of four parts, In the course of alice type heat exchanger weld department fract ure, cause burning exhaust transition pipe, radiant tube flange and radiant tube replacementrepair. After the exhaust system into a new radiation produced by the principle of tube heat exchangerin shell and tube heat exchanger, it is good to avoid these defects.

Key words: smoke extraction system, alice type heat exchanger, air import and export housing, exhaust transition pipe, hot air bellows, the radiation tube heat exchanger

辐射管燃烧系统的控制调节

张立宇

（本钢冷轧厂，辽宁本溪　117000）

摘　要：随着生产技术的进步，辐射管被广泛应用于连续退火炉中，作为辐射管加热炉的重要组成部分，烧嘴是加热系统控制的关键部件，在对辐射管炉燃烧系统的结构及主要组成特征了解的基础上，本文主要介绍了辐射管燃烧系统的工作原理，根据辐射管烧嘴的结构特点，确定其烧嘴的压力平衡调整方法，保证各个炉段的总体空气、煤气流量的分配平衡，然后又确定空燃比例调整的系数，利用空燃比例系数在分别调整各个烧嘴的空气、煤气流量配比，保证最佳燃烧效果。

关键词：辐射管，主烧嘴，压力平衡，空燃比

Control Regulation of Radiation Tube Combustion System

Zhang Liyu

(Cold Rolling Plant, Bensteel, Benxi 117000, China)

Abstract: With the progress of production technology, radiant tube is widely used in the continuous annealing furnace, as an important part of the radiation tube heating furnace burner are key components in heating system control, the radiant tube furnace combustion system on the basis of the structure and main characteristics of understanding, this article mainly introduces the working principle of radiant tube combustion system, according to the structure characteristics of the radiant tube burner, determine its burner of pressure balance adjustment method, ensure the overall air furnace section and the distribution of the gas flow balance, and then determine the coefficient of air-fuel ratio adjustment, use of air-fuel ratio in each burner adjusted respectively of air, gas flow ratio, guarantee the best burning effect.

Key words: radiant tube, the main burner, pressure balance, air-fuel ratio

数字化技术在高质量高精度钢材轧制中的应用

康永林[1]，朱国明[1]，陶功明[2]，张思勋[3]，汪水泽[4]

（1. 北京科技大学材料学院，北京　100083；2. 攀钢集团攀枝花钢钒有限公司，四川攀枝花　617062；
3. 山东钢铁股份有限公司，山东莱芜　271105；4. 武钢研究院，湖北武汉　430080）

摘　要：本文分析了板带材及型材热轧过程的特点，提出了数字化热轧系统的构成框图、板带热轧过程数值模拟平台和型材轧制数字化系统的基本架构。针对国内某热连轧产线典型低合金高强钢的实际生产过程进行了粗轧、精轧及轧后冷却残余应力形成的全过程模拟分析、组织转变模拟预测、数字化技术在百米重轨设计开发、尺寸精度控制、以及复杂断面型钢产品设计开发中应用的实例。

关键词：数字化，轧制，板带材，型材，重轨

Application of Digital Technology in High Quality and High Precision Steel Production Rolling

Kang Yonglin[1], Zhu Guoming[1], Tao Gongming[2], Zhang Sixun[3], Wang Shuize[4]

(1. School of Material Science and Engineering，University of Science and Technology Beijing，Beijing 100083,
China；2. Panzhihua Steel and Vanadium Co., Ltd., Pangang Group, Panzhihua 617062, China；
3. Technology Center of Laiwu Subsidiary, Shandong Iron and Steel Co., Ltd., Laiwu 271105, China;
4. Research and Development Center, Wuhan Iron & Steel (Group) Corp.,Wuhan 430080, China)

Abstract: A block diagram of hot rolling digital system, a simulation platform for the hot rolling process and a basic framework of section steel rolling digital system were established based on the analysis of hot rolling process. The whole rough and finish rolling process and the residual stress formation by the cooling after rolling of the typical high strength low alloy steel Q345B by a domestic hot continuous rolling line were simulated, and the microstructure transformation were predicted. Besides, the application of digital technology in the design & development and dimension precision control of 100-metre heavy rail was discussed, some instances about the digital technology application in high precision complex section steel rolling were also given.

Key words: digital technology, rolling, plate-strip, section steel, heavy rail

热轧无缝钢管在线控制冷却技术开发与工业应用

袁　国[1]，张忠铧[2]，王笑波[2]，杨为国[3]，刘耀恒[2]，王国栋[1]

（1. 东北大学轧制技术及连轧自动化国家重点实验室，辽宁沈阳　110819；2. 宝山钢铁股份
有限公司，上海　201900；3. 烟台宝钢钢管有限责任公司，山东烟台　265500）

摘　要：热轧无缝钢管传统生产流程中，管坯经热成形后采用空冷方式冷却至室温。由于缺乏有效的组织性能在线

调控手段，导致部分规格钢管的热轧态性能合格率较低，不合格产品不得不再进行离线热处理，生产成本高、能耗大、生产周期长。热轧钢管定径后进行控制冷却是有效的组织调控手段。本文介绍了控制冷却工艺在热轧无缝钢管组织调控方面的优势、国内外研究现状、技术开发难点，以及我国首台套兼具终冷温度精准控制和直接淬火功能的热轧无缝钢管在线控制冷却工艺装备开发与工业化应用的新进展。

关键词：钢管，热轧，控制冷却，直接淬火，组织调控，性能

Development and Industrial Application of On-line Control Cooling Technology for Hot Rolled Seamless Steel Tubes

Yuan Guo[1], Zhang Zhonghua[2], Wang Xiaobo[2], Yang Weiguo[3], Liu Yaoheng[2], Wang Guodong[1]

(1. State Key Laboratory of Rolling and Automation, Northeastern University, Shenyang 110819, China;
2. Baoshan Iron & Steel Co., Ltd., Shanghai 201900, China;
3. Yantai Baosteel Pipe Co., Ltd., Yantai 265500, China)

Abstract: In conventional producing process of hot rolled seamless steel tubes, steel pipe is air cooled to room temperature after hot forming. The qualified rate of as-rolled properties is low for lack of effective on-line microstructure and properties control measures. The unqualified product has to be subjected to off-line heat treatment, which results in increased cost, more energy consumption and longer production period. Controlled cooling of hot-rolled pipe after sizing mill is effective microstructure control method. Technical advantages of controlled cooling on microstructure control for hot rolled seamless steel tubes, research status at home and abroad, and technical development difficulties were introduced in this paper. Moreover, the development and application progresses of China's first control cooling technology and equipment, which possessed the capabilities of precise temperature control and direct quenching, were also presented.

Key words: steel tube, hot rolling, control cooling, direct quenching, microstructure control, mechanical property

耐海水腐蚀不锈钢/碳钢复合带肋钢筋轧制技术

余 伟，蔡庆伍，吴 伟，雷力齐

（北京科技大学工程技术研究院，北京 100083）

摘 要： 海洋工程用的带肋钢筋要求具有耐氯离子腐蚀能力，多采用奥氏体或奥氏体-马氏体类型不锈钢，导致材料成本居高不下。利用有限元方法模拟不锈钢/碳钢复合带肋钢筋的热轧过程，分析覆层在轧制过程尤其是成品孔型的变形规律。覆层采用 2205 双相不锈钢，基材为低合金钢 20MnSi，分析发现钢筋横肋根部的应变最大，覆层在此位置减薄明显。实验室采用焊接和真空处理制坯，1200℃加热，950℃终轧的四道次的轧制试验，制备了直径 16mm 的热轧复合钢筋。复合钢筋的覆层和基材界面有良好的冶金结合，Fe 和 Cr 的扩散层厚度 40μm。复合钢筋纵剖冷弯试验无覆层剥离现象，界面剪切强度达到 317.5MPa，复合钢筋的屈服强度 485MPa，抗拉强度 701MPa，断后延伸率约 37.1%，各项指标达到带肋钢筋 HRB400 级别。该生产工艺的复合带肋钢筋的生产成本较不锈钢降低 50% 以上。

关键词：热轧钢筋，复合，有限元，组织，力学性能

Rolling Technology for Stainless Steel/Carbon Steel Composite Rebar with Seawater Corrosion Resistance

Yu Wei, Cai Qingwu, Wu Wei, Lei Liqi

(Engineering Research Institute, University of Science and Technology, Beijing 100083, China)

Abstract: The ribbed reinforced bar for marine engineering requires the corrosion resistance ability of chloride ion. The austenitic or austenitic martensitic stainless steels are often used to make bar, which leads to high material cost. The finite element method is used to simulate the hot rolling process of stainless steel / carbon steel composite ribbed steel, 2205 duplex stainless steel as cladding, low-alloy steel 20MnSi as the substrate. It is found that the maximum strain exists at roots of transverse ribs of rebar, clad layer thinning significantly at the same place in finished pass. The hot rolled composite steel rebars with diameter 16mm were prepared by welding and vacuum treatment for billets, heating at 1200 degrees and four times final rolling at 950 degrees. The interface of cladding layer and substrate of the composite rebar has good metallurgical bonding and Fe and Cr element diffusion layer thickness of 40μm. There is no delamination in the cold bending test of the steel bars after longitudinal cutting. The interfacial shear strength reaches 317.5MPa. The composite steel yield strength is 485MPa, tensile strength 701MPa, elongation of about 37.1%. The index reaches requirement of the grade HRB400 in GB standard. The production cost of the composite ribbed steel bar of the production process is reduced by more than 50% than that of the stainless steel.

Key words: hot rolled bar, composite, finite element, microstructure, mechanical property

抗 H₂S 腐蚀无缝管线管的研制

赵　波[1,2]，王长顺[1,2]，解德刚[1,2]，陈克东[1,2]

（1. 海洋装备用金属材料及其应用国家重点实验室，辽宁鞍山　114009；
2. 鞍钢集团钢铁研究院，辽宁鞍山　114009）

摘　要： 依据美国石油协会 API Spec 5L《管线钢管规范》标准，通过合金成分设计、冶炼轧制工艺优化、热处理制度筛选等手段，研制了力学性能及抗 H₂S 腐蚀性能优良的无缝管线管。研究发现，该管线管热轧态组织晶粒细小均匀，珠光体片层间距较小，性能满足 X52 钢级管线管要求，经过 900℃淬火+600℃回火的调质工艺后，其综合力学性能最优，可用于 X65 钢级管线管使用。

关键词： 抗 H₂S 腐蚀，热处理，无缝管，管线管

Research of Anti-H₂S Corrosion Seamless Line Pipe

Zhao Bo[1,2], Wang Changshun[1,2], Xie Degang[1,2], Chen Kedong[1,2]

(1. State Key Laboratory of Metal Material for Marine Equipment and Application, Anshan 114009, China；
2. Iron & Steel Research Institute of Angang Group, Anshan 114009, China)

Abstract: According to API Spec 5L, the seamless line pipe with fine mechanical properties and H₂S corrosion resistance was researched through designing chemical composition, optimizing smelting and rolling processes, and screening heat treatment processes, etc. The results show that this hot rolled seamless line pipe has the uniform and fine crystal and small pearlite interlamellar spacing, and the mechanical properties meet X52 requirements. After 900℃ quenching and 600℃ tempering, the pipe performed better mechanical properties, and it can be used for X65 line pipe.

Key words: anti-H$_2$S corrosion, heat treatment, seamless, line pipe

钛合金无缝管热轧形变规律及试制

张　冰，孙　宇，赵苏娟，刘家泳，杨永昌

（天津钢管集团股份有限公司技术中心，天津　300301）

摘　要： 钛合金具有高强度、低密度、机械性能和韧性良好、抗腐蚀性能优异等特点，并广泛应用于航空航天领域，其新型的应用目标市场可用于含硫天然气井、地热海水钻井及其它抗腐蚀用管、钻井管道，为跟进国内外钛合金无缝管生产与研发的步伐，开拓无缝钢管的钛合金市场，天津钢管集团对高强度钛合金无缝管的轧制工艺设计及成型试验进行了研究，根据钛合金的材料特性选用典型材料进行热模拟试验，根据合金钢的轧制经验制定了钛合金无缝管的试制工艺以对轧制特性进行研究。通过产品试制及评价，成功的利用试验轧机完成了钛合金无缝管的穿孔与轧制。通过热处理制度的研究，得到了高强韧性的钛合金无缝管，为钛合金无缝管产品的进一步发展提供了理论及试制参考。

关键词： 钛合金无缝管，高强韧性，热模拟，形变规律

Titanium Seamless Pipe Rolling Design and Experimental Study

Zhang Bing, Sun Yu, Zhao Sujuan, Liu Jiayong, Yang Yongchang

(Tianjin Pipe Croup Co., Ltd., Technical Center, Tianjin 300301, China)

Abstract: Titanium alloys with high strength, low density, outstanding comprehensive mechanical properties and excellent corrosion resistance, have been extensively applied in aerospace industry. New application markets are sulfurous natural gas well, geothermal water drilling, anti-corrosion tube, drilling pipeline. In order to keep pace with production and development of titanium alloy seamless pipe at home and abroad, and further develop the market, rolling technology and forming test of high strength titanium alloy seamless pipe were studied by TPCO. According to the material properties of titanium alloy, some typical steel were studied by Gleeble 3500. The rolling process parameters were determined according to experience and the rolling characteristics of titanium alloy seamless pipe have been studied. Based on the trial process and evaluation test, piercing and rolling of titanium alloy seamless pipe were developed successfully by experimental rolling mill. Anti-corrosion titanium alloy seamless pipe with high strength and high fracture toughness were developed successfully by means of different heat-treatment processes. This provides a reference for further research of theirs.

Key words: titanium seamless pipe, high strength and toughness, thermal simulation, deformation law

钒在高强度钢中的析出动力学研究

胡友红[1]，张　毅[2]，谢　祥[1]，刘立德[3]，李　燚[3]，高长益[3]

（1. 首钢水城钢铁（集团）有限责任公司炼钢厂，贵州六盘水　553028；

2. 首钢水城钢铁（集团）有限责任公司，贵州六盘水　553028；

3. 首钢水城钢铁（集团）有限责任公司技术中心，贵州六盘水　553028）

摘　要： 为了优化 HRB500 钢的热轧工艺，使钢中的钒微合金化元素充分析出，达到提高钢材强度的目的，采用水钢生产的 $\phi25mm$ HRB500 钒微合金化钢筋，用 Gleeble-1500D 进行了压缩热模拟实验，用透射电镜对实验后的试样进行了微观分析，得出结论：钒氮微合金化钢筋的析出动力学曲线呈"C"形曲线，鼻尖温度随轧制变形量的增加而降低；当轧制变形量为 0.2~0.5 时，鼻尖温度为 850℃左右，在 3s 内便可完成析出过程；为了使加入钢筋中的钒微合金元素最大限度地以第二相质点的形式析出，充分发挥微合金元素的弥散强化与细晶强化作用，在钢筋的轧制过程中最后一道轧制的变形量应大于 0.2，同时轧后加速冷却的开始温度应低于 850℃，在终轧温度至 850℃的温度范围内慢冷。

关键词： HRB500，高强度钢筋，钒微合金化，析出动力学，强化措施

Study on Vanadium Precipitation Dynamics in High Strength Steel

Hu Youhong[1], Zhang Yi[2], Xie Xiang[1], Liu Lide[3], Li Yi[3], Gao Changyi[3]

(1. Steelmaking Plant, Shougang Shuicheng Iron and Steel Group Co., Ltd., Liupanshui 553028, China;

2. Shougang Shuicheng Iron and Steel Group Co., Ltd., Liupanshui 553028, China;

3. Technology Center, Shougang Shuicheng Iron and Steel Group Co., Ltd., Liupanshui 553028, China)

Abstract: In order to optimize hot rolling process of HRB500 to make vanadium microalloying element fully precipitate to improve strength of steel, compressive thermal simulative experiments are done by Gleeble-1500D with $\phi25mm$ HRB500 vanadium microalloying steel bar, samples after the thermal simulative experiments are analyzed by transmission electron microscope (TEM) and the following results are obtained: precipitation dynamics curves of vanadium-nitrogen microalloying steel appear as "C" shape, the nose temperature decreases as rolling reduction increases; when rolling reduction is 0.2~0.5, the nose temperature is 850℃ or so, precipitation finishes within 3 seconds; in order to make vanadium microalloying element in steel fully precipitate in form of second phase particles to make full use of dispersion strengthening and grain refinement strengthening of microalloying element, reduction of the last rolling process should be larger than 0.2, start temperature of accelerated cooling after rolling should be lower than 850℃ and slow cooling is needed from final rolling temperature to 850℃.

Key words: HRB500, high strength steel bar, vanadium microalloying, precipitation dynamics, strengthening measure

百米钢轨在线热处理生产线创新设计

黄　振，朱俊华，卓　见，戴江波

（中冶南方武汉钢铁设计研究院有限公司，湖北武汉　430080）

摘　要：介绍了目前国内外四种钢轨淬火工艺，设计开发百米钢轨在线热处理工艺，及其产品性能。采用新型淬火工艺，实现百米钢轨在线余热淬火，并使热处理钢轨产品质量达到指标要求；实现百米钢轨淬火形变控制；实现不同来料温度钢轨实时淬火控制；实现对卧式钢轨的翻转及精确定位输送功能。创新淬火工艺的自动识别与控制系统。

关键词：百米钢轨，在线，热处理，研发

Innovative Design of Online Heat Treatment Line for 100 - meter Rails

Huang Zhen, Zhu Junhua, Zhuo Jian, Dai Jiangbo

(WISDIR Wuhan Iron & Steel Design & Research Institute Co., Ltd., Wuhan 430080, China)

Abstract: This paper introduces four kinds of rail quenching process at home and abroad, and designs and develops the on-line heat treatment process of 100 meters rail and its product performance.Adopting the new quenching technology, the on-line waste heat quenching of 100 - meter rail is realized, and the quality of the heat treatment rail reaches the requirement of the index. To control the quenching deformation of 100 meters steel rails; Realize the real-time quenching control of the temperature rail of different materials; To realize the overturning and accurate positioning of the horizontal rail transport function. Automatic identification and control system of innovative quenching process.

Key words: 100 - metre steel rails, on – line, heat treatment, research and development

帘线钢加热过程中钛夹杂固溶行为研究

雷家柳[1]，赵栋楠[1]，徐先锋[1]，薛正良[2]

（1. 湖北理工学院材料科学与工程学院冶金工程系，湖北黄石　435003；
2. 武汉科技大学材料与冶金学院冶金工程系，湖北武汉　430081）

摘　要：本文通过热力学理论分析，结合试验钢中钛夹杂的高温扫描显微镜原位观察，探讨了帘线钢铸坯在加热过程中钛夹杂的固溶行为。结论如下：（1）根据 Ostwald 熟化规律碳氮化钛夹杂的粗化速率很小，加热保温阶段可以不考虑碳氮化钛夹杂的粗化行为。（2）钛夹杂在加热过程中存在较为明显的固溶行为。（3）通过适当控制轧钢加热炉温度和保温时间以及随后的冷却速率，有效控制高级别帘线钢中大颗粒钛夹杂的尺寸和数量，具有一定的可行性。

关键词：帘线钢，钛夹杂，CLSM，固溶

Research on the Solid Solution Behavior of Titanium Inclusion for the Tire Cord Steel during Heating Process

Lei Jialiu[1], Zhao Dongnan[1], Xu Xianfeng[1], Xue Zhengliang[2]

(1. School of Materials Science and Engineering, Hubei Polytechnic University Metallurgy Engineering Department, Huangshi 435003, China; 2. School of Material Science and Metallurgy, Wuhan University of Science and Technology Metallurgy Engineering Department, Wuhan 430081, China)

Abstract: Through the thermodynamic theory analysis, combined with the experiments of confocal laser scanning microscopy in situ observation, the solid solution behavior of titanium inclusion in tire cord steel slab during the heating process was discussed in this paper. The conclusions are as follows: (1)According to the law of Ostwald repening, the coarsening rate of titanium carbonitride inclusion is small, the coarsening behavior of titanium carbonitride inclusion canbe not considered in the heating and holding stage. (2)The solid solution behavior of titanium inclusion in the heating process is obviously existed. (3)Through the proper control of rolling furnace temperature, holding time and the subsequent cooling rate, It has certain feasibility to effectively control the size and quantity of large particle titanium inclusion in the high grade tire cord steel.

Key words: tire cord steel, titanium inclusion, CLSM, solid solution

低成本高强度螺纹钢生产工艺研究

周庆子，范鼎东，蒲雪峰，杭国龙，汪金林

（安徽工业大学冶金工程学院，安徽马鞍山　243032）

摘　要： 本课题联动转炉炼钢和轧后穿水冷却工艺，在少加或不加微合金元素 Nb、V、Ti 和控制 Mn 质量分数为 0.8~1.2 左右的条件下，通过喂碳线稳定提高碳质量分数、现场试验测得 20MnSiV 钢动态 CCT 曲线，提出合理钢种设计，调整开轧温度、利用 Marc 有限元软件对钢筋轧后冷却过程中钢筋截面温度变化进行数值模拟，优化轧后穿水冷却工艺，使III级热轧带肋钢筋边部和芯部获得良好性能组织，过渡层获得高强度组织，在保证棒材塑性指标的条件下提高棒材的强度，从而达降低合金成本提高钢筋综合性能的目的。

关键词： 低成本，HRB400 螺纹钢，穿水冷却，数值模拟，机械性能

Investigation of Production Process for Low Cost and High Strength Rebars

Zhou Qingzi, Fan Dingdong, Pu Xuefeng, Hang Guolong, Wang Jinlin

(School of School of Metallurgical Engineering, Anhui University of technology, Maanshan 243032, China)

Abstract: The linkage of converter steelmaking and rolling through water cooling process, without micro alloying elements Nb, V in less or Ti and control the mass fraction of Mn is about 0.8~1.2, the mass fraction of carbon increased by feeding carbon steel line stability, 20MnSiV dynamic CCT curve test, put forward the reasonable design of steel adjust the open, rolling temperature, numerical simulation of the change of temperature during the cooling section of reinforced steel after rolling by the finite element software Marc, water cooling after rolling process optimization, the hot-rolled ribbed steel bar edge and core has good performance, the transition layer to acquire the high strength organization, in order to increase the strength of steel bar bar plasticity index under the condition of lower cost and improve the comprehensive performance of the alloy steel.

Key words: low cost, HRB400 bar, water cooling, numerical simulation, mechanical propertiest

贝氏体型非调质钢大方坯加热横裂
分析及工艺改善

郝彦英，刘献达，白素宏

（河钢集团石钢公司，河北石家庄 050031）

摘 要：针对河钢集团石钢公司贝氏体型非调质钢大方坯轧制过程后形成"巢穴"状孔洞缺陷，结合贝氏体型非调质钢特点，分析了周向不同侧面开裂孔洞形貌、样品基体组织及连铸坯加热过程，认为加热速度过快及温度梯度过大产生的热应力、组织应力使连铸坯发生横裂造成轧制过程后孔洞缺陷。通过降低预热段加热温度，延长加热时间、降低温度梯度，有效遏制了孔洞缺陷的产生。

关键词：贝氏体型非调质钢，加热，横裂，热应力，组织应力，工艺措施

Analysis and Process Improvement for Heating Transverse Crack on Billet of Non-quenched and Tempered Bainite Steel

Hao Yanying, Liu Xianda, Bai Suhong

(Technique Center，Shijiazhuang Iron and Steel Company, Hebei Iron and Steel Group, Shijiazhuang 050031, China)

Abstract: Aiming at the "nest" shaped hole defects formed during the rolling of bainite quenched and tempered steel bloom of Hesteel Shisteel, and in combination with the characteristics of bainite quenched and tempered steel, the analysis of circumferential side hole cracking morphology, matrix microstructure of the sample and bloom reheating process are analyzed, it is considered that the thermal stress, structure stress due to too fast rolling speed and too large temperature gradient results in transverse crack of bloom and consequently the hole type defects during the rolling. The formation of hole defect is effectively suppressed by reducing the heating temperature of the preheating zone, prolonging the reheating time and decreasing the temperature gradient.

Key words: non-quenched and tempered bainite steel, heating, transverse crack, thermal stress, organizational stress, process improvement

连续共轭孔型在工字钢开发中的设计与应用

陶功明，朱华林，吴郭贤，陈 潇

（攀钢钒轨梁厂，四川攀枝花 617062）

摘 要：型钢产品开发中，共轭孔型是解决轧机辊身长度不足，或小轧机轧制大型材的一种孔型设计方法。共轭孔型设计较为复杂，三对连续共轭孔型设计则难度更大。某一中小型钢厂要利用 150 方坯开发 16 号工字钢，其辊身长度、电机功率均不满足常规孔型设计的要求。本文介绍了如何设计三对连续共轭孔型，如何采用新的共轭孔型配

置方法，在保证出钢平直度及轧制设备安全的情况下，一次性开发成功的过程，对小轧机轧制大型材具有一定的借鉴意义。

关键词：工字钢，连续共轭孔型，孔型配置，延伸系数，开发及应用

Design and Application of Continuous Conjugate Pass in the I-steel Development

Tao Gongming, Zhu Hualin, Wu Guoxian, Chen Xiao

(Rail & Beam Plant of Panzhihua Steel & Vanadium Co.,Ltd, Panzhihua 617062, China)

Abstract: In development of section steel , the conjugate pass is a method of   address the barrel length, or the small mill rolling middle-section steel.Conjugate groove design is relatively complex, three of continuous conjugate groove design is more difficult. A medium-sized mills to use 150 billet development No.16 beams, the roll barrel length, motor power does not meet the requirements of conventional groove design.This article introduced how to design the three pairs of continuous conjugate pass, how to adopt new methods of conjugate pass configuration for the safety of rolling equipment and the straightness of from mill , which is developed for small rolling mill rolling middle-section steel has a certain reference significance.

Key words: I-steel, continuous conjugate pass, pass elongation coefficient, pass configuration, application & development

V-N 微合金热轧 H 型钢形变奥氏体再结晶规律研究

邢 军[1]，吴保桥[1]，程 鼎[1]，潘红波[2]，许文喜[3]，夏 勐[1]

（1. 马鞍山钢铁股份有限公司技术中心，安徽马鞍山　243000；2. 安徽工业大学工程研究院，安徽马鞍山　243002；3. 马鞍山钢铁股份有限公司特钢公司，安徽马鞍山　243000）

摘　要： 通过热力模拟试验及光学显微组织分析，研究了 V-N 微合金化热轧 H 型钢形变奥氏体再结晶规律，分析了变形温度、变形量等工艺参数对形变奥氏体再结晶百分数的影响，绘制了实验钢变形奥氏体再结晶图。结果表明，在变形量 32%、轧制温度 1100℃和变形量 50%、轧制温度 950℃以上时发生完全再结晶；在变形量 14%、轧制温度 1100℃和变形量为 22%、轧制温度 800℃以下时未发生再结晶。V-N 微合金化热轧 H 型钢控轧工艺应为在 1050℃以上采用多道次轧制，累积变形量达 40%以上；未再结晶区开轧温度应低于 900℃，前几道次变形量应控制在 17%～22%。

关键词：再结晶，形变奥氏体，钒微合金化，热轧 H 型钢

Study on Recrystallization Rule of Deformed Austenite in V-N Micro-alloyed Hot-rolled H-beam Steel

Xing Jun[1], Wu Baoqiao[1], Cheng Ding[1], Pan Hongbo[2], Xu Wenxi[3], Xia Meng[1]

(1. Technology Center, Maanshan Iron and Steel Co., Ltd., Maanshan 243000, China; 2. School of Engineering Research Institute, Anhui University of Technology, Maanshan 243002, China; 3. Specialty steel Company, Ma'anshan Iron and steel Co., Ltd., Maanshan 243000, China)

Abstract: The recrystallization rule of deformed austenite in V-N micro-alloyed Hot-rolled H-beam was investigated by thermal-mechanical simulation experiment and optical microscope. The effects of process parameters, including rolling temperature and reduction, on the recrystallization fraction of deformed austenite were analyzed. The figure of deformed austenite recrystallization of tested steel was drawn. The results indicate that the complete recrystallization and non-recrystallization of tested steel occurs above of the reduction of 32% and the rolling temperature of 1100℃ or the reduction of 50% and under of the reduction of 14% and the rolling temperature of 1100℃ or the reduction of 22% the rolling temperature of 800℃. The controlled rolling technology of V-N micro-alloyed Hot-rolled H-beam should be above 1050℃, the multi-pass accumulative reduction is more than 40%; while the rolling temperature in non-recrystallization region should be below 900℃, a few times before deformation ranges from 17% to 22%.

Key words: recrystallization, deformed austenite, vanadium micro-alloyed, Hot-rolled H-beam steel

消除小规格弹簧钢矫直硬化的研究

张永兴，王会庆，刘丽果，谢文发

（河钢石钢轧钢厂，河北 石家庄 050031）

摘　要： 随着国内特钢行业的迅猛发展，以及钢铁市场形势的持续恶化，精整线具有提高、保证产品质量的功能，增加产品附加值，为钢铁行业扭亏增盈发挥着重要作用。矫直机是精整线中关键设备，具有满足用户和后续探伤设备平直度要求的功能。本项目研究了某矫直机各调整参数对小规格弹簧钢矫直硬化层的影响。结果说明：（1）矫直辊辊缝较钢材直径尺寸小于 0.5mm 以上时，开始产生矫直硬化现象；（2）上辊角度＜16°，下辊角度＜12°时，是钢材表面开始出现矫直硬化的临界值；（3）当矫直辊角度不变时，辊缝减小或速度增加加剧硬化，其中速度加剧程度较大。最终满足了国内高端弹簧钢用户的质量需求。

关键词： 精整线，平直度，硬化层，矫直

Research of Avoiding Straightening Hardening on Small Specification Spring Steel

Zhang Yongxing, Wang Huiqing, Liu Liguo, Xie Wenfa

(Hesteel Group Shisteel Company, Shijiazhuang 050031, China)

Abstract: With the fast development of the domestic special steel profession, and the circumstance keep on worsen in the steel market, the finishing line plays important role at increase profits , improve and promise product quality for the steel profession. The straightener is key equipment in finishing line, it can satisfy the requestion of customer and the leakage inspection equipment about the bar straightness. The item researched the influence of straightening hardening layer on small specification spring steel through adjust data about some straightener. As a result, (1) when the roll gap less than 0.5mm compare the diameter of steel, which produce hardening phenomenon; (2) the angel of up roll<16°, the angel of down roll <12°, which is the critical value of straightening harden on the surface of steel; (3) when the angel of straightener roll keeps constantly, the gap lessen or velocity increasing of straightener will deepen the hardening, especially velocity increased. Finally it is satisfied domestic customer's quality need on the high level spring steel.

Key words: finishing line, straightness, hardening layer, straightening

ER70S-6 热轧盘条氧化铁皮结构分析及工艺优化

朱传清，余坤洋，沈克非，周　锋

（日钢钢铁控股集团有限公司长材制造部，山东日照　276806）

摘　要：针对 ER70S-6 焊丝钢盘条氧化铁皮去除不净，拉拔丝表面发黑且模具损耗大问题，日钢对收集到的不同样品进行折弯实验和扫描电镜检测，经分析对比盘条表面氧化层较薄、结构异常是造成氧化铁皮去除不净的主要原因。通过优化钢坯加热工艺及控冷工艺，盘条表面氧化层厚度及结构得到明显改善，去除不净问题得到解决。

关键词：ER70S-6 盘条，氧化铁皮，工艺优化

热轧钢筋新标准下棒材轧机改造工艺技术分析

程知松，余　伟

（北京科技大学高效轧制国家工程研究中心，北京　100083）

摘　要：热轧带肋钢筋是我国产量最大的钢材产品，而提高钢筋产品质量并降低生产成本一直是技术创新的动力。为了适应我国即将出台的热轧带肋钢筋新标准，对于高强度钢筋轧制工艺和装备上如何适应将来的新标准要求，进行了控轧控冷工艺，以及对现有的棒材轧机进行改造的工艺进行了详细的分析，比较了切分轧制和高速轧制两种工艺的优缺点，提出了工艺布置新思路。新方法为新钢筋生产线工艺设计，老线改造设计，或者钢铁企业工程技术人员开发品种提供了独特的途径。

关键词：标准，钢筋，质量控制，轧机改造，工艺技术

Technique Analyses of Bar Mill Upgrading for Future New Hot Rolled Rebar Standard

Cheng Zhisong， Yu Wei

(NERCAR of University of Science and Technology Beijing, Beijing 100083, China)

Abstract: Hot rolled ribbed bar is the largest quantity steel products in our country. But the quality improvement and cost reduction of rebar production is always the motivation of technology innovation. In order to meet the new rebar standard which is going to come on in the near future by our country, the controlled rolling and controlled cooling technique for high strength rebar is discussed, the processes for upgrading existing bar mills are analyzed in details, the advantages and disadvantages for slit rolling and high speed rolling process are compared. The new idea of mill layout is put forward for controlled rolling and cooling process. The new thought gives an unique proposal for designing new rebar mill line or revamping old product line and developing new products for engineering and technical personnel in enterprise.

Key words: standard, rebar, quality control, mill upgrading, technique

环保型免酸洗 82B 盘条氧化铁皮控制研究

王海宾[1,2]，王宏斌[2]，李　娜[2]，刘雅政[1]，李　杰[3]，刘　毅[3]

（1. 北京科技大学，北京　100083；2. 河钢集团宣钢公司，河北宣化　075100；
3. 河钢集团钢研总院，河北石家庄　050000）

摘　要：本文结合河钢集团宣钢公司高速线材生产线装备与控制实践，研究了环保型免酸洗 82B 盘条氧化铁皮厚度与结构的控制。结果表明：吐丝温度是影响 82B 盘条表面氧化铁皮机械去除性能的主要参数，通过将吐丝温度由 900℃ 提高到 950℃，氧化铁皮总厚度由 6.26μm 增加到 12.54μm，Fe_3O_4 比例由 51.92% 提高到 70.98%，机械去除表面氧化铁皮效果较好。

关键词：82B，免酸洗，氧化铁皮，高速线材

Tin Oxide Practice of Environmental and No Picking 82B Wire Rod

Wang Haibin[1,2], Wang Hongbin[2], Li Na[2], Liu Yazheng[1], Li Jie[3], Liu Yi[3]

(1. Beijing University of Science and Technology, Beijing 100083, China; 2. HBIS Group Xuan gang Company, Xuanhua 075100, China; 3. HBIS Group Steel Research Institute, Shijiazhuang 050000, China)

Abstract: In this paper, combined with the equipment and control practice of high-speed wire rod production line of Hebei Steel Group Xuan gang Company, the control of the thickness and structure of environmental and no pickling 82B wire rod was studied. The results show that the temperature is the main parameter that affects the mechanical removal performance of 82B wire rod surface. The total thickness of the oxide scale is increased from 6.26μm to 12.54μm and the ratio of Fe_3O_4 is improved from 51.92 % to 70.98%, and the effect of surface oxide scale mechanical removal is better by raising the temperature from 900℃ to 950℃.

Key words: 82B, no pickling, tin oxide, high-speed wire

电渣重熔法制备双金属复合轧辊的工艺研究

曹玉龙[1]，姜周华[1]，董艳伍[1]，邓　鑫[2]，李传峥[1]，侯志文[1]

（1. 东北大学冶金学院，辽宁沈阳　110819；
2. 辽宁科技大学材料与冶金学院，辽宁鞍山　114051）

摘　要：本文基于 Fluent 软件及其自定义标量方程（UDS）、自定义函数功能（UDF）建立了电渣重熔法制备双金属复合轧辊过程的二维准稳态数学模型，研究分析了自耗电极-导电结晶器供电回路方案下的复合轧辊熔炼制备过程电磁场、流场及温度场的基本特征，同时，利用先进的导电结晶器技术开展了相应的实验室制备试验。结果表明：模拟结果得到了试验的良好验证，采用电极-导电结晶器供电回路方案因能有效调整渣池高温区位置、获得理想的

双金属界面温度而适合于冶炼双金属复合轧辊，其冶炼制备的双金属复合轧辊具有界面结合均匀、无明显缺陷等优点，在众多双金属复合轧辊制备工艺中具有一定的竞争力。

关键词：电渣重熔，复合轧辊，数值模拟，温度场，导电结晶器

Research on Electroslag Remelting Technology for Producing Bimetallic Composite Roll

Cao Yulong[1], Jiang Zhouhua[1], Dong Yanwu[1], Deng Xin[2],
Li Chuanzheng[1], Hou Zhiwen[1]

(1. School of Metallurgy, Northeastern University, Shenyang 110819, China;
2. School of Materials and Metallurgy, University of Science and Technology Liaoning, Anshan 114051, China)

Abstract: A 2D quasi-steady state mathematical model of the electroslag remelting technology for producing bimetallic composite roll was developed based on the Fluent software and the UDS and UDF function. Characteristics of the electromagnetic field, flow field and temperature field of the bimetallic composite roll system during the manufacturing process under the electrode-conductive mold power supply circuit program have been numerically simulated. At the same time, the laboratory scale experiments have also been developed by the advanced conductive mold technology. The results indicate that: the simulation results were well verified by the experiment, it is beneficial to improving the current density and Joule heat distributions of slag pool and keeping the high temperature zone away from the roll core surface with the using of conductive mold. It is beneficial to reducing the heat transfer from the slag bath to the roll core surface which can help us obtain a suitable temperature for the bimetallic bonding. Though the experiments, a composite roll with a uniform bonding interface and no obvious defects in the bonding interface has been manufactured. So, the electroslag remelting technology has a certain competitiveness among the lots of process technologies for producing the bimetallic composite rolls.

Key words: electroslag remelting, bimetallic composite roll, numerical simulation, temperature field, conductive mold

非对称射流 MILD 氧燃烧特性的研究

伍永福[1]，武殿斌[1]，刘中兴[2]，董云芳[2]

（1. 内蒙古科技大学能源与环境学院，内蒙古包头　014010；
2. 内蒙古自治区白云鄂博矿多金属资源综合利用重点实验室，内蒙古包头　014010）

摘　要：本文通过研究不同燃烧器结构，设计了一种非对称射流燃烧器，并通过实验研究了丙烷-氧气 MILD 氧燃烧在炉膛内的火焰现象、温度场和浓度场。结果表明从喇叭口射出的高速助燃气体可以将对称侧喷出的可燃气体迅速打散成诸多小体积的微团，气体流量越大，这种"打散"的效果越理想，温度波动比值越低，燃烧状态越来越接近于 MILD 燃烧。

关键词：MILD 燃烧，非对称射流，实验，低污染

Study on Oxygen Combustion Characteristics of Asymmetric Jet MILD

Wu Yongfu[1], Wu Dianbin[1], Liu Zhongxing[2], Dong Yunfang[2]

(1. School of Energy and Environment, Inner Mongolia University of Science and Technology, Baotou 014010, China;

2. Key Laboratory of Integrated Exploitation of Bayan Obo Multi-Metal Resources, IMUST, Baotou 014010, China，

Abstract: In this paper, an asymmetric jet burner was designed by studying the structure of different burners. The flame phenomena, temperature field and concentration field of propane oxygen MILD oxygen combustion in the furnace were studied experimentally. The results show that the high-speed combustion gas can be emitted from bellbottom combustible gas ejected quickly broken down into a symmetric side many small micelle volume, gas flow rate increases, the "scatter" effect is more ideal, the temperature fluctuation ratio is low, the combustion state is getting more and more close to the MILD combustion.

Key words: MILD combustion, asymmetric jet, experiment, low pollution

桥梁缆索钢丝用线材的研发

蒋跃东[1]，涂益友[2]，任安超[1]，鲁修宇[1]，夏艳花[1]，张　帆[1]

（1. 武汉钢铁有限公司研究院条材所，湖北武汉　430080；

2. 东南大学材料科学与工程学院，江苏南京　211189）

摘　要：介绍了武钢桥索钢线材 82MnQS、87MnQS 的工艺及质量难点，在研发过程中出现了镀锌钢丝扭转次数波动的情况，通过扫描电镜、透射电镜对扭转断口的分析，发现断口处的 MnS 夹杂和渗碳体片层球化是主要原因，相对应的措施是：（1）控制钢中的硫含量，达到控制硫化物夹杂物的目的。（2）连铸生产大断面的方坯轧制成 200mm×200mm 的钢坯时，优化轧制工艺技术措施。（3）优化线材的控冷工艺。试验结果表明，提高了线材的纯净度、索氏体含量和晶粒度级别，有利于改善镀锌钢丝的扭转性能。

关键词：桥索钢，硫含量，轧制，扭转

Research and Development of Wire Rope for Bridge Cable

Jiang Yuedong[1], Tu Yiyou[2], Ren Anchao[1], Lu Xiuyu[1],

Xia Yanhua[1], Zhang Fan[1]

(1. Research and Development Center，Wuhan Iron and Steel Corporation, Wuhan 430080, China;

2. College of Materials Science and Engineering, Southeast University, Nanjing 211189, China)

Abstract: Introduces the process and quality problems of WISCO bridge cable steel wire rod 82MnQS, 87MnQS, the galvanized steel wire twist number fluctuations occur in the development process, through transmission electron microscopy and scanning electron microscopy analysis of torsional fracture, MnS inclusions and cementite lamellar spheroidization fracture is the main reason, the corresponding measures are: (1) the control of sulfur content in steel and to control the purpose of sulfide inclusions. (2) optimizing the rolling technological measures when the large section billet is rolled into 200mm ×200mm billet. (3) optimize the control cooling process of wire rod. The test results show that the purity, the content of ferrite and the grain size of wire rod are improved, and the torsion property of galvanized steel wire is improved.

Key words: bridge cable rope wire, sulfur content, rolling, retortion

重轨钢坯的加热过程中脱碳层深度的有效控制途径

王晓晖，徐学良，吴保华

（河钢集团邯钢公司技术中心，河北邯郸　056015）

摘　要： 对加热炉内钢坯的加热过程在实验室进行了模拟实验，研究了钢材的脱碳与加热温度及加热时间的关系，分析了防氧化脱碳保护涂料对氧化脱碳的影响，提出防氧化脱碳保护涂料的使用是重轨钢坯加热过程中控制脱碳层深度的有效途径。

关键词： 重轨钢坯，有效控制途径，脱碳，涂层

Effective Way Control of Decreasing Decarbonization Percentage during Bloom for Heavy Rail Heating

Wang Xiaohui, Xu Xueliang, Wu Baohua

(Hesteel Group Hansteel Company，Technology Center, Handan 056015, China)

Abstract: Within the fumace and billet heating process was simulated in the aboratory experiments. Studied the relationship between the decarburization of steel and the heating temperature and time. Through the analysis of experimental results，protective coating on the impact of decarburization, the use of coatings is the effective way control of decreasing decarbonization percentage during bloom for heavy rail heating.

Key words: bloom for heavy rail, effective control of, decarbonization, coating

圆环链用钢 CG80 盘圆的开发实践

贾元海，张春雷，邓博赜

（河钢集团承钢公司钒钛工程技术研究中心，河北承德　067002）

摘　要： 介绍了以 CG80 为代表的链条钢高速线材的开发情况。通过制定合理的冶炼工艺、轧制温度制度及控制冷却制度，开发出了圆环链用钢 CG80 盘圆，并介绍了此种产品在客户使用过程中的质量情况及市场前景。

关键词： 圆环链，CG80，开发，实践

Development of CG80 Coil of Steel Ring Chain

Jia Yuanhai, Zhang Chunlei, Deng Boze

(Vanadium–Titanium Engineering Technique Research Center, Chengde Iron and Steel Company,
Hebei Iron and Steel Group, Chengde 067002, China)

Abstract: Introduced by CG80 chain development situation of high speed wire rod steel.By making reasonable smelting process, rolling temperature system and control cooling system, Developed a ring chain steel CG80 plates, And introduces the quality of this product is in the process of customers to use situation and market prospects.

Key words: round link chain, CG80, development, practice

美标系列线材的开发实践

贾元海，张春雷，王　林

（河钢集团承钢公司钒钛工程技术研究中心，河北承德　067002）

摘　要： 研究 C、Si、Mn 元素对钢材性能的影响，制定严格的内控成分范围，保证了性能的稳定性；利用全程保护浇注、LF 精炼、结晶器电磁搅拌（M-EMS）、稳定拉速等先进控制手段进行铸坯生产，降低了钢中气体含量、夹杂物，提高了铸坯低倍质量；采用控轧控冷技术，保证了产品的性能及组织均匀。

关键词： 美标系列，线材，开发实践，控轧控冷

The Development Practice of American Standard Series Wire

Jia Yuanhai, Zhang Chunlei, Wang Lin

(Vanadium–Titanium Engineering Technique Research Center, Chengde Iron and Steel Company,
Hebei Iron and Steel Group, Chengde 067002, China)

Abstract: The influence of C, Si and Mn elements on the properties of steel is studied, and strict internal control components are developed to ensure the stability of the performance. Used to protect all the pouring, LF refining, mold electromagnetic stirring (M - EMS), advanced control means such as stable speed for casting production, reduce the gas content, the inclusions in steel, improve the quality of the slab at low; The control cold technology is adopted to ensure the product's performance and organization.

Key words: american standard series, wire, development practices, the controlled rolling control cold

弹簧钢 60Si2MnA 盘条全脱碳层试验研究

陈志亮，韩　健，甄　玉，赵　阳，仝太钦

（中天钢铁集团公司技术中心，江苏常州　213000）

摘　要：对轧钢厂弹簧钢线材 60Si2MnA 表面全脱碳层进行研究，确定加热炉内气氛和加热温度对全脱碳层的影响及其形成机理。轧钢厂加热炉内钢坯加热时间较长为 2.0~2.5h，轧制高硅 60Si2MnA 盘条时其表面脱碳层较深且均有明显全脱碳层。调整加热炉加热温度和炉内气氛来控制全脱碳层，试验结果表明，适当增加炉内弱氧化气氛时间使钢坯在加热炉内表面氧化速度大于脱碳速度，表面全脱碳层可以逐步消失，加热炉内高温段 1000~1100℃采用弱氧化气氛 20~30min 时，开轧温度 970~1000℃，60Si2MnA 表面全脱碳层深度降为零，且总脱碳层平均深度降至 0.5%D。综合调节加热温度和气氛很好地解决了 60Si2MnA 表面全脱碳层问题，提高弹簧疲劳使用寿命。

关键词：弹簧钢，完全脱碳层，加热温度，弱氧化气氛

Experimental Study on Fully Decarburization Layer of Spring Steel 60Si2MnA Wire Rod

Chen Zhiliang, Han Jian, Zhen Yu, Zhao Yang, Tong Taiqin

(Technical Center Department of ZENITH Iron and Steel Group, Changzhou 213000, China)

Abstract: The influence of the atmosphere and the heating temperature in the reheating furnace on the total decarburization layer and its formation mechanism were studied in the spring steel wire rod 60Si2MnA of the rolling mill. The reheating time of steel billet in reheating furnace is longer than 2.0~2.5h. When rolling high silicon 60Si2MnA wire rod, its surface decarburization layer is deeper and all obvious decarburization layer exists. Adjust the gas furnace heating temperature and furnace to control decarburization layer, test results show that the appropriate increase in the furnace weak oxidizing atmosphere to billet in a heating furnace and surface oxidation rate is greater than the decarburization rate, surface decarburization layer can gradually disappear, the furnace temperature of 1000~1100℃ in the weak oxidizing atmosphere 20~30min, open rolling temperature of 970~1000℃, 60Si2MnA surface decarburization depth is reduced to zero, and the average depth of decarburized layer to 0.5%D. By adjusting the heating temperature and atmosphere comprehensively, the total decarburization layer on the surface of 60Si2MnA is solved, and the service life of the spring is improved.

Key words: spring steel, completely decarburization layer, heating temperature, weak oxidizing atmosphere

盘条离线盐浴索氏体化处理（QWTP）工业化生产实践

刘　澄，李　阳，朱　帅，孙　理

（青岛特殊钢铁有限公司中特研究院青钢分院，山东青岛　266409）

摘　要：本文介绍了盘条离线盐浴索氏体化处理（QWTP）工业化生产的工艺、产品性能和应用效果，以等温盐浴为淬火介质进行索氏体化处理，可以实现盘条近似等温相变。斯太尔摩线风冷盘条 QS87Mn 经离线盐浴索氏体化处理后，索氏体化率由89%提高至95%，片层间距由102.5nm 减小至78.8nm，消除了网状碳化物等不良组织，且显微组织更加均匀；盘条的抗拉强度由平均1280MPa 提高至1350MPa，波动范围由±40 MPa 降低至±15 MPa。以盐浴处理 PQS87Mn 盘条为原料制造的 Φ5mm-1960MPa 锌铝钢丝完全满足强度与扭转指标，已应用于虎门二桥工程。

关键词：离线盐浴，索氏体，工业化，桥梁缆索，环保

Industrial Production of the Off-line Salt Bath Patenting (QWTP) of Wire Rods

Liu Cheng, Li Yang, Zhu Shuai, Sun Li

(Qingdao Branch of Research Institute of CITIC Special Steel Group, Qingdao Special Iron and Steel Co., Ltd., Qingdao, 266409, China)

Abstract: The technology and product performance and application of the industrialized off-line salt bath patenting of steel wire rods (QWTP) are introduced in the paper. Isothermal phase transformation could be realized approximatively as the wire rod is quenched into the isothermal salt bath. Compared to Stelmor air-cooled QS87Mn wire rod, the sorbitic rate of QWTP product is increased from 89% to 95% while the lamellar space is decreased from 102.5 nm to 78.8 nm. Moreover, harmful microstructure such as network carbide is eliminated and the homogeneity of microstructure is increased in the QWTP product. As a result, the average tensile strength of QWTP QS87Mn wire rods is increased to 1350 MPa with a fluctuation of ± 15 MPa from 1280 MPa with a fluctuation of ± 40 MPa. In addition, 1960MPa grade hot dip galvanized steel wire made from the QWTP QS87Mn wire rods has been applied in Humen bridge engineering.

Key words: off-line salt bath patenting, sorbite, industrial production, bridge cable, environmental

环保型高强度桥梁缆索用盘条工业化热处理工艺的研发

刘 澄，朱 帅，李 阳，孙 理

（中信特钢研究院青钢分院，山东青岛 266409）

摘 要： 本文对比了传统铅浴工艺和创新环保型盐浴工艺对高端桥梁缆索盘条组织性能的影响，通过两种工艺对热轧盘条进行工业化处理和组织性能分析，结果显示采用盐浴处理后 87 级盘条索氏体化率从 89% 提高至 95%，索氏体片层间距从 105±5nm 降低至 75±5nm，索氏体片层更加平直，珠光体球团变得更加均匀，抗拉强度和面缩率分别提高 70MPa 和 4%，达到 1350MPa 和 40%。采用盐浴工艺处理的盘条可用于生产 1960MPa 级桥梁缆索用镀锌钢丝，钢丝强度达到 2000MPa 以上，扭转性能均值在 20 次左右，目前青岛特钢盐浴盘条已经向 1960MPa 强度级别桥梁项目批量供货。

关键词： 桥梁缆索，盐浴热处理，镀锌钢丝，扭转性能，抗拉强度，索氏体

An Environmental Protection Heat Treatment Process for Industrial Production on High-strength Bridge Cable Wire

Liu Cheng, Zhu Shuai, Li Yang, Sun Li

(Qingdao Steel Branch of Citic Special Steel Insititute, Qingdao 266409, China)

Abstract: Influence on high-end cable wire microstructure and mechanical properties of conditional lead patenting process and an innovation & environmental protection salt patenting process was contracted in this paper. The analysis result of salt patenting wire show that, sorbite rate was increased to 95% from 89% while lamellar spacing of sorbite decreased to 75±5nm from 105±5nm, sorbite lamellar was more straight and grain size was evidently refined and homogeneous. Tensile strength and area reduction was improved by 70MPa and 4%, to 1350MPa and 40% respectively. Salt patenting treated wire was used to made 1960MPa strength grade bridge cable galvanized wire, of which the strength and twisting number were above 2000MPa and around 20 times, now the salt patenting cable wire was used on 1960MPa grade bridge project.

Key words: cable wire, salt patenting, galvanized wire, twisting property, tensile strength, sorbite

线材中魏氏组织影响因素分析

马洪磊，刘　维，宋召勇，巩延杰

（邢台钢铁有限责任公司，河北邢台　054000）

摘　要： 对线材中的过冷度、扩散转变、新旧相的界面形式、惯习现象等影响魏氏组织等机理原因进行了分析。初步总结出产生魏氏组织的根本原因在于冷速过快导致的低温转变，进而导致扩散能力不足以及惯习现象的原因最终导致产生魏氏组织。同时并对生产现场中影响魏氏组织的吐丝温度、辊道速度、规格因素、头尾因素、位置因素等原因进行了调查分析。

关键词： 魏氏组织，过冷度，惯习现象，现场因素

Analysis of Influencing Factors of Widmanstatten Structure in Wire

Ma Honglei, Liu Wei, Song Zhaoyong, Gong Yanjie

(Xingtai Iron and steel Co., Ltd., Xingtai 054000, China)

Abstract: This paper analyzes the influence of the factors such as overcooling,diffusion transformation ,habit phenomenon，interface form of old and new phase, etc. The basic reason for the development of　Widmannstatten structure is that the low temperature change caused by the excessively fast speed leads to the lack of diffusion ability and the reason of the habit phenomenon which eventually leads to the production of Widmannstatten structure. At the same time, the paper analyses the factors in production field such as the temperature of spinning, the speed of the roller, the factors, the factors of the head, the position factors and so on.

Key words: Widmanstatten structure, overcooling, habit phenomenon, factors in production field

吐丝温度对 60 钢盘条组织和拉拔性能的影响研究

王文锋，王卫东，刘红兵，张卫强

（武钢集团昆明钢铁股份有限公司技术中心，云南昆明　650302）

摘　要： 本文采用了光学显微镜（OM）、拉伸试验机等设备针对 60 钢盘条吐丝温度对其组织性能的影响进行研究。结果表明：吐丝温度在 840～860℃范围内盘条组织及性能达到最优匹配，有利于保证 60 钢盘条良好的拉拔性能；

过高或过低的吐丝温度均对盘条拉拔性能造成不良影响。

关键词：60 钢盘条，吐丝温度，组织，拉拔性能

Research on the Effect of Laying Temperature on Microscopic Structure and Drawing Property of 60 Steel Wire Coil

Wang Wenfeng, Wang Weidong, Liu Hongbing, Zhang Weiqiang

(Technology Center of Wukun Steel Co., Ltd., Kunming 650302, China)

Abstract: The effect of laying temperature on microscopic structure and properties of 60 steel wire coiling has been studied by using optical microscope and tensile testing machine. Results show that the microscopic structure and the properties of wire coil achieved the optimal matching with the laying temperature in 840~860℃ temperature range. It is beneficial to ensure the good drawing property of 60 steel wire coil. Too high or too low laying temperature has bad influence on drawing property of wire coiling.

Key words: 60 steel wire coil, laying temperature, microscopic structure, drawing property

加热过程中弹簧钢 55SiCr 表面脱碳研究

张　凯[1]，陈银莉[1]，徐志军[2]

（1. 北京科技大学钢铁共性技术协同创新中心，北京　100083；

2. 北京科技大学工程技术研究院，北京　100083）

摘　要：利用真空管式加热炉对弹簧钢铸坯试样进行加热保温，控制炉气成分以及 $H_2O(g)$含量，采用金相显微组织法研究了加热温度、保温时间以及 $H_2O(g)$含量对弹簧钢表面脱碳的影响。结果表明，在混合气氛含有 $H_2O(g)$的条件下，600~950℃范围内总脱碳层随温度升高持续增厚，铁素体脱碳层于 850℃出现峰值；700~950℃时随保温时间延长，脱碳层增厚，脱碳速率衰减；表面总脱碳层、铁素体脱碳层以及氧化层厚度均随 $H_2O(g)$含量的提高而显著增加。控制气氛中 $H_2O(g)$含量对降低脱碳和减少氧化烧损具有重要意义。

关键词：弹簧钢，表面脱碳，水气，加热温度，保温时间，氧化层

Investigation on Surface Decarburization of Spring Steel 55SiCr during the Heating Process

Zhang Kai[1], Chen Yinli[1], Xu Zhijun[2]

(1. Collaborative Innovation Center of Steel Technology,University of Science and Technology Beijing, Beijing 100083, China; 2. Institute of Engineering Technology, University of Science and Technology Beijing, Beijing 100083, China)

Abstract: Controlling components of furnace gas and $H_2O(g)$ content, spring steel samples were heated by a tube vacuum furnace. Effect of heating temperature, holding time and $H_2O(g)$ content on surface decarburization of spring steel 55SiCr were investigated by metallographic method. The results show that in the mixed atmosphere containing H₂O (g) and under the temperature condition of 600~950℃, total decarburization layer increases with the temperature, and the depth of the decarburized ferrite layer at the temperature of 850℃ reaches the maximum. At temperature between 700 and 950℃, with the prolongation of the holding time, the decarburized layer continue to increase and decarburization rate has an attenuation. Total decarburized layer, decarburized ferrite layer and the oxide layer increases with the increasing of the H$_2$O(g) content. Decreasing H$_2$O(g) content in furnace gas is necessary to reduce the decarburization depth layer and the scale loss.

Key words: spring steel, surface decarburization, gaseous water, heating temperature, holding time, oxide layer

关于 50 穿孔机试车的探讨

银建华

（攀钢集团江油长城特殊钢有限公司能动中心，四川江油　621701）

摘　要： 根据 Φ50 穿孔机试车情况，结合在试车过程中采取的各种措施进行分析，探讨 Φ50 穿孔机在下一步试车的思路及应努力的方向。

关键词： 50 穿孔机，热轧参数，试车分析，改进措施

Abstract: According to Φ50 punch test conditions, combined with various measures in the process of test analysis, exploreΦ50 punch and efforts should be made to the next step on the way.

Key words: Φ50 punch, hot rolling parameters, test analysis, improvement measures

连铸方坯热送热装生产实践

谭武祥

（湖南华菱湘潭钢铁有限公司技术质量部，湖南湘潭　411101）

摘　要： 通过炼钢、轧钢现场跟踪分析，对各工序环节实现连铸方坯热送热装的可操作性提出改进意见。结合炼钢厂、轧钢厂的实际情况，主要针对设备改造、运输过程调整、温度控制、库房管理、信息传递等流程提出优化改进意见，通过改进前后能耗、产量、成材率等对比，得出了连铸方坯热送热装应用后的运行效果和经济效益。

关键词： 连铸方坯热送热装，能耗，产量

钢管张力减径过程的有限元模拟及验证

米 楠

（中冶京诚工程技术有限公司，北京　100176）

摘　要： 应用非线性有限元法建立三维热力耦合弹塑性有限元模型，运用该模型对实际张力减径过程进行了数值模

拟，得到了应力场和应变分布规律。通过节点参数，运用数学方法计算得到了各架出口壁厚。模拟得到的轧制力矩与现场实测吻合较好，可以为张减机的设备开发提供设计依据。

关键词：张力减径，壁厚分布，轧制力矩，有限元

Analysis on Stretch Reducing Rolling Process of Tube by FEM and Verify

Mi Nan

(MCC Capital Engineering & Research Incorporation Limited, Beijing 100176, China)

Abstract: A 3D thermal-mechanical coupled elasto-plastic finite element model had been established by using nonlinear finite element method. Actual stretch reducing rolling process was simulated using the model ,strain field and stress field were obtained ,as well as the wall thickness when tube billet passing every stand. Simulated rolling torgue are a good agreement with the field results, the model provides the methods for develop new stretch reducing machine.

Key words: stretch-reducing, wall thickness, rolling torgue, FEM

鞍钢免铅浴线材的开发和应用

尹 一，李建龙，梁 琪，胥 晶，赵学博

（鞍钢股份有限公司线材厂，辽宁鞍山 114042）

摘 要：EDC 高碳钢线材具有比斯太尔摩线材更好的组织性能，鞍钢线材厂通过对 EDC 控冷工艺的研究，在提高线材可拉拔性和组织性能两方面取得进展，实现了取消小规格半成品钢丝拉拔前和部分重要用途成品钢丝拉拔前的铅浴处理，可降低用户生产成本，减少或消除铅浴和随后的酸洗磷化等工序的环境污染。

关键词：EDC 线材，免铅浴，可拉拔性，组织性能，应用

Development and Application of Eliminating Lead Patenting Wire Rod of Ansteel

Yin Yi, Li Jianlong, Liang Qi, Xu Jing, Zhao Xuebo

(Wire Rod Plant of Angang Steel Company Limited, Anshan 114042, China)

Abstract: EDC high carbon wire rod has better microstructure and property than Stelmor wire rod, the wire rod plant of Ansteel performed technique research to the EDC control cooling process, and made progress on improving drawability and microstructure and property of the wire rods, realizes eliminating lead patenting before wire drawing of small size half-finished product wires and partial important uses finished product wires, reduces customer's production cost, reduces or eliminates environmental pollution of lead patenting and followed acid pickling and phosphating process.

Key words: EDC wire rod, eliminating lead patenting, drawability, microstructure and property, application

B-ULR1T 超低电阻导线用钢生产实践

田伟阳，吴东明，荆高扬

（本溪钢铁集团棒线材研究所，　辽宁本溪　117017）

摘　要： 介绍了本钢北营公司 B-ULR1T 超低电阻导线用钢的技术要求和生产试制。转炉采用双渣冶炼，终点 $w(C) \leqslant 0.05\%$；精炼采用 LF-RH 双路径，保证夹杂物的去除及深脱碳效果；连铸采用全程保护浇铸及电磁搅拌技术，拉速恒定控制 2.3m/min；加热炉均热段温度(1120±30)℃，开轧温度(1000±20)℃，吐丝温度(890±20)℃。热轧状态盘条抗拉强度 305~315MPa，伸长率 50%~53%，金相组织为单相 F、晶粒度 5.0~6.0 级。产品质量完全满足用户使用需要。

关键词： B-ULR1T，盘条，金相组织，非金属夹杂物，电导率

Production Practices of B-ULR1T Steel Wire Rod for Ultra-low Resistance Conductor

Tian Weiyang, Wu Dongming, Jing Gaoyang

(Bar & Wire Rod Research Department of Benxi Iron & Steel Group, Benxi 117017, China)

Abstract: The technical requirements and production test of B-ULR1T ultra-low resistance wire of this steel north battalion company are introduced. The converter adopts double slag smelting and the end point $w(C)$ is less than 0.05%; The LF - RH dual path is adopted to ensure the removal of inclusions and the deep decarbonization effect. In continuous casting, the whole process was used to protect the casting and electromagnetic stirring technology, and the speed constant control was 2.3m/min. Soaking zone heating furnace temperature (1120±30)℃, start rolling temperature (1000±20)℃, spinning temperature (890±20)℃. Hot rolling state phi rod tensile strength 305~315MPa, elongation ratio between 50% and 53%, metallographic group is single-phase F, grain size 5.0~6.0. The quality of the product meets the user's needs.

Key words: B-URL1T, wire rod, the microstructure, non-metallic inclusions, electrical conductivity

SD500 高强韩标钢筋开发

张春雷，王晓飞，贾元海，褚文龙

（河钢股份有限公司承德分公司，河北承德　067102）

摘　要： 河钢股份有限公司承德分公司（以下简称河钢承钢）为了应对国内建材市场严峻形势，积极开发韩标钢筋国际市场。在研制开发韩标 SD500 钢筋过程中，分析了 SD500 钢筋的质量特性，结合类似产品生产经历，设计了河钢承钢韩标钢筋的内控化学成分，优化了连铸坯的生产工艺，设计了合理的孔型方案和轧制工艺制度，成功地开发出质量达到 KS D 3504 要求的韩标钢筋产品，并顺利取得了产品认证证书，产品自销售以来无质量问题。

关键词： 韩标钢筋，SD500，生产工艺，机械性能，竹节肋

Development of High - strength South Korean Steel Bar SD500

Zhang Chunlei, Wang Xiaofei, Jia Yuanhai, Chu Wenlong

(HBIS Company Limited Chengde Branch, Chengde 067102, China)

Abstract: HBIS Company Limited Chengde Branch(simplified HeSteel Chengsteel)actively develop the international market for south Korean steel to deal with the severe situation of the domestic building materials market.Within the process ,refer to the similiar product procedure, Chengsteel make the internal control chemical composition , optimize the continuous casting billet productive technology and design the reasonable pass scheme and rolling process system. Finally sucessfully produce the steel bar which can reach the quality standard KS D 3504 and get the CRCC. The best result is that untill now there is no quality compliants from the customers.

Key words: south korean steel bar, SD500, productive technology, mechanical property, bamboo steel rib

减酸洗焊丝用线材的氧化铁皮控制

李 杰[1], 刘 毅[1], 杨边疆[2]

（1. 河钢集团钢研总院，河北石家庄 050023；2. 河钢集团唐钢公司，河北唐山 063016）

摘 要：开发了一种减酸洗实心焊丝高速线材生产工艺，分析了实心焊丝用高线产品氧化铁皮的结构和影响铁皮厚度和组织转变的因素，对斯太尔摩控冷线最后一台风机的布局位置和作用进行了分析；影响用户酸洗耗酸量和酸洗时间的主要因素为氧化铁皮的总厚度和 FeO 与 Fe_3O_4 的比例，针对 FeO 的共析反应温度区间，对控冷工艺进行了改进，降低了产品铁皮总厚度和其中难以酸洗的 Fe_3O_4 含量。

关键词：氧化铁皮，高速线材，实心焊丝，减酸洗，控制冷却

The Controlling of Reducing Acid Pickling Wire Rod for Welding Wire

Li Jie[1], Liu Yi[1], Yang Bianjiang[2]

(1. HBIS Group Technology Research Institute, Shijiazhuang 050023, China;

2. HBIS Group Tangsteel Company, Tangshan 063016, China)

Abstract: A new process has been developed of a reducing acid pickling wire rod for welding wire, The structure and influencing factors of scale thickness and organizational change for solid welding wire are analyzed, the last air blower 's position and role of Stelmor controlled cooling line is analyzed; Main factors affecting the washing time and the consumption of acid pickling is the scale's total thickness and the ratio of Fe_3O_4 and FeO. For the eutectoid reaction temperature range of FeO, the control cooling process has been improved, Reduced the total thickness of the iron scale and the Fe_3O_4 content which is difficult to be pickled.

Key words: iron scale, wire rod, solid welding wire, reducing acid pickling, controlled cooling

帘线钢表面异物压入及凹坑缺陷的控制

林　辉

（武钢条材厂，湖北武汉　430083）

摘　要： 帘线钢盘条表面存在凹坑或异物容易导致在拉拔过程中断丝，本文通关分析异物的组织和成分，总结产生该缺陷的原因，并通过采取有效的措施，使帘线钢表面异物压入及凹坑缺陷得到控制。

关键词： 帘线钢，凹坑，异物，断丝，控制

Foreign Body into the Cord Steel Surface Defect and Pit Defect Control

Lin Hui

(WISCO Limited Material Factory, Wuhan 430083, China)

Abstract: Cord steel wire rod surface there are pits or foreign bodies easily lead to broken wire in the drawing process, through the analysis of foreign body structure and composition, summarizes the causes of the defects, by taking effective measures, the foreign bodies in the steel cord and the control of the pit is controlled.

Key words: cord steel, pits, foreign bodies, broken wire, control

改善帘线钢氧化铁皮机械剥离性能的研究

苏　豪，赵　阳，刘　鹤，严冉冉

（中天钢铁集团有限公司，江苏常州　213011）

摘　要： 本文简要阐述了盘条表面氧化铁皮的形成机理以及氧化铁皮的机械剥离方法。并通过工业大生产试验，研究帘线钢吐丝温度和控冷工艺对盘条的力学性能、氧化铁皮厚度、重量及机械剥离性能的影响，确定了当吐丝温度为900℃，配合一定的控冷工艺，盘条具有良好的力学性能，同圈差在50MPa以内，氧化铁皮厚度可以达到10μm以上，并且具有良好的机械剥离性能。

关键词： 帘线，帘线钢，氧化铁皮，机械剥壳

Study on Improving Mechanical Stripping Performance of Oxide Skin on Cord Wire Steel

Su Hao, Zhao Yang, Liu He, Yan Ranran

(ZENITH Iron & Steel Group Co., Ltd., Changzhou 213011, China)

Abstract: This paper briefly describes the formation mechanism of wire rod surface oxide skin and the mechanical stripping of it. Study on how the spinning temperature and controlled cooling process influences the mechanical properties ,thickness of iron oxide, weight and mechanical peeling performance of cord steel wire rod by industrial production test. Get the following conclusion, when the spinning temperature was 900℃, with a certain cooling process, the wire rod has good mechanical properties, with the circle difference within 50MPa, the thickness of iron oxide can reach more than 10μm, and has good mechanical stripping performance.

Key words: cord wire, cord wire steel, iron oxide skin, mechanical shelling

20CrMnTiH 圆钢控轧控冷工艺研究

王 超，田文庆，柯加祥，肖 超，李祥才

（中信泰富特钢研究院青钢分院棒材研究所，山东青岛 266000）

摘 要：青特钢公司通过优化加热工艺、控轧控冷的方法，使得 20CrMnTiH 齿轮钢的带状组织控制在 1~1.5 级，满足了客户的使用要求。同时，分析了控轧控冷工艺对产品带状组织、晶粒度以及硬度的影响。通过不断优化改进齿轮钢 20CrMnTiH 的生产工艺参数，使产品质量更加稳定。

关键词：齿轮钢，带状组织，晶粒度，控轧控冷

Research on Controlled Rolling and Cooling Process of 20CrMnTiH Round Steel

Wang Chao, Tian Wenqing, Ke Jiaxiang, Xiao Chao, Li Xiangcai

(CITIC Pacific Special Steel Institute (Qinggang Branch) Institute of Bar Steel, Qingdao 266000, China)

Abstract: With measures of optimizing heating system, controlled rolling and controlled cooling, the banded structure of 20CrMnTiH pinion steel is controlled as level 1~1.5, and the steel can meet the requirements from different user. The parameters of controlled rolling and controlled cooling and their effects on banded structure, grain size and hardness of 20CrMnTiH were analysed. After optimizing the technology parameters, the rolled products with more consistent quality are obtained.

Key words: pinion steel, banded structure, grain size, controlled rolling and controlled cooling

50CrVA 弹簧圆钢的研发和生产

李祥才，田文庆，王 超

（青岛特殊钢铁有限公司棒材研究所，山东青岛 266400）

摘 要：根据客户需要，青岛特钢研发生产了 50CrVA 弹簧圆钢，生产和检验表明，其产品氧氮氢含量较低，钢质纯净，组织正常，奥氏体晶粒度不小于 11 级，晶粒较细，并且表面质量较好，其尺寸、夹杂物、脱碳层、低倍、

硬度和拉伸性能指标均满足相关国家标准要求，产品质量完全合格，并且用户反馈使用良好，青岛特钢已经完全具备了生产 50CrVA 弹簧圆钢的能力。

关键词：氧氮氢含量，奥氏体晶粒度，夹杂物，脱碳层，拉伸性能

Research and Production of 50CrVA Spring Round Bar

Li Xiangcai, Tian Wenqing, Wang Chao

(Round Bar Reseach Institute, Qingdao Special Iron and Steel Co., Ltd., Qingdao 266400, China)

Abstract: According to customer needs, 50CrVA spring round bar was developed and produced by Qingdao Special Iron and Steel Co., Ltd., Production and check showed the O/N/H contents were comparatively low,the quality of 50CrVA steel was pure and the microstructure was normal,the austenite grain size was not less than 11 grade,the grain was comparatively finer and the surface quality of 50CrVA round bar was good and the size、non-metallic inclusion、decarburized layer、macrostructure、hardness and tensile property all met the requirements of standard GB/T 1222—2007,the quality of 50CrVA spring round bar was fully qualified and the customer feedback the use of 50CrVA round bar was good.Qingdao Special Iron and Steel Co.,Ltd. has possessed the capability to produce the 50CrVA spring round bar fully.

Key words: O/N/H contents, austenite grain size, non-metallic inclusion, decarburized layer, tensileproperty

低 Si 焊接用钢热轧盘条表面氧化铁皮结构优化

董　强，亓奉友，李向春，黄　亮，唐　庆

（青岛特殊钢铁有限公司线材研究所，山东青岛　266400）

摘　要：针对 Φ6.5mm 的 EH14 盘条在下游客户使用过程中出现因机械剥壳不理想导致镀铜不良的问题。进行了检测分析，结果显示 EH14 焊丝与 ER50-6E 相比，Si 元素含量较少，盘条母材表面的氧化铁皮较厚，容易掉落，之后基体被锈蚀；同时存在爆皮的现象，在爆皮后，盘条表面出现二次氧化，主要是致密的 Fe_3O_4 层。机械剥壳都难以去除。通过优化轧制工艺，调节盘条的冷却温度和冷却速度，减小了盘条的氧化铁皮厚度，同时减轻了盘条母材的爆皮现象。解决了 EH14 出现的机械剥壳不理想的问题。

关键词：吐丝温度，氧化铁皮，冷却速度，低 Si 焊丝

Structure Optimization of Iron Oxide Sheet on Hot - rolled Wire Rod of Low - Si Welding Steel

Dong Qiang, Qi Fengyou, Li Xiangchun, Huang Liang, Tang Qing

(Qingdao Special Steel and steel Co., Ltd. Wire Research Institute, Qingdao 266400, China)

Abstract: EH14 wire rod（Φ6.5mm, produced by Qingdao Special Steel and Steel Co., Ltd.）　used in the downstream process of the customer, the peeling machine peeling is not ideal, leading to subsequent plating problems.The result of test and analysis shows, compared with ER50-6E,Si element content of EH14 is less, Iron oxide skin on the base metal of the wire rod are thicker easy to fall, After the substrate is rusted.;At the same time, the phenomenon of explosion cracking occurred. After the explosion, re-oxidation occurred on the wire rod surface,mainly dense Fe_3O_4. These are difficult to remove by mechanical descaling. By optimizing the rolling process, the cooling temperature and cooling rate of the wire rod are adjusted, and the phenomenon of explosion cracking is reduced, and the iron scale thickness of the wire rod is reduced.Thus,it is solved that the EH14 mechanical stripping is not ideal.

Key words: spinning temperature, iron scale, cooling rate, low Si wire

青特钢#1 高线 HotEye 热眼影像式表面检测设备在轧制过程的应用

曾祥洲，黄　波，闫　亮，陈强磊

（青岛特殊钢铁有限公司高线厂，山东青岛　266400）

摘　要：表面质量对于轧钢高端产品（高端冷镦钢、弹簧钢等）至关重要。随着影像式表面检测技术的发展和成熟，新型的检测设备已能对轧制过程进行在线表面质量检测，并实时传递最新的质量信息予轧制车间。传统上，轧钢产品的表面质量检测是以离线非破坏检测方式或等产品冷却后目测进行，新型的检测设备则无此限制，可于轧制时进行在线检测，并提供实时质量反馈，车间可利用此讯息进行轧制工艺或操作程序的调整与最佳化。本文将重点介绍在线检测数据的使用和应用，并利用实例解释如何运用数据提高产品质量和生产效率，并有助于解决表面缺陷、质量监控或客户异议等问题。

关键词：轧制，影像，缺陷，应用

不同冷速对 49MnVS3 钢过冷奥氏体转变的影响

刘艳丽[1]，安治国[1]，黄艳新[2]

（1. 河钢集团钢研总院，河北石家庄　050000；

2. 河北钢铁集团石钢公司技术中心，河北石家庄　050031）

摘　要：利用 DIL805L 型膨胀仪测定了 49MnVS3 钢的临界点、不同冷却速度下连续冷却时的转变曲线，并研究了冷却速度对其显微组织及硬度的影响规律。实验结果表明：49MnVS3 钢的临界点为 AC1=740℃，AC3=786℃；当冷却速度在 0.05~5℃/s 之间时，试验钢转变组织为较粗大的铁素体、珠光体；冷却速度在 10~40℃/s 之间时，试验钢转变组织为珠光体、贝氏体和马氏体的混合组织；冷却速度＞40℃/s 后，试验钢转变组织为单一的马氏体组织。

关键词：49MnVS3 非调质钢，连续冷却转变曲线，临界点，膨胀法，金相硬度法

Effects of Cooling Rate on Phase Transformation of Austenite in 49MnVS3 Steel

Liu Yanli[1], An Zhiguo[1], Huang Yanxin[2]

(1. Hesteel Group Central Iron and Steel Research Institute, Shijiazhuang, 050000, China;

2. Shijiazhuang Iron and Steel Company, Hebei Iron and Steel Group, Shijiazhuang, 050031, China)

Abstract: The critical points and the different dilatometric curves of continuous cooling transformation of 49MnVS3 steel were measured by DIL805L Quenching dilatometer, and the effect of cooling rate on structure and hardness was studied. Experimental results show that the critical point at AC1, AC3 of 49MnVS3 steel was，AC1=738℃，AC3=783℃. The products of austenite transformation are ferrite and pearlite with cooling rate 0.05~5℃/s; with 10~40℃/s the tested steel obtains mixed structure of pearlite, bainite and martensite; the structure will be single martensite when the cooling rate above 40℃/s.

Key words: 49MnVS3 unquench-tempered steel, continuous cooling transformation curve, critical point, dilatometric test, metallographic analysis-hardness measurement

低碳钢盘条表面氧化层颜色控制方法

刘鸢杰，庄　娜，石　敏，祝俊飞，范众维

（方大特钢科技股份有限公司，江西南昌　330029）

摘　要： 随着盘条材料使用行业的发展，用户对材料的质量要求更加精细和个性化，公司部分现有用户对表面氧化层颜色提出了具体要求。本文主要介绍了公司目前生产的低碳钢盘条表面氧化层的常见颜色，并对其轧制工艺控制提出了改进方法。

关键词： 盘条，低碳钢，氧化层，颜色，控制方法

Low Carbon Steel Wire Rod Surface Oxide Layer Color Control Method

Liu Yuanjie, Zhuang Na, Shi Min, Zhu Junfei, Fan Zhongwei

(Fangda Special Steel Technology Co., Ltd., Nanchang 330029, China)

Abstract: With the development of wire rod material industry, the quality requirements of materials are more sophisticated and personalized, some existing users of the company put forward specific requirements for the surface oxidation layer color. This paper mainly introduces the common color of the surface oxide layer of low carbon steel wire rod produced by the company, and propose improvement methods for its rolling process control.

Key words: wire, low carbon steel, oxide layer, color, control method

钢绞线用 82B 钢生产关键技术研究

于　浩，熊　浩，卢　军，宋成浩

（北京科技大学材料科学与工程学院，北京　100083）

摘　要：针对 SWRH82B 盘条在拉拔过程中断丝率高的问题，对生产采用的铸坯和拉拔盘条试样分别进行金相组织和力学性能分析，并系统地研究了 SWRH82B 的加热工艺和控轧控冷工艺。分析得出盘条断丝的原因是心部异常组织和 Al$_2$O$_3$·CaO·SiO$_2$ 的复合夹杂物的存在，并针对盘条心部的异常组织调整加热工艺和斯太尔摩风冷线冷却制度，最终得到性能优异的 SWRH82B 盘条。

关键词：盘条，加热工艺，控制冷却，复合夹杂物

The Key Technology Research of 82B Steel Wire

Yu Hao, Xiong Hao, Lu Jun, Song Chenghao

(USTB, Beijing 100083, China)

Abstract: The experiment research was carried out because of the high rate of breakage during SWRH82B drawing process. The causes were analyzed based on metallographic structure inspection, micro-hardness and mechanical properties of casting blank and steel wire rod. According to the systematic study of the billet thermal regulation and TMCP, the cause of the breakage were the abnormal structure and Al$_2$O$_3$·CaO·SiO$_2$ containing complex inclusions in the core of steel wire rod. The relative adjustment of heating and Stelmor cooling process were applied, making an improvement of SWRH82B property.

Key words: steel wire rod, thermal regulation, TMCP, abnormal structure, complex inclusions

改善轧制圆棒材头部弯曲实践

常鹏飞，刘建培，王绍丹

（河钢石钢，河北石家庄　050031）

摘　要：由于轧制过程中轧线对中、轧机弹跳、轧机速度、轧件温度、进口导卫位置、压下量等不对称的因素，造成棒材在轧制过程中头部产生弯曲，要彻底解决头部弯曲困难很大。通过对成品轧机出口导卫的改进，试验"缩小出口导卫内径、延长直线段长度"，轧件头部弯曲有较明显地改善，且不会带来表面缺陷和生产事故。

关键词：头部弯曲，出口导卫，改善方法

Practice of Improving the Head Bending of Rolled Round Bar

Chang Pengfei, Liu Jianpei, Wang Shaodan

(Hested Group Shisteel Company, Shijiazhuang 050031, China)

Abstract: The bar head bending is happened during the rolling process due to the alignment of rolling line, springing of rolling mill, rolling speed, rolling stock temperature, position of enter guide, asymmetry of reduction etc, which is very difficult to be settled completely. But the bending of rolling stock can be improved obviously through the improvement to the exit guide of finished rolling mill by "reducing the inner diameter and extending the length of the straight line of the exit guide", and the improvement does not result in any surface defect and production accident.

Key words: bar head bending, exit guide, improvement method

钢材在线倒棱生产中的应用实践

郑晓宁

（河钢集团石钢公司轧钢厂，河北石家庄　050031）

摘　要：研究钢材端部在线倒棱在生产中的应用，主要是对圆钢在线倒棱工艺制定、设备使用维护进行研究改进。石钢公司中型棒材线轧钢车间通过在钢材收集台架前安装倒棱机，实现圆钢规格 $\phi 60\sim\phi 130mm$ 的在线倒棱功能，完成圆钢在线倒棱，在线倒棱角度控制在 40°～60°，倒棱斜面宽度控制在 2～8mm；在线倒棱减少后部人工倒角处理量，节约了生产成本。本文简要介绍了石钢公司中型棒材线轧钢车间生产中圆钢在线倒棱机的应用。

关键词：柔性倒棱机，ST 螺面辊道，棕钢玉砂轮片

Application Practices in the Production of Steel Online Ream

Zheng Xiaoning

(Rolling Mill, Hesteel Group Shisteel Company, Shijiazhuang 050031, China.)

Abstract: Research steel end chamfering online application in the production, mainly for round steel online chamfering process improvement maintenance, equipment used for research.Steel co medium by steel bar line of steel rolling workshop collection platform before installing chamfering machine, round steel specification from 60 to 130 mm chamfering online function, complete online chamfering round steel, the online control in 40° ~ 60° chamfer angle, chamfering cant width control in 2~8 mm; online chamfering reduce the back of the artificial chamfering processing, saving the cost of production.This paper briefly introduces the steel co medium-sized bar production line of steel rolling workshop application of round steel online chamfering machine.

Key words: flexible chamfering machine, ST spiral surface roller, brown steel jade grinding wheel piece

小规格 60Si2MnA 弹簧钢棒材脱碳控制实践

王敬文，李　伟，刘建培，张志旺，袁　浩

（河钢集团石钢公司，河北石家庄　050031）

摘　要：河钢石钢轧钢厂为了降低小规格弹簧钢 60Si2MnA 的脱碳层深度，进行了一系列加热工艺试验，通过调整加热温度、加热时间、炉内气氛控制，控制钢温均匀性，从而提高了 60Si2MnA 脱碳检验的一次合格率，改善了产品质量。

关键词：弹簧钢 60Si2MnA，棒材，加热，脱碳

Practice to Control Decarburization of Small 60Si2MnA Spring Steel Round Bar

Wang Jingwen, Li Wei, Liu Jianpei, Zhang Zhiwang, Yuan Hao

(Hesteel Group Shisteel Company, Shijiazhuang 050031, China)

Abstract: In order to reduce the decarburized layer depth of small 60Si2MnA spring steel round bar，the Rolling Mill of Hesteel Group Shisteel Compang completed a series of technological tests. By adjusting heating parameter and atmosphere of the furnace and controlling the umiformity of the steel temperature, thereby increased the qualify rate of 60Si2MnA spring steel decarburization, and improved the product quality.
Key words: 60Si2MnA spring steel, hot rolled round bar, heating, decarburization

HRB400 盘螺生产中的工艺改进

余坤洋，沈克非，朱传清

（日照钢铁有限公司，山东日照 276806）

摘 要：本文介绍了盘螺生产中的工艺改进，分析了原有生产工艺在实际生产中存在的不足及问题。重点针对工艺改进的必要性进行分析，通过所采取改进工艺优化方案，从而达到稳定产品性能的目的。
关键词：HRB400 盘螺，工艺改进

Technology Improvement of HRB400 Rebar Coil Production

Yu Kunyang, Shen Kefei, Zhu Chuanqing

(Rizhao Steel Co., Ltd., Rizhao 276806 ,China)

Abstract: The Technology improvement of HRB400 rebar coil production is introduced, and the deficiencies and problems existed in the actual production by original production process are analyzed in this paper. The necessity of technology improvement is analyzed particularly, and the production performance is stabilized by the projection of technology improvement.
Key words: HRB400 rebar coil, technology improvement

高线粗轧机滚动导卫技术改进

刘庆雨

（日照钢铁有限公司，山东日照 276806）

摘　要：文章介绍了对日钢高速线材粗轧机组进口滚动导卫导辊的改造。通过改造，满足了轧钢生产需求，减少了因导卫轴承坏引发的堆钢事故，延长了导卫在线使用寿命。

关键词：滚动导卫，导辊，轴承，改进

Improvement of Rolling Defeng of Finishing Rolling of High Speed Wire Roughing Mill

Liu Qingyu

(Rizhao Steel Co., Ltd., Rizhao 276806, China)

Abstract: The article introduce the high-speed wire roughing mill guide rolling the import of ontology .Completion of the high import rolling defended the base unit and ontology of reform of the processing,the needs of the production of steel rolling,reduce the pile of steel defended bearing the cause of accidents and prolongde use online time.

Key words: rolling defended, guide roller, bearing, transformation

型钢矫直机与矫直工艺控制浅析

韩振琳，焦　峰

（日照钢铁控股集团有限公司，山东日照　276806）

摘　要：本文通过对等节距十辊悬臂式矫直机设备特点进行分析，综合 H 型钢在生产过程中易出现的质量缺陷，并针对质量缺陷通过有效利用冷却条件、改进工艺方式、改善管理水平达到提高矫直质量的目的。

关键词：H 型钢，矫直调整，缺陷控制

Bar Straightener and Straightening Quality Control Analysis

Han Zhenlin, Jiao Feng

(Rizhao Steel Holding Group Co., Ltd., Rizhao 276806, China)

Abstract: In order to improve the quality of the straightening quality through the analysis of the characteristics of the 10 - roll cantilever straightener, the quality defects easily appear in the production process, and the quality defects can improve the straightening quality through the effective use of cooling conditions, improving the process mode and improving the management level.

Key words: H-profiled bar, straightening adjustment, defect control

热轧无缝钢管在线冷却控制系统研究

李振垒，陈　冬，袁　国，王国栋

（东北大学轧制技术及连轧自动化国家重点实验室，辽宁沈阳　110819）

摘　要： 控制冷却技术在热轧产品开发过程中是改善轧件性能的有效途径，已广泛应用于板带材，但控制冷却技术尚未在热轧无缝钢管在线生产中广泛应用，本文对热轧无缝钢管在线冷却控制系统关键控制技术做了系统介绍。根据热轧无缝钢管定径后的生产工艺特征，采用合适的控制时序流程，实现钢管在线冷却多流程冗余控制；根据热轧无缝钢管圆形断面特征，建立热轧无缝钢管在线冷却过程温度计算数学模型，通过对控制系统功能的最优化设计，采用合适的冷却策略，热轧无缝钢管终冷温度控制精度在目标值±20℃范围之内，实现热轧无缝钢管在线冷却系统的稳定运行。投入使用后，系统具有高稳定性、高可靠性、高温度命中率，显著提高了热轧无缝钢管产品的质量和性能。

关键词： 热轧无缝钢管，在线快速冷却，数学模型，多流程控制，冷却策略

On-line Cooling Control System for Hot Seamless Tube

Li Zhenlei, Chen Dong, Yuan Guo, Wang Guodong

(State Key Laboratory of Rolling & Automation, Northeastern University, Shenyang 110819, China)

Abstract: Controlled cooling technique, which has been widely used in strip, is an effective approach to improve properties of product during hot rolling process. But the application of controlled cooling technique on hot rolled seamless tube has not been extensively proceed. In this paper, the key control technology of on-line cooling control system for hot rolled seamless tube is systemically introduced. According to the productive process characteristics of hot rolled seamless tube after sizing process, the on-line cooling process of seamless tube has implemented stabilization operation by using multi-process redundant control system. Considering the circular section feature of hot rolled seamless tube, temperature field calculation mathematical model during on-line cooling was established. Through the optimal design of control system function and proper cooling strategy application, the stable operation of on-line cooling system for hot rolled seamless tube has been realized. The control accuracy of final cooling temperature can obtain deviation of ±20℃ from target value. The on-line cooling control system for hot-rolled seamless possess high stability, high reliability and high temperature accuracy. The application of this system could obviously improve the quality and properties of hot rolled seamless tube production.

Key words: hot rolled seamless tube, on-line cooling system, mathematical model, multi-process control, process control strategy

控制模式和时效处理对热轧带肋钢筋拉伸性能的影响

宋祖峰[1]，徐　雁[1]，黄　飞[2]，孙志鹏[1]

（1. 马鞍山钢铁股份有限公司，安徽马鞍山　243000；

2. 国家钢铁及制品质量监督检验中心，安徽马鞍山　243000）

摘　要：本文对拉伸速度控制模式（B 法）和时效处理对热轧带肋钢筋拉伸性能的影响进行了研究，并通过比对两家实验室的试验结果，初步得出了不同拉伸控制模式（B 法）和不同时效处理方式对热轧带肋钢筋拉伸性能的影响规律。

关键词：控制模式，时效处理，热轧带肋钢筋，拉伸性能

Effect on Tensile Properties of Hot Rolled Ribbed Steel Bar of Tensile Control Modes and Aged Treatment

Song Zufeng[1], Xu Yan[1], Huang Fei[2], Sun Zhipeng[1]

(1. Maanshan Iron & Steel Co., Ltd., Maanshan 243000, China；
2. National Center for Quality Supervision & Test of Steel Material Products, Maanshan 243000, China)

Abstract: Effect on tensile properties of hot rolled ribbed steel bar of tensile control modes and aged treatment has been researched in this paper. The results from two laboratories have been compared. It was obtained preliminarily that different tensile control modes and aged treatment　influence tensile properties of hot rolled ribbed steel bar.

Key words: tensile control modes, aged treatment, hot rolled ribbed steel bar, tensile property

控轧控冷工艺对免退火冷镦钢 SWRCH35KM 组织与性能影响

周乐育[1,2]，何建中[1]，吕　刚[1]，赵晓敏[1]，董　捷[1]，谭晓东[1]，涛　雅[1]

（1. 包头钢铁集团公司技术中心，内蒙古包头　014010；2. 北京机电研究所，北京　100083）

摘　要：为了实现免退火冷镦钢 SWRCH35KM 在线软化处理，研究了控轧控冷工艺制度对中碳冷镦钢组织和性能的影响。结果表明，采用分段缓冷制度，850~750℃终轧后均获得 60%~63%铁素体加部分球化珠光体，随着终轧温度降低，铁素体量略有增加，晶粒尺寸从 15~16.9μm 细化至约 10μm，珠光体球化趋势趋于显著。冷却制度对于实验钢力学性能有显著影响。采用轧后分段缓冷制度后，不同终轧温度时抗拉强度约 490~510MPa，伸长率 36.5%~40.5%，洛氏硬度为 73~78HRB；与直接空冷工艺相比，不同终轧温度下抗拉强度降低约 30~40MPa，伸长率提高 1%~3%，洛氏硬度降低 2~3HRB。

关键词：免退火冷镦钢，在线软化，分段缓冷，珠光体球化

Effect of Controlled Rolling and Controlled Cooling Process on Microstructure and Mechanical Property of Non-annealed Cold Heading Steel SWRCH35KM

Zhou Leyu[1,2], He Jianzhong[1], Lv Gang[1], Zhao Xiaomin[1], Dong Jie[1], Tan Xiaodong[1], Tao Ya[1]

(1. Technology Center, Baotou Iron & Steel Group Co., Ltd., Baotou 014010, China;
2. Beijing Research Institute of Mechanical and Electrical Technology, Beijing 100083, China)

Abstract: In order to realize the on-line softening treatment of non-annealed cold heading steel SWRCH35KM, the influence of controlled rolling and controlled cooling process on the microstructure and mechanical properties of medium carbon cold heading steel was investigated. The results show that the microstructures of spherular pearlite dispersed in 60%~63% ferrite matrix are obtained under step slow cooling process. With decreasing of finish rolling temperature from 850℃ to 750℃, ferrite grain size refines from 15~16.9μm to about 10 μm, and spheroidization of pearlite tends to significantly. Tensile strength of tested steel with step slow cooling process is about 490~510MPa, elongation is 36.5%~40.5%, and hardness is 73~78HRB. Compared with the air cooling process, the tensile strength of tested steel decreases about 30~40MPa, elongation increased 1%~3% and hardness decreases about 2~3HRB.

Key words: non-annealed cold heading steel, on-line softening, step slow cooling, spheroidization of pearlite

ER70S-6 盘条拉拔断裂的研究

闫卫兵

（宣化钢铁集团有限责任公司，河北宣化　075100）

摘　要： 研究采用了宏观检测和微观检测相结合的方法，对断丝纵剖形貌，硬相成分，微裂纹、硬相及试样横截面的 C、Si、Mn 元素的偏析情况进行了研究。结果表明试样心部存在大量的马氏体组织是导致断裂的主要原因，通过严格控制成分、降低过热度、加强辊道保温等措施，可大幅减少马氏体的数量。

关键词： V 型裂纹，硬相，心部马氏体，偏析

Study on Fracture Section of ER70S-6 Wire Rods

Yan Weibing

(Steel Making Plant of Xuangang Steel, Xuanhua 075100, China)

Abstract: The study adopts the method of macroscopic detection and microcosmic detection. In this paper, microstructure of broken samples, energy spectrum of hard phase, and the partial analysis of the elements of C, Si and Mn of microcrack, hard phase and cross section was studied. The results show that a large number of martensite tissue is the main cause of fracture. We can reduces the amount of martensite significantly by composition controlling, lowering the superheat of liquidsteel, strengthening roller insulation.

Key words: type Vmicrocrack, hard phase, martensite in centre area, segregation

16mm 螺纹钢三切分工艺优化

杨章令，陈必胜，许颖波，紫建华，舒云胜，武天寿

（红河钢铁有限公司，云南蒙自　661100）

摘　要： 简述了红钢轧钢厂棒材在生产 16 螺纹钢三切分生产过程中出现中线表面折叠现象，分析了造成折叠的原因，通过对切分导卫尺寸优化、提高轧辊装配技能等措施解决了折叠现象，降低了生产中产生的废品率，提高了月产量。

关键词： 三切分，中线折叠，切分盒，同心度

16mm Three Segmentation Process Optimization

Yang Zhangling, Chen Bisheng, Xu Yingbo, Zi Jianhua, Shu Yunsheng, Wu Tianshou

(Honghe Iron & Steel Co., Mengzi 661100, China)

Abstract: The production of 16mm three segmentation folding line surface phenomena appeared in the process of Honghe Iron & Steel Co, analyses the causes of the folding, based on the segmentation guide size optimization, improve the measures to solve the folding phenomenon, such as roll assembly skill reduces the rejection rate of the production, increase the monthly output.

Key words: three-cut, middle-line folding, slitting box, concentricity

10.9 级高强冷镦钢 ML20MnTiB 的调质实验研究与分析

高　建，翟晓毅，吴蓓蓓，孙彩凤，李海斌

（河钢集团邯钢公司技术中心，河北邯郸　056015）

摘　要：本文主要研究不同的调质工艺，对高强冷镦钢 ML20MnTiB 组织、性能的变化影响规律。通过实验发现 ML20MnTiB 热轧盘条经 860~880℃在 0#柴油（25℃）中淬火，再经 420℃回火处理后，得到回火索氏体，抗拉强度达到 1014~1097MPa，断后伸长率达到 12%以上，具有良好的塑性和强度匹配，可用于制作 10.9 级高强度标准件；经 860~880℃淬火，200℃回火，可得到回火马氏体，抗拉强度达到 1127~1229MPa，断后伸长率达到 11%以上。如要达到 12.9 级以上强度要求，需对淬火介质及调质工艺进行调整。

关键词：调质处理, ML20MnTiB, 高强冷镦钢, 油淬

Research and Analyse on 10.9 Grade High Strength Cold Heading Steel ML20MnTiB Quenching and Tempering Experiment

Gao Jian, Zhai Xiaoyi, Wu Beibei, Sun Caifeng, Li Haibin

(Handan Iron & Steel Group Co., Ltd., Handan 056015, China)

Abstract: It was researched that the microscopic structure and performance variable regular of the high strength cold heading steel ML20MnTiB with different quenching and tempering processes in this page. Through experiment,it is discoveried that hot rolled ML20MnTiB coil rod quenched in 0# diesel oil（25℃）at 860~880℃,and tempered at 420℃,the microscopic structure of ML20MnTiB coil rod will turn to tempered sorbite,with better plasticity and hardness,the Rm is 1014~1097MPa,the A is above 12%,it can be used to make 10.9 grade high strength fastener;and it quenched at same craft ,tempered at 200℃, the microscopic structure of ML20MnTiB coil rod will turn to tempered martensite, with the Rm 1127~1229MPa,the A above 11%.To meet the request of 12.9 grade high strength it is needed to adjust quench bath and tempering process.

Key words: quendhing and tempering, ML20MnTiB, high strength cold heading steel, oiled quench

4　表面与涂镀

Effect of Dew Point during Heating on Selective Oxidation of TRIP Steel

Jiang Guangrui[1,2], Wang Haiquan[1,2], Liu Libin[1,2], Teng Huaxiang[1,2]

(1. Shougang Research Institute of Technology, Beijing 100043, China; 2. Beijing key Laboratory of Green Recyclable Process for Iron & Steel Production Technology, Beijing 100083, China)

Abstract: The influence of dew point during heating on the selective oxidation behavior of two kinds of transformation-induced plasticity (TRIP) steel was studied by means of a Hot Dip Process Simulator. The selective oxidation behavior of alloying elements on the surface was studied by glow discharge optical emission spectroscopy (GDOES) and scanning electron microscopy (SEM). For a Al-bearging TRIP steel, less Al-rich oxides were found on the steel surface and there was less Mn on the surface when the dew point during heating was lower. On the contrary, for a Si-bearging TRIP steel, more film-like Si-rich oxides were found on the steel surface and there was more Mn on the surface when the dew point during heating was higher.

Key words: selective oxidation, dew point, TRIP steel, oxides

宝钢 BMD 技术开发与应用

段明南[1]，李山青[1]，陈声鹤[2]，房 鑫[1]，杨向鹏[1]，徐江华[1]

（1. 宝山钢铁股份有限公司研究院，上海 201900；

2. 宝钢湛江钢铁有限公司，广东湛江 524000）

摘 要：金属材料热轧成形后的表面氧化物（俗称鳞皮）在后工艺处理之前通常需有效清除，称之为除鳞，目前常用的化学酸洗工艺始终存在污染大、成本高、质量不稳定等诸多缺陷。基于此，宝钢从 2009 年开始着手开发 BMD 技术（Baosteel Mechanical Descaling technology，简称 BMD 技术），该工艺采用水+磨料颗粒的混合射流方式，通过射流介质在金属表面的持续击打、磨削而实现除鳞后的表面质量达到 Sa3.0 级，充分满足下游用户后续冲压、辊压、涂装、冷轧等各项严苛工艺要求。本文对 BMD 工艺的前期研发历程、工艺特征以及近期的工业化推广实施与应用进行阐述，并对 BMD 技术的后续发展和应用前景进行了展望。

关键词：宝钢，BMD，开发

Development and Application of Baosteel BMD Technology

Duan Mingnan[1], Li Shanqing[1], Chen Shenghe[2], Fang Xin[1], Yang Xiangpeng[1], Xu Jianghua[1]

(1. Research Institute，Baoshan Iron & Steel Co., Ltd., Shanghai 201900,China;

2. Baosteel Zhanjiang Iron & Steel Co., Ltd., Zhanjiang 524000, China)

Abstract: The surface of metal oxide material after hot forming (commonly known as scaly skin) before postprocessing usually need to effectively remove, called descaling. At present, descaling technology is commonly used in the pickling process, the technology has high pollution and high cost and unstable quality defects Based on this, Baosteel began to develop BMD Technology from 2009 (Baosteel Mechanical Descaling technology, abbreviated as BMD).BMD is a process of using water and abrasive particles mixed jet to continue hitting, grinding the metal surface, and achiving the Sa3.0 level quality, it's quality fully meet the downstream users stamping, cold rolling and other stringent requirements In this paper, the development process, process characteristics, recent industrialization, implementation and application of BMD technology are expounded, and the subsequent development and application prospect of BMD technology are prospected.

Key words: Baosteel, BMD, development

冷轧马氏体钢退火炉氧化机理研究

李　研，蒋光锐，刘顺明，郑晓飞，王海全，巫雪松

（首钢技术研究院，北京　100043）

摘　要： 冷轧马氏体钢采用连续退火炉进行生产，为满足其性能要求，其退火温度较高，带速较慢。为满足其淬透性，该钢种加入较多的合金元素，Mn 元素的含量达到 1.7%。Mn 元素在还原性气氛中优先于 Fe 元素氧化，在带钢表面形成富集氧化层，以氧化物颗粒的形式存在。采用热镀锌模拟器对冷轧马氏体钢退火过程中的氧化行为进行了研究，分析了不同露点下带钢表面氧化物颗粒的形貌、尺寸及氧化物类型。在预热段及均热段进行露点控制，可以有效改善带钢表面氧化物颗粒的形貌，避免 Mn 元素的过度富集及大尺寸氧化物颗粒的形成。

关键词： 冷轧，退火，氧化，露点

The Oxidation Mechanism Research of the Cold Rolled Martensitic Steel

Li Yan, Jiang Guangrui, Liu Shunming, Zheng Xiaofei, Wang Haiquan, Wu Xuesong

(Shougang Research Institute of Technology, Beijing 100043, China)

Abstract: The cold rolled martensitic steel is produce by the continuous annealing furnace, to fulfill the mechanical property, the annealing temperature is high and the speed is slow. Some alloy elements are used to make sure the hardenability of the steel. The Mn content can reach 1.7%. The Mn element is preferred oxidized in the reduction atmosphere, the oxidation grain can be found on the surface of the strip. The galvanizing simulator was used to study the oxidation behavior of the cold rolled martensitic steel inside the annealing furnace, the morphology, size and type of the oxidation was analyzed. The morphology of the oxidation can be improved by controlling the dew point of the heating and soaking section which can avoid the large size oxidation grain.

Key words: cold rolling, annealing, oxidation, dew point

钢材表面除鳞方法探讨

徐言东[1]，顾　洋[2]，谢宝盛[1]，马树森[3]，李建强[1]，袁华朋[3]

（1. 北京科技大学工程技术研究院，北京　100083；2. 北京科技大学钢铁技术协同创新中心，北京　100083；3. 大连鑫永尚科技有限公司，辽宁大连　116021）

摘 要： 使用拉伸、冷轧等方法，利用 SEM、EDS、XRD 等手段对比分析钢材表面除鳞方法。结果表明：氧化铁皮由最外层红色 Fe_2O_3，中间层为磁性 Fe_3O_4，以及最内层的维氏体组成。以磷酸为基底的复合多元酸 GF2&GF1 可以进行酸洗以去除氧化铁皮，且效果好于盐酸。机械除鳞时压应力的除鳞效果显著好于拉应力，为了达到破鳞效果，必须迫使钢材产生较大变形。

关键词： 钢材，磷酸，盐酸，除鳞

Discussion on Phosphorous Removal Method of Steel Surface

Xu Yandong[1], Gu Yang[2], Xie Baosheng[1], Ma Shusen[3], Li Jianqiang[1], Yuan Huapeng[3]

(1. Engineering and Technology Research Institute, University of Science and Technology Beijing, Beijing 100083, China; 2. Collaborative Innovation Center of Steel Technology, University of Science and Technology Beijing, Beijing 100083, China; 3. Dalian Yongsun Co., Ltd., Dalian 116021, China)

Abstract: The method of phosphorous removal was analyzed by means of SEM, EDS, XRD, stretching and cold rolling. The results show: The oxide scale consists of the outermost red Fe_2O_3, the middle layer is magnetic Fe_3O_4, and the innermost layer is wustite. The phosphoric acid-based composite polyacid GF2 & GF1 can be acid washed to remove the oxide scale, and the effect is better than hydrochloric acid. Pressure stress descaling effect is significantly better than the tensile stress during mechanical descaling. In order to achieve the scale effect, the steel must be forced to produce a large deformation.

Key words: steels, phosphoric acid, hydrochloric acid, phosphorous removal

汽车用热镀锌先进高强钢 TRIP690 生产工艺探究

李大光

（本钢产品研究院，辽宁本溪 117000）

摘 要： 通过热模拟试验，利用金相显微技术，透射电镜检测方法，研究了两相区加热温度、贝氏体等温处理工艺对热镀锌 TRIP690 组织组成、室温时残余奥氏体含量及力学性能的影响规律；结果表明，基于镀锌产线的特殊性，合理选择两相区加热温度、贝氏体等温处理等重要工艺参数，可以获得良好的 TRIP 效应。

关键词： TRIP 钢，贝氏体等温处理，组织，性能

Study on the Production Technology of Advanced High Strength Steel TRIP690 for Hot Dip Galvanizing

Li Daguang

(Product research institute of Ben Xi Iron & Steel Co., Ltd., Benxi 117000, China)

Abstract: By means of thermal simulation test and metallographic microscopy,transmission electron microscopy, To study the two-phase zone heating temperature、the bainite isothermal treatment process for hot dip galvanized TRIP690 organization composition、retained austenite properties at room temperature and the effect of mechanical properties;the results show, based on the particularity of galvanizing line, by reasonable choice of two-phase zone heating temperature、the bainite isothermal treatment etc. important process parameters, be obtained good TRIP effect.

Key words: TRIP steel, bainite isothermal treatment, organization, performance

IF 钢合金化镀锌板相结构和抗粉化性能研究

周元贵，白会平

（武钢研究院，湖北武汉　430080）

摘　要： 采用扫描电镜（SEM）、聚焦离子束（FIB）、透射电镜（TEM）及弯曲试验研究了工业大生产条件下两种加热炉功率生产的 IF 钢合金化镀锌板（GA）镀层微观组织、相结构及抗粉化性能差异。结果表明两种 GA 板镀层表面均由致密的 δ 相和柱状 ζ 相组成；当合金化炉的功率由 464kW 提高到 672kW 时，镀层厚度保持不变；GA 板镀层均由 Γ 相、δ 相和 ζ 相组成，靠近基板区域主要为 Γ 相，镀层中心区域主要为 δ 相，镀层近表面主要为 ζ 相；抗粉化性能对比结果表明，镀锌层表面 δ 相更多，对应的镀层抗粉化性能更好。

关键词： 合金化镀锌板，镀层，粉化，相结构

Research on Phase Microstructure and Powdering-Resistance of Hot-dip Galvannealed Steel Sheets

Zhou Yuangui, Bai Huiping

(The Research and Development Center of WISCO, Wuhan　430080, China)

Abstract: The microstructure, phase morphology and powdering performance of two kinds of industrially produced hot-dip galvannealed (GA) coatings on interstitial free(IF) steels were investigated by means of SEM, FIB, TEM and 180 degree bending test. The results show that, the surface of these coatings is composed of δ phase and ζ .phase. The thickness of the coatings is not change as the power of furnace is enhanced from 464kW to 672kW . There are Γ phase、δ phase and ζ phase in both two GA coatings. Enhancement the GA furnace power is beneficial to forming more δ phase. The powdering resistance behavior is better when the coating surface is consisting of more δ phase.

Key words: hot-dip galvannealed steel sheet, coating, powdering, phase constitution

Si 对热轧酸洗平整板表面氧化铁皮的影响

周洪宝，王学伦，鲍生科，吴盛平，王　庆

（日照钢铁控股集团有限公司，山东日照　276806）

摘　要： 针对 ESP 生产 SPHC 系列产品时存在的氧化铁皮问题，发现随着 Si 含量的增加，氧化铁皮有减少趋势，本文对 Si 对氧化铁皮的影响作了简要分析。

关键词： 热轧酸洗，氧化铁皮，薄规格，以热代冷

The Effect of Si to the Scale of Hot Rolling Pickling Plate

Zhou Hongbao, Wang Xuelun, Bao Shengke, Wu Shengping, Wang Qing

(Rizhao Steel Holding Group Co., Ltd., Rizhao 276806, China)

Abstract: It is found that the scale is decreased following the increase of Si for the product of SPHC series steel. The effect of Si to the scale is analyzed in this paper.

Key words: hot rolling and picking, oxide scale, thin specification, hot rolling products take place of cold rolling products

Zn-Al-Mg 镀层热冲压前后组织比较

孙伟华，方　芳，邓照军，杜小峰，胡宽辉

（宝钢股份中央研究院武汉分院（武钢有限技术中心），湖北武汉　430080）

摘　要： 本工作采用 1500MPa 级热成形钢进行热浸镀试验，镀液成分是 Zn-7.1Al-1.6Mg(wt.%)，对镀后（热冲压前）和热冲压后的镀层组织进行了分析。结果表明，镀后组织为大块的富 Zn 相和细小的共晶组织。共晶组织中 Al 含量高。在高温奥氏体化和热冲压过程中，镀层转变为 Fe-Zn-Al 合金层。根据相图，该合金层是 α-Fe 固溶体。镀层中 Al 元素成分由里到外逐渐升高，而 Zn 元素逐渐降低，Mg 元素在镀层表面富集。镀层表面有氧化物生成，包括 MgO，Al_2O_3 和 ZnO。镀层局部有裂纹，但未扩展到钢基中。

关键词： 热成形钢，Zn-Al-Mg 镀层，扩散，氧化物，相图

Microstructures of the Zn-Al-Mg Coating Layers before and after Hot Stamping

Sun Weihua, Fang Fang, Deng Zhaojun, Du Xiaofeng, Hu Kuanhui

(Wuhan Branch Baosteel Central Research Institute (Technology Center of Wuhan Steel Ltd.),
Wuhan 430080, China)

Abstract: The hot dipping galvanizing experiment is carried out on the 1500MPa grade press hardening steel. The bath chemistry is Zn-7.1Al-1.6Mg (wt.%). The microstructures after galvanizing and hot stamping are analyzed. The results show that the microstructure of the coating layer after galvanizing consists of large Zn solid solution phase and fine eutectic structures. The content of Al in the eutectic structure is high. During the austenitizing and subsequent hot stamping process, the solidified microstructure transforms into Fe-Zn-Al layer. Based on the phase diagram, the layer is α-Fe solid solution phase. The content of Al in the coating layer is increased gradually from inside to outside, whereas that of Zn is decreased. Mg is concentrated at the surface. Oxides are formed at the surface, which include MgO, Al_2O_3 and ZnO. Cracks are formed in the layer locally, but are not expanded into the steel base.

Key words: press hardening steel, Zn-Al-Mg coating, diffusion, oxides, phase diagram

热浸镀铝锌液中添加 Ti 合金对金属间化合物形成的影响

丁志龙[1,2]，张　杰[1]，李　刚[3]，孙永旭[3]，江社明[1]，张启富[1]

（1. 钢铁研究总院，先进金属材料涂镀国家工程实验室，北京　100081；2. 宝钢研究院梅钢技术中心，江苏南京　210039；3. 上海梅山钢铁股份有限公司，江苏南京　210039）

摘　要： 为了研究含 Ti 铝锌合金添加对热浸镀铝锌锅中金属间化合物形成的影响，采用扫描电子显微镜（SEM）、能谱仪（EDS）对添加含 Ti 铝锌合金前后的铝锌锅中形成的的金属间化合物进行形貌、尺寸及成分、相组成分析。结果表明：添加含 Ti 合金前，铝锌锅中的铝锌液凝固态组织中生成 $FeAl_3$ 和 Fe_2SiAl_7，尺寸在 60-70μm 左右，数量较少；添加含 Ti 合金后，铝锌锅中的铝锌液凝固态组织中包括 $FeAl_3$、Fe_2SiAl_7 和尺寸为 20-40μm 的 $TiAl_3$ 和 $Al_{23}V_4$；且随着添加含 Ti 合金的时间增加，$TiAl_3$ 和 $Al_{23}V_4$ 组织有增大增多的趋势。

关键词： 铝锌镀层，Ti，合金，金属间化合物

Effect of Ti-added Alloy on Intermetallic Formation in the 55%Zn-Al-43.5%-Si Molten Pot

Ding Zhilong[1,2], Zhang Jie[1], Li Gang[3], Sun Yongxu[3], Jiang Sheming[1], Zhang Qifu[1]

(1. National Engineering Laboratory of Advanced Coating Technology for Metals, Central Iron and Steel Research Institute, Beijing 100081, China; 2. Meishan Technology Center of Shanghai BaoSteel Company, Nanjing 210039, China; 3. Meishan Iron and Steel Co., Ltd., Nanjing 210039, China)

Abstract: In order to investigate the Effect of Ti-added Al-Zn alloy on intermetallic formation in the 55%Zn-Al-43.5%-Si Molten Pot, SEM and EDS was used to study the morphology, size and composition of the intermetallic compound in the molten pot before and after Ti-added Al-Zn alloy added. The results showed that: before Ti-added Al-Zn alloy added, the intermetallic in the molten pot were little of $FeAl_3$ and Fe_2SiAl_7, the size was about 60-70μm; after Ti-added Al-Zn alloy added, the intermetallic compound in the molten pot were $FeAl_3$, $Fe2SiAl_7$ and $TiAl_3$, $Al_{23}V_4$, the size of $TiAl_3$, $Al_{23}V_4$ was about 20-40μm; with the time of Ti-added Al-Zn alloy added increase, the size and amount of $TiAl_3$, $Al_{23}V_4$ increased.

Key words: galvalume coating, Ti, Al-Zn alloy, intermetallic

开创彩涂板绿色生产新纪元

杨　力，刘　鹏，李绮屏，张　卫，焦时光，王　颖

（中冶京诚工程技术有限公司，北京　100176）

摘　要： 本文简述了彩涂板发展历史及目前我国彩涂板生产存在的问题，重点介绍了中冶京诚公司研发的新型绿色

环保的粉末彩涂生产技术，在对已投产机组生产数据分析基础上，将粉末彩涂与溶剂彩涂在成型机理、政策、环保、性能、成本以及市场方面的对比，提出了粉末彩涂环保绿色的技术特性及良好发展前景。

关键词：粉末彩涂，溶剂彩涂，绿色环保，VOCs

A New Era of Green Production for Color Coating Coil

Yang Li, Liu Peng, Li Qiping, Zhang Wei, Jiao Shiguang, Wang Ying

(MCC Capital Engineering & Research Incorporation Limited, Beijing 100176, China)

Abstract: This paper briefly introduces the development history and the existing problems in the production of color coated coil in China. It focused on the introduction of the new environmental powder coating production technologies developed by CERI. Based on the analysis of the production data, this paper contrasting the powder coating and solvent color coating in the molding mechanism, policy, environmental protection, performance, cost and market, the conclusion is drawn that powder coating production technology shows the green and environmental characteristics and also the vast potential for the future development.

Key words: powder color coating, solvent color coating, green and environmental, VOCs

纳米 TiO₂ 添加剂对 ZL101A 铝合金微弧氧化陶瓷涂层性能的影响

张 宇[1,2]，范 伟[1]，杜海清[1]，曾 丽[1]，宋仁国[3]

（1. 浙江工业职业技术学院机械工程学院，浙江绍兴 312000；2. 浙江工业大学特种装备制造与先进加工技术教育部/浙江重点实验室，浙江杭州 310014；3. 常州大学材料科学与工程学院，江苏常州 213164）

摘 要：本文通过在硅酸盐电解液中加入纳米 TiO₂ 添加剂，采用扫描电镜(SEM)、X 射线衍射仪(XRD)、电化学工作站、体视显微镜、摩擦磨损试验等，研究了纳米添加剂浓度的变化对 ZL101A 铸造铝合金微弧氧化陶瓷涂层耐蚀性能和耐磨性能的影响。结果表明，纳米添加剂进入到陶瓷涂层中，涂层表面变得更加致密。膜层硬度、厚度随着添加剂浓度的增加而增加。同时，摩擦磨损测试及磨损后 SEM 表明，随着纳米添加剂浓度的增加，磨损量逐渐较小，摩擦系数也随之减小，并在 20g/l 时具有最好的磨损性能。同时，动电位极化曲线测试和电化学阻抗测试表明，随着纳米添加剂的浓度增加，腐蚀电流密度不断减小，交流阻抗不断增大，表明涂层的耐蚀性有明显提高。

关键词：微弧氧化，铝合金，纳米二氧化钛，耐蚀性，耐磨性

Effects of Nano-additive TiO₂ on Properties of MAO Ceramic Coatings Formed on ZL101A Aluminum Alloy

Zhang Yu[1,2], Fan Wei[1], Du Haiqing[1], Zeng Li[1], Song Renguo[3]

(1. School of Mechanical Engineering, Zhejiang Industry Polytechnic College, Shaoxing 312000, China; 2. Key Lab.

of E&M, Ministry of Education & Zhejiang Province, Zhejiang University of Technology, Hangzhou 310014, China;

3. School of Materials Science and Engineering, Changzhou University, Changzhou 213164, China)

Abstract: TiO$_2$-containing ceramic coatings were prepared on the ZL101A aluminum alloy substrates by micro-arc oxidation (MAO) in silicate electrolytes doped with different concentrations. The effects of nano-additive concentration on corrosion resistance was investigated by means of SEM, XRD, stereo microscope, wear test and electrochemical workstation. The results show that some nano-particle were incorporated into the resulting coating during the MAO process, the ceramic coatings become more compact. The hardness and thickness of the ceramic coatings were increased with nano-additive concentration. Also the wear test and SEM test results showed that the friction coefficient was increased with nano-additive concentration, and when the coating was prepared at 20g/l nano-TiO$_2$, the wear behaviour of the coating was best. Potentiodynamic polarization test and Electrochemical impedance spectroscopy (EIS) test results showed that the corrosion current density was decreased, and the resistance was increased. It means the corrosion resistance of coatings was increased with increasing the TiO$_2$ concentration in the electrolyte.

Key words: micro-arc oxidation, aluminium alloy, nano TiO$_2$, corrosion resistance, wear resistance

无机自润滑热镀锌板电阻点焊性能评价及分析

赵晓非，苗雨芳，黎　敏，汪小培，伊日贵

（首钢技术研究院，北京　100041）

摘　要: 在热镀锌板表面涂覆无机自润滑涂层可能会对热镀锌材料电阻点焊性能造成明显影响，包括焊接窗口缩窄，焊接寿命下降等。本文针对上述现象，对无机自润滑处理后热镀锌板材料点焊性能下降的具体原因展开分析，结果显示，在焊接的高温过程中，热镀锌板表面自润滑膜与焊头、热镀锌板表面锌层发生化学反应生成复杂无机物涂层，该涂层附着在焊头表面严重阻碍焊接过程进行，进而造成焊接性能严重下降。

关键词: 自润滑涂层，热镀锌板，焊接性能，热分解反应

Welding Properties Investigation of Self-lubricated Film Coated Galvanized Autosheets

Zhao Xiaofei, Miao Yufang, Li Min, Wang Xiaopei, Yi Rigui

(Shougang Research Institute of Technology, Beijing 100041, China)

Abstract: Self-lubricated film were coated on the galvanized autosheets and then the welding properties were investigated. The results indicated that a thermal decomposition reaction of self-lubrication film was carried out under the high temperature during the welding process. Such a chemical reaction between the self-lubrication film, the zinc coating of autosheets and the surface of welding head had a negative effect on the final welding properties.

Key words: self-lubricated film, galvanized sheets, welding properties, thermal decomposition reaction

耐候钢表面锈层稳定涂层的耐腐蚀性能研究

于东云[1]，高立军，杨建炜

（首钢技术研究院，北京　100043）

摘　要：在耐候钢表面制备了一种锈层稳定涂层，并将涂覆和未涂覆涂层试样置于3.5%NaCl溶液中腐蚀不同时间后，通过扫描电镜和X-射线衍射分析试样表面的形貌和相组成，用失重腐蚀速率和电化学阻抗谱评价耐蚀性能。结果表明，锈层稳定涂层在腐蚀过程中促进了Cr元素的富集，并促使耐候钢表面生成连续致密的锈层，连续致密的锈层随着腐蚀时间的增加而逐渐增厚，为耐候钢基体提供保护作用。

关键词：耐候钢，锈层，稳定涂层，耐腐蚀

Study on Corrosion Resistance of Rust-Layer Stability Coating on Weathering Steel Surface

Yu Dongyun, Gao Lijun, Yang Jianwei

(Shougang Research Institute of Technology, Beijing 100043, China)

Abstract: A type of rust layer stability coating was prepared on the weathering steel, then the coated and uncoated samples were immersed in 3.5% NaCl solution corroded for different time. The surface morphology and phase of the samples were characterized by scanning electron microscopy, and X-ray diffraction, and their anti-corrosion property was assessed by weight loss rate, and electrochemical impedance spectroscopy. The results showed that the rust layer stability coating promoted the aggregation of Cr during the corrosion and the formation of continuous and compact rust layer on weathering steel. The rust layer become thicker with the corrosion time increased to improve protection for the weathering steel substrate.

Key words: weathering steel, rust-layer, stability coating, corrosion resistance

镀铝硅用无机类耐高温无铬钝化涂层的开发

杨家云，邹海霞

（上海凯密特尔化学品有限公司技术中心，上海　201305）

摘　要：镀铝硅无铬钝化钢板具有良好的耐热抗氧化、耐蚀性和热反射性，因此在汽车、家电及轻工领域得到了广泛应用。但常规镀铝硅无铬钝化钢板耐高温黄变性能较差，在高温环境中表面会迅速黄变，不仅影响其外观，还会降低其耐腐蚀性能，因此限制了其在烤炉烤箱内胆、汽车排气管等具有耐高温要求领域的应用。为解决该问题，本文设计开发了一种用于镀铝硅钢板的具有自修复能力的无机类耐高温无铬钝化涂层，经500℃高温烘烤测试、耐盐雾测试、扫描电镜分析及能谱分析可知，500℃高温烘烤后，耐高温无铬钝化涂层表面无龟裂、脱落等缺陷，且涂层表面氧元素含量无明显提高，这表明耐高温无铬钝化涂层耐热性良好且能隔绝氧气与铝硅镀层接触，防止铝硅镀层在高温下氧化。经耐高温无铬钝化涂层处理后，镀铝硅钢板500℃烘烤5min后表面色差<3，且烘烤前后耐盐雾

性能均能保证 72h 表面无锈蚀，因此特别适用于烤炉烤箱内胆、汽车排气管等具有耐高温要求的领域。

关键词：镀铝硅，耐高温，无铬钝化，高耐蚀

Development of High-temperature Resistant Cr-free Passivation for Al-Si Steel Strip

Yang Jiayun, Zou Haixia

(Technical Center, Shanghai Chemetall Chemicals Co., Ltd., Shanghai 201305, China)

Abstract: Al-Si steel strip with Cr-free passivation coating has good high-temperature resistance, oxidation resistance, anticorrosion property and heat reflection. Therefore, it has been widely used in automobile, home-appliance and light industry. However, the yellowing resistance of conventional Al-Si steel strip after high temperature heating was poor. In the high temperature environment, the surface of Al-Si steel strip would became yellow rapidly. This would not only affect the appearance, but also reduce the anticorrosion performance of the Al-Si steel strip. Thus, this limited the application in the fields which required high temperature resistance, such as the oven liner and automobile exhaust pipe. In order to solve this problem, a kind of inorganic Cr-free passivation coating with high temperature resistance and self-repairing property was introduced in this work. The inorganic Cr-free passivation coating was tested by high temperature heating test, salt spray test, SEM and EDX. The results showed that the inorganic Cr-free passivation coating had no cracks and peeling defects after heating at 500℃ for 5min, and the oxygen content of the coating did not improve obviously. This indicated that the coating had good high temperature resistance and could insulate oxygen from Al-Si steel strip to prevent the oxidation of Al at high temperature. The color difference of the Al-Si steel strip with inorganic Cr-free passivation coating was below 3 after heating at 500℃ for 5min, and the Al-Si steel strip with the coating had no rust after 72h salt spray test whether or not heating at 500℃ for 5min. Therefore, the Al-Si steel strip with inorganic Cr-free passivation coating was especially suitable for the fields which required high temperature resistance, such as the oven liner and automobile exhaust pipe.

Key words: Al-Si steel strip, high-temperature resistance, Cr-free passivation, high anticorrosion property

免涂装耐候钢锈层快速生成技术

高　鹏[1,2]，陈义庆[1,2]，钟　彬[1,2]，艾芳芳[1,2]，李　琳[1,2]，李　岩[1,2]

（1. 海洋装备用金属材料及其应用国家重点实验室，辽宁鞍山　114009；
2. 鞍钢集团钢铁研究院汽车与家电用钢研究所，辽宁鞍山　114009）

摘　要： 针对免涂装耐候钢锈层形成时间长，使用初期锈液流淌问题，开发了免涂装耐候钢锈层快速生成处理技术。研究结果表明，处理剂可有效促进内锈层中 α-FeOOH 的生成，使耐候钢表面快速生成致密、与钢基体附着良好的锈层。在大型建筑物上实际应用情况表明，处理剂可使免涂装耐候钢快速展现装饰性锈蚀外观并显著改善耐候钢使用初期的锈液流淌问题。

关键词： 免涂装，耐候钢，锈层，处理剂

Fast Formation Technology of Rust on Surface of Unpainted Weathering Steel

Gao Peng[1,2], Chen Yiqing[1,2], Zhong Bin[1,2], Ai Fangfang[1,2], Li Lin[1,2], Li Yan[1,2]

(1. State Key Laboratory of Metal Material for Marine Equipment and Application, Anshan 114009, China;

2. Iron & Steel Research Institute of Angang Group, Anshan 114009, China)

Abstract: A fast formation technology of rust on surface of unpainted weathering steel，was studied，which can quickly form rust layers and prevent liquid rust flowing. The results show that the treatment agent can expedite the formation of α-FeOOH in internal rust layers. Firmly and dense rust layers form on surface of weathering steel. The effects of practical application on large buildings indicates that, treatment agent can obviously suppress liquid rust flowing and promote formation of decorative appearances.

Key words: unpainted, weathering steel, rust, treatment agent

C-Mn-Si-Cr 系镀锌双相高强钢腐蚀行为研究

刘靖宝，杜明山，王嘉伟，魏焕君，李　勃

（河钢集团唐钢公司技术中心，河北唐山　063001）

摘　要: 通过盐雾腐蚀试验和扫描电镜研究了不同沉降量、不同腐蚀时间对双相高强钢失重速率和腐蚀行为的影响。试验表明，在不同的沉降量下，板材的腐蚀速率不同，随着沉降量的增加，失重增加；随着腐蚀时间增长，腐蚀产物覆盖在镀锌板表面，阻碍了纯锌层与腐蚀液的接触，减缓腐蚀的进行。但是，由于双相钢中 Mn 和 Cr 元素较高，造成纯锌层与基板的粘附性变弱，通过 72h 盐雾腐蚀后，带钢表面产生红锈的面积达到了 65%。通过显微组织观察，腐蚀产物的形貌为呈雪花状，主要成分是 Fe_3O_4。

关键词: 双相高强钢，腐蚀，失重，镀锌板

Study on Corrosion Behavior of C-Mn-Si-Cr Hot Dip Galvanized Dual Phase High Strength Steel

Liu Jingbao, Du Mingshan, Wang Jiawei, Wei Huanjun, Li Bo

(Technical Center, Hesteel Group Tangsteel Company, Tangshan 063001, China)

Abstract: Influence of the weight loss rate and corrosion behavior of the dual phase high strength steel at different settlement and corrosion times were investigated by the salt spray corrosion test and scanning electron microscopy. The results show that the corrosion rates of the sheets were different under different settlements. With the increase of the settlement, the weightlessness increases. With the increase of corrosion time, the corrosion products were covered on the surface of galvanized sheet, which hindered the contact between pure zinc layer and solution, and slows down the corrosion. However, due to the high content of Mn and Cr in the dual phase steel, the adhesion between the pure zinc layer and the substrate is weakened, and the area of the red rust on the surface of the steel strip was 65% after 72h salt spray corrosion.

Through the observation of the microstructure, the morphology of the corrosion products was a snowflake, and the main component was Fe_3O_4.

Key words: dual phase high strength steel, corrosion, weightlessness, galvanized sheet

高强 IF 钢热轧板边部细线形成机理研究

王　畅[1]，于　洋[1]，王　林[1]，刘文鑫[1]，陈　瑾[2]，焦会立[2]

（1. 首钢技术研究院，北京　100043；2. 首钢迁安钢铁有限责任公司，河北迁安　064404）

摘　要： 本文针对高强 IF 钢频发的热卷边部细线缺陷进行微观分析和数据统计，确定边部细线产生的机理以及提出相应的改善措施。微观分析发现热卷边部细线的微观特点为表面沿轧向裂口组成，截面可见裂纹双向延展性，表层组织有混晶形貌，距离边部位置在 15mm 左右，具有明显的冷轧遗传性，冷轧后表现为沿轧向凹坑和掉肉形貌。生产过程统计对比发现，缺陷产生与出炉温度，边部减宽量和立辊使用周期关系密切，由此制定相应的优化工艺。通过采用新的优化工艺跟踪实验发现高强 IF 钢边部细线缺陷得到明显的抑制，大生产下可控制细线发生位置在边部 10mm 以内。

关键词： 边部细线，高强 IF 钢，立辊，热轧

Research on Mechanism of High Strength IF Steel Edge Thin-wire Defects

Wang Chang[1], Yu Yang[1], Wang Lin[1], Liu Wenxin[1], Chen Jin[2], Jiao Huili[2]

(1. Shougang Research Institute of Technology, Beijing 100043, China;

2. Shougang Qian'an Steel Company, Qian'an 064404, China)

Abstract: The article research on the high strength IF steel edge thin-wire defects, by the way of microscopic analysis and data statistics, we can determine the mechanism of this defects and presented improvement measures. The defects contains lots of crack along the mill on surface, from cross section observation, we can find that the crack is two-way ductility, there are mixed crystal shape along the crack. The defects locate beside the edge about 15mm, after the cold rolling, steel plate surface exist pits and hole. Comparison of statistics in the production process, defects formation is closely to the heating furnace temperature, width reduction amount and the period of edger using. The problems in operation are studied and the new concept for process optimization is recommended. After tracking the defects occurrence position is less than 10 mm of edge position.

Key words: edge thin-wire defects, high strength IF, period, hot rolling

热轧 Parsytec 表面检测系统花纹板破豆缺陷的检测方法

于　洋[1]，焦会立[2]，王　畅[1]，徐文军[2]，王　林[1]，张嘉琪[2]

（1. 首钢技术研究院，北京　100041；2. 首钢迁安钢铁有限责任公司，河北迁安　064404）

摘　要：本文介绍了一种热轧 Parsytec 表面检测系统分析花纹板破豆缺陷的方法，包括通过配置文件增加花纹板材料组；设定花纹板材料组的边部检测参数；设定花纹板材料组的区域检测参数；设定花纹板敏感区域检测参数；收集花纹板破豆缺陷样本进行训练，最终达到对花纹板破豆缺陷 95% 以上的分类率。通过后处理文件对花纹板破豆缺陷进行操作界面报警。

关键词：破豆缺陷，花纹板，表面检测，热轧

DP780 热镀锌板表面点状缺陷分析

邓照军，林承江

（武汉钢铁有限公司研究院，湖北武汉　430080）

摘　要：对一种 DP780 双相钢热镀锌板表面的点状缺陷进行了分析，结果表明点状内部均存在漏镀，其内部和边缘存在 Zn 颗粒。离子束切割缺陷部位的截面观察发现，Zn 颗粒与钢基之间没有抑制层，为镀锌后嵌入漏镀区域形成的颗粒。萃取碳复型样品的透射电镜观察发现，漏镀区存在较多 30~100nm 的球形 Cr、Mn 氧化物颗粒，颗粒中部分元素被锌液中 Al 置换，光滑区域的薄膜状氧化物全部被置换形成 Al 的氧化物膜。

关键词：双相钢，热镀锌，漏镀，点状缺陷，氧化富集

Analysis on Spot Defects of a 780 MPa Hot Dip Galvanized Dual Phase Steel

Deng Zhaojun, Lin Cheng jiang

(Research and Development Center of Wuhan Iron and Steel Co., Ltd., Wuhan 430080, China)

Abstract: The spot defects on the surface of a dual phase steel were studied, it is showed that skip plating existed on the spot area, and there is Zn particles in the skip plating area. There is no inhibition layer between the Zn particle and the steel matrix after milled by Ga ion beam, the particles should be formed after hot dip process. Carbon replica sample of TEM showed that there is small particles with size from 30 to 100 nm covered on the skip plating surface, which was oxidation enrichment of Cr and Mn, and part of them was substituted by Al. Smooth area in the skip plating showed be thin film of Cr and Mn oxide, and the alloy atom of which was substituted by Al completely.

Key words: dual-phase steel, hot-dip galvanize, skip plating, spot defects, oxidation enrichment

镀铝锌机组热镀工艺研究及其应用

徐　勇

（宝钢股份上海梅山钢铁股份有限公司冷轧厂，江苏南京　210039）

摘　要： 本文通过对镀铝锌机组热镀工艺的研究，说明了目前影响热镀工艺的关键因素有锌锅的型式和尺寸规格、锌锅感应器型式、锌锅实际使用的功率大小、锌锅温度波动大小、锌锅内沉没辊系的布置、是否使用沉没辊系刮刀、温度制度是否合理、捞渣制度是否合理等方面，实际应用时应谨慎选择，慎重决定各项关键工艺和设备条件，采用合理的热镀工艺生产出高质量的镀铝锌产品。

关键词： 热镀铝锌，热镀，锌锅，锌渣

Research and Application of Hot-dip galvanizing Technology for GL Line

Xu Yong

(Cold Rolling Plant,Shanghai Meishan Iron & Steel Co., Ltd., Nanjing 210039, China)

Abstract: In this paper, the research on the hot-dip galvanizing process of the GL line shows that the key factors affecting the hot-dip galvanizing process are the type and size of the zinc pot, the type of the zinc pot electric inductor, the actual power of the zinc pot, Temperature fluctuations and the size of the installation of zinc pot sinking rolls system, the use of sink roll scraper, reasonable temperature system,reasonable slag system, etc., the practical application should be carefully chosen and determine the key process and equipment conditions, A reasonable hot-dip process should be chosen to produce high-quality GL products.

Key words: hot-dip 55%Al-Zn galvanizing, hot-dip, zinc pot, dross

湛江工程户外电气箱防盐雾腐蚀技术方案

朱政国

（中冶赛迪上海工程技术股份有限公司，上海　200940）

摘　要： 结合工程实例，分析盐雾腐蚀的危害成因及现状，阐述户外控制箱防盐雾腐蚀技术方案在钢厂中应用的意义，并提出防盐雾腐蚀技术在工程设计中具体实施措施。

关键词： 户外，控制箱，防盐雾腐蚀技术

The Technical Solution for Salt Spray Corrosion Resistance on Open-air Electrical Cabinet in Zhanjiang Project

Zhu Zhengguo

(Cisdi Shanghai Engineering Co.,Ltd.,Shanghai 200940, China)

Abstract: Combined with project examples, the hazardous reason and current situation on slat spray corrosion is analyzed. This paper also states the practical meaning of salt spray corrosion resistance technology on open-air control cabinet. Finally, the implementing measures of salt spray corrosion resistance technology are proposed in engineering design.

Key words: open-air, electrical cabinet, slat spray corrosion

两起冷轧板锈蚀客户抱怨的原因分析

王春喜，陈 斌，袁 群

（马鞍山钢铁股份有限公司制造部，安徽马鞍山 243000）

摘 要：本文简要阐述了冷轧板锈蚀机理和影响因素，并分析两起冷轧板锈蚀原因：（1）冲压后冷轧板表面锈蚀：原因为冲压过程中钢板表面与模具之间存在摩擦造成钢板表面油膜被破坏，新生的金属表面暴露于空气中，从而造成钢板生锈。（2）冷轧板点锈:原因为空气中的水分以灰尘颗粒为结露核，在灰尘处形成局部的水环境。微小的水环境可以提供腐蚀钢铁的关键介质氧，从而加速钢铁的局部域腐蚀，形成"点锈"。同时提出了防止锈蚀的解决方法,并成功地运用于钢厂冷轧板锈蚀技术服务。

关键词：锈蚀，缺陷，分析，服务

Cause Analysis of Corrosion of Cold Rolled Sheet

Wang Chunxi, Chen Bin, Yuan Qun

(Maanshan Iron & Steel Co., Ltd., Maanshan 243000, China)

Abstract: This paper briefly explains the mechanism and influence factors of the corrosion of cold rolled sheet, The cause of cold rolled plate corrosion is analyzed, (1) After drawing, the surface of cold rolled sheet is corroded. During the stamping process, the surface of the cold rolled sheet is rubbed against the die, and the oil film on the cold rolled sheet is destroyed, fresh metal surfaces are exposed to the air and oxygen rusts the steel plates. (2) Cold rolled sheet spot rust: the condensation of moisture in the air is bound up with dust particles, the small water environment can provide the key medium for the corrosion of iron, oxygen, which accelerates the local corrosion of steel, forming "spot rust". At the same time, the solutions of preventing corrosion are invented and successfully applied to the technical service of cold rolled plate corrosion.

Key words: corrosion, defect, analysis, service

钢材表面少酸除鳞工艺设备发展现状

徐言东[1]，顾　洋[2]，詹智敏[1]，马树森[3]，李建强[1]，袁华朋[3]

（1. 北京科技大学工程技术研究院，北京　100083；2. 北京科技大学钢铁技术协同创新中心，
北京　100083；3. 大连鑫永尚科技有限公司，辽宁大连　116021）

摘　要：长期以来多采用酸洗技术对热轧板、卷和浸镀件等表面进行除鳞处理。最终产物具有腐蚀性，一方面降低产品的表面质量和性能；另一方面对操作人员和设备造成损害，污染环境。随着国家对环境保护的重视，国内外均对除鳞工艺进行研究，探索最新的技术和工艺。本文综合以往的研究，指出目前除鳞工艺的发展方向为少酸或者无酸的机械式处理工艺和以磷酸代替盐酸的新工艺。

关键词：钢，除鳞工艺，酸洗，设备

Development Status of Sulfuric Acid Descaling Process Equipment on Steel Surface

Xu Yandong[1], Gu Yang[2], Zhan Zhimin[1], Ma Shusen[3], Li Jianqiang[1], Yuan Huapeng[3]

(1. Engineering and Technology Research Institute, University of Science and Technology Beijing, Beijing 100083, China; 2. Collaborative Innovation Center of Steel Technology, University of Science and Technology Beijing, Beijing 100083, China; 3. Dalian Yongsun Co., Ltd., Dalian 116021, China)

Abstract: The phosphorus removal process of hot-rolled plates, rolls and dregs was acid pickling technology for long-term. The final corrosive product could reduce the product surface quality and performance, also do damage to the operator and equipment, and pollute environment. With the national emphasis on environmental protection, both domestic and foreign researcher focus on phosphorus removal process to explore the latest technology and technology. This paper summarizes the previous research, pointing out the current development of the phosphorus removal process is less acid or acid-free mechanical process and phosphoric instead of hydrochloric.

Key words: steel, phosphorus removal process, acid pickling, equipment

低 Cr 管线钢抗 CO_2 腐蚀性能研究

郎丰军，黄先球，庞　涛，崔　雷

（武钢有限技术中心，湖北武汉　430080）

摘　要：本文介绍低 Cr 抗 CO_2 腐蚀管线钢实验室制备工作，探讨 Cr 含量和微观组织的对管线钢抗 CO_2 腐蚀性能的影响。Cr 在 0.02%~0.53%范围内，试验钢腐蚀速度先减小后增大。添加 0.13%Cr、铁素体+贝氏体组织的试验钢 CO_2 腐蚀速度较 X65 钢下降 54.4%。

关键词：管线钢，CO_2 腐蚀，低 Cr

Study on CO$_2$ Corrosion Resistance of Low Cr Pipeline Steel

Lang Fengjun, Huang Xianqiu, Pang Tao, Cui Lei

(Technical Center of Wuhan Iron and Steel Co., Ltd., Wuhan 430080, China)

Abstract: The development of low Cr anti-carbon dioxide corrosion pipeline steel was introduced, with the influence of Cr content on carbon dioxide corrosion. The corrosion rate of test steel decreased first and then increased with 0.02%~0.53%Cr. Compare to X65 steel, the corrosion rate of test steel with 0.13%Cr and ferrite + bainite decreased 54.4%.
Key words: pipeline steel, carbon dioxide corrosion, low Cr

连续热镀锌锌花不均影响因素及作用机理

李云龙，邓永刚，花福安，邸洪双，李建平

（东北大学轧制技术及连轧自动化国家重点实验室，辽宁沈阳 110819）

摘 要： 本研究使用共聚焦显微镜、扫描电镜、电子探针及化学定容方法研究了产生问题的原因及主要影响因素的作用机理。结果表明：热轧酸洗基板的表面粗糙度及钢中碳化物的分布以及带头带尾铁锌层增厚等是造成锌花不均匀的主要因素。随着基板粗糙度的增加，带钢在出锌锅冷却过程中形核点增加，表面积增大，在加热过程中吸热速率快，带头带尾厚度又大于带中导致头尾部含有的核心热增多，使头尾温度高于带中，进锌锅后铁锌反应剧烈且时间较长，铁锌合金层增厚；当钢中碳化物主要沿着铁素体晶界分布时，Fe$_2$Al$_5$ 的形成受到限制，使 Al 的消耗量增加，锌层附近的 Al 含量降低，使得锌液的流动性下降，导致锌液中的 Pb 和 Sb 有充足的时间向枝晶偏析，增加了枝晶的增长速度，形成了大锌花。
关键词： 热基镀锌，粗糙度，形核，锌花，铁锌合金

The Influence Factors and Mechanism of Continuous Hot Galvanizing Zinc Grain

Li Yunlong, Deng Yonggang, Hua Fuan, Di Hongshuang, Li Jianping

(State Key Laboratory of Rolling and Automation, Northeastern University, Shenyang 110819, China)

Abstract: This study used confocal microscope, scanning electron microscope, electron probe and chemical determination method to study the causes of the problems and the mechanism of the main influencing factors.The results show that the surface roughness of hot-rolled acid-washed substrate and the distribution of carbides in steel and the thickening of the lead belt are the main factors that cause the inhomogeneity of the zinc flowers.With the increase of substrate roughness, strip steel in the process of the zinc pot cooling nucleation point increase, increases surface area, absorption of heat faster, in the process of heating in the lead the thickness of the tail and larger than lead to increase in the number of core thermal head tail contains, in the end temperature is higher than the belt, iron zinc in zinc pot after violent reactions and for a long time, iron, zinc alloy layer thickening;When main along the ferrite grain boundary carbides distribution in steel, the formation of Fe$_2$Al$_5$ is restricted, increase the consumption of Al, zinc layer near the lower Al content, make the liquid zinc liquid drops, causes the Pb and Sb in the liquid zinc have plenty of time to the dendritic segregation, increased the dendrite growth, formed the big zinc grain.

Key words: hot-base galvanizing, roughness, nucleation, zinc grain, Fe-Zn alloy

连续退火时效工艺对镀锡板钢硬度的影响

魏宝民[1,2]，穆海玲[2]，白振华[1]

（1. 燕山大学国家冷轧板带装备及工艺工程技术研究中心、燕山大学亚稳材料制备技术与科学国家重点实验室，河北秦皇岛　066004；2. 上海梅山钢铁股份有限公司，江苏南京　210039）

摘　要： 本文在试验室模拟了 MR T-4CA 镀锡板连续退火不同时效工艺，研究时效工艺对其硬度(HR 30T，下同)的影响关系，并通过内耗等方法对硬度变化的机理进行分析，提出通过时效段工艺提高硬度的方案，并在现场得到了应用。现场的应用表明时效开始温度由 380℃降低至 340℃，时效时间由 150s 缩短至 50s，材料的硬度由 58.5 提高值 61.3,取到了预期的效果。

关键词： 时效工艺，硬度，内耗，MR T-4CA 镀锡板

Effect of Aging Process in Continuous Annealing on Hardness of Tinplate Steel

Wei Baomin[1,2], Mu Hailing[2], Bai Zhenhua[1]

(1. National Engineering Research Center for Equipment and Technology of C.S.R, State Key Laboratory of Metastable Materials Science and Technology, YanShan University, Qinhuangdao 066004, China; 2.Shanghai Meishan Iron & Steel Co., Ltd., Nanjing 210039, China)

Abstract: In this paper, the effects of aging process on hardness (HR 30T, the same) were studied in a laboratory by simulating MR T-4CA tinplate continuous annealing. and the mechanism of hardness change was analyzed by internal friction and other methods, Proposed by the aging process to improve the hardness of the program, and in the Production Line has been applied. The application of the Production Line shows that the aging start temperature is reduced from 380 ℃ to 340 ℃, the aging time is shortened from 150s to 50s, the hardness of the material increases from 58.5 to 61.3, and the expected effect is obtained.

Key words: aging process, hardness, internal friction, MR T-4CA tinplate

退火工艺对热镀锌双相钢力学性能的影响

李春诚，王鲲鹏，王　禹，李沈洋

（本钢产品研究院，辽宁本溪　117000）

摘　要： 本文采用连续退火模拟试验机模拟镀锌生产工艺，研究了退火工艺参数对热镀锌双相钢力学性能的影响。

试验结果表明：随着退火温度的提高，双相钢的屈服强度降低，抗拉强度增加，断后伸长率先增大后减小；随着退火时间的提高，双相钢的屈服强度、抗拉强度略有降低，断后伸长率变化不大；随着缓冷温度的提高，钢的屈服强度先降低后增加，抗拉强度增加，断后伸长率先增大后减小。

关键词：双相钢，热镀锌，力学性能，连续退火

Effect of Annealing Process on Mechanical Properties of Hot Dip Galvanized Dual Phase Steel

Li Chuncheng, Wang Kunpeng, Wang Yu, Li Shenyang

(Product Research Institute of BX Steel, Benxi 117000, China)

Abstract: Influence of hot dip galvanized process on mechanical properties of dual phase steel were studied by continuous annealing simulator. The results indicate that the yield strength decreased, tensile strength increased, the elongation increased and then decreased with the increasing of annealing temperature. The yield strength and tensile strength decreased, the elongation varied little with the increasing of soaking time. The yield strength decreased and then increased, tensile strength increased, the elongation increased and then decreased with the increasing of slow cooling temperature.

Key words: dual phase steel, hot-galvanize, mechanical property, continuous annealing

高铝镀锌双相钢表面点线状缺陷成因分析及成型影响

张亮亮[1,2]，于 洋[1,2]，高小丽[1,2]，李晓军[3]，王泽鹏[1,2]，李 欢[3]

（1. 首钢技术研究院，北京 100043；2. 绿色可循环钢铁流程北京市重点实验室，北京 100043；
3. 首钢京唐公司冷轧厂，河北唐山 063200）

摘 要：高 Al 镀锌双相钢表面存在两种典型缺陷——点状缺陷及线状有手感缺陷，造成冲压后脱锌影响产品美观及涂装。结合微观分析和产线工况发现，点状缺陷处含大量 Si、Mg、O，与镀后风机冷却引入杂质有关；间断线状缺陷基板存在起皮且含有大量 Mn、Al 的氧化颗粒质点，冷轧模拟轧制表明该缺陷与高 Al 成分导致的热轧细线类缺陷遗传有关。通过镀锌线镀后移动冷却设备投入，加强铸坯表面检查制度和火焰清理喷嘴更换等设备改进，点线状缺陷明显消除，成型试验模拟冲压无脱锌现象，满足客户使用要求。

关键词：高铝镀锌双相钢板，点线状缺陷，冷轧模拟，成型试验

Causes and Forming Effect of Spot and Linear Defects on the Surface of High Aluminum Galvanized Duplex Steel

Zhang Liangliang[1,2], Yu Yang[1,2], Gao Xiaoli[1,2], Li Xiaojun[3], Wang Zepeng[1,2], Li Huan[3]

(1. Shougang Research Institute of Technology, Beijing 100043, China; 2. Beijing Key Laboratory of Green Recyclable Process for Iron & Steel Production Technology, Beijing 100043, China; 3. Department of Cold rolling operation of Shougang Jingtang Iron & Steel Company, Tangshan 063200, China)

Abstract: The High Al-galvanized dual-phase steel has two typical defects --spot and linear defects, resulting in dezincification after forming, which affect the appearance and painting of the products. We found that, combined with micro-analysis and production line conditions, spot defects with a large number of Si, Mg, O, were due to the introduction of impurities in the air blown by the blower; The peeling substrate of non-continuous linear defects has oxidized particles with a large amount of Mn and Al, Cold rolling simulation shows that the hot-rolled thin-line defects were inherited. Spot and linear defects were eliminated through the use of locomotive cooling equipment in galvanized production line and the Execution of the surface inspection system and the nozzle to replace. There was no dezincification phenomenon after stamping simulation, which can meet customer requirements.

Key words: high Aluminum galvanized duplex steel, point and linear defects, cold rolling simulation, forming test

780MPa 镀锌高强 DP 钢表面沉没辊印缺陷控制优化

冯　岗，高兴昌

（本钢板材股份有限公司第三冷轧厂，辽宁本溪　117000）

摘　要： DP 钢具有较好的延展性和较高的强度，满足了加工前冲压和加工后强度的需求。沉没辊印缺陷严重的影响了 DP 钢产品的表面质量。通过对 DP 钢退火工艺的调整，从沉没辊印产生的机理上控制了锌液渣的产生，从而实现了 DP 钢表面沉没辊印缺陷的控制和消除，保证了产品的最终表面质量。

关键词： 双相钢，沉没辊印，镀锌工艺，表面质量

Control and Optimization of Sink Roll Mark of 780MPa Continuous Galvanizing Dual Phase Surface

Feng Gang, Gao Xingchang

(Ben Gang Steel Plates Co., Ltd., The Third Cold Rolling Mill, Benxi 117000, China)

Abstract: With the good ductility and high strength, the DP steel satisfy the requirement of drawing before produce and strength after produce. The sink roll mark seriously damages the surface quality of DP steel. By changing the CGL process parameters, based on the principle of the cause of sink roll mark, the goal that control and eliminate the sink roll mark and guarantee the product surface quality, is achieved.

Key words: DP, sink roll mark, continuous galvanizing process, surface quality

冷轧镀锡原板表面黑斑缺陷成因分析

吕家舜[1]，周　芳[1]，杨洪刚[1]，张福义[2]，李　锋[1]，刘仁东[1]

（1. 鞍钢集团钢铁研究院，辽宁鞍山　114009；2. 鞍钢股份冷轧厂，辽宁鞍山　114001）

摘　要： 表面黑斑缺陷常见于冷轧镀锡原板的表面，严重影响产品质量，本文利用激光共聚焦显微镜、扫描电镜、

X-射线光电子能谱仪等分析设备对黑斑缺陷进行了原位分析，结果显示，缺陷位置的粗糙度曲线显示异常，由钢板与轧辊之间的相对滑动引起，XPS 能谱分析发现缺陷处元素 C、O 的含量偏高，为轧制过程中的过热引起的氧化和随后退火过程中发生的乳化液的碳化，同时给出了解决措施。

关键词：冷轧镀锡原板，黑斑缺陷，激光共聚焦，光电子能谱仪

Analyze of Black Spot Defects in Black Plate Surface

Lv Jiashun[1], Zhou Fang[1], Yang Honggang[1], Zhang Fuyi[2], Li Feng[1], Liu Rendong[1]

(1. Research and development academe of Ansteel Group., Anshan 114009, China;
2. Cold rolling plant of Ansteel limited, Anshan 114001, China)

Abstract: This article introduces the usual flaw of carbon cold rolling steel strip surface that is pressing and oxidizing black spot defects characteristic and cause of its formation. O in this article the microstructure revolution of coating was discussed. Surface and cross-section microstructure, coating elements along depth, element surface and cross-section distribution and phases in coating were observed by SEM and LSM, GDOES (Glow discharge optical emission spectrometer), EPMA (electron probe micro analyzer) and X-ray diffract meter, respectively. The discussion results showed that: During two years' practice, the cold-rolling plant of Taishan steel controls black spot defects better by improving quality of pickled sheet and rolling degree of deformation in pass, and by controlling the temperature, thickness, pressure, outflow and cleanness of emulsion.

Key words: steel strip, black spot defects, LSM, XPS

红外线干涉法冷轧带钢表面涂层膜厚在线检测系统

施振岩，袁　焕，张　赢

（上海宝钢工业技术服务有限公司，上海　201900）

摘　要：冷轧带钢表面涂层膜厚在线检测需要，研发了红外线干涉法冷轧带钢表面涂层膜厚在线检测系统。系统主要由探头电控行走机构、探头装置、终端计算机装置等构成。系统采用红外线干涉法检测原理，来实现红外线干涉法冷轧带钢表面涂层膜厚在线实时检测。应用效果证明系统可检测涂层膜厚的种类较多，除彩涂板涂层以外，还可检测电镀镀/热镀锌/镀铝锌耐指纹膜厚和钝化膜厚和磷化膜厚和涂油量，同时还可以检测硅钢绝缘涂层膜厚。系统具有很好的可靠性和检测精度。

关键词：涂层膜厚，干涉法，在线检测系统，检测精度

Infrared Interference of Cold-Rolled Silicon Steel Strip Surface Coating Thickness Online Detection System

Shi Zhenyan, Yuan Huan, Zhang Ying

(ShangHai Baosteel Industry Technological Service Co., Ltd., Shanghai 201900, China)

Abstract: For the need for online detection of cold-rolled silicon steel strip surface coating thickness, we developed an infrared interference online detection system. The system is mainly composed of a probe control walking mechanism, probe device, and computer terminal device, etc. The system applies the method of infrared interference detection to detect cold-rolled silicon steel strip coating thickness in a real time online setting. Practical application results showed that the system can detect various types of coating thickness. Not only can it detect silicon steel insulation coating, it can also detect electroplating / galvanized fingerprint resistant film thickness, the film thickness and the phosphating film thickness，and also the film thickness of paint coated board. The system has been proved to have good reliability and detection accuracy.

Key words: insulation coating film thickness, interference method, the online detection system, the accuracy of detection

八钢 15 万吨热镀锌线卧式活套带钢跑偏问题浅析

任 轶，袁亚军

（八一钢铁股份有限公司轧钢厂冷轧分厂涂镀作业区，新疆乌鲁木齐 830022）

摘 要： 简述了八钢 15 万吨连续热镀锌机组带钢跑偏的原因及原理，经过现场试验找到了有效防止带钢跑偏的方法，同时取得了良好的效果，使八钢连续热镀锌产线入口卧式活套连续生产套量 45%提升至 80%。

关键词： 热镀锌，卧式活套，辊型，跑偏

150 Thousand Tons of hot Galvanized Line of Steel Strip Deviation of Horizontal Looper

Abstract: The reason of 150 thousand tons of continuous hot-dip galvanizing strip and the paper describes the principle of the field test to find the effective method to prevent strip deviation, and achieved good results.

热镀锌板表面锌灰缺陷的形成机理及控制机制

朱砚刚，彭 俭

（宝山钢铁股份有限公司，上海 201900）

摘 要： 锌灰缺陷是高等级质量要求热镀锌板难以接受的一类缺陷，其直接影响了板材的冲压后质量。本文结合近些年投产的热镀锌板线的生产实践，对锌灰缺陷的形成机理从构建理论模型的角度进行了推导，并对现场控制措施方面尝试建立模型，以期对实际控制水平进行进一步的提升。

关键词： 热镀锌，锌灰，露点，抽锌泵，溢流

The Formation and Control Mechanism of the Zinc Ash Defects on Galvanized Plate

Zhu Yangang, Peng Jian

(Baoshan Iron Steel Co., Shanghai 201900, China)

Abstract: The zinc ash defect is an unacceptable defect of hot galvanized plate according to high gr ade quality requirements, which has a direct impact on the quality of sheet metal upon stamping. This paper relates to the production of hot dip galvanized sheet line in recent years, deduces the zinc a sh defect formation mechanism from the perspective of constructing a theoretical model, and attem pts to establish the model of field control measures, in the hope of promoting the actual control level.

彩色涂层钢板涂层固化降低燃耗探讨

符新涛，刘佑爽

（宝钢集团八一钢铁股份有限公司冷轧薄板厂，新疆乌鲁木齐 830022）

摘　要：2015 年度八钢彩板生产线吨钢混合煤气消耗达到 145m^3/t，国内先进机组宝钢彩色钢板生产线吨钢混合煤气消耗 90m^3/t，对比展开原因分析，采取对策措施降低混合煤气吨钢消耗。

关键词：焚烧炉，固化炉，混合煤气，板温

Color Coating Steel Plate Solidification Reduce to Investigate the Mixed Gas Consumption

Fu Xintao, Liu Youshuang

(Baosteel Group Xinjiang Bayi Iron & Steel Co., Ltd., Urumchi 830022, China)

Abstract: Bayi Iron & Steel Color plate production line Per ton of steel Mixed gas consumption Is 145m^3/t, comparing the domestic advanced unit Baosteel Group color plate production line per ton of steel mixed gas consumption is 90m^3/t, Cause analysis on contrast take measures to per ton of steel mixed gas consumption.

Key words: incinerator, curing oven, mixed gas, plate temperature

连续热镀锌钢带边厚缺陷的浅析

艾厚波

（本溪钢铁(集团)有限责任公司冷轧薄板厂，辽宁本溪 117000）

摘　要：简述了连续热镀锌钢带边厚缺陷的产生的基本原理和技术情况。通过对原料、气刀参数、工艺速度等因素的分析，找出热镀锌钢带边厚缺陷解决方法。

关键词：连续热镀锌，边厚，气刀

Analysis of Edge Thickness Defect of Continuous Hot-dip Galvanizing Strip

Ai Houbo

(Benxi Iron and Steel (Group) Co., Ltd., Cold Rolling Sheet Plant，Benxi 117000, China)

Abstract: The continuous hot galvanized steel strip edge thickness defects resulting from the basic principle and technical condition. Based on the raw materials, air knife, the speed of the craft parameter analysis, find out the hot galvanized steel strip edge thickness defect solution.

Key words: continuous hot dip galvanized, edge thickness, air knife

镀铝锌板冲压发黑的原因分析

林　彬[1]，杨洪刚[2]

（1. 鞍钢股份冷轧厂，辽宁鞍山　114021；2. 鞍山钢铁研究院，辽宁鞍山　114021）

摘　要：高效耐蚀镀铝锌板越来越多的得到国内终端用户的认可和普及使用，但镀铝锌生产难度高于热镀锌，引进到国内的时间晚，镀铝锌工艺技术交流不充分，应用技术研究也有待完善和积累。针对鞍钢镀铝锌板在冲压中出现表面发黑现象，利用扫描电镜分析，结合 XPS 测量纳米层范围成分优势，分析冲压发黑的根本原因。得出铝锌镀层钢板冲压过程中，表面耐指纹膜破坏脱落，缺少耐指纹膜润滑的条件下，模具对钢板摩擦产生的热量使铝锌镀层产生剥落和氧化发黑的结论。从而得出改善耐指纹皮膜润滑性的改进方案，提高鞍钢镀铝锌产品应用性能。

关键词：镀铝锌，扫描电镜，能谱分析，X 射线光电子谱仪

Analysis of Galvalume Stamped to Blacken

Lin Bin[1], Yang Honggang[2]

(1. Cold Strip Work of Ansteel Company Ltd., Anshan 114021, China;
2. Academe of Steel of Ansteel, Anshan 114021, China)

Abstract: The galvalume are accepted to be used by more domestic End-User for high efficiency antirust. However GL are more difficult produced than GI. It was later introduced to domestic. Technology communion of GL is no more adequately. Study to applicability need to be improving and accumulate. For galvalume stamped to blacken, analysis the reason by SEM combine with XPS, that have advantage at measure rang of nanometer. Conclusion is galvalume layer flake off and oxidative to blacken because there is friction heat between mould and sheet, on condition of anti-finger film of surface flake away, lack of lubricate. Accordingly find out improve project that increase anti-finger film lubricating ability, thereby advance applicability of ansteel galvalume.

Key words: galvalume, SEM, EDS，XPS

热镀锌炉内钢板的表面氧化现象

刘军友，李 成，柳 军，张国强，何天庆

（鞍钢股份有限公司，辽宁鞍山 114021）

摘 要：本文在热镀锌连续退火炉内对硅锰含量相对较高的钢种进行了高温氧化试验。分别在氧化后和还原后的位置取样，并利用辉光光谱仪进行了测试，氧化后试样发现了硅和锰元素的内氧化现象，而且硅元素向内扩散较锰更为明显，表面已没有影响镀层质量的硅元素存在；还原后试样表面几乎没有氧存在，基本上为铁元素，为实现高表面高强热镀锌钢的生产可能提供了实践依据。

关键词：热镀锌，内氧化，铁硅锰合金

Oxidation of Steel in Hot-galvanizing Furnace

Liu Junyou, Li Cheng, Liu Jun, Zhang Guoqiang, He Tianqing

(Cold Strip Work Angang Steel Company Limited, Anshan 114021, China)

Abstract: The high temperature oxidation test based on hot dip galvanized steel in continuous annealing furnace of silicon manganese content were relatively high. Respectively after oxidation and reduction of the position after sampling, and tested by glow discharge spectrometer, oxidized specimens were found in the oxidation of silicon and manganese, silicon and manganese than inward diffusion is more obvious, the surface has no silicon influence the quality of the coating are also original; after the specimen surface is almost no oxygen basically, for iron, provides a practical basis for the realization of high surface strength galvanized steel production.

Key words: hot galvanized, internal oxidation, Fe-Mn-Si alloy

5　金属材料深加工

炼铁与原料
炼钢与连铸
轧制与热处理
表面与涂镀
★　金属材料深加工
钢铁材料
汽车钢
海洋工程用钢
轴承钢
电工钢
粉末冶金
非晶合金
高温合金
耐火材料
能源与环保
分析检测
冶金设备与工程技术
冶金自动化与智能管控
冶金技术经济

"绿色管带" 用镀锌钢丝绳深加工技术研究

冯　平，邵永清，施　瑾

（法尔胜泓昇集团江苏法尔胜特钢制品有限公司，江苏江阴　214434）

摘　要： 为了满足管带行业绿色化发展要求，对钢丝绳进行轻量化、结构多样化和包装环保化研究。

关键词： 钢丝绳，管带增强材料，深加工

Deep-processing Technology Research of Zinc-coated Steel Wire Ropes for "Greening Hose and Belt"

Feng Ping, Shao Yongqing, Shi Jin

(Jiangsu Fasten Steel Cord Co., Ltd., Jiangyin 214434, China)

Abstract: In order to meet the greening development requirements of hose and belt industry, light-weight、constructural diversification and packaging environmental protection of steel wire rope are studied.

Key words: steel wire rope, reinforcements for hose and belt, deep processing

钢帘线产业发展赋予上游企业的机遇与挑战

周志嵩[1,2]，姚利丽[1,2]，周　洁[1,2]，姜志美[1,2]，华　欣[1,2]

（1. 江苏兴达钢帘线股份有限公司，江苏泰州　225721；
2. 江苏省结构与功能金属复合材料重点实验室，江苏泰州　225721）

摘　要： 本文介绍了钢帘线的生产工艺流程和钢帘线用盘条的发展现状，并对钢帘线用国产盘条与进口盘条的化学成分、力学性能、微观组织等性能进行对比分析。结果表明，国产盘条的质量稳定性与进口盘条相比，仍存在一定差距，提高钢帘线用国产盘条的内部质量是目前我国钢铁行业亟待解决的重要问题之一。同时，近些年来环保、节能、安全意识高涨，对轮胎轻量化要求强烈。纤维帘线、复合帘线等高强度、新材料、新结构的帘线发展步伐加快，轮胎中的钢丝被完全取代后，势必会对钢丝的用量产生影响。对于线材企业来讲，这既是机遇又是挑战。

关键词： 钢帘线，盘条，国产，轮胎，骨架材料，纤维

The Development of Steel Cord Industry Gives the Opportunities and Challenges for Upstream Enterprises

Zhou Zhisong[1,2], Yao Lili[1,2], Zhou Jie[1,2], Jiang Zhimei[1,2], Hua Xin[1,2]

(1. Jiangsu Xingda Steel Cord Co., Ltd., Taizhou 225721, China;

2. Jiangsu Key Laboratory for Structural and Functional Metal Materials Composites, Taizhou 225721, China)

Abstract: The production process of steel cord and the developing situation of wire rod for steel cord were introduced in the paper. The chemical composition, mechanical properties, microstructure of domestic wire rod and imported wire rod used for steel cord were compared and analyzed. The results show that the quality stability of the domestic wire rod still has a certain gap compared with imported wire rod. Improving internal quality of domestic wire rod used for steel cord is one of the important problems that need to be solved urgently in the steel industry of our country. Meanwhile, with the increase of awareness for environmental protection and energy saving and safety, human beings have the strong demand for lightweight tires in recent years. The development of high strength, new material, and new structure of fiber cord or composite cord is accelerating. When the steel wire in the tire is completely replaced, the usage amount of steel wire will be affected inevitably. This is an opportunity as well as a challenge for the wire rod enterprises.

Key words: steel cord, wire rod, domestic, tire, framework material, fibre

轻量化功能化精密辊压型材技术和应用发展

晏培杰，毕若凌

（上海宝钢型钢有限公司，上海　201999）

摘　要：轻量化功能化精密辊压型材是一种通过以辊压型材产品为研究对象，进行其材料、结构轻量化以及功能化进行整合优化设计，并通过冲孔，辊压，焊接，切边等多工艺集成实现高效自动化制造的工艺，得到具有良好尺寸精度的近成品型材产品。本文对轻量化功能化精密辊压型材关键技术和特点进行了系统介绍，同时介绍该类型型材的使用现状和应用前景。

关键词：轻量化，功能化，辊压型材

Development and Technology of Lightweight Functional High Precision Roll Forming Profile

Yan Peijie, Bi Ruoling

（Shanghai Baosteel Section Steel Co., Ltd., Shanghai 201999, China）

Abstract: Lightweight functional high precision roll forming profile are nearly finished product profile products which has good dimensional accuracy. It has lightweight materials, structure and function of integrated optimization design and is manufactured by integrated manufacturing process including the punching, roll forming laser welding and cutting. In this paper, the design path, key technology, advantages and disadvantages of lightweight functional high precision roll Forming profile are introduced, and the application status and development are proposed.

Key words: Lightweight, functional, roll forming profile

冷成型工艺在汽车行业的应用

李文奇

（北京新光凯乐汽车冷成型件股份有限公司，北京　101102）

摘　要： 冷成型工艺具有多方面优势应用于汽车行业，汽车行业和冷成型件公司对钢材制造商及钢材质量有较高要求。

关键词： 冷成型工艺，冷镦钢，塑性成型，成型性件

非调质钢高强螺栓铁素体时效行为对性能的影响

罗志俊，李舒筎，孙齐松，王晓晨，马　跃，王丽萍

（首钢集团有限公司技术研究院特殊钢研究所，北京　100043）

摘　要： 采用热处理时效及透射电镜（TEM）观察的试验方法，分析了铁素体+马氏体双相（F+M）组织及铁素体+珠光体（F+P）组织中铁素体组织的时效行为及时效制度对力学性能、永久伸长的影响。结果表明：F+P 类型钢，在 200～400 ℃内时效，强度没有变化。而 F+M 类型非调钢，热轧状态与拉拔变形行为下变化规律相同，强度先增加后降低，在 300℃下达到最大值。而高强螺栓时效处理，强度先增加后减小，在 250～300 ℃时达到峰值，但均高于未时效处理螺栓。F+P 组织高强螺栓时效前永久伸长量达到 15 μm 以上，时效后大幅度下降至不高于 5 μm；F+M 类型非调质螺栓时效前后永久伸长量变化较小，均为 1～2 μm。螺栓制造过程中的拉拔变形诱导析出纳米尺寸的细小碳化物或氮化物阻碍位错运动和沉淀强化是时效后强度变化的原因。铁素体内部大量的相互缠结的位错以及胞状亚结构(亚晶粒)是铁素体-马氏体双相钢螺栓时效前后永久伸长变化较小的原因。

关键词： 非调钢，高强螺栓，铁素体-马氏体组织，铁素体-珠光体组织，时效处理

Effect of Ferritic Aging Behavior on Performance of Non Heat-Treated Steel High Strength Bolts

Luo Zhijun, Li Shujia, Sun Qisong, Wang Xiaochen, Ma Yue, Wang Liping

(Shougang Group Co., Ltd., Technology Research Institute Special Steel Research Institute, Beijing 100043, China)

Abstract: Effect of ferrite - martensite dual phase (F+M) structure and ferrite – pearlite (F+P) structure on high-strength bolts permanent deformation behavior under static load was investigated by utilizing transmission electron microscopy (TEM) and proof load. The results indicated that elongation of high-strength bolts with ferrite – pearlite (F+P) structure could be 15μm or even more before aging, while it declined to no more than 5μm after aging. However, in the case of non-heat-treated high-strength bolts with ferrite – martensite structure，the permanent elongation had small change before and after aging as only 1 ~ 2 μm. When aging treatment was conducted in the range of 200 ~ 400℃, the strength of both F+P and non-heat-treated F+M bolts increased first and decreased afterwards and reached a peak at 250 ~ 300℃, but were higher than that of the no aging treatment bolts. Reasons of small changes in permanent elongation before and after aging are cellular substructures (subgrains) and a lot of entangled dislocation inside ferrite. The strength variation after aging is caused by precipitation of fine disperse nitrides or carbides which impede dislocation motion in drawing deformation of the bolt manufacturing process.

Key words: non heat-treated steel, high strength bolts, ferrite-martensite, ferrite-pearlite, permanent elongation

非调质钢 46MnVS5 汽车发动机连杆胀断缺陷分析

张朝磊[1]，巴鑫宇[1]，樊世亮[2]，方　文[1]，刘雅政[1]

（1. 北京科技大学材料科学与工程学院，北京　100083；
2. 西宁特殊钢集团有限责任公司，青海西宁　810005）

摘　要： 通过合金成分、显微组织定量分析，以及断口观察和力学性能测定，对近期国内生产的非调质钢 46MnVS5 连杆胀断过程中的胀不断、断口变形大、断口不齐缺陷进行系统分析。结果表明：存在大量贝氏体异常组织、组织均匀性差、塑性过好是产生缺陷的主要直接原因。最后提出解决上述问题的策略。

关键词： 非调质钢，胀断加工技术，失效分析

Fracture Splitting Defect Analysis of Air-cooled Forging Steel 46MnVS5 Automotive Connecting Rod

Zhang Chaolei[1], Ba Xinyu[1], Fan Shiliang[2], Fang Wen[1], Liu Yazheng[1]

(1. School of Materials Science and Engineering, University of Science and Technology Beijing, Beijing 100083, China; 2. Xining Special Steel Co., Ltd., Xining 810005, China)

Abstract: Fracture splitting defects of air-cooled forging steel 46MnVS5 automotive connecting rod were analyzed by alloys composition, quantitative analysis of microstructure, fracture observation and mechanical propertie determination. The defects include not fracture, large deformation after fracture and fracture uneven. The results show that the defects are as a result of a large number of bainite, lack homogeneity of microstructure and excessive plasticity. Finally, the strategies for solving the above problems were proposed.

Key words: fracture splitting technology, air-cooled forging steel, failure analysis

罩式退火工艺对超高强热成形钢组织与性能的影响

赵征志[1,2]，陈伟健[1]，梁江涛[1]

（1. 北京科技大学钢铁共性技术协同创新中心，北京　100083；
2. 北京科技大学冶金工程研究院，北京　100083）

摘 要：本文研究了超高强热成形钢的罩式退火工艺对其性能的影响和组织演变规律，结合成分、相变、工厂的现状以及传统工艺，模拟了 680℃和 700℃的罩式退火工艺，保温时间分别为 1h、2h、5h 和 8h。实验结果表明：实验用钢的综合性能对罩式退火的温度比较敏感；保温时间对钢的力学性能和组织中的碳化物有影响；合理的设置罩式退火温度和保温时间，可以获得低强度高塑性的钢板，有利于热成形前基板的冲裁和切削，大幅度缩短生产周期，降低能耗和成本。

关键词：罩式退火，热成形钢，微观组织，力学性能，碳化物

Effect of Enclosure Annealing Process on Microstructure and Properties of Ultrahigh - strength Hot - formed Steel

Zhao Zhengzhi[1,2], Chen Weijian[1], Liang Jiangtao[1]

(1. Collaborative Innovation Center of Steel Technology,
University of Science and Technology Beijing, Beijing 100083, China;
2. Engineering Research Institute, University of Science and Technology Beijing, Beijing 100083, China)

Abstract: In this paper, the effect of bell-type annealing process on the properties and microstructure evolution of ultra-high strength hot forming steel has been studied, combining of components, phase change, the status of the factory and the traditional process, the temperature of the batch annealing process at 680℃ and 700℃ was simulated, and the holding time was 1h, 2h, 5h and 8h respectively. The results indicate that the comprehensive properties of the experimental steel are more sensitive to the temperature of bell-type annealing, holding time has effects on mechanical properties of steel and carbides in microstructure and reasonable setting bell-type annealing temperature and holding time can obtain low strength and high plasticity steel plate, which is beneficial to blanking and cutting of substrate before thermoforming, greatly shortening production cycle and reducing energy consumption and cost.

Key words: bell-type annealing, hot stamping steel, microstructure, mechanical property, carbides

高强钢 DP980 激光焊接接头组织性能研究

徐 梅，米振莉，徐亚鹏，李 龙

（北京科技大学工程技术研究院，北京 100083）

摘 要：利用 YLS-6000 型激光器对高强钢 DP980 进行激光拼焊试验，通过拉伸试验、焊接接头显微硬度测量、光学显微镜以及 EBSD 等方式来分析 DP980 焊接接头性能及显微组织。结果表明：焊接后的 DP980 抗拉强度受热输入影响不大，断后伸长率明显下降，并且随着焊接热输入的增大，断后伸长率呈现出不断下降的趋势；DP980 焊接接头组织分布非常不均匀，硬度变化较大，存在较为明显的软化区。

关键词：DP980 双相钢，激光焊接，热影响区，组织分析

Research on Microstructure and Properties of Laser Welding DP980 High Strength Steel Weld Joints

Xu Mei, Mi Zhenli, Xu Yapeng, Li Long

(Engineering Research Institute, University of Science and Technology Beijing, Beijing 100083, China)

Abstract: High strength steel DP980 was welded employing YSL-6000 laser. DP980 steel welded joint performance and microstructure of the study is through the tensile test, micro-hardness measurement of welded joints, optical microscopy and EBSD. The results show that the heat input has little effect on the tensile strength of DP980 after welding, and the elongation after fracture decreases obviously. With the increase of welding heat input, the elongation of DP980 shows a decreasing trend. The distribution of DP980 welded joint is very uneven, the hardness changes greatly, and there is obvious softening zone of welded joint.

Key words: DP980 steel, laser welding, heat affected zone, organizational analysis

DP600 高强钢多道次拉深旋压成形机理研究

石　磊[1,2]，肖　华[1,2]，夏琴香[3]

（1. 宝山钢铁股份有限公司研究院，上海　201900；2. 汽车用钢开发与应用技术国家重点实验室（宝钢），上海　201900；3. 华南理工大学机械与汽车工程学院，广东广州　510640）

摘　要： 为研究高强钢 DP600 多道次拉深旋压成形机理，本文基于杯形零件，采用 ABAQUS/Explicit 软件，对 DP600 高强钢多道次拉深成形过程建立数值模型，获得了其成形过程应力、应变及壁厚变化规律。研究结果表明，在 DP600 多道次拉深旋压成形中，材料始终处于径向与切向受压、轴向受拉的状态，其等效应力值在周向上分布均匀，而在轴向上差异明显，最大应力出现在旋轮接触区域及圆角区域。随着旋压进展，口部应力逐渐增加，而旋压完成后，圆角区域应力获得释放而减小。在每道次旋压成形后，其壁厚值在圆周方向上分布均匀，在轴向上先减小后增加的趋势。

关键词： 高强钢，DP600，杯形件，多道次拉深旋压，数值模拟

Research on Forming Mechanism of Muti-pass Drawing Spinning of the High Strength Steel DP600

Shi Lei[1,2], Xiao Hua[1,2], Xia QinXiang[3]

(1. Research Institute, Baoshan Iron & Steel Co., Ltd., Shanghai 201900, China;
2. State Key Laboratory of Development and Application Technology of Automotive Steels (Baosteel), Shanghai 201900, China; 3. School of Mechanical and Automotive Engineering, South China University of Technology, Guangzhou 510640, China)

Abstract: For studying the forming mechanism of muti-pass drawing spinning of the high strength steel DP600, numerical model of cup shaped part was established based ABAQUS/Explicit software, by which the stress, the strain and the wall thickness distribution was obtained. The results show that the material is always in the radial and tangential compression, axial tension, the equivalent stress value is distributed evenly in the circumferential direction, while the axial stress is obvious. The maximum stress appears in the roller contact area and the fillet region. With the development of spinning, the stress at the mouth increases gradually, but after the spinning, the stress in the fillet region is released and decreased. After each spinning pass, the wall thickness is distributed evenly in the circumferential direction, and decreases first and then increases in the axial direction.

Key words: high strength steel, DP600, cup, muti-pass deep drawing spinning, numerical simulation

中厚板电阻闪光焊线能量试验研究

黄治军，方要治，刘　斌

（武钢有限技术中心，湖北武汉　430080）

摘　要： 焊接线能量是常规电弧焊中的一个重要参数，对焊接接头性能有很重要的影响，但在电阻闪光焊中尚无此概念。本文分析了电阻闪光焊工艺特点，在焊接实验的基础上，提出了中厚板电阻闪光焊线能量概念及其计算方法总线能量指在预热、烧化、顶锻等焊接全过程的通电热输入，有效线能量则只指最终接头上承受热输入。在闪光焊过程中，大部分加热金属被烧化、挤出，少部分的加热金属锻挤成焊接接头，所以有效线能量只是总线能量的一部分。对于 8.0~18.0mm 钢板，闪光焊总线能量为 4.5~24.1kJ/mm，有效线能量为 1.0~4.6kJ/mm。结合 15mm 规格新型车轮进行了有效的试验分析，有趣的是其计算的有效线能量处于同规格钢种常用的埋弧焊单道线能量范围内。

关键词： 电阻闪光焊，线能量，车轮钢

Study on Heat Input of Electric Resistance Flush Welidng of Medium-Thick Steel

Huang Zhijun, Fang Yaozhi, Liu Bin

(Technology Center of Wuhan Iron & Steel Co.,Ltd., Wuhan 430080, China)

Abstract: Heat input is an key parameter for arc welding and has obvious effect on the properties of welded joint, but it is not mentioned in electric flush welding until now. In this article electric resistance flush welding was analysed, and heat input and its calculation mechod were proposed based upon welding experiments of medium thickness wheel steels. Total heat input is the electric resistance heat during the whole welding process starting from preheating, burning to squeezing, while effective heat input refers to heat input for the finished joint. During flush wedlng most heated metal were burnt and squeezed out, so effective heat input is just part of total heat input. For 8.0mm to 18.0mm thick steel plates, total heat inputs are from 4.5kJ/mm to 24.1kJ/mm, and effective heat inputs are from 1.0kJ/mm to 4.6kJ/mm. Focused on 15mm thick new type wheel steel the welding experiment and analysis were carried out in terms of heat input. Interestingly the effective heat input is in the normal range of single pass heat input of submerged welding for the similar thickness steel.

Key words: electric resistance flush welding, heat input, wheel steel

新一代超高强度热成形钢的强塑化技术

李光瀛[1]，马鸣图[2]，张宜生[3]，李　红[4]

（1. 中国钢研科技集团钢铁研究总院，北京　100081；2. 中国汽车工程研究院，重庆　401122；
3. 华中科技大学材料学院，湖北武汉　430074；4. 北京工业大学材料学院，北京　100124）

摘　要：随着世界各国汽车轻量化、节能减排环保、抗冲撞安全新材料的升级，超高强度级别（TS≥800~1800MPa）的先进高强度钢 AHSS 汽车板成为汽车业与钢铁业的开发热点。由于超高强度钢 UHSS 汽车板在冷冲压过程中存在严重回弹、起皱、开裂和成形极限等问题，热冲压以其高精度、无回弹、无缺陷、低冲压载荷等优点使超高强度 TS1500 级热成形钢构件在全球得到发展和应用，但其低塑性（总伸长率 TE≈6%）成为在冲撞应力超过屈服强度时（σ_I>YS1150MPa）发生脆性断裂的隐患。为提高热成形构件的抗脆断能力，一种方法是拼接式分区热成形（Tailored Hot Stamping）使构件局部区域强度降低、塑性提高；另一种方法是开发新一代热成形钢，提高整体构件的伸长率（TE≈15%）、强塑积（PSE≥20GPa·%）和抗脆断能力。本文介绍了国内外新一代热成形钢的强塑化方法及其 TS800、TS1000、TS1200、TS1500 级系列化超高强塑性钢（TE≥15%~25%）的先进热冲压处理 AHST 技术。与此同时，美国纳米钢公司 NanoSteel 开发了冷冲压用 NXG 1200 级超高强度钢，2016 年 4 月在 AK 钢公司生产线试制出冷轧带卷，性能达到 TS=1188MPa、YS=378MPa、TE=54.6%、UE=51.5%，具有竞争力和有益启示。

关键词：先进高强度钢，热冲压成形，超高强度钢，残余奥氏体，新一代热成形钢

Strengthening-Plasticizing Technology for New Generation of Ultra-High-Strength Hot-Stamping-Steel

Li GuangYing[1], Ma MingTu[2], Zhang YiSheng[3], Li Hong[4]

(1. Central Iron & Steel Research Institute, Beijing 100081, China; 2. Automotive Engineering Research Institute, Chongqing 401122, China; 3. Huazhong University of Science and Technology, Wuhan 430074, China; 4. Beijing University of Technology, Beijing 100124, China)

Abstract: With the upgrade of new auto-materials for weight-lightening, energy-saving, emission-control and collision-resistant safety, the ultra-high-strength grades of AHSS with TS≥800~1800MPa have been focused by auto & steel industry. Due to the serious spring back, wrinkling, tear crack and forming limit occured in cold stamping of UHSS, hot stamping technology has been developed globally with advantages in high-geometry-accuracy TS1500MPa grade auto-parts without any spring-back, wrinkling and forming-limit under low stamping load. While the low ductility TE≈6% is the hidden trouble in brittle fracture once high- collision-stress σ_I>YS1150MPa occurs. To increase the ductility of hot-stamped UHSS auto-parts, one method is tailored hot stamping to decrease the strength for better elongation in certain local area; another method is to develop a new generation of hot stamping steel with higher ductility TE≈15% for the whole components. In the present paper, the strengthening-plasticizing method for the new generation of hot-stamping-steel TS800、TS1000、TS1200、TS1500 with TE≥15%~25% is introduced through advanced-hot-stamping-treatment AHST. Meanwhile, NanoSteel company in USA developed NXG 1200 steel for cold stamping, and AK Steel made production trial of cold-rolled strip with property TS=1188MPa、YS=378MPa、TE=54.6%、UE=51.5% in April 2016. It is competitive and enlightening.

Key words: advanced high strength steel AHSS, hot stamping, ultra-high-strength-steel UHSS, retained austenite, new generation of hot stamping steels

热成形钢抗氢脆性能和冷弯性能研究

晋家春，谷海容，曹　煜，刘　飞，崔　磊，张　武

（马鞍山钢铁股份有限公司，安徽马鞍山　243000）

摘　要： 借助中试冶炼、轧制以及连退模拟等试验设备获得不同成分的实验用原材料，并通过平板模具淬火实验获得最终的实验钢。通过对比研究不同成分实验钢的三点弯曲性能、耐氢脆性能、微观组织、夹杂物以及析出物状态，分析了成分影响实验钢三点弯曲性能和耐氢脆性能的机理，并指出夹杂物的存在也会对实验钢耐氢脆性能产生影响。

关键词： Nb-V 微合金，热成形钢，夹杂物，氢脆，三点弯曲

Research on Hydrogen Embrittlement and Cold-bending Properties of PHS

Jin Jiachun, Gu Hairong, Cao Yu, Liu Fei, Cui Lei, Zhang Wu

(Maanshan Iron and Steel Co., Ltd., Maanshan 243000, China)

Abstract: The experimental materials with different composition were obtained by experimental equipments of smelting, rolling in pilot-scale and continuous annealing simulation, then the final test steel was obtained through quenching by plate die. The mechanism of the three-point bending performance of the test steel was analyzed by comparing the three-point bending performance and the microstructure of different composition. Meanwhile, the effects of the composition on the hydrogen embrittlement resistance were studied by comparing the hydrogen brittleness resistant and the precipitation status. Moreover, it was pointed out that the presence of inclusions would mask the effects of the composition on the hydrogen embrittlement resistance of the test steel.

Key words: Nb-V microalloyed, hot-stamped steel, inclusions, hydrogen-induced cracking, three-point bending

江淮汽车车身材料应用设计概述

孙启林，伏建博，黄家奇，鲁后国，张　龙

（安徽江淮汽车集团股份有限公司技术中心，安徽合肥　230601）

摘　要： 汽车车身包含白车身及四门两盖，在不降低性能的前提下，综合考虑轻质材料、结构优化和先进工艺，同时兼顾车身性能、重量和成本三个因素，建立起轻量化节能技术平台，为达到降低油耗目标提供有效的技术平台。本文介绍了钢质车身中涉及到的轻量化、安全性及防腐设计。

关键词： 白车身，轻量化，安全性，防腐

Application Design Summary of JAC Body Material

Sun Qilin, Fu Jianbo, Huang Jiaqi, Lu houguo, Zhang Long

(Jianghuai Motor Corporation Technical Center, Hefei 230601, China)

Abstract: The vehicle body includes the body in white and the four doors and two covers, without reducing the performance of the premise, considering the light material, structural optimization and advanced technology, consider the body performance, weight and cost, and establish a lightweight energy-saving technology platform, In order to achieve the target of saving fuel consumption to provide an effective technical platform. This article describes the steel body involved in the lightweight, security and corrosion protection design.

Key words: body in white, lightweight, safety, corrosion protection

不同原始组织状态对 SCM435 球化退火的影响

郑晓伟，左锦中，张剑锋

（江阴兴澄特种钢铁有限公司线材研究所，江苏江阴　214429）

摘　要： 为探索不同原始显微组织及状态对冷镦钢 SCM435 球化退火的影响，分别进行了珠光体+铁素体和贝氏体组织的两种显微组织的盘条球化退火试验，及分别经 27.75%减面率拉拔后的球化退火试验。研究表明：贝氏体组织比珠光体+铁素体组织易于球化，拉拔后再球化退火得到的组织球化率比热轧材直接球化退火更高；为获得更好的球化组织和更低的生产成本，采用球化退火前先经一定减面率拉拔或通过控轧控冷获得贝氏体组织的轧材直接球化退火的工艺。

关键词： SCM435，珠光体+铁素体，贝氏体，拉拔，球化退火，自发球化

Effect of Different Original Structure and State on the Spheroidizing Annealing of SCM435 Steel

Zheng Xiaowei, Zuo Jinzhong, Zhang Jianfeng

(Jiangyin Xingcheng Special Steel Co., Ltd. Wire Institute, Jiangyin 214429, China)

Abstract: To study the impact of different original structure and state before annealing on the spheroidizing annealing of SCM435 steel, the experiments were carried out on SCM435 steel of different original structure and state, including pearlite-feerite structure and bainite structure as well as their drawing samples with 27.75% reduction rate. The study results show that rate of spheroidization with original structure of bainite structure is better than rate of spheroidization with original structure of pearlite-feerite structure and it is better to anneal through drawing in a reduction rate. Therefore, it is better to use the annealing technology with as-drawed wire, but it is viable for cost reduction to anneal using as-rolled wire with original structure of bainite structure.

Key words: SCM435, pearlite-feerite, bainite, drawing, spheroidizing annealing, spontaneous spheroidizing

深冲钢板冲压开裂的实例分析

钱健清

（安徽工业大学冶金工程学院，安徽马鞍山　243002）

摘　要： 材料、模具以及现场生产工艺都可能影响的板材冲压开裂，本文在对现场冲压开裂进行基本分析后，对深冲钢板的性能、成分、组织进行全面的分析，并结合实际研究了解决该起冲压开裂的各方面原因，并且根据实际情况提出解决方案，解决该起冲压开裂问题。

关键词： 冲压，破裂，深冲钢板，实例分析

Case Study in Deep Drawing Steel Sheet Stamping Cracking

Qian Jianqing

(School of Metallurgical Engineering, Anhui University of Technology, Maanshan 243002, China)

Abstract: It can affect the sheet stamping cracking is materials, molds and the craft. The solution has been proposed for the stamping cracking, based on the basic analysis of stamping cracking and rounded analysis of deep drawing steel sheet. Solve the stamping cracking problem.

Key words: stamping, cracking, deep drawing steel sheet, case study

高强韧镁合金挤压型材的研制及其在
轨道交通上的应用

张　成[1,2,3]，谢　玉[1,2,3]，唐伟能[1,2,3]，徐世伟[1,2,3]

（1. 宝山钢铁股份有限公司研究院汽车用钢所，上海　201900；2. 汽车用钢开发与应用技术国家重点实验室（宝钢），上海　201900；3. 上海运输工具轻量化金属材料应用工程技术研究中心，上海　201900）

摘　要： 本文采用了宝钢自行优化开发的非稀土系镁合金，通过优化挤压模具、挤压加工工艺、热处理工艺等试制出了一批高铁座椅骨架用复杂截面镁合金挤压型材。其中编号为 BGZ8 的挤压态镁合金型材经过 T5 处理后，其屈服强度为 289 MPa，抗拉强度为 396 MPa，延伸率为 11.5%，机械性能完全达到高铁座椅骨架使用要求。显微组织观察表明，优化的挤压模具、挤压工艺参数和后续的热处理工艺，既保证了挤压型材截面不同部位上显微组织的均匀性，也获得了有效强化的基面织构、细晶组织和细小均匀分布的多种析出相，这种多尺度的显微组织保证了镁合金挤压型材获得较高的综合力学性能。针对某型号的高铁座椅骨架采用该高强韧镁合金挤压型材对现有铝型材进行了替代验证，结果显示宝钢开发的镁合金座椅骨架底座仅重 1.7 kg，相比原有的铝合金座椅骨架减重 25%，且各项力学性能优于客户要求。

关键词： 镁合金，挤压型材，工艺优化，高强韧

Research and Preparation of High Strength and Toughness Magnesium Alloys Extrusion Profiles and Their Application in Rail Transportation

Zhang Cheng[1,2,3], Xie Yu[1,2,3], Tang Weineng[1,2,3], Xu Shiwei[1,2,3]

(1. Research Institute, Baoshan Iron & Steel Co., Ltd., Auto Sheet Department, Shanghai 201900, China;
2. State Key Laboratory of Development and Application Technology of Automotive Steels (BaoSteel), Shanghai 201900, China; 3. Shanghai Engineering Research Center of Metals for Lightweight Transportation, Shanghai 201900, China)

Abstract: High strength and toughness magnesium alloy extrusion profiles with complex sections are prepared to apply in high-speed rail seat frame by extrusion mold design, extrusion process and heat treatment process optimization. A self-optimized non-rare earth magnesium alloy developed by Baosteel is used. Magnesium alloy extrusion profiles numbered BGZ8 have yield strength of 289 MPa, tensile strength of 396 MPa and elongation of 11.5% after T5 treatment, mechanical properties of which fully meet the application requirements of high-speed rail seat frame. Microstructure observation shows that not only the uniformity of microstructure at different parts of the extruded section have been guaranteed, but also effective bottom reinforced texture, fine grain size and uniform distribution of preparation phase have been obtained by optimized extrusion die, extrusion process and subsequent heat treatment process. Due to the muti-scale microstructure, the comprehensive mechanical properties of magnesium alloy extrusion profiles have been significantly improved. Furthermore, this paper also verifies the possibilities that the existing aluminum alloy profiles used in the seat frame of certain type high-speed rail are replaced by magnesium alloy profiles described in this paper. The result shows that the seat frame can achieve a weight loss of 25% compared to aluminium alloys with a weight of 1.7kg and the mechanical performance is better than customer requirements.

Key words: magnesium alloy, extrusion profile, process optimization, high strength and toughness

钢铁企业智慧生态物流系统的构建与实施

侯海云，王延明

（鞍山钢铁集团有限公司，辽宁鞍山　114000）

摘　要： 鞍山钢铁集团有限公司在供应物流、生产物流、销售物流、逆向物流和废弃物回收物流领域全面贯彻实施智慧生态物流理念和开发应用相关技术，实现了提高物流绩效，为企业及其供应链伙伴与股东创造价值的同时取得了良好的社会效益、生态效益。本文详细介绍了钢铁企业智慧生态物流系统的构建与实施的背景、主要内容和实施效果，为钢铁企业的物流运作模式升级和创新提供了较好的借鉴思路，而且，这种系统集成创新的技术和方法也可以为我国的钢铁企业和物流企业提供很好的借鉴。

关键词： 钢铁企业，智慧物流，生态物流，技术创新，物流模式创新

Construction and Implementation of Intelligent Ecological Logistics System in Iron and Steel Enterprises

Hou Haiyun, Wang Yanming

（Anshan Iron and Steel Group Co., Ltd., Anshan 114000, China）

Abstract: Anshan Iron and Steel Group Co. Ltd. fully implement the concept of ecological intelligent logistics, develop and apply related technology in the field of supply logistics, production logistics, distribution logistics, reverse logistics and recycling logistics. Improve the logistics performance, create value for the enterprise , supply chain partners and share holders ,and meanwhile achieve a good social and ecological benefits. This paper introduces detailedly the background , main content and implementation effect of construction and implementation of intelligent ecological logistics system in iron and steel enterprises, it provides a good reference and ideas for the upgrade and innovation of the logistics operation mode in iron and steel enterprises, and this kind of systemic technology and method of integrated innovation can also provide a good reference for Chinese iron and steel enterprises and logistics enterprises.

Key words: Iron and steel enterprises, intelligent logistics, ecological logistics, technological innovation, logistics model innovation

道次变形率对超低碳钢 Qst32-3 冷拉拔性能的影响

刘　超 [1,2]，陈继林 [1,2]，孟军学 [3]

（1. 邢台钢铁有限责任公司，河北邢台　054027；2. 河北省线材工程技术研究中心，
河北邢台　054027；3. 邢台新翔金属材料科技股份有限公司，河北邢台　054027）

摘　要： 本文以超低碳钢 Qst32-3 线材为研究对象，通过在冷拉拔加工工序设置不同的拉拔变形道次及道次变形率，研究其对材料力学性能的影响，利用透射电子显微镜(TEM)对材料变形后的位错组态进行分析，并进一步对比加工硬化态材料退火后的组织差异。结果表明，在总变形量一定的前提下，采用两道次拉拔，且首道次拉拔减面率占总减面率一半时，可得到最优的冷加工性能。经 TEM 分析，该工艺具有最低程度的位错塞积，且退火后再结晶不充分，证明形变储存能较低。这说明，拉拔变形道次及道次变形率对位错组态、畸变能及力学性能有直接影响。

关键词： Qst32-3，冷拉拔，道次变形率，位错，形变储能，力学性能

Effect of Pass Deformation Rate on Cold Drawing Property of Ultra-low Carbon Steel Qst32-3

Liu Chao[1,2], Chen Jilin[1,2], Meng Junxue[3]

(1. Xingtai Iron & Steel Co., Ltd., Xingtai 054027, China; 2. Hebei Engineering Research Center for Wire Rod, Xingtai 054027, China; 3. Xingtai Xinxiang Metallic Materials Technology Co., Ltd., Xingtai 054027, China)

Abstract: Ultra-low carbon steel wire rod Qst32-3 was researched in the present work. Different cold drawing passes and pass deformation rates were adopted in cold drawing process to analyze the effect on mechanical property. Meanwhile, the dislocation patterns after drawing process were also observed by TEM. Microstructures after annealing process were compared further. The results show that the optimum cold drawing property can be obtained in conditions of the total deformation was not changed when the two-passes drawing was adopted as well as the first pass deformation rate was set to be 50% of the total reduction rate. This kind of process can lead to the minimal dislocation pile up, and the inadequate recrystallization also indicates the lower distortion energy. The results show that the drawing pass and pass deformation rate have direct effect on dislocation pattern, distortion energy and mechanical property.

Key words: Qst32-3, cold drawing , pass deformation rate, dislocation, distortion energy, mechanical property

真空复合法生产 9Ni 超薄板实践

刘文飞，金百刚，周智勇，王鲁毅，张宏亮，王　亮

（鞍钢股份鲅鱼圈钢铁分公司，辽宁营口　115000）

摘　要： 用真空复合法生产 9Ni 超薄板中会出现诸多难题，比如初轧坯加工前的板型弓曲现象，剩磁高复合坯无法进行真空电子束焊接的问题，加热前复合坯鼓肚、轧制过程尾部放炮的现象以及分板后板厚不均匀等缺陷，本文针对这些问题，开展了一系列的研究工作，通过对 9Ni 初轧坯矫直方式优化，高剩磁坯消磁研究，复合坯内气体排除的探索以及成品厚度不均的分析等手段，有效的解决了此类制约生产的难题，顺利实现了厚度不大于 5mm9Ni 板的国产化，打破了国外对于 9Ni 超薄复合板的垄断。

关键词： 初轧坯矫直，消磁，板型弓曲，鼓肚

Practice of Vacuum Composite Technology on the 9Ni Steel Ultra Thin Sheet

Liu Wenfei, Jin Baigang, Zhou Zhiyong, Wang Luyi, Zhang Hongliang, Wang Liang

(Ansteel Bayuquan Iron & Steel Subsidary, Yingkou 115000, China)

Abstract: There are lots of problems in the production process of 9Ni thin slab by vacuum compound, such as the phenomenon of plate type bow before the initial billet processing; the problem of vacuum electron beam welding can not be carried out by the residual magnetic high compound blank; the phenomenon of the tail blasting in rolling process and compound billet bulging before heating; the defect of uneven thickness in the slab after separation, and so on. Several methods are put up in this paper.　These puzzles are solved through the optimization of straightening method for initial rolling, the research on the magnetic field of high remanence, the exploration of gas exclusion in composite billet and the anlysis of thickness of finished product. The domestic production of no more than 5mm 9Ni plate are successfully achieved. The abroad monopoly of the 9Ni steel production are broken.

Key words: straightening of cogged ingot, degaussing, plate type uneven, drum belly

数据时代的中国钢结构产业发展刍议

董树勇

（中国金属学会，北京　100711）

摘　要：中国钢结构产业是钢铁行业化解产能过剩和建筑行业转型升级的重要抓手。政府推动建筑绿色化、工业化发展战略，使钢结构产业迎来重要的发展机遇，中国钢结构产量到 2020 年将超过 1 亿吨，在 2030 年将达到 2 亿吨左右。钢铁行业要抓住历史机遇，创新发展，既要满足建筑行业的需求，研发生产高品质、高性能的钢结构用钢，又要努力打造客户驱动的全产业链的智能网络服务平台体系，将网络协同和数据智能双螺旋有机结合，为钢结构全产业链提供"精准服务"，使整个产业链逐步迈上中高端。

关键词：钢结构，钢铁行业，建筑行业，智能网络服务平台

On the Development of China Steel Construction Industry in the Data Age

Dong Shuyong

(The Chinese Society for Metals, Beijing 100711, China)

Abstract: China steel construction industry is an important starting point for the steel industry to resolve the overcapacity and the transformation and upgrading of the construction industry. Our government has put forward the development strategy of the greening and industrialization of construction. It will usher in an important development opportunity for the steel construction industry. The output of China steel construction will exceed 100 million tons by 2020, reaching about 200 million tons in 2030. China steel industry should seize the historical opportunities. We should make full use of our creativity. The high quality, high performance steel for steel construction should be researched and developed to meet the needs of the construction industry. Combined the double helix of network collaboration and data intelligent and driven by customer, the intelligent network service platform for the whole industry chain should be built. It will provide accurate service for the whole industry chain of the steel construction and make the whole industry chain step up to the medium and high level gradually.

Key words: steel construction, steel industry, construction industry, intelligent network service platform

45 圆钢拉拔断裂原因分析

齐晓峰，王晓春

（通化钢铁股份有限公司，吉林通化　134003）

摘　要：45 圆钢在拉拔六棱柱产品的过程中出现断裂问题，通过宏观形貌观察、化学成分分析、表面酸洗、金相分析和硬度检验，找出了该产品拉拔断裂的原因。结果表明，45 圆钢表面上的横向裂纹是导致其断裂的主要原因，

而横向裂纹是由于拉拔工艺不当造成的。

关键词：拉拔断裂，横向裂纹，金相分析，显微硬度

Research on the Dralling Fracture of 45 Steel

Qi Xiaofeng, Wang Xiaochun

(Tonghua Iron and Steel Group Co., Ltd., Tonghua 134003, China)

Abstract: 45 steel hexagonal products occurred breakup phenomenon during drawing, by adopting succession of macro appearance observation, chemical composition, surface pickling, metallurgical analysis and microhardness. The results show that the transversal cracks which produced during dralling on the products surface was the main reason for the 45 steel brittle fracture of the 45 steel products, improper operation leaged to the ansversal cracks.

Key words: drawing fracture, transversal crack, metallurgical analysis, microhardness

贝氏体-残余奥氏体钢冷拉拔过程中的相变模拟研究

安若维，米振莉，喻智晨

（北京科技大学工程技术研究院，北京　100083）

摘　要： 本文采用代表性体积单元法(RVE)构建二维贝氏体-残余奥氏体复相钢微观结构模型，使用 Fortran 语言进行 ABAQUS 用户材料子程序 VUMAT 进行二次开发，模拟分析了冷拉拔过程中残余奥氏体的相变行为。分析表明，拉拔过程中形变诱导相变首先发生在残余奥氏体的尖端和边缘附近，然后逐步向心部扩展，导致了较为明显的应力跃升现象。不同的拉拔参数对残余奥氏体的受力转变有较大的影响，随工作锥角的增大，残奥所受应力均逐渐增大，尖部受工作锥角变化的影响较大，边部和心部受工作锥角变化的影响较小。随道次压缩率的增大，应力峰值与微观应力都不断增大。压缩率对尖端处的应力影响较大，而当道次压缩率大于 10%时，对残余奥氏体的边部和心部所受应力影响较小。

关键词：贝氏体-残余奥氏体钢，冷拉拔，相变模拟，VUMAT

The Transformation Simulation Research of Bainite-Retained Austenitic Steel during Cold Drawing

An Ruowei, Mi Zhenli, Yu Zhichen

(Institute of Engineering Technology, University of Science and Technology Beijing, Beijing 100083, China)

Abstract: The 2D microstructure model of bainite-retained austenite steel was constructed by using a representative volume element (RVE). User-material subroutine (VUMAT) in ABUMQUS was implemented by Fortran to analyze transformation behavior of retained austenite during cold drawing. The analysis shows that the deformation induced transformation of phase first occurs near the tip and edge of the retained austenite, and then gradually extends to the center, resulting in a more obvious stress jump, during the drawing process. Different drawing parameters have a great influence on the transformation of residual austenite. With the increase of the die cone angle, the stress of the retained austenite is gradually increased, and the influence of the tip by die cone angle is larger, less affected by the edge and center of die cone angle. With the increase of the compression ratio, the stress peak and the micro stress increase. The compressive rate has a great influence on the stress at the tip, and when the compression ratio is more than 10%, the influence of the stress on the edge and the core of the retained austenite is small.

Key words: bainitic-retained austenitic steel, cold drawing, transformation simulation, VUMAT

基于电铸技术进行废旧结晶器再制造的探讨

定 巍[1]，李 岩[2]，王宝峰[1]

（1. 内蒙古科技大学材料与冶金学院，内蒙古包头 014010；2. 内蒙古科技大学，内蒙古自治区白云鄂博矿多金属资源综合利用重点实验室，内蒙古包头 014010）

摘 要： 铜质结晶器是连铸机心脏，也是冶金企业生产中的主要耗材之一。如何降低结晶器的消耗，提高其使用寿命是一个值得研究的话题。本文以小方坯结晶器为例，采用电铸技术对废旧结晶器进行了修复。对修复后结晶器经过不同温度热处理后的硬度进行了测试，测试结果表明电铸修复的结晶器具有硬度好，抗蠕变性能好的特点，能满足结晶器对铜质材料的性能要求。从电铸技术特点和电铸铜的性能来看，电铸技术用于废旧结晶器的再制造在技术上具有可行性。

关键词： 电铸，结晶器，铜，再制造

Discussion on Remanufactured of Abandoned Mold using Electroforming Technology

Ding Wei[1], Li Yan[2], Wang Baofeng[1]

1. School of Materials and Metallurgy, Inner Mongolia University of Science and Technology, Baotou 014010, China; 2. Bayan Obo Multimetallic Resource Comprehensive Utilization Key Lab., Inner Mongolia University of Science and Technology, Baotou 014010, China)

Abstract: Mold is the heart of continuous casting machine, and is one of the main supplies of metallurgical production. How to reduce the consumption of mold, improve its service life is a worthy topic. In this paper, a billet mold was remanufactured using the electroforming technology. The results show that the remanufactured mold has the characteristics of good hardness and good creep resistance, which can meet the performance requirements of the mold for the copper material. From the characteristics of electroforming technology and the performance of electroformed copper, electroforming technology for the remanufacture of abandoned mold is feasible.

Key words: electroforming, mold, copper, remanufacturing

室温高成形性镁合金板材制备及其性能

谢　玉[1,2,3]，张春伟[1,2,3]，唐伟能[1,2,3]，张　成[1,2,3]，徐世伟[1,2,3]

（1. 宝山钢铁股份有限公司研究院，上海　201900；
2. 汽车用钢开发与应用技术国家重点实验室（宝钢），上海　201900；
3. 上海运输工具轻量化金属材料应用工程技术研究中心，上海　201900）

摘　要：镁合金薄板用途广泛，可用于制备汽车座椅坐盆、高铁内部蒙皮、笔记本外壳等冲压件，实现显著的轻量化。本文针对上述市场领域目前急需的宽幅400mm以下的镁合金薄板，采用宝钢生产的镁合金挤压板材为原料，仅通过1~3个轧制道次制备出了可工业化批量生产的不同厚度规格（0.5~1.5mm）与宽幅（100~300mm）的系列AZ31B薄板。该系列薄板板型良好、无边裂、无需切边，且可以实现室温成卷和开卷。该系列板材，经退火调试后，在室温拉伸下具有较高的综合力学性能（屈服强度>250MPa，抗拉强度>300MPa，延伸率>20%），且在室温下可实现室温下反复折弯而无裂纹，显示出非常好的室温成形性。

关键词：镁合金，轧制，成形性，组织演变

Preparation and Properties of Magnesium Sheet with High Formability at Room Temperature

Xie Yu[1,2,3], Zhang Chunwei[1,2,3], Tang Weineng[1,2,3], Zhang Cheng[1,2,3], Xu Shiwei[1,2,3]

(1. Research Institute (R & D Centre), Baoshan Iron & Steel Co., Ltd., Shanghai 201900, China;
2. State Key Laboratory of Development and Application Technology of Automotive Steel　(Baosteel), Shanghai 201900, China; 3. Shanghai Transportation Lightweight Engineering
Research Center for Metal Materials, Shanghai 201900, China)

Abstract: Magnesium alloy sheet can be used to prepare car seat pots, high-speed rail internal skin, notebook shell and other stamping parts, to achieve significant lightweight effect. In this paper, the magnesium alloy sheet with a width of 100~300mm which is urgently needed in the above-mentioned market area, is prepared by using the magnesium alloy extruded sheet produced by Baosteel as raw material. Only 1~3 rolling passes was needed to prepare different thickness specifications of industrial mass production (0.5~1.5mm). The series of thin is of good shape with no crack on the edge. The tensile strength of the series plate, after annealing at room temperature has a high overall mechanical properties (yield strength>250MPa, tensile strength>300MPa, elongation> 20%), and can be easily bended at room temperature, showing excellent formability at room temperature.

Key words: magnesium alloys, rolling, formability, structure evolution

二元合金等轴枝晶生长的相场模拟

任　能，李宝宽，齐凤升，李林敏

（东北大学冶金学院，辽宁沈阳　110819）

摘　要：运用相场模型对二元合金过冷熔体中等轴晶的枝晶生长行为加以描述，模型中同时考虑了过冷度以及溶质再分配所产生的作用，通过有限差分法计算得到了凝固过程中等轴晶形貌和溶质分布的演变过程。计算所得到的枝晶形貌与枝晶生长理论契合。计算结果表明：晶内的溶质分布较为均匀，而熔体中形成了较大的溶质浓度梯度，溶质富集于相界面一侧；溶质在两个枝晶臂间的过冷熔体富集抑制了此处的凝固过程，造成了枝晶根部的"缩颈"现象；由于在凝固过程本身驱动了熔体中溶质元素的富集，因而枝晶臂末端以及缩颈处周围的相界面逐渐变得尖锐。

关键词：相场法，枝晶生长，凝固，数值模拟

Phase-field Simulation of Equiaxed Dendrite Growth of a Binary Alloy

Ren Neng, Li Baokuan, Qi Fengsheng, Li Linmin

(School of Metallurgy, Northeastern University, Shenyang 110819, China)

Abstract: Phase field method is employed to describe the equiaxed growth in undercooled binary alloy melt, where the effects of undercooling and solute redistribution is taken into account; finite difference method is used to get the evolution of grain morphology and t solute distribution. The calculated morphology is validated by the dendrite growing theory. The result shows: the solute distribution is uniformly in the grain, while a large concentrate gradient forms in the melt; the solute enrichment in melt between two dendrite arms delays the local solidification, leading to the "necking"; the solidification drives the solute enrichment in melt, so the end of dendrite arms and the phase interface around the necking get sharper.

Key words: phase-field, dendrite growth, solidification, numerical simulation

热轧平整机组挫伤缺陷控制的研讨

张明生，孙腾飞

（首钢京唐钢铁有限责任公司热轧部，河北唐山　063200）

摘　要：热轧平整机组主要针对常温下的耐候钢、冷轧料及马口铁等低强度的薄规格产品进行平整作业，以提高其表面质量、改善板形及机械性能等。但在生产过程中由于原料卷内圈松卷问题，易造成带钢中、尾部出现不同程度的挫伤缺陷，对带钢表面质量造成影响。本文主要介绍了热轧带钢挫伤缺陷的形貌特征与分布规律，通过研究挫伤产生的规律、设备功能的优化及固化操作人员的操作习惯等几个方面提出改进方案，达到了消除挫伤缺陷的目的。

关键词：平整机组，松卷，挫伤，开卷张力

Discussion on the Control of the Contusion Defect of Skin Pass Mill for Hot Strip

Zhang Mingsheng, Sun Tengfei

(Hot Rolling Department, Shougang Jingtang United Iron and Steel Co., Ltd., Tangshan 063200, China)

Abstract: Hot skin pass mill mainly deal with weathering steel, cold rolled material and tin and so on low strength of thin gauge products, in order to improve the surface quality, improve the plate shape and mechanical properties etc. However, in the process of production, the inner ring of the raw material is loose, which is easy to cause the different degree of bruise defects in the steel strip and the tail, which influences the surface quality of the strip. This paper mainly introduces the morphology characteristic and distribution law of contusion defect on hot rolling strip steel, By studying the law of the contusion, the optimization of the equipment function, the operation habit of the curing operator, the paper put forward the improvement plan, which can achieve the goal of eliminating the defect.

Key words: skin pass, loose coil, contusion, uncoil tension

不锈钢 304 带肋螺旋焊管提速生产实例

刘志龙

（山西智德安全公司冶金安全评价部，山西太原　030001）

摘　要： 不锈钢 304 带肋螺旋焊管主要用于煤矿的瓦斯抽放管，是最近两年才开发的新产品，以替代过去的 Q235 浸塑管。304 螺旋焊管用 MIG 焊接，由于最近煤矿需求量较大，提产成为主要任务。研究成果如下：焊接速度从 0.90m/min 提高到了 1.10m/min，解决了焊接参数和焊丝直径的优化问题。

关键词： 不锈钢螺旋焊管，MIG 焊接提速，焊缝

Production Example of Speed Increase of Ribbed Stainless Steel 304 Spiral Welded Pipe

Liu Zhilong

(Shanxi Zhide Safty Co., Ltd., Metallurgical Safety Evaluation Department, Taiyuan 030001, China)

Abstract: Ribbed stainless steel 304 spiral welded pipe is mainly used in the gas drainage pipe of coal mine. It is a new product developed in the last two years to replace the Q235 dipped plastic pipe in the past. 304 spiral welded pipe welding with MIG, because of the recent large demand for coal mine, production has become the main task. The research results are as follows: the welding speed is increased from 0.90m/min to 1.10m/min, and the optimization problem of welding parameters and welding wire diameter is solved.

Key words: stainless spiral welded pipe, MIG welding speed increase, weld line

电力高速热镀锌钢丝钢绞线清洁生产工艺及设备

王保元，卓　见，戴江波，刘云丹

（中冶南方武汉钢铁设计研究院有限公司，湖北武汉　430080）

摘　要： 近年来国家输电工程大力发展，作为输电导线加强芯和地线的热镀锌钢丝钢绞线也得到了越来越广泛的应用。针对国家环保要求的不断提高，实现清洁化生产已成为热镀锌钢丝企业持续发展的关键。本文以国内某厂新建 10 万吨镀锌钢丝钢绞线的工艺布置、设备选型、污染物排放、环保措施等设计为例，说明建设一个节能、环保、绿色低碳的先进热镀锌钢丝钢绞线生产车间的基本思路和方法。

关键词： 钢绞线，热镀锌，工艺，设备，环保

The Process and Equipment of Green Initiative about the High Speed Hot-dip Galvanized Steel Wire Strand of Electricity

Wang Baoyuan, Zhuo Jian, Dai Jiangbo, Liu Yundan

(WISDRI Wuhan Iron and Steel Design Research Institute Co., Ltd., Wuhan 430080, China)

Abstract: In recent years, the state power transmission engineering has been developing vigorously, and the hot-dip galvanized steel wire strand, which is the core and ground wire of transmission wire, has been widely used. Aiming at the continuous improvement of national environmental protection requirements, the realization of cleaner production has become the key to the continuous development of hot-dip galvanized steel wire enterprises. Based on the domestic process in newly built 100000 tons of galvanized steel wire strands plant layout, equipment selection, pollutants discharge and environmental protection measures, such as design, for example, illustrates the construction of an energy saving, environmental protection, green low carbon galvanized steel wire strands advanced production workshop of the basic ideas and methods.

Key words: steel strand, hot dip galvanized, process, equipment, green initiative

超高强双相钢 DP980 电阻点焊工艺研究

李 龙，徐 梅，徐亚鹏，米振莉

（北京科技大学工程技术研究院，北京 100083）

摘 要： 本文以超高强冷轧双相钢 DP980 为对象，通过调整焊接电流、焊接时间和电极压力，研究了不同的电阻点焊工艺参数对接头性能的影响。观察了焊接接头的微观组织，测试了接头的显微硬度。结果表明，DP980 点焊的最佳工艺参数为：焊接电流 7kA，焊接时间 360ms，电极压力 4kN，剪切力可达 20.8kN；焊接电流和焊接时间是影响接头性能的主要因素；接头由熔核区、热影响区和母材区构成，熔核区为粗大的柱状马氏体板条；热影响区内软化区的出现使该区显微硬度最低，成为了断裂发生的危险点。

关键词： 双相钢，电阻点焊，超高强，力学性能

Research on Resistance Spot Welding Process of Ultra High Strength Dual Phase Steel DP980

Li Long, Xu Mei, Xu Yapeng, Mi Zhenli

(Engineering Research Institute, University of Science and Technology Beijing, Beijing 100083, China)

Abstract: The resistance spot welding process of dual phase steel DP980 was studied by adjusting the welding current, welding time and electrode pressure.Microstructure of welded joint was observed and microhardness of joint was tested. The results show that the optimum process for DP980 spot welding is the welding current of 7kA, welding time of 300ms and electrode pressure of 4kN. The tensile shear force is 20.8kN under the optimum process conditions. The welding current and welding time are the main factors that influence the performance of the joint. The welded joint consists of nugget zone, heat affected zone and base metal, the nugget is coarse columnar martensite lath. The occurrence of softening zone in the heat affected zone makes the microhardness of the area lowest and becomes the dangerous point of fracture.

Key words: dual phase steel, resistance spot welding, ultra high-strength, mechanical properties

百足成形法中翼缘纵向应变仿真研究

苏　岚[1]，陈海斌[2]，杨永刚[2]，徐　梅[2]，丁士超[3]，米振莉[2]

（1. 北京科技大学钢铁共性协同创新中心，北京　100083；2. 北京科技大学工程技术研究院，
北京　100083；3. 昆士兰大学机械与采矿学院，澳大利亚圣卢西亚　4072）

摘　要： 本文针对高强钢一种新的成形方式——百足成形进行研究，分析了翼缘纵向应变产生的原因，并建立了翼缘纵向应变计算模型；随后，考察了翼缘纵向应变中下山量的影响。通过高强钢百足成形实验，采集翼缘纵向应变数据；并通过 ABAQUS 软件对百足成形过程的有限元模拟，分析下山量与板料翼缘纵向应变的关系。翼缘纵向应变仿真结果与实验、理论计算结果对比表明：有限元模拟的纵向应变与实验结果相符。

关键词： 百足成形，有限元，纵向应变，下山量

Study on Longitudinal Strain Simulation of Flange during Millipede Forming

Su Lan[1], Chen Haibin[2], Yang Yonggang[2], Xu Mei[2], Ding Shichao[3], Mi Zhenli[2]

(1. Collaborative Innovation Center of Steel Technology, University of Science and Technology Beijing, Beijing 100083, China; 2. Institute of Engineering Technology, University of Science and Technology Beijing, Beijing 100083, China; 3. School of Mechanical and Mining Engineering, University of Queensland, St Lucia 4072, Australia)

Abstract: This paper study a new forming of high strength steel-----Millipede forming, analyzing the cause of longitudinal strain on the flange and establishing the longitudinal strain calculating formula. Then, the effects of downhill on longitudinal strain are considered. According to the experiment of high strength steel of millipede forming, the data of the longitudinal strains on the flange was being collected. The forming process was simulated by ABAQUS software and the influences of downhill on longitudinal strain on the flange were analyzed. Comparing with the simulation results of the longitudinal strain of the flange and the experimental, theoretical calculation，the results show that the calculation formula of the longitudinal strain has certain reference significance and the result of finite element simulation consistent with the experimental results well.

Key words: millipede forming, FEA method, longitudinal strain, downhill

ML40Cr 奥氏体晶粒混晶的研究

肖　冬，李为龙

（湖南华菱湘潭钢铁有限公司技术质量部，湖南湘潭　411100）

摘　要：　ML40Cr 钢种用于冷镦制作 10.9 级紧固件，在成品紧固件奥氏体晶粒度检测过程中发现存在粗大晶粒现象。对钢厂原始盘条进行奥氏体晶粒度检测，原始盘条也同样存在粗大晶粒。通过对混晶机理研究和论证，钢厂采取将 Al 含量提高到 0.02%；增大连续累积变形量很好的解决了此问题。

关键词：混晶，AlN，连续累积变形量，粗化温度

Study on Mixed Grain of Austenite Grain Size of ML40Cr

XiaoDong, LiWeilong

(Hunan Valin Xiangtan iron and Steel Co., Ltd., Technical Quality Department, Xiangtan 411100, China)

Abstract: ML40Cr steel is used to make the 10.9 stage fastener for cold pier,and the coarse grain phenomenon is found in the austenite grain size test of the finished fastener.The austenite grain size of the original steel wire rod is detected,and the original wire rod also has coarse grain.Through the study and demonstration of the grain size mixing mechanism of austenite grain,the steel mill has solved the problem by increasing the content of Al to 0.02% and increasing the cumulative deformation.

Key words: mixed grain, AlN, continuous, cumulative deformation coarsening temperature

冬季 Q345B 低合金高强钢热轧板卷轧制工艺优化实践

李松波，柴　超，孙　媛

（通化钢铁股份有限公司技术质量部，吉林通化　134003）

摘　要：冬季生产的 Q345B 热轧板带易出现强度升高、塑性下降的问题，并可能导致力学性能不合。通过对比冬季不同时期的工艺、环境因素，确定气温及冷却水温剧烈降低是产生该问题的直接原因。在采取稳定水温、优化卷取温度及冷却工艺等手段后，有效解决了 Q345B 力学性能异常波动的问题。

关键词：冬季，Q345B，力学性能，冷却水温

Process Optimization Practice for Q345B Low Alloy High Strength Hot Rolled Strip in Winter

Li Songbo, Chai Chao, Sun Yuan

(Technology & Quality Department of Tonghua Iron & Steel Co., Ltd., Tonghua 134003, China)

Abstract: The Q345B hot rolled strip produced in winter exist some problem which the intensity will be increase and the plasticity will be reduce, and cause mechanical properties unsuitability possibilities. The reason of this problem is temperature of air and water fall acuity after contrast process and environment of different phase in winter. According to stabilize water temperature, and improve on coiling temperature and cooling process, solve the problem of mechanical properties fluctuate abnormity.

Key words: winter, Q345B, mechanical properties, cooling water temperature

浅析冷弯型钢应用

何　刚[1]，杨　海[2]

（1. 辽宁冶金职业技术学院冶金工程系，辽宁本溪　117000；

2. 本钢板材有限责任公司包装公司，辽宁本溪　117000）

摘　要： 冷弯型钢采用借助优化截面形状及改善材料性能来提高材料利用效果,这对于节约能源和资源具有十分重要的意义。冷弯型钢是通过顺序配置的多道次成型轧辊，把卷材、带材等金属板带不断的进行横向弯曲，以制成特定断面的型材，如圆管、槽钢、波形板、异型截面钢材等。冷弯型钢是经济断面型材，由于具有力学性能良好、断面均匀、形状任意、利用率高、能源消耗低、经济效益高等特点，在汽车、航空、建筑等国民经济各行各业应用前景广泛。方法是通过对我国目前情况的分析，例举了冷弯型钢的三种典型应用，对比与发达国家在品种、用途上的差距，提出冷弯型钢应该在产品品种、规格、质量以及应用领域等方面要有更大发展的结论。

关键词： 冷弯型钢，特点，应用，探讨

Analysis of Roll formed Sections Applications

He Gang[1], Yang Hai[2]

(1. Liaoning Vocational and Technical Metallurgy College, Benxi 117000, China;

2. Benxi Steel Plate Co., Ltd. Packaging Company, Benxi 117000, China)

Abstract: Cold bend section steel by optimal cross-section shape and improve the material performance is used to improve the effect of material use, for saving energy and resources is of great significance. Cold bend section steel is the multichannel time through the sequence configuration forming roll, the coil, the strip of metal plate with constant lateral bending, to make a specific section of the profile, such as pipe, channel steel, corrugated plate, single section steel, etc. Cold bend section steel is economical cross-section profiles, with good mechanical properties and uniform cross section, arbitrary shape, high efficiency, low energy consumption, high economic benefit, etc, in the national economy such as automobile, aviation, construction in all walks of life extensive prospect of application. Method is based on the analysis of the current situation in our country, illustrates the three typical application of cold bending steel, compared with the developed countries of the varieties, USES gap, put forward the cold bending steel should be in the field of product varieties, specifications, quality and application aspects to have greater development.

Key words: roll formed sections, characteristic, application, to discuss

6 钢铁材料

炼铁与原料
炼钢与连铸
轧制与热处理
表面与涂镀
金属材料深加工
★ 钢铁材料
汽车钢
海洋工程用钢
轴承钢
电工钢
粉末冶金
非晶合金
高温合金
耐火材料
能源与环保
分析检测
冶金设备与工程技术
冶金自动化与智能管控
冶金技术经济

流变应力曲线能量分析法确定动态再结晶临界条件

赵宝纯，李桂艳，马惠霞

（鞍钢股份有限公司技术中心，辽宁鞍山 114009）

摘　要：加工硬化率曲线能反映材料内部组织的变化特征，但通过加工硬化率曲线难于直接确定出材料发生动态再结晶的临界条件。Poliak 和 Jonas 提出可以通过加工硬化率 θ 与应力 σ 关系曲线上的拐点确定发生动态再结晶的临界应力，且 $-\partial\theta/\partial\sigma$ 的最小值处的应力也对应于临界应力。然而，Bambach 研究指出对一些模型在临界点处存在不可微性。本文应用能量的观点对流变应力曲线进行分析，通过曲线偏置得到能量增量曲线，该曲线用于确定动态再结晶发生的临界条件，避免了微分操作。将该方法和现有方法分别应用于 8 种试验钢在不同变形条件下获取的 22 条流变应力曲线，并将结果进行了比较。结果表明，两者具有较高的一致性，所提出方法的可信性得到了证实。

关键词：动态再结晶，临界条件，能量增量曲线，流变应力曲线

A New Analysis of Flow Stress Curve for Determining Critical Condition of Dynamic Recrystallization

Zhao Baochun, Li Guiyan, Ma Huixia

(Technology Center of Angang Steel Co., Ltd., Anshan 114009, China)

Abstract: The characteristic of microstructure evolution in materials can be reflected on strain hardening curve, but it is hard to identify the critical conditions for the onset of dynamic recrystallization (DRX) from the strain hardening curve directly. According to the approach by Poliak and Jonas, the critical conditions for the onset of DRX can be determined from the strain hardening rate as a function of flow stress. Moreover, the critical stress corresponds to the minimum of the differentiation of strain hardening rate with respect to stress. However, a flow stress model might suffer from insufficient differentiability at the critical point, which is found by Bambach. In the present work, the flow stress curve was analyzed in view of energy, which resulted in energy increment curve. The energy increment curve could be used to determine the critical strain for the onset of DRX in the absence of differential analysis. Twenty-two flow curves were studied by both the present analysis and the previous one to find out the critical strains and the results were compared. The results show that the resultant critical strains agree well with each other and the validity of the present analysis is confirmed.

Key words: dynamic recrystallization, critical condition, energy increment curve, flow stress curve

CT701 奥氏体热强不锈钢焊接接头组织与性能研究

蒋健博，及玉梅，付魁军，王佳骥，刘芳芳

（海洋装备用金属材料及其应用国家重点实验室，鞍钢钢铁研究院产品所，辽宁鞍山 114009）

摘　要：采用手工电弧焊接技术对 CT701 奥氏体热强不锈钢进行了焊接连接并通过金相显微镜、扫描电镜、万能

拉伸试验机等检测设备对焊接接头的微观组织和力学性能进行了检验分析。结果表明：采用手工电弧焊接技术成功实现了 CT701 奥氏体热强不锈钢的焊接连接，焊接过程稳定，焊缝连续均匀，无气孔、夹杂等缺陷存在。接头组织均由奥氏体构成，晶粒较粗大，晶界间存在（Fe/Cr）C 化合物析出。焊接接头常温抗拉强度达到 920MPa 以上，达到母材强度的 97%，失效位置为焊接热影响区。熔合区平均硬度为 225，较母材出现下降趋势。

关键词：奥氏体热强不锈钢，焊接接头，微观组织，力学性能

Study on Microstructures and Mechanical Properties of CT701 Heat-resistance Austenitic Stainless Steel Joint

Jiang Jianbo, Ji Yumei, Fu Kuijun, Wang Jiaji, Liu Fangfang

(State Key Laboratory of Marine Equipment made of Metal Material and Application, Iron & Steel Research Institute of Angang Group, Anshan 114009, China)

Abstract: CT701 heat-resistance austenitic stainless joint is produced by manual arc welding, the microstructures and mechanical properties of which is analyzed by universal tensile testing machine, optical microscopy (OM), scan electron microscopy (SEM) and energy spectrum analysis (EDS). The results show that through the manual arc welding, a well CT701 heat-resistance austenitic stainless steel joint is obtained. The welding process is steady, and the appearance of the welding joint is uninterrupted and smooth. There are no pores and lards in the welding joint. The microstructure of the welding joint is consisted of austenitic. The (Fe/Cr)C compound precipitated on the grain boundary. Room temperature tensile strength of the welding joint is all higher than 920MPa, which is 97% of the base metal. The broken position is all in heat infect zone. The Vickers hardness of the welding joint is 225, which is lower than base metal.

Key words: heat-resistance austenitic stainless steel, welding joint, microstructure, mechanical property

7Ni 钢与 9Ni 钢组织对比分析

朱莹光，侯家平，张宏亮

（海洋装备用金属材料及其应用国家重点实验室，辽宁鞍山　114009）

摘　要：采用常规生产方式制造 7Ni 钢，在力学性能、动态 CCT 曲线、微观组织和析出物等方面与 9Ni 钢展开全方位的对比分析。结果表明，7Ni 钢与 9Ni 钢的动态 CCT 曲线非常相似，7Ni 钢的 C 曲线向左、向上移动，7Ni 钢与 9Ni 钢在相同的生产工艺条件下可以获得相同种类的组织结构和相似的性能，只是组织构成的比例略有差异，而且 7Ni 钢屈强比低。7Ni 钢中 Cr、Mo 合金的加入，弥补了因 Ni 含量降低造成的强度的下降，采用两次淬火+高温回火工艺热处理后，虽然获得的奥氏体总量与 9Ni 钢相比略有差距，但是其尺寸更小，同样有利于对韧性的提高。

关键词：7Ni 钢，9Ni 钢，动态 CCT 曲线，热处理，奥氏体

Comparison of Microstructure of 9Ni and 7Ni Steel

Zhu Yingguang, Hou Jiaping, Zhang Hongliang

(State Key Laboratory of Metal Material for Marine Equipment and Application, Anshan 114009, China)

Abstract: A kind of 7Ni steel is made by conventional method, and it is compared with 9Ni steel in mechanical property, dynamic CCT curve, microstructure and precipitation. The results show that their dynamic CCT curves are similar, and the one of 7Ni steel moves left and up. 7Ni and 9Ni steel can obtain the same kind of structures and similar property, but there is a little difference on the composition and the yield ratio of 7Ni steel is lower. The addition of Cr and Mo makes up for the decline of strength for the absence of 2% Ni. The heat treatment process of two stage quenching and high temperature tempering is used to obtain austenite at the room temperature. Although the austenite in 7Ni steel is somewhat less than 9Ni steel, the smaller size is also beneficial to the toughness.

Key words: 7Ni steel, 9Ni steel, dynamic CCT curve, heat treatment, austenite

V-N 成分体系超低碳贝氏体钢组织及析出行为研究

李晓林[1,2]，崔　阳[1,2]，肖宝亮[1,2]，杨孝鹤[3]，金　钊[3]

（1. 首钢技术研究院，北京　100043；2. 绿色可循环钢铁流程北京市重点实验室，北京　100043；

3. 首钢京唐钢铁公司，河北曹妃甸　063200）

摘　要： 借助场发射扫描电镜(SEM)、透射电子显微镜(TEM)和能谱仪(EDS)等方法研究了 4 种不同 V、N 含量的低碳贝氏体钢在控轧控冷过程中的微观组织、力学性能及碳氮化钒的析出行为。结果表明：在 450℃卷取，四种试验钢的组织都为粒状贝氏体，V 含量由 0.05%增加到 0.15%，实验钢的屈服强度增加了 163 MPa，抗拉强度增加了 85 MPa。随着钢中 V、N 含量的增加，贝氏体组织明显细化；同时，在贝氏体铁素体基体中细小析出相的体积分数显著增加，析出相的平均尺寸在 4~6 nm 之间。尺寸在 10 nm 以下的析出颗粒呈细小碟片状，尺寸在 20 nm 左右的析出相，一般呈长条状或者椭球状。经 TEM 分析，尺寸小于 10 nm 的析出相为具有面心立方结构的 V(C,N)或(V,Cr)(C,N) 复合析出相。

关键词： 碳氮化钒, 粒状贝氏体, (V,Cr)(C,N), 面心立方结构

Study on Microstructure and Precipitation Behavior of Ultra-low Carbon Bainitic V-N-bearing Microalloyed Steel

Li Xiaolin[1,2], Cui Yang[1,2], Xiao Baoliang[1,2], Yang Xiaohe[3], Jin Zhao[3]

(1. Shougang research institute of technology, Beijing 100043, China; 2. Beijing key Laboratory of Green Recyclable Process for Iron & Steel Production Technology, Beijing 100043, China;

3. Shougang Jingtang steel company, Caofeidian 063200, China)

Abstract: The precipitation behavior of vanadium carbonitrides and its effect on the microstructure and mechanical property during the thermomechanical control processes (TMCP) in low carbon bainitic steel with four different vanadium and nitrogen content were studied by means of field emission scanning electron microscopy(SEM), transmission electron microscope(TEM) and energy disperse spectroscopy (EDS). The results show that the microstructure of the four tested steels coiled at 450℃ are granular bainite. When the vanadium content increased from 0.05% to 0.15%, the yield strength and tensile strength of the tested steel increased 163MPa, 85MPa, respectively. With the vanadium content increased, the bainite was apparently refined and volume fraction of very fine precipitated particles significantly increased. The average size of the precipitate is from 4nm to 6nm. The precipitated particles with 10 nm or less are in the form of fine pellets, and about 20 nm are generally elongated or ellipsoidal. The precipitated phase less than 10 nm is a complex precipitated phase

V(C,N) or (V,Cr)(C,N) with face-centered cubic structure by TEM analysis.

Key words: vanadium carbonitrides, granular bainite, (V,Cr)(C,N), face-centred cubic structure

基于渗碳体石墨化研究开发的亚共析结构钢

张永军，张鹏程，韩静涛

（北京科技大学材料科学与工程学院，北京　100083）

摘　要： 石墨具有特殊的简单六方点阵，其强度、硬度和塑性很低。在亚共析钢的成分范围内，被视为引起材质变坏的脆性相。然而，近年来人们利用钢中渗碳体的石墨化开发出了新型结构钢，如具有较高切削性能的石墨化易切削钢、具有良好成型性能的石墨化中、高碳钢板，其研制正是利用了石墨的质软、润滑等的作用。本文对这两类钢种的开发背景、开发思路、制备过程及其性能特点等进行了简要说明。

关键词： 渗碳体，石墨化，石墨化易切削钢，石墨化中、高碳钢板

Hypoeutectoid Constructional Steel Developmented by Graphitization of Cementite

Zhang Yongjun, Zhang Pengcheng, Han Jingtao

(School of Materials Science and Engineering, University of Science and Technology Beijing, Beijing 100083, China)

Abstract: Graphite has special simple hexagonal crystal structure, it's strength, hardness and plasticity are lower. In composition range of hypoeutectoid steel, graphite is considered to be brittle phase which deteriorate properties of material. In recent years, however, new kind of structural steel are developmented by graphitization of cementite，as graphitized free cutting steel with excellent machinability, graphitized medium or high carbon steel plate with good cold forming et al. Preparation of the steel It is using the characteristics of lower hardness and good lubrication of graphite. Development backgrounds , development idea , preparation process, and their characteristics in this paper are introduced.

Key words: cementite, graphitization, graphitized free cutting steel, graphitized medium or high carbon steel plate

客车方矩管用 700MPa 级热轧高强钢的研制与开发

陶文哲[1]，陈吉清[1]，刘志勇[1]，杨海林[2]，冯　佳[1]，陆在学[1]，徐进桥[1]

（1. 武汉钢铁有限公司研究院，湖北武汉　430080；
2. 武汉钢铁有限公司热轧总厂，湖北武汉　430083）

摘　要： 本文对比研究了不同 Mn 含量对 700MPa 级热轧高强钢组织的影响，确定了武钢热轧高强钢 WYS700 的成

分、工艺，并对 WYS700 的组织、析出、力学性能、焊接性能和成形性能进行了系统研究，结果表明，WYS700 具有良好的强度、低温冲击和焊接性能，铁素体的组织设计也保证了 WYS700 具有优良的成形性能，可满足客车行业多次变形的要求。

关键词：方矩管，铁素体，高强钢，成形性能

Research and Development of 700MPa Hot Rolled High Strength Steel for Rectangular Tube of Bus

Tao Wenzhe[1], Chen Jiqing[1], Liu Zhiyong[1], Yang Hailin[2], Feng Jia[1], Lu Zaixue[1], Xu Jinqiao[1]

(1. Research & Development Center of WISCO, Wuhan 430080, China;

2. Hot Strip Rolling Mill of WISCO, Wuhan 430083, China)

Abstract: The paper studied the effect of different Mn content on the microstructure of 700MPa hot rolled high strength steel. The component and hot holling technology of hot rolled high strength steel WYS700 were confirmed in WISCO. The metallographic structure, precipitation, mechanical properties, weldability and formability were studied. The results showed that WYS700 has good tensile property, low-temperature impact property and welding performance. Ferrite structure made WYS700 has excellent formability, and could meet the requirements of the deformation of bus industry.

Key words: rectangular tube, ferrite, high strength steel, forming property

Nb 微合金化对弹簧扁钢组织和性能的影响

丁礼权[1,2]，吴　润[2]，任安超[1]，丁文胜[3]，董水要[3]

（1. 武汉钢铁公司研究院，湖北武汉　430080；2. 武汉科技大学材料与冶金学院，湖北武汉　430081；

3. 武钢集团襄阳新材料产业有限公司，湖北襄阳　441103）

摘　要：通过光学显微镜、扫描电镜、万能试验机等测试研究了不同 Nb 含量（无 Nb，0.015%Nb，0.032%Nb）对中试工厂热轧 Si-Mn-Cr 系弹簧扁钢组织和性能的影响。结果表明，含 Nb 弹簧扁钢显微组织为细片 P ＋粗片 P ＋少量 F，Nb 合金元素的加入细化了珠光体片层间距，同时可降低弹簧扁钢表面脱碳敏感性，含 Nb 弹簧扁钢奥氏体晶粒度可达到 8.0 级以上，比不含 Nb 试验钢晶粒度高出 2.0 级以上。添加微量 Nb 的钢材强度和韧性指标均有明显提高，其中，屈服强度提高 2.8%以上，抗拉强度提高 2.5%以上；断后伸长率提高 25%以上，面缩率提高 14%以上。

关键词：Nb 微合金化，弹簧扁钢，组织，性能，晶粒细化

Effect of Nb Microalloying on Microstructure and Mechanical Properties of Spring Flat Steel

Ding Liquan[1,2], Wu Run[2], Ren Anchao[1], Ding Wensheng[3], Dong Shuiyao[3]

(1. Research and Development Center of WISCO, Wuhan 430080, China; 2. University of Science and Technology Wuhan, Wuhan 430081, China; 3. Wuhan Iron and Steel Group Xiangyang new materials

industry Co., Ltd., Xiangyang 441103, China)

Abstract: The effects of different Nb content (no Nb, 0.015%Nb, 0.032%Nb) on the microstructure and properties of hot-rolled Si-Mn-Cr spring flat steel in pilot plant were investigated by optical microscope, scanning electron microscope and universal testing machine. The results show that the content of Nb spring flat steel microstructure are fine pearlite, coarse pearlite and a small amount of ferrite. Nb alloy elements added can refine the interlamellar spacing of pearlite, and reduce the surface decarburization sensitivity of spring flat steel. The austenite grain size of Nb spring flat steel can reach more than 8, and the grain size is higher than 2 grade than that without Nb steel. The strength and toughness indexes of steel with addition of trace Nb were obviously improved. The yield strength is increased by more than 2.8%, and the tensile strength increased by more than 2.5%. The elongation increased by more than 25%, and the reduction of area increased by more than 14%.

Key words: Nb microalloying, spring flat steel, microstructure, properties, grain refining

16MnCr5 高温热塑性研究

王宁涛 [1,2]，阮士朋 [1,2]，王利军 [1,2]，李世琳 [1,2]，张　鹏 [1,2]

（1. 河北省线材工程技术研究中心，河北邢台　054027；

2. 邢台钢铁有限责任公司，河北邢台　054027）

摘　要： 采用 Gleeble 1500 热模拟试验机对 16MnCr5 的高温塑性进行了测定，并应用扫描电镜对拉断后试样的断口形貌进行了观察。16MnCr5 高温力学性能显示，钢种存在明显的两个脆性区间。16MnCr5 钢的第Ⅲ脆性温度区为 600-800℃，实际生产过程中应避开这一温度范围，在铸坯表面温度高于 800℃时进行矫直。

关键词： 16MnCr5，高温热塑性，断口形貌

High Temperature Hot Ductility of 16MnCr5 Steel

Wang Ningtao [1,2], Ruan Shipeng [1,2], WangLijun [1,2], Li Shilin [1,2], Zhang Peng [1,2]

(1. Hebei Engineering Research Center For Wire Rod, Xingtai 054027, China;

2. Xingtai Iron & Steel Co., Ltd., Xingtai 054027, China)

Abstract: The high temperature hot ductility of 16MnCr5 steel was determined on Gleeble 1500 thermal simulator and the morphology of the fractures was observed by scanning electron microscopy. The high temperature mechanical properties of 16MnCr5 show that there are three brittle zones. The results show that the third temperature range of brittleness of 16MnCr5 is in the range of 600 to 800℃. It is vital to avoid this temperature range for straightening in the actual production and the steel should be straightened when the temperature is above 800℃.

Key words: 16MnCr5, hot ductility, fracture morphology

奥氏体化工艺对 ZG9CrMoWVNbNB 钢组织与性能的影响

马煜林，刘 越，江 旭，张莉萍，刘春明

（东北大学材料科学与工程学院，辽宁沈阳 110819）

摘 要：奥氏体化工艺的温度为 1050℃、1100℃、1150℃和 1200℃，保温时间分别为 1h、2h 和 4h，采取空冷、水冷和油冷的冷却方式，之后进行 730℃回火 2h 处理。通过 OM、SEM、TEM 等方法观察材料的微观组织及析出相等。通过萃取的方法获得析出相粉末，再借助 SEM、TEM 和 XRD 等分析析出相成分。并对试样进行室温拉伸试验和硬度测试，对不同热处理条件下试样的性能进行对比分析。研究结果表明：试样在不同热处理条件下的组织均为马氏体组织。热处理工艺为 1200℃保温 4h 水冷，再进行 730℃保温 2h 的回火处理获得的马氏体板条更加致密，板条清晰，残余奥氏体较少组织均匀；$M_{23}C_6$ 和 MX 相尺寸较小并弥散分布；其抗拉强度、屈服强度分别为 910MPa 和 830MPa，伸长率和断面收缩率分别达 18.16% 和 57.49%，表现出具有较高的强度和良好的塑性及韧性。

关键词：铁素体耐热钢，热处理，奥氏体化工艺，微观组织

Effect of Austenitization on Microstructure and Properties of ZG9CrMoWVNbNB Steel

Abstract: The temperature of the austenitizing process is 1050℃, 1100℃, 1150℃ and 1200℃, the holding time is 1h, 2h and 4h, and then were taken by air cooling, water cooling and oil cooling method, followed by 730℃ tempered.Then observes the material microstructure and precipitation by OM, SEM, TEM and other methods. Precipitates powder obtained by extraction method then analysis the composition of precipitates by SEM, TEM, XRD and Raman spectra. And specimens at room temperature tensile test and hardness testing, the performance under different heat treatment conditions samples were analyzed. The results show that: the sample under different heat treatment conditions of the organization are martensite and a small amount of δ ferrite. Heat treatment process for the 1150 ~ 1200℃ heat 4h cooled, then tempered martensite laths 730℃ for 2h to obtain a more compact, clear lath residual austenite less homogeneous, $M_{23}C_6$ and MX phase are smaller and dispersed. Tensile strength, yield strength of 910MPa and 830MPa, elongation and reduction of 18.16% and 57.49%, respectively, exhibit high strength and good ductility and toughness.

Key words: ferritic heat-resistant steel, heat treatment, austenitizing process, microstructure

低碳贝氏体高强钢 Q550D 的开发

王新钢，吴尚超，檀丽静

（河钢集团邯钢公司 中板厂，河北邯郸 056015）

摘 要：邯钢中板厂宽厚板线生产车间采用低碳微合金化设计，在 3500 mm 中厚板轧机上采用 TMCP 工艺生产贝

氏体高强钢 Q550D，产品可以直接热轧交货。对其性能和组织进行分析，结果表明：采用 TMCP 工艺在两阶段控制轧制和加速冷却条件下生产的高强钢 Q550D 的性能全部符合客户的要求，Q550D 的屈服强度和抗拉强度，分别在 590MPa 和 700MPa 以上，延伸率也在 24%左右，冲击功在 150J 以上，综合力学性能良好。同时节约了大量昂贵合金的加入、降低了成本、减少了热处理环节，并且已经可以进行大批量生产，极有力地提升了邯钢高强钢产品的市场竞争力。

关键词：Q550D，热轧态，TMCP，屈服强度，延伸率

The Development of High Strength Steel Q550D Low Carbon Bainite

Wang Xingang, Wu Shangchao, Tan Lijing

(Hesteel Group Hansteel Company, Medium Plate Plant, Handan 056015, China)

Abstract: Based on the design of low C micro alloying, high strength steel Q550D with bainite are produced on the 3500 mm medium and heavy plate mill by TMCP at the heavy plate production line of Handan Iron & Steel, and the products can deliver as hot-rolled. The mechanical properties and micro-structure have been analyzed. The results show that the mechanical properties of the high strength steel Q550D produced with II phase controlled rolling and accelerated cooling by TMCP all satisfy the requirements of customers. The products have good comprehensive mechanical properties. the yield strength and tensile strength are respectively more than 590MPa and 700MPa, and elongation rate is around 24%, the impact energy is more than 150 J, thus saving a large quantity of expensive alloys, lowering the cost, canceling the heat treatment procedure, and having been able to mass produced. Effectively improving the market competitiveness of the high strength steel products of Handan iron steel.

Key words: Q550D, hot-rolled, TMCP, yield strength, elongation

高速铁路用 U75VG 钢轨断裂原因分析

王瑞敏，周剑华，朱　敏，郑建国，费俊杰，徐志东

（武汉钢铁有限公司研究院，湖北武汉　430080）

摘　要：采用断口宏观观察和微观金相组织观察，分析某线路 U75VG 钢轨发生异常断裂的原因。结果表明，钢轨轨头下颚边缘曾受到机械碰撞而导致外伤，引起加工硬化，产生白层有害组织，白层组织硬而脆，在外力作用下容易破碎产生微裂纹。在应力及应力集中共同作用下，裂纹不断扩展导致钢轨断裂失效，提出预防此类钢轨伤损发生的措施。

关键词：钢轨，断裂，金相组织，原因分析

Fracture Analysis of U75VG Rail of High Speed Railway

Wang Ruimin, Zhou Jianhua, Zhu Min, Zheng Jianguo, Fei Junjie, Xu Zhidong

(Research and Development Center of WISCO, Wuhan 430080, China)

Abstract: The cause of unusual fracture of U75VG rail was analyzed by facture surface observation and microstructure examination. The results indicate that collision once happened at the edge of rail base, and then results in work-hardening, produces the white layer. The hard and brittle white layer is fragile, resulting in transverse, the crack extends continuously to the whole cross section under the effect of stress and stress concentration, and the rail fractures completely in the end. Measures are also put forward to reduce the rail damages.

Key words: rail, fracture, metallurgical microstructure, cause analysis

TiN 对搪瓷性能的影响及其析出行为研究

汤亨强，张　宜，李　进

（马钢技术中心，安徽马鞍山　243000）

摘　要：第二相 TiN 粒子的析出对细化晶粒和提高搪瓷性能影响是不可忽略的。以某钢厂生产的两种钢板为研究对象，借助金相显微镜、扫描电镜以及计算模拟等手段，探讨了 TiN 粒子析出对钢板抗鳞爆性的影响。通过计算得到了不同析出温度下[Ti][N]的析出量的变化，并指出了在 TiN 的析出过程中，控制好 N 含量更有利于获得细小 TiN 粒子，以获得更加优异的综合性能。

关键词：TiN 粒子，合金成分，析出温度

Effect of TiN on the Properties of Enamel and Precipitation Behavior

Tang Hengqiang, Zhang Yi, Li Jin

(Institute of Home Appliance Board Research of Technical Center of Masteel, Maanshan 243000, China)

Abstract: The effects of the precipitation of the second phase TiN on the refinement of grain and the improvement of the performance of the enamel can not be negligible. Two steel plates are taken as the research object produced by a steel company, the effects of TiN particle precipitation on the scale resistance of steel plate were studied by means of metallographic microscope, scanning electron microscope and computer simulation. Through the calculation of different precipitation temperature [Ti][N] precipitation changes, control the content of N is more conducive to obtaining fine TiN particles, in order to obtain a more comprehensive performance is pointed out in the TiN evolution process.

Key words: TiN particle, alloy, precipitation temperature

40CrMoV4-6 高强度耐热冷镦钢盘条的生产实践

张　鹏[1,2]，阮士朋[1,2]，王利军[1,2]，李世琳[1,2]，王宁涛[1,2]，张素萍[2]

（1. 河北省线材工程技术研究中心，河北邢台　054027；
2. 邢台钢铁有限责任公司，河北邢台　054027）

摘　要：研究 40CrMoV4-6 的高温热塑性，为轧制冷却工艺的制定提供了理论依据，使邢钢成功生产出了 40CrMoV4-6 高强度耐热冷镦钢盘条。生产实践表明，该盘条显微组织良好，可以满足下游企业深加工要求。与

SCM435 相比，该盘条的高温性能良好，热强性优异、热稳定性显著。

关键词：40CrMoV4-6，高强度，耐热冷镦钢，高温热塑性，高温蠕变

The Production Practice of 40CrMoV4-6 High Strength Heat Resistant Cold Heading Steel Wire Rod

Zhang Peng[1,2], Ruan Shipeng[1,2], Wang Lijun[1,2], Li Shilin[1,2], Wang Ningtao[1,2], Zhang Suping[2]

(1. Hebei Engineering Research Center For Wire Rod ,Xingtai 054027, China;
2. Xingtai Iron & Steel Co., Ltd., Xingtai 054027, China)

Abstract: The hot plasticity research of 40CrMoV4-6 was applied to the cooling process engineering at XingTai steel, the 40CrMoV4-6 high strength heat resistant cold heading steel wire rod has been produced. The production practice has shown that, Its microstructure was good, which can meet the downstream enterprise deep processing requirements. High temperature performance, compared with SCM435, its thermal strength was remarkable, while its thermal stability was significantly.

Key words: 40CrMoV4-6, high strength, heat resistant cold heading steel, hot plasticity, high temperature creep

HSLA 船板钢在调质过程中富 Cu 相的析出行为

刘庆冬 [1,2]，顾剑锋 [1,2]，刘文庆 [3]，刘东升 [4]

（1. 上海交通大学材料科学与工程学院，材料改性与数值模拟研究所，上海 200240；2. 上海交通大学高新船舶与深海开发装备协同创新中心，上海 200240；3. 上海大学材料科学与工程学院，材料研究所，上海 200072；4. 江苏省（沙钢）钢铁研究院，江苏张家港 215625）

摘 要：高强度低合金（HSLA）船板钢在 900℃固溶处理 0.5h，水淬至室温，然后在不同温度等时回火 1 h，用显微硬度反映基体回火软化和富 Cu 相析出强化的作用效果,用(HR)TEM 和原子探针层析技术(APT)表征 500℃等温回火时富 Cu 相的析出特征。结果表明：马氏体组织在 500℃回火时，富 Cu 相优先在马氏体板条界面、位错等晶体缺陷处形核，呈弥散分布，出现硬化峰值。随着回火时间的延长，富 Cu 相的尺寸不断增加，数量逐渐减少，形态由球型向椭球型转变。纳米尺寸的富 Cu 团簇中，含有大量的 Fe 和一定量的 Ni 和 Mn，成分波动较大。随着富 Cu 相不断长大，Cu 含量逐渐升高，Fe 含量不断降低，Ni 和 Mn 含量趋于稳定.Ni 和 Mn 倾向在富 Cu 相/基体界面处偏聚，但在尺寸较大的富 Cu 相中，Mn 含量明显高于 Ni.随着回火时间延长，富 Cu 相的强化效果不断降低，与富 Cu 相的大小和数量的变化规律相对应。

关键词：高强度低合金钢，船板钢，富 Cu 相，析出反应，原子探针层析技术，强化机制

Precipitation Reactions of Cu-rich Phase and Its Strengthening Mechanism in a Quench-tempered HSLA Shipbuilding Steel

Liu Qingdong[1,2], Gu Jianfeng[1,2], Liu Wenqing[3], Liu Dongsheng[4]

(1. Institute of Materials Modification and Modelling, School of Materials Science and Engineering,

Shanghai Jiao Tong University, Shanghai 200240, China; 2. Collaborative Innovation Centre for Advanced Ship and Deep-Sea Exploration, Shanghai Jiao Tong University, Shanghai 200240, China; 3. Laboratory for Microstructures, School of Materials Science and Engineering, Shanghai University, Shanghai 200072, China; 4. Institute of Research of Iron and Steel, Jiangsu Province and Sha Steel, Zhangjiagang 215625, China)

Abstract: The nanometer sized Cu-rich phases contribute to precipitation strengthening in highstrength structural steels, and respond for the embrittlement in reactor pressure vessel steels under long-term neutron irradiation. The nature of precipitation reactions of Cu-rich phases in ferritic steels is of significant scientific interest. A quench-tempered high-strength low-alloyed (HSLA) steel was used to study the precipitation behaviors of Cu-rich phases in initial martensitic matrix by (high-resolution) TEM and atom probe tomography (APT). The micro-hardness varies with the combined effect of the softening of matrix for tempering and the precipitation strengthening of Curich phases. The maximum hardness, obtained when tempering at 500℃, was resulted from the dispersed Cu-rich phases that located primarily at crystallographic defects such as martensite lath boundaries and dislocations. As the tempering time prolonged, the Cu-rich phases increased in sizes and decreased in number density, and evolved from spheroidal to elliptical in aspect of morphology. The Cu-rich phases at the early nucleation stage contains a considerable high Fe and some Ni and Mn, but the concentrations were with high inaccuracy. The growing Cu-rich phases contains decreasing Fe and increasing Cu, as well as approximate 4 at.% Ni and Mn. Both Ni and Mn tended to segregate at the Cu-rich phase/matrix heterophase interfaces, but more Mn are detected in the core of Cu-rich phases especially that were with larger sizes. Modus strengthening was the dominated strengthening mechanism for Cu-rich phases in the present ferritic steel. In accordance with the evolution in sizes and number density of Cu-rich phases, the precipitation strengthening effects of Cu-rich phases decreased with prolonged tempering time.

Key words: high-strength low-alloy steel, shipbuilding steel, Cu-rich phase, precipitation reaction, atom probe tomography, strengthening mechanism

热轧厚规格 X80 管线钢静态再结晶过程以及 冷速对组织性能的影响

徐 勇，文小明

（本钢集团公司产品研究院，本溪 117000）

摘 要： 本文主要研究了本钢生产的厚度 21.4mm X80 管线钢静态再结晶过程以及冷速对组织性能的影响，通过利用 Gleeble2000 绘制了 X80 钢的静态 CCT 曲线和静态动力学曲线，变形温度是影响奥氏体静态再结晶的主要因素，在 960℃以下奥氏体静态再结晶体积分数为 30%以下，由此确定了非再结晶区轧制温度为小于等于 960℃；以及利用金相显微镜和拉伸试验机研究了显微组织和力学性能之间的关系，随着冷速的增加，组织由针状铁素体和块状 MA 转变为粒状贝氏体和板条贝氏体铁素体，研究表明，最佳冷却速率为 25℃/S，组织为针状铁素体中弥散分布细小的点状 MA，细小弥散分布的 MA 能阻碍位错运动和疲劳裂纹扩展，不易因应力集中而诱发裂纹，并使其长度小于裂纹失稳扩展的临界尺寸，可显著提高钢材的强度和 DWTT 值。通过对轧制工艺以及冷速的优化，生产的 X80 管线钢具有优良的力学性能，在-60℃下冲击功仍然在 300J 以上，在-20℃下 DWTT 值在 90%以上。

关键词： 管线钢，再结晶，冷速，组织

Static Recrystallization Process of Hot Rolled Thick Gauge X80 Pipeline Steel and the Effect of Cooling Rate on Microstructure and Properties

Xu Yong, Wen Xiaoming

(Research Institute of Benxi Iron & Steel Co., Ltd., Benxi 117000, China)

Abstract: This paper mainly studies the static recrystallization process of 21.4mm X80 pipeline steel that is produced by Benxi Steel and the effect of cooling rate on microstructure and properties, we draw the static recrystallization curve of X80 steel by using Gleeble2000, the deformation temperature is the main factor that affects the austenite static recrystallization. At 960℃, the austenite static crystalline volume fraction is below 30%, and the non recrystallization zone rolling temperature is less than or equal to 960℃; And it studies the relationship between microstructure and mechanical properties by using metallographic microscope and tensile testing machine. with the increase of cooling rate, microstructure consisted of acicular ferrite and massive MA turns into granular bainite and lath bainite ferrite, research shows that the optimal cooling rate is 25 ℃/s, the microstructure of dot MA is dispersed in acicular ferrite, which is small MA. Because the dispersed and small MA can prevents dislocation moving and prevent cracks growing, it is not easy to induce the crack due to stress concentration and the length of the crack is less than the critical size of the crack instability expansion, which can significantly improve the strength and DWTT value of the steel. Through the optimization of the rolling process and cooling rate, the production of X80 pipeline steel has excellent mechanical properties, the impact energy is still above 300J at –60 ℃, the DWTT value is above 90% at –20℃.

Key words: pipeline steel, recrystallization, cooling rate, microstructure

工艺参数对二次冷轧镀锡板组织性能的影响

方　圆[1,2], 吴志国[1], 孙超凡[1], 宋　浩[1], 王雅晴[1], 石云光[1]

（1. 首钢集团有限公司技术研究院，北京　100043；2. 绿色可循环钢铁流程北京市重点实验室，北京　100043）

摘　要：采用洛氏硬度计、拉伸试验机、光学显微镜、场发射电镜以及制盖设备分析了退火温度和二次冷轧压下率对二次冷轧镀锡板组织性能和成形性的影响规律。研究结果表明，退火温度提高将粗化二次冷轧镀锡板的组织晶粒使得硬度和强度降低，延伸率略有改善。在适当范围内提高退火温度有助于制耳减小。二次冷轧压下率提高将会使得硬度和强度提高，但会导致各向异性增加，制耳增大，塑性降低，且{111}织构密度下降。

关键词：二次冷轧，镀锡板，工艺参数，组织性能

Effect of Processing Parameters on Microstructure and Properties of Double Cold-reduced Tinplate

Fang Yuan[1,2], Wu Zhiguo[1], Sun Chaofan[1], Song Hao[1],

Wang Yaqing[1], Shi Yunguang[1]

(1. Technology Research Institute of Shougang Group Co., Ltd., Beijing 100043, China; 2. Beijing Key

Laboratory of Green Recyclable Process for Iron & Steel Production Technology, Beijing 100043, China)

Abstract: The effects of annealing temperature and double cold-reduced rate on the microstructure and properties of double cold-reduced tinplates were analyzed by Rockwell hardness tester, tensile tester, optical microscope, field emission electron microscope and canning system. The results show that the increase in annealing temperature will coarsen the microstructure of the double cold-reduced tinplate so that the hardness and strength are reduced and the elongation is slightly improved. Increasing the annealing temperature in the proper range contributes to the reduction of the ear. The increase in double cold-reduced rate will increase the hardness and strength, but will lead to anisotropy increase, the ear increases, the plasticity decreases, and the {111} texture density decreases.

Key words: Double cold-reduced, Tinplate, Processing parameters, Microstructure and properties

V-N 微合金化 HRB400E 钢筋的研究与生产

刘金花，向浪涛，汪 涛，廖 明

（重庆钢铁股份有限公司，重庆 401258）

摘 要： 钒微合金化是高强度热轧钢筋的主要合金化方式。在含钒钢筋中增氮，能促进 V（CN）的析出，显著提高钒元素的沉淀强化效果。研究表明，采用 V-N 微合金化，在同等强度水平下可显著降低 V 含量。本文通过生产实践，经大生产数据表明：与 V 微合金化相比，采用 V-N 微合金化，能降低 V 含量 50% 左右，大大降低合金成本。

关键词： 低成本，V-N 微合金化，钢筋，性能

Research and Production of Vanadium- nitrogen Microalloying in HRB400E Reinforcing Steel Bar

Liu Jinhua, Xiang Langtao, Wang Tao, Liao Ming

(Chongqing Iron & Steel Co., ltd., Chongqing 401258, China)

Abstract: Vanadium microalloying is the main process of high strength hot rolled steel bars. Improve the content of nitrogen in vanadium containing steel bars, The precipitation of Vanadium carbonitride was promoted, The precipitation strengthening effect of vanadium was improved remarkably, Research shows, vanadium- nitrorgen microalloying is adopted, Under the same mechanical condition, the content of vanadium can be reduced .Production data show: vanadium- nitrogen microalloying technology is adopted, Compared with vanadium containing reinforcement, vanadium content decreased by about 50%,greatly reduce the cost of the alloy.

Key words: low cost, vanadium-nitrogen microalloying, steel bar, properties

钢铁冶金过程多尺度架构的研究概述

林腾昌[1]，曾加庆[1]，朱 荣[2]，李士琦[2]，王 杰[1]，董 凯[2]

（1. 钢铁研究总院冶金工艺研究所，北京 100081；2. 北京科技大学冶金与生态工程学院，北京 100083）

摘　要：将时空多尺度理论引入到钢铁冶金过程研究中，以炼钢洁净度控制为切入点，简述了钢的洁净度控制过程存在时空多尺度结构。在国家"互联网+"与智能制造战略背景下进一步深化思考钢铁冶炼工艺与产品质量间多尺度关系研究的重要性。分析认为基于我国钢铁研究发展需要，未来钢铁冶炼过程中的现象与本质间多尺度关系基础研究可为钢铁行业智能化发展提供数据支撑。

关键词：冶金过程，时空多尺度，洁净度，互联网+，智能制造

Overview of Multiscale Architecture for Iron and Steel Metallurgy Process

Lin Tengchang[1], Zeng Jiaqing[1], Zhu Rong[2], Li Shiqi[2], Wang Jie[1], Dong Kai[2]

(1. Institute for Metallurgical Proeess, Central Iron and steel Research lnstitute, Beijing 100081, China;
2. School of Metallurgical and Ecological Engineering, University of Science and Technology Beijing, Beijing 100083, China)

Abstract: It is suitable for introducing the theory of space-time multiscale to the study on metallurgy process. Based on the controlling of cleanliness, the space-time multiscale structure of cleanliness controlling is illuminated. The importance of multi-scale study of iron and steel smelting process and product quality is deep considered in the context of national "Internet +" and intelligent manufacturing. Based on the needs of China's iron and steel research development, it considers that the basic research of multi-scale relationship of phenomenon and nature in the process of iron and steel smelting can provide data support for the intelligent development of steel industry.

Key words: steelmaking process, space-time multiscale, cleanliness, internet +, intelligent manufacturing

Q235 系列结构钢柔性制造技术生产实践与探讨

王运起，胡恒法，郭园园

（上海梅山钢铁股份有限公司制造管理部，江苏南京　210039）

摘　要：论文以 Q235 系列结构钢为例介绍了微合金结构钢的柔性制造技术，即采用同一种化学成分的板坯轧制不同性能要求的产品，梅山钢铁公司通过该项技术的实施，减少了钢种数量，减少了混浇坯和合同余材，大幅度提升了生产效率和原辅料利用率，降低了生产制造成本，缩短了产品制造周期，提升了合同交货及时率。

关键词：热轧，结构钢，探讨，柔性制造

Practice and Discussion on the Flexible Manufacturing Technology of Q235 Series Structural Steel Products

Wang Yunqi, Hu Hengfa, Guo Yuanyuan

(Shanghai Meishan Iron & Steel Co., Ltd., Nanjing 210039, China)

Abstract: The Q235 series of structural steel for example introduced micro alloy steel structure of flexible manufacturing technology, which uses the same chemical composition of slab rolling with different performance requirements of products,

Meishan Iron and steel company through the implementation of the technology, reducing the number of steel, reduce the mixed casting billet and contract more than wood, greatly enhance the production efficiency and raw material utilization ratio, reduce the cost of manufacturing, shorten the product manufacturing cycle, enhance the contract on time delivery rate.

浅论不锈钢复合管在市政管网中的应用前景

王小勇[1]，罗家明[2]，黄乐庆[1]，马长文[1]，王海宝[2]

（1. 首钢集团有限公司技术研究院宽厚板所，北京　100043；
2. 北京市能源用钢工程技术研究中心，北京　100043）

摘　要：概述了国内外市政管网的现状。在总结国内城市管网问题与借鉴国外经验的基础上，分析认为不锈钢复合管在市政管网领域具有广阔的应用前景。
关键词：不锈钢，轧制复合管，市政管网，应用前景

Discuss the Application Prospects of Stainless Steel Clad Pipe in Municipal Pipe Network

Wang Xiaoyong[1], Luo Jiaming[2], Huang Leqing[1], Ma Changwen[1], Wang Haibao[2]

(1. Plate Technology Department of Shougang Research Institute of Technology, Beijing 100043, China;
2. Beijing Engineering Research Center of Energy Steel, Beijing 100043, China)

Abstract: The status quo of municipal pipeline network at home and abroad was summarized. It showed that the stainless steel clad pipe had a broad application prospect in the area of municipal pipe network, on the basis of summarizing the problems of the domestic urban pipe network and drawing on the experience of foreign countries.
Key words: stainless steel, rolled clad pipe, municipal pipeline network, application prospects

热变形工艺对 430 不锈钢组织和织构的影响

李　娜[1]，杨博威[2]，詹　放[3]，倪　偲[1]

（1. 辽宁科技大学材料与冶金学院，辽宁鞍山　114051；2. 东北大学材料科学与工程学院，辽宁沈阳　110819；3. 昆明理工大学材料科学与工程学院，云南昆明　650093　）

摘　要：本文利用热模拟单轴压缩实验研究了热变形工艺对 430 铁素体不锈钢组织均匀性和织构的影响。实验过程中分别采用不同的变形温度、道次和热轧压下率，利用金相显微镜、扫描电子显微镜（SEM）对样品进行了组织观察和 EDX 能谱测试，利用 XRD 进行了织构测试，实验结果表明在较低的变形温度下，单道次大变形的样品组织更加细小均匀；并且热加工织构强度都较弱。
关键词：铁素体不锈钢，热变形，组织，织构

Effects of Hot Deformation Technology on the Microstructure and Texture of 430 Stainless Steel

Li Na[1], Yang Bowei[2], Zhan Fang[3], Ni Cai[1]

(1.School of Material and Metallurgy, University of Science and Technology Liaoning, Anshan 114051, China; 2. School of Materials Science and Engineering, Northeastern University, Shenyang 110819, China; 3. School of Materials Science and Engineering, Kunming University of Science and Technology, Kunming 650093, China)

Abstract: The effects of hot rolling technology on the microstructure and texture of 430 stainless steel were studied in this work by using the thermal simulation experiment of uniaxial compression to simulate hot rolling process. Samples were prepared with different reduction rate, deformation temperatures and rolling passes during the experiment. The microstructures were observed with optical microscope (OM) and scanning electronic microscope (SEM), the energy spectrum was obtained by energy dispersive X-ray detector (EDX) and the textures were examined by XRD. The results show that, the microstructures are finer and more uniform under the condition of lower deformation temperature, single rolling pass and larger deformation; and the texture strength after hot rolling is weak.

Key words: 430 stainless steel, thermal simulation, microstructure, texture

重载化铁路用钢轨生产技术研究

智建国，赵桂英，宋　海，梁正伟，高明星

（包头钢铁（集团）有限责任公司，内蒙古包头　014010）

摘　要： 为了应对重载铁路发展，减少钢轨伤损，选择合金化的途径，开发出轧态钢轨抗拉强度 $Rm \geqslant 1080MPa$、伸长率 $A \geqslant 9\%$、踏面硬度 $\geqslant 300HB$ 的高强 U76CrRE 钢轨，与 U75V 相比低温冲击性能改善、耐磨性能、耐腐蚀性能提高，钢轨焊接性能优良，已经批量出口和国内上线使用。通过新轨头廓形钢轨 60N1-2 优化轮轨接触形面，接触位置明显趋向于轨头踏面中心，最大接触应力和最大 Mises 应力显著减小，铺设使用明显改善曲线上股钢轨的轮轨接触关系，可较好地平衡钢轨斜裂纹、剥离掉块和核伤等疲劳伤损与钢轨磨耗尤其是侧面磨耗之间的矛盾，使小半径曲线上股钢轨获得最大的使用寿命。

关键词： 重载铁路，高强钢轨，U76CrRE 钢轨，新廓形钢轨，60N1-2 钢轨

Study on Production Technology of Heavy Haul Railway Rail

Zhi Jianguo, Zhao Guiying, Song Hai, Liang Zhengwei, Gao Mingxing

(Baotou Iron & Steel(Group) Co., Ltd., Baotou 014010, China)

Abstract: In order to cope with the heavy haul railway development, reduce the damage of rail, choose the method of alloying, developed the high strength U76CrRE, with tensile strength $Rm \geqslant 1080MPa$, elongation $A \geqslant 9\%$, foot surface hardness $\geqslant$ 300HB, compared with U75V, raising of low temperature impact properties, wear resistance, corrosion resistance, excellent performance of rail welding already, batch export and domestic on-line use. Through the new railhead profile rail 60N1-2 optimization of wheel rail contact surface. The contact position of the apparent tread center in the head, the maximum contact stress and maximum Mises stress decreased significantly. The obvious change of laying The wheel rail contact relationship between good curve rail, rail oblique crack can balance better, off the block and nuclear injury fatigue damage and wear of rail side wear especially the contradiction between, so that the small radius curve rail for maximum service life.

Key words: heavy haul railway, high-strength rail, U76CrRE rail, New profile rail, 60N1-2 rail

耐腐蚀不锈钢覆层钢筋的开发和应用前景

向　勇[1,2]，黄　玲[1,2]，曾麟芳[1,2]，谢扬平[1]，谢昭昭[1,2]

（1. 湖南三泰新材料股份有限公司，湖南娄底　417009；2. 湖南省双金属钢基复合材料工程技术研究中心，湖南娄底　417009）

摘　要：基础设施的破坏是当今世界的重大问题之一，而钢筋腐蚀被确认为是影响其耐久性的第一因素。然而纯不锈钢钢筋价格昂贵，环氧树脂涂层钢筋及镀锌钢筋也存在着一系列影响使用的问题；新型耐腐蚀钢筋的开发势在必行。采用净界面复合、形性协同变形工艺制备的不锈钢覆层钢筋具有强度高、耐腐蚀性能优越、性价比高的优势，有望在桥梁、公路及海洋工程等领域得到广泛推广及应用。

关键词：钢筋混凝土，不锈钢覆层钢筋，腐蚀，净界面复合，形性协同变形

Development and Application Prospect of Corrosion Resistant Stainless Steel Cladded Rebars

Xiang Yong[1,2], Huang Ling[1,2], Zeng Linfang[1,2], Xie Yangping[1], Xie Zhaozhao[1,2]

(1. Hunan 3T New Materials Co., Ltd., Loudi 417009, China; 2. Hunan Technology and Engineering Research Center for Steel Matrix Bimetal Composites, Loudi 417009, China)

Abstract: The destruction of infrastructure is one of the major issues in the world today, and the corrosion of rebarsis recognized as the first factor affecting its durability. As pure stainless steel bars are expensive, epoxy resin coated steel bars and galvanized steel bars also exist a series of problems affect use; the development of new corrosion resistant steel bars is imperative. The stainless steel clad bar was maded by using pure interface composite and collaborative deformation of shape and property technologies has the advantages of high strength, superior corrosion resistance and cost-effective, it's expected to be widely extension and used in highways, bridges, marine engineering and other fields.

Key words: reinforced concrete, stainless steel clad rebar, corrosion, pure interface composite, collaborative deformation of shape and property

42CrMo 钢在高浓度 Cl⁻溶液中的腐蚀行为研究

周　文，兰　伟，曹献龙，邓洪达

（重庆科技学院冶金与材料工程学院，重庆　401331）

摘　要： 采用浸泡实验、电化学测试技术研究 42CrMo 在不同质量浓度 NaCl 溶液中的平均腐蚀速率和电化学特性，并结合 SEM 扫描电镜对浸泡 58 h 后的挂片试样进行分析。结果表明，随着 Cl⁻ 浓度的不断增大，42CrMo 钢的平均腐蚀速率呈现先增大后减小的趋势，在 Cl⁻ 浓度为 60 g/L 时，平均腐蚀速率达到最大值为 0.124 g/(m²·h)。在常温、静态环境下，42CrMo 在高浓度 Cl⁻ 溶液中以局部点蚀为主，电化学腐蚀过程主要受阴极去氧极化的控制。

关键词： 42CrMo，Cl⁻，点蚀，电化学腐蚀

Study on Corrosion Behavior of 42CrMo Steel in Chloride Solution

Zhou Wen, Lan Wei, Cao Xianlong, Deng Hongda

(School of Metallurgy and Materials Engineering, Chongqing University of Science and Technology, Chongqing 401331, China)

Abstract: The average corrosion rate and electrochemical characteristics of 42CrMo in different mass concentrations of NaCl solution were studied by immersion experiment and electrochemical testing technology, and combined with the scanning electron microscope(SEM) to analyze specimen after 58h immersion. Results show that the average corrosion rate of 42CrMo steel increases firstly and then decreases with the increasing of chloride ion concentration, and the concentration of Cl⁻ is 60 g/L, the average corrosion rate reached a maximum of 0.124 g/(m²·h). Under normal temperature and static environment, 42CrMo is mainly local pitting in high concentration Cl⁻ solution, and the electrochemical corrosion control mainly by the cathodic oxygen depolarization control.

Key words: 42CrMo, chloride ion, pitting, electrochemical corrosion

180t 转炉大方坯+控轧控冷工艺开发高等级 FD42CrMoA 风电用钢

张　群，赵千水

（本钢板材股份有限公司产品研究院，辽宁本溪　117000）

摘　要： 采用 180t 转炉大方坯连铸连轧工艺开发高等级风电用钢 FD42CrMoA。转炉冶炼采用高拉碳法，精炼采用白渣操作，化学成分实行窄带严格控制，连铸过程投入结晶器电磁搅拌、末端电磁搅拌、动态轻压下技术，控制过热度在≤20℃，氧含量控制在≤14×10⁻⁶，轧制过程采用控轧控冷技术。统计的 30 炉的检验结果，成品球磨钢材晶粒度为 7 级，平均氧含量为 11×10⁻⁶，各项物理性能均达到高等级球磨用钢要求。

关键词： 大方坯，连铸连轧，物理性能

High Class FD42CrMoA Wind Power Steel Produced by 180t LD Big Billet+Continuous Casting and Rolling Process

Zhang Qun，Zhao Qianshui

(Product Research Institute，Bengang Steel Plates Co., Ltd., Benxi 117000, China)

Abstract: Controlling technology on high class FD42CrMoA wind power steel produced by 180t LD Big billet continuous casting and rolling process was adopted. High tapping carbon content was adopted during converter steelmaking process，and white slag operation during refining processes. Chemical composition was strictly controlled, MEMS, FEMS and dynamic soft reduction technology were invested and the superheat temperature was no more than 25℃ during concasting process. the oxygen content was no more than 14×10^{-6}, and the technology of controlled rolling and controlled cooling were adopted during rolling. The result of statistic 30 furnaces was showed that: the grain size was grade 7,the average oxygen content of finished ball mill steel was 11×10^{-6}, and the requirements of high class ball mill steel were met at all the physical properties.

Key wordss: big billet, continuous casting and rolling process, physical properties

优质石油钻头用钢 EX30、SN2025 的生产实践

阚　开，卢秉军，王德勇

（本钢板材股份有限公司产品研究院，辽宁本溪　117000）

摘　要：论述了采用"50t 电炉→LF→VD→模铸 3.16 t 钢锭→轧制"工艺流程试制生产优质石油钻头用钢 EX30、SN2025 的过程。通过对关键化学元素成分控制，试制 6 炉钢，晶粒度 7 级。EX30 钢力学性能 Rel 值 803～1070MPa，Rm 值 1075～1311MPa。SN2025 钢力学性能 Rel 值 1088～1143MPa，Rm 值 1340～1401MPa，各项技术指标均满足标准要求。

关键词：石油钻头，EX30，SN2025，电炉，模铸

Production Practice of Premium Petroleum Drill Steel EX30 and SN2025

Kan Kai, Lu Bingjun, Wang Deyong

（Products Research Institute of Benxi Steel Plates Co., Ltd., Benxi 117000, China）

Abstract: Introducing "50tEAF-LF-VD-molded3.16t ingot-rolling" process to produce EX30 and SN2025 petroleum drill steel, the grade of grain size is 7 produce by key chemical elements control of 6 furnace. The yield strength of EX30 is 803 ~ 1070 MPa, and tensile strength is 1075 ~ 1075MPa. The yield strength of SN2025 is 1088~1143MPa, and tensile strength is 1340~1340MPa.Various technical indicators meet the standard requirements.

Key words: petroleum drill, EX30, SN2025, eaf, ingot casting

矿用耐磨高碳钢奥氏体晶粒长大行为的研究

吉　光[1,2]，高秀华[1]，王子健[2]，郑力宁[2]，顾志文[3]

（1. 东北大学轧制技术及自动化国家重点实验室，辽宁沈阳　110819；

2. 江苏沙钢集团淮钢特钢股份有限公司总工办，江苏淮安　223002；

3. 常熟非凡新材股份有限公司生产部，江苏常熟　215557）

摘　要：利用高温激光共聚焦显微镜，将矿用耐磨高碳钢 75MnCr 试样在 950～1200℃之间经 0～3600s 等温奥氏体化处理，通过测量奥氏体晶粒尺寸对其奥氏体晶粒长大行为进行了深入研究，建立了奥氏体晶粒长大的动力学模型。结果表明：随着加热温度的升高和保温时间的延长，75MnCr 钢的奥氏体晶粒尺寸逐渐增大；且加热温度对奥氏体晶粒长大过程的影响要高于保温时间的影响；奥氏体晶粒粗化温度为 1100℃。当温度低于 1100℃时，试验钢中第二相粒子 AlN 起到了阻碍奥氏体晶粒长大的作用；建立了 75MnCr 钢等温加热过程的奥氏体晶粒长大的动力学模型 $D^{5.26}-D_0^{5.26}=9.20\times10^{28}t\cdot\exp(-524796/RT)$。

关键词：耐磨高碳钢，高温激光共聚焦显微镜，晶粒长大，第二相粒子，动力学模型

Behavior of Austenite Grain Growth in Wear-Resistant High Carbon Steel for Mine

Ji Guang[1,2], Gao Xiuhua[1], Wang Zijian[2], ZhengLining[2], GuZhiwen[3]

(1. State Key Laboratory of Rolling and Automation, Northeastern University, Shenyang 110819, China;

2. Jiangsu Shagang Group Huaigang Special Steel Co., Ltd., Huaian 223002, China;

3. Changshu Feifan Metalwork Co., Ltd., Changshu 215557, China)

Abstract: In order to investigate austenite grain growth behavior of 75MnCr wear-resistant steel for mine at various temperatures ranging from 950℃ to 1200℃ with subsequent isothermal holding time of 3600s by using confocal laser scanning microscope(LSCM). And kinetics model of austenite grain growth was established. The results show that the austenite grain diameter of the 75MnCr steel increases with the increasing heating temperature and holding time. The significant effect of heating temperature on the growth of austenite grain is observed compared with that of holding time. Coarsening temperature for the austenite grains is 1100℃. When the temperature is below 1100℃, the AlN particles has played a role in hindering growth of austenite grain; On the basis of the nonlinear fitting of experimental data, the kinetics model of austenite grain growth in the steel during isothermal heat treatment was established using regression analysis: $D^{5.26}-D_0^{5.26}=9.20\times10^{28}t\cdot\exp(-524796/RT)$.

Key words: wear-resistant high carbon steel, CLSM, grain growth, second phase particles, dynamical model

25Mn 低温钢力学性能研究

罗　强[1]，孙　超[2]，李东晖[2]，吴开明[1]，王红鸿[1]

（1. 省部共建耐火材料与冶金国家重点实验室，高性能钢铁材料及其应用湖北省协同创新中心，武汉科技大学，湖北武汉　430081；2. 南京钢铁集团有限公司研究院，江苏南京　210035）

摘　要：本文研究了 25Mn 低温钢的屈服/抗拉强度和冲击韧性值等力学性能。采用光镜，扫描电镜（SEM）和 EBSD 技术对 25Mn 钢在室温不同拉伸变形量后形貌组织进行了观察和测量。结果表明 25Mn 低温钢具有优异的低温力学性能，在 0℃到−196℃范围内没有观察到其韧脆转变温度。室温拉伸过程中有形变孪晶，ε-马氏体和 α′-马氏体产生，拉伸开始形变孪晶只出现在少数有利取向晶粒中，形变量加大取向因素作用降低，大多晶粒内都产生形变孪晶。ε-马氏体优先出现在奥氏体晶界上，拉伸形变量加大晶内也会产生 ε-马氏体，α′-马氏体出现在 ε-马氏体内部或者边缘且非常细小。

关键词：高锰钢，拉伸形变，形变孪晶，ε-马氏体，α′-马氏体

Investigation of Mechanical Properties of 25Mn Cryogenic Steel

Luo Qiang[1], Sun Chao[2], Li Donghui[2], Wu Kaiming[1], Wang Honghong[1]

(1. The State Key Laboratory of Refractories and Metallurgy, Hubei Collaborative Innovation Center for Advanced Steels, Wuhan University of Science and Technology, Wuhan 430081, China; 2. Research Institute of Nanjing Iron and Steel Co., Ltd., Nanjing 210035, China)

Abstract: The yield strength, tensile strength and the toughness of 25Mn cryogenic steel were investigated. Microstructure and morphology of experiment steel after tension process have been characterized using optical microscope, Scanning Electron Microscope (SEM) and Electron Backscattered Diffraction (EBSD). Results showed that experimental steel has excellent mechanical properties at cryogenic temperature. The ductile-brittle transition temperature was not observed range 0 to $-196℃$. There would be ε- martensite and α'- martensite produced during tension process. ε- martensite first appeared on the austenitic grain boundary and then appeared in the grain when the experiment steel sustained larger deformation. Furthermore, α'- martensite would be generated during the process of the collision of ε- martensite.

Key words: high manganese cryogenic steel, tensile strain, deformation twin, ε- martensite, α'- martensite

TiC 颗粒强化型马氏体耐磨钢的性能研究

吴建鹏[1], 梁小凯[2]

（1. 江阴兴澄特钢有限公司，江苏江阴 214400；2. 钢铁研究总院，北京 100081）

摘 要：采用模铸、连铸两种工艺工业化试制一种 TiC 颗粒强化型马氏体耐磨钢，分析了 TiC 颗粒的析出规律特点，对比研究了试验钢与传统马氏体耐磨钢的组织、力学性能及耐磨性能。试验结果表明：凝固速度越大，TiC 析出相越细；轧制压缩比越大，颗粒分布越均匀；TiC 颗粒强化马氏体钢强度与传统马氏体钢相当，韧性有所降低；微米级的 TiC 可以有效提高材料的磨粒磨损性能，试验钢磨损失重仅为同等硬度传统马氏体钢的 70%；耐磨性能的提高主要是因为在磨粒磨损条件下，微米级 TiC 硬质点可以破碎磨砺、钝化尖角、阻断磨痕。

关键词：颗粒强化型马氏体耐磨钢，微米级 TiC，耐磨性能

Study of Performance of a TiC Particle Reinforced Martensite Wear-resistant Steel

Wu Jianpeng[1], Liang Xiaokai[2]

(1. Division of Special Plate of Xingcheng Special Steel Co., Ltd., Jiangyin 214400, China;
2. Division of Engineering Steels，Central Iron and Steel Research Institute, Beijing 100081, China)

Abstract: TiC particle reinforced martensite wear-resistant steels with a hardness of HB450 were fabricated by mold casting and continuous casting processes. The principal of precipitation of TiC in experimental steels was studied. The microstructures, mechanical and wear-resistant properties were analyzed compared with traditional martensitic wear-resistant steels. The results showed that a higher solidification rate of continuous casting resulted in finer micron-sized TiC precipitates. In addition, a larger deformation ratio resulted in a more homogeneous distribution of TiC particles. Uniformly distributed micron-sized TiC particles could passivate the tip of the grains and stop scratches, which improved the wear resistant property of steels significantly, and reduced the weight loss to 70% compared with the traditional martensitic wear-reissitant steels.

Key words: particle reinforced martensitic wear-resistant steel, micron-sized TiC particles, wear-resistant properties

高强度建筑钢筋质量分析及标准修改建议

陈雪慧，杨才福，王瑞珍

（钢铁研究总院工程用钢研究所，北京　100081）

摘　要： 深入调查了中国 HRB400/400E 和 HRB500/500E 高强度热轧钢筋的质量现状。从不同地区的钢筋生产厂和使用场所取样分析了高强度热轧钢筋成分、宏观金相、微观组织和截面硬度。按照钢筋化学成分范围，目前中国高强度热轧钢筋可归为 4 类，即含钒钢筋、低钒钢筋（<0.02%钒）、20MnSi 钢筋和 C-Mn 钢筋。宏观金相和微观组织观察结果表明，只有含钒钢筋能完全满足热轧钢筋的组织要求，其他 3 类钢筋截面基圆外围均出现非铁素体－珠光体组织的表面硬化层。基于高强度热轧钢筋的质量分析结果，提出了热轧带肋钢筋国家标准 GB1499.2-2007 的修订建议。

关键词： 高强度热轧钢筋，余热处理，钒微合金化，回火马氏体，钢筋国家标准

Quality Assessment and Suggestion of Standard Revision for High Strength Rebars in China

Chen Xuehui, Yang Caifu, Wang Ruizhen

(Department of Structure Steel, Central Iron and Steel Research Institute, Beijing 100081, China)

Abstract: The quality status of HRB400/400E and HRB500/500E high strength hot-rolled reinforcing rebars was thorough investigated. The samples of hot-rolled reinforcing rebars were taken from steel companies and construction sites throughout the country. Their chemical composition，macrograph, microstructure and section hardness of the samples were analyzed. Based on their chemical composition, the high strength hot-rolled reinforcing rebars can be classified into four categories: V microalloyed reinforcing rebars, low V reinforcing rebars (<0.02%V), 20MnSi reinforcing rebars and C-Mn reinforcing rebars. The macrograph and microstructure observation show that, only vanadium microalloyed reinforcing rebars can completely meet the microstructure requirements of hot-rolled reinforcing rebars while other three kinds of reinforcing rebars have hardening layer with tempering matensite on the surface. Based on quality analysis results of high strength hot rolled reinforcing rebars, it is necessary to revise further the current national standard for hot rolled ribbed rebars (GB1499.2—2007)

Key words: high strength hot-rolled reinforcing rebars, tempcore, V microalloying, tempering matensite, GB1499.2—2007

合金元素 Cu、Ni、Ti 对中碳高强度弹簧钢耐腐蚀性能的影响

王　轩[1]，陈银莉[2]

（1. 北京科技大学工程技术研究院，北京　100083；

2. 钢铁共性技术协同创新中心，北京　100083）

摘　要： 采用盐雾腐蚀实验，研究了 Cu、Ni、Ti 对中碳高强度弹簧钢耐腐蚀性能的影响。结果表明，在腐蚀时间 24h 之后，各组试验钢腐蚀速率均随时间的增加而逐渐降低，并且趋于稳定，腐蚀时间越长，随合金元素 Ti、Cu、Ni 的提高，实验钢就越表现出较好的耐腐蚀性能；并且不同周期的实验钢腐蚀坑的最大深度降低；结果显示，在盐雾腐蚀环境下，Cu、Ni、Ti 元素对基体起了很好的保护作用。

关键词： 合金元素，高强度弹簧钢，耐腐蚀性能，腐蚀坑

Effect of Alloying Elements Cu, Ni and Ti on Corrosion Resistance of Medium Carbon High Strength Spring Steel

Wang Xuan[1], Chen Yinli[2]

(1. Institute of Engineering Technology, University of Science and Technology Beijing, Beijing 100083, China;

2. Collaborative Innovation Center of Steel Technology, Beijing 100083, China)

Abstract: The effects of alloying elements on corrosion resistance of medium carbon high strength spring steel have been investigated by salt spray corrosion test. The results indicated that the corrosion rates of all the tested steels decreased and tended to be stable with the corrosion time increasing after 24h. The tested steels showed a better corrosion resistance and the maximum depth of corrosion pits of the tested steels in different period decreased with the contents of the alloying elements increased. The results showed that Cu, Ni and Ti elements played a significant protective effect on the matrix in the salt spray environment.

Key words: alloying elements, high strength spring steel, corrosion resistance, corrosion pit

10Cr21Mn16NiN 奥氏体不锈钢中氮化物析出行为

张寿禄

（先进不锈钢材料国家重点实验室，山西太原　030003）

摘　要： 研究了 500~1050℃时效处理过程中 10Cr21Mn16NiN 高锰氮奥氏体不锈钢中氮化物的析出规律。结果表明：氮化物析出温度范围在 650~1000℃之间，鼻尖温度约为 850℃。在 700~950℃时效观察到了 $\gamma \rightarrow \gamma' + Cr(M)_2N$ 胞状析出的类珠光体组织，分布极不均匀，850℃下数量最多，类珠光体的片层间距随着时效温度的提高而加大变宽。这些氮化物富 Cr 且含 V、Mn 和 Fe，是一种 Cr_2N 型合金氮化物。

关键词： 高锰氮奥氏体不锈钢，10Cr21Mn16NiN，时效处理，氮化物，胞状析出

Study of Precipitation Behavior of a High Nitrogen Austenitic Stainless Steel 10Cr21Mn16NiN during Aging Process

Zhang Shoulu

(State Key Laboratory of Advanced Stainless Steels, Taiyuan 030003, China)

Abstract: This paper studied the precipitation behavior of 10Cr21Mn16NiN high nitrogen austenitic Stainless steel during the aging process at 500~1050℃. The results showed that nitrides were precipitated between 650~1000℃, the nose temperature of precipitation was determined to be about 850℃. Between 700~950℃, cellular precipitation of nitrides were observed. Lamella-like nitride was grown from grain boundary into grains. The most amount of cellular structure were observed at 850℃. These nitrides, which were rich in Cr and contained Mn and a small amount of V, were Cr_2N-type alloy.

Key words: high-nitrogen austenitic stainless steel, 10Cr21Mn16NiN, aging treatment, nitride, cellular structure

3Cr13 线材耐蚀性改善与研究

孙增淼，姜　方，白李国，张荣兴，王　刚，田付顺

（邢台钢铁责任有限公司不锈钢公司，河北邢台　054000）

摘　要： 3Cr13 是 400 系马氏体不锈钢的代表钢种之一，本文为解决用户反馈的耐蚀性差的问题，对 3Cr13 产品进行了小炉成分调整试验，通过极化曲线试验的方法验证了成分调整后的耐蚀性。试验结果表明，提高 Cr 元素含量并添加 Mo、Ni 元素的 3Cr13 试样通过极化曲线试验所表现出的耐蚀性明显优于低 Cr 的试样。成分调整后的 3Cr13 产品的耐蚀性得到改善，得到了下游用户的认可。

关键词： 马氏体不锈钢，3Cr13，耐蚀性，极化曲线

The Improvement and Research on Corrosion Resistance of 3Cr13 Wire Rod

Sun Zengmiao, Jiang Fang, Bai Liguo, Zhang Rongxing, Wang Gang, Tian Fushun

(Xingtai Iron and Steel Co., Ltd., Stainless Steel Company, Xingtai 054000, China)

Abstract: 3Cr13 is one of the 400 series martensitic stainless steel. In order to solve the problem of the poor corrosion resistance of user feedback, in this paper, the adjustment test of small furnace was carried out on the 3Cr13 products, and the corrosion resistance was tested by polarization curve test. The test results show that, the corrosion resistance of the 3Cr13 sample, which improves the content of Cr element and adds Mo and Ni elements, is obviously better than that of low Cr sample by polarization curve test. The corrosion resistance of 3Cr13 products has been improved and has been recognized by downstream users.

Key words: martensitic stainless steel, 3Cr13, corrosion resistance, polarization curve

浦项高锰钢的研究与应用进展

郑　瑞，代云红，刘　行

（首钢集团有限公司技术研究院，北京　100043）

摘　要： 综述了浦项高锰钢产品系列的研究进展和应用情况，浦项已实现利用熔融锰铁合金的高锰钢生产技术的商用化，高锰钢在汽车、LNG 储罐、油砂浆管、建筑等行业已实现商业化应用。分析了浦项在高锰钢产品开发和推广应用中值得借鉴的特点，包括以标准推动应用、持续推进解决方案式营销活动等。建议国内钢铁企业根据市场应用潜力确定重点开发高锰钢成分体系，重视用户使用解决方案的研究。

关键词： 高锰钢，汽车，高强度，高塑性，LNG，减噪，原料

Research and Application Progress of Pohang High Manganese Steel

Zheng Rui, Dai Yunhong, Liu Hang

(Shougang Research Institute of Technology, Beijing 100043, China)

Abstract: The research progress and application of POSCO high manganese steel products are reviewed, POSCO commercializes high manganese steel production technology based on molten FeMn alloy, high manganese steel products have been commercially applied in automobile, LNG storage tank, oil sands slurry pipe and building industry. The characteristics of Pohang in the development and popularization of high manganese steel products are analyzed. It is suggested that the domestic iron and steel enterprises should focus on the development of high manganese steel composition system according to the potential of market application, and pay attention to the study of users' solutions.

Key words: high manganese steel, automobile, high strength, high plasticity, LNG, noise reduction, raw material

适用于热处理升级高强 HFW 油井套管用钢研究

缪成亮[1]，张宏艳[1]，王志鹏[1]，程　政[2]，朱滕威[1]，陈　一[2]

（1. 首钢技术研究院薄板所，北京　100042；

2. 首钢京唐钢铁联合有限责任公司，河北唐山　063200）

摘　要： 针对出口用户对高强油井套管的需求，以及低强钢卷经高频电阻焊（HFW）制管后再整管快速热处理升级到不同强度钢管的工艺路线，本文介绍了适用于热处理升级的低成本 C-Mn-Cr 系套管用钢，讨论其不同冷速下的相变动力行为特征，以及不同热处理工艺处理后升级到 N80/P110 的可行性，同时介绍了实际工业生产的性能和组织特征。此外，对该类套管在工业生产中的探伤缺陷原因和钩状裂纹主因进行统计说明。

关键词： 石油套管，热处理性能，钩状裂纹，夹杂物

Research on the Steel for High Strength HFW Oil Casing by Heat-treatment Upgrading

Miao Chengliang[1], Zhang Hongyan[1], Wang Zhipeng[1],
Cheng Zheng[2], Zhu Tengwei[1], Chen Yi[2]

(1. Strip Technology Department, Shougang Research Insititute of Technology, Beijing 100042, China;
2. Shougang Jingtang United Iron & Steel Co., Ltd., Tangshan 063200, China)

Abstract: Aiming to the requirements of high strength oil casing from export customer, and new process route that different high strength casings are upgraded by different fast on-line heat-treatments after HFW pipe-making using low strength coils, this paper introduces a new low cost steel material with C-Mn-Cr chemical compostions which can apply to heat-treatment upgrading, its phase transformation characteristic is researched under different cooling rate, in addition, the upgrading feasibility of N80/P110 is disscussed based on the properties after different heat-treatment processes, and it also is shown that properties and microstructure of actual industrial products. Furthermore, it is analyzed and illustrated that the reasons for the detection defects and the main reason of hook cracks in industrial production, and the optimization direction of the steel material is discussed.

Key words: oil casing, heat-treatment proerty, hook crack, inclusions

逆相变退火制备高强度高塑性纳米/超细晶钢工艺研究

类承帅，邓想涛，王昭东

（东北大学轧制技术及连轧自动化国家重点实验室，辽宁沈阳 110819）

摘 要： 本文以 Fe-17Cr-6Ni 奥氏体钢不锈为材料，研究了"冷轧+逆相变退火"制备纳米/超细晶奥氏体不锈钢工艺。研究了冷轧及逆相变退火过程中组织演变规律。研究发现在650~750℃退火时获得了平均晶粒尺寸为210~400nm的奥氏体组织，同时实现了高强度高塑性组合，屈服强度达到 790~1041MPa，抗拉强度达到 1023~1093MPa，断裂延伸率达到 26.6%~40.5%。研究发现，相比于其他纳米超细晶金属材料，其较高的塑性主要是由于变形过程中产生了形变诱导马氏体相变及形变孪晶，增强了其加工硬化能力，延迟颈缩的产生，从而改善塑性。

关键词： 纳米/超细晶钢，马氏体逆相变，应变硬化，高强度，高塑性

Fabrication of Phase Reversion-induced High Strength-high Ductility Nano/ultrafine Grained Steel

Lei Chengshuai, Deng Xiangtao, Wang Zhaodong

(State Key Laboratory of Rolling and Automation, Northeastern University, Shenyang 110819, China)

Abstract: A novel process consisted of heavy cold rolling and phase reversion annealing was used to fabricate nano/ultrafine grained Fe-17Cr-6Ni austenitic stainless steel. The microstructure evolution during the cold rolling and annealing process was investigated. The results indicated that nano/ultrafine grained steels with average grain size of 210~400 nm were obtained after ~75% cold reduction and annealing at 650~750 ℃. Yield strength of 790~1041 MPa, ultimate tensile strength of 1023~1093 MPa and fracture tensile elongation of 26.6%~40.5% were obtained. Investigations indicated that the high ductility was due to the comprehensive effect deformation induced martensite transformation and deformation twining, which increased the strain hardening ability and delayed necking.

Key words: nano/ultrafine grained steel, phase reverse transformation, strain hardening, high strength, high ductility

成分及力学性能对耐磨蚀性能的影响

宋凤明，温东辉，杨阿娜

（宝山钢铁股份有限公司研究院，上海　201900）

摘　要： 在围海造陆、航道疏浚等作业中，大量泥沙通过管道以浆体的形式输送。管体不仅发生磨损，还承受不同程度的腐蚀，从而使得管体产生严重的磨蚀失效。采用旋转型试验装置研究了成分及力学性能对钢铁材料在模拟疏浚工况条件下耐磨蚀性的影响。试验结果表明较多耐蚀元素的添加不利于钢种耐磨蚀性能的改善，钢的磨蚀失重与抗拉强度呈线性关系，强度越高耐磨蚀性能越好。在抑制腐蚀的基础上提高强度是改善耐磨蚀性能的有效途径。

关键词： 磨损，腐蚀，疏浚，交互作用，钢

Effect of Composition and Mechanical Properties on Resistance of Erosion Corrosion

Song Fengming, Wen Donghui, Yang Ana

(Research Institute, Baoshan Iron and Steel Co., Ltd., Shanghai 201900, China)

Abstract: Large-scale solid particles are transported by pipeline in form of slurry in dredging works, the pipe wastage occurs due to erosion corrosion. The effect of composition and mechanical properties on resistance of erosion corrosion in dredging was investigated using the rotation method. The experimental results indicate that the addition of resistant corrosion elements in steels is invalid for erosion corrosion resistance. The erosion corrosion weight loss increased with tensile strength according to a nearly linear relationship. The resistance of erosion corrosion can be greatly improved as strength increased considering corrosion resistance.

Key words: erosion, corrosion, dredging, synergism, steel

桥梁钢 Q370qNH 耐腐蚀性研究

石晓伟，郑文超，刘宝喜，王云阁，尹绍江，郝　鑫

（河钢唐钢技术中心，河北唐山　063000）

摘　要： 采用实验室加速腐蚀实验对唐钢中厚板公司研发的耐候桥梁钢 Q370qNH 的耐腐蚀性能进行了研究，并与

普通桥梁钢 Q370q 的耐腐蚀性能进行了对比，对 Q370qNH 的耐腐蚀机理进行了研究，对耐腐蚀实验结果进行了分析。结果表明，Q370qNH 的耐腐蚀性能明显优于普通的 Q370q，其年腐蚀速率为 0.0008mm/a，耐腐蚀性能满足用户对耐候桥梁钢的要求。

关键词：桥梁钢，耐腐蚀性能，加速腐蚀实验，Q370qNH

Study on Corrosion Resistance of Bridge Steel Q370qNH

Shi Xiaowei, Zheng Wenchao, Liu Baoxi, Wang Yunge, Yin Shaojiang, Hao Xin

(Technical Center of HBIS Tangsteel Company,Tangsteel 063000, China)

Abstract: A study on corrosion resistance of newly developed Q370qNH by Tangshan heavy plate was carried out in laboratory by means of wet-dry alternated accelerated corrosion and electro-chemical tests, and the corrosion resistance of Q370qNH was compared with bridge steel Q370q.The corrosion resistance mechanism of Q370qNH was studied and the results of test was analyzed. The results show that the corrosion resistance of Q370qNH was better than Q370q. The annual corrosion rate is 0.0008mm/a and the corrosion resistance meets requirements of the users for the weathering bridge steel.
Key words: bridge steel, corrosion resistance, accelerated corrosion test, Q370qNH

质子辐照对 RPV 材料组织和性能的影响

罗 聪

（华北电力大学能源动力与机械工程学院，北京　100026）

摘　要： 反应堆压力容器（RPV）是反应堆最大的不可更换的核心部件，其使用寿命直接决定了整个核电站的服役寿命。本文采用质子辐照模拟中子辐照，探索国产 A508-3 钢在不同注量辐照条件下微观组织和力学性能的变化。在室温下，进行 190keV 的电子辐照。利用金相显微镜和透射电子显微镜观察了辐照前后样品的显微组织，质子辐照对板条数量密度及分布、碳化物尺寸均没有明显影响。随着辐照注量增加，位错环的尺寸及数量密度均增加。利用纳米压痕实验测试辐照前后 A508-3 钢的力学性能，结果表明，A508-3 钢辐照后弹性模量无显著改变，但纳米硬度值随辐照注量增加而增加。本文用幂函数拟合了位错环引起的强化和脆化关系式，也表征了屈服强度增量和转变温度增量随辐照注量变化的关系式。

关键词：反应堆压力容器钢，质子辐照，微结构损伤，位错环，辐照硬化和脆化

Influence of Proton Irradiation to Microstructure and Properties of the Reactor Pressure Vessel

Luo Cong

(North China Electric Power University, Beijing 100026, China)

Abstract: The reactor pressure vessel (RPV) is the core component of the largest nuclear reactor, and its service life directly determines the service life of the nuclear power station. Irradiation embrittlement is the main failure mode of RPV materials, however, the experimental data of RPV material in high irradiation is less. It is necessary to study the structure and properties of the domestic RPV material under high irradiation. In this paper, proton irradiation is used to simulate neutron irradiation to explore the microstructure and mechanical properties of A508-3 steel under different irradiation.

In the present work, 190 keV proton are conducted on a A508-3 steel. The microstructure of the samples before and after irradiation was observed by optical microscope. The result indicates that the microstructure of the domestic A508-3 steel is granular bainite. In this paper, the relationship between the reinforcement and embrittlement caused by dislocation ring is fitted by power function, and the relationship between yield strength increment and temperature increment of the increment of the change of the irradiation dose is also characterized. At last, nano indentation experiment was conducted to express the mechanic properties. The hardness value increased with the increase of the irradiation dose whose growth rate tends to ease. In this paper, the relationship between the reinforcement and embrittlement caused by dislocation ring is fitted by power function, and the relationship between yield strength increment and temperature increment of the increment of the change of the irradiation dose is also characterized.

Key words: reactor pressure vessel steel, proton irradiation, microstructure damage, dislocation loops, irradiation hardening and embrittlement

输电铁塔耐候试验钢耐大气腐蚀性能分析

张瑞琦[1]，郭晓宏[1]，刘志伟[1]，钟　彬[1]，渠秀娟[2]，李　琳[1]

（1. 鞍钢股份有限公司技术中心，辽宁鞍山　114009；2. 鞍钢股份有限公司产品发展部，辽宁鞍山　114021）

摘　要： 对实验室冶炼的 4 种输电铁塔耐候试验钢，采用周期浸润腐蚀试验、中性盐雾腐蚀试验、电化学腐蚀试验，使用 Q345B 作为对比试样，对其在模拟海洋大气环境、工业大气环境（SO$_2$）的耐候性能进行评估。研究表明：4 种输电铁塔耐候试验钢无论是在模拟海洋大气还是工业大气环境中，耐大气腐蚀性能均优于 Q345B；在模拟工业大气（SO$_2$）环境中，输电铁塔耐候试验钢的耐大气腐蚀性能更加突出，约为对比试样 Q345B 的 3 倍；当 Cr 含量小于 1.0%时，随着 Cr 含量的增加，对于耐大气腐蚀性能的影响不是非常明显；综合来看 3#试验钢的耐大气腐性能和成本更适合工业试制。

关键词： 输电铁塔，耐候钢，周期浸润腐蚀试验，中性盐雾腐蚀试验，电化学腐蚀试验

Analysis of Atmospheric Corrosion Resistance of the Transmission Tower Weathering Tested Steel

Zhang Ruiqi[1], Guo Xiaohong[1], Liu Zhiwei[1], Zhong Bin[1], Qu Xiujuan[2], Li Lin[1]

(1. Technology Center of Angang Steel Company Limited, Anshan 114009, China; 2. Product Development Department of Angang Steel Company Limited, Anshan 114021, China)

Abstract: Evaluating atmospheric corrosion-resistance performance of 4 kinds of transmission tower weathering tested steels which are smelted and rolled in lab, by alternate immersion and dry corrosion test, neutral salt spray corrosion test and electrochemical corrosion test in simulated oceanic atmosphere and industrial atmosphere(SO$_2$). Q345B will be used as the contrast sample. The research shows that the corrosion-resistance of 4 transmission tower tested steels is better than Q345B in two different atmosphere. In simulated industrial atmosphere (SO$_2$), the corrosion-resistance is about three times compared to the sample Q345B. With the increase of Cr content, the effect on atmospheric corrosion-resistance performance is not prominent when Cr content is less than 1.0%. According to the atmospheric corrosion-resistance performance and cost, 3# is more suitable for industrial trial.

Key words: transmission tower, weathering steel, alternate immersion and dry corrosion test, neutral salt spray corrosion test, electrochemical corrosion test

60Si₂CrVAT 货车转向架弹簧失效分析

李绍杰[1]，樊一丁[1]，寇　鑫[2]，孙晓明[1]

（1. 河钢集团石家庄钢铁有限责任公司，河北石家庄　050031；2. 天津市大港汽车配件弹簧厂，天津　300270）

摘　要： 采用断口分析、电子显微分析、金相分析和硬度分析等分析方法，对铁路货车转向架弹簧 60Si2CrVAT 在疲劳试验中断裂的原因进行了分析。结果表明，弹簧支撑圈与工作圈之间存在点接触产生硌伤而导致应力集中是弹簧早期疲劳断裂的主要原因。另外，弹簧局部存在异常下贝氏体组织也对弹簧疲劳寿命产生不良影响。

关键词： 弹簧，疲劳断裂，点接触，应力集中，下贝氏体

Failure Analysis of 60Si₂CrVAT Van Spring

Li Shaojie[1], Fan Yiding[1], Kou Xin[2], Sun Xiaoming[1]

(1. HBIS Group Shijiazhuang Iron & Steel Co., Ltd., Shijiazhuang 050031, China; 2. Da gang Automobile Accessories & Spring Factory, Tianjin 300270, China)

Abstract: Fracture analysis,electronic microscopy analytical, metallurgical structure analysis and hardness test were used to analysis the reason of van spring fracture. The results indicate that stress concentration at the effective coils of spring was the mail reason of fracture, which created by point-contact between the stay coil and effective coil.In addition, lower bainite of spring was disadvantage.

Key words: spring, fatigue fracture, point-contact, stress concentration, lower bainite

汽车齿轮钢及生产控制技术

李冠军，王信康

（河钢石钢公司，河北石家庄　050031）

摘　要： 研究了汽车变速器用齿轮对齿轮钢质量要求，并对齿轮钢的窄淬透性控制技术、夹杂物控制技术、细晶控制技术和偏析控制技术等进行了实践，通过制定合适的冶炼、连铸及轧制工艺参数，生产出满足高端汽车变速器用的齿轮钢。

关键词： 窄淬透性，夹杂物，晶粒度，偏析

Automobile Gear Steel and Its Production Control Technology

Li Guanjun, Wang Xinkang

(Hebei I & S Group Shijiazhuang I & S Co., Ltd., Shijiazhuang 050031, China)

Abstract: The requirements on the quality of gear steel for automobile transmission gear is researched, and control technology of narrow hardenability, inclusions, fine grains and segregation is tested. The gear steel for high-end automobile transmission is successfully produced through the setting of proper process parameters for steelmaking, continuous casting and rolling.

Key words: narrow hardenability, inclusions, grain size, segregation

运用失效分析成果服务客户，实现产品质量和服务质量双提升

高建华

（石家庄钢铁有限责任公司技术中心，河北石家庄　050031）

摘　要：通过对某汽车配件厂生产的转向节表面结瘤件进行金相、硬度和能谱分析，结合生产工艺进行研究，最终找到了造成结瘤废品的原因是由于锻造时加热温度较高造成氧化铁皮熔化，吹撒的石墨膏水溶液浓度较高，熔融的氧化铁皮与熔化的石墨膏瞬间发生合金反应生成亚共晶融滴，粘结在工件表面形成瘤体和气孔，结瘤体硬度较基体硬度高，造成钻孔困难而产生批量废品。通过采取降低加热温度、稀释石墨膏水溶液，加强吹扫等改进措施，避免了该缺陷产生，成功将失效分析成果服务于客户，实现了产品质量和服务质量双提升。

关键词：氧化铁皮，石墨膏，亚共晶，结瘤

Customer Service with Failure Analysis, Promoting Product Quality and the Service Quality

Gao Jianhua

(Shijiazhuang Iron & Steel Co., Ltd. Technical Center, Shijiazhuang 050031, China)

Abstract: Failure sample is the joint nodulation part from one of the automobile parts factory, considering the process details, do the analysis by the Metallographic, Hardness and EDS, find out the failure reason is that the high heating temperature during the forging process make the oxide scale melting, alloyed with the graphite paste which from the sprayed high concentration graphite paste water solution, generating the hypoeutectic tear, pasted on the part surface, formed the nodulation and gas hole, the hardness of the nodulation part is higher than the base part which leading the drilling process is much more difficult and producing the batch rejected parts. Suggested to the customer that reducing the heating temperature, dilute the graphite paste concentration, reinforce the cleaning process will avoid the rejection, and finally we succeed customer service with failure analysis, promoted the product quality and the service quality.

Key words: oxide scale, graphite paste, hypoeutectic, nodulation

X80 级 1422mm 大口径 UOE 焊管开发

章传国[1]，郑　磊[1]，张清清[1]，孙磊磊[1]，谢仕强[2]，王　波[3]

（1. 宝钢股份中央研究院，上海　201900；2. 宝钢股份钢管条钢事业部，上海　201900；
3. 宝钢股份制造管理部，上海　201900）

摘　要： 为满足大输量管道建设需求，开发试制了 X80 级 1422mm 大口径 UOE 焊管用厚规格钢板及焊管。采用低碳铌微合金化成分设计，通过实验轧制和热模拟试验方法，得到优化的再结晶区变形工艺和合适的制管焊接热输入范围，并成功进行 X80 钢板及 1422mm 焊管试制。开发的宽至 4350mm 厚度为 21.4mm、25.7mm、30.8mm 规格 X80 管线钢板具有高的低温冲击韧性及 DWTT（Drop weigh tear test）性能；在 UOE 机组上试制的 X80 级 1422mm 大口径直缝埋弧焊管管体、焊缝及热区 −10℃冲击功分别达到 340J、200J 及 270J 以上，管体 DWTT 的 85%FATT（Fracture appearance transition temperature）低于 −15℃，同时具有良好的椭圆度、焊高、错边等管型尺寸。进一步比较分析了小批量生产的板-管性能变化规律，钢管所有性能均满足技术要求，部分规格焊管已在大输量管道工程中得到应用。

关键词： 1422mm 口径，UOE 焊管，组织性能，焊接

Development of X80 Grade 1422mm Large Diameter UOE Welded Pipe

Zhang Chuanguo[1], Zheng Lei[1], Zhang Qingqing[1], Sun Leilei[1],
Xie Shiqiang[2], Wang Bo[3]

(1. Research Institute, Baoshan Iron and Steel Co., Ltd., Shanghai 201900, China; 2. Products & Technique Management Department, Baoshan Iron and Steel Co., Ltd., Shanghai 201900, China; 3. Tube, pipe and bar business unit, Baoshan Iron and Steel Co., Ltd., Shanghai 201900, China)

Abstract: To meet the requirements of larger transmission capacity pipeline construction, the X80 grade UOE longitudinal submerge arc welding (LSAW) pipe with Φ1422mm outside diameter (O.D.) and its raw plate were developed. The low carbon and niobium micro-alloy composition was designed. The optimized deformation process in rough rolling and suitable range of heat input for submerged arc welding were investigated by using the way of pilot rolling and thermo-simulation, which promote the trial production of X80 plate and O.D.1422mm welded pipe. The trial produced 21.4mm, 25.7mm, 30.8mm plate with width up to 4350mm characterized by high impact toughness and good DWTT performance. The X80 welded pipes with dimension in Φ1422×21.4mm、Φ1422×25.7mm and Φ1422×30.8mm were manufactured at UOE mill. The charpy impact absorbed energy of pipe body, weld seam and heat affected zone is over 300J, 200J and 270J, respectively. The DWTT 85% FATT for pipe body is lower than −15℃. Meanwhile, the out of roundness, weld height and misalignment are well meeting the requirements. Furthermore, the mechanical properties changes from plate to pipe were analyzed for the produced pipe in small quantity. A part of produced O.D.1422mm X80 UOE pipe have been used in large transmission gas pipeline.

Key words: O.D.1422mm, UOE welded pipe, microstructure and property, welding

超快冷条件下 Q420 桥梁钢焊接热影响区组织性能研究

刘宝喜[1]，胡　军[2]，王云阁[1]，高彩茹[2]，高秀华[2]，杜林秀[2]，李双江[3]

（1. 河北钢铁集团唐钢技术中心，河北唐山　063001；2. 东北大学轧制技术及连轧自动化国家重点实验室，辽宁沈阳　110819；3. 河钢集团钢研总院，河北石家庄　050023）

摘　要： Q420q 桥梁钢需要具有优异的强韧性的同时拥有良好的可焊性，焊缝(WM)两侧的热影响区(HAZ)的显微组织和性能直接影响焊接构件接头质量。在未采用焊前预热和焊后热处理的条件下，在实验室对采用超快冷技术生产的 20mm 控轧控冷 Q420q 钢板进行了气体保护焊焊接试验。当线能量为 15kJ/cm 时，焊接接头的屈服强度、抗拉强度、断后伸长率达到国家标准。焊接接头各部分-20℃的冲击功均≥279J，冲击韧性远超国家标准。焊接接头各区域断口均由韧窝组成，呈现韧性断裂模式。焊接接头区域并未出现明显的软硬化现象。WM 显微组织多以针状铁素体(AF)为主，能很好地阻碍裂纹的扩展。熔合线(FL)显微组织包含粒状贝氏体(GB)、侧板条铁素体(FSP)、AF 和多边形铁素体(PF)。粗晶区(CGHAZ)的显微组织为 GB、板条贝氏体(LB)、AF 及少量的 PF 的混合组织。细晶区(FGHAZ)的显微组织为大量的 PF、珠光体(P)及少量的渗碳体(C)。

关键词： 超快冷，桥梁钢，焊接热影响区，组织性能

Study on Microstructure and Properties of Heat Affected Zone of Q420 Bridge Steel under Ultra-fast Cooling Condition

Liu Baoxi[1], Hu Jun[2], Wang Yunge[1], Gao Cairu[2], Gao Xiuhua[2], Du Linxiu[2], Li Shuangjiang[3]

(1. Technology Center of Tangsteel, Hebei Iron & Steel Group Co., Ltd., Tangshan 063001, China; 2. The State Key Laboratory of Rolling and Automation, Northeastern University, Shenyang 110819, China; 3. HBIS Group Technology and Research Institute, Shijiazhuang 050023, China)

Abstract: Q420q bridge steel require not only excellent strength and toughness but also great weldability. The microstructure and properties of the heat affected zone(HAZ) on both sides of the weld metal (WM) directly affect the performance of welding joint in welded components. Under the condition of no preheating before welding and post-weld heat treatment, the gas shielded welding was carried out for 20mm thick plates of Q420q bridge steel. The plates were produced by ultra-fast cooling technology on the principle of TMCP. When heat input is 15kJ/cm, the yield strength, tensile strength and elongation of welding joint meet national standard. The impact energy of each zone in welding joint is ≥279J, impact toughness greatly exceed the national standard. The impact fracture in each zone of welding joint are composed of dimples, exhibiting ductile fracture behavior. There is no obvious softening and hardening behavior in the welding joint. The microstructure of the WM is mostly acicular ferrite(AF), which can resist the propagation of cracks. The microstructure of fusion line(FL) contains granular bainite(GB), ferrite side plates (FSP), AF, and polygonal ferrite(PF). The microstructure of the coarse grain heat affect zone(CGHAZ) is the mixed structure of GB, lath bainite(LB), AF, and a small amount of PF. The microstructure of the fine grain heat affect zone (FGHAZ) is a large number of PF, pearlite(P)and a small amount of cementite(C).

Key words: ultra-fast cooling, bridge steel, heat affected region, microstructure and properties

建筑抗震用极低屈服点钢板制备工艺研究

陈振业[1,2]，孙　力[2]，刘宏强[2]，李睿昊[1]

（1. 东北大学轧制技术及连轧自动化国家重点实验室，辽宁沈阳　110819；

2. 河钢集团钢研总院，河北石家庄　050023）

摘　要： 通过合理的成分设计及轧制、热处理工艺选择，开发出满足 GB/T28905—2012 要求的 40mm 和 80mm 厚 160MPa 级建筑抗震用低屈服点钢。所开发的 40mm 钢板的微观组织为 F+B，晶粒度为 6.5～7 级，屈服强度在 157～168MPa 之间，抗拉强度在 291～304MPa 之间，0℃下冲击功在 289～311J 之间；所开发的 80mm 钢板微观组织为 F，晶粒度为 5～5.5 级，屈服强度在 150～163MPa 之间，抗拉强度为 280～285MPa，0℃下冲击功在 73～92J 之间。试制的 LY160 钢板具有良好的高应变低周疲劳性能和焊接性能，为 160MPa 级低屈服点钢板的工业化生产提供了理论依据及工艺基础。

关键词： 抗震用钢，低屈服点钢，力学性能，高应变低周疲劳

Study on Production Process of Low Yield Point Steel Used For Earthquake Resistant

Chen Zhenye[1,2], Sun Li[2], Liu Hongqiang[2], Li Ruihao[1]

(1. State Key Laboratory of Rolling Technology and Automation, Northeastern University, Shenyang 110819, China;

2. The Centre Iron and Steel Technology Research Institute of HBIS, Shijiazhuang 050023, China)

Abstract: The 160MPa low-yield strength steel with 40mm and 80mm thicknesses for building aseismicity was developed through reasonable composition design and rolling technology. The LY160 steel could meet the requirements of GB/T28905—2012. The microstructure of the LY160 steel plate (40mm) was ferrite with very little bainite. The grain size was from grade 6.5 to grade 7. Yield strength is from 157MP to 168MPa and tensile strength is from 291MPa to 304MPa. The impact property below 0℃ ranged between 289J and 311J. The microstructure of the LY160 steel plate (80mm) was ferrite. The grain sizes was from grade 5 to grade 5.5. In addition, the yield strength is from 150MP to 163MPa, the tensile strength is from 280MPa to 285MPa. The impact property below 0℃ ranged between 73J and 92J. The trail production had favorable high adaptability to the low-cycle fatigue property and welding property, and it was reference for the industrial trail of the 160MPa low point steel.

Key words: steels for earthquake resistant, low yield point steel, mechanical properties, high strain low cycle fatigue

新型高铁车轮材料的组织和性能调控研究

李昭东[1]，周世同[1,2]，崔银会[3]，王群娣[4]，雍岐龙[1,2]，杨才福[1]

（1. 钢铁研究总院工程用钢研究所，北京　100081；2. 昆明理工大学材料科学与工程学院，

云南昆明　650093；3. 马鞍山钢铁股份有限公司，安徽马鞍山　243000；

4. 太原重工轨道交通设备有限公司，山西太原　030024）

摘　要：通过显微组织调控和夹杂物改性设计了两种新型中碳珠光体高铁车轮钢，并研究了其组织、非金属夹杂物与力学性能的关系。与 ER8 车轮钢相比，V-Si 合金化的新车轮材料的奥氏体晶粒尺寸和珠光体片层间距得到细化，强度和韧性同时提高；通过硫化物包裹氧化物的夹杂物改性后，车轮材料的韧性进一步优化，疲劳性能也有所改善。

关键词：高铁车轮，显微组织，非金属夹杂物，韧性

Study on the Controlling of Microstructure and Mechanical Properties of High-speed Railway Wheels

Li Zhaodong[1], Zhou Shitong[1,2], Cui Yinhui[3], Wang Qundi[4],
Yong Qilong[1,2], Yang Caifu[1]

(1. Department of Structural Steels, Central Iron and Steel Research Institute, Beijing 100081, China;
2. Department of Materials Science and Engineering, Kunming University of Science and Technology, Kunming 650093, China; 3. Maanshan Iron & Steel Co., Ltd., Maanshan 243000, China;
4. Taiyuan Heavy Industry Transit Equipment Co., Ltd., Taiyuan 030024, China)

Abstract: Two new types of medium-carbon pearlitic steels for high-speed railway wheel were developed by microstructure control and inclusion modification. The effects of microstructure and non-metallic inclusions on mechanical properties were studied. Compared to steel ER8, the austenite grain size and pearlite interlamellar spacing of new wheel steels were refined by V-Si alloying, leading to a simultaneous improvement of strength and toughness. The toughness was found to be further improved through inclusion modification by formation of oxide-sulfide duplex inclusion (sulfide-encapsulated oxide) in the steel, and the fatigue properties were also found to be slightly improved.
Key words: high-speed railway wheel, microstructure, non-metallic inclusion, toughness

奥氏体气阀钢 21-4N 炸裂原因分析

曹美姣

（攀钢集团江油长城特殊钢股份有限公司技术中心不锈钢研究室，四川江油　621700）

摘　要：本文通过阐述 21-4N 工艺试验过程、分析结果及微观组织分析，分析了 21-4N 钢初轧方坯轧后水冷后炸裂原因，同时提出了有效的改进措施。

关键词：气阀钢，21-4N，炸裂

Analysis of Reasons for Cracking of Austenitic Valve Steel 21-4N

Cao Meijiao

(Technology Center, Sichuan Changcheng Special Steel Co., Ltd., Pangang Group, Jiangyou 621700, China)

Abstract: Based on the analysis of process test, test results and micro-structure of the steel 21-4N, the reasons for cracking of the 21-4N billet after rolling followed by water cooling were found out and effective improvement measures were put forward.

Key words: valve steel, 21-4N, cracking

齿轮轴开裂原因分析

谢金鹏，杨　春，钟振前

（钢铁研究总院失效分析中心，北京　100081）

摘　要： 某公司齿轮轴加工过程中发生开裂。本文对齿轮轴的材料力学性能、金相组织、裂纹及断口形貌进行了分析和检验。结果表明，齿轮轴存在过烧现象，部分晶界出现熔化，在锻造过程中材料内部形成大量沿晶裂纹。

关键词： 齿轮轴，20CrMoH，过热

Failure Analysis of Fractured Gear Shaft

Xie Jinpeng, Yang Chun, Zhong Zhenqian

(Failure Analysis Center, Centre Iron & Steel Research Institute, Beijing 100081, China)

Abstract: Some gear shaft has cracked in the processing. Mechanical properties, microstructure and fracture morphology were analyzed to determine failure reason. The results showed that the material has been over-burned in heat treatment. The grain boundary was melted and cracked in the flowing forge process.

Key words: gear shaft, 20CrMoH, over-burning

现场工程师学术化促成特殊钢成果高效产业化

殷　匠[1,2]，裴顺福[2]

（1.亚星锚链，江苏靖江　214531；2.（原）宝钢特钢，上海　201900）

夹杂物不同晶面诱导点蚀机理与功函数的关系

曹羽鑫[1,2]，李光强[1,2]，侯延辉[1,2]

（1. 武汉科技大学耐火材料与冶金省部共建国家重点实验室，湖北武汉　430081；2. 武汉科技大学钢铁冶金及资源利用省部共建教育部重点实验室，湖北武汉　430081）

摘　要：以 Al-Ti 脱氧钢中夹杂物为研究对象，采用基于密度泛函理论第一性原理的计算方法，计算比较了 Al_2O_3，Ti_2O_3，$CaTiO_3$ 和钢基体的多个晶面的电子功函数，分析了夹杂物和钢基体在不同晶面以及同一晶面不同终端暴露在最外层时，本征电势差的大小，以此分析了不同夹杂物不同晶面、同一晶面不同终端面诱导点蚀的机理。

关键词：夹杂物，第一性原理，电子功函数，点蚀

Correlation between Work Function and Inducing Pitting Corrosion Mechanism of Inclusion Different Surface

Cao Yuxin[1,2], Li Guangqiang[1,2], Hou Yanhui[1,2]

(1. State Key Laboratory of Refractories and Metallurgy, Wuhan University of Science and Technology, Wuhan 430081, China; 2. Key Laboratory for Ferrous Metallurgy and Resources Utilization of Ministry of Education, Wuhan University of Science and Technology, Wuhan 430081, China)

Abstract: Inclusions of Al - Ti deoxidizing steel was researched, based on density functional theory and first principles calculation, compared electron work function of Al_2O_3, Ti_2O_3, $CaTiO_3$ and steel matrix multiple crystal, and analyzed potential difference of inclusion and the steel substrate in various crystal planes and terminal, which exposed to the outer layer. It was illuminated that the mechanism of inclusion different surface and diverse terminal on pitting corrosion.

Key words: inclusion, first principle, electron work function, pitting corrosion

美标 AISI4140 钢种的开发

任树洋，孟庆勇，李　行，尹绍江，秦　坤，马欣然

（河钢唐钢技术中心，河北唐山　063016）

摘　要：通过制定 4140 钢种成分、工艺路线、生产过程中工艺参数进行组织生产，获得了较为理想的板材性能数据，板材的外形质量未发生因应力不均而造成的瓢曲现象。

关键词：4140，硬度，CCT 曲线，高温回火

Development of AISI4140 Steel

Ren Shuyang, Meng Qingyong, Li Hang, Yin Shaojiang, Qin Kun, Ma Xinran

(HBIS Group Tangsteel Company, Tangshan 063016, China)

Abstract: Applying appropriate composition design and process route, excellent property of AISI4140 steel is obtained. Plate profile is fine, which is uninfluenced by residual stress.

Key words: 4140, hardeness, CCT curve, high-temperature tempering

烘烤时间对低碳钢烘烤硬化性能影响内耗研究

金晓龙[1]，王　旭[1]，刘仁东[1]，李伟娟[2]

（1. 鞍钢集团钢铁研究院，辽宁鞍山　114009；2. 辽宁科技大学材料与冶金学院，
辽宁鞍山　114051）

摘　要：测试了大晶粒尺寸低碳钢不同烘烤时间的烘烤硬化性能的内耗曲线。结果表明，预变形量为 5%，烘烤温度为 170℃时，不同烘烤时间的大晶粒低碳钢的烘烤硬化值（BH）均为负值；BH 值和 SKK 峰随着烘烤时间的增加先增加后不变，BH 值随着 SKK 峰的变化而变化，即 BH 值主要受 Cottrell 气团强化大小影响；当烘烤时间为 1000min 时，BH 值达到最大值，烘烤时间继续增加，BH 值变化不大；烘烤时间为 20min 时，SKK 峰值最大，当烘烤时间为 1000min 以上，SKK 峰高变化不大。

关键词：烘烤硬化，内耗，Cottrell 气团强化，预变形

Internal Friction on Effect of Baking Time on Bake Hardening in Low Carbon Steel

Jin Xiaolong[1], Wang Xu[1], Liu Rendong[1], Li Weijuan[2]

(1. Iron & Steel Research Institute of Angang Group, Anshan 114009, China; 2. School of Materials and Metallurgy, University of Science and Technology Liaoning, Anshan 114051, China)

Abstract: Bake hardening property and internal friction curves in pre-deformed 5% and baking temperature 170℃ for different baking time of coarse grained low carbon steel were measured. The results show that the bake hardening properties are negative in different baking time; the BH value and SKK peak are from increasing to steady with baking time's increasing, BH value vary with SKK peak, that is to say, the BH values are affected by Cottrell atmosphere`s density; BH value is the maximum when baking time is 1000min, BH values are not varied when the baking time continue increasing; SKK peak is the maximum when the baking time is 20min, SKK peak are not varied when the baking time greater than 1000min.
Key words: bake-hardening, internal friction, Cottrell atmosphere strengthening, pre-deformation

模焊制度对大厚度 14Cr1MoR 钢板组织和性能的影响

李样兵，吴艳阳，柳付芳，牛红星

（舞阳钢铁有限责任公司，河南舞阳　462500）

摘　要：本文研究了模拟焊后热处理制度对 14Cr1MoR 钢板组织和性能的影响。结果表明：在相同的模焊保温时间、装出炉温度和升降温速率下，三次分段保温 20h 与一次性保温 20h 两种模焊制度下组织中均有大量颗粒状碳化物析出，不同之处在于三次分段保温 20h 还在晶界上明显有碳化物析出；相应的室温拉伸、高温拉伸的屈服和强度与一次性保温 20h 相比有一定程度下降，冲击韧性明显变差。

关键词：模焊制度，14Cr1MoR 钢板，组织，性能

Effect of PWHT Process on Microstructure and Properties of Large Thickness 14Cr1MoR Plate

Li Yangbing, Wu Yanyang, Liu Fufang, Niu Hongxing

(HBIS Group Wuyang Iron and Steel Co., Ltd., Wuyang 462500, China)

Abstract: This paper mainly studies the effect of PWHT system on Microstructure and Properties of large thickness 14Cr1MoR Plate. The results shows that under the same PWHT soaking time, loading & unloading temperature and rate of heating &cooling, three stages heat preservation 20h and one stage heat preservation 20h，There are numerous granular carbides in the metallographic structure. The difference is that the three stages heat preservation 20h has obvious carbide precipitation on the grain boundaries,accordingly,the yield and strength at room temperature tensile test and high temperature tensile test are decreased to some extent compared with that one stage heat preservation 20h, and the impact toughness is obviously worse.

Key words: PWHT, 14Cr1MoR plate, microstructure, properties

气雾罐顶盖冲压开裂原因分析及改进措施

张　飞[1]，李建设[2]，白丽娟[1]，任振远[2]，董伊康[1]，刘　需[1]，张　鹏[1]

（1. 河钢集团钢研总院，河北石家庄　050023；2. 唐山不锈钢有限责任公司，河北唐山　063010）

摘　要： 冲压开裂是气雾罐顶盖生产过程中常见的缺陷。利用扫描电镜、能谱仪对气雾罐顶盖冲压开裂断口形貌进行分析；利用拉伸试验机对试验材料力学性能进行分析；对顶盖冲压过程进行计算机模拟，为工艺优化提供理论依据。研究表明，镀锡板轧向 r 值偏低是造成冲压开裂的直接原因，基板中的大颗粒夹杂物增加了开裂几率。为提高冲压性能，镀锡板的 r 值应该控制在 1.3 以上。采取优化措施后，气雾罐顶盖冲压开裂率由 1% 降低至 0.03% 以下，满足了用户需求，实现了稳定供货。

关键词： 镀锡板，冲压开裂，夹杂物，r 值

Cause Analysis and Improvement Measures on Stamping Crack of Aerosol Tank Top Cover Steel Sheet

Zhang Fei[1], Li Jianshe[2], Bai Lijuan[1], Ren Zhenyuan[2], Dong Yikang[1], Liu Xu[1], Zhang Peng[1]

(1. Hesteel Group Technology Research Institute, Shijiazhuang 050023, China; 2. Tangshan Stainless Steel Co., Ltd., Tangshan 063010, China)

Abstract: The stamping cracking is a common defect in the production process of the aerosol can top cover. The fracture morphology of the stamping cracking of the aerosol can was analyzed by SEM and EDS. The mechanical properties of the test materials were analyzed by tensile testing machine. The stamping process of the top cover was simulated by computer software, which can provide theoretical basis for the process optimization. The results show that the low r value of tinplate rolling direction is the direct cause of the stamping cracking, and the large particle inclusions in the substrate increase the cracking probability. In order to improve the stamping performance, the r value of tin plate should be controlled more than 1.3. After taking optimization measures, the cracking rate of the top cover of aerosol is reduced from 1% to less than 0.03%, which meets the needs of users and achieves stable supply.

Key words: tinplate steel sheet, stamping crack, inclusion, r value

循环热处理对马氏体时效钢组织和性能的影响

谢　东，林明新，张　萌，程含文，石　莉，刘印子

（舞阳钢铁有限责任公司，河南舞阳　462500）

摘　要： 本文对 T250 马氏体时效钢进行 820℃循环热处理，并得出其组织和性能的变化。结果表明：循环热处理可以明显细化晶粒；循环热处理不能提高钢的强度和硬度，却能提高钢的塑性。

关键词： 马氏体时效钢，循环热处理，细化，塑性

Effect of Circulating Heat Treatment on Microstructure and Properties of Martensitic Aging Steel

Xie Dong, Lin Mingxin, Zhang Meng, Cheng Hanwen, Shi Li, Liu Yinzi

(Ministry of Science and Technology of HBIS Group Wu Steel Company, Wuyang 463500, China)

Abstract: In this paper, T250 martensitic aging steel was subjected to cyclic heat treatment at 820 ℃, and its microstructure and properties were obtained. The results show that the heat treatment can obviously refine the grain size. The cyclic heat treatment can not improve the strength and hardness of steel, but it can improve the plasticity of steel.

Key words: martensitic aging steel, circulating heat treatment, refinement, plasticity

7 汽车钢

炼铁与原料
炼钢与连铸
轧制与热处理
表面与涂镀
金属材料深加工
钢铁材料
★ 汽车钢
海洋工程用钢
轴承钢
电工钢
粉末冶金
非晶合金
高温合金
耐火材料
能源与环保
分析检测
冶金设备与工程技术
冶金自动化与智能管控
冶金技术经济

铝元素对中锰钢不同热处理下组织性能的影响

徐海峰[1,2]，周天鹏[1,3]，王存宇[4]，曹文全[1]

（1. 钢铁研究总院特殊钢研究所，北京　100081；2. 华中科技大学材料科学与工程学院，湖北武汉　430074；3. 重庆大学材料科学与工程学院，重庆　400045；4. 钢铁研究总院华东分院，北京　100081）

摘　要： 本文研究了铝元素对中锰钢的组织性能的影响规律，结果发现铝元素基本不改变中锰钢热处理过程中组织演变，添加铝元素抑制退火时碳化物析出，从而提高了奥氏体的碳含量。尽管含铝中锰钢的抗拉强度有所下降，但是其屈服强度明显增加。含铝中锰钢经675℃奥氏体化淬火及短时间退火后获得最佳力学性能，即抗拉强度880MPa，总延伸率39.5%，强塑积34.5GPa%。

关键词： 含铝中锰钢，残余奥氏体量，碳含量，微观组织，强塑积

Effects of Aluminium on the Microstructure and Mechanical Properties in the Medium Manganese Steel under Different Heat Treatment Conditions

Xu Haifeng[1,2], Zhou Tianpeng[1,3], Wang Cunyu[4], Cao Wenquan[1]

(1. Institute for Special Steels, Central Iron and Steel Research Institute, Beijing 100081, China; 2. School of Materials Science and Engineering, Huazhong University of Science and Technology, Wuhan 430074, China; 3. School of Materials Science and Engineering, Chongqing University, Chongqing 400045, China; 4. East China Branch, Central Iron and Steel Research Institute, Beijing 100081, China)

Abstract: The effects of aluminium on microstructure and mechanical properties in medium manganese steels were investigated in this study. It is found that addition of aluminium don't significantly change the microstructure evolution during heat treatment process, aluminium can retard the precipitation of carbides in the following annealing, resulting in higher carbon content in several austenite. Though aluminium addition leads to a certain decline of tensile stress, the yield stress and elongation are neither decreasing significantly, especially the former is improved after heat treatment. Furthermore, the Al-bearing medium manganese steels can achieve optimal product of strength and ductility with austenization at 675°C and short time (5min) annealing, the excellent mechanical properties are strength 880MPa, total elongation 39.3% and product of tensile strength to total elongation (Rm*AT) 34.6GPa%.

Key words: Al-beating medium manganese steel, retained austenite fraction, carbon concentration, microstructure, product of tensile strength to total elongation

一步法工艺对热轧 Q&P 钢组织和性能的影响

胡　俊，刘　洋，梁　文，刘　斌，彭　周

（武钢研究院，湖北武汉　430080）

摘　要： 研究了一步法工艺对热轧 Q&P 钢的组织和力学性能的影响，利用单轴拉伸试验获得了不同热处理条件下的力学性能，利用金相显微镜，扫描电镜和透射电镜对热轧 Q&P 钢的微观组织进行了表征，利用 XRD 衍射测量了热轧 Q&P 钢中的残余奥氏体含量。结果表明：采用一步法 Q&P 工艺，可以获得抗拉强度超过 980MPa，屈服强度超过 600MPa，延伸率超过 15%的热轧 Q&P 钢。随着淬火温度从 290℃升高至 410℃，Q&P 钢的微观组织由马氏体逐渐转变为马氏体+残余奥氏体，并且抗拉强度逐渐降低，延伸率逐渐升高。随着淬火温度升高，热轧 Q&P 钢中残余奥氏体含量也逐渐升高。Q&P 钢中的残余奥氏体在拉伸形变过程中通过 TRIP 效应，使得 Q&P 钢不仅具有高强度而且具有良好的塑性，强塑积可以达到 19.41GPa%。

关键词： 热轧 Q&P 钢，一步法工艺，残余奥氏体，TRIP 效应，强塑积

The Effect of One-Step Process on the Microstructure and Mechanical Properties of Hot-Rolled Q&P Steel

Hu Jun, Liu Yang, Liang Wen, Liu Bin, Peng Zhou

(Research and Development Center of WISCO, Wuhan 430080, China)

Abstract: The effect of one-step process on the microstructure and mechanical properties of hot-rolled Q&P steel was investigated. The mechanical properties were studied by uniaxial tensile test. The microstructure of Q&P steel was characterized by optical microscope (OM), scanning electron microscope (SEM) and transmission electron microscope (TEM). The volume fraction of retained austenite in the hot rolled Q&P steel was measured by XRD diffraction. The results show that the hot-rolled Q&P steel with tensile strength over 980MPa, yield strength over 600MPa and total elongation over 15% can be obtained by one-step Q&P process. With the quenching temperature increasing from 290℃ to 410℃, the microstructure of Q&P steel gradually changed from martensite to martensite + retained austenite and the tensile strength decreased while the elongation increased gradually. As the quenching temperature increased, the amount of retained austenite in the hot rolled Q&P steel increased gradually. The TRIP effect of the retained austenite in the Q&P steel during deformation makes Q&P steel not only high strength but also good plasticity, in which the product of tensile strength to total elongation exceeded 19.41GPa%.

Key words: hot-rolled Q&P steel, one-step process, retained austenite, TRIP effect, product of tensile strength to total elongation

热成形钢 AC1500HS 热变形行为的研究

时晓光，董　毅，韩　斌，张　宇，孙成钱

（鞍钢股份有限公司技术中心，辽宁鞍山　114009）

摘　要： 本文主要采用鞍钢生产的热成形钢 AC1500HS，进行了连续冷却静态 CCT 曲线测试。采用 Gleeble3800 热模拟试验机进行了热成形模拟试验，揭示了热成形钢的流变行为及组织特点，给出了热成形钢 AC1500HS 模拟热成形的最佳工艺参数，并进行了实际热成形汽车零件应用制造。

关键词： 热成形钢，AC1500HS，流变行为，汽车

The Investigation of the Thermal Transformation Behavior of AC1500HS Hot Stamping Steel

Shi Xiaoguang, Dong Yi, Han Bin, Zhang Yu, Sun Chengqian

(Technology Centre, Angang Steel Co., Ltd., Anshan 114009, China)

Abstract: The CCT curves of AC1500HS hot stamping steel of Ansteel has been determined by conductive in this paper. The simulation tests of hot stamping with a Gleeble 3800 has been determined for the flow behavior and microstructures characteristics of hot stamping steels, pointing out the excellent processes parameters of hot stamping, and the actual automobile parts were produced by hot stamping processes.

Key words: hot stamping steel, AC1500HS, flow behavior, automobile

Al 元素含量对高锰钢力学行为的影响

郭金宇，刘仁东，徐荣杰，孟静竹，王　旭，王科强，金晓龙，孙健伦

（鞍钢钢铁研究院，辽宁鞍山　114009）

摘　要：以具有不同 Al 含量的 Fe-18Mn-（0, 1.5, 3.5, 6, 8, 11）Al-0.3Si-0.8C 热轧高锰钢为研究对象，通过室温拉伸、金相组织观察和透射电镜观察等研究手段，研究 Al 元素对高锰钢力学性能、显微组织、变形方式的影响。研究结果表明，随着 Al 元素含量的增加，实验钢的屈服强度增加、抗拉强度先减小后增加、延伸率和应变硬化能力下降。力学性能的改变与 Al 元素对实验钢层错能的影响有关。

关键词：高锰钢，显微组织，力学性能，变形方式，层错能

Influence of Al Element Content on the Mechanical Behavior of High Manganese Steel

Guo Jinyu, Liu Rendong, Xu Rongjie, Meng Jingzhu, Wang Xu, Wang Keqiang, Jin Xiaolong, Sun Jianlun

(Iron & Steel Research Institute of Angang Group, Anshan 114009, China)

Abstract: This research is focused on the high Mn steel with different Al element content. By means of tensile tests, OM and TEM observation, the influence of Al element content on the mechanical behavior, microstructure and deformation mechanism of high Mn steel was studied. The experimental results showed that the yield strength increased, the tensile strength decreased firstly then increased, the elongation and strain hardening exponent decreased. The change of high Mn steel mechanical behavior is relevant to the affection of Al element content on the stacking fault energy.

Key words: high manganese steel, microscopic structure, mechanical property, deformation behaviors, stacking fault energy

中锰 TRIP 钢 ART 工艺处理的组织性能

邹　英，许云波，胡智评，韩仃停，陈树青

（东北大学轧制技术及连轧自动化国家重点实验室，辽宁沈阳　110819）

摘　要：研究了奥氏体逆相变（ART）退火工艺参数对中锰 TRIP 钢微观组织和力学性能的影响。结果表明，显微组织主要由超细晶的交替分布的板条状铁素体及残余奥氏体组成，并存在少量块状铁素体及奥氏体，残余奥氏体与临近铁素体保持 N-W 取向关系。随着退火温度的升高及保温时间的延长，残余奥氏体体积分数稍有下降。经两相区 700℃退火保温 6h 的实验钢残余奥氏体含量达到 40.8%，并获得了屈服强度 740MPa、抗拉强度 960MPa、断后延伸率接近 40%、强塑积超过 38GPa·%的优异力学性能。

关键词：中锰钢，奥氏体逆相变，残余奥氏体，TRIP 效应，强塑积

Microstructure and Mechanical Properties of Medium-Mn TRIP Steel Treated by ART Process

Zou Ying, Xu Yunbo, Hu Zhiping, Han Dingting, Chen Shuqing

(State Key Laboratory of Rolling and Automation, Northeastern University, Shenyang 110819, China)

Abstract: In this study, the effects of austenite-reverted transformation (ART) annealing process on microstructure and mechanical properties of medium-Mn transformation induced plasticity (TRIP) steel were studied. Results show that the microstructure is mainly composed of ultra-fine grained lath-like ferrite and retained austenite, as well as a small amount of blocky ferrite and austenite. The retained austenite and adjacent ferrite keeps the N-W orientation relationship. With the increase of annealing temperature and time, the volume fraction of retained austenite slightly deceases. Excellent mechanical properties with yield strength of 740 MPa, tensile strength of 960 MPa, total elongation of ~40% and product of tensile strength and elongation (PSE) over 38GPa·% can be obtained when intercritical annealed at 700℃ for 6 h. At the same time, the volume fraction of retained austenite reaches up to 40.8%.

Key words: medium-Mn steel, austenite-reverted transformation, retained austenite, TRIP effect, product of strength and elongation

980MPa 级热镀锌复相钢研究及生产实践

邱木生 [1,2]，张环宇 [3]，韩　赟 [1,2]，王勇围 [1,2]，邝　霜 [1,2]

（1. 首钢技术研究院薄板研究所，北京　100043；2. 绿色可循环钢铁流程北京市重点实验室，北京　100043；3. 首钢京唐钢铁联合有限责任公司，河北唐山　063200）

摘　要：基于首钢京唐连续热镀锌产线的工艺配置开发了一种低碳含铌钛的 980MPa 级复相钢，采用连退模拟器研究了不同退火温度对实验钢性能的影响，基于连退模拟器结果制定了冷轧镀锌工业试制关键工艺参数，并对工业试制产品的力学性能及显微组织进行了分析，研究结果表明模拟不同退火工艺下实验钢应力应变曲线为连续屈服特征，

并且屈服强度随退火温度的提高而增加。工业试制产品屈服强度 776MPa，抗拉强度 1039MPa，断后延伸率 12%，显微组织为铁素体+贝氏体及细化分布的马氏体，产品力学性能满足客户要求并成功应用于某畅销合资车型地板梁零件。

关键词：980MPa 级复相钢，热镀锌，力学性能，生产实践

Research and Industrial Practice of Hot Dip Galvanizing MP980 Steel

Qiu Musheng[1,2], Zhang Huanyu[3], Han Yun[1,2], Wang Yongwei[1,2], Kuang Shuang[1,2]

(1. Sheet Metal Research Institute, Shougang Research Institute of Technology, Beijing 100043, China; 2. Beijing Key Laboratory of Green Recyclable Process for Iron & Steel Production Technology, Beijing 100043, China; 3. Shougang Jingtang United Iron & Steel Co., Ltd., Tangshan 063200, China)

Abstract: Based on the process configuration of Shougang Jingtang continuous hot dip galvanizing production line, a low carbon and Nb/Ti bearing galvanizing multiphase steel of 980MPa grade was developed. The continuous annealing simulator was used to investigate the influences of annealing temperature on mechanical properties of a hot-dip galvanizing MP980 steel. Base on these simulation process above, the key process for industrial production of hot-dip galvanizing multiphase steel of 980MPa grade has been established. The results shown that the stress-strain curve of experiment steels exhibited continuous yielding feature after different annealing process, the yielding strength rises as the annealing temperature increases, The industrial trial products has the comprehensive performance with the tensile strength of 1039MPa, yield strength of 776MPa, elongation of 12%, the microstructures of the products consisted of ferrite , bainite and fine dispersed distribution martensite, the mechanical properties of products meets request of customers and has been successfully applied to a joint venture vehicle type of floor beam parts.

Key words: 980MPa grade multiphase steel, hot dip galvanizing, properties, industrial practice

短流程薄规格热轧汽车结构钢 QStE340TM 的开发与研究

魏　兵，刘永前，刘　斌，朱万军，郑海涛，王伟波

（武汉钢铁有限公司研究院，湖北武汉　430080）

摘　要： 针对武钢 CSP 产线特点，利用 Nb-Ti 复合微合金化技术，采取相应的生产控制，成功开发出 1.2~2.0mm 规格热轧汽车结构用钢 QStE340TM。分析结果表明：开发的含 Nb-Ti 的钢带头尾性能控制在 30MPa 以内，波动小，性能稳定，具有良好的冷冲压成形性能和良好的韧性，与相应的冷轧产品性能相当，能够实现产品的"以热代冷"，具有显著的经济社会效益和良好的市场前景。

关键词：短流程，QStE340TM，热轧汽车结构钢，成形极限

The Research of Hot-rolled Automobile Thin Plate QStE340TM By CSP

Wei Bing, Liu Yongqian, Liu Bin, Zhu Wanjun, Zheng Haitao, Wang Weibo

(Research and Development Center of WISCO, Wuhan 430080, China)

Abstract: The Nb-Ti micro-alloyed QStE340TM hot rolled steel with thickness from1.2 to 2.0mm was successfully developed for the CSP production line of WISCO through corresponding processing control. The analysis results showed that mechanical properties of the Ni-Ti micro-alloyed sheet steel were controlled within 30MPa for the head and tail, and the fluctuation was small. Moreover the cold forming properties and toughness were also very good. The performance of the hot rolled sheet steel was equivalent to that of the cold rolled product so that it can realize the "hot rolled product on behalf of the cold rolled product" and it can produce significant economic and social benefits with good market prospect.

Key words: CSP, QStE340TM, hot rolled structural steel, FLC

形变温度和应变速率对基于马氏体温变形的高锰 TRIP 钢的影响

郭志凯 [1,2]，李龙飞 [2]

（1. 河钢唐钢技术中心，河北唐山　063016；
2. 北京科技大学新金属材料国家重点实验室，北京　100083）

摘　要： 利用 Gleeble 1500 热模拟试验机和箱式电阻炉进行单轴热压缩实验及两相区退火实验，研究了形变温度和应变速率对基于马氏体温变形的高锰 TRIP 钢的组织与力学性能的影响。结果表明：马氏体温变形过程中发生的马氏体分解、铁素体动态再结晶和奥氏体逆转变等过程均为热激活过程，提高形变温度或降低应变速率可促进上述过程的进行，在相同退火条件下有利于获得铁素体基体更加均匀和残余奥氏体含量更高的最终组织，改善高锰 TRIP 钢的加工硬化能力，使其获得更好的强度-塑性配合。

关键词： 高锰 TRIP 钢，马氏体温变形，形变温度，应变速率，力学性能

Effects of Deformation Temperature and Strain Rate on High Manganese TRIP Steel Based on Warm Deformation of Martensite

Guo Zhikai[1,2], Li Longfei[2]

(1. Technology Center of Tangsteel of Hesteel, Tangshan 063016, China; 2. State Key Laboratory for Advanced Metals and Materials, University of Science and Technology Beijing, Beijing 100083, China)

Abstract: The effects of deformation temperature and strain rate on microstructure and mechanical properties of high manganese TRIP steel based on warm deformation of martensite were investigated by uniaxial hot compression tests on a Gleeble 1500 simulation test machine and intercritical annealing tests in a muffle furnace. The results indicate that during warm deformation of martensite the decomposition of martensite, dynamic recrystallization of ferrite and the reverse transformation of austenite took place, all of which are thermal activation processes. The increasing in the deformation temperature or the decreasing in the strain rat of warm deformation could promote the kinetics of these processes during warm deformation and be of benefit to obtaining the final microstructure with more uniform ferrite matrix and more retained austenite by intercritical annealing, thus improving the work-hardening capability of high manganese TRIP steel and resulting in the well balance of strength and ductility of high manganese TRIP steel.

Key words: high manganese TRIP steel, warm deformation of martensite, deformation temperature, strain rate, mechanical property

临界区退火温度对 0.1C-7Mn-1.1Si 钢组织性能影响

韩仃停，许云波，邹　英，胡智评，陈树青

（东北大学轧制技术及连轧自动化国家重点实验室，辽宁沈阳　110819）

摘　要：研究了临界区退火温度对 0.1C-7Mn-1.1Si（wt.%）中锰钢组织演变及力学性能的影响。利用 EBSD、EPMA 和 XRD 等组织表征手段以及拉伸试验，通过调整及控制退火温度，热轧后得到全马氏体组织，退火后为超细晶奥氏体和铁素体组织，最终得到高强塑积。结果表明，当退火温度大于 640℃时，组织为板条状和块状奥氏体+板条铁素体；当临界区退火温度小于 640℃时，组织由板条相间的奥氏体+铁素体组成。由 XRD 测得退火后实验钢的残余奥氏体含量，得到随着退火温度的升高，残余奥氏体晶粒尺寸增大，稳定性降低，体积分数先上升后缓慢下降（42%~64%~60%）。在 620℃退火时，实验钢性能最佳，此时残余奥氏体的体积分数为 45%，屈服强度为 780MPa，抗拉强度 1060MPa，延伸率为 46%，强塑积达到 43GPa·%。

关键词：中锰钢，临界区退火，残余奥氏体，组织演变，力学性能

Effect of Annealing Temperature on Microstructure and Properties of 0.1C-7Mn-1.1Si Medium Manganese Steel

Han Dingting, Xu Yunbo, Zou Ying, Hu Zhiping, Chen Shuqing

(The State Key Laboratory of Rolling and Automation in Northeastern University, Shenyang 110819, China)

Abstract: The effect of annealing temperature on microstructure evolution and mechanical properties of 0.1C-7Mn-1.1Si medium Mn steel was investigated. EBSD, EPMA and XRD are used to study the test steel and tensile test for mechanical properties through adjusting and controlling the temperature, microstructure composed of martensite after hot rolling and UFG ferrite and austenite, high product of strength and elongation are obtained. When the temperature was higher than 640℃, both and lath retained austenite was discovery rather than lath retained austenite only. And the width of the lath retained austenite is wider with increase the annealing temperature. With the annealing temperature increasing from 600℃ to 660℃, the tensile strength increased and the total elongation decreased. The volume of the fraction of retained austenite before and after fracture were identified by means of X-ray diffraction(XRD), it was found that with increasing the annealing temperature the retained austenite volume fraction increased firstly and then gradually decline（42%~64%~60%）, but the stability also have a gradual downtrend. When the annealing temperature reached at 620℃, get the best mechanical yield strength 780MPa, tensile strength 1060MPa, the ductility reached 46%, and the product of strength and elongation is 43GPa%.

卷取温度对铁素体区热轧 Ti-IF 钢析出物和织构的影响

李　浪[1]，邱晨阳[1]，王建功[2]，周　旬[2]，夏银峰[2]，康永林[1]

（1. 北京科技大学材料科学与工程学院，北京　100083；

2. 首钢京唐钢铁联合有限责任公司热轧作业部，河北唐山　063200）

摘　要：铁素体区轧制 Ti-IF 钢能显著提高其深冲性能。本文主要研究了卷取温度对 Ti-IF 钢热轧板金相组织、析出物及织构的影响。结果表明：随着卷取温度提高，热轧板铁素体晶粒变得粗大均匀，TiC 粒子尺寸增大，对深冲性能有利的 {111}//ND 组分织构强度先增大后减小，而不利的 {001}<110>织构强度显著减小。

关键词：Ti-IF 钢，铁素体区轧制，组织，析出物，织构

Effect of Coiling Temperature on Precipitation and Texture of Hot Rolled Ti-IF Steel Sheet in Ferrite Region

Li Lang[1], Qiu Chenyang[1], Wang Jiangong[2], Zhou Xun[2], Xia Yinfeng[2], Kang Yonglin[1]

(1. University of Science and Technology Beijing, Materials Science and Engineering, Beijing 100083, China;
2. Shougang Jingtang United Iron and Steel Co., Ltd., Tangshan 063200, China)

Abstract: The deep drawing property of Ti-IF steel could be obviously improved by ferritic-region hot rolling technology. Effect of coiling temperature on microstructure、precipitate and texture of the hot rolled Ti-IF steel sheet has been studied in this paper . The results show that high coiling temperature is beneficial to the growth and homogeneity of ferritic grains. High coiling temperature is beneficial to the growth of TiC particles. With the increasing of coiling temperature, the intensity of {111}//ND component increases first and decreases later, while the {001} <110> texture is dramatically weakened.

Key words: Ti-IF steel, ferrite rolling, microstructure, precipitate, texture

热镀锌高强钢点焊工艺及电极失效分析

刘东亚，张　武，计遥遥，成昌晶，王伟峰

（马鞍山钢铁股份有限公司技术中心，安徽马鞍山　243000）

摘　要：本文分别使用 CrZrCu 电极和 Al₂O₃ 弥散强化铜电极（简称氧化铝电极）对 800MPa 级热镀锌高强汽车用钢进行了电阻点焊实验研究。探索了不同电极材料对汽车板电阻点焊焊接性的影响，并采用扫描电子显微镜（SEM）和 EDS 分析了不同电极材料的失效机理。结果表明：采用氧化铝电极时点焊可焊性范围是采用 CrZrCu 电极时的 1.5 倍，且焊点拉剪力较 CrZrCu 电极时高；氧化铝电极寿命较 CrZrCu 电极寿命明显提高；焊接过程中镀锌板表面的锌层与电极中 Cu 元素的合金化是造成电极失效的内因，氧化铝电极由于具有良好的耐热性能，避免电极端面镦粗，且弥散氧化铝强化相可以抑制 Cu 与 Zn 的合金化，从而提高电极的寿命。

关键词：电极材料，镀锌高强钢，电阻点焊，失效分析

Research on Spot Welding Process and Electrode Failure Mechanism of Hot-Dip Galvanized High Strength Steel

Liu Dongya, Zhang Wu, Ji Yaoyao, Cheng Changjing, Wang Weifeng

(Automotive Sheet Research Institute of Maanshan Iron & Steel Co., Ltd., Maanshan 243000, China)

Abstract: In this paper, the Chromium-Zirconium-copper electrode (CrZrCu electrode) and the Al_2O_3 dispersion strengthened copper electrode (Al_2O_3 electrode) are used to study the resistance spot welding of 800MPa high-strength galvanized steel. The effects of different electrode materials on the spot welding resistance of automobile plates were investigated. The failure mechanism of different electrode materials was analyzed by scanning electron microscopy (SEM) and Energy Disperse Spectroscopy (EDS). The results show that the range of weldability when using Al_2O_3 electrode is 1.5 times that of CrZrCu electrode and the tensile-shear load of the joints is a little higher than that of CrZrCu electrode as well. The Al_2O_3 electrode life is much higher than that of CrZrCu electrode under the same experimental conditions. The internal cause of the electrode failure is the reaction between zinc layer on the steel face and Cu in the electrode when the galvanized high strength steel for welding. The Al_2O_3 electrode exhibits better heat-resistance. As a barrier, the Al_2O_3 phase in the Al_2O_3 electrode can suppress the alloying of Cu and Zn, thereby extending the electrode life.

Key words: electrode material, galvanized high strength steel, resistance spot welding, failure analysis

短时效与缓时效烘烤硬化钢热镀锌板材的自然时效性

佟皑男，孙　彦，邓　军，张长伟

（鞍钢蒂森克虏伯汽车钢有限公司，辽宁大连　116600）

摘　要： 乘用车用烘烤硬化钢热镀锌板材的主要作用，就是利用钢的特定成分和工艺使钢板在变形后，提升晶粒内应力，在一定温度作用下，改变其机械性能的钢种。很多专家已经非常深入地探讨了这方面的规律性。由于这个钢种是亚稳形态，在无应变的自然条件下也具有时效属性。经过不同时间长度自然时效后对其烘烤硬化属性影响，人们一般采用替代方法："快速人工时效"或者"时效指数"来表征。本文用自然时效的实际验证方法，探讨不同时长的自然条件下的时效性以及其与烘烤硬化叠加对烘烤硬化值的影响。

关键词： 烘烤硬化钢，短时效，缓时效，自然时效与烘烤硬化叠加，热镀锌

Nature Aging Characteristics of Bake Hardening Steel that Respectively are Shorter Aging or Postponement Aging One for Hot Dip Galvanizing Strip

Tong Ainan, Sun Yan, Deng Jun, Zhang Changwei

(TKAS Auto Steel Company Limited, Dalian 116600, China)

Abstract: Bake hardening steel is with the possibility to change mechanical properties when forming and heating. It already was, researched enough to bake hardening characteristic of BH steel by many experts. But nobody has researched its regulation of natural aging in the mechanical properties, even though some body use substitution with "fast aging" and "aging index". This text will discuss the regularity of aging characteristics on natural condition as well as complex characteristics both natural aging and bake hardening.

Key words: bake hardening steel, short aging, postponement aging, complex characteristics both natural aging and bake hardening, hot dip galvanizing

过时效温度对980MPa级冷轧双相钢组织及力学性能的影响

肖洋洋，詹　华，刘永刚，郑笑芳，景宏亮

（马鞍山钢铁股份有限公司技术中心，安徽马鞍山　243000）

摘　要： 利用 ULVAC-CCT-AY-II 型多功能连续退火模拟器模拟了 980MPa 级冷轧退火双相钢的生产过程，采用扫描电镜和透射电镜进行组织观察，采用拉伸试验机进行力学性能检测，研究了不同过时效温度条件下实验钢的组织形貌和力学性能特点。结果表明：随着过时效温度的升高，试验钢组织中马氏体含量基本不变，马氏体微观形态由板条状逐渐转变为岛状，过时效温度提高到 340℃ 后，实验钢的马氏体出现分解，有部分颗粒状碳化物析出。随着过时效温度的增加，实验钢抗拉强度呈单调降低的趋势，屈服强度先降低后增大。试验钢在过时效温度为 290℃ 时力学性能最佳，其强塑积（AT×σb）可达到 13.8GPa·%。

关键词： 双相钢，连续退火，过时效温度，组织性能

Effect of Overaging Temperature on Microstructure and Properties of 980MPa Grade Cold-rolled Dual-phase Steel

Xiao Yangyang, Zhan Hua, Liu Yonggang, Zheng Xiaofang, Jing Hongliang

(Auto Sheet Strategic Business Unit of Maanshan Iron and Steel Company, Maanshan 243000, China)

Abstract: The ULVAC-CCT-AY-II annealing simulator was used to simulate the production process of 980MPa grade dual phase (DP) steel. The characteristics of microstructure in experimental steel under different overaging temperature were observed by using scanning electron microscopy and transmission electron microscopy, and mechanical properties were measured by uniaxial tensile tests. With the increase of the overaging temperature, the martensite content changes little, and the microstructure of the martensite is gradually changed from plate to island.After the overaging temperature increased to 340℃, the martensite begins to decompose and some carbides precipitate. With overaging temperature increase, the tensile strength of the experimental steel decreases monotonically and the yield strength decreases first and then increases. The experimental steel has an optimum mechanical properties when the overaging temperature is 290℃, and it's PSE can reach 13.8GPa ·%.

Key words: dual-phase steel, continuous annealing, overaging temperature, microstructure and properties

连续退火工艺条件下中锰 TRIP 钢的热处理工艺研究

定　巍[1]，李一磊[1]，李　岩[2]，龚志华[1]

（1. 内蒙古科技大学材料与冶金学院，内蒙古包头　014010；2. 内蒙古科技大学，内蒙古自治区白云鄂博矿多金属资源综合利用重点实验室，内蒙古包头　014010）

摘　要：对 0.2C-5Mn 中锰钢的相变规律、微观组织以及力学性能进行了研究，在此基础上分析了临界退火温度和临界退火时间对微观组织和力学性能的影响。研究结果表明：当临界退火温度为 700℃时，在冷却过程中发生明显的马氏体相变；在本文所设计工艺条件下，由于临界退火时间短其微观组织主要由铁素体、残余奥氏体以及渗碳体所构成；试样在 650 ～ 700℃之间退火能获得良好性能，但是不同临界退火温度出现良好力学性能所需要的时间不一样，温度越高，所需要时间越短；675℃退火 5 min 时有最佳的力学性能，抗拉强度约为 980 MPa，断后延伸率约为 30%，强塑积接近 30 GPa•%。

关键词：连续退火，中锰 TRIP 钢，相变规律，微观组织，力学性能

Heat Treatment Process of a Medium Mn TRIP Steel under Continuous Annealing Line

Ding Wei[1], Li Yilei[1], Li Yan[2], Gong Zhihua[1]

(1. School of Materials and Metallurgy, Inner Mongolia University of Science and Technology, Baotou 014010, China; 2. BayanObo Multimetallic Resource Comprehensive Utilization Key Lab., Inner Mongolia University of Science and Technology, Baotou 014010, China)

Abstract: The transformation, microstructure and mechanical properties of the 0.2C-5Mn TRIP steel with different intercritical annealing temperatures and time are investigated. The result shows when the intercritical annealing temperature is 700 ℃, an obvious martensitic transformation occurs during the cooling process, the microstructure of the steel was mainly comprised of ferrite, retained austenite and some of cementite, the cementite remained after intercritical annealing. The steel has good mechanical properties when the intercritical annealing temperature between 650 ～ 700 ℃, and with increasing the annealing temperature from 650 to 700 ℃, the time required to achieve the optimal mechanical property is decreased. The optimal mechanical properties of investigated steel with tensile strength of 980 MPa and total elongation of 30% can be obtained after annealing at 675 ℃ when the intercritical annealing time is 5 min.

Key words: continuous annealing, medium Mn TRIP steel, transformation, microstructure, mechanical properties

稀土对 TRIP/TWIP 钢中夹杂物形态分布和弯曲性能的影响

王立辉，刘祥东，潘利波，魏　琼，龚　涛

（宝武集团武汉钢铁有限公司研究院，湖北武汉　430080）

摘　要：对比分析了 3 种 RE 含量 TRIP/TWIP 钢中夹杂物的形态分布及冷弯性能。结果表明，试验钢 01（含 0% RE）中夹杂物为长条状硫化锰和硒化物，夹杂物的直径与试验钢截面横向负半轴的夹角为 82.9°；添加稀土后钢中夹杂物为稀土硫化物和稀土硫氧化物，试验钢 02（含 0.035%RE）夹杂物呈类椭球形，夹杂物的直径与试验钢截面横向负半轴的夹角为 91.6°；试验钢 03（含 0.061%RE）的夹杂物呈椭球形，夹杂物的直径与试验钢截面横向负半轴的夹角为 138.1°。扫描电镜观察结果显示，夹杂物均分布在奥氏体晶内，随着稀土含量的增加夹杂物的长轴直径方向呈与试验钢横向负半轴角度增大的趋势且夹杂物逐渐球化。试验钢 02 在弯心直径达到钢板厚度 1/2 时，未出现裂纹，弯曲性能理想。

关键词：稀土，TRIP/TWIP 钢，夹杂物，弯曲性能

Effect of RE on the Inclusion Morphology, Distribution and Bending Properties of the TRIP/TWIP Steels

Wang Lihui, Liu Xiangdong, Pan Libo, Wei Qiong, Gong Tao

Abstract: The inclusions morphology, distribution and cold bending properties of TRIP/TWIP(transformation induced plasticity/twinning induced plasticity) steels with 3 kinds of RE (rare earth) content were analyzed. The results showed that the inclusions were manganese sulfide and selenium compounds of strip shape in test steel 01 (0% RE), and it was at an included angle of 82.9° to the horizontal negative half axis; the inclusions were rare earth sulfide and rare earth oxysulfide compounds after the RE were added in the steels composition, the inclusions shape was ellipsoid, and it was at an included angle of 91.6° to the horizontal negative half axis in the steel 02 (containing 0.035%RE); the inclusions shape was spheroidicity, and it was at an included angle of 138.1° to the horizontal negative half axis in the steel 03 (containing 0.061%RE); According to the SEM (scanning electronic microscopy) experimental results, these inclusions distributed in austenite grains, with the increasing of RE content, the angle of the inclusions long axis diameter and the horizontal negative half axis of the test steels was bigger and bigger, the inclusions gradually were spheroidized. The bending property of test steel 02 was ideal as there was no crack phenomenon when the bending core diameter reached to one half of the thickness of steel.

Key words: rare earth, TRIP/TWIP steel, inclusion, bending property

中锰第三代汽车钢及其先进性

王存宇，曹文全，董　瀚

（钢铁研究总院，北京　100081）

摘　要： 介绍了中锰第三代汽车钢的发展思路与过程，中锰钢经逆相变退火处理后具有超细晶粒的铁素体+亚稳奥氏体组织，具有高强度和高塑性的力学性能特点，提高了零件的冷成形能力。基于中锰钢提出了温成形工艺，研究表明，在获得 1500MPa 汽车零件时，中锰钢温成形工艺的加热温度为 800℃，较传统热成形钢降低 150℃，表明几乎不发生脱碳现象，微观组织明显细化，断后伸长率提高到 10%。

关键词： 第三代汽车钢，中锰钢，强度，塑性

The Third Generation Automobile Steel of Medium Manganese and Its Advantages

Wang Cunyu, Cao Wenquan, Dong Han

(Central Iron and Steel Research Institute, Beijing 100081, China)

Abstract: The idea and progress of the third generation automobile steel of medium manganese are introduced. The duplex microstructure of ferrite and metastable austenite with ultrafine grain size are obtained after medium manganese steel being

ART-annealing heat treatment, as well as the mechanical properties of high strength combine with good ductility, thus improves the cold forming ability. Warm stamping technology is proposed based on the medium manganese steel, the heat temperature for medium manganese steel to obtain automobile parts with tensile strength of 1500MPa level is 800℃, which is lower by 150℃ than that of traditional hot stamping, and almost no decarburization occur during heating process, the microstructure of warm stamping parts is much finer than that of hot stamping, and the total elongation teach 10%.

Key words: the third generation automobile steel, medium manganese steel, strength, ductility

几种汽车用马氏体超高强钢的氢致延迟断裂性能研究

熊林敞[1]，周庆军[1]，Andrej Atrens[2]

（1. 宝山钢铁股份有限公司，上海 201900；

2. 昆士兰大学材料系，澳大利亚圣卢西亚 4072）

摘 要： 采用两点弯曲加载和线性应力加载（LIST）试验方法，研究了四种抗拉强度 1000MPa 以上的汽车用马氏体超高强钢在（1）0.1M HCl 溶液和（2）浓度为 3.5%的 NaCl 溶液和充氢条件下的延迟开裂性能和氢脆敏感性。试验结果表明，在 0.1M 的 HCl 溶液、0.8TS（抗拉强度）和 1.0TS 张应力条件下，四种马氏体超钢试样浸泡 300h 浸泡均没有发生延迟断裂；在 3.5%的 NaCl 溶液和充氢条件下，四种马氏体钢均没有表现出明显的氢脆敏感性。

关键词： 马氏体钢，汽车钢，氢，延迟断裂

Study of Hydrogen Induced Delayed Fracture of Some Ultrahigh Strength Martensitic Steels

Xiong Linchang[1], Zhou Qingjun[1], Andrej Atrens[2]

(1. Baoshan Iron & Steel Co., Ltd., Research Institute, Shanghai 201900, China;

2. The University of Queensland, Division of Materials, St. Lucia 4072, Australia)

Abstract: The delayed fracture behavior and hydrogen embrittlement of four kinds of martensitic steel with tensile strength above 1000MPa were studied with two-point load and linearly increasing stress method in (1) 0.1M HCl solution and (2) 3.5% NaCl solution and hydrogen charging conditions. The results indicated that, there were no delayed fracture occurred for the four kinds of tested materials after immersed 300 hours in 0.1M HCl solution. All the tested materials didn't exhibit obvious hydrogen embrittlement sensitivity.

Key words: martensitic steel, auto sheet steel, hydrogen, delayed fracture

冷轧中锰钢(0.1C-5Mn)不同变形温度下
奥氏体稳定性与力学性能之间的关系

周天鹏[1,2]，徐海峰[1,3]，王存宇[4]，曹文全[1]

（1. 钢铁研究总院特殊钢研究所，北京　100081；2. 重庆大学材料科学与
工程学院；重庆　400044；3. 华中科技大学材料科学与工程学院，
湖北武汉　430074；4. 钢铁研究总院华东分院，北京　100081）

摘　要： 冷轧中锰钢经过奥氏体逆转变退火，组织中形成了大量的亚稳奥氏体，在变形过程中发生形变诱导马氏体相变进而获得了优异的力学性能。而奥氏体的稳定性受到多方面的影响，对力学性也产生了很大影响作用。本文主要针对变形温度对奥氏体稳定性的影响，通过对冷轧中锰钢在不同温度下进行拉伸实验，研究残余奥氏体在不同变形温度条件下的微观组织状态以及对奥氏体的稳定性进行分析，同时结合不同变形温度下的力学性能，探究奥氏体稳定性与力学性能之间的关系。

关键词： 中锰钢，变形温度，奥氏体稳定性，力学性能

Relationship Between Austenite Stability and Mechanical Properties of Cold Rolled Medium Manganese Steel (0.1C-5Mn) at Different Deformation Temperatures

Zhou Tianpeng[1,2], Xu Haifeng[1,3], Wang Chunyu[4], Cao Wenquan[1]

(1. Institute for Special Steels, Central Iron and Steel Research Institute, Beijing 100081, China;
2. School of Materials Science and Engineering, Chongqing University, Chongqing 400044, China;
3. School of Materials Science and Engineering, Huazhong University of Science and Technology, Wuhan 430074, China; 4. East China Branch, Central Iron and Steel Research Institute, Beijing 100081, China)

Abstract: A large amount of metastable austenite has been formed in cold rolled medium manganese during the austenite reverted transformation (ART) annealing. The deformation induced martensite transformation occurs, and excellent mechanical properties are obtained during the deformation process, but the stability of austenite is affected by many factors, and it also has a great influence on the mechanical properties. In this paper, the effect of deformation temperatures on the stability of austenite was studied. The tensile test of medium manganese steel under different temperatures was carried out and the microstructure and the stability of retained austenite were studied. The relationship between the stability of austenite and mechanical properties was investigated by combining the mechanical properties at different deformation temperatures.

Key words: medium manganese steel, deformation temperatures, stability of austenite, mechanical properties

成形极限图不同应变路径对冷轧 TRIP 钢残余奥氏体转变量的影响

王德宝，宋祖峰，刘志军，程志远

（马鞍山钢铁股份有限公司技术中心，安徽马鞍山　243000）

摘　要：以冷轧 TRIP600 为研究对象，通过试验建立成形极限图（FLD），同时采用金相显微镜、扫描电镜对未变形和变形后显微组织进行观察，并利用 X 射线衍射方法测定了经历不同应变路径下的残余奥氏体含量。试验结果表明：在平面应变状态下极限应变值（FLD0）为 0.397。随着应变路径由拉伸至平面应变、再到胀形，残余奥氏体转变量逐渐增加。与 DP600 相比，TRIP600 较高的 FLD0 值是由于 TRIP 效应的存在，变形过程中，缩颈区域较宽。

关键词：TRIP 钢，相变诱发塑性，残余奥氏体，FLD，应变路径

Effect of Different Strain Paths Based on Forming Limit Diagram on Retained Austenite Transformation of Cold-rolled TRIP Steel

Wang Debao, Song Zufeng, Liu Zhijun, Cheng Zhiyuan

Abstract: With cold-rolled TRIP600 steel as the object of research, forming limit diagram (FLD)was established by experiments. The undeformed and deformed sample microstructures were examined by metalloscopy and scanning electron microscopy, at the same time the content of retained austenite which experienced different strain paths were measured by X-ray diffraction. The results showed that the ultimate strain under plane strain state (FLD0) is 0.397.With the strain path from tensile to plane strain and the bulging, the transformation contents of residual austenite gradually increase. Compared with DP600,the higher FLD0 values of TRIP600 were ascribed to the TRIP effect, the necking area becoming wider during deformation.

DP 钢拉伸成形影响因素的比较

贾生晖[1]，张　杰[2]，李洪波[2]，孔　政[2]

（1. 武汉钢铁股份有限公司，湖北武汉　430083；2. 北京科技大学机械工程学院，北京　100083）

摘　要：基于有限元仿真，选择了三个有代表性的拉伸成形影响因素，从冲压力、截面厚度分布和最大减薄率的角度对各影响因素进行了比较和评价。发现摩擦系数、压边力和 r 值对冲压力和板料减薄率的影响规律。提出应重视不同生产线间 DP 钢性能的差异性，丰富和补充 DP 钢的性能标准，以提高 DP 钢冲压成形稳定性的建议。

关键词：DP 钢，拉伸，有限元，r 值，摩擦系数

Comparison of Influence Factors on Deep Drawing of Dual Phase Steel

Jia Shenghui[1], Zhang Jie[2], Li Hongbo[2], Kong Zheng[2]

(1. Wuhan Iron & Steel Co., Ltd., Wuhan 430083, China; 2. School of Mechanical Engineering, University of Science and Technology Beijing, Beijing 100083, China)

Abstract: Three representative influence factors of deep drawing are compared and evaluated in terms of punch force, thickness distribution and maximum thinning rate based on finite element analysis. The effects of coefficient of friction, blank-holder force and Lankford coefficient r on the punch force and maximum thinning rate are obtained. It is suggested that the mechanical property difference of DP steel from different production lines should be focused and the property criteria should be revised to improve the stamping manufacture stability.

Key words: dual phase steel, deep drawing, finite element analysis, anisotropy coefficient, coefficient of friction

车轮钢氧化铁皮的机理研究和控制策略

李成亮，李爱民，陈志桐

（河钢集团邯钢公司，河北邯郸　056015）

摘　要：为了解决热轧卷板生产中出现的表面氧化铁皮缺陷，尤其是厚规格车轮钢，进行了热轧氧化铁皮形成机理的简单探讨，并以此为依据，进行了成分和轧制工艺的工业性试验。结果表明，降低 Si 含量，降低出炉温度和精轧入口温度，合适的卷取温度可以有效的控制氧化铁皮的性质和结构，从而减轻甚至抑制红锈氧化铁皮的生成。

关键词：车轮钢，氧化铁皮，Si 含量，出炉温度，精轧入口温度，卷取温度

Controlling Strategies on Scale of Thick Wheel Steel

Li Chengliang, Li Aimin, Chen Zhitong

(Hot Strip Mill of Handan Iron and Steel Group Co.,Ltd., Handan 056015, China)

Abstract: To solve the scale defects on the surface of hot strip, especially thick Wheel Steel, authors had discussed the mechanism of scale formation. According to the mechanism, the experiments of changing the component content and hot-rolling process had been tested. The results shows that the percentage of the scale decreases with the decrease of Si content, reheating temperature, FET (FM entry temperature) and appropriate coiling temperature.

Key words: wheel steel, scale, Si, reheating temperature, FM entry temperature, coiling temperature

近年汽车用热镀锌钢板新增产线分析

郑　瑞，代云红，周谊军

（首钢集团有限公司技术研究院，北京　100043）

摘　要：综述了近年国内外重点钢铁企业汽车用热镀锌钢板新增产线的建设进展和新产品开发情况，总结了热镀锌高强钢、超高强钢和铝硅镀层热成形钢等重点、热点产品的发展情况。建议国内重点钢铁企业在未来向高端、细分及差异化领域发展，保持对高性能涂镀产品的研发投入，加快新产品、新工艺的研制与应用，推动国内汽车热镀锌钢板品种结构优化和下游应用进展。

关键词：汽车，热镀锌，钢板，腐蚀，镀层，高强钢，热成形钢

Analysis on New Production Lines of Hot-Dip Galvanized Steel Sheet for Automobile in Recent Years

Zheng Rui, Dai Yunhong, Zhou Yijun

(Shougang Research Institute of Technology, Beijing 100043, China)

Abstract: The construction progress of new production lines and new product development of hot-dip galvanized steel sheet for automobile for key steel enterprises in recent years in China and abroad are reviewed. The development of hot-dip galvanized high-strength steel, ultra-high strength steel and al-si coating hot forming steel are summarized. It is suggested that the domestic key iron and steel enterprises will develop to high-end, subdivision and differentiation fields in the future, maintain the R & D investment in high performance coating products, to promote the optimization and downstream application of domestic hot - dip galvanized steel sheet for automobile.

Key words: automobile, hot galvanized, steel sheet, corrosion, coating, high strength steel, hot forming steel

汽车大梁用钢 B510L 延伸不合原因分析与改进

赵　亮，张爱梅

（宝钢集团八钢公司制造管理部，新疆乌鲁木齐　830022）

摘　要：本文通过金相分析等手段对汽车大梁用钢 B510L 出现了批量延伸率不合的问题进行了分析，指出热轧工序卷取温度的波动造成部分钢卷局部位置卷取温度较低，进入贝氏体转变区，从而带钢局部区域出现一定量贝氏体组织以及由于粗轧温度过高，带钢四分之一断面处出现混晶是造成 B510L 的延伸率偏低的原因。在后续改进生产中，通过增加 Nb 含量及提高中间坯厚度及提高卷取温度精度等技术措施后，B510L 延伸率不合的情况有了明显改善，大大提高了性能合格率。

关键词：汽车大梁钢，卷取温度，显微组织，力学性能

The Analysis of Elongation Unqualified and Improvement on B510L Grade Sheet for Automobile Crossbeam

Zhao Liang, Zhang Aimei

(Manufacturing Management Department, Bayi Iron & Steel Co., Baosteel Group, Urumchi 830022, China)

Abstract: The problem of the batch elongation unqualified in B510 automotive beam plate is analyzed by meana through the microstructure analysis. It indicated that coiling temperature fluctuations in hot rolled process caused coiling temperature lower in part steel coils or local position in one steel. Organization phase change into the bainite transition zone, and the strip surface formed a certain thickness bainite organization layer is caused the elongation lower in B510L.In production, by increasing the Nb content, improve intermediate slab thickness and coiling temperature technical measures,the elongation unqualified situation has obvious improvement in B510, greatly improved qualified rate.

Key words: automotive beam plate, coiling temperature, microstructure, mechanical property

590MPa 级高强汽车车轮用热卷的开发

田　鹏[1,2]，康永林[1]，耿立唐[2]，刘　伟[2]，陆凤慧[2]

（1. 北京科技大学材料科学与工程学院，北京　100083；2. 河钢承钢，河北承德　067102）

摘　要： 通过对影响高强汽车车轮用钢性能的分析，认为采用铌钛微合金化低碳成分生产 590MPa 级车轮用钢是可行的。结合成分、温度和板形控制，生产的 C590CL 具有理想的显微组织、力学性能和抗冲击性能。用户制作轮辋的质量反馈表明该 C590CL 成形性能和焊接性能优异，可替代进口产品和满足以薄代厚的汽车轻量化要求，推动了高强汽车车轮用钢的热卷生产及其在轮辋的应用。

关键词： 590MPa 级，高强车轮用钢，铌钛微合金化，轻量化

Development 590 MPa Grade of High Strength Hot Rolled Coil for Automotive Wheels

Tian Peng[1,2], Kang Yonglin[1], Geng Litang[2], Liu Wei[2], Lu Fenghui[2]

(1. School of Materials Science and Engineering, University of Science and Technology Beijing, Beijing 100083, China; 2. HBIS Group Chengsteel Company, Chengde 067102, China)

Abstract: Through the analysis of the properties of high strength automobile wheels steel, it is considered feasible to produce 590MPa grade wheel steel with Nb-Ti microalloyed and low-carbon components. Combined with the control of composition, temperature and plate shape, C590CL has the ideal microstructure, mechanical properties and impact resistance. The quality feedback from the user's made rim indicated that the C590CL was excellent in formability and weldability, which can replace imported products and meet the requirements of automotive lightweighting. It promoted the hot coil production of high strength automotive wheel steel and its application in rims.

Key words: 590 MPa grade, high strength wheel steel, Nb-Ti microalloyed, lightweight

含残余奥氏体的冷轧 DP780 的组织及其性能的研究

王卫卫[1]，李光瀛[1]，刘　浏[1]，张　建[2]，刘永刚[2]，詹　华[2]

（1. 钢铁研究总院冶金工艺研究所，北京　100081；
2. 马钢股份有限公司技术中心，安徽马鞍山　243000）

摘　要：本文通过成分工艺优化，在传统冷轧铁素体和马氏体双相钢 DP780 的显微组织上引入了一定含量的残余奥氏体，研究了不同退火温度条件下残余奥氏体的比例、尺寸、形貌、分布及不同的变形条件下残余奥氏体的稳定性，同时也研究了连退工艺对力学性能的影响。结果表明，通过引入 5~7% 的残余奥氏体，不仅可以获得 YS/TS≤0.5 的超低屈强比型冷轧 DP780，而且在提高伸长率的同时也改善了成型性能。
关键词：冷轧双相钢，残余奥氏体，连续退火，工艺参数，超低屈强比，成型性能

Study on Microstructure and Properties of Cold Rolled DP780 Containing Retained Austenite

Wang Weiwei[1], Li Guangying[1], Liu Liu[1], Zhang Jian[2], Liu Yonggang[2], Zhan Hua[2]

(1. Metallurgical Technology Institute of Central Iron & Steel Research Institute, Beijing 100081, China;
2. Technology Center, Maanshan Iron & Steel Co., Ltd., Maanshan 243000, China)

Abstract: In this paper, by means of the optimization of the composition and the process parameters, the influence of cold rolling process parameters on Microstructure and mechanical properties of dual phase steel DP780 were investigated and a certain amount of retained austenite were added in the microstructure of the conventional cold rolled ferrite and martensite dual phase steel DP780. The proportion, size, morphology, distribution and the stability of retained austenite under different deformation conditions were researched. The influence of the process on the mechanical properties was also studied. The results show that the microstructure with 5~7% of retained austenite in traditional cold-rolled ferrite and martensite dual phase steel, not only can be obtained more than 0.5 ultra low yield ratio, but also improve the elongation while improving the formability of cold-rolled DP780.
Key words: cold rolled dual phase steel, retained austenite, continuous annealing, the process parameters, ultra low yield ratio, formability

冷轧压下率对铁素体轧制 Ti-IF 钢组织性能的影响

邱晨阳[1]，李　浪[1]，王建功[2]，周　旬[2]，夏银峰[2]，康永林[1]

（1. 北京科技大学材料科学与工程学院，北京　100083；
2. 北京首钢京唐钢铁联合责任有限公司热轧部，河北唐山　063200）

摘　要：以工业生产的 Ti-IF 钢热轧板为原料，进行不同冷轧压下率的试验，然后在连退模拟机上模拟退火过程，得到不同冷轧压下率的退火板。通过微观组织观察、力学性能以及 XRD 宏观织构检测等实验，分析冷轧压下率对

Ti-IF 钢组织性能的影响。试验的结果表明：在相同的热轧及退火工艺下，高的冷轧压下率会使晶粒更加粗大，且还会在退火过程中形成较强的{111}织构，使其强度达到了 10 以上，从而极大的提高了退火板的 r 值。

关键词：Ti-IF 钢，冷轧压下率，织构，r 值

Effect of Cold Rolling on Texture and Property of Ti-IF Steel

Qiu Chenyang[1], Li Lang[1], Wang Jiangong[2], Zhou Xun[2],

Xia Yinfeng[2], Kang Yonglin[1]

(1. University of Science and Technology Beijing, School of Materials Science and Engineering, Beijing 100083, China; 2. Shougang Jingtang United Iron and Steel Co., Ltd., Tangshan 063200, China)

Abstract: The hot rolled plate of commercial Ti-IF steel were used for trials which the plate was reduced with different rate, and the annealed plate with different cold rolling reduction after annealed on the simulated annealed machine was gotten. The result show that under the same hot rolled and annealed process, the high cold rolling reduction rate makes the grains coarser, and the strong {111} texture which the average intensity reaches 10 is formed during the annealing process, which greatly improves the r-value of annealing plates.

Key words: Ti-IF steel, cold reduction, texture, r-value

热镀锌 DP780 辊压开裂原因分析及对策

刘华赛[1,2]，陈凌峰[1,2,3]，邝　霜[1,2]，韩　赟[1,2]

（1. 首钢技术研究院，北京　100043；2. 绿色可循环钢铁流程北京市重点实验室，北京　100043；
3. 北京科技大学材料科学与工程学院，北京　100083）

摘　要：辊压成形工艺被越来越多的应用到超高强钢汽车零件的成形中。某车型左右门槛使用热镀锌双相钢 DP780 作为原材料，在辊压过程中在圆角半径外侧发生开裂。本文采用宏观形貌观察、断口观察、化学成分分析、力学性能分析、金相观察等方法对其辊压开裂的原因进行了分析和讨论。结果表明：辊压件的断裂为韧性断裂，其开裂原因主要是开裂位置存在大量的带状组织，从而导致该处均匀变形能力被削弱，最终造成辊压开裂；通过降低热轧终轧温度和卷取温度可以有效地减轻热镀锌 DP780 中的带状组织，改善热镀锌 DP780 的组织均匀性，降低辊压开裂发生风险。

关键词：热镀锌，DP780，辊压开裂，带状组织

Roll-forming Cracking Mechanism of Hot Dip Galvanized DP780 Steel

Liu Huasai[1,2], Chen Lingfeng[1,2,3], Kuang Shuang[1,2], Han Yun[1,2]

(1. Shougang Research Institute of Technology, Beijing 100043, China; 2. Beijing Key Laboratory of Green Recyclable Process for Iron & Steel Production Technology, Beijing 100043, China; 3. School of Materials Science and Engineering, University of Science and Technology Beijing, Beijing 100083, China)

Abstract: Roll-forming are used more and more for the forming process of ultra-high strength steel during automobile component production. A hot-dip galvanized DP780 steel is used for rolled formed left and right rear rocker. During roll forming process, cracking can be found on the outside of the bending angle. In this study, macroscopic observation, fracture observation, chemical composition analysis, mechanical properties testing and microstructure analysis were used to analyze the roll-forming cracking mechanisms. The results shows that it is ductile fracture of the roll-formed automobile component. There are a lot of band structure at the fracture area, therefore the uniform deformation ability was weaken at the band structure area during roll-forming process. Finish rolling temperature and coiling temperature plays an important role in controlling band structure, and band structure can be reduced by decreasing finish rolling temperature and coiling temperature which helps to reduce the risk of roll-forming cracking.

Key words: hot dip galvanized, DP780, roll-forming cracking, band structure

合金结构钢 40CrNiMo 的研制开发

田志国，徐瑞军

（湖南华菱湘潭钢铁有限公司技术质量部，湖南湘潭　411101）

摘　要： 采用 240×240mm 的方坯轧制 90mm 规格合金结构钢 40CrNiMo 圆钢。采用 LF 精炼精准控制化学成分，经 VD 炉真空脱气，其力学性能 ReL≥835MPa，Rm≥980MPa，A≥25%，Z≥50%，KU2C/J≥75J，满足截面较大和载荷较重的重要零件的要求。

关键词： 合金结构钢，40CrNiMo

Research and Development of the Alloy Structure Steels 40CrNiMo

Tian Zhiguo, Xu Ruijun

(Xiangtan Iron & Steel Co., Ltd., Hunan Valin Group, Xiangtan 411101, China)

Abstract: Using the 240 × 240mm square billet rolled 90mm round steel of alloy structural steel 40CrNiMo. Using LF refining, precise control of chemical composition, Vacuum degassing by VD furnace. Its mechanical properties is ReL≥835MPa, Rm≥980MPa, A≥25%, Z≥50%, KU2C/J≥75J.Meet the requirements of larger parts and heavy load of important parts.

Key words: alloy structure steels, 40CrNiMo

冷轧 CR550/980DP 组织性能分析

邝春福，郑之旺

（攀钢集团研究院钒钛钢研究所，四川攀枝花　617000）

摘　要： 双相钢（DP 钢）具有低屈服强度、高抗拉强度和良好成形性能等优点，广泛应用于汽车结构件和加强件，成为轿车用首选高强钢，预计在汽车用先进高强钢用量中将超过 70%。西昌钢钒板材厂 2#连退线为高强汽车板生产线，机组于 2015 年初投产后即开展了 390~980MPa 系列冷轧双相钢的工业试制，本文简述了冷轧 CR550/980DP 的工业试制情况，重点进行了组织性能分析。

汽车轮圈断裂原因分析及建议

肖　亚，黄微涛，马建国，汪　涛，陈　锐，舒遗一

（重庆钢铁股份有限公司技术中心，重庆　400084）

摘　要： 对汽车轮圈在加工过程中发生断裂失效进行了分析。结果表明，轮圈断裂原因系由焊接工艺不当，使焊缝处出现裂纹、疏松、孔洞及焊渣卷入等缺陷，在轮圈扩圆过程中，形变拉应力极大，使其在焊缝缺陷处应力集中导致出现裂纹或断裂。

关键词： 轮圈断裂，焊接，缺陷，应力集中

DC06 冷轧深冲钢的动态力学性能测试及研究

王亚芬，赵广东，李志伟，杨　振

（本钢板材股份有限公司产品研究院，辽宁本溪　117000）

摘　要： 采用常温拉伸试验机和高速拉伸试验机对实验钢 DC06 在不同应变速率下分别进行准静态及高速拉伸实验，应变速率范围为 0.001~1000s^{-1}，试验测得工程应力-应变曲线，并通过建立本构方程拟合外推其真塑性应力-应变曲线，对其力学性能特征进行研究。结果表明：与准静态拉伸相比，高速拉伸下 DC06 钢的强度、塑性等力学性能均发生明显的变化，屈服强度和抗拉强度均大于准静态。在高应变率拉伸条件下，由于加工硬化、应变速率强化及绝热软化等综合因素，钢板表现出强烈的应变速率敏感性。

关键词： 高速拉伸，应变速率，本构方程，动态力学

Test and Study on Dynamic Mechanical Properties of DC06 Cold Rolled Deep Drawing Steel

Wang Yafen, Zhao Guangdong, Li Zhiwei, Yang Zhen

(Research of Product Research Institute of Benxi Steel Plates Co., Ltd., Benxi 117000, China)

Abstract: The tensile testing machine and tensile testing machine of high speed steel DC06 experiments at different strain rates were conducted under quasi-static and high strain rate tensile test, the range of 0.001~1000s^{-1}. The stress-strain curves of DC06 steel were measured and the mechanical properties were studied through the establishment of the constitutive equation fitting extrapolation true stress-strain curves. The results show that the mechanical properties such as strength and plasticity of DC06 steel are obviously changed at elevated speed compared with quasi-static tension, and the yield strength and tensile strength were all greater than quasi static. Under the condition of high strain rate tensile, the steel exhibits a strong strain rate sensitivity due to the factors such as work hardening, strain rate strengthening and adiabatic softening.

Key words: high speed tensile, strain rate, constitutive equation, dynamic mechanics

不同厚度热轧双相钢的成形极限分析

吴青松

（日照钢铁控股集团有限公司，山东日照　276806）

摘　要： 采用基于应变速率变化的测试方法测试了不同厚度双相钢的成形极限图，分析了双相钢的成形极限图随着厚度的变化规律，随着厚度的减薄，双相钢的成形能力下降。组织中铁素体晶粒大小、马氏体含量及大小是影响双相钢成形能力的主要原因。

关键词： 双相钢，应变速率，成形极限图

Forming Limit Analysis of Hot-rolled Dual Phase Steel with Different Thickness

Wu Qingsong

(Rizhao Steel Holding Group Co., Ltd., Rizhao 276806, China)

Abstract: The forming limit diagram of dual-phase steel with different thickness was tested by the test method based on the change of strain rate. The variation law of forming limit diagram of dual phase steel with thickness is analyzed. The results show that the forming ability of dual phase steel decreases with the decrease of thickness. Ferrite grain size, martensite content and size are the main reasons that affect the forming ability of dual phase steel.

Key words: dual phase steel, strain rate, forming limit diagram

8　海洋工程用钢

炼铁与原料
炼钢与连铸
轧制与热处理
表面与涂镀
金属材料深加工
钢铁材料
汽车钢
★ 海洋工程用钢
轴承钢
电工钢
粉末冶金
非晶合金
高温合金
耐火材料
能源与环保
分析检测
冶金设备与工程技术
冶金自动化与智能管控
冶金技术经济

EH47 高止裂钢的组织和性能

周 成，严 玲，张 鹏，朱隆浩

（鞍钢集团钢铁研究院，辽宁鞍山 114009）

摘 要： 借助 SEM 等显微组织分析手段及力学性能测试，研究了集装箱船用 EH47 级别高止裂韧性钢的显微组织与力学性能的关系。实验结果表明，为满足集装箱船用 EH47 级别高止裂韧性钢对止裂韧性性能的严格要求，组织设计可采用针状铁素体+贝氏体的组织方法。晶粒尺寸细小的针状铁素体+贝氏体相互配合具有较高的强度、止裂韧性性能；增大针状铁素体晶粒尺寸或引入多边形铁素体过多则会降低材料强度；而贝氏体体积分数过多、晶粒尺寸过大时，材料塑性和止裂韧性性能显著降低。

关键词： 高强高韧，止裂韧性，针状铁素体，粒状贝氏体，集装箱船用止裂钢

Study on Microstructures and Mechanical Properties of EH47 High Strength Brittle Crack Arrest Steel

Zhou Cheng, Yan Ling, Zhang Peng, Zhu Longhao

(Iron & Steel Research Institute of Angang Group, Anshan 114009, China)

Abstract: Relationship between microstructure and mechanical properties of the thick steel plates EH47 with high crack arrest toughness for container ships were investigated by SEM and mechanical property tests. The results indicate that the microstructure design can use acicular ferrite and bainite microstructure to satisfy the strict requirements of the high crack arrest toughness of the EH47 High Strength Brittle Crack Arrest Steel. Microstructure consisting of granular bainite and acicular ferrite in the tested steels is a guarantee to obtain optimum match of strength and toughness. The strength can decrease with the volume fraction of polygonal ferrite increasing in the duplex microstructure. While enlarging the acicular ferrite grain size can also decrease the strength. The excessive volume fraction and too large grain size of bainite can decrease the plasticity and crack arrest toughness.

Key words: high strength and toughness, arrest toughness, acicular ferrite, granular bainite, brittle crack arrest steel for container ship

特厚板坯皮下横裂纹原因浅析

闫晓旭

（秦皇岛首秦金属材料有限公司炼钢部，河北秦皇岛 066001）

摘 要： 本文阐述 3#铸机皮下裂纹形成机理。改善结晶器振动，提高矫直温度和铸机二冷段设备精度，选择合适保护渣，有利于消除皮下横裂纹。

关键词： 直弧型连铸机，板坯，皮下横裂纹，影响因素，预防措施

Analysis of the Causes of Transverse Crack of Extra Thick Slab

Yan Xiaoxu

(Steelmaking Department of Qinhuangdao Shouqin Metal Materlals Co., Ltd., Qinhuangdao 066001, China)

Abstract: This paper discusses the causes to form subsurface transversal crack No.3 caster. Subsurface transversal crack can be reduced by improving mould oscillation, raising the straightening temperature , improving the precision of segments as well as by using suitable mold powder.

Key words: straight mold type caster, slab, subsurface transversal crack, influencing factor, precaution

淬火工艺对大厚度 690MPa 级海工钢组织性能影响

张　鹏，严　玲，周　成，朱隆浩

（鞍钢集团钢铁研究院，海洋装备用技术材料及其应用国家重点实验室，辽宁鞍山　114021）

摘　要: 本文对比了淬火+回火（QT），一次两相区淬火+一次淬火+回火（LQT），一次淬火+一次两相区淬火+回火（QLT）三种热处理工艺对大厚度超高强度 690MPa 级海洋工程用钢组织性能的影响，结果表明，三种不同淬火+回火工艺对 690MPa 级海洋工程用钢的低温韧性影响不同，其中采用一次淬火+回火工艺不能保证大厚度海洋工程钢板的低温韧性，尤其是不能保证钢板心部低温韧性，采用一次两相区淬火+一次淬火+回火（LQT）工艺能够一定程度提升钢板低温韧性，一次淬火+两相区二次淬火+回火（QLT）工艺结果最理想，能够大幅度提高钢板的低温韧性，同时，还能够获得最好的强韧匹配，其中细化晶粒及高密度位错是决定钢板优良低温韧性的关键因素。

关键词: 两相区淬火，大厚度，海洋工程用钢，低温韧性，细化晶粒，高密度位错

Effect of Quenching Process on Microstructure and Properties of Large Thickness 690 Grade Marine Engineering Steel

Zhang Peng, Yan Ling, Zhou Cheng, Zhu Longhao

(Ansteel Group Research institute, State Key Laboratory of Metal Material for Marine Equipment and Application, Anshan 114021, China)

Abstract: Treatment of QT、LQT and QLT was contrasted in this paper to research the mechanical properties of 690MPa grade ship hull steel, the results show that three different heat treatment processes have different effects on the low temperature impact toughness of 690MPa grade ship hull steel, using traditional quenching and tempering process can not guarantee steel with good low temperature impact toughness, especially it can not guarantee the low temperature impact toughness at the 1/2 position of steel plate. Using LQT technology to a certain extent to enhance the ow temperature impact toughness of the steel plate. the result of using QLT technology is the best , this process can greatly improve the low temperature toughness of steel plate. At the same time also can get the best match of strength and toughness. Grain refinement and high density dislocation are the key factors to determine the excellent low temperature toughness of steel plate.

Key words: two-phase quenching, heavy thickness, marine engineering steel, low temperature toughness, refine grain size, high density dislocation

Mg 脱氧钢中夹杂物对晶内针状铁素体形成的影响

徐龙云[1,2]，杨　健[2]，王睿之[3]，王万林[1]

（1. 中南大学冶金与环境学院，湖南长沙　410083；2. 上海大学省部共建高品质特殊钢冶金与制备国家重点实验室，上海　200072；3. 宝钢股份研究院炼钢技术研究所，上海　201999）

摘　要：通过焊接热模拟试验和扫描电镜观察研究了 Mg 脱氧钢中夹杂物对焊接热影响区（Heat-Affected Zone, HAZ）晶内针状铁素体（Intragranular Acicular Ferrite, IAF）形成的影响。研究结果表明，钢中尺寸在 2 μm 左右的 Mg-(Al)-Ti-O+MnS 氧硫化物复合夹杂能够有效地促进 IAF 形成，IAF 体积分数为 55.4%；具备诱发 IAF 形核的有效夹杂物中 MnS 含量明显低于无效夹杂物。采用 Mg 脱氧的 50 mm 厚试验钢板经 400 kJ/cm 大线能量焊接后 HAZ –20 ℃冲击功在 160 J 以上。

关键词：氧化物冶金，热影响区，Mg 脱氧，夹杂物，晶内针状铁素体

Effect of Inclusions on the Formation of Intragranular Acicular Ferrite in Steel with Mg Deoxidation

Xu Longyun[1,2], Yang Jian[2], Wang Ruizhi[3], Wang Wanlin[1]

(1. School of Metallurgy and Environment, Central South University, Changsha 410083, China; 2. State Key Laboratory of Advanced Special Steel, Shanghai University, Shanghai 200072, China; 3. Steelmaking Research Department, Research Institute, Baoshan Iron and Steel Co., Ltd., Shanghai 201999, China)

Abstract: Effect of inclusions on the formation of intragranular acicular ferrite (IAF) in heat-affected zone (HAZ) of steel plate after high heat input welding has been investigated employing welding thermal cycle simulation experiments and SEM observation. The results showed that the formation of IAF in HAZ was effectively promoted by Mg-(Al)-Ti-O+MnS oxysulfide complex inclusions with sizes around 2 μm. The volume fraction of IAF was 55.4%. The content of MnS in effective inclusion acting as nucleant of IAF was obviously lower than that in ineffective inclusion. After welding thermal cycle simulation with the heat input of 400 kJ/cm, the HAZ impact energy at –20 ℃ of 50 mm thickness experimental steel plate with Mg deoxidation reached more than 160 J.

Key words: oxide metallurgy, heat-affected zone, Mg deoxidation, inclusion, intragranular acicular ferrite

集装箱船用高强度超宽特厚钢板生产工艺开发

李新玲，丛津功，王若钢，肖青松，刘　源，应传涛

（鞍钢股份有限公司鲅鱼圈钢铁分公司，辽宁营口　115007）

摘　要：基于多相组织调控原理获得高止裂性钢板，通过实施高洁净化与高均质冶金技术、低温加热原始奥氏体晶粒细化、临界温度区间大厚度截面钢板奥氏体超细化再结晶控制及驰豫过程中先共析铁素体析出形态与比例等全流程控制关键技术，获得了具有合适软硬相配比、晶粒细化的多相组织。

关键词：止裂钢，多相组织，晶粒细化

The Production Process of High Strength and Overwide Steel Plate for Container Ships

Li Xinling, Cong Jingong, Wang Ruogang, Xiao Qingsong, Liu Yuan, Ying Chuantao

(Bayuquan Iron & Steel Subsidiary Company of Angang Steel Co., Ltd., Yingkou 115007, China)

Abstract: Based on the principle of multi-phase microstructure control, high fracture steel plate was obtained. Through the implementation of high clean and high homogeneous metallurgical technology, low temperature heating of the original austenite grain refinement, critical temperature range large thickness section steel austenite superfine recrystallization control and relaxation process of the first eutectoid ferrite precipitation morphology and proportion of the whole process control key technology, with a suitable soft and hard match ratio, grain refinement of the multi - phase structure.

Key words: cracked steel, multiphase structure, grain refinement

深海大壁厚 X65MOS 海底管线钢关键技术研究

丁文华[1,2]，张　海[1,2]，李少坡[1,2]，李　群[3]，王雪松[3]

（1. 首钢集团有限公司技术研究院宽厚板所，北京　100043；2. 北京市能源用钢工程技术研究中心，北京　100043；3. 秦皇岛首秦金属材料有限公司，河北秦皇岛　066326）

摘　要：基于深海大壁厚 X65MOS 海管低温断裂韧性、屈强比和抗腐蚀性等技术难点，本文开展了钢坯粗轧不同变形工艺下低温断裂韧性、工业制管拉伸性能变化与实验室"预拉伸+弯曲"方法模拟制管抗酸性能变化等方面的研究工作，最终确定了钢坯粗轧极限变形工艺要求、板-管圆棒样拉伸性能变化规律与钢板 HIC 性能内控判定标准。

关键词：海底管线钢，X65MOS，低温断裂韧性，屈强比，抗腐蚀性

Research of Heavy Wall X65MOS Submarine Pipeline Steel Key Technology

Ding Wenhua[1,2], Zhang Hai[1,2], Li Shaopo[1,2], Li Qun[3], Wang Xuesong[3]

(1. Shougang Group Research Institute of Technology, Beijing 100043, China; 2. Beijing Engineering Research Center of Energy Steel, Beijing 100043, China; 3. Qinhuangdao Shouqin Metal Materials Co., Ltd., Qinhuangdao 066326, China)

Abstract: Based on the technical difficulties of heavy wall X65MOS submarine pipeline steel, such as low temperature fracture toughness, yield ratio, corrosion resistance. This paper reports on the research of the low temperature toughness under different rough rolling process, the strength properties change during pipe making and the corrosion resistance change by "pre-stretching + bend" deformation simulate pipe making in the lab. Eventually determine the limit deformation process of rough rolling, the tensile properties change of round bar sample and the HIC decision criteria of plate.

Key words: submarine pipeline, X65MOS, low temperature fracture toughness, yield ratio, corrosion resistance

自升式海洋平台用大厚度齿条钢的开发

陆春洁，杨 汉，镇 凡，曲锦波

（江苏省（沙钢）钢铁研究院板带材研究室，江苏张家港 215625）

摘 要： 沙钢依靠先进的装备与技术优势，联合国内重机和冶金领域优势力量，开发出自升式海洋平台用 152.4mm、177.8mm 厚齿条钢 ASTM A514 GrQ Mod / ASTM A517 GrQ Mod / EQ70。对钢板进行各项性能检验和焊接性能评定，结果表明：钢板整板长度和厚度方向性能均匀，具有良好的强塑性、抗层状撕裂性、低温韧性和焊接性能。满足海洋平台用齿条钢的技术要求，通过了 ABS、CCS、DNV GL 三家船级社认证。

关键词： 海洋平台，齿条钢，强塑性，抗层状撕裂性，低温韧性，焊接性能

Development of Heavy Thickness Steel Plates for Racks of Jack-up Rigs

Lu Chunjie, Yang Han, Zhen Fan, Qu Jinbo

(Institute of Research of Iron and Steel (IRIS), Shasteel, Zhangjiagang 215625, China)

Abstract: In association with heavy machinery and metallurgical advantages, Sha-steel has developed 152.4mm and 177.8mm heavy thickness steel plates for racks of jack-up rigs by using advanced equipment and technology. Various mechanical property and welding tests were carried out and the result indicates that the plates have a good combination of strength and plasticity, resistance to lamellar tear, low temperature toughness and weldability, which meet the technical requirement of racks of jack-up rigs, and have been approved by ABS, CCS and DNV.GL ship classification societies.

Key words: offshore platform, rack steel, strength and plasticity, resistance to lamellar tear, low temperature toughness, weldability

低压缩比 120mm 海洋工程用 EH36 钢板研制及开发

刘朝霞，李经涛，赵 孚，刘 俊，宁康康，高 俊

（江阴兴澄特种钢铁有限公司研究院，江苏江阴 214400）

摘 要： 为了使用连铸坯生产低压缩比 120mm 海洋工程用 EH36 钢板，进行成分设计、低倍分析以及对精轧开始轧制温度以及热处理关键工艺进行探索，随后进行现场批量试制验证研究。结果表明，本研究成分体系下，轧制工

艺、热处理工艺窗口较大，适用于中厚板生产。小批量试制结果表明，钢板典型组织主要为细小均匀铁素体和弥散分布其周围的珠光体组织及碳化物颗粒组成。拉伸、冲击、时效、弯曲、落锤等性能优异，满足 EH36 级海洋工程用钢船级社规范认证要求。

关键词：海洋工程用钢板，连铸坯，低压缩比，EH36

The Development of Low Compression Ratio 120mm EH36 Offshore Steel Plate

Liu Zhaoxia, Li Jingtao, Zhao Fu, Liu Jun, Ning Kangkang, Gao Jun

(Research Institute, Jiangyin Xingcheng Special Steel Works Co., Ltd., Jiangyin 214400, China)

Abstract: To produce low compression ratio 120 mm EH36 offshore steel plate by continues casting slab, the design of chemical composition, the observation of slab macrostructure and the key process parameters as start rolling temperature of final rolling and normalizing temperature and holding time are studied in this paper. The results show that rolling process and normalizing process windows are big and easy for Middle heavy plate at the composition system. The typical trial product microstructure is mainly composed of fine and uniform ferrite dispersed by pearlite and fine carbides. Tensile properties, impact properties, aging properties, bending properties, NDT are excellent performed; meet the social classification requirements of EH36 offshore steel plate.

Key words: offshore steel plate, continuous casting slab, low compression ratio, EH36

热处理工艺对超纯铁素体不锈钢
组织性能的影响

梁轶杰，李　阳，姜周华，杨　光，杜鹏飞

（东北大学冶金学院，辽宁沈阳　110819）

摘　要：通过光学金相显微镜，扫描电镜，硬度测试，动电位极化和电化学阻抗谱等分析测试手段，研究了不同的退火温度和保温时间对铌钛稳定化的超级铁素体不锈钢组织、力学性能以及其在 3.5%NaCl 溶液中的耐腐蚀性能的影响。研究结果表明，超纯铁素体不锈钢热处理后，钢中主要析出相为微米级的铌钛的碳氮化物和细小 χ 相；退火温度为 1050℃，保温 5min 后水冷的超纯铁素体不锈钢再结晶程度最好、点蚀电位、钝化膜阻抗最高，耐腐蚀性能最佳。

关键词：超纯铁素体不锈钢，热处理，硬度，点蚀电位

Effect of Heat Treatment on Microstructure and Properties of Super Ferritic Stainless Steel

Liang Yijie, Li Yang, Jiang Zhouhua, Yang Guang, Du Pengfei

(School of Metallurgy, Northeast University, Shenyang 110819, China)

Abstract: The effect of heat treatment on microstructure, macro hardness and corrosion behaviour in 3.5%NaCl solution under 30℃ of super ferritic stainless steel have been investigated by scanning electron mircroscopy (SEM), optical microscopy (OM), potentiodynamic polyrization technique and electrochemical impedance spectrum (EIS). The results show that (Nb, Ti) (C, N) and χ phase have been observed after heat treatment; The recrystallization, ptting potential, and impedance of passivation film of super ferritic stainless steel after annealing at 1050℃ for 5min followed by water quenching are the best.
Key words: super ferritic stainless steel, heat treatment, hardness, pitting corrosion

大厚度集装箱船用止裂钢的研制

陈 华[1]，王 华[1]，严 玲[1]，韩 鹏[1]，陈军平[1]，亢淑梅[2]，刘 源[1]

（1. 鞍钢股份有限公司，辽宁鞍山 114021；2. 辽宁科技大学，辽宁鞍山 114051）

摘 要：介绍鞍钢 EH47BCACOD 集装箱船用止裂钢板开发情况，通过采用低碳+微合金化，复合添加 Cr、Mo、V 等合金成分设计，结合高质量钢坯的冶炼、多阶段 TMCP 轧制及超快速冷却工艺，成功开发符合 CCS、DNVGL 等多国船级社认证规范要求的 90mm 厚度 EH47BCACOD 止裂钢板。该钢板从表面至心部组织均匀，晶粒度达到 10.5 级，具有优异的抗层状撕裂性、低冷裂纹敏感性、可焊性、低温韧性及止裂性能。
关键词：多阶段 TMCP 轧制，抗层状撕裂性，可焊性，低温韧性，止裂性能

Research of Heavy Steel Plates with Brittle Crack Arrest Properties for Container Ships

Chen Hua[1], Wang Hua[1], Yan Ling[1], Han Peng[1], Chen Junping[1], Kang Shumei[2], Liu Yuan[1]

(1. Angang Steel Company Limited, Anshan 114021, China; 2. Liaoning University of Science and Technology, Anshan 114051, China)

Abstract: The present development status was introduced about heavy steel plates with brittle crack arrest properties for container ships. The 90mm thickness EH47BCACOD plate with brittle crack arrest properties conforming to the certification of CCS, DNVGL was produced by reasonable micro alloying design, the steel making of high quality steel billet, multi-stage TMCP rolling and ultra-fast cooling process. The steel plate is evenly distributed from the surface to the core and has a grain size 10.5. It has good lamellar tearing resistance, low cold crack sensitivity, good weldability and low temperature toughness.
Key words: multi-stage TMCP rolling, lamellar tearing resistance, weldability, low temperature toughness, brittle crack arrest properties

回火工艺对中锰钢显微组织和力学性能的影响

李军辉[1]，孙 超[2]，李 丽[2]，李东晖[2]，吴开明[1]，王红鸿[1]

（1. 武汉科技大学高性能钢铁材料及其应用湖北省协同创新中心，湖北武汉 430081；
2. 南京钢铁集团有限公司研究院，江苏南京 210035）

摘　要：通过热处理工艺实验及显微组织观察，研究了正火+回火和淬火+回火工艺中，回火工艺对中锰钢显微组织和力学性能的影响。结果表明，在正火+回火工艺中，650℃回火，主要为贝氏体以及少量的逆转变奥氏体；680℃回火，主要为贝氏体，同时存在少量二次马氏体及逆转变奥氏体；710℃回火，主要为贝氏体+二次马氏体+残余奥氏体，二次马氏体板条束特征明显。在两种热处理工艺中，屈服强度和冲击吸收功都随着回火温度的升高而降低。在淬火+回火工艺中，屈服强度和抗拉强度随着回火保温时间的延长而增大，冲击吸收功随着回火保温时间的延长有升高的趋势。

关键词：回火工艺，中锰钢，贝氏体，二次马氏体，逆转变奥氏体，力学性能

Effect of Tempering Process on Microstructure and Mechanical Properties of Medium Manganese Steel

Li Junhui[1], Sun Chao[2], Li Li[2], Li Donghui[2], Wu Kaiming[1], Wang Honghong[1]

(1. Hubei Collaborative Innovation Center for Advanced Steel, Wuhan University of Science and Technology, Wuhan 430081, China; 2. Research Institute Nanjing Iron and Steel Group Corp., Nanjing 210035, China)

Abstract: The effects of tempering process on microstructure and mechanical properties of medium manganese steel during normalizing + tempering and quenching + tempering process were studied by heat treatment process and microstructure observation. The results show that in the normalizing + tempering process, , the microstructure was mainly bainite and a small amount of reversed austenite after 650 ℃ tempering;, it was mainly bainite, while the presence of a small amount of secondary martensite and reveersed austenite by 680 ℃ tempering; 710 ℃ tempering, mainly for bainite + secondary martensite + reversed austenite, secondary martensite plate bundle characteristics are obvious. In both heat treatment processes, yield strength and impact absorbed energy decreased with increasing tempering temperature. In the quenching and tempering process, the yield strength and the tensile strength increased with the increase of the tempering holding time, and the impact absorbed energy increased with the increase of the tempering holding time.

Key words: tempering process, medium manganese steel, bainite, secondary martensite, reversed austenite, mechanical properties

高铝 δ 铁素体钢的软化和再结晶行为研究

李建赫[1]，徐翔宇[2]，王学敏[2]，张　威[1]，刘倩男[2]

（1. 北京科技大学材料科学与工程学院，北京　100083；
2. 北京科技大学钢铁共性技术协同创新中心，北京　100083）

摘　要：本文采用了等温双道次压缩和单道次压缩试验研究 Fe-4Al-2Ni 的 δ 铁素体钢的软化和再结晶行为。与奥氏体相同：铁素体的软化率和再结晶率随变形温度和等温时间的升高而增大；新再结晶晶粒优先在原形变晶界处形核；再结晶可以帮助均匀和细化晶粒结构。不同的是，铁素体钢中再结晶行为前的软化率较高（大于 0.5）。该铁素体可采用再结晶区和非再结晶区轧制配合离线回火的方式进行加工。

关键词：δ 铁素体，软化，静态再结晶，结构细化

Softening and Recrystallization Behavior of Al-bearing Ferritic Steel

Li Jianzhe[1], Xu Xiangyu[2], Wang Xuemin[2], Zhang Wei[1], Liu Qiannan[2]

(1. School of Materials Science and Engineering, University of Science and Technology Beijing, Beijing 100083, China; 2. Collaborative Innovation Center of Steel Technology, University of Science and Technology Beijing, Beijing 100083, China)

Abstract: In this study, softening and recrystallization behavior of Fe-4Al-2Ni δ-ferrite steel was investigated by interrupted and single-pass compression test. Similar to the austenite steel, the fractional softening and recrystallization fraction were enhanced with the increasing deformation temperature and holding time. The preferred nucleation sites of recrystallized grains were the original grain boundaries and homogeneous and fine grained structure can obtained by recrystallization progress. Somewhat differently, recrystallization in the ferritic alloy commenced after a significant degree of softening (0.5). The two-stage rolling process (recrystallization controlled rolling and non-recrystallization-zone rolling) and annealing process could be carried out to obtain refined microstructure.

Key words: δ-ferrite, softening, static recrystallization, microstructure refinement

固溶温度对高 Cr、Mo 双相不锈钢组织和冲击性能的影响

朱隆浩，严 玲，周 成，张 鹏

（鞍钢集团钢铁研究院，辽宁鞍山 114001）

摘 要： 对实验室试制的 S31803 双相不锈钢进行 1050、1100、1150℃固溶处理，应用光学显微镜、XRD、TEM 和全自动冲击试验机，探究固溶温度对高 Cr、Mo 双相不锈钢组织结构及冲击性能的影响，获得了高 Cr、Mo 双相不锈钢析出相随固溶温度的变化规律。结果表明：经 1050、1100、1150℃固溶处理的 S31803 双相不锈钢，其组织均由 α-Fe，γ-Fe 和 σ 相组成。σ 相主要分布在 α/γ 晶界以及 α/α 晶界处。随着固溶温度的增加，σ 相晶粒尺寸逐渐减小，并且发生局部溶解，晶粒形状从具有明显晶体学特征的尖角型转变为局部球型。S31803 双相不锈钢的冲击性能随固溶温度的升高呈现先升后降的规律，在 1100℃固溶处理后，冲击性能最优。

关键词： 双相不锈钢，固溶温度，组织，冲击性能

Influence of Solute Temperature on Microstructure and Impact Property of Duplex Stainless Steel with Higher Cr, Mo Contents

Zhu Longhao, Yan Ling, Zhou cheng, Zhang Peng

(Iron & Steel Research Institute of Angang Group, Anshan 114001, China)

Abstract: The duplex stainless steel used in this experiment is the first production of Ansteel group corperation. Solid solution treatment temperature is 1050, 1100, 1150℃. The analysis was carried out by means of optical microscope, XRD

(X-ray diffraction) and TEM (transmission electron microscope) and auto impact testing machine. Explore the effect of different solid solution temperature on microstructure and impact properties of duplex stainless steel with higher Cr and Mo content, and find out the relationship between precipitated phase and solution temperature. The results show that the microstructure of S31803 duplex stainless steel under different solid solution temperature is composed of α-Fe, γ-Fe and σ phase . The σ phase is mainly distributed in α/γ grain boundaries and α/α grain boundaries. With the increase of solution temperature, the grain size decreases and the local dissolution occurs. The shape of the crystal is changed from a sharp angle with a distinct crystallographic feature to a local sphere. The impact property of S31803 duplex stainless steel increases with the increase of solid solution temperature, which is first and then decreased, and the impact property is the best after the solution treatment at 1100℃.

Key words: duplex stainless steel, solution temperature, microstructure, impact property

深海用 X70 厚壁钢管多丝埋弧焊接工艺研究

毕宗岳[1,2]，牛爱军[1,2]，牛　辉[1,2]，黄晓辉[1,2]，刘海璋[1,2]

（1. 国家石油天然气管材工程技术研究中心，陕西宝鸡　721008；2. 宝鸡石油钢管有限责任公司 钢管研究院，陕西宝鸡 721008）

摘　要： 针对深海用高强度、高韧性及良好焊接性能的厚壁管线钢管，采用低 C、高 Mn-Nb-Ni-Mo 的合金化成分体系及洁净钢冶炼技术和控轧控冷工艺，开发出了 X70 厚度 36.5mm 深海用管线钢。试验研究了厚壁管线钢多丝埋弧焊接工艺，提出并建立了 X70 厚壁 36.5mm 直缝埋弧焊接钢管的内焊四丝、外焊五丝的多丝埋弧焊接工艺参数，以此成功开发出深海用 X70ϕ1016×36.5mm 大口径厚壁直缝埋弧焊管。检测结果表明，钢管焊缝强度达 645MPa，–40℃冲击韧性平均 121J，0℃下 CTOD 值为 0.54mm，各项性能指标优于标准要求。该多丝埋弧焊接工艺完全满足深海厚壁管线的苛刻要求。

关键词： 深海管线，壁厚，直缝埋弧焊管，多丝埋弧焊接，焊接工艺

Study on Multi-wire Submerged Arc Welding Technical Process of X70 Steel Pipe with Heavy Wall Used for Deep-sea Pipeline

Bi Zongyue[1,2], Niu Aijun[1,2], Niu Hui[1,2], Huang Xiaohui[1,2], Liu Haizhang[1,2]

(1. Chinese National Engineering Research Center for Petroleum and Natural Gas Tubular Goods, Baoji 721008, China; 2. Research Institute of Steel Pipe, Baoji Petroleum Steel Pipe Co., Ltd., Baoji 721008, China)

Abstract: According to the high strength, high toughness and good welding performance of thick wall pipe used deep-sea pipeline, the X70　pipeline steel with thickness of 36.5mm used in deep sea has been developed by using low C and high Mn-Nb-Ni-Mo alloy composition system and clean steel smelting technology and controlled rolling and controlled cooling technology.The multi-wire submerged arc welding process parameters of the internal welding with four wire and the external welding with five wire applied to X70 LSAW steel pipe with thickness of 36.5mm is put forward and established by experimental study the multi-wire submerged arc welding process on thick wall pipeline steel.And then the X70 phi 1016×36.5mm LSAW pipe with large diameter and thick wall used for deep-sea pipeline has been developed successfully. The test results show that the weld strength of the steel pipe is 645MPa, the impact toughness average value is 121J, –40℃ and the CTOD average value is 0.54mm, 0℃on the weld. Each performance index is superior to the standard requirements. The multi-wire submerged arc welding process can fully meet the harsh requirements of the thick wall pipeline for deep-sea.

Key words: deep-sea pipeline, heavy wall thickness, longitudinal submerged arc welded steel pipe (LSAW pipe), multi-wire submerged arc welding, welding technical process

高性能船用锻件工艺与成分研究

高家悦

（石钢京诚装备技术有限公司，辽宁营口 151000）

摘 要：近年来，我公司 20Mn-C 船用锻件产品性能合格率不高，非常规、性能指标要求高的船件（指 20Mn-C 碳锰钢抗拉强度大于 560MPa）性能合格率为 62.5%，多次船检出现抗拉值低于要求指标的情况，甚至易导致产品的报废。经过近一年多的努力，通过实践解决了这一问题，本文研究了船用锻件的成分、工艺的设计与结果的关系，提出加入固溶强化元素可以获得比理论计算值更高的强度结果，对理论强度计算公式可以进行局部修正。

关键词：船用锻件，20Mn-C，成分设计

Study on Process and Composition of High Performance Marine Forgings

Gao Jiayue

(ShiGangJingCheng Equipment Development and Manufacturing Co., Ltd., Yingkou 151000, China)

Abstract: In recent years, our company 20Mn-C marine forgings product quality rate is not high, the very rules, the performance requirements of high (20Mn-C carbon steel tensile strength greater than 560Mpa) performance of the qualified rate is 62.5%, the ship appears repeatedly tensile value is lower than the index requirements, even easily lead scrap caused by products. After nearly a year of hard work, through the practice of solving this problem, this paper studies the relationship of ship design and the forging process, composition, added to the solid solution strengthening elements can be obtained more than the strength calculation theory, can be a local correction formula for calculating strength.

Key words: Marine forgings, 20Mn-C, composition design

抗硫管线钢 KS30 炼钢生产特点及实践

魏 巍，李 虹

（石钢京诚装备技术有限公司，辽宁营口 115100）

摘 要：抗硫管线钢主要用于加工石油、天然气的输送管道，对钢的强度、韧性、抗氢致裂纹(HIC)、抗硫应力腐蚀裂纹(SSC)和焊接性能等要求很高。本文结合石钢京诚生产抗硫管线钢的实践情况，对各成分的作用及影响进行了分析，并对抗硫管线钢 KS30 生产实践情况进行介绍。

关键词：抗硫管线钢，成分的作用及影响，生产实践

Production Characteristics and Practice of Sulfur Resistant Pipeline Steel KS30

Wei Wei, Li Hong

(Shi Gang Jing Cheng Equipment Development and Manufacturing Co., Ltd., Yingkou 115100, China)

Abstract: Sulfur resistant pipeline steel is mainly used in the processing of petroleum and natural gas pipeline. It requires high strength and toughness of steel, anti hydrogen induced cracking(HIC), anti sulfide stress corrosion cracking(SSC) and welding performance. Combining with the practice of Shi Gang Jing Cheng Equipment Development and Manufacturing Co., Ltd. Analyzed the role and influence of each component. Introduced the production practice of sulfur resistant pipeline steel KS30.

Key words: sulfur resistant pipeline steel, role and influence of each component, production practice

深海服役用复合贝氏体型链钢和链的性能边界

殷　匠 [1,2,3,5]，李松杰 [2]

（1. 亚星锚链，江苏靖江　214531；2. 郑州大学，河南郑州　450001；3. （原）宝钢特钢，
上海　200940；4. 兴澄特钢，江苏江阴　214400；5. 茵矩公司）

9 轴承钢

炼铁与原料
炼钢与连铸
轧制与热处理
表面与涂镀
金属材料深加工
钢铁材料
汽车钢
海洋工程用钢
★ 轴承钢
电工钢
粉末冶金
非晶合金
高温合金
耐火材料
能源与环保
分析检测
冶金设备与工程技术
冶金自动化与智能管控
冶金技术经济

基于质量大数据平台的高洁净度轴承钢钛含量控制

张海宁，席军良，孙玉春，赵瑞华，李双居

（石家庄钢铁有限责任公司，河北石家庄 050031）

摘　要：论述了高品质轴承钢中钛对轴承钢使用寿命的影响，利用大数据分析高品质轴承钢生产过程中钛含量变化，结合热力学分析，确定了影响轴承钢钛含量变化的主要因素。分析了钢中钛与原材料中钛、初炼炉终点碳、炉渣中钛等方面的对应关系，制定并实施了钢中钛的控制措施，结果显示，高品质轴承钢利用大数据系统的数据资源进行分析控制，可稳定实现钛含量在 0.0015% 以下，可达到 0.0009% 的水平。

关键词：大数据，洁净度，轴承钢，钛含量

The Control of Titanium Content of High Clean Bearing Steel Based on Quality Data Platform

Zhang Haining, Xi Junliang, Sun Yuchun, Zhao Ruihua, Li Shuangju

(Shijiazhuang Iron & Steel Co., Ltd., Shijiazhuang 050031, China)

Abstract: This paper discusses the high quality bearing steel in titanium influence on service life of bearing steel, through the analysis of high quality bearing steel production by using big data in the process of titanium content changes, combined with the thermodynamics analysis, determined the main factors influencing the bearing steel titanium content changes. The corresponding relationship between titanium in steel and titanium in raw materials、convertor end point carbon and titanium in slag is analysised, formulation and implementation measures for the control of the titanium steel, results show that high quality bearing steel titanium content is stable under 0.0015%, the lowest can reach 0.0009% by using data resources of big data platform to analyze and control.

Key words: big data, cleaness, bearing steel, Ti content

高氮轴承钢 SV30 过冷奥氏体冷却相变研究

张艳君[1]，陈鹏飞[2]，方志波[1]，闵永安[1]，顾家铭[2]

（1. 省部共建高品质特殊钢冶金与制备国家重点实验室，上海大学材料工程与科学学院，
上海　200072；2. 上海天安轴承有限公司，上海　200233）

摘　要：利用 DIL805A 相变仪测定 SV30 高氮轴承钢 1030℃×10min 奥氏体化后的过冷奥氏体连续冷却转变及其第二相高温析出后的冷却转变热膨胀曲线，在光学显微镜及扫描电子显微镜下观察微观组织，并测定其硬度及残余奥氏体含量。结果表明：随着过冷奥氏体冷速的增大，SV30 钢的淬火硬度先增大后减小再增大；冷速为 30℃/s 左右时，奥氏体出现明显的稳定化，淬火后残奥的增加降低了淬火硬度。过冷奥氏体在 1000～800℃ 不同温度下等温，

在900℃等温过程中碳化物、氮化物析出最多，过冷奥氏体稳定性显著下降，淬火获得了58.6HRC最高硬度。

关键词：相变，轴承钢，高氮钢，奥氏体稳定化

Cooling Phase Transformation of Undercooled Austenite in High Nitrogen Alloyed Bearing Steel SV30

Zhang Yanjun[1], Chen Pengfei[2], Fang Zhibo[1], Min Yongan[1], Gu Jiaming[2]

(1.　State Key Laboratory of Advanced Special Steel, School of Materials Science
and Engineering, Shanghai University, Shanghai 200072, China;
2. Shanghai Tianan Bearing Co., Ltd., Shanghai 200233, China)

Abstract: High nitrogen alloyed bearing steel SV30, austenitized at 1030℃ for 10min, was studied with DIL805A dilatometer, to determine the continuous cooling transformation CCT characteristics, as well as the cooling transformation after the high temperature precipitation of second phases. The microstructure was observed by OM and SEM, while hardness and retained austenite were also measured. The results show that the hardness of SV30 steel increases firstly, then decreases and increases again along with the cooling rate increases. When the cooling rate rises to about 30℃/s, austenite is obviously stabilized, quenching hardness decreases due to the increasing of retained austenite. When undercooled austenite of SV30 steel keep at temperature range of 1000~800℃, the carbide/nitride precipitates most and fastest at about 900℃. And then SV30 steel obtains the highest hardness of 58.6HRC due to the weakest stabilization of undercooled austenite.

Key words: phase transformation, bearing steel, high nitrogen alloyed steel, austenite stabilization

一种新型纳米贝氏体渗碳轴承钢研究

杨志南[1,2]，秦羽满[2]，武东东[2]，纪云龙[2]，戴力强[1]，张福成[1,2]

（1. 燕山大学，国家冷轧板带装备及工艺工程技术研究中心，河北秦皇岛　066004；
2. 燕山大学，亚稳材料制备技术与科学国家重点实验室，河北秦皇岛　066004）

摘　要： 本文系统研究了一种新型的纳米贝氏体渗碳轴承钢表层和心部组织、性能，以及耐磨性。结果显示新材料经过低温等温并回火后表面硬度在60HRC以上，心部硬度可以达到38HRC，并且心部韧性达到131J，明显高于经典的G20Cr2Ni4A钢。同时，与G20Cr2Ni4A钢相比，其耐磨性提高30%以上。纳米贝氏体钢更优异的强韧性匹配，以及组织中更高含量的残余奥氏体保证了优异的性能。随着纳米贝氏体钢的研究不断深入，纳米贝氏体轴承钢必将有更广阔的应用前景。

关键词： 轴承钢，纳米贝氏体，磨损，组织，性能

Research on a Novle Nanobainite Carburized Bearing Steel

Yang Zhinan[1,2], Qin Yuman[2], Wu Dongdong[2], Ji Yunlong[2],
Dai Liqiang[1], Zhang Fucheng[1,2]

(1. National Engineering Research Center for Equipment and Technology of Cold Strip Rolling,

Yanshan University, Qinhuangdao 066004, China; 2. State Key Laboratory of Metastable Materials
Science and Technology, Yanshan University, Qinhuangdao 066004, China)

Abstract: The microstructure, mechanical properties and wear property of a novelnanobainite carburized bearing steel was studied in this paper. Results show that a surface hardness of 60HRC and an interior hardness of 38HRC can be obtained on the novel steel after austempering treatment at low temperature of 190℃. As compared with the conventional bearing steel, G20Cr2Ni4A, the interior toughness of the novel steel is notably improved, and the wear resistance is improved 30% for the novel steel. The excellent strength and toughness of nanobainite steel, and the high content of retainedaustenite in microstructure of nanobainite steel, ensure the excellent comprehensive mechanical properties of the novel steel. The study on the nanobainite steel becomes more and more in-depth, which makes sure the wide application prospect of the steel.
Key words: bearing steel, nanobainite, wear, microstructure, property

VD 流程生产轴承钢夹杂物演变研究

徐迎铁[1]，杨宝权[2]，陈兆平[1]

（1. 宝山钢铁股份有限公司研究院，上海　201900；

2. 宝山钢铁股份有限公司钢管条钢事业部，上海　201900）

摘　要： 通过对 VD 流程冶炼轴承钢过程各个工序夹杂物的分析检测，研究了轴承钢 VD 流程冶炼过程夹杂物成分及大小的演变，得出 LF 处理结束后就出现大量含氧化钙的复合夹杂物，VD 处理结束后夹杂物的钙含量会进一步升高，而在浇铸过程中夹杂物中部分 CaO 会逐步演变成 CaS；VD 处理前小颗粒（2～5μm）低熔点 CaO-Al$_2$O$_3$ 夹杂和 CaO-MgO-Al$_2$O$_3$ 夹杂越多，VD 处理后此类大颗粒夹杂（>5μm）会增多，此类夹杂物碰撞融合长大是大颗粒夹杂增多的主要原因；还发现连铸中包水口堵塞与中包内镁铝尖晶石数量有对应关系，对于轴承钢冶炼，高熔点镁铝尖晶石对水口堵塞的影响超过了氧化铝夹杂。

关键词： 轴承钢，Ds 夹杂，VD，炉渣

Study of Inclusions Evolution of Bearing Steel Through LF-VD-CC Processes

Xu Yingtie[1], Yang Baoquan[2], Chen Zhaoping[1]

(1. Research Institute, Baoshan Iron & Steel Co., Ltd., Shanghai 201900, China; 2. Tube, Pipe and Bar Business Unit, Baoshan Iron & Steel Co., Ltd., Shanghai 201900, China)

Abstract: The inclusions evolution of bearing steel through LF-VD-CC processes were investigated by inductions observation on the specimens sampling at each process end. It was found that large amount of CaO containing inclusions formed in the steel after LF treatment, the contents of CaO in the inclusions would be higher after VD treatment，and the CaO in the inclusions would change CaS in the casting process.The more of little small particles (size in 2～5μm) of CaO-Al$_2$O$_3$ and CaO-MgO-Al$_2$O$_3$ inclusions having lower melting point existed before VD treatment, the number of the large inclusions (>5μm) of this sort would be increased in the steel. The reason is the collision of these inclusions resulting in the size larger. It was also found that there is relationship between the nozzle clogging and and the number of the MgO-Al$_2$O$_3$ spinal inclusions in steel sampling in tundish. The effects of these spinal inclusions on clogging is greater than the Al$_2$O$_3$ inclusions in the bearing steel.
Key words: bearing steel, Ds type inclusion, VD, slag

高温轴承钢 Cr4Mo4V 的晶粒度研究及讨论

马永强，李 涛，孙大利

（东北特钢集团抚顺特殊钢股份有限公司技术处，辽宁抚顺 113001）

摘 要：采用电子显微镜、扫描电镜、能谱分析分析了高温轴承钢 Cr4Mo4V 的退火原始晶粒度、钢材退火后基体中的块状碳化物及晶格内、晶界上析出相成分；通过不同热处理制度标定了 Cr4Mo4V 钢晶粒度长大情况。研究结果表明：退火后 Cr4Mo4V 钢的晶粒度大于 10 级，基体析出相及晶界上析出相的成分均为（Mo、Fe、V、Cr）C 的混合碳化物，但晶界上、晶粒中析出的 Mo_xC_y 成分低于基体中块状碳化物的成分，晶界上、晶粒中 V_xC_y 成分高于基体中块状碳化物的成分。（Mo、V、Fe、Cr）C 在晶界上析出，起到了钉扎在晶界有助于晶粒细化的作用。在 1050~1130℃保温相同时间（15min），随着温度的提高晶粒在不断长大；在 1100℃随着保温时间的延长，晶粒在不断长大。通过结果分析及讨论，认为通过均质化技术控制、热处理温度、保持时间的控制及进一步研究"晶界钉扎效应"将有助于控制热处理后的 Cr4Mo4V 钢晶粒度及控制残余奥氏体含量，进而达到提高高温轴承的使用寿命及产品稳定性。

关键词：高温轴承钢，Cr4Mo4V，轴承钢，晶粒度

Research and Discussion on Grain Size of Cr4Mo4V for High Temperature Bearing Steel

Ma Yongqiang, Li Tao, Sun Dali

(Technology Department of Fushun Special Steel Co., Ltd.,
Dongbei Special Steel Group, Fushun 113001, China)

Abstract: The optical microscope, scanning electron microscopy (SEM) and energy dispersive spectroscopy (EDS) were used to analyze the original grain size and the contents of precipitated phase in grain boundary of Cr4Mo4V steel. The heat treatment system calibrates the grain growth of Cr4Mo4V steel. The results show that the grain size of Cr4Mo4V steel is more than 10 after annealing, and the precipitated phase of the precipitated phase and the grain boundary are all mixed carbides of (Mo, Fe, V, Cr) C, but the grain boundary The Mo_xC_y component precipitated in the granules is lower than that of the block carbides in the matrix. The grain boundary has a higher V_xC_y component in the grains than the block carbides in the matrix. (Mo, V, Fe, Cr) C in the grain boundary precipitation, played a nail rolling in the grain boundary contribute to the role of grain refinement. The results show that the grain size is increasing at 1050 ~1130 ℃ in the same holding time of 15min, and the grain size is growing at 1100 ℃ with the increase of the holding time. Through the analysis and discussion of the results, it is believed that the control of the heat treatment temperature, the holding time and the further study of the "grain boundary pinning effect" through the homogenization technology will help to control the grain size of the Cr4Mo4V steel after heat treatment and control the retained austenite, And thus to improve the service life of high temperature bearings and product stability.

Key words: high temperature bearing steel, Cr4Mo4V, bearing steel, grain size

穿水冷却对 GCr15 轴承钢低倍酸蚀孔洞的影响

韩怀宾 [1,2]，赵宪明 [1]，王　维 [2]，万长杰 [2]，虞学庆 [2]

（1. 东北大学轧制技术及连轧自动化国家重点实验室，辽宁沈阳　110819；

2. 河南济源钢铁（集团）有限公司，河南济源　459000）

摘　要：通过 4 种穿水冷却工艺生产 ϕ20mm 热轧 GCr15 轴承钢棒材，研究了穿水冷却对轴承钢低倍酸蚀孔洞组织的影响。结果表明：终轧前后均采用穿水冷却，控制终轧温度在 800~860℃，上冷床返红温度为 620℃，能明显减少热轧及热处理轴承钢中的酸蚀孔洞。同时也验证了酸蚀孔洞的形成与热轧轴承钢中网状先共析碳化物的形成密切相关。

关键词：GCr15 轴承钢，穿水冷却，热轧，先共析碳化物，酸蚀孔洞

Effect of Through-water Cooling on Macrostructure Acid Holes of GCr15 Bearing Steel

Han Huaibin[1,2], Zhao Xianming[1], Wang Wei[2], Wan Changjie[2], Yu Xueqing[2]

(1. The State Key Laboratory of Rolling and Automation, Northeastern University, Shenyang 110819, China;

2. Henan Jiyuan Iron and Steel Group Co., Ltd., Jiyuan 459000, China)

Abstract: The ϕ20mm hot-rolled GCr15 bearing steels were obtained by four through-water cooling processes, the effect of through-water cooling on macrostructures acid hole of hot-rolled bearing steel was investigated. The results showed that the through-water cooling both before and after finish rolling was carried out to reach the finish rolling temperature 800~860℃ and the re-reddening temperature 620℃, then the acid holes in hot-rolled and heat-treated bearing steels after acid pickling both decreased. It is also proved that the formation of acid holes is closely related to the formation of proeutectoid carbides in hot rolled bearing steels.

Key words: GCr15 bearing steel, through-water cooling, hot-rolling, proeutectoid carbide, acid holes

GCr15 热轧圆钢低倍中心孔洞形成机理的研究

肖　超，王　超，柯加祥，袁长波，张　虎，田文庆

（青岛特殊钢铁有限公司中特研究院青钢分院，山东青岛　266409）

摘　要：240mm×300mm 断面 GCr15 连铸坯轧制 Φ80mm 以上规格热轧圆钢，低倍中心有孔洞出现，经过对热轧圆钢以及连铸坯横、纵截面低倍和金相组织进行系统观察，成功找到了中心孔洞的形成机理，并分两个方面采取相应措施，一是对连铸参数进行优化，中心缩孔的尺寸成功控制到<1mm，最大 C 含量仅为 1.08%，二是对加热炉的加热制度进行优化，高温段温度控制到 1200～1240℃，扩散时间>3h，碳化物和显微孔隙与优化前控制水平相同，200 和 500 倍下显微孔洞的数量却明显减少，采用优化完成之后的参数，轧制多批 Φ80mm～Φ90mm 规格的圆钢，低倍

合格率达到100%。

关键词：轴承钢，低倍，中心孔洞，连铸，加热炉

Study on Formation Mechanism of the Central Holes in Macrostructure for Hot Rolled Round Steel GCr15

Xiao Chao, Wang Chao, Ke Jiaxiang, Yuan Changbo, Zhang Hu, Tian Wenqing

(Qingdao Special Iron and Steel Co., Ltd. CITIC Pacific Special Steel Institute Qingdao Branch, Qingdao 266409, China)

Abstract: Hot rolled round steel whose size was more than Φ80mm, produced by using bloom GCr15 with the cross section 240mm×300mm，in which some holes were found on the macrostructure, through systematically observing the cross and longitudinal sections of round steel and bloom, the formation mechanism of central holes were found successfully, in the following, measures from two aspects were taken, the first that the continuous casting parameters were optimized, the sizes of the central shrinkage cavities were steadily controlled to less than 1mm, the maximum carbon content was only 1.08%, the second that heating systems of heating furnace were optimized, temperature range of high temperature zone was controlled to 1200-1240℃,diffusion time more than 3h, the control level of carbide and micro-void was the same as before, microscopic holes had reduce significantly at the 200 and 500 time, multi-batch round steels with the sizes of Φ80mm-Φ90mm were rolled using the optimized parameters, the qualified percent reached to 100%.

Key words: bearing steel, macrostructure, central hole, continuous casting, heating furnace

石钢轴承钢结晶器液面波动控制

杨立永，郝增林，陈良勇，王翠亮

（河钢石钢炼钢厂，河北石家庄　050031）

摘　要：通过采用结晶器自动加保护渣系统、调整水口插入深度、控制中包液面等措施，轴承钢结晶器液面波动范围从±5mm降低到±3mm，液面波动超标甩坯率从1.34%降低到0.3%以下。

关键词：轴承钢，液面波动，中包液面，自动加渣

Control of Liquid Level Fluctuation in Mould of Bearing Steel

Yang Liyong, Hao Zenglin, Chen Liangyong, Wang Cuiliang

(HBIS Group SHISteel Company, Shijiazhuang 050031, China)

Abstract: Through the use of mold automatic protective slag adding system and adjust the immersion depth of nozzle, control tundish liquid level, bearing steel mold level fluctuation range decreased from ±5mm to ±3mm, exceed the standard rejection level fluctuation rate from 1.34% reduced to less than 0.3%.

Key words: bearing steel, liquid level fluctuation, tundish liquid level, automatic slag adding

石钢 GCr15 轴承钢连铸坯偏析现状研究

高　晗，李反辉，范俊硕

（河北钢铁集团石钢公司，河北石家庄　050031）

摘　要：本文采用化学成分分析的方法对石钢轴承钢连铸小方坯横截面和纵切面上不同部位的碳含量分布进行分析，研究得出宏观偏析的规律，以求找出影响石钢轴承钢中心偏析的工艺因素。这对提高石钢控制轴承钢连铸小方坯中的宏观碳偏析水平，具有重要的实际意义。

关键词：轴承钢，连铸小方坯，中心偏析

Study on Macrosegregation of Billets in Continuous Casting of GCr15 Bearing Steel

Abstract: This paper uses the method of chemical analysis of carbon content in different parts of the distribution of Shi Steel Bearing Steel Billet cross section and longitudinal section analysis, research that macro segregation laws, in order to find out the influence of process factors on segregation of steel bearing center stone. It has important practical significance to improve the macro carbon segregation in continuous casting billet of controlling bearing steel.

Key words: bearing steel, continuous casting billet, central segregation

高品质轴承钢钛含量控制分析与实践

张海宁，孙玉春，李双居，赵瑞华

（河钢集团石钢公司，河北石家庄　050031）

摘　要：论述了高品质轴承钢中钛对轴承钢使用寿命的影响，通过分析高品质轴承钢生产过程中钛含量变化，结合热力学分析，确定了影响轴承钢钛含量变化的主要因素。分析了钢中钛与原材料中钛、初炼炉终点碳、炉渣中钛等方面的对应关系，制定并实施了钢中钛的控制措施，结果显示，石钢公司高品质轴承钢钛含量稳定在 0.0015% 以下，最低可达到 0.0009% 的水平。

关键词：轴承钢，钛含量，控制

The Research and Practice of the Controlling of Titanium Content in the High Quality Bearing Steel

Zhang Haining, Sun Yuchun, Li Shuangju, Zhao Ruihua

(HBIS SHISTEEL, Shijiazhuang 050031, China)

Abstract: This paper discusses the high quality bearing steel in titanium influence on service life of bearing steel, through the analysis of high quality bearing steel production in the process of titanium content changes, combined with the thermodynamics analysis, determined the main factors influencing the bearing steel titanium content changes. The corresponding relationship between titanium in steel and titanium in raw materials、convertor end point carbon and titanium in slag is analysised, formulation and implementation measures for the control of the titanium steel, results show that the steel company's high quality bearing steel titanium content is stable under 0.0015%, the lowest can reach 0.0009%.

Key words: bearing steel, Ti content, control

邢钢轴承钢 Al 含量的控制

霍志斌，曹晓思，赵彦岭，王建林，王秋坤

（邢台钢铁有限责任公司炼钢厂，河北邢台　054000）

摘　要：本文分析了炼钢厂生产轴承钢各工序铝含量的变化情况，并通过热力学计算指出影响钢水铝含量的因素，根据出钢氧、铝块加入量的实际数据回归出计算关系式，使钢水中酸溶铝含量得以稳定控制。

关键词：轴承钢，出钢氧，铝脱氧，酸溶铝

Aluminum Content Controlling of Bearing Steel in Xingtai Steel

Huo Zhibin, Cao Xiaosi, Zhao Yanling, Wang Jianlin, Wang Qiukun

Abstract: This paper analyzes the aluminum content changing of every process in the production of bearing steel, and through the thermodynamic calculation gives factors that affect the aluminum content, according to the actual data of tapping oxygen and aluminum addition, the calculation formula is regressed, which provides the stable controlling of acid soluble aluminum content in molten steel.

Key words: bearing steel, tapping oxygen content, aluminium deoxidization, acid soluble aluminium

GCr15 钢的球化退火及碳化物粒度控制的研究

江运宏，周　靖，马宝国

（宝钢股份特殊钢事业部，上海　200940）

摘　要：本文结合国内外相关的研究成果和实际生产经验，探讨了 GCr15 钢球化机制和球化退火工艺；从退火质量稳定性等角度出发，对比分析了 GCr15 钢棒线材的三种不同炉型的球化退火方式及退火工艺；进而对影响球化退火质量的多种因素作了深入研究，实现了碳化物粒度的有效控制，从而满足了用户不同的使用要求。

关键词：GCr15 钢，棒线材，球化退火，碳化物粒度

Discussion on Spheroidizing Annealing of GCr15 Bearing Steel

Jiang Yunhong, Zhou jing, Ma Baoguo

(Baoshan Iron & Steel Co., Ltd. Special Steel Business Unit, Shanghai 200940, China)

Abstract: The mechanism of spheroidzation and the selection of annealing process for GCr15 bear steel are discussed in the paper, according to the correlative research civil and abroad and the experience of the productive practice. In the same, analyze and compare the three main spheroidizing annealing mode of GCr15 bearing steel in the paper, based on the production quality stability after annealing. Then, explain the important effect imposed on the spheroidizing annealing quality improvement of bearing steel by the continuous advancement of annealing equipment condition. In the end, bring up the further development direction for the quality of bearing steel in forms of bar and wire civil after spheroidizing annealing.

Key words: GCr15 bearing steel, bar and wire, spheroidizing annealing, carbide particle size

石钢 GCr15 氧含量及大型夹杂物控制实践

马玉强，孙玉春，张 力，武 森

（石家庄钢铁有限责任公司，河北石家庄 050031）

摘 要：本文介绍了河钢石钢在生产 GCr15 轴承钢氧含量及大型非金属夹杂物控制的部分实践，通过过程工艺的不断优化改进，降低了 GCr15 轴承钢氧含量及钢中大型非金属夹杂物的级别和数量，提高了轴承钢的实物质量，满足了用户需求。

关键词：轴承钢，氧含量，非金属夹杂物

The Control of Oxygen Content and Large Inclusion of GCr15 Clean Bearing Steel at Shijiazhuang Iron & Steel Co., Ltd.

Ma Yuqiang, Sun Yuchun, Zhang Li, Wu Sen

(Shijiazhuang Iron & Steel Co., Ltd., Shijiazhuang 050031, China)

Abstract: This paper discusses Shijiazhuang Iron & Steel Co.,Ltd in the production of GCr15 bearing steel oxygen content and large non-metallic inclusion control part of the practice, through the process of optimization to improve continuously, reduces the GCr15 bearing steel oxygen content and the number of levels and the largest non-metallic inclusions in steel, improve the physical quality of bearing steel, meets the requirements of users.

Key words: bearing steel, oxygen content, non-metallic inclusion

10　电工钢

炼铁与原料

炼钢与连铸

轧制与热处理

表面与涂镀

金属材料深加工

钢铁材料

汽车钢

海洋工程用钢

轴承钢

★　电工钢

粉末冶金

非晶合金

高温合金

耐火材料

能源与环保

分析检测

冶金设备与工程技术

冶金自动化与智能管控

冶金技术经济

相变法制备 Fe-0.43Si-0.5Mn 电工钢时工艺参数对组织结构和磁性能的影响

杨　平，夏冬生，王金华，毛卫民

（北京科技大学材料学院，北京　100083）

摘　要：相变法可显著粗化无取向电工钢成品板晶粒尺寸和提高有利织构{100}、{110}的强度，是提高低牌号无取向钢磁性能的有效方法。本文在实验室条件下研究了 Fe-0.43Si-0.5Mn 电工钢在利用相变法中工艺参数对组织、织构和磁性能的影响规律。结果表明，增大氢气流量可粗化柱状晶尺寸和完善柱状晶组织，减低{111}织构的强度；奥氏体晶粒尺寸的增大使铁素体晶粒也增大，但柱状晶特征减弱。Σ3 关系晶界的出现多数情况下对{110}或{100}取向晶粒的形成不利。大的奥氏体晶粒和较快的冷却速率均有利于相变过程中 Σ3 关系晶界的形成。此电工钢中可得到 $P_{1.5/50}$=3.791W/kg，B_{50}=1.764T 的磁性能。

关键词：相变，电工钢，织构，磁性能，柱状晶

Influences of Processing Parameters on Microstructures, Textures and Magnetic Properties in a Fe-0.43Si-0.5Mn Electrical Steel Subjected to Phase Transformation Treatment

Yang Ping, Xia Dongsheng, Wang Jinhua, Mao Weimin

(School of Materials Science and Engineering, University of Science and Technology Beijing, Beijing 100083, China)

Abstract: The treatment through phase transformation can increase grain sizes and enhance favorite {100} or {110} textures in non-oriented steels significantly, thus it is an effective method for improving magnetic properties in low grade electrical steels. This work investigates the influence of processing parameters on microstructures, textures and magnetic properties in a laboratory-melt Fe-0.43Si-0.5Mn electrical steel by means of phase transformation treatment. The results show that the increase of hydrogen flow rates will coarsen columnar grain sizes and improve columnar grain morphology in addition to the decrease of {111} texture. The increase of austenitic grain sizes by longer time holding will increase transformed ferrite grain sizes, but weaken the feature of columnar grain structure. The occurrence of Σ3 grain boundary generally deteriorates the strengthening of {100} or {110} texture, and it is favored by large austenitic grain size and high cooling rate. The magnetic properties of $P_{1.5/50}$=3.791W/kg, B_{50}=1.764T are obtained in this steel.

Key words: phase transformation, electrical steel, texture, magnetic property, columnar grain

常化温度对薄规格无取向硅钢织构和高频磁性能影响

施立发，王立涛，何志坚，张乔英

（马鞍山钢铁股份有限公司技术中心，安徽马鞍山　243000）

摘　要： 本文研究了不同常化温度对含 3.0%Si 的 0.30mm 薄规格无取向硅钢的组织、织构和高频磁性能的影响规律。结果表明：随着常化温度的提升，常化态和再结晶退火态的晶粒尺寸增大，P_e 占铁损比例增大，$P_{1.0/400}$ 降低、B_{50} 升高；常化态的立方织构{100}<001>、Goss 织构、α 纤维织构组分逐步增强，对应的再结晶退火态有利织构组分相应增强；常化温度为 950℃时，退火态晶粒尺寸为 121μm、立方织构{100}<001>、Goss 织构比例最大，$P_{1.0/400}$ 达到 14.60W/kg。

关键词： 常化温度，高频磁性能，织构，薄规格无取向硅钢

Effect of Normalizing Temperature on the Texture and High Frequency Magnetic Property of Thin Non-oriented Silicon Steel

Shi Lifa, Wang Litao, He Zhijian, Zhang Qiaoying

(Technology Center, Maanshan Iron and Steel Co., Ltd., Ma'anshan 243000, China)

Abstract: this paper, the influence of different normalizing temperature on microstructure, texture and high frequency magnetic property of thin non-oriented silicon steel containing 3.0%Si has been studied. The results show that with increasing normalizing temperature, the average grain size of the normalizing and recrystallization annealed plate increases, and the ratio of Pein the total iron loss rises. The cubic texture{100} <001>, Goss texture, α-fiber texture of the normalized plate and thetextures of the corresponding recrystallization annealed plate are enhanced; In addition, when the hot band normalizedis 950℃, the grain size of annealed plate is 121μm and the proportion of favorable components of cubic texture {100} <001> andGoss texture is the largest, and the minimum of high frequency iron loss $P_{1.0/400}$ is up to 14.60W/kg.

Key words: silicon steel, iron loss, magnetic induction, texture, property of motor

Cu 及退火工艺对 6.5% Si 高硅钢组织及磁性能的影响

程朝阳[1,2]，刘　静[1]，骆忠汉[2]，向志东[1]

（1. 武汉科技大学　省部共建耐火材料与冶金国家重点实验室，湖北武汉　430081；

2. 国家硅钢工程技术研究中心，湖北武汉　430080）

摘　要： 本文研究了 Cu 含量及退火工艺对 6.5%Si 高硅钢组织及磁性能的影响。通过热轧、温轧工艺制备了厚度约为 0.3 mm 的不同 Cu 含量的 6.5%高硅钢薄板。经不同退火工艺处理后，研究了不同试样的组织及磁性能，试验结果表明：当 Cu 含量为 0.3%~0.5%时，可提高 6.5%Si 高硅钢的综合磁性能，而当 Cu 含量达到 1.0%时，会恶化其各项磁性指标而降低其综合磁性能。6.5%Si 高硅钢经低温短时间退火 $P_{15/50}$ 降低量约为 35%，经高温长时间退火 $P_{15/50}$ 降低量约为 43%，两种退火工艺对磁感影响均较小。添加适量 Cu（0.3%~0.5%）可促进 6.5%Si 高硅钢退火过程中晶粒长大，有利于降低其铁损；而当 Cu 含量过高时，晶粒长大受阻，导致铁损增大。

关键词： 6.5% Si 高硅钢，Cu 合金化，退火工艺，磁性能，组织

Effects of Cu and Annealing Treatment on the Microstructure and Magnetic Properties of 6.5%Si High Silicon Steel

Cheng Zhaoyang[1,2], Liu Jing[1], Luo Zhonghan[2], Xiang Zhidong[1]

(1. The State Key Laboratory of Refractories and Metallurgy, Wuhan University of Science and Technology, Wuhan 430081, China; 2. National Engineering Research Center for Silicon Steel, Wuhan 430080, China)

Abstract: The effects of Cu and annealing treatment on the microstructure and magnetic properties of 6.5%Si high silicon steel were investigated in this paper. 0.3 mm thickness 6.5%Si silicon steel sheets with various Cu contents were produced by hot-rolling and warm-rolling process. The microstructure and magnetic properties of the specimens after different annealing treatments were studied. It was found that the magnetic properties of 6.5%Si high silicon steel would be improved by adding 0.3%~0.5% Cu, while the Cu content was 1.0%, the magnetic properties of the steel would be deteriorated. The iron loss $P_{15/50}$ of the 6.5%Si high silicon steel could be reduced by 35% after annealing at low temperature for short time, while it would be reduced by 43% for annealing at high temperature for long time. The effect of annealing treatment on magnetic induction was very little. The addition of a suitable amount of Cu (0.3%~0.5%) could promote the grain growth of 6.5%Si high silicon steel during annealing treatment, which is beneficial for the reduction of iron loss. However, when the Cu content was too high, it had opposite effect.

Key words: 6.5% Si high silicon steel, Cu alloyed, annealing treatment, magnetic properties, microstructure

薄板坯无取向硅钢表面'夹渣'缺陷研究

乔浩浩

（本钢产品研究院，辽宁本溪　117000）

摘　要： 针对薄板坯无取向硅钢表面'夹渣'缺陷，成立攻关组进行技术研究，通过对缺陷形貌的检验分析以及对各工艺参数的调整，明确了缺陷的实质和成因，并成功消除了产品缺陷，提高了产品质量。

关键词： 薄板坯连，无取向硅钢，'夹渣'缺陷

Study On 'Entrapped Slag' Defect of the Thin Slab Non-oriented Silicon

Qiao Haohao

(The Product Research Institute of BX Steel, Benxi 117000, China)

Abstract: Against to the 'Entrapped Slag' Defect From The Surface of Non-oriented Silicon Steel Produced By Thin Slab, BX steel set up a working group for technology research, through analysis on defect morphology and adjusting the process parameters, the nature and causes of the defects have be cleard, and successfully eliminated the product defects ultimately, improved the quality of the products.

Key words: thin slab, oriented silicon steel, 'entrapped slag' defect

冷轧无取向硅钢 50BW470 生产实践

乔浩浩，李德君

（本钢产品研究院硅钢所，辽宁本溪　117000）

摘　要： 在没有一贯制技术和电磁搅拌及常化设备的前提下，利用薄板坯连铸连轧工艺生产硅钢的优势，开发生产冷轧无取向硅钢 50BW470。通过对化学成分和生产工艺的调整及优化，较好地避免了产品出现明显的瓦楞缺陷，产品电磁性能满足标准要求，达到同行业中上水平，为后续的研发工作奠定了基础。

关键词： 电磁搅拌，薄板坯连铸连轧，冷轧无取向硅钢，瓦楞缺陷，电磁性能

Production Application of the Cold-rollde Non-oriented Silicon Steel 50BW470

Qiao Haohao, Li Dejun

(The Product Research Institute of BX Steel, Benxi 117000, China)

Abstract: In the absence of consisten technique and electromagnetic stirring and constant equipment,Use the advantage of thin slab continuous casting and rolling process of silicon steel production, develop and product the cold-rolled non oriented silicon steel 50BW470. By adjusting and optimizing the chemical composition and production process,the products can avoid the obvious corrugated defects,the electromagnetic performance of the product meets the standard requirements, reaching the upper level of the industry, which lays the foundation for the follow-up research and development.

Key words: electromagnetic stirring, thin slab continuous casting and rolling, cold-rolled non oriented silicon steel, corrugated defect, electromagnetic properties

氧化铁粉中 CaO 杂质的分析与控制

李　慧，韩　晋，吴大银，张文康

（山西太钢不锈钢股份有限公司，山西太原　030003）

摘　要： 氧化铁粉是铁氧体磁性材料的主要原料。本文介绍了太钢冷轧硅钢厂盐酸再生生产工艺，分析了影响酸再生副产品氧化铁粉质量的因素，结合现场实际对其影响因素进行逐一排查。结果表明，CaO 是目前影响太钢氧化铁粉质量的主要因素，且新酸为铁粉中 Ca 元素的主要来源，通过调整补酸模式、控制酸损失率及增加酸循环次数等措施后，降低了铁粉中 CaO 的含量，其合格品率由原来的 77% 提高到 90%，有效提高了太钢氧化铁粉质量，使其达到生产软磁用粉的标准。

关键词： 磁性材料，酸再生，氧化铁粉，CaO

Analysis and Control of CaO in the Ferric Oxide Powder

Li Hui, Han Jin, Wu Dayin, Zhang Wenkang

(Cold-rolled Silicon Steel Mill, Shanxi Taigang Stainless Steel Co., Ltd., Taiyuan 030003, China)

Abstract: Ferric oxide powder is the main raw material for ferrite magnetic material. This article introduces the process of hydrochloric acid regeneration used by Taigang cold-rolled silicon steel mill. The influence factors of ferric oxide powder quality are analyzed and investigation was performed on these possible causes based on the production practice. The result shows CaO is the main factor affecting the quality of ferric oxide powder and the clean acid is the main source. The content of CaO is reduced by adjusting the acid supplement, controlling the acid loss, increasing times of acid cycle and some other measures. The eligibility of CaO in ferric oxide powder increases from 77% to 90%. The quality of the ferric oxide powder reaches the standard production of soft magnetic material.

Key words: magnetic material, hydrochloric acid regeneration, ferric oxide power, CaO

关于电工钢测量标准体系的思考

向　前[1,2]，吴开明[1]，黄　双[2]

（1. 武汉科技大学，湖北武汉　430081；2. 武汉钢铁有限公司质量检验中心，湖北武汉　430080）

摘　要：从电工钢的交、直流磁特性、表面涂层性能、机械性能及应用性能的测量标准出发，对当前相关测量标准在满足产品及应用方面存在的问题进行分析，形成了电工钢性能测量标准体系完善的相关建议：要关注 SST 单片试样交直流性能测量方法、0.10mm 以下薄带的交流磁性能测量方法以及变温状态或高频（10kHz）条件下的磁性能测量的研究，监控新制修订标准的实施情况以及正在实施标准的适应性，探索涂层附着性测量的定量化方法，探讨建立涡流磁性测量标准测量涂层厚度的必要性，开展体现电工钢应用性能的测量标准研究。

关键词：电工钢，测量标准，交、直流磁特性，表面涂层性能

Thoughts on Measuring Standard System of Electrical Steel

Xiang Qian[1,2], Wu kaiming[1], Huang Shuang[2]

(1. Wuhan University of Science and Technology, Wuhan 430081, China;

2. Quality Inspection Center of Wuhan Iron and Steel Co., Ltd., Wuhan 430080, China)

Abstract: Based on the measurement standard of AC and DC magnetic properties, surface coating performance, mechanical properties and application performance of electrical steel, the existing problems of measurement standards relevant to the product and application are analyzed, and formed some recommendations to improve the electrical steel performance measurement standard system: It is necessary to pay attention to study the measurement method of AC/DC performance of SST single sheet samples, and the AC magnetic properties of stripes below 0.10mm, and the magnetic properties under changed temperature or high-frequency (10kHz) conditions, monitor the implementation of new revision standards and the adaptability of standards being implemented, then explore the quantitative method of coating adhesion measurement and discuss the necessity of establishing a eddy current magnetic measurement standard to measure the thickness of the coating, and develop some research on measuring standard of application performance of electrical steel.

Key words: Electrical steel, measurement standards, AC and DC magnetic properties, surface coating performance

CSP 流程生产高牌号无取向硅钢析出物演变规律研究

夏雪兰，裴英豪，王立涛，刘青松

（马鞍山钢铁股份有限公司技术中心，安徽马鞍山　243000）

摘　要： 研究了 CSP 流程生产高无取向硅钢各工序间析出物的演变规律。结果表明，热轧过程是析出物最快速析出的工序，后续工序中析出物数量减小，析出物分布密度减小，析出物所占体积分数增大。从尺寸和数量上看，铸坯中尺寸较小的析出物数量较多，热轧板中析出物数量达到顶峰，且更加细小弥散，而常化、退火过程，析出物粗化长大。

关键词： CSP，高牌号无取向硅钢，析出物

Research on Evolution of Precipitates in High Grade Non- oriented Silicon Steel by CSP Process

Xia Xuelan, Pei Yinghao, Wang Litao, Liu Qingsong

(Technical Center, Maanshan Iron and Steel Co., Ltd., Maanshan 243000, China)

Abstract: The evolution of precipitates in high grade non-oriented silicon steel by CSP process was studied in this paper. The results show that the hot rolling process is the fastest precipitation process, and the number of precipitates decreases, the distribution density of precipitates decreases, and the volume fraction of precipitates increases. In terms of size and quantity, the smaller amount of precipitates in the slab, the number of precipitates in the hot-rolled plate reaches the peak, and is more fine and dispersed. The precipitates is coarsening and growing in the process of normalizing and annealing.

Key words: CSP, high grade non-oriented electrical steel, precipitates

电工钢单片检测标准方法中有效磁路相关问题解析

周　星，沈　杰，李建龙，唐　灵

（宝山钢铁股份有限公司制造管理部检化验中心，上海　200941）

摘　要： 本文解析了电工钢单片检测标准方法有效磁路与测量结果的关系及问题的背景，涉及国际电工委员会（IEC）磁性合金和钢技术委员会（TC68）对相关问题的讨论，提出了修改现行标准方法中有效磁路长度的建议，以及由此对测量结果的改善情况。期望对相关技术问题的综述和解析有助于对相关领域国内外标准化活动更好的了解。

关键词： 电工钢，单片法，标准化

Summary and Analysis of the Relative Problem about the Effective Magnetic Path Length in Standardizing Single Sheet Testing Methods for Electrical Steel

Zhou Xing, Shen Jie, Li Jianlong, Tang Ling

(Test, Inspection and Analysis Center, Dept. of Manufacture Mgmt., Baosteel Co.. Ltd., Shanghai 200941, China)

Abstract: The relative problem about the effective magnetic path length in standardizing single sheet testing methods for electrical steel and its effects on the testing results have been described in the view of practice. It's related to the discussions about this problem in Technical Committee n°68 (TC68, Magnetic Alloys and Steels), International Electrotechnical Commission (IEC). One proposal for changing the effective magnetic path length and the advantage changes on the testing results have been described. These will help to comprehend the international and national standard developing activities in the concerning area.

Key words: electrical steel, single sheet testing method, standardization

锆对无取向电工钢组织、织构及性能的影响

石文敏[1]，杨　光[1]，冯大军[1]，李秀龙[2]

（1. 国家硅钢工程技术研究中心，湖北武汉　430080；

2. 武钢有限公司硅钢事业部，湖北武汉　430080）

摘　要： 本文研究了 Zr%对无取向电工钢组织，织构，磁性能和机械性能的影响。结果表明 0.06wt.%Zr 即可对常化和退火后的组织产生明显抑制作用，过多的 Zr%对成品退火组织影响不大，而 $ZrFe_2$ 是成品中 Zr 的主要析出物形式。Zr%对{100}，{111}织构的影响最为显著，Zr%≤1.3wt.%时随着 Zr%的增加{100}显著增强，Zr%=0.6wt.%时{111}最强，不同 Zr%对特定取向晶粒的长大和团聚作用各不相同。 随着 Zr%的增加，铁损和磁感均呈恶化趋势，而屈服强度在 Zr%=0.6wt.%时大幅上升，硬度在 Zr%=0.6wt.%时略微下降后保持稳定。

关键词： 无取向电工钢，组织，织构，析出物，磁性能，机械性能

Effect of Zirconium on the Microstructure, Texture and Properties of Nonoriented Electrical Steel

Shi Wenmin[1], Yang Guang[1], Feng Dajun[1], Li Xiulong[2]

(1. National Research and Engineering Center for Silicon Steel, Wuhan　430080, China;

2. WuHan Limited company of iron and steel, Wuhan　430080, China)

Abstract: The effect of Zir conium on the microstructure,texture and properties of nonoriented electrical steel were investigated.The results show that the microstructure after normalizing and annealing are restrained by 0.06%Zr,the microstructure after finishing annealing are not affected by too much Zr. The $ZrFe_2$ is the main precipitate in the specimen after finishing annealing. The {100}, {111}texture are affected greatly by Zr, the {100}texture increase significantly when

Zr%≤1.3 wt.% while the {111}texture is strongest when Zr%=0.06wt.%. The effect of different Zr% on the growth and reunion of specific oriented grain is different. The iron loss and magnetic induction are deteriorated with the increase of Zr content, the tensile strength rises sharply and the vickers hardness decreases slightly when Zr%=0.06wt.%，then they remain stable.

Key words: non oriented electrical steel, microstructure, texture, precipitate, magnetic property, mechanical property

国内薄板坯连铸连轧生产电工钢的现状及研究进展

项　利[1]，陈圣林[2]，裴英豪[3]，骆忠汉[2]，仇圣桃[1]

（1. 钢铁研究总院连铸技术国家工程研究中心，北京　100081；2. 国家硅钢工程技术研究中心，湖北武汉　430080；3. 马鞍山钢铁股份有限公司技术中心，安徽马鞍山　243000）

摘　要： 文中介绍了国内薄板坯连铸连轧流程生产电工钢的生产现状，在此基础上提出了利用该流程生产电工钢存在的技术难点及其解决思路。然后对国内薄板坯连铸连轧流程生产电工钢的研究进展进行了概述，为该项技术在国内薄板坯连铸连轧企业中的应用提供参考。

关键词： 取向电工钢，无取向电工钢，薄板坯连铸连轧，技术难点

Domestic Current Status and Research Progress of Producing Electrical Steel by Thin Slab Casting and Rolling Process

Xiang Li[1], Chen Shenglin[2], Pei Yinghao[3], Luo Zhonghan[2], Qiu Shengtao[1]

(1. National Engineering Research Center of Continuous Casting Technology, Central Iron and Steel Research Institute, Beijing 100081, China; 2. National Engineering Research Center for Silicon Steel, Wuhan Iron & Steel (Group) Corp, Wuhan 430080, China; 3. Technical Center, Maanshan Iron and Steel Co., Ltd., Ma'anshan 243000, China)

Abstract: Basing on the domestic current status of producing electrical steel by thin slab casting and rolling process introduced in the paper, the technology difficulties and solving ideas were put forward.　Then, domestic research progress of producing electrical steel by thin slab casting and rolling process was described for providing a reference to the iron and steel company.

Key words: grain oriented electrical steel, non-oriented electrical steel, thin slab casting and rolling process, technology difficulties

供西门子高速冲床用 50BW600 生产工艺研究

李毅伟

（本钢产品研究院，辽宁本溪　117000）

摘　要：本文研究了高速冲床 50BW600 冷轧无取向硅钢的生产工艺技术，对于影响产品的力学性能和电磁性能的各项因素进行分析，通过对金相组织和织构等组织结构检验分析，找出生产工艺对显微组织和织构的影响。通过生产工艺的优化，使产品的力学性能和电磁性能达到最佳，保证产品的冲压性能和电机特性。

关键词：显微组织，织构，力学性能，电磁性能

Study on Production Process of 50BW600 for High Speed Punch

Li Yi wei

(Product research institute of BX Steel, Benxi 117000, China)

Abstract: In this paper, we study the Cold rolled non oriented silicon steel production technology of 50BW600 for high speed punch, through the analysis of microstructure and texture of organizational structure inspection, find out effects of production process on the microstructure and texture. By optimizing the production process, the mechanical properties and electromagnetic properties of the product to achieve the best, ensure the characteristics of stamping performance and Motor characteristics.

Key words: macrostructure, texture, mechanical properties, electromagnetic properties

激光划线（刻痕）技术在电工钢表面的应用

Rauscher P.[1]，Hauptmann J.[1]，Müller E.[2]，丁元东[3]

（1. 激光烧蚀与切割部门，弗劳恩霍夫研究所 IWS 分所，德累斯顿 德国；2. Rofin-Sinar 激光公司，汉堡 德国；3. 德商罗芬激光技术(上海)有限公司，上海 中国）

摘　要：本文介绍了对取向硅钢的磁畴进行细化以提高磁性能的激光划线技术。此技术以工业化应用为背景，选用高功率连续波激光器。为满足工业化应用需求，尤其是很高的钢带运行速度，开发了一种基于扫描振镜技术的激光束传导系统。

关键词：激光划线，磁畴，软磁性材料

Aspects of Application of Laser Scribing in Industry

Rauscher P.[1], Hauptmann J.[1], Müller E.[2], Ding Yuandong[3]

(1. Laser Ablation and Cutting, Fraunhofer IWS, Dresden Germany; 2. Rofin-Sinar Laser GmbH, Hamburg Germany; 3. Rofin-Baasel China Co., Ltd., Shanghai China)

Abstract: In this article we report about the improvement of the magnetic properties of grain oriented electrical steel by using laser scribing as one possible method to refine the magnetic domains. The method is discussed against the background of using high power continuous wave laser beam sources and the possibility to transfer it into the industrial environment. Therefore, a laser beam deflection system based on so-called galvanometer scanner was developed for continuous moving material that fulfills the demands of the industrial application, especially the high coil speed.

Key words: laser scribing, magnetic domains, soft magnetic materials

浅析抗氧化剂对中温炭套及结瘤的影响

何明生[1]，谢国华[2]，龚学成[2]，周旺枝[1]，张　敬[2]，许　建[2]

（1. 武汉钢铁有限公司研究院，湖北武汉　430080；

2. 武汉钢铁有限公司硅钢部，湖北武汉　430080）

摘　要： 硅铝和磷酸盐类抗氧化剂是国内外炭套厂家生产中温炭套普遍使用的抗氧化剂。通过对中温炭套表面粘附瘤子和嵌入瘤子的微观结构和成分分析，结果表明瘤子的主要成分来自于硅钢本身，其次为抗氧化剂和石墨粉。采用浸泡法对磷酸盐处理中温炭套的耐水性能进行了研究，结果表明其耐水性能较差。瘤子的形成与抗氧化剂、炉内高温以及特殊气氛条件有关。

关键词： 抗氧化剂，结瘤，中温炭套，硅钢

Effect of Antioxidant on Forming Buildups of Medium-temperature Carbon Sleeve

He Mingsheng[1], Xie Guohua[2], Gong Xuecheng[2], Zhou Wangzhi[1],

Zhang Jing[2], Xu Jian[2]

(1. R&D Center of Wuhan Iron & Steel Co., Ltd., Wuhan 430080, China; 2. Silicon Steel Division of Wuhan Iron & Steel Co., Ltd., Wuhan 430080, China)

Abstract: Carbon sleeve is the most important kind of hearth rolls. Aluminosilicate and Phosphate is widely used as an antioxidant in the production of medium-temperature carbon sleeve at home and abroad. Based on the investigation and study of the microstructure, topography and composition of buildups adhered to the surface of the medium-temperature carbon sleeve and buildups embedded in the surface of the sleeve, the results show that the main component of the buildup comes from silicon steel strip, the secondary component from the antioxidant and graphite powder of carbon sleeve. The water resistance of medium-temperature carbon sleeve treated with aluminosilicate and phosphates is examined by immersion method, and the results show that the resistance to water is poor. The forming buildup in or on the surface of the carbon sleeve is because of antioxidant, high temperature and special atmosphere in continuous annealing furnace.

Key words: antioxidant, forming buildup, medium-temperature carbon sleeve, silicon steel

高温隧道退火炉在取向电工钢生产中的应用及效果研究

刘鹏程，刘宝志

（包头市威丰稀土电磁材料股份有限公司，内蒙古包头　014000）

摘　要：企业降低能耗,创新设备技术,为降低生产成本、优化工艺技术、降低能源介质、提高产品质量,本文重点介绍高温隧道退火炉的研发过程及实践。

关键词：高温隧道退火炉,基本结构,研发过程,工作流程及生产工艺,主要特点及创新生产技术

The Application and Effect Research of High Temperature Tunnel Annealing Furnace in the Production of Grain-oriented Silicon Steel

Liu Pengcheng, Liu Baozhi

(Baotou Weifeng Rare Earth Electromagnetic Material Limited Liability Company, Baotou 014000, China)

Abstract: In order to reduce production cost, optimizing process technology, lowering the energy medium, improving the quality of products, the company has reduced energy consumption and made equipment technology innovation. This paper focused on introducing the R&D process and practice of the high temperature tunnel annealing furnace.

Key words: high temperature tunnel annealing furnace, the R&D process of basic structure, workflow and manufacturing technique, key feature and innovative production technology

制氮机富氧尾气回收在电工钢的加热炉、退火炉酸再生炉中的应用及研究

杨　松，王洪彬

（大庆开发区三春节能技术有限公司，黑龙江大庆　163316）

摘　要：通过对不同氧浓度和燃烧火焰及温度场的试验研究,并结合富氧燃烧技术机理分析了富氧燃烧技术的可行性和经济合理性；而富氧燃烧技术一直受制氧成本高和运行费用高等因素限制,制氮机富氧尾气回收系统的研发成功解决了富氧来源及运行费用高的问题,将制氮机富氧尾气回收后用于加热炉、退火炉、酸再生炉中进行富氧燃烧可实现增产节能的作用。

关键词：制氮机富氧尾气回收,加热炉,退火炉,酸再生炉,富氧燃烧技术,节能

The Oxygen Rich Tail Gas of the Nitrogen Making Machine is Recycled in the Heating Furnace and the Annealing Furnace of the Electric Steel Application and Research of Acid Regeneration Furnace

Yang Song, Wang Hongbin

(Daqing trispring Energy Saving Technology Co., Ltd., Daqing 163316, China)

Abstract: The feasibility and economic rationality of oxygen-enriched combustion technology were analyzed by the experiment of different oxygen concentration and combustion flame and temperature field. And rich oxygen combustion technology has been restricted by high oxygen generation cost and operation cost higher factor, oxygen enriched nitrogen making machine tail gas recovery system developed to resolve the sources of oxygen enrichment and high operation cost, after oxygen-enriched exhaust gas recycling nitrogen making machine is used for heating furnace, annealing furnace and the regenerative furnace acid oxygen-enriched combustion increase productivity and energy saving can be realized.

Key words: The oxygen combustion technology of fuoxic acid regeneration furnace is energy saving

我国高磁感取向电工钢生产技术及供给现状的研究

陈　卓

（中国金属学会电工钢分会，湖北武汉　430080）

摘　要： 高磁感取向电工钢是我国电力工业节能减排和提升变压器产品能效的重要原材料产品之一。近年来，一直受到国家和电工钢产业以及变压器行业发展的关注。笔者重点就我国高磁感取向电工钢生产技术现状、现有生产能力、国内需求等方面进行研究并提出了未来的工作方向及目标。

关键词： HIB 钢特点，生产技术，能力，需求现状分析

The Research on the Production Technology and the Current Supply Situation of High Magnetic Induction Grain-oriented Silicon Steel

Chen Zhuo

(Electrical Steel of Chinese Society for Metal, Wuhan 430080, China)

Abstract: High Magnetic Induction Grain-oriented Silicon Steel （HIB steel） is one of the important raw materials for energy conservation of the power industry in our country and promoting energy efficiency of a transformer. In recent years, it has been brought into focus by the nation and the electrical steel industry, as well as the transformer industry. This paper is focus on the research on the production technology status of the HIB steel in China, the existing production capacity, the domestic demand, and so on. The future work direction and objective are also proposed.

Key words: characteristics of HIB steel, production technology, capacity, demand status analysis

IE4 超超高效率三相异步电动机系列产品的开发

黄　坚[1]，顾德军[1]，李光耀[1]，孙茂林[2]，胡志远[2]，王付兴[2]

（1. 上海电器科学研究所，上海　200063；2. 北京首钢股份有限公司，河北迁安　064404）

摘　要： IE4 为目前全球统一的最高的电动机效率等级，本文所介绍的 YE4 系列三相异步电动机为目前国际上首个符合 IE4 效率等级的全封闭（IP55）自扇冷散嵌绕组三相笼型感应电动机整马力系列产品。其效率、功率因数和噪声指标优于德国 Siemens 公司和巴西 Weg 公司同类产品，居国际领先水平。

关键词： IE4 超超高效率，三相异步电动机

The Development of the Three-phase Asynchronous Motor Series Products with IE4 Ultra-super High Efficiency

Huang Jian[1], Gu Dejun[1], Li Guangyao[1], Sun Maolin[2], Hu Zhiyuan[2], Wang Fuxing[2]

(1. Shanghai electrical sicence institute, Shanghai 200063, China; 2. Shougang Ltd., Qian'an 064404, China)

Abstract: IE4 is the highest motor efficiency grade in the world at present. The YE4 series three-phase asynchronous motor introduced in this paper is the first three-phase cage induction motor integral horse-power series product of totally-enclosed (IP55) self-fan cooled dispersion who meets IE4 efficiency grade of worldwide now. Its efficiency, power factor and noise index are better than the similar products of German Siemens company and Brazil Weg company, it has reached international advanced level.

Key words: IE4 ultra-super high efficiency, three-phase asynchronous motor

无取向电工钢环保半无机涂层的绝缘规律研究

段 辉[1], 李国明[2], 吴红兵[3]

（1. 淄博高新区耐火材料工程研究院，山东淄博 250086；2. 海军工程大学理学院，
湖北武汉 430033；3. 湖北武洲新材料科技有限公司，湖北鄂州 43600）

摘 要： 半无机涂层被广泛地应用于电工钢片的涂敷，由于温度和压力等因素，使服役电机的绝缘性能下降，严重地影响了电机的使用效率。本文研究了电阻率与聚合物、无机物和涂层组成结构之间的关系，为控制涂层电性能提供了理论依据，其成果也将为电力能源工业的可持续化发展提供高性能绝缘材料技术支撑。

关键词： 半无机涂层，电学性能

The Study on the Law of the Insulation of Semi-inorganic Coating for Electrical Steel

Duan Hui[1], Li Guoming[2], Wu Hongbing[3]

(1. Zibo hi-tech Zone Refractory Material Engineering Research Institute, Zibo 250086, China;
2. School of science, Naval University of Engineering, Wuhan 430033, China;
3. Hubei Wu Zhou New Materials Technology Ltd, Ezhou 436000, China)

Abstract: The semi-inorganic coating is widely used in electrical steel sheet of motor for its performance of insulation. However due to higher operating temperature and powder lead to insulation decreased of the motor, the efficiency of the motor is seriously affected. The project studies the microstructure and resistivity were investigated, the quantitative relationship between the insulation resistance and components of coating will be established. These studies will provide theoretical and applied basis for the semi-inorganic coating used in the large motors, the results will also provide sustainable development of high performance insulating materials for the electric energy industry.

Key words: semi-inorganic coating, electrical property

APL 机组使用陶瓷辊道代替耐热合金钢辊的研究

陈天龙，陈圣贤

（浙江中硅新材料有限公司，浙江嘉兴　314419）

摘　要：电工钢和不锈钢 APL 机组在生产中合金钢炉底辊的结瘤严重影响生产效率品质以及成本。为解决合金钢炉底辊结瘤和使用寿命短的问题，开发使用陶瓷辊道。经过设计和使用验证，陶瓷辊道的材料特性使其在高温下保持表面光滑，不结瘤，使用寿命是合金钢炉底辊的 4 倍长。

关键词：冷轧不锈钢钢带，退火酸洗线，炉底辊

Research on Fused Silica Rollers Replacing Alloy Rollers in APL

Chen Tianlong, Chen Shengxian

(Cencera Corporation, Jiaxing 314419, China)

Abstract: The build-up on the surface of the roller in APL causes pimples in the surface of the strip. The only contact of the steel strip with pick-up formation drives to a modification of the electrical characteristics. The introduction of fused silica rollers tremendously reduces the pick-up defects on the surface of the steel strip.

Key words: cold rolling strip, anneal-pickling line, roller hearth furnace

以氮化硅为抑制剂的取向电磁钢板氮化均匀性对磁性能的影响

井上博贵，新垣之启等（特开 2015-172223）

（JFE，日本）

王　杰，潘　妮（译、摘、编）

（原武汉钢铁股份有限公司，湖北武汉　430081；武汉科技大学，湖北武汉　430080）

摘　要：采用以氮化硅为主体抑制剂，氮化处理后的氮化量均匀性也是一项主要的指标，其均匀性的评价方法采用 SEM 观察钢板的断面，计算出最大长度大于 $0.1\mu m$ 析出物的个数。被观察的区域的析出物数量分别记为"N 表面"和"N 中心"（N 代表析出物的个数），并且沿钢板轧制方向选择 5 个点加以观测并计算出平均数，并对 N 表面和 N 中心加以比较，即 N 表面/N 中心的比值在 10 以下的区间内，其高温退火后测定的 B800（T）都很高，磁性能良好。

关键词：氮化硅抑制剂，氮化处理，均匀性评价，高磁感

The Nitriding Homogeneity of the Oriented Silicon Steel with Silicon Nitride Inhibitor

Wang Jie, Pan Ni (translate)

(Wuhan Steel Limited Liability Company, Wuhan 430081, China; Wuhan University
of Science and Technology, Wuhan 430080, China)

Abstract: As silicon nitride is the main inhibitor, the homogeneity of the amount of nitrogen after nitrogen treatment is a key index, its evaluation methodology is through observing the cross section of the steel by SEM, and counting the number of the precipitates whose the longest length are more than 0.1 μm. The number of the precipitates of different areas is labelled as "$N_{surface}$" and "N_{center}" (N is the number of the precipitates). There were five points observed along the rolling direction of the steel, and the average number of the precipitates was counted. Comparing $N_{surface}$ with N_{center}, when the ratio of $N_{surface}/N_{center}$ was less than 10, the B_{800} of the steel was very high, the magnetic property of the steel was excellent.

Key words: silicon nitride inhibitor, nitrogen treatment, homogeneity evaluation, high magnetic induction

油电混合/电动汽车驱动马达用硅钢片

(日本钢铁研究所 加工技术研究开发中心 藪本政男 開道力 脇坂岳顕等)
李军锋（译、摘、编）

（深圳市荣昌盛金属科技有限公司，广东深圳　518100）

摘　要： 硅钢片用于油电混合汽车(HEV)及电动汽车(EV)驱动马达铁芯，对 HEV/EV 的性能造成影响。为了实现驱动马达的小型轻量化、高效率,硅钢片中有很多的要求。满足这些要求，在进行最佳性能硅钢片开发的同时，优质的技术应用也是必要的。

关键词： 驱动马达，硅钢片

Electrical Steel Sheet for Traction Motor of Hybrid/Electric Vehicles

Yabumoto Masao, Kado Chkara, Wakisaka Tkakeaki, Kubota Takeshi, Suzuki Noriyuki
Li Junfeng (translate)

Abstract: Electrical steel sheet is used for the core of traction motors of hybrid electric vehicles (HEV)andelectric vehicles(EV), and affects performance of HEV/EV. In order to make motors to be small, light, powerful and efficient, there are many demands to electrical steel sheet. To realize these demands, development of electrical sheets with suitable qualities and suitable application techniques of electrical steel sheet are required as well.

Key words: HEV/HV motor, electrical steel sheet

2%Si 无取向硅钢组织研究

刘书明[1]，樊立峰[1,2]，陆　斌[2]，何建中[2]，张　昭[2]

（1. 内蒙古工业大学材料科学与工程学院，内蒙古呼和浩特　010051；
2. 内蒙古包钢钢联股份有限公司，内蒙古包头　014010）

摘　要： 研究了 2%Si 无取向硅钢各工艺阶段的组织和夹杂物。采用光学显微镜观察各工艺阶段组织，扫描电镜观察夹杂物。结果表明：（1）提高热轧后两道次压下率和终轧温度使热轧板等轴晶区所占比例增多，从 29%增多到 44%，中心纤维状组织减少，从 19%减少到 16%；（2）常化温度从 950℃提高到 1000℃使成品板平均晶粒尺寸从 106μm 增大到 111μm；（3）钢中主要夹杂物为 AlN 和 MnS，平均尺寸达到 4.8μm。

关键词： 无取向硅钢，组织，夹杂物，热轧，退火

Study on Microstructure of 2% Si Non-oriented Silicon Steel

Liu ShuMing[1], Fan LiFeng[1,2], Lu Bin[2], He JianZhong[2], Zhang Zhao[2]

(1. School of Materials Science and Engineering, Inner Mongolia University of Technology, Hohhot 010051, China; 2. Bao tou Steel (Union) Co., Ltd., Baotou 014010, China)

Abstract: The microstructures and inclusions in the different process stages of 2% Si non-oriented silicon steel were investigated, using optical microscope observing microstrctures and SEM observing inclusions. The results show that (1) the proportion of equiaxed grains increased from 29% to 44% with the increasing of the hot rolling reduction ratio in the last two passes and finishing rolling temperature, meanwhile the proportion of center fibrous tissues decreased from 19% to 16%; (2) the average grain size of the finished plate increased from 106μm to 111μm with the increasing of normalizing temperature from 950℃ to 1000℃; (3) The major inclusions observed in the test steel are AlN and MnS, the average size reached 4.8μm.

Key words: non-oriented silicon steel, organization, inclusions, hot rolling, annealing

0.18mm 取向硅钢不同角度高频磁性能研究

沈侃毅，马长松

（宝山钢铁股份有限公司中央研究院，上海　201900）

摘　要： 本文利用 500×500mm 大单片研究了 0.18mm 取向硅钢刻痕产品不同角度高频磁性能。磁感应强度选择三个水平 0.5T（低），1.0T（中），1.5T（高）和不同频率下（100Hz，200Hz，400Hz，1000Hz 和 2000Hz）测定磁场强度和损耗，结果表明，不同角度的磁化特性和损耗特性有明显不同，这主要根源于磁化过程中磁畴的变化。

关键词： 取向硅钢，高频性能，各向异性

Research on High-frequency Magnetic Properties in Different Angles of 0.18mm Grain Oriented Electrical Steels

Shen Kanyi, Ma Changsong

(Central Research Institute, Baoshan Iron & Steel Co., Ltd., Shanghai 201900, China)

Abstract: The high-frequency magnetic properties of 0.18mm grain oriented silicon steel products in different angles were studied by using 500×500mm single sheet test. The magnetic field strength and iron loss were measured under magnetic induction intensity at three levels of 0.5T (low), 1.0T (middle), 1.5T (high), respectively and at different frequencies 100Hz, 200Hz, 400Hz, 1000Hz, 2000Hz, respectively. The results show that the magnetization and iron loss characteristics are quit different, which mainly due to irreversible movement and rotation of magnetic domains.

Key words: grain oriented electrical steels, high-frequency magnetic properties, anisotropy

炉底辊对硅钢表面质量及生产的影响

何明生[1]，彭守军[2]，金 犁[2]，龚学成[2]，周旺枝[1]，张 敬[2]

（1. 武汉钢铁有限公司研究院，湖北武汉 430080；
2. 武汉钢铁有限公司硅钢部，湖北武汉 430080）

摘 要： 结合硅钢连续退火炉内炉底辊的运行环境和条件（如温度、露点、气氛等），对硅钢实际生产过程中炉底辊的破损形式及原因进行了分析，探讨了不同种类炉底辊在使用过程中存在的主要问题，提出了硅钢连退炉内炉底辊的选用原则。

关键词： 炉底辊，炭套，陶瓷辊，金属辊，硅钢，连续退火

Effect of Hearth Roll on Surface Quality of Silicon Steel Strip

He Mingsheng[1], Pen Shoujun[2], Jin Li[2], Gong Xuecheng[2], Zhou Wangzhi[1], Zhang Jing[2]

(1. R&D Center of Wuhan Iron & Steel Co., Ltd., Wuhan 430080, China;
2. Silicon Steel Division of Wuhan Iron & Steel Co., Ltd., Wuhan 430080, China)

Abstract: The quality of hearth roll in continuous annealing furnace is one of the most important factors that affect production efficiency and surface quality of silicon steel products. According to the operating environment and conditions (such as temperature, dew point, atmosphere, etc.) in continuous annealing furnace for silicon steel, the breakage forms and causes of the hearth roll during production are analyzed, and the main problems of the different types of hearth roll in the process of use are discussed. Some selection principles of hearth roll in continuous annealing furnace for silicon steel are proposed.

Key words: hearth roll, carbon sleeve, ceramic roller, metal roller, silicon steel, continuous annealing

磁轭钢直流磁性能检测技术的探讨

李关仁，周新华

（长沙天恒测控技术有限公司，湖南长沙　410100）

摘　要： 磁轭钢板材是用于大型水力发电机转子的关键部件，随着水电机组的大型化发展，具有良好磁性能的磁轭钢有着快速增长的市场需求。而由于磁轭钢具有在很大的磁化场下才会饱和磁化的特点，目前关于其饱和磁极化强度的检测方法还没有完整的国标与国际标准。本文主要探讨采用在 GB/T 3655—2008 的基础上使用爱泼斯坦方圈以及在 GB/T 13012—2008 的基础上使用 A 类磁导计测量其饱和磁极化强度的方法。

关键词： 磁轭钢，饱和磁极化强度，爱泼斯坦方圈，A 类磁导计

Technical Discussion on DC Magnetic Properties of Yoke Steel

Li Guanren, Zhou Xinhua

(Tunkia Co.,Ltd., Changsha 410100, China)

Abstract: Yoke steel plate is suitable for the key components of large - scale hydraulic generator rotor. With the increasing rapidly development of hydropower unit, the yoke steel with good magnetic performance has a huge market demand. Because the yoke steel has the characteristics of saturation magnetization in the large magnetization field, there is no full GB standard and IEC standard for the measurement method of its saturation magnetic polarization. This paper focuses mostly on discussing the methods of using Epstein frame based on GB/T 3655—2008 and using Class A permeameter based on GB/T 13012—2008 to measure its saturation magnetic polarization.

Key words: yoke steel, saturation magnetic polarization, epstein frame, class a permeameter

冷连轧前滑模型应用及其调整的研究

曹　静，范正军，刘玉金，陈　伟，刘　磊

（北京首钢股份有限公司，河北迁安　064400）

摘　要： 冷连轧数学模型是影响工控效果的关键因素，数学模型的精度与产品质量密切相关。冷连轧过程中前滑的波动会引起轧制速度的变化和机架间的张力波动，使得整个连轧过程处于不稳定的状态，存在轧制风险。为实现大张力下稳定的连续轧制控制，在保证秒流量相等的原则下，根据各机架不同负荷分配，给出匹配的精确设定速度对自动化控制有重要指导意义，而前滑的预算精度将直接影响速度设定的精度。本文介绍在实际生产中如何利用前滑设定模型、前滑自适应实现不同批次、不同规格带钢前滑的准确设定，并对前滑设定模型进行优化，解耦与摩擦相关的前滑补偿系数，实现摩擦系数设定改变时，对前滑设定的迅速调整。

关键词： 冷连轧，前滑模型，补偿系数

Study of Forward Slip Model Application and Adjustment for Tandem Cold Mill

Cao Jing, Fan ZhengJun, Liu YuJin, Chen Wei, Liu Lei

(Beijing Shougang Co., Ltd., Qian'an 064400, China)

Abstract: The mathematical model of tandem cold mill is the critical factor, which has an effect on industrial control system observably. The accuracy of the mathematical model is closely related to the quality of the products. In the process of the Tandem Cold Rolling, the fluctuation of the forward slip will lead to the rolling speed changing and the fluctuation of the tension between neighbouring stands, which makes the whole rolling process unstable and having rolling risks. In order to achieve stable continuous rolling control, under the large tension, ensuring in the principle of the Mass Flow is equally. According to the different load distribution of each stand, matching the accurate setting speed is important to the automatic control system. While the accuracy of the forward slip pre-calculation will affect the speed setting accuracy directly. In this paper we will introduce how to use the forward slip model and self-adaptive model to make different batches or specifications materials having more accurately setting value. Then we optimize the forward slip model to decouple the friction compensation coefficient. Once changing the friction coefficient, the forward slip setting will be set quickly in the end.

Key words: tandem cold mill, forward slip model, compensation coefficient

冷轧无取向电工钢变形抗力模型研究

裴英豪，夏雪兰，王立涛，杜　军

（马鞍山钢铁股份有限公司技术中心，安徽马鞍山　243000）

摘　要：针对薄板坯连铸连轧流程生产的3种不同成分无取向硅钢，利用 MTS880 拉伸试验机对冷轧变形抗力进行了实验测定，回归出相应的冷轧变形抗力公式，并对公式判定系数分析和回归计算结果与实际测试结果进行比较，结果表明此模型具有良好的曲线拟合特性。

关键词：变形抗力，无取向硅钢，冷轧，薄板坯连铸连轧，数学模型

Research on Deformation Resistance Model of Cold Rolled Non-oriented Electrical Steel

Pei Yinghao, Xia Xuelan, Wang Litao, Du Jun

(Technical Center, Maanshan Iron and Steel Co., Ltd., Maanshan 243000, China)

Abstract: The deformation resistance model of three different content of Si+Al non-oriented electrical steel produced by thin slab casting and rolling were investigated in this paper by use of MTS880 cupping machine. The formula of cold rolling deformation resistance is regressed accordingly. Through compare the coefficient analysis and regression calculation results of this formula with the actual test results, The results show that this model was proved to have good curve fitting characteristics.

Key words: deformation resistance, non-oriented electrical steel, cold rolling, thin slab casting and rolling, mathematic model

11 粉末冶金

炼铁与原料
炼钢与连铸
轧制与热处理
表面与涂镀
金属材料深加工
钢铁材料
汽车钢
海洋工程用钢
轴承钢
电工钢
★ 粉末冶金
非晶合金
高温合金
耐火材料
能源与环保
分析检测
冶金设备与工程技术
冶金自动化与智能管控
冶金技术经济

CNTs 与 Al₂O₃ 颗粒协同增强铜基复合材料

周　川，路　新，贾成厂，刘博文，吴　超，洪　逍

（北京科技大学　新材料技术研究院，北京　100083）

摘　要： 采用高能球磨、分子水平混合、行星球磨结合放电等离子体烧结的工艺制备了 Cu-CNTs-Al₂O₃ 复合材料。研究了单独添加和同时添加的 CNTs、Al₂O₃ 颗粒在 Cu 基体中的分布状态，以及对复合材料致密度、硬度、强度和导电性能的影响。结果表明：单独添加 CNTs、Al₂O₃ 颗粒，复合材料力学性能改善不显著，且电导率下降较明显。Cu-1.0CNTs-0.5Al₂O₃ 复合材料，综合性能最佳，其硬度、抗拉强度和断后伸长率分别为 131 HV、314 MPa、7.5%，电导率为 47.9 MS/m。复合材料性能的提高主要归因于同时添加的 CNTs、Al₂O₃ 颗粒显著改善了 Cu 基体中强化相的分布，从而提高了对位错的钉扎作用。

关键词： 混杂增强，铜基复合材料，碳纳米管，氧化铝

刹车速度对铜基粉末冶金摩擦材料性能影响的研究

刘联军[1]，李　利[1]，吴其俊[2]，车明超[1]

（1. 西安航空制动科技有限公司粉末冶金厂，陕西兴平　713106；

2. 中国舰船研究设计中心，湖北武汉　430064）

摘　要： 以粉末冶金法制备铜基粉末冶金摩擦材料。采用洛氏硬度计和夏比冲击试验机对摩擦材料的力学性能进行表征，利用 MM-3000 型摩擦磨损性能试验台研究了刹车速度对其摩擦磨损性能的影响，并借助电子扫描显微镜（SEM）表征了摩擦材料的微观形貌。研究表明：铜基粉末冶金摩擦材料的摩擦磨损性能与刹车速度密切相关，随着刹车速度的增大，摩擦吸收功率近似线性增长，而摩擦系数呈先增大后减小的趋势。并且在高速刹车条件下，铜基体自身发生软化会破坏摩擦材料表面形成的氧化膜，降低了分子键的抗剪切强度，从而增大了磨损量。

关键词： 粉末冶金，摩擦材料，刹车速度，摩擦磨损

Effects of Braking Velocity on Friction Properties of the Cu-based Powder Metallurgy Friction Material

Liu Lianjun[1], Li Li[1], Wu Qijun[2], Che Mingchao[1]

(1. Xi'an aviation brake technology Co. Ltd., Xingping 713106, China;

2. China Ship Development and Design Center, Wuhan 430064, China)

Abstract: Cu-based powder metallurgy friction material is prepared by powder metallurgy method. Mechanical properties are characterized using the hardness tester and charpy impact test machine, the effects of braking velocity on friction and wear properties are researched by the machine of MM3000, the micromorphology of friction material is characterized by scanning electron microscopy (SEM). The results show that the friction and wear properties of Cu-based powder metallurgy friction material are related to braking velocity, and friction absorption powers present almost linear growth and friction coefficients increase at frist and then decrease with increase of the braking velocity. Softening of copper matrix can frature oxide film forming on the surface of friction material, which reduces the shear strength of molecular bond and increases wear loss.

Key words: braking velocity, powder metallurgy, friction material, friction and wear

纳米 AlN 颗粒弥散增强铜基复合材料

刘佳思[1]，纪　箴[1]，贾成厂[2]，张一帆[1]，刘博文[2]，周　川[2]

（1. 北京科技大学，材料科学与工程学院，北京　100083；

2. 北京科技大学，新材料技术研究院，北京　100083）

摘　要： 本文采用高能球磨结合放电等离子烧结方法制备了不同 AlN 质量分数的 AlN/Cu 复合材料。探究了 AlN 含量对 AlN/Cu 复合材料微观形貌、致密度、显微维氏硬度、拉伸及导电性能的影响。结果表明：当 AlN 质量分数＜1.0%时，随着 AlN 质量分数的提高，复合材料的硬度、抗拉强度提高，断后伸长率、电导率降低。但当 AlN 质量分数为 1.0%时，AlN/Cu 复合材料致密度为 97.8%，显微硬度和抗拉强度分别达到了 119.5HV 和 259.7MPa，电导率为 49.30 MS/m，综合性能达到最优。

关键词： 放电等离子烧结，弥散增强 Cu 基复合材料，维氏硬度，抗拉强度，电导率

Nano-AlN Particle Dispersion Strengthened Cu-matrix Composite

Abstract: Cu-matrix composite dispersion strengthened by AlN with different AlN mass fraction were prepared by mechanical ball milling and spark plasma sintering. The effects of different AlN contents on the microstructure, density, hardness, tensile properties and electrical conductivity of AlN/Cu composites were investigated. The results show that as AlN mass fraction is less than 1.0% the hardness and tensile strength were improved, while the elongation and the conductivity decrease with the increase of AlN mass fraction. The densities, hardness, tensile strength and electrical conductivity of 1.0% AlN/Cu composites were 97.8%, 119.5HV and 259.7 MPa, and 49.30MS/m, respectively.

Key words: spark plasma sintering, dispersion strengthened Cu-matrix composite, Vickers-hardness, tensile strength, electrical conductivity

不同含量掺杂对 Al₂O₃ 陶瓷显微结构及介电性能的影响

洪　逍[1]，纪　箴[1]，贾成厂[2]，张一帆[1]，刘博文[2]，周　川[2]

（1. 北京科技大学材料科学与工程学院，北京　100083；

2. 北京科技大学新材料技术研究院，北京　100083）

摘　要： 利用放电等离子烧结制得了不同含量 Y₂O₃ 单独掺杂及不同含量 Y₂O₃/MgO 共同掺杂的 Al₂O₃ 微波介电陶瓷。本文研究发现掺杂含量的变化会影响 Al₂O₃ 陶瓷的显微结构，从而进一步影响其介电性能。Al₂O₃ 陶瓷的显微

结构如致密度和第二相的含量是影响其介电性能的主要因素。单独掺杂 0.25 wt%Y$_2$O$_3$ 时，Al$_2$O$_3$ 陶瓷得到最优的介电性能，介电常数 ε_r =11.3±0.2，介质损耗角 tanδ 稳定在 10^{-3} 数量级以内。同时掺杂 Y$_2$O$_3$ 和 MgO 能进一步改善其介电性能，当两者的添加量均为 0.25wt%时，得到最优值，介电常数 ε_r =10.2±0.2，介质损耗角 tanδ 稳定在 8×10^{-4} 以下。

关键词： Y$_2$O$_3$ 单独掺杂，Y$_2$O$_3$/MgO 共同掺杂，Al$_2$O$_3$ 陶瓷，显微结构，介电性能

烧结金属含油轴承的耐磨性

陈璐璐

（江苏海安赢球集团，江苏海安　226600）

摘　要： 耐磨性是烧结金属含油轴承的重要性能之一，对该类轴承的使用状况与寿命有着至关重要的影响。提高烧结金属含油轴承的耐磨性的主要措施是材料成分的优化。例如增加耐磨相，添加有用的合金元素，使用稀土氧化物等。还有制备工艺的优化，例如严格控制成形工艺参数，烧结温度与保温时间，使用合适的烧结气氛等。

关键词： 烧结金属含油轴承，耐磨性，成分，工艺

用于电子驻车制动系统的粉末冶金斜齿轮

丁　霞，陈　迪，彭景光，吴增强

（1. 上海汽车粉末冶金有限公司，上海　200072；
2. 上海粉末冶金汽车材料工程技术研究中心，上海　200072）

摘　要： 本文所开发的粉末冶金斜齿轮为电子驻车制动系统中涡杆传动中的关键粉末冶金零件，其内孔为花键非圆结构，成型难度相当大，本文论述了其结构和性能特点、材料选择和生产工艺等。通过合理的工艺制定，开发出符合实际工况要求、高精度、高强度、形状复杂的内孔为花键的斜齿轮。

关键词： 粉末冶金，斜齿轮，主动旋转，阴模旋转，烧结硬化

Powder Metallurgy Helical Gear for the Electronic Parking Brake System

Ding Xia, Chen Di, Peng Jingguang, Wu Zengqiang

(1. Shanghai Automotive Powder Metallurgy Co.,Ltd. , Shanghai 200072, China;
2. Shanghai Powder Metallurgy Automobile Material Engineering Research Center, Shanghai 200072, China)

Abstract: The powder metallurgy helical gear developed in this paper is the key part of the turbine drive in the electronic parking brake system. The inner hole of the part is non-circular structure with the spline, Which is very difficulty to form. The structure, performance, material selection and production process of it is introduced in this paper. By rational technology, the helical gear would be produced with high-accuracy, high-strength, complex-shaped and meeting the demand of working condition.

Key words: powder metallurgy, helical gear, active rotation, die rotation, sinter hardening

原始粉末颗粒边界对 FGH96 热变形再结晶的影响

王梦雅[1]，纪　箴[1]，田高峰[2]，贾成厂[1]，傅　豪[1]

（1. 北京科技大学材料科学与工程学院，北京　100083；
2. 北京航空材料研究院先进高温结构材料重点实验室，北京　100095）

摘　要： 通过 Gleeble-1500 热模拟试验机对热等静压成形制成的 FGH96 高温合金，在 1075℃、变形速率为 $0.001s^{-1}$，进行变形量为 70% 的热压缩试验。采用场发射扫描电镜、透射电镜和电子能谱等表征方法，研究合金原始颗粒边界（PPB）的组成，以及它对合金热变形再结晶的影响。结果表明：热等静压成形 FGH96 高温合金中的 PPB 主要由 Ti 的碳化物和 γ′ 相共同组成；PPB 对合金热变形再结晶形核起促进作用，对晶粒长大起阻碍作用，但在 PPB 密度较低的区域，再结晶晶界向前迁移时不受 PPB 影响，最终穿过 PPB 完成再结晶。

关键词： FGH96 高温合金，热压缩，PPB，再结晶

The Effect of Prior Particle Boundary on Recrystallization of FGH96 Superalloy duringhot Deformation

Wang Mengya[1], Ji Zhen[1], Tian Gaofeng[2], Jia Chengchang[1], Fu Hao[1]

(1. School of Materials Science and Engineering, University of Science and Technology Beijing, Beijing 100083, China; 2. Beijing Institute of Aeronautical Materials, Beijing 100095, China)

Abstract: Hot compression test of FGH96 superalloy formed by hot isostatic pressing was prepared at 1075 ℃ and a strain rate of $0.001s^{-1}$ with the deformation of 70% on a Gleeble-1500 thermo-simulation machine.The composition of the prior particle boundary (PPB) and its effect on the recrystallization of the alloy during hot deformationwere studied by field emission scanning electron microscopy, transmission electron microscopy and electron spectroscopy. The results show that the PPB of FGH96 superalloy is composed mainly of Ti carbide and γ' phase. The PPB plays a promotingrole in recrystallization nucleationand hinders the grain growth. But in the region with low PPB density, the grain boundarymigration is not affected by PPB, and finally through the PPB to complete the recrystallization.

Key words: FGH96 superalloy, hot compression, PPB, recrystallization

铁-硅烷软磁复合材料的电磁性能研究

李发长，李　一，李　楠，柳学全，李金普，丁存光

（安泰科技股份有限公司，北京　100081）

摘　要： 本文研究了铁-硅烷软磁复合材料的结构和软磁性能。SEM、EDS 分析、元素化学分析和面分布结果表明，硅烷包覆工艺能在铁粉表面生成一层很薄的完整均匀的包覆层。磁性能测量结果表明，硅烷添加量为 10wt% 时，所

得到的包覆铁粉磁芯具有优异的综合软磁性能。与传统的磷化铁粉相比，铁-硅烷软磁复合材料具有更优异的耐热性能，550℃热处理后，磁损耗无明显增大。当热处理温度从 500℃提高至 550℃，铁-硅烷软磁复合材料电阻率降低约 20%，但磷化铁粉磁芯电阻率降低超过 80%。

关键词：软磁复合材料，铁粉，硅烷，软磁性能

Study of the Magnetic and Electrical Properties of Iron-silane Soft Magnetic Composites

Li Fachang, Li Yi, Li Nan, Liu Xuequan, Li Jinpu, Ding Cunguang

(Advanced Technology & Materials Co., Ltd., Beijing 100081, China)

Abstract: This paper investigates the magnetic and structural properties of iron-silane soft magnetic composites(SMC). Scanning electron microscopy, energy dispersive spectrometer analysis, element chemical compostion and distribution maps show that the iron particles surface layer contains a thin layer of silane coating with high coverage of powders surface. Magnetic measurements show that when the addition level of silane was 10wt.%, the soft magnetic composites has excellent comprehensive electromagnetic properties. The silane insulation coating has a greater heat resistance than conventional phosphate insulation, which enables stress reliving annealing at higher temperature (550℃) without a large increase in magnetic loss. When the annealing temperature increased to 550℃ from 500℃, the electrical resistivity of the iron-silane soft magnetic composites decreased by about 20%, while the resistivity of the phosphated samples decreased by more than 80%.

Key words: soft magnetic composites, iron powder, silane, soft magnetic properties

镍基粉末高温合金 γ′相体积分数的成分敏感性研究

陈　阳，田高峰，王　悦，马国君，杨　杰，邹金文

（中国航发北京航空材料研究院先进高温结构材料重点实验室，北京　100095）

摘　要：研究了各合金成分对镍基高温合金γ′相体积分数的影响。结果表明，γ′相体积分数主要敏感于 Al、Ti、Nb、Ta、Cr 元素。对二十种镍基高温合金的成分含量及γ′相体积分数进行分析，使用 Matlab 软件，利用线性回归法，建立了一个γ′相体积分数与合金成分之间的数学关系式，并与已公开的其他经验关系式结果进行了比较分析，此数学关系式的结果更加精确，关系式为：$V\gamma' = 21.06 + 5.76Al + 0.41Co - 0.95Cr + 0.84Hf + 2.15Nb + 0.82Ta + 3.06Ti$。相关系数 $r = 95.5\%$。

关键词：镍基高温合金，γ′相体积分数，合金化学成分

Effect of Alloying Elements on Volume Fraction of γ′ Phase in Powder Metallurgy Nickel-base Superalloy

Chen Yang, Tian Gaofeng, Wang Yue, Ma Guojun, Yang Jie, Zou Jinwen

(Science and Technology on Advanced High Temperature Structural Materials Laboratory, Beijing Institute of Aeronautical Materials, Beijing 100095, China)

Abstract: This paper studied the effect of alloy chemical component on the volume fraction of γ′ phase at aging temperature. The results indicated the volume fraction of γ′ was mainly sensitive to Al, Ti, Nb, Ta, and Cr. By analyzing the chemical component and the volume fraction of γ′ phase of 20 kinds of nickel-base superalloys and linear regression method in Matlab software, a mathematical formula between the volume fraction of γ′ phase and alloy chemical component was established, the results of which were much more accurate compared with the three public empirical formulas. The formula is: $V\gamma'=21.06+5.76Al+0.41Co-0.95Cr+0.84Hf+2.15Nb+0.82Ta+3.06Ti$. And correlation coefficient $r=95.5\%$.

Key words: nickel-base superalloy, volume fraction of γ′ phase, alloy chemical component

新型纳米硼酸盐粉体制备及其介电特性

梁　栋[1]，贾成厂[2]

（1. 北京京东方光电科技有限公司 Panel 平台开发部，北京　100176；

2. 北京科技大学材料科学与工程学院，北京　100083）

摘　要： 采用溶胶-凝胶法结合表面活性剂制备了 $Mn_3B_7O_{13}Cl$ 纳米粉体，讨论了纳米颗粒的形成过程。用 XRD，TEM 等技术手段对其进行了表征，并探讨了利用纳米 $Mn_3B_7O_{13}Cl$ 制得陶瓷基片的介电性能。结果表明，干凝胶在 550℃下焙烧可以获得粒径分布较为均匀、平均粒径在 50nm 左右的 $Mn_3B_7O_{13}Cl$ 纳米粉体。纳米 $Mn_3B_7O_{13}Cl$ 有较好的介电性能，可用于微波介质陶瓷介质衬底材料。

关键词： 溶胶凝胶法，$Mn_3B_7O_{13}Cl$，介电性能

Preparation and Dielectric Properties of $Mn_3B_7O_{13}Cl$ Nano-powders

Liang Dong[1], Jia Chengchang[2]

(1. Beijing BOE Optoelectronics Technology Co., Ltd., Panel Platform Development Department,Beijing 100176, China; 2. University of Science and Technology Beijing, Beijing 100083, China)

Abstract: $Mn_3B_7O_{13}Cl$ nano-powders were prepared by sol-gel process. X-ray diffraction (XRD) and Transmission electron microscope (TEM) were utilized to characterize the structure, shape, size of the obtained products. The formation of $Mn_3B_7O_{13}Cl$ nanoparticles and the dielectric properties of $Mn_3B_7O_{13}Cl$ were discussed. The results show that $Mn_3B_7O_{13}Cl$ nano-powders which obtained after xerogel roasted at 550℃ are uniform. Their average particles size is about 50nm. $Mn_3B_7O_{13}Cl$ nano-powders have good dielectric properties that can be used as microwave dielectric ceramic dielectric substrate material.

Key words: sol-gel method, $Mn_3B_7O_{13}Cl$, dielectric properties

注射成形用金属粉末粒度分形维数与粉末性能的关系

王聪聪，王迎馨，饶晓雷，胡伯平

（北京中科三环高技术股份有限公司研究院，北京　102200）

摘　要：粉末的粒度分布于粉末的流动性密切相关，但是细粉特别是 20 μm 以下的粉末很难采用传统的方式表征其性能，本文通过引入粒度分形维数来表征细粉粒度分布差异，同时结合 FT4 粉末流变仪对 3 种原料粉末测试压缩性、透气性、剪切性、动力学和充气流动性测试进行了实验研究，以确定注射成形粉末的主要性能参数，为注射成形用粉的选择提供理论参考。结果表明：粉末粒度分布窄，粒度分形维数较小，粉末的基本流动能和比流动能大，对于流速敏感，流动函数也大，粉末流动稳定性好。粒度分形维数和流动函数是判断注射粉末流动性好坏的重要参数。

关键词：注射成形，流动性，FT4 粉体流变仪，流动函数

Relationship between Metal Powder Size Distribution Fractal Dimension and Powder Properties of Powder Injection Molding

Wang Congcong, Wang Yingxin, Rao Xiaolei, Hu Boping

(Beijing Zhong Ke San Huan Hi-Tech Co., Ltd. Research, Beijing 102200, China)

Abstract: The flowability of metal powder is an important parameter that can affect the powder properties, and it also has closely relationship with powder size distribution. However, the properties of powder are difficult to characterize by traditional methods, especially for the powder of particle size of less than 20 μm. In this paper, we introduced size distribution fractal dimension to characterize the difference between these powders, five main properties including compressibility, permeability, shear property, dynamic test and variable flow rate as well as aeration were characterized by FT4 powder rheometer for the 3 kinds of raw powders in order to disclose the main properties of metal injection molding, to provide theoretical reference of powder selection. The results indicated that particle size distribution is narrower, the fractal dimension is smaller. And then the powder has higher basic flowability energy and higher specific energy. The powder with narrow size distribution has sensitive to flow rate, the greater the flow function is, the more stable the powder flowability. Powder size distribution fractal dimension and flow function are the key parameters of judgement of powder performances.

Key words: metal injection molding, flowability, FT4 powder rheometer, flow function

蒙乃尔 400 粉末轧制复合板工艺技术研究

白云波，李　红，高　楠，田玉婷

（鞍钢重型机械设计研究院有限公司，辽宁鞍山　114031）

摘　要：蒙乃尔 400 以其具有较强的耐腐蚀性在海洋建设中得到了广泛应用。目前，该材料价格昂贵，致使许多海洋工程项目成本居高不下。本文提出一种采用蒙乃尔 400 粉末与 IF 汽车板复合，经冷轧制成复合板坯，再经加热烧结，热轧成复合板，以替代纯蒙乃尔 400 钢板的工艺制备技术。实验结果表明，蒙乃尔 400 粉末能够与 IF 汽车板进行复合轧制。复合后的板坯既可保证蒙乃尔 400 的原有性能，还可降低材料成本。

关键词：蒙乃尔 400，粉末轧制，耐腐蚀，复合板

Study on the Technology of Monel 400 Powder Rolling Composite Plate

Bai Yunbo, Li Hong, Gao Nan, Tian Yuting

(Angang heavy machinery designing and research institute Co., Ltd., Anshan 114031, China)

Abstract: Monel 400,with its strong orrosion resistant properties,has been widely used in Marin construction.It costs too much in ocean engineering project construction.This paper presents a method which combines monel 400 powder with Interstitial-Free Steel by cold rolling,sintering and hot rolling to replace the pure technological process of monel 400 plate.The experimental result shows that monel 400 can compound with Interstitial-Free Steel by rolling,and the composite slab can not only guarantee the original performance of monel 400, but reduce the cost of material as well.

Key words: monel 400, powder rolling, orrosion resistant, composite plate

热处理对 SiC-Al-Mg 复合材料性能的影响

吴　超，孙爱芝，贾成厂，刘博文，周　川

（北京科技大学新材料技术研究院，北京　100083）

摘　要： 本文采用放电等离子体烧结（Spark Plasma Sintering，以下简称 SPS）制备了含纳米 SiC$_p$（平均粒径为 40 nm）质量分数分别为 0%、0.5% 和 1% 的 SiC$_p$/Al-Mg 复合材料，并对制备的复合材料进行了退火和固溶时效热处理；通过扫描电镜（SEM）对复合材料进行微观组织形貌的观察以及力学性能测试，研究了两种不同热处理方式对复合材料力学性能的影响规律。研究表明：固溶时效处理后质量分数为 1% 的 SiC/Al-Mg 复合材料的硬度和抗拉强度分别达到了 70.6 HB 和 297 MPa，比未处理前的复合材料提高了 13.3% 和 15.7%，比相同温度下退火处理的复合材料的硬度和强度也要高，固溶时效的热处理效应要优于退火的热处理效应。

关键词： 复合材料，热处理，退火，固溶时效，力学性能

Abstract: In this paper，SiCp / Al-Mg composites with nano-SiCp (average particle size 40nm) mass fraction of 0%、0.5% and 1% were prepared by spark plasma sintering (hereinafter referred to as SPS).The effects of two different heat treatment methods on the prepared composites . The microstructure and mechanical properties of the composites were investigated by scanning electron microscopy (SEM). The results show that the hardness and tensile strength of SiC / Al-Mg composites with the mass fraction of 1% are 70.6HB and 297MPa respectively, which is 13.3% and 15.7% higher than that of the untreated composites. The hardness and strength of the composites annealed at the same temperature are higher than those of the annealing materials. The heat treatment effect of the solution is better than that of the annealing.

Key words: composite, heat treatment, annealing, solid solution aging, mechanical properties

12　非晶合金

低维非晶合金材料

汪卫华

（中国科学院物理研究所）

摘　要：低维材料已经成为物理、化学、电子的核心研究对象，在解决信息、能源、医药等领域内的重大科学和技术问题上扮演关键角。本报告介绍准二维、零维非晶合金材料表现出的丰富的独特特性和行为，包括非晶合金的表面动力学行为、晶化行为，超稳定性，超薄非晶膜的独特物理、化学性能。纳米级非晶颗粒的动力学行为和性能。介绍这些低维特性和非晶的特征、特性的密切相关，对调控非晶的性能的作用。研究表面低维非晶的深入研究可能提供认识和解决非晶传统难题的新途径。

超纳双相材料研究新进展：原理、工艺、性能及应用

吕　坚

（香港城市大学，先进结构材料研究中心）

摘　要：发展高强高韧材料是材料科学家永恒的梦想，中国金属材料的制造和使用都名列世界第一，但很多关键结构材料的技术都还需要引进消化。金属非晶材料是新型材料中发展最快的具有广阔前景的高强材料，但是它们工业应用的最大障碍是它们的低断裂韧性，其在结构领域的应用一直受到韧性差，制造工艺复杂昂贵，工业合金稀少的制约。本报告将系统地介绍总结我们及合作者近年来通过金属玻璃断裂机制及组织结构的关系和增韧制备方法的研究进展[1]。和其他金属材料一样，金属玻璃的力学性能和断裂机制和材料的组织结构是紧密相关的，我们研究了不同的多尺度力学实验方法用来表征金属玻璃的断裂性能及机制，我们又研究了金属玻璃原子尺度的组织结构的规律性及其和力学性能的相关性[2~4]，提出并发展了动态微墩及快 beta 弛豫测量等实验以表征金属玻璃的原子尺度的组织结构缺陷用以研究破坏过程的原子尺度结构演变过程。我们通过热处理的方式得到了超小尺度的纳米晶混杂在大体积的非晶基体里，后来我们通过前期的模拟计算[7]得出超小尺度双相非晶材料只需很少力学性能差别就可以形成多剪切带形成的增韧机制[1,7]. 最近我研究组成功地使用磁控溅射获得超纳尺度（<10nm）镁基非晶-纳米晶非同质的镶嵌结构的新材料，由于该材料的纳米晶晶粒约 6nm，在加载过程中几乎不产生位错，变形能力由非晶部分产生非灾难性多剪切带形成的新变形机制，材料强度接近材料的理论值（E/20），这种超纳新结构为材料科学家带来超小晶粒间的有别于晶界的新型材料家族，因而会改变金属材料所有和超小晶粒及晶界有关所有物理及化学性能。最后介绍超纳材料在微型结构，生物医学等领域的应用前景及材料集成先驱结构设计的案例。

关键词：超纳，非同质双相，高强，镁合金

非晶合金的功能性能及应用前景

姚可夫，陈双琴，施凌翔，贾蓟丽，邵洋

（清华大学材料学院）

摘　要：非晶合金的结构特征是原子呈长程无序、短程有序堆垛排列，这使其具有完全不同于晶态合金材料的多种性能。非晶合金不仅具有高强度、高硬度、大弹性变形量等力学性能，还具有优异的耐蚀性能、高活性、优异的软磁性能等物理化学性能。在非晶合金研究领域，非晶合金的功能性能一直受到研究人员和工业界的高度关注。在最近 10 年间，铁基非晶合金的软磁研究取得了一些列重要进展，研发出了具有高饱和磁感强度的铁基非晶/纳米晶合金，并在不断改进相关制造与工艺性能，以期实现工业化应用。在铁基非晶合金生产中，中国已成为主要生产和应用基地。最近，研究人员还发现，非晶合金具有远优于传统工业铁粉的污水降解净化性能，一些铁基非晶合金工业带材呈现出极为优异的有机染料的降解净化性能，使非晶合金在污水降解净化处理中展现出很好的应用前景。本文将介绍非晶合金软磁性能研究和化学活性研究中的主要进展，以飨读者。

高饱和磁感应强度非晶和纳米晶软磁合金开发

王新敏[1]，王安定[1]

（中科院宁波材料所中科院磁性材料与器件重点实验室）

摘　要：由于软磁性能优异，损耗低，非晶合金具有诱人的应用前景，吸引了广泛的关注和研究。然而，由于非晶形成能力的要求，通常需要添加较高比例的非磁性非晶形成元素，导致合金的饱和磁感应强度降低。其次，非晶合金还具有磁致伸缩系数大，噪音高，热处理后脆性大等问题，这都不利于非晶软磁合金的更广泛推广应用。因此，开发高饱和磁感应强度的非晶合金，解决脆性和噪音问题，成为近年来的研究热点，迫切需要开发适合现有带材生产设备和工艺的兼具高饱和磁感应强度、优异软磁性能和大非晶形成能力的新型铁基非晶和纳米晶软磁合金，并进行配套工艺研究。

最近，本团队围绕这些重大问题，开展了高饱和磁感应强度和良好软磁性能的合金成分和热处理工艺开发、性能调控工艺及机理研究，性能与微观结构关联机制分析等工作，取得了重要突破。本报告中，将系统介绍 FeSiBPC(Cu) 系高 Bs 非晶合金的设计方法、获得兼具最佳软磁性能和韧性的热处理技术，以及工业原料制备、熔体净化工艺等成果，并基于工艺性对未来的合金成分设计展开分析。本工作将为未来高性能非晶和纳米晶软磁合金的开发及应用提供重要参考。

非晶合金：一类高效处理印染废水的新型材料

张海峰，朱正旺，秦鑫冬，王爱民

（中国科学院金属研究所沈阳材料科学国家(联合)实验室，辽宁沈阳　110016）

摘　要：非晶合金的原子结构呈长程无序、短程有序状态，每个短程有序区域可以成为催化活性中心，使其具有高催化活性中心密度，在某些化学反应中表现出优异的催化活性、高的选择性和稳定性。本文系统研究了 Fe 基等非晶合金作为催化剂在降解染料废水—除色中的特性，包括合金成分、表面状态、结构、添加量、助剂、溶液性质等与降解—除色之间的相关性。

非晶合金的 3D 打印：机遇与挑战

柳　林

（华中科技大学材料学院，材料成形与模具技术国家重点实验室）

摘　要：非晶合金，特别是块体非晶合金因其具有长程无序的原子结构，表现出一系列优于晶态合金的物理、化学和力学性能，近 30 年来受到科学和工程界的广泛重视，并成为当前凝聚态物理与材料科学领域的前沿热点。然而，非晶合金固有的室温脆性和本征超高强度特征，使其难以通过常规手段进行加工成形，这成为非晶合金工程应用的瓶颈。近年来发展起来的激光 3D 打印技术，为大尺寸非晶合金的制备及非晶零件的成形提供了新途径，也为非晶合金的工程应用带来了新机遇。激光 3D 打印技术是利用精确控制的高能量激光束将非晶粉末逐层堆积成形，理论上可以制备任意尺寸和任意复杂形状的非晶合金构件。本报告重点介绍华中科技大学非晶态研究室近年来在激光 3D 打印非晶合金领域的研究进展，包括：非晶合金粉末制备、3D 打印过程中温度场与热应力场分布、3D 打印非晶合金的结构演变以及性能。最后，介绍激光 3D 打印非晶合金面临的挑战，包括：非晶合金粉末的制备困难与选择、缺陷的产生以及晶化的发生。

高损伤容限 $Zr_{61}Ti_2Cu_{25}Al_{12}$ 块体金属玻璃的断裂行为

徐　坚[1]，贺　强[1]，宋贞强[1]，Ma Evan[2]

（1.中国科学院金属研究所；2. The Johns Hopkins University, Baltimore, USA）

摘　要：块体金属玻璃（BMGs）具有高强度、大弹性应变极限（~2%）等特点，在过去 20 多年里得到了广泛的关注与研究，被期待作为新一类工程材料得到应用。研究表明，在无约束的载荷作用下（如单向拉伸或压缩），BMG 的形变与失效受控于高度局域化的剪切带，几乎没有宏观的塑性变形能力。然而，BMG 却可拥有反常的性能结合，诸如拥有高断裂韧性（或损伤容限）。我们的研究发现，在 Zr-Ti-Cu-Al 四元合金中，$Zr_{61}Ti_2Cu_{25}Al_{12}$（ZT1）BMG 具有高断裂韧性（$K_{JIC}=133\,\mathrm{MPa}\sqrt{m}$），其高韧性源于在裂纹尖端可形成由剪切带交互作用构成的数百微米量级的塑性区。同时可引起裂纹的偏转，使得局域载荷模式由初始阶段的 I-型，演化为 I-型与 II-型相混合的模式。而且断裂过程中出呈现"裂纹阻力曲线（R-curve）"的特征，即随着裂纹的扩展应变能释放速率（J）增加。ZT1 BMG 不仅实现了高强度（$\sigma_y=1600\mathrm{MPa}$）与高断裂韧性的结合，而且也具有良好的疲劳抗力，疲劳极限 σ_{fat} 和疲劳极限与断裂强度之比（$\sigma_{fat}/\sigma_{UTS}$）分别达到 440 MPa 和 0.27 的水平。在扭转载荷的作用下，ZT1 BMG 的剪切屈服强度 $\tau_y=950\,\mathrm{MPa}$、剪切弹性应变极限 $\gamma_C=3.0\%$，其断裂亦由单一主剪切带控制，无发生明显的宏观塑性形变。微观上，扭转断裂过程由若干事件构成，包括剪切带起始阶段的冷剪切、软化的剪切带由于圆柱试样横断面上应力梯度的约束而发生的短时稳定扩展，以及最终因形成空洞和微裂纹导致失稳断裂。进一步对 ZT1 BMG 反平面剪切（III 型）断裂韧性的研究显示，ZT1 的 III 型断裂韧性（表达为塑性应变强度因子）$\Gamma_{\text{IIIC}} \approx 29 \ \mu m$，转换为应力强度因子为 $K_{\text{IIIC}} \approx 51 \ \text{MPa} \sqrt{m}$，对应 III 型载荷下裂纹尖端塑性区的尺寸约为 0.54 mm。在反平面剪切载荷下，ZT1 在发生灾难性断裂之前会发生亚临界裂纹扩展。扩展过程中裂纹前沿微观上呈现之字形，从而在裂纹尖端产生局域的 II/III 混合型载荷。这种微观的裂纹扩展路径降低了裂纹尖端有效的扩展驱动力，从而在一定程度上提高了 ZT1 的 III 型断裂韧性。ZT1 的 $K_{\text{IIIC}}/K_{\text{IC}}$ 与 $G_{\text{IIIC}}/G_{\text{IC}}$ 值分别为 0.39 和 0.24，表明相对于 I 型载荷，ZT1 在 III 型载荷模式下具有更低的裂纹扩展阻力。与传统工程材料相比，ZT1 的 III 型断裂韧性（G_{IIIC}）与高强度铝合金相当，为某些钢铁材料的四倍。但相对于 I 型张开型载荷，材料中的裂纹对反平面剪切载荷更加敏感。因此，III 型断裂韧性可以代替 I 型断裂韧性作为金属玻璃构件的保守设计基准。这些对 BMG 在各种载荷模式下断裂行为的理解对于 BMG 作为结构件的工程应用具有重要意义。

关键词：金属玻璃，力学行为，断裂韧性，疲劳，剪切带

块体非晶合金的塑韧性研究

惠希东

（北京科技大学新金属材料国家重点实验室，北京　100083）

摘　要：非晶合金由于在结构上是长程无序而短程有序，所以具有高强度、高弹性极限、优异的耐腐蚀和软磁性能，已经在电力和电子行业获得了广泛应用。但是，在结构材料应用方面，非晶合金的脆性已经成为亟待解决的难题。研究证明，非晶合金的各种性能与其短程和中程有序团簇结构密切相关，因此非晶合金结构与性能的相关性就成为认识非晶合金塑性和韧性的关键所在。本报告以非晶合金原子排列结构、团簇结构与其弛豫规律、相分离和纳米晶化的关联性，非晶合金的结构与其塑性变形能力之间的关系等科学问题为主线。从非晶合金局域有序程序、团簇结构、软硬区对合金塑性变形能力、剪切带特征和自由体积的影响；非晶合金的结构对合金结构弛豫、相分离和纳米晶形成的影响规律等方面，阐述近年来在高强高韧非晶材料研究成果，以期为非晶合金在结构材料中的应用提供参考。

高 B_s Fe-Si-B-Cu 系纳米晶合金的组织结构和磁性能研究

张伟

（大连理工大学材料科学与工程学院，辽宁大连　116024）

摘　要：随着电力、电子设备和电子器件向节能化、小型化、高效化方向发展，需研制具有高饱和磁感应强度（B_s）、低矫顽力（H_c）、高磁导率（μ_e）、低铁损的新型软磁合金。近年，Makino 等人报道的高 Fe 含量 Fe-Si-B-P-Cu 系纳米软磁合金不仅显示出 1.80 T 以上的 B_s 值，还具有和 FINEMET 相当的优异软磁性能，因而受到广泛关注。但该合金含有 P 元素且纳米晶化热处理需高升温速率，限制了其在实际生产中的应用。

　　最近，我们调查了不含 P 元素的 Fe-Si-B-Cu 系液态急冷合金的结构、热处理结晶化组织及其磁性能，研究了纳米晶合金成分—组织结构—磁性能间的关系。结果表明：Fe-Si-B-Cu 液态急冷合金在较宽的成分范围内为非晶态，经高升温速率热处理晶化可形成由 α-Fe 和残余非晶相组成的微细纳米晶组织，显示出优异的软磁性能；高 Cu 含

量（＞1.5 at.%）的液态急冷合金的非晶基体中分散着高密度的平均尺寸约为6nm的 α-Fe 晶核，经低升温速率热处理也能获得 α-Fe 的平均尺寸小于20nm的纳米晶组织，具有 10 A/m 以下的 H_c、1.75 T 以上 B_s 以及高 μ_e 值。这是由于高 Cu 量可在 Fe 基非晶基体产生较强的成分起伏，促使富 Cu 团簇形成而诱发高密度 α-Fe 晶核在非晶前驱体内形成；这种既存晶核和热处理结晶化生成的 α-Fe 晶核间的竞争生长能抑制 α-Fe 晶粒的不均衡生长，从而得到微细、均匀的纳米组织，获得低 H_c 值。此外，我们还研究了微量合金元素添加对高 Cu 含量 Fe-Si-B-Cu 液态急冷合金的结构、纳米晶化组织及其磁性能影响。

关键词：纳米晶软磁合金，软磁性能，Cu 含量，α-Fe 晶核，高 B_s

我国块体非晶产业化发展

李扬德

（宜安科技）

摘　要：非晶合金已经开始应用在汽车、消费电子、声学和航空航天等领域。宜安科技在非晶合金的产业化发展方面做了系统的工作，涉及满足工业需求的材料成分开发、材料批量熔炼设备与工艺、先进成型工艺、成型设备、专用后处理工艺和模具设计与优化等一整套产业化技术及相关知识产权，具有全球领先地位。通过控制美国 LQMT，实现知识产权的平行授权，拥有了完整的知识产权布局，并成功将中国非晶合金产业延伸到全球市场，同时将美国的先进技术引进消化，进一步拓展了中国非晶合金的应用空间和市场。宜安科技正在开展从原材料到最终产品开发的多方位合作，已经成立了专门生产先进非晶合金压铸设备的真空压铸机公司。宜安科技已经建立起高水平的非晶合金产业化平台，能够满足工业生产和高校研究机构的产品和打样需求。宜安科技可以实现非晶合金产品的批量制造、个性化定制和深度合作研发。

关键词：产业化，真空压铸，非晶合金

Ce-Ni 合金熔体的互扩散行为研究 Ce-Ni 合金熔体的互扩散行为研究

张　博

（合肥工业大学材料科学与工程学院&非晶态物质科学研究所）

摘　要：熔体的互扩散行为在材料科学中具有重要的意义，但相关的研究却非常稀少。利用滑动剪切技术，我们对 Ce80Ni20 二元金属合金熔体的互扩散行为进行了测量和研究。

　　测量结果表明在此 Ce-基熔体体系，熔体的互扩散行为严重偏离标准的 Darken 关系，且随着温度的降低，偏离倾向更加明显。另外，动力学因子 S 因子明显小于1。进一步的分析表明 S 因子的本质来源于原子间的化学相互作用。在此 Ce-Ni 二元合金熔体中，原子间的化学相互作用使 Ni 原子倾向于与 Ce 原子一起运动，而避免与同类的 Ni 原子一起，从而使得 Ni-Ni 及 Ni-Ce 之间的扩散偏离理想扩散行为，并最终导致 Darken 公式的失效。

Multi-stage Relaxations in Metallic Glasses during Isothermal Annealing

Wang Junqiang

(Ningbo Institute of Materials Technology & Engineering, Chinese Academy of Sciences)

Abstract: A comprehensive description of the relaxation dynamic relationship of glasses is an interesting and challenging problem. In this work, the enthalpy relaxation kinetics of a Au-based metallic glass is studied in wide temperature range between 0.6 Tg and 1.1 Tg using a high-precision calorimeter. Striking multiple relaxations was confirmed during isothermal annealing. Especially, the fast beta' relaxation was firstly found in heat flow trace using DSC. The quantitative relation between the glass enthalpy and relaxation activation energy is studied. These results challenge the protocol that slow relaxation happens at high temperatures while fast relaxations happen at low temperatures; it also suggests that the role of different relaxations on other properties is worthy of further study.

Molecular Dynamics Simulations of Dynamical Mechanical Spectroscopy in Metallic Glasses: Linear and Nonlinear Responses

Yu Haibin

(Wuhan National High Magnetic Field Center, Huazhong University of
Science and Technology, China)

Abstract: Relaxation dynamics are the central topics in glassy physics. They are usually measured by the experimental techniques of dynamical spectroscopies such as the dielectric and mechanical spectra. Here, we put the protocols of the dynamical mechanical spectroscopy into molecular dynamics simulations. It reproduces the salient features of relaxation spectra asmeasured experimentally. Moreover, combining it with detailed studies atomic movements and structural analysis, we clarified the mechanisms of internal friction of metallic glasses[1]. We observed the strain driven glass transition and a pronounced fragility transition as induced by nonlinear strain [2]. We find that the primary (α) relaxation always takes place when the most probable atomic displacement reaches a critical fraction (~20%) of the average interatomic distance, irrespective of whether the relaxation is induced by temperature (linearresponse) or by mechanical strain (nonlinear response)[3].We show the time dependent viscoelastic moduli of metallic glasses are mainly contributed from the most probable atomic displacement, irrespective of the dynamical heterogeneity [4]. Moreover, by studying the relaxation dynamics up-to time scales of microseconds, we demonstrate that the secondary relaxation (Johari-Goldstein, β)originates from the string-like cooperative atomic motions.[5]

从非晶合金断裂到工程材料性能定量联通

张哲峰

（中国科学院金属研究所沈阳材料科学国家（联合）实验室，辽宁沈阳 110016）

摘 要：基于报告人十余年关于非晶合金断裂与强度理论方面的研究工作，提出了脆性材料统一拉伸断裂准则，并进一步发展了复杂应力条件下断裂准则与强度理论，通过揭示非晶合金韧-脆转变机制，建立了非晶合金弹性参数与拉伸/压缩强度及断裂行为的定量关联。通过高强度材料剪切与解理断裂的竞争关系，进一步建立了工程材料（钢铁、铝合金与钛合金）冲击/断裂韧性与拉伸性能、疲劳强度与拉伸性能的定量关系，为实现工程材料力学性能之间的定量联通奠定基础。

关键词：非晶合金，断裂机制，强度理论，拉伸性能，疲劳强度

非晶态材料与物理的计算模拟研究

管鹏飞

（北京计算科学研究中心）

摘 要：非晶合金，即金属玻璃，是一类特殊的由基本化学元素组成的非晶态物质，由于其独特的微观组织结构，展现出了不同于传统晶态合金材料的特殊物性，而成为高性能材料应用领域的重要一员。由于非晶合金的结构无序性，相应的理论模型也不完善，人们对非晶合金中一些基础物理问题的认识尚且不足，无法形成基本的理论框架来精确地描述其物性产生的微观机理。因而，当前非晶合金研究的核心问题可以概括为：如何建立以微观特征或结构为基础的基本理论框架，准确地概括非晶合金物性的微观机理。由于非晶合金微观结构的复杂性和实现精确实验表征的手段缺乏，计算模拟成为了研究非晶合金结构、物性及其关联的重要手段。基于计算模拟，我们通过对非晶合金体系的结构及其对外场的响应的系统研究，尝试理解非晶体系中的两个重要的物理过程：玻璃转变与剪切形变，并期望基于非晶合金中不均匀性的特征，理解和建立微观特征与物性之间的关联。本次报告我们将重点介绍非晶合金中短程有序结构与物性之间的退耦合关系，而中程序或更大尺度上的不均匀性可能将帮助我们更准确地理解非晶合金物性机理。

Pressure-induced Elastic Anomaly in a Polyamorphous Metallic Glass

Zeng Qiaoshi[1], Zeng Zhidan[1], Lou Hongbo[1], Kono Yoshio[4], Zhang Bo[5], Curtis Kenney-Benson[4], Park Changyong[4], Mao Wendy L.[2,3]

(1. Center for High Pressure Science and Technology Advanced Research (HPSTAR), Shanghai 201203, China; 2. Geological Sciences, Stanford University, Stanford, CA 94305, USA; 3. Stanford Institute for Materials and Energy Sciences, SLAC National Accelerator Laboratory, Menlo Park, CA 94025, USA; 4. HPCAT, Geophysical

Laboratory, Carnegie Institution of Washington, 9700 South Cass Avenue, Argonne, IL 60439, USA; 5. Institute of Amorphous Matter Science, School of Materials Science and Engineering & Anhui Provincial Key Lab of Functional Materials and Devices, Hefei University of Technology, Hefei 230009, China.)

Abstract: The pressure-induced transitions discovered in metallic glasses (MGs) have attracted considerable research interest offering an exciting opportunity to study polyamorphism in densely packed systems. Despite the large body of work on these systems, the elastic properties of the MGs during polyamorphic transitions remain unclear. In this work, using an in situ high-pressure ultrasonic sound velocity technique integrated with synchrotron radiation x-ray radiography and x-ray diffraction in a Paris-Edinburgh cell, we precisely determined both the compressional and shear wave sound velocities of a $Ce_{68}Al_{10}Cu_{20}Co_2$ metallic glass up to 5.8 GPa through its polyamorphic transition. Then, its elastic moduli and Poisson's ratio versus pressure were all derived. For the first time, we observed elastic anomalies with minima (at ~1.5 GPa) in the sound velocities, bulk modulus, and Poisson's ratio of a metallic glass. Compared with the compression behaviors of the silica glass and the crystalline Ce in detail, we conclude that the elastic anomaly is closely associated with the pressure-induced transitions. In addition, sound velocities have been extensively used to calculate the density of glasses under pressure. We found that the density of $Ce_{68}Al_{10}Cu_{20}Co_2$ metallic glass calculated from sound velocities obviously deviated from the directly measured density when a polyamorphic transition occurs. Our results suggest that the density-sound velocity relationship may be invalid in glasses with a polyamorphic transition, although the transition is continuous.

Preparation of Ductile FeNbB Bulk Metallic Glass with Excellent Magnetic and Mechanical Properties after Annealing Optimization

Gao Zhikai[1, 2], Wang Anding[2, 3], Chen Pingbo[2], Li Fushan[1], Chang Chuntao[2, 3], He Aina[1], Wang Xinmin[2, 3], Liu Chain-Tsuan[4]

(1. School of Materials Science and Engineering, Zhengzhou University, Zhengzhou 450001, China;

2. Key Laboratory of Magnetic Materials and Devices, Ningbo Institute of Materials Technology and Engineering, Chinese Academy of Sciences, Ningbo 315201, China;

3. Zhejiang Province Key Laboratory of Magnetic Materials and Application Technology, Ningbo Institute of Materials Technology and Engineering, Chinese Academy of Sciences, Ningbo 315201, China;

4. Center for Advanced Structural Materials, Department of Mechanical and Biomedical Engineering, College of Science and Engineering, City University of Hong Kong, Hong Kong, China)

Abstract: Annealing induced magnetic and mechanical property changes for the $Fe_{71}Nb_6B_{23}$ bulk metallic glass were systematically investigated. It is interesting that this ternary alloy can combinedly exhibit outstanding magnetic and mechanical properties, especially good ductility, after optimally annealing in structure relaxation stage for eliminating the internal stress and homogenizing the microstructure. The alloy exhibits low coercive force of 1.6 A/m, high effective permeability of 15×10^3, high fracture strength of 4.2 GPa and good plastic strain of 1.8%. It is also found that the responses of mechanical and magnetic properties to structure relaxation are asynchronous. The glass transition and crystallization will greatly deteriorate the magnetic and mechanical properties. Here we propose a physical picture and demonstrate that the primary structure factors determining magnetic and mechanical properties are different. This work will bring a promising material for application and a new perspective to study the effect of annealing induced structure relaxation on mechanical and magnetic properties.

Key words: bulk metallic glass, annealing, mechanical property, magnetic property, structure relaxation

Pressure Effects on Structure and Dynamics of Metallic Glass-forming Liquid with Miscibility Gap

Cheng Y., Wang P. F., Peng C. X., Jia L. J., Wang Y. Y., Wang L.

(School of Mechanical, Electrical & Information Engineering, Shandong University (Weihai), Weihai 264209, China.)

Abstract: The metallic liquid with miscibility gap has been widely explored recently because of theincreasing plastic deformation abilityof phase-separated metallic glass. While the poor glass formation ability limits its application of the structural materials due to the positive mixing enthalpy of the two elements of all of them. Since the high pressure is in favor of the formation of the glass, in the paper, the effect of pressure on the structural and dynamical heterogeneity of phase-separated $Cu_{50}Ag_{50}$ liquid is investigated by molecular dynamics simulation in the pressure range of 0~16GPa.Our results clearly show that the pressure promotes the formation of metallic glass by increasing the number of five-fold symmetry cluster and dynamical relaxation time, meanwhile the liquid-liquid phase separation is also enhanced, the homogenous atom pairs show more stronger interaction than that of heterogeneous atom pairs with increasing pressure.The dynamical heterogeneity is related to the formation of five-fold symmetry clusters. The lower growing rate of W at higher pressure with decreasing temperature corresponds to the slow increasing of dynamical heterogeneity.The pressured glass with miscibility gap may act as a candidate glass with improving plastic formation ability.Our results explore the structural and dynamics heterogeneity of phase separated liquid at atomic level.

Keywords: glass-forming liquid with miscibility gap, high pressure, structural and dynamical heterogeneity

一种测量液态金属原子扩散的多层平动剪切技术

钟浪祥，胡金亮，张　博

（合肥工业大学材料科学与工程学院非晶态物质科学研究所，安徽合肥　230009）

摘　要： 液态金属扩散系数是描述质量传输现象的核心参数。液态金属原子扩散行为的研究对于人们了解液态金属的原子结构和原子迁移的物理本质有重大意义。本文介绍了一种自主研发的测量液态金属原子扩散的多层平动剪切技术，并测量了铁基金属熔体（FeSi、FeB 和 FeSiB 等）扩散系数，并研究了铁基金属熔体扩散系数的温度依赖关系。基于铁基非晶合金的广泛用途，铁基熔体原子扩散行为的研究对铁基非晶材料的开发和利用有重要意义。多层滑动剪切技术也能够应用于其他高温合金熔体扩散系数的测量，具有广泛的工程应用价值。

晶化对 Gd-Er-Al-Co 金属玻璃磁热性能的调控

罗　强，沈　军

（同济大学材料科学与工程学院，上海　201804）

摘　要： 用高真空电弧熔炼炉铜模吸铸法制备了直径为 3mm 大块金属玻璃，并对该合金及其晶化后样品的磁化过程和磁热性能做了仔细研究。铸态非晶样品随着温度降低，先后经历顺磁到铁磁转变和再入自旋玻璃的冻结行为。随着晶化程度的增加，类似自旋玻璃冻结行为越明显，表现在低温下场冷和零场冷的磁化曲线之间的分叉特征越来越明显。同时晶化后样品的居里温度和磁化强度均降低。铸态金属玻璃样品在 5T 外场变化的条件下展现出最大的磁熵变为 9.9 (J·kg⁻¹K⁻¹)，该值与 Gd 金属单质以及 Gd 基的金属玻璃相当。样品晶化后，磁熵变降低，在研究的一系列晶化样品中，最大磁熵变的最小值为 6.7 (J·kg⁻¹K⁻¹)，该值仍然大于 Fe 基和 Co 基的非晶合金。该类合金优异的磁热性能、高电阻、良好的软磁性能、优异的力学性能以及工作温度方便可调的特点表明这类材料在低温区作为磁制冷材料有良好的应用前景。

关键词： 金属玻璃，磁热效应，自旋玻璃，晶化，结构

Tune the Magnetocaloric Effect of Gd-Er-Al-Co Metallic Glass Through Crystallization

Luo Qiang, Shen Jun

(School of Materials Science and Engineering, Tongji University, Shanghai 201804, China)

Abstract: Magnetization and magnetocaloric effec (MCE) of Gd-based Bulk Metallic Glass (BMG) $Gd_{24}Er_{32}Co_{24}Al_{20}$ and the crystallized samples are investigated. Reentraned spin glass like behavior is observed in the as-cast sample. With increasing degree of crystallization, the magnetic frustration increases, which can be seen from the increased divergence of the zero field cooled and field cooled magnetization behaivor(ie.the spin glass like freezing behavior). The Curie temperature and magnetization decrease after crystallization. The as-cast BMG shows large maximum magnetic entropy change $(-\Delta S_m)$ of 9.9 $(J \cdot kg^{-1}K^{-1})$ under a field change of 5 T, which is comparable with that of Gd metal and other Gd-based metallic glasses. The maximum magnetic entropy reduces to 6.7 $(J \cdot kg^{-1}K^{-1})$ after crystallization, which is still superior to the Fe-/Co-based amorphous alloys. Their good MCE combining with high electrical resistivity, outstanding mechanical properties, tunable nature, and sufficiently soft magnetic property make them to be attractive candidate for magnetic refrigerants in the low temperature range.

Key words: metallic glass, magnetocaloric effect, spin glass, crystallization, microstructure

锆铜铝熔体中液-液相变的原位散射研究

兰　司[1,2]，董蔚霞[1]，任　洋[3]，王循理[2,4]

（1.南京理工大学材料科学与工程学院/格莱特研究所，江苏南京　210094；2. 香港城市大学物理

系，中国香港；3.美国阿贡国家实验室 X 射线科学部，阿贡，伊利诺伊州　60439；
4.香港城市大学深圳研究院中子散射中心，广东深圳　518057）

摘　要：本文利用高能 X 射线原位研究了具有不同玻璃形成能力的锆铜铝合金熔体（$Zr_{56}Cu_{36}A_{18}$ 和 $Zr_{46}Cu_{46}A_{18}$）在冷却过程中的原子结构变化。$Zr_{46}Cu_{46}A_{18}$ 相比 $Zr_{56}Cu_{36}A_{18}$ 具有较优良的玻璃形成能力。结果表明，$Zr_{56}Cu_{36}A_{18}$ 在熔点以上并未发生明显异常结构变化，但 $Zr_{46}Cu_{46}A_{18}$ 在熔点以上发生了液-液相变，证据为散射结构因子的主衍射峰的强度随着温度的变化在特定温度处 T_C 发生了突变，表明合金熔体发生了由低有序度向高有序度熔体转变的相变过程。实空间分析结果表明，$Zr_{46}Cu_{46}A_{18}$ 合金熔体的中程有序度的结构变化更加显著，暗示其内在的原子团簇的协同重组及连接性在液-液相变的过程中发挥了作用，使其在冷却过程中具有更强的抵抗结晶的能力从而具有更优异的玻璃形成能力。我们的研究结果将有助于指导新型非晶合金体系的研发。

关键词：液-液相变，同步辐射散射，玻璃形成能力，原子机制

In-situ Scattering Research on Liquid-liquid Phase Transition in Zr-Cu-Al Melts

Lan Si[1,2], Dong Weixia[1], Ren Yang[3], Wang Xunli[2,4]

(1. Herbert Gleiter Institute of Nanoscience, School of Materials Science and Engineering, Nanjing University of Science and Technology, Nanjing 210094, China;

2. Department of Physics, City University of Hong Kong, Hong Kong, China.

3. X-ray Science Division, Argonne National Laboratory, Argonne, Illinois 60439, USA;

4. City University of Hong Kong Shenzhen Research Institute Neutron Scattering Center, Shenzhen Hi-Tech Industrial Park, Shenzhen 518057, China)

Abstract: In this paper, we in-situ studied the atomic structural evolutionof two Zr-Cu-Al molten liquids, i.e. $Zr_{56}Cu_{36}A_{18}$ and $Zr_{46}Cu_{46}A_{18}$, with different glass forming abilities during cooling by synchrotron high-energy X-ray scattering. $Zr_{46}Cu_{46}A_{18}$ has better glass-forming ability than that of $Zr_{56}Cu_{36}A_{18}$. Our resultssuggested that there is no anomalousstructural change for $Zr_{56}Cu_{36}A_{18}$ liquid. However, a liquid-liquid phase transition occurred in Zr46Cu46Al8 alloy above the melting point, which is evident by the abrupt intensity increase of the first diffraction peak in structural factor at a critical temperature, T_C, upon cooling. The transition is from a low ordered liquid to a higher ordered liquid. The real space analysis suggested that the structural change in a medium-range length scale for $Zr_{46}Cu_{46}A_{18}$ is more pronounced, indicating that the collective atomic arrangement and connectivity of the clusters may play important rolesduring the liquid-liquid phase transition. As a result, the melt with a liquid-liquid phase transition may possess a stronger ability to resist crystallization during cooling, and thus has better glass forming ability. Our results may shed light on research and development of new amorphous alloys.

Key words: liquid-liquid phase transition, synchrotron scattering, glass-forming ability, atomic mechanism

Development of Ultra-thin Nanocrystalline Ribbons with Excellent Soft Magnetic Properties

Li Zongzhen, Zhou Shaoxiong, Zhang Guangqiang, Zhang Qiang

(China Iron & Steel Research Institute Group, Advanced Technology & Materials Co., Ltd., Beijing 100081, China)

Abstract: A new ultra-thin amorphous ribbon with thickness less than 15μm was successfully prepared by using traditional planar flow melt spinning (PFMS) and novel puddle embedded nozzle in the Fe-Si-B-P-Nb-Cu alloy system.The effect of planar flow melt spinning (PFMS) parameters on the continuity and surface quality of ribbons has been investigated in detail and a window of process parameters for obtaining continuous ribbons with good surface quality has been evaluated. It is found that the stability of puddle plays a key role in a wide range of processing operations. The introduction of melt puddle embedded nozzle can reduce the lateral disturbance of the melt in the puddle and the instability of puddle, reducing the roughness of the free surface and the depth of scratches, therefore the quality of free surface is efficiently improved.After isothermal annealing, the FeSiBPNbCu nanocrystalline ribbons showed a typical nanocomposite microstructure composed of α-Fe grains of approximately 13nm in diameter randomly dispersing in the residual amorphous matrix, exhibiting a comparatively high saturation magnetization of 1.74 T and a lower coercivity of 3.3 A/m. Based on the results, a better understanding of the influential mechanisms of the process parameters on the surface roughness can be obtained. The strategy for the achievement of ultra-thin nanocrystalline ribbons with excellent soft magnetic properties is also discussed.

Fabrication and Magnetic Properties of $Fe_{60}Co_{13}Ni_4Si_{11}B_{12}$ Amorphous Soft Magnetic Powder Cores Using Ball-milled Powders

Liu T., Lu K., Si J.J., Wang T., Zhu J., Wu Y.D., Hui X.D.

(State Key Laboratory for Advanced Metals and Materials, University of Science and Technology Beijing, Beijing 100083, China)

Abstract: Toroidal shape $Fe_{60}Co_{13}Ni_4Si_{11}B_{12}$(at. %) alloy amorphous powder core has been firstly prepared using ball-milled powders and its magnetic properties has been investigated. The powders were obtained after ball-milling for 12h and annealing at 350℃ for 1h.FeCoNiSiB amorphous powder cores (outer diameter = 20 mm, inner diameter =16mm, height =4mm) were made of amorphous alloy powder below 150μm by cold pressing. The core loss of the cores increase greatly with the increasing of frequency in frequency range of 30-400 kHz. The core for Bm= 30 mT shows core loss of about 110W/kg at 100 kHz. The quality factor Q decreases with increasing frequency in frequency range of 30-400 kHz. The core for Bm= 30mT shows quality factor about 30 at 100 kHz. Initial permeability in the frequency range up to 500 kHz are approximately constant and the value is about 24. The present Fe-based amorphous magnetic powder cores with excellent soft magnetic properties are a potential candidate for a variety of electronic system and industrial applications.

Key words: amorphous powder core, annealing, low core loss, magnetic properties

具有拉伸塑性和加工硬化能力的大尺寸 TiZr 基非晶复合材料

朱正旺，刘丁铭，李　伟，张海峰

（中国科学院金属研究所沈阳材料科学国家(联合)实验室，辽宁沈阳　110016）

摘　要：非晶复合材料兼具高强度和塑性，是一类先进结构材料。形成尺寸、拉伸塑性和加工硬化能力等均是限制其实际应用的关键问题。本工作研究了内生枝晶相的 TiZr 基非晶复合材料的形成能力、微观结构和力学性能。结果发现，$Ti_{44.3}Zr_{32.7}Cu_{6.2}Ni_{3.2}Be_{13.6}$（原子百分比）合金具有超高非晶复合材料形成能力，最大形成尺寸超过 50mm；当尺寸不超过 30mm 时，样品具有良好的拉伸塑性（大于 6%）和显著的加工硬化能力，最大强度达到 1620MPa。机制研究表明，基体非晶存在合适的二十面体局域有序度是其具有高形成能力的主要原因，同时增强了凝固过程中亚稳 β 相的稳定性；而亚稳 β 相发生变形诱发马氏体转变，致使材料具有优异的力学性能。该发现对于研制新型高性能非晶复合材料、促进其实际应用具有重要作用。

关键词：内生型非晶复合材料，加工能力行为，拉伸塑性，形变诱发马氏体转变

Large-sized in-situ Ti-based Bulk Metallic Glass Composite with Work Hardening Ability and Tensile Plasticity

Zhu Zhengwang, Liu Dingming, Li Wei, Zhang Haifeng

(Shenyang National Laboratory for Materials Science, Institute of Metal Research, Chinese Academy of Sciences, Shenyang 110016, China)

Abstract: Bulk metallic glass composites (BMGCs) with high strength and ductility have been recognized as a class of advanced structural materials. Large formation size, tensile ductility and work hardening ability have become the challenging issues for their practical application. In this work, the $Ti_{44.3}Zr_{32.7}Cu_{6.2}Ni_{3.2}Be_{13.6}$at.% alloy presents the ultrahigh forming ability of BMGC and superior mechanical properties. The maximum size of BMGC is above 50mm. The sample with a diameter less than 30mm exhibits large tensile ductility and work hardening behaviors. It is indicated that the existence of icosahedral short range ordering contributes to high glass forming ability in the matrix, which favors stabilizing the metastable β-phase in the dendrites during solidification process. Superior mechanical properties are attributed to deformation-induced martensitic transformation in the dendrites. These findings not only promote its application but also shed light on the preparation of other high-performance BMGCs.

Key words: in-situ bulk metallic glass composite, work hardening ability, tensile ductility, deformation-induced martensitic transformation

生物医用器件用 $Zr_{55.8}Al_{19.4}(Co_{1-x}Cu_x)24.8$ (x = 0-0.8) 块体金属玻璃研究

韩凯明，羌建兵，王英敏，Haeussler Peter

（大连理工大学三束材料改性教育部重点实验室，辽宁大连　116024）

摘　要：不含有毒元素 Zr-Al-Co-Cu 系金属玻璃有望用作生物医用器件材料。本文从 $Zr_{55.8}Al_{19.4}Co_{24.8}$ 三元基础成分出发，探讨了 Cu 的添加对基础合金的玻璃形成能力(GFA)及相关金属玻璃的热稳定性、室温力学与耐蚀性能等的影响作用，并用普通铜模铸造法制备出 $Zr_{55.8}Al_{19.4}(Co_{1-x}Cu_x)_{24.8}$ (x = 0-0.8)系列块体金属玻璃（BMGs）。其中，$Zr_{55.8}Al_{19.4}Co_{17.36}Cu_{7.44}$ 成分处合金的 GFA 最优，其形成玻璃的临界直径 d_{max} 达 12 mm，并且，该金属玻璃具有良好的综合室温力学与耐蚀性能：杨氏模量为 89.8 GPa，室温压缩塑性应变达 4.0%，断裂强度约 2.0 GPa，断裂韧性近 120 MPa m$^{1/2}$；在 37 ℃磷酸盐缓冲液（PBS）中 $Zr_{55.8}Al_{19.4}Co_{17.36}Cu_{7.44}$ 金属玻璃显示出优良的耐蚀性，其自腐蚀电流密度为 5.0×10^{-9} A/cm^2，比同条件下 Ti6Al4V 合金低一个数量级。这种兼具大 GFA 与优良物理化学性能的

Zr-Al-Co-Cu 金属玻璃适合用作生物医用器件材料。

关键词：Zr-Al-Co-Cu，金属玻璃，力学性能，生物医用器件材料

$Zr_{55.8}Al_{19.4}(Co_{1-x}Cu_x)_{24.8}$ (x = 0-0.8)Bulk Metallic Glasses for Biomedical Devices Applications

Han Kaiming, Qiang Jianbing, Wang Yingmin, Haeussler Peter

(Key Laboratory of Materials Modification by Laser, Ion, and Electron Beams (Ministry of Education), Dalian University of Technology, Dalian 116024, China)

Abstract: Toxic element-free Zr-Al-Co-Cu bulk metallic glasses (BMGs) are probable candidates as surgical devices materials. In the present work, the quaternary compositions are produced by taking $Zr_{55.8}Al_{19.4}Co_{24.8}$ as a basic alloy, and Cu as an alloying element. Then the effect of Cu addition on glass formation,thermal glass stability, mechanical behaviors and corrosion resistance of a series of $Zr_{55.8}Al_{19.4}(Co_{1-x}Cu_x)_{24.8}$ (in at.%, x=0-0.8) bulk metallic glasses (BMGs) prepared by copper mold casting method wereinvestigated. The $Zr_{55.8}Al_{19.4}Co_{17.36}Cu_{7.44}$ BMG alloy is found to combine a centimeter scale glass formation size (d_{max} = 12 mm) and with high Young's modulus (E =89.8 GPa),ultrahigh strength (σ_y = 2.04 GPa), large room-temperature plasticity (ε_P=4.0%) and high fracture toughness about 120 MPa m$^{1/2}$. And the BMG exhibitsgood corrosion resistance in PBS open to air at 37 °C.The passive current density is 5.0×10^{-9} A/cm^2, which is one order of magnitude lower than Ti6Al4V alloy at same condition.The combination of high GFA and excellent physicochemical properties suggests that the $Zr_{55.8}Al_{19.4}Co_{17.36}Cu_{7.44}$ BMG can be a promising surgical devices material.

Key words: Zr-Al-Co-Cu, bulk metallic glasses, mechanical property, biomedical device materials

超音速火焰喷涂 Fe 基非晶涂层的组织与腐蚀性能

荣　震，李宝明，徐世霖

（北京中机联供非晶科技股份有限公司）

摘　要： 采用超音速火焰喷涂系统（HVAF）在 45 钢表面制备了 Fe 基非晶涂层。利用金相显微镜表征了涂层的显微形貌，X 射线衍射仪（XRD）分析涂层相组成，差热分析仪（DTA）测试了涂层的热稳定性，并测试了涂层的显微硬度、结合强度和抗腐蚀性能。结果表明：超音速火焰喷涂 Fe 的基非晶涂层，非晶含量可高达 97%，孔隙率为 0.49%，显微硬度 785HV，结合强度 49MPa。Fe 基非晶涂层在 HCl、NaCl 和稀 H_2SO_4 溶液中表现出较优的腐蚀性能。

关键词：超音速火焰喷涂，Fe 基非晶，腐蚀性能

Microstructures and Corrosion Resistance of Amorphous Coating by HVAF

Abstract: Fe-based amorphous coatings were prepared on 45 steel by High Velocity Air-Fuel (HVAF). The coating microstructure is characterized by metallographic microscope and X-ray diffraction (XRD).The thermal stability was tested by Differential Thermal Analysis (DTA). The hardness, bond strength and corrosion resistance were tested too. The results show that the amorphous content of the coating can be at 97 percent, the porosity is 0.49%, the hardness is 785HV, and the bond strength is 49MPa. Fe-based amorphous coatings showexcellent corrosion resistance in HCl, NaCl and H_2SO_4.

Key words: HVAF, Fe-based amorphous, corrosion resistance

国产非晶合金带材工程应用的研究

凌 健，顾雪军

（上海置信电气股份有限公司， 上海 200335）

摘 要：通过介绍国产非晶合金带材的 X 射线衍射（XRD）和差示扫描量热法（DSC）实验数据，得出目前国产非晶合金带材的非晶化结构的 XRD 图谱和起始晶化温度；通过分析非晶铁心单框成形过程受力仿真情况，提出了改进模具插入方式和移动方式的设想；通过以 120 只非晶铁心实际应用在 15 台非晶配变为样机的空载损耗和声压级实测值验证了国产非晶合金带材应用在非晶配变中是可以满足国标、国网企标的要求。

关键词：国产非晶合金带材，非晶铁心，非晶配变，空载损耗，声压级

An Analysis of Engineering Application of Domestic Amorphous Alloy Ribbon

Ling Jian, Gu Xuejun

(Shanghai Zhixin Electric Co., Ltd., Shanghai 200335, China)

Abstract: According to the experimental data of X-ray diffraction (XRD) and differential scanning calorimetry (DSC) of domestic amorphous alloy ribbon, the amorphous structure and crystallization temperature of this material were obtained. Through simulation analysis of the force of amorphous metal core in its formation process, this paper puts forward improvement method of mold movement. It is proved that the domestic amorphous alloy ribbon can meet the sound limitation requirement of China national standards by testing the no load loss and sound level of 15 sets of amorphous metal distribution transformers in which 120 amorphous metal cores are used.

Key words: domestic amorphous alloy ribbon, amorphous metal core, amorphous distribution transformer, no-load loss, sound level

铁基非晶热处理后软磁性能优化的结构起源

曹成成，朱 力，孟 洋，夏功婷，王寅岗

（南京航空航天大学材料科学与技术学院，江苏南京 210016）

摘 要：本文研究了退火温度、深冷处理、预退火时间等热处理工艺对铁基非晶合金微结构和磁性能的影响。退火和深冷处理研究的结果表明，虽然两种工艺使得非晶合金结构朝着两个不同的方向演变，但非晶合金的矫顽力均有

所降低。穆斯堡尔谱研究显示，矫顽力的降低是由于合金带材的易磁化轴在热处理过程中朝着带面的方向翻转导致的。对于非晶合金在不同预退火时间下的研究结果表明，弛豫过程中非晶基体 Cu 原子倾向于从 Fe 原子的第一壳层排出，并有与 P 原子结合形成 Cu-P 团簇的趋势。这些 Cu-P 团簇可以为非晶形核提供更多异质形核位置，使合金具有更为均匀细小的纳米晶结构，从而其软磁性能更为优异。

关键词：非晶带材，退火处理，深冷处理，磁软化，穆斯堡尔谱，纳米晶化

Linking the Optimization of Soft Magnetic Properties to the Structure Response during Thermal Treatment in Fe-based Metallic Glasses

Cao Chengcheng, Zhu Li, Meng Yang, Xia Gongting, Wang Yingang

(College of Materials Science and Technology, Nanjing University of Aeronautics
and Astronautics, Nanjing 210016, China)

Abstract: Effects of annealing temperature, cryogenic treatment and pre-annealing time on structure and magnetic properties of amorphous ribbons were investigated. Annealing causes relaxation while cryogenic treatment induces rejuvenation that the structure transfers to the opposite direction relative to that of annealing, but they both enhance soft magnetic properties of samples. The magnetic softening is attributed to the easy axis rotating toward the ribbon plane. For pre-annealed samples, Cu atoms tend to separate from the nearest shell of Fe atoms and form the Cu-P clusters. These clusters can provide more heterogeneous nucleation sites for α-Fe and promote grain refinement and thus an enhancement of soft magnetic properties of nanocrystalline alloy is thus obtained.

Key words: amorphous ribbon, annealing treatment, cryogenic treatment, magnetic soften, Mössbauer spectra, nanocrystalline

快速凝固 Al-Ca-Zn 合金共晶组织及形成机理

张慧贤[1]，陈子潘[2]，徐向棋[2]，段宗银[2]，孙　建[2]，陈守东[2]

（1. 钢研纳克检测技术有限公司；2. 铜陵学院机械工程学院）

摘　要：为了研究 Al-Ca-Zn 合金在非平衡凝固条件下的组织特征，采用水冷铜模吸铸法制备出直径为 1 mm 的合金试样。利用 X 射线衍射仪（XRD）进行结构检测，并通过扫描电镜（SEM）观察组织形貌特征，系统地分析了 $Al_{100-2x}Ca_xZn_x$(x=3.5，4，4.5)合金的铸态组织。结果表明，$Al_{100-2x}Ca_xZn_x$(x=3.5，4，4.5)合金的铸态组织主要由α-Al 和 Al_3CaZn 两相组成，由于冷却速度较快，当 x=3.5 时，合金由先共晶α-Al 相和纳米伪共晶组织构成；当 x=4 时，合金由完全纳米伪共晶组织构成；当 x=4.5 时，合金由先共晶 Al_3CaZn 相和纳米伪共晶组织构成。本文研究的快速凝固 Al-Ca-Zn 合金还未见报道，可为开发新型 Al-Ca-Zn 合金材料提供参考。

关键词：Al-Ca-Zn 合金，非平衡凝固，伪共晶组织，微观组织

Eutectic Structure and Its Formation Mechanism in Rapid Solidified Al-Ca-Zn Alloys

Zhang Huixian[1], Chen Zipan[2], Xu Xiangqi[2], Duan Zongyin[2], Sun Jian[2], Chen Shoudong[2]

(1. NCS Testing Technology Co., Ltd.；2. School of Mechanical Engineering, Tongling University)

Abstract: In order to study the microsturcture of Al-Ca-Zn alloys under non-equilibrium solidified condition, rod samples with a diameter of 1 mm were produced by water-cooled copper mold method. The microstructure of casting $Al_{100-2x}Ca_xZn_x$(x=3.5, 4, 4.5) alloys was systematically investigated by X-ray diffraction (XRD) and scanning electron microscope (SEM). The results show that the microstructure of as-cast of $Al_{100-2x}Ca_xZn_x$(x=3.5, 4, 4.5) alloys is mainly composed of α-Al and Al_3CaZn phases. As the cooling rate increasing, when x is 3.5, the alloy is composed of proeutectic α-Al phase and nano-pseudoeutectic structure; when x is 4, the alloy is completely consisted of nano-pseudoeutectic structure; when x is 4.5, the alloy contains proeutectic Al_3CaZn phase and nano-pseudoeutectic structure; Presently, reports about rapid solidified Al-Ca-Zn alloys haven't been found, therefore, in this work, it can provide guidance for exploring novel Al-Ca-Zn alloys.

Key words: Al-Ca-Zn alloy, non-equilibrium solidification, pseudoeutectic structure, microstructure

大原子半径元素对 Al-Ca-Ni 合金非晶形成能力的影响

陈子潘[1]，张慧贤[2]，徐向棋[1]，陈守东[1]，孙　建[1]

（1. 铜陵学院，安徽铜陵　244061；2. 钢研纳克检测技术有限公司，北京　100081）

摘　要：通过替换具有相似性质的大半径元素，系统地研究了 Na、Eu、Sr 元素替换对 Al-Ca-Ni 合金非晶形成能力的影响。采用单辊旋淬法制备出铝基非晶薄带，并通过 X 射线衍射仪（XRD）进行结构分析，示差扫描量热仪（DSC）进行热稳定性分析。采用楔形铜模吸铸法制备出楔形试样，通过扫描电镜（SEM）进行组织分析，确定合金的最大非晶形成尺寸。结果表明，当 Sr 的原子分数为 1% 时，合金的晶化温度增大，液相线温度降低，其非晶形成能力最好，合金的最大非晶形成尺寸为 410 μm。研究还发现，随着替换元素原子半径的增加，合金的非晶形成能力不断增加。

关键词：铝基非晶，非晶形成能力，热稳定性，大原子半径元素

Effect of Large Atomic Size Elements on the Glass Formation of Al-Ca-Ni Amorphous Alloys

Chen Zipan[1], Zhang Huixian[2], Xu Xiangqi[1], Chen Shoudong[1], Sun Jian[1]

(1. Tongling University, Tongling 244061, China;

2. China NCS Testing Technology Co., Ltd., Beijing 100081, China)

Abstract: Effects of different large atomic radius elements (i.e.,Na, Eu and Sr) with similar properties on the glass forming ability of Al-Ca-Ni alloy system were systematically investigated.Al-based amorphous belts were prepared by single roller melt-spinning, the structure was analysed by X-ray diffraction (XRD), the thermal properties were inverstigated by differential scanning calorimetry (DSC). Wedge shaped specimens were obtained by injecting the alloy melt into a copper-mold with a wedge cavity. The critical dimension for glass formation of the wedge samples was examined by Scanning electron microscopy (SEM). The results show that the crystallization temperature increased and the liquidus temperature decreased when the alloy contains 1 at.% Sr. Alloy has the best glass formng ability (GFA) and the maximal attainable thickness for glass formation is 410μm in Al-Ca-Ni alloy with 1 at.% Sr. Furthermore, the results also demonstratewd that the glass forming ability increased with increase of atomic size of adding elements.

Key words: Al-based amorphous alloy, glass forming ability, thermal stability, large atomic radius elements

超硬 Cr 基块体非晶合金

斯佳佳，惠希东

（北京科技大学，新金属材料国家重点实验室，北京　100083）

摘　要：开发了一种具有超高硬度和模量的四元 Cr 基块体非晶合金。测试表明，$Cr_xCo_{79-x}Nb7B_{14}$ (x = 45, 50, 55 at. %) 块体非晶合金的维氏显微硬度均高于 1300 HV，尤其是 $Cr_{55}Co_{24}Nb_7B_{14}$ 块体非晶合金的维氏硬度高达 1605 HV，仅低于已报道的少数富 B 的 Co 基块体非晶合金。纳米压痕测试表明，该系列合金的纳米压痕硬度均超过 15 GPa，纳米压痕模量均高于 250 GPa；超声模量测试表明，该系列铬基块体非晶的泊松比均较低（≤0.28），弹性模量高于 250 GPa。其中，$Cr_{55}Co_{24}Nb_7B_{14}$ 块体非晶合金的弹性模量为 278 GPa，比模量高达 37×10^6 Nm/kg，为目前所有块体非晶合金中的最高值。该合金的高硬度主要来源于合金内金属元素和硼元素之间的强烈的相互作用，该系列合金含有高含量 Cr 元素，因此在耐蚀耐磨领域具有很好的应用前景。

新型铁基非晶涂层抗辐照性能研究

吕　旷，惠希东

（北京科技大学，北京　100083）

摘　要：非晶结构失效将对材料性能产生巨大影响，因此研究非晶涂层的抗辐照能力对保证材料安全服役十分重要。本文采用 Xe 重离子对制备的 Fe-Cr-Mo-Zr-B 非晶合金进行辐照强度分别为 2dpa、6dpa 和 12dpa 的辐照试验，利用 XRD、SEM、SKP、纳米压痕等方法研究了 Xe 离子辐照对非晶合金涂层在结构和耐腐蚀性能的影响。结果表明，对比商业 Fe-Cr-Mo-C-B 非晶涂层，新型 Fe-Cr-Mo-Zr-B 非晶涂层在 Xe 重离子辐照下的抗辐照性能优于商业涂层的耐蚀性能和抗辐照性能。该结果对设计抗辐照非晶合金具有一定的指导意义，该新型非晶涂层有希望应用于各种辐照和耐蚀多重复杂环境中。

Effects of Cu Addition on the Glass Forming Ability and Corrosion Resistance of Ti-Zr-Be-Ni Alloys

Gu Jialun, Yao Kefu

(School of Materials Science and Engineering, Tsinghua University, Beijing 100084, China)

Abstract: The Ti-based bulk metallic glasses (BMG)with high Ti content (>50 at.%) have attracted wide attention for their outstanding mechanical properties, despite their glass forming ability (GFA)is still relatively poor. Here the $(Ti_{55}Zr_{15}Be_{20}Ni_{10})_{100-x}Cu_x$(x=2, 4, 6, 8 and 10 at.%) BMGs were designed and prepared by copper mold casting. The critical diameter of the original alloy could be enhanced from 5 mm to 10 mm with the Cu addition of 4 at.%, which is the first reported centimeter-sized Ti-based BMG with Ti content more than 52 at.%. In addition, all the as-prepared BMGs exhibited higher yield strength (above 2000 MPa). And the supercooled liquid range of the as-prepared Ti-based BMGs were enlarged from 35 K to 74 K by Cu addition. Electrochemical measurements showed that the BMGs with Cu addition of 0 ~ 6 at.% exhibited much higher pitting potentials (over 573 mV/SCE) than that of 304 stainless steel (304SS)in 3.5 wt.% NaCl aqueous solution. And The XPS analysis revealed that the Cu addition might lead to the enrichment of Cu and the deficiency of Ti, Zr, Ni elements in passive film, which could induce the worse corrosion resistance of the Ti-based BMGs with high Cu content. The results indicate that the developed Ti-based BMGs possess good GFA, mechanical properties and corrosion resistance.

TiZrNbMoV 高熵合金的相组成与固溶强化机理

吴一栋，惠希东

（新金属材料国家重点实验室，北京科技大学，北京 100083）

摘要： 使用铜模铸造的方法制备了 $TiZrNbMo_xV_y$ 高熵合金系，其中 x = 0 ~ 2，y = 1 和 0.3。本文系统研究了该高熵合金系的相组成和相稳定性。实验结果表明，低 V 含量的合金（y = 0.3）组织由单一体心立方结构相组成，并且该组织在 1273K 下具有优异的稳定性。升高 V 和 Mo 的含量，合金组织由两种体心立方相组成，并且组织稳定性降低，在 1273K 下保温 72 小时生成了 Laves 相。本文基于文献中报道的预测判据详细研究了合金系的相形成规律。铸态和退火态合金的压缩试验结果表明 Mo 在该合金体系中起到强烈的固溶强化作用，尤其是 $TiZrNbMo0.3V0.3$ 合金的压缩屈服强度达到 1312Mpa，压缩塑性超过 50%。该合金在 1273K 下保温 72 小时依然保持同样优异的压缩性能。

Influence of Similar Atom Substitution on Glass Formation and Property in (Fe,Co)-B-Si-Zr/Hf Bulk Glassy Alloys

Geng Y.X.[1], Lin X.[2], Jie L[1], Fan S.M.[1], Xu J.H.[1], Yu L.H.[1]

(1.School of Materials Science and Engineering, Jiangsu University of Science and Technology, Zhenjiang 212003, China; 2. State Key Laboratory of Solidification Processing, Northwestern Polytechnical University, Xi'an 710072, China)

Abstract: In this work, the alloying effects of similar atom substitution of Co for Fe on the glass-forming ability and the various properties of $Fe_{71.7-x}Co_xB_{16.7}Si_{8.3}Zr_{3.3}$ and $Fe_{72.5-y}CoyB_{16.7}Si_{8.3}Hf_{2.5}$ (x, y = 0, 5, 10, 15, 20 and 25) metallic glass are investigated. Our experiments show that the substitution of Co for Fe causes a decrease in the glass-forming ability. By increasing the content of Co, the critical glass formation size gradually decreases from 2.5 mm to 1 mm. Nano-indentation results indicate that the elastic modulus and hardness intensify with increasing Co content in $Fe_{71.7-x}Co_xB_{16.7}Si_{8.3}Zr_{3.3}$ bulk glassy alloys. $Fe_{51.7}Co_{20}B_{16.7}Si_{8.3}Zr_{3.3}$ bulk glassy alloy exhibits super-high elastic modulus and hardness of over 240 GPa and 19 GPa, respectively. All of the glassy alloys exhibit good soft magnetic properties with high saturation magnetization (1.12~1.33 T) and low coercive force (0.6~   5.2 A/m).

Key words: (Fe,Co,)-B-Si-Zr/Hf bulk glass alloys, glass-forming ability, mechanical property, magnetic property

Influences of the Preparation Techniques on Glass Forming Ability of Fe–P–B–Si–C Amorphous Alloys

Zhang Jijun[1,2,3,4], Dong Yaqiang[1,2], Bie Luyang[4], Li Qiang[4], Li Jiawei[1,2] Wang Xinmin[1,2]

(1. Key Laboratory of Magnetic Materials and Devices, Ningbo Institute of Materials Technology & Engineering, Chinese Academy of Sciences, Ningbo 315201, China;

2. Zhejiang Province Key Laboratory of Magnetic Materials and Application Technology, Ningbo Institute of Materials Technology & Engineering, Chinese Academy of Sciences, Ningbo 315201, China;

3. University of Chinese Academy of Sciences, Beijing 100049, China;

4. School of Physics Science and Technology, Xinjiang University, Urumqi 830046, China)

Abstract: In contrast to the ultrahigh cooling rate that is required to for glass formation, larger glass forming ability (GFA) is achieved at lower cooling rate in $Fe_{76}P_5(B_{0.5}Si_{0.3}C_{0.2})19$ synthesized in argon atmosphere. Cylindrical rods with diameters of 1 to 2 mm are prepared by water quenching without flux treatment, Cu-mold injection casting and Cu-mold suction casting, respectively. The influences of the preparation techniques with different cooling rates on GFA, thermal property and nucleation/growth behavior are examined. The critical diameter of $Fe_{76}P_5(B_{0.5}Si_{0.3}C_{0.2})_{19}$ amorphous alloys is 1.7 mm for water quenchingwhile smaller than 1 mm for injection casting. Microstructures analysis indicates that the crystallization and solidification processes are quite different between the water-quenched and the injection casted rods.This finding provides fundamental understanding on the relationship between the cooling rate, techniques and glass forming ability of Fe-based amorphous alloys.

Key words: Fe-based amorphous alloys, preparation techniques, glass forming ability, crystallization behaviors

Correlation between Sructure and Gass-forming Ability of Al$_{86}$Ni$_{14-x}$La$_x$(x=3,5,9) Metallic Gasses: An ab Initio Molecular Dynamics Study

王芳茹

（中国人民大学）

Abstract: The correlation of atomic and electronic structures with glass-forming ability (GFA) in Al$_{86}$Ni$_{14-x}$La$_x$ (x=3,5,9 at. %) metallic glasses (MGs) has been systematically elucidated via ab initio molecular dynamics simulations. The local atomic structures in Al$_{86}$Ni$_{14-x}$La$_x$ (x=3,5,9 at. %) metallic liquids and glasses change systematically with adjusting the concentration of solute atoms. It is found that GFA in Al-based MGs is strongly correlated with the chemical short-range order (CSRO) around Al atoms. Furthermore, the good glass-forming composition of Al$_{86}$Ni$_9$La$_5$ MG can be obtained when the Fermi surfaces nearly touch the quasi-Brillouin boundaries, and the Fermi level also lies at a minimum of total electronic density of states essentially caused by the local minima of La-5d and Ni-3d electron bands, which is consistent with the prediction of nearly-free-electron model. This finding provides new insights into the origin of GFA in Al-based MGs.

钴基非晶合金丝磁性能及传感技术研究

满其奎，王新敏，李润伟

（浙江省磁性材料及应用技术重点实验室，中国科学院磁性材料与器件重点实验室，
中国科学院宁波材料技术与工程研究所，浙江宁波　315201）

摘　要：非晶态合金丝，以其高强度、高弹性模量、良好的耐磨耐蚀性以及优异的磁电性能等优点在磁性功能材料和结构材料领域具有应用前景。采用 CoFeBSiNb 非晶合金体系，获得了高质量直径约 50μm 的非晶细丝，研究了钴基非晶丝软磁特性、巨磁阻抗效应和弱磁检测传感器。非晶丝的饱和磁化强度约 0.55T，矫顽力为 2 A/m 左右，起始磁导率32000。研究了驱动条件、退火条件、应力状态、排布方式对巨磁阻抗效应的影响规律和形成机制。

研究发现：等温退火热处理能改善铸态非晶合金丝的软磁性能，通过调控钴基非晶丝数量和间距获得了巨磁阻抗效应的增强，巨磁阻抗变化率由35%增加到205%，该效应的增强源于非晶丝间的磁交互作用。通过对非晶丝进行硅烷处理来调控 Co 基非晶丝/树脂复合材料的界面，发现硅烷分子与非晶丝表面形成 Si—O—Si 和 Fe—O—Si 共价键，导致感生磁各向异性增强，巨磁阻抗变化率由85%增加到310%。通过调控非晶丝的面内弹性应变，有效调控了非晶丝的磁畴结构和磁阻抗。采用压电衬底材料施加应力，实现磁畴壁的移动和部分磁畴反转。0.052%的弹性变形导致巨磁阻抗变化率增加了 2 倍。基于巨磁阻抗效应研究结果，开发了应用于车辆检测系统的弱磁检测传感器和应力传感器样机。

关键词：非晶合金，软磁性能，巨磁阻抗效应，传感技术

Ta 丝增强 Zr 基非晶合金复合材料变形行为研究

陈　森[1,2]，付华萌[2]，李正坤[2]，张　龙[2]，朱正旺[2]，张宏伟[2]，

李　宏[2]，王爱民[2]，张海峰[2]

（1.东北大学材料科学与工程学院，辽宁沈阳　110819；

2.中国科学院金属研究所沈阳材料科学国家（联合）实验室，辽宁沈阳　110016）

摘　要： 本文利用液态浸渗法成功制备了一种新型结构形式的 Ta 丝增强 Zr 基非晶合金复合材料，Ta 丝以螺旋密排方式均匀分布于非晶基体中。复合材料在压缩时经历了两次屈服过程，样品最终被压缩成薄饼状。基体和 Ta 丝之间的界面层首先萌生垂直于 Ta 丝的微裂纹，随后扩展到非晶基体诱发剪切带。剪切带相互作用形成剪切裂纹，并导致 Ta 丝发生剪切变形。随加载进行，Ta 丝产生严重的挤压变形而进入裂纹内部。螺旋形式的 Ta 丝改变了复合材料的内部应力场，使其变形方式和机制不同于传统的纤维增强复合材料，为非晶合金复合材料的应用提供了新思路和新方法。

关键词： 非晶合金复合材料，Ta 丝，螺旋，剪切带，变形行为

Deformation Behaviour of Ta-wire-reinforced Zr-based Bulk Metallic Glass Composites

Chen Sen[1,2], Fu Huameng[2], Li Zhengkun[2], Zhang Long[2], Zhu Zhengwang[2],

Zhang Hongwei[2], Li Hong[2], Wang Aimin[2], Zhang Haifeng[2]

(1. School of Materials and Metallurgy, Northeastern University, Shenyang 110819, China;

2. Shenyang National Laboratory for Material Science, Institute of Metal Research,

Chinese Academy of Sciences, Shenyang 110016, China)

Abstract: The Ta wire-reinforced Zr-based metallic glass composites with a new type of structure were prepared successfully by the method of liquid metal infiltration. Ta wires distributed uniformly in the matrix in the form of spiral. The composites experienced two yield stages under the compressive stress, and had been compressed into the pancake shape. The microcracks originated perpendicular to the interface between the Ta wire and the matrix, and then induced the shear bands in the matrix due to the stress concentration. The shear cracks formed with the interaction of shear bands during the continued loading process, and the shear deformation of Ta wires occurred on the impact of shear bands. The stress fields in the composites had been changed obviously due to the introduction of spiral formed reinforcements. The investigation of the deformation behaviour and mechanism suggested a new method for the application of bulk metallic glass composites in the structural materials.

Key words: metallic glass composites, Ta wire, spiral, shear band, deformation behaviour

Zr 基块体非晶合金在形变中的相分离

王　璐，王　拓，惠希东

（北京科技大学，新金属材料国家重点实验室，北京　100083）

摘　要：由于非晶合金没有晶界，不会像晶体材料一样具有位错。因此大多数块体非晶合金往往没有或只有很少的塑性。提高非晶合金的塑性是近年来研究的热点。本文通过成分设计，制备出了具有 20%以上压缩塑性的 Zr 基非晶合金。我们通过高分辨透射电镜（HRTEM）观察了铸态以及发生变形后的合金组织。研究发现合金在原始状态是典型的迷宫状组织，而经过变形后的合金在剪切带的周围明显发生了相分离。这表明变形给合金注入了能量，促进了具有正混合焓元素的非晶合金在压缩变形中发生相分离。

关键词：块体非晶，Zr 基，塑性，相分离

亚快速凝固/长周期结构增强的高强度 $Mg_{94}Zn_{2.4}Y_{3.6}$ 合金

朱　建

（北京科技大学，北京　100083）

摘　要：为了得到高强度的镁合金，我们用亚快速凝固/水冷铜模喷铸的方法制备了 $Mg_{94}Zn_{2.4}Y_{3.6}$ 合金。我们全面地研究了该合金的微观组织及力学性能，并将其与常规铸造的 $Mg_{94}Zn_{2.4}Y_{3.6}$ 合金作比较。我们发现，亚快速凝固 $Mg_{94}Zn_{2.4}Y_{3.6}$ 合金具有远远偏离平衡凝固的组织特点，导致它含有 1～3μm 的细晶α-Mg、不同寻常的 Y、Zn 溶质的超固溶度以及网状的长周期相和细小的 $Mg_{24}Y_5$ 相。长周期相与α-Mg 基体之间呈半共格关系，具有如下取向 $[0002]_\alpha//[1120]_{LPSO}$, $(1010)_\alpha//(0002)_{LPSO}$。亚快速凝固 $Mg_{94}Zn_{2.4}Y_{3.6}$ 合金具有该合金具有高硬度(112HV)、高抗拉强度(335MPa)和良好的耐热及耐蚀性能。该合金的强化机制可以归结为：细晶强化、长周期相强化、固溶强化和 $Mg_{24}Y_5$ 相的弥散强化。

W 元素对 Fe 基非晶合金热稳定性和电化学行为的影响

梁丹丹，魏先顺，江浩然，沈　军

（同济大学材料科学与工程学院，上海　201804）

摘　要：本文研究了添加难熔金属 W 元素对 $Fe_{47-x}Cr_{20}Mo_{10}W_xC_{15}B_6Y_2$ (x=0, 2, 4, 6 at.%)非晶合金的热稳定性和电化

学行为的影响。结果表明，适量添加 W 元素可抑制 $Fe_{23}B_6$ 晶体相的形成，$\Delta T_x(T_x-T_g)$ 和 γ [$T_x/(T_g+T_l)$] 均随着 W 含量的增加先增加后降低。采用工业纯度的原材料可制备出直径为 8 mm 的 $Fe_{43}Cr_{20}Mo_{10}W_4C_{15}B_6Y_2$ 非晶棒材。通过 Mott-Schottky 测试和角分辨 XPS 分析可知，$Fe_{47-x}Cr_{20}Mo_{10}W_xC_{15}B_6Y_2$ 非晶合金的钝化膜均随着外界干扰电位呈 p-n 半导体性质，且 W 元素的添加会降低受体浓度(N_A)和供体浓度(N_D)。此外，W 元素添加可促进钝化膜中的 Fe^{3+}/Cr^{3+}-氧化物的形成，从而提高钝化膜的厚度和稳定性，改善其耐蚀性。

关键词：非晶合金，热稳定性，钝化膜，Mott-Schottky 测试，电子结构，角分辨 XPS

The Effect of W Addition on the Thermal Stability and Electrochemical Behavior of Fe-based Amorphous Alloys

Liang Dandan, Wei Xianshun, Jiang Haoran, Shen Jun

(School of Materials Science and Engineering, Tongji University, Shanghai 201804, China)

Abstract: The effects of refractory element W addition on the thermal stability and electrochemical behavior of $Fe_{47-x}Cr_{20}Mo_{10}W_xC_{15}B_6Y_2$ (x=0, 2, 4, 6 at.%) amorphous alloys were investigated. An appropriate substitution of Fe by W (4 at. %) could suppress the formation of the $Fe_{23}B_6$ phase and thus improve the GFA and thermal stability. The supercooled liquid region $\Delta T_x(T_x- T_g)$ and criterion γ [$T_x/(T_g+T_l)$] of $Fe_{47-x}Cr_{20}Mo_{10}W_xC_{15}B_6Y_2$ amorphous alloys increased from 55 K and 0.387 to 62 K and 0.390, respectively, when x increased from 0 at. % to 4 at. %. The GFA and thermal stability declined when x increased to 6 at. %. The $Fe_{43}Cr_{20}Mo_{10}W_4C_{15}B_6Y_2$ bulk metallic glass (BMG) rod with an 8 mm diameter could be fabricated by using commercial purity materials. The electrochemical measurements showed that the substitution of W with Fe element in Fe-based amorphous alloy could enhance the corrosion resistance and all passive films with different W addition exhibited dipolar (p-n) semiconducting characteristics with varied applied potential. Both the acceptor density (N_A) and donor density (N_D) reduced with the incensement of W content, which demonstrated that W addition modified the outer and inner oxide layers of passive films and improved the stability of passive films. Moreover, it can be obtained from the angle-resolved XPS quantitative analysis results that adding W element promotes the formation of Fe^{3+}- and Cr^{3+}-oxides and thickens the passive film. The existence of W ions in the passive film has a beneficial effect on corrosion resistance by forming a thicker and more stable passive film to inhibit the dissolution of Fe and Cr elements.

Key words: amorphous alloy, thermal stability, passive film, Mott-Schottky measurement, electronic structure, angle-resolved XPS

HVAF 喷涂铁基非晶合金涂层的
组织和电化学行为研究

高　涵，魏先顺，沈　军

（同济大学材料科学与工程学院，上海　201804）

摘　要： 本文采用超音速火焰喷涂(HVAF)技术，在 45#碳钢基体上制备了成分为 $Fe_{49.7}Cr_{18}B_{15.2}Mo_{7.4}C_{3.8}Si_{2.4}Mn_{1.9}W_{1.6}$ (at.%)的非晶合金涂层，Fe 基非晶合金涂层结构致密，其孔隙率仅为 0.5%，硬度高达 1025 ± 74 $HV_{0.1}$。通过电化学方法对非晶合金涂层在 3.5% NaCl，0.5M H_2SO_4 和 1M HCl 溶液中的腐蚀行为进行探究。动态极化测试结果表明，与不锈钢相比，涂层可以形成更宽的钝化区间和稳定的钝化膜来抵御局部腐蚀。EIS 曲线反映了在 NaCl、HCl 环境中，高的孔隙度缺陷可作为氯离子的扩散通道，使得溶液渗入内部，诱发产生腐蚀产物，从而降低涂层性能。此外，涂层较高的非晶相含量使其在含氢离子的 HCl 和 H_2SO_4 溶液中表现出极佳的耐腐蚀性能。HVAF 非晶合金涂层在严苛环境的工业生产中具有广阔应用前景。

关键词：铁基非晶合金涂层，HVAF，热喷涂，耐腐蚀性能

Structure and Corrosion Behavior of Fe-based Amorphous Alloy Coatings Produced by HVAF

Gao Han, Wei Xianshun, Shen Jun

(School of Materials Science and Engineering, Tongji University, Shanghai 201804, China)

Abstract: Fe-based amorphous coatings with the composition of $Fe_{49.7}Cr_{18}B_{15.2}Mo_{7.4}C_{3.8}Si_{2.4}Mn_{1.9}W_{1.6}$(at.%) were deposited on the ASTM 1045 carbon steel substrate by high velocity air fuel (HVAF) thermal spraying process. A dense amorphous alloy coating with low porosity of 0.5% and microhardness of 1025 ± 74 $HV_{0.1}$ was obtained. The corrosion behaviors of as-deposited coatings were studied by electrochemical test in 3.5% NaCl, 0.5M H_2SO_4 and 1M HCl solutions. Potentiodynamic polarization tests revealed that amorphous coatings have wider passive region and stable passivation film to resist localized corrosion than the 304 stainless steel. This amorphous alloy coating exhibited outstanding corrosion resistance due to the formation of stable passive films. Moreover, the porosity defects are the diffusion channels of chloride ions immersed in NaCl and HCl solution, the clogging of the corrosion product easily occurs in coatings with higher porosity reflected by EIS curves, which can also palliate the corrosion process, when the solution penetrated into the defects of the coatings in Cl^- solution. The better corrosion resistancein hydrogen-containing HCl and H_2SO_4 solution was attributed to higher amorphous content of coatings. The results demonstrate that the superior corrosion resistance of amorphous coatings in the aggressive corrosive environments is attributed to dense coating structure and high amorphous content. The amorphous alloy coatings fabricated by HVAF have a promising potential as the corrosionresistant coating in the industrial environment.

Key words: Fe-based amorphous alloy coating, HVAF, thermal spray, corrosion behavior

封孔处理对电弧喷涂 Fe 基非晶涂层耐蚀性的影响

颜月梅，魏先顺，沈　军

（同济大学材料科学与工程学院，上海　201804）

摘　要： 本文采用磷酸铝和环氧树脂封孔剂对电弧喷涂 Fe 基非晶涂层进行封孔处理。利用 X 射线衍射仪（XRD）和扫描电镜（SEM）对涂层的组织结构进行了表征，通过电化学方法对封孔前后的涂层在中性（3.5 wt.%NaCl）和酸性（0.5 mol/L H_2SO_4）环境下的耐腐蚀性能进行研究。结果表明：两种封孔剂均可以渗透到涂层内部并填充涂层孔隙，封孔后涂层的耐腐蚀性能得到明显提升。在 3.5 wt.%NaCl 溶液中，磷酸铝封孔涂层的钝化电流密度显著低于环氧树脂封孔涂层，表明在中性环境中磷酸铝封孔剂的封孔效果更好。在 0.5 mol/L H_2SO_4 溶液中，二者表现相当。封孔处理可以阻止或减缓外部腐蚀介质对涂层的渗透作用，从而起到有效的防护作用。

关键词：电弧喷涂，非晶合金涂层，封孔处理，耐腐蚀性能

Effect of Sealing Treatment on Corrosion Resistance of Arc Sprayed Fe-based Amorphous Metallic Coatings

Yan Yuemei, Wei Xianshun, Shen Jun

(School of Materials Science and Engineering, Tongji University, Shanghai 201804, China)

Abstract: Arc sprayed Fe-based amorphous metallic coatings were sealed with aluminum phosphate and epoxy resin sealants. The morphology and microstructure of the as-sprayed coating and sealed coatings were characterized by X-ray diffraction (XRD) and scanning electron microscopy (SEM). The corrosion behavior of the coatings in neutral (3.5 wt.% NaCl) and acid (0.5 mol/L H_2SO_4) solutions was examined by potentiodynamic polarization and electrochemical impedance spectroscopy. The microstructure analysis suggests that both aluminumphosphate and epoxy resin sealants can penetrate into the coating and fill the structural defects. The electrochemical results show that the sealed coatings have better corrosion resistance than the unsealed coating in both neutral and acid solutions. The passive current density of coatingssealed with aluminum phosphate sealant is obviously lower than that of the coatings sealed with epoxy resinse alant in 3.5 wt.% NaCl solution, implying that the aluminum phosphate sealant is more efficient for the arc sprayed amorphous coating in neutral environment. The coatings sealed with aluminum phosphate and epoxy resin sealants perform similarly in 0.5 mol/L H_2SO_4 solution. The improved corrosion behavior of the sealed coatings is attributed to the blocking effect of sealants which prevents the penetration of corrosive solution.

Key words: arc spray, amorphous coating, sealing treatment, corrosion resistance

HVAF 喷涂 Fe 基非晶合金涂层在模拟燃料电池环境下的腐蚀行为研究

魏先顺，梁　丹，沈　军

（同济大学材料科学与工程学院，上海　201804）

摘　要: 采用空气激发超音速火焰喷涂(HVAF)技术制备了成分为 $Fe_{40}Cr_{23}Mo_{14}C_{15}B_6Y_2$ (at. %)的 Fe 基非晶合金涂层材料，采用电化学动电位极化曲线和阻抗谱研究了 Fe 基非晶合金涂层在不同浓度 SO_4^{2-} 和 Cl^- 的模拟燃料电池环境中的腐蚀行为。电化学动电位极化曲线结果表明 Fe 基非晶合金涂层在含有 SO_4^{2-} 和 Cl^- 的酸性介质中能够发生自发钝化，并且具有较宽的钝化区和较低的钝化电流密度，瞬态电流测试结果证明非晶合金涂层可以在很短的时间内形成钝化膜，I-t 曲线表明涂层具有很低的钝化电流，表明非晶合金涂层在腐蚀介质环境下可以保持钝化膜稳定。Fe 基非晶合金涂层的阻抗曲线呈现单一的时间常数，表明具有高自发钝化能力。温度对电化学反应有显著的影响，虽然非晶涂层在 25~80℃温度下都可以保持稳定钝化，并形成较宽的钝化区，但随着温度的升高，Fe 基非晶合金涂层的耐蚀能力和耐点蚀能力都随之降低。

关键词: 涂层，HVAF，热喷涂，腐蚀行为，PEM 燃料电池

Corrosion Behavior of HVAF Sprayed Fe-based Amorphous Alloy Coating in Simulated PEM Fuel Cell Environments

Wei Xianshun, Liang Dan, Shen Jun

(School of Materials Science and Engineering, Tongji University, Shanghai 201804, China)

Abstract: Fe-based amorphous alloy coatings with nominal composition of $Fe_{40}Cr_{23}Mo_{14}C_{15}B_6Y_2$ (at.%) were fabricated by high velocity air fuel (HVAF) process. Corrosion behavior of HVAF sprayedFe-basedamorphous alloy coatings was investigated by potentiodynamic polarisation and electrochemical impedance spectroscopy (EIS) method in simulated proton exchange membrane (PEM) fuel cell environments with different concentration of SO_4^{2-} and Cl^-. Potentiodynamic polarization results indicatethat Fe-based amorphous coatings can passivate spontaneously with wide passive region and low passive current density in the SO_4^{2-} and Cl^- acidic environment.Transient current measurements show that the passive film of Fe-based amorphous coatings can be formed in in a short time.Low current density was observed in I-t curves, which indicates the stability of the passive film.The impedance curve of Fe-based amorphous alloy coatingsshows a single time constant, representing a good spontaneous passivation behavior.Temperature has a significant effect on the corrosion behavior of Fe-based amorphous alloy coatings. Although the amorphous coating can be passivated stably with a wide passivation region at 25℃~80℃, the corrosion resistance and pitting resistance were reduced with the increase of temperature.

Key words: coating, HVAF, thermal spray, corrosion behavior, PEM fuel cell

Fe-Co-Si-B-P-Cu 纳米晶条带软磁性能研究

苗宝雯，罗 强，沈 军

（同济大学材料科学与工程系，上海 201804）

摘 要：本文采用单辊旋淬法制备了 FeCoSiBCuP 系非晶/纳米晶合金,利用 X 射线衍射仪和差示扫描量热法表征了该软磁合金的结构特征，热力学性能以及晶化过程。然后利用振动样品磁强计、阻抗分析仪、交/直流 B-H 仪等测定了该合金的软磁性能，系统研究了 Co 元素的添加以及退火条件对 FeSiBPCu 体系的纳米晶结构和软磁性能影响。获得了具有。研究表明，该系列纳米晶软磁合金材料具有高饱和磁感应强度、低矫顽力和高磁导率。$Fe_{80}Co_xB_{14-x}Si_2P_3Cu_1$ 合金中 Co 和 B 相对含量的改变对该合金的微观结构和磁性能均有显著的影响，在 Co 含量为 4 at.%时该合金的软磁性能最佳。随着退火工艺的改变，非晶态合金晶化后将得到不同结构和性能的纳米晶合金。在最佳退火条件下，该合金的饱和磁感应强度可达 1.84T，矫顽力为 5.3A/m,磁导率为 12601（1kHz）。

关键词：纳米晶，单辊旋淬法，饱和磁感应强度，矫顽力，磁导率

Effect of Co Addition and Annealing Condition on Magnetic Properties of Nanocrystalline FeCoSiBPCu Ribbons

Miao Baowen, Luo Qiang, Shen Jun

(Tongji University, Shanghai 201804, China)

Abstract: We report on the effect of substituting Co for metalloid element on glass forming ability, thermal property, and magnetic properties of $Fe_{80}Co_xB_{14-x}Si_2P_3Cu_1$(x=0,2,4,6) alloy ribbons. The experimental results showed that addition of Co decreased the thermal ability against crystallization of the amorphous phase, and thus enlarged the region of heat treatment temperature of this alloy. This series of alloys exhibit high saturation magnetic flux density, low magnetic coercivity and high effective permeability. Among them, the $Fe_{80}Co_4B_{10}Si_2P_3Cu_1$ nanocrystalline alloy after optimized heat treatment exhibits the best soft magnetic performance with a high saturation magnetic flux density B_s of 1.84 T, a low coercivity H_c of 5.3 A/m, and a high effective permeability μ_e of 12601 at 1 kHz.

Key words: nanocrystalline alloy, soft magnetic properties, saturation magnetization, coercivity, core loss

$Cu_{60}Zr_{30}Ti_{10}$ 块体金属玻璃的纳米化及性能研究

蔡 晋[1]，刘 辉[2]，王 征[3]，朱继宏[2]

（1. 沈阳航空航天大学航空航天工程学部，辽宁沈阳　110136；2. 西北工业大学机电学院，陕西西安　710072；3. 中国航发沈阳黎明航空发动机有限责任公司技术中心，辽宁沈阳　110043）

摘　要：本文研究了铜基块体金属玻璃的力学性能，并且对于热松弛的程度和不同的晶体体积分数进行了分析。对样品以恒定加热速率的差示扫描量热法、宏观压缩和断口性能进行测试。样品在加热到 781 K，随着退火过程时间的增加导致了第一个峰值转移到较低的温度处，且峰的焓值降低。膨胀测量显示了结晶的两个阶段，且第一阶段的体积效应约为 2.4%。宏观压缩测试表明，断裂强度和杨氏模量随着热处理(退火)时间的增加而提高。然而，当长时间退火处理时，材料会发生一定的脆化。退火后的非晶样品中发现了晶粒尺寸为 1~5 nm 的纳米晶。

关键词：金属玻璃，力学性能，热松弛，纳米晶

Study of Nanocrystalline and Properties of $Cu_{60}Zr_{30}Ti_{10}$ Bulk Metallic Glass

Cai Jin[1], Liu Hui[2], Wang Zheng[3], Zhu Jihong[2]

(1. Faculty of Aerospace Engineering, Shenyang Aerospace University, Shenyang 110136, China；
2. School of Mechanical Engineering, Northwestern Polytechnical University, Xi'an 710072, China；
3. Technical Center, AECC Shenyang Liming Aeroengine Co., Ltd., Shenyang 110043, China)

Abstract: The effect of thermal relaxation degree and different crystalline volume fractions on the mechanical response of a Cu-based bulk metallic glass has been investigated. The different degree the thermal relaxation was characterized by constant rate heating differential scanning calorimetry (DSC) and the mechanical tests including the macroscopic compression and fracture morphology analysis. When the sample heated to 781K, the first peak shift to the lower temperature and the enthalpy value of the peak decrease. Dilatometry measurement showed that two stages of the crystallization, the volume effect of the first stage about 2.4%. Macroscopic compression tests showed that the fracture strength and Young's modulus increase with increasing annealing time. However, embrittlement of the material was also found for the long term annealing. Moreover, the grian size of 1~5 nm nanocrystals were observed after annealing.

Key words: metallic glass, mechanical property, thermal relaxation, nanocrystalline

钨/低活化钢钎焊用 Fe 基非晶钎料研究

刘天鸷[1]，羌建兵[1]，王英敏[1]，Haeussler Peter[1]，王建豹[2]，练友运[2]

（1.大连理工大学三束材料改性教育部重点实验室，辽宁大连 116024）；

（2.核工业西南物理研究院聚变所，四川成都 610041）

摘　要：在 DEMO 聚变示范堆的偏滤器中，钨/低活化钢之间的有效连接至关重要。本文运用团簇合金设计理论结合单辊急冷技术开发出 $Fe_{60}Mn_{15}B_{16.67}Si_{6.33}Sn_2$ 箔状新非晶，并将其作为钎料应用于钨/低活化钢的连接，以此为基础研究了钎焊温度（1180-1220℃）与保温时间（15~45min）对相关钎焊接头组织与性能的影响规律。最终结果表明：当钎焊温度为 1220℃、钎焊时间为 30min 时，所获得的钎焊接头质量最优，其界面结合良好，无孔洞、裂纹等宏观缺陷，焊缝具有较高的剪切强度。

关键词：偏滤器，Fe-B-Si-Mn-Sn，钎焊温度，钎焊时间，钎焊接头

The Research on Fe-based Amorphous Filler Metal for Joining Tungsten/Reduced Activation Steels

Liu Tianzhi[1], Qiang Jianbing[1], Wang Yingmin[1], Haeussler Peter[1], Wang Jianbao[2], Lian Youyun[2]

(1.Key Laboratory of Materials Modification by Laser,Ion,and Electron Beams (Ministry of Education),Dalian University of Technology, Dalian 116024, China;

2.Fusion Laboratory,Southwestern Institute of Physics,Chengdu 610041, China)

Abstract: The effective joining of tungsten and reduced activation steels is crucial for the divertor of developing demonstration fusion reactor (DEMO). The amorphous alloy composition of $Fe_{60}Mn_{15}B_{16.67}Si_{6.33}Sn_2$ has been designed as a new filler metal in the present work by using the so-called cluster-formula alloy design method, andribbon sampleswere prepared through melt-spinning technique, then the amorphous filler metal was used to braze the tungsten and reduced activation steels, and the influences of welding temperature (1180~1220℃) and brazing time（15~45min）on the microstructure and property of brazed joints were investigated. The experimental results revealed that theholes and cracks-free brazing joint with high shear strength can be achieved at the welding temperature of 1220℃ with 30 minutes duration.

Key words: amorphous filler, Fe-B-Si-Mn-Sn, tungsten, reduced activation steels, brazing joint

铁基(Fe, Ni, Mo, Cr)-P-C-B 系非晶合金的玻璃形成能力和耐蚀性

王思雯，贾行杰，李艳辉，王雪威，张　伟

（大连理工大学材料科学与工程学院，辽宁大连 116024）

摘　要：铁基非晶合金因其高强度、优异的耐磨性、良好的耐蚀性和较低的成本而具有广阔的应用前景。利用非晶合金过冷液体的粘性流动特性，可采用热喷涂技术将铁基非晶合金粉体在器件表面制成致密的涂层，提高器件的耐磨耐腐蚀性能。适用于热喷涂技术的非晶合金需兼备低玻璃化转变温度（T_g）、大过冷液相区间（ΔT_x）、高玻璃形成能力、优异的力学性能和耐蚀性。本工作中，我们研究了耐蚀元素 Mo、Cr 和 Ni 的复合添加对具有低 T_g 的 Fe-P-C-B 系非晶合金热稳定性、玻璃形成能力、力学性能和耐蚀性的影响。

采用感应熔炼制备母合金锭；采用铜模铸造法制备棒状样品；采用 X 射线衍射仪对样品进行组织结构表征；采用差示扫描量热仪和差热分析仪测定合金的热性能；采用电化学工作站评价合金在 0.5 mol/L 硫酸溶液和 1.0 mol/L 盐酸溶液中耐蚀性；采用万能实验机测定合金在压缩条件下的力学性能。

在 $Fe_{80}P_{12}C_4B_4$ 合金中添加适量的 Mo、Cr 和 Ni 元素不仅显著提高了合金过冷液体热稳定性和玻璃形成能力，还降低了合金的 T_g。其中，$Fe_{55}Ni_{15}Cr_7Mo_3P_{12}C_4B_4$ 非晶合金的 T_g、ΔT_x 和非晶形成临界直径分别为 690 K、60 K 和 2.5 mm。Mo、Cr 和 Ni 的添加还显著提高了合金的耐蚀性，其在硫酸和盐酸溶液中优于 SUS316L 不锈钢。该系块体非晶合金具有高达约 3000 MPa 的屈服强度，并具有良好的塑性变形能力。这种新型 Fe 基块体非晶合金有望在耐磨和耐腐蚀涂层领域获得应用。

关键词：铁基块体非晶合金，玻璃形成能力，过冷液相区间，耐蚀性

Mo 元素添加对 Fe-Si-B-P-Cu 系合金非晶形成能力和软磁性能的影响

贾行杰，李艳辉，张　伟

（大连理工大学材料科学与工程学院，辽宁大连　116024）

摘　要：铁基 Fe-Si-B-P-Cu 系纳米晶软磁合金因其高饱和磁感应强度（Bs）、低矫顽力（Hc）、高磁导率和低铁损和而备受关注。但该系合金的非晶形成能力较低，并且获得具有优异软磁性能的纳米晶合金的所需升温速率较高，可操作热处理温度区间较窄，使其难以获得工业化应用。本工作中，我们研究了微量 Mo 元素的添加对 Fe-Si-B-P-Cu 系合金非晶形成能力、结晶化组织和磁性能的影响。

采用感应熔炼炉制备母合金锭；采用单辊甩带设备制备条带样品；采用差示扫描量热仪测定合金的热性能；采用真空等温退火对样品进行热处理，升温速率为 50~400K/min；采用 X 射线衍射仪和透射电子显微镜对急冷和热处理样品进行结构表征；分别采用振动样品磁强计和直流磁滞回线仪测量合金的 B_s 和 H_c。

实验结果表明，添加微量 Mo 元素能提高合金的非晶形成能力，非晶条带的临界厚度可由 14μm 提高至 20μm。Mo 元素的添加也使合金可在低升温速率下和宽热处理温度区间内获得均匀细微的纳米晶组织和良好的软磁性能。经低升温速率的热处理后的 Fe-Si-B-P-Cu-Mo 系合金的 α-Fe 晶粒平均尺寸、B_s 和 H_c 分别为~25nm、~1.75T 和~20A/m。此外，我们从晶化激活能的角度探讨了 Mo 元素添加影响合金结构和磁性能的机理。

关键词：纳米晶软磁合金，非晶形成能力，微合金化，升温速率

CuZr 基金属玻璃液体的液液相变现象

赵　茜[1]，田泽安[2]，胡丽娜[1]

（1. 山东大学千佛山校区，材料液固结构演变与加工教育部重点实验室，山东济南　250061；
2. 湖南大学，物理与微电子学院，湖南长沙　410082）

摘　要：液液相变是指在特定的温度、压力条件下，同种成分的液体会存在两种不同密度或结构的液相的现象。本论文利用高温熔体粘度测量仪和差示扫描量热仪，探索了 CuZr 基高温熔体在远高于液相线温度的动力学及热力学行为，得到了液液相变存在的直接证据，发现 CuZr 二元合金普遍具有液液相变现象。结合流体团簇模型和过热熔体脆性概念，提出了表征高温熔体液液相变程度的参数 F，建立了参数 F 与玻璃形成能力之间的关系，从而确定了 CuZr 基高温熔体的液液相变有利于玻璃形成。结合经典动力学模拟，探索了高温熔体动力学转变过程中的团簇演变规律。本文对探究液液相变与固体性质的关系，提高非晶合金制备过程中的标准化程度具有重要的意义。

关键词：非晶合金，液液相变，玻璃形成

有关 Fe 基非晶玻璃形成能力的液体结构演变性质研究

任楠楠[1]，管鹏飞[2]，胡丽娜[1]

（1. 山东大学材料液固结构演变与加工教育部重点实验室，山东济南　250061；

2. 北京计算科学研究中心，北京　100193）

摘　要：Fe 基金属玻璃具有优异的软磁性能且成本较低，近年来受到广泛关注，但相对较差的玻璃形成能力成为限制其发展和应用的一大瓶颈。因此，研究与玻璃形成能力有关的结构演变特点非常重要。本文选取二元玻璃形成液体 Fe-Ni 和 Fe-P 作为研究对象，以玻璃形成能力较差的纯 Fe 液体作为对比，利用分子动力学模拟的方法，研究了 Fe 基金属玻璃形成液体的结构演变规律。计算结果表明，玻璃形成液体在高于 T_g 的某温度区间内，双体分布函数第二峰位置急剧偏移，对应中程序结构的变化。Voronoi 多面体统计结果表明，玻璃形成液体中，类二十面体团簇与变形的 bcc 团簇之间的竞争程度远高于纯 Fe 液体。本文为揭示 Fe 基非晶的玻璃形成能力提供了新的思路。

关键词：Fe 基玻璃形成液体，玻璃形成能力，结构演变

高 B 含量铁基非晶合金涂层制备与性能研究

王钦佳，斯佳佳，惠希东

（北京科技大学，新金属材料国家重点实验室，北京　100083）

摘　要：开发出具有较高非晶形成能力的高 B 含量 Fe-Cr-B-Mo-Mn 铁基非晶合金，采用超音速雾化制粉技术和超音速火焰喷涂技术，在 304 不锈钢基体上制备非晶涂层，获得了优化的工艺参数。实验研究了该涂层的结构、结合强度、硬度和摩擦磨损性能，结果表明，涂层以非晶相为主，呈片状或带状粒子相互搭接堆积而成的层片状结构，有少量层间孔隙，无明显裂纹，整体结构较为致密；与基体之间的结合强度大于 55 MPa；显微硬度约为 1150HV$_{0.1}$；耐磨性显著优于 80+商用 Fe 基非晶涂层和 304 不锈钢基体。良好的综合性能表明该涂层有望作为耐磨材料应用于工程实际。

The Influence of Element Segregation on the Viscosity of Fe-based Amorphous Alloys and Its Solutions

Dong B.S.[1], Zhou S.X.[1], Wang Y.G.[2]

(1.Advanced Technology & Materials Co., Ltd., Beijing Key Laboratory of Energy Nanomaterials,
China Iron & Steel Research Institute Group, Beijing 100081, China;

2.Institute of Physics, Chinese Academy of Sciences, PO Box 603, Beijing 100080, China)

Abstract: The properties of initial alloy melt have shown great influence on its following solidification state for the preparation and property control of Fe-based amorphous alloys. The phenomenon that the viscosity of alloy melt is unable to keep continuous with changing temperatures, i.e.,a sudden drop occurred during the heating process in the viscosity-temperature plots is reported in Fe-based amorphous alloys[1-3]. But the nature of this transition is still an open question at present. Our group has investigated the viscosity-temperature-ingot relation in $Fe_{78}Si_9B_{13}$ and $Fe_{85}B_{15}$ systems by high rotational vibration type high-temperature melt viscometer and scanning electron microscope (SEM). Experimental results have shown that, the typical phenomenon of viscosity sudden drop can be shown only for ingots having element segregation in Fe-based amorphous alloys. This element segregation needs more energy to eliminate. So the segregation zone is disintegrated with the increasing heating temperature toward homogeneous melt and the viscosity will drop suddenly. And the viscosity-temperature curve is completely reversible when the heating temperature is lower than the needed disintegrated energy. However, for the homogeneous Fe-based ingots, the viscosity-temperature is almost reversible at all the investigated temperatures. So, a homogeneous Fe-based alloy melt can be obtained by a proper heat treatment at a temperature higher than the needed disintegrated energy. These findings probably offer the internal causes of the viscosity's sudden dropin viscosity-temperature plot, and the significant reference for the preparation and property control of amorphous alloys in the fields of materials science and metallurgy.

Research and Development of Iron-based Amorphous and Nanocrystalline Powder Cores

Li Xiantao[1, 2], Zhou Shaoxiong[1], Kuang Chunjiang[1],
Gao Hui[1], Song Yanyan[2]

(1. Advanced Technology & Materials Co., Ltd., Beijing 100094, China;
2. Central Iron and Steel Research Institute, Beijing 100081, China)

Abstract: Metallic magnetic powder cores, which are made from metallic soft magnetic powders subjected to processes such as insulation coating, compression molding, and heat treatment, have incomparable advantages over other soft magnetic materials. With the rapid development of electronic technology, mass production of electronic devices for high-frequency and high-power-density applications, and increasing demand for miniaturization devices and anti-electromagnetic interference materials, the demand for magnetic powder cores has increased. Magnetic powder cores materials consist of a basic metallic soft magnetic powder and inorganic or organic electrically insulating additives and binders. These materials may be divided into some main classes according to the composition and microstructure. The relevant main classes are pure iron, iron-silicon, iron-silicon-aluminium, iron-nickel, iron-nickel-molybdenum, iron-based amorphous and nanocrystalline powder core materials. The silicon and boron content of iron-based amorphous powder are 5~8 wt.% and 2~3 wt.%, respectively. Remainder is iron. Typical compositions of iron-based nanocrystalline powder are $Fe_{73.5}Cu_1Nb_3Si_{13.5}B_9$. The range of the initial permeability of the amorphous and nanocrystalline powder cores is from 19 to 125. These materials are used for manufacturing the powder cores with ring shape. Iron-based amorphous powder cores are used for inductive components in switching power, filter and sensor, etc. Nanocrystalline powder cores are used for booster rings and filter inductors in uninterrupted power supply, filter inductors and energy storage inductors in military switching power supply, filter inductors in high power supply, power factor correction inductors, etc.

Key words: magnetic powder cores, soft magnetic materials, amorphous, nanocrystalline

热处理不脆的高性能非晶软磁合金开发

王安定[1]，高志开[1,2]，陈平博[1]，李福山[2]，王新敏[1]

（1.中科院宁波材料技术与工程研究所磁材事业部，浙江宁波 315201；

2.郑州大学材料科学与工程学院，河南郑州 450001）

摘 要： 因为特殊的长程无序原子排列结构，非晶合金具有完全不同于晶态合金材料的多种优异性能，例如，良好软磁性能、频率特性、力学性能和耐腐蚀性等。然而，当前商业化的非晶合金存在明显的热处理脆化问题，严重影响实际工程应用，并制约其应用领域的推广。因此，开发热处理后仍具有良好韧塑性的非晶软磁合金，并探究热处理过程对非晶合金韧塑性和磁性能的影响机理具有重要意义。

通过梳理合金成分设计方法和研究元素对性能的影响，我们成功制备了两种具有高非晶形成能力、优异软磁性能和塑韧性的非晶合金。其中，$Fe_{83}(Co_x, Ni_y)(B_{11}Si_2P_3C_1)_{1-x,y/17}$ (x, y=1~3) 非晶软磁合金的铁含量超过高硅硅钢，该合金 B_s 达到1.72T，H_c 低至 5 A/m，μ_e 达到 10×10^3，最重要的是该带材样品热处理达到最佳软磁性能后依然可以弯曲对折，非常适合带材生产和应用要求，具有优异的应用前景。通过研究 $Fe_{71}Nb_6B_{23}$ 三元块体非晶合金的结果弛豫和晶化过程，我们成功制备出了热处理后仍保有较高塑性及优良软磁性能的样品，该合金非晶形成能力达到1mm，塑性达到1.8%，H_c 低至 1.6 A/m，μ_e 达到 15×10^3。采用这两种模型材料，我们探讨了热处理对其磁性及塑韧性的影响机理，非晶合金原子排列短程有序，长程无序，在宏观尺度表现出结构的高度均匀性，在微观尺度存在大量的自由体积、类液体区等"缺陷"，正是这些结构缺陷和基体结构在热处理过程中的呈现不同衍化过程，并对磁性能和力学性能的产生不同的影响，使获得兼具优异韧脆性和磁性能的高性能非晶软磁合金成为可能。通过不同热处理阶段对其性能影响，我们推演出不同阶段的非晶合金结构衍变模型，从而探究了不同的热处理阶段对非晶合金性能的影响机理，解释了在热处理后非晶合金不脆的原因。本研究的提供了两种具有优异应用前景的合金材料，并可为获得不脆的高性能非晶软磁合金提供理论参考，从而进一步促进非晶软磁合金在推广和应用。

Thin and Flexible Fe–Si–B/Ni–Cu–P Metallic Glass Multilayer Composites for Efficient Electromagnetic Interference Shielding

Zhang Jijun[1,2,3], Li Jiawei[1,2], Tan Guoguo[1,2], Hu Renchao[1,2], Wang Junqiang[1,2], Chang Chuntao[1,2], Wang Xinmin[1,2]

(1.Key Laboratory of Magnetic Materials and Devices, Ningbo Institute of Materials Technology & Engineering,

Chinese Academy of Sciences, Ningbo 315201, China;

2.Zhejiang Province Key Laboratory of Magnetic Materials and Application Technology, Ningbo Institute of Materials Technology & Engineering, Chinese Academy of Sciences, Ningbo 315201, China;

3.University of Chinese Academy of Sciences, Beijing 100049, China)

Abstract: Thin and flexible materials that can provide efficient electromagnetic interference (EMI) shielding are urgently needed, especially if they can be easily processed and withstand harsh environments. Herein, layer structured Fe–Si–B/Ni–Cu–P metallic glass composites have been developed by simple electroless plating Ni–Cu–P coating on commercial Fe–Si–B metallic glasses. The 0.1-millimeter-thick composite shows EMI shielding effectiveness of 40 dB over the X-band frequency range, which is higher than that of traditional metals, metal oxides and their polymer composites of larger thickness. Most of the applied electromagnetic waves (EMWs) are proved to be absorbed rather than bounced back. This performance originates from the combination of superior permeability, excellent electrical conductivity, and multiple internal reflections from multilayer composites. In addition, the flexible composites also exhibit good corrosion resistance, high thermal stability, and excellent tensile strength, making them suitable for EMI shielding in harsh chemical or thermal environments.

Key words: Fe-based metallic glasses, electroless Ni–Cu–P, multilayer composites, electromagnetic shielding, harsh environments

Microstructure and Properties of CoCrFeNi(WC) High-entropy Alloy Coatings Prepared by Mechanical Alloying and Hot Pressing Sintering

Xu Juan, Shang Caiyun, Wang Yan

(School of Materials Science and Engineering, University of Jinan, Jinan, 250022 China.)

Abstract: The CoCrFeNi high-entropy alloy coatings (HEACs) with different weight ratio (10 and 30%) of WC addition have been prepared by mechanical alloying (MA) and vacuum hot pressing sintering (VHPS) technique on Q235 steel substrate. The microstructures, microhardness and wear resistance of HEACs were studied. The CoCrFeNi HEA powders were fabricated by MA process, which contain single FCC solid solution. The coating powder mixtures of CoCrFeNi(WC) HEAs were obtained by milling the CoCrFeNi HEA powders and WC powders for 120h. The products of VHPS-ed CoCrFeNi(WC) HEACs are still composed of FCC solid solution and WC phase. CoCrFeNi HEACs with 10 and 30% WC are 850 and 700 μm in thickness respectively, and all bearthe good metallurgical bonding to the substrate. The average microhardness values of HEACs with 10 and 30% WC reach 475 and 525 HV respectively, which far exceed the substrate (160 HV). Both coatings exhibit superior wear resistance than substrate under the same wear conditions.

Key words: high entropy alloy coating, mechanical alloying, vacuum hot pressing sintering, wear resistance, corrosion resistance

Nonlinear Fragile-to-strong Transition in a Magnetic Glass System Driven by Magnetic Field

Huo Juntao[1,2], Luo Qiang[3], Wang Junqiang[1,2], Xu Wei[1,2], Wang Xinmin[1,2]
Li Runwei[1,2], Yu Haibin[4]

(1.Key Laboratory of Magnetic Materials and Devices, Ningbo Institute of Materials Technology, and Engineering, Chinese Academy of Sciences, Ningbo 315201, China;

2.Zhejiang Province Key Laboratory of Magnetic Materials and Application Technology, Ningbo Institute of Materials Technology, and Engineering, Chinese Academy of Sciences, Ningbo 315201, China;

3.School of Materials Science and Engineering, Tongji University, Shanghai 201804, China;

4.Wuhan National High Magnetic Field Center, Huazhong University of Science and Technology, Wuhan 430074, China.)

Abstract: Relaxation dynamics in nonlinear response regime has become an emerging novel tool to study the dynamics and structure of glassy materials. It provides more insights relative to the standard linear response experiments. However, limited by inherent endurance of the materials to external fields, up to now, almost all the probed nonlinear effects are very weak. Here, strong nonlinear effects are observed in magnetic systems with disordered spins (i.e. magnetic glass). In particular, we report a pronounced fragility transition as driven by the external magnetic field as a result of nonlinear dynamic response. Such model systems provide a new platform to study the glassy dynamics with large and tuneable nonlinearity.

Cu-Ni-P 非晶钎料及其钎焊行为研究

张明海[1]，羌建兵[1]，王英敏[1]，Haeussler Peter[1]，王建豹[2]

（1.大连理工大学三束材料改性教育部重点实验室，辽宁大连 116024；

2.核工业西南物理研究院聚变所，四川成都 610041）

摘 要： Cu-Ni-P 合金常用作 Cu/CuCrZr 材料连接钎料，但其钎焊接头中易出现 Cu_3P 脆性相连续析出，导致相关接头性能不佳。针对这一问题，本文基于 Cu-Ni-P 三元相图特征，利用非晶合金团簇理论设计出 $Cu_{69}Ni_{15}P_{16}$ 成分，利用感应熔炼法和急冷甩带技术制备了上述成分的合金铸锭与箔带，分析了两种合金样品的凝固组织和熔化行为，并研究了 $Cu_{69}Ni_{15}P_{16}$ 箔带钎焊 Cu/CuCrZr 合金的接头组织与性能。结果表明：通过急冷甩带技术可以获得完全非晶态的 $Cu_{69}Ni_{15}P_{16}$ 箔带，由其钎焊获得的 Cu/CuCrZr 合金接头中的 Cu_3P 相被 Cu 基固溶体所割裂，钎缝性能大为改善。

关键词： Cu-Ni-P 合金，非晶态，Cu/CuCrZr 连接，钎焊

Newly Cu-Ni-P Amorphous Fillers for Brazing Cu/CuCrZr Alloys

Zhang Minghai[1], Qiang Jianbing[1], Wang Yingmin[1], Haeussler Peter[1], Wang Jianbao[2]

(1.Key Laboratory of Materials Modification by Laser, Ion, and Electron Beams (Ministry of Education), Dalian University of Technology, Dalian 116024, China;

2.Southwest Institute of physics and nuclear fusion industry, Chengdu 610041, China)

Abstract: Cu-Ni-P alloys are usually utilized as filler metals for brazing Cu/CuCrZr. However, the continuous brittle Cu_3P phases are easily found in the relevant joint, which will worsen the property of the joint significantly. To overcome the shortcomings, the composition of $Cu_{69}Ni_{15}P_{16}$ is design by the cluster theory for amorphous alloy in the present work basing on ternary Cu-Ni-P phase diagram, and the botton and ribbon samples were prepared by the induction smelting and melt-spinning method respectively. And the solidification structure and melting behavior of the above samples were investigated systematically, then the $Cu_{69}Ni_{15}P_{16}$ melt-ribbon was used as filler metal to braze the Cu/CuCrZr alloys, and the micro-structure and properties of brazing joint were also examined. The results showed that fully amorphous-state $Cu_{69}Ni_{15}P_{16}$. ribbons can be achieved by the melt-spun technique, and btain total amorphous substance ribbon, and the composition is And in the joint, that obtained by brazing the Cu/CuCrZr, have Cu_3P brittle compound is refined and separated by Cu based solid solution phase in the Cu/CuCrZr joint brazed by the $Cu_{69}Ni_{15}P_{16}$ amorphous ribbon, which enhances the properties of the resultant joint obviously.

Key words: Cu-Ni-P alloy, amorphous, brazing, Cu/CuCrZr joint

球磨时间对铁基 FeSiBPCu 合金粉体结构和电磁性能的影响

李亚楠，李艳辉，张　伟

（大连理工大学材料科学与工程学院，辽宁大连　116024）

摘　要： 现代电子工业和信息产业的高速发展所引起的电磁辐射危害不容忽视，探索高效的电磁屏蔽材料已成为目前迫切需要解决的问题。高的磁导率（μ）和饱和磁感应强度（B_s）是高效能电磁屏蔽材料需具备的先决条件。近年发展的铁基 FeSiBPCu 系纳米晶合金具有高 B_s、高μ和低损耗等优异的软磁性能，但关于其电磁屏蔽性能的研究甚少。本文通过高能球磨制备 $Fe_{83.3}Si_4B_8P_4Cu_{0.7}$ 合金粉体，调查了球磨时间和热处理工艺对粉体的结构、形貌和电磁性能的影响，获得了具有优异电磁屏蔽性能的纳米晶合金粉体。

采用感应熔炼制备母合金锭；采用行星式球磨机制备合金粉体；采用真空等温退火对粉体进行热处理；分别采用 X 射线衍射仪和扫描电镜表征合金粉体的结构和形貌；分别采用差示扫描量热仪和振动样品磁强计评价粉体的热性能和磁性能；将合金粉体与石蜡按质量比 3:1 混合均匀，使用矢量网络分析仪测试其电磁参数，并计算反射损耗。

结果表明，随着球磨时间的延长，合金粉体的颗粒尺寸逐渐降低，非晶相含量不断升高，α-Fe 晶粒尺寸（$D_{\alpha\text{-Fe}}$）逐渐减小。合金经球磨 100 h 后可形成平均颗粒尺寸为 7 μm、$D_{\alpha\text{-Fe}}$ 为 5.4 nm 的纳米晶粉体。粉体的 B_s 和电磁屏蔽性能随着球磨时间的延长而逐渐提高。球磨 100 h 后，粉体的 B_s 达到 1.74 T，此时对应厚度为 2 mm 的样品的反射损耗（RL）< -10 dB 的频带宽度达到 5.8 GHz，并在 11.5 GHz 的频率下达到峰值-44 dB。经 300 ℃热处理后，粉

体屏蔽性能进一步改善，RL＜–10 dB 的频带宽度达到 6.2 GHz，峰值增加到–54 dB，对应频率为 11.7 GHz。

关键词：纳米晶软磁合金，FeSiBPCu 合金，合金粉体，球磨，电磁屏蔽性能

Fe-Pt-B 非晶合金脱合金化制备 Pt 基纳米多孔材料及其性能研究

马殿国[1]，欧淑丽[1]，王英敏[1]，汤蓋邦夫[2]，李艳辉[1]，张伟[1]

（1. 大连理工大学材料科学与工程学院，辽宁大连　116024；

2. 东北大学金属材料研究所，日本仙台　980-8577）

摘　要：纳米多孔金属由三维双连续的纳米孔隙和韧带组成，兼具纳米材料、多孔材料和金属材料的特性，在催化、传感和驱动等领域具有广阔的应用前景。其中，铂基纳米多孔金属由于其突出的电催化性能有望在燃料电池上实现应用，目前已成为该领域的研究热点之一。脱合金化法是制备纳米多孔金属的有效方法，目前采用的脱合金化前驱体主要集中在连续固溶体合金体系。非晶合金的组织结构均匀、组成元素种类多、成分范围宽，且各组元间存在明显的电化学活性差异，适合作为脱合金化前驱体来制备纳米多孔金属。利用非晶合金脱合金化已经制备出了金基、钯基等纳米多孔金属，但用非晶合金来制备铂基纳米多孔金属的研究甚少。本工作研究了 Fe-Pt-B 非晶合金脱合金化后的组织结构、成分与形貌特征，并调查了其催化性能与磁性能。

分别采用电弧熔炼和单辊甩带法制备母合金锭和非晶合金条带；利用电化学工作站在 0.1 mol/L H_2SO_4 溶液中对合金条带进行脱合金化制备纳米多孔合金；采用 X 射线衍射和透射电子显微镜表征脱合金化前后合金样品的组织结构；利用扫描电子显微镜和能谱仪观察纳米多孔合金的形貌并进行成分分析；采用电化学法评价其对甲醇的电催化性能并利用振动样品磁强计测试其磁性能。

结果表明，$Fe_{60}Pt_{10}B_{30}$ 非晶合金经脱合金化后可形成由面心立方 FePt 相组成的纳米多孔合金，其中 Fe 与 Pt 的含量分别约为 35.8 at.% 和 64.2 at.%。该合金具有三维双连通的纳米多孔结构，孔径分布均匀，平均孔径大小及孔壁厚度分别为 5 nm 和 7 nm。纳米多孔合金具有优异的甲醇电催化性能，其催化活性和效率均高于商用 Pt/C 催化剂。该合金还呈现软磁特性，饱和磁化强度为 19.09 emu/g。

关键词：非晶合金，Pt-Fe 合金，脱合金化，纳米多孔金属，电催化性能，磁性能

Composition Designof Novel Fe-based Amorphous Alloys with Good Corrision Resistance

Zhang Qian[1], Hui Xidong[1], Li Zongzhen[2], Zhou Shaoxiong[2]

(1. State Key Laboratory for Advanced Metals and Materials, University of Science and Technology Beijing, Beijing 100083, China;

2. Central Iron and Steel Research Institute, Beijing100081,China)

Abstract: FeCBSiP alloys with high Bs and amorphous-forming ability as well as good softmagnetic properties were developed via component design and composition adjustment. In order to extend the applicability of FeCBSiP, it is significant to improve its corrosion resistance through adding other element into this alloy.Here element Cr is chosen based on Cr is a corrosion resistance element and many papers have been reported that the Cr containing Fe-based glassy alloy possess high corrosion resistance so far.The influence of replacing Fe by Cr on the corrosion resistance of $Fe_{83}Si_2B_{11}P_3C_1$ amorphous alloys produced by melt-spun method under an argon atmosphere were studied by immersion test and potentiodynamic polarisation in 1M HCl and 0.6M NaCl solutions. the corrosion rate of almost the present Fe based amorphous alloys in both 1M HCl and 0.6M NaCl solutions are on the order of 10-1 mm/year and the corrosion current density in these two solutions decreased from 10^{-3} to10^{-6} A cm^{-2} and 10^{-5} to 10^{-6} A cm^{-2}, respectively, with increasing Cr content from x=0 to 4.The minor-addition of Cr improved the corrosion resistance through the increase in Ecorr value, which makes easy to reach passive state, and the suppression of pitting corrosion.The SEM/EDS analysis shows that the high corrosion resistance is due to the formation of chromium-rich passive films during immersion in HCl asolutions. The study also focused on the modification of the magnetic properties in the material as a result of exposure to an aggressive medium and Cr addition.

Key words: amorphous alloys, corrosion resistance, magnetic properties

铁基非晶复合磁粉芯的制备及性能研究

董亚强[1]，郭俊江[1,2]，刘　珉[1]，常春涛[1]，李　强[2]，王新敏[1]

（1.浙江省磁性材料及应用技术重点实验室，中国科学院磁性材料与器件重点实验室，
中国科学院宁波材料技术与工程研究所，浙江宁波　315201；
2.新疆大学物理科学与技术学院，新疆乌鲁木齐　830046）

摘　要：本文系统研究了微细 FeSi 合金粉末的含量对 FeSiBPNb 非晶复合磁粉芯性能的影响规律，所采用的 FeSi 合金粉末平均粒径约为 5 μm，铁基非晶合金粉末的平均粒径约为 50μm，采用冷压的方法压制成外径为 27 mm，内径为 14.8 mm，厚度为 6.5 mm 的环形样品，为了提高复合磁粉芯的软磁性能，所有的样品均在 673 K 热处理 1h，以消除在压制过程中磁粉芯内部产生的应力。研究发现，当 FeSi 粉末含量为 10% 时，复合磁粉芯表现出优异的软磁性能，其有效磁导率达到 60，且在 10 MHz 频率范围内保持恒定，呈现出优异的频率特性；在 $B_m = 0.1$ T，$f = 50$ kHz 条件下，其损耗仅为 159 W/kg；同时，在外加直流偏置磁场为 100 Oe 时，磁粉芯的磁导率仅下降 20%，呈现出良好的抗直流偏置能力。这种性能优异的非晶复合磁粉芯符合电力电子器件向高频化、小型化和大电流方向发展的趋势，具有广泛的应用前景。

关键词：磁粉芯，非晶合金，磁导率，直流偏置，损耗

绝缘包覆工艺对铁基非晶磁粉芯性能的影响

黄柯瑜[1,2]，董亚强[2]，刘　珉[2]，赵占奎[1]，常春涛[2]，王新敏[2]

（1.长春工业大学材料科学与工程系，吉林长春　130012；
2.浙江省磁性材料及应用技术重点实验室，中国科学院磁性材料与器件重点实验室，
中国科学院宁波材料技术与工程研究所，浙江宁波　315201）

摘　要：本文以气雾化法制备的球状$(Fe_{0.76}Si_{0.09}B_{0.1}P_{0.05})_{99}Nb_1$ 非晶粉末为对象，首先利用正硅酸乙酯的水解反应，在非晶粉末表面原位生成 SiO_2 绝缘包覆层，通过调节正硅酸乙酯的含量，可以控制非晶粉末表面 SiO_2 绝缘层的厚

度。随后，在非晶粉末表面包覆 2 wt%的环氧树脂，形成均匀的双层核壳结构包覆层，利用液压成型机，在 1800 MPa 的压强下，将复合粉末压制成外径 20.3 mm，内径 12.7 mm，厚度 6.0 mm 的环形磁粉芯，并系统研究了正硅酸乙酯含量对 FeSiBPNb 磁粉芯性能的影响规律。研究发现，当正硅酸乙酯含量为 2 ml/g 时，制备的磁粉芯综合性能最好，其磁导率达到 49，且在 10 MHz 频率范围内保持恒定；在 $B_m = 0.1$ T，$f = 100$ kHz 条件下，其损耗仅为 320 W/kg。

关键词：绝缘包覆，非晶磁粉芯，磁导率，损耗，气雾化

核壳结构的 FeB@SiO₂ 非晶颗粒制备的金属基磁性液体

于孟春，边秀房

（山东大学，山东济南　250100）

摘　要：首次制备了核壳结构的 FeB@SiO₂ 非晶颗粒并将其均匀分散在液态 $Ga_{85.8}In_{14.2}$ 合金中，制备了金属基非晶磁性液体。FeB 非晶颗粒的形貌是规则的球形，平均尺寸大约是 190nm。SiO_2 包覆层稳定的包覆在 FeB 非晶颗粒的表面，包覆层的厚度大约是 40nm。VSM 结果表明 FeB 非晶颗粒的饱和磁化强度大约是 131.5emu/g，约是 Fe_3O_4 的 2 倍多。由于非磁性的 SiO_2 包覆层的缘故，FeB@SiO₂ 非晶颗粒的饱和磁化强度比 FeB 非晶颗粒的饱和磁化强度降低了 18.7%，约为 106.9emu/g。旋转震荡粘度仪测试结果表明 FeB 非晶颗粒制备的金属基磁性液体表现出良好的高温性能，在高温应用方面具有一定的应用前景。

片状 FeSiBPNbCu 纳米晶粉体吸波材料的制备与性能研究

陈淑文 [1,2]，谭果果 [1,2]，顾习胜 [1,2]，满其奎 [1,2]，王新敏 [1,2]，李润伟 [1,2]

（1. 中国科学院材料技术与工程研究所，磁性材料与器件重点实验室，浙江宁波　315201；
2. 中国科学院宁波材料技术与工程研究所 浙江省磁性材料与应用技术重点实验室，
浙江宁波　315201）

摘　要：相比于铁磁性合金粉末，铁基纳米晶软磁粉末由于铁磁交换耦合作用能解决传统吸波材料面密度大、频带窄、吸收弱的缺点，在高频吸波材料领域展现巨大应用潜力。本研究选用自主研发的强非晶形成能力和高饱和磁化强度的 FeSiBPNbCu 合金成分体系，首先将粉体片状化处理，再进行热处理，通过与高分子材料复合制备得到高频吸波材料。实验结果表明：当热处理工艺为 530℃，保温 15mim 时，在 2.45GHz 处，吸波复合材料最低反射损耗值（RL）可达到–45.7dB，–10dB 频宽为 1.63GHz，适用于无线局域网抗电磁干扰吸波材料。

关键词：纳米晶，高频，吸波材料，反射损耗

Preparation and Microwave Absorption Properties of FeSiBPNbCu Nanocrystalline Composites

Chen Shuwen[1,2], Tan Guoguo[1,2], Gu Xisheng[1,2], Man Qikui[1,2],
Wang Xinmin[1,2], Li Runwei[1,2]

(1. Key Laboratory of Magnetic Materials and Devices, Ningbo Institute of Materials Technology & Engineering, Chinese Academy of Sciences, Ningbo 315201, China;

2. Zhejiang Province Key Laboratory of Magnetic Materials and Application Technology, Ningbo Institute of Materials Technology & Engineering, Chinese Academy of Sciences, Ningbo 315201, China)

Abstract: Due to the ferromagnetic exchange coupling, Fe-based nanocrystalline soft magnetic absorbing material can solve the problems on high density, narrow frequency band and weak absorption compared to ferromagnetic alloy powder. Thus it shows great potential for high frequency absorbing materials. In this research, the self-developed high glass forming ability and saturation magnetization of FeSiBPNbCu alloy system was used to treated as the absorber, the spherical nanocrystalline powders were flaked by milling, then were perpared to polymer composite absorbing materials after the heat treatment process. The experimental results show that the minimum reflection loss value (RL) can reach −45.7dB at 2.45GHz, and the bandwidth is 1.63GHz for RL≤10dB under the condition of 530 ℃ for 15min, which is suitable for wireless LAN anti-electromagnetic interference absorbing materials.

Key words: nanocrystalline, high frequency, absorbing material, reflection loss

合金相图与高温熔体动力学行为之间的联系

商继祥，胡丽娜

（山东大学材料液固结构演变与加工教育部重点实验室）

摘　要： 探究合金相图与高温熔体粘度之间的关系，对揭示液液相变现象以及经快冷后冻结的金属玻璃的弛豫、焓异常现象有重要的指导意义，对进一步探究玻璃转变的本质，拓宽金属玻璃的应用领域,加强人们对金属玻璃认识有着重要的现实意义。通过对 CuZr 二元合金相图、CuZrAl 三元合金相图及其高温熔体粘度的观察，我们发现存在粘度突变的合金体系其对应合金相图比较复杂，相图中有多种中间化合物生成。我们进一步探究了与 CuZr 合金相图相似的 CuAl、CuSn 合金相图，其高温熔体的粘度也存在突变行为，以及没有中间化合物生成的 GaSn、GaZn 等共晶相图，其高温熔体的粘度符合阿伦尼乌斯方程，不存在突变行为。这一发现证明了高温熔体动力学行为与其合金相图存在着一定的关联。

关键词： 合金相图，高温熔体，粘度，中间化合物，液液相变

CeNi 二元合金 EAM 势的构建

雷亚威[1]，孙晓锐[1]，周如龙[1]，张　博[1]

（合肥工业大学，材料科学与工程学院非晶态物质科学研究所，安徽合肥　230009）

摘　要：根据 CeNi 二元合金实验数据以及第一性原理计算的数据，我们构建了一个新的 CeNi 原子嵌入势。这个原子嵌入势能够很好的计算平衡晶格常数，结合能，未驰豫的空位形成能，弹性常数，$Ce_{80}Ni_{20}$ 熔体径向分布函数以及自扩散系数，$Ce_{80}Ni_{20}$ 非晶玻璃形成温度。使用这个势能，我们研究了深度共晶成分 $Ce_{80}Ni_{20}$ 的玻璃形成能力以及动力学行为。

铁基非晶合金的玻璃形成能力

赵云波，胡丽娜

（山东大学材料液固结构演变与加工教育部重点实验室）

摘　要：非晶合金以优越的热力学性能、电磁学性能、化学性能等被广泛应用于各个领域，但由于其生产尺寸的局限性，限制其发挥优良的性能，因此，生产大尺寸的非晶合金和预测其玻璃形成能力是我们首要的任务。基于以上背景，本工作测量了 Fe-Ga-P-C-B 系列非晶合金的粘度，发现该系列非晶合金也会有微弱的液液相变现象，通过 Arrhenius 公式拟合得到表征液液相变系数 F。通过分析实验数据可以发现：当 Ga 代替 Fe 的含量为百分之五时，非晶合金的玻璃形成能力最大，此时的 F 值最大。

关键词：铁基非晶合金，液液相变，玻璃形成能力，粘度

Ni 元素替换与等温退火对于铁基非晶条带性能的影响

刘鹏飞，张　博

（合肥工业大学材料科学与工程学院非晶态物质科学研究所，安徽合肥　230009）

摘　要：在此工作中，我们开发了一系列由工业纯度合金制备的 $Fe_{75}Co_{4-x}B_{13}Y_4Nb_4Ni_x$（$x$=0、0.5、1、2、3 和 4)非晶合金带。此外，系统地研究了 Ni 元素替换和等温退火对饱和磁化强度(Ms)和内禀矫顽力(jHc)的影响。根据 squid-vsm 数据指出，随着 Ni 元素的添加，饱和磁化强度最大的增加了 24%，从 114 emu/g 增加到 141 emu/g。条带在最佳温度区域快速退火后，Ms 达到 200 emu/g，jHc 从 40 Oe 下降到 25 Oe。DSC 测量显示，该系列的非晶条带在加热过程中存在多步结晶。在非晶形成能力不降低的前提下，两个结晶峰之间的间距随着 Ni 元素的增加而先减小后增大,宽的晶化峰意味着退火处理中更容易形成 α-Fe 纳米晶，这也与 X 射线衍射仪的结果相符合。Scherrer 公式计算结果表明等温退火对纳米晶的晶粒尺寸有着很大作用，从而对饱和磁化强度和内禀矫顽力产生影响。退火温度和时

间不仅影响纳米晶的析出，还能控制晶粒尺寸。因此，适当的退火工艺对于提高这种铁基纳米晶磁性材料的性能是非常重要的。综上所述，研究结果将有助于我们了解镍元素和等温退火对 $Fe_{75}Co_{4-x}B_{13}Y_4Nb_4Ni_x$ 非晶态合金磁性、微观结构与形成能力等性能的影响。

La-Ce-Ni-Cu-Al 高熵非晶合金的性能

吴 林，张 博

（合肥工业大学材料科学与工程学院非晶态物质科学研究所，
安徽合肥　230009）

摘　要： 我们报道一种新型的高熵非晶合金材料，它具有很低的玻璃转变温度（$T_g = 416K$）、较宽的过冷液相区间（$\triangle T_x = 66\ K$)和好的非晶形成能力(临界尺寸至少为 3 mm)。通过 Kissinger 公式计算，发现我们的高熵非晶合金材料在已报道的高熵非晶合金材料中，具有最低的玻璃转变激活能（$E_g = 120\pm8kJ/mol$)。而低的玻璃转变激活能表明材料具有良好的热塑性成形能力，这表明我们的材料具有很好的热塑性成形能力和广泛的应用前景。进一步探究其内在因素，确定玻璃转变温度的高低是决定玻璃转变激活能大小的关键性因素。

CuZr 二元非晶合金高温拉伸蠕变行为

洪　凯，张　博

（合肥工业大学材料科学与工程学院非晶态物质科学研究所，安徽合肥　230009）

摘　要： 我们研究 CuZr 非晶带材在玻璃转变温度点 Tg 以下的高温拉伸蠕变行为，通过多次独立实验和控制变量法得到不同温度（$T/Tg = 0.967\text{-}1$），应力（100，145，190MPa），成分（$Cu_{40}Zr_{60}$，$Cu_{50}Zr_{50}$，$Cu_{60}Zr_{40}$）对 CuZr 金属玻璃蠕变变形的影响。基于蠕变应变-时间曲线得到应变速率，计算了蠕变表观激活能与应力指数，发现 $Cu_{40}Zr_{60}$，$Cu_{50}Zr_{50}$，$Cu_{60}Zr_{40}$ 非晶带材三种成分 Tg 以下的蠕变激活能与应力指数不同。

LaCe 基块体非晶态高熵合金合成及性能研究

蒋　伟，张　博

（合肥工业大学材料科学与工程学院非晶态物质科学研究所，安徽合肥　230009）

摘　要： 成功开发了一种含有 LaCe 稀土元素的块体非晶态高熵合金，块体高熵非晶，是指在成分上具有较大混合熵的，结构上为非晶态的一类新型金属材料，同时临界尺寸超过 1 mm，其组分包含 5 个以上的组元，并且单一组元的含量不低与 5 at.%且不高于 35 at.%，组元的成分特征为 5 个及 5 个以上。对其形成能力进行了探讨，得到最大直径为 4mm 的块体高熵非晶。其压缩应力和弯曲应力能达到 800MPa 明显大于 La 基和 Ce 基的应力值，并且有小的塑性。尤其在动态力学分析仪（DMA）的温度谱和频率谱上发现其都具有明显的β弛豫，这是在其他高熵非晶里很少发现的，这为研究β弛豫的起源提供了新思路。

用正电子湮灭技术探究 Ce-基大块非晶合金的局域原子结构和晶化行为

赵 勇[1,2]，张 博[1]，Sato K.[2]

（1.合肥工业大学材料科学与工程学院非晶态物质科学研究所，安徽合肥 230009；
2.东京学芸大学环境科学系，东京 184-8501）

摘 要：Ce-基大块非晶合金具有低玻璃化转变温度（T_g）、优异的玻璃形成能力、良好的塑性精密成型能力、强的过冷液体行为及高压下非晶多形态现象等特性。但目前对其原子结构几乎一无所知，严重限制了人们对于 Ce-基非晶合金优异性能的进一步理解。

时效温度对 $Fe_{78}Si_9B_{13}$ 非晶合金活化过硫酸钠降解偶氮染料橙黄 G 的影响

李俊俊，张 博

（合肥工业大学材料科学与工程学院非晶态物质科学研究所，安徽合肥 230009）

摘 要：基于硫酸根自由基（$SO_4^{-}\cdot$）的高级氧化技术是近些年发展起来的、具有发展潜力的、降解难降解有机污染物的一项新技术。有研究表明热力学亚稳态的铁基非晶合金可以活化过硫酸盐快速降解有机污染物，具有很好的应用前景。本文基于此对不同时效温度处理的 $Fe_{78}Si_9B_{13}$ 非晶合金条带所得到不同结构的 Fe-Si-B 合金条带活化过硫酸钠（PS）降解偶氮染料橙黄 G（OG)进行了研究，并探索其降解机理，这些问题的解决有助于将基于硫酸根自由基的高级氧化技术推广应用到实际工业废水处理中，改善水污染现状。同时也为铁基非晶合金的应用提供了一个可行的方向。

本文研究了将 $Fe_{78}Si_9B_{13}$ 非晶合金条带分别在 573K,623K,673K,723K,773K,823K 等温退火 10min 活化 PS 降解 OG，并进行了动力学分析，结果表明降解过程均符合一级动力学模型。退火温度 623K 以下的 Fe-Si-B 合金均具有非常快的降解速率，20ppm 橙黄 G 水溶液在温度为 25℃、合金条带用量为 0.5g/L、PS 投加量为 1.0mmol/L 的条件下，经过 10 min 的降解，OG 脱色率达 97%。各时效温度的 Fe-Si-B 合金在活化 PS 降解 OG 时具有不同的降解速率，随退火温度升高，降解速率总体呈下降趋势，但 773K 退火得到的 Fe-Si-B 合金降解 OG 速率高于在 673K,723K 及 823K 退火得到的 Fe-Si-B 合金。通过扫描电子显微镜分析合金条带反应过后的表面微观形貌，发现不同结构的 Fe-Si-B 合金微观形貌存在明显差异，推测 773K 退火合金异常的降解速率与反应过程中合金表面形成微米级的规则四角星状坑蚀有关。

成功开发具有优异析氢电催化性能的
FeCo 基非晶合金

张发宝，吴继礼，蒋　伟，胡青卓，张　博

（合肥工业大学材料学院非晶态物质科学研究所，安徽合肥　230009）

摘　要：氢气作为一种清洁、可再生能源，有望替代化石燃料成为一种理想的能源载体。水电解产生氢气是一种有效的获取氢能源的方法。析氢反应作为电解水反应的一个重要半反应，是当今人们研究的热点[1,2]。然而在电解水反应中，催化剂起到了至关重要的作用。非晶合金，由于在腐蚀介质中具有很好的稳定性以及高催化活性，因而被认为是一种潜在的电催化剂。FeCo 基纳米材料也已被大量实验证实符合低的成本效益，并且具有良好的耐腐蚀特性与高的电催化活性[3]。

利用中子散射技术探究 Ce-Ga-Cu
熔体体系的自扩散行为

胡金亮[1]，Embs J. P.[2]，张　博[1]

（1.合肥工业大学材料科学与工程学院非晶态物质科学研究所，安徽合肥　230009；
2.Laboratory for Neutron Scattering, Paul Scherrer Institute,
WHGA/112, CH-5232 Villigen PSI, Switzerland）

摘　要：作为表征系统动力学的重要参数，金属熔体的扩散不仅是熔体理论的重要组成部分，而且对玻璃化转变、非晶合金的制备及成型都具有重要影响。利用准弹性中子散射技术(QNS)，我们在液相线温度以上一个很宽的温度范围(750~1150K)系统研究了 $Ce_{70}Ga_xCu_{30-x}(x=4; 10; 14)$ 这一典型 Ce-基金属玻璃形成体系熔体的自扩散行为。研究结果表明，在高温部分熔体的扩散系数的温度依赖行为能很好的符合 Arrhenius 关系，并且随着 Ga 元素的增加，扩散激活能越大。但随着温度的降低，特别是温度接近 T_l 时，扩散行为明显偏离 Arrhenius 关系，我们认为温度的降低会导致熔体的结构变紧密，堆积密度增加，从而使扩散行为急剧降低。

液态金属铈的结构及动力学行为

孙晓锐，周如龙，张　博

（合肥工业大学材料学院非晶态物质科学研究所，安徽合肥　230009）

摘　要：利用第一性原理分子动力,本工作直接证明低密度铈中存在两种不同价态的铈原子。并发现在低密度铈中,相同价态的铈原子相互聚集，不同价态的铈原子相互分离。在高价态铈原子周围容易形成二十面体结构，而低价态的铈原子周围很少有二十面体结构。并且高价态的铈原子扩散比低价态铈原子慢。本工作提供了新的视角和方法来深入理解液态铈及其合金。

$Ce_{70}Ga_xCu_{30-x}$ (x=6, 10, 13) 非晶合金形成能力的分子动力学方法研究

陈　恒，张　博

（合肥工业大学材料学院非晶态物质科学研究所，安徽合肥　230009）

摘　要：$Ce_{70}Ga_xCu_{30-x}$ (x=6, 10, 13)非晶合金作为一种典型的金属塑料材料，目前常规的 XRD 等实验方法仅能确定其非晶态的结构特征，并不能获得其微观局域结构，本工作结合同步辐射精细结构吸收谱（XAFS）和第一性原理的方法对其微观局域结构进行研究。首先，利用第一性原理分子动力学获得的原子坐标进行 XAFS 模拟计算，得到的吸收谱线与实验测得的吸收谱线符合的很好。然后我们进一步的分析了 $Ce_{70}Ga_xCu_{30-x}$ (x=6, 10, 13)三种成分熔体原子结构。结果表明 x=10 时，Ga 和 Cu 原子周围具有最强的二十面体有序性，而二十面体短程有序性与晶体结构有序性不相容，从而有利于非晶的形成。除此之外，我们发现，当 x=10 时具熔体有最大的堆积密度，而高的堆积密度同样与好的非晶形成能力有关。最后通对 $Ce_{70}Ga_{10}Cu_{20}$ 的电子结构分析发现，在熔体中存在 Ga-Ce 和 Ga-Cu 的共价键。而共价键具有方向性从而阻碍熔体的晶化，在一定程度上解释了 CeGaCu 非晶合金具有很好的非晶形成能力。

激光表面熔融对块体非晶合金微观结构和性能的影响

张文武，陶平均，涂　其，李东阳，杨元政

（广东工业大学材料与能源学院，广东广州　510006）

摘　要：通过铜模铸造法在氩气气氛中成功制备了尺寸为 $10 \times 3 \times 80$ mm 的完全致密的无裂纹板状 $Cu_{46}Zr_{42}Al_7Y_5$ 块状金属玻璃（BMG）样品，并在不同激光频率下进行表面激光表面熔融（LSM）。X 射线衍射图显示：晶化程度由热影响区和熔融区中的结晶相决定，且在相同的激光频率下热影响区比熔融区更容易晶化。熔融区的显微硬度在激光频率为 16 Hz 时达到最大值 508.2 HV。然而，热影响区的显微硬度随着激光频率的增加而急剧下降。此外，随着滑速的增加，平均摩擦系数先急剧增加后降低，其值在滑动速度为 500 mm/min 时达到最大值。在激光频率为 12、14 和 16 Hz 下，材料的耐磨性得到适当增加。

关键词：块体非晶，激光表面融化，显微组织，显微硬度，摩擦磨损

Effect of Laser Surface Melting on Bulk Metallic Glass: Investigation of Microstructure and Properties

Zhang Wenwu, Tao Pingjun, Tu Qi, Li Dongyang, Yang Yuanzheng

(School of Materials and Energy, Guangdong University of Technology, Guangzhou 510006, Chian)

Abstract: $10\times3\times80$ mm^3 fully dense crack-free plate-like $Cu_{46}Zr_{42}Al_7Y_5$ bulk metallic glass (BMG) samples were successfully prepared by a copper mold casting method in an argon atmosphere and subjected to surface laser surface melting (LSM) under different laser frequencies. X-ray diffraction patterns showed that the degree of crystallization is determined by the crystalline phase in the heat-affected zone and the welding fusion zones, in the heat-affected zone is easier to crystallize than the welding fusion zones at the same laser frequency. The microhardness of the welding fusion zones reached the maximum of 508.2 HV at the laser frequency of 16 Hz. However, the microhardness of the heat-affected zone declined drastically with increasing laser frequency. In addition, the average frictional coefficient dramatically increased and then dropped with the increase of sliding velocity, and the values reached the maximum at the sliding velocity of 500 mm/min. The property of wear resistance is moderately improved at laser frequencies of 12、14 and 16 Hz.

Key words: bulk amorphous alloy, laser surface melting, microstructure, microhardness, friction and wear,

Novel Multicomponent Ti-Zr-Cu-Ni-Co-Fe-Al-Sn(-Si) Amorphous Brazing Filler Alloys for High-strength Joining of Titanium Alloys

Liu Ying[1], Sun Lulu[1], Xiong Huaping[2], Pang S.J.[1], Zhang Tao

(1. Key Laboratory of Aerospace Materials and Performance (Ministry of Education), School of Materials Science and Engineering, Beihang University, Beijing 100191, China;
2. Welding and Plastic Forming Division, Beijing Institute of Aeronautical Materials, Beijing 100095, China)

Abstract: Ti-based brazing filler metals (BFMs), which are widely used for the brazing of titanium alloys, usually contain high contents of Cu and Ni elements to achieve a relatively low melting temperature. However, high contents of Cu and Ni in Ti-based BFMs usually cause the formation of brittle intermetallic compounds in the joints during brazing, which is detrimental to the mechanical properties of the joints. In comparison to the crystalline BFMs, amorphous brazing filler alloys exhibit advantages such as excellent homogeneity, high purity and easy forming ability of foils with superior flexibility and ductility, etc. Therefore, design and investigation of novel amorphous Ti-based BMFs with low contents of Cu and Ni is of great importance to acquire improved mechanical properties of titanium alloy brazed joints.

In our recent work, novel multicomponent Ti-Zr-Cu-Ni-Co-Fe-Al-Sn(-Si) amorphous filler alloys with low Cu and Ni contents were developed for brazing titanium alloys. These Ti-based metallic glasses possess high glass-forming ability and relatively low melting temperatures, and can be prepared into ribbons with good flexibility by melt spinning. The Ti-6Al-4V joints brazed with the present Ti-based amorphous BFMs exhibited high shear strength up to 460 MPa, which is mainly attributed to the reduced amount of intermetallics in the braze zone due to the low contents of Cu and Ni in the BFMs. This presentation reports the formation and thermal properties of the Ti-based amorphous filler alloys and the resultant microstructure and mechanical properties of Ti-6Al-4V brazed joints. The mechanisms for the formation of the amorphous alloys and the improvement of the joint mechanical properties will be discussed. It is suggested that multicomponent composition design of amorphous alloys would be an effective approach for tailoring novel BFMs satisfying various composition and property requirements.

Key words: brazing, metallic glass, filler metal, Ti-based alloy, mechanical property

TaMoNiVTiAl 高熵合金的高温力学性能探究

葛绍璠[1,2]，张　龙[2]，付华萌[2]，张海峰[2]

（1. 中国科学技术大学材料科学与工程学院，辽宁沈阳　110016;

2. 中国科学院金属研究所沈阳材料科学国家（联合）实验室，辽宁沈阳　110016）

摘　要：高熵合金自诞生以来由于极高的混合熵使金属间化合物的大量减少的特性引起了极大的关注，TaMoNiVTiW 体系耐高温高熵合金中的系列合金已经证明了其在超高温条件下优于常规材料的强度和塑性。为了得到更为优异的性能，本文通过计算开始研究 TaMoNiVTiAl 合金，首先通过 CALPHAD 对合金进行了相组成的分析，接着系统研究了该合金经不同退火温度后的微观结构以及在 500、700、900℃下的压缩力学性能。讨论了该合金相的变化与力学性能的关系，以及其在高温核材料方面的应用前景。

关键词：高熵合金，CALPHAD，高温压缩，固溶强化

High Temperature Mechanical Property of TaMoNiVTiAl High Entropy Alloy

Ge Shaofan[1,2], Zhang Long[2], Fu Huameng[2], Zhang Haifeng[2]

(1. School of Materials Science and Engineering, University of Science and Technology of China, Shenyang 110016, China; 2. Shenyang National Laboratory for Material Science, Institute of Metal Research, Chinese Academy of Sciences, Shenyang, 110016, China)

Abstract: High entropy alloys have gathered remarkable concerns due to its specially property about high mixing entropy induced reduction of number of phases in multicomponent. The TaMoNiVTiW series system refractory high entropy alloys have proved better mechanical properties than current refractory structural materials under super high temperature. In order to get more progress, based on rigorous calculation, TaMoNiVTiAl alloy was chosen and analyzed the phase components by calculation of phase diagram(CALPHAD). Then, the compression mechanical properties of homogenously annealed alloys under 500、700 and 900℃　were investigated. The phase segregation and microstructure and were obtained by SEM, EDX and XRD. In the end, we discussed the potential application prospect on refractory nuclear material.

Key words: high entropy alloy, CALPHAD, high temperature compression, solidification strengthening

具有高形成能力的 ZrCu 基非晶合金的制备及性能研究

耿铁强[1]，朱正旺[2]，李　文[1]，张海峰[2]

（1. 沈阳理工大学材料科学与工程学院，辽宁沈阳　110159；2. 中国科学院金属研究所沈阳材料科学国家（联合）实验室，辽宁沈阳　110016）

摘　要：ZrCu 基非晶合金因其具有优异的机械、物理、化学性能而受到诸多研究学者的广泛关注。本文以非晶合金 $Zr_{52}Cu_{33}Ni_4Al_{8.5}Ag_{1.5}Nb_1$ 为研究对象，使用核级锆和海绵锆制备两种合金锭，利用铜模翻转方法获得直径分别为 5mm、10mm、20mm 的样品。采用 DSC、XRD、SEM、万能材料试验机等检测分析手段表征了材料的热力学行为、相组成、微观组织、压缩力学性能等，研究了冷却速度及原材料纯度对合金的微观组织和力学性能的影响。试验结果表明，原材料纯度越高，合金锭析出物越少，直径为 5mm、10mm 样品为纯非晶结构，直径为 20mm 为晶态+非晶相复合结构。冷却速度越低，晶态相含量越高，但非晶相的热力学参数如玻璃转变温度、晶化温度等基本一致。晶态相的析出对合金的压缩强度影响不大。

关键词：非晶合金，冷却速度，原材料纯度，强度

Preparation and Properties of ZrCu - based Amorphous Alloy with High Formation Ability

Geng Tieqiang[1], Zhu Zhengwang[2], Li Wen[1], Zhang Haifeng[2]

(1. School of Materials Science and Engineering, Shenyang University of Technology, Shenyang 110159, China; 2. Shenyang National Laboratory for Material Science, Institute of Metal Research, Chinese Academy of Sciences，Shenyang 110016, China)

Abstract: ZrCu-based amorphous alloys have attracted much attention because of their excellent mechanical, physical and chemical properties. In this paper, $Zr_{52}Cu_{33}Ni_4Al_{8.5}Ag_{1.5}Nb_1$ amorphous alloys were prepared using different-scale raw materials. Samples with diameters of 5 mm、10 mm and 20 mm were obtained using a copper mold tilting casting method. Phase composition, thermal and mechanical properties were measured by differential scanning calorimetry (DSC), X-ray diffraction (XRD), scanning electron microscope (SEM), universal material testing machine, respectively. The results show that the higher the purity of the raw material, the less the ingot precipitate, samples for the 5 mm and 10 mm diameter is pure amorphous structure, 20 mm diameter is amorphous matrix composite with crystalline phase. The thermal properties including glass transition temperature and onset-crystallization temperature are almost similar. Compression tests showed the little influence of the formation of crystalline phase on the strengths for the present alloy.

Key words: amorphous alloy, cooling rate,　raw material, strength

非晶合金 $Fe_{78}Si_8B_{14}$ 活化过硫酸钠快速降解酸性橙 II

李海龙[1,2]，朱正旺[2]，付华萌[2]，张宏伟[2]，李　宏[2]，王爱民[2]，张海峰[2]

（1.东北大学材料科学与工程学院，辽宁沈阳　110819；

2. 中国科学院金属研究所沈阳材料科学国家（联合）实验室，辽宁沈阳　110016）

摘　要：铁基非晶合金不仅在活化过硫酸钠得到有较高氧化电位的硫酸根自由基（E_o=2.5~3.1V）方面有着优越的性能，而且在偶氮染料废水除色方面表现出色。本工作采用正交设计实验 $L_{16}(4^3)$ 探究过硫酸钠浓度、反应温度、条带量等因素对降解反应的影响，同时，对水溶液中的酸性橙降解反应机理以及动力学过程也做了深入研究。结果表明，从高到低影响降解效率的因素依次是条带量、过硫酸钠浓度、温度。降解过程主要分为两个反应，第一个反应为铁基非晶合金中的零价铁对酸性橙的直接降解生成小分子有机物和 Fe^{2+}，第二个反应是 Fe^{2+} 活化过硫酸钠生成硫酸根自由基降解酸性橙，两个反应协同作用于降解酸性橙。该规律揭示将促进非晶合金作为催化剂在污水处理

领域的应用。

关键词：金属玻璃，过硫酸钠，降解，偶氮染料

Efficient Degradation of Acid Orange II Using Persulfate Oxidation Activated By $Fe_{78}Si_8B_{14}$ Metallic Glass

Li Hailong[1,2], Zhu Zhengwang[2], Fu Huameng[2], Zhang Hongwei[2], Li Hong[2],
Wang Aimin[2], Zhang Haifeng[2]

(1. School of Materials and Metallurgy, Northeastern University, Shenyang 110819; 2. Shenyang National Laboratory for Material Science, Institute of Metal Research, Chinese Academy of Sciences, Shenyang 110016)

Abstract: Iron-based metallic glasses have excellent performance not only in activating sodium persulfate for sulfate radicals ($SO_4^-\cdot$) with a high oxidative potential (E_o=2.5~3.1V), but also in the removal of azo dye wastewater. In this work, orthogonal design experiment L_{16} (4^3) was used to investigate the parameters influencing this reaction, including persulfate concentration, reaction temperature and ribbon dosage. The mechanism and kinetics of orange II (AOII) dye degradation in aqueous solution was also investigated in this study. The results showed that the factors affecting the degradation efficiency from high to low are ribbon dosage, sodium persulfate concentration and temperature .The degradation reactions in solution mainly contains two reactions, one is the direct degradation by ZVI（zero valent iron）of Fe based amorphous alloy in acid orange with products of small organic molecules and Fe^{2+}, the other reaction is that Fe^{2+} activates sodium persulfate with products of sulfate radical which can destroy the structure of acid orange. This work will accelerate the application of amorphous alloys as catalysts in the field of wastewater treatment.

Key words: metallic glass, persulfate oxidation, degradation, azo dye

内生 β 相 TiZr 基非晶复合材料的两相平衡

李正坤，秦鑫冬，刘丁铭，张海峰

（中国科学院金属研究所沈阳材料科学国家(联合)实验室，辽宁沈阳 110016）

摘 要：为设计和调控非晶复合材料的力学性能，本文以内生 β 相型 TiZr 基非晶复合材料为研究对象，在 $Ti_xZr_yNi_{2.65}Cu_{4.5}Be_{11.35}$ 合金体系中，通过调控 Ti 和 Zr 的原子比，开发了一系列具有两相平衡特征的内生非晶复合材料。通过成分分析发现，该体系在 $Ti\text{-}Zr\text{-}Ni_{14.3}Cu_{24.3}Be_{61.4}$ 伪三元相图中具有相平衡特征，其中内生相可在 β 相和 α+β 相间连续变化。在载荷作用下，内生相中形成密集的滑移带，并且 β 相可发生应力诱发马氏体相变，提高复合材料塑性。

关键词：非晶复合材料，内生，β 相，两相平衡

Phase Equilibrium in TiZr-based Metallic Glass Composites Containing In-situ β Phase

Li Zhengkun, Qin Xindong, Liu Dingming, Zhang Haifeng

(Shenyang National Laboratory for Materials Science, Institute of Metal Research, Chinese Academy of Sciences, Shenyang 110016, China)

Abstract: In order to design the metallic glass composites with improved mechanical properties, this work studies the TiZr-based in-situ β phase composites. By regulating the percentages of Ti and Zr in $Ti_xZr_yNi_{2.65}Cu_{4.5}Be_{11.35}$ alloy, a series of in-situ β phase composites with phase equilibrium are developed. The component analysis indicates that the phase equilibrium can be established in the pseudo-ternary phase diagram of Ti-Zr-$Ni_{14.3}Cu_{24.3}Be_{61.4}$. The precipitations transfer from β to $\alpha+\beta$ continuously. Besides, intense slip bands appear in the in-situ phases under loading, and the stress-induced martensite phase transformation of β phase contributes to the improvement of the plasticity.

Key words: metallic glass composite, in-situ, β phase, phase equilibrium

具有高玻璃形成能力的钛基金属玻璃

林师峰[1,2]，朱正旺[2]，李　丹[2]，付华萌[2]，张宏伟[2]，

李　宏[2]，王爱民[2]，张海峰[2]

（1. 东北大学材料科学与工程学院，辽宁沈阳　　110819；

2. 中国科学院金属研究所沈阳材料科学国家（联合）实验室，辽宁沈阳　　110016）

摘　要： 在 ZT3（$Ti_{32.8}Zr_{30.2}Ni_{5.3}Cu_9Be_{22.7}$）的基础上，保持 Cu、Ni 和 Be 的原子百分比不变，通过提高钛的百分比含量、降低锆的百分比含量来制备新型钛基金属玻璃 T-X（$Ti_{X.8}Zr_{63-X.8}Ni_{5.3}Cu_9Be_{22.7}$（X=35,38,41,44,47））。结果表明，该系列合金的玻璃形成能力（GFA）与二十面体局域有序密切相关。当 X≤41，合金临界非晶形成尺寸大于 20mm，竞争晶化产物均为二十面体准晶相(I-Phase)；随着钛含量的进一步提高，晶化产物变得复杂，I-Phase 的量变少，导致对应体系的合金过冷液相的稳定性降低，GFA 降低。力学性能试验表明，该系列钛基金属玻璃的断裂强度相近，约为 1800 MPa。以上结果表明，该类钛基金属玻璃是一类潜在的、性能优异的结构材料。

关键词： 钛基金属玻璃，晶化产物，玻璃形成能力，断裂强度

New Ti-base Bulk Metallic Glasses with High Glass Forming Ability

Lin Shifeng[1,2], Zhu Zhengwang[2], Li Dan[2], Fu Huameng[2], Zhang Hongwei[2],

Li Hong[2], Wang Aimin[2], Zhang Haifeng[2]

(1. School of Materials and Metallurgy, Northeastern University, Shenyang 110819, China;

2. Shenyang National Laboratory for Material Science, Institute of Metal Research,

Chinese Academy of Sciences, Shenyang 110016, China)

Abstract: Ti-base bulk metallic glasses (BMGs) of T-X ($Ti_{X.8}Zr_{63-X.8}Ni_{5.3}Cu_9Be_{22.7}$) were prepared by optimizing the alloy composites through increasing the content of titaniumin a ZT3 ($Ti_{32.8}Zr_{30.2}Ni_{5.3}Cu_9Be_{22.7}$) alloy. The results presented that glass forming ability of the present alloys is very closely related to icosahedral short-range ordering which is indicated by the formation of I-phase. For the alloys with $X\leq41$, the primary crystallization product consisted of the icosahedral quasicrystal phase (I-phase).With further raising the content of titanium, it can be found that the crystallization products become complex that resulting in the decreasing of the content of I-Phase. It probably leads to reducing the stability reduction of the supercooled liquid phase. Mechanical tests show that the present alloys possess high compressive fracture strength of about 1800 MPa. These findings indicated that these new Ti-based BMGs are potentially used as structural materials.

Key words: Ti-base bulk metallic glass, crystallization products, glass forming ability, fracture strength

利用 Ti6Al4V 制备内生枝晶相非晶复合材料及性能研究

刘丁铭[1,2]，朱正旺[2]，李正坤[2]，张 龙[2]，付华萌[2]，

李 宏[2]，张宏伟[2]，王爱民[2]，张海峰[2]

(1. 东北大学材料科学与工程学院，辽宁沈阳 110819；)

2. 中国科学院金属研究所沈阳材料国家实验室，辽宁沈阳 110016)

摘 要： 本文通过异质复合思想制备了一系列内生枝晶相非晶复合材料，具体策略是将非晶合金 $Ti_{32.8}Zr_{30.2}Ni_{5.3}Cu_9Be_{22.7}$ 和传统钛合金 Ti6Al4V 按照不同的比例复合得到一系列复合材料。通过 XRD、SEM、TEM 及力学性能测试研究了这一系列复合材料的微观结构和力学性能。结果表明，随着添加的 Ti6Al4V 质量百分数的增加，复合材料中析出的枝晶相的尺寸和体积分数增大。这一系列复合材料的压缩力学性能也表现出明显的变化，当 Ti6Al4V 的质量百分数为 50% 时复合材料力学性能最佳，屈服强度、断裂强度和塑性分别达到 1551MPa、2365MPa 和 12.9%。并且由于压缩过程中发生了应力诱发马氏体相变，复合材料表现出明显的加工硬化行为。

关键词： 非晶合金，复合材料，Ti6Al4V，加工硬化，马氏体相变

Preparation and Mechanical Properties of a Series of in-situ Ductile Dendrite Reinforced Bulk Metallic Glass Composites using Ti6Al4V

Liu Dingming[1,2], Zhu Zhengwang[2], Li Zhengkun[2], Zhang Long[2], Fu Huameng[2],

Li Hong[2], Zhang Hongwei[2], Wang Aimin[2], Zhang Haifeng[2]

(1. School of Material Science and Engineering, Northeastern University, Shenyang 110819, China; 2. Shenyang National Laboratory for Materials Science, Institute of Metal Research, Chinese Academy of Sciences, Shenyang 110016, China)

Abstract: A series of in-situ ductile dendrite reinforced bulk metallic glass composites using Ti6Al4V was prepared successfully by a new design strategy. The nominal composition of these composites was obtained by mixing a metallic glass ($Ti_{32.8}Zr_{30.2}Ni_{5.3}Cu_9Be_{22.7}$) and a traditional Ti alloy (Ti6Al4V) with different ratios. The microstructure and mechanical properties of these composites were studied by SEM, XRD, TEM and compression test. The results show that the dendrite size and volume fraction in these composite increase with the increasing of the percentage of Ti6Al4V. The optimal property with the yielding strength of 1551MPa, fracture strength of 2365MPa and plasticity of 12.9% was obtained when the weight percentage of Ti6Al4V is 50%. A significant work-hardening phenomenon was observed during the deformation due to the stress-induced martensitic transformation.

Key words: metallic glass, composites, Ti6Al4V, work-hardening, martensitic transformation

Ti 元素对 Zr 基非晶合金在含 Cl⁻溶液中抗腐蚀性的影响

邱张维佳[1, 2]，付华萌[2]，张海峰[2]

（1. 中国科学技术大学材料科学与工程学院，安徽合肥　230026；

2. 中国科学院金属研究所沈阳材料科学国家（联合）实验室，辽宁沈阳　110016）

摘　要：本实验制备了合金成分为（$Zr_{51.3}Al_{10}Ni_6Cu_{31.8}Ag_{0.1}Y_{0.8}$）$_{100-x}Ti_x$（x=2，4，6，8 at%）的非晶棒材与条带。DSC 与 XRD 结果表明各成分在 φ3 的冷却速率下均可形成纯非晶；利用失重法和电化学方法测试了各成分在含 Cl⁻溶液中的抗腐蚀性，结果表明随着 Ti 含量的提升，合金的抗腐蚀性提升；通过 SEM 观察了恒电位极化后各成分非晶棒材的腐蚀形貌，随着 Ti 的添加，点蚀现象减少。分析表明，Ti 作为一种更具阀金属特性的元素，对 Zr 基非晶合金在含 Cl⁻溶液中的抗腐蚀性有较大提升。

关键词：Zr 基非晶合金，Ti，抗腐蚀性，电化学，腐蚀行为

Effect of Ti Addition on Corrosion Resistance of Glassy Zr-Based Alloy In Chloride Medium

Qiu Zhangweijia[1, 2], Fu Huameng[2], Zhang Haifeng[2]

(1. School of Materials Science and Engineering, University of Science and Technology of China, Hefei 230026, China; 2. Shenyang National Laboratory for Material Science, Institute of Metal Research, Chinese Academy of Sciences, Shenyang 110016, China)

Abstract: In this paper cylindrical bulk samples and melt-spun ribbon of $(Zr_{51.3}Al_{10}Ni_6Cu_{31.8}Ag_{0.1}Y_{0.8})_{100-x}Ti_x$(x=2，4，6，8 at%) alloys were prepared. The amorphous nature of melt-spun ribbon and both bulk samples with 3 mm diameter using confirmed by X-ray diffraction (XRD) and Differential Scanning Calorimetry (DSC). Weight loss method and electrochemical method were applied to characterize the corrosion resistance in chloride-contained solution. The results showed that the corrosion resistances increase with increasing Ti content. The corrosion morphology of sample surface were observed by scanning electron microscope (SEM), it shows that pitting corrosion was inhibited when more Ti were added into amorphous matrix. Analysis shows that Ti as an element with valve-metal nature benefit the corrosion resistance of Zr-based glassy alloys in chloride-contained solution.

Key words: Zr-based amorphous alloy, Ti, corrosion resistance, electrochemical method, corrosion behaviour

预变形对 Ti 基非晶复合材料力学性能的影响

赵子彦，年　娟，王沿东

（东北大学材料各向异性与织构教育部重点实验室，辽宁沈阳　110819）

摘　要：非晶合金具有高强度、高硬度等优异的性能，但较差的塑性变形能力严重制约着非晶合金的应用。本文主要从预变形的角度，研究了室温下其对 TiZrNiCuBe 块状非晶合金复合材料的准静态压缩性能的影响。结果表明，轴向预压缩、预拉伸，横向预压缩处理都会诱发马氏体相变以及形成剪切带。轴向预压缩使塑性降低，而横向预压缩和轴向预拉伸使塑性提高。不同预加载处理对塑性的影响归因于预加载阶段萌生的马氏体与剪切带，两者具有明显的择优取向，并对之后压缩过程中马氏体相变和剪切带的萌生以及分布产生强烈影响。由于多重剪切带的萌生以及应力的分散，使得复合材料的力学性能显著改善。这种预变形处理方法为内生非晶合金复合材料的研究提供了新思路。

关键词：非晶复合材料，预变形，力学性能

Influence of Pre-deformation on Mechanical Properties in the Ti-based Bulk Amorphous Alloy Composites Containing in-situ Formed Dendrites

Zhao Ziyan, Mu Juan, Wang Yandong

（Key Laboratory for Anisotropy and Texture of Materials (Ministry of Education), School of Material Science and Engineering, Northeastern University, Shenyang 110819, China）

Abstract: Bulk amorphous alloys (BAAs) exhibit superior mechanical properties including high strength, high hardness, etc., but poor tensile plastic deformation capacity at room temperature, which largely limits their practical application. In this work, the Ti-Zr-Cu-Ni-Be bulk amorphous alloy matrix composites (BAACs) containing in-situ formed dendrites with deformation-induced martensitic transformation were selected, and the influence of pre-deformation on the quasi-static mechanical properties was studied at room temperature. The plastic pre-deformation in compression and tension along the axial and lateral directions induce martensitic transformation and the formation of shear bands. In the subsequent mechanical tests, the different pre-deformation result into different behaviors. The axial pre-deformation reduces the plasticity of BAACs, and contrarily, the lateral tensile/compressive pre-deformations significantly improve the plasticity of BAACs. The contrary effects are attributed to the preferentially orientated martensitic variants and shear bands formed in the pre-deformation stages, which show a large influence on martensitic transformation behaviors and the nucleation and propagation of shear bands. This strategy for enhancing the plastic deformation capacity of BAACs will shed light on developing new BAACs of high performance.

Key words: bulk amorphous alloy composite, pre-deformation, mechanical property

Hf 元素对 Zr 基非晶合金形成、性能影响

朱玉辉[1,2]，朱正旺[2]，王爱民[2]，张海峰[2]

（1. 中国科学技术大学材料科学与工程学院，辽宁沈阳　110016；

2. 中国科学院金属研究所沈阳材料科学国家（联合）实验室，辽宁沈阳　110016）

摘　要：本文以 ZrNbCuNiAl 为基础，研究了 Hf 部分取代基体中 Zr 对于非晶合金形成能力、热力学和力学性能的影响规律。实验结果表明：随着 Hf 含量的增加，合金临界尺寸增加，非晶形成能力得到了提高，过冷液相区 ΔT 增大、约化玻璃转变温度 Trg 减小，对比合金临界尺寸的变化规律可发现，ΔT 可以更好地反映本合金系非晶形成能力的强弱。合金的压缩力学性能变化不明显，屈服强度 σ_s、抗压强度 σ_{bc} 及塑性应变均先增大后减小，在 Hf 含

量约为 0.5 at%时，综合压缩塑性达到最佳，压缩试样以剪切方式断裂。冲击韧性由不含 Hf 的 3.3J/cm², 增至大于 12J/cm²，得到了明显改善，裂纹由试样表面产生并快速向内部扩展导致试样断裂。

关键词：非晶合金，非晶形成能力，压缩性能，冲击韧性

Effects of Hf Addition on the Glass-forming Ability and Mechanical Properties of Zr-based Amorphous Alloys

Zhu Yuhui[1,2], Zhu Zhengwang[2], Wang Aimin[2], Zhang Haifeng[2]

(1. School of Materials Science and Engineering, University of Science and Technology of China, Shenyang 110016, China;

2. Shenyang National Laboratory for Material Science, Institute of Metal Research, Chinese Academy of Sciences, Shenyang 110016, China)

Abstract: In this work, the effects of which added Hf partially replaces Zr as basis in the ZrNbCuNiAl metallic glasses on the glass-forming ability (GFA), thermal and mechanical properties were studied. The results showed that the addition of Hf enhances the glass-forming ability and enlarges the critical size of the alloy gradually. And at the same time, the super-cooled liquid region ΔT increased and the reduced glass transition temperature Trg decreased.Compared with the critical size of alloy, it can be found that ΔT could better reflect the relative strength trend of the glass-forming ability.The fluctuation of compressive mechanical properties of the alloy within a small range, the yield strength σ_s, the compressive strength σ_{bc} and the plastic strain increased first and then decreased. When the Hf content was about 0.5 at%, the comprehensive performance was the best. And the fracture mode of all samples are shear fracture.The impact toughness increased from 3.3J/cm² which without Hf to greater than 12J/cm². And the crack was generated from the surface of the sample and expanded rapidly to the inside.

Key words: metallic glass, the glass-forming ability, compression performance, impact toughness

Bond Length Deviation in CuZr Metallic Glasses

Peng Chuanxiao[1], Şopu Daniel[2,3], Song Kaikai[1], Zhang Zhenting[1], Wang Li[1], Eckert Jürgen[3,4]

(1. School of Mechanical, Electrical & Information Engineering, Shandong University (Weihai), Weihai 264209 China;

2. Division of Materials Modeling, Institute of Materials Science, TU Darmstadt, Jovanka-Bontschits-Straße 2, D-64287 Darmstadt, Germany;

3. ErichSchmid Institute of Materials Science, Austrian Academy of Sciences, Jahnstraße 12, A-8700 Leoben, Austria;

4. Department Materials Physics, Montanuniversität Leoben, Jahnstraße 12, A-8700 Leoben, Austria)

Abstract: We define a structural parameter, called atomic bond length deviation (BLD_i), to characterize structural heterogeneity of CuZr melt and metallic glass (MG). Molecular dynamics (MD) simulations have been performed to explore the average BLD_i of the system evolution with temperature during $Cu_{64}Zr_{36}$ and $Cu_{50}Zr_{50}$ MGs formation and the correlation between BLD_i and thermal relaxation/local atomic shear strain upon compressive loading. The results indicate that *BLD*$_i$ contains both symmetrical characteristic and volumetricinformation of the short-range order clusters while symmetry seems to play more important role in relaxation and deformation events; the fast decreasing of average *BLD*$_i$ near above T_g with decreasing temperature corresponds to the sharp increase of the number of full icosahedra (FI) while the shear transformation zones or single jump events have a high propensity to originate from those regions with the higher *BLD*$_i$ and less FI cluster.Additionally, the system average *BLD*$_i$ can also be accessed experimentally, through the radial distribution function (RDF).

Key words: atomic bond length deviation, structural heterogeneity of metallic glass, shear transformation zones

Atomic Structure Evolution of Cu$_{50}$Ag$_{50}$ Phase Separated Glass under Compression Deformation

Jia L. J., Peng C. X., Cheng Y., Wang P. F., Wang Y. Y., Wang L.

(School of Mechanical, Electrical & Information Engineering, Shandong University (Weihai), Weihai 264209, China)

Abstract: Although the phase-separated metallic glass has been reported in literature, its plastic deformation mechanism is still open to date. In this paper, molecular dynamics (MD) simulation are performed to explore the plastic deformation mechanism of phase-separated glass Cu$_{50}$Ag$_{50}$ metallic glass during the compression deformation. Our finding indicates that the homogeneous atom pairs show stronger interaction than that of heterogeneous atom pairs in Cu$_{50}$Ag$_{50}$ glass, which also shows higher plastic deformation ability exhibited in stress-strain curves. The degree of phase separation increases and the glass becomes more ordering with increasing deformation. Ag atom shows higher Voronoi volume and potential energies, localize larger local shear strain than Cu atom, which undergo flow deformation preferentially, Voronoi volume and potential energy of Ag atom in the entire region keep unchanged during plastic deformation period because of the release of elastic energy and atom arrangement; while it increases for Cu atom due to its resistance to the shear transformation in this case.

Key words: phase–separated metallic glass, compression deformation, local shear strain, voronoi volume

Formation of Nanoglassy Alloy in Al-Ni-Mg-Ce System

Wang S. H., Li D. F., Yuan X. D, Song K. K., Li X. L., Wang L.

(School of Mechanical, Electrical & Information Engineering, Shandong University (Weihai), Weihai 264209, China)

Abstract: The present letter aims to report the formation of nanoglassy alloy in the Al-Ni-Mg-Ce alloy system. The nanoglassy alloys exhibit a unique combination of primary crystallization of fcc-Al preceded by a clear glass transition phenomenon. It is found that the proper Mg additions can promote the formation of nanoglassy alloy in the Al-Ni-Mg-Ce alloy system. As proved in the Al$_{84}$Ni$_{10-x}$Mg$_x$Ce$_6$ alloy system, the nanoglassy alloys are formed where $2 \leqslant x \leqslant 4$ at. %; and for the Al$_{83}$Ni$_{10-x}$Mg$_x$Ce$_7$ and Al$_{82}$Ni$_{10-x}$Mg$_x$Ce$_8$ alloy system, the nanoglassy alloys are formed where $2 \leqslant x \leqslant 4$ and $4 \leqslant x \leqslant 5$ at. %, respectively. The formation of nanoglassy alloy can be ascribed to the small negative heat of mixing between Mg and other constituents.

Key words: metallic glass, microstructure, nanoglassy alloy, primary crystallization

Structure and Thermodynamic Characteristics of Miscible and Immiscible Binary Liquids

Cui W. C., Wang L., Wang S. H.

(School of Mechanical, Electrical & Information Engineering, Shandong University (Weihai), Weihai 264209, China)

Abstract: Molecular dynamics (MD) based on the embedded atom method (EAM) simulation has been performed to investigate the structural and thermodynamic difference between miscible and immiscible liquid mixture. For fully miscible Cu-Zr liquid mixture, there is stronger interatomic interaction between heterogeneous particles: The pair correlation functions (PCF) of heterogeneous pairs in liquid $Cu_{50}Zr_{50}$ show highest peak among the three partial PCFs (PPCF); the coordination number (CN) of Cu-Zr pairs is much higher than that of Cu-Cu and Zr-Zr pairs. In contrast to that, for Fe-Cu melt with miscibility gap, there is stronger ineratomic interaction of homogeneous atom pairs: Bhatia-Thornton (B-T)structure factor Scc(q) of $Fe_{50}Cu_{50}$ melts increases sharply at q→0; PPCF of homogeneous particles show the higher peak than that of heterogeneous pairs; CN of Cu-Cu and Fe-Fe is much higher than that of Fe–Cu. For the miscible systems, enthalpy of mixing is positive while it is negative for the immiscible systems. Further investigation indicates that the excess volume is related to the microscopic interaction between particles, and it could be negative and positive for both miscible and immiscible mixtures.The present work provides an understanding of atomic-scale structure and thermodynamic properties in miscible and immiscible liquid mixtures.

Key words: miscible and immiscible binary liquids, structural and thermodynamics characteristics, molecular dynamics simulation

Form of Precursor Material on Laser Clad Al-based Amorphous Composite Coating

Zhang Chunzhi, Chen Shu, Wu Yumei, Li Huiping, Kong Lingliang

(School of Materials Science and Engineering, Shandong University of Science and Technology, Qingdao 266590, China)

Abstract: Al-Ni-La amorphous composite coatings were fabricated by laser cladding(LC) technology on Q235 steel substrates with two forms of precursor materials. Properties of the LC specimens in single clad tracks were compared in terms of the microstructure, the elemental composition distribution, the microhardness and the corrosion resistance. The morphological observation indicates that satisfying metallurgical bonding can be realized between the cladding layer and the substrate. Both cladding layers exhibit much higher microhardness than that of the substrate due to the high solidification rate of the laser irradiation. The cladding layer obtained with the hyperquenched (HQ) ribbons possesses the higher micohardness and better corrosion resistance than that with the ball-milled powders. Such a result maybe attribute to the lower dilution and the oxygen participation in the subsurface of the former. The present study shows that the adoption of the HQ ribbons is a feasible method to obtain a LC layer with the desired performance on the carbon steel substrate.

Key words: coating, hardness, corrosion resistance, laser cladding

利用超声提高块体非晶合金的屈服强度

王　拓，惠希东

摘　要： 非晶合金在热力学上属于亚稳态材料，在外场作用下会发生结构的弛豫。本文研究了 ZrCuFeAlNb 块体非晶合金在经过不同时间超声以后的机械性能。结果表明，合金经过超声后的塑性变化不大，但是屈服强度随着超声的时间增加而呈现增加的趋势。通过高分辨透射电镜（HRTEM）观察发现合金在经过 3h 超声后析出了几个纳米尺度均匀的纳米晶，同时在剪切带附近发生了调幅分解。这表明超声给合金注入了能量，同时发生了结构的弛豫。通过超声的手段形成的纳米晶和调幅分解在保证合金塑性不变的情况下有助于提高合金屈服强度。

高压下 $Ce_{70}Al_{10}Cu_{20}$ 熔体结构和扩散行为的分子动力学研究

李冬冬，陈　恒，孙晓锐，屈冰雁，周如龙，张　博

（合肥工业大学材料学院，非晶态物质科学研究所）

摘　要： 我们采用第一性原理方法，对高压下 $Ce_{70}Al_{10}Cu_{20}$ 熔体的结构和扩散行为进行了理论研究。研究结果表明，在加压的条件下，熔体的短程有序性和扩散系数发生的异常的变化。通过结构分析发现，在熔体中存在两种不同半径的 Ce 原子：大半径的 Ce 和小半径的 Ce。随着压力的增大，大半径的 Ce 原子逐渐转化为小半径的 Ce。当压力超过 4.6GPa 的时候，小半径的 Ce 原子占主导。在常压下，Al 原子周围具有较强的五次对称性或二十面体有序性而 Cu 原子有序性很弱。随着 Ce 半径减少，Cu 原子周围团簇的二十面体有序性增强，甚至强于 Al 原子周围的二十面体有序性。这种增强的二十面体有序性大大地迟滞了 Cu 原子的扩散。我们的理论计算结果能够帮助我们理解，加压的条件下结构和动力学行为的关联，同样能够解释实验上观察到的 Ce 基非晶合金在加压条件下的结构转变。

13 高温合金

新一代航空发动机对高温合金材料的需求

董志国

（中国航发沈阳发动机研究所，辽宁沈阳　110015）

摘　要：航空发动机是航空武器装备发展建设的关键，是衡量一个国家装备水平、科技工业实力和综合国力的重要标志。高温合金是航空发动机用量最大的材料，也是关键重要部件的首选材料，对于保障发动机的性能、可靠性及安全性起着极大作用，新一代发动机的性能提高对于高温合金的发展提出了更高的要求，需要更多的先进技术支持发展。

关键词：航空发动机，高温合金

高合金化合金均质化控制技术研究

付　锐

（钢铁研究总院，北京　100081）

摘　要：高合金化合金是广泛应用于航空、航天、舰船、机械、汽车、石化等领域的关键材料，是衡量一个国家工业实力的重要标志。均质化是高合金化合金生产永恒的追求目标，包括成分均匀和组织均匀两个方面，成分均匀通过降低铸锭凝固过程的成分偏析来实现，而组织均匀通过热加工实现。先进装备的不断发展，对材料性能的要求不断提高，材料性能的欲度不断减小，因此对材料性能的一致性提出了更高的要求，进一步提高材料均质化水平，开发均质化检测技术并建立材料应用大数据库成为材料研究工作者奋斗的重要方向之一。

关键词：高合金化合金，均质化，检测技术

CALPHAD 技术及其在高温合金设计中的应用

郭翠萍

（北京科技大学，北京　100083）

摘　要：CALPHAD 技术是目前利用传统实验相图和其他所有热化学实验信息建立现代相图最成熟的一种技术。可以用来计算真实多元材料体系的成分、相组成，模拟组织演变及相转变过程等，在合金设计中起到了至关重要的作用。而高精度的热力学以及动力学数据库的建立是利用 CALPHAD 技术进行高温合金设计必不可少的组成部分。

关键词：CALPHAD 技术，高温合金设计，数据库

超超临界电站用高温材料的发展

江　河

（北京科技大学材料科学与工程学院，北京　100083）

摘　要：随着现代工业的发展，能源和环境危机日益严峻，为此需提高火电站发电效率以实现节能减排。世界各国均大力发展超超临界电站技术，然而超超临界电站需在高温高压环境下长期安全服役，对材料的性能和稳定性提出了极高的要求。传统铁素体、奥氏体钢无法满足如此严苛的条件，超超临界电站的发展为高温材料提出了新的要求和挑战。

关键词：超超临界，高温材料

先进航空材料和复杂构件的钎焊技术

静永娟

（中航工业集团北京航空制造工程研究所，北京　100024）

摘　要：随着航空科学技术的发展，在航空装备中越来越多地采用新材料、新结构，对其焊接技术也提出了更高的要求。钎焊技术作为特种连接技术的一种，在耐温材料的连接和复杂精密结构的制造方面上具有独特优势，开展高温合金、钛合金、金属间化合物材料以及陶瓷等材料及构件的钎焊技术研究已成为目前航空领域焊接技术的主要发展方向。本研究主要介绍了钎焊金属蜂窝夹层结构件、大间隙钎焊修复涡轮叶片类零件、钎焊多孔层板以及焊料设计等相关研究工作，这些研究结果为新型航空材料的应用和复杂航空构件的设计制造提供了技术支撑。

关键词：航空材料，复杂构件，钎焊

Nb-Si 基超高温合金及其精密铸造技术研究

康永旺

（中国航发北京航空材料研究院先进高温结构材料重点实验室，北京　100095）

摘　要：Nb-Si 基超高温合金具有高熔点(>1750℃)、低密度(6.6~7.2g/cm^2)和良好的高温强度，承温能力可达1200~1400℃，是最有希望替代现有 Ni 基高温合金的新型高温结构材料之一。但是，Nb-Si 基超高温合金的应用仍然面临三大挑战：材料综合力学性能的提高、高温抗氧化性的提升以及超高温成形工艺。目前的研究工作主要集中在材料成分优化、超高温定向凝固组织控制、超高温热处理组织优化及显微组织对力学和氧化行为的影响，并开发适合 Nb-Si 基超高温合金使用的高温抗氧化涂层，同时发展 Nb-Si 基超高温合金熔炼和超高温定向凝固成形相关技术，为未来 Nb-Si 基超高温合金在 1200℃以上服役环境的使用奠定基础。

关键词：Nb-Si 基超高温合金，超高温精密铸造

新型γ′相强化 Co-Al-W 基高温合金及其蠕变机理

李龙飞，路　松，李文道，薛　飞，周海晶，冯　强

（北京科技大学新金属材料国家重点实验室，北京　100083）

摘　要：与镍基高温合金相比，钴基高温合金具有良好的抗热腐蚀、抗热疲劳和焊接性能，非常适用于地面燃气轮机和海航用航空发动机的热端部件；传统钴基合金由于缺少γ′相强化，其高温强度显著低于镍基合金，阻碍了其在高温条件下的使用。近年来，高温稳定的γ′相(Co_3(Al,W))的发现使发展综合性能优异的高温合金成为可能，并使发展新型钴基高温合金迅速成为国际高温合金界研究的前沿。但是，目前该类合金的研发仍处于初级阶段，且对其蠕变行为和机理知之甚少，严重阻碍了其工程应用。本文从 Co-Al-W 基三元合金出发，研究了 Ta、Ti、Nb、Mo 和 V 等元素对该合金系γ′相溶解温度以及组织稳定性的影响。在此基础上，研究并开发出了一种目前国际上在 1000℃以上具有最好组织稳定性和蠕变性能的 Co-Al-W-Ti-Ta 五元单晶高温合金，其蠕变性能介于第一、二代镍基单晶高温合金之间。该合金作为模型合金，为未来具有优异综合性能的新型钴基高温合金的开发奠定了基础。针对该合金的特殊蠕变行为（最小蠕变速率小于稳态蠕变速率），系统研究了合金在蠕变不同阶段γ/γ′两相微观组织演变规律以及亚微观组织（位错、层错和反相畴界等晶体缺陷）的形成与演变机制，建立了"错配度-微观组织-亚微观组织-蠕变行为"之间的关系。上述研究揭示了具有正错配度的新型钴基单晶高温合金高温蠕变机理，为新型钴基高温合金的强度设计理论提供了物理冶金依据，也为丰富和发展γ/γ′两相合金的强韧化理论奠定了基础。

热等静压近净成形粉末高温合金制备技术

刘建涛 [1,2]

（1.钢铁研究总院高温材料研究所，北京　100081；
2.高温合金新材料北京市重点试验室，北京　100081）

摘　要：粉末高温合金是先进航空发动机涡轮盘等热端关键转动部件用首选材料。采用热等静压近净成形工艺制备的粉末高温合金具有组织和性能一致性好，材料利用高的显著特点，是高性价比粉末高温合金盘、轴、环等制件制造的解决方案。零原始颗粒边界的合金设计和精度达毫米数量级的包套设计是粉末高温合金近净成形技术的关键。
关键词：粉末高温合金，热等静压近净成形，原始颗粒边界，包套设计

脉冲电流对 GH4169 合金组织及变形行为的影响

刘　杨，王　磊，安金岚，宋　秀，刘恢弘，张　伟

（东北大学　材料各向异性与织构教育部重点实验室/材料科学与工程学院，辽宁沈阳　110819）

摘　要：在 GH4169 合金的时效过程、拉伸变形过程施加脉冲电流，研究了脉冲电流对合金强化相析出、再结晶行为及拉伸变形行为的影响，探究了脉冲电流对合金组织演化、变形行为的作用机理。结果表明，脉冲电流时效处理

后合金强度和塑性可实现同时增加，脉冲电流可在时效过程中诱发合金缺陷密度增大，促进 δ 相于晶界析出、γ″相长大，部分 γ″相在位错附近呈链状特征形貌析出。塑性变形过程中施加脉冲电流，GH4169 合金流变应力显著下降，断裂方式由沿晶断裂转变为穿晶断裂，合金塑性变形能力提高。脉冲电流诱导原子振动加剧导致 Peierls 力降低是引起合金塑性变形过程中流变应力降低的本质，同时脉冲电流促进合金动态再结晶亦是流变应力降低的主要原因。

关键词：GH4169 合金，脉冲电流，显微组织，电塑性，变形行为

航空发动机用 Ti₂AlNb 合金的研制进展

马　雄，张建伟，梁晓波，程云君

（钢铁研究总院高温材料研究所，北京　100081）

摘　要：Ti₂AlNb 合金由于具有低密度、高比强度、良好的抗氧化性及无磁性等优势，成为能在 600~750℃ 使用的最具潜力的轻质高温结构材料之一，并可替代现役高密度镍基、铁基高温合金，通过减轻自重以提高发动机性能。本文主要介绍了 Ti₂AlNb 合金的研究进展和在航空发动机领域的应用。通过合金化和组织设计的系统研究，形成了具有自主知识产权的 Ti-22Al-25Nb 合金，改善了传统 Ti₂AlNb 合金的塑性，突破了合金制备、热加工及成型工艺等技术，具有了一定的批量供货能力。近年来，应用这类合金针对我国高推比航空发动机的部件研制工作进展顺利。

关键词：Ti₂AlNb 合金，合金化，组织设计，制备技术

液态金属冷却定向凝固中 Sn 与高温合金铸件的反应与控制

申　健，卢玉章，徐正国，郑　伟，张　健

（中国科学院金属研究所，辽宁沈阳　110016）

摘　要：提高燃气初温是燃气轮机发展的最重要途径之一，燃气初温的提高首先要求热端部件具有足够高的承温能力，为了提高涡轮叶片的承温能力，燃气轮机用涡轮叶片广泛使用定向凝固技术制备的定向柱晶叶片或单晶叶片。液态金属冷却（Liquid Metal Cooling, LMC）定向凝固技术相对国内广泛使用的高速凝固法（High Rate Solidification, HRS）具有更高的温度梯度和冷却速率，尤其适合制备大尺寸定向柱晶和单晶叶片。LMC 定向凝固法中采用低熔点金属 Sn 作为冷却介质，在定向凝固过程中复杂形状型壳可能开裂造成 Sn 与高温合金在高温下发生反应，导致铸件报废。本文系统研究了低熔点金属 Sn 与高温合金在定向凝固过程中的反应现象。通过研究发现：当温度达到 500℃时，高温合金和 Sn 之间开始发生反应，并且随着温度的提高，反应产物变得复杂。当温度达到 750℃以上时，1min 之内将导致严重的表面反应；通过实验观察和数值模拟相结合的方法获得了高温合金铸件接触 Sn 的临界时间。如果型壳在此之前开裂，就会发生表面反应。

关键词：液态金属冷却，定向凝固，高温合金，铸件

第四代单晶高温合金的发展

史振学，刘世忠，李嘉荣，王效光，岳晓岱

摘　要：针对未来高推重比航空发动机高压涡轮叶片的优异高温综合性能要求，发展了第四代单晶高温合金研究。借助计算机辅助设计和试验相结合，研究 Ru、Re、Cr、Co、W、Y 等合金元素在新一代单晶合金中的作用机理，设计了第四代单晶高温合金 DD15 的化学成分，建立了 DD15 合金的定向凝固工艺和热处理工艺。DD15 合金具有良好的组织稳定性，优良的力学性能，其拉伸性能与持久性能达到了美国第四代单晶高温合金 EPM-102 的水平；该合金具有良好的抗氧化性能、铸造工艺性能和焊接性能；DD15 合金可用于制造具有复杂结构的薄壁涡轮叶片。

关键词：第四代单晶高温合金，DD15

第三代镍基单晶高温合金设计及其组织稳定性

苏海军，张　军，王　博，李卓然，黄太文，刘　林，傅恒志

（西北工业大学凝固技术国家重点实验室，陕西西安　710072）

摘　要：从合金组织稳定性出发，设计了系列具有不同 Re、W、Mo、Cr 含量的第三代镍基单晶高温合金，系统考察了合金组织稳定性及其影响因素。研究结果表明：合金经固溶和时效处理后，γ'相均呈典型的立方状形貌，体积分数均在 75% 左右。γ'相尺寸随合金元素含量不同呈现出不同的变化，增加 Re 含量能够显著降低 γ'相尺寸，Cr 和 Mo 含量增加会小幅度降低 γ'尺寸，而 W 含量对 γ'尺寸没有影响。大部分合金经 1000℃/1000 h 热暴露后，析出了针状，棒状和块状的 TCP 相。Re 或 Mo 含量的增加会显著促进 TCP 相的析出，Cr 也会促进 TCP 相的析出，但 W 含量的增加对 TCP 相的含量无明显影响。合金中的 TCP 相主要为 σ 相、μ 相和 R 相。

关键词：镍基单晶高温合金，难熔元素，γ'相，TCP 相，组织稳定性

激光熔化沉积增材制造镍基高温合金微细柱晶组织研究

汤海波，王华明，田象军，李　佳

（北京航空航天大学大型金属构件增材制造国家工程实验室，北京　100082）

摘　要：镍基高温合金是航空、航天、核电等重大装备高温部件的重要制造材料，具有定向凝固柱晶/单晶组织和复杂内腔结构高效冷却的镍基高温合金涡轮叶片，是高推重比航空发动机、重型燃气轮机等研制生产的关键材料/制造技术和瓶颈之一，其采用最先进的液态金属冷却定向凝固技术制备仍不可避免存在（1）凝固组织粗大和疏松严重导致力学性能低；（2）偏析严重导致初熔温度和使用温度低；（3）叶片生产合格率低三大难题。激光原位冶金/快速凝固"高性能金属材料制备"与"大型、复杂构件成形制造"一体化的金属激光熔化沉积增材制造技术，为涡轮叶片高性能高品质制造提供了新的技术途径。本文通过激光约束熔化沉积定向凝固工艺，建立了微细柱晶凝

固组织控制模型，成功制备出组织细小均匀、无 γ/γ′共晶 、无显微疏松、无明显偏析、低发散度的微细柱晶高温合金试件，发现其一次枝晶间距可达 7.6~9.1μm（二次枝晶不明显）、较现有定向凝固叶片显著细化 1~2 个数量级，实现其晶粒取向的主动调控，获得优异的晶体取向一致性、EBSD 定量测试晶体取向差小于 6°。测试结果表明，激光约束熔化沉积微细柱晶高温合金(LMD-408)初熔温度较传统定向凝固的提高 8℃、室温/高温拉伸、高温持久等关键性能优异。

关键词：激光熔化沉积，增材制造，高温合金，显微组织

先进航空发动机用粉末高温合金涡轮盘技术

王晓峰，邹金文

（先进高温结构材料重点实验室，北京航空材料研究院，北京　100095）

摘　要：涡轮盘是航空发动机中安全系数要求最高的核心部件，工作时承受高温、高应力、高转速及复杂的燃气腐蚀环境，服役条件异常苛刻。涡轮盘的性能决定了发动机的整体性能。采用粉末高温合金材料制造的涡轮盘是推重比 8 以上先进航空发动机涡轮盘的首要选择。粉末盘技术是设计—材料—制造—考核与验证的系统化集成技术，本文重点介绍了国内粉末高温合金材料研究进展及粉末盘制造技术难点及创新，以及未来粉末盘技术发展趋势。

关键词：粉末高温合金，涡轮盘技术，先进航空发动机

The P/M Superalloy Turbine Disc Technology in Advanced Aero-engine Manufacturing

Wang Xiaofeng, Zou Jinwen

(Science and Technology on Advanced High Temperature Structural Materials Laboratory, Beijing Institute of Aeronautical Materials, Beijing 100095, China)

Abstract: As the most safety demanding component in aero-engine, the turbine disc extreme working conditions including high temperature, high stress, high rotation rate and erosion from a complex burning atmosphere. The service conditions for a turbine disc are extremely harsh, and therefore the ultimate performance of it determines the overall performance of the aero-engine. The best way for manufacturing the turbine disc of an aero-engine with a thrust-weight ratio approaching eight is the P/M superalloy turbine disc technology.

The P/M superalloy turbine disc technology is an integrated systematic technology based on designing-materials-manufacturing-assessment and evaluation. The attention of this work is focused on (1) the current research progress (both domestic and abroad) on developing P/M superalloy, (2) existing difficulty, innovation and possible future development in designing and manufacturing the P/M superalloy turbine disc.

Key words: P/M superalloy, turbine disc technology, advanced aero-engine

电极缩孔对IN718合金电渣重熔过程影响的数值模拟研究

王资兴[1,2]，李青[3]，王磊[1]

（1. 东北大学新材料技术研究院，辽宁沈阳 110819；2. 宝钢特钢有限公司，上海 200940
3. 宝钢股份中央研究院，上海 201900）

摘　要：利用自主开发的ESR过程仿真软件，针对直径430 mm的IN718合金铸锭，通过设计不同形状尺寸的电极缩孔，进行重熔过程的数值模拟计算和分析。结果表明，电极中缩孔的存在改变了电极与渣池的接触面积，从而显著影响渣池的焦耳热和电磁力分布，而缩孔沿电极轴向尺寸的变化对二者分布的影响则很少。在恒熔速条件下，当缩孔半径小于0.025 m时，缩孔对熔炼过程的影响可以忽略；当缩孔尺寸继续增大时，渣池流态发生改变，流速相对减弱，渣池温度逐步升高，而熔池形状及两相区尺寸变化不明显；缩孔尺寸对电压、电流和功率的影响呈非线性关系，当缩孔半径大于0.05 m时，缩孔尺寸的变化将显著影响功率、电流和电压等输入参数。从工艺控制稳定性角度而言，电极缩孔半径应尽量控制在0.05 m以下。

关键词：电渣重熔(ESR)，数值模拟，电极缩孔

高温合金真空电弧重熔过程中白斑缺陷的成因研究

于　腾，王志刚，赵长虹，吴贵林

（抚顺特殊钢股份有限公司技术中心，辽宁抚顺 113001）

摘　要：真空电弧重熔（VAR）是在真空下利用直流电弧产生的热量将电极熔化的工艺。金属电极上滴落的金属熔滴滴入到水冷铜结晶器，当熔炼继续进行时，在结晶器里就逐渐地形成了一支新的锭料，合金锭上方覆盖一层金属熔池。采用真空电弧重熔工艺主要是为了获得纯净度和化学成分均匀的钢锭。对于高温合金和钛合金，真空电弧重熔通常是最后一步冶炼工艺。采用真空电弧重熔工艺冶炼的材料通常应用于航空发动机以及机身上的转动和承力部件。

真空电弧重熔合金锭的常见缺陷有超声波探伤单显（本质为夹杂物聚集导致内部开裂）、黑斑、白斑等，其中黑斑和白斑可以通过成品材的低倍检验进行控制。超声波探伤可以发现由于夹杂物聚集导致成品材出现内部裂纹而造成的单显，但超声波探伤并不能使夹杂物聚集缺陷得到完全控制。夹杂物聚集造成的内部裂纹长度只有0.8～2 mm，甚至仅有0.4 mm，如果聚集的夹杂物在热加工变形过程未发生开裂，超声波探伤将无法识别该类缺陷，对于材料的后期加工和与服役存在致命的风险。因此，本文对真空电弧重熔夹杂缺陷的产生原因进行了详细的分析。

本研究对真空电弧重熔过程中的白斑缺陷进行了分类，并对其成因进行了详细的分析。白斑缺陷可分为离散型、枝晶型和凝固型。偏弧和过长的弧长是白斑缺陷形成的主要原因，稳定的漫散电弧可以避免白斑缺陷的形成，对提高真空电弧重熔铸锭的冶金质量起着至关重要的作用。

关键词：高温合金，真空电弧重熔，白斑，冶金缺陷

国内高温合金大型锻件的研制现状和发展

袁士翀

（中国第二重型机械集团德阳万航模锻有限责任公司，四川德阳　618000）

摘　要： 高温合金因具有较高的高温强度，良好的抗氧化和抗腐蚀性能，良好的疲劳性能、断裂韧性等综合性能，被广泛用于航空、航天、石油、化工、舰船等领域。又因其合金化程度高，对变形高温合金锻件来说，锻造窗口窄，变形抗力大，锻造工艺复杂，制造难度大。近年来，随着国内新建一批以 8 万吨模锻压机为代表的大型模锻设备，我国在高温合金大型锻件的研制生产上取得了较大突破，并逐步向国际先进水平迈进。

关键词： 高温合金，大型锻件

SiC 纤维增强 GH4169 复合材料界面分析

张国兴，王玉敏，杨　青，张　旭，杨丽娜，杨　锐

（中国科学院金属研究所，辽宁沈阳　110016）

摘　要： 连续 SiC 纤维增强高温合金复合材料能够显著提高比强度和比刚度，降低构件重量，提高航空发动机效率。SiC 纤维与高温合金基体的界面相容性是发展该类复合材料的主要障碍。本文采用溅射方法在纤维表面沉 Ti 合金涂层作为反应阻挡涂层，再制备 SiC$_f$/GH4169 复合材料，并采用 SEM 及 EDS 等方法对复合材料的界面进行分析。在复合材料制备过程中，Ti 合金与 GH4169 之间发生元素扩散；Ti 涂层厚度达到 5μm 能够有效减缓纤维与基体之间的界面反应，提高界面相容性。

关键词： SiC 纤维，GH4169 基复合材料，反应阻挡涂层，界面反应

Interface Analysis of SiC Fiber Reinforced GH4169 Composite

Zhang Guoxing, Wang Yumin, Yang Qing, Zhang Xu, Yang Lina, Yang Rui

(Institute of Metal Research, Chinese Academy of Sciences, Shenyang 110016, China)

Abstract: Continuous SiC fiber reinforced Ni based alloys composite is a promising candidates for high temperature applications owing to its high specific stiffness and specific strength. The serious interfacial reaction between fiber and matrix decreased the mechanical properties of composite. A Ti alloy coated SiC fiber reinforced GH4169 matrix composite was fabricated by vacuum hot pressing. The interfacial reaction zone was analyzed by scanning electron microscopy and energy dispersive spectrometer. The inter-diffusion between Ti alloy and GH4169 matrix occurred during the fabrication of composite; the Ti coating with a thickness of 5μm can retard the interfacial reaction and increase compatibility of interface.

Key words: SiC fiber, GH4169 matrix composite, reaction barrier coating, interfacial reaction

定向凝固 DZ125 涡轮叶片服役后显微组织损伤研究

陈亚东，郑运荣，冯　强，郑为为

（北京科技大学新金属材料国家重点实验室，北京　100083）

摘　要：航空发动机涡轮叶片长期在高温、高压的环境下服役，易发生各类组织损伤退化，甚至导致叶片失效。了解涡轮叶片的服役损伤行为进而对其进行服役损伤评价对于叶片的应用及维修具有重要指导意义。涡轮叶片在服役后，横截面的解剖分析相对简单易行，能快速反映叶片的服役损伤状态。而纵截面γ'相的演变则与叶片服役的温度和应力密切相关。本文利用扫描电镜对服役不同时间的定向凝固 DZ125 合金高压涡轮叶片从榫头到叶尖不同截面以及同一截面上不同位置，包括进气边、排气边、叶盆、叶背四个部位的显微组织进行了定量表征和分析。研究结果发现，300h 服役后，叶片中部截面进气边枝晶干γ'相即出现了明显的退化现象，γ'相已显著溶解和连接，体积分数显著降低至 57%，γ'相已发生筏排，形成垂直于应力方向的筏形组织，γ通道宽度已明显增加；至 600h 时，γ'相发生十分明显的聚集连接和回溶，已形成近似"孤岛状"γ'相组织，出现较大面积的"无γ'相区"，体积分数下降至 55%以下；服役 900h 后，γ'相已严重聚集连接和溶解，形成类似"孤岛状"不规则γ'相，原始立方状的γ'相已不复存在，体积分数已下降至 40%左右，γ'相的粗化也很明显，形成垂直于应力方向的不规则"短棒状"γ'相，相邻"短棒状"γ'相之间的间距已很大，即γ通道宽度很宽，已大于γ'相厚度。同时叶片中上部的枝晶间的初生 MC(1) 型碳化物发生了一定程度的分解，M6C 碳化物的相对面积分数由原始的 5%~10%增加至 15%~20%，晶界γ'相膜的厚度增加。

　　通过对比服役前后叶片不同部位枝晶干γ'相、碳化物及晶界组织的演变行为发现 3 个服役不同周期叶片的组织损伤规律一致：即损伤退化最先在叶身中部进气边发生，依次是叶身中部叶盆及排气边。叶片其余部位基本没有明显损伤，特别是靠近叶根的截面。

　　依据服役前组织均匀一致、对服役环境敏感、与合金性能相关以及易于量化等原则，确定了 DZ125 合金枝晶干γ'相的体积分数（Vf）和筏形完善程度（Ω）为主要服役表征参量，γ'相厚度（D）、γ通道宽度（W）、M6C 型碳化物的相对面积分数（R）及晶界γ'相膜的厚度为辅助服役表征参量。通过对枝晶干γ'相形貌、体积分数和筏形完善程度的定量表征，提出了叶片服役损伤程度分级的方法。

Al、Ti、Nb 含量对 GH600 合金组织和性能的影响

刘丰军[1]，孙文儒[2]，张　滨[1]

（1. 东北大学，辽宁沈阳　110819；2. 中国科学院金属研究所，辽宁沈阳　110016）

摘　要：本文研究了 Al、Ti、Nb 含量对 GH600 合金组织和性能的影响。结果表明，在试验合金成分范围内，随着 Al、Ti、Nb 含量的提高，铸锭铸态组织的枝晶间析出相增多、枝晶间距减小，但对铸锭低倍组织的影响不明显。提高 Al、Ti、Nb 的含量还可以细化合金棒材的晶粒组织，增大其室温抗拉强度(σ_b)和屈服强度($\sigma_{0.2}$)，但对合金的拉伸塑性影响不大。坯料经均匀化退火后，Al、Ti、Nb 含量增加对合金的晶粒细化效果减弱，棒材室温拉伸强度值及其升高幅度都要略低于未经均匀化处理的棒材。

关键词：GH600 合金，Al、Ti、Nb 含量，组织与性能

Effects of Al, Ti and Nb Content on Microstructure and Properties of GH600 Alloy

Liu Fengjun[1], Sun Wenru[2], Zhang Bin[1]

(1. Northeastern University, Shenyang 110819, China;
2. Institute of Metal Research, Chinese Academy of Sciences, Shenyang 110016, China)

Abstract: The effects of Al, Ti, Nb content on microstructure and properties of GH600 alloy were studied in this paper. The results showed that along with the increase of Al, Ti, Nb content, the interdendritic precipitation phase and the number of eutectic increasing, the dendrite spacing decreases. Howerer, due to the absolute amount of Al, Ti, and Nb content are very low, the changes on macrostructure of the alloy were not obvious. In the meantime, as Al, Ti, Nb content increasing, the grains get finer, tensile strength (σ_b) and yield strength ($\sigma_{0.2}$) increases, but made little contribution on the tensile plasticity of the alloy. After homogenization annealing, the effect of Al, Ti, Nb content on microstructure and properties of GH600 alloy become weaker. Both the room temperature tensile strength and its rising amplitude of the homogenizing treatmented bar are less than that the bar which the billet without treated by homogenization treatment.
Key words: GH600 alloy, Al, Ti, Nb content, microstructure and mechanical properties

GH4169 合金时效过程残余应力演化的原位中子衍射研究

秦海龙[1]，毕中南[1,2]，张瑞尧[3]，董洪标[3]，Lee Tung-lik[4]，

杜金辉[1,2]，张　继[1,2]

（1. 钢铁研究总院高温材料研究所，北京　100081；2. 高温合金新材料北京市重点实验室，北京　100081；3. 莱斯特大学工程学系，英国　LE1 7RH；4. ENGIN-X ISIS，英国　OX11 0DE）

摘　要： 中子衍射技术是一种测量材料或工程部件内部的三维应力状态的方法。本文采用原位中子衍射方法研究时效过程中 GH4169 合金圆盘部件内部残余应力的演化规律。研究结果表明，在时效过程中，强化相的析出行为会受到具体位置应力状态的影响。基于修正后的基准试样晶格参数变化，可以分析得出时效过程中残余应力的演化规律：在时效前的升温过程中，材料逐渐软化，超过 720℃下屈服强度（400 MPa）的残余应力会通过塑性变形进行释放；在时效保温过程中，剩余残余应力会通过蠕变变形进行部分释放。采用高分辨中子衍射谱仪来分解强化相与基体的衍射峰，补充测试分析残余应力，其实验结果与工程谱仪测量结果的量级保持一致。
关键词： 原位中子衍射，残余应力，时效析出行为，高分辨中子衍射

Study on In-situ Neutron Diffraction for Residual Stress Evolution during Aging in GH4169 Alloy

Qin Hailong[1], Bi Zhongnan[1,2], Zhang Ruiyao[3], Dong Hongbiao[3], Lee Tung-lik[4], Du Jinhui[1,2], Zhang Ji[1,2]

(1.Central Iron and Steel Research Institute, Beijing 100081, China;
2. Beijing Key Laboratory of Advanced High Temperature Materials, Beijing 100081, China;
3. University of Leicester, UK LE1 7RH; 4. ENGIN-X ISIS, UK OX11 0DE)

Abstract: The evolution of residual stress during aging was studied by using in-situ neutron diffraction for a GH4169 alloy disk. The results show that precipitation behavior is strongly affected by local stress condition in the workpiece. Based on the modified lattice parameter of stress-free sample, the evolution law of residual stress during aging is analyzed: residual stress relaxation is a result of plasticity due to the reduction of the yield strength in elevated temperature and creep during further isothermal treatment. The separation of γ'' peaks from the matrix is studied by high resolution powder diffraction to analysis residual stress.

Key words: in-situ neutron diffraction, residual stress, precipitation behavior, high resolution powder diffraction

氧化钇增强超细晶高温合金中弥散相析出行为的研究

夏 天[1,2], 曾 巍[1,2,3], 谢跃煌[1,2], 祝国梁[1,2], 张德良[4]

（1. 上海交通大学，材料科学与工程学院，上海 200240；
2. 上海交通大学，上海市先进高温材料及其精密成形重点实验室，上海 200240；
3. Department of Chemical Engineering & Materials Science, University of California, Irvine 92607；
4. 东北大学，材料科学与工程学院，辽宁沈阳 110819）

摘 要： 本研究使用高能球磨和粉末冶金方法制备了含有 5 vol.% Y_2O_3 纳米颗粒的 FGH4096、FGH4097 和 GH4169 超细晶高温合金，使用 XRD、SEM 和 TEM 分析了它们的晶粒组织、相组成、弥散颗粒和 γ'/γ'' 析出相。研究表明，高能球磨和粉末冶金方法是有效地制备超细晶高温合金的工艺方法，晶粒尺寸与粉末固结工艺有关；高能球磨和粉末固结过程中，Al、Ti、Nb 等活性元素容易与 O 和 N 反应，形成 Y-Al-O 氧化物和 $(Ti, Nb)_xN_y$ 氮化物；超细晶组织及基体中 Al、Ti、Nb 含量的变化对 γ' 及 γ'' 相的形成和长大有影响。

关键词： 高温合金，弥散强化，超细晶，Y_2O_3

Dispersoids Precipitation Behavior in Ultrafine Grained Superalloy with Y₂O₃ Nanoparticles Reinforcement

Xia Tian[1,2], Zeng Wei[1,2,3], Xie Yuehuang[1,2], Zhu Guoliang[1,2], Zhang Deliang[4]

(1. School of Materials Science and Engineering, Shanghai Jiao Tong University, Shanghai 200240, China;
2. Shanghai Key Lab of Advanced High-temperature Materials and Precision Forming, Shanghai Jiao Tong University, Shanghai 200240, China;

3. Department of Chemical Engineering & Materials Science, University of California, Irvine 92607, USA;

4. School of Materials Science and Engineering, Northeastern University, Shenyang 110819, China)

Abstract: In this study, ultrafine grained FGH4096-5 vol.% Y_2O_3, FGH4097-5 vol.% Y_2O_3 and GH4169-5 vol.% Y_2O_3 alloys were fabricated by high energy mechanical milling and powder metallurgy consolidation process, and XRD, SEM and TEM were used to analyze the microstructures, phases, dispersoids and γ'/γ" precipitates. The results shows that it is effective to fabricate ultrafine grained superalloys with Y_2O_3 nanoparticles addition by using high energy mechanical milling and powder metallurgy consolidation process. During the process, some active elements as Al, Ti and Nb would react with O and N, forming Y-Al-O oxides and $(Ti, Nb)_xN_y$ nitrides. The ultrafine grain microstructures and the contents variation of Al, Ti and Nb would effect on the γ' and γ" precipitates formation.

Key words: superalloy, dispersion strengthening, ultrafine grained microstructure, Y_2O_3

Al 对 HR3C 奥氏体耐热钢高温抗氧化性能的影响

王　剑[1]，权　鑫[1]，董　楠[1]，方旭东[2]，韩培德[1]

（1. 太原理工大学材料科学与工程学院，山西太原　030024；

2. 山西太钢不锈钢股份有限公司技术中心，山西太原　030003）

摘　要： 含铝奥氏体耐热钢由于具有良好的抗高温氧化性能正受到人们的广泛关注。本文通过在 HR3C 合金成分的基础上添加 1.5、2.5、3.5 wt%的 Al 元素，研究了 HR3C 钢和含铝奥氏体耐热钢的抗高温氧化性能。结果表明：含铝奥氏体耐热钢高温氧化后的增重明显小于 HR3C 钢，在 800℃下具有优异的抗高温氧化性能。进一步采用 SEM、EDS、XRD、GDS 等分析手段对含铝 HR3C 钢的抗高温氧化机理进行了研究，发现 800℃氧化过程中，含铝 HR3C 钢基体中的 Al 通过向表面扩散，并在高温下同 O_2 结合形成了一层连续致密的 Al_2O_3 保护膜，从而阻碍了金属基体的进一步氧化，提高了其高温抗氧化性能。

关键词： HR3C，高温氧化，Al_2O_3 膜，新型含铝奥氏体耐热钢

Effect of Al on High Temperature Oxidation Behavior of HR3C Austenitic Stainless Steel

Wang Jian[1], Quan Xin[1], Dong Nan[1], Fang Xudong[2], Han Peide[1]

(1. School of Materials Science and Engineering, Taiyuan University of Technology, Taiyuan 030024, China;

2. Technology Center, Shanxi Taiyuan Stainless Steel Co., Ltd., Taiyuan 030003, China)

Abstract: A new alumina-forming austenitic stainless steel with excellent high-temperature oxidation resistance is widely concerned. This paper investigates oxidation behavior of HR3C before and after adding 1.5, 2.5, 3.5 %Al. Compared with HR3C steel, the oxidation resistance property of the steels containing aluminum is markedly better at 800℃ due to lesser weight gain. The high temperature oxidation mechanism of the steels containing aluminum was analyzed by using scanning electron microscopy (SEM) with energy-dispersive spectrum (EDS) system, X-ray Diffraction (XRD) and glow discharge optical emission spectroscopy (GDS) techniques. Experimental results show that Al element in HR3C steel substrate diffuses to the surface combines with O_2 at high temperature and forms a continuous and compact Al_2O_3 protection film, which hinders the further oxidation of the base metal, improves the high temperature oxidation resistance.

Key words: HR3C, high temperature oxidation, Al_2O_3 film, austenitic heat-resistant steel

GH4738 合金持久性能与取向演变的关联性分析

韦　康，艾卓群，陈少华，朱琳烨，张　琪，张麦仓

（北京科技大学材料科学与工程学院，北京　100083）

摘　要： 利用背散射电子衍射（EBSD，Electron Back Scattered Diffraction）对 GH4738 合金经不同持久试验后取向的变化进行了系统分析，并从晶体学角度分析了晶粒取向变化与持久性能的关联性。结果表明：持久变形过程中，不同试验条件的样品呈现不同的取向演化规律，且 GH4738 合金的持久性能与取向演化密切相关。持久试验温度从 700 ℃升高到 725 ℃，试样的延伸率提高；相同的 725 ℃温度下，较低的应力对应更高的延伸率和持久寿命。此外，700 ℃/450 MPa 和 725 ℃/380 MPa 试样断口附近晶粒取向相对于试验前的无明显变化，仍均匀分布，但延伸率高达 19.36% 的 725 ℃/350 MPa 试样中晶粒发生明显转动，大部分晶粒处在 [001]、[111] 和 [001]-[111] 取向，持久性能优异。

关键词： GH4738，EBSD，持久性能，晶粒取向

Investigations on the Correlation between Stress Rupture Properties and Orientation Evolution of GH4738 Alloy

Wei Kang, Ai Zhuoqun, Chen Shaohua, Zhu Linye, Zhang Qi, Zhang Maicang

(School of Materials Science and Engineering, University of Science and Technology Beijing, Beijing 100083, China)

Abstract: The orientation evolution of GH4738 alloy under different stress rupture tests were systematically investigated by electron back scattered diffraction (EBSD) method. Correlation between the orientation evolution and the changes of stress rupture properties were also studied based on the theory of crystallography. The results showed that the samples presented different orientation evolution under various stress loading processes, and the stress rupture properties of GH4738 alloy were closely related to the orientation changes. Elongation was improved as the test temperature increased from 700℃ to 725℃. At the same temperature of 725℃, the lower stress corresponded to higher elongation and longer life. Compared to the pre-experiment sample, there was no significant change in grain orientation near the fracture of 700℃/450 MPa and 725℃/380 MPa sample. However, the crystal of 725℃/350 MPa sample, with the elongation of up to 19.36%, occurred significantly rotation. In addition, most of the grains in 725℃/350 MPa sample were in [001], [111] and [001]-[111] orientation, and presented excellent lasting performance.

Key words: GH4738, EBSD, stress rupture properties, grains orientation

热力学平衡计算和相鉴别在 FGH4097
原始颗粒边界研究中的应用

朱琳烨，陈少华，姬忠硕，薛　渊，黄一君，张麦仓

（北京科技大学材料科学与工程学院，北京　100083）

摘　要： 原始颗粒边界（PPB）是粉末高温合金中三大缺陷之一。本文通过热力学平衡计算和相鉴别的方法研究了 FGH4097 中的原始颗粒边界微观形成机理。热力学计算结果表明，随着温度的降低，MC 有向 M_6C 过渡并且转化为 $M_{23}C_6$ 的趋势，MB_2 有向 M_3B_2 转化的趋势；并且析出了 μ 相 σ 相，析出量随着温度的降低而不断增加。相鉴别结果表明，PPB 组成为包含或者不包含氧化物核心的（Nb, Ti, Hf）C 型碳化物，分别对应两种形成方式，一种是直接在颗粒边界形成，另一种是在氧化物质点周围形核长大，并给出了后一种的示意图。

关键词： FGH4097，原始颗粒边界，热力学计算，相鉴别

Application of Thermodynamic Equilibrium Calculation and Phase Identification in FGH4097 Prior Particle Boundary Study

Zhu Linye, Chen Shaohua, Ji Zhongshuo, Xue Yuan, Huang Yijun, Zhang Maicang

(School of Materials Science and Engineering, University of Science and Technology Beijing, Beijing 100083, China)

Abstract: The prior particle boundary (PPB) is one of the three major flaws in the powder metallurgy superalloys. Thermodynamic equilibrium calculation and phase identification are carried out to investigate the microscopic formation mechanism of the prior particle boundary in FGH4097. The thermodynamic equilibrium calculation indicates that with the reduction of temperature, MC has a tendency to transition to M_6C and eventually convert to $M_{23}C_6$ and MB_2 has a tendency to convert to M_3B_2. Meanwhile, μ and σ start to precipitate and the amount of both is getting much as temperature decreases. Phase identification results show that PPB is composed of (Nb, Ti, Hf) C-type carbides with or without oxide particles, corresponding to two formation modes, respectively; one is to directly form in the powder grain boundary, the other is to gather and grow around the nucleation of oxides, and a diagram is provided for the latter formation form.

Key words: FGH4097, prior particle boundary, thermodynamic equilibrium calculation, phase identification

激光焊接 DZ125L 和 IN718 异质合金接头组织及拉伸力学性能的研究

梁涛沙，王　磊，刘　杨，宋　秀

（东北大学材料科学与工程学院，辽宁沈阳　110819）

摘　要： 研究了激光焊接 DZ125L 和 IN718 异质合金接头组织及拉伸力学行为。结果表明，接头熔合区主要由柱状

树枝晶组成，枝晶间分布着大量尺寸细小的 Laves 相和汉字状 MC 型碳化物。较高的焊接温度使 DZ125L 一侧热影响区 γ'相部分溶解，并使两侧热影响区晶界液化。接头经过双时效热处理后，在各区域分别析出细小的 γ''和 γ'相，接头力学性能恢复。拉伸试验结果表明，热处理后接头的强度高于 DZ125L 母材合金强度但是低于 IN718 母材合金强度，断裂发生在强度较低的 DZ125L 母材合金。熔合区可以观察到明显滑移带痕迹，但是并未观察到明显裂纹。DZ125L 合金的断裂主要由于尺寸较大的碳化物在塑性变形初期发生碎裂，在随后变形过程中成为裂纹源，微裂纹的长大和连接以及主裂纹的扩展最终导致基体断裂。

关键词：激光焊接，镍基高温合金，焊后热处理，拉伸性能

交变磁场对单晶高温合金凝固组织的影响

刘承林，苏海军，张　军，刘　林，傅恒志

（西北工业大学凝固技术国家重点实验室，陕西西安　710072）

摘　要：本文利用定向凝固技术，通过改变石墨套厚度研究了感应线圈交变磁场对 DD90 单晶高温合金凝固组织的影响规律，同时结合 Ansys 有限元分析软件对合金熔体内磁场、温度场以及流场分布进行了模拟，探讨了磁场对组织影响的机制。结果表明：当石墨套厚度为 10mm、15mm 时 DD90 单晶生长性受到破坏而形成杂晶，一次枝晶间距减小，合金元素的偏析降低；进一步增大石墨套厚度为 20mm、25mm、30mm 时，单晶性保持完好，一次枝晶间距随厚度增加而变大。Ansys 有限元模拟表明，合金熔体内磁场的分布，磁场大小和实测的磁场大小基本吻合，由于熔体内出现强的磁拉力和磁压力，使得枝晶被打断，破坏枝晶的生长。交变磁场产生的感应电流使枝晶尖端产生较大的焦耳热，促进枝晶尖端重熔，从而抑制枝晶生长。熔体内产生的电磁力会促进液体流动，使得枝晶附近的元素偏析减轻。

关键词：单晶高温合金，有限元，温度场，磁场

不同锻造工艺下 Ti-6Al-4V 合金热处理后组织和取向演化

姬忠硕，陈少华，薛　渊，黄一君，张　琪，张麦仓

（北京科技大学材料科学与工程学院，北京　100083）

摘　要：在快锻液压机上对 Ti-6Al-4V 合金进行了锻造变形，在锻坯的不同位置取样进行热处理，采用光学显微镜和背散射电子衍射技术研究不同锻造工艺下 Ti-6Al-4V 合金热处理后组织和取向的演化规律。在单向镦拔、换向镦拔两种不同锻造方式下，大变形区、小变形区及难变形的组织均匀性基本一致，两种锻造方式下锻坯不同区域的应变积累稍有差别，进而对 α 相织构的组分和分布产生显著影响。进一步研究 α 相的织构组分及分布可知：单向镦拔的小变形区与难变形区应变积累少，主要以 {0001} 基面滑移，形成基面织构，大变形区应变积累多，主要为 {11$\bar{2}$0} 和 {10$\bar{1}$0} 等柱面织构；而换向镦拔的不同区域织构组分较多且分布随机性较大。此外，两种锻造方式均能改善 Ti-6Al-4V 合金中 α 相的取向均匀性和织构集中，但换向镦拔优于单向镦拔。

关键词：钛合金，锻造，微观组织，取向均匀性

Microstructure and Orientation Evolution of Ti-6Al-4V Alloy after Heat Treatment under Different Forging Processes

Ji Zhongshuo, Chen Shaohua, Xue Yuan, Huang Yijun, Zhang Qi, Zhang Maicang

(School of Materials Science and Engineering, University of Science and Technology Beijing, Beijing 100083, China)

Abstract: The Ti-6Al-4V alloy was subjected to forging process on a fast forging hydraulic press, samples taken from different locations of forging billet were subsequently carried out heat treatment. The microstructure and texture evolution of Ti-6Al-4V alloy after heat treatment under different forging processes were investigated by Optical Microscopy and Electron Backscattered Diffraction. It turns out that the microstructures of Ti-6Al-4V alloy under different forging processes, unidirectional forging and cross upsetting and stretching, are nearly the same in the large deformation zone, small deformation zone and the difficult deformation zone. The difference of strain in different forging regions has significant effect on the composition and distribution of the α texture. Further observations indicate that, strain is less in the small deformation zone and the difficult deformation zone of the unidirectional forging, the slip system is mainly the {0001} basal plane and forms the basal texture, the strain is high in the large deformation zone and results in the {11$\bar{2}$0} and {10$\bar{1}$0} prismatic textures; but the texture components are more and distribute randomly in different regions of the cross upsetting and stretching forging. In addition, the two forging processes can both improve the orientation uniformity and texture concentration of α phase, while cross upsetting and stretching is better than unidirectional forging.

Key words: titanium alloys, forging, microstructure, uniformity

镍在新型钴基高温合金高温氧化行为中的作用

高　博，王　磊，刘　杨，宋　秀

（东北大学 材料科学与工程学院，辽宁沈阳　110819）

摘　要： 研究了新型钴基高温合金 800℃ 和 900℃ 下的高温氧化行为。研究表明，高温下新型钴基高温合金具有优良的高温抗氧化行为。900℃ 新型钴基高温合金的抗氧化性能优于传统镍基高温合金 K417G 和钴基高温合金 DZ40M。增加合金 Ni 含量有利于形成富 Cr 氧化层。利用第一原理计算了 Co-Cr-Ni (111)表面与氧的吸附行为。计算表明，低 Ni 含量 Co-Cr-Ni 模型对氧的吸附作用强于高 Ni 模型，即氧易进入基体与 Cr 相结合，导致 Cr 原子扩散受阻，阻碍了富 Cr 氧化层的形成，进而影响新型钴基高温合金的高温氧化行为。

深海油气开采用高等级 In718 合金棒管材开发

代朋超，王资兴，孙　敏

（宝山钢铁股份有限公司研究院，上海　201900）

摘　要：随着油气开采不断向深海发展，镍基合金，特别是 In718 合金，以其高强度、高韧性以及良好的抗腐蚀性能在井下工具中获得了广泛应用。宝钢特钢公司凭借近 30 年航空 In718 合金开发和生产技术积累，成功开发出 120KSI 等级的 In718 棒、管材。材料组织均匀、性能优异，完全满足 API 6A718 规范要求。特别是 In718 管材的成功开发，可以替代井下工具制造时从棒材掏孔的传统制造方式，大大提高材料利用率，降低工具制造成本。

关键词：深海油气开采，高等级，In718，棒管材

The Development of High Strength Grade In718 Bar and Tube for Deep-sea Oil and Gas Exploitation

Dai Pengchao, Wang Zixing, Sun Min

(Research Institute, Baoshan Iron & Steel Co., Ltd., Shanghai 201900, China)

Abstract: With the development of oil and gas exploitation to the deep sea, nickel based alloys, especially In718 alloy, is widely used in the downhole tools area, because of its high strength, high toughness and good corrosion resistance performance. On the basis of more than 30 years' experience in In718 development and technology accumulation, Baosteel Special steel company has successfully developed 120KSI grade In718 bar and tube for oil and gas. With the uniform micro-structures and good properties, the material can well meet the API 6A718 standard requirements. Deserve to be mentioned, this extruded In718 tube can be a good way to reduce the material cost, especially for the downhole tools manufacture. As we know, the traditional way, digging a hole in the bar, is quite a waste of material.

Key words: oil and gas exploitation, high grade, In718, bar and tube

14　耐火材料

首钢耐火材料应用现状与发展趋势

祝少军[1]，张卫东[1]，邵俊宁[2]，沙远洋[3]，张启东[3]

（1. 首钢集团有限公司技术研究院，北京　100041；2. 首钢股份公司迁安钢铁公司，河北迁安　064404；3.首钢京唐钢铁联合有限责任公司，河北唐山　063200）

摘　要：本文简述了首钢迁钢公司和京唐公司钢铁生产板块中现有主要冶炼设备，对采取的耐火材料区域承包模式进行了简要说明，比较详细地介绍了出铁口、鱼雷罐、铁水包、转炉、钢包、中间包和连铸用耐火材料的品种、性能和应用效果，从降耗、节能、环保、安全和应用技术等方面，展望了"十三五"期间首钢耐火材料的发展趋势。

关键词：耐火材料应用，耐火材料品种，区域承包，发展趋势

Application Status and Development Trend of Refractories in Shougang

Zhu Shaojun[1], Zhang Weidong[1], Shao Junning[2], Sha Yuanyang[3], Zhang Qidong[3]

(1. Shougang Research Institute of Technology, Beijing 100041, China; 2. Shougang Qian'an Iron & Steel Co., Ltd., Qian'an 064404, China; 3. Shougang Jingtang United Iron & Steel Co., Ltd., Tangshan 063200, China)

Abstract: This paper describes the main equipments of Shougang iron and steel-making department in Jingtang and Qian'an. The refractory regional contracts of Shougang are introduced briefly. The classifications, selected properties and application effects of refractories for taphole, torpedo tanks, molten iron ladle, converter, ladle, tundish and continuous casting are introduced in detail. From the aspects of the energy saving, environmental protection, safety and so on, and application of technology, the development trend of Shougang refractories are expected during "13th Five-Year".

Key words: refractory application, refractory variety, regional contract, development trend

湿法喷注技术在炼铁系统方面的应用

章荣会，徐吉龙，刘贯重，邓乐锐，孙赛阳

（北京联合荣大工程材料股份有限公司，北京　101400）

摘　要：本文介绍了湿法喷注工艺的技术优势。联合荣大湿法喷注工艺技术在炼铁系统方面的一些技术成果，其中包括：高炉内衬湿法喷注技术，湿法喷注技术在高炉煤气封罩、上升管、下降管及重力除尘器的应用，湿法喷注技术在铁沟上利用出铁间隔对铁沟进行喷注修补的开发及实践应用。

关键词：湿法喷注技术，高炉内衬，C沉积，可伸缩喷枪，溶胶结合，铁沟喷补

VOD 炉冶炼超纯铁素体不锈钢关键部位耐火材料的开发

柳　军，李红霞，冯海霞

（中钢集团洛阳耐火材料研究院有限公司先进耐火材料国家重点实验室，河南洛阳　471039）

摘　要：分析超纯铁素体不锈钢的冶炼钢种特点及所用耐火材料的损毁机理。对 VOD 用后镁钙砖进行电镜分析发现：C-A-S 系熔渣对镁钙砖渗透侵蚀严重，渗透区域可达 90% 以上，残砖表面呈现明显梯度变化依次为重度变质层、轻度变质层、渗透层和原砖层。熔渣中 CaO、Al_2O_3、SiO_2、Fe_2O_3 等成份与原砖反应后形成的 C_3S 和 C_4AF 是主要硅（铝）酸盐矿物相。C_3S 分解产生的 C_2S 和 CaO 中低温产生体积膨胀效应，造成耐火砖表面出现粉化和大面积剥落。富 Al_2O_3 的熔渣增加了镁钙砖中的 CaO 向熔渣中熔解的趋势，CaF_2 能够明显降低熔渣的粘度，加剧对镁钙砖的渗透侵蚀。分析 VOD 采用低碳镁碳砖作为炉衬的理论基础，并在 VOD 渣线进行现场使用，比较现用 20 烧成镁钙砖，低碳镁碳砖的抗渣性及抗热震效果取得明显改善。

关键词：超纯铁素体不锈钢，低碳镁碳砖，熔渣，真空精炼 VOD 炉

钢冶炼过程炉衬耐火材料对钢水洁净度的影响

王雅杰，黄　奥，顾华志，邹永顺，付绿平，连朋飞

（武汉科技大学省部共建耐火材料与冶金国家重点实验室，湖北武汉　430081）

摘　要：为满足越来越严格的钢材市场需求，高品质钢越来越受到重视，提高钢洁净度具有重要意义。钢冶炼过程炉衬耐火材料对钢中氧、氮、碳、硫、磷等以及合金元素有重要影响，既可以吸附去除夹杂，也会产生夹杂，是钢中非金属夹杂物的主要来源之一。耐火材料的材质对钢水中元素含量有较大的影响，其机制各异；除耐火材料组分与钢液之间发生反应形成夹杂外，在钢水冲蚀作用下，耐火材料剥落到钢水中形成夹杂物颗粒也是主要因素；同时，钢包釉也是钢中夹杂物的主要来源，受到广泛关注，研究表明铝酸钙质隔离层不仅能延长耐火材料寿命，还可净化钢液，为高品质洁净钢生产提供了更为完善的指导作用。

关键词：耐火材料，精炼，钢水，杂质元素，非金属夹杂物

Effect of Lining of Steel Smelting Process Refractories on Cleanliness of Molten Steel

Wang Yajie, Huang Ao, Gu Huazhi, Zou Yongshun, Fu Lvping, Lian Pengfei

(The State Key Laboratory of Refractories and Metallurgy, Wuhan University of Science and Technology, Wuhan 430081, China)

Abstract: In order to meet the increasing demand of steel market, high quality steel is paid more and more attention, it is necessary to raise the cleanliness of steel. The lining of steel smelting process refractories have great effects on the elements in steel, such as, oxygen, nitrogen, carbon, sulfur, phosphorous and other alloying elements. It can not only remove inclusions but also produce inclusions, which is one of the main sources of non-metallic inclusions in steel. Refractory materials have great influence on the content in metal, and the mechanism is different. Expect for the reaction between the components of refractory material and steel liquid can form inclusions, one more major factor is that the grain from refractory was entered into liquid steel under the action of molten steel erosion. Simultaneously, ladle glaze is also the main source of inclusions in steel, which has been received extensive attention, experiment results show that the calcium aluminate insulating layer can not only prolong the life of refractory materials, but also purify molten steel and provided more perfect guidance for the production of high quality steel.

Key words: refractory, refining, molten steel, impurity element, non-metallic inclusion

新型复合结构钢包透气元件的设计与应用

陈 卢，张 晖，禄向阳，尹洪丽，郭 鹏

（中钢集团洛阳耐火材料研究院有限公司先进耐火材料国家重点实验室，河南洛阳 471039）

摘 要： 通过再现钢包狭缝型透气元件顶端工作层横向断裂的发展过程，深入剖析了狭缝型透气元件的失效机理。揭示出出钢和吹氩精炼时，传统狭缝型透气元件工作面横向断裂和热震稳定性差是导致吹通失效的主要原因。详细阐述了芯板型复合结构透气元件的设计理念，分析了该元件的结构特点和功能特性。列举了在典型冶炼工况下的应用实例，展示了该元件长寿命、高吹通率、易维护等优异的使用效果。

关键词： 透气元件，复合结构，芯板型，损毁机理，工业应用

Design and Application of Novel Composite Structure Purging Plug for Bottom Argon Blowing Ladle

Chen Lu, Zhang Hui, Lu Xiangyang, Yin Hongli, Guo Peng

(Sinosteel Luoyang Institute of Refractories Research Co., Ltd., State Key Laboratory of Advanced Refractories, Luoyang 471039, China)

Abstract: The failure mechanism of purging plug for bottom argon blowing ladle is analyzed by means of reconstructing development process of the transverse fracture of the top working layer. During tapping or refining process, argon blowing refining cannot performed normally due to poor thermal shock stability of slit type plug, which is result of transverse fracture of the top working layer. The design idea of segment purging plug is elaborated in detail, the structural features and functional characteristics of the plug are analyzed. The application examples under the typical smelting conditions are given. Excellent using effects of this plug，such as long service life , high air permeability stability and maintenance friendly, are exhibited.

Key words: purging plug, composite structure, segment type, damage mechanism, industrial application

转炉挡渣氧化锆滑板热震性能的改善与应用

余同署[1]，齐庆俊[1]，闫磊鑫[1]，余西平[2]，梁保青[3]，张　晖[1]，禄向阳[1]

（1. 中钢集团洛阳耐火材料研究院有限公司先进耐火材料国家重点实验室，河南洛阳　471039；

2. 马鞍山利尔开元新材料有限公司，安徽马鞍山　243041；

3. 河南熔金高温材料股份有限公司，河南卫辉　453100）

摘　要：本工作针对目前氧化锆滑板使用过程中较易发生开裂、剥落等热震性能差的问题，进行了工艺方案的优化、物相组成的调控和显微结构的优化等改进措施，改进后氧化锆滑板在水急冷—裂纹判定法试验中由先前的 3 次提升到 6 次，热震性能显著提高。改进后氧化锆滑板在某 120t 和 150t 转炉出钢口批量使用，使用寿命稳定在 20 炉附近，下线板面平整、光滑，没有出现先前常见的开裂、剥落等现象，应用结果证实了改进措施的有效性。

关键词：转炉挡渣，氧化锆，滑板，热震性能

RH 炉用无铬耐火材料性能优化研究

洪建国，管　灿，张　炳

（上海梅山钢铁股份有限公司技术中心，江苏南京　210039）

摘　要：RH 插入管内衬一直用镁铬砖作为工作衬，因存在 Cr^{6+} 的公害问题，近年来尝试用 Al_2O_3-尖晶石浇注料取而代之，为此要求 Al_2O_3-尖晶石浇注料的性能须满足苛刻的使用要求，对其组成、显微结构和性能进行优化十分必要。研究了结合体系、尖晶石的粒度和 CaO 的引入方式对 Al_2O_3-尖晶石浇注料性能的影响，进行了浇注料的性能优化和应用试验。

选取烧结氧化铝、富铝尖晶石、活性氧化铝超细粉等为主要原料，制备了 Al_2O_3-尖晶石浇注料。研究了水硬性氧化铝加入量（分别为 0%，1%，2% 和 3%），尖晶石微粉加入量（分别为 0%，2%，4% 和 6%）和纳米级 $CaCO_3$ 加入量（分别为 0%，0.3%，0.6% 和 0.9%）对 Al_2O_3-尖晶石浇注料 110℃、1000℃、1400℃ 和 1550℃ 烧后试样的永久线变化率、常温抗折强度、常温耐压强度，1400℃ 热态抗折强度、荷重软化温度等性能的影响。采用静态坩埚法研究了上述三因素对 Al_2O_3-尖晶石浇注料抗渣性的影响，并借助 SEM 和 EDAX 对部分抗渣后的试样做了显微结构分析。

对 Al_2O_3-尖晶石浇注料的性能进行了优化。性能优化后的浇注料在宝钢集团梅山钢铁公司炼钢厂的 RH 现场实炉进行了替代镁铬砖的使用试验，取得了预期的效果。

关键词：Al_2O_3-尖晶石，浇注料，性能，结合方式

Performance Optimization of Al₂O₃-Spinel Castables for RH Liner Lining

Hong Jianguo, Guan Can, Zhang Bing

(Shanghai Meishan Iron and Steel Co., Ltd., Nanjing 210039, China)

Abstract: Considering the Cr^{6+} pollution of magnesia-chrome bricks for seconclary refining and ladle lining[1], Al₂O₃-spinel castable is a potential choice to replace magnesia-chrome bricks, but the thermal shock stability needs to be optimized.

Adopting tabular alumina[2], fused spinel and high purity spinel and fused zirconium corundum (ZrO_2-Al_2O_3, hereinafter referred as ZA) as the main raw materials, Al₂O₃-spinel castables have been prepared with the introduction of ZA (respectively adding 0, 2%, 4%, 6%, 8%, 10%). Thermal shock stability has been investigated by residual strength ratio. Slag resistance has been investigated by static crucible method and microstructure of specimens after slag test has been analyzed by SEM and EDAX. Permanent linear change (PLC), cold modulus of rupture (CMOR) and cold compression strength (CCS) of the specimens pre-heated at 110℃, 1000℃, 1400℃ and 1550℃ respectively , hot modulus of rupture (HMOR) at 1400℃ were also tested.

Results indicate that the castables with fused zirconium corundum cements (ZA) exhibit an excellent thermal shock stability and slag resistance. In this condition, the HMOR gets worse with the replacement of tabular alumina by ZA, all specimens tested are above 20 MPa. And PLC remains between 0 to +0.5%. Based on the achievements by this work, properties of the Al₂O₃-spinel castable was optimized and an industrial trail using the improved castable as RH snorkel working lining at Baosteel's Meishan Steel Plant was carried out with positive result as desired.

Key words: Al₂O₃-spinel, castables, fused zirconium corundum, slag resistance

钢液对刚玉质耐火材料冲蚀实验研究

杨　梦，张美杰，顾华志，黄　奥

（武汉科技大学耐火材料与冶金省部共建国家重点实验室，湖北武汉　430081）

摘　要： 在钢铁冶炼过程中，耐火材料会受到钢液的冲蚀。本文采用静止试样浸液通气法研究了钢液对刚玉质耐火材料冲蚀。结果表明，冲蚀首先从耐火材料的基质部分开始，基质部分被冲蚀后留下孔洞与缝隙，钢液沿着孔洞与缝隙向基质内部渗透，且渗透深度随着时间的延长而逐渐增加；当骨料周围的基质被钢液冲蚀掉后，骨料与基质之间的结合强度降低，骨料剥落，耐火材料损毁。

关键词： 钢液，刚玉质耐火材料，气相搅动，冲蚀

Experiment of Erosion of Corundum Refractory by Molten Steel

Yang Meng, Zhang Meijie, Gu Huazhi, Huang Ao

(The State Key Laboratory of Refractories and Metallurgy,
Wuhan University of Science and Technology, Wuhan 430081, China)

Abstract: In the steelmaking process, refractory are subject to erosion by molten steel. In this paper, the erosion of corundum refractory by molten steel was studied with using immersion sample immersion method. The results showed that the erosion started from the matrix part of the refractory, the matrix part was eroded and left after the pores and the gap, molten steel penetrated through the pores and the gap to the inside of the matrix, and the depth of penetration increased with the extension of time. When the matrix around the aggregate was eroded by molten steel, the bonding strength between the aggregate and the matrix was reduced, the aggregate was peeled off and the refractory was damaged.

Key words: molten steel, corundum refractory, gas phase agitation, erosion

用后 Al$_2$O$_3$-SiC-C 砖在鱼雷罐衬砖中的应用

蔡长秀，刘　毅，董童霖

（武钢耐火材料有限责任公司，湖北武汉　430080）

摘　要：用后 Al$_2$O$_3$-SiC-C 砖经过整形、均化处理后获得 Al$_2$O$_3$-SiC-C 再生料，并作为原料引入鱼雷罐衬砖中，研究了 Al$_2$O$_3$-SiC-C 再生料加入量和碳化硅微粉加入量对试样性能的影响。结果表明：（1）通过整形、均化处理，可减少再生料中的假颗粒，稳定再生料的质量；（2）添加 Al$_2$O$_3$-SiC-C 再生料可研制出性能优良的鱼雷罐衬砖，但当再生料加入量大于 45% 后，试样的各项性能较差；（3）添加碳化硅微粉可以降低试样的气孔率，改善高温性能，且碳化硅微粉最佳加入量为 6%；（4）研制的鱼雷罐衬砖在某钢厂 320t 鱼雷罐铁水区试用，使用寿命超过 1200 炉。

关键词：Al$_2$O$_3$-SiC-C 再生料，碳化硅微粉，高温抗折强度，抗侵蚀性

Application of After-used Al$_2$O$_3$-SiC-C Brick in Torpedo Tank Lining Brick

Cai Changxiu, Liu Yi, Dong Tonglin

Abstract: After-used Al$_2$O$_3$-SiC-C brick were plasticized and homogenized to obtain Al$_2$O$_3$-SiC-C recycled material, introducing into the torpedo tank lining brick as a raw material. The effects of the Al$_2$O$_3$-SiC-C recycled material and silicon carbide power additions were studied. The results reveal that: (1) By plasticizing and homogenizing treatment to reduce the fake particle in the recycled material and stable quality of recycled material. (2) The excellent performance torpedo tank lining bricks are prepared by adding Al$_2$O$_3$-SiC-C recycled material,when the amount of Al$_2$O$_3$-SiC-C recycled material more than 45%, the performance of the sample is poor. (3) Adding silicon carbide powder can reduce the porosity of the sample and inprove the high temperature performance, and the optimum addition of silicon carbide powder is 6%. (4) The prepared products used in a 320 tons torpedo cans,the service life is more than 1200 times.

Key words: Al$_2$O$_3$-SiC-C recycled material, silicon carbide power, flexural strength, corrosion resistance

电磁场下熔渣对镁钙耐火材料的侵蚀和渗透行为

李世明 [1,2]，马北越 [1,2]，李红霞 [2]，刘国齐 [2]，赵世贤 [2]，于景坤 [1]

（1. 东北大学冶金学院，辽宁沈阳　110819；2. 中钢集团洛阳耐火材料研究院有限公司
先进耐火材料国家重点实验室，河南洛阳　471039）

摘　要：为了研究电磁场作用下熔渣对镁钙耐火材料的侵蚀和渗透行为，配置 R=4 的熔渣，采用静态坩埚法，在电阻炉和多因素抗渣炉（电磁场、真空、惰性气氛 N₂）中于 1600 ℃下保温 2 h 和 1 h 对镁钙耐火材料进行侵蚀试验。测量试验后试样的侵蚀和渗透深度，并利用 SEM、EDS、XRD 对试样进行表征。结果表明：（1）在电磁场环境下镁钙耐火材料经碱性渣侵蚀的渣线部位存在方镁石、硅酸三钙、CaO 和硅石相；在无电磁场环境下渣线部存在方镁石、硅酸三钙和方石英相。（2）在电磁场环境下碱性渣对镁钙耐火材料侵蚀和渗透相对于无电磁场环境下碱性渣对镁钙耐火材料侵蚀和渗透较为严重。（3）在电磁场环境下镁钙耐火材料中的 Fe₂O₃ 中的铁离子会发生扩散。

关键词：电磁场，镁钙耐火材料，熔渣，侵蚀，渗透

Corrosion and Penetration Behaviors of Molten Slag on the Magnesia Calcia Refractories under Electromagnetic Field

Li Shiming[1,2], Ma Beiyue[1,2], Li Hongxia[2], Liu Guoqi[2], Zhao Shixian[2], Yu Jingkun[1]

(1. Northeastern University, School of Metallurgy, Shenyang 110819, China; 2. State Key Laboratory of Advanced Refractories, Sinosteel Luoyang Institute of Refractories Research Co.,Ltd, Luoyang 471039, China)

Abstract: In order to study the corrosion and penetration behaviors of molten slag on the magnesia calcia refractories under the action of electromagnetic field, the slag with R = 4 was prepared, and the corrosion test was conducted using the static crucible method, in a resistance furnace and a multi-factor anti-slag furnace (electromagnetic field, vacuum, inert atmosphere N₂), and at 1600 ℃ for 2 h and 1 h, respectively. The corrosion and penetration depths of the samples after test were measured, and the samples were characterized by SEM, EDS and XRD. The results show: (1) There exist magnesite, tricalcium silicate, CaO and silica phases in the slag line part of magnesia calcia refractories corroded by alkaline slag under the electromagnetic field. However, magnesite, tricalcium silicate, and cristobalite phase can be detected in the slag line parts without the electromagnetic field. (2) In the electromagnetic field environment, the corrosion and penetration degrees of alkaline slag on the magnesia calcia refractories are relatively serious compared to none of electromagnetic. (3) In the electromagnetic field environment, the iron ions in Fe₂O₃ from magnesia calcia refractories will spread.

Key words: Electromagnetic field, Magnesia calcia refractories, molten slag, corrosion, penetration

环保型"水基"转炉大面自流料应用技术

董战春，邓乐锐，章荣会

（北京联合荣大工程材料股份有限公司，北京　101400）

摘　要：本文对比了环保型"水基"大面自流料与传统型大面补炉料使用特性及应用效果。实践结果表明：环保型"水基"大面料烧结时间更短，使用寿命更高，可以有效提升转炉维护效果，有利于转炉生产平稳顺行，提高了转炉生产效率，并降低了炉前劳动强度，烧结过程绿色环保，具有很好的经济、社会效益优势。

关键词：转炉大面料，无碳，环保，烧结时间短，长寿

Application Technology of Environmental Protection No Carbon "Water" Bonded Self-flowing Magnesia Castable

Dong Zhanchun, Deng Lerui, Zhang Ronghui

(Beijing Allied Rongda Engineering Material Co., Ltd., Beijing 101400, China)

Abstract: In this paper, the characteristics and application effects of environmental protection no carbon "water" bonded Self-flowing magnesia castable and traditional large fabrics are compared. Practice results show that: environmental protection no carbon "water" bonded Self-flowing magnesia castable sintering time is shorter, the service life is higher, can effectively improve the converter maintenance effect, is conducive to the smooth operation of the production, improve the production efficiency of the converter, and reduce the labor intensity of the green environmental protection furnace, sintering process, has good economic and social benefits.

Key words: fefractory for converter bedding face, free carbon, environmental protection, short sintering time, long life

电铝热法冶炼 FeV80 合金用炉衬侵蚀过程分析

谢毓敏 [1,2]，宋明明 [1,2]，宋　波 [3]，曹　敏 [3]，薛正良 [1,2]，徐润生 [1,2]

（1. 武汉科技大学耐火材料与冶金国家重点实验室，湖北武汉　430081；

2. 武汉科技大学钢铁冶金与资源综合利用省部共建教育部重点实验室，湖北武汉　430081；

3. 北京科技大学冶金与生态工程学院，北京　100083）

摘　要： 电铝热法生产 FeV80 渣线部位炉衬分为三层：第一个是挂渣层，主要物相有 $MgO \cdot Al_2O_3$、Al_2O_3 和 $12CaO \cdot 7Al_2O_3$；第二是反应层，主要包括 $MgO \cdot Al_2O_3$ 和 Ca-Al-O；最后一个是渗透层，其主要组成物相是 MgO、Ca-Al-O 和 Ca-Al-Mg-O。反应层中液态炉渣与 MgO 反应形成了大块状的 $MgO \cdot Al_2O_3$ 及较大的间隙，使液态炉渣在反应层中的传递变得容易。在渗透层中有少量的液态渣渗入，镁砂颗粒保持着很紧密的接触，液态炉渣沿镁砂颗粒呈网状分布。MgO 进入炉渣主要是由于反应层 $MgO \cdot Al_2O_3$ 间填充的液态炉渣造成了炉衬的软化，大大降低了镁砂捣打炉衬的强度，在热应力、机械力等作用时，炉衬中大量的高熔点 $MgO \cdot Al_2O_3$ 相很容易通过挂渣层进入渣中，恶化 FeV80 冶炼工艺。

关键词： 电铝热法，FeV80 合金，炉衬侵蚀，物相组成，镁砂捣打炉衬

Erosion Mechanism of Lining during Electro-aluminium Heating Melting Process for FeV80 Alloy Producing

Xie Yumin[1,2], Song Mingming[1,2], Song Bo[3], Cao Min[3],

Xue Zhengliang[1,2], Xu Runsheng[1,2]

(1. Key Laboratory for Ferrous Metallurgy and Resources Utilization of Ministry of Education, Wuhan University of Science and Technology, Wuhan 430081, China; 2. The State Key Laboratory of Refractories and Metallurgy, Wuhan University of Science and Technology, Wuhan 430081, China; 3. School of Metallurgical

and Ecological Engineering, University of Science and Technology Beijing, Beijing 100083, China)

Abstract: The lining of the furnace producing FeV80 alloy is divided into three layers in the parts near slag line: The first layer is the slag adhesion, in which the main contents of the slag layer are $MgO \cdot Al_2O_3$, Al_2O_3 and $12CaO \cdot 7Al_2O_3$. The second is the reaction layer mainly including $MgO \cdot Al_2O_3$ and Ca-Al-O. And the last is the permeation layer mainly component of MgO, Ca-Al-O and Ca-Al-Mg-O. In the reaction layer, the liquid slag can react with MgO to form the large $MgO \cdot Al_2O_3$ lumpy and the larger clearance, which makes the transfer of liquid slag in the reaction layer easily. In the permeable layer, there is a small amount of liquid slag. The magnesia particles contacted closely, and the slag along the magnesia particles is distributed in a network. MgO get into the slag easily, it is mainly caused by the thermal stress and mechanical force on the loose lining with large MgO. Al_2O_3 lumpy filled by liquid slag, which reduce the strength of magnesia ramming furnace lining obviously and deteriorating the smelting process of FeV80 alloy.

Key words: electro-aluminium heating melting, FeV80 alloy, lining erosion, composition, magnesite ramming lining

高炉炉缸溶胶结合浇注料的研究与应用

唐勋海，高　栋，薛乃彦，徐自伟，张君博

（中国京冶工程技术有限公司　中冶建筑研究总院有限公司，北京　100088）

摘　要：本文通过对溶胶浓度（百分比溶度 30%、40%）、促凝剂、微粉加入量（3%、5%、8%）及微粉复合等试验分析，设计溶胶结合炉缸浇注料，研究其施工性能（流动性能、硬化时间、烘烤性能）、硬化性能、物理化学性能和耐侵蚀性能等，结果表明：溶度 40%溶胶做结合剂，加 5%活性氧化铝和 0.04%复合促凝剂，溶胶结合浇注料的施工性能优良，其物理化学性能最佳。经实践应用，溶胶结合浇注料可代替高炉炉缸中的陶瓷杯和风口组合砖，通过支模浇筑工艺，可形成整体密封的高炉炉缸结构，该技术克服传统结构成百上千砖缝的设计缺陷，缩短施工维修时间，减少施工人为因素和维修施工的二次损害，延长炉缸使用寿命，提高安全系数。

关键词：溶胶，炉缸浇注料，研究，应用

Studies and Application of Nanosol Bonded BF Hearth's Castables

Tang Xunhai, Gao Dong, Xue Naiyan, Xu Ziwei, Zhang Junbo

(Central Research Institute of Building and Construction Co., Ltd., Beijing 100088, China)

Abstract: In order to design nanosol bonded castables and its application, The effect of nanosol concentration (percentage solubility: 30%, 40%), Coagulant, activated alumina addition (3%, 5%, 8%) on construction performance (flow performance, hardening time, baking property), hardening time, physical properties, physical properties and corrosion resistance of the properties of castables was studied. The results show that the solubility of 40% nanosol as binder, adding 5% activated alumina and 0.05% composite coagulant, nanosol bonded castables' performance is excellent, The physical and chemical properties reach or even exceed the performance of blast furnace hearth with ceramic cup. This paper studies the nanosol bonded castables by formwork pouring process, can replace the hearth in the design of ceramic cup and tuyere combined bricks, forming an integral sealing structure, and overcome the defects of traditional hearth structure of thousands of bricks, greatly improves the safety performance of the hearth. To ensure high efficiency, energy saving and long life operation of blast furnace.

Key words: nanosol, BF hearth's castables, studies, application

镁碳砖产品配方的剖析

崔园园，钟　凯，曹　勇，祝少军

（首钢集团有限公司技术研究院，北京　100041）

摘　要：为加强耐火材料质量管控，满足首钢一业多地钢铁业的安全生产、工艺技术和产品质量要求，特对一些关键耐火材料产品进行配方剖析。本文以钢包渣线镁碳砖为例，介绍含碳耐火材料产品常用的剖析方法和剖析步骤。采用氧化脱碳和碱液浸泡解体两种方式解体镁碳砖，通过各粒级的化学成分分析、扫描电镜、X衍射分析和粒度分析等对产品的粒度级配、可能的配料组元、组元比例进行推测，结合文献等资料对检测结果进行一系列推算，最终给出接近合理的产品配料方案。

关键词：镁碳砖，产品配方，剖析

The Analysis of Magnesia Carbon Brick Product Formula

Cui Yuanyuan, Zhong Kai, Cao Yong, Zhu Shaojun

(Shougang Group Co., Ltd. Research Institute of Technology, Beijing 100041, China)

Abstract: To strengthen the quality control of refractories, meet the safety production, process technology and product quality requirements of the multi-steel industry in Shougang, a number of key refractory products were analyzed. In this paper, the steel slag line magnesia carbon brick as an example, introduced carbon refractory products commonly used analysis methods and steps. Using oxidation decarburization and alkali immersion disintegration in two ways to disintegrate magnesium carbon bricks. The particle size, the possible ingredient group, the component proportion of speculation were determined by chemical composition analysis, SEM, XRD and particle size analysis, combined with literature and other information on the test results for a series of projections, and ultimately give a reasonable product ingredients.

Key words: magnesia carbon brick, product formula, analysis method

化学成分对 $3Al_2O_3·2SiO_2$ 莫来石流钢砖耐火度的影响

柯　超

（大冶特殊钢股份有限公司特冶厂，湖北黄石　435001）

摘　要：莫来石流钢砖是一种典型的硅铝系无机非金属材料，在硅铝系二元相图中流钢砖的液相点出现温度为1840℃。流钢砖的耐火度不光与其复合矿相的熔点有关，还与物理性质有关。研究流钢砖中各种氧化物的分子量以及高温下发生的共晶反应，可推算出各种低熔物的化学式。统计进厂莫来石流钢砖的化学成分可知主成分的变化：铝氧最小值为62.94%，最大值为73.61%；而二氧化硅平均值为23.11%。流钢砖中各氧化物的熔点和质量比的乘积之和称为流钢砖的液相线温度，流钢砖中低熔点的碱性氧化物RO起到降低液相线的作用。另外，流钢砖的工作面熔洞等各类缺陷，会对耐火度产生不利的影响。

关键词：化验成分，莫来石流钢砖，耐火度，氧化物熔点

The Influence of Composition to the Refractoriness of Mullite-refractory which is $3Al_2O_3 \cdot 2SiO_2$

Ke Chao

(Forging Business Division, Hubei Xinyegang Steel Co., Ltd., Huangshi 435001, China)

Abstract: Mineralogical phase of mullite-refractory brick is $xAl_2O_3 \cdot ySiO_2$, the temperature of liquidus arrives 1840℃. To increasing thermal shock stability of refractory brick. In actual production, temperature of baking house could be higher than 80℃. Studing the oxide molecular weight in refractory brick and eutectic reaction at high temperature, can calculate molecular formula of various low fusant. Inspecting chemical composition of mullite-refractory brick, could find the change of aluminum oxide (minimum value is 62.94%, maxiumum value is 73.61%), and average value of silicon dioxide is 23.11%. Basic oxide achieve 1.15% by contrast.

Key words: testing component, mullite-refractory brick, refractoriness, melting point of oxide

碱性熔渣和钢水对镁钙材料的侵蚀和渗透行为

李世明[1,2]，马北越[1,2]，杨文刚[2]，钱　凡[2]，刘国齐[2]，于景坤[1]

（1. 东北大学冶金学院，辽宁沈阳　110819；2. 中钢集团洛阳耐火材料研究院有限公司
先进耐火材料国家重点实验室，河南洛阳　471039）

摘　要： 为了研究碱性熔渣（钢水和熔渣）对镁钙耐火材料的侵蚀和渗透行为，配置 R=4 的熔渣，采用静态坩埚法，在埋碳气氛中1600℃下保温2h对镁钙耐火材料进行侵蚀试验。测量试验后试样的侵蚀和渗透深度，并利用SEM、EDS、XRD对试验后试样进行表征。结果表明：（1）在 1600 ℃下，镁钙耐火材料有较好的抗碱性渣（R=4）侵蚀和渗透能力，亦有很好的抗熔渣和钢水共同侵蚀和渗透能力。（2）从 SEM 及 EDS 图谱分析可得，在 1600 ℃下碱性渣（R=4）对镁钙耐火材料的侵蚀和渗透程度都大于在同温度下钢水和熔渣的侵蚀和渗透程度。从 XRD 图谱可得，在1600℃下碱性渣（R=4）侵蚀作用下，镁钙耐火材料侵蚀层中的方石英衍射峰强高于相同温度下钢水和熔渣对镁钙耐火材料侵蚀层中的方石英衍射峰，且硅酸三钙和 Al_2O_3 相的衍射强度几乎一致。

关键词： 镁钙材料，熔渣，钢水，侵蚀，渗透，埋碳气氛

Corrosion and Penetration Behaviors of Alkaline Slag and Molten Steel on the Magnesia-Calcia Materials

Li Shiming[1,2], Ma Beiyue[1,2], Yang Wengang[2], Qian Fan[2],
Liu Guoqi[2], Yu Jingkun[1]

(1. Northeastern University, School of Metallurgy, Shenyang 110819, China; 2. State Key Laboratory of Advanced Refractories, Sinosteel Luoyang Institute of Refractories Research Co., Ltd., Luoyang 471039, China)

Abstract: In order to study the corrosion and penetration behaviors of alkaline slag (molten steel and slag) on the magnesium calcia materials, the molten slag of $R = 4$ was designed, a static crucible method was used to conduct the corrosion test on the magnesia-calcia materials at 1600 ℃ for 2 h under the condition of embedded carbon. The corrosion and penetration depths of the samples after the corrosion test were measured, and the samples were characterized by SEM, EDS and XRD. The results show: (1) Magnesium calcia material have good corrosion and penetration resistances to alkaline slag (R=4), as well as slag and molten steel at 1600 ℃. (2) From the analyses of SEM and EDS, the corrosion and penetration degrees of alkaline slag (R=4) on the magnesium calcia materials at 1600 ℃ are greater than those of molten steel and slag at the same temperature. From the XRD pattern, the diffraction peaks of cristobalite in the corrosive layer of magnesium calcia materials under the corrosion of alkaline slag ($R = 4$) at 1600 ℃, are higher than those of in the corrosive layer under the corrosion of molten steel and slag at the same temperature. The diffraction intensities of dicalcium silicate and alumina are almost unchanged.

Key words: magnesia-calcia refractories, molten slag, molten steel, corrosion, penetration, carbon-embedded atmosphere

连铸低碳尖晶石碳塞棒棒头的研究与应用

钱　凡，李红霞，杨文刚，刘国齐，郑　卫，闫广周

（中钢集团洛阳耐火材料研究院有限公司，河南洛阳　471039）

摘　要： 在连铸特殊钢种如高氧钢、高锰钢时广泛使用尖晶石碳质塞棒，但随着冶金技术的发展，对其性能要求也进一步提高：浇铸时间长，适用钢水种类多变，对钢水增碳少等。为此在试验室开展了不同基质组成的尖晶石碳材料与现有尖晶石碳材质性能对比研究，特别是抗热震性能与抗侵蚀性能，根据结果进一步优化方案，并在连铸中间包中获得了应用，与之前的产品进行对比，新型低碳尖晶石碳棒头材料的抗热震性满足现场要求，同时抗钢液冲刷性能的增强，使用寿命获得进一步提高。

关键词： 塞棒，尖晶石碳，抗热震性，抗冲刷性

The Research and Application of Low Carbon Spinel-C Refractories for Monolithic Stopper

Qian Fan, Li Hongxia, Yang Wengang, Liu Guoqi, Zheng Wei, Yan Guangzhou

(Sinosteel Luoyang Institute of Refractories, Luoyang 471039, China)

Abstract: Spinel-C refractories for monolithic stopper is widely used in special steel casting such as high oxygen steel, high manganese steel, which is requested that long service life, low carbon content and various kinds of steel casting with the development of metallurgical technology. This paper tries to make a comparative research on properties between low spinel-c refractories with different composition and SG100 in the laboratory, especially some properties such as thermal shock resistance and erosion resistance, at last obtain optimization scheme of low carbon content and had been tested in continuous casting. The result shows that thermal shock resistance of the low carbon spinel carbon refractories could satisfy the field thermal shock requirement, at the same time, the erosion resistance by liquid steel was enhanced, and the service life was further enhanced compared with SG100.

Key words: monolithic stopper, spinel-c refractories, thermal shock resistance, erosion resistance

热风炉陶瓷燃烧器的研究与应用

张伯鹏[1]，李富朝[2]，王潘峰[2]，李贯朋[2]

（1. 北京戈尔登科技开发有限责任公司，北京　100010；

2. 郑州安耐克实业有限公司，河南郑州　452370）

摘　要：本文阐述了各种热风炉用燃烧器的特点与工作原理，通过对各种热风炉用燃烧器的特点与工作原理的研究，分析产生弊病的原因，并运用各种技术手段，对各种燃烧器结构，拱顶空间空气动力学特性进行了研究，在此基础上研发出具有自主知识产权的，三维混合非预混型陶瓷燃烧器，并详细介绍了该燃烧器的研究内容与方法，以及新型燃烧器结构在顶燃式热风炉工业化应用中取得的良好应用效果，展示了其广阔的发展前景。

关键词：新型顶燃式热风炉，陶瓷燃烧器，高风温，工业应用

Research & Application of Ceramic Burner for Hot Blast Stove

Zhang Bopeng[1], Li Fuchao[2], Wang Panfeng[2], Li Guanpeng[2]

(1. Beijing Golden Science and Technology Development Co., Ltd., Beijing 100010, China;

2. Zhengzhou Annec Industrial Co., Ltd., Zhengzhou 452370, China)

Abstract: This article expound the characteristics and working principle of burners for various hot blast stove(hereinafter called HBS). Through the study of the characteristics and working principle of various HBS burners, analysis the causes of the malady, and study the aerodynamic characteristics of vault space for various burner structures by using various techniques, we have developed a three-dimensional mixed non premixed ceramic burner with independent intellectual property. The research contents and methods of the burner are introduced in detail. And the application of the new type burner structure in the industrial application of the top combustion type HBS has shown its vast potential for future development.

Key words: new top combustion type HBS, ceramic burner, high blast temperature，industrial application

15 能源与环保

炼铁与原料
炼钢与连铸
轧制与热处理
表面与涂镀
金属材料深加工
钢铁材料
汽车钢
海洋工程用钢
轴承钢
电工钢
粉末冶金
非晶合金
高温合金
耐火材料
★ 能源与环保
分析检测
冶金设备与工程技术
冶金自动化与智能管控
冶金技术经济

钢铁企业自发电率评价模型及提升路径研究

桂其林，周佃民

（宝山钢铁股份有限公司能源环保部，上海 200941）

摘 要：钢铁企业自发电率是评价企业节能综合水平的重要指标。国内学者和钢铁企业围绕自发电率做了大量工作。本文从自发电率的定义出发，分析其影响因素，建立了一套用于评价钢铁企业自发电率水平的指标群和基于此的评价模型，并对提升自发电率的技术进行了综述。

关键词：钢铁企业，自发电率，评价模型，提升路径

Research of Evaluation Models and Promotion Paths of Self Generation Rate in Iron & Steel Complex

Gui Qilin, Zhou Dianmin

(Energy & Environment Department, Baoshan Iron and Steel Co., Ltd., Shanghai 200941, China)

Abstract: The self generating rate of iron and steel complex is an important index to evaluate the comprehensive level of energy saving. Domestic scholars and steelmakers have done a lot of works around it. Based on the definition of self generating rate, this paper analyzes its influencing factors. A set of index groups and evaluation models for evaluating the self generating rate of iron and steel enterprises are established, and the technologies to improve the self generating rate are summarized.

Key words: iron & steel complex, self generation rate, evaluation models, promotion paths

中国钢铁工业能源消耗概述

宋强建，宁晓钧，张建良，焦克新，王 翠

（北京科技大学冶金与生态工程学院，北京 100083）

摘 要：20 世纪末，世界发达国家如日本、美国等相继走上了钢铁工业的去产能、兼并重组和绿色发展的道路。中国作为最大的发展中国家，过去几十年钢铁工业取得了巨大的成就，同时由于粗放式发展带来的弊端也日益显现出来。2015 年国家推行供给侧改革，全国粗钢产量首次出现下滑，吨钢综合能耗降低到 573.72 千克标煤/吨。进入 2016 年，全国粗钢产量为 8.08 亿吨，比去年出现小幅增长，吨钢综合能耗比去年增加 11.94 千克标煤/吨，增长率为 2.08%。这种反弹的现象尤其应该引起注意，改革的力度不能放松，进一步的深入改革必须抓住重点，对症下药。本文通过对 2015 年、2016 年全国重点钢铁企业的能源消耗进行对比分析，从而为下一步释放钢铁工业活力，建立资源节约型钢铁工业提供数据借鉴。

关键词：钢铁工业，粗钢产量，能源消耗，概述

Summary on Energy Consumption of Chinese Iron and Steel Industry

Song Qiangjian, Ning Xiaojun, Zhang Jianliang, Jiao Kexin, Wang Cui

(School of Metallurgical and Ecological Engineering, University of Science and
Technology Beijing, Beijing 100083, China)

Abstract: At the end of the 20th century, Developed countries such as Japan, the United States and so on have entered the road of cutting overcapacity, merger and reorganization and green development. China as the largest developing country, the past few decades the steel industry has made great achievements, and the drawbacks are increasingly apparent because of the extensive development.In 2015, the government carried out the supply-side structural reform, the countrywide crude steel capacity fell for the first time. the overall energy consumption per ton steel company dropped to 573.72 kgce/t. Another small capacity increase come in 2016 which turn into 808 million tons. The overall energy consumption per ton steel company increased 11.94 kgce/t, the growth rate is 2.08%. The resilient phenomenon should be pay attention especially. The intensity of reform can not be relaxed, further in-depth reform must focus on primary issues. In this paper, the national key iron and steel enterprises of energy consumption and environmental indicators have been analysed in 2015 and 2016. So as to release the iron and steel industry vitality,and offer data reference for establish resource-saving steel industry.

Key words: iron and steel, crude steel capacity, energy consumption, summary

高热值燃烧技术在 M701S(D)型 CCPP 机组上的应用

薛晓金，刘　伟，詹守权

（鞍钢股份鲅鱼圈钢铁分公司能源动力部，辽宁营口　115007）

摘　要： 文章主要对冬季 CCPP 机组使用高热值运行方式，即热值由 4396kJ/m^3 提升至 4605kJ/m^3 后，机组负荷的增加情况，以及燃机高温部件的运行、空压机的运行、煤压机等各主机设备运行情况进行了探索试验。

关键词： 高热值，压气机，压比

Application of High Calorific Combustion Technology on M701S(D) CCPP Power Plant

Xue Xiaojin, Liu Wei, Zhan Shouquan

(The Energy and Power Department of Ansteel Bayuquan Iron & Steel Subsidiary, Yingkou 115007, China)

Abstract: This paper mainly on the winter CCPP unit using high calorific value operation mode, namely the calorific value increased from 4396kJ/m^3 to 4605kJ/m^3, the unit load increased, explored the testingresults of each host and combustion engine high temperature parts operation, operation of air compressor, gas compressor etc..

Key words: high calorific value, compressor, pressure ratio

球团焙烧回转窑温度分布数值模拟

张义奇，徐天骄，董 辉

（东北大学热能工程系，辽宁沈阳 110819）

摘 要：对球团焙烧回转窑内的温度分布进行了研究，建立了回转窑内一维轴向传热数学模型，应用数值计算方法对该模型求解，得到了回转窑内轴向温度分布。分析了回转窑热工参数对于球团焙烧质量的影响，结果表明：球团在刚进入窑内快速升温，随后升温过程减缓，且球团温度受窑内多种热工参数的综合影响。

关键词：回转窑，传热，数值模拟，温度分布

Numerical Study of Temperature Distribution in the Pellet Roasting Rotary Kiln

Zhang Yiqi, Xu Tianjiao, Dong Hui

(Northeastern University, Shenyang 110819, China)

Abstract: Researched distribution of temperature in pellet roasting rotary kiln. The heat transfer model is built and utilize numerical method to obtain the solution. Thermal parameter is studied for effecting roasting quality. Results show: Pellet heat up sharply at first, and then slow down. Pellet's temperature is influenced by thermal parameter of the kiln.

Key words: rotary kiln, heat transfer, numerical simulation, temperature distribution

煤粉锅炉全烧高炉煤气改造技术措施

刘 婕

（上海梅山钢铁股份有限公司热电厂，江苏南京 210039）

摘 要：高炉煤气与煤粉燃料特性相差较大，在煤粉锅炉全烧高炉煤气改造实践中，得出如下的结论：合理调整过热器受热面防止过热器超温，有效地增加省煤器受热面降低排烟温度，保证煤改气后锅炉安全稳定高效。

关键词：煤粉锅炉，全烧高煤锅炉，改进技术措施

Technical Measures for Changing Powdered-coal Boiler to BFG Fired Boiler

Liu Jie

(Thermal Power Plant of Meishan Iron & Steel Co., Ltd., Nanjing 210039, China)

Abstract: There is great difference between powdered coal and BFG. In the boiler improvement,a conclusion is reached with efforts that reasonable heating surface design can prevent overtemp risk and guarantee boiler efficiency.

Key words: powdered coal boiler, improvement, BFG fired boiler

钢铁厂烧结主抽风机变频节电改造节能量校核方法的研究

朱红兵

（上海宝钢节能环保技术有限公司，上海　201900）

摘　要： 本文分析了导致烧结主抽风机变频改造后节能量波动的各种独立变量，在独立变量劣化趋势不能改变的前提下，提出了采用动态电耗基准的节能量计算校核方法，可对主抽风机变频改造后的节能效果有一个客观准确的评价。同时提出了多种节能措施下的节能分成比例的确定方法，在负荷变化范围有限和变化周期不太频繁的工艺系统中，风机设备提效改造的节能贡献度远大于变频调节的贡献度，此时决定节能效果优劣的主要在于主体设备风机本身的运行效率。

关键词： 烧结，主抽风机，节电，变频，效率，电耗基准

The Research of the Energy Check Method of the Main Sintering Fan in Iron & Steel Plant

Zhu Hongbing

(Shanghai Baosteel Energy Saving Technology Co., Ltd., Shanghai 201900, China)

Abstract: This paper analyzes independent variables that causing quantity of energy saving rise and fall of the sinter fan after invertor reform. Under the premise of the degradation trend of independent variables can't change, the dynamic energy consumption is applied to energy saving calculation and checking, so have an objective and accurate evaluation for energy saving of the main sinter fan. Simultaneously, a method is put forward using for determining the proportion of energy of various energy-saving measures. At the process system of the limited load range and less frequent change, fan equipment effect of energy-saving than frequency conversion, the decision lies mainly in the fan itself.

Key words: sintering, main sintering fan, power saving, frequency conversion, efficiency, energy consumption baseline

高炉喷吹石灰窑尾气对碳素消耗的影响

周建安，陶　迅，谢剑波，王　宝

（钢铁冶金与资源化教育部重点实验室，省部共建耐火材料与冶金国家重点实验室，武汉科技大学，湖北武汉　430081）

摘　要：基于目前石灰窑尾气不能够充分利用，提出将石灰窑尾气通入高炉风口回旋区内，达到降低理论燃烧温度和高炉碳素消耗的目的。在保证高炉正常工作的前提下，研究石灰窑尾气喷吹量对理论燃烧温度以及极限石灰窑尾气喷吹量对碳素消耗的影响，理论上证明了对于较高理论燃烧温度的高炉可以达到降低碳耗消耗的目的，这为石灰窑尾气新的利用找到一条途径。结果表明：基于一座 3000 m³ 的生产高炉，在不影响高炉炉况正常运行的前提下，降低理论燃烧温度 200℃，喷吹石灰窑尾气最大的碳素降低约 8.6kg/tFe；与未喷吹高炉相比，新工艺不仅可以降低碳耗，而且还可以对废弃资源进行二次利用，增加综合效益。

关键词：石灰窑尾气，高炉，理论燃烧温度，碳耗

Effect of Injecting Lime Kiln Exhaust Gas into Blast Furnace on Carbon Consumption

Zhou Jian'an, Tao Xun, Xie Jianbo, Wang Bao

(Key Laboratory of Ferrous Metallurgy and Resources Utilization, Ministry of Education, State Key Laboratory of Refractories & Metallurgy, Wuhan University of Science and Technology, Wuhan 430081, China)

Abstract: Based on present scarce utilizations of lime kiln exhaust gas, we inject the exhaust gas into the raceway of blast furnace in order to reduce theoretical flame temperature and carbon consumptions.The effect of injecting amount on carbon consumption of blast furnace is explored.And the effects of limited injecting amount on carbon consumption and theoretical flame temperature are investigated on the basis of the normal working condition of blast furnace. It is theoretically proved that gas injection can achieve reduced carbon consumption for the blast furnace with high theoretical flame temperature,which finds a new way for the utilization of exhaust gas. The results show that,under reduced 200℃ of the theoretical flame temperature and the normal working condition of blast furnace, the maximum carbon consumption can be reduced by 8.6kg/tFe after exhaust gas is injected at a 3000m³ blast furnace.In contrast to non-injection, this new process can reduce carbon consumptions, reuse abandon resources and increase combined profits.

Key words: lime kiln exhaust gas, blast furnace, theoretical flame temperature, carbon consumption

2017 年上半年中钢协会员单位能源利用评述

王维兴

（中钢金属学会，北京　100711）

摘　要：2017 年上半年比上年同期相比中钢协单位吨钢综合能耗下降，焦化、烧结、高炉和电炉工序能耗在升高；转炉、钢加工工序能耗得到降低。钢铁企业之间，能源利用水平差距较大，各工序均有一批指标较先进、进步较大的企业，但也有一批指标落后、退步的企业；说明钢铁企业尚有一定节能降耗的潜力。部分企业能耗统计有误差，没有完全执行国家标准，出现能耗指标与生产指标的对应性差，没有真实反映出企业实际生产情况，应急时改进。

关键词：17 年上半年，重点企业，能源利用，评述

Abstract: In the first half of 2017, compared with the same period in the first half of 2017, the total energy consumption of tonnage is reduced, and the energy consumption of coking, sintering, blast furnace and electric furnace is increasing. The energy consumption of converter and steel processing is reduced. Among steel enterprises, there is a large gap in energy utilization. Each process has a group of enterprises with advanced and advanced indicators, but there are also a number of enterprises which are backward and backward. It shows that steel enterprises have the potential to reduce energy consumption. Some enterprises have energy consumption statistical error, not fully implement the national standard, energy consumption index and production indexes of the correspondence is poor, there is no real reflect the enterprise actual production conditions, improve emergency.

Key words: first half of 2017, key enterprises, energy use, review

基于两化的能源管理中心在泰钢的研究与应用

李　涛，朱　林，唐巧梅，王宝军，祁建伟，高玉树

（山东泰山钢铁集团有限公司自动化部，山东莱芜　271100）

摘　要： 随着泰钢的发展，能源消耗量不断增大，能源数据采集点分散、管理粗放等问题愈发显露出来，在一定程度上制约了泰钢循环经济的持续整体推进。企业能源管理中心是一项整合自动化和信息化技术的管控一体化节能新技术，是通过对企业能源生产、输配和消耗实施动态监控和管理，改进和优化能源平衡，从而实现系统性节能降耗，为企业增加经济效益，提升能源现代管理水平。

关键词： 能源管理中心，节能新技术，管控一体化

Research and Development of Energy Management Center Based on Automation Information Fusion in Taigang

Li Tao, Zhu Lin, Tang Qiaomei, Wang Baojun,

Qi Jianwei, Gao Yushu

(Shandong Taishan Steel Group Co., Ltd., Laiwu 271100, China)

Abstract: With the development of Taishan steel, the energy consumption is increasing, the scattered and extensive management of energy data collection point is more and more exposed, which restricts the sustainable development of the circular economy of Taishan steel at a certain extent. Enterprise energy management center is a new and integrated energy-saving technology integrating automation and information technology. it can improve and optimize energy balance through dynamic monitoring and management of enterprise energy production, transmission and distribution and consumption, so as to achieve systematic energy saving and consumption reduction, increase economic benefits and improve energy modern management level for enterprises.

Key words: energy management center, new energy saving technology, integration of control and control

鞍钢热风炉高风温及节能技术进步

孟凡双[1]，刘德军[2]，郝　博[2]

（1. 鞍钢股份有限公司炼铁总厂，辽宁鞍山　114021；
2. 鞍钢股份有限公司技术中心，辽宁鞍山　114021）

摘　要： 介绍了鞍钢高炉热风炉高风温及其相应的节能技术的进步。重点叙述了鞍钢针对本企业热风炉长期坚持使用低热值煤气烧炉的特点，开展了热风炉结构形式的改造和热风炉自预热、前置炉及辅助热风炉等根本性改造；继而开展了针对热风炉的多项综合节能技术的研究与应用，实现了热风温度的大幅提高和热风炉烧炉煤气消耗的大幅降低，取得了良好的效果，极大地推动了鞍钢高炉热风炉技术的进步。

关键词： 高炉，热风炉，技术，发展

150MW 燃气–蒸汽轮机联合循环发电技术改进与优化

王贺敏，陈　兵，任　强

（河钢集团邯钢公司，河北邯郸　056015）

摘　要： 本文介绍了邯钢东区现有发电机组状态，分系统说明了 150MW 燃气–蒸汽联合循环发电项目，包括设备主机参数、制造厂家、设备配备等内容。简单介绍了整个系统燃气蒸汽联合循环工艺流程原理，详细阐述了燃气轮机技术的主要关键技术点。通过不断吸收及消化机组技术要点，进行优化，保设备运行稳定，发电效率高效，创造了巨大的经济效益和社会效益，为钢铁企业充分利用低热值燃料发电技术的应用提供了可靠的实践依据。

关键词： 燃气轮机，喷嘴，离心式压缩机，改进，效益

150 MW CCPP Generator Technical Digestion and Absorption

Wang Hemin, Chen Bing, Ren Qiang

(Hesteel Group Hansteel Company, Technology Center, Handan 056015, China)

Abstract: In this paper, Han Steel was introduced the east area the existing state of generator set, partial system shows 150 MW combined cycle power generation project, including equipment parameters, manufacturers, equipment equipped etc. Brief introduce the whole system combined cycle process principle, expounds the main key technology points of the technology of gas turbine. Through continuous technology main points, absorption and digestion unit is optimized, the Unit operation is stable, power efficiency and efficient, has created a huge economic and social benefits, for iron and steel enterprises make full use of the application of low calorific value of fuel power generation technology provides a reliable practical basis.

Key words: gas turbine, nozzle, centrifugal compressor, improvement, benefits

环冷机内气固传热过程仿真及优化

张　晟，董　辉

（东北大学国家环境保护生态工业重点实验室，辽宁沈阳　110819）

摘　要： 在先前理论和实验研究的基础上，选取环冷机一二段为研究对象，借助多物理场仿真软件 COMSOL 的自定义函数功能，将烧结矿沿周向的移动速度定义到物理模型中，藉此建立环冷机二维稳态数值模型。以此作为开展环冷机内气体流动和气固换热研究的基础。研究结果表明：国内某 360 m^2 烧结机对应的环冷机适宜的热工参数为：冷却风表观流速为 1.5m/s，料层高度为 1.5m，进口风温 424K，余热回收段长度 48m。

关键词： 环冷机，余热回收，传热，数值模拟

Simulation and Optimization of Gas-Solid Heat Transfer in Sinter Annular Cooler

Zhang Sheng, Dong Hui

(SEPA Key Laboratory on Eco-industry, Northeastern University, Shenyang 110819, China)

Abstract: According to the previous theoretical and experimental researches, the first and second stage of sinter annular cooler was taken as the research subject. The two-dimensional steady-state numerical model of sinter cooler was established. At the same time, The speed along the moving direction of sinter trolley was defined in the model, with the help of user-defined functions in COMSOL. This physical model was taken as the foundation to research the law of gas flow and gas-solid heat transfer. The result shows that, for the sinter annual cooler with 360m^2 sintering machine, the suitable thermal parameters are that the superficial velocity of cooling air is 1.8m/s, the height of sinter is 1.5m, the temperature of inlet air is 424K, the length of heat recovery stage is 48m.

Key words: sinter annular cooler, waste heat recovery, heat transfer, numerical simulation

转炉煤气"单柜双厂"回收的研究与应用

韩新萍，左建平

（宝钢股份武钢有限能源环保部，湖北武汉　430080）

摘　要： 通过对影响转炉煤气"单柜双厂"回收因素进行研究，采取可行措施加以改进，实现单座 10 万立方米气柜回收两个炼钢厂煤气，确保了安全生产的运行，并创造可观的经济效益。

关键词： 转炉煤气，10 万立方米气柜，单柜双厂，研究与应用

Reserch and Application of "One Gasholder Double Steel Plant" LDG Recovery

Han Xinping, Zuo Jianping

(Energy and Environmental Protection Department of Baosteel Wuhan Iron and Steel Co., Ltd., Wuhan 430080, China)

Abstract: The complications were analyzed of "One Gasholder Double steel plant"and important measures were taken.LDG from two steel plant can be reconvered by One gasholder of 100000 m^3 in more safe and economical way.

Key words: LDG, 100000m^3 gasholder, one gasholder double steel plant, anlyze and application

高炉鼓风机叶片加级改造及运行效果分析

胡 翰

（武汉钢铁有限公司，湖北武汉 430080）

摘 要： 针对高炉扩容改造后鼓风机性能不匹配造成运行效率降低、高炉炉况波动、风机叶片受损的问题，对其原因进行了分析，并采取措施对转子叶片进行加级改造，改变安全运行曲线设置方式。机组恢复后进行了性能试验，并与改造前机组效率进行了对比。通过改造后的运行效果分析给出解决方案及提高高炉鼓风系统能源利用效率及安全运行的合理建议。

关键词： 高炉鼓风机，叶片加级改造，特性曲线，高炉鼓风系统效率

Blade Stage Increase Modification and Operation Effect Analysis of Blast Furnace Blower

Hu Han

(Wuhan Iron & Steel Co., Ltd., Wuhan 430080, China)

Abstract: For the blast furnace capacity expansion, the blower performance does not match operation efficiency, blast furnace condition fluctuation, blower blade damaged. Analyzes the reasons, and to take measures on the rotor blades stage increase transformation, change the security operation curve setting scheme. The performance test was carried out after the unit was recovered and compared with the efficiency of the former unit. Through the analysis of the operation effect after modification, the solutions and suggestions for improving the energy efficiency and safety operation of blast furnace blower system are given.

Key words: blast furnace blower, modification of blade stage increase, characteristic curve, efficiency of blast furnace blowing system

预燃式空燃比检测装置设计

何　苗，曹　凯，徐劲林，马　珺

（武汉钢铁有限公司计控厂，湖北武汉　430080）

摘　要： 武钢一热轧加热炉各段均装有残氧分析仪，根据实时残氧量进行空燃比的调节，但当燃气热值波动时，设定的空燃比无法实现炉内气氛的动态修正。因此，进行了预燃式空燃比检测装置的设计，该装置采用燃气预燃烧分析法，计算出设定空燃比，和原炉段残氧控制系统一起，实现空燃比的精确控制。装置投运后运行稳定，有效避免了燃气热值波动对空燃比的影响，钢坯氧化烧损和吨钢能耗均降低，经济效益显著。

关键词： 预燃烧，空燃比，燃气热值，氧化烧损，吨钢能耗

The Design for Air-fuel Ratio Detection Device Based on Pre-combustion

He Miao, Cao Kai, Xu Jinlin, Ma Jun

(Instrumentation and Control Company of Wuhan Iron and Steel Co., Ltd., Wuhan 430080, China)

Abstract: Oxygen analyzer is installed in each section of the heating furnace in WISCO No.1 hot rolling mill, which can adjust the air-fuel ratio in real time. But the setting air-fuel ratio cannot realize the dynamic correction of atmosphere in the furnace when the fuel calorific value is fluctuating. Therefore, device based on pre-combustion for air-fuel ratio detection is designed. It can calculate the setting ratio and control the ratio accurately with the original oxygen control system. The device runs steadily after production which can avoid the influence of calorific value fluctuating on air-fuel ratio effectively. Economic benefits are obvious with the reduction of slab oxidation burning loss and energy consumption per ton steel.

Key words: pre-combustion, air-fuel ratio, fuel calorific value, oxidation burning loss, energy consumption per ton steel

钢铁企业智慧能源研究

周佃民，桂其林，汤晓帆

（宝山钢铁股份有限公司能源环保部，上海　200941）

摘　要： "智慧能源"作为一个比较新的概念，是指通过信息系统与能源系统的高度融合，提升能源系统环境资源友好水平。自这个概念诞生起，因其蕴含了人们对未来能源系统的各种美好期待于一身而吸引了大量业者的关注。"智慧能源"的具体内涵因其讨论对象不同而又有区别，本文主要从钢铁企业的具体工业能源系统背景出发，研究"智慧能源"的理念与技术在钢铁企业中的具体实现及相关问题，包括内涵、驱动、价值、架构、关键问题及建设路径等方面，并提出了在理论研究方面亟待解决的关键问题。

关键词： 智慧能源，钢铁企业，能源系统，能源互联网

Research on the Smart Energy in Iron & Steel Complex

Zhou Dianmin, Gui Qilin, Tang Xiaofan

(Energy & Environment Department, Baoshan Iron and Steel Co., Ltd., Shanghai 200941, China)

Abstract: As a new concept, smart energy refers to the deep integration of information technology and energy technology to enhance environmental friendliness. Since the birth of this concept, it contains all the people's expectations for the future energy systems, and attracts many engineering researcher. From the background of energy system in iron & steel complex, smart energy and its connotation, architecture, and other aspects are discussed in this paper .

Key words: smart energy, iron & steel complex, energy system, energy internet

煤气混合站掺烧天然气的方案技术研究

李　佳，刘孝清，邓　灿

（中冶南方工程技术有限公司钢铁分公司，湖北武汉　430023）

摘　要： 为优化煤气能源供应，适应煤气供应变动情况，整体考虑全厂煤气平衡，煤气混合站原采用高焦混合煤气，现采用掺烧天然气实现高焦天混合煤气。同样热值情况下，对基准气和置换气的华白数、燃烧势、AGA 判定三指数进行了计算，华白数偏差在允许范围 10%以内，AGA 三指数判定在合适范围内。最后综合考虑用户燃烧喷嘴等情况，得出天然气掺烧的合适比例。

关键词： 混合站，天然气，互换性指标，AGA 指数

Research on Interchangability of Natural Gas Mixed Fuel Gas

Li Jia, Liu Xiaoqing, Deng Can

(WISDRI Engineering & Research Incorporation Limited，Wuhan 430023, China)

Abstract: In oder to optimize gas energy supply and adapt to gas energy variation, the gas mixing station mixes natural gas into former BFG-COG mixed gas, overall considering the gas balance. On base of the same calorific value, Wobbe index, combustion potential, AGA index of the substitution gas and the reference gas are calculated and compared. The Wobbe index deviation is within 10% allowed by the specification, while the AGA indexs are judged in the adequate range. Finally, considering factors such as user combustion nozzle, appropriate natural gas mixing volume ration is concluded.

Key words: gas mixing station, natural gas, interchangeabiliy index, AGA index

韩国钢铁企业的节能减排举措

罗 晔[1]，王 超[2]

（1. 宝钢股份中央研究院武汉分院（武钢有限技术中心），湖北武汉　430080；2. 武汉杉舍环保科技有限公司利福康有害生物防治服务中心，湖北武汉　430050）

摘　要：本文主要介绍了韩国三大钢铁企业在节能减排方面的实践。为了积极应对国内碳排放权交易制度，实现减排目标，浦项制铁建立了副生煤气发电站，将高炉煤气、FINEX 炉尾气及焦炉煤气作为燃料使用；与化工企业合建光阳-丽水海底隧道项目，实现副产物的回收再利用；自主开发的 FINEX-CEM 技术已投入商业化应用。下属的研发机构目前主要研究的相关课题包括：氨水吸收捕集二氧化碳，变压吸附分离副产煤气中一氧化碳和二氧化碳，炉渣显热回收，利用中低温余热发电，氢气还原炼铁，利用熔融盐回收余热等。现代制铁的节能减排成果已经获得了 CDP 的高度评价和 VCS 的权威认证，利用余热开展热配送业务，在节能减排的同时为地区降低了燃料采购成本。东国制钢引进 Eco-Arc 电炉和间接热装轧制工艺，通过提升装备技术达到节能减排的效果。

关键词：韩国，钢铁，企业，尾气，余热

The Measures about Energy-saving and Emission-reduction in South Korean Iron & Steel Enterprises

Luo Ye[1], Wang Chao[2]

(1. Research and Development Center of Wuhan Iron & Steel Co., Ltd., Wuhan 430080,China; 2. Lifuk ang Pest Control Service Center of Wuhan Sunse Environmental Protection Technology Co., Ltd., Wuhan 430050, China)

Abstract: This paper mainly introduces the practice of energy saving and emission reduction in three major iron and steel enterprises in South Korea. In order to actively respond to domestic carbon emissions trading system and achieve the target of emission reduction,POSCO has constructed gas power stations, the blast furnace gas and FINEX furnace gas, coke oven gas are used as fuel and built subsea tunnel between Kwangyang and Yosu in cooperation with chemical enterprises,to realize the recycling of by-products; the self-developed FINEX-CEM technology has been put into commercial applications. The on-going projects of its affiliated relevant research institution include: trapping carbon dioxide by the absorption of ammonia, carbon monoxide and carbon dioxide gas separating by way of pressure swing adsorption, slag sensible heat recovery, power generation by use of low temperature waste heat, hydrogen reduction for ironmaking, waste heat recovery using molten salt etc. The achievements of energy saving and emission reduction of Hyundai Steel have been highly appraised by Carbon Disclosure Project and authorized by the Verified Carbon Standard. Heat distribution has been carried out using waste heat, and fuel purchasing costs can been reduced for the region while saving energy and reducing emissions. The Eco-Arc electric furnace and hot charging process have been introduced in Dongkuk Steel, and the effect of energy saving and emission reduction has been greatly improved by upgrading equipment technology.

Key words: Korea, iron and steel, enterprises, off-gas, waste heat

峰谷电价条件下钢铁企业副产煤气调度规律与优化策略

郝聚显，赵贤聪，韩玉召，白 皓

（北京科技大学冶金与生态学院，北京 100083）

摘 要： 钢铁行业是目前中国最大的电力消费者之一。为了减少电力采购成本，提高电力供应的可靠性，钢铁企业建立了大量的自备电厂，而钢铁生产过程产生的副产煤气是其主要燃料。近年来，随着峰谷电价的实施，自备电厂和煤气柜柜位之间的协调调度越来越受到重视。当电价在峰段区间向自备电厂分配更多的副产煤气以产生更多的电力，同时在谷段区间向自备电厂分配较少的副产煤气，可以减少电力采购成本。本文通过构建基于峰谷电价的混合整数线性规划调度模型，讨论了在不同的柜容条件下煤气柜总标准偏移量和电力采购成本之间的关系。同时对具有不同配置的两个钢铁企业实施案例研究，并讨论了最优运营策略。

关键词： 调度模型，副产煤气，峰谷电价，自备电厂，极限柜容

Scheduling and Optimization Strategy of Byproduct Gas in Iron and Steel Enterprises under Time-of-Use Power Price

Hao Juxian, Zhao Xiancong, Han Yuzhao, Bai Hao

(School of Metallurgical and Ecological Engineering, University of Science and
Technology Beijing, Beijing 100083, China)

Abstract: The steel industry is one of the largest electricity consumers in China. In order to reduce the cost of electricity procurement and improve the reliability of power supply, iron and steel enterprises established a large number of on-site power plants (OSPPs), and the by-product gas produced during steel production process is its main fuel. In recent years, with the implementation of Time-of-use power price in China (TOU), the collaboration between OSPPs and gasholders is paid more[1] and more attention. Through allocating less by-product gas to reducing the electricity generation during the valley price period (VPP) and distributing more by-product gas increasing the electricity generation during the peak price period (PPP), the electricity purchasing cost can be reduced. In this paper, the relationship between the sum of standard deviation volume (SSDV) and the electricity purchasing cost under different limit cabinet capacity is discussed by building a mixed integer linear programming (MILP)-based scheduling model based on Time-of-use power price. A case study is conducted on two steel enterprises with different configurations of OSPPs, and the optimal operation strategy is also discussed.

Key words: scheduling model, byproduct gas, time-of-use power price, on-site power plants, limit capacity

浅谈煤气管网不同气源并网运行的风险控制及措施

柴晓慧，尹 林

（宝钢集团八钢公司能源中心，新疆乌鲁木齐 830022）

摘　要：通过对煤气管网不同气源并网运行的风险因素分析，从系统运行安全控制出发，提出了管理措施和技术措施，实现在线调度的系统安全。

关键词：管网，不同气源，并网，风险，措施

钢铁工业分布式能源及其在宝钢的开发实践

周佃民，桂其林

（宝山钢铁股份有限公司能源环保部，上海　200941）

摘　要：分布式能源作为建设在用户端的能源供应方式，已经成为能源系统的重要发展方向并得到国家大力提倡。钢铁工业是典型的高耗能行业，其生产过程伴随着大量而复杂的多种能源的转换与利用，大力发展分布式能源具有十分重要的节能减排意义。钢铁企业本身已开发或正在开发有一定数量的分布式能源。参照分布式能源的定义及国际认可的分类方式，本文对钢铁企业内部的多种分布式能源进行了分类和说明，并以此为基础，介绍宝钢在分布式能源方面的开发实践。

关键词：分布式能源，钢铁企业，能源系统

Distributed Energy System in Iron & Steel Complex and Its Application in Baosteel

Zhou Dianmin, Gui Qilin

(Energy & Environment Department, Baoshan Iron and Steel Co., Ltd., Shanghai 200941, China)

Abstract: Distributed energy system, as one form of energy at the demand side, has become an important development direction of energy system, and has been vigorously promoted by the state. Iron and steel industry is a typical energy-intensive industry, and its producing process is accompanied by a large number of complex energy conversion and utilization. It is very important to develop distributed energy resources to save energy and reduce emission. The steel enterprise itself has developed or is developing a certain number of distributed energy sources. According to the definition of distributed energy, this paper classifies and explains the distributed energy within the iron and steel enterprises, and then introduces the practical application of Baosteel in the distributed energy.

Key words: distributed energy system, iron & steel industry, energy system

钢铁企业二次能源回收利用评述

王维兴

（中国金属学会，北京　100711）

摘　要：钢铁企业二次能源占总能耗的 70%（包括副产煤气和余热、余能），可回收利用的二次能源量（不包括副产煤气）约占钢铁企业总用能的 15%左右。文章指出了余热余能的转换、回收和利用基本原则；文章还评述了我国

钢铁工业二次能源利用现状；同时指出了钢铁生产各工序节能技术和回收二次能源的技术装备，及相关指标。

关键词：钢铁工业，二次能源，回收利用，评述

Abstract: Iron and steel enterprises secondary energy accounted for 70% of total energy consumption (including the by-product gas and remaining heat and can), the amount of secondary energy recycling (not including by-product gas) accounted for about 15% of the total energy of iron and steel enterprise. The paper points out the transformation, recovery and utilization of the residual heat energy. The paper also analyzes the current situation of secondary energy utilization of steel industry in China. The technical equipment and related indexes of energy saving technology and recycling secondary energy are also pointed out.

Key words: iron and steel industry, secondary energy, recycling, review

轧钢加热炉采用热装工艺的蓄热式燃烧
技术的应用前景

张强国

（华菱湘钢有限公司五米板厂，湖南湘潭　411101）

摘　要：近年来，公司轧钢加热炉实施了一系列节能改造，如汽化冷却和烟气余热利用、热送热装技术的应用，远红外高温辐射涂料的应用等，取得了可观的节能效果。随着五米板和宽厚板的热装率达到85%左右的水平，热装节能接近极限水平，再提高热装率对节能的贡献有限。本人研究国内蓄热式燃烧技术的发展现状，认为高热装水平下的蓄热式燃烧技术的应用，将是公司轧钢加热炉大幅节能的另一个途径。

关键词：加热炉，蓄热式燃烧技术，热装率

Application Prospects of Regenerative Combustion Technology with Hot Charging Process in Steel Rolling Heating Furnace

Zhang Qiangguo

(Valin Xiangtan Iron and Steel Co., Ltd., Xiangtan 411101, China)

Abstract: In recent years, the company has implemented on a series of energy saving transformation on reheating furnace. Such as the application of vaporization cooling and waste heat utilization, technology of HCCR and technology of high temperature far infrared radiant coating, has achieved considerable energy saving. With The ratio of HCCR approached 85% in Five-meters wide heavy plate mill plant and Wide and heavy plate plant, the thermal energy is close to the limit level, and then increase the contribution of ratio of HCCR to energy saving. I study the development status of regenerative combustion technology in China, that the application of regenerative combustion technology in high ratio of HCCR level will be another way for the company to save energy.

Key words: the reheating furnace, the regenerative combustion technology, the ratio of HCCR

烧结余热发电技术在泰钢的研究与应用

许海涛，高玉树，祁建伟，孙庆霞

（山东泰山钢铁集团有限公司，山东莱芜　271100）

摘　要： 烧结余热发电技术是一项将烧结废气余热资源转变为电力的节能技术。钢铁生产过程中，烧结工序的能耗约占总能耗的 10%，仅次于炼铁工序而位居第二。原来由于种种原因，没有进行余热回收利用，使这些具有较高热焓的烟气全部通过烟囱排放至大气中，浪费了能源，同时也污染了环境。本文结合烧结余热回收系统国内外发展概况，对泰钢利用的烧结余热工艺技术时影响回收利用的主要因素进行分析，然后通过相应的技术改造、工艺控制和设备改进，使之得到改善，取得了良好的经济效益。

关键词： 余热发电，烧结

Research and Application of Sintering Waste Heat Power Generation Technology in the Taishan Steel

Xu Haitao, Gao Yushu, Qi Jianwei, Sun Qingxia

(Shandong Taishan Steel Group Co., Ltd., Laiwu 271100, China)

Abstract: Sintering waste heat power generation technology is a sintered exhaust heat resources into electricity saving technology. Iron and steel production process, sintering energy consumption accounts for about 10% of total energy consumption of ironmaking process, second only in second. The original due to various reasons, no waste heat recovery and utilization, which has high enthalpy of flue gas through the chimney emissions to the atmosphere, resulting in a waste of energy, but also pollutes the environment. The development of this combination of sintering waste heat recovery system at home and abroad, the main factors affecting the sintering waste heat recycling technology Taigang used is analyzed, and then through technical transformation, the corresponding process control and improvement of equipment, which has improved, and achieved good economic benefit.

Key words: waste heat power generation, sintering

引进低温技术　降低能源消耗
——化厂化产车间硫铵低温技术工艺改造

戴乐龙

（江苏沙钢集团有限公司，江苏张家港　215600）

摘　要： 针对企业的喷淋式饱和器处理过程中蒸汽消耗大，饱和器出口煤气温度较高，影响粗苯回收生产的实际，在交流学习国内先进硫铵生产工艺，结合本厂生产实际，车间决定引进硫铵低温生产工艺，降低蒸汽消耗，同时降低母液出口煤气温度，在不影响硫铵产量的前提下，进一步增加粗苯回收率。组织现场工艺设备改造，保证硫铵产品质量符合一级品标准。

关键词： 喷淋式饱和器，低温技术，节能

热轧平流池气浮降油泥含油率技术与工业应用

杨大正[1]，于 洋[2]，徐鹏飞[1]，马光宇[1]，杨立军[3]，李 静[4]

（1. 鞍钢技术中心环境与资源研究所，辽宁鞍山 114009；2. 鞍山钢铁技改部，
辽宁鞍山 114009；3. 鞍钢股份热轧带钢厂，辽宁鞍山 114000；
4. 辽宁科技大学材料与冶金学院，辽宁鞍山 114000）

摘 要：本文简要介绍了传统的冶金企业热轧带钢厂平流池产出的热轧油泥处理方法，提出一种在热轧平流池增加气浮设备系统，降低热轧油泥含油率的新技术。就油泥除油原理进行了分析，对鞍钢 1780 线平流池完成工业化实施，简要介绍了工艺、设备情况、除油效果。结果表明：热轧油泥的含油率由 16.1% 降至 2.16%，已直接作为烧结原料使用；回收的废油量约是改造前的 12 倍；平流池水质得到改善。该项目具有投资少、见效快、成本低、生产过程环保、减排二氧化碳、经济效益、环境效益和社会效益显著的特点。在国内外具有很大的推广价值。

关键词：热轧油泥，气浮，除油，平流沉淀池

New Technology and Commercial Application of Air Flotation in Horizontal Flow Sedimentation Tank to Reduce Oil Content in Hot Rolling Sludge

Yang Dazheng[1], Yu Yang[2], Xu Pengfei[1], Ma Guangyu[1], Yang Lijun[3], Li Jing[4]

(1. Environment and Resources Research Institute of AISC Technology Cnter, Anshan 114009, China;
2. Transformation Department of AISC, Anshan 114009, China; 3. Hot Strip Rolling Mill of AISC,
Anshan 114000, China; 4. School of Material & Metallurgy of University of Science &
Technology Liaoning, Anshan 114000, China)

Abstract: Traditional treatment methods of hot rolling sludge from horizontal flow sedimentation tank have been briefly introduced, a new technique for decreasing the oil content in hot rolling sludge by adding air flotation equipment in horizontal flow sedimentation tank was put forwarded. The process, equipment and oil removal principle and oil removal efficiency of the new technique are introduced and analyzed, and carried out in the 1780 level horizontal flow sedimentation tank in AISC . Result show that the oil content of hot rolling sludge decreased from 16.1% to 2.16%, which can be used directly as sintering material, the amount of waste oil recovered is about 12 times as much as before reformation, the water quality of the leveling tank has been improved at the same time. The new technique has the characteristics of low investment, quick results, low cost, environmental protection and reduce carbon dioxide emission. The economic, social and environmental benefits of this technology are remarkable, which has great popularization value in metallurgical enterprises at home and abroad.

Key words: hot rolling sludge, air flotation, oil removal, horizontal flow sedimentation tank

高温钢渣在线工业化改性研究与实践

张亮亮，郭　冉，卢忠飞

（中冶建筑研究总院有限公司环保事业部，北京　100088）

摘　要： 针对钢渣水硬胶凝活性差，难以实现在水泥和混凝土中的大规模应用等现状，分析了钢渣水硬胶凝活性差的原因，研究了添加钙质和硅铝质材料对钢渣矿物组成和活性改善的影响，设计了适用于现有钢渣辊压处理生产线的熔融钢渣在线改性螺旋给料试验装置，并进行了熔融钢渣在线改性，取得了 28d 活性指数提高 10%的良好改性效果。

关键词： 钢渣，辊压，活性，改性

Research and Practice of On-line Modification of Molten Steel Slag

Zhang Liangliang, Guo Ran, Lu Zhongfei

(Central Research Institute of Building and Construction, MCC Group, Co., Ltd., Beijing 100088, China)

Abstract: As steel slag has poor cementitious property, this makes it difficult to achieve large-scale application in cement and concrete. The paper analyzed the reason of poor cementitious property for steel slag and studied the effect of calcareous material and aluminosilicate material to improve mineral composition and cementitious property of steel slag. Test equipments by screw-feed that apply to the existing product line of roll-in cracking of steel slag was designed, by which tests were carried out about on-line modification of molten steel slag and 28d activity index was creased 10%.

Key words: steel slag, roll-in cracking, cementitious property, modification

FeO 调控 $CaO\text{-}SiO_2\text{-}MgO\text{-}Al_2O_3\text{-}Cr_2O_3$ 体系尖晶石析出的热力学分析

余　岳，王　迪，李建立，朱航宇，薛正良

（武汉科技大学省部共建耐火材料与冶金国家重点实验室，钢铁冶金及资源利用
省部共建教育部重点实验室，湖北武汉　430081）

摘　要： 不锈钢渣是冶炼不锈钢的副产品，Cr_2O_3 的存在对钢渣的资源化利用造成了困难；而以尖晶石矿物赋存的铬是稳定的，可有效抑制不锈钢渣中铬的溶出行为。本文基于熔体的非平衡凝固理论，采用热力学数据库 FactSage 7.0 研究了 FeO 含量对 $CaO\text{-}SiO_2\text{-}MgO\text{-}Al_2O_3\text{-}Cr_2O_3$ 体系中尖晶石晶体析出行为的影响，得到了尖晶石晶体析出温度、析出量和化学组成及铬元素赋存状态的变化规律。计算结果表明尖晶石晶体为高温析出相，主要由 $MgCr_2O_4$ 和 $FeCr_2O_4$ 组成；随着 FeO 含量增加，析出温度逐渐降低，最终析出量逐渐增加。FeO 含量的增加促进了尖晶石固溶体中 $FeCr_2O_4$ 组元的增加而抑制了 $MgCr_2O_4$ 析出，但对铬元素的赋存状态未产生影响，仍以尖晶石形式存在。

关键词： 不锈钢渣，FeO，尖晶石，FactSage，非平衡凝固

Thermodynamic Calculation of FeO Effect on Precipitation of Spinel in CaO-SiO₂-MgO-Al₂O₃-Cr₂O₃ System

Yu Yue, Wang Di, Li Jianli, Zhu Hangyu, Xue Zhengliang

(Key Laboratory for Ferrous Metallurgy and Resources Utilization of Ministry of Education, The State Key Laboratory of Refractories and Metallurgy, Wuhan University of Science and Technology, Wuhan 430081, China)

Abstract: Stainless steel slag (SSS) is a by-product during stainless steel-making process, and contains a certain content of Cr_2O_3, which impedes the utilization of SSS as a secondary resource. The spinel mineral is known as a stable phase and effectively controls the elution of chromium. Based on the Scheil-Gullive equation, the thermodynamic software FactSage was employed to simulate the solidification process of CaO-SiO₂-MgO-Al₂O₃-Cr₂O₃ system, and then discuss the effect of FeO on the precipitating temperature, sediment amount and component of spinel phase and the existence of chromium. The calculation results show that the spinel precipitated at high temperature, and mainly consists of $MgCr_2O_4$ and $FeCr_2O_4$. When the content of FeO increases, the precipitation temperature gradually decreases, and the amount grows finally. FeO could promote the formation of $FeCr_2O_4$ component, and restrain the precipitation of $MgCr_2O_4$, but have few effect on the existing occurrence of chromium in the solid slag.

Key words: stainless steel slag, FeO, spinel, FactSage, non-equilibrium solidification

连续流分段进水 A/O 工艺在焦化废水中的应用

杨红军，杨庆彬，董凤杰，王　逊

（唐山首钢京唐西山焦化有限责任公司，河北唐山　063200）

摘　要：本文以连续流分段进水生物脱氮工艺为理论依据，在原有 AOO 工艺的基础上，通过改变进水方式，完善原系统的脱氮功能，使产水达到总氮达标的环保指标要求。同时，保证整个酚氰废水处理系统处于良好、有序的运转状态，达到提高环保效益，提高企业社会竞争力的目的。连续流分段进水的工艺理论成功地应用于处理高毒、高氨氮的焦化废水，也为焦化废水的处理技术提供了有益的参考和借鉴。

关键词：连续流分段进水，AO 工艺，焦化废水，生物脱氮

浅谈稳定化处理的钢渣作混凝土骨料应用技术

胡天麒，杨景玲，郝以党，吴　龙

（中冶建筑研究总院有限公司，北京　100088）

摘　要：我国是钢铁生产大国，钢渣堆存问题困扰着许多钢铁企业。利用钢渣代替天然砂石作骨料配制混凝土既可以节约混凝土生产成本，又可以快速消纳堆存的钢渣。本文介绍了钢渣热闷自解稳定化处理技术原理和特点。稳定化处理后的钢渣由于具有颗粒级配好、强度高、金属铁含量低、无有害物质、水化硬化等特点，是作混凝土骨料的

理想材料。

关键词： 钢渣，混凝土骨料，稳定化，热闷

冷轧废水污泥减量化实践

姚治国，袁留锁，孙贵锁

（北京首钢冷轧薄板有限公司，北京　101304）

摘　要： 通过对冷轧污泥产源的分析，在强化现有工艺设备的基础上，生化过程采用了 OSA 污泥减量化工艺，含酸系统采用了 PH、DO 双指标控制工艺，实现了污泥的治理的源头化。同时也进一步实践了，在保证出水水质前提下，通过提高生化进水有机物浓度减少预处理含油污泥产生量的可操作性。

关键词： 环保，冷轧，源头治理，污泥减量化

Practice of Sludge Reduction in Cold Rolling Wastewater

Yao Zhiguo, Yuan Liusuo, Sun Guisuo

(Beijing Shougang Cold Rolled Sheet Co., Ltd., Beijing 101304, China)

Abstract: Through the analysis of the cold rolling sludge source, based on strengthening the existing process equipment, biochemical process using the OSA sludge reduction process, acid system by PH and DO double index control technology, the source of sludge treatment. At the same time, under the premise of ensuring the quality of the effluent, the operability of reducing the amount of oily sludge generated by pretreatment is improved by improving the concentration of organic matters in the influent.

Key words: environmental protection, cold rolling, source treatment, sludge reduction

钢铁企业水处理专家管理系统

高康乐，李　红，逯博特

（中冶节能环保有限责任公司，北京　100088）

摘　要： 钢铁企业水处理专家管理系统，通过将企业工艺节水技术、污水处理技术和计算机技术、物联网技术以及大数据技术相结合，利用信息化建设实现钢铁企业水处理的精细化和智能化管理，提高水处理的效率，保障水处理的安全、稳定的生产运营；通过钢铁企业全流程水质水量平衡优化设计，实现企业内部供排水系统的最优化，达到钢铁企业节水减排的目标。

关键词： 钢铁企业，水处理专家系统，循环水，智能管理

Water Treatment Expert System of Iron and Steel Enterprise

Gao Kangle, Li Hong, Lu Bote

(Energy-saving and Environmental Protection Co., Ltd., MCC, Beijing 100088, China)

Abstract: The water treatment expert system of iron and steel enterprise, through combination of process water-saving technology, wastewater treatment technology and computer technology, internet technology, big data technology. Using information construction to realize the accurate and intelligent management of water treatment of iron and steel enterprises, improve the efficiency of water treatment, and guarantee safe and stable operation of water treatment. Through the balance optimization design of quality and quantity of water, achieve the optimization of water supply and drainage system inside the enterprise, and finally achieve the goal of water saving and emissions reduction for iron and steel enterprises.

Key words: iron and steel enterprise, water treatment expert system, circulating water, intelligent management

转炉一次烟气高温膜管干法除尘及余热回收系统研究开发

杨 倩[1]，杨慧斌[1]，杨印东[2]，李 宇[3]，麦克琳[2]

（1. 秦皇岛同力达环保能源股份有限公司，河北秦皇岛 066004；2. 加拿大多伦多大学材料科学与工程系，184 College Street, Toronto, Canada M5S 3E4；3. 北京科技大学钢铁冶金国家重点实验室，北京 100083）

摘 要： 在中试取得的数据基础上，对转炉一次除尘工艺做出了全新的干法工艺设计，把传统的喷雾冷却塔设计成汽化冷却高温换热降尘塔，回收余热，同时粗除尘。粗除尘后的烟气进入高温膜除尘器精除尘，除尘器内置汽化冷却骨架梁，在除尘过滤的同时回收转炉煤气里的显热。与传统湿法除尘工艺相比，可多回收蒸汽 50～70kg/t 钢，系统运行阻力是传统湿法除尘系统的 1/3，一次风机电机耗电降低 2/3，实现纯干法回收煤气、除尘。这项技术已经被京津冀钢铁行业节能技术创新联盟推荐为先进适用技术。

关键词： 高温膜管，高温换热降尘塔，煤气换热冷却塔，回收蒸汽，回收煤气，干法除尘

Gas Cleaning and Heat Recovery System for Flue Gas of Converter Using High Temperature Membrane

Yang Qian[1], Yang Huibin[1], Yang Yindong[2], Li Yu[3], McLean A.[2]

Abstract: On the basis of the data obtained by the pilot test, a new dry gas cleaning process was developed for the primary dust removal of converter flue gas. The traditional spray-cooling tower was changed to the high temperature tower with vaporization cooling for heat exchange and dust removal, in which the waste heat was recovered and coarse dust was collected. Fume gas after deducting was then introduced into the high temperature membrane dust collector. A vaporization cooling skeleton beam was built in the dust collector. During dust collecting through the filter, recovery of the sensible heat in the converter gas was carried out. Comparing with the traditional dust recovery process, steam recovery increased by 50 ~70kg/t steel; Running resistance is 1/3 of the traditional wet dust removal system; A fan motor power consumption reduced by 2/3, and a pure dry recovery of gas and dust was achieved. This technology has been recognized as advanced and applicable technology in the field.

Key words: high temperature membrane tube, waste heat recovery, gas cleaning, dust removal

转炉钢渣的理化特性分析及其磷元素的富集技术分析

唐卫军，张德国

（北京首钢国际工程技术有限公司，北京市冶金三维仿真设计工程技术研究中心，北京　100043）

摘　要： 本文依据钢铁冶金理论论述了转炉钢渣的产生机理，综合分析了转炉钢渣的物理性质、化学性质以及矿物组成，特别分析了转炉钢渣中磷元素的分布及其在转炉钢渣中的矿物特征。在此基础上，综合评价了当前转炉钢渣降磷的富集分离技术，为钢渣在冶金企业内部循环利用提供了技术依托。

关键词： 转炉钢渣，理化特性，磷，富集

Physical and Chemical Characteristics of Converter Slag and the Concentration Technology for Phosphorus from Converter Slag

Tang Weijun, Zhang Deguo

(Beijing Shougang International Engineering Technology Co., Ltd., Beijing Metallurgical 3-D Simulation Design Engineering Technology Research Center, Beijing 100043, China)

Abstract: Based on the theory of iron and steel metallurgy, the occurrence of converter slag was investigated, and its physical properties, chemical properties and mineral composition were analyzed. Especially, the distribution of phosphorus in converter slag and its mineral characteristics were discussed. Accordingly, the concentration and separation technology for dephosphorization of converter slag was assessed, which provides a technical support for the internal recycling of steel slag in metallurgical enterprises.

Key words: converter slag, physical and chemical characteristics, phosphorus, concentration

粗灰回炉在安钢转炉干法除尘中的应用

曹树卫，谷　洁，李军强

（安阳钢铁股份有限公司第二炼轧厂，河南安阳　455004）

摘　要： 安阳钢铁股份有限公司（以下简称安钢）第二炼轧厂将三座 150t 转炉湿法除尘系统改造为干法电除尘。改造完成后，转炉一次烟气达标排放，能源消耗降低。蒸发冷却器捕集的粗灰回炉，有效利用固废资源、降低生产

成本、减少二次污染。为实现安钢"绿色转型、生态发展"的战略目标做贡献。

关键词：干法除尘，粗灰回炉，机械输灰，气力输送

The Application of Thick Dust Melt Down in Dry Process Dust Precipitation for Converter in Anyang Iron and Steel Group Co., Ltd.

Cao Shuwei, Gu Jie, Li Junqiang

(Anyang Iron and Steel Group Co., Ltd., No.2 Steel Making & Rolling Mill，Anyang 455004, China)

Abstract: It will be remake three 150t converter from wet dust precipitation to dry process dust precipitation to Anyang Iron and Steel CO.,Ltd (hereinafter referred to as Angang) No.2 Steel Making & Rolling Mill.After the remake,primary exhaust gas of converter meets the standard discharge,reduced energy consumption.Melt down evaporative cooler gathered thick dust melt down for converter can effective use of solid waste resources, lower production cost and reduce pollution two times.To realize the strategic goal of "green transformation and ecological development" of Anyang company.

Key words: dry process dust precipitation, thick dust melt down for converter, mechanical transportation dust, pneumatic transmission

雾化喷淋除尘技术在翻包区域的除尘设计应用

张　明，王兴敏

（武钢集团昆明钢铁股份有限公司安宁分公司）

摘　要： 新区炼钢厂在进行翻包倒渣作业时，产生阵发性粉尘，现有收集罩收集效果不佳，造成大量粉尘逸散，对环境保护及操作人员危害较大，基于炼钢厂翻包区域的现有装备和实际生产情况，采用雾化喷淋除尘技术对翻包区域除尘系统进行优化设计改造，通过生产及应用实践，改造取得明显效果，有效改善翻包区域的工作环境。

关键词： 新区炼钢厂，粉尘，雾化喷淋除尘设计，降尘效果

我国钢铁企业固体废弃物资源化处理模式和发展方向

张寿荣[1]，张卫东[2]，姜　曦[3]

（1. 武汉钢铁（集团）公司，湖北武汉　430083；2. 首钢技术研究院，北京　100043；
3. 中国钢铁工业协会，北京　100711）

摘　要： 本文论述了我国钢铁企业可持续发展的建设理念和绿色环保的新特点，讨论了钢铁厂从焦化、球团、烧结、炼铁、炼钢和轧钢及配套系统全工序产生的固废产物种类和相关冶金特性，研究不同种类固废产物的处理模式。重点针对工程堆放和外销类固体废弃资源，深入探讨"含有害元素除尘灰处理方式"、"危险废弃物处置"和"冶金渣高

温转换与余热回收方式"的难题，并围绕这些难题创造性地提出了钢铁企业固废产物大平台处理的总体技术发展路线，希望能为今后我国的钢铁企业固体废弃物实现"零排放"起到促进作用。

关键词：钢铁企业，固体废弃物，再利用，零排放

Solid Waste Resources Treatment Mode and Development Tendency of Iron and Steel Enterprises in China

Zhang Shourong[1], Zhang Weidong[2], Jiang Xi[3]

(1. Wuhan Iron and Steel (Group) Corp., Wuhan 430083, China; 2. Shougang Research Institute of Technology, Beijing 100043, China; 3. China Iron and Steel Association, Beijing, 100711, China)

Abstract: This paper elaborated the sustainable development concepts and new environment-friendly features of iron and steel enterprises in China. As to a major iron and steel plant, material types and metallurgical properties of solid wastes produced from coking, pellet, sinter, iron-making, steel-making, roll and affiliated processes in a certain steel plant were discussed comprehensively, while the corresponding treated measures were also researched. Focused on the solid waste of engineering disposal and export sale from steel plant, this paper also discussed the major bottleneck tasks of dusting methods of containing harmful elements, hazardous waste disposal and high temperature slag waste heat recovery, and represented a new solid waste treatment technology development route for iron and steel enterprises. It was expected to make a further contribution to the zero emission of solid waste treatment for iron and steel enterprises in China.

Key words: iron and steel enterprises, solid waste, reuse, zero emission

不锈钢渣中铬的深度还原实验研究

吴　拓，张延玲，袁　方，张康晖

（钢铁冶金新技术国家重点实验室，北京　100083）

摘　要：针对某企业的不锈钢渣的分析表明，（Fe+Cr+Mn）含量为 12.8%，其中 Cr 主要存在于含铬尖晶石相中。在小型感应炉中进行了不锈钢渣的高温还原实验，重点考察了还原剂种类以及加入方式、终点渣碱度以及 Al_2O_3 含量、温度对渣中残铬以及铬还原率的影响。结果表明，1550℃下，加入固态还原剂后渣中铬显著降低，Al 和过量的硅铁可将渣中的铬还原至 0.12%；以 C 为主要还原剂时，配加少量的硅铁可以促进铬的还原。本实验条件下，最佳渣碱度为 1.2，Al_2O_3 含量对铬的还原影响不大，温度越高越有利于降低渣中的铬含量。在 1650℃下，以 Fe-2.0C 为金属熔池，按 n_C：n_O=1.0 和 n_{Si}：n_O=0.1 配加还原剂，调节渣碱度为 1.2 后，得到的 Fe-7.5%Cr-C 合金可以作为不锈钢冶炼的原料使用。

关键词：不锈钢渣，铬回收，深度还原

Experimental Investigation on the Deep Reduction of Cr in Stainless Steel Slag

Wu Tuo, Zhang Yanling, Yuan Fang, Zhang Kanghui

(State Key Laboratory of Advanced Metallurgy, Beijing 100083, China)

Abstract: An analysis for the stainless steel slag from an enterprise showed that the total content of (Fe+Cr+Mn) is 12.8%, and Cr mainly existed in the Cr-containing spinel phase. A small induction furnace was used to carry out the high temperature reduction experiment on the stainless steel slag. The effects of reductants type and addition way, slag basicity, Al_2O_3 content and temperature on the residual Cr and chromium reduction rate were investigated. The results indicated that solid reductants significantly reduced the chromium in the slag, especially aluminum and excess SiFe could reduce the chromium to 0.12% at 1550℃. If C was chosen as the main reductant, additional small amount of ferrosilicon promoted the reduction of chromium. Under the current experimental conditions, the optimum slag basicity was 1.2, and the content of Al_2O_3 had little effect on the reduction of chromium. The increase in temperature was beneficial for reducing residual Cr in the slag. At 1650℃, with Fe-2.0C as the metal pool, and with reductants by $n_C : n_O = 1.0$ and $n_{Si} : n_O = 0.1$, adjusting slag basicity to 1.2 , the Fe-7.5% Cr-C alloy can be obtained and will be used as a raw material for stainless steel smelting.

Key words: stainless steel slag, recovery of Cr, deep reduction

散料输送喷雾除尘

宋会江

（斯普瑞喷雾系统（上海）有限公司，上海　201612）

摘　要：本文介绍了喷雾除尘机理，探讨了喷雾除尘技术在散料输送的皮带输送机、料仓落料点的应用，湿式除尘液体的配制。

关键词：喷雾除尘，转运点，落料点

The Application of Spraying De-dust Technology in Bulk Material Transfer Process

Song Huijiang

(Spraying Systems (Shanghai) Co., Ltd., Shanghai 201612, China)

Abstract: This paper introduces spraying de-dust mechanism, the typical application of spraying de-dust technology in bulk material transfer process, such as belt transferring point and material bin downloading point.

Key words: spraying de-dust, downloading point, transfer point

[Omim][BF₄]离子液体萃取铬渣酸浸液中 Cr(VI)

魏君怡，李　勇，何金桂

（东北大学冶金学院，辽宁沈阳　110819）

摘　要：离子液体作为一种新型的绿色溶剂，在重金属离子萃取分离方面较传统的有机溶剂有显著的优势。本文讨论了 1-辛基-3-甲基咪唑四氟硼酸盐离子液体（[Omim][BF₄]）萃取 Cr(VI) 的影响因素，包括相比（O/A）、萃取时间、萃取温度、pH。研究结果表明，最佳工艺条件为：相比 O/A≥0.2、时间为 4.0 min、温度为 25.0℃、pH=2.00~5.38，

并且[Omim][BF$_4$]对于铬渣酸浸液中的 Cr(VI)具有较好的萃取能力，为铬渣的绿色资源化利用开辟了新的途径。

关键词：铬渣酸浸液，离子液体，六价铬，萃取

Extraction Chromium(VI) from the Chromium Slag Acid Leaching Solution by 1-octyl-3-methylimidazolium Tetrafluoroborate

Wei Junyi, Li Yong, He Jingui

(School of Metallurgy, Northeastern University, Shenyang 110819, China)

Abstract: As a new type of green solvent，ionic liquid shows remarkable advantages in the extraction of heavy metal ions comparing with traditional organic solvents. This review discussed the influencing factors of 1-octyl-3-methylimidazolium tetrafluoroborate([Omim][BF$_4$]) for extraction process, which included phase ratio(O/A), extraction time, extraction temperature, pH value. The results showed that the best experimental conditions are on the extraction phase ratio(O/A)≥0.2, shaking time 4.0 min，temperature on the extraction 25.0 ℃, the leaching solution pH=2.00~5.38. The ionic liquid has good effect on the extraction efficiency for the extraction of chromium(VI) with [Omim][BF$_4$] and presents a new approach of its green recycling application.

Key words: chromium slag acid leaching solution, ionic liquid, chromium(VI), extraction

浅谈生物生化法处理焦化废水及常见问题的处理措施

翟伟光

（本钢集团北营公司焦化厂，辽宁本溪　117017）

摘　要：介绍了本钢集团北营公司焦化厂三区污水的可生化性，通过预处理、生化处理、混凝沉淀、污泥处理四个方面来叙述生物生化法的处理系统，并在本文中详细论述了在污水处理的过程中所涉及到的关键因素和在运行过程中所遇到的问题原因及解决措施。

关键词：废水，生物，污泥，措施

烧结炼铁协同处置含铬污泥的应用研究

张　垒，刘尚超，张道权，李　军，冯红云，罗之礼

（武汉钢铁有限公司，湖北武汉　430080）

摘　要：通过烧结炼铁规模化试验考察了配入含铬污泥后对烧结工况、产品和烧结、炼铁环保指标的影响。试验结果表明，当铬泥配入比例低于 0.05%，对烧结工况、烧结产品和炼铁工序环保无明显影响；对铬平衡进行了分析，表明铬泥中的铬元素是通过烧结矿流向铁水，说明烧结炼铁协同处置含铬污泥不会对烧结和炼铁环保带来二次污染。

关键词：含铬污泥，烧结炼铁，协同处置，铬平衡

Application Research on Sintering and Ironmaking Co-Disposal of Chromium-Containing Sludge

Zhang Lei, Liu Shangchao, Zhang Daoquan, Li Jun, Feng Hongyun, Luo Zhili

(Wuhan Iron & Steel Co.,Ltd., Wuhan 430080, China)

Abstract: The effects of chromium-containing sludge on sintering conditions, products, sintering and ironmaking environmental protection influences were investigated by scale test of sintering and ironmaking. The results showed that when the proportion of chromium-containing sludge was less than 0.05%, there were no obvious influences on the sintering condition, sintering products and the environmental protection of ironmaking process. Chromium balance was analyzed, indicating that chromium in chromium-containing sludge flowed from sintering section to ironmaking section, indicating that the sintering and ironmaking co-disposal of chromium-containing sludge would not bring secondary pollution to sintering and ironmaking processes.

Key words: chromium-containing sludge, sintering and ironmaking, co-disposal, chromium balance

全膜法在焦化废水深度处理及回用技术中的应用

姜　楠[1]，邢　磊[2]

（1. 中冶京诚工程技术有限公司，北京　100176；2. 唐山港陆钢铁有限公司，河北唐山　063016）

摘　要： 以实际工程为例，通过采用"二级砂滤+微滤+纳滤+反渗透"为核心的焦化废水深度处理及回用技术，提高厂区废水回收率，减少厂区废水排放量，降低新水耗水量，同时通过 PDMS 三维技术，进一步降低成本，缩短设计工期。为焦化废水深度处理及回用的设计提供指导参考。

关键词： 全膜法，焦化废水，深度处理，回用，PDMS

Application of Membrane Process in Advanced Treatment and Reuse of Coking Wastewater

Jiang Nan[1], Xing Lei[2]

(1. Capital Engineering and Research Incorporation Limited, MCC, Beijing 100176, China;

2. Tangshan Ganglu Iron and Steel Co.,Ltd., Tangshan 063016, China)

Abstract: In practical engineering, for example, through the adoption of "sand filter + microfiltration + nanofiltration+ reverse osmosis" as the core of the coking wastewater advanced treatment and reuse technologyto improve plant wastewater recovery, reduce plant wastewater discharge, reduce water consumption. At the same time through the PDMS three-dimensional technology to further reduce costs and shorten the design duration. In order to provide reference for coking wastewater advanced treatment and reuse.

Key words: membrane process, coking wastewater, depth treatment, reuse, PDMS

焦化废水低成本深度处理回用技术

张功多，安路阳，张立涛，宋迪慧，孟庆锐，尹健博

（中钢集团鞍山热能研究院有限公司环境工程院士专家工作站，辽宁鞍山　114044）

摘　要：采用电化学和催化湿式过氧化氢氧化(CWPO)组合技术作为深度处理工艺，处理不同地区焦化废水，表现出良好的处理效果和较低的运行成本。其中 COD 的去除率可达 66.1%~82.7%，TOC 的去除率在 65.3%~81.1%，悬浮物去除率 100%，色度的去除率在 93.8%~96.87%。该组合工艺吨水处理成本不高于 5 元，出水基本符合工业循环冷却水或再生水标准，可用于生化稀释水或循环冷却水补充水，实现了工业废水的循环利用。

关键词：电化学，CWPO，深度处理，循环利用

Study on Low Cost Advanced Treatment and Reuse Technology of Coking Wastewater

Zhang Gongduo, An Luyang, Zhang Litao, Song Dihui, Meng Qingrui, Yin Jianbo

(Academician Experts Workstation, Sinosteel Anshan Research Institute of
Thermo-energy Co., Ltd., Anshan 114044, China)

Abstract: Electrochemical and catalytic wet peroxide oxidation (CWPO) combined process has been applied to advanced treatment of coking wastewater with different water quality. Showed excellent treatment effect and lower operating cost. The removal rate of COD was 66.1%~82.7%, the TOC removal rate was 65.3%~81.1%, the suspended matter removal rate was 100% and the chroma removal rate was 93.8%~96.87%. Treatment of a ton of water costs less than 5 yuan and the effluent could be compliant with the standard for industrial circulating cooling water or the reclaimed water, and could be used for the water supplement of the biochemical dilution water or circulating cooling water, and the recycling utilization of industrial waste water was realized.

Key words: electrochemistry, CWPO, advanced treatment, recycling

烧结烟气污染物处理技术的发展趋势与展望

王　永[1]，金增玉[2]，张天赋[1]，袁　玲[1]，王　飞[1]

（1. 鞍钢股份有限公司技术中心，辽宁鞍山　114009；
2. 鞍山市环境保护研究所，辽宁鞍山　114009）

摘　要：随着环保的要求越来越严格，烧结烟气除了脱硫还需要脱硝，目前仅仅脱硫工艺已经不适用当前的环保要求。文章首先对燃煤电厂烟气与烧结烟气的特点进行分析，并对目前应用在烧结烟气脱硫工艺进行分析，指出其工艺的优点及不足，并对未来最适合烧结烟气脱硫脱硝的工艺提出一些看法。

关键词：烧结烟气，脱硫，脱硝，脱硫脱硝

The Development Trend and Prospect of Sintering Flue Gas Pollutants Treatment Technology

Wang Yong[1], Jin Zengyu[2], Zhang Tianfu[1], Yuan Ling[1], Wang Fei[1]

(1. Technology Center of Angang Steel Co., Ltd., Anshan 114009, China; 2. Anshan City Institute of Environmental Protection, Anshan 114009, China)

Abstract: According to the striking environment requirement, sintering flue gas treatment is not only including desulfurization, but DeNO$_x$ and most of existing desulfurization processes does not meet the requirement. In this paper, the analyze is made on the nature of flue gas both in coal-fired power plant and sinter and existing sintering flue gas desulfurization process, the advantages and shortcoming are put forward, and a few suggestions are given on future suitable desulfurization process in sintering process.

Key words: sintering flue gas, desulfurization, DeNO$_x$, desulfurization and DeNO$_x$

Fenton 氧化-焦粉吸附深度处理焦化废水的研究

陈 鹏，胡绍伟，王 飞，刘 芳

（鞍钢集团钢铁研究院环境保护研究室，辽宁鞍山 114009）

摘 要：以实际焦化废水经 A^2/O 工艺处理后的出水为研究对象，试验采用 Fenton 氧化-焦粉吸附工艺深度处理焦化废水。探讨了 Fenton 氧化阶段 H$_2$O$_2$ 加入量、Fe^{2+} 加入量、进水 pH 值、反应时间以及吸附阶段焦粉投加量、废水 pH 值和吸附时间等因素对焦化废水 COD 和色度处理效果的影响。在综合考虑经济性和去除效果的前提下，提出工艺的最佳操作条件：Fenton 氧化阶段 H$_2$O$_2$ 加入量为 260mg/L、Fe^{2+} 加入量为 230mg/L、进水 pH 值为 3.5、反应时间为 30min，吸附阶段焦粉投加量为 90g/L、废水 pH 值为 9.0、吸附时间为 1h。在最佳操作条件下，最终 COD 去除率可达 81.6%，色度降为 22 倍，工艺出水水质稳定，水质可以达到《辽宁省污水综合排放标准》(DB21/1627—2008)，同时试验结果为该集成工艺的工业化应用提供了实验依据。

关键词：Fenton 氧化，焦粉，深度处理，焦化废水

Research on Advanced Treatment of Coking Wastewater by Fenton Oxidation and Coke Powder Adsorption

Chen Peng, Hu Shaowei, Wang Fei, Liu Fang

(Iron & Steel Research Institute of Ansteel Group, Anshan 114009, China)

Abstract: The article mainly investigates the Fenton oxidation and coke powder adsorption technology advanced treat coking wastewater after A^2/O technology, the integration technology is adopted by Fenton oxidation and coke powder adsorption. Studying on influence factors of Fenton oxidation and coke powder adsorption. on the COD removal and chroma, it contains H$_2$O$_2$ dosage, Fe^{2+} dosage, pH of wastewater, reaction time and coke powder dosage, adsorption time and so on. The experiment results show that discharge of waste water is stable and water quality can meet local emission standards of Liaoning Province (DB21/1627—2008) when H_2O_2 dosage is 260mg/ L, Fe^{2+} dosage is 230mg/ L, pH of wastewater is 3.5, reaction time is 30min, coke powder dosage is 90g/L, pH of wastewater is 9.0, adsorption time is 1h, removal efficiency of COD is 81.6% and chroma is reduced to 22 times. The experimental result provides basis for industrial application of the process.

Key words: Fenton oxidation, coke powder, advanced treatment, coking wastewater

冶金煤气干法除尘技术创新与成就

张福明

（首钢集团有限公司总工程师室，北京　100041）

摘　要： 高炉炼铁和转炉炼钢过程中所产生的高炉煤气和转炉煤气是钢铁厂重要的二次能源，冶金煤气干法除尘是钢铁制造绿色发展的关键共性技术。首钢京唐钢铁厂采用了高炉煤气和转炉煤气干法除尘技术，攻克了 5500m³ 特大型高炉和 300t 转炉煤气干法除尘国际性技术难题，实现了世界首套 5500m³ 特大型高炉煤气纯干法除尘、首套铁水"脱硫-脱硅-脱磷"工艺条件下 300t 转炉煤气干法除尘重大技术突破和工业化稳定应用。本文结合首钢京唐钢铁厂工程设计研究，论述了特大型高炉和转炉煤气干法除尘的技术创新及其成就。

关键词： 高炉，转炉，煤气干法除尘，煤气余压发电，节能减排，循环经济

Achievement and Innovation of Metallurgical Gas Dry De-dusting Technology

Zhang Fuming

(Shougang Group Co., Ltd., General Engineer Office, Beijing 100041, China)

Abstract: Metallurgical gas includes blast furnace gas (BFG) and basic oxygen furnace gas (BOFG), produced by blast furnace ironmaking and basic oxygen furnace steelmaking. Metallurgical gas is an important secondary energy in steel plant, the BFG and BOFG dry dedusting process are the key common technology of steel manufacturing green development. The BFG and BOFG dry dedusting technologies were applied in Shougang Jingtang steel plant, the international technological difficult problems of 5500m³ huge blast furnace and 300t converter metallurgical gas dry dedusting technology have been solved. The major breakthrough and stable industrialization application of the world's first 5500m³ huge blast furnace gas dry dedusting technology and the first example under the conditions of hot metal desulfurization-desilication-dephosphorization pretreatment 300t converter gas dry dedusting technology have been achieved. In this paper, combine with the study on engineering design of Shougang Jingtang steel plant, the technological innovation and achievement of BFG and BOFG dry dedusting process are discussed.

Key words: blast furnace, converter, gas dry dedusting, top gas recovery turbine, energy saving and emission reducing, circulating economy

自动化控制系统在环境除尘中的应用

何志龙

（北京首钢自动化信息技术有限公司自动化事业部，北京　100141）

摘　要：完善的自动化系统在工业生产中的应用作用越来越重要，工业生产中的环境保护越来越受到重视，本文探讨的就是自动化控制系统在环境除尘中的典型应用。本文还介绍了一些自动化系统中用到的小技巧。

关键词：自动化控制，电除尘，施耐德，二进制

Application of Automatic System in the Environmental Dust Removal

Abstract: Good automatic system is more important in the industrial production. The environmental protect is more seriously in the industrial production. The paper discusses the typical application of automatic system in the environmental dust removal. Some skill of automatic system are also introduced in this article.

Key words: automatic control, electrostatic precipitator, Schneider, binary system

半干法烧结烟气脱硫灰回配烧结技术研究

李小丽，金勇成

（鞍钢股份有限公司炼铁总厂，辽宁鞍山　114021）

摘　要：半干法脱硫工艺从 2000 年以来已成为火电机组、烧结烟气脱硫的主导方向，脱硫灰的利用问题已经制约半干法脱硫工艺的使用及推广。本篇文章介绍了鞍钢炼铁总厂旋转喷雾半干法烧结脱硫工艺（SDA）脱硫灰的产生、特性以及污染现状，总结了当前国内外对脱硫灰综合利用的现状和利用过程中存在的问题，提出了烧结脱硫灰加配烧结的新途径，从而使之变废为宝，并降低烧结成本。

关键词：半干法，烧结，脱硫灰，回配烧结

Study on Sintering Technology of Semi Dry Sintering Flue Gas Desulphurization Ash

Li Xiaoli, Jin Yongcheng

(Ironmaking Plant of Anshan Steel Co., Ltd., Anshan 114021, China)

Abstract: Semi dry desulphurization process has become the main direction of thermal power unit and sintering flue gas desulphurization since 2000. The use of desulphurization ash has restricted the use and promotion of semi dry desulphurization process. This article introduces the general iron making plant of Angang rotary spray semi dry sintering desulphurization (SDA) generation, characteristics and pollution status of desulphurization ash, summarizes the problems existing in the process of the current situation of comprehensive utilization of desulphurization ash and utilization at home and abroad, puts forward new ways of sintering FGD ash with sintering, thereby turning waste into treasure, and reduce the sintering cost.

Key words: semi dry, sintering, desulphurization ash, back sintering

钢铁厂酸洗废液处理技术及邯钢研究现状

王晓晖，朱文玲

（河钢集团邯钢公司技术中心，河北邯郸　056015）

摘　要： 在对钢材表面进行加工处理时，酸洗工序可以改善钢材表面结构，利用强酸的腐蚀作用对钢材表面进行清洗而生成酸洗废液。钢铁酸洗废液具有腐蚀性，其中含有可回收利用的大量酸和铁资源。按酸洗废液的回收方法的不同，简要介绍了目前酸洗废液的回收和利用的研究进展。

关键词： 酸洗废液，资源回收，研究进展

Wastewater Treatment of Steel Pickling Waste Liquors and Research Development in Hangang

Wang Xiaohui, Zhu Wenling

(Hesteel Group Hansteel Company, Technology Center, Handan 056015, China)

Abstract: For improving the structure of the steel surface or the surface processing，the strong acid were used to cleaning the steel surface. Steel picklingwaste acid liquor，a kind of bad corrosion contaminant，contains a great deal of reproducible acid and ferrous ion. According to the difference of recycling method of steel hydrochloric acid pickling waste liquid，the article gives brief introduction of recent research development.

Key words: waste acid liquor, recycle, research development

浅析钢铁企业烧结烟气脱硫几种适用方法

程　岗，刘　鑫

（陕西钢铁集团龙钢公司，陕西韩城　715400）

摘　要： 烧结烟气脱硫是目前环保治理的一个重点工作，排放不达标，对大气污染尤为严重，针对目前钢铁企业炼铁烧结烟气脱硫的类别及适用，例举三种不同的烧结烟气脱硫方法，通过对比，分析，走访，学习探索出我公司烧结烟气脱硫适合于烧结烟气—石灰石—石膏湿法的烧结烟气脱硫处理办法，通过近几年的运行，不断改造完善，在实际操作运行中不断提高，改进，使钢铁企业烧结烟气脱硫更高效，环保。

关键词：钢铁企业，烧结烟气，废气处理，脱硫效率，脱硫类别

浅谈镍铁渣综合回收利用的方法

余　杰，王万林，周乐君

（中南大学冶金与环境学院，湖南长沙　410000）

摘　要：镍铁渣为我国继铁渣、钢渣、赤泥之后第四大冶炼工业废渣，其综合利用已迫在眉睫。目前，国内对镍铁渣综合利用研究的主要方向多数集中在建材方面，在别的方面也有一定研究。例如：镍铁渣在生产微晶玻璃、无机纤维、保温砖、无机盐等方面也有相关应用。笔者在查阅近年来国内外有关镍铁渣综合利用研究的大量文献之后，总结出本文。

关键词：镍铁渣，综合回收，水泥，微晶玻璃，无机纤维，保温砖

The Comprehensive Recovery and Utilization Ways of Ferronickel Slag

Yu Jie, Wang Wanlin, Zhou Lejun

(School of Metallurgy and Environment in Central South University, Changsha 410000, China)

Abstract: Ferronickel slags has become the fourth-largest industry waste residues after iron slags, steel slag, red mud in China, and its comprehensive utilization is imminent. At present, the main direction of the comprehensive utilization of domestic ferronickel slags are concentrated on building materials, but there are other aspects of the research. For example, there are some applications in producing glass ceramics, inorfil, insulating brick, inorganic salt and so on. After consulting a lot of literatures in recent years at home and abroad on the comprehensive utilization of nickel slag, the auther summarized this paper.

Key words: ferronickel slags, comprehensive recovery and utilization, cement, glass ceramics, inorfil, insulating brick

电炉粉尘中湿法回收锌时铁酸锌的有效处理方法研究

李　进[1]，张　梅[2]

（1. 西安建筑科技大学冶金工程学院，陕西西安　710055；
2. 北京科技大学冶金与生态工程学院，北京　100083）

摘　要：炼钢电炉的粉尘中含有较多的锌。由于粉尘中大量的锌以稳定的铁酸锌形式存在，使得锌不易浸出、回收。此外，残余锌的粉尘污染环境，处理代价昂贵。近年来，国内外开发了许多有效的新工艺从电炉粉尘中浸出回收锌。本文通过归纳分析的方法对这些处理铁酸锌以浸出锌的新工艺和新方法进行了全面的总结、分类，指出浸出锌的关键问题是活化铁酸锌中的锌，可以通过物理方法活化或通过焙烧把铁酸锌转化为易溶结构的化学方法。同时，对这些浸锌工艺的特点进行了分析，为经济、环保地回收电炉粉尘中的锌提供了更全面的解决思路。

关键词：电炉粉尘，铁酸锌，锌的浸出，碱浸，酸浸

Study on Effective Zinc Ferrite Processing Methods for Zinc Leaching Recovery from EAFD

Li Jin[1], Zhang Mei[2]

(1. School of Metallurgical Engineering, Xi'an University of Architecture and Technology,
Xi'an 710055, China; 2. School of Metallurgical and Ecological Engineering, Beijing University
of Science and Technology, Beijing 100083, China)

Abstract: Electric arc furnace dust (EAFD) contains a high level of zinc. However, zinc can hardly be extracted because of the stable structure of zinc ferrite in the dust. This makes the residue difficult and costly to be dealt with. Some new routes involving leaching are effective to recover zinc from EAFD. The new zinc ferrite processing methods are inductively analyzed in this paper and the effective ways are classified. It is pointed that there are two kinds of effective methods to destroy the stable structure of zinc. One is physical activation method and the other one is chemical method in which zinc ferrite is converted to other soluble chemical form. Moreover, the characteristics of the processing methods are analyzed and discussed to give comprehensive information to design more economic and environmental friendly, more effective recovery route of zinc from EAFD.

Key words: EAF dust, zinc ferrite, zinc leaching, alkaline leaching, acid leaching

焦化厂 VOCs 综合治理新技术研究

杨邵鸿[1]，孙利军[2]

（1. 本溪北营钢铁（集团）股份有限公司焦化厂，辽宁本溪　117017；2. 武汉科林精细化工有限公司，湖北武汉　430000）

摘　要： 本溪北营钢铁（集团）股份有限公司焦化厂（以下简称北营焦化厂）冷鼓、脱硫、硫铵、粗苯、油库等区域焦油各类储槽、苯各类储槽以及氨水各类储槽对空排放尾气，尾气中含有氨、苯类、酚类、氰化物等有害气体，属于挥发性有机化合物(volatile organic compounds，即 VOCs)，危害人体健康，不符合《炼焦化学工业污染物排放标准》（GB16171—2012）。因此，必须对该部分污染气体配置处理装置进行综合治理。

关键词： VOCs，综合治理，新技术，研究

Research on New Technology of Comprehensive Treatment of VOCs in Coking Plant

Yang Shaohong[1], Sun Lijun[2]

(1. Benxi Beiying Steel (Group) Co., Ltd., Coking Plant, Benxi 117017, China; 2. Wuhan Colin Fine Chemical Co., Ltd., Wuhan 430000, China)

Abstract: Beiying coking plant, such as cold drum, desulfurization, ammonium sulfate, crude benzene, oil depot and other regional tar various types of storage tanks, benzene and various types of storage tanks and ammonia storage tanks on the exhaust emissions, exhaust gas containing ammonia, Phenols, cyanide and other harmful gases, are volatile organic compounds (ie VOCs), endangering human health, does not meet the "coking chemical industry pollutant discharge standards"（GB16171—2012）,so it must study the comprehensive management.

Key words: VOCs, comprehensive management, new technology, research

强化管理和治理，提高焦化企业环保水平

郭晓强

（本溪北营钢铁（集团）股份有限公司，辽宁本溪　117017）

摘　要：焦化企业生产过程中因其污染源排放点多、面大，环保工作压力相应地较大。强化环保工作管理和治理，提高焦化企业的环保管理水平，是当前焦化企业谋求长久发展的关键性因素。

关键词：管理，治理，增强，检查考核，改进

Strengthening Management and Governance and Improving the Environmental Protection Level of Coking Enterprises

Guo Xiaoqiang

(Coke-making Plant of Benxi Beiying Iron and Steel (Group) Co., Ltd., Benxi 117017, China)

Abstract: In the process of coking enterprise production, the pressure of environmental protection work is correspondingly larger because of the large number of emission sources and large surface. Strengthening the management and improvement of environmental protection work and improving the environmental protection management level of coking enterprises are the key factors for the long-term development of coking enterprises.

Key words: management, governance, enhancement, inspection and assessment, improvement

MBBR 改进型焦化废水两级 A/O 处理试验研究

赵二华，余云飞，杨世辉

（中冶赛迪技术研究中心有限公司，重庆　401122）

摘　要：针对焦化废水处理工艺抗冲击能力差，总氮、氨氮、COD 去除效率低，运行费用高等问题，采用流化床生物膜反应器（Moving Bed Biofilm Reactor, MBBR）改进型焦化废水低成本组合型两级 AO 工艺开展试验研究。研究表明该工艺在不加消泡水或稀释水情况下，可实现 COD 从 4000~4600 mg/L 降低到 250mg/L；通过 Fenton 氧化工艺，提高废水生化性，废水 B/C 值从 0.08~0.1 提高至 0.4~0.6，从而减少外加碳源（葡萄糖）投加量至 100mg/L，实现出水总氮≤20mg/L，氨氮≤5.2mg/L，运行费用 5.96 元/m³。

关键词：焦化废水，MBBR-活性污泥法，两级 A/O，Fenton 氧化

Study on Decontaminating Coking Wastewater Adopting Two Stage Anoxic and Aerobic Processes Using MBBR

Zhao Erhua, Yu Yunfei, Yang Shihui

(CISDI Research & Development Co., Ltd., Chongqing 401122, China)

Abstract: As the coking wastewater treatment processes encounter problems including unstable removal rate of total nitrogen (TN), ammonia (NH$_4$-N) and chemical oxygen demand (COD) and high running cost, this paper conducted a lab study on decontaminating coking wastewater adopting a two stage anoxic and aerobic (A/O) processes using MBBR and activated sludge. The results showed that COD of coking wastewater decreased from 4000~4600 mg/L to 250 mg/L after this treatment. In order to improve the biodegradability of the bio-treated coking wastewater after the first A/O treatment, a Fenton process was applied and the B/C of the coking wastewater evidently increased following this step from 0.08~0.1 to 0.4~0.6, which indicated the significant enhancement of biodegradability. This in turn resulted in less extra carbon (i.e. glucose) added in the second A/O process which was essential for the denitrification process. The concentration of TN and NH$_4$-N in the effluent was less than 20 and 5.2 mg/L, respectively, and the operational cost of this proposed technology was 5.96 RMB/m^3.

Key words: coking wastewater, MBBR-activated sludge, two stage A/O, Fenton

焦化厂废水综合深度处理工艺研究

周稳华[1]，叶青保[1]，马广伟[2]

（1. 铜陵泰富特种材料有限公司生产质量部；2. 上海申昱环保新材料有限公司）

摘　要： 本文研究了焦化厂废水的深度处理方法，通过采用新的吸附催化材料，探索了处理工艺条件对降低废水中 COD 含量，氨氮、废水含油量、废水中苯并芘含量的影响。研究表明，采用分子筛吸附催化材料 CZR-ST-64，可同时降解焦化废水中的 COD 含量，氨氮、废水含油量、废水中苯并芘含量、总氰、总氮含量。当废水中苯并芘含量 2.4μg/L 时，COD 为 286mg/L，处理后的废水中苯并芘含量达到 0 μg/L，COD 降低到 86mg/L，当进水重量空速在 4h^{-1} 以下时，出口废水中苯并芘均低于 0.02μg/L，COD 低于 120mg/L。提高废水的进口温度，由 20℃到 60℃和 80℃，苯并芘出口浓度和 COD 浓度基本不变，均在 0.016μg/L 以下，提高废水的氧化环境，废水出口的苯并芘和 COD 含量基本没有影响，运行一个月后再生催化剂，再生后继续使用，苯并芘和 COD 出口浓度基本达到吸附剂初始活性时的浓度。上述研究表明，采用吸附催化材料 CZR-ST-66，在进水温度 30~40℃，进水重量空速在 2h^{-1}，不需要其他氧化环境下，废水中 COD、氨氮、总氰、总氮、油含量和苯并芘的降解效果最佳，催化剂再生后能反复使用。

关键词： 焦化厂，废水，苯并芘，COD，氨氮，吸附，催化降解

焦化厂含苯并芘废水处理工艺研究

叶青保[1]，马广伟[2]

（1. 中信泰富特钢集团铜陵泰富特种材料有限公司；2. 上海申昱环保新材料有限公司）

摘　要：本文研究了焦化厂含苯并芘废水的处理方法，通过采用新的催化材料，探索了处理工艺条件对降低苯并芘的影响。研究表明，采用催化材料 ST-66 对废水中的苯并芘处理，当废水中苯并芘含量 2.4μg/L 时，处理后的废水中苯并芘含量达到 0 μg/L。提高废水重量空速，当进水重量空速在 5h⁻¹ 以下时，出口废水中苯并芘均低于 0.02μg/L，提高废水的进口温度，由 30℃到 50℃和 70℃，苯并芘出口浓度基本不变，均在 0.02μg/L 以下，向废水中加入少量的双氧水，或者通入空气，提高氧化环境，废水出口的苯并芘基本没有影响，运行一个月后再生催化剂，再生后继续使用，苯并芘出口浓度基本未变，维持原来的浓度。上述研究表明，采用催化材料 ST-66，在进水温度 30~40℃，进水重量空速在 2h⁻¹，不需要其他氧化环境下，废水中苯并芘的降解效果较佳，催化剂能反复再生后使用。

关键词：焦化厂，废水，苯并芘，催化降解，ST-66 催化剂

燃机发电用煤气净化系统创新性改造

孙广亿

（中信泰富特钢集团铜陵泰富特种材料有限公司）

摘　要：原有的燃机发电所采用的传统的煤气净化技术，采用四个脱硫塔，合计约 1000m³ 氧化铁脱硫剂脱硫；采用四个脱萘塔，合计 640m³ 活性炭吸附剂脱萘；采用四个脱苯塔，合计 480m³ 活性炭吸附剂脱苯；一共使用 12 个吸附塔，共约 2000m³ 吸附剂依次净化煤气。净化后的煤气经煤压机压缩后，送到燃机发电，吸附剂一到两年更换一次。升级改造后的煤气净化工艺，采用目前国际上尖端的纳米疏水分子筛吸附剂技术，同时选择性地吸附苯、萘、焦油，其他重质芳烃、硫化氢、有机硫、氨气，仅采用 4 个吸附塔，400m³ 吸附剂，同时完成脱焦油、硫化氢和有机硫、苯、萘、氨气、氢氰酸等混合气体的工作。吸附塔使用个数减少了 8 个，吸附剂的使用量仅为原来的 20%。以前的 8 个塔中，两两循环再生，现在减少为 4 个塔单循环再生，换热器、加热器、冷却器也相应减少，再生周期由以前的 7 天缩短为 3 天，蒸汽使用量减少，吸附剂使用效率显著提高。吸附剂 5~7 年更换一次，不仅显著降低了环保危废的处理量，而且节约了大量人力物力，装置运行成本也大大降低。

关键词：燃机发电，煤气净化，综合吸附，脱硫，脱苯，脱萘

150t 转炉除尘浊环水系统的优化

侯铭新，陈胜喜，耿　淼

（河钢承钢能源事业部，河北承德　067002）

摘　要：转炉在吹炼时产生大量含有 CO 和氧化铁类粉尘的高温烟气，为了防止污染，保护环境，河钢承钢 150t 三座转炉分别设置一套洗涤塔＋环缝装置（新 OG 系统）全湿烟气净化系统，对烟气进行净化处理并回收煤气。但系统存在强结垢现象，且在除尘过程中烟囱有冒黄烟的现象。为达到除尘效果，对发现问题进行分析，对洗涤塔喷嘴进行改造，及对水质进行优化控制，减少用水量，减少黄烟，减少煤气粉尘，节约能源降低消耗，减少排放，提高煤气回收质量及效率。

关键词：转炉除尘，浊环水，结垢，水质控制

Optimization of Dedusting and Circulating Water System for 150t Converter

Hou Mingxin, Chen Shengxi, Geng Miao

(Energy Center of Hebei Iron & Steel Group Chengde Steel Corp, Chengde 067002, China)

Abstract: A large number of converter high-temperature flue gas containing CO and iron oxide dust in the process, in order to prevent pollution, protect the environment, River steel bearing steel 150t three converter are respectively arranged a washing tower + ring joint device (OG system) all wet flue gas purification system, processing of flue gas recovery and purification. But there is strong scaling phenomenon, and the chimney in the process of dust removing out yellow phenomenon。To achieve the effect of dust, the discovery of the problem analysis, the transformation of the washing tower nozzle, and optimal control of water quality, reduce water consumption, reduce tobacco, reduce gas dust, reduce the energy consumption, reduce emissions, improve the quality and efficiency of gas recovery.

Key words: dust removal of converter flue gas, turbid circulating water, scaling, water quality control

酒钢炼轧厂除尘系统运行分析

朱青德[1]，慕进文[2]，常全举[2]，郝会斌[2]，魏国立[1]

（1. 酒泉钢铁（集团）宏兴股份公司，甘肃嘉峪关　735100；

2. 酒泉钢铁（集团）宏兴股份公司炼轧厂，甘肃嘉峪关　735100）

摘　要：介绍了酒钢炼轧厂转炉一次除尘工艺流程及除尘系统存在的问题，根据冶炼工艺参数核算烟气量，与现有除尘风机参数进行对比，结合国内转炉一次除尘技术的发展方向及炼轧厂转炉冶炼工艺，探讨转炉一次除尘的改进方向，提出了改进建议。

关键词：除尘，环保，鼓风机，烟气

Operation Analysis of Dust Removal System of Hongxing Stock Company of Jiuquan Iron

Zhu Qingde[1], Mu Jinwen[2], Chang Quanju[2], Hao Huibin[2], Wei Guoli[1]

(1. Hongxing Stock Company of Jiuquan Iron, Jiayuguan 735100, China;

2. Hongxing Stock Company of Jiuquan Iron and Lianzhachang, Jiayuguan 735100, China)

Abstract: The wine process dust removal system of steel rolling plant and existing problems, according to the technology parameters calculation of dust blower air volume, compared with the existing dedusting fan parameters, combined with the development direction of a dust removal technology and domestic converter steelmaking and rolling plant of smelting process and explore the improvement direction of primary dedusting of converter, puts forward some suggestions for improvement.

Key words: remove dust, environmental protection, blower, smoke

完全平衡法——除尘管路阻力平衡计算方法的探讨

张秀江

（中钢集团工程设计研究院有限公司公辅设计部，北京　100080）

摘　要：对常用的除尘管路几种阻力平衡计算方法进行分析比较，指出它们存在的不足，介绍了完全平衡法的基本思路和具体做法，举例说明了完全平衡法的设计计算方法，总结了完全平衡法的优势。

关键词：阻力平衡，完全平衡法，管路，计算，设计

The Full Balance Method - the Discussion of the Resistance Balance Calculation Method of Dedusting Duct

Zhang Xiujiang

(Sinosteel Engineering Design & Research Institute Co., Ltd., Beijing 100080, China)

Abstract: This paper analyzes several common resistance balance calculation methods of dedusting duct, and points out the flaws in them. This paper introduces the basic idea and practice of the full balance method, and use examples to illustrate the design calculation of the full balance method. The advantages of the full balance method are summarized in the end.

Key words: resistance balance, full balance method, dedusting duct, calculation, design

电炉粉尘-深共晶溶剂选择性浸出锌的实验研究

杨　莹[1]，李佳兴[1]，许继芳[1]，郭恒睿[1]，邹长东[2]

（1. 苏州大学沙钢钢铁学院，江苏苏州　215021；
2. 江苏省沙钢钢铁研究院，江苏张家港　215625）

摘　要：电炉粉尘中富含大量的铁、锌等元素，高效率回收利用电炉粉尘中有价元素具有重要意义。本文主要研究了超声波对电炉粉尘在氯化胆碱/尿素离子液体中的溶解性的促进作用以及 ChCl/urea 离子液体电沉积 ZnO 制备金属 Zn 过程的电化学行为，结果表明：在相同时间内，超声波处理后的电炉粉尘中锌元素的浸出率高于机械搅拌后的电炉粉尘中锌元素的浸出率，超声波可以促进电炉粉尘中的锌元素在离子液体中的溶解；ChCl/urea-ZnO 离子液体中 Zn（II）的还原是准可逆过程，在 ChCl/urea 离子液体中可以通过恒电流电沉积法制备金属 Zn。

关键词：电炉粉尘，超声波，离子液体，电沉积

Study on Selective Leaching of Zinc in EAF Dust - Deep Eutectic Solvent

Yang Ying[1], Li Jiaxing[1], Xu Jifang[1], Guo Hengrui[1], Zou Changdong[2]

(1. Shagang School of Iron and Steel, Soochow University, Suzhou 215021, China;
2. Institute of Research of Iron and Steel (IRIS), Shasteel, Zhangjiagang 215625, China)

Abstract: Electric furnace dust contains iron, zinc and other elements. Efficient recycling of valuable elements in EAF dust is of great significance. In this paper, the effect of ultrasonic wave on the solubility of EAF dust in ChCl/ urea ionic liquid has been investigated. The electrochemical behaviors of Zn in ChCl/urea-ZnO ionic liquid have also been investigated. The following conclusions have been drawn; during the same time, the leaching rate of zinc of EAF dust after ultrasonic treatment is higher than that after mechanical agitation. Ultrasonic wave can promote the dissolution of zinc in the EAF dust in ionic liquids; The reduction of Zn(II) in ChCl/urea-ZnO ionic liquid is quasi-reversible process. Zn can be prepared by constant current electrodeposition in ChCl/urea-ZnO ionic liquid.

Key words: EAF dust, ultrasonic wave, ionic liquid, electrodeposition

国内烟气脱硝工程概述

李丽坤，卢丽君，康凌晨，刘　瑛，张　垒，韩　斌

（武钢研究院环保所，湖北武汉　430080）

摘　要： 今年，环保部起草的《钢铁烧结、球团工业大气污染物排放标准》（GB 28662—2012）修改单中将钢铁烧结烟气氮氧化物限值降低至 100 mg/Nm²。现阶段国内钢铁企业只有太钢、宝钢少数几家企业进行了烧结烟气脱硝，全国烧结烟气脱硝任务迫在眉睫，以往，钢铁烧结烟气脱硫多数复制电厂脱硫工程，但钢铁脱硝工程不可能简单复制电厂脱硝技术。本文综述了国内近年来烟气治理增设的脱硝工程，总结了国内钢厂现有的脱硝技术，最后对于武钢烧结烟气如何开展适合钢铁行情的脱硝选择方案提出了建议。

关键词： 烟气，脱硝，工程

Overview of Domestic Flue Gas Denitration Engineering

Li Likun, Lu Lijun, Kang Lingchen, Liu Ying, Zhang Lei, Han Bin

(Institute of Environmental Protection, Wuhan Research Institute of WISCO, Wuhan 430080, China)

Abstract: This year, the Ministry of environmental protection drafted the "iron and steel sintering and pelletizing industry emission standards for atmospheric pollutants" (GB 28662—2012) revision, the steel sintering flue gas nitrogen oxide limit lowered to 100 mg/Nm². Nowaday only a few domestic iron and steel enterprises like TISCO, Bao steel have sintering flue gas denitrification. From now on denitrification is a formidable task throughout the country, in the past, the flue gas desulfurization in the iron and steel industry referred to flue gas desulfurization of power plants, however the steel denitration project can not simply duplicate the power plant denitration technology. This paper summarized the domestic denitrification engineering of flue gas treatment added in recent years, summarizes the denitration technology of existing domestic mills, finally in this paper Wuhan Iron and Steel sintering flue gas denitration project for the steel market was proposed.

Key words: flue gas, denitrification, engineering

焦油深加工废水预处理实验研究

付本全[1]，张　云[2]，刘　霞[3]，王丽娜[1]

（1. 武钢有限公司研究院，湖北武汉　430080；2. 武钢有限公司条材厂，湖北武汉　430086；

3. 武汉平煤武钢联合焦化有限责任公司，湖北武汉　430082）

摘　要：焦油深加工废水采用适当的预处理，可以有效降低废水水量和后续处理成本。实验室对某焦油加工过程中的酚盐分离水开展了酸化混凝气浮和钙法吸附沉淀实验。得出酸化混凝气浮的最佳处理工艺参数为：废水稀释 10 倍，pH 7.2~7.4，搅拌时间 10min，沉降时间选择为 120 min。PAM 投加量为 2mg/L，此时 COD 去除率为 19.57%。钙法吸附沉淀处理焦油深加工废水的最佳投药量为 20kg/t 废水，搅拌和沉降时间分别为 15min，此时废水的 COD 去除率为 8.29%。

关键词：酚盐分离水，酸化气浮，预处理，减量

Research Progress on Pretreatment of Coal Tar Deep Processing Wastewater

Fu Benquan[1], Zhang Yun[2], Liu Xia[3], Wang Lina[1]

(1. Research and Development Center of Wuhan Iron & Steel Limited, Wuhan 430080, China;

2. Strip Mill of Wuhan Iron & Steel Limited, Wuhan 430086, China;

3. WISCO-Pingmei Joint Coking Co., Ltd., Wuhan 430082, China)

Abstract: Wastewater limited and subsequent treatment costs were reduced by the pretreatment of coal tar deep processing wastewater. The experiment of acidification, coagulation flotation and calcium adsorption precipitation had been carried out on the separation of phenolic salt in a tar processing process. The best parameters of the experiment of acidification, coagulation flotation were 10 times diluted, 7.2-7.4 of pH, stirred for 10 minutes, precipitated in 120 minutes and dropped 2 mg/L PAM. The removal rate of COD was 19.57%. The best parameters of the calcium adsorption precipitation were dropped 20kg calcium oxide in the 1ton wastewater, stirred for 15 minutes and precipitated in 15minutes. The removal rate of COD was 8.29%.

Key words: separate water of phenate, acidification and flotation, pretreatment, reduction

焦化烟气脱硝技术研究现状与进展

郭凯岳，周景伟，樊　响，邓志鹏，许艳梅，闫武装

（北京中冶设备研究设计总院有限公司能源与环保分院，北京　100029）

摘 要： 在烟气治理领域焦炉烟气脱硝一直是时下关注的重点，特别是国家颁布了最新的《炼焦化学工业污染物排放标准》之后，对焦化烟气脱硝技术提出了更高的要求。介绍了传统 SCR 烟气脱硝技术的研究现状和进展。同时也对目前重点研究的脱硫脱硝一体化技术包括臭氧氧化法、吸附法、等离子体法、液相氧化吸收法等技术做了详细分析。分类阐述了各种脱硝技术的机理与技术特点，展望了未来脱硝技术的发展方向。

关键词： 焦化烟气，催化还原，催化剂，烟气脱硝，一体化

Research Status and Progress of Coke Flue Gas Denitration Technology

Guo Kaiyue, Zhou Jingwei, Fan Xiang, Deng Zhipeng, Xu Yanmei, Yan Wuzhuang

(Beijng Metallurgical Equipment Research & Design Corporation Limited of MCC Group Environment and Energy Branch, Beijing 100029, China)

Abstract: In the field of flue gas of coke oven flue gas denitration has always been the focus of nowadays, especially the state issued the latest after coking chemical industrial pollutants emission standards, the coking flue gas denitration technology put forward higher requirements. The research status and progress of traditional SCR denitration technology were introduced. At the same time, the integration techniques of desulfurization and denitration in the present study include ozone oxidation method, adsorption method, plasma method and liquid phase oxidation absorption method. The mechanism and technical characteristics of various denitrifying technology are described and the development direction of denitration technology is prospected.

Key words: coking flue gas, catalytic reduction, catalyst, flue gas denitrification, integration

高温气体除尘技术的发展与应用

魏 嵩

（天津钢铁集团有限公司 动力厂燃气作业区，天津 300301）

摘 要： 高温气体除尘技术是在高温条件下，直接进行气体与固体的分离从而实现气体净化，它可以最大程度利用气体的物理显热、化学潜热和动力能等有用资源。基于此本文将对高温气体除尘技术的发展与引用进行研究，其目的在于让更多人对目前高温气体除尘技术的发展现状以及应用领域有更为明确的了解，希望此文的研究对促进该项技术的应用和发展有所助益。

关键词： 高温气体除尘技术，发展现状，研究进展，陶瓷过滤材料

Development and Application of High Temperature Gas Dust Removal Technology

Wei Song

(Tianjin Iron & Steel Group Co., Ltd., Gas Operation Area of Power Plant, Tianjin 300301, China)

Abstract: High-temperature gas dust removal technology is the direct separation of gas and solid in high temperature conditions to achieve gas purification, it can maximize the use of gas physical sensible heat, chemical latent heat and dynamic energy and other useful resources. Based on this paper, the development and reference of high temperature gas dust removal technology will be studied. The purpose of this paper is to make more people understand the current situation and application of high temperature gas dust removal technology. It is hoped that this paper The application and development of technology is helpful.

Key words: high temperature gas dust removal technology, development status, research progress, ceramic filter material

铁矿烧结过程二噁英排放特征研究

钱立新 [1]，龙红明 [1,2]，吴雪健 [1]，春铁军 [1]，王毅璠 [1]

(1. 安徽工业大学冶金工程学院，安徽马鞍山　243002；2. 冶金减排与资源利用教育部
重点实验室（安徽工业大学），安徽马鞍山　243002)

摘　要： 铁矿烧结过程是二噁英排放的主要污染源之一，为了查明烧结各个工序二噁英的排放特性，对国内某钢铁企业 3 台烧结机静电除尘灰、脱硫灰和脱硫装置进出口烟气中二噁英及其同系物排放情况进行了检测分析。结果表明：Cl 元素含量较高的烧结原料会促进二噁英的生成；脱硫系统对 3 台烧结机产生的二噁英减排率分别为 59.6%、93.5% 和 75%；在 4#烧结机检测结果中，脱硫前后气相中的二噁英同系物主要以 PeCDFs 为主，电场除尘灰和脱硫灰中的 PeCDFs 和 HxCDFs 含量之和分别占总排放量的 84% 和 85%。另外，脱硫灰和脱硫装置进出口烟气中 PCDFs/PCDDs 比值分别为 9.88、9.14 和 6.29，说明烧结过程二噁英的合成机理以从头合成为主。

关键词： 铁矿烧结，二噁英，脱硫灰，除尘灰，同系物

Study on the Characteristics of Dioxin Emission in Iron Ore Sintering Process

Qian Lixin[1], Long Hongming[1,2], Wu Xuejian[1], Chun Tiejun[1], Wang Yifan[1]

(1. School of Metallurgical Engineering, Anhui University of Technology, Maanshan 243002, China;
2. Key Laboratory of Metallurgical Emission Reduction & Resources Recycling (Anhui University
of Technology), Ministry of Education, Maanshan 243002, China)

Abstract: The iron ore sintering process is one of the main emission of dioxin. In order to identify the emission characteristics of dioxin in each sintering process, the dioxin and homologues emission of sintering dust, desulphurization ash and sintering flue gas were analyzed. The results show that the sintering raw with high Cl elements will promote the formation of dioxin. The dioxin emission reduction rates for the three sintering machines in the desulfurization system were 59.6%, 93.5% and 75%, respectively. No.4 sintering machine test result shows that PeCDFs were the main homologues in the flue gas of desulphurization device import and export. The sum of the PeCDFs and HxCDFs contents in sintering dust or desulfurization ash accounts for 84% and 85% of the total emissions, respectively. In addition, the PCDFs/PCDDs ratios in desulfurization ash and flue gas of desulphurization device import and export were 9.88, 9.14 and 6.29, respectively. So it is generally believed that the de novo synthesis is the main generation mechanism of dioxin in sintering process.

Key words: iron ore sintering, dioxin, desulfurization ash, sintering dust, homologues

改性钢渣对土壤中 NO_3^--N 淋失及土壤性质的影响

杨毛毛，杨丽韫，钱晓明，白　皓，李　宏

（北京科技大学冶金与生态工程学院，北京　100083）

摘　要： 本研究对钢渣改性制备出一种新型廉价的土壤 NO_3^--N 吸附保持剂，并通过土柱淋溶实验分析了原钢渣和改性钢渣对土壤淋溶液中 NO_3^--N 淋失以及添加改性钢渣对土壤性质的影响。结果表明：灭菌和未灭菌组的土壤 NO_3^--N 淋失量随改性钢渣和原钢渣用量的增加而减少，与对照组相比降低幅度范围分别为 12.28%~36.43% 和 11.33%~34.93%，且改性钢渣效果优于原钢渣。这说明土壤中添加改性钢渣可有效减少土壤 NO_3^--N 的淋失。添加原钢渣和改性钢渣的土壤淋溶液 pH 值和 ORP 值与对照组相比均存在显著性差异（p<0.05），pH 值增加幅度在 28.2%~48.1% 之间，ORP 值降低幅度在 19.39%~97.52% 之间。因此，在实际应用中应根据当地土壤的用途和理化性质来确定改性钢渣的添加量。

关键词： 改性钢渣，土壤，NO_3^--N，淋失

Effect of Modified Steel Slag on NO_3^--N Leaching and Changes of Properties in Soil

Yang Maomao, Yang Liyun, Qian Xiaoming, Bai Hao, Li Hong

(School of Metallurgical and Ecological Engineering, University of Science and Technology Beijing, Beijing 100083, China)

Abstract: In this study, a new and inexpensive adsorbent for NO_3^--N in soil was prepared by modification of steel slag. The effects of original steel slag and modified steel slag on the leaching loss of NO_3^--N in soil leaching solution and the change of soil properties after adding two kinds of steel slag were analyzed by soil column leaching experiment. The results showed that the leaching loss of NO_3^--N in the sterilized and unsterilized soils decreased with the increase of the amount of original steel slag and modified steel slag, which were 12.28%~36.43% and 11.33%~34.93% respectively compared with the control group. And the effect of modified steel slag was better than that of original steel slag. This indicated that the addition of modified steel slag in soil could effectively reduce the leaching of NO_3^--N in soil. The pH and ORP value of soil leaching solution after adding original steel slag and modified steel slag were significantly different (p<0.05) from those of the control group, which the increase of pH value was 28.2%~48.1% and the decrease of ORP value was 19.39%~97.52%. Therefore, the amount of modified steel slag added in soil should be determined according to specific use and physical and chemical properties of local soil.

Key words: modified steel slag, soil, NO_3^--N, leaching

MOFs 在钢铁企业 VOCs 吸附、选择性还原脱硝和焦炉煤气制氢领域的应用

陈 琛

（中冶京诚工程技术有限公司，北京 100176）

摘 要：MOFs 材料是一种高效多孔材料在吸附领域和催化领域具有很广阔的前景。VOCs 和 NOx 污染是钢铁企业废气排放污染主要源头之一，严重地威胁人类健康和影响社会的可持续发展，有效地治理 VOCs 的环境污染已迫在眉睫。选择性还原脱硝（SCR）工艺是一种高效的烟气脱硝技术，焦炉煤气制氢技术是廉价氢气的来源之一。吸附性能和反应活性是 VOCs 治理、烟气 SCR 脱硝和 PSA 制氢的关键技术。本文主要探索和总结 MOFs 作为一种高效吸附剂和催化剂材料在 VOCs NOx 和 PSA 制氢的应用前景。

关键词：金属有机框架材料，挥发性有机物，吸附性能，选择性催化还原脱硝

The Application of Metal-Organic Frameworks (MOFs) on VOCs Treatment, SCR and Hydrogen Separation from Coke Oven Gas in Metallurgical Industry

Chen Chen

(Capital Engineering & Research Incorporation Limited, Beijing 100176, China)

Abstract: Pollution of VOCs and NOx is one of the main sources of emission pollution of iron and steel enterprises. It seriously threatens human health and impacts the sustainable development of society. Therefore, it is urgent to effectively control the environmental pollution of VOCs. Selective Catalytic Reduction (SCR) Process is a high-efficient flue gas denitrification (De-NOx) technology. Pressure Swing Adsorption (PSA) of Hydrogen production from coke oven gas is one of the low price sources of hydrogen. Adsorption performance and catalysts reaction activity are the key technologies for VOCs treatment, flue gas SCR De-NOx and PSA hydrogen production. This paper mainly explores and summarizes the application prospects of MOFs as an efficient adsorbent and catalyst material for the treatment of VOCs, De-NOx and PSA hydrogen produciton.

Key words: metal-organic frameworks (MOFs), volatile organic compounds (VOCs), pressure swing adsorption (PSA), selective catalytic reduction (SCR)

微波强化冰乙酸浸出高炉渣中钙的行为

李 峰，马国军，张 翔，刘孟珂

（武汉科技大学钢铁冶金及资源利用省部共建教育部重点实验室，湖北武汉 430081）

摘 要：本实验探究了浸出剂种类及初始浓度、搅拌速度、微波功率、固液比和高炉渣颗粒粒径对高炉渣中钙的浸

出行为的影响。实验结果表明：采用冰乙酸浸出效果明显优于氯化铵和乙酸铵，后两者高炉渣中 Ca^{2+} 的浸出率均未超过 6%；增加浸出剂冰乙酸的初始浓度，可促进 Ca^{2+} 的浸出，当初始浓度为 2mol/L 时，Ca^{2+} 的浸出率可达 70%；当搅拌速度在 200r/min 以下时，提高转速能显著提高 Ca^{2+} 的浸出率，当转速超过 200r/min 时，转速对 Ca^{2+} 的浸出率影响不大；Ca^{2+} 最终浸出率随微波功率的增加呈递减趋势，但微波浸出较常规浸出高；Ca^{2+} 的浸出率随固液比减小而增大，当固液比为 1:30 时，Ca^{2+} 的浸出率可达 90%；高炉渣颗粒粒径越小，浸出反应速度越快，且 Ca^{2+} 的浸出率越高。

关键词：冰乙酸，高炉渣，微波，浸出

Microwave Strengthening Leaching of Blast Furnace Slag By Glacial Acetic Acid

Li Feng, Ma Guojun, Zhang Xiang, Liu Mengke

(Key Laboratory for Ferrous Metallurgy and Resources Utilization of Ministry of Education, Wuhan University of Science and Technology, Wuhan 430081, China)

Abstract: In this paper, the effects of leaching agents, initial acetic acid concentration, stirring speed, microwave power, liquid-solid ratio and particle size of blast furnace slag on Ca^{2+} leaching behavior of blast furnace slag were investigated. The results show that the leaching rate of Ca^{2+} using acetic acid was significantly higher than that using ammonium chloride or ammonium acetate, the leaching rate of Ca^{2+} using both ammonium chloride and ammonium acetate was less than 6%. Increasing the concentration of acetic acid and improving liquid-solid ratio benefit to leach calcium from blast furnace slag. At the optimum conditions of initial acetic acid concentration of 2mol/L, solid-liquid ratio of 1:30, the leaching rate of Ca^{2+} was up to 90%. When the stirring speed is less than 200r/min, increasing rotational speed can significantly improve the leaching rate of Ca^{2+}, and when the speed exceeds 200r/min, the stirring speed of has no obvious effect on the leaching rate of Ca^{2+}. The final leaching rate of Ca^{2+} reduces withthe improvement of microwave power, and it's higher than the conventional leaching process. The leaching rate of Ca^{2+} increases with the decrease of solid-liquid ratio, and the leaching rate of Ca^{2+} can reach up to 90% with the solid-liquid ratio of 1:30. Decreasing the particle size of blast furnace slag is helpful to accelerate the leaching reactions and increase the leaching rate of Ca^{2+}.

Key words: glacial acetic acid, blast furnace slag, microwave, leaching

源头有效控制 末端高效治理 实现焦炉烟气达标排放

李　超，王明登，尹　华，郑亚杰，孙刚森，康　婷

（中冶焦耐（大连）工程技术有限公司，辽宁大连　116085）

摘　要： 随着大气环境的恶化和环保法规的日益严格，焦炉烟囱排放烟气中的 SO_2 和 NO_X 因其产生机理复杂、对环境危害大、治理成本高，已成为炼焦工业面临的重大技术难题。本文从焦炉烟气中 SO_2 和 NO_X 的来源及产生机理入手，量化分析了各因素的影响效果，提出了各种源头消减和过程控制方法，定性分析了国内主流脱硫脱硝工艺的优缺点。提出了综合应用源头控制方法和末端治理技术，实现焦炉烟气达标排放的技术路线。

关键词：焦炉烟气，脱硫脱硝，源头减排，末端治理

Effective Source Control and Terminal Treatment Realizing Discharge of Coke Oven Flue Gas in Qualified Standard

Li Chao, Wang Mingdeng, Yin Hua, Zheng Yajie, Sun Gangsen, Kang Ting

(ACRE Coking & Refractory Engineering Consulting Corporation (Dalian), MCC, Dalian 116085, China)

Abstract: With the deterioration of the atmospheric environment and the increasingly stringent environmental regulations, discharge of SO_2 and NO_x in coke flue gas has become a serious technical problem against coke industry all around the world, which is characteristic for its complex creation mechanism, environment hazards, and high cost of terminal treatment. In this paper, the source and creation mechanism of SO_2 and NO_x in coke flue gas are analyzed, and the effect of various factors is discussed quantitatively. Various sources reduction and process control methods are put forward, and the advantages and disadvantages of domestic mainstream desulfurization and denitration process are qualitatively analyzed. The technical route of comprehensive application of source reduction and terminal treatment to realize the discharge of coke flue gas is also provided.

Key words: coke oven flue gas, desulfurization and De-NOx, source reduction, terminal treatment

石煤改质对转炉钢渣矿相变化的影响研究

林　超，王　珏，钟娜娜，吴六顺

（安徽工业大学 冶金工程学院，安徽马鞍山　243032）

摘　要：转炉钢渣作为一种冶金二次资源，微粉化处理是实现其大宗量综合利用的重要前提。但是转炉钢渣硬度高、易磨性差，实际微粉化困难。而利用转炉钢渣中 $2CaO \cdot SiO_2$ 相（简称 C_2S）晶型转变时体积膨胀产生的内应力可实现钢渣自粉化，因此钢渣中 C_2S 相含量高低将影响钢渣的自粉化效果。本文根据实际钢渣成分配制了合成转炉钢渣，研究了石煤改质对转炉钢渣矿相变化及 C_2S 优势析出的影响。实验结果表明：利用石煤改质钢渣，当碱度较高时，渣中主要生成 C_2S 相；当碱度小于 2 时，$CaO \cdot SiO_2$ 相（简称 CS）开始生成；当碱度小于 1.6 时，$CaO \cdot MgO \cdot SiO_2$（简称 CMS）开始生成，$C_2S$ 相逐渐消失；碱度降低至 1.2 时，钢渣矿相以 CMS 相为主，含有少量 CS 相。为使 C_2S 相优势析出，应控制石煤改质转炉钢渣的碱度大于 2.4。本论文实验结果为石煤应用和转炉钢渣微粉化提供了参考。

关键词：转炉钢渣，石煤，矿相，硅酸二钙，碱度

Study on Influence of Stone Coal Modification on Mineral Phase Change in Converter Steel Slag

Lin Chao, Wang Jue, Zhong Nana, Wu Liushun

(School of Metallurgical Engineering, Anhui University of Technology, Ma'anshan 243032, China)

Abstract: As one of metallurgical secondary resources, pulverization of converter steel slag is an important precondition for its mass utilization. However, the high hardness and poor grindability of converter steel slag makes it very difficult to be pulverized. While the converter steel slag can be self-pulverized by volume expanding internal stress when $2CaO \cdot SiO_2$ (C_2S)

in steel slag changed from β-C$_2$S to γ-C$_2$S during cooling process. So the change of C$_2$S contents in converter steel slag will influence its self-pulverization effect. In this paper, according to chemical compositions of practical steel slags, a converter steel slag has been synthetized as experimental slag. The influence of stone coal modification on mineral phase change in conveter steel slag has been studied. The experiments results show that for stone coal modification, when basicity of steel slag is high, its main mineral phase in slag is C$_2$S; when basicity is lower than 2, CaO·SiO$_2$ (CS) phase appears in the slag; when basicity is lower than 1.6, CaO·MgO·SiO$_2$ (CMS) phase appears and C$_2$S phase disappears gradually; when basicity reduces to 1.2, the main mineral phase in steel slag is CMS phase as well as small amount of CS phase. To make C$_2$S phase precipitate dominantly in converter steel slag, its basicity should be controlled higher than 2.4. The results of the experiment provide scientific reference for stone coal application and pulverization of converter steel slag.

Key words: converter steel slag, stone coal, mineral phase, dicalcium silicate, basicity

昆钢新区二次资源综合利用实践

惠世谷，李信平

（武钢集团昆明钢铁股份有限公司安宁分公司）

摘　要：新区将各生产工序间产生的二次资源根据其价值高低，一部分参与混匀造堆生产；另一部分附加值较高的二次资源进行招标外卖处理，对实现企业可持续发展具有重要的现实意义。

关键词：钢铁企业，二次资源，循环利用，效益

安钢炼钢工序大气污染防治

曹树卫，李军强，谷　洁

（安阳钢铁股份有限公司第二炼轧厂，河南安阳　455004）

摘　要：目前，我国冶金企业污染严重，尤其是炼钢产能经过一轮扩张后严重过剩，污染物总量攀升，突出表现在大气的污染方面。近年来，雾霾成了最热的词汇之一。随着应对雾霾对策的不断升级，国家环保部门已在 2012 年对一系列环保法和标准进行修订，在以"壮士断腕"的决心，加强对建设项目和专项规划的环境评价、严格控制污染物排放总量和排放浓度标准，重罚和累计处罚甚至关停不达标排放企业等一系列手段，减少工业企业对大气造成的污染。钢铁企业成了重灾区，环保重压成了企业生死抉择。

关键词：炼钢工序，大气污染，防治，节能减排

The Prevention and Control of Atmospheric Pollution in the Steel-Making Process in Anyang Steel and Iron Group Co., Ltd.

Cao Shuwei, Li Junqiang, Gu Jie

(Anyang Iron and Steel Co., Ltd., No.2 Steel Making & Rolling Plant, Anyang 455004, China)

Abstract: The prevention and control of atmospheric pollution in the steel-making process in Anyang Steel and Iron Group

Com. Ltd. The metallurgy industry has had surplus production capacity and has been making a lof of pollution in the present years, especially after the new round of expansion. The total amount of pollutants is increasing and the air pollutants occupies the most. Recently, "haze" has been one of the top words. The national environment departments has revised a series of environmental law and standards and upgraded the policies of preventing haze. In order to decrease the atmospheric pollution, they also strengthened the environmental assessment in construction projects and subject plan, made strict control of the total emission amount and concentration, implemented heavy fine on, even closed the companies which had excessive emissions and so on. The iron and steel enterprise has the life or death choice due to the environmental stress.

Key words: steel-making process, atmospheric pollution, prevention and control, energy conservation and emission reduction

16　分析检测

炼铁与原料
炼钢与连铸
轧制与热处理
表面与涂镀
金属材料深加工
钢铁材料
汽车钢
海洋工程用钢
轴承钢
电工钢
粉末冶金
非晶合金
高温合金
耐火材料
能源与环保
★ 分析检测
冶金设备与工程技术
冶金自动化与智能管控
冶金技术经济

中国材料与试验标准体系建设

杨植岗

（中关村材料试验技术联盟）

摘 要：针对目前材料与试验标准体系存在的标准供给不足、标准重复、标准体系不够完备等问题，提出建设材料与试验标准共享平台的设想，通过平台汇集各类标准资源、标准需求信息、新材料研发信息并进行梳理，从而将标准资源甄别比对、协调统一，并与汇集到的标准需求信息，新材料研发信息相匹配以解决标准供给不足和重复的问题。通过标准以及信息的共享完善标准体系建设，补全目前体系中的缺失。

在此基础之上建设具有系统性、先进性、适用性、时效性、多元性、包容性、动态性的中国材料与试验团体标准体系。并充分发挥市场在标准制修订的引领作用。CSTM 从材料属性和材料应用两个维度，设立不同的专业技术委员会，以实现材料基本属性与应用属性的标准协调和各类材料指标-试验-评价标准体系的统筹。

为说明 CSTM 的组织结构，报告中详细介绍了 CSTM 专家委员会、CSTM 标准委员会的职责以及领域委员会的设置、运作模式、以及拟立项的标准项目等信息。

Construction of CSTM Standards System

Yang Zhigang

(Chinese Society for Testing and Materials)

Abstract: Nowadays, the several deficiencies existed in the testing & material standard system，for example, the standards are in short supply, so many standards are repeated, and the system is not integrated enough. An idea constructing a testing & materials standard sharing platform is thus put forward here. Through the platform, all kinds of standard, including the standard demand information, new materials research & development information would be gathered and integrated. The standard resources would be discriminated and harmoniously unified, and further matched with the standard requirement information and the new materials research & development information. The above problem would thus be resolved and the missing part of the system would also be complemented by the sharing of standards and information.

On this basis, CSTM standards system would be established with following obvious characteristics: systematicness, advancement, suitability, timeliness, pluralism, inclusiveness, and dynamics. Letting the market itself to play the leading role in the standard system development is also very important. CSTM established different field committees from two dimensions, namely, from the material basic attribution and application attribution, so that the coordination of these two attributions and the overall planning of index - test - evaluation various materials standard systems would be realized.

To illustrate the CSTM organization structure, the report details the duties of CSTM expert committee and CSTM standards committee, the field committee setting, operation pattern, the proposed standard project information, and so on.

多元合金组合材料的微波高通量微制造与表征的新方法研究

赵　雷，陈学斌，冯　光，王海舟

（钢铁研究总院，金属材料表征北京市重点实验室）

摘　要： 基于材料基因组思想的材料研究新方法是通过一次实验获取大量数据和结果，即通过量变实现质变的全新路径。采用传统迭代方法的合金研究一次只能尝试一种成分配比的合金材料，这种方法已严重滞后于新技术的发展，成为新技术发展的主要瓶颈问题。微波特种能场被认为是人类的"第二团火焰"，其具有加热的快速性，能够大幅度的缩短制备周期和降低烧结温度，从而使工艺过程简化，为实现材料的快速制备奠定了基础，而且独有的加热特性可以制备出显微组织均匀、晶粒细小的材料，同时体积加热、选择性加热以及非热效应加热等特性是其它加热方式和手段所不具备的。本研究中采用微波特种能场进行金属块体材料的高通量制备，能够一次制备一批含有多种成分变化的样品，并结合相应的分析表征手段，实现短时间内大量样品的制备与表征。本报告以模具钢 H13 金属粉为基体材料，通过添加不同的纯元素粉末，再以微波制备手段一次性合成梯度组分变化的系列样品，并对其成分均匀性、组织状态和力学性能等参数进行高通量表征，初步探索性能和组织与组分变化的相关性，为多元合金的高通量制备提供必要的基础信息，提出一种块体材料高通量制备的新原理和新技术。

A New Approach to High Throughput Micro Synthesis and Characterization of Multi-element Alloy Combinatorial Materials by Microwave Heating

Zhao Lei, Chen Xuebin, Feng Guang, Wang Haizhou

(China Iron & Steel Research Institute, Beijing Key Laboratory of Metal Material Characterization)

Abstract: The new approach of material genome research is to obtain large amounts of data and results through an experiment, that is, to achieve a new path of qualitative change through quantitative change. The conventional iterative method of alloy research can only test one composition ratio of the alloy material. This method has seriously lagged behind the development of new technology and become the main bottleneck of new technology. Microwave heating, a special energy, is regarded as the "Second flame" of human and has the advantages of rapid heating. It can greatly shorten the preparation cycle, reduce sintering temperature and simplify the process procedure. Microwave heating laid the research foundation of the rapid synthesis of materials. And the unique heating can be used for materials synthesis with the characteristics of homogeneous microstructure, fine grains, volume heating, selective heating and non-thermal effect heating. These advantages are not available for the other heating methods. In this paper, microwave heating was used for high throughput synthesis of metal bulk materials. This approach can produce a batch of small samples with different composition once time. Meanwhile different characterization methods were used to realize the synthesis and characterization of a number of samples in a short time. This report took the mold steel H13 powder as the matrix material and the different elemental powders were added into the matrix material. Then the mixture powder was sintered into a series samples with different gradient composition by microwave energy. The composition uniformity, metallographic structures and mechanical properties were characterized by high throughput methods. This research preliminarily explored the correlation of performance, metallographic structures and compositions. This research provides the necessary basic information for high-throughput synthesis of multicomponent alloy and put forward a new principle and technique for high-throughput synthesis of bulk materials.

钢中超低氢分析的研究

朱跃进，李素娟，邓　羽，郝士一

（中国科学院金属研究所，辽宁沈阳　110016）

摘　要：钢中氢含量低至 0.Xμg/g 时测不准问题突显，首先表面微量油脂污染单纯浸泡清洗不净，必须在溶剂中超声波清洗。在此应用感应热抽取法分析稀土钢样品，观察到清洗前后的明显差异。石墨坩埚与钢样的相互作用同样可以影响到超低氢分析的准确性，通过加料空白可以观察到 0.1μg/g 的正干扰。两种石墨坩埚两种分析功率对同一标样的氢分析结果表明：高功率（高温）联测条件下，必须使用套坩埚。如果使用单坩埚，氢将会产生严重偏离。钢中氢常规分析出现"零峰值"异常样，对现行脉冲熔融法和现行仪器提出了挑战。改用基于感应热抽取法的仪器可以解决问题。日常分析过程中，异常样偶尔可见，本文例举了一组稀土钢。稀土元素挥发污染可以造成"零峰值"现象曾被试验证实。标样中也有类似的异常样，用两种方法分析结果不一致。感应热抽取法为目前判断和解决异常样的有效手段，使用石英坩埚，钢与石墨相互作用的正干扰完全不存在；样品保持固态，合金元素挥发造成的负干扰微乎其微；感应热抽取法完全适合于钢中超低氢分析的方法。而脉冲熔融法则不同，脉冲石墨电极炉无法摆脱由石墨坩埚引起的 0.Xμg/g 数量级的正干扰，熔融法无法摆脱合金元素挥发引起的负干扰；在正负干扰的双重作用下，难于准确测定钢中超低氢，特别是 6μg/g 以下值。应用感应热抽取法及相关仪器分析测定钢中超低氢才可能保障其测量的准确性。

关键词：超低氢，感应热抽取法，脉冲熔融法，石墨套坩埚，石墨单坩埚，石英坩埚，加料空白

Research on Determination of Ultralow Hydrogen in Steel

Zhu Yuejin, Li Sujuan, Deng Yu, Hao Shiyi

(Institute of Metal Research, Chinese Academy of Sciences, Shenyang 110016, China)

Abstract: The analysis uncertainty problem was appeared when the hydrogen content lower to 0.Xμg/g for steel samples. First it is caused by the sample surface. It is not cleaned completely for the sample surface by the dipping in solution. The ultrasonic cleaning has to be used for the sample preparation. The obvious difference of the hydrogen analysis was shown for the example-samples compared with ultrasonic cleaning or not. The reaction between graphite crucible with steel sample can affect the accuracy of ultralow hydrogen analysis also. 0.1μg/g positive interference was observed by the filling-blank test. The designed crucible tests show that the duplex-crucible must be applied for the H/O/N multi-element analysis. If the single-crucible be applied instead of the duplex-crucible, the results of hydrogen analysis would be deviated from the normal. The "zero peak" sample, which had been appeared on daily analysis, has challenged the current methods and the current apparatuses. The problem is solved by the outdated apparatus based on the induction-heating hot extraction method. This kind of abnormal sample can be met occasionally. One group rare earth-steel samples is listed as example here. It was proved that the pollution of the rare earth element can result in "zero peak" phenomenon during hydrogen analysis. To calibration steel samples, The abnormal sample is also be found whose results differ from each other by the two different methods. The induction-heating hot extraction method is the best method to solve the abnormal sample problem. The positive interferences caused by the reaction of steel sample with graphite crucible does not exist by quartz crucible. The negative interferences caused by the volatilization of the elements exist little by solid state keeping. The induction-heating hot extraction method is the most suitable method for the determination of ultralow hydrogen in steel. But the impulse-heating fusion method is different entirely. The positive interferences caused by the reaction of steel sample with graphite crucible exist about 0.Xμg/g by using of the graphite crucible. The negative interferences caused by the volatilization of the elements exist because of the fusion. It is time to use the induction-heating hot extraction instead of the impulse-heating fusion for ultralow hydrogen analysis especially lower than 6μg/g.

Key words: ultralow hydrogen, induction-heating hot extraction, impulse-heating fusion, duplex graphite crucible, single graphite crucible, quartz graphite, filling-blank

微波消解-电感耦合等离子体原子发射光谱法测定钢中化合铌

宋鹏心，杨志强，鞠新华，刘卫平

（首钢集团有限公司技术研究院，北京　100043）

摘　要： 电解提取钢中析出物，以王水、硫酸作为溶剂对聚碳酸酯滤膜和提取物进行微波消解处理，加入酒石酸作为络合剂抑制铌的水解，以电感耦合等离子体原子发射光谱仪测定钢中化合铌含量。研究表明：通过微波消解仪可以将聚碳酸酯滤膜和提取物完全溶解，加入 5 mL 酒石酸即可达到完全络合铌的目的。以干扰较少的 Nb316.340 nm 为分析谱线，并使校准曲线和试样在处理方式上保持一致以消除物理干扰等不利因素。校准曲线线性相关系数大于 0.9997，方法检出限为 0.0062 mg/L。对 3 个钢种的化合铌含量进行测定，相对标准偏差在 0.12%以内，精密度良好，并与国家标准方法 GB/T 223.40—2007 进行比对，结果相符，准确度良好。该方法可以满足对钢中化合铌的提取与测定，结果可靠。

关键词： 电解提取，微波消解，电感耦合等离子体原子发射光谱法，析出物，化合铌

Determination of Niobium in Precipitation of Steel by Microwave Digestion and Inductively Coupled Plasma Atomic Emission Spectrometry

Song Pengxin, Yang Zhiqiang, Ju Xinhua, Liu Weiping

(Shougang Group Co., Ltd., Research Institute of Technology, Beijing 100043, China)

Abstract: The precipitation was extracted from the steel by electrolysis, the polycarbonate filter and the precipitation were dissolved with aqua regia and sulfuric acid as the solvent. The tartaric acid was added as complexing agent to inhibit the hydrolysis of niobium. The content of niobium in precipitated phase was determined by inductively coupled plasma atomic emission spectrometry. The results showed that the polycarbonate filter and the precipitated phase could be completely dissolved by microwave digestion, and 5mL tartaric acid can achieve the the purpose of niobium complexation completely. Nb316.340nm was choosed to be the analysis line for its little interference, and the calibration curve and the dissolution of the sample should be operated by the same steps to eliminate physical interference and other unfavorable factors. The linearity correlation coefficient of calibration curve is more than 0.9997, and the detection limit is 0.0062mg/L. The relative standard deviations were within 0.12%, so the precision of the method was good. The results were compared with the national standard method GB/T 223.40—2007. they matched well. The method can satisfy the extraction and determination of niobium in the precipitated phase of steel, and the result is reliable.

Key words: electrolytic extraction, microwave digestion, inductively coupled plasma atomic emission spectrometry, precipitation, niobium

手持式 X 射线荧光光谱仪在铜矿品位测定中的应用

屈华阳，史玉涛，杜　效，王舒冉，张金伟，沈学静

（钢研纳克检测技术有限公司，北京　100081）

摘　要：本研究通过使用手持式 X 射线荧光光谱仪（Port X-200,钢研纳克检测技术有限公司）对 Cu 矿含量进行分析，建立了矿山现场测定 Cu 矿中 Cu，Pb，Zn，Mo，Fe 等元素的分析方法。优化了电流、电压等条件对测定的影响，比较了不同参比范围、分段曲线范围的测试效果，采用经验系数法进行计算和校正，确定了最佳干扰元素克服了激发增强效应。该方法对矿山现场实际样品进行分析，结果准确度高，重现性好，能实现对 Cu 矿的快速定量检测，满足现场选矿的需要，已经应用于国内某些大型铜矿采矿基地。

关键词：手持，X 射线荧光光谱，铜矿，现场，便携

Application of Hand Held X-ray Fluorescence Spectrometer in Determination of Copper Ore

Qu Huayang, Shi Yutao, Du Xiao, Wang Shuran, Zhang Jinwei, Shen Xuejing

(NCS Testing Technology Co., Ltd., Beijing 100081, China)

Abstract: In this study, the content of Cu, Pb, Zn, Mo, Fe in Cu ore was measured by using a handheld X-ray fluorescence spectrometer (Port X-200, NCS Testing Technology Co., Ltd.). The influence of current, voltage and other conditions on the measurement was optimized. The results of different reference range and segmented curve range are compared. The empirical coefficient method is used to calculate the optimal interference element, which overcomes the excitation enhancement effect. The method has high accuracy and good reproducibility. The instrument and method have been applied to some domestic copper mining fields due to meet the needs of quickly analyzing the actual samples in site.

Key words: hand-held, X-ray fluorescence spectroscopy, copper mine, in site

电感耦合等离子体原子发射光谱法测定硅锰、锰铁中磷

闫　丽

（北京首钢股份有限公司质量检验部，河北迁安　064400）

摘　要：本文建立了电感耦合等离子体原子发射光谱法测定硅锰、锰铁中磷含量的方法，通过对称样量的确定、溶样条件的选择、仪器参数的优化，仪器分析谱线的选择、干扰及消除等研究，最终制定了一套切实可行的分析方法，该方法精密度 RSD 小于 5%，相关系数＞0.9990，回收率为 97%~103%，标准样品分析值小于标准中给定的允许差；国标方法与本文方法的检验结果不存在差异。

关键词：电感耦合等离子体原子发射光谱法，硅锰，锰铁，磷

Determination of Phosphorus in Silicomanganese and Ferromanganese by Inductively Coupled Plasma Atomic Emission Spectrometry

Yan Li

(Beijing Shougang Co., Ltd., Quality Inspection Department, Qian'an 064400, China)

Abstract: In this paper, a method for the determination of phosphorus content in silicon manganese and ferromanganese by inductively coupled plasma atomic emission spectrometry was established. Through the determination of symmetry volume, the selection of sampling conditions, the optimization of instrument parameters, the selection of spectrum, And the correlation coefficient is more than 5%, the correlation coefficient is more than 0.9990, the recovery is 97% -103%, the standard sample analysis value is less than the allowable difference given in the standard. GB method and the method of this test results there is no difference.

Key words: inductively coupled plasma atomic emission spectrometry, silico manganese, ferromanganese, phosphorus

60CrMnBA 弹簧扁钢低倍裂纹分析与改进措施

袁长波

（青岛特殊钢铁有限公司中信特钢研究院青钢分院，山东青岛　266409）

摘　要：对 60CrMnBA 弹簧扁钢低倍裂纹进行了系统检验分析，连铸坯中碳化物在轧钢加热过程中没有充分溶解是产生低倍裂纹的主要原因，通过优化轧钢加热温度基本消除了该缺陷，同时对钢中硼含量控制目标以及连铸工艺提出了改进建议。

关键词：低倍裂纹，合金碳化物，加热溶解，硼含量，连铸工艺

Analyze and Improvement Measure of Crack from Macrostructure for Spring Steel Flat Bar 60CrMnBA

Yuan Changbo

(Qingdao Special Iron and Steel Co., Ltd., CITIC Pacific Special Steel Institute Qingdao Branch, Qingdao 266409, China)

Abstract: Cracks in macrostructure for spring steel flat bar 60CrMnBA were analyzed systematically, the main cracking reason was that carbide could not adequately dissolve during rolling process, though optimizing the heating temperature of heating furnace, the defects were completely eliminated, meanwhile, for control objective of boron content and continuous casting parameters the suggestion was given.

Key words: crack of macrostructure, alloy carbide, dissolve in heat, boron content, continuous casting parameters

电感耦合等离子体原子发射光谱法测定液态高纯硫酸锰中的钙镁

吴 菡[1,2]，宫小艳[1,2]，袁文东[1,2]，杨超彬[1,2]，夏钟海[2]，赵温冬[2]

（1. 钢铁研究总院，北京 100081；2. 钢研纳克检测技术有限公司，北京 100094）

摘 要：液态高纯硫酸锰溶液是制备硫酸锰粉末的半成品，需对其钙镁含量进行严格的把控。液态高纯硫酸锰中锰含量高达 160g/L，选择标准加入法消除测定过程中的基体效应。综合考虑各条谱线的谱图、背景轮廓和强度值，最终选择了干扰较少的 Ca315.887nm 和 Mg285.213nm 作为待测元素的分析线。钙和镁校准曲线的相关系数 R^2 分别为 0.9999 和 0.9997，线性关系良好，方法中钙和镁的测定下限分别为 0.0117μg/mL 和 0.0063μg/mL，结果相对标准偏差(RSD，n=11)为 0.70%和 0.89%，回收率为 98.2%和 90.0%，测定结果准确可靠；并提出了适合生产企业批量快速测定的变异系数法，与标准加入法的测定值基本一致，可一次测定多个样品，工作效率得到明显提高，已应用于实际样品分析，结果满意。

关键词：电感耦合等离子体原子发射光谱法(ICP-OES)，硫酸锰，标准加入法，变异系数法

Determination of Calcium and Magnesium in High - purity Manganese Sulfate Solution by Inductively Coupled Plasma Atomic Emission Spectrometry

Wu Han[1,2], Gong Xiaoyan[1,2], Yuan Wendong[1,2], Yang Chaobin[1,2], Xia Zhonghai[2], Zhao Wendong[2]

(1.Central Iron & Steel Research Institute, Beijing 100081, China; 2. NCS Testing Technology Co., Ltd., Beijing 100094, China)

Abstract: High-purity manganese sulfate solution is the preparation of manganese sulfate powder semi-finished products, its content of calcium and magnesium need strict control, high purity manganese sulfate solution content of manganese up to 160g/L, the standard addition method was selected to removed matrix effect, according to the spectra account, the background profile and the intensity values, the experiment selected spectral lines Ca315.887nm and Mg285.213nm as the analytical lines. The linear correlation coefficients of the calibration curves for calcium and magnesium were 0.9999 and 0.9997, the linear relationship is good, the lower limits of quantification of calcium and magnesium in the method were 0.0117μg/mL and 0.0063μg/mL, the relative standard deviations (RSD, n=11) were 0.70% and 0.89%, the recoveries were 98.2% and 90.0% respectively, and the results are accurate and reliable. And also the coefficient of variation method was used for mass determinations of batch production, its measurement results were consistent with standard addition method, in this method multiple samples can be measured at one time, and the working efficiency were obviously improved. In the actual sample analysis, the results were satisfactory.

Key words: inductively coupled plasma atomic emission spectrometry (ICP-OES), manganese sulfate, standard addition method, coefficient of variation method

用立式管式炉红外吸收法测定硫铁矿中有效硫

王学华，黄小峰，吴利民，张长均，王 蓬

（钢研纳克检测技术有限公司，北京 100081）

摘 要： 本方法采用立式管式炉，在氧气流中，850℃条件下对硫铁矿样品进行加热，其中有效硫转化为 SO_2，由载气携带进入红外检测器进行测定。本方法对样品称样量、分析时间进行了考察，确定最优称样量为 0.1g。本方法分析时间在 120～200s 范围内，较传统燃烧-中和滴定法时间大为缩短，且本方法操作简单，没有传统分析方法繁杂的化学操作流程，易于掌握。对于不同有效硫含量的铁矿石样品，本法分析精度相对标准偏差（RSD，n＝7）0.80% 左右，满足生产需求。采用不同含量梯度的参考物质对仪器进行校准，建立工作曲线，实现了待测样品中有效硫含量的准确测定，与传统燃烧-中和滴定法进行比对，结果一致。

关键词： 红外吸收法，硫铁矿，有效硫，称样量，分析时间

Determination of Effective Sulfur in Pyrite by Using Vertical Tube Furnace Infrared-absorption Method

Wang Xuehua, Huang Xiaofeng, Wu Limin, Zhang Changjun, Wang Peng

(NCS Testing Technology Co., Ltd., Beijing 100081, China)

Abstract: This method uses vertical tube furnace to heat Pyrite sample in oxygen flow at 850℃，and the effective sulfur is tranformed into SO_2, which is carried into the infrared cell by carrier gas and measured. The sample weight and the analysis time is discusssed, and the optimal weight is found 0.1g. The analysis time of this method is between 120～200s, which is much shorter than traditional combustion-neutralized titration method, moreover the operation of this method is simple and easy without complex chemical operation procedures like the traditional method has. The RSD (n=7) of this method is around 0.80% for different content of pyrite samples, which meets the need of production. The working curve is made to calibrate the instrument by using different content certificated reference materials (CRM). The content of effective sulfur in testing samples is determinated accurately, and the results are in accord with the ones obtained by traditonal combustion-neutralized titration method.

Key words: infrared absorption method, pyrite, effective sulfur, sample weight, analysis time

电感耦合等离子体原子发射光谱法测定铜合金中铬

罗岁斌，范小芬，张 霞

（钢研纳克检测技术有限公司 国家钢铁材料测试中心，北京 100081）

摘 要： 本文关于铜合金中铬的测定，研究比较了硝酸-盐酸和硝酸-硫酸-氢氟酸等几种分解铜合金试样的方法以及共存元素对电感耦合等离子体原子发射光谱法（ICP-AES）测定铜合金中铬的影响。结果表明，硝酸-盐酸分解试样

出现的残渣中含铬，导致测试结果偏低；采用硝酸-硫酸-氢氟酸分解，测定结果具有很好的稳定性，其分析结果和JB/T 9552.2—1999（过硫酸铵氧化容量法测定铬）吻合，常见的共存元素对铬的干扰可以忽略，铬的加标回收率在97.6%~101%。此分析方法简便快速，分析准确，成本低，测定铜合金中的铬取得了满意效果。

关键词：电感耦合等离子体原子发射光谱法，铜合金，铬的测定，共存元素影响，试样分解方法

Determination of Chromium in Copper Alloy by Inductively Coupled Plasma Atomic Emission Spectrometry(ICP-AES)

Luo Suibin, Fan Xiaofen, Zhang Xia

(NCS Testing Technology Co., Ltd., National Analysis Center for Iron and Steel, Beijing 100081, China)

Abstract: The influence of the sample decomposition methods and the coexisting elements on the determination of chromium by inductively coupled plasma atomic emission spectrometry is studied in this paper. The test results demonstrate that the residue of the HNO_3-HCl decomposition method contains chromium which leads to lower test results. The HNO_3-H_2SO_4-HF sample decomposition method has good stability and the results are consistent with those obtained by ammonium peroxydisulfate oxidization volumetric method(JB/T 9552.2—1999). The spiked recovery is between 97.6%~101%. This analysis method is simple and accurate, the determination of which obtains satisfied effects.

Key words: inductively coupled plasma atomic emission spectrometry, copper alloy, determination of chromium, influence of coexisting elements, sample decomposition methods

含硼冷镦钢 10B21 镦头成型开裂原因分析

罗新中，林晏民，李富强，章玉成，朱祥睿

（宝武集团广东韶关钢铁有限公司检测中心，广东韶关 512123）

摘 要： 含硼冷镦钢 10B21 在拉拔后镦头成型过程中多次出现开裂，为分析造成冷镦开裂的主要原因，对冷镦开裂试样进行宏观观察、金相组织、非金属夹杂物、扫描电镜及能谱等检测分析。结果表明：10B21 镦头成型开裂的主要原因有连铸卷渣、铸坯裂纹、轧制折叠、轧制划伤、拉拔划伤等五类。

关键词：10B21，镦头开裂，卷渣，裂纹，划伤，折叠

Analysis on the Cracking of the 10B21 Cold -heading Steel Containing Boron in the Cold Heading Forging Process

Luo Xinzhong, Lin Yanmin, Li Fuqiang, Zhang Yucheng, Zhu Xiangrui

(Baowu Steel Group Guangdong Shaoguan Iron and Steel Co., Ltd., Shaoguan 512123, China)

Abstract: After drawing, the 10B21 boron-containing cold heading steel cracks several times in the cold heading forging process. In order to study the main cause of the cracking of the cold heading steel, the cracking cold heading samples were analyzed by macroscopical observation, the microstructure, the nonmetallic inclusions , scanning electron microscopy and energy spectrum detection. The results show that the main reasons for the cracking of 10B21 cold heading forging are steel rolling, slab cracking, rolling folding, rolling scratches and drawing scratches.

Key words: 10B21, cracking of head, slag, crack, folding, scratches

ICP-AES 测定保护渣中的铁、铝、锰、钙、钛、硅和镁

张　军，冯晴晴，张燕茹

（唐山钢铁集团公司技术中心，河北唐山　063016）

摘　要：探讨了电感耦合等离子体发射光谱法（ICP-AES）测定保护渣中铁、铝、锰、钙、钛、硅和镁的分析条件。试样用碳酸钠-硼酸混合熔剂熔融，用盐酸溶解浸出冷却后的熔块，低温加热使之分解，稀释到规定体积。分取部分溶液，将溶液雾化导入电感耦合等离子体发射光谱仪，于所推荐的波长处，测量溶液中待测元素对钇内标元素的相对强度，根据标准溶液制作的校准曲线计算出各待测元素的质量分数。对该方法进行精密度试验，相对标准偏差（RSD，n=8）均小于 3.0%。测定了样品的加标回收率均在 97%～101%之间。本测试方法简单、快速、精度高。将该方法用于保护渣中铁、铝、锰、钙、钛、硅和镁含量的测定，取得满意效果。

关键词：电感耦合等离子体发射光谱法，保护渣，钇，内标法

Determination of Iron, Aluminium, Manganese, Calcium, Titanium, Silicon and Magnesium in Mold Fluxes by ICP-AES

Zhang Jun, Feng Qingqing, Zhang Yanru

(Technology Center of Tangshan Iron and Steel Group Company, Tangshan 063016, China)

Abstract: Analytical conditions of iron，aluminium, manganese, calcium, titanium, silicon and magnesium in mold fluxes by ICP-AES were discussed. The sample was melted with sodium carbonate - mixing boric acid flux, and then was dissolved with hydrochloric acid. Dispensing a portion of the solution, the solution was atomized into the inductively coupled plasma atomic emission spectrometer. Using the standard solution for calibration curve and the internal standard method, the relative intensity of analyte elements were determined at the recommended wavelength. In the precision test, the relative standard deviation (RSD, n = 8) were less than 3.0%. In this way, the paper determined that the standard sample addition recovery rate at 97%～101% .The method is simple, fast and high-precision. It achieved satisfactory results.

Key words: inductively coupled plasma atomic emission spectrometry, mold fluxes, yttrium, internal standard method

中、高镍合金光谱分析用内部控制样品的研制

刘　洁，葛晶晶，任玲玲，禹青霄，禹继志

（河钢集团钢研总院理化检测中心，河北石家庄　050000）

摘　要：本方法根据相关标准研制了包含 C、Si、Mn、P、S、Ni、Cr、Cu、Mo、Co 共 10 个元素的 Hasteloy、Inconel、Monel、Invar 和 Kovar 五种典型的镍合金光谱分析用内部控制样品。采用 50 kg 真空感应炉冶炼，冶炼过程利用感应磁力搅拌，铝粒脱氧，小截面锭型 (130 mm×130 mm×350 mm) 金属模浇铸，从而保证了钢液成分均匀、良好的脱氧效果和元素的较高回收率，避免了成分偏析和夹杂物的形成。铸锭经锻造后进行了均匀性和稳定性检验，均符合控制样品的要求。实验室采用多种不同原理不同方法，采用相应的国家标准、行业标准进行分析检测，总结出准确、可靠的分析方法进行定值分析，对分析结果进行统计和处理，得到了被定值元素的认定值和不确定度。

关键词：中、高镍合金，光谱分析，内部控制样品，研制

Development of the Internal Control Samples for Spectroanalysis of Medium and High Nickel-base Alloys

Liu Jie, Ge Jingjing, Ren Lingling, Yu Qingxiao, Yu Jizhi

(HBIS Group Technology Research Institute, Shijiazhuang 050000, China)

Abstract: According to the related standards, five typical medium and high nickel-base alloys internal control samples for spectroanalysis including Hasteloy, Inconel, Monel, Invar and Kovar containing 10 elements (C, Si, Mn, P, S, Ni, Cr, Cu, Mo, Co) were developed. A 50 kg vacuum induction furnace was used for smelting, and the induction magnetic stirring was used in smelting process. The deoxy-genation was conducted using aluminum particles and casting was used small cross section ingot (130 mm×130 mm×350 mm). The uniform liquid steel composition, good deoxygenation effect and high recovery of elements were guaranteed, avoiding composition segregation and the formation of inclusions. After forging, the statistic test and stability test were conducted. The results comply with the requirements of the internal control sample. The certified value was obtained in laboratories using accurate and reliable analytical methods with different principles. The analytical results were statisticed and processed. The certified values and expanded uncertainty of testing elements were obtained.

Key words: medium and high nickel-base alloy, spectroanalysis, internal control sample, development

汽车板埃里克森杯突试验能力验证的设计实施

张慧贤[1]，佟艳春[2]，张　亮[2]

（1. 钢研纳克检测技术有限公司，北京　100094；
2. 北京中实国金国际实验室能力验证研究中心，北京　100081）

摘　要：随着社会进步和国民生活水平的提高，汽车已经成为日常生活中不可或缺的交通工具，汽车制造业高速发展，汽车用钢检测行业越来越受到重视。其中汽车用钢良好的冲压成型性是其重要的性能指标之一，而埃里克森杯突试验则是该性能的常规检验方法，埃里克森杯突值能够直接反映出汽车板材的优劣程度。能力验证是认可机构和管理机构判定实验室能力的重要技术手段，也是实验室内部质量控制的重要补充。实验室通过参加有效的能力验证活动，并获得满意结果，能够增加客户以及相关方对实验室的信任。本文介绍了汽车板埃里克森杯突试验能力验证计划的设计实施过程及结果分析等情况。汽车板埃里克森杯突试验能力验证项目的设计实施，有助于帮助实验室发现日常检测存在的问题，为提高实验室的测试水平提供了依据，具有十分重要的现实意义。

关键词：埃里克森杯突试验，埃里克森杯突值，汽车板，能力验证

Design and Implementation Proficiency Testing Scheme of Automobile Sheet-Erichsen Cupping Test

Abstract: Along with the social progress and the improvement of the national standard of living, the car has become an indispensable means of transportation in daily life. With the rapid development of automobile manufacturing industry, more and more attention has been paid to automobile steel testing industry. Good formability is one of the most important performance indexes of automobile steel, Erickson cupping test is a routine test method of this performance. Erickson cupping index can directly reflect the quality and the degree of Automobile Sheet. Proficiency Testing is an important technical means to determine the ability of the authorized institutions and management organizations, and is also an important supplement to the laboratory internal quality control. Laboratory to participate in effective Proficiency Testing, and obtain satisfactory results, can increase the trust of the customer and related parties. This paper introduces the design and implementation process, results analysis of proficiency testing scheme of Automobile Sheet-Erichsen cupping test. Design and implementation proficiency testing scheme of Automobile Sheet-Erichsen cupping test, may help the lab to find the problems in daily testing, provides a basis in order to improve the level of laboratory, has very important practical significance.

Key words: erichsen cupping test, erickson cupping index, automobile sheet, proficiency testing

火焰原子吸收测定 GH4169 合金中锰的干扰消除研究

范小芬，罗岁斌，齐 荣，刘庆斌，孟子敬，李美慧

（钢研纳克检测技术有限公司国家钢铁材料测试中心 北京 100081）

摘 要： 目前在采用火焰原子吸收（FAAS）测定 GH4169 高温合金中的锰元素时，存在共存元素的干扰问题，使测定结果偏低。本文采用王水和氢氟酸来溶解 GH4169 高温合金试样，加入 1.5 mL 的磺基水杨酸（0.5 mol/L）作为掩蔽剂消除共存元素对锰测定的干扰，测定结果与标准值以及电感耦合等离子体原子发射光谱（ICP-AES）测定值相吻合。此方法的检出限（3 s）为 0.036 μg/mL，采用标准加入法进行回收率实验，测得回收率在 92.8%~95.6%之间。

关键词： 火焰原子吸收，GH4169，锰，磺基水杨酸，共存元素干扰

The Elimination of Interference on the Determination of Manganese of GH4169 by Flame Atomic Absorption Spectrometry

Fan Xiaofen, Luo Suibin, Qi Rong, Liu Qingbin, Meng Zijing, Li Meihui

(NCS Testing Technology Co., Ltd., National Analysis Center for Iron and Steel, Beijing 100081, China)

Abstract: The coexisting elements interfere the determination of manganese in the high temperature alloy GH4169 by flame atomic absorption spectrometry (FAAS), which lead to lower results. In this paper, GH4169 was digested by aqua regia and hydrofluoric acid, 1.5 mL of sulfosalicylic acid was added(0.5 mol/L) as a masking agent to eliminate the interference of coexisting elements, the results (FAAS) are consistent with those obtained by inductively coupled plasma atomic emission spectrometry (ICP-AES) and also consistent with certified value. The detection limit (3S) of this method is 0.036 g/mL, and the recovery rate is between 92.8%~95.6%.

Key words: FAAS, GH4169, manganese, sulfosalicylic acid, coexisting element interference

直读光谱法快速测定易切削钢中高硫含量

张世欢，李世晶

（宝钢集团广东韶关钢铁有限公司检测中心，广东韶关　512123）

摘　要： 本文试验了用火花源原子发射光谱法测定易切削钢（1215MS）中的硫含量。本法通过实验选择检测该钢种的光谱仪最佳预燃时间、最佳分析强度等和工作曲线的延伸、扩展，同时对易切削钢试样进行光谱法和红外碳硫分析进行比较，结果表明火花源原子发射光谱法测定易切削钢中的硫含量效果明显，满足生产的要求。

关键词： 火花源发射光谱法，易切削钢，硫含量

Photoelectric Direct Reading Spectrometry Determination of Free Cutting Steel Sulfur Content

Zhang Shihuan, Li Shijing

(BaoSteel Group Shaogang Quality Inspection Center, Shaoguan 512123, China)

Abstract: This article introduces the tests by spark source atomic emission spectrometry (1215MS) content in free-cutting steel. This method is tested by experimental test of the steel grade spectrometer best precombustion time and the best intensity and the extension of work curve, extend, at the same time, we draw a comparison between spark source atomic emissio spectrometry and infrared carbon sulfur for free-cutting steel, the results show that the spark source atomic emission spectrometry makes the obvious effects of sulfur content in free cutting steel, meet the requirements of production.

Key words: photoelectric direct reading spectrometry, free-cutting steel, sulfur content

2Cr12NiMo1W1V 叶片钢高温拉伸性能的分析探讨

王丽英

（山西太钢不锈钢股份有限公司，山西太原　030003）

摘　要： 通过对四种不同温度、四种不同拉伸应变速率下叶片钢 2Cr12NiMo1W1V 的拉伸变形行为进行分析研究，结果表明：温度升高，2Cr12NiMo1W1V 叶片钢强度下降，塑性升高，300℃时塑性反而略有下降；随着变形的增大，各温度下应变硬化率都持续下降，900℃时整个变形过程中应变硬化率几乎为零；600℃以上拉伸变形流变应力可以用 Zener-Hollomon 方程表征。

关键词： 叶片钢，高温拉伸，应变速率，强度

Analysis and Discussion of Elevated Temperature Tensile Properties of 2Cr12NiMo1W1V Blade Steel

Wang Liying

Abstract: The tensile deformation behavior was investigated of 2Cr12NiMo1W1V blade steel under four different temperatures and four different strain rates.The results indicate that ,as the temperature rises ,the tensile strength of 2Cr12NiMo1W1V blade steel decreases ,and the plasticity increase, on the contrary the plasticity have a little decrease at 300℃. When the strain increase,the tensile strain hardening rate always decrease , the tensile strain hardening rate is almost zero during all deformation at 900℃.The tensile flow stress may be expressed by Zener-Hollomon equation when test temperature is higher than 600℃.

Key words: blade steel, evevated temperature tensile, strain rate, strength

轧机轴承外套断轧机轴承外套断裂失效分析

马惠霞，钟莉莉，李文竹，严平沅，王晓峰

（鞍钢股份有限公司技术中心，辽宁鞍山　114009）

摘　要： 采用断口分析技术、金相分析技术，以及表面硬度测定，结合森吉米尔多辊轧机辊系特征，对一背衬轴承外套早期断裂失效原因进行检验分析。结果表明：套圈外表面因摩擦而产生的机械热损伤导致套圈出现挤压裂纹。裂纹的存在，显著地降低了断裂强度，而且在裂纹尖端产生应力集中。当应力达到或超过材料强度极限时，套圈发生了早期断裂。

关键词： 20辊轧机，轴承外套，早期断裂，表面热损伤

Failure Analysis of Broken Bearing Cover of Rolling Mill

Ma Huixia, Zhong Lili, Li Wenzhu, Yan Pingyuan, Wang Xiaofeng

(Technology Center, Angang steel Co., Ltd., Anshan 114009, China)

Abstract: The premature fracture of bearing cover was analysed, with the fracture analysis, the metallographic analysis, the surface hardness determination and characters of the 20 high sendzimir mill. The results showed that mechanical thermal damage cause extruded crack of the bearing cover. The crack reduced the fracture strength and stress concentration occurred on the crack tip. The bearing cover broke when the stress greater than or equal to the breaking point.

Key words: 20 high rolling mill, bearing cover, premature fracture, thermal damage

统计分析技术在冶金实验室质量管理中的应用

杜士毅，王贵玉，金　伟，沈　涛，孙　娟，闫　丽

（首钢股份公司质量检验部，河北迁安　064406）

摘　要：本文阐述了统计分析技术在实验室质量管理的应用方法，以实例的方式说明了配对 T 检验、方差分析、筛选实验设计、田口设计、统计过程控制在冶金实验室应用的方法及效果。根据不同的质量管理要求，有针对性地采用统计质量管理方法，可有效提高质量管理工作的效率和可靠性，对于冶金实验室质量管理水平的提高、完善质量管理体系有重要的借鉴意义。

关键词：质量，统计分析，冶金实验室，管理措施

Application of Statistical Analysis Technology in Quality Management of Metallurgical Laboratory

Du Shiyi, Wang Guiyu, Jin Wei, Shen Tao, Sun Juan, Yan Li

(Shougang Co., Qian'an Steel Corp., Qian'an 064406,China)

Abstract: This article describes the statistical analysis methods applied quality management in laboratory. The following examples illustrate paired T test, ANOVA, filtration of DOE, Taguchi design, statistical process control in the application of the method and effect of the metallurgical laboratory. According to the different requirements of the quality management, the quality management by statistical methods, can effectively improve the efficiency and reliability of quality management, the quality management level of the metallurgical laboratory to enhance and improve the quality management system has important significance.

Key words: quality, statistical analysis, metallurgical laboratory, management measures

一种基于线激光的目标深度测量方法

李文浩，贾　同，薛　宇，侯明亚

（东北大学信息科学与工程学院，辽宁沈阳　110000）

摘　要：本文通过线激光和摄像头构建测量系统，利用三角测距原理获取目标物体的深度信息。首先，将线激光投射到被测物体表面，利用摄像头获取线激光的图像；然后采用光条中心提取技术计算线激光的位置；最后，基于三角测量原理获得物体的深度信息。通过实验表明，本方法近距离（小于 1 m）测量误差在 5 mm 以内，远距离（最远 6 m）最大测量误差 80 mm。相比较同类，本方法测量精度更高。

关键词：线激光，深度测量，光条中心提取，三角测量

A Method of Target Depth Measurement Based on Line Laser

Li Wenhao, Jia Tong, Xue Yu, Hou Mingya

(College of Information Science and Engineering, Northeastern University, Shenyang 110000, China)

Abstract: In this paper, the measurement system is constructed by line laser and camera ,and the target depth information is acquired on the principle of triangulation measurement. Firstly, the line laser is projected onto the target surface, and then the camera is used to obtain the line laser image. Secondly, the center position of the line laser image is calculated by the stripe center extraction technique. Finally, the depth information of the target is obtained based on the triangulation principle. The experimental results show that the measurement error of this method is less than 5mm at short range (less than 1m), and the maximum measurement error is 80mm at long distance (more than 6m). Compared with the related measurement method, it has higher measurement accuracy.

Key words: line laser, depth measure, light center extraction, triangulation measurement

全谱直读光电光谱法测定中低合金钢中的痕量硼

史玉涛，屈华阳

（钢研纳克检测技术有限公司技术中心，北京　100081）

摘　要： 钢铁中 B 的加入量多少对材料的性能具有重要影响，采用全谱火花光谱仪 SparkCCD6000 建立了准确快速测定钢铁中痕量 B 的方法：优化了光源激发条件，比较了不同的基体参比线对曲线线性和测定结果稳定性的影响，最终选择 B 的分析线为 182.6nm，参比线为 182.1nm。建立的校准曲线线性相关系数 R^2 为 0.99624，用该曲线分析未知样品 10 次，考察方法精密度，结果的相对标准偏差 RSD 为 3.31%（测试平均值为 0.0024%）。同时，考察了方法的准确度，用该曲线分析标准样品 GBW01395~GBW01400，测定结果与标样认定值一致。

关键词： 火花源原子发射光谱法，CCD 全谱，中低合金钢，痕量硼

Determination of Trace Boron in Mid & Low Alloy Steel by Whole Spectrum Optical Emission Spectroscopy

Shi Yutao, Qu Huayang

(Technique Centre of NCS Testing Technique Co., Ltd., Beijing 100081, China)

Abstract: The quantity of boron added into steel has an important influence on the properties of the material. The determination method of trace boron in steel was established by Spark CCD 6000, a whole spectrum optical emission spectrometer. Source parameters of the spectrometer were optimized and the influence of different reference lines was compared on the linear correlation coefficient of the calibration curve and precision of the final results. As a result, the analytical line of boron was selected as 182.6nm and the reference line was selected as 182.1nm. The Linear correlation coefficient R^2 of the calibration curve is 0.99624. Precision test of the method was conducted by analyzing an unknown sample for 10 times, the relative standard deviation (RSD) of the result is 3.31%(with a mean value of 0.0024%). Accuracy tests were also performed by analyzing a set of certified reference materials(CRMs) GBW01395~GBW01400. The found results were consistent with the certified values of the CRMs.

Key words: spark source optical emission spectroscopy (spark-OES), charge coupled device(CCD) whole spectrum, mid & low alloy steel, trace boron

厚规格 X80 管线钢显微组织与力学性能

王建钢[1]，何建中[1]，樊立峰[2]，王　皓[1]，袁晓鸣[1]

（1. 内蒙古包钢钢联股份有限公司，内蒙古包头　014010;

2. 内蒙古工业大学材料科学与工程学院，内蒙古呼和浩特　010051）

摘　要: 对 22 cm 厚壁 X80 管线钢的显微组织、第二相析出物及力学性能进行了分析研究，结果表明：(1) X80 管线钢组织主要为针状铁素体，部分粒状贝氏体和 M/A 岛，晶粒度均大于 10 级; (2) 第二相主要为 Nb、Ti 的碳氮复合析出，尺寸主要集中在 30~50nm; (3) 热轧板沿 30°方向平均屈服、抗拉强度分别为 583MPa 和 681MPa，DWTT85%为–33℃，CVN50%低于–60℃，均满足 X80 管线钢的质量要求。

关键词: X80，针状铁素体，析出物，力学性能

Microstructure and Mechanical Properties of X80 Pipeline Steel

Wang Jiangang[1], He Jianzhong[1], Fan Lifeng[2], Wang Hao[1], Yuan Xiaoming[1]

(1. Baotou Steel (Union)Co., Ltd., Baotou 014010, China; 2. School of Materials Science and Engineering, Inner Mongolia University of Technology, Hohhot 010051, China)

Abstract: The microstructure, precipitates and mechanical properties of X80 pipeline steel were investigated, the results show that: (1) the microstructure of X80 was composed of acicular ferrite and a little granular bainite and M/A islands, the grain size is greater than 10. (2) the precipitates were mainly carbonitride of Nb and Ti, and the size is 30~50nm. (3)the average yield strength and tensile strength of hot rolled plate with the 30°deviation of rolling direction were 583 MPa and 681 MPa, the CVN50% was under –20℃, the DWTT85% was –33℃, which can meet the X80 requirements.

Key words: X80, acicular ferrite, precipitate, mechanical property

河钢石钢高端钢纯净度分析及进一步改善措施

王殿峰，丁　辉，席军良，任鹏飞

（河钢集团石家庄钢铁有限责任公司，河北石家庄　050031）

摘　要: 水浸高频超声检测设备可以检测和评估产品宏观夹杂物，由于检测速度快、检测体积大，所以具有快速、准确评估产品纯净度的能力，检测结果可直观反映钢中缺陷的形状、位置和分布，并可通过计算量化纯净度指数，进而得到一段时间内纯净度变化的趋势，指导炼钢工艺完善和技术进步。河钢石钢利用水浸高频超声设备对产品进

行了长期的跟踪检测和总结分析，为河钢石钢产品迈向高端架起了一座桥梁。本文通过对河钢石钢高端钢纯净度检测分析，指出影响产品纯净度的主要夹杂物类型，为进一步改善部分产品的纯净度，提出了工艺改进措施并验证实施效果。同时探讨通过高频超声设备检测微观夹杂或较小宏观夹杂物的可能性。

关键词：纯净度，水浸高频超声，评估，纯净度指数，趋势

Hesteel Shisteel High Level Product Cleanness Analysis and Further Improvement Action

Wang Dianfeng, Ding Hui, Xi Junliang, Ren Pengfei

(Hesteel Group Shijiazhuang Iron and Steel Co., Ltd., Shijiazhuang 050031, China)

Abstract: Water immersion high-frequency ultrasonic testing machine can test and evaluate the macroscopic inclusion, the testing speed is fast, testing volume is large, the testing have the fast and accurate ability to evaluate the product's cleanness, testing results will directly show the defect profile, position and distribution, and after calculation will get the cleanness index, collecting some period cleanness index data will get the cleanness change trend, this help to improve the steel making process and technical progress. Hesteel Shisteel use this testing machine made the long terms trace test , summary and analysis, build up a bridge for Hesteel Shisteel products go to high level. In this thesis, we use the cleanness index data, along with the inclusion type, give the instruction steel making process to promoting some product's cleanness and verify its results, meanwhile make a discussion for the possibility to use this testing machine to check the microscopic inclusion or the small macroscopic inclusion.

Key words: cleanness, high-frequency ultrasonic, evaluation, cleanness index, trend

X 射线荧光光谱法测定钢铁企业中返生产利用资源主次成分

李世晶，马秀艳，张世欢，吴超超，邢文青

（宝武集团广东韶关钢铁有限公司，广东韶关　512123）

摘　要：本文通过熔融制样低温预处理技术和高温熔融制样，建立了 X 射线荧光光谱法测定成分复杂的水渣铁粒、富集磁粉、氧化铁皮等返生产利用资源中的主次成分的工作曲线，探讨了不同氧化剂、熔融温度、熔融时间对检测结果的影响，解决了部分试样中金属颗粒物腐蚀铂金坩埚的难题，实现了对返生产利用资源中的全铁、S、SiO_2、Al_2O_3、CaO、MgO 等元素的快速检测。试验采用（1：14）的试样熔剂稀释比，采用氧化钴作内标，在 1020℃ 熔融处理试样，样品熔解完全，且对铂金坩埚无明显腐蚀。试验结果表明，采用熔融法制样—X 射线荧光光谱法测定返生产利用资源样品，该方法有效消除了基体效应和粒度效应对检测结果的影响，具有良好的精密度和正确度，已经应用于生产检测。

关键词：熔融法制样，X 射线荧光光谱法，返生产利用资源，主次成分

Determination of Major and Minor Components in the Utilized Resources of Back Production for Iron and Steel Enterprises by X-ray Fluorescence Spectrometry

Li Shijing, Ma Xiuyan, Zhang Shihuan, Wu Chaochao, Xing Wenqing

(Baowu Steel Group Shaoguan Iron and Steel Co., Ltd., Shaoguan 512123, China)

Abstract: Through the sample melt pretreatment of low temperature and high temperature melting technology, established the working curve by X ray fluorescence spectrometry determination, which can measure the major and minor components of complex composition of slag iron particles, enrichment of magnetic powder, iron oxide scale etc for the utilized resources of back production. And also discusses the influence of different oxidants, melting temperature, melting time of detection the results, to solve the problem of corrosion of metal particles in the platinum crucible part of the sample and to achieve rapid detection of backward production by using total iron, S, SiO_2, Al_2O_3, CaO, MgO and other elements in the resources. In the experiment, the flux dilution ratio of the sample (1:14) was used, and the cobalt oxide was used as the internal standard. The samples were melted at 1020℃, and the samples were completely melted. Experimental results show that the method-the determination of production resource utilization by return sample melting method—X ray fluorescence spectrometry, which can effectively eliminate the matrix effect and the effect of particle size on the test results with good precision and accuracy, has been applied in production testing.

Key words: fusion sample preparation, X-ray fluorescence spectrometry, the utilized resources of back production, major and minor components

钢铁材料中氧氮的测定分析研究

刘文玖，李建红，刘　静

（陕钢集团汉中钢铁有限责任公司，陕西汉中　724200）

摘　要： 钢铁材料中的氮含量超过一定限值，在加热时便会出现"兰脆"，钢的韧性下降，脆性增加；氧含量决定夹杂物的类型、分布，破坏钢基体连续性而造成干裂[1]，影响钢铁材料的力学及加工性能和再加工拉拔处理。汉钢公司在冶炼新品钢（如 77B、82B 钢绞线用母材）时，精炼操作，要求控制氧含量≤30ppm，氮含量≤60ppm。氧氮氢分析仪应用于钢中氧氮氢检测。本文就氧氮氢分析仪分析金属材料中氧氮含量，进行研究，探讨出能够准确分析金属材料中氧氮含量的有效方法，收到良好效果，科学指导精炼生产。

关键词： 钢铁材料，氧氮含量，红外吸收热导法

Determination Analysis of Oxygen and Nitrogen in Steel Materials

Liu Wenjiu, Li Jianhong, Liu Jing

(Shaanxi Steel Group Hanzhong Iron and Steel Co., Ltd., Hanzhong 724200, China)

Abstract: When nitrogen in steel materials exceeds a certain limit, the toughness of steel will decrease, brittleness increases;

and it will appear "blue brittle" when heated. The oxygen content determines the type and distribution of inclusions, when it exceeds a certain limit, it will damage the continuity of steel substrate and cause cracking[1], which will affect the properties of steel and iron and reprocessing. Our company have a set of strict production requirements, that requires to control oxygen content less or equal than 30ppm, nitrogen content less or equal than 60ppm when smelt new type of iron and steel (such as base metal of 77B, 82B steel strand). Oxygen and nitrogen analyzer is mainly used in detecting steel's oxygen and nitrogen contents. This article is about how oxygen and nitrogen analyzer detects oxygen and nitrogen contents, carries on research and explores the effective way of precisely analyzing oxygen and nitrogen contents in metal materials, which has received good effects and used in refining production.

Key words: steel materials, the oxygen and nitrogen contents, infrared absorption and thermal conductivity method

碱熔络合置换滴定法测定铝基钢包渣中总铝含量

吕 琦，沈 克，郭 芳，张小凡，陈高莉

（宝钢股份武钢股份质检中心，湖北武汉　430080）

摘　要： 基于铝的还原性及两性特征，采用 HAc-NaAc 缓冲液-碱溶法对炼钢用铝基钢包渣中总铝进行提取，并采用络合置换滴定法对铝含量进行测定，建立了一种铝基钢包渣中总铝含量快速、准确测定的分析方法。实验结果表明，总铝含量的滴定结果标准偏差小于 0.28%，加标回收率为 99.21%～100.46%，实现了样品的高准确度定量分析。

关键词： 铝基钢包渣，总铝，络合置换滴定

An Alkali Fusion Complexometric Displacement Titration Analysis the Metallic Al in the Al-Based Ladle Slag

Lv Qi, Shen Ke, Guo Fang, Zhang Xiaofan, Chen Gaoli

(Quality Inspect Center of Baowu Steel Group Corporation Limited, Wuhan 430080, China)

Abstract: Based on the characteristics of Al, reducibility and amphotericity, the Al metal and total Al in series of steelmaking Al-based deoxidizer are respectively extracted with iodine-ethanol-extraction and HAc-NaAc buffer-alkali dissolving method, whose content are determined by the complex displacement titration, developing a analysis method to determine the content of Al in the Al-based deoxidizer and realizing a rapid and accurate determination. It shows that, the standard deviation of results using complexing-displacement titration method in metal-Al determination is less than 0.45%, while the recovery being 98.8%～102.96%, realizing a high accuracy and high parallel quantitative analysis.

Key words: Al-based ladle slag, total Al, complexometric replacement titration

激光诱导击穿光谱技术分析高含氧量的合金样品

王 辉[1]，邵慧琪[2]，郭飞飞[1]，侯红霞[1]，冯 光[1]，陈吉文[1]

（1. 钢研纳克检测技术有限公司技术中心，北京　100081；
2. 北京科技大学化学与生物工程学院，北京　100083）

摘　要：采用激光诱导击穿光谱（LIBS）技术分析火花光谱无法正常激发的高含氧量合金样品。选择能涵盖分析试样各元素含量和强度的标准样品绘制工作曲线，采用工作曲线分别分析待测的分析试样和标准样品，由标准样品的认定值与其测定值绘制趋势线，得到趋势线计算式，计算分析试样的元素含量。标准样品工作曲线、标准样品认定值与测定值趋势线线性良好，碳元素和硫元素分析结果与碳硫分析仪分析结果有良好的对应。

关键词：激光诱导击穿光谱，高含氧量，扩散放电，工作曲线

Analysis of High Oxygen Content Alloy Samples by Laser Induced Breakdown Spectroscopy

Wang Hui[1], Shao Huiqi[2], Guo Feifei[1], Hou Hongxia[1], Feng Guang[1], Chen Jiwen[1]

(1. Technology Center, The NCS Testing Technology Co., Ltd., Beijing 100081, China; 2. School of Chemistry and Biological Engineering, University of Science and Technology Beijing, Beijing 100083, China)

Abstract: Laser induced breakdown spectroscopy (LIBS) was used to analyze the high oxygen content alloy samples which could not be excited normally by spark OES. Standard samples which could cover each element concentration and intensity of analytical samples were selected to draw the working curve. Analytical samples and standard samples were analyzed respectively by the working curve. The trend lines of certified values and measured values of standard samples were drew. The element concentration of analytical samples was calculated by the trend lines. The working curves and the trend lines of certified values and measured values of standard samples were satisfied. The analysis results of carbon and sulfur were in good agreement with those of the carbon and sulfur analyzer.

Key words: laser induced breakdown spectroscopy, high oxygen content, diffuse discharge, working curve

钢帘线用钢盘条 C82D2 质量研究

李桂英

（中信特钢青岛特殊钢铁有限公司，山东青岛　266409）

摘　要：通过试验和检验，分析研究了盘条化学成分、连铸坯中心偏析、非金属夹杂物、钢中 O、N 含量、轧后控冷工艺等 5 种因素对 C82D2 盘条质量的影响。研究表明，应严格控制 C82D2 盘条化学成分的均匀性和有害元素的含量；连铸坯中心偏析严重，易使盘条在拉拔时断裂，产生杯锥状断；盘条中的非金属夹杂物，使盘条在拉拔和捻制变形时，因应力作用而造成钢丝断裂；钢中 O、N 含量过高，会使钢的强度和硬度升高，塑韧性下降；合理的吐丝温度和冷却速度，能保证盘条获得理想的细索氏体组织。为了进一步提高 C82D2 盘条的质量，提出了质量改进建议。

关键词：钢帘线，C82D2，中心偏析，夹杂物，质量研究

Investigation of C82D2 Wire Rod for Steel Cord Quality

Li Guiying

(Qingdao Special Iron and Steel of the Citic Special Steel, Qingdao 266409, China)

Abstract: To analyze and investigate the effect of such factors as chemical composition，central segregation in continuous casting billet, non-metallic inclusions, oxygen and nitrogen contents in steel, cooling process after rolling on quality of C82D2 wire rod. The investigation indicates that the homogeneous of chemical composition and deleterious　elements　is strictly controled, serious carbon segregation in the center of wire rod is easy to make wire rod break in drawing and bring on cup-cone shape fracture, non-metallic inclusions in wire rod cause steel wire rupture in wire drawing and stranding for stress，the high weight of oxygen and nitrogen improve the hardness and intensity of steels，reasonable rolling and cooling process can get fine microscope sorbite.The suggestions of improving quality of C82D2 wire rod are put forward.

Key words: wire rod, C82D2, central segregation, non-metallic inclusions, investigation quality

钢中析出相的透射电子显微镜表征方法及应用

马家艳，关　云，邓照军

（武汉钢铁（集团）公司研究院，湖北武汉　430080）

摘　要： 本文阐述了钢中析出相的透射电子显微镜表征方法。以硅钢等为例，显示了采用该方法对钢中析出相进行表征而得到的析出相的形态、分布、尺寸、成分、类型、结构、数量等定性及半定量分析结果，并讨论了定量结果的误差来源。结果表明此方法是表征纳米级析出相的最直观的有效途径。

关键词： 钢，析出相，透射电子显微镜，方法

Transmission Electron Microscope Characteristic Method and Application of the Precipitates in Steels

Ma Jiayan, Guan Yun, Deng Zhaojun

(Research and Development Center of Wuhan Iron and Steel (Group) Co., Wuhan 430080, China)

Abstract: Transmission electron microscope characteristic method of the precipitates in steels was elaborated in this paper. As examples, the precipitates in the silicon steel and other steels were analyzed by this method and the qualitative and quantitative results about shape, distribution, size, composition, type, structure and number were obtained. The error origin of quantitative results was discussed. The above results appear that this method is the most intuitive and effective way of characterizing the nano-size precipitates.

Key words: steel, precipitates, transmission electron microscope, method

电感耦合等离子体原子发射光谱法测定土壤中 8 种有效态元素

罗剑秋[1,2]，吴　菡[1,2]，袁文东[1,2]，杨超彬[1,2]，夏钟海[2]，隗立晶[2]

（1. 钢铁研究总院，北京　100081；2. 钢研纳克检测技术有限公司，北京　100094）

摘　要：土壤有效态元素是指在植物生长期内土壤中能够被植物根系所吸收的元素，重金属进入土壤后对土壤的生态结构，植物的生长以及人体健康均会产生危害。本文根据环境新标准 HJ 804—2016 《土壤 8 种有效态元素的测定 二乙烯三胺五乙酸浸取-电感耦合等离子体发射光谱法》，准确称取 10.0 g 土壤样品置于 100 mL 三角瓶中，加入 20.0 mL DTPA 浸提液，将瓶塞盖紧，在 20℃±2℃室温下，以 180r/min±20r/min 的振荡频率振荡 2h，将浸提液缓慢导入离心管中，300r/min 离心 10min，上清液经定量滤纸重力过滤后于 48h 内进行测定分析。综合考虑各条谱线的谱图、背景轮廓和强度值，最终选择了 Cu 327.396 nm、Fe 238.204 nm、Mn 257.610 nm、Zn 213.856 nm、Cd 214.438 nm、Co 228.616 nm、Ni 231.604 nm、Pb 220.353 nm 作为分析线，各元素校准曲线的相关系数 R_2 均大于 0.999，线性关系良好，测定下限为 0.003~0.129 μg/g，测定结果的相对标准偏差（RSD, n=10）均小于 2.0%，回收率为 90.69%~104.60%，测定结果与标准认定值相吻合。

关键词：电感耦合等离子体原子发射光谱法（ICP-OES），有效态元素，DTPA

Determination of Eight Available Element Contents in Soils by Inductively Coupled Plasma Optical Emission Spectroscopy

Luo Jianqiu[1,2], Wu Han[1,2], Yuan Wendong[1,2], Yang Chaobin[1,2],
Xia Zhonghai[2], Wei Lijing[2]

(1. Central Iron & Steel Research Institute, Beijing 100081, China;
2. NCS Testing Technology Co., Ltd., Beijing 100094, China)

Abstract: Soil available elements refer to the elements that can be absorbed by the plant roots in the soil during the growing period. After the heavy metals enter the soil, the soil ecological structure, plant growth and human health will be harmful. In this paper, according to the new environmental standard HJ 804–2016 《Determination of 8 available elements in soils - DTPA leaching - inductively coupled plasma optical emission spectroscopy》, 10.00 g (accurate to 0.0001 g) of soil was weighted and extracted in the 100 mL flask by 20.0 mL DTPA extracts. Then the samples was oscillated by 180r/min±20r/min under 20℃±2℃. Then the extractions was poured into a centrifuge tube and centrifuged 10 min in 300r/min to get supernatant and it was measured within 48 hours. According to the spectra account, the background profile and the intensity values, the experiment selected spectral lines Cu 327.396 nm, Fe 238.204 nm, Mn 257.610 nm, Zn 213.856 nm, Cd 214.438 nm, Co 228.616 nm, Ni 231.604 nm, Pb 220.353 nm as the analytical lines. The linear correlation coefficients of each element were more than 0.999, the linear relationship is good, the low limit of elements were between 0.003μg / g and 0.129μg / g, the relative standard deviations (RSD, n=10) were less than 2.0%, the recoveries were between 90.69% and 104.60%, the determination of the results consistent with the standard value.

Key words: inductively coupled plasma atomic emission spectrometry (ICP-OES), available element, DTPA

冷轧镀锌线带钢表面质量检测系统开发与应用

李志锋[1]，李连成[1]，秦大伟[2]，宋宝宇[2]，曹忠华[2]

（1. 鞍钢集团钢铁研究院，辽宁鞍山　114009；2. 鞍钢未来钢铁研究院，北京　102209）

摘　要：目前国内外许多钢厂在冷轧带钢生产线上配置了表面质量检测系统，但多数系统都是进口的，自主研发的国有冷轧带钢表面质量检测系统并投入在线使用的显有报道。论述了冷轧镀锌线生产工艺的技术特点，镀锌带钢表面检测系统技术需求，设计开发了一套冷轧镀锌带钢表面质量在线检测系统，成功应用于某冷轧镀锌线上。分别从硬件组成、软件设计、应用效果进行了详细论述。

关键词：冷轧镀锌线，表面质量检测，图像识别，照明光源

Development and Application of Surface Quality Inspection System for Cold Rolled Galvanized Strip

Li Zhifeng[1], Li Liancheng[1], Qin Dawei[2], Song Baoyu[2], Cao Zhonghua[2]

(1. Iron & Steel Research Institute of Angang Group, Anshan 114009, China;

2. Ansteel Beijing Research Institute, Beijing 102209, China)

Abstract: Currently many domestic and foreign steel mills in cold-rolled strip steel production line configuration of the surface quality inspection system, but most of the system are imported, the independent research and development of the state-owned cold-rolled strip surface quality detection system and put into use online are reported. Analysis of the technical characteristics of production technology of cold-rolled galvanized plant, galvanized steel strip surface inspection system technical requirements, design and development of a cold rolled galvanized steel strip surface quality on-line detection system is successfully applied in a cold-rolled galvanized plant. Separately from the hardware composition, the software design, the application effect has carried on the detailed elaboration.

Key words: galvanized plant of cold rolled、surface quality inspection, image recognition, Illumination light source

X 射线荧光光谱法分析高速工具钢

夏鹏飞，牛国锋，王立军，苏建民，张文华

（邢台钢铁有限责任公司质量控制部，河北邢台 054001）

摘　要：本法是采用真空取样器插入钢水中取球拍样，使用南京和澳铣样机直接铣制出分析面，试样表面应平整、光洁，不能有沙眼、夹杂、裂纹等。采用 X 荧光光谱法分析其中的 Si、Mn、P、S、V、Cr、W、Mo 等元素。本方法建立的校准曲线线性良好，测定结果准确、快速的指导大规模生产。

关键词：X 荧光光谱仪，高速工具钢，检测

超低碳检验在生产中的分析应用

林 丽

（本钢集团北营分公司质量管理中心，辽宁本溪　117017）

摘　要：本文主要叙述了使用红外碳硫分析仪对超低碳、硫进行分析的实验条件,分析方法及相关的实验过程，列举了大量的实验数据。试验数据表明采用红外吸收法可对超低碳、硫进行分析测定的精密度和准确度满足 ASTME1019 的要求，超低碳检验在实际生产中切实可行。

关键词：超低碳、硫，试验条件，精密度，准确度

The Analysis and Application of Ultra-low Carbon Inspection in the Production

Lin Li

(BX Steel Beiying Group, Benxi 117017, China)

Abstract: The text mainly discusses the experimental conditions and analysis method for ultra-low carbon and sulfur which are tested by the infrared carbon sulfur analyzer. In the experimental process, listed a lot of data, The data show that the accuracy and precision of determination of the ultra-low carbon and sulfur which is by the infrared absorption method meet the requirements of standard ASTME1019.Ultra-low carbon inspection is feasible in the production.

Key words: ultra-low carbon、sulfur, experimental conditions, accuracy, precision

钢铁材料 XPS 面分析时仪器参数的优化研究

王文昌，蔡　宁，鞠新华

（首钢集团有限公司技术研究院检测中心，北京　100043）

摘　要： XPS 是一种表面纳米级深度的分析手段，研究钢铁材料 XPS 面分析可以为表面纳米级元素分布提供研究指导。通过 XPS 研究了不同步长对镀锡板表面面分析结果的影响。结果表明，随着步长降低，XPS 面分析分辨率提高，元素含量结果更加准确。当步长为 60μm 时，XPS 面分析可以清晰地展现出镀锡板表面元素的分布状况。继续降低步长，分析时间增加，元素含量结果变化不大。步长 60μm 是钢铁材料 XPS 面分析的最佳分析条件。

关键词： XPS，镀锡板，面分析，步长

Study on Optimization of Instrument Parameter of XPS Mapping for Steel Materials

Wang Wenchang, Cai Ning, Ju Xinhua

(Shougang Research Institute of Technology, Beijing 100043, China)

Abstract: XPS is an analysis method of nanometer depth resolution. The research on XPS mapping for steel materials could provide guidance for researching the distribution of elements in nanometer depth of materials. The influence of step length on tin plate mapping results was studied via XPS. It indicated that, with the decrease of step length, the resolution of XPS mapping improved and the results of element content were more accurate. When the step length was 60μm, the distribution of different elements on tin plate surface was explicitly presented by XPS mapping. With the further decrease of step size, the analyzing time increased, but the element contents had little change. The step size of 60μm was the optimal parameter of XPS mapping.

Key words: XPS, tin plate, mapping, step length

红外碳硫分析仪在线测定硫系易切削钢中的高硫含量

邢　武

（湖南华菱湘潭钢铁有限公司技术质量部，湖南湘潭　411100）

摘　要： 本文应用红外碳硫分析仪对测定易切削钢中的高硫含量进行了研究，获得了满意的测定结果，并通过优化流程，利用红外碳硫仪分析的硫元素结果与直读光谱仪分析其他元素相结合的方式在线指导炼钢生产，确保了硫系易切削钢在冶炼过程中对关键元素快速和准确获得的基本要求。

关键词： 红外法，易切削钢，高硫

无损检测对于圆棒材的应用

周淮林，李桂元

（南京钢铁联合有限公司特钢事业部精整厂，江苏南京　210035）

摘　要： 无损检测技术，是利用声，光，磁和电等特性在不损害或不影响被检对象使用性能的前提下，检测被检对象中是不是存在缺陷或不均匀性，并给出缺陷的大小、位置、性质和数量的所有技术手段的总称。由于不影响被检对象的使用性能，无损检测技术的应用在这些年得到了飞速的发展。光电检测技术是光电信息技术的主要检测技术之一，它是以激光红外光纤等现代光电子器件作为基础通过对被检测成品物体的光辐射经光电检测器接受光辐射并转换为电信号，再经过后续的处理，获取有用信息的技术。

　　本文无损检测对于圆钢的应用指出光电检测技术与无损检测技术的结合对于圆钢的应用，可以取两者之中的优点，得到越来越广泛的圆钢检测应用，在本文中将对常用的基于光电技术的无损检测技术进行简要概述。主要论述超声波检测，漏磁检测，磁粉检测，涡流检测等几种无损检测技术对于圆钢的应用。对他们的原理和适用范围都做了详细的论述，并举例说明了每一种技术在圆钢中的应用。

关键词： 无损检测，超声波，涡流，漏磁，圆钢

The Use of Nondestructive Testing in Bar Steel

Zhou Huailin, Li Guiyuan

(Nanjing Iron & Steel Co.,Ltd., Nanjing 210035, China)

Abstract: Nondestructive testing technology, is the use of sound, light, magnetic and electrical properties of the detected object does not affect the performance without damage or detection, the detected object is not defective or not uniformity, and gives the general defect size, location, nature and quantity of all technical means. The application of nondestructive testing technology has been developing rapidly in recent years because it does not affect the performance of the tested object. Photoelectric detection technology is one of the main detection technology of optoelectronic information technology, it is based on the laser infrared optical fiber and other modern optoelectronic devices as a basis by receiving radiation radiation and converted into electrical signal by photoelectric detector on the finished object is detected by light, and then after the follow-up treatment, to obtain useful information technology.

In this paper, the application of nondestructive testing for steel pointed out that the photoelectric detection technology and nondestructive testing technology for steel combined application can take advantages of the two, has been more and more widely used bar detection, this paper makes a brief overview of nondestructive detection technology based on Optoelectronic Technology in common use. This paper mainly discusses the application of nondestructive testing technology such as ultrasonic testing, magnetic flux leakage testing, magnetic particle testing, eddy current testing and so on. The principle and scope of application are discussed in detail, and an example is given to illustrate the application of each technique in round steel.

Key words: nondestructive testing, ultrasonic, vortex, magnetic leakage, bar steel

X-射线荧光光谱熔片法同时测定
锰硅合金中锰、硅、磷

李 京

（江苏沙钢集团淮钢特钢股份有限公司理化检测中心，江苏淮安 223002）

摘 要：研究了锰硅合金试样的熔融制片方法，确定了在铂黄坩埚中以四硼酸锂挂壁、氢氧化锂预熔试样的方法，避免损伤铂黄坩埚，同时，加入 Co 元素为 Mn 元素内标、溴化钾为脱模剂制备熔片。将此锰硅合金熔片应用于 X-荧光分析，可同时测定锰硅合金中锰、硅、磷。用该方法对不同生产单位的标准样品进行测定，测定值与标准值相吻合，各元素测定结果的相对标准偏差（n＝11）为 0.26%～1.79%。

关键词：X-射线荧光光谱法，熔融，锰硅合金，内标

Simultaneous Determination of Mn, Si and P in Silicomanganese Alloy by X-Ray Fluorescence Spectrometry

Li Jing

(Physics and Chemistry Testing Centre of Huaigang Special Steel Ltd.,

Shagang Iron &Steel Group, Huaian 223002 , China)

Abstract: Method of fusion of Silicomanganese Alloy sample for XRFS analysis was studied, and the sample was fluxed with LiOH in Pt-Au crucible with the walls of the crucible covered with fluxed $Li_2B_4O_7$ to avoid cauterization, while the element Co had been added as internal standard for Mn and KBr as demoulding agent. Mn, Si, and P in the sample were determined by XRFS simultaneously with the glassy disc-shaped melt. In applying the method to the analysis of 3 standard samples of different manufacturers, the results obtained were in consistency with the standard values, and values of RSD's(n=11) found were in the range of 0.26%～1.79%.

Key words: XRFS, fusion, silicomanganese alloy, internal standard

浅谈引伸计在盘螺生产检测应用中存在问题

娄　颖

（本钢北营钢铁集团质量管理中心，辽宁本溪　117017）

摘　要： 盘螺生产中存在的主要问题是对生产出的钢筋进行力学检验时，钢筋应力－应变曲线图上无明显的屈服平台，这就必须使用引伸计进行测试，为保证使用引伸计检测盘螺规定非比例延伸强度 $R_{p0.2}$ 结果的准确性，引伸计的选择、使用及引伸计使用中产生误差、误差原因分析将是至关重要的，引伸计在力学性能测试中，是应变测试的必要手段，是非常重要的，它的准确度直接影响所测数据的准确度。

关键词： 引伸计，规定非比例延伸强度，使用，准确度

The Extensometer Problems in Plate Production Detection Application

Lou Ying

(Benxi Beiying Iron & Steel Group Test Center, Benxi 117017, China)

Abstract: The main problems existing in plate production is to produce the mechanical testing of reinforced, reinforced the stress-strain curve without obvious yield platform, it must use the extensometer test, in order to ensure the accuracy of extensometer detection using half the specified non proportional extension strength $R_{p0.2}$ results, extensometer, selection use and error, error analysis will be crucial in the use of extensometer, extensometer in the test of mechanical properties, is a necessary means of strain testing, is very important, it directly affects the accuracy of the measured data accuracy.

Key words: extensometer, prescribed non proportional elongation strength, use, accuracy

干熄焦机械强度及筛分组成在线测定技术研究

李馨泉，王剑云，祁　威，徐　昱，陈　勇

（宁波钢铁有限公司制造管理部理化检验中心，浙江宁波　315807）

摘　要： 在国家鼓励创新政策的有力支持与引导下，中国已成为世界上使用干熄焦技术最多的国家。而与干熄焦配套的质量检测技术却没有同步发展，干熄焦物理性能检测过程中的筛分、手穿孔等需要人直接与焦炭粉尘接触，检测时扬起的干灰给从业人员身体健康带来严重危害。本文旨在 ISO13909《硬煤 焦炭机械采样》基础上，研究一种焦炭全自动采制样系统工艺，在实现焦炭机械采样的同时，同步实现焦炭筛分组成测定及焦炭机械强度在线测定。在线检验焦炭平均粒径以快速反映焦炭质量波动，更好指导炼焦配煤及高炉炼铁生产。解决了国内推广干熄焦技术以来存在的取样困难及手工物理检测困难的问题。填补国内外空白。

关键词： 煤化工，干熄焦，焦炭机械采制样，焦炭物理检测，环保

Technical Study of Coke Dry Quenching Mechanical Sampling and Test Mechanical Strength of Coke Online

Li Xinquan, Wang Jianyun, Qi Wei, Xu Yu, Chen Yong

(Shanghai Baosteel Industry Technological Service Co., Ltd., Ningbo Branch, Ningbo 315807, China)

Abstract: The State encourages innovation powerful policy support and guidance, China is the country with the largest number of the use of Coke Dry Quenching technique .But the technique of the test on the Coke Dry Quenching do not synchronous develop. Physical performance test of the Coke Dry Quenching would make a lot of inhalable particles .It will do harm to human body. This paper studies a technology of coke automatic sampling system. The sample preparation of coke total moisture、proximate analysis, coke reactivity index (CRI) and coke strength after reaction (CSR) was finished and on-line measurement of coke size distribution and coke mechanical strength was realized while cokemechanical sampling was achieved. It can be quick response quality fluctuation,resolve the difficulties of the popularization of the technique. To fill the domestic and abroad blank.

Key words: coal chemical industry, CDQ, mechanical sampling and sample preparation of coke, mechanical strength of coke, environmental protection

酸溶法测定高碳铬铁中钛含量

张延新，吕艳艳，高立杰，刘　爽，韩长花

（中信特钢研究院青钢分院试验检测所，山东青岛　266409）

摘　要：高碳铬铁中的碳化铬极其稳定不易被酸分解，一般分析检验是利用镍锅加过氧化钠碱熔分解，整个过程中引入了大量的钠离子和镍离子，样品分析时基体影响大，钛含量低，背景扣除对结果测定有很大的影响。本文利用酸溶方法溶解样品，测定结果稳定。

关键词：高碳铬铁，混合酸，电感耦合等离子体发射光谱仪，钛

Determination of Ti in High Carbon Ferrochrome by Acid Dissolution

Zhang Yanxin, Lv Yanyan, Gao Lijie, Liu Shuang, Han Changhua

(Laboratory in the Qingdao special steel Attached to the Citic special steel research institute, Qingdao 266409, China)

Abstract: The carbonized chromium in high carbon ferrochrome is extremely stable and is not easily decomposed by acid, The general of the test is to use the nickel pot to add sodium peroxide decomposition, A large amount of sodium and nickel ions are introduced throughout the process. There is a lot of interference, resulting in inaccurate measurement results. In this paper, the sample was dissolved by acid melting method and the results were stable.

Key words: high carbon ferrochrome, mixed acids, ICP-AES, Ti

预浓缩-气相色谱/质谱法测定空气中有机硫

凌　冰，黄　晓，高巨鹏

（宝钢环境监测站，上海　201900）

摘　要： 建立了苏玛罐预浓缩-气相色谱/质谱联用定量分析空气中有机硫化合物的分析方法，该方法的回收率在 84%~103% 之间，方法的检出限在 0.003~0.008 mg/m³ 之间，低于 EPA TO15 的要求。通过对实际样品的分析表明，本方法操作简便，重复性好，能满足日常分析的要求。

关键词： 预浓缩，有机硫，气质联用

Determination of Organic Sulfur Compounds in Air by GC-MS

Ling Bing, Huang Xiao, Gao Jupeng

(Baosteel Environment Monitoring Center, Shanghai 201900, China)

Abstract: A method for the determination of Organic Sulfur Compounds in air by GC-MS coupled with concentrator was established. The recoveries of every compounds was between 84% and 103%. The Method detection limit was between 0.003~0.008 mg/m³. This method has been used for the determination of Organic Sulfur Compounds in waste gas. The results were satisfactory.

Key words: preconcentration, organic sulfur, GC-MS

高频红外燃烧法测定钢铁中高硫含量

蔺　菲，文元梅，汪秀珍，杨国荣

（钢研纳克检测技术有限公司，北京　100081）

摘　要： 本文以硫酸钾溶液配制校准曲线，采用钨做助熔剂，在自动分析模式下，建立高频红外燃烧法测定钢铁中高硫含量的方法。探索了用硫酸钾标准溶液，采用标准物质作为样品，对实验进行了验证，同时，对比了采用 ISO13902: 1997 中用硫酸钡做标准曲线的实验结果。结果表明，采用硫酸钾标准溶液做校准曲线，得到的实验结果，能够满足实验室的需求。

关键词： 高频红外，燃烧法，硫酸钾，高硫

Determination of High Sulfur Content in Iron and Steel by High Frequency Infrared Combustion Method

Lin Fei, Wen Yuanmei, Wang Xiuzhen, Yang Guorong

(NCS Testing Technology Co., Ltd., Beijing 100081, China)

Abstract: The calibration curve was prepared by using potassium sulfate solution and tungsten was used as flux. In the automatic analysis mode, high frequency infrared combustion method was established for the determination of high sulfur content in iron and steel. The calibration curve was prepared with standard solution of potassium sulfate. The standard material was used as a sample to verify the experiment. At the same time, the experimental results of using barium sulfate as standard curve in ISO13902: 1997 were compared. The results show that the standard curve of potassium sulfate can be used as calibration curve, and the experimental results can meet the needs of the laboratory.

Key words: high frequency infrared, combustion method, potassium sulfate, high sulfur

氮化硅锰中低氮的快速测定法

蔺　菲，杨国荣

（钢研纳克检测技术有限公司，北京　100081）

摘　要：硅锰合金中的氮化物主要是 Si_3N_4 和 MnN。Si_3N_4 中 N 的热导检测法普遍采用镍铂助熔,高温熔融; MnN 中 N 的热导检测法普遍采用锡或镍包裹来防止 Mn 挥发造成的干扰。这两种化合物中的 N 含量都高于 1%，称样量都低于 0.1g。在此基础上，开发了以铁基 CRM 校准，镍箔包裹助熔，高温套坩埚熔融，热导池检测为手段的氮化硅锰中低 N 的快速测定法。N 含量测试推荐范围 0.01%～0.66%。

关键词：氮化硅锰，低氮

Rapid Determination of Silicon Nitride of Low Nitrogen

Lin Fei, Yang Guorong

(NCS Testing Technology Co., Ltd., Beijing 100081, China)

Abstract: Nitride silicon manganese alloy is mainly Si_3N_4 and MnN. In Si_3N_4, the thermal conductivity detection method of nitrogen generally adopts the nickel platinum assistant melting and high temperature melting. In MnN, the thermal conductivity detection method of nitrogen generally adopts tin or nickel package to prevent the interference caused by Mn volatilization. The content of nitrogen in these two compounds is higher than 1%, and the sample weight is lower than 0.1g. In this experiment, the curve is calibrated by Fe-base standard material. Nickel foil as flux wrap samples. Melt in a crucible at high temperature. Low nitrogen content was measured by thermal conductivity method in silicon nitride. Recommended the range of nitrogen content test is 0.01%～0.66%.

Key words: manganese silicon nitride, low nitrogen

超声 C 扫描检测技术在高炉煤气管道局部腐蚀检测和评价中的应用

顾素兰，罗云东，何　磊

(宝钢技术上海金艺检测技术有限公司，上海　201900)

摘　要：针对工业高炉煤气管道局部腐蚀缺陷，采用超声 C 扫描检测技术进行检测，检测数据直观、准确，可为高炉煤气管道维修和安全评价提供有效的技术支撑。

关键词：超声 C 扫描，煤气管道，局部腐蚀

The Application of Ultrasonic C-scan Detection Technique on Detection and Evaluation of Partial Corrosion in Industrial Blast-furnace Gas Pipeline

Gu Sulan, Luo Yundong, He Lei

（Shanghai Jinyi Inspection Technology Co., Ltd., Shanghai 201900, China）

Abstract: Aiming at the defect of partial corrosion in the industrial blast-furnace gas pipeline,ultrasonic C-scan works as a direct and accurate way of data detection.As a scanning detection technique,ultrasonic C-scan provides effective technological support for the maintenance of blast-furnace gas pipeline as well as safety evaluation.

Key words: ultrasonic C-scan, blast-furnace gas pipeline, partial corrosion

标准测量方法在冶金分析仪器计量性能评价中的应用研究

毕经亮

（钢研纳克检测技术有限公司，北京　100081）

摘　要：研究了标准测量方法规定的测量范围，精密度 r 和 R 在冶金分析仪器计量性能评价中的应用，包括检出限、重复性、稳定性、示值误差等，可以作为仪器无计量技术规范或现有规范规定的参数无法满足实验室要求时的参考。

关键词：标准测量方法，计量性能，检出限，重复性，示值误差

Application of Standard Measurement Method in Metrological Performance Evaluation of Metallurgical Analytical Instruments

Bi Jingliang

(NCS Testing Technology Co. ,Ltd., Beijing 100081, China)

Abstract: The measurement scope of the provisions of a standard measurement method, application and evaluation of instrument performance in the precision of R and R in metallurgical analysis, including the detection limit, repeatability, stability, error, can be used as instruments without parameter specification or the existing provisions of the metrology standard cannot meet the requirements of the reference laboratory.

Key words: standard measurement method, measurement performance, detection limit, repeatability, indication error

氮化合金中氮含量的测定

禹青霄，禹继志，刘　洁，葛晶晶，任玲玲

（河钢集团钢研总院理化检测中心，河北石家庄　050000）

摘　要：本文采用脉冲加热惰气熔融热导法测定氮化铬铁、氮化锰铁等氮化合金中氮元素的含量，对方法的准确度和精密度进行了研究与讨论。通过分析确定了脉冲加热惰气熔融热导法分析氮化合金时所必要的分析参数，其中包括：分析功率的选择、样品称样量的选择、助熔剂的选择等，可得到较高的准确度和精密度。测定结果的相对标准偏差小于 1.29%，完全满足氮化合金中氮元素的分析要求。

关键词：氮化合金，氮含量，脉冲加热惰气熔融热导法，检测

Determination of Nitrogen Content in Nitride Alloys

Yu Qingxiao, Yu Jizhi, Liu Jie, Ge Jingjing, Ren Lingling

(HBIS Group Technology Research Institute, Shijiazhuang 050000, China)

Abstract: The content of nitrogen in accuracy and precision of the method were studied and discussed in this paper, using the pulse heating inert gas fusion thermal conductivity method determination of chromium iron nitride, nitride ferromanganese alloy nitride. Pulse heating was determined by the analysis of inert gas fusion thermal conductivity method nitride alloy, can get high accuracy and precision, the determination results, when necessary to the analysis of the parameters, including: analysis of the choice of power, sample selection, the selection of flux to the sample weight, etc. The relative standard deviation less than 1.29%, completely satisfy the requirement of the nitrogen in the nitride alloy elements analysis.

Key words: nitride alloy, nitrogen content, pulse heating and inserting heat conduction method, detection

万能材料试验机检定/校准不确定度的讨论

冯冉，宋洋，杨东

（钢研纳克检测技术有限公司，钢铁研究总院，北京　100081）

摘　要：万能材料试验机在金属/非金属检测中发挥着重要的作用，对万能材料试验机校准时的测量准确与否直接影响着被测材料结果的准确性。本论文对万能材料试验机校准时所产生的不确定度进行分析。

关键词：万能材料试验机，检定/校准，不确定度

Discussion of Uncertainty of Measurement in the Universal Material Experiment Machine Verification/Calibration

Feng Ran, Song Yang, Yang Dong

(NCS Testing Technology Co., Ltd., Central Iron & Steel Research Institute, Beijing 100081, China)

Abstract: Universal material testing machine in the metal / non-metallic detection plays an important role. The accuracy on the measurement of the calibration of universal material experiment machine will directly affect the accuracy of the results of the measured material. In this paper, the uncertainty of the universal material testing machine is analyzed.

Key words: universal material experiment machine, verification/calibration, uncertainty of measurement

电感耦合等离子体原子发射光谱法
测定硼铁中硼铝磷

吴超超，王　岩，梁小红，叶玉锋

（宝武集团广东韶关钢铁有限公司，广东韶关　512123）

摘　要：采用王水溶解试样，过滤残渣，灰化，用碳酸钠进行碱熔，试样溶解完全后，用 ICP 光谱法进行了测定。试验进一步探讨了溶样方法、分析谱线的选择、基体效应等因素对测定结果的影响，建立了 ICP 光谱法对硼铁中硼、铝、磷等元素的测定方法。该方法替代了传统的手工湿化学分析方法，实现了对韶钢外购硼铁中硼铝磷等元素的快速测定，提高了检测效率，缩短了检测周期。

关键词：ICP 光谱法，硼铁，测定，硼铝磷

Determination of Niobium in Steel by Inductively Coupled Plasma Atomic Emission Spectrometry

Wu Chaochao, Wang Yan, Liang Xiaohong, Ye Yufeng

(Baowu Steel Group Shaoguan Iron and Steel Co., Ltd., Shaoguan 512123, China)

Abstract: The samples were dissolved in aqua regia, filtered and ashed, and the alkali was melted with sodium carbonate. After the samples were dissolved, the samples were measured by ICP. The influence of the factors such as the method of dissolution, the selection of the spectral line and the matrix effect on the determination results were discussed. The determination of boron, aluminum and phosphorus in boron iron by ICP spectroscopy was established. This method replaces the manual wet chemical method, and realizes the rapid determination of boron, aluminum and phosphorus in boron iron and the detection efficiency was improved with the method.

Key words: ICP spectrometry, boron iron, determination, boron and aluminum phosphate

电感耦合等离子体原子发射光谱法测定
氮化增强剂、氮化硅锰中硅含量

曾海梅，章祝雄，孙肖媛，苏　宁，欧阳光

（武钢集团昆明钢铁股份有限公司技术中心，云南昆明　650302）

摘　要： 提出了用电感耦合等离子体原子发射光谱法（ICP-AES）测定氮化增强剂、氮化硅锰中硅含量的分析方法。试样分别采用碱融与酸溶的方法溶解，经过对比，样品经碳酸钠—硼酸熔融，用盐酸浸出分析效果更佳。在选用的最佳光谱线和合适的工作条件下测定，优化了仪器工作条件，确定了最佳实验条件。结果表明，该方法准确、快速、简便，相对标准偏差小于 2%，回收率为 98.33%～105.0%，样品分析结果满意。

关键词： 电感耦合等离子体原子发射光谱法，氮化增强剂，氮化硅锰，硅

Determination of Silicon in Nitride Enhancer, Silicon Manganese Nitride by Inductively Coupled Plasma Atomic Emission Spectrometry

Zeng Haimei, Zhang Zhuxiong, Sun Xiaoyuan, Su Ning, Ouyang Guang

(Technology Center of Wukun Steel Co., Ltd., Kunming 650302, China)

Abstract: Analysis method of determination of silicon in nitride enhancer, silicon manganese nitride by inductively coupled plasma atomic emission spectrometry introduced in this paper. dissolved methods of the sample, elimination of the interference and the selection of element spectrum lines were discussed. We optimized instrument conditions and determined the optimal experimental conditions. The results showed that this method is accurate, fast and simple. Precision, recovery and sample analysis have achieved satisfactory results.

Key words: inductively coupled plasma atomic emission spectrometry, nitride enhancer, silicon manganese nitride, silicon

高炉热风炉炉壳裂纹检测、原因分析及处理建议

李贵文，罗云东，李　明，茅晨昊，陈融融

（宝武集团上海金艺检测技术有限公司，上海　201900）

摘　要： 热风炉是为高炉加热鼓风的设备，是现代高炉不可缺少的重要组成部分，炉壳钢板和焊接质量的任何缺陷都会影响热风炉的安全运转和使用寿命，本文针对某钢厂高炉 4#热风炉蓄热室炉壳标高 20.21m 环焊缝处的大量裂纹，利用多种技术手段，进行了深入细致的研究，明确了热风炉炉壳焊缝开裂的详细情况，指出了裂纹产生的真实原因，提出了改进建议，为大修工作奠定了良好的技术基础，同时也形成了一项有特色的大型炉壳综合检测评价技术。

关键词： 热风炉，裂纹，检测，分析

Inspection、Analysis and Solutions of Blast Furnace Hot-blast Stove Furnace Shell Crack

Li Guiwen, Luo Yundong, Li Ming, Mao Chenhao, Chen Rongrong

(Shanghai Jinyi Inspection Technology Co., Ltd., Shanghai 201900, China)

Abstract: Hot-blast stove(HBS) is an indispensable part of modern blast furnace(BF), which heats the BF and provides ventilation, any defects of furnace shell plate and welding quality may jeopardize the equipment and its life span. This article mainly discussed a large of cracks that discovered at elevation of 20.21 m of regenerator's shell, and this regenerator belongs to a certain steel plant's 4# HBS. In order to make clear the details of weld cracking, author used many types of technical measures, made intensive study, and pointed out the reasons and gave suggestions to solve the problem. Therefore, the study laid solid foundation for overhaul, and concluded an comprehensive inspection evaluation technique for stove shell.

Key words: hot-blast stove, crack, inspection, analysis

火花源原子发射光谱法测定无取向硅钢中超低碳硫元素的含量

杨　琳，王　鹏，刘步婷

（武钢有限质量检验中心，湖北武汉　430083）

摘　要： 通过高频燃烧红外分析法对无取向硅钢中的超低碳硫试样进行定值并作为内控标样，在 ARL-4460 光谱仪上分析该内控标样，通过测量得出的强度值，与标样的定值之间建立相应的数学关系，从而在火花源原子发射光谱仪建立分析超低碳硫元素的方法。实验结果表明，该方法用于超低碳硫分析周期短，化验分析成本低，样品的分析准确度与红外气体法相比无明显差异，分析精密度均满足日常生产的需求。

关键词： 火花放电原子发射光谱法，超低碳硫，内控标样，高频燃烧红外分析法

Determination of Ultra - low Carbon and Sulfur in Non - Oriented Silicon Steel by Spark Discharge Atomic Emission Spectrometry

Yang Lin, Wang Peng, Liu Buting

(Quality Inspection Center of WISCO Limited Company, Wuhan 430083, China)

Abstract: The ultra-low carbon and sulfur samples in non-oriented silicon steel were calibrated by high frequency combustion infrared analysis method and were used as internal control standards,This internal control standards are analyzed on ARL-4460 spectrometer.Through the analysis of the strength value obtained, and the standard value of the establishment of the corresponding mathematical relationship between the discharge atomic emission spectrometer to establish a method for the analysis of ultra-low carbon and sulfur.The results of the experiment show that the method can shorten the ultra low carbon and sulfur analysis cycle and reduce the cost of laboratory analysis.The accuracy of the analysis of the sample is not significantly different from that of the infrared analysis method, and the analytical precision satisfies the daily production demand.

Key words: spark discharge atomic emission spectrometer, the ultra-low carbon and sulfur, internal control standards, high frequency combustion infrared analysis method

ICP-AES 法测定低合金钢中锑和锡元素含量

亢德华，王　莹，于媛君，邓军华，王一凌

（鞍钢集团钢铁研究院化学室，辽宁鞍山　114009）

摘　要：建立了利用电感耦合等离子体原子发射光谱法（ICP-AES）同时测定低合金钢中锑和锡元素含量的分析方法。通过试验选择了合适的溶样酸体系和酸度，优选了适宜的仪器测定参数和分析谱线，采用基体匹配法进行基体效应的校正，并通过干扰系数校正试验研究了共存元素的干扰情况。实验结果表明，各元素分析谱线线性相关系数大于 0.999，方法检出限为 Sn 0.02 μg/mL 和 Sb 0.03 μg/mL。该方法应用于国家或国际标准物质分析，测定值与认定值相符，测定结果的相对标准偏差均小于 7.0%。

关键词：电感耦合等离子体原子发射光谱法（ICP-AES 法），低合金钢，锑，锡，基体匹配法，干扰系数校正

Determination of Antimony and Tin Contents in Low-alloy Steel by Inductively Coupled Plasma Atomic Emission Spectrometry

Kang Dehua, Wang Ying, Yu Yuanjun, Deng Junhua, Wang Yiling

(Iron & Steel Research Institute of Ansteel Group, Anshan 114009, China)

Abstract: A method for simultaneous determination of antimony and tin contents in low-alloy steel by inductively coupled plasma atomic emission spectrometry was established. The optimum dissolution system and acidity were selected by experiment, the optimum instrument working conditions and the analytical spectrum lines were optimized. Besides, the matrix match method was used to eliminate the matrix effect and the coexist elements interference was corrected by interfering coefficients. The results showed that the linear correlation coefficient of all the elements were lager than 0.999, the method detection limits were Sn 0.02 μg/mL and Sb 0.03 μg/mL. The determination of certified reference materials with similar matrix showed that the measured results were in good agreement with the certified values, and the relative standard deviations were all lower than 7.0%.

Key words: inductively coupled plasma atomic emission spectrometry (ICP-AES method), low-alloy steel, antimony, tin, matrix match method, interfering coefficients correction

沉淀分离 EDTA 滴定法测定镁铝合金中镁量

沈　真

（湖南华菱湘潭钢铁有限公司，湖南湘潭　411101）

摘　要：采用硝酸溶解试料，以氢氧化钠沉淀镁使其与干扰元素分离。盐酸溶解沉淀后，以氨水和氨性缓冲溶液控制 pH=10，以铬黑 T 作指示剂，建立了 EDTA 标准溶液滴定法测定镁铝合金中镁含量的分析方法。镁的精密度为 0.17%；加标回收率为 99.9%。

关键词：镁铝合金，镁，EDTA 标准溶液滴定

Determination of Magnesium Content in Magnesium Aluminum Alloy by Precipitation Separation EDTA Titration

Shen Zhen

(Xiangtan Iron and steel Co., Ltd., of Hunan Valin, Xiangtan 411101, China)

Abstract: The sample is dissolved by nitric acid and sodium hydroxide to precipitate magnesium separating it from interfering elements.Dissolved in hydrochloric acid after precipitation with ammonia and ammonia buffer solution of pH=10 control, with chrome black T as indicator, analysis method was established for determination of magnesium content in magnesium aluminum alloy in EDTA standard solution titration. The precision of magnesium is 0.17%, and the recovery rate is 99.9%.

Key words: magnesium aluminum alloy, magnesium, EDTA standard solution titration

X-荧光分析法在原料质量检验上的推广

李桂东

（本钢北营钢铁集团质量管理中心，辽宁本溪　117017）

摘　要：通过对 X-荧光分析原料可能存在的影响结果的因素进行分析，探索克服及解决的办法，使 X-荧光分析炼铁、炼钢原料的分析方法得以建立，并进一步推广，最终将该方法用于炼铁、炼钢原料分析。以提高原料检验效率，降低分析成本。

关键词：X-荧光分析，原料分析，推广

X- ray Analytic Method in Raw Material Performance Test Promotion

Li Guidong

(Benxi Beiying Iron & Steel Group Test Cente, Benxi 117017, China)

Abstract: Based on X-ray fluorescence analysis raw material possible affect results analyzes the factors, explore overcome and the solution, make the X-ray fluorescence analysis of raw materials, steel-making iron-making analysis method to establish, and further promotion, finally the method is applied in iron and steel material analysis. In order to improve the raw material inspection efficiency, reduce the cost analysis.

Key words: X-ray fluorescence analysis, raw material, analysis promotion

梅钢选用连铸保护渣的熔化温度评估方法

刘 为

（上海梅山钢铁股份有限公司制造部，江苏南京　210039）

摘　要：为了保证梅钢在连铸过程能够稳定生产，本文针对结晶器内关键性的功能材料连铸保护渣的熔化温度进行了评估检测，并分析了实际检测过程中遇到的问题，探讨各种因素对连铸保护渣性能检测带来的影响，为今后梅钢连铸保护渣的选用提供帮助。

关键词：连铸保护渣，熔化温度，检测方法，脱碳

Evaluation Method of Melting Temperature for Mold Flux Used in Meishan Iron and Steel Company

Liu Wei

(Manufacturing Department of Meishan Iron & Steel Co.,Ltd., Nanjing 210039, China)

Abstract: In order to ensure the stable production of Meishan steel during the continuous casting process, the melting temperature of mold flux is evaluated and analyzed. The problems encountered in the actual testing process are analyzed . The influence of various factors on the performance test of continuous casting mold flux was discussed, which was helpful for the selection of mold flux.

Key words: mold flux, melting temperature, detection method, decarburization

TD-16 高温拉伸塑性探究

徐俊凯，赵　磊，唐秀艳

（中信泰富青岛特殊钢铁有限公司青钢研究分院试验检测所，山东青岛　266000）

摘　要：铁路道钉是铁路轨道用金属道钉的简称，这种道钉是固定在铁路轨枕上并扣压住钢轨的关键元件，并使得铁路轨道更好的压紧在铁路路基上。TD-16 主要应用于 5.6 级道钉螺栓，青岛钢铁自 2013 年于老厂区即生产该钢种，新厂区生产规格为Φ24、Φ25 的线材，为探究我厂 2016 年 4 月份生产的 TD-16 钢高温力学性能指标，在 800℃-1100℃范围内（50℃一个温度点）对 TD-16 进行高温拉伸试验，试验结果表明，屈服强度和抗拉强度均随温度升高而降低，但断面收缩率在 900℃骤降，且在 950℃塑性指标（断面收缩率、断后延伸率）为最低点，TD-16 钢在 900~1000℃存在高温脆性区。

关键词：铁路道钉，TD-16 钢，高温脆性，断面收缩率，断后延伸率

Study on Plasticity of TD-16 under High Temperatures

Xu Junkai, Zhao Lei, Tang Xiuyan

(Citic Qingdao Special Iron and Steel Co., Ltd., Testing Institute, Qingdao, 266000, China)

Abstract: A railroad spike is used for the railway track to fix the railway tightly on the railway sleeper as the key component. TD-16 is a new steel material mainly used in spike bolt of 5.6 level. Qingdao iron and steel Co.Ltd products this material at the old factory since 2013, and now mot of the newproductionsare wires with the diameter is 24 and 25 mm.In order to research the mechanical performance parameters under high temperatures of TD-16 steel producted during April2016, the high temperature tensile tests are carried out between the temperatures from 800℃ to 1100℃ (every 50 ℃ is selected for a test point).The results show that the yield strength and tensile strength decrease with the temperature increasing, while the rate of reduction in area drops suddenly at the temperature of 900℃, meanwhile, the percentage reduction of area and the elongation after fracture behave the smallest at the temperature of 950℃. That is to say, TD-16 steel exits an high temperature brittleness area at the temperature of 900~1000℃.

Key words: railway spike, TD-16 steel, high temperature brittleness, percentage reduction of area, percentage elongation after fracture

影响螺纹钢力学性能因素的分析

李 鹏，孙振理，徐 伟，陈 宇，张 强，孙 宏

（陕西龙门钢铁有限责任公司能源检计量中心，陕西韩城 715405）

摘 要：钢材力学性能的稳定性直接影响质量的好坏，在当下日新月异的发展速度下，市场对外观以及性能的挑剔已经达到了新的高度，对钢材的质量要求越来越高，尤其是对性能不合格采取零容忍。针对钢材性能偏低或者偏高的原因，经过长时间检验，采用金相分析以及外观抽检得到了答案。最后表明：成分差异、魏氏组织、夹杂物、表面缺陷与不同的外观尺寸均会导致性能出现异常。

关键词：性能，成分分析，魏氏组织，夹杂物，外观尺寸

Analysis of Factors Affecting the Mechanical Properties of Threaded Steel

Li Peng, Sun Zhenli, Xu Wei, Chen Yu, Zhang Qiang, Sun Hong

(Shaanxi Longmen Iron and Steel Company Energy Measurement Center, Hancheng 715405, China)

Abstract: The stability of mechanical properties of steel directly affects the quality, In the current rapid pace of development, the market's attention to appearance and performance has reached new heights, The quality of the steel is getting higher and higher, especially for performance failure to take zero tolerance. Reasons for low or high performance of steel, After a long inspection, metallographic analysis and appearance sampling were used to get the answer. It is finally showed that compositional difference、Widmanstatten structure inclusion surface defects and different appearance sizes can lead to abnormal performance.

Key words: performance, component analysis, widmanstatten structure, inclusions, appearance size

SCM435 冷镦钢镦头开裂原因分析

朱祥睿，罗新中，李富强，章玉成

（宝武集团广东韶关钢铁有限公司检测中心，广东韶关　512123）

摘　要： 使用 SCM435 冷镦钢盘条镦制螺钉过程中部分螺钉头部出现开裂。对开裂试样进行各种理化试验分析，最终确认螺钉头部开裂的主要原因是材料内部近表面区域存在大尺寸夹渣，次要原因是材料表面存在裂纹，两者共同作用最终导致螺钉镦头开裂。

关键词： 冷镦钢，螺钉，开裂，夹渣

Analysis of Cracking Reason of Upsetting Head of SCM435 Cold Heading Steel

Zhu Xiangrui, Luo Xinzhong, Li Fuqiang, Zhang Yucheng

(Baosteel Group Guangdong Shaoguan Iron and Steel Co., Ltd., Shaoguan 512123, China)

Abstract: Part of heads of SCM435 screws cracked in the pier process from cold heading steel wire rod to screws. To analyze the cracked sample by a variety of physical and chemical tests, we finally confirm the main reason for the cracking of the screw head is that there is a large size slag in the near surface area of the material. The secondary reason is that there is a crack on the surface of the material, two factors together eventually lead to screw headset cracking.

Key words: cold heading steel, screw, crack, slag

鞍钢矿业球磨机轴承状态监测与故障诊断

于广宇

（鞍钢集团矿业有限公司装备制造分公司，辽宁鞍山　114001）

摘　要： 基于对鞍钢矿业有限公司现有选矿运行设备球磨机情况的掌握，使用先进的检测系统对球磨机进行状态监测与故障诊断工作。结合球磨机常见故障特点，以一台球磨机轴承故障为典型案例，剖析了球磨机小齿轮轴承的劣化过程振动变化规律及谱图特征，总结了在一般工况下球磨机轴承早期故障特点。

关键词： 球磨机轴承，状态监测，故障诊断，维修管理模式

Anshan Iron and Steel Group Mining Co., Ltd., Ball Bearing Condition Monitoring and Fault Diagnosis

Yu Guangyu

(Equipment Manufacturing Branch of Anshan Iron and Steel Group Mining Co., Ltd., Anshan 114001, China)

Abstract: Based on the mastery of the existing processing equipment and the ball mill in Anshan Iron and Steel Group Mining Co. Ltd. the state monitoring and fault diagnosis of the ball mill is carried out by using advanced detection system. According to common fault features of ball mill, take one bearing fault for a typical example, analyzes the change rule of the deterioration process of vibration ball mill pinion bearings and its spectrum characteristics, summarizes the general conditions of the ball mill bearing early fault features.

Key words: ball mill bearing, condition monitoring, fault diagnosis, maintenance management model

汽车用热轧板的残余应力分析

薛　峰[1]，张晓蓉[2]，海　岩[1]，白丽娟[1]，邢承亮[1]，梁爱国[1]

（1. 河钢集团钢研总院，河北石家庄　050000；2. 山东凯文科技职业学院，山东济南　250200）

摘　要： 采用 X 射线衍射法测量了汽车用热轧板表面残余应力的分布情况，结果表明，沿钢板宽度方向上残余应力分布不均匀，钢板上下表面的残余应力也不相同，上表面主要表现为拉应力，最高为 218.3MPa，最小为–24.7MPa；下表面主要表现为压应力，最大为–285.0MPa，最小为–25.7MPa。

关键词： 残余应力，汽车钢，热轧板，X 射线衍射法

Residual Stress Analysis of Hot Rolled Automobile Steel Sheet

Xue Feng[1], Zhang Xiaorong[2], Hai Yan[1], Bai Lijuan[1], Xing Chengliang[1], Liang Aiguo[1]

(1. Hesteel Group Technology Research Institute, Shijiazhuang 050000, China;

2. Shandong Kaiwen College of Science & Technology, Ji'nan 250200, China)

Abstract: Distribution of residual stress on the surface of hot rolled automobile steel sheet was measured by X-ray diffraction (XRD), it was showed that distribution of residual stress was not uniform not only along the width of steel sheet but also on the upper and lower surfaces of steel sheets. The upper surface was mainly shown as tensile stress with a maximum of 218.3MPa and a minimum of –24.7MPa, while the lower surface is mainly shown as compressive stress with a maximum of –285.0MPa and a minimum of –25.7MPa.

Key words: residual stress, automobile steel, hot rolled sheet, X-ray diffraction

SWRCH18A 线材拉拔断裂分析

祝小冬， 何春根

（方大特钢科技股份有限公司检测中心, 江西南昌 330012）

摘 要：本文阐述了 SWRCH18A 线材拉拔断裂的两种形态，分别对笔尖状断口和斜茬状断口进行分析，指出了线材拉拔断裂的原因。

关键词：笔尖状断口，斜茬状断口

Analysis of Drawing Fracture of SWRCH18A Wire Rod

Zhu Xiaodong, He Chungen

(Fangda Special Steel Technology Co., Ltd., Nanchang 330012, China)

Abstract: In this paper, analysis on the two kinds of SWRCH18A wire rod drawing fracture, the fracture of pen point shape fracture and inclined fracture are analyzed. It is pointed out that the reason of the drawing fracture.

Key words: pen point shape fracture, inclined fracture

火花光谱法测定稀土镁合金中稀土和非稀土共 11 种元素的含量

王 洋，冯 光，李小佳

（钢研纳克检测技术有限公司，北京 100081）

摘 要：本文研究了采用高能固态预火花光源新型火花发射光谱仪测定稀土镁合金中 Al、Mn、Zn、Ni、Fe、Ag、Zr、La、Nd、Ce、Pr 等元素的方法。考察了各元素的线性相关性，选择了最佳分析谱线；对受到干扰的元素进行了干扰校正，提高了线性相关性。该法简便快速，精密度好，准确度高，标准样品的测定值与认证值一致。

关键词：火花，发射光谱，镁合金，稀土，成分分析

Determination of 11 Kinds of Rare Earth and Non-rare Earth in Mg-RE Alloys

Wang Yang, Feng Guang, Li Xiaojia

(Central Iron & Steel Research Institute, Beijing 100081, China)

Abstract: Magnesium alloy, as the lightest metal structure material in engineering application field, has a great application prospect in the aerospace, military industry, electronic communication, transportation and other fields, because of its low density, high specific strength, good shock absorption, and easy processing and recycling, and many other advantages. Recently studies have shown that rare earth elements added into magnesium alloy can effectively improve its microstructure. This paper studied the mothed by using Spark-OES to determine the concentration of the 11 elements in Mg Rare alloy. The mothed mentioned in this paper used CCD detectors to choose the best analysis spectrum lines and added interference correction model to calibrate the curve. That made the RE element testing results very close to the certified value.

Key words: spark, optical spectrometer, Mg alloy, rare earth, concentration determination

钼铁中硅、磷、铜、锡、锑的联合测定

齐　兵，周淑新

（河钢石钢技术中心，河北石家庄　050031）

摘　要：研究了采用等离子体发射光谱仪测定钼铁中微量元素和残余元素的检测方法。主要研究了几种元素不同国标方法和采用一种方法溶样方式后，不同元素同时测定。由于国产钼铁标样没有锡、锑的定值，在极少量的进口标样中虽含有这两个元素的值，但含量也很低，采用定量加入法加入锡、锑的标液方法，利用 ICP 光谱法同时测定钼铁中的硅、磷、铜、锡、锑元素的含量，提高工作效率。

关键词：微量元素，残余元素，联合测定，定量加入，标液

In Silicon Molybdenum Iron, Phosphorus, Copper, Tin, Antimony, Combined Determination

Qi Bing, Zhou Shuxin

(HBIS Shisteel Technology Center, Shijiazhuang, 050031, China)

Abstract: The use of plasma emission spectrometer was developed for the determination of trace elements and residual elements in molybdenum iron test methods. Mainly studies several elements of different national standard method and adopting a method after dissolving sample method, determination of different elements at the same time. Because domestic molybdenum iron sample without fixed value of tin, antimony, although in very small amounts of imported sample containing these two elements of value, but the content is low, quantitative join method is used to join the liquid method of tin, antimony, using ICP spectrometry determination of silicon in molybdenum iron at the same time, the content of phosphorus, copper, tin, antimony elements, improve the work efficiency.

Key words: trace elements, residual elements, combined determination, quantitative add, the liquid

影响悬浮物测定结果的因素分析

李　琳

（昆钢动力能源分公司，云南昆明　650000）

摘　要：通过实际具体日常工作中的实际分析情况判断并进行总结，对水中 SS 测定结果的影响因素做出简要分析，作为指导今后实验条件的参数，从而提高分析结果的准确度。
关键词：悬浮物，恒重，滤纸，精密性，质量控制

Factors Analysis Affecting the Determination of Suspended Substance

Li Lin

(Kunming Steel Power Energy Branch, Kunming 650000, China)

Abstract: Through the judgement and summarization to the actual situation of the daily analysis work, the affecying factors of SS determination results in water were briefly analyzed.　Guidance to the future experimental conditions was thus provided. and the accuracy of analysis results were improved at the same time by using these parameters.
Key words: suspended substance, constant weight, filter paper, precision, quality control

奥氏气体分析仪在铁合金煤气分析中的应用

李　琳

（昆钢动力能源分公司，云南昆明　650000）

摘　要：本文主要是以提高奥氏分析仪测定铁合金煤气含量的准确度为出发点。预先设定定量的煤气成分含量，运用控制变量法，通过实验寻找与设定值最为匹配的结果，记下该结果所对应的实验条件，作为指导今后实验条件的参数，从而提高实验的准确度。
关键词：奥氏气体分析，铁合金煤气，含量测定

Application of Austenite Gas Analyzer in the Analysis of Ferroalloy Gas

Li Lin

(Kunming Steel Power Energy Branch, Kunming 650000, China)

Abstract: The paper aims at improving the accuracy of the determination of ferroalloy gas content by the austenite analyzer. By preseting quantitative gas component content and using the control variable method, try to find the most matching results with the set-value through experiment. The relative the experimental conditions were recorded.Guidance to the future experimental conditions was thus provided and the accuracy of analysis results were improved at the same time by using these parameters.
Key words: austenite gas analysis, ferroalloy gas, content determination

电感耦合等离子发射光谱法测转炉渣中三氧化二铬含量

刘炼伟，曹兴旺，马文广，张文龙

（日照钢铁控股集团有限公司技术规划发展处，山东日照　276806）

摘　要： 试样经沸水溶解，残渣碱熔，硝酸溶解盐类，用电感耦合等离子体发射光谱法测定转炉渣中三氧化二铬的含量。在选定条件下，三氧化二铬的含量在 0.073%～1.462% 范围内线性关系良好，相关系数大于 0.9990，相对标准偏差（RSD）小于 1.5%，回收率在 97%～104% 之间。本方法适用于转炉渣中三氧化二铬的测定。

关键词： 转炉渣，铬，ICP-AES 法

The Exaimination of Content for Chromium Oxide in Converter Slag by ICP

Liu Lianwei, Cao Xingwang, Ma Wenguang, Zhang Wenlong

(Technical Planning and Development Branch Rizhao Steel Holding Group Co., Ltd., Rizhao 276806, China)

Abstract: The sample was dissolved by boiling water, residual was fused by alkali, and dissolve the salt with nitric acid, the content of chromium oxide in converter slag was examined by ICP. Under the selected conditions, The content of chromium trioxide in the range of 0.073% ~ 1.462% was good, the correlation coefficient was greater than 0.9990, the relative standard deviation (RSD) was less than 1.5%, and the recovery rate was between 97% and 104%. This method is applicable to the determination of chromium trioxide in the furnace slag.

Key words: converter slag, chrome, ICP-AES

高温环境下拉伸速率对材料强度的影响

倪佳俊，杨　斌

（钢研纳克检测技术股份有限公司上海分公司，上海　200231）

摘　要： 采用夹头分离速率控制方式对高温合金 GH738 在不同位移速率下进行高温拉伸试验，探讨 800℃ 温度下不同位移速率对该材料屈服强度、抗拉强度的影响。发现：高温条件下材料对拉伸速率较敏感，位移速率的改变会引起高温拉伸试验测试力学性能结果的不确定性，危及到高温拉伸试验技术的可靠性。

关键词： 高温拉伸，拉伸速率，屈服强度，抗拉强度

The Influence of Tensile Test Rate on Material Strength in High Temperature Environment

Ni Jiajun, Yang Bin

(NCS Tseting Technology Co., Ltd., Shanghai Branch, Shanghai 200231, China)

Abstract: The high temperature tensile tests designed by control of grip separate rate to the same high temperature alloy material under different displacement rate, and discussed the influence of different tensile test rate on yield strength and tensile strength of the material in temperature of 800℃. It is found that the material is sensitive to the tensile rate at high temperature, and the change of the displacement rate will lead to the uncertainty of mechanical properties at high temperature tensile test and endanger the reliability of the high temperature tensile test technique.

Key words: elevate temperature tension, tensile rate, yield strength, tensile strength

ICP-AES 法测定硅铁合金中铌、钒和锆量

英江霞，宋祖峰，陆　尹，王昔文

（马鞍山钢铁股份有限公司，安徽马鞍山　243000）

摘　要：使用硝酸、氢氟酸和高氯酸冒烟溶解样品。选择 Nb322.548 nm、V 310.230 nm、Zr 349.621 nm 为分析线并设置合适的背景扣除位置，采用基体匹配法绘制校准曲线可消除基体效应的影响，利用电感耦合等离子体原子发射光谱法（ICP-AES）同时测定铌、钒和锆，建立了硅铁合金中铌、钒和锆的测定方法。各待测元素校准曲线的线性相关系数均大于 0.9995；各元素的检出限分别为 6 mg/kg，5 mg/kg 和 5 mg/kg。方法应用于硅铁合金样品中铌、钒和锆的测定，结果的相对标准偏差（RSD，n=10）为 2.28%~4.73%。回收率为 95.0%~105.0%。

关键词：电感耦合等离子体原子发射光谱法，硅铁合金，检出限，光谱干扰，相对标准偏差

Determination of Nb、V and Zr Elements in Ferro-Silicon Alloy by ICP-AES

Jia Jiangxia, Song Zufeng, Lu Yin, Wang Xiwen

(Maanshan Iron & Steel Co., Ltd., Maanshan 243000, China)

Abstract: The sample was processing by using nitric acid, hydrofluoric acid and perchloric acid. Nb322.548 nm, V 310.230 nm, Zr 349.621 nm was selected as the analysis line and the appropriate background subtraction position was setted. The effection of calibration curve can be eliminated by matrix matching.The method for determination of Niobium, Vanadium and Zirconium in ferro-silicon alloy was established by using Inductively Coupled Plasma Atomic Emission Spectrometry (ICP-AES) simultaneous.The linear correlation coefficients of the calibration curves of each element to be measured are all greater than 0.9995. The detection limits of each element are 6mg/kg, 5 mg/kg and 5 mg/kg, respectively. The method was applied to the determination of Nb, V and Zr in ferrosilicon alloy samples. The Relative Standard Deviations (RSD, n=10) were between2.28% and 4.73%. The recovery rate was between 95.0% and 105.0%.

Key words: inductively coupled plasma atomic emission spectrometry, ferrosilicon alloy, detection limit, spectral interference, relative standard deviation

力学网改造项目的开发与实施

袁晓辉

（舞阳钢铁有限责任公司，河南舞阳　462500）

摘　要： 本文通过深入讨论力学网现状，指出了改造的必要性，采用新的数据库和开发工具，优化了服务器结构。并阐述了力学网改造思路，详细说明改造的实施过程，对力学系统的功能模块及设计方案进行了描述，针对原力学网中存在的疑难问题和不能实现的功能，提出了解决办法和改进措施。最后总结了力学系统在公司生产经营中发挥的重要作用。

关键词： 力学网，现状，改造思路，实施过程

The Development and Implementation of Mechanical Power Network Reconstruction Project

Yuan Xiaohui

(Wuyang Iron and Steel Co., Ltd., Wuyang 462500, China)

Abstract: In this paper, the necessity of the transformation is pointed out by discussing the present situation of the mechanical network, and the new database and development tools are used to optimize the structure of the server. And describes the mechanical network transformation ideas, detailed description of the implementation process of transformation, function module and the design scheme of mechanical system are described, aiming at the problems and can-not-realize founction in the original mechanical network, put forward the solutions and measures for improvement. Finally, the important role of the mechanical system in the production and operation of the company is summarized.

Key words: mechanical power network, present situation, thought of reforming, implementation process

常用玻璃量具项目相关建设

李金生，董萍萍

（河北钢铁股份有限公司唐山分公司信息自动化部，河北唐山　063000）

摘　要： 本文适用于唐钢企业最高标准玻璃量具项目的相关建设，叙述了常用玻璃量具建立的相关背景，技术要求，相关标准器设备配备，项目实施及售后培训等内容。该项目的建立不仅提高了唐钢在玻璃量具检定能力，节约了大量外检成本，提高了产品质量，也为唐钢做大做强实验室，应对来自各个方面的认证，打下坚实的基础，具有十分重要的意义。

关键词： 唐钢，玻璃量具，不确定度，测量

工作用辐射温度计计量标准项目的建设

万 利

（河钢集团唐钢校准实验室，河北唐山 063000）

摘 要：本文适用于唐钢企业内部最高计量标准工作用辐射温度计项目的相关建设，叙述了工作用辐射温度计建立的相关背景，技术要求，相关标准设备配备，项目实施及售后培训等内容。该项目的建立不仅提高了唐钢校准实验室的检定能力，为企业节约大量外检成本，而且保证测量设备提供数据的准确性，提高产品质量，为公司带来更多的经济效益。

关键词：唐钢，工作用辐射温度计，数据，经济效益

沉淀分离 EDTA 滴定法测定镁铝合金中镁量

沈 真

（湖南华菱湘潭钢铁有限公司，湖南湘潭 411101）

摘 要：采用硝酸溶解试料，以氢氧化钠沉淀镁使其与干扰元素分离。盐酸溶解沉淀后，以氨水和氨性缓冲溶液控制 pH=10，以铬黑 T 作指示剂，建立了 EDTA 标准溶液滴定法测定镁铝合金中镁含量的分析方法。镁的精密度为 0.17%；加标回收率为 99.9%。

关键词：镁铝合金，镁，EDTA 标准溶液滴定

Determination of Magnesium Content in Magnesium Aluminum Alloy by Precipitation Separation EDTA Titration

Shen Zhen

(Xiangtan Iron and Steel Co., Ltd., of Hunan Valin, Xiangtan 411101, China)

Abstract: The sample is dissolved by nitric acid and sodium hydroxide to precipitate magnesium separating it from interfering elements.Dissolved in hydrochloric acid after precipitation with ammonia and ammonia buffer solution of pH=10 control, with chrome black T as indicator, analysis method was established for determination of magnesium content in magnesium aluminum alloy in EDTA standard solution titration. The precision of magnesium is 0.17%, and the recovery rate is 99.9%.

Key words: magnesium aluminum alloy, magnesium, EDTA standard solution titration

17　冶金设备与工程技术

炼铁与原料

炼钢与连铸

轧制与热处理

表面与涂镀

金属材料深加工

钢铁材料

汽车钢

海洋工程用钢

轴承钢

电工钢

粉末冶金

非晶合金

高温合金

耐火材料

能源与环保

分析检测

★ 冶金设备与工程技术

冶金自动化与智能管控

冶金技术经济

切向旋风除尘器在 POSCO 浦项 2 号和 3 号高炉的应用

潘铁毅，胡雪萍，单 良，吴晓东，李 晔，沈 军，田莎丽

（中冶南方工程技术有限公司钢铁公司，湖北武汉 430223）

摘 要：POSCO 浦项 2 号和 3 号高炉在第三次大修时将重力除尘器改造为单管路切向旋风除尘器，结构简单、流速不高、内衬寿命长、生产维护工作量小。设计中，采用了一系列的新工艺、新技术、新装备，克服了一系列的技术难点，进行了一系列的技术创新。高炉在投产后旋风除尘系统运行良好，旋风除尘器除尘效率高达 85%~88%，比肖夫湿法煤气洗涤系统作业负荷、电耗和维修工作量均大幅下降，延长了煤气清洗系统寿命。该除尘器便于回收利用干燥的瓦斯灰中的 Fe、C，二次除尘更利于锌的回收处理。

关键词：单管路，切向旋风除尘器，应用，除尘效率，瓦斯灰回收

Application of Tangential Cyclone Separators for No. 2 and No. 3 Blast Furnaces at POSCO Pohang Steel Works

Pan Tieyi, Hu Xueping, Shan Liang, Wu Xiaodong,

Li Ye, Shen Jun, Tian Shali

(Iron & Steel Incorporation of WISDRI Engineering & Research Incorporation Limited, Wuhan 430223, China)

Abstract: During the 2 blast furnaces' 3rd relining the gravity dust catchers of No. 2 and No. 3 blast furnaces at POSCO Pohang Steel Works are revamped to single inlet tangential cyclone separators, whose flow velocity isn't high, lining has a long lifetime and maintenance work during operation is less. During the engineering, a series of new process, new technology and new equipment were adopted; a series of technologic difficulty was conquered; a series of technology was innovated. After the blast furnaces' blow-in, the cyclone de-dusting system works well. The dust collection efficiency of the tangential cyclone reaches 85%~88%. The work load, power consumption and maintenance work of Bischoff wet scrubber system decrease a lot. The lifetime of gas cleaning system is prolonged. The Fe and C in the gas dust are easily recovered by the separators. Zinc's more easily being recovered and treated in the 2nd de-dusting system.

Key words: single inlet, tangential cyclone separator, application, dust collection efficiency, recovery of gas dust

滚切式定尺剪剪切机理与剪刃断裂分析

郝建伟，胡典章，陈玉柏

（中冶京诚工程技术有限公司，北京 100176）

摘 要：研究了滚切剪剪切钢板的过程，分析了影响剪切质量的主要因素，对剪刃使用寿命进行了详细分析。针对

某现场 5m 滚切剪剪刃断裂事故进行了深入分析，避免再次发生类似事故。

关键词：滚切剪，剪刃，剪刃间隙，磨损，寿命，断裂

Fracture Analysis and Shearing Mechanism for Dividing Shear

Hao Jianwei, Hu Dianzhang, Chen Yubai

(Capital Engineering & Research Incorporation Limited, Beijing 100176, China)

Abstract: The main factors influencing the shear quality were analyzed in cutting process, and the service life of the cutting edge was analyzed in detail. A detailed investigation was carried out to investigate the accident of a 5m roll cutting shear fracture in this paper.

Key words: rolling shear, shear, knife gap, wear, life, fracture

二十辊轧机辊系稳定性动力学研究

刘海超[1]，陈　兵[2]，李　琰[2]，张利杰[2]，刘　昌[2]，刘　磊[3]

（1. 北京科技大学材料科学与工程学院，北京　100083；2. 北京科技大学机械工程学院，北京　100083；3. 北京首钢股份有限公司硅钢事业部，河北迁安　064400）

摘　要：二十辊轧机广泛应用于冷轧取向硅钢和高牌号无取向硅钢，我国由国外引进的 HZ21 型二十辊轧机具有更优秀的厚度控制及板形控制能力。为了提高轧辊的服役周期，提高生产效益，本文以 HZ21 型轧机为研究对象，利用 ADAMS 多体动力学软件建立了二十辊轧机辊系-轧件一体刚柔耦合模型，分别研究了轧制力与窜辊量对辊系稳定性的影响，研究表明：轧制力的增大、窜辊量的不足都会影响辊系受力及运动规律，并且导致轧辊的磨损加剧，影响辊系稳定性。

关键词：森吉米尔轧机，辊系稳定性，动力学仿真

The Study on Stability of Roll System of Twenty Rolling Mill

Liu Haichao[1], Chen Bing[2], Li Yan[2], Zhang Lijie[2], Liu Chang[2], Liu Lei[3]

(1. University of Science and Technology Beijing, Beijing 100083, China; 2. School of Mechanical Engineering, University of Science and Technology Beijing, Beijing 100083, China; 3. Shougang Co., Ltd., Qian'an 064400, China)

Abstract: 20-roll mill is widely used in cold-rolled oriented silicon steel and high-grade non-oriented silicon steel, China introduce the HZ21 20-roll mill with better thickness control and shape control. In order to improve the service life of the roll and improve the production efficiency, this paper takes the HZ21 mill as the research object, and uses the ADAMS multi-body dynamics software to establish the roll-to-roll rigid-flexible coupling model of the 20-roll mill. The results show that the increase of the rolling force and the lack of the amount of the roll will affect the force and movement law of the roller system, and the wear of the roll will increase, which will affect the stability of the roller system.

Key words: sendzimir mill, stability of rolls system, dynamic simulation

非对称工艺条件对热连轧精轧机组振动及轧制稳定性的影响

任天宝[1]，陈　斌[1]，郜志英[2]，孔　宁[2]

（1. 马鞍山钢铁股份有限公司，安徽马鞍山　243003；

2. 北京科技大学机械工程学院，北京　100083）

摘　要： 热连轧机精轧机组的振动问题是引起板带厚差与表面振纹等产品质量问题的主要原因，也是造成轧辊频繁更换与零部件疲劳破坏的根源，不同的轧件规格和轧制工艺直接影响振动的剧烈程度，尤其是生产高强度薄规格产品时振动问题更为突出，甚至会导致轧制失稳而危害系统运行安全。本文针对某热轧机组现场出现的典型振动现象并结合实际生产工艺，讨论上下辊面非对称工艺条件工艺参数对振动问题的影响机理与规律，为提出抑制振动的工艺措施提供依据。

关键词： 热连轧机，振动，稳定性，非对称，工艺条件

Influence of Asymmetrical Technical Condition on the Vibration and Stability in the Process of Hot Rolling

Ren Tianbao[1], Chen Bin[1], Gao Zhiying[2], Kong Ning[2]

(1. Maanshan Iron & Steel Company Ltd., Maanshan 243003, China; 2. School of Mechanical Engineering, University of Science and Technology Beijing, Beijing 100083, China)

Abstract: The vibration of hot tandem rolling mill is the main cause for thickness fluctuation and marks on the rolling strip surface. It also leads to roll changing and fatigue of parts. The rolling schedule and process directly affect the vibration degree. The rolling mill is more easily to oscillate violently when the rolled strip is thin. It even causes the destabilization of rolling mill and unsafe operation. For the typical vibration phenomenon in hot tandem rolling mill, we studied the effects of asymmetric rolling process parameters on the vibration mechanism, so that the vibration suppress strategies can be put forward finally.

Key words: hot tandem rolling mill, vibration, stability, asymmetric, technical condition

高精度高表观质量大型冷轧铝板带生产及防划痕控制技术

李谋渭，杨海波，张少军，张大志，边新孝，刘国勇，朱冬梅

（北京科技大学机械工程学院，北京　100083）

摘　要：本文简要介绍高精度高表观质量大型铝板带生产及防划痕关键技术，该技术已使 2450mm 六辊 CVC 板带生产线生产出高质量的铝罐料，划痕损伤率已从 2012 年 2 月的 35.2%下降到 2013 年 7 月的 0.7%。

关键词：冷轧机，铝带，划痕，润滑，振动

The Control Technology of High Accuracy High Surface Quality Large Aluminium Cold Mill Production and Avoiding Scratch

Li Mouwei, Yang Haibo, Zhang Shaojun, Zhang Dazhi, Bian Xinxiao,
Liu Guoyong, Zhu Dongmei

(Mechanical Engineering School, USTB, Beijing 100083, China)

Abstract: The key technology of high accuracy high surface quality large aluminiun cold mill production and avoiding scratch will be introduced in this paper. The high quility aluminnium can materials had been produced in 2450mm six-high rolling mill CVC strip produce line by this technology. The scratch damage rate had been lowered from 35.2% in February, 2012 to 0.7% in July, 2013.

Key words: cold rolling mill, aluminium strip, scratch, lubrication, vibration

大型转炉 OG 风机故障原因分析及改进优化

黄建东，龚年生

（宝山钢铁股份有限公司炼钢厂，上海　200941）

摘　要：宝钢股份一炼钢 3 座 300t 转炉的国产 OG 风机一度故障频繁，主要表现在叶片焊缝撕裂、振动大频繁停机、轴瓦频繁损坏等，严重影响转炉生产的安全顺行。通过对国产风机的一系列优化改进，取得了理想的效果。

关键词：OG 风机，腐蚀，水膜，轴向力

The Analysis and Optimizing about the Breakdown of OG Fan on Big BOF

Huang Jiandong, Gong Niansheng

(BaoSteel Steel Making Plant, Shanghai 200941, China)

Abstract: The accidents happyed frequently on OG fans of three BOFs in Baosteel's first steel making plant.These faults include tearing of blade welding seam,vibration increasing frequent and bearing busing damaging ,etc,which effects the safe run of BOFs badly.The perfect result is achieved by a series of optimizing steps.

Key words: OG-fan, corrosion, moisture-film, axial-force

Cognex 表面检测系统在冷轧薄板产线应用能力优化

王　林[1]，于　洋[1]，孙　海[2]，王　畅[1]，张　栋[1]，王明哲[3]

（1. 首钢技术研究院薄板研究所，北京　100043；2. 北京首钢冷轧薄板有限公司技术质量部，北京　101304；3. 首钢股份公司迁安钢铁公司，河北迁安　064404）

摘　要：首钢冷轧厂定位于高端汽车板和家电板生产，全流程工序引入了基于机器视觉的 Cognex 表面质量在线检测系统，通过高速摄像头以线扫描方式检测缺陷，实现了连续的表面质量在线检测和表面缺陷的分类判级。通过对产线设备和检测系统进行硬件和软件两方面的优化，解决了应用过程中出现的寻边问题和 GA/GI 板面检测差异等问题，大大提升了 Cognex 系统的检测效率和准确性，提高了薄板生产的可控性和生产效率。

关键词：冷轧薄板，寻边，GA/GI，缺陷检测

Application Capacity Optimization of Cognex Surface Inspection System on the Production of Cold Rolled Strip

Wang Lin[1], Yu Yang[1], Sun Hai[2], Wang Chang[1], Zhang Dong[1], Wang Mingzhe[3]

(1. Shougang Research Institute of Technology, Sheet Metal Research Institute, Beijing 100043, China;
2. Shougang Cold Rolling Sheet Co., Ltd., Technology and Quality Division, Beijing 101304, China;
3. Shougang Group: Shougang Qian' an Iron & Steel Company, Qian'an 064404, China)

Abstract: Shougang cold rolling mill is located in high-quality automobile and household sheet production, Cognex surface quality on-line detection system based on machine vision development has been brought in and applied in the whole process. Continuous surface quality on-line detection and surface grade classification is achieved through high-speed camera online scanning detection on defects. By the optimization of hardware and software on the production equipment and detection system, problems such as edge-searching and detection difference between GA/GI is solved, detection efficiency and accuracy of the Cognex system is enhanced greatly, and the controllability and production efficiency is improved.

Key words: cold rolled strip, edge-searching, GA/GI, defects detection

高炉无料钟加压水冷齿轮箱新技术及应用

于成忠，张荣军，徐宝军

（鞍钢股份公司炼铁总厂，辽宁鞍山　114021）

摘　要：鞍钢 11 号高炉 1990 年在国内首次引进卢森堡 Paul Wurth 公司的高炉无料钟炉顶装料技术，标志国内进入高炉无料钟装料工艺时代。随着高炉冶炼技术不断的发展，高炉的中心布焦的高强度冶炼和低成本节能炼铁的需求，为高炉布料的关键设备水冷齿轮箱提出如何能够适应高炉高顶温布料环境，实现低成本节能运行的新课题。加压水冷齿轮箱的设计理念就是基于这样的思维应运而生的，简要介绍高炉无料钟加压水冷齿轮箱的原理、优点和在

鞍钢 11 号高炉的应用。

关键词：加压水冷传动齿轮箱，S-杯，水冷站

New Technology Application of Bell Less Top-CTGW-PC

Yu Chengzhong, Zhang Rongjun, Xu Baojun

(General Ironmaking Plant of Angang Steel Co., Ltd., Anshan 114021, China)

Abstract: BF11 of Angang Steel Co., Ltd applied the bell less top charging technology of Luxemburg Paul Wurth for the first time in China in 1990, marking the beginning the bell less top charging process in China. With the development of blast furnace smelting technology, high smelting intensity and low cost energy saving is requested the blast furnace center feeding, as the key equipment of blast furnace, water-cooling gear box, how to adapt to the high temperature of blast furnace top material environment, realize low cost & energy saving operation is a new topic. The design concept of CTWW-PC is based on this thinking. This article is briefly introduced the advantages and application of CTGW-PC on Angang BF11 site.

Key words: CTGW-PC, S-cup, cooling skid

电磁场下保护渣性能测试系统的设计与搭建

赵　立，王　雨，赵潞明

（重庆大学材料科学与工程学院，重庆　400044）

摘　要：目前国内外尚未有成熟设备研究交变磁场对熔融保护渣性能的影响，为了探究不同电磁参数对熔融保护渣性能的影响规律及影响机理。本文设计并搭建了电磁场下保护渣性能测试系统，并完成了设备材质的选择、设备指导加工、装配及工作效果测试整个过程，所设计的测试系统的磁场强度、频率等参数可以实现一定范围的自由调节。该测试系统有坩埚炉、电磁搅拌器、水冷机等主要设备，并具有价格便宜、电磁铁与坩埚炉能同时工作、操作方便等特点。

关键词：测试系统，坩埚炉，电磁搅拌装置，保护渣

Design and Construction of Performance Test System for Mold Flux under Electromagnetic Field

Zhao Li, Wang Yu, Zhao Luming

(College of Materials Science and Engineering, Chongqing University, Chongqing 400044, China)

Abstract: Nowadays, there are no mature equipment at home and abroad to study the influence of alternating magnetic field on the performance of molten slag. In order to study the influence of different electromagnetic parameters on the performance of molten slag and the influence mechanism. This paper designs and builds the slag performance test system of electromagnetic field, and completed the crucible material selection, processing, assembly and work effect test to guide the whole process. The testing system design of the magnetic field intensity, frequency and other parameters can achieve a range of free adjustment. The test system has the main equipment, such as crucible furnace, electromagnetic stirrer, water cooling machine, etc. It has the advantages of cheap price, electromagnetic and crucible furnace working at the same time, easy operation and so on.

Key words: test system, crucible furnace, electromagnetic stirring device, mold flux

承钢 120t 板坯连铸机蒸汽抽风机改造

张翼斌，王平刚，胥　阳，张　跃

（河钢集团承钢公司维护检修中心，河北承德　067102）

摘　要：120t 板坯连铸二冷室蒸汽抽风机使用过程中因产生剧烈振动、噪声大、风机轴承和叶轮损坏、软连接破裂造成蒸汽外泄，影响浇钢工人操作，同时存在一定安全隐患。改变风机传动形式和机体结构，延长风机使用周期和检修时间。

关键词：离心通风机，传动形式，振动，断裂，联轴器

Modification of Steam Exhauster for Chengde Steel 120t Slab Caster

Zhang Yibin, Wang Pinggang, Xu Yang, Zhang Yue

(HBIS Group Chengsteel Company Maintenance Center, Chengde 067102, China)

Abstract: 120t slab continuous casting two cold chamber steam exhaust fan in use process because of vibration, noise, bearing and fan impeller damage, soft connection rupture caused by steam leakage, influence of cast steel workers to operate, at the same time, there are some hidden danger. Change fan drive type and airframe structure, prolong fan cycle and overhaul time.

Key words: centrifugal fan, transmission type, vibration, fracture, coupling

青岛特钢 2 号高炉无料钟炉顶冷却系统设计与优化

戴建华[1]，闫树武[1]，毛庆武[1]，耿云梅[1]，赵　强[2]

（1. 北京首钢国际工程技术有限公司炼铁事业部，北京市冶金三维仿真设计工程技术研究中心，北京　100043；2. 青岛特殊钢铁有限公司炼铁部，山东青岛　266400）

摘　要：青岛特钢 2 号高炉炉顶装料设备采用了首钢国际新型串罐无料钟炉顶装料设备。本文主要介绍该高炉的炉顶设备结构以及气密箱冷却系统的设计与水冷系统管路优化改进实施情况。炉顶气密箱采用氮气密封，采用工业水开路循环冷却。在水冷系统调试过程中，发现管路中水流不稳定，特别是在炉内升压到一定压力时，容易发生断流、击穿水封甚至往炉内溢水的现象。经过分析找出原因后优化改进系统管路布置，解决了上述问题，高炉投产后系统运行稳定，冷却效果很好，满足了生产要求。

关键词：高炉，串罐无料钟炉顶，气密箱，冷却

Design and Optimization of Bell-less Top Cooling System on No. 2 BF of Qingdao Special Steel Co.,Ltd.

Dai Jianhua[1], Yan Shuwu[1], Mao Qingwu[1], Geng Yunmei[1], Zhao Qiang[2]

(1. Beijing Shougang International Engineering Technology Co., Ltd. Iron-making Division, Beijing Metallurgy 3D Simulation Design Engineering Technology Research Center, Beijing 100043, China;

2. Qingdao Special Steel Co., Ltd. Iron-making Plant, Qingdao 266400, China)

Abstract: A serial hoppers bell-less top charging equipment supplied by BSIET is used on Qingdao Special Steel No. 2 BF. And the structure of the top charging equipment, as well as the design and optimization of the cooling system of the gas-sealing gear box, is introduced on this paper. Normally the gear box is sealed with nitrogen and cooled with opening circuit loop industrial water. During the commissioning phase of the water cooling system, it was found that the water flow in the pipeline was unstable, especially when the pressure in the furnace was risen up on a high level, the water flow broke down, the water seal broke up and even some water spilled into the furnace. After analyzing the reasons and optimizing the piping layout of the system, the above problems have been solved. Since the blast furnace blow-in, the cooling system runs stably and the cooling effect meets the production requirements well.

Key words: blast furnace, serial hoppers bell-less top, gas-sealing gear box, cooling system

六辊轧机三维振动特性

郑永江，李一耕，申光宪

（燕山大学机械工程学院，河北秦皇岛　066004）

摘　要：冷轧板带六辊轧机是装备技术现代化水平最高的轧机，但轧机制造企业未能给出轧机振动特性指标。轧机按机构学静定理论可分为稳定和不稳定轧机两种，其区别在于辊系结构中有无辊间偏移距设置和轧辊轴承座与机架窗口立柱间侧隙。已有的轧机振动特性分析均按照稳定轧机进行分析。在重载及高速工况下，考虑轧机辊系的偏移距和轴承座侧隙影响的两种六辊轧机，用作者开发的修正多物体传递矩阵法进行轧辊垂直、水平、轴向位移及扭转、交叉及摇摆等6个自由度的三维振动行为解析，给出六辊轧机三维振动特性。由此深度区分不稳定轧机振动特性劣于稳定轧机辊系，指明了偏移距和轴承座与机架的侧隙放任的危害性，阐明轧机更新换代的必要性。

关键词：六辊轧机，三维振动，传递矩阵法，稳定轧机

Three-dimensional Vibration Characteristics of 6-H Mill

Zheng Yongjiang, Li Yigeng, Shen Guangxian

(College of Mechanical Engineering, Yanshan University, Qinhuangdao 066004, China)

Abstract: The cold rolling mills are equipments with a high level of modernization. However, the chatter vibration the vibration characteristics parameters cannot be given by the manufacturing enterprise. The cold rolling mills can be divided into two kinds by the statically determinate theory of mechanism, namely the stable and unstable mills. The main difference is that, in the roll system, whether the offset distance set and the side clearance between the chock and side surface of housing window exists. The existing mill vibration analysis all aims at the stable mill. Considering the effects of offset distance set and side clearance, two kinds of six-high mill with heavy load and high speed rolling condition are set as examples to analysis the three-dimensional vibration behavior involving 6 degree of vertical, horizontal, axial, torsional, cross and swinging vibration by the modified multi-body transfer matrix method. The results of three-dimensional vibration characteristics of two kinds of six-high mill can be used for distinguishing the good and bad vibration characteristics to demonstrate the perniciousness of the offset distance set and uncontrolled side clearance, and to clarify the essential of upgrading of the mill.

Key words: 6-h mill, three-dimensional vibration, transfer matrix method, stable mill

首例智能化新能源钢卷运输系统的创新及应用

韦富强 [1,2]，杨金光 [3]，杨建立 [1]，郑江涛 [1]，吕冬梅 [1]，毕国龙 [1]

（1. 北京首钢国际工程技术有限公司智能运输设备研究所，北京　100043；

2. 北京市冶金三维仿真设计工程技术研究中心，北京　100043；

3. 山东钢铁集团日照有限公司，山东日照　276806）

摘　要：本文概述了 10 年来冶金企业钢卷运输技术的发展，介绍了超级电容器及其在公交行业的应用，详细说明了首台智能化新能源钢卷运输车及其创新点，对比了某工程采用的智能化新能源钢卷运输系统方案和传统的托盘式运输方案，指出了该技术的优点及在其他领域、其他行业的可能应用。该技术符合中国制造 2025 的要求，为智能重载物流提供了一种可靠的运输手段解决方案。

关键词：智能化，新能源，重载，钢卷运输，创新

Innovation & Application of the First Intelligent New Energy Steel Coil Transportation System

Wei Fuqiang[1,2], Yang Jinguang[3], Yang Jianli[1], Zheng Jiangtao[1],

Lv Dongmei[1], Bi Guolong[1]

(1. Institute of Intelligent Transportation Equipment of BSIET Co., Ltd., Beijing 100043, China;

2. Beijing Metallurgical 3-D Simulation Design Engineering Technology Research Center, Beijing 100043, China;

3. Rizhao Co., Ltd. of Shandong Steel, Rizhao 276806, China)

Abstract: This paper summarized the development in this 10 years of the steel coil transportation technology of metallurgical enterprises, introduced the super capacitor and its application in the public transport industry, described in detail of the first intelligent new energy coil car and its innovations, compared two technical proposals of a project that using the new transportation system and the traditional system, points out the advantages and the possible applications of this technology in other fields and other industries. The technology meets the requirements of 2025 made in China, and provides a reliable transportation solution for intelligent heavy load logistics.

Key words: intelligent, new energy, heavy load, steel coil transportation, innovation

基于嵌入式图像处理热轧辊系平行度的检测

陈盼盼，苏兰海

（北京科技大学机械工程学院，北京　100083）

摘　要：热轧辊系平行度检测系统是用来检测离线状态下热轧辊系中支撑辊轴线相对于支撑辊轴承座上的两支撑点连线的平行度，得到辊系的装配误差，判断辊系是否合适用于生产线。本文提出利用嵌入式图像处理的方法来进行检测系统的视觉检测模块的设计，得到支撑辊轴承座两支撑点连线的绝对角度，作为检测系统检测时的测量基准，然后结合倾角传感器对轧辊轴线角度的测量数据，得到轧辊轴线相对于测量基准的相对角度，完成辊系平行度的检测。实验结果表明，便携的嵌入式设备加上视觉检测的方法，使得高度检测精度小于 0.5mm，角度检测精度为 0.009 度，满足系统的检测要求，且检测算法的鲁棒性较好，准确可行。

关键词：嵌入式，视觉检测，热轧辊系，中心点提取

Detection of Parallelism of Hot Rolling Roll System Based on Embedded Image Processing

Chen Panpan, Su Lanhai

(School of Mechanical Engineering, University of Science and Technology Beijing, Beijing 100083, China)

Abstract: The system of measuring the degree of parallelism of the roll system is used to detect the parallelism between the supporting roll axis in the hot roll system and the two support line on the support roller bearing seat in the hot roll system, getting the assembly error of roll system and judge whether the roll system is suitable for production line. In this paper, the method of embedded image processing is used to design the visual inspection module of the detection system, the coarse positioning of the first images contour using correlation algorithm, then use the method of sub-pixel accurate positioning, to get the absolute angle support two support line of roller bearing seat, the measurement standard of detection system at, then, the relative angle of the roll axis relative to the measuring reference is obtained by measuring the angle of the roll axis with the angle sensor, and the parallelism of the roll system is measured. The experimental results show that portable embedded device with visual inspection method, and the height detection accuracy is less than 0.5mm, and the angle detection accuracy is 0.009 degrees, which meets the detection requirements of the system, and the robustness of the detection algorithm is better, accurate and feasible.

Key words: embedded system, the visual detection, hot roll system, center point extraction

磷化钢氧化铁皮压入缺陷描述与剥离机理研究

陈　斌[1]，孟庆军[1]，杨　迪[2]，孔　宁[2]

（1. 马鞍山钢铁股份有限公司，安徽马鞍山　243003；
2. 北京科技大学机械工程学院，北京　100083）

摘　要：本文对现场酸洗后的磷化钢表面存在的黑点进行了微观分析与成分测定，结果判断为压入氧化铁皮缺陷。为了有效去除氧化铁皮保证带钢表面质量，采用拉矫-酸洗试验模拟了磷化钢表面氧化铁皮的剥离过程，并以延伸率表征酸洗前一道工序拉矫破鳞的程度，通过表面微观检测可知随着带钢延伸率的增加，试样表面裂纹数目及宽度不断增加；在进行不同时间的盐酸酸洗后，发现与无裂纹相比，裂纹的存在使得酸洗机理发生了改变，无裂纹酸洗以很慢的表面溶解为主，有裂纹以化学溶解和机械剥离相结合的方式进行酸洗，大大加快了酸洗程度。最后通过酸洗动力学曲线进一步验证了机械破鳞对于酸洗效率的提高起着关键性的作用，使得拉矫机破鳞与酸洗除鳞的结合更加有意义，为氧化铁皮的高效剥离创造了条件。

关键词：压氧缺陷，酸洗，化学溶解，机械剥离

Defect Description and Peeling Mechanism of Oxide Scales in Phosphating Steel

Chen Bin[1], Meng Qingjun[1], Yang Di[2], Kong Ning[2]

(1. Maanshan Iron & Steel Company Ltd., Maanshan 243003, China; 2. School of Mechanical Engineering, University of Science and Technology Beijing, Beijing 100083, China)

Abstract: This paper analysis and determination of micro components spots on steel surface after pickling phosphating site, the judgment is pressed into the scale defect. In order to effectively remove the oxide scale to ensure strip quality, the straightening and pickling tests on steel surface oxidation phosphating tin stripping process, and to extend the characterization of a pickling rate process of straightening and descaling degree by surface micro analysis showed that with the increase of strip elongation, the number and width of surface crack is increasing; in different time after compared with hydrochloric acid, no crack, crack due to the existence of pickling mechanism changed, no crack on the surface of the acid dissolved very slowly the main crack combined with chemical dissolution and mechanical stripping method for pickling, greatly accelerate the pickling degree. Finally further verified the mechanical kinetic curve of pickling Scale breaking plays a key role in the improvement of acid washing efficiency, which makes it more meaningful to combine the descaling of the straightening machine with the descaling process, thus creating the conditions for the complete stripping of the oxide scales.

Key words: oxide scale indentation defect, acid pickling, chemolysis, mechanical peeling

冷轧薄板高速轧制时的振动问题及抑振措施

王康健[1]，王　欣[1]，郜志英[2]，张清东[2]

（1. 宝山钢铁股份有限公司，上海　201900；2. 北京科技大学机械工程学院，北京　100083）

摘　要：高强薄带钢在高速冷连轧生产过程中，设备安全、生产效能和产品质量都受到轧机振动及轧制失稳的影响和制约，随产品规格和工艺条件的不同，振动失稳现象表现为随机和不确定性，但随着速度提升振动又表现为失稳的必然性，产品越薄问题越突出，严重影响生产效率和产品质量。本文针对某冷轧机生产现场中的不同产品规格，通过对轧件等效刚度和临界速度的计算，分析了轧件厚度和变形抗力对轧机振动及失稳的影响，并提出通过间断升速、降速以及改善润滑等工艺措施实现对振动的抑制。

关键词：冷轧，薄板，轧机振动，临界轧制速度，抑振

Mill Vibration and Suppression in the High-speed Cold Rolling Process of Thin Strip

Abstract: In the production process of cold rolling of thin steel sheets, such as automobile plate and tin plate，Equipment safety, production efficiency and product quality are affected and restricted by rolling mill vibration and rolling instability. By focusing on different production standards of a certain cold rolling machine on production site, the equivalent stiffness of the strip and the critical velocity is calculated, and the influence of rolling piece thickness and strain resistance on the vibration and instability of rolling machine is analyzed. Finally some technology strategies for suppressing vibration are presented, such as discontinuous speed lift, lowing speed and improving lubrication condition.

Key words: cold rolling, thin strip, mill vibration, critical rolling speed, vibration suppression

基于遗传算法的智能加渣机器人运动规划

刘景亚，张燕彤，万小丽，刘向东，彭晓华

（中冶赛迪技术研究中心有限公司装备系统研究所，重庆　401122）

摘　要： 工业机器人在钢铁冶金行业中的应用，有利于企业降低人力资源成本、降低生产运行维护成本、提高生产效率、提高产品质量和生产稳定性。由于钢铁冶金流程工作环境复杂恶劣，对工业机器人的运动规划提出了更高的要求。本文基于遗传算法对智能加渣机器人的运动规划问题进行了研究。首先介绍了智能加渣机器人的系统组成及工作原理；其次建立了运动规划模型，以关键点和障碍点为约束，以最短运动行程和不超过各关节预定最大扭矩值为目标利用遗传算法进行了优化；最后以一个平面三自由度工业机器人为例对模型进行了验证。结果表明，该模型可以使规划的路径以最短运动行程经过关键点并实现避障，同时各关节的扭矩不超过预定值。

关键词： 智能加渣机器人，遗传算法，运动规划，避障

Motion Plan of Intelligent Slag Addition Robot Based on Genetic Algorithm

Liu Jingya, Zhang Yantong, Wan Xiaoli, Liu Xiangdong, Peng Xiaohua

(Equipment System Research Division, CISDI Research & Development Co., Ltd., Chongqing 401122, China)

Abstract: The application of industry robot in metallurgical area is benefit for reducing manpower cost, production cost and accidents, and also for improving the productivity, process stability and product quality. The requirement for the motion plan of the industry robot is more and more higher because of the complicated and harsh working environment. Based on genetic algorithm the motion plan of intelligent slag addition robot is studied. Firstly, the system composition and the working principle of the intelligent slag addition robot is introduced; secondly, the model of the motion plan is established and optimized using genetic algorithm which is constrained by key points and obstacle points, and aimed at the shortest motion distance, while not exceeding a maximum joint torque; finally, the model is verified by a planar 3-dof industry robot. The results reveal that the model can plan the motion passing through the key points with avoiding the obstacle, while not exceeding the pre-defined joint torque.

Key words: intelligent slag addition robot, genetic algorithm, autonomous motion plan, autonomous obstacle avoidance

热连轧工作辊冷却喷嘴布置的优化设计

陈　斌[1]，毛　鸣[1]，宋　鸣[2]，刘国勇[2]，朱冬梅[2]，张少军[2]

（1. 马鞍山钢铁股份有限公司，安徽马鞍山　243003；

2. 北京科技大学机械工程学院，北京　100083）

摘　要：在热轧带钢生产中，冷却喷嘴的布置是影响工作辊温度分布的重要因素。为了改善工作辊温度分布，本文利用有限元软件 ANSYS 建立了工作辊二维非稳态温度场计算模型，对冷却喷嘴布置的优化进行了研究。研究发现：当 F1~F7 机架工作辊冷却水的水凸度分别为：33%、33%、33%、38.7%、38.7%、41.2%、41.2%时，工作辊的中部与边部的温差将降低 7~8℃。采用两种方式对喷嘴布置进行了优化，一种方式是：每排的喷嘴间距是相同的，通过改变喷嘴流量大小，从轧辊中部向两侧按照喷嘴流量从大到小进行梯度布置。另一种方法是：每排喷嘴型号、流量等都相同，通过改变喷嘴间距，从轧辊中部向两侧按照喷嘴间距从小到大进行布置。

关键词：热连轧，工作辊，冷却，温度分布优化，喷嘴布置

Optimized Arranging Form of Cooling Nozzles of Hot Rolling Work-roll

Chen Bin[1], Mao Ming[1], Song Ming[2], Liu Guoyong[2], Zhu Dongmei[2], Zhang Shaojun[2]

(1. Maanshan Iron & Steel Company Ltd., Maanshan 243003, China; 2. School of Mechanical Engineering, University of Science and Technology Beijing, Beijing 100083, China)

Abstract: The arranging form of cooling nozzles plays an important role in affecting temperature of work roll during hot strip rolling. The 2D model for simulating the temperature field of work roll was established with ANSYS software. In order to improve the temperature distribution of work roll, the arranging form of nozzles was researched. It is found that when the water crowns of cooling water are 33% of F1~F3 stands, 38.7% of F4~F5 stands, and 41.2% of F6~F7 stands respectively, the temperature differences between middle and edge of work roll will reduce 7~8℃. Then two arranging forms of nozzles were proposed. One arranging form is changing water flow of nozzles that nozzle spacing is consistent. The other arranging form is changing nozzle spacing that water flow is consistent.

Key words: hot strip mill, work roll, cooling, temperature distribution optimization, arranging form of nozzles

高温耐热不锈钢工艺罩失效机理分析

廖礼宝，邓　聪，吴荣杰

（宝钢股份设备部，上海　200941）

摘　要：某高温加热炉内部布置一种大型圆柱体隔离工艺罩，其目的是用来将某种钢卷实施高温热处理。该工艺罩结构形状类似倒扣的水杯。罩内放置钢卷，同时充入一定比例的氢气和氮气等混合气体，然后，将罩口四周用特定材料密封。该工艺罩由日本进口，使用约 11 个月后，主要发现如下失效现象：顶部陆续出现塌陷变形，罩体拼接

焊缝出现开裂，罩口附近出现大小不同、深浅不一的点状圆形麻坑，材料明显脆化等。针对失效现象，文章对罩体失效宏观形貌、材料成份、金相组织、脆性相析出及形成和焊接接头组织形貌等分别进行了试验与分析。并依此提出了结论和相应的建议。

关键词：高温，耐热不锈钢，工艺罩，失效机理，分析

2230 酸轧活套车转向轮掉落浅析

李子俊，刘　森，张森建

（首钢京唐钢铁联合有限责任公司冷轧作业部，河北唐山　063200）

摘　要：2230 酸轧机组运行过程中多次发生活套车转向轮掉落故障，每次停车均超过十个小时，对连续生产形成较大影响，轴承损坏是根本原因。在选取活套最大速度和最大张力的情况下，通过计算轴承使用寿命，得出轴承损坏原因为在线超期使用，提出轴承更换周期，彻底解决了该问题，对实际工作有很大的指导意义。

关键词：活套，转向轮，轴承损坏，寿命计算

Analysis on the Steering Wheel Drop of 2230 PLTCM Loop Car

Li Zijun, Liu Sen, Zhang Senjian

(Shougang Jingtang United Iron & Steel Co., Ltd., Cold Rolling Department, Tangshan 063200, China)

Abstract: In the operation process of the 2230 PLTCM, the steering wheel drop fault occurred several times, every stop for more than ten hours, which has a great impact on continuous production, bearing damage is the root cause. Under the condition of selecting the maximum speed and maximum tension of the loop, through calculating the bearing service life, the cause of bearing damage is online overdue use, the bearing replacement cycle is proposed, thoroughly solved the problem, and has great guiding significance to the actual work.

Key words: live sleeve, steering wheel, bearing damage, life calculation

3 号空分离心式氮压机漏油故障分析及处理

张建刚

（昆钢动力能源分公司）

摘　要：3 号空分氮压机为ЦKOH-320/0.8-16YXπ4 型离心式氮压机，该氮压机在运行中存在着漏油现象，不仅影响了氮气的产品质量，增加了生产和维护的成本，而且污染了生产现场，对现场操作、检修人员的工作造成了较大的影响。氮压机漏油现象主要反映在排放管口、高低压室、出口氮气露点上。经过我们现场分析和论证，导致漏油（烟）的故障原因主要有油封密封不严、油冷却器和气体冷却器本体通洞造成水、气、油三者互窜，排放管满足不了大量油烟的排放要求。因此我们在故障分析的基础上针对漏油（烟）原因进行了检查、排除和相应的处理：(1) 保

证密封效果，阻断油（烟）的泄漏途径：检查了油封、冷却器。(2) 提高降温效率，减少油（烟）的产生：清理了油冷却器、气体冷却器。(3) 对油（烟）进行疏导，解决了油（烟）的排放：改造了排放管，加装了油气分离器、抽油烟机等设施。从而圆满地解决了进口设备在使用过程中的缺陷，以较低的成本获得了较好的经济效益。

关键词：氮压机，漏油，故障分析

光源质量对冷轧表面缺陷检测仪效果的影响

蒋　渝

（宝山钢铁股份有限公司冷轧厂，上海　201900）

摘　要：针对生产现场热镀锌表面缺陷检测仪发生的锌灰缺陷漏检问题，逐项排查了缺陷检测仪的成像系统，对视场、照明、信号参数等关键影响因素分别进行了分析验证，并结合现场实际情况，寻找到了问题的原因，明确了检测仪光源照明的不均匀会严重影响缺陷检测仪的检测效果。

关键词：冷轧，表面缺陷，检测仪，光源，均匀性，检测效果

Influence of the Quality of Light Source on the Surface Defects Detector for Cold Rolled Strips

Jiang Yu

(Cold Rolling Mill, Baoshan Iron & Steel Co., Ltd., Shanghai 201900, China)

Abstract: Aiming at the problem of the surface defect detector not detected of zinc ash in hot-dip galvanizing line, the imaging system of the detector is investigated. The key factors, such as field of view, illumination and signal parameters, are analyzed and validated respectively. Combined with the actual situation in the field, the reasons for the problem are found, the uneven illumination of light source will seriously affect the detection effect of the detector.

Key words: cold rolling, surface defect, detector, light source, uniformity, detection effect

斗轮机斗轮轴连接装置的故障分析及改进

王明江

（武汉钢铁有限公司炼铁厂原料分厂，湖北武汉　430082）

摘　要：斗轮取料机由于斗轮轴连接装置上存在的问题，导致斗轮无法运转，针对存在的问题，围绕现场实际对其连接装置的故障进行了分析和改进。该文设计思路清晰，对斗轮轴连接装置的故障制定了切实可行的改造方案。通过改造实施，效果明显。

关键词：斗轮机，斗轮轴，法兰连接器，改进

Failure Analysis and Improvement of Bucket Wheel Axle Connection Device of Bucket Wheel Machine

Wang Mingjiang

(Raw Material Branch of Ironmaking Plant of Wuhan Iron and Steel Co., Ltd., Wuhan 430082, China)

Abstract: Because the problems existing in the connecting device of the bucket wheel and bucket wheel cause the bucket wheel to be unable to run, according to the existing problems, the fault of the connecting device is analyzed and improved according to the field practice. The design concept is clear, and the feasible scheme for the failure of the bucket wheel shaft connection device is worked out. Through the implementation of the reform, the effect is obvious.

Key words: bucket wheel machine, bucket wheel shaft, flange connector, improvement

---

新型 CPE 法无缝钢管连轧设备—拉管机结构与性能

丁　军，徐　刚，张海军，贾庆春

（中冶京诚工程技术有限公司轧钢工程技术所，北京　100176）

摘　要： 针对目前 CPE 的生产工艺，提出用"拉"替代"推"的方式来进行惰辊轧管生产的新型 CPE 热轧拉拔制管工艺设备—拉管机，其传动布置、受力方式和工具设计等都与顶管机有显著的区别，电动机通过减速机驱动齿条，拉动芯棒拔制得荒管。沿轧制线方向，在床身内串列设置若干个辊模，芯棒带着毛管依次穿过各个辊模，从而实现荒管轧制。此拉管机提供更为高效经济的大延伸率轧制的无缝钢管生产方式，辊模贴近均布，同时咬入辊模数大为增加，床身系数更小，芯棒直径更小，可生产荒管直径更小，且没有压杆稳定性问题，大量节省了扣瓦工具和产品换规格的更换时间。

关键词： CPE，拉管机，辊模，芯棒，毛管，轧制，荒管

Construction and Performance of New CPE Pipe Tandem Mill with Pull Benches

Ding Jun, Xu Gang, Zhang Haijun, Jia Qingchun

(Capital Engineering & Research Incorporation Limited (CERI), Beijing 100176, China)

Abstract: Pull benches, which use "pull" method to replace traditionally "push" on CPE process is a new hot-rolling tube-drawing equipment, whose drive layout, force mode and facility design have significant differences. The motor drive the reducer, then move the gear rack to drag mandrel bar for drawing steel shell. Along the center of rolling line, mandrel bar with hollow bloom sequentially through the benches which are set in the ring bed to roll the shell. Pull benches provide a high-performance methods of production for seamless pipe, which keep arrange closely, more roll stand working simultaneously, less bed length ratio and mandrel bar, then less shell diameter, and with no problem of pressure lever stability, and save a lot of time for changing tools and tube size or specifications.

Key words: CPE, pull benches, roll stand, mandrel bar, hollow bloom, rolling, shell

板形检测辊非概率区间可靠性设计方法研究

吴海淼[1]，王　凯[1]，孙建亮[2]

（1. 河北工程大学机械与装备工程学院，河北邯郸　056038；
2. 燕山大学国家冷轧板带装备及工艺工程技术研究中心，河北秦皇岛　066004）

摘　要：针对整辊内嵌式板形检测辊，应用灰度理论确定了带钢张力的不确定区间，采用非概率区间可靠性设计方法推导了板形检测辊与带钢之间不打滑的可靠度函数及板形检测辊不发生强度失效的可靠度函数，在此基础上建立了板形检测辊内径的非概率可靠性优化设计模型。采集现场实验数据进行计算并试制板形检测辊样机进行工业试验，结果表明，当外径取 400mm、内径取 260mm 时，板形检测辊满足设计要求，工作可靠。

关键词：板形检测辊，非概率可靠性设计，不确定区间，张力，内径

The Non-probability Interval Reliability Design Method on Entire Roller Embedded Shapemeter Roll

Wu Haimiao[1], Wang Kai[1], Sun Jianliang[2]

(1. College of Mechanical and Equipment Engineering, Hebei University of Engineering, Handan 056038, China;
2. National Engineering Research Center for Equipment and Technology of Cold Strip Rolling,
Yanshan University, Qinhuangdao 066004, China)

Abstract: For entire roller embedded shapemeter roll, the uncertain range of the strip tension was determined by the gay degree theory, the reliability function that there was no slip between strip and shapermter roll and the reliability function that strength failure of the shapemeter roll did not occurred were deduced by the non-probability interval reliability theory. On this basis, the non-probability interval reliability optimization design model of shapermeter roll was built, and the shapemeter roll diameter was calculated by field experiment data. The result showed that the shapemeter roll can normally work, when the outer diameter and inner diameter of shapemeter roll were 400mm and 260mm respectively.

Key words: shapemeter roll, non-probability interval reliability design, the uncertain range, strip tension, inner diameter

旋转过滤并联组合式高炉炉前渣水分离技术的研究

游　涛

摘　要：针对国内外研究现状和各种渣处理渣水分离方法存在的缺陷，提出了一种渣水分离创新技术——旋转过滤并联组合式高炉炉前渣水分离技术。介绍了该创新装置的结构，对其可靠性进行了探讨。根据模拟装置试验数据，与现有各种高炉渣处理机械装置进行了对比。探讨分析了创新技术装置，从根本上克服现有各种渣处理方法不能适应细小颗粒渣、棉渣、浮渣的缺陷。

关键词：旋转过滤并联组合式，高炉渣处理，渣水分离，创新

A Study on Slag Water Separation Technology of Rotary Filter Parallel Combination Type in Front of Blast Furnace

You Tao

Abstract: In view of the research status at home and abroad and the existing deficiencies of various slag water separation methods in slag treatments, a new technology for slag water separation was proposed—slag water separation technology of rotary filter parallel combination type in front of blast furnace. The structure of the innovative device was introduced, and its reliability was discussed. According to the experimental data of simulation device, an comparison with various blast furnace slag treatment machines was carried out. The innovation technology device was discussed and analyzed, it can fundamentally overcome the defects of existing slag treatment methods that can not be adapted to small particle slag, cotton residue and dross.

Key words: rotary filter parallel combination type, blast furnace slag treatment, slag water separation, innovation

采用电动势对实验高炉炉缸液面测量

李　映，魏　涵，于要伟

（省部共建高品质特殊钢冶金与制备国家重点实验室、上海市钢铁冶金新技术开发应用
重点实验室和上海大学材料科学与工程学院材料工程系，上海　200072）

摘　要： 实时掌握高炉炉缸渣铁液面高度和炉缸热状态信息可以辅助炼铁工作者更好地操作高炉，保证高炉生产稳定和延长高炉寿命，但是目前，对炉缸热状态监控主要通过操作工经验判断、热电偶和水冷（或风冷），这些提供渣铁液面信息很有限。本文研究了电动势（electromotive force，EMF）测量用于监测炉缸中熔融锡液面变化实验。实验结果表明 ΔEMF 信号和液面变化有直接联系，垂直方向的 ΔEMF 值随着液面下降而减小，耐火材料（炉墙）厚度和 ΔEMF 信号强弱相关，同一圆周方向上 ΔEMF 值反映液面和死焦柱状态。

关键词： 高炉炉缸，炉缸侵蚀，EMF，液面测量

Monitoring Liquid Level by Electromotive Force (EMF) Signal in a Laboratory Scale of Blast Furnace Hearth

Li Ying, Wei Han, Yu Yaowei

(State Key Laboratory of Advanced Special Steel, Shanghai Key Laboratory of Advanced Ferrometallurgy, School of Materials Science and Engineering, Shanghai University, Shanghai 200072, China)

Abstract: The ability to exactly measure slag and iron liquid levels in the hearth creates more stable cast house operations and prohibits blast furnace from fluctuations. In practice, status of the hearth liquid and thermal state is evaluated mainly by the experience of operators thermocouple and cooling water. In present article, Electromotive force (EMF) is used to monitor liquid level of a laboratory scale of blast furnace hearth using molten tin. Experimental results show ΔEMF signal can characterize the liquid level in the hearth of blast furnace. ΔEMF value in the vertical direction decreases as the liquid level decreases. The thickness of graphite crucible only decreases the ΔEMF magnitude. Circumferential difference of ΔEMF explains the status of liquid surface and the status of dead man during tapping.

Key words: blast furnace hearth, hearth erosion, electromotive force, liquid level measurement

高线减定径机在线监测与诊断技术研究

蔡正国

（上海金艺检测技术有限公司诊断部，上海　201900）

摘　要： 高速线材减定径机的运行状态好坏直接影响到轧制稳定性，一旦出现故障有可能造成停机事故。本文提出了一种减定径机运行状态的在线振动监测与诊断方法，通过采集设备的振动参数和工艺过程量参数，建立了监视减定径机状态劣化趋势的分类指标和在线预警方法，避免因齿轮箱轴承故障、轴系安装不当、啮合表面状况不良、齿轮啮合异常等导致的非计划停机，支撑高线轧机的正常生产。

关键词： 减定径机，在线监测，诊断

Research on Vibration On-line Monitoring & Diagnosis Technology for Reducing Sizing Mill

Cai Zhengguo

(Shanghai Jinyi Industry Technology Co., Ltd. Diagnosis Department, Shanghai 201900, China)

Abstract: Condition of reducing sizing mill (RSM) definitely impacts on rolling stability while gearbox and bearing faults occur furthermore result in production breakdown. The condition on-line monitoring and fault diagnosis technology for RSM is established, vibration parameters and process data are collected to reconstruct the classified criterions forecasting RSM abnormal such as rolling bearing, misalignment of shaft system, gearbox surface flakes and broken tooth etc. The practical application shows that the multiple parameters diagnosis technology can effectively monitor and forecast the typical faults in the process of RSM running.

Key words: RSM, condition on-line monitoring, smart diagnosis

窄搭接焊机质量控制分析及管理要素

任予昌，张东方，欧阳娜，田志辉

（宝武钢铁武汉有限公司冷轧厂，湖北武汉　430000）

摘　要： 窄搭接电阻焊机是将带尾、带头以一定的搭接量搭接在一起，上下电极轮夹紧搭接处，转动电极轮从操作侧移至传动侧，即完成焊接过程。本文介绍了窄搭接焊机的常发问题，介绍了窄搭接焊机的工作原理，分析了焊接电流、电极压力、焊接速度、搭接量、搭接补偿量、焊轮状态及带钢表面状况等因素对窄搭接焊接质量的影响，并总结了实际生产过程中及焊接试验时焊缝质量判断的方法。

关键词： 窄搭接焊机，焊接原理，焊接质量，影响因素，质量判断

Analysis on Quality Control and Management Factors of Narrow Lap Welding Quality

Ren Yuchang, Zhang Dongfang, Ouyang Na, Tian Zhihui

Abstract: Narrow lap welding machine is with tail and lead to a certain amount of lap together, the upper and lower electrode wheel clamping joints, the rotating electrode wheel from the operating side to the transmission side to complete the welding process. This paper introduces the common problems of narrow lap welding machine, introduces the working principle of narrow lap welding machine, analyzes the influence of welding current, electrode pressure, welding speed and the amount of overlap, the amount of compensation, the lap welding wheel and strip surface condition on welding quality of narrow lap, and summarizes the actual production process and welding when the test method of weld quality judgment.

Key words: narrow lap welding machine, welding problems, welding quality, influencing factor, quality judgement

Formastor-F 全自动相变仪的典型应用

刘文艳，陈叶清，黄治军

（宝钢股份中央研究院武汉分院，湖北武汉　430080）

摘　要: Formastor-F 全自动相变仪是目前世界上较先进的研究钢材相变特征的试验装置之一。为了更好地了解和掌握 Formastor-F 设备特点和使用，本文主要介绍了 Formastor-F 全自动相变仪的系统组成、功能及其典型应用包括钢材临界点测试、奥氏体晶粒度测定、CCT 曲线测试、TTT 曲线测试及其他模拟实验。

关键词: Formastor-F，临界点，CCT，TTT

Typical Application of Formastor-F

Liu Wenyan, Chen Yeqing, Huang Zhijun

(Wuhan Branch of Baosteel Central Research Institute, Baoshan Iron & Steel Co., Ltd., Wuhan 430080, China)

Abstract: Formastor-F full automatic transformation measuring instrument is one of the most advanced simulating equipments to study phase transformation in the world. In this paper, The systems, functions and typical application of Formastor-F including the determination of critical points, austenitic grain size, CCT, TTT and other simulation experiments were demonstrated for the sake of understanding the characteristic of Formastor-F and knowing well how to use it.

Key words: Formastor-F, critical points, CCT, TTT

方坯在线称重系统的技术改进

王　欣，郑　虎

（河钢唐钢二钢轧厂）

摘　要：原有的铸坯在线称重系统采用了传感器-称重仪表组合结构，秤体处于铸坯输送辊道下方，使用过程中存在着高温高湿、环境恶劣、现场传感器多、算法繁琐的问题，导致称体故障点多、维修困难、事故频发等问题，不利于生产的顺利进行。对此，重新设计新的称重系统并进行了改进，如：采用称重模块代替称重仪表，称体安装位置由辊道下方改为直供辊道上方，优化系统检测点和检测算法。通过以上措施，定重率（控制精度±0.2%）由原先的75%提高到85%-90%。

关键词：定重供坯，在线称重，方坯，连铸

Technological Optimization of the Online Weighing for Billet

Wang Xin, Zheng Hu

(HBIS Tangsteel Company No.2 Steel Making and Rolling Plant)

Abstract: The original online-weighing system adopted in the sensor - weighing instrument structure, the body of online-weighing system was used under the roller. There were some disadvantages such as high temperature, high humidity environment, lots of sensor, complicated algorithm, maintenance difficulties, accidents and other problems. These were not conducive to smooth production. So the new online-weighing system were designed, such as using the weighing module, installation position being above the direct supply roller above, optimization system detection and detection method. Through the above measures, weight ratio (control accuracy ±0.2%) increase from the original 75% to 85%-90%.

Key words: fixed weight for billet, online weighing, billet, continuous casting

组合式电搅改善大方坯齿轮钢内部质量的试验研究

张维宽，杨文中

（攀枝花钢钒有限公司提钒炼钢厂，四川攀枝花　617000）

摘　要：针对2#方坯在2014年进行断面改造、新增凝固末端电磁搅拌后，对360mm×450mm断面品种钢的组合式电磁搅拌效果未实质性的进行优化问题，相应的开展了不同工艺参数下的现场试验，摸索最佳组合电搅参数，充分发挥电磁搅拌的冶金功能。通过试验表明：当结晶器电搅投运600A、2.4Hz，末端电搅投运200A、7Hz时，铸坯等轴晶率最高，偏析最小。

关键词：大方坯，齿轮钢，组合式电搅，偏析

Experimental Study on Improvement of Internal Quality
of 2# Billet Gear Steel by Combined Electric Stirring

Zhang Weikuan, Yang Wenzhong

(Panzhihua Steel Vanadium Co., Ltd., Panzhihua 617000, China)

Abstract: For the 2# billet in 2014 to transform the section, the new solidification of the end of electromagnetic stirring, the 360mm×450mm cross-section steel varieties of electromagnetic stirring effect is not substantive to optimize the problem, the corresponding process parameters carried out under different field test, to explore the best combination of electric stirring parameters, give full play to the metallurgical function of electromagnetic stirring. The results show that: when the mold electric stirring operation 600A, 2.4Hz, the end of electrical stirring operation 200A, 7Hz, the billet equiaxed the highest rate, the smallest segregation.

Key words: billet, gear steel, combined electric stirring, segregation

阀控缸位置伺服控制液压系统的分析与仿真

王广吉，宗　旭

（鞍钢股份有限公司冷轧厂，辽宁鞍山　114000）

摘　要： 阀控缸液压位置控制系统是液压伺服系统中重要的一种类型，本文通过以米巴赫焊机夹板台移行系统为对象，完成液压控制系统的建模分析，利用 Matlab 软件进行参数仿真。通过这一过程，重现出生产现场的设备模型，以利于现场故障的排除与参数的优化调整以及生产工艺参数的优化。

关键词： 液压系统，阀控缸，位置控制，仿真

The Analysis and Simulation of theValve-Cylinder
Position Control System

Wang Guangji, Zong Xu

(Cold Strip Works of Ansteel, Anshan, 114000, China)

Abstract: Valve-Cylinder position control System is a typical hydraulic servo-control system, the thesis researches mainly about the position control system based on the clamping device of the Miebach laser welding machine. And then , we use the tools of Matlab software to simulate the real system. The model of the system can truly describe the real working procedures of the clamping device. It makes great sense for the operators to adjust the parameters of the production control.

Key words: hydraulic servo-control system, valve-cylinder system, position control, simulation

宝钢原料跨吊车梁断裂原因分析及设计优化

黄建东[1]，龚年生[1]，庄继勇[2]

（1. 宝山钢铁股份有限公司炼钢厂，上海　200941；2. 宝钢技术，上海　200941）

摘　要：2015 年初，宝钢一炼钢原料跨吊车梁发生撕裂下沉的险兆事故。本文分析了事故发生的原因和抢修方案，并进一步介绍了后续改进方案和吊车梁更换的实施过程，对相关问题做了一些讨论。

关键词：吊车梁，变截面，疲劳，优化

The Analysis and Optimizing about Break Cause of Crane Beam in Baosteel Raw Material Span

Huang Jiandong[1], Gong Niansheng[1], Zhuang Jiyong[2]

(1. Baosteel Steel Making Plant, Shanghai 200941, China; 2. Baosteel Engineering, Shanghai 200941, China)

Abstract: In the 2015'beginning, an accident of crane beam break in Baosteel raw material span happened. This article analyes the cause and plan of urgent repair, the next optimization and replace of the crane beam is described, relevant discuss is also involved.

Key words: crane beam, variable cross-section, fatigue, optimizing

伺服阀故障检测及其在热轧机上的应用维护

吴建荣，王　强，孟　钢，杨举宪，葛明毅

（鞍钢股份有限公司热轧带钢厂，辽宁鞍山　114000）

摘　要：本文主要介绍了应用现场建立伺服阀检测系统及检测方法，以现场实例说明伺服阀性能测试及修复过程，通过对伺服阀性能进行定量的分析，判断其是否满足运行要求，并基于此分析总结了热轧机上伺服阀的故障及原因，以现场多年实践经验提出了宝贵的预防与维护措施，通过对伺服阀的故障检测能够对现场伺服阀进行更好的应用维护，降低备件维护成本，保证生产稳定运行，获得了可观的经济效益，具有重要的推广价值。

关键词：伺服阀，检测，修复，热轧机，维护

Fault Detection and Maintenance of Servo Valve in the Hot Rolling Mill

Wu Jianrong, Wang Qiang, Meng Gang, Yang Juxian, Ge Mingyi

(Ansteel Co., Ltd., The Hot Strip Mill, Anshan 114000, China)

Abstract: This paper described the system structure、working principle and detection method of the testing platform of the servo vavle, it also introduced the functional test and repair process of servo vavle with a practical example, by quantitative analysis on the performance of the servo valve, to determine whether it can meet working requirements; the fault types and causes of servo vavle were analyzed and summarized in the hot rolling mill, it also put forwarded the series valuable prevention and maintenance measures. With many years of practical experience, the practice proves that maintenance costs of spare parts was obviously reduced and the production operated steadily, and the better economic benefits have been achieved, so that it has the important spreading value.

Key words: servo valve, detection, repair, hot rolling mill, maintenance

首钢迁钢 4 号板坯连铸机电磁搅拌电流保护故障分析与处理对策

成建峰

（首钢迁钢公司炼钢作业部，河北迁安 064400）

摘　要：本文主要对首钢股份公司迁安钢铁公司 4 号板坯连铸机电磁搅拌系统的典型故障报警—电流保护的故障原因进行了分析，并提出了排查处理方法及解决对策，降低了电流保护报警造成的系统停机和铸机断浇故障，保障了首钢迁钢高品质硅钢冶炼电磁搅拌系统的高效、稳定运行。

关键词：电磁搅拌，电流保护报警，IGBT，处理对策

Analysis and Solutions of No.4 Continuous Caster EMS Current Protection of Shougang Qiangang Steel

Cheng Jianfeng

(Shougang Qian'an Steel Corp, Steelmaking Department, Qian'an 064400, China)

Abstract: This article mainly analyzed the reasons of the typical fault alarm-current protection of an electromagnetic stirring system (EMS) of 4# slab continuous caster of Shougang Qian'an Steel Corporation. And it also put forward the troubleshooting methods and countermeasures. It reduced the system shut down and casting fault caused by current protection alarm, and guarantees the efficient and stable operation of the EMS of high quality silicon steel smelting of Shougang Qian'an Steel Corporation.

Key words: EMS, current protection fault-alarm, IGBT, countermeasures

ABB 变频器在冶金铸造起重机主起升机构的应用

李理，王赓，董园

（鞍钢股份炼钢总厂，辽宁鞍山 114021）

摘　要：330/80t 铸造起重机是炼钢总厂三分厂 5#线转炉高产达标的关键设备之一，担负着为 5#线 D、E 两座转炉

吊运重铁水罐和重钢水罐的工作，其作业率高，工作环境恶劣。一旦 330t 起重机主起升机构发生故障，将严重影响转炉及连铸机生产。

330t 起重机主起升机构电气系统选用 ABB 公司 ACS800 变频器进行电气控制，电气元件大幅减少，其故障诊断系统不仅可以最大限度的保护电机，而且通过故障代码的方式查找故障原因，易于快速判断故障，缩短了判断设备故障的时间，实现快速排查及处理故障，减少设备故障率和缩短了设备故障处理时间，提高了起重机工作效率。为应急控制和维修管理提供准确、可靠的依据，避免生产系统的临时性非计划停机、停产，从而保证生产连续进行。

关键词：冶金铸造起重机，主起升机构，变频器，应用

The Application of AAB Frequency Converter to Metallurgical Casting Crane Main Lifting Mechanism

Li Li, Wang Geng, Dong Yuan

(Anshan iron & steel Steelmaking plant, Anshan 114021, China)

Abstract: Anshan iron & steel Steelmaking plant 3 operating area 5 # line, 330/80 t ladle crane is One of the key equipment for high production standards,converter steelmaking factory high standards, Responsible for the 5# line D, E two converter lifting heavy iron and steel tank tank work, its operation efficiency is high, poor working conditions. Once the 330t crane main lifting mechanism fails, it will seriously affect the production of converter and continuous casting machine.

330t main hoisting mechanism of a crane electrical system with ACS800 converter ABB, electrical control, electrical components significantly reduced, the fault diagnosis system can not only protect the motor to the maximum, and the reasons to find fault fault code, easy to quickly determine the fault, shorten the judgment equipment failure time, achieve rapid investigation and handling of faults. Reduce the equipment failure rate and shorten the fault processing time, improve the work efficiency. For crane emergency control and maintenance management to provide accurate and reliable basis, to avoid the temporary production system of unplanned downtime, production, to ensure continuous production.

Key words: metallurgical casting crane, main lifting mechanism, converter, application

ABB 新型变频器在冶金铸造起重机上的应用与实践

李 理，王 赓，董 园

（鞍钢股份炼钢总厂吊车三作业区，辽宁鞍山 114021）

摘 要：变频调速技术具有调速效率高、调速范围宽、机械特性较硬等优点。近年来，随着电力电子技术的快速发展，变频调速技术日趋成熟，变频器广泛应用在工业及民用的各个领域。

本文主要介绍 ABB 公司的 ACS880 系列变频器，在鞍钢股份炼钢总厂 330/80t 铸造起重机上的实际应用及变频器的日常维护、常见故障处理，并在实际使用过程中的设计优化。

ACS880 系列变频器在 330/80t 铸造起重机上的使用，使起重机在设备操作稳定性、设备使用寿命、设备故障快速排查、节能降耗环保等方面各项重要技术指标得到了本质的提升，确保了连铸机实现高效生产，为鞍钢股份炼钢总厂大连铸生产工序顺行提供了有力地保证。

关键词：冶金铸造起重机，ABB 变频器，应用与实践

Application and Practice of ABB Frequency New Converter on Metallurgical Casting Crane

Li Li, Wang Geng, Dong Yuan

(Anshan Iron & Steel Steelmaking Plant, Anshan 114021, China)

Abstract: The frequency converter has high efficiency speed, wide speed range, stronger mechanical characteristic and so on. In recent years, with the rapid development of power electronic technology, frequency conversion technology matures, the inverter is widely used in industrial and civil fields.

This paper mainly introduces ACS880 series ABB inverter, in Angang steelmaking plant maintenance 330/80t casting and application of converter on the crane, trouble shooting, design optimization in the actual use of the process.

ACS880 series inverter in 330/80t casting using crane, the crane in the equipment operation stability, the service life of the equipment, equipment quick troubleshooting, energy saving and environmental protection and other aspects of the important technical indicators are the essential promotion, to ensure the continuous casting machine to realize the high efficiency of production, the production process for Angang Steel casting direct steelmaking plant of Dalian provides strong guarantee.

Key words: metallurgical casting crane, ABB series inverter, application and practice

专用吊运废钢槽旋转吊具的设计与研究

董　园，李　理，王　赓

（鞍钢股份炼钢总厂吊车三作业区，辽宁鞍山　114021）

摘　要： 鞍钢炼钢总厂三分厂 5#线由于厂房各跨横向空间狭小，厂房内屋面空间低等原因限制，通过汽车运输来的废钢槽进入现场，造成废钢槽槽口方向与转炉口方向垂直成90°，无法实现加废钢作业。

为了实现在现场将废钢槽旋转90°，确保能够正常加入废钢，我们根据现场实际情况，研究设计了一套具有四个吊点强制平衡的废钢槽专用旋转吊具。在吊卸废钢槽作业时，通过操作吊车主小车起升机构完成废钢槽升降，通过旋转启动按钮控制旋转吊具正向或反向进行旋转，可实现废钢槽水平方向360°范围内，任意角度摆放，从而实现给转炉加入废钢作业。

关键词： 旋转吊具，吊运废钢槽，设计研究

Design and Research of Special Rotary Lifting Gear for Hoisting Scrap Steel Tank

Dong Yuan, Li Li, Wang Geng

(Anshan Iron & Steel General Assembly Plant 3 Operating Area Liaoning Province, Anshan 114021, China)

Abstract: Anshan iron and steel factory three branch line # 5 because factory all across the horizontal space is narrow, in low roof space reasons limit of the plant, through the motor transport to scrap steel tank into the field, causing scrap steel slit direction with the converter mouth direction vertical slot into 90°, can't add scrap steel work.

In order to realize the scrap of the tank rotate 90° at the scene, to ensure the normal join scrap, we according to the actual situation, study design a set of has four lifting point force balance of scrap steel tank dedicated rotating hook. When crane unloading the scrap steel tank, by manipulating the owner carriage complete scrap steel tank lifting hoist, by rotating start button control rotating hook forward or reverse rotation, which can realize scrap steel tank level within the scope of the direction of 360°, arbitrary Angle, so as to realize to join scrap of converter operation.

Key words: rotary lifting gear, hoisting waste steel troughs, design and research

高炉布料溜槽故障分析及长寿化研究应用

张荣军，于成忠

（鞍钢股份炼铁总厂，辽宁鞍山 114021）

摘　要： 布料溜槽是高炉无料钟炉顶装料设备的关键部件，其使用寿命直接影响高炉正常生产。通过对鞍钢 3200m³ 以上高炉炉顶布料溜槽故障原因进行了深入调查分析研究，在鞍钢现有的高炉布料溜槽进行一系列的优化改进，经过鞍钢高炉实践表明，改进的布料溜槽完全满足 3200m³ 高温高强度的冶炼需求，达到预期使用寿命目标，取得良好的经济效益，是一种耐高温长寿型高炉布料溜槽。

关键词： 高炉，布料溜槽，改进

Failure Analysis and Lifetime Research of Blast 3200m³ Blast Furnace Distribution Chute

Zhang Rongjun, Yu Chengzhong

(General Ironmaking Plant of Angang Steel Co., Ltd., Anshan 114021, China)

Abstract: Distribution chute is the key component of bell less top, the lifetime of distribution chute will affect the routine operation of blast furnace directly. After the deep-going research to the failure of blast 3200m³ furnace distribution chute which happened on Angang site, a series of improvement has been carried out. It has been proved by operation that the improved distribution chute could not only meet the high temperature and high intensity working condition of 3200m³ blast furnace, but also the expected lifetime, and good economic benefit has been obtained from the improvement as well. The improved distribution chute is a kind of high temperature resistant and long lifetime distribution chute.

Key words: blast furnace, distribution chute, improvement

活套在带材连续机组的应用

吴安民，苏　华，魏春颖

（中冶京诚工程技术有限公司，北京 100176）

摘　要：介绍了带材连续机组中活套应用的三种形式，并详细介绍了坑式活套、卧式活套和塔式活套的结构特点及在相关连续机组中的应用情况。

关键词：活套，卧式活套，塔式活套

Application of Looper in Continuous Strip Processing

Wu Anmin, Su Hua, Wei Chunying

(Capital Engineering & Research Incorporation Limited, Beijing 100176, China)

Abstract: The article describes the type of looper used in continuous strip processing and detail the construction and application of Pit type looper, horizontal looper and tower type looper.

Key words: looper, horizontal looper, tower type looper

铁道车辆车体钢结构防腐处理的研究推广应用

宋正学

（鞍钢集团铁路运输公司机车厂，辽宁鞍山　114021）

摘　要：介绍了鞍钢集团铁运公司厂内铁道车辆车体钢结构存在的问题及其铁道部铁道车辆车体钢结构用钢经历三个阶段和发展方向。提出了解决鞍钢集团铁运公司厂内铁道车辆车体钢结构防腐问题的四种方案，并根据鞍钢实际及性价比确定最终方案：铁道车辆进厂检修时板材表面喷涂一层 1.5mm 厚防腐涂料，在涂料上面加装一层较薄的（3mm 厚）TCS 不锈钢板，在三辆车上作实验取得良好效果，并陆续推广应用。

关键词：钢结构，防腐，应用

点检定修制在唐钢的推广应用

韩一杰

（河钢唐钢自动化公司，河北唐山　063000）

摘　要：本文介绍了设备点检定修制的基础理论和特点，结合唐钢实际情况就点检定修制的实施状况和技术方面的推广应用进行详细分析，提出深化推行点检定修制在设备管理中应采取的方法和措施，以提高唐钢设备综合管理水平，提升企业竞争力。

关键词：点检定修制，设备管理，推广应用

Spot Inspection and Customized Repairing System in Application of Tangshan Iron and Steel Group

Han Yijie

(Tangshan Iron and Steel Group Automation Co., Ltd., Tangshan 063000, China)

Abstract: This paper introduces the basic theory and characteristics of the equipment spot inspection and customized repair system. The implementation status and technical aspects of spot inspection and customized repair system are analyzed in detail according to the actual situation of Tangshan Iron and Steel Group. Implementation methods and measures of equipment inspection and maintenance in equipment management should be put forward in order to improve the equipment comprehensive management level of Tangshan Iron and Steel Group and enhance the competitiveness of enterprises.

Key words: spot inspection and customized repairing system, device management, popularization and application

武钢热轧总厂一分厂翻钢机的设计及应用

李 威

（中冶南方工程技术有限公司，湖北武汉 430223）

摘 要： 翻钢机作为钢卷运输线的重要组成设备，其结构特点可适应立式和卧式两种钢卷运输方式。本文依托武钢热轧总厂一分厂运输链改造项目，对该改造所使用的翻钢机的技术参数、结构特点、自动化控制以及联锁进行了阐述，对其他同类设备的改造设计具有一定的指导意义和实用价值。

关键词： 翻钢机，联锁，卧式运输

Design and Application of Steel Turning Machine in No. 1 Branch of Hot Rolling Mill of WISCO

Li Wei

(WISDRI Engineering and Research Incorporation Limited, Wuhan 430223, China)

Abstract: As an important component of the steel coil transportation line, the steel turning machine can adapt to two kinds of steel coil transportation modes, vertical and horizontal. This paper is based on the hot rolling factory total transportation chain transformation project, the use of the transformation of the technical parameters, tilter structure characteristics, automatic control and interlock were described, which has a certain guiding significance and practical value for other similar equipment design.

Key words: steel turning machine, interlocking, horizontal transportation

鞍钢热轧厂 1780mm 精轧机架辊常见故障及处理

李　林

（鞍钢股份有限公司热轧带钢厂，辽宁鞍山　114000）

摘　要：精轧机组是对钢板进行成型加工的最重要的设备，鞍钢热轧精轧机组采用日本三菱设计，三菱与一重合作制造。本文通过对精轧机组中机架辊常见故障的总结，分析故障的原因，提出相应的解决措施。

关键词：机架辊，故障，润滑装置，处理方法

Failure Analysis of Frame Roll Its Solution in Anshan Hot Strip Steel 1780mm Unit

Li Lin

(Hot Strip Steel of Ansteel Co., Ltd., Anshan 114000, China)

Abstract: The Finishing Mill Group is the most important equipment for Hot Strip Steel of Ansteel, The Ansteel Hot Strip Steel Mill Group is designed by MITSUBISHI of Japan, made by MITSIBISHI and China First Heavy Machinery Co. Ltd. Based on the Summary of the common faults of the Finishing Mill Stand, This paper analyzes the causes of failure and puts forward the corresoponding solutions.

Key words: frame roll, fault, the lubrication device, processing methods

气动三联件减压阀压力调节保持器的应用与研究

赵永强

（河北钢铁集团承钢板带事业部，河北承德　067000）

摘　要：气动三联件减压阀部分由塑料制成，工作环境温度较高时，随着气动三联件减压阀塑料部分老化以及减压阀内的高强弹簧压力的双重作用下，气动三联件减压阀就会在应力集中的排气口处发生断裂失效，从而导致气动三联件失效。我们设计制造了气动三联件减压阀压力调节保持器。安装在已失效气动三联件减压阀损坏部位，从而用极其低廉的成本继续实现其功能。

关键词：气动三联件，减压阀，失效，压力调节保持器，低成本，实现功能，节约大量成本

淮钢 VD 炉机械真空泵应用

裴王敏

（江苏沙钢集团淮钢特钢股份有限公司，江苏淮安 223002）

摘　要：主要介绍了机械泵在淮钢 100tVD 炉的应用实践，重点阐述了机械真空泵的运行情况和冶金效果。生产时间表明，使用机械真空泵的 VD，其冶金效果达到或由于使用多级蒸汽喷射泵真空系统的 VD 炉，脱[H]率达≤1.5ppm，能源介质消耗成本较多级蒸汽喷射泵真空系统低 3.5/t 钢。机械真空泵以电能为介质，生产组织灵活。

关键词：淮钢，机械真空泵，VD，应用

转炉横移烟道强制循环冷却改造实践

裴王敏

（江苏沙钢集团淮钢特钢股份有限公司，江苏淮安 223002）

摘　要：介绍了转炉横移烟道强制循环改造所选择的主要技术参数、涉及的设备改造。

关键词：转炉，烟道，强制循环，改造

梅钢 12 万 m³ 转炉煤气柜橡胶膜运行状态分析及改进

吴先吉，王自龙，孙光华

（上海梅山钢铁股份有限公司能源环保部，江苏南京 210039）

摘　要：介绍了梅钢 12 万 m³ 转炉煤气柜的橡胶膜运行情况，描述了煤气柜的橡胶膜情况，重点针对煤气柜内部密封橡胶膜破损原因进行分析，针对存在的问题采取相应的对策措施，使气柜恢复正常运行，为以后的生产运行提供参考。

关键词：煤气柜，橡胶膜，破损

Analysis of Operation State of Sealing Rubber Film of 120000m³ Converter Gas Holder in Shanghai Meishan Steel Cooperation

Wu Xianji, Wang Zilong, Sun Guanghua

(Energetic-Environmental Protection Department in Shanghai Meishan
Iron & Steel Co., Ltd., Nanjing 210039, China)

Abstract: The article introduces the operation state of sealing rubber of 120000m³ Converter Gas Holder in Shanghai Meishan Steel Cooperation, and describes the condition of converter gas holder. It mainly focuses on the analysis of damage cause of interior sealing rubber film, and adopts relevant solutions to the existing issues, which guarantee its normal operation state and offer references to its subsequent running.

Key words: gas holder, rubber film, damage

消除 1700 冷轧连退平整机组带钢表面亮印缺陷

王占军

（首钢京唐钢铁联合有限责任公司冷轧部，河北唐山　　063200）

摘　要： 首钢京唐公司 1700 冷轧连退平整机组生产过程中经常在带钢上表面出现亮印缺陷的问题。经过现场试验与研究发现平整液系统中含有大量细菌。细菌经常堵塞平整液喷梁、喷嘴。为了解决带钢表面亮印缺陷的问题，首先把消毒剂加入到平整液中，然后把平整液温度提高至 40~45℃，杀死平整液中的细菌，最后用压缩空气清理管道及喷梁、喷嘴中的细菌，有效的缓解了亮印缺陷的产生。为消除带钢表面亮印缺陷，在平整机上支持辊增加一道平整液喷梁，增加了对轧辊的润滑，避免了轧辊表面粗糙度局部不匀。同时制订了上支持辊喷梁调节标准，最终成功解决了带钢上表面亮印缺陷问题。

关键词： 亮印，喷梁，喷嘴，细菌

The Elimination of the Bright Printing Defects on the Surface of Strip in Skin Pass Mill in 1700 Cold Rolling Continuous Annealing

Wang Zhanjun

(Shougang Jingtang Steel Co., Ltd., Tangshan 063200, China)

Abstract: The problem of bright printing defects on the surface of strip is often occurred in the process of production in skin Pass Mill in 1700 cold rolling continuous annealing line in Shou Gang Jing Tang United Iron&Steel Co. Ltd. Through field tests and studies, it is found that a lot of bacteria are contained in the wet skin passing system. The bacteria often block up spray beam and nozzle. In order to solve the problem of the bright printing defects on the surface of strip, Fist we add the disinfectant to oil and raise the oil temperature to 40~45 degree Celsius, next kill the bacteria, and then clean the pipe and spray beam and nozzle using compressed air, Finally effectively alleviate the bright printing defects. In order to eliminate the bright printing defects on the surface of strip, top BUR is added with a spray beam which increases the lubrication of roll and avoids unevenness of roll surface roughness. At the same time, we develop regulative standard of spray beam of top BUR, finally we successfully solve the problem of the bright printing defects on the surface of strip.

Key words: bright printing, spray beam, nozzle, bacteria

关于旋流井自动抓渣的探讨

墙新奇

（宝钢集团新疆八钢有限公司，新疆乌鲁木齐　830022）

摘　要：钢铁企业普遍采用旋流井工艺处理富含氧化铁皮的浊循环污水，通过旋流沉淀实现固液分离，用抓斗清除氧化渣，污水再进行处理后实现循环利用。旋流井清渣作业，通常由人工操作单梁桥式起重机，通过电动葫芦控制抓斗抓渣，简单重复性工作，人员劳动效率不高。通过对旋流井抓渣过程和相关设备的研究，探索基于对现有设备的改造来完成自动抓渣的活动途径。

关键词：钢铁企业，旋流井，单梁桥式起重机，抓斗，自动抓渣

Discussion of the Automatic Catch Slag of Whirlpool

Qiang Xinqi

(Baosteel Group Xinjiang Baiyi Iron & Steel Co., Ltd., Urumqi 830022, China)

Abstract: The widespread adoption of the iron and steel enterprise whirlpool process turbid circulating water, rich in iron oxide by vortex sedimentation solid-liquid separation, grab to remove oxide slag, sewage water for processing after recycling. Whirlpool slag removal operations, usually by manual single-beam bridge grab crane, control grab catch slag, simple and repetitive work, personnel labor efficiency is not high. Through the study of the process and related equipment of the catch slag process, research and explore a new way to be used on the basis of the existing equipment to complete the activities of automatically catch slag.

Key words: the iron and steel enterprise, whirlpool, single beam bridge grab crane, grab, automatic catch slag

射水箱溢流水回收装置

冯友祥

（铜陵泰富特种材料有限公司动力分厂，安徽铜陵　244000）

摘　要：本文研究了发电厂汽轮机射水箱溢流水回收装置，主要解决以往技术中溢水箱回水系统能耗高，冷却效率低，操作繁琐，智能化程度度差的技术问题。本文通过采用一种汽轮机射水箱溢流水回收装置，包括汽轮机冷却器，射水箱，溢流水箱。冷却水管进口和冷却水管出口和汽轮机冷却器相连，射水箱通过冷凝管和汽轮机冷却器相连，溢流水箱通过溢流管和射水箱相连，溢流水箱回水管连接到冷却水管出口的技术方案，较好地解决了溢流水浪费问

题，该设备用于汽轮机发电，起到了节能降耗的效果。

关键词：射水箱，溢流水，回收装置，节能降耗

Abstract: This paper studies the overflow water recovery unit of steam turbine water tank in power plant, which mainly solves the technical problems of high energy consumption, low cooling efficiency, cumbersome operation and poor intelligence of the overflow system in the past. This paper uses a steam turbine water tank overflow water recovery device, including the steam turbine cooler, water tank, overflow water tank. The cooling water pipe inlet and the cooling water pipe outlet are connected with the steam turbine cooler. The water tank is connected with the steam turbine cooler through the condensing pipe and the overflow water tank is connected through the overflow pipe and the water tank. The overflow water tank is connected with the technical scheme of the cooling water pipe outlet, a better solution to the problem of overflow water, the equipment used for steam turbine power generation, has played a saving effect.

Key words: water tank, overflow water, recovery device, energy saving

振动除渣系统的动力学分析

李子俊，刘　森

（首钢京唐钢铁联合有限责任公司冷轧作业部，河北唐山　063200）

摘　要： 针对开发振动除渣系统的实际需求，本文设计了一台双频组合振动除渣机构。首先，通过简化得到该偏心激励振动系统的动力学模型，推导了其多自由度动力学方程，经过数值仿真分析了系统的动力学特性。然后，研究了弹簧刚度组合、激振频率及激励位置对除渣机刀具运动轨迹的影响。

关键词：偏心激励，多刚体，多自由度，运动轨迹

Dynamics Analysis of Low-noise Vibration Mining System

Li Zijun, Liu Sen

(Shougang Jingtang United Iron & Steel Co.,Ltd., Cold Rolling
Department, Tangshan 063200, China)

Abstract: In this paper, to meet the actual need, we design a dual-frequency low-noise vibrating mining machine. First, we observe the dynamical model of eccentric excitation through simplifying vibrating system; obtain dynamical equations of multi-degree-of-freedom; analyze dynamical character of system after numerical simulation. And then we study the influencing factor to tool's trajectory such as the combination of spring stiffness, vibrating frequency and the location of excitation.

Key words: eccentric excitation, multi-rigid-body, multi-degree-of-freedom, trajectory

2230 酸轧入口防皱辊使用与控制策略

孙　抗

（首钢京唐公司冷轧部，河北唐山　063200）

摘　要：本文介绍了热轧卷在开卷时"横折"的产生机理及预防措施，继而引入了解决此问题的防皱辊设备。以京唐公司 2230 酸轧机组入口防皱辊为基础，就此设备的主要功能，设备结构，液压系统以及电气系统的控制策略进行了详细介绍，重点描述了防皱辊压下量的计算与控制方式，为防皱辊的使用提供了新的思路。

关键词："横折"缺陷，防皱辊，位置控制，压力限幅控制

Function and Control Strategies of Anti-Coil Break Roll on 2230 PLTCM#2

Sun Kang

(Shougang Iron and Steel Corporation Cold Roll Department, Tangshan 063200, China)

Abstract: This paper introduces the mechanism and preventive measures of "cross break" defect when hot-roll coil unwinding, then introduce Anti-Coil Break Roll that can solve this defect. Based on Anti-Coil Break Roll on 2230 PLTCM#2, introduces mainly function, equipment structure, hydraulic system and electrical control strategies in detail, focus on discussing bending level calculation and control strategies, bring some new ideas for usage.

Key words: coil break defect, anti-coil break roll, position control, pressure limitation

320t 筒型混铁车切轴断裂原因分析及防范措施

项克舜

（武汉钢铁有限公司运输部，湖北武汉　430083）

摘　要：对 320t 筒型混铁车切轴断裂事故进行了分析，从断轴的断口形状分析判断为冷切轴断裂。分析了车轴发生冷切轴断裂的原因，提出了防止车轴切轴断裂的措施，取得了明显效果。

关键词：筒型混铁车，切轴断裂，超声波探伤，分析，措施

Analysis and Prevention of Cutting-axle Fracture of 320t Cylindrical Torpedo Ladle Car

Xiang Keshun

(Transportation Department of Wuhan Iron and Steel Co., Ltd., Wuhan 430083, China)

Abstract: This article analyzes the accident of cutting-axle fracture of 320t cylindrical torpedo ladle car. According to the shape of the fracture surface, the fracture was defined as a cold cutting-axle fracture. This paper further analyzes the causes of the cold cutting-axle, and puts forward the preventive measures, and has obtained the obvious effect.

Key words: cylindrical torpedo ladle car, cutting-axle fracture, ultrasonic flaw detect, analysis, preventive measures

冷轧厂中间库起重机供电扩容改造

曾进东，周景革

（北京首钢冷轧薄板有限公司设备部，北京　101304）

摘　要： 冷轧厂新建罩式退火生产线，将原有中间库⑬/©跨延长 72.5m，并增加 1 台 42t 桥式起重机。原有中间库⑬/©跨与新建厂房的起重机分属 1#和新建 2#低压供电中心供电，因而，延长后的中间库⑬/©跨起重机的供电需考虑整合与统一。经研究，决定采取只由 1#低压供电中心对全部中间库⑬/©跨的起重机进行供电，并对系统进行相应改造。确定系统的计算电流、尖峰电流，校核供电系统的电压损失，并提出供电系统的扩容措施；通过对滑触线的载流量及稳定性进行校验，确定不需更换滑触线，但需局部敷设并行电缆。改造后，系统已正常稳定地运行 3 年了，说明改造是成功的。

关键词： 起重机，滑触线，供电，阻抗，电压损失，短路电流

Power Supply Expansion of Intermediate Storehouse Crane in Cold Rolling Mill

Zeng Jindong, Zhou Jingge

(Beijing Shougang Cold Rolling Co.,Ltd., Beijing 101304, China)

Abstract: The new bell type annealing production line in cold rolling mill will extend the original span ⑬/© to 72.5 meters, and increase a 42t bridge cranes. The original warehouse ⑬/© spans the crane with the new plant and is powered by the 1# and the new 2# low voltage power supply center. Therefore, the power supply of the extended intermediate ⑬/© crane shall be considered integration and unification. It has been decided that the power supply of all intermediate ⑬/© 1# cranes will be powered by the low-voltage power supply center only, and the system will be transformed accordingly. To determine the current and peak current calculation system, voltage loss of power supply system checking, and puts forward the measures of expansion of power supply system; through the calibration of the load flow and stability of sliding contact line, need not replace sliding contact line, but local parallel cable laying. After the transformation, the system has been running normally and stably for 3 years, which shows that the transformation has been successful.

Key words: crane, sliding contact wire, power supply, impedance, voltage loss, short circuit current

大型重载起重机服役期内的系统综合安全评估技术

李贵文，罗云东，陈　琳，姜文吉，陈融融，张　炯

（中国宝武集团上海金艺检测技术有限公司特种检验部机电检验室，上海　201900）

摘　要： 国内目前很少有针对大型起重机服役期内的系统综合安全评估技术分析的案例，文章介绍了起重机系统综合安全评估技术的发展历程，就目前在役大型起重机实施系统综合安全技术路线进行了详细的叙述。文章同时结合

现场测试的经典案例，对在役起重机综合安全评估技术中先进测试技术进行了描述，该先进测试技术的现场应用事实证明其相对于以往传统的测试办法的数据准确率为100%，现场测试时间缩短50%以上，有效提升了测试技术水平和应用效率，保障了生产运行的安全可靠，有很强的推广应用意义，起到了为我们的安全生产工作添砖加瓦和保驾护航的作用。

关键词：起重机，安全，评估

连铸闭路事故水阀非事故状态下开启原因的分析

金　辉

（本溪钢铁集团公司炼钢厂，辽宁本溪　117021）

摘　要：通过对连铸机械闭路水系统的电脑记录参数曲线的分析和计算，找出事故水阀在非事故状态下开启的理论原因，并以此为依据对系统事故水阀开启设定压力进行调整，从而彻底解决这一问题。

关键词：事故水，事故水阀，开启设定压力，膨胀罐，水位曲线，压力曲线，内漏

Energy-saving Practices Compressor Running Buck

Jin hui

(Benxi Iron and Steel Group Steel Plant, Benxi 117021, China)

Abstract: The use of on-site investigations approved by the wind pressure point, reasonably determine the pipe network pressure control value, transforming the way the individual points with the wind, low pressure air compressor to run the process to develop the system, reduce equipment power consumption, achieved tremendous economic benefits.

Key words: air compressor with air pressure surveys approved for wind buck save energy

浅谈在线棒材探伤设备中心线校准技术

王会庆，张永兴，刘丽果，赵争喜

（河钢集团石钢公司，河北石家庄　050031）

摘　要：随着钢铁市场高端用户的需求，在线探伤设备越来越得到广泛的应用，在线自动无损探伤更是特钢发展不可缺少的质量检测和保证手段，在钢铁行业发挥着无可替代的重要作用。然而辊道或机械部件出现磨损、变形进行更换后，如何保证探伤设备±0.5mm 甚至±0.2mm 的对中精度，成为一大难题。石钢探伤技术人员根据探伤设备机械对中高精度的特点，与相关厂家和其他特钢探伤技术人员交流，经过不断总结实践，辅以标准样棒、模板等工具对中心线进行校准测量，摸索出一套探伤设备中心线校准方法，为探伤设备维护积累了宝贵经验。

关键词：在线探伤，对中精度，标准样棒，校准

Introduction the Calibration Technology of the On-line Bar Detection Device

Wang Huiqing, Zhang Yongxing, Liu Liguo, Zhao Zhengxi

(Hesteel Group, Shisteel, Shijiazhuang 050031, China)

Abstract: Along with the need of customer from high level in steel market, the detection device on-line is more and more widely used, online automatic NDT is an indispensable quality examination and assurance means, t plays important unexampled role in steel profession. However, after the roller or machine parts was replaced because which appears wear away, transform etc, how to guarantee the accuracy ±0.5mm even ±0.2mm of the equipment which became a difficult problem. Shisteel technical personnel according to the characteristics of the testing equipment in the machinery of high precision, explored a set of testing method about centerline measurement, through exchanged with the related manufacturers and other technology personnel in special steel detection, supplemented by standard sample, templates, tools to measure centerline with constant practice, it accumulated valuable experience for the detection equipment maintenance.

Key words: online inspection, medium accuracy, standard bar, calibration

浅谈冶金压缩空气系统节能技术措施

步　彬，吴　滨

（河钢邯钢集团邯宝能源中心，河北邯郸　056015）

摘　要： 随着环保要求日益严格，冶金行业已进入减量发展、清洁生产的时代，生产原料面临着供给侧结构性改革。压缩空气作为工业生产的第二大动力源，在企业总电耗量中占据着相当大的比重，其节能技术措施的应用与推广，对冶金行业生产能源成本的降低有着较为显著的作用。

关键词： 冶金，压缩空气系统，节能技术措施

Technical Measures for Energy Conservation in Metallurgical Compressed Air System

Bu Bin, Wu Bin

(Hanbao Company, Hegang Iron and Steel Group, Handan, 056015, China)

Abstract: With the increasing requirements of environmental protect,metallurgical industry has entered the era of reducing development and cleaner production, and raw materials are facing structural reforms in the supply side.The compressed air,as the second largest power source of industrial production, occupies a large proportion in the total power consumption of enterprises. The application and popularization of energy conservation have a significant effect on the reduction of energy cost in metallurgical industry.

Key words: metallurgical industry, compressed air system, energy conservation measures

设备检修平台的各种形式与细节

李 贝

（中冶南方工程技术有限公司炼钢分公司，湖北武汉 430223）

摘 要：在钢厂里，需定期对设备进行检修、维护，为了确保工人的人身安全，就需要设计检修平台。本文就介绍各种各样的设备检修平台，包括固定式检修平台、折叠式检修平台，以及活动栏杆、安全止挡、安全插销等设计细节，供设计、制造、维护时参考。

关键词：检修平台，固定式，折叠式，安全止挡，安全插销

Forms and Details of Equipment Maintenance Platform

Li Bei

(Steelmaking Branch, WISDRI, Wuhan 430223, China)

Abstract: In steel plant, equipment needs to be regularly checked and maintained. In order to ensure the safety of workers, maintenance platform is required. This paper describes various kinds of equipment maintenance platform, including fixed type platform and folding type platform, etc. Design details such as movable railing, safety stop and safety pin, etc. are also discussed. This paper can be consulted for design, fabrication and maintenance of equipment maintenance platform.

Key words: maintenance platform, fixed type, folding type, safety stop, safety pin

高炉炉缸气隙的判定方法

张 科，宋彦坡，周 萍，朱蓉甲

（中南大学能源科学与工程学院，湖南长沙 410083）

摘 要：在实际的生产过程中，高炉炉缸经常会出现气隙，危害高炉安全。本文通过传热学原理分析高炉炉缸冷却壁与碳砖间填料层出现气隙对炉衬温度场的影响，找到炉衬温度的变化规律，提出一个判断气隙程度的方法，为炼铁高炉的实际生产及高炉长寿提供重要的参考依据。

关键词：高炉，炉缸，气隙，方法

Determination Method of Gas Gap in Hearth Wall of Blast Furnace

Zhang Ke, Song Yanpo, Zhou Ping, Zhu Rongjia

(School of Energy Science and Engineering, Central South University, Changsha 410083, China)

Abstract: Gas gap was one of the biggest threats to BF hearth in the practical production process. In this paper, according to the theory of heat conduction, the influence of temperature field by linings gas gap of BF hearth is analyzed, finding the change rules of temperature, and the judgment method of BF hearth gas gap is given. Providing an important reference for the long campaign life of the blast furnace.

Key words: blast furnace, hearth, gas gap, method

免酸洗高强钢氧化铁皮控制策略研究

张文洋[1]，游慧超[1]，王一博[2]，孔　宁[2]

（1. 马鞍山钢铁股份有限公司，安徽马鞍山　243003；

2. 北京科技大学机械工程学院，北京　100083）

摘　要：氧化铁皮粉状剥落对产品后续冷弯和冲压加工有不利影响，是目前制约免酸洗高强钢品质提升的关键因素。本文针对氧化铁皮粉状剥落缺陷，结合钢厂热连轧实际生产工艺，提出控制氧化铁皮粉状剥落的三方面措施：（1）"高温快轧"和"低温卷取"，将卷取温度尽量设定在靠近共析温度的区间；（2）在进行低硅生产的同时，可以兼顾考虑硅元素对氧化铁皮剥落的改善效果，提高氧化铁皮的附着力和韧性；（3）优化工艺水流量，提高冷却能力，防止轧辊氧化膜裂纹扩展引起的粉状氧化铁皮剥落。

关键词：氧化铁皮，免酸洗，硅酸亚铁，轧辊，控制策略

Study on the Control Strategy of the Oxide Scale of High Strength Steel without Pickling

Zhang Wenyang[1], You Huichao[1], Wang Yibo[2], Kong Ning[2]

(1. Maanshan Iron&Steel Company Ltd., Maanshan 243003, China; 2. School of Mechanical Engineering, University of Science and Technology Beijing, Beijing 100083, China)

Abstract: The oxide scale powder, which is the key factor to restrict the quality improvement of high strength steel without pickling, has adverse effect on the subsequent processing such as cold bending and punching press. In this paper, three measures to control the powder flaking of oxide scale are put forward. (1) "High temperature fast rolling" and "low temperature coiling". The coil temperature should be set near the eutectoid temperature of Fe_3O_4; (2) In the process of low silicon production, the improvement effect of silicon elements on the oxide scale powder should be considered to improve the adhesion and toughness of oxide scale. (3) Optimizing the flow of process water to improve the cooling ability so as to prevent the powder flaking of oxide scale which caused by the crack propagation of thick roll oxide film.

Key words: oxide scale, pickling free, ferrous silicate, roll, control strategy

热连轧工作辊温度场影响因素研究

司小明[1]，游慧超[1]，宋　鸣[2]，刘国勇[2]，朱冬梅[2]，张少军[2]

（1. 马鞍山钢铁股份有限公司，安徽马鞍山　243003；

2. 北京科技大学机械工程学院，北京　100083）

摘　要：在热轧带钢生产中，工作辊温度是影响带钢板形和凸度的重要因素。利用有限元软件 ANSYS 建立了工作辊的二维非稳态温度场计算模型，对精轧工作辊的温度场进行了数值模拟，研究了不同因素等对工作辊辊温变化规律的影响。研究发现：下工作辊的冷却效果要好于上工作辊的冷却效果；出口侧水量越大，工作辊表面处于高温状态的时间就越短，表面温度下降的越快，有利于抑制工作辊氧化，减少带钢表面氧化铁皮缺陷；带钢温度越高，工作辊在轧制过程中的最高表面温度就越大。

关键词：热连轧，工作辊冷却，二维非稳态温度场，水量分配比例

Research on the Influence Factors of Work Roll Temperature in Hot Mill

Si Xiaoming[1], You Huichao[1], Song Ming[2], Liu Guoyong[2],
Zhu Dongmei[2], Zhang Shaojun[2]

(1. Maanshan Iron&Steel Company Ltd., Maanshan 243003, China; 2. School of Mechanical Engineering, University of Science and Technology Beijing, Beijing 100083, China)

Abstract: The temperature of work roll is an important factor affecting profile and flatness of strips during hot strip rolling. The 2D model for simulating the temperature field of work roll was established with ANSYS software. The influences of different parameters on the evolution and distribution of work roll temperature were analyzed. It is found that the cooling effect of lower work roll is better than that of top work roll. Increasing water flow rate of strip exit is a reasonable method for dropping surface temperature of work roll soon and enhancing cooling effect. In addition, the maximum surface temperature of work roll will increase with increasing strip temperature.

Key words: hot strip mill, work roll cooling, two dimensional unsteady temperature field, cooling water flow rate distribution proportion

粗氩塔氮塞的分析及处理

王　富

（昆钢动力能源分公司云南安宁　650302）

摘　要：简介粗氩塔氮塞的情况及处理方法，分析了影响粗氩塔精馏工况的主要因素，并谈了氩馏分含氮量发生变化的原因及调整措施。

关键词：粗氩塔，氩馏分，氮塞，原因，措施

工艺氩出氩干燥器含水量超标原因与处理措施

陈　吕

（武昆股份公司能源分公司制氧车间，云南安宁　650300）

摘　要： KDON-10000/20000-I 型空分设备氩纯化系统 1#、2#干燥器加热再生温度达不到工艺要求，1#、2#使用时 AIA2103 水份分析仪表超标；且 V2105 期间逐渐开大；影响了整个制氩系统的稳定生产。在对现象观察一段时间及原因分析后，发现当 1#、2#干燥器切换使用时，水份分析仪表 AIA2103 均达到较高值，且较之前上升，故障越来越严重，随即停运制氩系统，对氩干燥器 GD 的 4A 分子筛进行更换。空分设备氩纯化系统恢复正常运行。

关键词： 制氩系统，干燥器，分子筛，含水量

The Causes of Excessive Moisture Content in Argon Dryer and Treatment Measures

Chen Lv

(Wukun Steel Co., Ltd., Energy Branch Oxygen-making Plant, Anning 650300, China)

Abstract: The heating regeneration temperature of dryer 1#、2# of KDON-10000/20000-I air separation unit argon purification system can not reach the technological requirements. The water analyzers of dryer 1#、2# in use show that the moisture content is over-standard. In the meantime, V2105 is getting up gradually. What mentioned above have affected the production stability of the whole argon-generating system. After long-time careful observation and thorough analysis, we find that when switching between 1# and 2#, the moisture content reaches a higher level and the problem is getting worse and worse. Then shut down the argon-generating system and replace the 4A molecular sieves of the argon dryer GD, the air separation unit argon purification system quickly resume normal operation.

Key words: argon- generating system, dryer, molecular sieves, moisture content

宽厚板轧机压下位移传感器改造

王朱涛

（江苏沙钢集团有限公司，江苏张家港　215625）

摘　要： 本文主要描述了宽厚板轧机压下 Sony 位移传感器频繁报错故障和改造方法。改造主要是在原先 Sony 位移传感器的外型基础上，将其磁头及磁杆部分改造为 MTS 的 SSI 绝对值传感器，来提高系统运行稳定性，降低维护费用。

关键词： SONY 位移传感器，MTS 位移传感器，轧机压下

The Screwdown Position Transducer of Plate Mill Reform

Wang Zhutao

(Jiangsu Shasteel Group Co.,Ltd., Zhangjiagang 215625, China)

Abstract: This article mainly describes fault messages of the screwdown position transducer and propose solutions.Keep the shape structure of the Sony position transducer, replace magnetic head and magnetic rod by MTS absolute transducer. It can improve stability of the system and reduce cost of maintenance through the reform

Key words: SONY position transducer, MTS position transducer, screw system of plate mill

设备检修平台的各种形式与细节

李 贝

（中冶南方工程技术有限公司炼钢分公司，湖北武汉 430223）

摘 要：在钢厂里，需定期对设备进行检修，维护，为了确保工人的人身安全，就需要设计检修平台。本文就介绍各种各样的设备检修平台，包括固定式检修平台、折叠式检修平台，以及活动栏杆、安全止挡、安全插销等设计细节，供设计、制造、维护时参考。

关键词：检修平台，固定式，折叠式，安全止挡，安全插销

Equipment Overhaul Platform in Various Forms and Details

Li Bei

(Steelmaking Branch, WISDRI, Wuhan 430223, China)

Abstract: In steel mill, equipment needs to be regularly checked and maintained. In order to ensure the safety of the workers, it is necessary to design and repair the platform. This paper introduces the equipment maintenance platform of all kinds, including fixed platform, folding and maintenance platform, movable rails, safety stop, the safety latch design details, to provide reference for design, manufacture, maintenance.

Key words: inspection platform, stationary, folding, safety stopper, safety pin

冷却塔风机控制系统节电优化

董庆玮，周恩会

（江苏沙钢集团有限公司，江苏张家港 215625）

摘 要：沙钢某车间内有大量的冷却塔风机，一直以来在轧线正常生产时均处于全部开启状态，总功率可达1000kW以上。在节能降耗的大背景下，现将其改造为由水温控制的自动控制系统。为了增加程序的可移植性和可读性，温度自动控制部分划分成三个模块分别进行编写，包括运行时间累积模块、风机选择模块和温度自动控制模块等。

关键词：冷却塔，自动控制，节能降耗

Energy Saving Optimization of Cooling Tower Fan Control System

Dong Qingwei, Zhou Enhui

(Jiangsu Shasteel Group Co., Ltd., Zhangjiagang 215625, China)

Abstract: A large number of cooling tower fans in a workshop in Shasteel have been in the fully open state when the rolling line is normal, and the total power is more than 1000kW. In the context of energy saving, it is intended to be transformed into an automatic control system controlled by water temperature. In order to increase the program's portability and readability, the temperature automatic control part is divided into three modules were prepared, including the run-time accumulation module, fan selection module and temperature automatic control module.

Key words: cooling tower, automatic control, energy saving

事故缺陷的修复在热轧板带轧辊上的实践应用

田吉祥，席江涛，罗北平

（河钢集团邯钢公司 连铸连轧厂，河北邯郸　056015）

摘　要：本文根据现场出现的事故缺陷轧辊种类、形式及产生原因进行了分析，通过采用各种不同人工修磨操作方法代替磨床磨削修复处理工艺，同时加以无损检测操作的辅助应用，经现场实践，达到了从源头上控制轧辊使用成本，延长轧辊使用寿命的目的，取得了较好的经济效益。

关键词：轧辊，修复，超声波，辊耗

Defect Repair of Accident in Hot Rolled Strip Roll in Practical Application

Tian Jixiang, Xi Jiangtao, Luo Beiping

(Hesteel Group Hansteel Company, CSPMill, Handan 056015, China)

Abstract: According to the scene of accident roll defects types，forms and causes are analyzed，through the adoption of various artificial grinding operation method instead of grinding machine for grinding repair process, the operation of auxiliary application to nondestructive testing at the same time, the field practice, achieved from the source control roll use cost, prolong the service life of roll and achieved good economic benefits.

Key words: roll, repair, ultrasound, roll consumption

CCPP 锅炉模块 1 紧固装置的改造应用

薛晓金

（鞍钢股份鲅鱼圈钢铁分公司能源动力部，辽宁营口　115007）

摘　要：文章主要针对 CCPP 余热锅炉模块 1 投产后由于振动造成锅炉多次爆管原因进行分析，然后对模块 1 实施了增加紧固梁、紧固支架等紧固装置，结合鞍钢 CCPP701(D)型锅炉实际情况，实施了可靠的改造后锅炉检查方案，避免了传统的复杂吹管工艺。

关键词：CCPP 锅炉，紧固装置，吹管

Reform and Application of CCPP Boiler Module One Fastening Device

Xue Xiaojin

(The Energy and Power Department of Ansteel Bayuquan Iron & Steel Subsidiary, Yingkou 115007, China)

Abstract: This paper mainly focused on CCPP HRSG module 1 after the operation due to vibration caused by the boiler tube explosion several causes are analyzed, then fastening devices such as adding fastening beam and fastening bracket are applied to module 1, combined with the actual situation of CCPP701 (D) HRSG, the implementation of a reliable reconstruction of boiler inspection program, to avoid complex the traditional blowing process.

Key words: CCPP boiler, fastening device, blowpipe

VEGAPULS64 雷达液位计在湘钢高炉铁水罐液位测量上的应用

阳 丹

（湖南华菱湘潭钢铁有限公司炼铁厂）

摘 要：通过对 VEGAPULS64 型雷达液位计工作原理的介绍，分析了其性能特点，介绍了其在湘钢高炉铁水罐液位测量的应用。

关键词：铁水罐液面，雷达液位计，连续调频

鞍钢 1#高炉 1-4 带炉皮更换炉体加固方案

李焕龙，张荣军，李金辉

（鞍钢股份炼铁总厂，辽宁鞍山 114021）

摘 要：为解决炉缸炉皮开裂问题，采用更换 1-4 带炉皮方案。炉体加固采用外加固形式，经计算炉体上部荷载约为 3200t。炉壳上新增环梁，利用新增环梁将 3200t 荷载传递到 35.400m 的炉体框架梁上，为将炉壳的荷载传递到框架梁上，需设置支座和"八卦"梁。再对框架进行加固，通过框架将力传递到基础上。计算加固梁的强度、刚度、稳定性及沉降确定钢结构截面尺寸。

关键词：炉皮开裂，外加固，强度，刚度，沉降

Reinforcement Scheme for Replacing Furnace Body with 1-4 BF Shell in Ansteel 1# BF

Li Huanlong, Zhang Rongjun, Li Jinhui

(General Lronmaking Plant of Angang steel Co.Ltd., Anshan 114021, China)

Abstract: In order to solve the cracking of BF shell,By replacing the 1-4 leather belt furnace program. The external reinforcement form is adopted for the reinforcement of the furnace body, By calculation, the load on the upper part of the furnace is about 3200t. New ring beam is added to the shell, The 3200t load is transferred to the frame frame of the 35.400m furnace with the new ring beam. The pedestal and the Eight Diagrams beam should be set, Reinforce the frame again. Calculate the strength, rigidity, stability and settlement of the strengthened beam, and determine the dimension of the steel structure.

Key words: bf shell breach, external reinforcement, strength, inflexibility, sink

微尺度静定大型四辊轧机研制

申光宪[1]，郑永江[1]，余　超[1]，申妍红[2]，沈　光[3]

（1. 燕山大学机械工程学院，河北秦皇岛；2.无锡市卫生及计划生育委员会，江苏无锡；
3.南京果逸康通信有限公司，江苏南京）

摘　要： 针对国产一重 1200 四辊精密轧机存在的"憋劲"和"乱逛荡"弊端，按微尺度静定设计理论创新设计微尺度静定大型四辊轧机。总括起来轧辊弯曲挠度约 0.3mm 及轧辊轴承座与机架窗口立柱之间 1mm 间隙的微尺度量，机构学静定化设计和控制。取消原偏移距设置理论创新设置智能衬板，完成严格控制无间隙状态的四辊平行工况轧制过程，确保其四列短圆柱滚动轴承不被烧损。防止两端支反力不等、板带跑偏，镰刀弯生成以及"叼板窜辊"等怪像。还提高轧机振动固有频率，跳跃现场 10—20Hz 低频干扰。最终，强化同类机构学不静定大型四辊轧机，释放轧机"高档次"轧制能力，降本绩效，迎接大压下轧制新工艺的采用，生产高强度新品种精品的新时代。

关键词： 大型四辊轧机，微尺度参量，机构学静定化，高档轧机能力

Research of Micro-dimensional Statically Determinate Large 4-H Rolling Mill

Shen Guangxian[1], Zheng Yongjiang[1], Yu Chao[1], Shen Yanhong[2], Shen Guang[3]

(1.College of Mechanical Engineering, Yanshan University, Qinhuangdao, China; 2.Wuxi City Health and Family Planning Committee, Wuxi, China; 3.Nanjing GuoYi Kang Communication Technology Co.,Ltd., Nanjing, China)

Abstract: For the disadvantages of block and movement of the roll system in the 1200 4-h mill designed by First Heavy Industry, the micro-scale statically determinate large 4-high mill is designed according to the micro-scale statically determinate design theory. The new mill is mean to design and control two parameters involving the roll bending deflection about 0.3mm and the 1mm clearance between chock and housing window. The traditional offset distance theory should be abandoned and the new intelligent lining plate is designed to maintain the roll parallel without clearance in the roll system and to protect the four-row roller bearing. In addition, the problems involving unequal roll reaction force on both ends of back up roll, lateral movement and camber surface of strip and clamping and lateral movement of the rolls, can also be solved. The natural frequency of rolling mill can be improved as well to avoid the interfering frequency of 10-20Hz. Finally, intensified large statically determinate mill can make full use of high-grade rolling mill capacity to achieve the goal of "low cost to increase efficiency". The large reduction rate technology can be used to produce high-strength new varieties strip.

Key words: large 4-h rolling mill, micro-scale parameters, statically determinate theory of mechanism, high-grade rolling mill capacity

新型环冷机的应用

于成忠，张荣军，徐福玉，董津博

（鞍钢股份炼铁总厂，辽宁鞍山　114021）

摘　要：烧结矿目前广泛采用强制通风冷却。强制通风冷却分为抽风式和鼓风式两种，目前鼓风式冷却机被广泛采用。烧结矿冷却设备按大的流程划分，有机上冷却和机外冷却两种，用于机外冷却的设备以环冷机、带冷机为主。鞍钢股份炼铁总厂烧结作业区采用的是环式鼓风冷却机。炼铁总厂二烧作业区在传统环式鼓风冷却机基础上进行了升级改造。本文介绍了原环冷机存在的问题以及改造后环冷机的优点和效果。

关键词：烧结矿，环冷机，改造

Application of A New Type of Annular Cooler

Yu Chengzhong, Zhang Rongjun, Xu Fuyu, Dong Jinbo

(General Ironmaking Plant of Angang Steel Co., Ltd., Anshan, 114021, China)

Abstract: Forced ventilation cooling is widely adopted in sinter .Forced draft cooling is divided into two type: air suction type and blast type. At present, blower type cooling machine is widely adopted. The sinter cooling equipment is divided into two kinds according to the big process, organic cooling and outside machine cooling. The equipment used for the outer cooling is mainly the ring cooler and belt cooler. The ring type blast cooling machine is adopted in the sintering operation area of Angang iron and steel general factory. The two firing operation area of iron smelting plant has been upgraded and upgraded on the basis of the traditional ring type blast cooling machine. This paper introduces the existing problems of the primary ring cooler and the advantages and effects of the ring cooler after modification.

Key words: sinter, ring cooler, modification

钢卷吊装定位自动检测与应用

李雁志，何　可，汤　璞，程　曦，易　钊，杨　松

（武汉钢铁有限公司设备管理部，湖北武汉　430080）

摘　要：进行行车吊装作业数字化模拟实现。使用激光检测装置对大车和两个小车的移动位置进行精确的在线跟踪

测量，指导司机精确走位，形成对吊装作业的数字化处理，提高吊装的准确性。保证圆心与炉台的圆心在同一垂直于炉台的直线上，这样才能保证钢卷退火的均衡性和科学性。

关键词：钢卷，吊，定位

Steel Coil Hoisting Positioning

Li Yanzhi, He Ke, Tang Pu, Cheng Xi, Yi Zhao, Yang Song

(Wuhan Iron and Steel Limited Equipment Management Department, Wuhan 430080, China)

Abstract: Realize the digital simulation of driving hoisting. The use of laser detection device on the cart and the two carts of the mobile location of the precise online tracking and measurement, to guide the driver to accurately walk, the formation of the lifting operation of the digital processing to improve the accuracy of lifting. To ensure that the center of the heart and the center of the furnace in the same vertical line directly on the furnace, so as to ensure the balance of steel annealing and scientific.

Key words: steel coil, hoisting, positioning

35000m³/h 制氧机精氩储槽含氮高分析与改造

刘　赫

（本钢集团北营制氧厂，辽宁本溪　117017）

摘　要： 本文简要介绍了北营制氧厂生产一区内压缩工艺林德 35000 m³/h 制氧机在投产后，针对精氩储槽精氩产品纯度不合格含氮量超标的情况，采取改造仪表管线与密封气系统进行了分析与处理，并改造成功，获得了合格的精氩产品。

关键词：密封气，精氩纯度，改造

35000m³/h Oxygen High Nitrogen Analysis and Transformation Contain Pure Argon Storage Tank Purity

Liu He

(Benxi Iron and Steel Group North Camp Oxygen Plant, Benxi 117017, China)

Abstract: This paper briefly introduces lind 35000 m³/h oxygen making machine in oxygen plant production area is put into production, the pure argon storage tank of pure argon product unqualified purity nitrogen content exceed the standard situation, reform of instrument piping and sealing gas system were analyzed and processed, and the successful transformation, get the qualified refined argon.

Key words: argon purity improvement of sealing gas

降低 50000m³/h 空分机组变负荷对制氩系统影响的实践

张　猛

（本钢北营制氧厂，辽宁本溪　117017）

摘　要：介绍了 50000 m³/h 空分机组在变负荷时，由于系统波动对制氩系统产生三个方面的影响，具体分析了操作问题影响的根源，制定了从三个方面解决制氩系统波动的详细措施，制氩系统波动能很好的控制。

关键词：空分机组，变负荷，制氩系统

Practice of Reducing Air Separation Unit Load Change on the Impact of Argon Producing System

Zhang Meng

(Beiying Iron-making Plant of BX Steel,Benxi 117017, China)

Abstract: This paper introduces the 50000 m³/h air separation unit under variable load, because the system, fluctuations produce three effects on argon producing system, specific analysis of the root of problems.

Key words: air separation unit, variable load, argon producing

激光跟踪仪精度仿真与实验分析

雷振尧，陈伟刚，陈文礼，任海峰

（首钢京唐钢铁联合有限责任公司，河北唐山　063200）

摘　要：随着现代工业技术的迅猛发展，制造业对设备尺寸及空间位置精度要求越来越严苛，已达微米级。激光跟踪仪作为一种高精密，便携测量工具被越来越广泛地应用于工业测量项目中。本文在误差传递理论及空间几何学基础上，推算激光跟踪仪在空间测量中定向误差与测程和测角之间的关系，并通过 MATLAB 软件进行定性与定量仿真分析，最终设计实验对结果加以验证。所得结论为构建大尺寸空间测量体系提供了理论依据并对今后工业现场测量作业起到指导性作用。

关键词：激光跟踪仪，误差传播理论，定向误差，仿真分析

Analysis of Simulation and Experiment on Accuracy of Laser Tracker

Lei Zhenyao, Chen Weigang, Chen Wenli, Ren Haifeng

（Shougang Jingtang United Iron and Steel Co., Ltd., Tangshan 063200, China）

Abstract: With the rapid development of modern industrial technology, the manufacturing industry on the equipment size and positional precision requirements more stringent, even it has reached the micron level. Laser tracker as a high precision, portable measurement tools are widely used in industrial measurement project. Aiming at such issues，calculate the relationship between the specific direction error with the line-angle measurement when measured with a laser tracker was studied in this paper, analysis by theory of error propagation and space geometry, then simulated by using MATLAB. Finally, the results were verified experiments. The conclusion provides a theoretical basis for the construction of large-scale space measurement system and plays a guiding role in the future industrial field measurement technology.

Key words: laser tracker, theory of error propagation, specific direction error, simulation analysis

优化 20000m³/h 空分设备操作提高氩纯度的实践

焦兆文

（本钢北营制氧厂，辽宁本溪　117017）

摘　要： 针对 20000 m³/h 空分氩产品含微量氧、微量氮较高的情况，经过对空分系统的各道工序分析后，通过工况的物料及冷量动态平衡、减小系统波动，稳定精馏工况、改进粗氩塔控制，降低氩中含氧量、改进精氩塔控制，降低氩中含氮量四个方面优化操作并缩小控制限，精氩纯度达到高纯氩的指标。

关键词： 空分设备，提氩系统，优化操作，高纯氩

Optimization of 20000m³/h Air Separation Equipment Operation to Improve the Practice of Argon Purity

Jiao Zhaowen

(Beiying Iron-making Plant of BX Steel,Benxi 117017, China)

Abstract: For 20000 m³/h air separation equipment of argon product containing trace oxygen and trace nitrogen is higher, after system the procedure of analysis of air separation, by working material and cold quantity dynamic balance, reduce the fluctuation of system, stable condition, improvement of the crude argon column control, reduce argon oxygen content, improve the refining argon tower control, reduce nitrogen argon in four aspects of optimized operation and to reduce the control limit, the pure argon purity reached standard of high purity argon.

Key words: air separation equipment, optimizing the operation system, argon extraction, high purity argon

干熄焦旋转密封阀故障案例分析与思考

王　宁，李红伟

（唐山首钢京唐西山焦化有限责任公司，河北唐山　063200）

摘　要： 本文结合典型的事故案例和生产实践，对处理干熄焦旋转密封阀卡阻过程中存在的问题进行了探讨，优化

了处理过程，改善了处理措施的有效性和安全性，找出了在这些问题上存在的认识误区，并提出了相应的改进方法和建议。

关键词：干熄焦，旋转密封阀

Analysis and Thinking of Rotary Seal Valve in Dry Quenching of Coke

Wang Ning, Li Hongwei

(Shougang Jingtang Xishan Coking Co., Ltd., Tangshan 063200, China)

Abstract: Combined with the typical case of accidents and production practice, the problems in the process of jamming of the CDQ rotary seal valve were discussed, the process was optimized, the effectiveness and safety of the treatment measures were improved. And the misunderstanding of these problems has been found. The corresponding improvement methods and suggestions were presented.

Key words: dry quenching of coke, rotary seal valve

基于高频伺服阀的轧机三倍频程振动消除方法

吴守民[1]，顾廷权[1]，齐 鹏[2]，张 伟[2]，白振华[2]

（1. 宝山钢铁股份有限公司，上海 201900；2. 燕山大学，河北秦皇岛 066004）

摘 要：轧机振动历来是困扰轧钢行业发展的复杂问题，振动会导致板带精度和板型问题，严重时甚至会损坏轧机设备。因此弄清轧机振动原因，掌握其机理，并采取有效措施抑制振动，是十分紧迫和必要的。本次研究主要针对生产中常见的轧机垂直方向三倍频程振动，做出了将在高频伺服阀并联在 AGC 伺服油缸来主动抑制三倍频程振动的对策，通过建立 Amesim 模型，验证了此方案对于解决轧机振动问题的有效性。

关键词：三倍频程振动，高频伺服阀，AGC 液压缸，抑制振动

Abstract: The vibration of cold mill machine has been a complex problem.This will make great impact on strip thickness and flatness ,what's more,it will damage the mill equipment.So it is important to find out the mechanism of vibration and take effective methods to prevent it from happening.In industrial production, 3-octave vibration is common,so this research will pay more attention to this type of vibration,and try to suppress it. A brand new method has been presented,which is to put high frequency servo valve and AGC piston in parallel.By Amesim simulation,this method is proved to be effective to suppress vibration.

Key words: 3 - octave vibration, high frequency servo valve, AGC, suppression

钢铁工业煤气储配系统的运行诊断探索

黄永红[1]，金宇晖[2]

（1. 中冶南方钢铁工程技术有限公司技术研究院，湖北武汉 430223；

2. 华中科技大学能源与动力工程学院，湖北武汉 430074）

摘　要：设备诊断作为一门学科，有其比较完整的理论和方法。但工业流程系统运行问题的诊断具有复杂性，相关研究不多。本文根据流程制造工业的运行特点，创立了一种根据分层诊断的方法。并就钢铁工业煤气储配系统的组成及运行特点，明确了各组成子系统诊断的关键要素。结合煤气储存系统的诊断实践，探索了工业复杂系统的诊断思路和方法。该诊断实践，为将来的智能化诊断作了有益地探索。最后提出了钢铁工业 4.0 智能化升级设备及系统智能化的诊断相关建议。

关键词：钢铁产业，工业煤气，储配系统，故障诊断

Exploration of Operation Diagnosis for Gas Storage and Transportation System in Iron and Steel Industry

Huang Yonghong[1], Jin Yuhui[2]

(1. Technology Research Institute ,WISDRI Steel Engineering Co., Ltd., Wuhan 430223, China;
2. School of Energy and Power Engineering, Huazhong University of Science and
Technology, Wuhan 430074, China)

Abstract: As a subject, equipment diagnosis has its more complete theory and methods. However, the diagnosis for the operation of industrial process system is complicated and the related research is limited. Considering the operation characteristics of process manufacturing industry, this paper established a method based on hierarchical diagnosis. The composition and operation characteristics of the steel industry gas storage system are analyzed, and the key factors for the diagnosis of the subsystems are clarified. Combined with the diagnosis practice of gas storage system, the ideas and methods of complex industrial system diagnosis are explored. This diagnostic practice is a useful exploration for future intelligent diagnosis. Finally, some suggestions for intelligent upgrading equipment and intelligent system of Steel Industry 4.0 are proposed.

Key words: iron and steel industry, industrial gas, storage and transportation system, operation diagnosis

高炉探尺电气控制研究及故障分析处理

岑家贵

（宁波钢铁有限公司）

摘　要：目前钢铁冶金行业，炼铁高炉操作炉内的情况，其依据就是靠高炉炉顶的探尺来判断炉内的料面情况；高炉炉顶探尺根据高炉的大小数量也不一样，本论点就依据 2500 m³ 的高炉来研讨；目前国内炼铁高炉 2500m³ 的高炉，炉顶探尺一般设计为 3 台探尺（3 台机械探尺或 2 台机械 1 台雷达探尺等组合）；电气部分：变频电机、主令控制器、绝对值编码器、译码器、西门子变频器、增量型编码器（1024 转）、西门子 DTI 放大板等电气设备组成。本论文的重点就是研讨探尺的电气控制，传统的高炉探尺是没有以上复杂的电气设备，最简单电气控制的就是用接触器、电阻等来控制；现在炼铁高炉探尺是以上变频器控制为主；本论文研讨的控制方式以西门子变频器矢量控制带编码器闭环控制方式（6SE70 系列）及其控制方式的缺陷（日常难点故障分析处理）。

关键词：高炉探尺电气控制

RYL-02 型自动熔样炉中硅碳棒的日常使用方法的改进

毕 欣，齐 兵，薛 松

（河钢集团石家庄钢铁有限责任公司，河北石家庄）

摘 要：通过日常使用中自动熔样炉的硅碳棒使用情况，主要研究了硅碳棒电阻率对其使用寿命的影响。结果表明：在相同情况下，炉中一组硅碳棒的电阻率越接近其使用寿命越长。可见炉中硅碳棒电阻率相同可以较好地延长其使用寿命，降低使用成本。

关键词：硅碳棒，电阻率，改进，使用寿命，使用成本

The Improvement of the Daily Use of Silicon Carbide Rod in Type RYL-02 Automatic Melting Furnace

Bi Xin, Qi Bing, Xue Song

(HBIS Group Se Company, Shijiazhuang, China)

Abstract: Through the use of silicon carbide rod in the daily use, the influence of the resistivity of silicon carbide on its service life was studied. The results showed that, in the same situation, a group of silicon carbon robs with the same resistivity in the furnace had a longer useful-working life than it with the different resistivity. It is shown that the same resistivity of silicon carbide in the furnace can prolong its service life and reduce the cost of use.

Key words: silicon carbide rod, electric resistivity, improvememts, useful-working life, use cost

法孚卧磨液压缸用蓄能器国产化及应用

高 健

（日照钢铁循环经济部）

摘 要：本文在进口蓄能器国产化改造过程中，通过对蓄能器国产化的目的、应遵循的原则、方法的详细分析，总结出进口备件、材料国产化的重要意义、应遵循的原则和一般方法，为公司降本增效而提出进口备件国产化的建议。

关键词：蓄能器国产化，目的，遵循原则，方法，降本增效，国产化建议

Abstract: Localization transformation process in the import accumulator through a detailed analysis of the purpose of localization of the accumulator, should follow the principles, methods, summed up the import of spare parts, the significance of the localization of the material, should follow the principles and general methods, efficiency and cost reduction for the company to put forward the proposal of the localization of imported spare parts.

Key words: localization of the accumulator, purpose, follow the principle, method, cost reduction and efficiency, localization of recommendations

10kV 电缆终端头故障原因分析及制作工艺要求

代俊杰

（攀钢集团西昌钢钒公司维修中心）

摘　要：本文通过对西昌钢钒余热发电六次 10kV 电力电缆故障事件介绍，并结合实际电缆故障原始照片及资料，详细阐述 10kV 交联聚乙烯电缆电力电缆终端头故障产生原因及电缆冷缩终端头制作工艺，并纠正作业区作业人员常见的错误的 10kV 交联聚乙烯电缆冷缩终端头的制作方法。

关键词：交联聚乙烯电缆，电缆冷缩终端头，制作工艺

浅谈影响产品氧纯度的因素及调整

李天青

（昆明钢铁集团动力能源分公司制氧车间，云南安宁　650300）

摘　要：氧气是高炉生产中不可或缺的能源，对高炉的生产起着重要作用；而其中氧气的纯度是极为关键；目前在全低压空分制氧过程中，对氧气纯度的影响因素较多，如氧气产量过大、下塔液空纯度过低、进上塔的膨胀空气量过大、精馏工况异常、冷凝蒸发器液氧面过高、加工空气量变化等；为此，对日常生产就提出了较高要求，应根据实际工况的变化及时作出调整，在满足氧气产量的同时也要保证氧气纯度。

关键词：氧气纯度，空分，精馏，物料平衡，回流比

35kV 制氧变电站供电系统改造

王治军

（昆钢动力能源分公司制氧车间）

摘　要：针对供电系统故障引起的系统全面停车事故，为满足车间今后不断发展新增负荷用电和消除 35 kV 供电存在的安全隐患，同时为公司、车间连续稳定的安全生产提供安全稳定的供电，及时、有序的调整，尽快组织恢复生产，降低事故引起的损失、减少事故对设备的损伤，因此对 35 kV 制氧变电站供电系统进行安全改造。

关键词：35kV 制氧变电站，供电系统，改造

Transformation of Power Supply System for 35kV Oxygen Producing Substation

Abstract: According to the system of power system fault caused by comprehensive parking accident, in order to meet the future development of the new workshop load and eliminate safety hazards exist in 35 kV power supply power supply, safe and stable production safety for the company at the same time, the workshop continuous stable, timely and orderly adjustment of the organization to resume production as soon as possible, to reduce the losses caused by the accident reduce the number of accidents, damage to the equipment, so the security transformation for 35kV oxygen generator substation power supply system.

Key words: 35 kV oxygen generating substation, power supply system, transformation

变工况操作及其对制氩系统工况的影响

余成志

（昆钢动力能源分公司制氧车间，云南安宁　650300）

摘　要： 简单介绍液体空分设备变工况操作的要点以及变工况操作对制氩系统工况的影响，并阐述了变工况操作时氮塞发生的原因，调整手段及预防措施。

关键词： 变工况操作，制氩系统，氩馏分，粗氩，氮塞

Variable Operation and Its Influence on the Working Condition of Argon Production System

Yu Chengzhi

(Kunming Power Branch Oxygen Plant, Anning 650300, China)

Abstract: The off design operation is simple introduction of liquid air separation equipment and the effect of variable condition operation on argon recovery system, and expounds the causes of nitrogen block during the off design operation, adjustment methods and preventive measures.

Key words: variable operation, argon recovery system, argon fraction, crude argon, nitrogen plug

浅谈一起水环机分离器堵塞的原因和处理措施

恭元芳

（昆钢动力能源分公司制氧车间，云南安宁　650300）

摘　要：KDON-10000/20000-I 型空分设备氩纯化系统进入水环机粗氩流量、进气压力、分离器出口压力等不断波动，液氩产量不断减少，影响了整个制氩系统的稳定生产。在对故障现象原因分析及短暂观察后，发现故障越来越严重，停运精氩塔，对堵塞的水环机分离器进行清理，空分设备氩纯化系统恢复正常运行。

关键词：制氩系统，水环式压缩机，水分离器，堵塞

免拆 LGA 闪存测试台的制作和使用

陈久升，陈　新

（山东泰山钢铁集团有限公司自动化部，山东莱芜　271100）

摘　要：随着集团公司现代化的发展，闪存类信息化以及工业设备越来越多。本文主要综述了免拆闪存测试台的制作以及使用方法，希望大家从中得到些收获。

关键词：基板，拖焊，恒温烙铁，焊料，焊接温度，测试

Without Removing the Production and Use of LGA Flash Memory Test Bench

Chen Jiusheng, Chen Xin

(Shandong Taishan Steel Department of Automation, Laiwu 271100, China)

Abstract: With the modern development of the group company, flash class information and industrial equipment more and more. This paper summarizes the production and application of the flash memory test stand, and hopes that we can get some harvest from it.

Key words: substrate, tow welding, constant temperature iron, solder, welding temperature, test.

18 冶金自动化与智能管控

炼铁与原料

炼钢与连铸

轧制与热处理

表面与涂镀

金属材料深加工

钢铁材料

汽车钢

海洋工程用钢

轴承钢

电工钢

粉末冶金

非晶合金

高温合金

耐火材料

能源与环保

分析检测

冶金设备与工程技术

★ 冶金自动化与智能管控

冶金技术经济

基于深度学习的板坯低倍质量评级系统

贾永坡

（河钢集团邯钢公司自动化部，河北邯郸　056015）

摘　要：低倍检验方法可以快速直接对板坯内部偏析和疏松等影响质量的关键问题进行判断，进而可以对板坯质量进行评级。在实际生产过程中，采用硫印和酸洗的方法对板坯表面进行腐蚀，然后人工根据表面腐蚀程度，通过经验来判断板坯内部缺陷严重程度。为实现质量自动评级，我们独立研发了基于深度学习的板坯低倍质量评级系统。本文介绍了该系统的研发过程以及主要系统构架和功能。该系统采用图像处理技术对板坯低倍检验后的表面进行处理，然后基于深度学习方法通过标定数据来学习判定模型。该系统的判定准确率大于 93%，综合 F1-score 高达 95%，系统稳定可靠，并且具备质量判定、信息管理以及自学习功能。

关键词：板坯质量评级，低倍检验，深度学习，图像处理，人工智能

A Deep Learning based Steel Slab Quality Grading System

Jia Yongpo

(Automation Department, HBIS Hansteel, Handan 056015, China)

Abstract: Low magnificence test is a widely adopted method to inspect the internal defects, including internal segregation and loose, of slab and based on which to measure its quality level. A traditional procedure is doing sulphur printin or pickling to slab surface, and then technician will use experience to judge the slab's internal defects' intensity based on surface image. In order to do this automatically, we devised a slab quality grading system based on deep learning. In this paper, we introduce the developing procedures and the architecture as well as functions of our system. This system adopted image processing techniques to handle the slab surface pictures taken after low magnificence test, and then based on deep learning and tagged ground truth data we learned a model for quality grading. This system has achieved 93% and 95% in terms of precision and F1-score, respectively. Our system is stable and robust, besides it contains quality grading, data management as well as self-learning modules to make it powerful enough for different application scenairos.

Key words: slab quality grading, low magnificence test, deep learning, image processing, artificial intelligence

基于主成分分析的连铸坯质量预测研究

陈恒志，杨建平，余相灼，刘　青

（北京科技大学钢铁冶金新技术国家重点实验室，北京　100083）

摘　要：为提高连铸坯质量预测模型的预测精度，提出了将主成分分析与 GA-BP 神经网络相结合的连铸坯质量预测方法。使用主成分分析法对多个影响连铸坯质量的因素进行降维处理或重新组合，将处理后所得较少的主成分变量作为样本输入 GA-BP 神经网络进行训练而得到连铸坯质量预测模型，对方大特钢的 60Si2Mn 连铸坯中心疏松和

中心偏析缺陷进行预测，并与以未经处理的影响因素作为输入变量的 GA-BP 神经网络连铸坯质量预测模型进行对比分析。结果表明：基于主成分分析与 GA-BP 神经网络相结合的连铸坯质量模型的预测精度较高，对连铸坯中心疏松和中心偏析缺陷的预测准确率分别为 85%、80%，且模型的运算速度有了显著的提升。

关键词：连铸坯，BP 神经网络，遗传算法，主成分分析，质量预测

Concentrate Quality Prediction of Continuous Casting Strand Based on Principal Component Analysis

Chen Hengzhi, Yang Jianping, Yu Xiangzhuo, Liu Qing

(State Key Laboratory of Advanced Metallurgy, University of Science and Technology Beijing, Beijing 100083, China)

Abstract: In order to improve the prediction accuracy of the quality prediction model of continuous casting bloom，a data processing method based on the combination of principal component analysis and BP neural network was presented. By using principal component analysis，the amount of variables which affect the quality of continuous casting will be reduced. Then the principal components are employed to train the BP neural network in order to obtain the quality prediction model of continuous casting bloom. As the number of inputs is reduced, the train process can be faster and the iteration time can be reduced. In comparison with the network model which uses the original variables as the inputs to predict the degree of the center porosity and central segregation of 60Si2Mn continuous casting bloom produced by Fangda Special Steel, the results show that GA-BP neural network prediction modeling method based on principal component analysis method is improved to 85% and 80% in the center loose and central segregation defects respectively, and this modeling method has both high efficiency in calculation.

Key words: continuous casting bloom, BP neural network, genetic algorithm, PCA, quality prediction

钢厂生产运行模式的仿真研究

郑　忠[1]，徐兆俊[1]，高小强[2]，范计鹏[1]，黄世鹏[1]，张　开[1]

（1. 重庆大学材料科学与工程学院，重庆　400045；
2. 重庆大学经济与工商管理学院，重庆　400044）

摘　要：为分析钢厂复杂生产组织行为的运行特征，基于钢厂生产流程的生产运行仿真研究，寻求生产运行模式及其主要影响因素之间的调控关系。论文建立了以炼钢生产流程网络的制造单元作为节点、以生产调度规则作为模型演化规则的元胞自动机仿真模型；通过现实钢厂的生产运行实例验证了基于该模型进行炼钢生产过程仿真的可靠性；通过对钢厂生产系统运行状态有直接影响的主要因素的变化情况进行仿真研究，以了解不同生产运行模式下的钢厂生产系统特征。研究结果可以为生产组织策略优化提供决策支持。

关键词：炼钢，生产运行模式，元胞自动机，仿真

Simulation-based Research on the Production Operation Mode in Steel Plant

Zheng Zhong[1], Xu Zhaojun[1], Gao Xiaoqiang[2], Fan Jipeng[1],

Huang Shipeng[1], Zhang Kai[1]

(1. College of Material Science and Engineering, Chongqing University, Chongqing 400045, China;
2. College of Economics and Business Administration, Chongqing University, Chongqing 400044, China)

Abstract: In order to analyse the operation characters of the complex production behaviors in steel plant, a simulation model based on cellular automata is built to dig the relationship between the production operation mode and its main influencing factors. In the model, manufacturing units of the steelmaking process network is regarding as the nodes, and the production scheduling rules are abstracted as the model evolution rules. The simulation experiments with real process data shows that the model can effectively simulate the real steelmaking process. Based on the simulation of the change of factors which mainly influenced the system status, the operation system characteristics under different production operation mode is rigorously studied. The research results can provide decision support for the optimization of production organization strategy.

Key words: steelmaking, production operation mode, cellular automata, simulation

PROFIBUS-DP 及 G130 变频器在焦罐车控制系统中的应用

王　宁，孙得维，敖轶男，丁　彪，何凤伟

（唐山首钢京唐西山焦化有限责任公司，河北唐山　063200）

摘　要：本文主要对以 PROFIBUS-DP 及 G130 系列变频器为核心的首钢京唐焦罐车控制系统进行介绍，并着重对该系统的基本构成、各主要功能模块结构以及 G130 主从变频器间的通讯、控制软件的结构和系统自动对位等问题进行了相关阐述。

关键词：PROFIBUS-DP，G130，变频器

The Application of the PROFIBUS-DP and G130 Inverter in ACBC Control System

Wang Ning, Sun Dewei, Ao Yinan, Ding Biao, He Fengwei

(Shougang Jingtang Xishan Coking Co., Ltd., Tangshan 063200, China)

Abstract: Shougang jingtang ACBC control system based the core of PROFIBUS-DP and G130 series inverter are introduced. The basic structure of the system, main control modules and G130 inverter communication between master and slave section, system software and automatic positioning and other issues are emphatically described.

Key words: PROFIBUS-DP, G130, inverter

热轧窄带钢二级模型应用

孟　新，谢新亮，申铁强，马小军，易湘德，伍延平

（北京冶自欧博科技发展有限公司工程部，北京　100071）

摘　要：本文介绍了热轧窄带钢二级模型系统的应用方案，重点阐述设定计算功能、计算数据流程、应用效果三个方面。模型具备完整的轧制规程计算、设定参数计算和轧辊状态计算，自学习功能具有短期和长期自学习，分别对轧制力、扭矩、温度、厚度及宽度进行学习。目前二级模型的应用使热轧窄带产品质量大幅提升，节能，降低成本，提升热轧窄带钢生产的现代化、智能化水平。

关键词：热轧窄带钢，二级模型，自学习

Application of Level 2 Model Technology for Hot Rolled Narrow Strip Mills

Meng Xin, Xie Xinliang, Shen Tieqiang, Ma Xiaojun, Yi Xiangde, Wu Yanping

(Beijing Ablyy Technology Development Co., Ltd., Beijing 100071, China)

Abstract: This paper introduces the level2 model technology application solutions for Hot Rolled Narrow Strip Mills, and the model calculation function, the data flow, the application effect are given emphasis introduction. Learning function consists of short term and long term, the learning items including the rolling force, torque, temperature, thickness and width. Now the application of the level2 model significantly improved the performance of the product, energy efficiency, reduce the cost, and promote the modernization and intelligent level of hot rolled narrow strip production.

Key words: hot rolled narrow strip mills, level2 model, learning

热轧微合金钢力学性能预报模型研究

李维刚[1,2]，刘　超[1]，杨　威[1]，严保康[1]

（1. 武汉科技大学信息科学与工程学院，湖北武汉　430081；2. 武汉科技大学冶金自动化与检测技术教育部工程研究中心，湖北武汉　430081）

摘　要：提出一种融合工业大数据与冶金机理的钢材力学性能建模方法，借助冶金机理和实际生产数据对各要素的影响进行剖析、把复杂问题拆分成若干子问题。首先，综合运用随机森林、贝叶斯网、因果图等数据挖掘方法，并结合冶金机理与人的先验知识筛选模型的影响因素；接着，根据确定的影响因素，构建热轧带钢力学性能预报模型，包括数据清洗、子模型建立、子模型验证等环节。最后，针对某热连轧机组进行了微合金钢力学性能建模实验，采用三次光滑样条非参数估计方法求得各自变量的单变量光滑函数，从而获得成分、碳氮化物及工艺参数对力学性能的影响曲线。实际预测实践表明，抗拉强度、屈服强度的预测误差分别为2.54%和3.34%，新模型具有计算精度高、

适应能力强等优点，可为微合金钢产品的优化设计提供参考。

关键词：力学性能预报，抗拉强度，屈服强度，析出，可加模型

Prediction Model of Mechanical Properties of
Hot-rolled Micro-alloyed Steel

Li Weigang[1,2], Liu Chao[1], Yang Wei[1], Yan Baokang[1]

(1. School of Information Science and Engineering, Wuhan University of Science and Technology, Wuhan 430081, China; 2. Engineering Research Center for Metallurgical Automation and Detecting Technology of Ministry of Education, Wuhan University of Science and Technology, Wuhan 430081, China)

Abstract: A modeling method for mechanical properties of steel via industrial big data and metallurgical mechanism is proposed. Based on metallurgical mechanism and actual production data, the influence of each factor is analyzed and the complex problem is divided into several sub-problems. First of all, combined with metallurgical mechanism and the priori knowledge, the influence factors of the model were selected by the comprehensive utilization of data mining methods, such as random forests, bayesian network and causality diagram. Then, according to the determined influencing factors, the property prediction model of hot-rolled strip was constructed, the steps include data cleaning, sub-model building, sub-model validation and so on. Then, a modeling experiment was carried out for a hot strip mill, the univariate function of each independent variable is estimated by the cubic smoothing spline nonparametric estimation method, the relationship curves of the composition, the carbon and nitrogen precipitates and the process parameters on the mechanical properties are given. The practical prediction case shows that the prediction errors of tensile strength and yield strength were 2.54% and 3.34%, respectively. The new model has the advantages of high precision and strong adaptability, which can provide reference for the optimal design of micro-alloyed steel.

Key words: mechanical property prediction, tensile strength, yield strength, precipitation, additive model

基于数据驱动的轧机振动预测研究

彭 艳，张 明，刘宣亮，崔金星

（燕山大学国家冷轧板带装备及工艺工程技术研究中心，河北秦皇岛　066004）

摘　要：轧机振动是影响轧制过程稳定性的关键因素。本文利用数据挖掘技术研究轧机振动问题，提出基于 BP-AdaBoost 的轧机振动预测模型和 POS-SVM 的轧机振动预测模型，并通过某钢厂热连轧系统振动实测数据对两种预测模型进行训练和检验，证明了利用数据挖掘技术能够实现轧机振动的预测；通过比较验证数据，得出 POS-SVM 轧机振动预测模型预测结果优于 BP-AdaBoost 轧机振动预测模型；最后分析了轧制工艺参数对轧机振动的影响规律，提出抑振措施，达到了抑振效果。

关键词：轧机振动，数据挖掘，预测模型，BP 神经网络，支持向量机

Prediction of Rolling Mill Vibration Based on Data Driven

Peng Yan, Zhang Ming, Liu Xuanliang, Cui Jinxing

(National Engineering Research Center for Equipment and Technology of Cold Strip Rolling, Yanshan University, Qinhuangdao 066004, China)

Abstract: Rolling mill vibration is the key factor affecting the stability of rolling process.In this paper, the data mining technology is used to study the vibration of rolling mills,and two kinds of rolling mill vibration prediction models were established, respectively the vibration mill prediction model based on BP-AdaBoost strong predictor and the rolling mill vibration prediction model based on PSO-SVM algorithm.The two prediction models are trained and tested by the vibration data of a hot rolling mill. It is proved that the vibration prediction of rolling mill can be realized by using data mining technology.By comparison, we can see that the POS-SVM algorithm has better prediction effect.And at the end,by analyzing the influence law of rolling parameters on rolling mill vibration, vibration suppression measures were put forward,and the vibration suppression effect was achieved.

Key words: rolling mill vibration, data mining, prediction model, BP neural network, support vector machine

钢铁生产智能制造顶层设计的探讨

李鸿儒[1]，封一丁[2]，杨英华[1]，庞哈利[1]，贾明兴[1]

（1. 东北大学信息科学与工程学院，辽宁沈阳　110819；

2. 河钢集团有限公司研究总院，河北石家庄　050021）

摘　要：本文以信息物理系统（CPS）为基础，探讨了未来钢铁生产的智能制造架构，强调全流程数据的应用，突出了横向集成的概念。该架构包括工程数据中心、物理系统、信息系统和虚拟系统四部分。从铁前、炼钢、精炼、连铸、热轧、冷轧、产成品多个生产工序出发，分别构建适合的 CPS 作为钢铁行业智能制造的基本单元形成物理系统；进而建立工厂数据中心为基础的信息系统，从一体化生产管控、全产业供应链系统、工厂设备资产全生命周期等多维度构建组织管理功能出发进行设计，其核心特点是基于统一的大数据平台，统筹管理企业运营。虚拟系统面向钢铁企业制造与管理过程的全生命周期进行仿真和优化。

关键词：智能制造，钢铁生产，信息物理系统，工业数据中心

Discussion on Top Level Design for of Smart Manufacturing in Iron and Steel Production

Li Hongru[1], Feng Yiding[2], Yang Yinghua[1], Pang Hali[1], Jia Mingxing[1]

(1. College of Information Science and Engineering, Northeastern University, Shenyang 110819, China;
2. General Research Institute of Hebei Iron Steel Group Co., Ltd., Shijiazhuang 050021, China)

Abstract: Based on the Cyber-Physical System (CPS), a framework of smart manufacturing for future iron and steel production is discussed, which emphasizes the application of full process data, highlighting the concept of horizontal integration. The architecture consists of four parts: engineering data center, physical system, information system and virtual system. Starting from iron, steelmaking, refining, continuous casting, hot rolling, cold rolling and finished products, some appropriate CPSs are constructed as the basic unit of smart manufacturing in iron and steel industry, which named as physical system . And then establish the information system based on the data center of the factory. The information system construct the organization management functions from the integration of production control, the whole industry supply chain system, product lifecycle management, its core characteristic is based on the unified big data platform, and the overall management and operation of enterprise. The virtual system aims at the simulation and optimization of the whole life cycle of the steel manufacturing and management process.

Key words: smart manufacturing, iron and steel production, cyber-physical system (CPS), industrial data center

宝钢宝山基地合金购决策系统的研发与应用

杜　斌[1]，谢树元[1]，林　云[1]，赵　科[2]，贾树晋[1]，刘舒良[3]

（1. 宝钢股份研究院，上海　201900；2. 宝钢股份原料采购中心，上海　201900；
3. 宝钢股份财务部，上海　201900）

摘　要： 通过分析炼钢合金消耗机理，在合金最优控制模型和生产数据的基础上，将统计与机理建模相结合，研制出了钢记号与合金消耗的对应智能模型群；再根据一定生产周期的合同与销售计划信息，开发了炼钢合金需求预测模型；以此为基础结合库存理论，综合考虑订购费、存贮费、缺货损失费等，以总费用最小为目标，研制出合金综合管理与采购智能优化决策支持系统。该系统在宝钢股份宝山基地获得应用，经济效益明显。

关键词： 冶金机理，采购智能优化，合金综合管理，库存论，过程控制模型

Development and Application of Management and Purchasing Decision System for Steel Making Alloys

Du Bin[1], Xie Shuyuan[1], Lin Yun[1], Zhao Ke[2], Jia Shujin[1], Liu Shuliang[3]

(1. Research Institute, Baoshan Iron and Steel Co., Ltd., Shanghai 201900, China; 2. Raw Materials Purchase Center, Baoshan Iron and Steel Co., Ltd., Shanghai 201900, China; 3.Finance Department, Baoshan Iron and Steel Co., Ltd., Shanghai 201900, China)

Abstract: The decision support system on alloys management and purchasing is composed of three sub-models. The first part is a set of models on alloy consumption of different steel grades. These models are based on sophisticated steal-making experience as well as alloy consumption knowledge. The second model is built to predict demand using the output from the first part as well as information on orders and sales. The last model on comprehensive management and purchasing is aimed to minimize total cost on purchasing, inventory and unfilled orders. This decision support system has been applied at Bao steel Corp.

Key words: metallurgy mechanism, purchasing optimization, alloy management, inventory theory, process control model

机房设备远程监控系统的建立与应用

李　青

（宝钢工程集团上海金艺检测技术有限公司，上海　201900）

摘　要：本文介绍了机房设备远程监控系统的建立和在宝钢股份各产线过程控制机房的应用。重点介绍 UPS，温湿度传感器和网络设备数据监控的方案及系统在机房人员进行设备状态把握和故障查询分析等方面的应用及效果。

关键词：远程监控，UPS，环境温湿度，设备状态

Establishment and Application of Remote Monitoring System for Machine Room Equipment

Li Qing

(Baosteel Engineering Group Shanghai JinYi Detection Technology Co., Ltd., Shanghai 201900, China)

Abstract: This article introduces the establishment of remote monitoring system for machine room equipment and its application in process control computer rooms of production lines all over BaoSteel. It mainly introduces the remote online monitoring scheme of UPS, temperature and humidity sensor and network equipment. The effective use of the monitoring system in equipment state control and fault query analysis is also introduced.

Key words: romote monitoring, UPS, evironment temperature and humidity, equipment state

冶金工艺先进过程控制系统定义高品质钢材生产

李立勋，田　勇，史页殊，王鲁毅，赵文涛，金百刚，徐国义

（鞍钢股份有限公司鲅鱼圈钢铁分公司，辽宁营口　115007）

摘　要：冶金工业发展至今对于工艺与智能测控系统的集成越来越重视，智能测控系统与冶金工艺是相辅相承的关系，一方面冶金工艺的进步和发展是冶金智能测控系统技术进步的主要动力，另一方面冶金智能测控技术的进步可以引发和促进冶金工艺的实现与进步。将现代冶金工艺原理及前沿研发理论应用于生产且适合冶金工作者的工程平台就是各类冶金先进过程控制系统（Advanced Process Control）及系统的 APC 模型接口。冶金 APC 模型及测控系统为生产高品质钢材提供机理、装备、工艺、质量的保证。提高企业的核心竞争力。

关键词：钢铁行业，发展，冶金工艺，先进过程控制，工程平台，核心竞争力

The Metallurgical Advanced Process Control System Defines High Quality Steel Production

Li Lixun, Tian Yong, Shi Yeshu, Wang Luyi, Zhao Wentao, Jin Baigang, Xu Guoyi

(Bayuquan Subsidiary of Ansteel Co., Ltd., Yingkou 115007, China)

Abstract: Iron and Steel Industry developed to now, put more and more important on embedding Metallurgical–Process and intellectualized control system. Metallurgical Process and intellectualized control system has closed relationship. Firstly, improvement of Metallurgical–Process drives control system improvement. Secondary, Improvement of Metallurgical intellectualized control system supports Metallurgical–Process come true. The suitable Engineering-workbench for Process worker is Advanced Process Control system and its interface. Metallurgical APC model and control system provide assurance for producing high quality steel. Enhance enterprise kernel competition.

Key words: iron and steel industry, develop, metallurgical -process, advanced process control, engineering-workbench, kernel competition

能源管控系统中的设备状态实时统计功能的研究

张长开，解 凯，姜 彬

（南京南瑞继保电气有限公司研究院，江苏南京 211102）

摘 要： 钢铁企业能源管控系统利用采集到的设备运行状态的信息，进行分析和处理，可以准确的对现场设备的运行时长进行实时统计。统计结果不仅方便技术人员了解设备的运行状态，也为提高生产调度能力、保证生产顺利进行提供了重要支撑。本文主要研究了能源管控系统中针对设备运行状态的实时统计功能，阐述了实时统计功能的对象、方法，并介绍了实时统计功能带来的有益效果。

关键词： 能源管控系统，钢铁企业，设备运行状态，实时统计

Research on Real Time Statistics of Equipment Running Status Based on Energy Management System

Zhang Changkai, Xie Kai, Jiang Bin

(System Software Institute of Nanjing NR Electric Co., Ltd., Nanjing 211102, China)

Abstract: Iron and steel enterprise can get accurate real time statistical results of equipment running status based on the analysis and process of the data acquired from the equipment. It not only makes convenient that the technical staff can understand the equipment running status in time, and also helps improving dispatch and guaranteeing production. This article mainly researches the real time statistics of equipment running status based on energy management system, and describes the object and method, and introduces the benefit using the method.

Key words: energy management system, iron and steel enterprise, equipment running status, real time statistics

MES 标准化平台概述和应用前景

殷娟娟

（马钢自动化信息技术有限公司软件研发事业部，安徽马鞍山 243000）

摘　要： 本文参考 ISA、AMR、MESA 等国内外研究机构的研究成果，首先阐述了在实现马钢整体信息化过程中，马钢人研发的 MES 经过逐步发展，形成了标准化的 MES 平台；接着介绍了 MES 标准化平台的设计模式、技术路线、平台组成架构和标准化功能；最后结合国家"十三五"的智能制造规划，描绘了 MES 标准化平台发展趋势和广阔前景。

关键词： MES，标准化，平台，智能

MES Standardized Platform Overview and Prospects

Yin Juanjuan

(Ma Steel Automation & Information Technology Co., Ltd., Software R & D Division, Maanshan 243000, China)

Abstract: In this paper, the research results of ISA, AMR, MESA and other domestic and foreign research institutions, Firstly, MES developed by Ma steel group is developed step by step in the process of realizing the overall informationization of Ma steel, and then MES Standardized platform design pattern, technical route, platform composition structure and standardization function. Finally, it combined with the national "thirteen-five" intelligent manufacturing planning, depicting the MES standardization platform development trend and broad prospects.

Key words: MES, standardization, platform, intelligent

板形测量输出信号处理方法的研究与应用

刘佳伟，文继刚，王　勇，孙喜龙

（鞍钢股份有限公司，辽宁鞍山　114001）

摘　要： 深入研究了某厂 1780 冷连轧机板形控制系统的板形检测原理及控制方法。板形测量值作为板形控制系统的反馈值，它的准确性直接关系到板形控制的效果。本文通过对数据处理方法的研究，应用通道数确定、插值和滤波等方法对检测数据进行处理并成功应用于现场，取得明显的效果。

关键词： 板形检测，带材信号处理，插值，滤波

Shape of Measuring Output Signal Processing Method Research and Application

Liu Jiawei, Wen Jigang, Wang Yong, Sun Xilong

(Anshan Iron and Steel Company, AnShan 114001, China)

Abstract: A throughout investigation has been carried out on the flatness measurement system and flatness control system of a 1780 tandem cold mill. The shape of strip measured value as the shape of strip control system's feedback value, its accurate direct relation shape of strip control's effect. This paper through to the data processing method's research, using the channel number determined that methods and so on interpolation and filter carry on to the examination data process apply successfully in the practical, makes the tangible progress.

Key words: shape of strip examination, strip signal processing, interpolation, filter

基于电磁出钢系统中铁碳合金添加量的设计

史纯阳，王 强，朱晓伟，何 明，刘兴安，赫冀成

（东北大学材料电磁过程研究教育部重点实验室，辽宁沈阳 110819）

摘 要： 在电磁出钢系统中，铁碳合金的添加量会直接影响到出钢时间，并且不同钢种填加量具有一定的差异。针对上述问题，为缩短开浇时间，优化碳合金添加量，本文针对某钢厂 110t 钢包电磁出钢过程设计了一种将模糊逻辑和专家系统相结合数学模型的评估诊断方法。并在专家系统中设计了长、短周期自学习模型功能，提高铁碳合金添加量的控制精度。本文所介绍的设计和实现方法对电磁开浇专家系统的研发具有一定的指导意义。

关键词： 电磁开浇，开浇时间，碳合金添加量，电磁开浇专家系统

Abstract: During the electromagnetic automatic steel-teeming system, the addition amount of Fe-C will directly affect teeming time and different steel grades has different filling amount of Fe-C alloy. In order to shorten teeming time and optimize additions amount of Fe-C alloy, This article design a evaluation and diagnosis method for combining fuzzy logic with expert system mathematical model according to a 110t ladle electromagnetic automatic steel-teeming system. And also design the function of long period self -learning model and short period self- learning model for the additions amount of Fe-C alloy expert system to improve the precision for addition amount of Fe-C alloy. This paper describes the design and implementation methods for electromagnetic automatic steel-teeming expert system, which has the universal guiding significance.

Key words: electromagnetic automatic steel-teeming, teeming time, additions amount of Fe-C alloy, electromagnetic automatic steel-teeming expert system

热轧带钢轧后冷却实测温度智能滤波算法开发

李旭东[1]，王淑志[2]，龚 波[1]，刘子英[1]，张长利[1]，王凤琴[1]

（1. 首钢集团有限公司技术研究院，北京 100043；

2. 首钢股份公司迁安钢铁公司，河北迁安 064404）

摘　要： 针对热轧带钢实测卷取温度的滤波算法进行了系统研究，结合中位值滤波算法的数值分析结果，阐明其在低温卷取工艺条件下的滤波效果无法满足工艺需求。同时，开发出适于生产实际的实测温度智能滤波算法，并结合生产数据，对滤波算法参数进行了优化。应用效果表明，新开发的滤波算法能够满足不同工艺条件下对温度实测值可靠性的需求，为卷取温度的高精度、智能化、自适应控制提供了有利支撑。

关键词： 热轧带钢，轧后冷却，智能控制，卷取温度，滤波算法

Development of Intelligent Filtering Algorithm of Measured Temperature for Cooling Control System of HSM

Li Xudong[1], Wang Shuzhi[2], Gong Bo[1], Liu Ziying[1],
Zhang Changli[1], Wang Fengqin[1]

(1. Shougang Research Institute of Technology, Beijing 100043, China; 2. Shougang Qian'an Iron & Steel Company, Qian'an 064404, China)

Abstract: The filtering algorithm of measured coiling temperature of hot rolling strip steel was researched systematically. Combined with the numerical analysis results, the deficiencies of median average filter were illuminated under the process of low temperature coiling. Meanwhile, the intelligent algorithm for measured temperature was developed to meet the different requirement of process. Furthermore, the parameters of the intelligent algorithm were optimized based on the production data. The application results indicated that the new developed filtering algorithm could satisfy the demands of reliability of measured temperature under different conditions, which provided a great support for high-precision, intelligence, self-adaption control of the coiling temperature.

Key words: hot rolled strip, cooling after rolling, intelligent control, coiler temperature, filtering algorithm

西门子厚度自动控制系统的研究与改进

张喜榜[1]，王凤琴[1]，刘子英[1]，范建鑫[2]，王　伦[2]

（1. 首钢技术研究院过程所，北京　100043；

2. 河北首钢迁安钢铁有限责任公司热轧厂，河北迁安　064404）

摘　要： 厚度自动控制系统是实现热轧高精度轧制的重要手段。本文介绍了首钢2160热连轧西门子厚度控制模型，并结合生产过程中出现厚度控制异常的问题，综合运用GM-AGC、MN-AGC，利用西门子CFC程序进行自主开发，完成压力AGC分钢种分规格增益控制功能和监控AGC分层动态增益控制功能的开发。投入使用后，进一步提高了带钢厚度控制精度，取得了较好效果。

关键词： 厚度控制，压力AGC，监控AGC，AGC功能优化

Research and Improvement on Siemens Automatic Gauge Control

Zhang Xibang[1], Wang Fengqin[1], Liu Ziying[1], Fan Jianxin[2], Wang Lun[2]

(1. Process Technology Department, Shougang Research Institute of Technology, Beijing 100043, China;
2. Qian'an Iron and Steel Company of Shougang Steel Corporation, Qian'an 064404, China)

Abstract: The automatic gauge control system is an important method to realize the high precision hot rolling. In this paper, the SIEMENS thickness control model of Shougang 2160 hot-rolling line is introduced. Combined with the problem of abnormal thickness control in production process, GM-AGC and MN-AGC are used synthetically. The SIEMENS CFC program was used for independent development. The development of gain control for pressure AGC in terms of steel grade and specification and monitoring AGC with layered dynamic gain control function was finished. After application, the strip thickness control precision has been further improved to get the good control results.

Key words: thickness control, rolling force agc, monitor agc, agc function optimization

鞍钢 2150mm 生产线加热炉智能装钢的应用

车 兰

（鞍钢热轧带钢厂，辽宁鞍山 114000）

摘 要：概述了鞍钢 2150mm 生产线加热炉智能装钢的模型和功能，以及系统的开发和实现过程，形成了完整的加热炉装钢管理系统，适合在其它类似生产线上推广应用。

关键词：智能装钢，模型，轧机作业计划，跟踪

Intelligence Charging of Heating Furnace Applicated on 2150mm Production Lines of Ansteel

Che Lan

(Hot Strip Mill of Ansteel, Anshan 114000, China)

Abstract: Have summed up the intelligence charging model and function of heating furnace on 2150mm production lines of Anshan Iron and Steel Plant, and systematic development and course of realizing, have formed integrated management system of intelligence charging, suitable for popularizing and applying on other similar production lines.

Key words: intelligence charging, model, rolling plan, track

高压变频器常见故障分析处理

张　伟，陈　欣，张　东，段　利

（宁波钢铁有限公司，浙江宁波　315807）

摘　要： 介绍了主排变频系统的故障分析处理。通过高压变频器的工作及节能原理并对变频器的故障现象、故障原因、处理措施的描述。总结同步电机变频实际应用经验，保证高压变频器长周期稳定运行，有效发挥节能的作用，节能效果显著。

关键词： 高压变频器，变频器电流波动，变频调速节能

Common Failure Analysis of High Voltage Inverter

Zhang Wei, Chen Xin, Zhang Dong, Duan Li

Abstract: This paper introduces the failure analysis of the main row frequency conversion system. Through the work and energy saving principle of high voltage inverter, the fault phenomenon, fault cause and treatment measures of inverter are described. In this paper, the practical experience of synchronous motor frequency conversion is summarized to ensure the stable operation of the high voltage inverter, and the energy saving effect can be effectively played.

Key words: high voltage inverter, inverter current fluctuation, variable frequency regulation energy saving

论单件小批量机械加工企业信息化建设

田　煜

（本钢信息自动化公司，辽宁本溪　117000）

摘　要： 主要阐述以订制产品为主，单件小批量生产的机加企业，如何通过信息化建设改善企业的生产经营环境，通过信息化手段将离散制造的各个环节有机的联系起来，使管理更清晰更流畅。对机械加工行业整体信息化以及 ERP 层面的简单概述，着重通过销售系统、生产系统、采购系统的对接过程，阐述如何通过信息化手段实现相关业务的联动，反映企业 ERP 系统如何在不同系统间建立整体管控的核心。

关键词： ERP，销售管理，生产管理，采购管理

The Information Construction of the Single Small Batch Machining Enterprise

Tian Yu

(Benxi Steel Information & Automation Co.,Ltd., Information Development Department, Benxi 117000, China)

Abstract: This paper mainly describes the production of customized products based, single piece small batch production of enterprises, how to improve the production and operation of enterprises through information technology, through the use of information technology will be integrated into the various links of discrete manufacturing, so that the management is more clear and more smooth. In this paper, the overall level of ERP management is outlined, focusing on the sales system, production system, procurement system docking process, to explain the relevant business implementation approach.

Key words: ERP, sales management, production management, procurement management

铁水智能扒渣系统在梅山炼钢厂的应用

郭海滨

（宝钢梅山炼钢厂，江苏南京 210039）

摘　要：介绍了铁水智能扒渣系统在宝钢集团梅山钢铁公司第二炼钢厂脱硫扒渣系统中的应用，介绍了系统的结构、系统的工作原理、主要控制对象及实现控制的方法。该系统运行稳定可靠，提高了工作效率，自动化程度和经济效益。

关键词：脱硫扒渣，摄像系统，信息处理系统，分析软件

Abstract: The present paper introduces the PCS7 in the desulphurisation system in Meishan Steelmaking Plant，involving the structure and function of the system and control objects as well as how to control it .The system ,reliable and stable ,can improve the labouring efficiency ,automation level and economic benefits.

Key words: desulphurization and slag skimming, camera system, information processing system, analysis software

以合同为中心、工程为主线的中小企业信息化系统建设的研究和应用

赵路征

（本溪钢铁（集团）信息自动化有限责任公司，辽宁本溪 117021）

摘　要：为了规范本钢信息自动化公司的业务流程和提高管理水平，设计研发以合同为中心、工程为主线的中小企业信息化系统。信息系统与实际业务相结合，以销售合同为中心，辅以工程为主线，从自动化公司的管理实际出发，以满足应用为前提，对销售、工程、采购、库房、财务等信息统一管理，形成闭环工作流，通过图形展示工作进度和业务数据。系统用户具有自己的角色和权限，按照级别和权限进行相关的业务操作和信息浏览。系统采用 B/S 模式，主要采用 C#语言和当前比较流行的前台技术，较高版本的操作系统和数据库。信息化系统的应用使得自动化公司实现了管理流程之间、部门之间环环相扣，消灭了信息孤岛，实现了数据共享，达到透明、实时和网络化管理，操作简便，提高了工作效率。

关键词：合同，工程，管理

Research and Application of Information System Construction of Small and Medium Sized Enterprises Based on Contract as the Center and Engineering as Main Line

Zhao Luzheng

Abstract: In order to standardize the business process and improve the management level of Benxi Iron and Steel information automation company, the design and development of engineering information systems for SMEs. The actual business information system and combined with the sales contract as the center, with the project as the main line, from the automation company's management practice, to meet the practical needs of the premise, the sales, engineering, purchasing, warehouse, financial information management, to form a closed loop workflow, display work and business data through graphics. System users have their own roles and privileges, according to the level and authority of the relevant business operations and information browsing. The system adopts B/S mode, mainly adopts C# language and current popular foreground technology, higher edition operating system and database. Interlocking between the application of information system makes the automation between the company and realize the management process of departments, eliminate the information island, realize data sharing, achieve transparent, real-time and network management, simple operation, improve work efficiency.

Key words: contract, engineering, administration

热连轧管轧机主传动动态速降研究

周　涛

（中冶赛迪工程技术股份有限公司，重庆　400013）

摘　要：分析了热连轧管轧机主传动咬钢冲击动态速降的动力学过程，提出了两种改善动态速降的方法，分别是冲击速降阶跃切换补偿和冲击速降微斜坡切换补偿，通过实际工程应用对比，发现冲击速降微斜坡切换补偿策略简单可行，可有效减小动态速降影响并保证速度切换时电流输出的稳定，补偿效果良好。

关键词：热连轧管，动态速降，补偿

Research on Dynamic Speed Drop of Main Drive in Pipe Mill

Zhou Tao

(CISDI Engineering Co.,Ltd., Chongqing 400013, China)

Abstract: Dynamic speed drop when biting impact load is added in pipe mill is described, two ways to improve dynamic speed drop are presented, these method can be described as below: Step Switch Impact Setpoint Compensation (SSISC) and Micro-ramp Switch Impact Setpoint Compensation (MRSISC). The result of SSISC and MRSISC is compared in real project. MRSISC is simple and feasible，moreover，is better to reduce speed drop and keep current stable when switching.

Key words: pipe mill, dynamic speed drop, compensation

烧结配料闭环控制系统开发与应用

秦雪刚

（北京首钢自动化信息技术有限公司自动化研究所，北京　100041）

摘　要： 烧结过程燃料及灰石配比对烧结矿的质量影响很大，尤其是燃料配比的变化将改变烧结过程的气氛性质和温度水平，直接影响烧结矿的成分和强度。传统烧结配料系统具有计算精度差、调整滞后、受人为因素干扰大等缺陷。本系统从精准控制理念出发，运用先进算法，实现了对原料批次及成分的动态追踪，运用"前馈+后馈"的控制方式，制定符合生产实际的碱度、含碳自动调整方法及规则，实现了烧结碱度、含碳的自动调整和烧结配料过程的闭环控制。烧结配料闭环控制系统的成功开发，有效的解决了传统操作的种种缺陷，稳定了主机参数，各项经济技术指标稳步提升。

关键词： 烧结配料，闭环，控制

Development and Application of Closed - loop Control System for Sinter Mix Blending

Qin Xuegang

(Beijing Shougang Automation Information Technology Co.,Ltd., Institute Automation, Beijing 100041, China)

Abstract: The ratio of fuel and limestone in the sintering process has a great influence on the quality of the sinter, especially the change of the fuel ratio will change the atmosphere properties and temperature level of the sintering process, which directly affect the composition and strength of the sinter. The traditional sinter mix blending system has the characteristics of poor precision, lag of adjustment and large interference by human factors. The system from the precise control concept, the use of advanced algorithms to achieve the raw material batches and components of the dynamic tracking, the use of "feed forward + feedforward" control, to develop in line with the actual production of alkalinity, carbon automatic adjustment method and rules to achieve the sintering alkalinity, carbon-containing automatic adjustment and sintering process of closed-loop control. The successful development of the closed-loop control system of sinter mix blending effectively solved the defects of the traditional operation, stabilized the host parameters, and steadily improved the economic and technical indexes.

Key words: sintering ingredients, closed loop, control

WinCC 在温轧实验轧机中的应用

甄立东，李建平，牛文勇，孙　涛

（轧制技术及连轧自动化国家重点实验室，辽宁沈阳　110819）

摘　要： 温轧实验轧机是一种新型实验装备，可以对金属板带施加张力并进行在线加热的温轧工艺实验。WinCC

是西门子的监控软件，WinCC 通过与基础自动化 PLC 和过程机通讯，完成数据采集、数据记录和实验操作。

关键词：温轧实验机，WinCC 软件，实时监控，HMI，PLC

Application of WinCC in Experimental Warm Rolling Mill

Zhen Lidong, Li Jianping, Niu Wenyong, Sun Tao

(The State Key Laboratory of Rolling Technology and Automation, Shenyang 110819, China)

Abstract: The experimental warm rolling mill was a new type of experimental equipment. The workpiece could be rolled with tensions exerted at its two ends, and be heated online.Siemens WinCC monitoring software, WinCC can complete data acquisition, data logging and experimental operation by automation PLC and process computer communication.

Key words: experimental warm rolling mill, WinCC, real time monitoring function, HMI, PLC

基于钢铁自动化设备故障诊断的专家知识库系统的研究与设计

刘富春，管锋利

（北京首钢自动化信息技术有限公司首迁运行事业部，河北迁安　064400）

摘　要： 基于故障诊断的规则配置专家知识库故障诊断方法是一种能以人类专家级水平进行故障诊断的计算机程序，它利用人工智能技术、现代设备诊断技术、信息传感与通信技术等，摆脱了传统的诊断技术对数学模型的强依赖性，将事先输入的经验知识以某种或几种方法表示出来，再应用不同的推理技术及诊断策略结合现场来的数据及事实进行诊断。在电子设备维护保障中应用基于规则的专家知识库故障诊断软件，可以降低故障检测、故障诊断的难度，提高系统的可靠性、有效性，保障设备发挥最大设计能力，延长使用寿命，降低寿命周期费用；通过检测监视、故障统计分析及性能评估等，为设备优化设计提供大量数据和信息以及经验。

关键词：专家系统，规则推理，模糊决策，故障诊断，交互式知识获取

Universal Fault and Diagnosis Method of Expert KB on Regulation

Liu Fuchun, Guan Fengli

(Shougang Automatic Information Technology Co.,Ltd., Qian'an 064400, China)

Abstract: In order to overcome contradiction of practicality and generality for fault diagnosis expert system , the design of general fault diagnosis method based on user and regular characterization is presented af ter the expert system is int roduced concisely. Rule reasoning and fuzzy decision algorithm are merged, and the step fault diagnosis method is accomplished. For different object , when necessary knowledge properties former is given, a special fault diagnosis method on regulation will be automatically produced , and the diagnosis result will be given.

Key words: expert system, rule reasoning, fuzzy decision, fault diagnosis

基于 Hybrid App 的企业信息化系统移动应用的研究与实现

刘富春

（北京首钢自动化信息技术有限公司首迁运行事业部，河北迁安　064400）

摘　要：随着智能终端的普及和企业内部信息化程度的完善，企业级应用与移动终端的对接，企业系统移动信息化已发展为一种趋势。随时随地把握处理工作信息，打破空间地域限制，企业信息综合管理移动应用的建立将作为一种重要的行业应用和管理应用。

基于 Hybrid App 的企业信息化系统移动应用的研究与实现是在企业现有的办公系统的基础上，整合公司信息基础资源、设备管理系统的权限管理和基础数据，面向移动办公和各系统的移动化需求，通过对现有各系统的无缝集成和移动开发，实现移动门户、移动办公、设备查询、知识库查询、信息发布、领导审批、员工交流等多种功能，实现数据管理、业务处理的移动化。

关键词：Hybrid App，移动设备，移动应用，企业信息化，办公管理

Research and Implementation of Mobile Application of Enterprise Information System Based on Hybrid App

Liu Fuchun

(Shouqian Division, Beijing Shougang Automation & Information Technology Co., Ltd., Qian'an 064400, China)

Abstract: With the popularization of intellective terminal and completion of internal enterprise informatization extent as well as the connection between enterprise application and mobile terminal. To master and handle working information anytime and anywhere, to break the space and region limit, the establishment of enterprise's general management application of operation and maintenance will be deemed as an important industrial application and management application.

Shougang Qiangang's operation and maintenance general management mobile application system is based on Android which is developed from the existing company office system. to integrate the company's operation and maintenance basic resource management system and equipment management system auothority management as well as basic data, to realize multi-functions of mobile information-exchange platform、daily officing、incident inquery、knowledge base inquery、information publishment、leaders' approval、staffs'communication, etc.

Key words: Hybrid App, mobile equipment, mobile application, enterprise informatization, operation and maintenance management

基于数据驱动的加热炉钢温预报模型

杨英华[1]，杨　茹[2]，石　翔[1]，秦树凯[1]

（1. 东北大学信息科学与工程学院，辽宁沈阳　110819；

2. 沈阳鼓风机集团股份有限公司，辽宁沈阳　110869）

摘　要：为实现钢坯出炉温度精准检测与控制的目的，采用基于数据驱动的研究方法建立加热炉钢坯出炉温度预报模型。首先分析加热炉过程工艺，由于单个钢坯彼此间的被加热过程的差异性，采用间歇过程分析方法进行研究。鉴于钢坯在炉内生产节奏不同导致各单体钢坯加热时间不等长的情况，运用改进的动态时间规整（DTW）方法将长度不等的数据进行等长处理，该方法能够最大限度的保持原始数据的分布。最后采用多向偏最小二乘（MPLS）方法将处理后的数据展开成二维形式，并建立钢坯出炉温度的回归模型。基于上述方法对加热炉钢坯出炉温度进行预报，预报误差能够满足工业应用的精度要求，同时也为钢铁工业智能化研究提供了支持。

关键词：钢温预报，加热炉，数据驱动，多向偏最小二乘法，动态时间规整

Billet Temperature Predication Model of Reheating Furnace Based on Data-driven

Yang Yinghua[1], Yang Ru[2], Shi Xiang[1], Qin Shukai[1]

(1. College of Information Science and Engineering, Northeastern University, Shenyang 110819, China;
2. Shenyang Blower Works Group Corporation, Shenyang 110869, China)

Abstract: In order to measure and control billet outlet temperature for reheating furnace precisely, a prediction model can be established based on data-driven method. The process engineering of reheating furnace industrial processes is analyzed at first, the heating process is defined as the batch process because of the difference among different billets. Considering that the variational production rhythm, data length are difficult to keep consistent among different batches. The dynamic time warping (DTW) method is used to process uneven-length data, which can keep the original data distribution. The multi-way partial least squares (MPLS) is employed to unfold three-way data matrix and build billet steel outlet temperature prediction model. Temperature prediction of billet outlet temperature based on above method insures that the predicting error can satisfy the demands of industrial application, which also provides a support for intelligent development of steel industry.

Key words: prediction of billet temperature, reheating furnace, data-driven, multi-way kernel partial least squares, dynamic time warping

钢铁行业 MES 系统进程管理模块分析

刘毅斌

（江西新余钢铁集团有限公司，江西新余　338001）

摘　要：本文通过总结本企业 MES 系统中的订单、物料的进程状态分析过程与做法，得出了清晰、合理、适用的进程状态体系。从介绍进程和状态命名规则开始，结合业务模型，完整介绍了这一体系在企业的应用效果。对类似的钢铁企业规划与设计适合企业自身特点的 MES 系统有一定的参考价值。对钢铁企业信息化系统服务提供商的方案优化也有一定的借鉴作用。

关键词：钢铁行业，MES 系统，订单进程，物料进程

Analysis on MES System Process Management Module for Iron and Steel Industry

Liu Yibin

(Xinyu Iron and Steel Co., Ltd., Group, Xinyu 338001, China)

Abstract: In this paper, through summarizing the analysis and practices of MES orders, material process state in enterprise, a clear, reasonable and applicable system of process status are obtained. Starting from the introduction of process and state naming rules, combined with business model, the application effect of this system are introduced completely in the enterprise, It has a certain reference value for planning and design in similar steel enterprise MES system suited to the characteristics of the enterprise itself. And it also has a certain reference function for service providers offering the optimization scheme of iron and steel enterprise information system.

Key words: iron and steel industry, MES system, order process, material process

冶金实验室大型分析仪器的绿色维修应用

王明利，崔全法，费书梅，夏碧峰，司洪珲

（北京首钢股份有限公司质量检验部，河北迁安　064404）

摘　要：绿色维修是综合考虑环境影响和资源利用效率的现代维修模式，通过先进的技术和工艺设备实现资源的高效利用和污染的最小排放，对企业的可持续发展和环境的保护都有着十分重要的意义。本文以冶金实验室最常用的 ARL4460 直读光谱仪和 ARL9900 荧光光谱仪电路板故障的芯片级检查和处理为例，介绍绿色维修技术在冶金实验室分析仪器维护中的应用。

关键词：绿色维修，冶金实验室，ARL4460，ARL9900，电路板，芯片级维修，仪器维护

Application of Green Maintenance for Large Analytical Instruments in Metallurgical Laboratory

Wang Mingli, Cui Quanfa, Fei Shumei, Xia Bifeng, Si Honghui

(Quality Inspection Department of Beijing Shougang Co., Ltd., Qian'an 064404, China)

Abstract: Green maintenance is a modern maintenance mode with comprehensive consideration of environmental impact and resource utilization efficiency. Through advanced technology and process equipment achieving efficient use of resources and minimum emissions of pollution. It is of great significance for the development of enterprises and the protection of the environment. In this paper, taking examples of circuit board fault chip level examination and treatment of ARL4460 spectrometer and ARL9900 X-ray fluorescence spectrometer which most commonly used in metallurgical laboratory, introduces the application of green maintenance technology in the maintenance of analytical instruments in metallurgical laboratories.

Key words: green maintenance, metallurgical laboratories, ARL4460, ARL9900, chip level maintenance, circuit board, maintenance of analytical instrument

首钢京唐 2 号热轧线传动系统技术改进与创新

刘海云，高文华，吕　梁，齐明光，肖胜亮，王新朋

（首钢京唐钢铁联合有限责任公司钢轧项目筹备组，河北唐山　063200）

摘　要： 首钢京唐热轧公司 2 号热轧生产线传动系统，引进国外直接转矩控制模式的变频调速系统，直接转矩控制模式为目前交流变频调速技术领域的最新技术，具有动态响应快等优势，但在我公司的应用实践中也发生了很多问题。针对系统运行后逐步暴露出的各种问题，在对这一新技术进行消化吸收的基础上，对系统硬件构成和软件参数进行了多项改造与创新。

关键词： 热轧，变频调速，直接转矩控制，开关频率，直流母线过电压

Shougang Jingtang United Iron & Steel Co., Ltd., No. 2 Hot Rolling Mill Drive System Technology Updated and Innovated

Liu Haiyun, Gao Wenhua, Lv Liang, Qi Mingguang,

Xiao Shengliang, Wang Xinpeng

(Shougang Jingtang United Iron & Steel Co., Ltd., Tangshan 063200, China)

Abstract: Shougang Jingtang United Iron & Steel Co., Ltd., No. 2 Hot Rolling Mill drive systems are imported foreign AC inverter of direct torque control mode. It is the newest technology in the AC inverter area. And it has the advantage of quickly dynamic response for this direct torque control mode. But it has been taken place many problems in the apply practice in our company. Against these problems Wemade many updating and innovating on hardware and software based on digestive and absorption of this new technology.

Key words: hot rolling, frequency control, direct torque control, switching frequency, DC over voltage

轧辊管理模块设计与应用

王志军，吴有生

（中冶南方工程技术有限公司技术研究院，湖北武汉　430223）

摘　要： 轧辊管理模块是轧机过程控制系统的重要组成部分。轧辊管理水平直接影响产品质量与生产成本。本文设计了轧辊管理模块的各项功能，包括常数管理、辊形管理、参数管理、信息监视与管理、报表；同时设计了轧辊管理模块与其他系统的通讯接口。基于本设计的轧辊管理模块应用于多个生产现场，应用实践表明，轧辊管理模块能提高生产机组的管理水平，降低管理成本，对控制产品板形，预防生产事故起到积极作用。

关键词： 轧辊管理模块，过程控制系统，通讯接口，辊形

Design and Application of Roller Management Module

Wang Zhijun, Wu Yousheng

(R&D Institute, WISDRI Engineering & Research Incorporation Limited, Wuhan 430223, China)

Abstract: The roller management module is an important part of the mill process control system. Roller management level directly affects product quality and production costs. This paper designs the functions of the roller management module, including constant management, roller shape management, parameter management, information monitoring and management, and report. At the same time, the communication interface between roller management module and other systems is designed. The roller management module designed by this paper has been applied to many production sites, application practice shows that the roller management module can improve the management level of production units, reduce management costs, and play a positive role in controlling the product shape, preventing production accidents.

Key words: roller management module, process control system, communication interface, roller shape

求解铝液配载问题的混合整数规划模型

汪恭书，白校源

（东北大学工业与系统工程研究所，辽宁沈阳 110819）

摘 要： 针对铝生产过程中的电解槽调配和出铝调度问题，将配铝计划和天车调度进行集成决策，以最小化天车行驶路程和抬包间化验元素含量差异为目标建立了混合整数规划模型。对以往文献给出算例进行测试，结果显示本文建立的模型能够获得更短天车行驶路程，明显优于文献提出的遗传算法。进一步考虑电解车间的双通廊布局情形，建立了新型混合整数规划模型，突破单通廊布局的限制，具有更广泛的扩展性。

关键词： 铝工业，配铝计划，天车调度，混合整数规划模型

A Mixed Integer Programming Model for Integrated Molten Aluminum Blending and Crane Scheduling Problem

Wang Gongshu, Bai Xiaoyuan

(Institute of Industrial & Systems Engineering, Northeastern University, Shenyang 110819, China)

Abstract: To solve the problem of electrolyzer blending and aluminum tapping in the electrolytic aluminum production process, we develop a mixed integer programming model in which the decisions of molten aluminum blending and crane scheduling are integrated. The objective is to minimize travelling distance of the crane used to lift ladle and diversity of impurity elements in different ladles. We test our model on instances collected from an existing article, and the computational results show that the model can get shorter travelling distance of the crane and outperforms the genetic algorithm proposed in the existing article. Moreover, we consider the double corridor layout case of the electrolytic workshop based on which a novel mixed integer programming model is developed. The new model breakthroughs the limitation of the single corridor layout, and can be extended to more generalized cases.

Key words: aluminum industry, molten aluminum blending, crane scheduling, mixed integer programmingmodel

连铸机大包下渣检测智能机器人的设计

汪　洋，傅爱玲，占时光

（宝武集团武钢有限计控公司，湖北武汉　430080）

摘　要：针对目前国内钢厂使用的下渣检测产品存在的不足，设计了一套测量可靠性高、信号稳定、且安装和维护成本低、自动化程度高的大包下渣检测智能机器人。该机器人检测装置采用电磁原理进行渣信号检测，测量方式是在大包水口处测量，机器人机械动作采用电机伺服控制，机器人的工作方式根据生产工艺设定。

关键词：下渣检测，电磁原理，机器人，中央控制单元

Abstract: Aiming at the problems existing in the current domestic steel mills using slag detection products, design a set of measure robot has high reliability, stable signal, and the installation and low maintenance cost, high degree of automation of large slag detection products. Electromagnetic principle of signal detection by the design of the product, the measurement is measured at gate ladle, the mode of operation with hydraulic drive, robot manipulator uses motor servo control. The robot's working mode is set according to the production process.

Key words: slag detection, electromagnetic principle, robot, the central controller

宝武武新自动下料 PLC-ERP 计量控制系统

冉光超，胡志群，吴　超，冯　喆

（宝武集团武钢有限计控公司，湖北武汉　430080）

摘　要：该控制系统由武宝计控公司自主研发，实现了现场操作室司机自主下料，中控室集中监控，ERP 通过主干网获取相关计划信息。下料控制方式有远程自动下料、中控室控制箱手动和中控室 HMI 手动控制，可以在 HMI 上实现三种控制方式的切换。PLC、监控和 ERP 可实时数据交换和控制。在下料过程中，PLC 和 ERP 数据传输不可中断，以确保相关步序不被锁定。其运行稳定、技术先进，在宝武武新得到广泛应用，值得推广。

关键词：PLC，计量，自动控制，数据

Abstract: The control system is designed independently by instrument and control company of WISCO, which can be loaded by driver in operator room and monitored in the central control room, and ERP receives the relevant plan information through the backbone network. There are three kinds of loading control methods: remote automatic loading, loading through control box in the control room manually and loading on HMI in the control room manually, you can achieve to switch between three kinds of control mode on the HMI. PLC, monitoring and ERP can be exchanged and controlled data in real-time. In the process of controlling, PLC and ERP data transmission cannot be interrupted, otherwise the relevant steps will be locked. It is controlled stably with advanced technology and widely used in it is in Baowu, it is worth promoting.

Key words: PLC, metering, automatic control, data

煤压机防喘控制技术研究

汪文婷

（武钢计控公司，湖北武汉　430080）

摘　要：本文主要介绍煤压机喘振的现象，分析了煤压机喘振的原因及危害，并对煤压机防喘控制系统的控制原理、喘振曲线及 PLC 控制系统等进行研究，解决了煤压机的喘振问题，对提高整个发电机组运行的质量和效率有重要意义，在同类型机组中有较大的推广应用价值。

关键词：离心式压缩机，喘振，喘振曲线，防喘控制，控制系统

Research of Anti-surge Control Technology for Gas Compressor

Wang Wenting

(Wugang Instrument and Control Company, Wuhan 430080, China)

Abstract: This paper mainly introduces the phenomenon of surge in coal press, analyzes the causes and hazards of gas compressor surge, and studies the control principle, surge curve and PLC control system of gas compressor antisurge control system. Solving the gas compressor surge problem that to improve the quality of the generator operation quality and efficiency is of great significance , it has a greater value in the promotion and application in the similar unit.

Key words: centrifugal compressor, surge, surge curve, antisurge control, control system

热连轧出钢节奏优化控制的应用

熊　创[1]，宗胜悦[2]，张　飞[2]，梁　勤[3]

（1. 北京科技大学设计研究院有限公司，北京　100083；2. 北京科技大学高效轧制国家工程研究中心，北京　100083；3. 广西柳州钢铁股份有限公司热轧厂，广西柳州　545000）

摘　要：基于某带钢热连轧生产线加热炉出钢节奏方法传统，已成为其影响生产线生产效率的关键瓶颈，结合其生产现状和需求，依据加热炉和粗轧区域的工艺过程和控制系统框架，提出出钢节奏的控制优化解决方案，利用 PLC 对其方案成功实现和应用，实现了区域间实时数据交换，提高了生产线节奏控制，降低人工劳动强度，具备同类生产线上使用和推广，并为工厂自动化和智能制造提供有益参考。

关键词：热连轧，粗轧机，节奏优化，PLC

Application of Optimum Control for Tapping Rhythm in Hot Strip Rolling

Xiong Chuang[1], Zong Shengyue[2], Zhang Fei[2], Liang Qin[3]

(1. Design and Research Institute of University of Science and Technology Beijing, Ltd., Beijing 100083, China;
2. National Engineering Research Center of high efficiency rolling, University of Science and Technology Beijing,
Beijing 100083, China; 3. Liuzhou Iron & Steel Limited by Share Ltd Hot Rolling Mill, Liuzhou, 545000, China)

Abstract: Method based on a production line for HSM furnace tapping rhythm tradition, has become the key to affect production line production efficiency bottleneck, combined with its production status and needs, according to the areas of the heating furnace and roughing process and control system framework, tapping rhythm control optimization solution is put forward, using PLC to its successful implementation and application, has realized the real-time data exchange between areas, improve the production rhythm control, reduce labor intensity of labor, have similar production line to use and promotion, and provide beneficial reference for factory automation and intelligent manufacturing.

Key words: hot strip mill, roughing mill, rhythm optimization, PLC

数据融合技术在热连轧温度测量值处理中的应用

宋　勇，荆丰伟，孙树萌

（北京科技大学工程技术研究院，北京　100083）

摘　要： 在热连轧的产品质量过程控制中，轧件温度是一个至关重要的参数，其测量值的准确性尤为重要。在实际生产中，轧件温度测量一般采用的是非接触式的红外高温计，轧制过程的恶劣测量环境有可能对其测量值造成干扰。目前主要采用硬件冗余方式来保证轧件温度测量值可靠性，但只能解决温度测量值异常的问题，而无法有效应对现场环境偶然因素对测量值精度所造成的影响。本文利用聚类分析得到历史轧件生产的典型工况和温度测量数据，通过对当前轧件工况与历史典型工况的匹配分析，根据大量历史工况的测量数据分布规律，采用数据融合技术来排除外界干扰因素对温度测量精度的影响。通过测试数据对所构建的融合模型及融合系数确定方法进行了试验，分析了该方法的有效性。

关键词： 热连轧，温度测量，数据融合，聚类分析

Application of Data Fusion Technology in Temperature Measurement of Hot Rolling

Song Yong, Jing Fengwei, Sun Shumeng

(Engineering Technology Research Institute of University of Science and
Technology of Beijing, Beijing 100083, China)

Abstract: In the process of product quality control of hot rolling, the rolling temperature is a critical parameter, and the accuracy of the measured value is particularly important. In actual production, the non-contact infrared pyrometer is usually used in the strip temperature measurement. Therefore, rolling process of the poor measurement environment may cause interference to its measured value. At present, hardware redundancy is used to ensure the reliability of the temperature measurement of the rolling parts, but only the problem of abnormal temperature measurement can only be solved, and the influence of the accidental factors on the accuracy of the measurement cannot be effectively dealt with. In this paper, the typical working conditions and temperature measurement data of historical rolling production are obtained by cluster analysis. Through the matching analysis of the current rolling conditions and typical working conditions, according to the distribution of the measured data in a large number of historical conditions, to eliminate the external interference factors on the temperature measurement accuracy. The fusion model and the fusion coefficient method are tested by the test data, and the effectiveness of the method is analyzed.

Key words: hot strip, temperature measurement, data fusion, cluster analysis

深度学习在中厚板表面缺陷识别中的应用

贺　笛，徐　科，孙金胜

（北京科技大学钢铁共性技术协同创新中心，北京　100083）

摘　要：中厚板表面覆盖着大量氧化铁皮，传统基于特征提取的表面缺陷识别方法产生大量的误报。本文将深度学习方法应用于中厚板表面缺陷识别，使识别率得到了很大提高。通过采用基于分层的训练、差异化学习率的训练方法，将卷积神经网络应用于小样本集合的中厚板表面缺陷识别中。通过将本文所提方法应用于某钢厂的中厚板表面缺陷检测，取得了90%的缺陷识别率，相对于传统方法有了较大的提升。

关键词：中厚板，表面检测，深度学习，卷积神经网络

Application of Deep Learning to Surface Defect Recognition of Medium and Heavy Plates

He Di, Xu Ke, Sun Jinshen

(Collaborative Innovation Center of Steel Technology, University of Science and Technology Beijing, Beijing 100083, China)

Abstract: Traditional surface defect recognition of medium and heavy plates based on feature extraction prone to a large number of false positives, as there are a lot of scales on the surface of medium and heavy plates. In this paper, the deep learning method was applied to the surface defect recognition of medium and heavy plates, and the recognition rare of surface defects was improved greatly. By using layer-wised training and different learning rates in different convolutional layer, the convolution neural network was applied to surface defect recognition of small sample sets of medium and heavy plates. The method achieved a recognition rate of 90% at the production line of medium and heavy plates, which was greatly improved compared with the traditional methods.

Key words: medium and heavy plates, surface inspection, deep learning, convolutional neural networks

基于物联网的吊车梁全生命周期智慧监测

李贵文，罗云东，陈　琳，陈融融，张　炯，姜文吉

（中国宝武集团上海金艺检测技术有限公司，上海　201900）

摘　要：文章主要围绕吊车梁全生命周期的智慧监测进行阐述，包括原有的吊车梁寿命后期的应力状态监测的实现、新吊车梁从设计计算及制造过程的在线监测介入，以及新吊车梁运行状态的远程智慧监测等。

从宝钢股份一炼钢 AB 跨吊车梁在 2015 年 1 月 31 日事故开始介绍，主要包括吊车梁事故抢险时实施的临时的应力状态在线监测、依据临时在线技术数据做出吊车梁进行更换、更换新吊车梁的材料选型、新吊车梁应力状态远程在线监测实现、现场数据采样及分析、吊车梁疲劳寿命分析预测，结合吊车梁关键部位应力谱和影响吊车梁疲劳寿命的影响因素，评估整个厂房内吊车梁的群体疲劳状态，为厂房内吊车梁的维护和管理提供依据。

希望通过文章的详尽记录，展现大型钢结构吊车梁全生命周期远程在线监测技术研究与应用实践，为今后类似项目的实施和该项技术的应用推广奠定一些基础。

关键词： 物联网，吊车梁，全生命，监测

提高冷连轧轧制力设定精度的方法研究

赵文姣，闫洪伟，裴勇梅，温玉莲，温　杰

（首钢京唐钢铁联合有限责任公司冷轧部，河北唐山　063200）

摘　要： 轧制力计算模型是冷轧过程控制计算的核心模型，其精度对其他参数的设定至关重要。经对轧制力设定模型进行研究，提高轧制力设定精度的方法主要包括轧制力不断自学习进行提高，优化变形抗力参数和钢种轧制力系数补偿进行提高。本文主要对这三种优化方法进行了详细研究，实际现场应用表明通过这三种方式优化后的轧制力设定精度完全能够满足现场越来越高的速度和精度需求。

关键词： 轧制力，设定精度，自学习，变形抗力，系数补偿

Research on Improving the Setting Precision of Rolling Force of Cold Rolling

Zhao Wenjiao, Yan Hongwei, Pei Yongmei, Wen Yulian, Wen Jie

(Cold Metal Dept., Shougang Jingtang United Iron and Steel Company, Tangshan 063200, China)

Abstract: The calculation model of rolling force is the key model of cold rolling process control calculation, and its precision is very important to the setting of other parameters. The method of improving the rolling force setting accuracy by studying the rolling force setting model mainly includes the continuous adaptive of the rolling force, the optimization of the deformation resistance parameters and the compensation of steel rolling force coefficient. In this paper, the three optimization methods are studied in detail. The actual field application shows that the precision of rolling force after these three methods can meet the requirements of increasing speed and precision.

Key words: rolling force, setting precision, adaptive, deformation resistance, coefficient compensation

炼钢工艺精细化成本管理系统设计实现

张志良，吕培培

（北京首钢自动化信息技术有限公司，北京　100043）

摘 要：MES(Manufacturing Execution System)系统在炼钢行业的应用已十分普遍和成熟,但是仍有很多钢铁企业需要耗费大量的人力来完成基于实际生产层面的成本核算和统计分析工作。炼钢工艺精细化成本管理系统充分结合了Extjs 框架的优越性，利用 Oracle 数据库的接口技术获取实时数据，以及多维报表适应多种用户需求的特性。该系统的设计实现不仅能实时查询工艺物料消耗和成本消耗，还能够站在更高层面上查看整个炼钢工艺的实际成本情况，进行多角度成本分析，有效解决了炼钢工艺生产成本的核算问题，保证数据的实时性和准确性，减少人为工作量，便于实时监控，为企业提高经济效益。

关键词：成本管理系统，MES，Extjs 框架,吨钢成本

Design and Implementation of Refining Cost Management System for Steelmaking Process

Zhang Zhiliang, Lv Peipei

Abstract: The application of MES (Manufacturing Execution System) in steelmaking industry is very general and matural, however, there are still many iron enterprises need to spend a lot of manpower to complete the cost accounting and statistical analysis based on the actual production level. Refining cost management system for steelmaking process combines the advantages of the Extjs framework, and uses the interface technology of Oracle database to obtain real-time data and multi-dimensional reports to adapt to the characteristics of a variety of user needs.This system can help not only query the process material consumption and cost in real time,but also look over the actual cost of the whole steelmaking process at a higher level,and make multi angel cost analysis. The system can effectively solve the accounting problem of steelmaking process cost, ensure the real-time and accuracy of data, reduce human workload, facilitate real-time monitoring and improve economic efficiency for enterprises.

Key words: cost management system, MES, Extjs framework, unit steel cost

唐钢智能化能源系统数据采集的升级改造与应用

郑熙华

（河钢集团唐山钢铁公司，河北唐山　063000）

摘 要：工业以太网技术和先进的控制技术应用于钢铁能源数据采集系统中,通过与企业生产网络平台的对接，实现信息快速传递、共享、管理和应用。再者，通过预测和预警、数据仓库和数据挖掘等理论方法和技术对有关数据进行深入的加工处理及分析，以提高监控数据的应用水平。

关键词：能源，数据采集，以太网，接口

The Data Collection Upgrade and Application of the Intelligent Energy System in Tangsteel

Zheng Xihua

(HBIS Group Tangsteel Company, Tangshan 063000, China)

Abstract: Industry Ethernet and advanced control technology are applied to the steel energy data collection system. The quick sending, sharing, management and application of information can be realized via connecting with the production network of companies. Besides, the related data are deeply processed and analyzed by the theories of forecast and early warning, data warehouse and data mining in order to improve the level of the application of monitoring data.
Key words: energy, data acquisition, ethernet, interface

基于 Android 的生产日报 APP 的设计与实现

汪　洋

（河钢集团宣钢公司，河北宣化　075100）

摘　要：在客户端与服务器间通过 WebService 进行通信，包含 Android 应用程序开发的完整过程，服务器端基于 C#语言，利用 WebService 实现接口的访问，Android 客户端通过 Ksoap2 调用 WebService，并以网格化管理为例，最终实现了 Android 客户端与服务器灵活的信息交互。以 Android 作为运行平台，将数据传输到 Android 客户端，进行解析，处理分析，提交上传等工作。最终实现了生产日报的预览功能。
关键词：WebService，Android，Ksoap2，移动办公，生产日报

Design and Implementation of Production Daily APP Based on Android

Wang Yang

(Hebei Iron and Steel Group in Xuanhua, Xuanhua 075100, China)

Abstract: Through the WebService in the communication between client and server, contains the complete process of Android application development, server based on the C# language interface using WebService access Android client through the Ksoap2 call WebService, and the grid management as an example, finally realizes the information interaction between Android client and server flexible. Android is used as the operating platform to transfer data to the Android client for analysis, processing, submission, uploading and so on. Finally, the preview function of production daily is realized.
Key words: WebService, Android, Ksoap2, mobile office, production daily

热轧炉卷轧机层流冷却系统改进

葛居娜[1]，尹丽琼[2]

（1. 昆钢技术中心，云南昆明　650302；2. 安宁公司板带厂，云南昆明　650302）

摘　要：本文介绍了安宁公司板带厂热轧炉卷轧机层流冷却控制系统硬件改造及模型的调整与优化。从硬件改造及模型优化方面出发，详细阐述了层流冷却控制系统存在的问题、产生的原因、改造的方案、模型的优化、改进及优化后的效果等。实际运行效果表明，该系统改进及优化后，不仅达到了预期的效果，提高了带钢卷取温度控制精度，保证了热轧带钢机械性能均匀性和产品综合质量，同时还降低了热轧的各项成本及维修费用，直接达到了"降本增

效、节能减排"的目的。

关键词：炉卷轧机，层流冷却，系统改造，模型调整与优化

Improvement of Laminar Flow Cooling System for Hot Rolling Mill

Ge Juna[1], Yin Liqiong[2]

(1. Kungang Technology Innovation Center, Technical Center Kunming 650302, China;
2. Quiet Company Board Factory, Kunming 650302, China)

Abstract: This paper introduces the hardware modification and optimization of laminar cooling control system of hot rolling mill of the company.From hardware modification and optimization model, expounds the problems existing in the laminar cooling control system, the reasons, the transformation of the scheme, optimization of the model, the effect after improvement and optimization, etc.Practical operation effect shows that the system after improvement and optimization, not only achieve the expected effect, improved the precision of strip coiling temperature control, to ensure the uniformity of the mechanical properties of hot rolling steel strip and comprehensive quality of products, but also reduces the cost of hot rolling and maintenance costs, directly to the "authors efficiency, energy conservation and emission reduction" goal.

Key words: furnace rolling mill, laminar cooling, system transformation, model adjustment and optimization

矿山选矿工艺 OPC 实时数据采集系统的研究

许业华

（本溪钢铁（集团）信息自动化有限责任公司信息化部，辽宁本溪　117000）

摘　要：实时数据采集系统要求汇集生产过程中大量的实时数据和信息，通过对大量实时数据的采集和在新型数据库技术的支持下进行存储、分析、提炼和发掘，为上层三级、四级系统的操作和管理人员提供所需的数据和结果、运行指导和决策依据。

关键词：OPC，OLE/COM，实时数据采集，C#

Research on OPC Real Time Data Acquisition System of Mine Beneficiation Technology

Xu Yehua

(Benxi Steel Information & Automation Co.,Ltd., Information Development Department, Benxi 117000, China)

Abstract: Real time data acquisition system requires a large amount of real-time data and information in the production process，Through the collection of a large number of real-time data and the support of the new database technology for storage, Analyze, refine, and excavate，Provide the required data and results, operation instructions and decision basis for the operation and management of the upper three and four level systems.

Key words: OPC, OLE/COM, real-time data acquisition, C#

制造执行系统（MES）在矿山选矿生产过程中的应用

史晓奇

（本溪钢铁（集团）信息自动化有限责任公司信息化部，辽宁本溪　117000）

摘　要： 目前，我国大部分矿山企业已经进行了不同程度的信息化建设，但限于实施的时间、规模、厂家等不同，存在部分信息孤岛，不能对企业的资源进行有效的整合。本文主要叙述了以本溪钢铁（集团）矿业有限责任公司歪头山铁矿厂马耳岭选矿车间为背景的制造执行系统（MES）的应用，并重点介绍了计划管理的设计思路及实现过程。通过 MES 的应用，消除信息孤岛，打通信息壁垒，最终形成横向贯通、纵向集成的制造执行系统，帮助企业实现生产管理的信息资源化、传输网络化、管理科学化的现代企业目标。

关键词： MES，矿山，选矿生产，计划管理

Application of Integrated Production Management and Control System (MES) in Mine Beneficiation Process

Shi XiaoQi

(Benxi Steel Information & Automation Co.,Ltd., Benxi, 117000, China)

Abstract: At present, most of the mining enterprises in China have carried out information construction in different degree, but limited to the implementation of the time scale, different manufacturers, there are some isolated information islands, and can not effectively integrate the resources of the enterprise. This paper mainly describes the Benxi iron and steel (Group) mining limited liability company Waitoushan Iron ore plant MaErLing dressing plant for integrated production control system (MES) application background, and introduces the design and Realization of project management process. Through the application of MES, eliminate isolated information islands, open information barriers, and ultimately the formation of lateral and vertical integration of the integrated production control system, help the enterprise to realize the production and management of information resources, transmission network, scientific management of modern enterprise target.

Key words: MES, mine, mineral processing production, plan management

宁钢焦化集中管控系统的整合与优化

李　慧，吴　成

（宁波钢铁有限公司，浙江宁波　315807）

摘　要： 论文研究中，主要是对以下模块的整合和编程，包括运煤管带机 PLC 系统模块、圆形料场 PLC 系统模块、自动配煤装置模块、备煤 PLC 系统模块、焦炉 DCS 系统模块、煤塔秤系统模块、自动放散点火装置模块等。在整合过程中，对焦化厂工艺性能的需求进行了分析，并利用 C/S 结构对于整合后的系统进行了设计，并实现了系统整合的功能。通过编程及网络连接，将各个模块最终通过中控室 HMI 画面进行监控，实现模块整合。

通过系统的整合，实现了焦化厂各个子系统的统一控制，达到了硬件、软件统一管理的目的，生产管理层面上实现了集中的管理。

关键词：大集中控制系统，整合，OPC，HMI

The Integration and Optimization of Coking Centralized Control System of Ningbo Steel

Li Hui, Wu Cheng

基于机器视觉技术的钢板表面质量检测系统

高　冰[1]，贾吉祥[1]，郭庆涛[1]，柴明亮[1]，费　静[1]，董咨雨[2]

（1. 鞍钢股份有限公司技术中心，辽宁鞍山　114009；

2. 哈尔滨工业大学计算机科学与技术学院，黑龙江哈尔滨　150001）

摘　要：要介绍了自主研发的可应用于冷轧钢板生产线上的基于机器视觉的钢板表面质量检测系统的硬件构成，包括了高速 CCD 相机的选择，光源的设计，系统照明方式的设计，嵌入式图像处理系统的设计以及分析如何基于千兆以太网实现高速线阵 CCD 相机的图像实时采集与传输。实验验证通过千兆以太网对采集的图像进行实时传输。该系统易于升级，可扩展到其他相关领域。

关键词：阵 CCD 相机，嵌入式处理器，千兆以太网

The System Based on Machine Vision in Cold Rolled Steel Sheet Surface Quality Inspection

Gao Bing[1], Jia Jixiang[1], Guo Qingtao[1], Chai Mingliang[1], Fei Jing[1], Dong Ziyu[2]

(Technology Center of Angang Steel Co.,Ltd., Anshan 114009, China)

Abstract: Briefly introduced the independent research can be applied on the cold rolled steel sheet production line surface quality detection, which is based on machine vision. It includes the design of system hardware, the design of the light, embedded image processing system design and analysis of how to realize high speed image acquisition and transmission based on gigabit Ethernet. This system is easy to upgrade, it also can spread to other related fields.

Key words: linear array CCD camera, embedded processor, gigabit Ethernet

浅析 IMS 厚度测量系统

刘嘉庆

（天津太钢天管不锈钢有限公司，天津　300450）

摘　要：本文主要介绍了 IMS 公司的厚度测量系统的控制过程及结构的组成，分析了厚度测量系统中各个元件的工作原理，主要包括了 X 射线、shutter、高压发生控制器、离子室，并且简单的描述了一些故障的处理方法。
关键词：X 射线，系统构成，离子室

The IMS Shallow Thickness Control System

Liu Jiaqing

(Tianjin TISCO&TPCO Stainless Steel Co., Ltd., Tianjin 300450, China)

Abstract: This paper mainly introduces the control process and the structure thickness measurement system of IMS company, analyzes the working principle of the various components of thickness measurement system, including X-ray, shutter, high pressure controller, ion chamber and a simple description of some methods to handle faults.
Key words: X-ray, System structure, Ion chamber

浅谈电子汽车衡防过载计量技术实例分析

田晋斌，马　欣，金小刚，田　栋

（陕西龙门钢铁有限责任公司，陕西韩城　715405）

摘　要：现阶段，随着我国工业产业的不断发展，电子汽车衡在工业生产中的应用越来越广泛。但是随着企业的发展，电子汽车衡的超吨位计量已经成为限制汽车衡应用的重要因素，本文对电子汽车衡在运行中承载力不足的问题进行技术研讨，通过增加秤体受力点提高汽车衡的承载力，有效地解决了电子汽车衡承载力不足的问题，降低了企业重建电子汽车衡的成本。
关键词：电子汽车衡，传感器，承载力

工业 4.0 背景下硅钢生产计划的设计

胡盛凯，韩震超

(安徽工业大学冶金工程学院，安徽马鞍山　243002)

摘　要：介绍了工业 4.0 产生的背景和工业 4.0 的主要内容，并点明了工业 4.0 的重要性。将工业 4.0 和工业 3.0 作比较，凸显工业 4.0 更为先进和智能化，介绍了工业 4.0 一特点——大规模定制生产，表明其科学性和合理性。根据大规模定制生产与硅钢生产工艺的特点，建立数学模型对炼钢—连铸批量计划的炉次计划和浇次计划进行设计。
关键词：工业 4.0，大规模定制，硅钢，数学模型

Design of the Production Plan of Silicon Steel in Industry 4.0

Hu Shengkai, Han Zhengchao

(Metallurgical Engineer School,Anhui University of Technology, Maanshan 243002,China)

Abstract: This paper introduces the background of industry 4.0 and the main contents of industry 4.0, and points out the importance of industry 4.0.Comparing with industry 4.0 and industry 3.0, the industry 4.0 is more advanced and intelligent, introducing the industrial 4.0 characteristic -- mass customization production, indicating its scientific nature and rationality.According to the characteristics of mass customization production and silicon steel production process, a mathematical model is established to design the furnace plan and casting plan of steelmaking and continuous casting batch plan.

Key words: industry 4.0, mass customization, silicon steel, mathematical model

矿山选矿工艺 MES 系统的研究

史晓奇

（本溪钢铁（集团）信息自动化有限责任公司信息化部，辽宁本溪　117000）

摘　要：本文主要叙述的是以本溪钢铁（集团）矿业有限责任公司歪头山铁矿厂马耳岭选矿车间为背景的矿山选矿工艺 MES 系统的设计与实现，主要包括逻辑模型的分析与建立和应用程序的设计与实现。本系统围绕破碎、磨矿、磁选、筛分、过滤等主要工艺环节，对生产、质量、设备运行、能源介质等信息统一管理，主要包括计划管理、生产实绩管理、质量管理、调度管理、设备运行管理、能源介质管理、跟踪管理等功能。

关键词：MES，矿山，选矿工艺，计划管理

Research on MES System of Mine Ore Dressing Process

Shi Xiaoqi

(Benxi Steel Information & Automation Co.,Ltd., Benxi 117000, China)

Abstract: This paper describes to design and implement the MES system of mine ore dressing process in the background of Maerling ore dressing workshop of Waitoushan iron ore plant of Benxi Iron & Steel (Group) Mining Co., Ltd.. It mainly includes the analysis and establishment of the logical model, and the design and implementation of the application. This system unified manage production, quality, equipment operation, energy medium and other information, around the crushing, grinding ore, magnetic separation, screening, filtering and so on main technological aspects. It mainly includes plan management, production performance management, quality management, scheduling management, equipment operation management, energy medium management, tracking management and other functions.

Key words: MES, mine, ore dressing process, plan management

软化水站 PLC 控制系统设计

尹成宇

（本溪钢铁(集团)信息自动化有限责任公司，辽宁本溪　117000）

摘　要： 通常把钙、镁离子含量高的水称为硬水。硬水对工业设备危害极大。降低水中钙、镁离子的含量，就叫做硬水的软化。本文对本钢供水厂软化水站的软化水 PLC 控制系统的应用进行介绍，分析了控制难点、解决方案及系统工艺特点。该系统自投入运行以来，取得了良好的控制效果。

关键词： 硬水，过滤器，钠床

基于流程的企业仓储配送管理信息中心解决方案

初海鹏，白如雪，万峻竹，吕　宁

（首钢京唐钢铁联合有限责任公司焦化作业部设备工程室，河北唐山　063200）

摘　要： 生产设备在企业运行的生产力构建过程中发挥着不可或缺的重要作用，备件材料等物资的合理储备更是设备维修工作的物质基础，做好材料备件的供应管理是提高设备投产率、完好率的必要保证。基于流程管控思维和依托信息技术的应用为企业仓储的现代化、规范化、集成化和智能化提供了良好的保障，有利于形成统一高效的仓储配送管理体系。

关键词： 备件材料管理，仓储，流程，信息化

Solution of Enterprise Storage and Distribution Management Information Center Based on Process

Chu Haipeng, Bai Ruxue, Wan Junzhu, Lv Ning

(Shougang Jingtang United Iron & Steel Co.,Ltd. Coking Operation department equipment Engineering Room, Tangshan 063200, China)

Abstract: Production equipment construction plays an important role in the process of enterprise operation in productivity, reasonable reserve spare parts materials is the material basis for equipment maintenance, spare parts supply management to material is needed to improve the operation rate and equipment intact rate guarantee. The application of process control thinking and relying on information technology based on the standard for enterprise warehouse modernization, provide good protection, integrated and intelligent, is conducive to the formation of a unified and efficient warehousing and distribution management system.

Key words: spare parts management, storage, technological process, promotion of information technology

基于 PLC400 的自动配煤系统改造浅析

王　宁，孙得维，敖轶男，丁　彪

（唐山首钢京唐西山焦化有限责任公司，河北唐山　063200）

摘　要： 京唐焦化自动配煤系统自 2008 年交付使用以来，由于固有的软件和硬件缺陷导致系统反应迟钝，重启严重，不能很好的满足生产需求，本文对自动配煤系统进行了相关改造升级。

关键词： 自动配煤，PLC400

7m 顶装焦炉余煤提升机自动控制系统改造

程　飞，胡中平，严　斌，袁本雄

（铜陵泰富特种材料有限公司，安徽铜陵　244000）

摘　要： 本文介绍了 7 m 顶装焦炉余煤提升机在铜陵泰富的应用情况，对余煤提升机存在的问题进行了详细的分析，采用 PLC 结合变频器编码器通讯技术对原控制系统进行改造，取得了良好效果，大大增强了余煤提升机设备在运行过程中的安全性能，提高了生产效率，实现了无人操作。

关键词： 余煤提升机，编码器，变频器，料位计

Improvement of Automatic Control System of 7 Meter Top Mounted Coke Oven Remaining Coal Hoist

Cheng Fei, Hu Zhongping, Yan Bin, Yuan Benxiong

(Tongling Pacific Special Materials Co., Ltd., Coking Plant, Tongling 244000, China)

Abstract: This paper introduces 7 meters more than the top charging coke oven coal lifting machine used in the Tongling Pacific, the remaining coal hoist problems are analyzed in detail, using PLC to transform the original control system with frequency conversion technology, and achieved good results, greatly enhanced the safety performance of the remaining coal hoist the equipment in the operation process, improve the production efficiency.Unmanned operation is realized.

Key words: coal mine hoist, encoder, frequency converter, material level meter

TCS 自动辊缝控制在日钢大 H 型钢应用

孙　岗

（日照钢铁有限公司电控处，山东日照　276806）

摘　要： 介绍日照钢铁股份有限公司万能精轧机 TCS 系统自动辊缝控制系统结构和分布及功能简介，TCS 系统控制原理，自动辊缝控制方式及选择，重点介绍系统标定-轧制线标定和辊缝标定。

关键词： TCS 系统，　HGC，AGC，线性编码器，系统标定

TCS Automation Gap Control in H-beam RiZhaosteel Application

Sun Gang

（Rizhao Iron and Steel Co., Ltd. Electric Control Office, Rizhao 276806, China）

Abstract: This paper profiles for RiZhaosteel Co., Ltd., Universal mill TCS system automatic gap control system structure and distribution，function introduction, TCS system control principle, the automatic gap control mode and choice, emphasis on the system calibration - rolling pass-line and the gap calibration.

Key words: TCS system, HGC, AGC, the linear encoder, system calibration

大 H 型轧线二级新增辊缝修改功能应用

孙　岗，刘贵超，官朋涛

（日照钢铁有限公司电控处，山东日照　276806）

摘　要： 为缩短轧制规格调整时间，避免人工操作失误，新增 TCS 离线修改、工作辊径离线修改和轧辊辊系离线修改功能，该功能基于 .NET 平台开发，利用 visual studio 2012 开发环境，用 C# 设计界面和后台控制，这三项功能实现提高轧制节奏，确保生产稳定。

关键词： TCS，工作辊径，轧辊辊系

H-Beam Level2 of Rolling Line New Rollgap Adjustion Function Application

Sun Gang, Liu Guichao, Guan Pengtao

(Electric Control Department, Rizhao Iron and steel Co., Ltd., Rizhao 276806, China)

Abstract: For reducing the time of rolling adjustment scheme and avoiding operation error, new TCS off-line modification, work roll diameter changes and roll offline modification function, the function is based on. NET platform development, using the visual studio 2012 development environment, using c # design and backcontrol, these three functions to improve the rolling rhythm, to ensure the production stable.

Key words: TCS, roll work diameter, rollerSet

GOR 二级服务器与 Excle 服务器的连接技术

张　丽，吴荣刚

（山东泰山钢铁集团有限公司，山东莱芜　271100）

摘　要： GOR（Gas Oxygen Refining）车间是不锈钢炼钢厂安全高效生产的重要环节。为改进 GOR 车间的管理水平，需要将存放在二级服务器中的生产数据连接到 Excle 服务器中。而传统操作需手动录入的方式，效率低、易出错。为解决这一问题我们采取了使用 ODBC 数据的转换方式，将二级服务器 Oracle 数据库中的数据快速准确的连接到 Excle 服务器 SQL 数据库中来。实际使用证明从二级服务器中直接提取数据，速度快、及时性强、数据准确，满足实际生产需要。

关键词： GOR 二级服务器，Excle 服务器，Oracle，SQL，ODBC

Research on the Connection Technology between the Secondary Server of GOR and the Excel Server

Zhang Li, Wu Ronggang

(Shandong Taishan Iron and Steel Group Co., Ltd., Laiwu 271100, China)

Abstract: Gas oxygen refining (GOR) workshop is an important process of the safe and high-efficiency production of the steelmaking plant. To improve the management of the GOR workshop, the production data stored in the secondary server of the GOR workshop need to be transport into the Excel server. The traditional method, handled by an operator manually, is inefficient and easy to make mistakes. To solve this problem, a data transport method based on ODBC is developed to transport the production data stored in Oracle database of the secondary server into the SQL database in Excel server. The application in practical system illustrates that the proposed method can obtain data from secondary server with high speed, on-time, and high accuracy, which meets the practical requirements.

Key words: the secondary server of the GOR, Excel server, Oracle, SQL, ODBC

钢铁行业工作用辐射温度计建标意义及所需标准器选型

毕宏图

（唐钢信息自动化部，河北唐山　063000）

摘　要： 工作用辐射温度计广泛应用于钢铁生产过程的温度控制，对生产过程的重要温度点进行监控，是保障产品质量的重要测量设备，为保证工作用辐射温度计提供的数据准确可靠，必须通过定期检定/校准的方式对其溯源。本文主要介绍建立工作用辐射温度计计量标准的意义，目前唐钢工作用辐射温度计数量及溯源情况及建立工作用辐射温度计计量标准的要求，主要检定设备黑体炉如何选择。

关键词： 钢铁黑体辐射源，辐射温度计，建标，黑体炉

宁波钢铁有限公司智能化物流管理的实践

马　琳

（宁波钢铁有限公司物流部，浙江宁波　315807）

摘　要： 介绍了宁波钢铁有限公司多年来实施自动化、信息化，探索物流管理智能化的工作实践，分析了针对钢铁行业所面临的市场形势，企业应如何借鉴"中国制造2025"，德国"工业4.0"等先进理念，通过推进实施物流管理智能化，建设智慧钢厂，提升企业市场竞争力和可持续发展能力。

关键词： 智能化，物流管理，中国制造2025，实践

Ningbo Iron& Steel Co., Ltd Intelligent Logistics Management Practice

Ma Lin

(Logistics Department, Ningbo Iron& Steel Co., Ltd., Ningbo 315807, China)

Abstract: The present paper take the Ningbo steel and iron Limited company as the object of study, in has the foundation which to the Iron and steel enterprise physical distribution management correlation theories grasps, introduced the exploring of logistics management that Ningbo Iron&steel Co., Ltd., have practiced on automation, informationization and intellectulalization for many years, analyzed how should the enterprise use "made in China 2025", Germany "industrial 4.0" and other advanced ideas to promote the logistics management intelligently and construct the factory wisely. Accordingly, elevate the market competitiveness and sustainable development ability for the enterprise.

Key words: intelligentize, logistics mangement, made in china 2025 strategy, practice

19 冶金技术经济

炼铁与原料

炼钢与连铸

轧制与热处理

表面与涂镀

金属材料深加工

钢铁材料

汽车钢

海洋工程用钢

轴承钢

电工钢

粉末冶金

非晶合金

高温合金

耐火材料

能源与环保

分析检测

冶金设备与工程技术

冶金自动化与智能管控

★ 冶金技术经济

钢铁企业物流财务共享 2.0 价值创造服务体系建设

侯海云，王　静

（鞍山钢铁集团有限公司，辽宁鞍山　114000）

摘　要：鞍钢股份公司以提供用户最佳服务体验和挖掘价值创造潜力为目标，以稳健的风险控制为核心，以先进的云计算、大数据、物联网等智能化信息技术为支撑，依托互联网，实施物流财务共享 2.0 价值创造服务体系建设，整合财务资源、规范财务运作、优化业务流程、实施创新驱动、提供专业服务，提升信息整合力、价值洞察力、决策支持力、风险管控力四项财务核心能力，走出一条"定位准、服务好、效率高、流程简、绩效优"的创新路，这种系统集成创新的技术和方法也可以为我国的钢铁企业和物流企业提供很好的借鉴。

关键词：物流财务共享 2.0，价值创造

建立安全隐患排查系统，全面提升安全管理水平

孟令书，吴　鹏

（河北钢铁集团邯钢公司）

摘　要：邯钢安全生产管理系统，依托公司现有网络资源，以车间为最小单元的网络终端，实现安全生产信息化办公。邯钢安全生产管理系统核心模块"安全隐患排查"将实现全程信息化管理隐患整改工作，这标志着邯钢安全管理工作正式步入了由传统管理向智能化、信息化管理的转变。

关键词：安全生产管理系统，安全隐患排查

Abstract: Safety management system of Handan Iron and steel, relying on the company's existing cyber source, the workshop is the smallest unit of network terminal, realization of safe production information office. Handan Iron and steel production safety management system core module "safety hazards investigation" will achieve the full information management hazards rectification work, this marks the Handan Iron and steel safety management work officially entered the change from traditional management to the intelligence and information management.

Key words: safety management system，safety hazards investigation

钢铁行业上市公司
2016 年年报分析及投资策略研究

张旭中，郭乃祺，王晶莹，王凌晗

（冶金工业规划研究院经济处，北京　100711）

摘　要：2016 年，钢铁行业大力化解过剩产能，各项政策措施陆续出台，行业运行走势稳中趋好，行业困境有所缓解；但产能过剩基本面没有改变、钢材价格大幅波动、资金链紧张的形势还将持续一段时间。面对充满机遇与挑战的 2016 年，代表着钢铁行业优质资产的各上市钢企，在提升盈利能力、保障股东权益、完善法人治理结构、控制成本、优化资金来源与结构、调整资产配置等方面做了大量工作。本文基于钢铁行业上市公司 2016 年年报数据，从发展规模与成长性、盈利能力、成本控制能力、偿债能力与流动性、社会责任等方面展开分析，并结合行业基本面与发展趋势，研究钢铁行业上市公司投资策略，进而提出钢铁行业上市公司的价值投资建议。

（注：统计范围共计 30 家上市公司，分别是：大冶特钢、河钢股份、韶钢松山、本钢板材、新兴铸管、太钢不锈、鞍钢股份、华菱钢铁、首钢股份、沙钢股份、三钢闽光、中原特钢、永兴特钢、包钢股份、宝钢股份、山东钢铁、西宁特钢、杭钢股份、凌钢股份、南钢股份、酒钢宏兴、抚顺特钢、方大特钢、安阳钢铁、八一钢铁、新钢股份、马钢股份、柳钢股份、重庆钢铁、武进不锈。此外与 2015 年年报相比，武进不锈为 2016 年新上市公司；宝武重组，宝钢股份 2016 年年报不含原上市公司武钢股份资产；攀钢钒钛因主营业务变更不予考虑。）

关键词：年报分析，投资策略，上市钢企，基本面，投资价值

制造协同初探

贾生晖，黄　勇

（宝钢股份武钢有限冷轧厂，湖北武汉　430080）

摘　要：本文结合武钢和宝钢实施联合的实际案列，通过快赢项目实施，分析了协同理论在制造过程实施的具体方法，为宝钢股份四山基地今后的融合发展，创造更大价值进行了有益的尝试。

关键词：联合，项目，协同，制造

Study on Collaborative Manufacturing

Jia Shenghui, Huang Yong

(Baosteel Wuhan Steel & Iron Company Ltd.Cold Rolling Mill, Wuhan 430080, China)

Abstract: in this paper, the implementation of the actual case of WISCO and Baosteel combined column, through the implementation of quick win project, analyzed the collaborative implementation method theory in the manufacturing process, Baosteel four mountain base in the development of fusion, and attempts to create greater value.

Key words: joint, project, synergy, manufacturing

建章立制　规范管理
构建企业制度流程管理体系

李　妞，李　刚，李振江

（山东钢铁集团永锋淄博有限公司，山东淄博　256400）

摘　要：在企业战略重组下，转型过程中，面对突如其来的巨大内外部环境变化，为实现企业的规范运作、高效运营，保障公司各项管理工作稳定有序开展，理顺工作流程，按照三级管理体制，结合公司实际，从制度的适用性、可塑性、合法性、操作性、全面性五个方面，实施制度流程管理体系的建设，并搭建开发制度流程信息化管理平台，实现公司制度内容电子文档汇编、网上审批、分类检索查阅等功能。

关键词：制度流程体系，信息化平台，战略重组，体系框架

Establishing Rules and Systems Standardizing Management Set Up the Management System of Institutional and Process about Enterprise

Li Niu，Li Gang, Li Zhenjiang

(Shandong Steel Group Yong Feg Zibo Co., Ltd., Zibo 256400, China)

Abstract: Under the strategic reorganization and transformation of the enterprise,in the face of sudden large internal and external environmental changes, to achieve the standardized and efficient operation, ensure all management work can be worked steady and orderly,according to the three levels of management system, combining the reality of the company, from the system's five aspects about applicability, plasticity, legality, operability, comprehensiveness,implementing the system construction,and building information platform,to realize the functions of electronic document compilation, online approval,classification retrieva and so on.

Key words: the system of institutional and process, information platform, strategic reorganization, system framework

钢铁企业期货风险管理策略探析

周　勋，赵　峰

（冶金工业规划研究院经济处，北京　100711）

摘　要：在钢铁行业产能严重过剩、市场竞争激烈的背景下，钢铁企业的抗风险能力亟待加强，随着我国期货市场快速发展，越来越多的钢铁企业运用期货工具进行风险管理，本文结合我国钢铁期货市场运行情况和钢铁企业参与期货的现状，分析企业利用期货进行风险管理存在的问题和进行期货风险管理的意义，并阐述企业如何进行期货风险管理，提出实施路径和建议。

关键词：期货，风险管理，套期保值，价格

Analysis on the Strategy of Futures Risk Management in Steel Enterprises

Zhou Xun, Zhao Feng

(China Metallurgical Industry Planning and Research Institute Economic Department, Beijing 100711, China)

Abstract: Under the background of severe overcapacity and fierce market competition in the iron and steel industry, the anti-risk ability of steel enterprises needs to be strengthened. With the rapid development of China's futures market, more and more steel enterprises use the futures tools to manage risk. This paper tells the operation condition of China's iron and steel futures market and the condition of steel enterprises participating in futures. By analyzing the problems existing in the risk management of steel enterprises using futures and the significance of futures risk management, it expounds how to carry on the futures risk management, and puts forward the implementation paths and suggestions.

Key words: futures, risk management, hedging, price

首钢国际工程公司构建科技创新体系的研究与创新

颉建新[1,3]，张福明[2,3]，李　欣[1,3]，张艾峰[1,3]，王　璇[1,3]，刘　然[1,3]

（1. 北京首钢国际工程技术有限公司，北京　100043；2. 首钢集团有限公司，北京　100041；
3. 北京市冶金三维仿真设计工程技术研究中心，北京　100043）

摘　要： 本文分析了首钢国际工程公司构建科技创新体系的目的与意义，针对公司"十二五"以前科技创新体系主要存在的问题，创造性地提出了公司构建科技创新体系的主要理论依据及目标方案，从技术创新体系、机制、能力三个方面入手，实施技术创新工程，其理论、措施、方法具有突出的管理创新性，实施后取得了较高的经济效益，具有广泛的推广与应用价值。

关键词： 首钢国际工程公司，构建，科技创新，体系，研究与创新

The Research and Innovation of Establishing system of Scientific and Technological Innovation among Shougang International Engineering Co., Ltd

Xie Jianxin[1,3], Zhang Fuming[2,3], Li Xin[1,3], Zhang Aifeng[1,3], Wang Xuan[1,3], Liu Ran[1,3]

(1. Beijing Shougang International Engineering Co., Ltd., Beijing 100043, China; 2. Shougang Group Co., Ltd., Beijing 100041, China; 3. Beijing Metallurgy Three-dimensional Simulation Design Engineering Technology Research Centre, Beijing 100043, China)

Abstract: The aim and significance of establishing system of science and technology innovation among Shougang International Engineering Company is analyzed in this essay. Targeting at the main existing problem concerning the system of science and technology innovation before the 12th Five-Year Plan, the main theoretical basis and plans of establishing the system of science and technology innovation among Shougang International Engineering Company is creatively put forward, Starting with 3 such aspects as technology innovation system, mechanism, capacity, the engineering of technical innovation is implemented, its theory, methods and measures are innovative and creative, gaining relatively high economic benefits after the measures are taken, meaning that this system can be further promoted and applied.

Key words: Shougang International Engineering Company, establishment, scientific and technological innovation, system, research and innovation

现场紧急采购模式在总承包工程中的运用及分析

杨　帆

（中冶南方工程技术有限公司，湖北武汉　430023）

摘 要：目前冶金行业总承包工程呈现出项目建设工期"与日俱短"的趋势，形成"设备订货时间严重缩短、现场增补设备需求时间紧迫"的局面，如按正常采购招标流程进行采买，将无法满足工期需求。本文结合了本人在总承包工程现场采买设备工作中的体会，讨论了如何适应目前总承包项目特点，如何解决总承包工程现场工期采购任务急、采购量小、时间紧迫的问题，以及如何避免常规采买流程带来的项目进度影响情况发生，保证项目正常实施。

关键词：现场紧急采购，总承包工程，运用，分析

比较优势理论在钢企跨地区经营中的应用研究

乐 洋

（马钢（集团）控股有限公司资本运营部，安徽马鞍山 243000）

摘 要：跨地区经营是不少特大型钢铁企业面临的经营状态。不同生产基地之间的基础条件、生产成本、地理位置等各不相同。比较优势理论通常应用于国际贸易领域，随着当前企业的跨地区经营规模不断扩大，一些实际经营中的问题也可以在微观层面用比较优势理论进行分析。本文拟以李嘉图模型和 H-O 模型为基础，从成本约束的角度进行研究。

关键词：跨地区经营，比较优势理论，成本约束，李嘉图模型，H-O 模型

Application Research of Comparative Advantage Theory in Trans-Regional Steel Enterprises

Yue Yang

(Magang (Group) Holding Co., Ltd., Maanshan 243000, China)

Abstract: Trans-regional operation is the operating status of many steel enterprises. The basic conditions, production cost and geographical location of different production bases are different. Comparative advantage theory is usually applied to international trade. With the expansion of the scale of enterprises operating trans regions, some problems in actual operation can also be analyzed with comparative advantage theory at the micro level. Based on the Ricardian and h-o model, this paper has carried out research from the perspective of cost constraint.

Key words: trans-regional enterprise, comparative advantage theory, cost constraints, Ricardian model, H-O model

我国粉末冶金产业现状和市场前景

王泽群，刘 琦，肖邦国

（冶金工业规划研究院市场研究中心，北京 100711）

摘 要：粉末冶金技术由于具有传统熔铸压延加工方法所不具备的特点，在现代制造业中发挥着不可替代的作用。随着汽车、机械、工具制造、消费电子产品等产业的发展，粉末冶金的市场规模不断扩大。本文介绍了我国粉末冶

金行业发展现状，包括企业分布和原材料粉末的产量。分析了粉末冶金零件、硬质合金和粉末注射成型制品三种粉末冶金产品的市场情况。对粉末冶金产业未来的市场前景进行了展望。

关键词：粉末冶金，产业现状，市场

鲅鱼圈分公司 4038m³ 大型高炉优化配煤研究与应用

王宝海，张延辉，唐继忠，蒋　珅，冯宝泽，姜彦冰

（鞍钢股份有限公司鲅鱼圈钢铁分公司，辽宁营口　115007）

摘　要：在分析鲅鱼圈分公司所用喷煤煤粉理化性能指标基础上，采用数学优化设计，确定了不同煤种，不同配入水平下配煤方案，并在 2 座高炉实际喷吹应用中，取得了提高喷煤比 4.44kg/t（注 1# 高炉）和 5.78kg/t（注：2# 高炉），降低燃料比 14.94kg/t 和 13.20kg/t 等良好效果，对比传统配煤技术，可以有效地减少燃料消耗，改善高炉技术指标，降低炼铁生产成本。

关键词：数学优化，配煤，喷煤比，燃料比

A Reasonable Injecting Coal Proportion Study for 4083m³ Big Blast Furnace and Its Application in Bayuquan Filiale

Wang Baohai, Zhang Yanhui, Tang Jizhong, Jiang Shen, Feng Baoze, Jiang Yanbing

(Bayuquan Iron&Steel Subsidiary Company of Angang Co., Ltd., Yingkou 115007, China)

Abstract: Having taken the coal as factors and proportion as level, a mathematical orthoplan has been carried out based on the analysis of coal physcial-chemical properties, the relative practice in 2 blast furnace of Ansteel has been conducted with the resonable coal plan, and the coal injection has been up by 4.44Kg caol per ton hot metal in 1# BF and 5.78 in 2# BF, the fuel rate has been decreased by 14.94 and 13.20kg fuel per ton hot metal. Finally this method can reduce the fuel consumption and improve BF performance compared with traditonal ones.

Key words: mathematical optimization, injecting coal proportion, coal ratio, fuel ratio

实现绿色铁路运输方式的实践与探索

赵征红

（鞍山钢铁集团有限公司铁路运输分公司运输总站，辽宁鞍山　114021）

摘　要：本文以鞍山钢铁集团有限公司路运输分公司为例，阐述了铁路运输过程中一些不当的行为产生多余的污染物对环境造成破坏以及如何消除污染，实现绿色铁路运输。如：运输过程中不必要的燃油消耗加重空气污染、物料飘浮产生的空气污染、机车运行控制不当产生的大量噪音污染、车辆漏料造成环境污染、铁路道口发生撞车事故造成物料倾泻污染环境等，以及如何通过合理调配和使用机车、选用添加剂改善燃油品质、加强道口管理减少路外事

故、规范车辆装载防止物料散落等措施，降低环境污染，实现绿色铁路运输。

关键词：铁路，货物运输，降低环境污染

The Practice and Exploration to Attain the Mode of Green Railway Transportation

Zhao Zhenghong

(Anshan Iron and Steel Group Co., Ltd., Railway Transport Branch Transport terminus, Anshan 114021, China)

Abstract: Based on Anshan Iron and Steel Group Co., Ltd., Railway transport branch Transport terminus, this paper suggests that the produce excess pollution by some improper behavior of the railway transport and the way to eliminate pollution to realizing the green transportation finally. We summarize the source of pollution such as adding unnecessary fuel consumption in the process of transport lead to air pollution, material floating air pollution, improper control of locomotive running a lot of noise pollution, the environment pollution by traffic material produced leak, railway crossing accidents cause the material produced leak which besmear the environment, etc. And we list some method to achieve the green transportation such as reasonable allocation and use of the locomotive，choosing additives for improving the quality of crude oil, strengthening the management of intersections to reduce accidents, standard vehicle load to prevent material scattered.

Key words: railway, transport of goods, reduce the environment pollution

浅议中厚板企业集约化生产措施

王东柱，王　新，万　潇，师大兴

（秦皇岛首秦金属材料有限公司制造部，河北秦皇岛　066326）

摘　要： 本文主要论述中厚板企业集约化生产措施，包括异钢种连浇过渡坯处理，钢坯牌号整合集约化生产和订单合并规则优化等，满足了当前订单生产的需求。

关键词：集约化，过渡坯，钢坯牌号，合并规则

Intensive Production Measures of Plate Mill

Wang Dongzhu, Wang Xin, Wan Xiao, Shi Daxing

(Qinhuangdao Shouqin Metal Materials Co., Ltd., Qinhuangdao 066326, China)

Abstract: This paper especially discusses the intensive measures of plate mill, including treatment of intermixing slab, Slab steel grades intensive production, and production orders combination. These measures meet the demands of orders.

Key words: intensification, intermixing slab, slab steel grades, combinationg rules

基于粗钢产量的全球钢铁发展分析

王晓燕，潘开灵，马云峰

（武汉科技大学管理学院，湖北武汉　430065）

摘　要： 工业化是国家走向现代强国的必由之路，而钢铁业是一国工业化、城镇化的支柱产业、先导产业，也是一国立于世界之林的战略资源。按照工业化各项评价指标，我国处于工业化中期阶段，并已出现工业化后期阶段的明显特征。按国际钢铁大国和钢铁强国的经验，这一阶段正是钢铁业快速发展期。但我国钢铁业却出现各种相悖现象：盈利水平持续下降，产量越来越高；大力推进重组，行业集中度不升反降；产能调控越调越高；规模经济的钢铁业却日益规模不经济。政府、行业、企业各方的目标与行为结果日趋背离。对全球钢铁发展历程进行分析和研究，以为我国钢铁发展准确定位并吸取经验。

关键词： 全球钢铁，区域转移，粗钢产量

Present Situation and Analysis of World Steel Development

Wang Xiaoyan, Pan Kailing, Ma Yunfeng

(College of Management, Wuhan University of Science and Technology, Wuhan 430065, China)

Abstract: Industrialization is the only way to a powerful country, while the steel industry is a pillar and a leading industry in the industrialization and urbanization of a country. It is also a strategic resource for a country to stand in the world's forests. According to the evaluation indexes of industrialization, China is in the middle stage and has obvious characteristics to the late stage of industrialization. This stage is a rapid development period in iron and steel industry in those power countries. However, Chinese steel appears all kinds of contrary phenomenon: profitability continued to decline while the yield is higher and higher; The government vigorously promote the restructuring of industry while the concentration declined; production capacity regulation tune higher; scale economy of iron and steel industry has become increasingly diseconomies of scale. The objectives and actions of the government, industry and enterprise are increasingly deviating from each other. Analyze and study of the process of world steel development, in order to accurately locate the development status of China's steel industry and absorb the successful experience of other countries.

Key words: the world iron and steel, zone transfer, crude steel output

用雁行理论分析厚料层烧结优化技术推广问题

翟江南

（湖南广播电视大学网络工程职业学院，湖南长沙　410004）

摘　要： 借用工程仿生学的分析方法，通过对雁行理论与厚料层烧结优化技术两者进行比较，将宝钢烧结比拟为大雁迁徙的领头雁，说明宝钢烧结技术的发展在中国该领域具有举足轻重的地位。用宝钢各期烧结机演变的详实数据，

揭示了宝钢在转化厚料层烧结优化技术的同时，未能抓住机遇打造核心竞争力，错失了引领中国烧结技术沿着集约发展方向前行。为了重整厚料层烧结优化技术，必须破除固有观念，开展在理论创新方面引入经济学的生产函数概念，在技术创新方面引入熊彼特的技术创新解释，在制度创新方面加强行业宏观调控的制度保障能力和管理水平。

关键词：雁行理论，厚料层烧结，优化技术，路径依赖，理论创新，技术创新，制度创新

Analysis of Thick-bed Sintering Optimization Techniques with the Flying Geese Paradigm

Zhai JiangNan

(Hunan Radio & Television University Net Work Engineering Vocational College, Changsha 410004, China)

Abstract: borrowing methods of analysis of engineering bionics, based on theories of the Flying Geese Paradigm，it is compared a Flying Geese's behaviour with popularizing of thick-bed sintering optimization technology. Through taking Baoshan iron sintering as personificated to been the leader of the wild geese migration, it indicates that an important role of Baosteel's in this area in China. With the evolution of the sinter machine at Baosteel and detailed data, the essay reveals the Baosteel unconscious while using thick-bed sintering optimization technology, failed to carry forward this technology as its core competitiveness, missed leading historical opportunities of development of sintering technology in China. In order to restructure the thick-bed sintering optimization techniques, it must break the stereotypes and concepts in theoretical innovation of Economics production function, lead into technological innovation in Schumpeter explained in technical innovation, and reinforce macro-control in institutional innovation and management levels.

Key words: Flying Geese Paradigm, thick-bed sintering, optimize technology, path dependency, theoretical innovation, technological innovation, institutional innovation